Milestones
in Biotechnology

BIOTECHNOLOGY

JULIAN E. DAVIES, *Editor*
Pasteur Institute
Paris, France

BIOTECHNOLOGY SERIES

Milestones in Biotechnology: Classic Papers on Genetic Engineering

Edited by

Julian Davies
University of British Columbia
Vancouver, British Columbia
Canada

William S. Reznikoff
University of Wisconsin
Madison, Wisconsin

Butterworth–Heinemann
Boston   London   Oxford   Singapore   Sydney   Toronto   Wellington

Recognizing the importance of preserving what has been written, it is the policy of Butterworth–Heinemann to have the books it publishes printed on acid-free paper, and we exert our best efforts to that end.

Library of Congress Cataloging-in-Publication Data
Milestones in biotechnology : classic papers on genetic engineering /
edited by Julian Davies and William Reznikoff.
p.   cm. — (Biotechnology series : 24)
Includes bibliographical references and index.
ISBN 0–7506–9251–0 (alk. paper : pb)
1. Microbial genetic engineering.   2. Microbial biotechnology.
I. Davies, Julian E.   II. Reznikoff, William S.   III. Series:
Biotechnology (Reading, Mass.) : 24.
TP248.6.M55   1992
660'.65—dc20
92–4598

                                                                CIP

British Library Cataloguing-in-Publication Data. A catalogue record for this book is available from the British Library.

Butterworth–Heinemann
80 Montvale Avenue
Stoneham, MA 02180

10 9 8 7 6 5 4 3 2 1

Printed in the United States of America

CONTENTS

Biotechnology, the use of living organisms or their components to provide processes or products useful to man, has been known for a long time. The Old Testament refers to products obtained by yeast-based fermentations, and in literature from other ancient cultures one can find many references to food and pharmaceutical use of micro-organisms and products obtained from plants and domesticated animals. Since these times countless practical applications have found their way into all aspects of industry. One of the landmarks was the Weizmann process for the production of acetone, which proved to be of considerable importance during the First World War (1914–1918). Biotechnology has been so generally assimilated into practice that many of its products are not recognized to be microbial metabolites; the emergence of the antibiotic era since 1945 is a good example of this.

Nonetheless, the term "biotechnology" has taken on a new meaning in the past 20 years, as if the word and the subject had been recently discovered. What has changed, since the demonstration of recombinant DNA methodology in the early 1970s, is the *technology* of biotechnology. We are no longer dependent on natural sources for the most effective processes; it is now possible to engineer genes and organisms for more convenient, efficient and safe production, modification and isolation. This should not be taken to imply that we can dispense with what nature may provide! Nothing can be further from the truth since everything we do in biotechnology depends on natural processes, and it is extremely difficult to improve on nature. But it can be done, and we shall make reference to the striking advances made in protein engineering to provide useful commercial enzymes with improved characteristics for particular processes. If, however, what we want is a thermostable protease, we can often do no better than to isolate such an enzyme from a natural source, a thermophilic microbe, rather than trying to create an enzyme with such characteristics by protein engineering. In a similar way the future of the degradation of xenobiotics and the production of antibiotics will rely on our ability to take natural processes and manipulate them by genetic engineering to obtain new hybrid products and processes.

In the medical field, advances in modern biotechnology have been nothing short of miraculous. Some 15-20 years ago we could make small amounts of crude interferon and similar products for pharmaceutical applications. At that time key cytokines such as interleukin-1, interleukin-2, and the multiple colony-stimulating factors were barely recognized. Hormones such as erythropoietin were known but unidentified; now in the 1990s we have all of these pharmacologically active proteins available and in quantity at purities greater than 99%! Advances in our understanding of human physiology in the past 10-20 years have been more profound and more rapid than anyone could have imagined, purely as a result of the ability to clone and express heterologous genes in simple microbes. Plant biotechnology, to mention but one, is another area where considerable conceptual and technical advances have been realized. Plants resistant to herbicides and vegetables resistant to physical damage from packing or transport are examples of the utilization of modern methods of gene identification and advanced techniques of gene transfer and expression.

Almost all fields of biology, both fundamental and practical, have benefitted from the new discipline that we call biotechnology. It is not a new science, but represents a fusion of aspects of genetics, molecular biology, biochemistry, microbiology, enzymology, and immunology leading to developments in biotechnology. The last two decades have been an exciting period in the field, and we have tried to communicate this excitement by providing a compendium of key publications that, in our opinion, describe the essentials of the development of modern biotechnology. It, of course, obvious that the choice we have made is very personal; when we began to plan this collection we had at least 150 papers, which would have been overly comprehensive. Our final selection is here, and we believe that it covers the key findings that have led to what we all now accept as biotechnology. This does not mean that other work should be considered less important and therefore be excluded: additional references to work of significance are provided in each section. We are sure that as readers you will enjoy our selection, and that you will be able to appreciate what has been accomplished by looking through this collection; we would appreciate your comments on our choice.

Julian Davies
William S. Reznikoff

Enzymology

Providing the Tools for Genetic Engineering.

Modern biotechnology was born because scientists learned to manipulate (engineer) DNA. This capability was derived largely from the field of nucleic acid enzymology. For over twenty-five years before anyone contemplated cloning a gene or constructing a transgenic animal, basic scientists were trying to define the biochemical reactions that underlie fundamental biological processes—such as, how do genes duplicate? How does genetic recombination occur? How do organisms exclude foreign genetic material? Remarkable progress was made. However, in addition to the critical basic science information that was being derived, these studies were also generating the instruments or tools—DNA metabolizing enzymes—which allowed scientists starting in the mid 1970s to contemplate and initiate genetic engineering.

All recombinant DNA research developed from the ability to cut DNA molecules at defined sequences. In other words, it was based upon the discovery of type II restriction enzymes. The isolation of the first enzymes (HindII and HindIII) was a result of a serendipitous discovery by Ham Smith and his coworkers that *Hemophilis influenza* extracts contained activities that cut large DNAs into defined fragments. The consequences of this finding were enormous. Recombinant DNA research was born. In addition, an entire industry developed with the prime purpose being discovery, characterization, purification and marketing of over 100 different site-specific restriction enzymes.

The bringing together of DNA fragments to form covalently linked chimeric molecules is the basis of recombinant DNA research; it is crucial in genetic engineering. This is accomplished by ligation catalyzed by DNA ligase, an enzyme whose discovery predated that of restriction enzymes. The Sgaramella and Khorana paper presents a case study of the use of DNA ligase to link cohesive ends together. We also know that T4 encoded DNA ligase can be used to bring together blunt or flush ends. An alternative means of bringing together two DNA fragments is to generate homopolymeric, single stranded "tails" at the 3' ends of each molecule, but with the tails on one molecule being complementary to the tails on the second. Such a tailing procedure, which uses deoxynucleotidyl terminal transferase, is described in the Jackson et al. paper.

Before 1970 the existing dogma in molecular biology was that genetic information transfer occurred from DNA→DNA, and from DNA→RNA→protein. The existence of RNA containing viruses which appeared to transfer their genetic information from RNA→DNA after cell infection (we now call these retroviruses) was a basic challenge to this model. The proof that RNA-to-DNA information transfer did occur rested on the discovery and characterization of reverse transcriptase by Temin and Baltimore. The applications of this enzyme have been of great significance. For example, reverse transcriptase allowed the direct comparison of mRNAs to their genes and thence led to the discovery of introns. Reverse transcriptase permits the direct sequence analysis of RNA molecules. But most important is that reverse transcriptase allows us to generate

DNA copies of mRNA subsequent to cloning. The generation of cDNA, containing direct protein coding information cleansed of introns, is the usual first step in cloning that leads to expression of eukaryotic proteins in bacteria. A major portion of Section 4 will be devoted to cDNA cloning.

Perhaps the most revolutionary and in many respects the most simplistic molecular biological technical development is PCR (polymerase chain reaction). PCR is (in retrospect) a straightforward application of DNA polymerase to permit the test tube binomial amplification of specific DNA sequences (see Mullis et al., 1986). The sequences which are amplified are defined by specific oligonucleotide primers localized in opposing directions, usually a few thousand base pairs away. Through the use of PCR one has an alternative to cloning for the purification and amplification of specific sequences. Having recognized the immense power of PCR, molecular biologists have busily sought to use it in combination with other technologies; e.g., it is used in cloning (the product DNA is introduced into a vector DNA molecule), in the amplification of cDNA molecules and in site specific or randomized mutagenesis protocols. PCR can also be used to detect specific genes (and thus organisms); e.g., whether the AIDS virus is present in a given sample. It opens the opportunity of diagnosing Abraham Lincoln's inherited disease or of performing molecular evolution studies (how closely related are mammoth and elephant DNA sequences?).

An important technical advancement which made PCR a useful procedure was the isolation and characterization of thermal stable DNA polymerases from thermophylic bacteria (Innis et al., 1988). The critical problems which this resolved were that at each cycle the denaturation of the product DNAs to yield new templates resulted in denaturation of the enzyme, and the temperature limitation of previously used DNA polymerases restricted the stringency (accuracy) of primer hybridization and length of polymerization. Of course some of the properties of thermal stable DNA polymerases which make them so useful in PCR (allowing stringent primer annealing conditions and highly processive polymerization) also make them ideal for other applications such as DNA sequencing. It is likely that the ever-increasing usefulness of thermostable DNA polymerases will encourage interest in a variety of "unusual" microorganisms if only as sources of enzymes with very special desirable properties.

The discovery of type II restriction enzymes demonstrated the immense power and utility of site specific DNA cleavage reagents. Two papers referred to in the supplemental readings (Schultz and Dervan, and Dreyer and Dervan) lay the foundation for a likely new generation of sequence specific cleavage reagents. This work describes approaches for giving specificity to the nonspecific DNA cleavage reagent EDTA-Fe(II).

Discoveries are still being made (and hopefully still will be made) which upset our dogmatic applecarts. Certainly the discovery by Tom Cech and his coworkers that RNA molecules can act as enzymes is in this category. Not only was this a fundamental discovery in its own right, but it has also led to the development of a new class of sequence specific RNA cleavage reagents. In this section we will present a paper from Cech's laboratory, and in Section 5 we will present an extension of this discovery (Haseloff and Gerlach) leading to the development of RNA based sequence specific endoribonuclease which may become a tool of the future for downregulating specific gene expression *in vivo*.

Finally, in the supplemental reading section, we refer the reader to communications which describe what may become a rich source of completely new protein based enzymes in the future; catalytic antibodies or "abzymes." Chemists and biochemists have for years wondered, can one create a biological catalyst from scratch? The answer is yes, and although there is a long way to go before abzymes approach the efficiency (turnover number, Km) of enzymes, advances in the last few years engender promise. We will present in the Immunology Section (12) two additional papers which transfer the abzyme technology to *E. coli* and thus allow far greater genetic manipulation flexibility and low cost production (see Ward et al. and Huse et al.). Perhaps in ten years we will wonder how we managed in the absence of abzymes.

Viral RNA-dependent DNA Polymerase

D. Baltimore

Two independent groups of investigators have found evidence of an enzyme in virions of RNA tumour viruses which synthesizes DNA from an RNA template. This discovery, if upheld, will have important implications not only for carcinogenesis by RNA viruses but also for the general understanding of genetic transcription: apparently the classical process of information transfer from DNA to RNA can be inverted.

RNA-dependent DNA Polymerase in Virions of RNA Tumour Viruses

DNA seems to have a critical role in the multiplication and transforming ability of RNA tumour viruses[1]. Infection and transformation by these viruses can be prevented by inhibitors of DNA synthesis added during the first 8–12 h after exposure of cells to the virus[1-4]. The necessary DNA synthesis seems to involve the production of DNA which is genetically specific for the infecting virus[5,6], although hybridization studies intended to demonstrate virus-specific DNA have been inconclusive[1]. Also, the formation of virions by the RNA tumour viruses is sensitive to actinomycin D and therefore seems to involve DNA-dependent RNA synthesis[1-4,7]. One model which explains these data postulates the transfer of the information of the infecting RNA to a DNA copy which then serves as template for the synthesis of viral RNA[1,2,7]. This model requires a unique enzyme, an RNA-dependent DNA polymerase.

No enzyme which synthesizes DNA from an RNA template has been found in any type of cell. Unless such an enzyme exists in uninfected cells, the RNA tumour viruses must either induce its synthesis soon after infection or carry the enzyme into the cell as part of the virion. Precedents exist for the occurrence of nucleotide polymerases in the virions of animal viruses. Vaccinia[8,9]—a DNA virus, Reo[10,11]—a double-stranded RNA virus, and vesicular stomatitis virus (VSV)[12]—a single-stranded RNA virus, have all been shown to contain RNA polymerases. This study demonstrates that an RNA-dependent DNA polymerase is present in the virions of two RNA tumour viruses: Rauscher mouse leukaemia virus (R-MLV) and Rous sarcoma virus. Temin[13] has also identified this activity in Rous sarcoma virus.

Incorporation of Radioactivity from ³H-TTP by R-MLV

A preparation of purified R-MLV was incubated in conditions of DNA polymerase assay. The preparation incorporated radioactivity from ³H-TTP into an acid-insoluble product (Table 1). The reaction required Mg^{2+}, although Mn^{2+} could partially substitute and each of the four deoxyribonucleoside triphosphates was necessary for activity. The reaction was stimulated strongly by dithiothreitol and weakly by NaCl (Table 1). The kinetics of incorporation of radioactivity from ³H-TTP by R-MLV are shown in Fig. 1, curve 1. The reaction rate accelerates for about 1 h and then declines. This time-course may indicate the occurrence of a slow activation of the polymerase in the reaction mixture. The activity is approximately proportional to the amount of added virus.

For other viruses which have nucleotide polymerases in their virions, there is little or no activity demonstrable unless the virions are activated by heat, proteolytic enzymes or detergents[8-12]. None of these treatments increased the activity of the R-MLV DNA polymerase. In fact, incubation at 50° C for 10 min totally inactivated the R-MLV enzyme as did inclusion of trypsin (50 μg/ml.) in the reaction mixture. Addition of as little as 0·01 per cent 'Triton N-101' (a non-ionic detergent) also markedly depressed activity.

Table 1. PROPERTIES OF THE RAUSCHER MOUSE LEUKAEMIA VIRUS DNA POLYMERASE

Reaction system	pmoles ³H-TMP incorporated in 45 min
Complete	3·31
Without magnesium acetate	0·04
Without magnesium acetate + 6 mM MnCl₂	1·59
Without dithiothreitol	0·38
Without NaCl	2·18
Without dATP	< 0·10
Without dCTP	0·12
Without dGTP	< 0·10

A preparation of R-MLV was provided by the Viral Resources Program of the National Cancer Institute. The virus had been purified from the plasma of infected Swiss mice by differential centrifugation. The preparation had a titre of 10^4·88 spleen enlarging doses (50 per cent end point) per ml. Before use the preparation was centrifuged at 105,000g for 30 min and the pellet was suspended in 0·137 M NaCl–0·003 M KCl–0·01 M phosphate buffer (pH 7·4)–0·6 mM EDTA (PBS–EDTA) at 1/20 of the initial volume. The concentrated virus suspension contained 3·1 mg/ml. of protein. The assay mixture contained, in 0·1 ml., 5 μmoles Tris-HCl (pH 8·3) at 37° C, 0·6 μmole magnesium acetate, 6 μmoles NaCl, 2 μmoles dithiothreitol, 0·08 μmole each of dATP, dCTP and dGTP, 0·001 μmole [³H-methyl]–TTP (708 c.p.m. per pmole) (New England Nuclear) and 15 μg viral protein. The reaction mixture was incubated for 45 min at 37° C. The acid-insoluble radioactivity in the sample was then determined by addition of sodium pyrophosphate, carrier yeast RNA and trichloroacetic acid followed by filtration through a membrane filter and counting in a scintillation spectrometer, all as previously described[12]. The radioactivity of an unincubated sample was subtracted from each value (less than 7 per cent of the incorporation in the complete reaction mixture).

Characterization of the Product

The nature of the reaction product was investigated by determining its sensitivity to various treatments. The product could be rendered acid-soluble by either pancreatic deoxyribonuclease or micrococcal nuclease but was unaffected by pancreatic ribonuclease or by alkaline hydrolysis (Table 2). The product therefore has the properties of DNA. If 50 μg/ml. of deoxyribonuclease was

BALTIMORE, D.
Viral RNA-dependent DNA polymerase. Reprinted by permission from *Nature* v. 226: pp. 1209-1211.

added to a reaction mixture there was no loss of acid-insoluble product. The product is therefore protected from the enzyme, probably by the envelope of the virion, although merely diluting the reaction mixture into 10 mM $MgCl_2$ enables the product to be digested by deoxyribonuclease (Table 2).

Table 2. CHARACTERIZATION OF THE POLYMERASE PRODUCT

Expt.	Treatment	Acid-insoluble radioactivity	Percentage undigested product
1	Untreated	1,425	(100)
	20 μg deoxyribonuclease	125	9
	20 μg micrococcal nuclease	69	5
	20 μg ribonuclease	1,361	96
2	Untreated	1,644	(100)
	NaOH hydrolysed	1,684	100

For experiment 1, 93 μg of viral protein was incubated for 2 h in a reaction mixture twice the size of that described in Table 1, with ³H-TTP having a specific activity of 1,133 c.p.m. per pmole. A 50 μl. portion of the reaction mixture was diluted to 5 ml. with 10 mM $MgCl_2$ and 0·5 ml. aliquots were incubated for 1·5 h at 37° C with the indicated enzymes. (The sample with micrococcal nuclease also contained 5 mM $CaCl_2$.) The samples were then chilled, precipitated with trichloroacetic acid and radioactivity was counted. For experiment 2, two standard reaction mixtures were incubated for 45 min at 37° C, then to one sample was added 0·1 ml. of 1 M NaOH and it was boiled for 5 min. It was then chilled and both samples were precipitated with trichloroacetic acid and counted. In a separate experiment (unpublished) it was shown that the alkaline hydrolysis conditions would completely degrade the RNA product of the VSV virion polymerase.

Localization of the Enzyme and its Template

To investigate whether the DNA polymerase and its template were associated with the virions, a R-MLV suspension was centrifuged to equilibrium in a 15–50 per cent sucrose gradient and fractions of the gradient were assayed for DNA polymerase activity. Most of the activity was

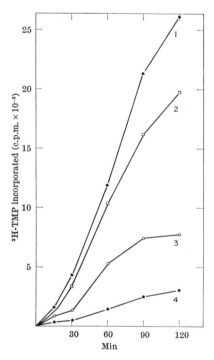

Fig. 1. Incorporation of radioactivity from ³H-TTP by the R-MLV DNA polymerase in the presence and absence of ribonuclease. A 1·5-fold standard reaction mixture was prepared with 30 μg of viral protein and ³H-TTP (specific activity 950 c.p.m. per pmole). At various times, 20 μl. aliquots were added to 0·5 ml. of non-radioactive 0·1 M sodium pyrophosphate and acid insoluble radioactivity was determined[12]. For the preincubated samples, 0·06 ml. of H_2O and 0·01 ml. of R-MLV (30 μg of protein) were incubated with or without 10 μg of pancreatic ribonuclease at 22° C for 20 min, chilled and brought to 0·15 ml. with a concentrated mixture of the components of the assay system. Curve 1, no treatment; curve 2, preincubated; curve 3, 10 μg ribonuclease added to the reaction mixture; curve 4, preincubated with 10 μg ribonuclease.

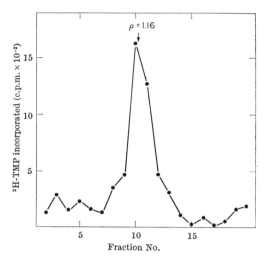

Fig. 2. Localization of DNA polymerase activity in R-MLV by isopycnic centrifugation. A preparation of R-MLV containing 150 μg of protein in 50 μl. was layered over a linear 5·2 ml. gradient of 15–50 per cent sucrose in PBS–EDTA. After centrifugation for 2 h at 60,000 r.p.m. in the Spinco 'SW65' rotor, 0·27 ml. fractions of the gradient were collected and 0·1 ml. portions of each fraction were incubated for 60 min in a standard reaction mixture. The acid-precipitable radioactivity was then collected and counted. The density of each fraction was determined from its refractive index. The arrow indicates the position of a sharp, visible band of light-scattering material which occurred at a density of 1·16.

found at the position of the visible band of virions (Fig. 2). The density at this band was 1·16 g/cm³, in agreement with the known density of the virions[14]. The polymerase and its template therefore seem to be constituents of the virion.

The Template is RNA

Virions of the RNA tumour viruses contain RNA but no DNA[15,16]. The template for the virion DNA polymerase is therefore probably the viral RNA. To substantiate further that RNA is the template, the effect of ribonuclease on the reaction was investigated. When 50 μg/ml. of pancreatic ribonuclease was included in the reaction mixture, there was a 50 per cent inhibition of activity during the first hour and more than 80 per cent inhibition during the second hour of incubation (Fig. 1, curve 3). If the virions were preincubated with the enzyme in water at 22° C and the components of the reaction mixture were then added, an earlier and more extensive inhibition was evident (Fig. 1, curve 4). Preincubation in water without ribonuclease caused only a slight inactivation of the virion polymerase activity (Fig. 1, curve 2). Increasing the concentration of ribonuclease during preincubation could inhibit more than 95 per cent of the DNA polymerase activity (Table 3). To ensure that the inhibition by ribonuclease was attributable to the enzymic activity of the added protein, two other basic proteins were preincubated with the virions. Only ribonuclease was able to inhibit the reaction (Table 3). These experiments substantiate the idea that RNA is the template for the reaction. Hybridization experiments are in progress to determine if the DNA is complementary in base sequence to the viral RNA.

Ability of the Enzyme to Incorporate Ribonucleotides

The deoxyribonucleotide incorporation measured in these experiments could be the result of an RNA polymerase activity in the virion which can polymerize deoxyribonucleotides when they are provided in the reaction mixture. The VSV RNA polymerase and the R-MLV DNA polymerase were therefore compared. The VSV RNA polymerase incorporated only ribonucleotides. At its pH optimum of 7·3 (my unpublished observation), in the presence of the four common ribonucleoside triphosphates, the enzyme incorporated ³H-GMP extensively[12]. At this pH, however, in the presence of the four deoxyribonucleoside triphosphates, no ³H-TMP incorporation was demonstrable (Table 4). Furthermore, replacement of even a single ribonucleotide by its homologous deoxyribonucleotide led to no detectable synthesis (my unpublished observation). At pH 8·3, the optimum for the R-MLV DNA polymerase, the VSV polymerase catalysed much less ribonucleotide incorporation and no significant deoxyribonucleotide incorporation could be detected.

Table 3. EFFECT OF RIBONUCLEASE ON THE DNA POLYMERASE ACTIVITY OF RAUSCHER MOUSE LEUKAEMIA VIRUS

Conditions	pmoles ³H-TMP incorporation
No preincubation	2·50
Preincubated with no addition	2·20
Preincubated with 20 μg/ml. ribonuclease	0·69
Preincubated with 50 μg/ml. ribonuclease	0·31
Preincubated with 200 μg/ml. ribonuclease	0·08
Preincubated with no addition	3·69
Preincubated with 50 μg/ml. ribonuclease	0·52
Preincubated with 50 μg/ml. lysozyme	3·67
Preincubated with 50 μg/ml. cytochrome c	3·97

In experiment 1, for the preincubation, 15 μg of viral protein in 5 μl. of solution was added to 45 μl. of water at 4° C containing the indicated amounts of enzyme. After incubation for 30 min at 22° C, the samples were chilled and 50 μl. of a 2-fold concentrated standard reaction mixture was added. The samples were then incubated at 37° C for 45 min and acid-insoluble radioactivity was measured. In experiment 2, the same procedure was followed, except that the preincubation was for 20 min at 22° C and the 37° C incubation was for 60 min.

Table 4. COMPARISON OF NUCLEOTIDE INCORPORATION BY VESICULAR STOMATITIS VIRUS AND RAUSCHER MOUSE LEUKAEMIA VIRUS

Precursor	pH	Incorporation in 45 min (pmoles) Vesicular stomatitis virus	Mouse leukaemia virus
³H-TTP	8·3	< 0·01	2·3
³H-TTP (omit dATP)	8·3	N.D.	0·06
³H-TTP (omit dATP; plus ATP)	8·3	N.D.	0·08
³H-GTP	8·3	0·43	< 0·03
³H-GTP	7·3	3·7	< 0·03

When ³H-TTP was the precursor, standard reaction conditions were used (see Table 1). When ³H-GTP was the precursor, the reaction mixture contained, in 0·1 ml., 5 μmoles Tris-HCl (pH as indicated), 0·6 μmoles magnesium acetate, 0·3 μmoles mercaptoethanol, 9 μmoles NaCl, 0·08 μmole each of ATP, CTP, UTP; and 0·001 μmole ³H-GTP (1,040 c.p.m. per pmole). All VSV assays included 0·1 per cent 'Triton N–101' (ref. 12) and 2–5 μg of viral protein. The R-MLV assays contained 15 μg of viral protein.

The R-MLV polymerase incorporated only deoxyribonucleotides. At pH 8·3, ³H-TMP incorporation was readily demonstrable but replacement of dATP by ATP completely prevented synthesis (Table 4). Furthermore, no significant incorporation of ³H-GMP could be found in the presence of the four ribonucleotides. At pH 7·3, the R-MLV polymerase was also inactive with ribonucleotides. The polymerase in the R-MLV virions is therefore highly specific for deoxyribonucleotides.

DNA Polymerase in Rous Sarcoma Virus

A preparation of the Prague strain of Rous sarcoma virus was assayed for DNA polymerase activity (Table 5). Incorporation of radioactivity from ³H-TTP was demonstrable and the activity was severely reduced by omission of either Mg^{2+} or dATP from the reaction mixture. RNA-dependent DNA polymerase is therefore probably a constituent of all RNA tumour viruses.

These experiments indicate that the virions of Rauscher mouse leukaemia virus and Rous sarcoma virus contain a DNA polymerase. The inhibition of its activity by ribonuclease suggests that the enzyme is an RNA-dependent DNA polymerase. It seems probable that all RNA tumour viruses have such an activity. The existence of this enzyme strongly supports the earlier suggestions[1-7] that genetically specific DNA synthesis is an early event in the replication cycle of the RNA tumour viruses and that DNA is the template for viral RNA synthesis. Whether the viral DNA ("provirus")[2] is integrated into the host genome or remains as a free template for RNA synthesis will require further study. It will also be necessary to determine whether the host DNA-dependent RNA polymerase or a virus-specific enzyme catalyses the synthesis of viral RNA from the DNA.

Table 5. PROPERTIES OF THE ROUS SARCOMA VIRUS DNA POLYMERASE

Reaction system	pmoles ³H-TMP incorporated in 120 min
Complete	2·06
Without magnesium acetate	0·12
Without dATP	0·19

A preparation of the Prague strain (sub-group C) of Rous sarcoma virus[14] having a titre of 5×10^7 focus forming units per ml. was provided by Dr Peter Vogt. The virus was purified from tissue culture fluid by differential centrifugation. Before use the preparation was centrifuged and the pellet dissolved in 1/10 of the initial volume as described for the R-MLV preparation. For each assay 15 μl. of the concentrated Rous sarcoma virus preparation was assayed in a standard reaction mixture by incubation for 2 h. An unincubated control sample had radioactivity corresponding to 0·14 pmole which was subtracted from the experimental values.

I thank Drs G. Todaro, F. Rauscher and R. Holdenreid for their assistance in providing the mouse leukaemia virus. This work was supported by grants from the US Public Health Service and the American Cancer Society and was carried out during the tenure of an American Society Faculty Research Award.

DAVID BALTIMORE

Department of Biology,
Massachusetts Institute of Technology,
Cambridge,
Massachusetts 02139.

Received June 2, 1970.

[1] Green, M., *Ann. Rev. Biochem.*, **39** (1970, in the press).
[2] Temin, H. M., *Virology*, **23**, 486 (1964).
[3] Bader, J. P., *Virology*, **22**, 462 (1964).
[4] Vigier, P., and Golde, A., *Virology*, **23**, 511 (1964).
[5] Duesberg, P. H., and Vogt, P. K., *Proc. US Nat. Acad. Sci.*, **64**, 939 (1969).
[6] Temin, H. M., in *Biology of Large RNA Viruses* (edit. by Barry, R., and Mahy, B.) (Academic Press, London, 1970).
[7] Temin, H. M., *Virology*, **20**, 577 (1963).
[8] Kates, J. R., and McAuslan, B. R., *Proc. US Nat. Acad. Sci.*, **58**, 134 (1967).
[9] Munyon, W., Paoletti, E., and Grace, J. T. J., *Proc. US Nat. Acad. Sci.*, **58**, 2280 (1967).
[10] Shatkin, A. J., and Sipe, J. D., *Proc. US Nat. Acad. Sci.*, **61**, 1462 (1968).
[11] Borsa, J., and Graham, A. F., *Biochem. Biophys. Res. Commun.*, **33**, 895 (1968).
[12] Baltimore, D., Huang, A. S., and Stampfer, M., *Proc. US Nat. Acad. Sci.* **66** (1970, in the press).
[13] Temin, H. M., and Mizutani, S., *Nature*, **226**, 1211 (1970) (following article).
[14] O'Conner, T. E., Rauscher, F. J., and Zeigel, R. F., *Science*, **144**, 1144 (1964).
[15] Crawford, L. V., and Crawford, E. M., *Virology*, **13**, 227 (1961).
[16] Duesberg, P., and Robinson, W. S., *Proc. US Nat. Acad. Sci.*, **55**, 219 (1966).
[17] Duff, R. G., and Vogt, P. K., *Virology*, **39**, 18 (1969).

DNA Sequencing with *Thermus Acquaticus* DNA Polymerase and Direct Sequencing of Polymerase Chain Reaction-amplified DNA

M.S. Innis, K.B. Myambo, D.H. Gelfand and M.A.D. Brown

ABSTRACT The highly thermostable DNA polymerase from *Thermus aquaticus* (*Taq*) is ideal for both manual and automated DNA sequencing because it is fast, highly processive, has little or no 3'-exonuclease activity, and is active over a broad range of temperatures. Sequencing protocols are presented that produce readable extension products >1000 bases having uniform band intensities. A combination of high reaction temperatures and the base analog 7-deaza-2'-deoxyguanosine was used to sequence through G+C-rich DNA and to resolve gel compressions. We modified the polymerase chain reaction (PCR) conditions for direct DNA sequencing of asymmetric PCR products without intermediate purification by using *Taq* DNA polymerase. The coupling of template preparation by asymmetric PCR and direct sequencing should facilitate automation for large-scale sequencing projects.

DNA sequencing by the Sanger dideoxynucleotide method (1) has undergone significant refinement in recent years, including the development of additional vectors (2), base analogs (3, 4), enzymes (5), and instruments for partial automation of DNA sequence analysis (6–8). The basic procedure involves (*i*) hybridizing an oligonucleotide primer to a suitable single- or denatured double-stranded DNA template; (*ii*) extending the primer with DNA polymerase in four separate reaction mixtures, each containing one α-labeled dNTP, a mixture of unlabeled dNTPs, and one chain-terminating ddNTP; (*iii*) resolving the four sets of reaction products on a high-resolution polyacrylamide/urea gel; and (*iv*) producing an autoradiographic image of the gel, which can be examined to infer the DNA sequence. The current commercial instruments address nonisotopic detection and computerized data collection and analysis. The ultimate success of large-scale sequencing projects will depend on further improvements in the speed and automation of the technology. These include automating the preparation of DNA templates and performing the sequencing reactions.

One technique that appears to be ideally suited for automating DNA template preparation is the selective amplification of DNA by the polymerase chain reaction (PCR) (9). With this method, segments of single-copy genomic DNA can be amplified >10 million-fold with very high specificity and fidelity. The PCR product can then either be subcloned into a vector suitable for sequence analysis or, alternatively, purified PCR products can be sequenced (10–13).

The advent of *Taq* DNA polymerase greatly simplifies the PCR procedure because it is no longer necessary to replenish enzyme after each PCR cycle (14). Use of *Taq* DNA polymerase at high annealing and extension temperatures increases the specificity, yield, and length of products that can be amplified and, thus, increases the sensitivity of PCR for detecting rare target sequences. Here we describe other properties of *Taq* DNA polymerase that pertain to its advantages for DNA sequencing and its fidelity in PCR.

MATERIALS

Enzymes. Polynucleotide kinase from T4-infected *Escherichia coli* cells was purchased from Pharmacia. *Taq* DNA polymerase, a single subunit enzyme with relative molecular mass of 94 kDa (specific activity, 200,000 units/mg; 1 unit corresponds to 10 nmol of product synthesized in 30 min with activated salmon sperm DNA), was purified from *Thermus aquaticus*, strain YT-1 (ATCC no. 25104), according to S. Stoffel and D.H.G. (unpublished data). More recently, *Taq* DNA polymerase (GeneAmp) was purchased from Perkin–Elmer Cetus Instruments. The polymerase (5–80 units/μl) was stored at $-20°C$ in 20 mM Tris·HCl, pH 8.0/100 mM KCl/0.1 mM EDTA/1 mM dithiothreitol/autoclaved gelatin (200 μg/ml)/0.5% Nonidet P-40/0.5% Tween 20/50% (vol/vol) glycerol.

Nucleotides, Oligonucleotides, and DNA. 2'-Deoxy-, and 2',3'-dideoxynucleotide 5'-triphosphates (dNTPs and ddNTPs) were obtained from Pharmacia. 7-Deaza-2'-deoxyguanosine 5'-triphosphate (c⁷GTP) was from Boehringer Mannheim. dATP[α-³⁵S] (650 Ci/mmol; 1 Ci = 37 GBq) was from Amersham, and [γ-³²P]ATP was from New England Nuclear. Oligonucleotide primers for sequencing and PCR were synthesized on a Biosearch 8700 DNA Synthesizer. Oligonucleotide primers were 5'-end-labeled (3×10^6 cpm/pmol) with [γ-³²P]ATP and T4 polynucleotide kinase (15). Single-stranded M13 DNA templates were prepared as described (16).

SEQUENCING METHODS

Annealing Reaction. Single annealing and labeling reactions were performed for each set of four sequencing reactions. The annealing mixture contained 5 μl of oligonucleotide primer (0.1 pmol/μl) in 6× *Taq* sequencing buffer (10 mM MgCl₂/10 mM Tris·HCl, pH 8.0, at room temperature), and 5 μl of template DNA (0.05–0.5 pmol). The mixture was heated to 90°C for 3 min, incubated at 42°C for 20 min, cooled to room temperature, and briefly spun to collect the fluid at the bottom of the tube.

Labeling Reaction. To the 10-μl annealing reaction mixture were added 2 μl of labeling mix (10 μM dGTP/5 μM dCTP/5 μM TTP in 10 mM Tris·HCl, pH 8.0), 2 μl of dATP[α-³⁵S] (5 μM in 10 mM Tris·HCl, pH 8.0), 2 μl of *Taq* DNA polymerase (5 units/μl in dilution buffer: 10 mM Tris·HCl, pH 8.0/0.5% Tween 20/0.5% Nonidet P-40), and 4 μl of H₂O. The labeling reaction mixture was incubated for 1 min at 37°C (see Fig. 3). Note: for sequencing with 5'-labeled primers, the addition of

---

Abbreviations: c⁷GTP, 7-deaza-2'-deoxyguanosine 5'-triphosphate; PCR, polymerase chain reaction.

INNIS, M.S., MYAMBO, K.B., GELFAND, D.H. and BROWN, M.A.D.
DNA sequencing with *Thermus aquaticus* DNA polymerase and direct sequencing of polymerase chain reaction-amplified DNA. *Proc. Natl. Acad. Sci. U.S.A.* 85:9436-9440 (1988). Reprinted by permission from the authors.

dNTP[α-^{35}S] and the labeling reaction step were omitted, and the volume was made up with 10 mM Tris·HCl (pH 8.0).

Extension–Termination Reaction. Four separate extension–termination reactions were performed in 96-well microtiter plates (Falcon 3911) for each labeled template, using concentrated deoxy/dideoxy termination mixes: "G-mix" (30 μM each dNTP, 0.25 mM ddGTP, 0.37 mM MgCl$_2$); "A-mix" (30 μM each dNTP, 1.0 mM ddATP, 1.12 mM MgCl$_2$); "T-mix" (30 μM each dNTP, 1.5 mM ddTTP, 1.62 mM MgCl$_2$); and "C-mix" (30 μM each dNTP, 0.5 mM ddCTP, 0.62 mM MgCl$_2$). Aliquots (4 μl) from the labeling reaction mixtures were added at room temperature to wells containing 2 μl of the appropriate termination mix. Reaction mixtures were overlaid with 10 μl of mineral oil to prevent evaporation and then incubated at 70°C for 1–3 min. Reactions were stopped by the addition of 2 μl of 95% deionized formamide with 0.1% bromophenol blue, 0.1% xylene cyanol, and 10 mM EDTA (pH 7.0). Samples were heated at 80°C for 3 min before loading 1–2 μl onto a buffer gradient sequencing gel (17).

Asymmetric PCRs. The template for PCRs was single-stranded M13mp10 DNA containing a 400-base insert in the *Eco*RI site of the polylinker. Oligonucleotides (20-mers) were synthesized to flank the polylinker, immediately outside of the universal "−20" and "Reverse" sequencing primer binding sites, and these were designated RG05 (5′-AGGGTTTTCCCAGTCACGAC-3′) and RG02 (5′-GTGTGG-AATTGTGAGCGGAT-3′), respectively. Each PCR contained 20 pmol of one primer and 0.2 pmol of the other, 20 μM each dNTP, 1–10 ng of DNA, 1× modified PCR buffer (10 mM Tris·HCl, pH 8.0/3.0 mM MgCl$_2$), 0.05% each of Tween 20 and Nonidet P-40, and 2.5 units of *Taq* DNA polymerase in a total vol of 100 μl. Reactions were performed in 0.5-ml microcentrifuge tubes with the Perkin–Elmer Cetus Thermal Cycler. The thermal profile involved 35 cycles of denaturation at 93°C for 30 sec, primer annealing at 50°C for 1 min, and extension at 72°C for 1 min.

Sequencing of PCR Products. Aliquots of the PCRs were directly incorporated into dideoxy chain-termination sequencing reaction mixtures. A set of four base-specific chain-termination mixes was made up, each in 1× modified PCR buffer and 20 μM each dNTP. The individual mixes contained 250 μM ddGTP, 1.28 mM ddATP, 1.92 mM ddTTP, or 640 μM ddCTP. For each PCR product to be sequenced, four wells on a 96-well microtiter plate were labeled G, A, T, and C, and each well received 2.5 μl of the appropriate termination mix. A 20-μl aliquot of each PCR mixture was mixed with 0.5 μl of fresh *Taq* DNA polymerase (48 units/μl), 1 μl of the appropriate ^{32}P-labeled M13 "forward" or "reverse" sequencing primer (5′-GTAAAACGA-CGGCCAGT-3′, 5′-AACAGCTATGACCATG-3′, respectively; 1.2 pmol/μl) and 10.5 μl of 1× modified PCR buffer. The PCR/primer preparation was immediately dispensed in 7.5-μl aliquots into the wells containing the termination mixes and mixed with the pipette. The reactions were incubated at 70°C for 2 min, and stopped by the addition of 4 μl of 91% formamide with 20 mM EDTA (pH 8.0) and 0.05% each of xylene cyanol and bromophenol blue. Aliquots (5μl) of these reaction mixtures were heated to 75°C for 5 min, and 1–2 μl was loaded on a buffer gradient sequencing gel.

RESULTS

***Taq* DNA Polymerase Is Fast and Very Processive.** The experiments shown in Fig. 1 involved extending a 5′ ^{32}P-labeled 30-mer primer hybridized to M13mp18 single-stranded DNA with an equimolar amount of *Taq* DNA polymerase at various temperatures. Aliquots were taken over time and analyzed as described. Within 2 min at 70°C the entire 7.25-kilobase template was replicated; this corre-

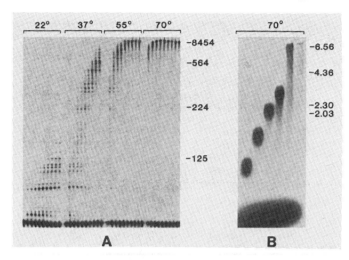

Fig. 1. Autoradiographs of a polyacrylamide/urea gel (*A*) and an alkaline agarose gel (*B*) comparing the extension rate of *Taq* DNA polymerase at different temperatures. Time points are as follows: (*A*) 0 (no enzyme), 15, 30, and 45 sec, and 1, 2, 3, 5, 7, and 10 min; (*B*) 15, 30, and 45 sec, and 1, 2, and 5 min. M13mp18 template DNA (2 pmol) and 5′ ^{32}P-labeled primer DG48 (4 pmol) (5′-GGGAAGGGC-GATCGGTGCGGGCCTCTTCGC-3′, calculated t_m = 78°C in 0.1 M Na$^+$) were annealed in 40 μl of 10 mM Tris·HCl, pH 8.0/5 mM MgCl$_2$, as described. The reaction mixtures were adjusted to 200 μM each dNTP, 0.05% each Tween 20 and Nonidet P-40, 10 mM Tris·HCl (pH 8.0), 50 mM KCl, and 2.5 mM MgCl$_2$, in a total vol of 80 μl, then brought to the desired temperature in the absence of enzyme. *Taq* DNA polymerase (2 pmol) was added to start the reactions, and 8-μl aliquots were removed and added to 8 μl of a stop solution containing 100 mM NaOH, 2 mM EDTA, 5% Ficoll, and 0.1% each bromophenol blue and xylene cyanol. The aliquots were further diluted to 40 μl with half-strength stop solution. Aliquots (5 and 20 μl) of the time points were denatured at 80°C for 3 min and loaded onto a buffer gradient sequencing gel (17) and a 0.8% alkaline agarose gel (18), respectively. Reduction in the signal of full-length product observed at the 5-min time point (*B*) is consistent with the presence of significant polymerization-dependent 5′ exonuclease activity associated with the enzyme. Markers refer to the number of bases incorporated in nucleotides (*A*) or in kilobases (*B*).

sponds to an extension rate in excess of 60 nucleotides per sec. *Taq* DNA polymerase retained significant activity at lower temperatures with calculated extension rates of 24, 1.5, and 0.25 nucleotides per sec at 55°C, 37°C, and 22°C, respectively. At 70°C and at substantial substrate excess (0.1:1 molar ratio of polymerase to primer/template; data not shown) most of the initiated primers were completely extended prior to reinitiation on new primer/template substrate. These results showed *Taq* DNA polymerase to be highly processive.

Factors Affecting the Sequencing Reactions. The buffer (14) for *Taq* DNA polymerase PCRs was modified for DNA sequencing. Each component was investigated individually by using a 5′ ^{32}P-labeled M13 forward sequencing primer (17-mer) and an M13 single-stranded DNA template. Sequencing reactions were performed as described above except that the labeling step was omitted. KCl was included at 0–300 mM. The best extensions occurred in the absence of KCl; at 50 mM KCl there was slight inhibition of enzyme activity, and at ≥75 mM KCl, the activity of *Taq* DNA polymerase was significantly inhibited. The presence of gelatin, which acts as an enzyme stabilizer in PCRs, did not affect the sequencing reactions *per se*; however, it produced distortions during electrophoresis. Addition of nonionic detergents (final concentrations, 0.05% Tween 20 and 0.05% Nonidet P-40) both stimulated the activity of the *Taq* DNA polymerase and reduced the background caused by false terminations from the enzyme (data not shown).

Taq DNA polymerase is sensitive to the free magnesium ion concentration. Accordingly, stock dNTPs and ddNTPs contained equimolar amounts of $MgCl_2$. We varied all four deoxynucleotide triphosphate concentrations between 1 and 20 μM. At concentrations of <5 μM each, or when the concentration of one dNTP was low relative to the other dNTPs, a high background of incorrect termination products was seen because of misincorporation of both dNTPs and ddNTPs. Thus, the optimum concentration for each ddNTP was empirically determined with all four dNTPs at 10 μM. We found that *Taq* DNA polymerase incorporated the four ddNTPs with varying efficiencies, and much less efficiently than the corresponding dNTPs. Ratios that generated optimal distributions of chain-termination products were [dGTP/ddGTP (1:6), dATP/ddATP (1:32), TTP/ddTTP (1:48), and dCTP/ddCTP (1:16)]. *Taq* DNA polymerase concentration was varied between 1 and 20 units per set of four reactions containing 0.2 pmol of single-stranded DNA template, 0.5 pmol of primer, and the dNTP/ddNTP concentrations just described. The signal intensity increased up to 10 units of polymerase per reaction set, representing approximately a 2.5-fold molar excess of enzyme over template/primer.

Developing a Two-Step Labeling and Extension Protocol. We then sought to develop a protocol for incorporation of labeled nucleotide during the sequencing reaction. A "Klenow-type" protocol, in which one labeled nucleotide is present at low concentration relative to the other three during the synthesis reaction, was impractical because of misincorporation of dNTPs and ddNTPs. We estimate the apparent K_m values for each of the four dNTPs to be between 10 and 20 μM. When the concentration of the labeled nucleotide was significantly below K_m (i.e., $\approx$1 μM), ddNTPs present at 80–500 μM were inappropriately incorporated at high frequency (data not shown). Concentrations higher than 1 μM for an α-^{35}S-labeled dNTP are not practical. Also, because the enzyme lacks 3'-exonuclease (proofreading) activity, misincorporated dNTPs induced chain termination. Fig. 2 shows a sequencing ladder generated in the absence of ddNTPs by

A   B   C
GATC GATC GATC

Fig. 2. Autoradiograph of a polyacrylamide/urea gel demonstrating base-specific chain termination due to misincorporation of dNTPs. The sequencing ladder generated with standard dideoxy chain terminations (A) is shown beside ladders generated by limiting one of the four dNTPs, before (B) and after (C) chasing with concentrated balanced dNTP mix. The standard dideoxy reactions were carried out as described for sequencing with a ^{32}P-labeled primer. In the other reactions, the primer and template were annealed in 10 μl of 10 mM Tris·HCl, pH 8.0/6 mM $MgCl_2$. Diluted *Taq* DNA polymerase (2 μl) was added and the reaction was brought to 20 μl with 10 mM Tris·HCl (pH 8.0). The sample was divided into four aliquots, identified by the nucleotide to be limited in that reaction. The "G" and "A" aliquots were brought to 0.5 μM in the limiting nucleotide and to 30 μM in the other three dNTPs; the "T" and "C" reactions were similar, with the limiting nucleotide increased to 1.5 μM. All reaction mixtures were incubated for 10 min at room temperature and then chased by addition of 0.25 vol of 10 mM Tris·HCl, pH 8.0/0.1 mM EDTA (B) or 250 μM (each) dNTP mix (C). The samples were overlaid with mineral oil and incubated at 70°C for 2 min before addition of 4 μl of formamide/EDTA stop solution. The products were denatured at 75°C for 5 min and resolved on a buffer gradient sequencing gel (17).

forcing misincorporation of dNTPs with imbalanced dNTP concentrations. These reactions produced a doublet at most base positions, and chasing these reactions revealed that the upper band of each doublet likely represents molecules that have misincorporated a base, while the lower band represents a pause in the polymerization. Accordingly, the lower bands disappeared when the reaction was chased. Misincorporated bases appeared to be inefficiently extended by the chase.

To circumvent these problems, we developed a two-step procedure similar in concept to one published by Tabor and Richardson for sequencing with a modified bacteriophage T7 DNA polymerase (5): an initial low-temperature labeling step using low concentrations of all four dNTPs (one of which is labeled) followed by a processive extension in the presence of higher dNTP and ddNTP concentrations. To read the sequence next to the primer, it was necessary to use both low temperature and limiting dNTP concentrations to generate an array of extension products ranging in size from a few to >100 nucleotides long. Minimum concentrations of 0.5 μM each dNTP were necessary in this step to generate signals on an overnight exposure, and increasing one of the unlabeled dNTPs to 1.0 μM made the signals very easily readable (data not shown). This effect was seen regardless of which nucleotide was increased, but increasing more than one did not provide additional benefit. The effects of temperature and incubation time on the labeling reaction are shown in Fig. 3. Termination reactions were incubated at either 55°C or 70°C using high dNTP concentrations to ensure maximum processivity and fidelity. The reactions performed at 55°C occurred at a slower rate, but there was no detectable difference in fidelity as compared with 70°C experiments. Using these conditions, we found remarkable uniformity in the band intensities, and we have not detected any idiosyncrasies in the band patterns. In addition, the same reaction conditions cover both short and long gel runs. Fig. 3 includes an autoradiograph of an extended electrophoresis, which yields DNA sequence information in excess of 1000 nucleotides from the priming site.

Using Base Analogs and High Temperature to Sequence Through G+C-Rich DNA and to Eliminate Band Compressions. Band compressions resulting from abnormal gel migration of certain sequences are frequently encountered with G+C-rich DNA templates. Substitutions of dITP (3), or the base analog c^{7}GTP (4), for dGTP have been particularly useful in resolving compression artifacts. We compared incorporation of these nucleoside triphosphates by *Taq* DNA polymerase using either an M13mp18 template or a G+C-rich insert in M13, which contains several regions of strong dyad symmetry (Fig. 4). We found that *Taq* DNA polymerase incorporated c^{7}GTP with essentially the same kinetics as dGTP and that a combination of high reaction temperature and c^{7}GTP was very efficient for resolving difficult sequences.

In contrast, inosine-containing reaction mixtures required a 4-fold higher level of dITP as compared to dGTP, the labeling reaction needed 4 min, and the ratio of ddGTP to dITP was reduced by a factor of 20 compared to dGTP. As shown in Fig. 4, dITP appears to promote frequent terminations during the extension reaction. Terminations caused by inosine result both from a higher rate of misincorporation with dITP as compared to the other dNTPs, and because *Taq* DNA polymerase lacks sufficient 3'-exonuclease activity for editing misincorporated bases. Terminations induced by dITP are reduced if the reactions are initiated at 70°C.

Coupling DNA Sequencing to the PCR. The PCRs were performed with one of the oligonucleotide primers present in a 100-fold greater concentration than the other. In this type of reaction, termed "asymmetric" PCR (13), one of the two PCR primers is depleted during the earlier thermal cycles,

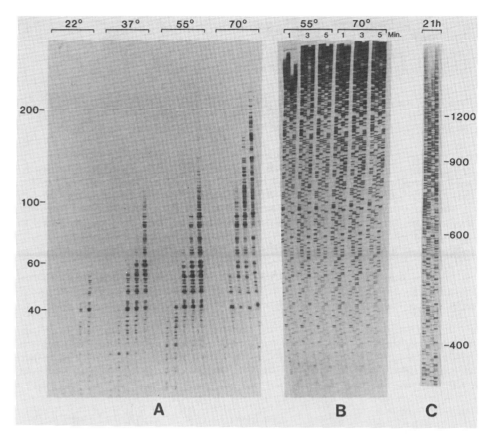

FIG. 3. Autoradiographs of polyacrylamide/urea gels showing the products of labeling reactions (A), extension–termination reactions performed at various temperatures (B), and sequencing reaction products resolved during extended electrophoresis (C). The labeling reactions were performed as described, except the reactions were brought up to temperature before the addition of the enzyme. Aliquots were removed at 0.5, 1, 3, 5, 7, and 10 min. The extension–termination reactions were performed as described for sequencing. Reactions were stopped and resolved on a buffer gradient sequencing gel as described in Fig. 2. Extended electrophoresis (C) was performed on the products of a 70°C 3-min extension–termination sequencing reaction. Samples were run at 15 W for 21 hr on a 7% acrylamide gel (18 × 50 cm × 0.4 mm) (24:1 cross-linking) with 7 M urea and 1× TBE (90 mM Tris/64.6 mM boric acid/2.5 mM EDTA, pH 8.3). Markers indicate the distance in nucleotides from the beginning of the primer. Reaction sets were loaded G, A, T, C.

and the reaction generates single-stranded product with the remaining primer.

Sequencing of asymmetric PCR-generated templates did not require purification of the product. Based on an estimated yield of 1 μg of total product, we calculate that one-third to one-half of the dNTPs initially added were used up during the PCR cycles. In addition, the stability of the dNTPs during PCR was determined to be ≈50% after 60 cycles of PCR (Corey Levenson, Cetus Corporation; personal communication). Accordingly, the termination mixes were formulated to boost the dNTPs to a final concentration of ≈10 μM in the sequencing reaction, to supply specific ddNTPs at appropriate concentrations as determined above, and to provide

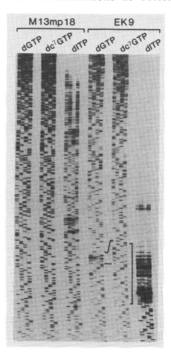

FIG. 4. Autoradiograph of a polyacrylamide/urea gel comparing extension products generated with base analogs. The effects of replacing dGTP with c7GTP (dc7GTP) or dITP are shown in sequencing reactions performed on M13mp18 single-stranded DNA or on a partially palindromic clone, EK9. Reaction conditions and electrophoresis were as described. Lanes are loaded G, A, T, C. Lines between the EK9 dGTP and c7GTP reaction sets align the same positions upstream and downstream of the compressed region. The bracket indicates the limits of the palindrome. The correct sequence of the region is 5'-CCAT<u>GTGACCCTGCCCGA-CTTCGACGGGAATTCCC-GTCGAAGTCGGGCAGGGT-CAC</u>CATA-3'. The complementary bases are underlined and the bases compressed in the dGTP reactions are boldface.

additional DNA polymerase. We used a ^{32}P-labeled sequencing primer to avoid purifying the PCR product and to simplify the sequencing protocol to a single extension/termination step. It is obvious that fluorescent-labeled sequencing primers could also be used, allowing the products to be analyzed on an automated DNA sequencing instrument.

The gel presented in Fig. 5 compares the DNA sequence obtained with Taq DNA polymerase using either an asymmetric PCR-generated template, or the same DNA insert cloned in M13mp18 as template. The resulting sequence ladders show the clarity and uniformity of signal characteristic of Taq-generated sequences. Any degradation of enzyme or dNTPs that may have occurred during the PCR thermal cycling did not seem to affect the generation of clean sequence data. Synthesis of single-stranded DNA template during 35 cycles of PCR was largely independent of the initial DNA concentration. Asymmetric PCRs performed with 0.1 to 100 ng of M13mp10 single-stranded DNA, or 10 μl of an M13 phage plaque picked directly into 100 μl of water, sequenced equivalently.

DISCUSSION

In this paper, we present convenient and efficient protocols for sequencing with Taq DNA polymerase. This enzyme worked equally well with either 5'-labeled primers or by incorporation of label in a two-step reaction protocol. Both approaches generated DNA sequencing ladders that were characteristically free of background bands or noticeable enzyme idiosyncrasies, were uniform in intensity, and were readable over long distances. These protocols also gave very clean results with alkali-denatured double-stranded DNA templates (data not shown).

Our results suggest that Taq DNA polymerase has advantages for many sequencing applications. Sequencing results obtained with the Taq enzyme were clearly superior to either Klenow or avian myeloblastosis virus reverse transcriptase and were often better (on G+C-rich templates) than results

FIG. 5.   Autoradiograph of a polyacrylamide/urea gel comparing the extension products from an M13-based single-stranded template (A) and an asymmetric PCR template of the same sequence (B). The sequencing of the M13 clone was carried out as described with a ^{32}P-labeled primer. The asymmetric amplification, DNA sequencing, and electrophoresis were performed as described. Reaction sets were loaded G, A, T, C.

obtained by using modified T7 DNA polymerase (data not shown). Unlike any of these polymerases, *Taq* DNA polymerase works over a broad temperature optimum centered around 75°C. Regions of DNA structure (hairpins) are commonly encountered that strongly hinder polymerases and cause premature termination bands across all four sequencing lanes. The ability of *Taq* DNA polymerase to operate at high temperature and low salt allows heat destabilization of hairpins during the sequencing reaction, permitting the enzyme to read through such structures. The concomitant use of a structure-destabilizing dGTP analog, c^{7}GTP, yields sequencing products from G+C-rich templates that are fully resolved upon electrophoresis.

We attribute the absence of background bands and the uniformity of signal to our observations that *Taq* DNA polymerase is highly processive, has a high turnover number, and has very little or no proofreading activity. Such properties of the enzyme are ideal for sequencing because they reduce pausing and premature termination at sequences with secondary structure and diminish discrimination against dideoxy nucleotide analogs (5).

Under certain circumstances, the absence of significant *Taq*-associated 3'-exonuclease activity causes chain-termination due to misincorporated bases. The misincorporation rate is enhanced when one or more of the dNTPs are well below K_m and/or when the concentration of one dNTP is very low relative to the other dNTPs. Because dITP base pairs promiscuously, we observed frequent chain termination near regions of high secondary structure with dITP and do not recommend it for sequencing with *Taq*. We do not observe misincorporation of bases if the concentration of all four dNTPs is similar and/or if they are present at ≥10 µM each. Sequence analysis of cloned PCR products generated with *Taq* DNA polymerase suggests that the fidelity of PCR using 50–200 µM each dNTP is quite respectable (approximately one mistake in 4000–5000 base pairs sequenced after 35 cycles of PCR and cloning of the products; unpublished results) and is comparable with that observed using other DNA polymerases for PCR. In addition, our data suggest that misincorporation errors that occur during the PCR promote chain termination (presumably because of significantly higher K_m values for mismatch extension), thus attenuating amplification of defective molecules and maintaining fidelity.

Several methods, with varying degrees of speed and reliability, have been published for sequencing PCR products (10–13). The remarkable sequencing properties demonstrated by *Taq* and its use in PCRs suggest it as the ideal enzyme for directly analyzing PCR products. Here, the protocols for sequencing with *Taq* were successfully used to sequence asymmetric PCR products without prior purification, and the results compared favorably with sequencing the same insert using M13 single-stranded DNA template.

While this approach has been developed for sequencing inserts in M13 or pUC-based vectors, it is applicable to direct sequencing of clones in λ phage and other cloning vectors. Some variability in the single-stranded DNA yield of the PCR has been observed with different primer pairs and ratios (13), and the reaction conditions for each amplification system will need to be adjusted for optimal results. Some investigators

have increased the homogeneity of their PCR products from genomic DNA by electrophoretic separation and reamplification of eluate from a selected gel slice (19). Our direct sequencing method is easily applied to this "secondary" PCR. Direct sequencing of PCR products from DNA by any method produces a "consensus" sequence; those bases that occur at a given position in the majority of the molecules will be the most visible on an autoradiograph and any low-frequency errors will be undetectable. In a coupled experiment of this kind, the resulting sequence data will be only as clean as the amplified product, and heterogeneous products will naturally produce mixed ladders.

The ability to couple template preparation by asymmetric PCR with direct sequencing by using the *Taq* enzyme opens the possibility of automating both DNA template preparation and the performance of the sequencing reactions in a manner that should be compatible with current DNA sequencing instruments.

We thank Susanne Stoffel for providing *Taq* DNA polymerase; Corey Levenson, Laurie Goda, and Dragan Spasic for preparation of synthetic oligonucleotide primers; Ulf Gyllensten for sharing data prior to publication; members of the Cetus PCR Group for their continued interest in this work; and Eric Ladner and Sharon Nilson for artwork.

1. Sanger, F., Nicklen, S. & Coulson, A. R. (1977) *Proc. Natl. Acad. Sci. USA* **74**, 5463–5467.
2. Yanisch-Perron, C., Vieira, J. & Messing, J. (1985) *Gene* **33**, 103–119.
3. Mills, D. R. & Kramer, F. R. (1979) *Proc. Natl. Acad. Sci. USA* **76**, 2232–2235.
4. Barr, P. J., Thayer, R. M., Laybourn, P., Najarian, R. C., Seela, F. & Tolan, D. R. (1986) *BioTechniques* **4**, 428–432.
5. Tabor, S. & Richardson, C. C. (1987) *Proc. Natl. Acad. Sci. USA* **84**, 4767–4771.
6. Smith, L. M., Sanders, J. Z., Kaiser, R. J., Hughes, P., Dodd, C., Connell, C. R., Heiner, C., Kent, S. B. H. & Hood, L. E. (1986) *Nature (London)* **321**, 674–679.
7. Prober, J. M., Trainor, G. L., Dam, R. J., Hobbs, F. W., Robertson, C. W., Zagursky, R. J., Cocuzza, A. J., Jensen, M. A. & Baumeister, K. (1987) *Science* **238**, 336–341.
8. Ansorge, W., Sproat, B., Stegemann, J., Schwager, C. & Zenke, M. (1987) *Nucleic Acids Res.* **15**, 4593–4602.
9. Saiki, R. K., Scharf, S., Faloona, F., Mullis, K. B., Horn, G. T., Erlich, H. A. & Arnheim, N. (1985) *Science* **230**, 1350–1354.
10. Engelke, D. R., Hoener, P. A. & Collins, F. S. (1988) *Proc. Natl. Acad. Sci. USA* **85**, 544–548.
11. Wong, C., Dowling, C. E., Saiki, R. K., Higuchi, R. G., Erlich, H. A. & Kazazian, H. H. (1987) *Nature (London)* **330**, 384–386.
12. Stoflet, E. S., Koeberl, D. D., Sarkar, G. & Sommer, S. S. (1988) *Science* **239**, 491–494.
13. Gyllensten, U. B. & Erlich, H. A. (1988) *Proc. Natl. Acad. Sci. USA* **85**, 7652–7656.
14. Saiki, R. K., Gelfand, D. H., Stoffel, S., Scharf, S. J., Higuchi, R., Horn, G. T., Mullis, K. B. & Erlich, H. A. (1988) *Science* **239**, 487–491.
15. Maxam, A. & Gilbert, W. (1980) *Methods Enzymol.* **65**, 499–560.
16. Zinder, N. D. & Boeke, J. D. (1982) *Gene* **19**, 1–10.
17. Biggin, M. D., Gibson, T. J. & Hong, G. F. (1983) *Proc. Natl. Acad. Sci. USA* **80**, 3963–3965.
18. Maniatis, T., Fritsch, E. F. & Sambrook, J. (1982) in *Molecular Cloning: A Laboratory Manual* (Cold Spring Harbor Lab., Cold Spring Harbor, NY).
19. Higuchi, R., von Beroldingen, C. H., Sensabaugh, G. F. & Erlich, H. A. (1988) *Nature (London)* **332**, 543–546.

Biochemical Method for Inserting New Genetic Information Into DNA of Simian Virus 40: Circular SV40 DNA Molecules Containing Lambda Phage Genes and the Galactose Operon of *Escherichia Coli*

D.A. Jackson, R.H. Symons and P. Berg

ABSTRACT We have developed methods for covalently joining duplex DNA molecules to one another and have used these techniques to construct circular dimers of SV40 DNA and to insert a DNA segment containing lambda phage genes and the galactose operon of *E. coli* into SV40 DNA. The method involves: (a) converting circular SV40 DNA to a linear form, (b) adding single-stranded homodeoxypolymeric extensions of defined composition and length to the 3′ ends of one of the DNA strands with the enzyme terminal deoxynucleotidyl transferase (c) adding complementary homodeoxypolymeric extensions to the other DNA strand, (d) annealing the two DNA molecules to form a circular duplex structure, and (e) filling the gaps and sealing nicks in this structure with *E. coli* DNA polymerase and DNA ligase to form a covalently closed-circular DNA molecule.

Our goal is to develop a method by which new, functionally defined segments of genetic information can be introduced into mammalian cells. It is known that the DNA of the transforming virus SV40 can enter into a stable, heritable, and presumably covalent association with the genomes of various mammalian cells (1, 2). Since purified SV40 DNA can also transform cells (although with reduced efficiency), it seemed possible that SV40 DNA molecules, into which a segment of functionally defined, nonviral DNA had been covalently integrated, could serve as vectors to transport and stabilize these nonviral DNA sequences in the cell genome. Accordingly, we have developed biochemical techniques that are generally applicable for joining covalently any two DNA molecules.‡ Using these techniques, we have constructed circular dimers of SV40 DNA; moreover, a DNA segment containing λ phage genes and the galactose operon of *Escherichia coli* has been covalently integrated into the circular SV40 DNA molecule. Such hybrid DNA molecules and others like them can be tested for their capacity to transduce foreign DNA sequences into mammalian cells, and can be used to determine whether these new nonviral genes can be expressed in a novel environment.

---

* Present address: Department of Microbiology, University of Michigan Medical Center, Ann Arbor, Mich. 48104.

† Present address: Department of Biochemistry, University of Adelaide, Adelaide, South Australia, 5001 Australia.

‡ Drs. Peter Lobban and A. D. Kaiser of this department have performed experiments similar to ours and have obtained similar results using bacteriophage P22 DNA (Lobban, P. and Kaiser, A. D., in preparation).

MATERIALS AND METHODS

DNA. (a) Covalently closed-circular duplex SV40 DNA [SV40(I)] (labeled with [³H]dT, 5×10^4 cpm/μg), free from SV40 linear or oligomeric molecules [but containing 3–5% of nicked double-stranded circles—SV40(II)] was purified from SV40-infected CV-1 cells (Jackson, D., & Berg, P., in preparation). (b) Closed-circular duplex λ*dvgal* DNA labeled with [³H]dT (2.5×10^4 cpm/μg), was isolated from an *E. coli* strain containing this DNA as an autonomously replicating plasmid (see ref. 3) by equilibrium sedimentation in CsCl–ethidium bromide gradients (4) after lysis of the cells with detergent. A more detailed characterization of this DNA will be published later. Present information indicates that the λ*dvgal* (λ*dv–120*) DNA is a circular dimer containing tandem duplications of a sequence of several λ phage genes (including C_I, O, and P) joined to the entire galactose operon of *E. coli* (Berg, D., Mertz, J., & Jackson, D., in preparation). DNA concentrations are given as molecular concentrations.

Enzymes. The circular SV40 and λ*dvgal* DNA molecules were cleaved with the bacterial restriction endonuclease RI (Yoshimori and Boyer, unpublished; the enzyme was generously made available to us by these workers). Phage λ-exonuclease (given to us by Peter Lobban) was prepared according to Little *et al.* (5), calf-thymus deoxynucleotidyl terminal transferase (terminal transferase), prepared according to Kato *et al.* (6), was generously sent to us by F. N. Hayes; *E. coli* DNA polymerase I Fraction VII (7) was a gift of Douglas Brutlag; and *E. coli* DNA ligase (8) and exonuclease III (9) were kindly supplied by Paul Modrich.

Substrates. [α-³²P]deoxynucleoside triphosphates (specific activities 5–10 Ci/μmol) were synthesized by the method of Symons (10). All other reagents were obtained from commercial sources.

Centrifugations. Alkaline sucrose gradients were formed by diffusion from equal volumes of 5, 10, 15, and 20% sucrose solutions with 2 mM EDTA containing, respectively, 0.2, 0.4, 0.6, and 0.8 M NaOH, and 0.8, 0.6, 0.4, 0.2 M NaCl. 100-μl samples were run on 3.8-ml gradients in a Beckman SW56 Ti rotor in a Beckman L2-65B ultracentrifuge at 4° and 55,000 rpm for the indicated times. 2- to 10-drop fractions were collected onto 2.5-cm diameter Whatman 3MM discs, dried without washing, and counted in PPO–dimethyl POPOP–toluene scintillator in a Nuclear Chicago Mark II

JACKSON, D.A., SYMONS, R.H. and BERG, P.
Biochemical method for inserting new genetic information into DNA of Simian Virus 40: circular SV40 DNA molecules containing lambda phage genes and the galactose operon of *Escherichia coli.*
Proc. Natl. Acad. Sci. U.S.A. 69:2904-2909 (1972). Reprinted with permission from the authors.

scintillation spectrometer. An overlap of 0.4% of ^{32}P into the ^{3}H channel was not corrected for.

CsCl–ethidium bromide equilibrium centrifugation was performed in a Beckman Type 50 rotor at 4° and 37,000 rpm for 48 hr. SV40 DNA in 10 mM Tris·HCl (pH 8.1)–1 mM Na EDTA–10 mM NaCl was adjusted to 1.566 g/ml of CsCl and 350 μg/ml of ethidium bromide. 30-Drop fractions were collected and aliquots were precipitated on Whatman GF/C filters with cold 2 N HCl; the filters were washed and counted.

Electron Microscopy. DNA was spread for electron microscopy by the aqueous method of Davis *et al.* (11) and photographed in a Phillips EM 300. Projections of the molecules were traced on paper and measured with a Keuffel and Esser map measurer. Plaque-purified SV40(II) DNA was used as an internal length standard.

Conversion of SV40(I) DNA to Unit Length Linear DNA [$SV40(L_{RI})$] *with* R_I *Endonuclease.* [^{3}H]SV40(I) DNA (18.7 nM) in 100 mM Tris·HCl buffer (pH 7.5)–10 mM MgCl$_2$–2 mM 2-mercaptoethanol was incubated for 30 min at 37° with an amount of R_I previously determined to convert 1.5 times this amount of SV40(I) to linear molecules [SV40(L_{RI})]; Na EDTA (30 mM) was added to stop the reaction, and the DNA was precipitated in 67% ethanol.

Removal of 5′-Terminal Regions from $SV40(L_{RI})$ *with* λ *Exonuclease.* [^{3}H]SV40(L_{RI}) (15 nM) in 67 mM K-glycinate (pH 9.5), 4 mM MgCl$_2$, 0.1 mM EDTA was incubated at 0° with λ-exonuclease (20 μg/ml) to yield [^{3}H]SV40(L_{RI}exo) DNA. Release of [^{3}H]dTMP was measured by chromatographing aliquots of the reaction on polyethyleneimine thin-layer sheets (Brinkmann) in 0.6 M NH$_4$HCO$_3$ and counting the dTMP spot and the origin (undegraded DNA).

Addition of Homopolymeric Extensions to $SV40(L_{RI}exo)$ *with Terminal Transferase.* [^{3}H]SV40(L_{RI}exo) (50 nM) in 100 mM K-cacodylate (pH 7.0), 8 mM MgCl$_2$, 2 mM 2-mercaptoethanol, 150 μg/ml of bovine serum albumin, [α-^{32}P]dNTP (0.2 mM for dATP, 0.4 mM for dTTP) was incubated with terminal transferase (30–60 μg/ml) at 37°. Addition of [^{32}P]dNMP residues to SV40 DNA was measured by spotting aliquots of the reaction mixture on DEAE-paper discs (Whatman DE-81), washing each disc by suction with 50 ml (each) of 0.3 M NH$_4$-formate (pH 7.8) and 0.25 M NH$_4$HCO$_3$, and then with 20 ml of ethanol. To determine the proportion of SV40 linear DNA molecules that had acquired at least one "functional" (dA)$_n$ tail, we measured the amount of SV40 DNA (^{3}H counts) that could be bound to a Whatman GF/C filter (2.4-cm diameter) to which 150 μg of polyuridylic acid had been fixed (13). 15-μl Aliquots of the reaction mixture were mixed with 5 ml of 0.70 M NaCl–0.07 M Na citrate (pH 7.0)–2% Sarkosyl, and filtered at room temperature through the poly(U) filters, at a flow rate of 3–5 ml/min. Each filter was washed by rapid suction with 50 ml of the same buffer at 0°, dried, and counted. Control experiments showed that 98–100% of [^{3}H]oligo(dA)$_{125}$ bound to the filters under these conditions. When the ratio of [^{32}P]dNMP to [^{3}H]DNA reached the value equivalent to the desired length of the extension, the reaction was stopped with EDTA (30 mM) and 2% Sarkosyl. The [^{3}H]SV40(L_{RI}exo)-[^{32}P]dA or -dT DNA was purified by neutral sucrose gradient zone sedimentation to remove unincorporated dNTP, as well as any traces of SV40(I) or SV40(II) DNA.

Formation of Hydrogen-Bonded Circular DNA Molecules. [^{32}P]dA and -dT DNAs were mixed at concentrations of 0.15 nM each in 0.1 M NaCl–10 mM Tris·HCl (pH 8.1)–1 mM EDTA. The mixture was kept at 51° for 30 min, then cooled slowly to room temperature.

Formation of Covalently Closed-Circular DNA Molecules. After annealing of the DNA, a mixture of the enzymes, substrates, and cofactors needed for closure was added to the DNA solution and the mixture was incubated at 20° for 3–5 hr. The final concentrations in the reaction mixture were: 20 mM Tris·HCl (pH 8.1), 1 mM EDTA, 6 mM MgCl$_2$, 50 μg/ml bovine-serum albumin, 10 mM NH$_4$Cl, 80 mM NaCl, 0.052 mM DPN, 0.08 mM (each), dATP, dGTP, dCTP, and dTTP, (0.4 μg/ml) *E. coli* DNA polymerase I, (15 units/ml) *E. coli* ligase, and (0.4 unit/ml) *E. coli* exonuclease III.

RESULTS

General approach

Fig. 1 outlines the general approach used to generate circular, covalently-closed DNA molecules from two separate DNAs. Since, in the present case, the units to be joined are themselves circular, the first step requires conversion of the circular structures to linear duplexes. This could be achieved by a double-strand scission at random locations (see *Discussion*) or, as we describe in this paper, at a unique site with R_I restriction endonuclease. Relatively short (50–100 nucleotides) poly(dA) or poly(dT) extensions are added on the 3′-hydroxyl termini of the linear duplexes with terminal transferase; prior

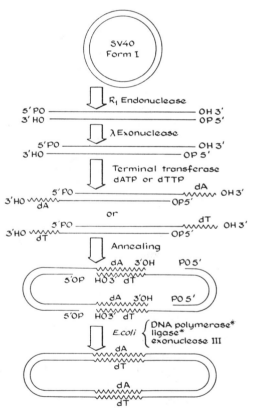

Fig. 1. General protocol for producing covalently closed SV40 dimer circles from SV40(I) DNA.

* The four deoxynucleoside triphosphates and DPN are also present for the DNA polymerase and ligase reactions, respectively.

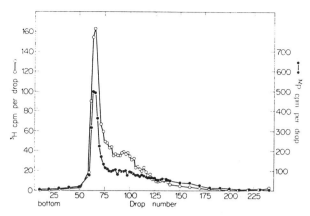

FIG. 2. Alkaline sucrose gradient sedimentation of [³H]SV40-(L_{RI}exo)–[³²P](dA)_{80} DNA. 0.16 μg of DNA was centrifuged for 6.0 hr.

removal of a short sequence (30–50 nucleotides) from the 5′-phosphoryl termini by digestion with λ exonuclease facilitates the terminal transferase reaction. Linear duplexes containing (dA)_n extensions are annealed to the DNA to be joined containing (dT)_n extensions at relatively low concentrations. The circular structure formed contains the two DNAs, held together by two hydrogen-bonded homopolymeric regions (Fig. 1). Repair of the four gaps is mediated by *E. coli* DNA polymerase with the four deoxynucleosidetriphosphates, and covalent closure of the ring structure is effected by *E. coli* DNA ligase; *E. coli* exonuclease III removes 3′-phosphoryl residues at any nicks inadvertently introduced during the manipulations (nicks with 3′-phosphoryl ends cannot be sealed by ligase).

Principal steps in the procedure

Circular SV40 DNA Can Be Opened to Linear Duplexes by R_I Endonuclease. Digestion of SV40(I) DNA with excess R_I endonuclease yields a product that sediments at 14.5 S in neutral sucrose gradients and appears as a linear duplex with the same contour length as SV40(II) DNA when examined by electron microscopy [(18); Jackson and Berg, in preparation; see Table 1]. The point of cleavage is at a unique site on the SV40 DNA, and few if any single-strand breaks are introduced elsewhere in the molecule (18); moreover, the termini at each end are 5′-phosphoryl, 3′-hydroxyl (Mertz, J., Davis, R., in preparation). Digestion of plaque-purified SV40 DNA under our conditions yields about 87% linear molecules, 10% nicked circles, and 3% residual supercoiled circles.

Addition of Oligo(dA) or -(dT) Extensions to the 3′-Hydroxyl Termini of SV40 (L_{RI}). Terminal transferase has been used to generate deoxyhomopolymeric extensions on the 3′-hydroxyl termini of DNA (7); once the chain is initiated, chain propagation is statistical in that each chain grows at about the same rate (12). Although the length of the extensions can be controlled by variation of either the time of incubation or the amount of substrate, we have varied the time of incubation to minimize spurious nicking of the DNA by trace amounts of endonuclease activity in the enzyme preparation; we have so far been unable to remove or selectively inhibit these nucleases (Jackson and Berg, in preparation).

Incubation of SV40(L_{RI}) with terminal transferase and either dATP or dTTP resulted in appreciable addition of mononucleotidyl units to the DNA. But, for example, after addition of 100 residues of dA per end, only a small proportion of the modified SV40 DNA would bind to filter discs containing poly(U) (13). This result indicated that initiation of terminal nucleotidyl addition was infrequent with SV40(L_{RI}), but that once initiated those termini served as preferential primers for extensive homopolymer synthesis.

Lobban and Kaiser (unpublished) found that P22 phage DNA became a better primer for homopolymer synthesis after incubation of the DNA with λ exonuclease. This enzyme removes, successively, deoxymononucleotides from 5′-phosphoryl termini of double-stranded DNA (15), thereby rendering the 3′-hydroxyl termini single-stranded. We confirmed their finding with SV40(L_{RI}) DNA; after removal of 30–50

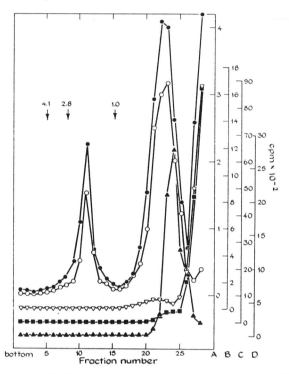

FIG. 3. Alkaline sucrose gradient sedimentation of [³H]-SV40(L_{RI}exo)–[³²P](dA)_{80} and –(dT)_{80} DNA incubated 4 hr with and without addition of *E. coli* DNA polymerase I (*P*), ligase (*L*), and exonuclease III (*III*). Conditions are described in *Methods*. 8-Drop fractions were collected. Samples *A*, *C*, and *D* were centrifuged for 60 min, sample *B* for 90 min. Line *A*, dA-ended, plus dT-ended SV40 linears, plus (P+L+III) (³²P, ●; ³H, ○); line *B*, dT-ended, SV40 omitted, plus (P+L+III) (³²P, ▼); line *C*, dA-ended SV40 omitted, plus (P+L+III) (³²P, ■); line *D*, dA-ended plus dT-ended SV40 linears, without (P+L+III) (³²P, ▲). ³H profiles are not shown for lines *B*, *C*, and *D*, but all show that the SV40 DNA sediments in its normal monomeric position. The ³²P and ³H profiles in line *A* are shifted to a faster-sedimenting position with respect to the ³²P profile in line *D* because SV40 strands are covalently linked to one another through (dA)_{80} or (dT)_{80} bridges in most of the molecules, whether or not covalently closed-circles are formed. Very little ³²P remains associated with the SV40 DNA in lines *B* and *C* because tails that remain single-stranded are degraded to 5′-mononucleotides by the 3′- to 5′-exonuclease activity of *E. coli* DNA polymerase I (7).

The *arrows* indicate the position in the gradient of different size supercoiled marker DNAs; the *number* is the multiple of SV40 DNA molecular size (1.0).

nucleotides per 5′-end (see *Methods*), the number of SV40(L_{RI}) molecules that could be bound to poly(U) filters after incubation with terminal transferase and dATP increased 5- to 6-fold. Even after separation of the strands of the SV40(L_{RI}exo)-dA, a substantial proportion of the [3H]-label in the DNA was still bound by the poly(U) filter, indicating that both 3′-hydroxy termini in the duplex DNA can serve as primers.

The weight-average length of the homopolymer extensions was 50–100 residues per end. Zone sedimentation of [3H]-SV40(L_{RI}exo)-[32P](dA)$_{80}$ (this particular preparation, which is described in *Methods*, had on the average, 80 dA residues per end) in an alkaline sucrose gradient showed that (*i*) 60–70% of the SV40 DNA strands are intact, (*ii*) the [32P](dA)$_{80}$ is covalently attached to the [3H]SV40 DNA, and (*iii*) the distribution of oligo(dA) chain lengths attached to the SV40 DNA is narrow, indicating that the deviation from the calculated mean length of 80 is small (Fig. 2). SV40(L_{RI}exo), having (dT)$_{80}$ extensions, was prepared with [32P]dTTP and gave essentially the same results when analyzed as described above.

Hydrogen-Bonded Circular Molecules Are Formed by Annealing SV40(L_{RI}exo)-(dA)$_{80}$ and SV40(L_{RI}exo)-(dT)$_{80}$ Together. When SV40(L_{RI}exo)-(dA)$_{80}$ and SV40(L_{RI}exo)-(dT)$_{80}$ were annealed together, 30–60% of the molecules seen by electron microscopy were circular dimers; linear monomers, linear dimers, and more complex branched forms were also seen. If SV40(L_{RI}exo)-(dA)$_{80}$ or -(dT)$_{80}$ alone was annealed, no circles were found. Centrifugation of annealed preparations in neutral sucrose gradients showed that the bulk of the SV40 DNA sedimented faster than modified unit-length linears (as would be expected for circular and linear dimers, as well as for higher oligomers). Sedimentation in alkaline gradients, however, showed only unit-length single strands containing the oligonucleotide tails (as seen in Fig. 2).

Covalently Closed-Circular DNA Molecules Are Formed by Incubation of Hydrogen-Bonded Complexes with DNA Polymerase, Ligase, and Exonuclease III. The hydrogen-bonded complexes described above can be sealed by incubation with the *E. coli* enzymes DNA polymerase I, ligase, and exonuclease III, plus their substrates and cofactors. Zone sedimentation in alkaline sucrose gradients (Fig. 3) shows that 20% of the

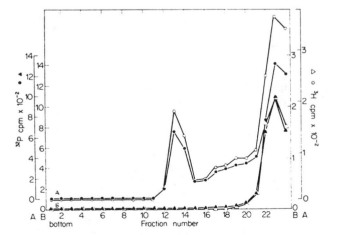

Fig. 4. CsCl–ethidium bromide equilibrium centrifugation of the products analyzed in Fig. 4. Line *A*, dA-ended, plus dT-ended SV40 linears, plus (P+L+III) (32P, ●; 3H, ○); line *B*, the same mixture without (P+L+III) (32P, ▲; 3H, △).

TABLE 1. *Relative lengths of SV40 and λdvgal-120 DNA molecules*

DNA species	Length ± standard deviation in SV40 units*	Number of molecules in sample
SV40(II)	1.00	224
SV40(L_{RI})†	1.00 ± 0.03	108
(SV40–dA dT)$_2$	2.06 ± 0.19	23
λdvgal–120(I)	4.09 ± 0.14	65
λdvgal–120(L_{RI})	2.00 ± 0.04	163
λdvgal–SV40	2.95 ± 0.04	76
λdv-1	2.78 ± 0.05	13

* The contour length of plaque-purified SV40(II) DNA is defined as 1.00 unit.

† Data supplied by J. Morrow.

input 32P label derived from the oligo(dA) and -(dT) tails sediments with the 3H label present in the SV40 DNA, in the position expected of a covalently closed-circular SV40 dimer (70–75 S). About the same amount of labeled DNA bands in a CsCl–ethidium bromide gradient at a buoyant density characteristic of covalently closed-circular DNA (Fig. 4).

DNA isolated from the heavy band of the CsCl–ethidium bromide gradient contains primarily circular molecules, with a contour length twice that of SV40(II) DNA (Table 1) when viewed by electron microscopy. No covalently closed DNA is formed if either one of the linear precursors is omitted from the annealing step or if the enzymes are left out of the closure reaction. We conclude, therefore, that two unit-length linear SV40 molecules have been joined to form a covalently closed-circular dimer.

Covalent closure of the hydrogen-bonded SV40 DNA dimers is dependent on Mg^{2+}, all four deoxynucleoside triphosphates, *E. coli* DNA polymerase I, and ligase, and is inhibited by 98% if exonuclease III is omitted (Lobban and Kaiser first observed the need for exonuclease III in the joining of P22 molecules; we confirmed their finding with this system). Exonuclease III is probably needed to remove 3′-phosphate groups from 3′-phosphoryl, 5′-hydroxyl nicks introduced by the endonuclease contaminating the terminal transferase preparation. 3′-phosphoryl groups are potent inhibitors of *E. coli* DNA polymerase I (14) and termini having 5′-hydroxyl groups cannot be sealed by *E. coli* ligase (8). The 5′-hydroxyl group can be removed and replaced by a 5′-phosphoryl group by the 5′- to 3′-exonuclease activity of *E. coli* DNA polymerase I (7).

Preparation of the Galactose Operon for Insertion into SV40 DNA. The galactose operon of *E. coli* was obtained from a λ*dvgal* DNA; λ*dvgal* is a covalently closed, supercoiled DNA molecule four times as long as SV40(II) DNA (Table 1). After complete digestion of λ*dvgal* DNA with the R_I endonuclease, linear molecules two times the length of SV40(II) DNA are virtually the exclusive product (Table 1). This population has a unimodal length distribution by electron microscopy and appears to be homogeneous by ultracentrifugal criteria (Jackson and Berg, in preparation). The R_I endonuclease seems, therefore, to cut λ*dvgal* circular DNA into two equal length linear molecules. Since one R_I endonuclease cleavage per λ*dv* monomeric unit occurs in the closely related λ*dv-204* (Jackson and Berg, in preparation), it is likely that λ*dvgal* is cleaved at the

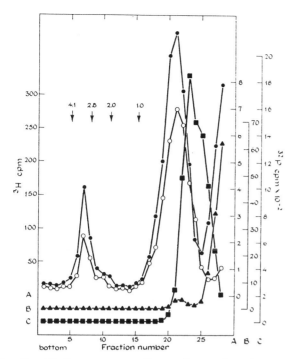

FIG. 5. Alkaline sucrose gradient sedimentation of annealed [³H]SV40(L_RIexo)-[³²P](dA)_80 and [³H]λdvgal-120) (L_RIexo)-[³²P](dT)_80 incubated for 3 hr with and without (P+L+III). Centrifugation was for 60 min. Line A, dA-ended SV40, plus dT-ended λdvgal-120 linears, plus (P+L+III)(³²P, ●; ³H, ○); line B, dT-ended λdvgal-120 linears, plus dT-ended SV40 linears, plus (P+L+III) (³²P, ▲); line C, dA-ended SV40 linears, plus dT-ended λdvgal-120 linears, without (P+L+III) (³²P, ■).

The *arrows* indicate the position in the gradient of supercoiled marker DNAs having the indicated multiple of SV40 DNA molecular size.

same sites and, therefore, that each linear piece contains an intact galactose operon.

The purified λdvgal (L_RI) DNA was prepared for joining to SV40 DNA by treatment with λ-exonuclease, followed by terminal transferase and [³²P]dTTP, as described for SV40-(L_RI).

Formation of Covalently Closed-Circular DNA Molecules Containing both SV40 and λdvgal DNA. Annealing of [³H]-SV40(L_RIexo)-[³²P](dA)_80 with [³H]λdvgal(L_RIexo)-[³²P]-(dT)_80, followed by incubation with the enzymes, substrates, and cofactors needed for closure, produced a species of DNA (in about 15% yield) that sedimented rapidly in alkaline sucrose gradients (Fig. 5) and that formed a band in a CsCl-ethidium bromide gradient at the position expected for covalently closed DNA (Fig. 6). The putative λdvgal–SV40 circular DNA sediments just ahead of λdv-1, a supercoiled circular DNA marker [2.8 times the length of SV40(II)DNA], and behind λdvgal supercoiled circles [4.1 times SV40(II)DNA] in the alkaline sucrose gradient. Electron microscopic measurements of the DNA recovered from the dense band of the CsCl-ethidium bromide gradient showed a mean cóntour length for the major species of 2.95 ± 0.04 times that of SV40(II) DNA (Table 1). Each of these measurements supports the conclusion that the newly formed, covalently closed-circular DNA contains one SV40 DNA segment and one λdvgal DNA monomeric segment.

Omission of the enzymes from the reaction mixture prevents λdvgal–SV40 DNA formation (Figs. 5 and 6). No covalently closed product is detectable (Fig. 5) if λdvgal and SV40 linear molecules with identical, rather than complementary, tails are annealed and incubated with the enzymes. This result demonstrates directly that the formation of covalently closed DNA depends on complementarity of the homopolymeric tails.

We conclude from the experiments described above that λdvgal DNA containing the intact galactose operon from *E. coli*, together with some phage λ genes, has been covalently inserted into an SV40 genome. These molecules should be useful for testing whether these bacterial genes can be introduced into a mammalian cell genome and whether they can be expressed there.

DISCUSSION

The methods described in this report for the covalent joining of two SV40 molecules and for the insertion of a segment of DNA containing the galactose operon of *E. coli* into SV40 are general and offer an approach for covalently joining any two DNA molecules together. With the exception of the fortuitous property of the R_I endonuclease, which creates convenient linear DNA precursors, none of the techniques used depends upon any unique property of SV40 and/or the λdvgal DNA. By the use of known enzymes and only minor modifications of the methods described here, it should be possible to join DNA molecules even if they have the wrong combination of hydroxyl and phosphoryl groups at their termini. By judicious use of generally available enzymes, even DNA duplexes with protruding 5'- or 3'-ends can be modified to become suitable substrates for the joining reaction.

One important feature of this method, which is different from some other techniques that can be used to join unrelated DNA molecules to one another (16, 19), is that here the joining is directed by the homopolymeric tails on the DNA. In our protocol, molecule A and molecule B can only be joined to each other; all AA and BB intermolecular joinings and all A and B intramolecular joinings (circularizations) are prevented. The yield of the desired product is thus increased, and subsequent purification problems are greatly reduced.

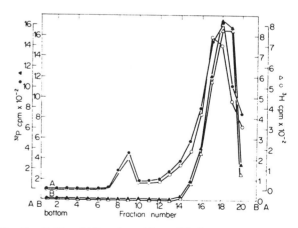

FIG. 6. CsCl-ethidium bromide equilibrium centrifugations of joined [³H]SV40(L_RIexo)-[³²P](dA)_80 and [³H]λdvgal-120(L_RIexo)-[³²P](dT)_80 DNA. The samples were those referred to in Fig. 5. Line A, dA-ended SV40 linears, plus dT-ended λdvgal-120 linears, plus (P+L+III) (³²P, ●; ³H, ○); line B, the same mixture without (P+L+III) (³²P, ▲; ³H, △).

For some purposes, however, it may be desirable to insert λ*dvgal* or other DNA molecules at other specific, or even random, locations in the SV40 genome. Other specific placements could be accomplished if other endonucleases could be found that cleave the SV40 circular DNA specifically. Since pancreatic DNase in the presence of Mn^{2+} produces randomly located, double-strand scissions (17) of SV40 circular DNA (Jackson and Berg, in preparation), it should be possible to insert a DNA segment at a large number of positions in the SV40 genome.

Although the λ*dvgal* DNA segment is integrated at the same location in each SV40 DNA molecule, it should be emphasized that the orientation of the two DNA segments to each other is probably not identical. This follows from the fact that each of the two strands of a duplex can be joined to *either* of the two strands of the other duplex $\left(\text{e.g.,}\ \begin{smallmatrix}W\frown W\\C\smile C\end{smallmatrix}\ \text{or}\ \begin{smallmatrix}W\frown C\\C\smile W\end{smallmatrix}\right)^{\S}$. What possible consequences this fact has on the genetic expression of these segments remains to be seen.

We have no information concerning the biological activities of the SV40 dimer or the λ*dvgal*–SV40 DNAs, but appropriate experiments are in progress. It is clear, however, that the location of the R_I break in the SV40 genome will be crucial in determining the biological potential of these molecules; preliminary evidence suggests that the break occurs in the late genes of SV40 (Morrow, Kelly, Berg, and Lewis, in preparation).

A further feature of these molecules that may bear on their usefulness is the $(dA \cdot dT)_n$ tracts that join the two DNA segments. They could be helpful (as a physical or genetic marker) or a hindrance (by making the molecule more sensitive to degradation) for their potential use as a transducer.

The λ*dvgal*–SV40 DNA produced in these experiments is, in effect, a trivalent biological reagent. It contains the genetic information to code for most of the functions of SV40, all of the functions of the *E. coli* galactose operon, and those functions of the λ bacteriophage required for autonomous replication of circular DNA molecules in *E. coli*. Each of these

§ The symbols W and C refer to one or the other complementary strands of a DNA duplex, and the "connectors" indicate how the strands can be joined in the closed-circular duplex.

sets of functions has a wide range of potential uses in studying the molecular biology of SV40 and the mammalian cells with which this virus interacts.

We are grateful to Peter Lobban for many helpful discussions. D. A. J. was a Basic Science Fellow of the National Cystic Fibrosis Research Foundation; R. H. S. was on study leave from the Department of Biochemistry, University of Adelaide, Australia and was supported in part by a grant from the USPHS. This research was supported by Grant GM-13235 from the USPHS and Grant VC-23A from the American Cancer Society.

1. Sambrook, J., Westphal, H., Srinivasan, P. R. & Dulbecco, R. (1968) *Proc. Nat. Acad. Sci. USA* 60, 1288-1295.
2. Dulbecco, R. (1969) *Science* 166, 962-968.
3. Matsubara, K., & Kaiser, A. D. (1968) *Cold Spring Harbor Symp. Quant. Biol.* 33, 27-34.
4. Radloff, R., Bauer, W., Vinograd, J. (1967) *Proc. Nat. Acad. Sci. USA* 57, 1514-1521.
5. Little, J. W., Lehman, I. R. & Kaiser, A. D. (1967) *J. Biol. Chem.* 242, 672-678.
6. Kato, K., Goncalves, J. M., Houts, G. E., & Bollum, F. J. (1967) *J. Biol. Chem.* 242, 2780-2789.
7. Jovin, T. M., Englund, P. T. & Kornberg, A. (1969) *J. Biol. Chem.* 244, 2996-3008.
8. Olivera, B. M., Hall, Z. W., Anraku, Y., Chien, J. R. & Lehman, I. R. (1968) *Cold Spring Harbor Symp. Quant. Biol.* 33, 27-34.
9. Richardson, C. C., Lehman, I. R. & Kornberg, A. (1964) *J. Biol Chem.* 239, 251-258.
10. Symons, R. H. (1969) *Biochim. Biophys. Acta* 190, 548-550.
11. Davis, R., Simon, M. & Davidson, N. (1971) in *Methods in Enzymology*, eds. Grossman, L. & Moldave, K. (Academic Press, New York), Vol. 21, pp. 413-428.
12. Chang, L. M. S. & Bollum, F. J. (1971) *Biochemistry* 10, 536-542.
13. Sheldon, R., Jurale, C. & Kates, J. (1972) *Proc. Nat. Acad. Sci. USA* 69, 417-421.
14. Richardson, C. C., Schildkraut, C. L. & Kornberg, A. (1963) *Cold Spring Harbor Symp. Quant. Biol.* 28, 9-19.
15. Little, J. W. (1967) *J. Biol. Chem.* 242, 679-686.
16. Sgaramella, V., van de Sande, J. H. & Khorana, H. G. (1970) *Proc. Nat. Acad. Sci. USA* 67, 1468-1475.
17. Melgar, E. & Goldthwait, D. A. (1968) *J. Biol. Chem.* 243, 4409-4416.
18. Morrow, J. F. & Berg, P. (1972) *Proc. Nat. Acad. Sci. USA* 69, in press.
19. Sgaramella, V. & Lobban, P. (1972) *Nature*, in press.

Specific Enzymatic Amplification of DNA
In Vitro: The Polymerase Chain Reaction

K. Mullis, F. Faloona, S. Scharf, R. Saiki, G. Horn and H. Erlich

The discovery of specific restriction endonucleases (Smith and Wilcox 1970) made possible the isolation of discrete molecular fragments of naturally occurring DNA for the first time. This capability was crucial to the development of molecular cloning (Cohen et al. 1973); and the combination of molecular cloning and endonuclease restriction allowed the synthesis and isolation of any naturally occurring DNA sequence that could be cloned into a useful vector and, on the basis of flanking restriction sites, excised from it. The availability of a large variety of restriction enzymes (Roberts 1985) has significantly extended the utility of these methods.

The de novo organic synthesis of oligonucleotides and the development of methods for their assembly into long double-stranded DNA molecules (Davies and Gassen 1983) have removed, at least theoretically, the minor limitations imposed by the availability of natural sequences with fortuitously unique flanking restriction sites. However, de novo synthesis, even with automated equipment, is not easy; it is often fraught with peril due to the inevitable indelicacy of chemical reagents (Urdea et al. 1985; Watt et al. 1985; Mullenbach et al. 1986), and it is not capable of producing, intentionally, a sequence that is not yet fully known.

We have been exploring an alternative method for the synthesis of specific DNA sequences (Fig. 1). It involves the reciprocal interaction of two oligonucleotides and the DNA polymerase extension products whose synthesis they prime, when they are hybridized to different strands of a DNA template in a relative orientation such that their extension products overlap. The method consists of repetitive cycles of denaturation, hybridization, and polymerase extension and seems not a little boring until the realization occurs that this procedure is catalyzing a doubling with each cycle in the amount of the fragment defined by the positions of the 5' ends of the two primers on the template DNA, that this fragment is therefore increasing in concentration exponentially, and that the process can be continued for many cycles and is inherently very specific.

The original template DNA molecule could have been a relatively small amount of the sequence to be synthesized (in a pure form and as a discrete molecule) or it could have been the same sequence embedded in a much larger molecule in a complex mixture as in the case of a fragment of a single-copy gene in whole human DNA. It could also have been a single-stranded DNA molecule or, with a minor modification in the technique, it could have been an RNA molecule. In any case, the product of the reaction will be a discrete double-stranded DNA molecule with termini corresponding to the 5' ends of the oligonucleotides employed.

We have called this process polymerase chain reaction or (inevitably) PCR. Several embodiments have been devised that enable one not only to extract a specific sequence from a complex template and amplify it, but also to increase the inherent specificity of this process by using nested primer sets, or to append sequence information to one or both ends of the sequence as it is being amplified, or to construct a sequence entirely from synthetic fragments.

MATERIALS AND METHODS

PCR amplification from genomic DNA. Human DNA (1 μg) was dissolved in 100 μl of a polymerase buffer containing 50 mM NaCl, 10 mM Tris-Cl (pH 7.6), and 10 mM $MgCl_2$. The reaction mixture was adjusted to 1.5 mM in each of the four deoxynucleoside triphosphates and 1 μM in each of two oligonucleotide primers. A single cycle of the polymerase chain reaction was performed by heating the reaction to 95°C for 2 minutes, cooling to 30°C for 2 minutes, and adding 1 unit of the Klenow fragment of *Escherichia coli* DNA polymerase I in 2 μl of the buffer described above containing about 0.1 μl of glycerol (Klenow was obtained from U.S. Biochemicals in a 50% glycerol solution containing 5 U/μl). The extension reaction was allowed to proceed for 2 minutes at 30°C. The cycle was terminated and a new cycle was initiated by returning the reaction to 95°C for 2 minutes. In the amplifications of human DNA reported here, the number of cycles performed ranged from 20 to 27.

Genotype analysis of PCR-amplified genomic DNA using ASO probes. DNA (1 μg) from various cell lines was subjected to 25 cycles of PCR amplification. Aliquots representing one thirtieth of the amplification mixture (33 ng of initial DNA) were made 0.4 N in NaOH, 25 mM in EDTA in a volume of 200 μl and applied to a Genatran-45 nylon filter with a Bio-Dot spotting apparatus. Three replicate filters were prepared. ASO probes (Table 1) were 5'-phosphorylated with [λ-^{32}P]ATP and polynucleotide kinase and purified by spin dialysis. The specific activities of the probes were between 3.5 and 4.5 μCi/pmole. Each filter

MULLIS, K., FALOONA, F., SCHARF, S., SAIKI, R., HORN, G. and ERLICH. H.
Specific enzymatic amplification of DNA *in vitro*: the polymerase chain reaction. Reprinted by permission from *CSH Symp. Quant. Biol.* 51:263-273 (1986), Cold Spring Harbor Lab.

was prehybridized individually in 8 ml of 5× SSPE, 5× DET, and 0.5% SDS for 30 minutes at 55°C. The probe (1 pmole) was then added, and hybridization was continued at 55°C for 1 hour. The filters were rinsed twice in 2× SSPE and 0.1% SDS at room temperature, followed by a high stringency wash in 5× SSPE and 0.1% SDS for 10 minutes at 55°C (for 19C) or 60°C (for 19A and 19S) and autoradiographed for 2.5 hours at −80°C with a single intensification screen.

Cloning from PCR-amplified genomic DNA using linker primers.

An entire PCR reaction was digested at 37°C with PstI (20 units) and HindIII (20 units) for 90 minutes (for β-globin) or BamHI (24 units) and PstI (20 units) for 60 minutes (for HLA-DQα). After phenol extraction, the DNA was dialyzed to remove low-molecular-weight inhibitors of ligation (presumably the dNTPs used in PCR) and concentrated by ethanol precipitation. All (β-globin) or one tenth (DQα) of the material was ligated to 0.5 μg of the cut M13 vector under standard conditions and transformed into approximately 6 × 10⁹ freshly prepared competent JM103 cells in a total volume of 200 μl. These cells (10–30 μl) were mixed with 150 μl of JM103 culture, plated on IPTG/X-Gal agar plates, and incubated overnight. The plates were scored for blue (parental) plaques and lifted onto BioDyne A filters. These filters were hybridized either with one of the labeled PCR oligonucleotide primers to visualize all of the clones containing PCR-amplified DNA (primer plaques) or with a β-globin oligonucleotide probe (RS06) or an HLA-DQα cDNA probe to visualize specifically the clones containing target sequences. Ten β-globin clones from this latter category were sequenced by using the dideoxy extension method. Nine were identical to the expected β-globin target sequence, and one was identical to the homologous region of the human δ-globin gene.

PCR construction of a 374-bp DNA fragment from synthetic oligodeoxyribonucleotides.

100 pmoles of TN10, d(CCTCGTCTACTCCCAGGTCCTCTTCAA-GGGCCAAGGCTGCCCCGACTATGTGCTCCTCAC-CCACACCGTCAGCC), and TN11, d(GGCAGGGGC-TCTTGACGGCAGAGAGGAGGTTGACCTTCTCCT-GGTAGGAGATGGCGAAGCGGCTGACGGTGTGG), designed so as to overlap by 14 complementary bases on their 3′ ends, were dissolved in 100 μl of buffer containing 30 mM Tris-acetate (pH 7.9), 60 mM sodium acetate, 10 mM magnesium acetate, 2.5 mM dithiothreitol, and 2 mM each dNTP. The solution was heated to 100°C for 1 minute and cooled in air at about 23°C for 1 minute; 1 μl containing 5 units Klenow fragment of E. coli DNA polymerase I was added, and the polymerization reaction was allowed to proceed for 2 minutes. Gel electrophoresis on 4% NuSieve agarose in the presence of 0.5 μg/ml ethidium bromide indicated that eight repetitions of this procedure were required before the mutual extension of the two primers on each other was complete (Fig. 5, lane I). A 2-μl aliquot of this reaction without purification was added to 100 μl of a second-stage reaction mixture identical to the one above except that 300 pmoles each of oligonucleotides LL09, d(CCTGGCCAATGGCATGGATCTGAAAGATAACC-AGCTGGTGGTGCCAGCAGATGGCCTGTACCTCG-TCTACTCCC), and LL12, d(CTCCCTGATAGATGG-GCTCATACCAGGGCTTGAGCTCAGCCCCCTCTG-GGGTGTCCTTCGGGCAGGGGCTCTTG), were substituted for TN10 and TN11. LL09 and LL12 were designed so that their 3′ ends would overlap with 14 complementary bases on the 3′ ends of the single-stranded fragments released when the 135-bp product of the previous reaction was denatured (see Fig. 4, no. 2). The cycle of heating, cooling, and adding Klenow fragment was repeated 15 times in order to produce the 254-bp fragment in lane II of Figure 5; 2 μl of this reaction mixture without purification was diluted into a third-stage reaction mixture as above but containing 300 pmoles of TN09, d(TGTAGCAAACCATCAAGTT-GAGGAGCAGCTCGAGTGGCTGAGCCAGCGGGC-CAATGCCCTCCTGGCCAATGGCA), and TN13, d(GATACTTGGGCAGATTGACCTCAGCGCTGAGT-TGGTCACCCTTCTCCAGCTGGAAGACCCCTCCC-TGATAGATG). After 15 cycles of PCR, the 374-bp product in lane III of Figure 5 was evident on gel electrophoresis. After gel purification, restriction analysis of the 374-bp fragment with several enzymes resulted in the expected fragments (data not shown).

PCR amplification with oligonucleotide linker primers.

DNA was amplified by mixing 1 μg of genomic DNA in the buffer described above with 100 pmoles of each primer. Samples were subjected to 20 cycles of PCR, each consisting of 2 minutes of denaturation at 95°C, 2 minutes of cooling to 37°C, and 2 minutes of polymerization with 1 unit of Klenow DNA polymerase. After amplification, the DNA was concentrated by ethanol precipitation, and half of the total reaction was electrophoresed on a gel of 4% NuSieve agarose in TBE buffer. The ethidium-bromide-stained gel was photographed (see Fig. 6A), and the DNA was transferred to Genatran nylon membrane and hybridized to a labeled probe (RSO6) specific for the target sequence (Fig. 6B). The blot was then washed and autoradiographed. For the amplification of β-globin, the starting DNA was either from the Molt-4 cell line or from the globin deletion mutant GM2064 For lane 5, the starting material was 11 pg of the β-globin recombinant plasmid pBR328::βᴬ, the molar equivalent of 5 μg of genomic DNA. For lane 6, the reaction was performed as in lane 1 except that no enzyme was added. To increase the efficiency of amplification of the longer HLA-DQα segment, DMSO was added to 10% (v/v), and the polymerization was carried out at 37°C for 27 cycles. The starting DNA for the amplification of DQα was either from the consanguineous HLA typing cell line LG-2 or from the HLA class II deletion mutant LCL721.180.

DISCUSSION

Extraction and Amplification

Figure 1 describes the basic PCR process that results in the extraction and amplification of a nucleic acid sequence. "Extraction" is used here in the sense that the sequence, although contained within a larger molecule that may in fact be a heterogeneous population of larger molecules, as in the case of a chromosomal DNA preparation, will be amplified as a single discrete molecular entity, and thus extracted from its source. This feature of the chain reaction can be, in some cases, as important as the amplification itself. The source DNA is denatured and allowed to hybridize to an excess of primers that correspond to the extremities of the fragment to be amplified (Fig. 1A). The oligonucleotide primers are employed in micromolar concentrations, and thus the hybridization is rapid and complete but not particularly stringent. A DNA polymerase is added to the reaction, which already contains the four deoxynucleoside triphosphates, and the primers are extended (Fig. 1B,C). After a short time, the reaction is stopped, and the DNA is denatured again by heating. On cooling, the excess of primers again hybridizes rapidly, and this time there are twice as many sites for hybridization on sequences representing the target. Sequences not representing the target, which may have been copied in the first polymerization reaction by virtue of an adventitious interaction with one of the primers, will only very rarely have generated an additional site for either of the primers. The unique property of the targeted sequence for regenerating new primer sites with each cycle is intrinsic to the support of a chain reaction, and the improbability of its happening by chance accounts for the observed specificity of the overall amplification.

In Figure 1D, the discrete nature of the final product becomes evident. Whereas initial extension of the primers on templates with indefinite termini results in products with definite 5' ends but indefinite 3' ends (Fig. 1C), their extension on a template that is itself an extension product from a previous cycle results in a DNA strand that has both ends defined (Fig. 1D). It is the number of these DNA fragments, with both ends defined by primers, that increases exponentially during subsequent cycles. All other products of the reaction increase in a linear fashion. Thus, the reaction effectively amplifies only that DNA sequence that has been targeted by the primers.

Nested Primer Sets

Despite its intrinsic specificity, the amplification of fragments of single-copy genes from whole human DNA sometimes produces a molecule exclusively representative of the intended target, and sometimes it does not. Given the complexity of the human genome and the tendency for many genes to have some sequences in common with other related genes, this lack

of complete specificity is not surprising. In many applications, extreme specificity is not required of the amplification process because a second level of specificity is invoked by the process of detecting the amplified products, e.g., the use of a labeled hybridization probe to detect a PCR-amplified fragment in a Southern blot. However, in other applications, such as attempting to visualize an amplified fragment of the β-globin gene by the ethidium bromide staining of an agarose gel, further specificity than that obtained by a simple amplification protocol was required. We achieved this by doing the amplification in two stages (Fig. 2). The first stage amplified a 110-bp fragment; the second stage employed two oligomers that primed within the sequence of this fragment to produce a subfragment of 58 bp (Table 1). By thus employing the specificity inherent in the requirement of four independent but coordinated priming events, we were able to amplify exclusively a fragment of the human β-globin gene approximately 2,000,000-fold (Mullis and Faloona 1986).

Addition and Amplification

Figure 3 depicts a PCR in which one of the oligonucleotides employed, primer B, has a 5' sequence that is not homologous to the target sequence. During cycle 1 of the amplification, priming occurs on the basis of the homologous 3' end of the oligomer, but the nonhomologous bases become a part of the extension product. In later cycles, when this extension product is copied by virtue of the fact that it contains a site for primer A, the entire complement to primer B is incorporated in this copy; thereafter, no further primer-template interactions involving nonhomologous bases are required, and the product that increases exponentially in the amplification contains sequence information on one end that was not present in the original target.

A similar situation results if both primers carry an additional 5' sequence into the reaction, except that now both ends of the product will have been appended in the course of amplification. We have employed this embodiment of the PCR to insert restriction site linkers onto amplified fragments of human genomic sequences to facilitate their cloning; it also underlies the strategy described below for construction of DNA from oligonucleotides alone.

If one or both primers are designed so as to include an internal mismatch with the target sequence, specific in vitro mutations can be accomplished. As in the case above, the amplified product will contain the sequence of the primers rather than that of the target.

Construction and Amplification

The process whereby extrinsic sequences are appended to a fragment during the PCR amplification can be utilized in the stepwise construction of a totally synthetic sequence from oligonucleotides. The advan-

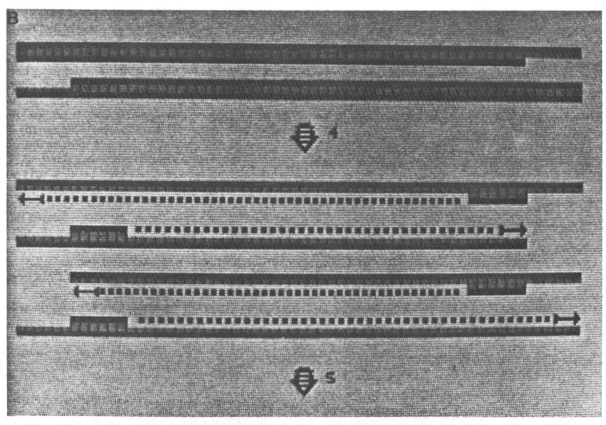

Figure 1. Three complete cycles of the polymerase chain reaction resulting in the eightfold amplification of a template sequence defined by the 5′ ends of two primers hybridized to different strands of the template are depicted above. The first cycle is shown in detail as reactions 1, 2, and 3. The second cycle is depicted in less detail as reactions 4 and 5, and the third cycle, as reactions 6 and 7. By cycle 3, a double-stranded DNA fragment is produced that has discrete termini, and thus the targeted sequence has been extracted from its source as well as amplified.

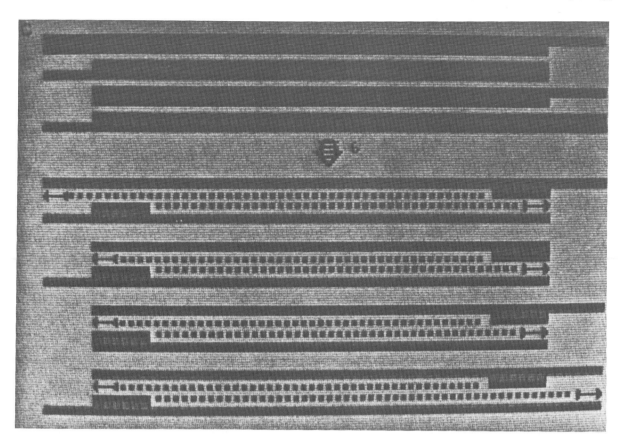

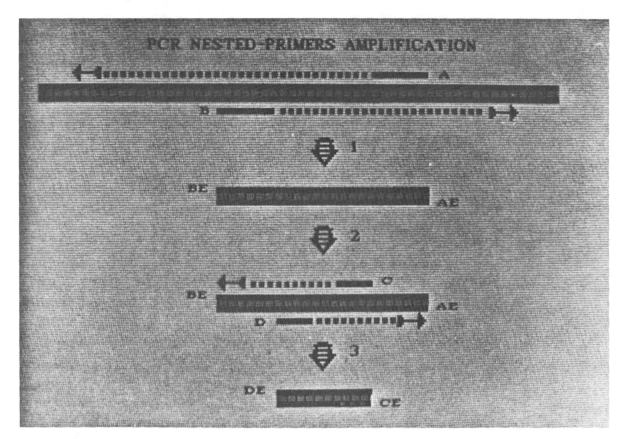

Figure 2. A PCR amplification employing oligonucleotides A and B results after a number of cycles in the fragment depicted as AE:BE. A subfragment, CE:DE, of this product can be extracted and further amplified by use of oligonucleotides C and D. This overall process requires four independent and coordinated primer-template interactions and is thereby potentially more specific than a single-stage PCR amplification.

tages of this approach (Figs. 4 and 5) over methods currently in use are that the PCR method does not require phosphorylation or ligation of the oligonucleotides. Like the method employed by Rossi et al. (1982), which involves the mutual extension of pairs of oligonucleotides on each other by polymerase (Fig. 4, no. 1), the PCR method does not require organic synthesis of both strands of the final product. Unlike this method,

however, the PCR method is completely general in that no particular restriction enzyme recognition sequences need be built into the product for purposes of accomplishing the synthesis. Furthermore, the PCR method offers the convenience of enabling the final product, or any of the several intermediates, to be amplified during the synthesis or afterwards to produce whatever amounts of these molecules are required.

Table 1. Oligodeoxyribonucleotides

	Sequence	Use
PC03	ACACAACTGTGTTCACTAGC	produce a 110-bp fragment
PC04	CAACTTCATCCACGTTCACC	from β-globin 1st exon
PC07	CAGACACCATGGTGCACCTGACTCCTG	produce a 58-bp subfragment
PC08	CCCCACAGGGCAGTAACGGCAGACTTCTCC	of the 110-bp fragment above
GH18	CTTCTGCAGCAACTGTGTTCACTAGC	as PC03 and PC04, but with *Pst*I
GH19	CACAAGCTTCATCCACGTTCACC	and *Hin*dIII linkers added
GH26	GTGCTGCAGGTGTAAACTTGTACCAG	242-bp from HLA-DQα
GH27	CACGGATCCGGTAGCAGCGGTAGAGTTG	with *Pst*I and *Bam*HI linkers
RS06	CTGACTCCTGAGGAGAAGTCTGCCGTT ACTGCCCTGTGGG	probe to central region of 110-bp fragment produced by PC03 and PC04
19A	CTCCTGAGGAGAAGTCTGC	ASO probe to β^A-globin
19S	CTCCTGTGGAGAAGTCTGC	ASO probe to β^S-globin
19C	CTCCTAAGGAGAAGTCTGC	ASO probe to β^C-globin

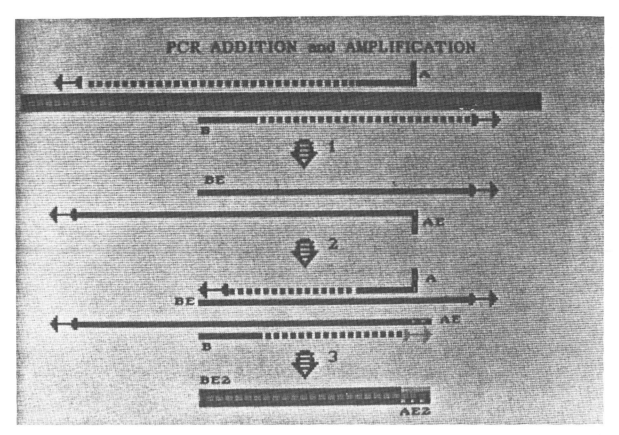

Figure 3. Addition of the nonhomologous sequence on the 5′ end of oligonucleotide A results, after the first PCR cycle, in the incorporation of this nonhomologous sequence into extension product AE, which, in a later cycle as template for B, directs the synthesis of BE2 that together with AE2 comprises the double-stranded DNA fragment AE2:BE2. This fragment will accumulate exponentially with further cycles.

Genetic Analysis

In addition to its use as an in vitro method for enzymatically synthesizing a specific DNA fragment, the PCR, followed by hybridization with specific probes, can serve as a powerful tool in the analysis of genomic sequence variation. Understanding the molecular basis of genetic disease or of complex genetic polymorphisms such as those in the HLA region requires detailed nucleotide sequence information from a variety of individuals to localize relevant variations. Currently, the analysis of each allelic variant requires a substantial effort in library construction, screening, mapping, subcloning, and sequencing. Here, using PCR primers modified near their 5′ ends to produce restriction sites, we describe a method for amplification of specific segments of genomic DNA and their direct cloning into M13 vectors for sequence analysis. Moreover, the cloning and sequencing of PCR-amplified DNA represents a powerful analytical tool for the study of the specificity and fidelity of this newly developed technique, as well as a rapid method of genomic sequencing. In addition, allelic variation has been analyzed using PCR amplification prior to hybridization with allele-specific oligonucleotide probes in a dot-blot format. This is a simple, general, and rapid method for genetic analysis and recently has been demonstrated in crude cell lysates, eliminating the need for DNA purification.

PCR Cloning and Direct Sequence Analysis

To develop a rapid method for genomic sequence determination and to analyze the individual products of PCR amplification, we chose the oligonucleotide primers and probes previously described for the diagnosis of sickle cell anemia (Saiki et al. 1985). These primers amplify a 110-bp segment of the human β-hemoglobin gene containing the Hb-S mutation and were modified near their 5′ ends (as described above in Fig. 3) to produce convenient restriction sites (linkers) for cloning directly into the M13mp10 sequencing vector. These modifications did not affect the efficiency of PCR amplification of the specific β-globin segment. After amplification, the PCR products were cleaved with the appropriate restriction enzymes, ligated into the M13 vector, and transformed into the JM103 host, and the resulting plaques were screened by hybridization with a labeled oligonucleotide probe to detect the β-globin clones. The plaques were also screened with the labeled PCR oligonucleotide primers to identify all of the clones containing amplified DNA (Table 2). In-

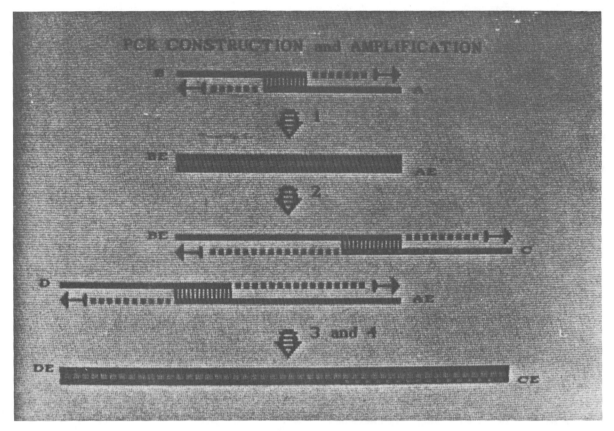

Figure 4. PCR construction and amplification. Oligonucleotides with complementary overlapping 3′ ends are extended on each other using DNA polymerase. The double-stranded DNA product of this reaction is extended on both ends by a second stage of several PCR cycles depicted by reactions 2, 3, and 4.

dividual clones were then sequenced directly by using the dideoxy primer-extension method.

Over 80% of the clones contained DNA inserts with the PCR primer sequences, but only about 1% of the clones hybridized to the internal β-globin probe. These nonglobin fragments presumably represent amplifications of other segments of the genome. This observation is consistent with the gel and Southern blot analysis of the PCR-amplified DNA from a β-globin deletion mutant and a normal cell line (Fig. 6). The similarity of the observed gel profiles reveals that most of the amplified genomic DNA fragments arise from nonglobin templates. Sequence analysis of two of these nontarget clones showed that the segments between the PCR primer sequences were unrelated to the β-globin gene and contained an abundance of dinucleotide repeats, similar to some genomic intergenic spacer sequences.

When ten of the clones that hybridized to the β-globin probe were sequenced, nine proved to be identical to the β-globin gene and one contained five nucleotide differences but was identical to the δ-globin gene. Each β-globin PCR primer has two mismatches with the δ-globin sequence. Each of these ten sequenced clones contains a segment of 70 bp originally synthesized from the genomic DNA template during the PCR amplification process. Since no sequence alterations were seen in these clones, the frequency of nucleotide misincor-

poration during 20 cycles of PCR amplification is less than 1 in 700.

To analyze the molecular basis of genetic polymorphism and disease susceptibility in the HLA class II loci, this approach has been extended to the amplification and cloning of a 242-bp fragment from the second exon, which exhibits localized allelic variability, of the HLA-DQα locus. In this case, the primer sequences (based on conserved regions of this exon) contain 5′-terminal restriction sites that have no homology with the DQα sequence. The specificity of amplification achieved by using these primers is greater than that achieved with the β-globin primers, since gel electrophoresis of the PCR products reveals a discrete band at 240 bp, which is absent from an HLA deletion mutant (Fig. 6). In addition, hybridization screening of the M13 clones from this amplification indicates that about 20% are homologous to the DQα cDNA probe (data not shown), an increase of 20-fold over the β-globin amplification. At this time, the basis for the difference in the specificity of amplification, defined here as the ratio of target to nontarget clones, is not clear, but may reflect the primer sequences and their genomic distribution. As described above, in some cases, the specificity of the PCR amplification can be significantly enhanced by using nested sets of PCR primers (Fig. 2).

Three HLA-DQα PCR clones derived from the

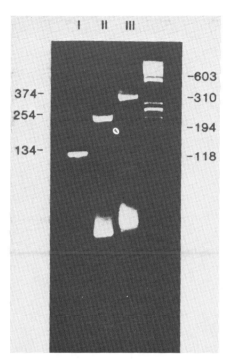

Figure 5. PCR construction of a 374-bp DNA fragment from synthetic oligodeoxynucleotides. (Lane I) 134-bp mutual extension product of TN10 and TN11; (lane II) 254-bp fragment produced by polymerase chain reaction of LL09 and LL12 with product in lane I; (lane III) 374-bp fragment produced by polymerase chain reaction of TN08 and TN13 with product in lane II; extreme right-hand lane is molecular-weight markers.

homozygous typing cell LG2 were subjected to sequence analysis. Two clones were identical to an HLA-DQα cDNA clone from the same cell line. One differed by a single nucleotide, indicating an error rate of ap-

Table 2. Cloning β-Globin Sequences from PCR-amplified DNA

Category	Number	Frequency (%)
Total plaques	1496	100
White plaques	1338	89
Primer plaques	1206	81
Globin plaques	15	1

proximately 1/600, assuming the substitution occurred during the 27 cycles of amplification. We are also currently using this procedure to analyze sequences from polymorphic regions of the HLA-DQα and DRβ loci. These preliminary studies suggest that the error rate over many cycles of amplification appears to be sufficiently low so that reliable genomic sequences can be determined directly from PCR amplification and cloning. This approach greatly reduces the number of cloned DNA fragments to be screened, circumvents the need for full genomic libraries, and allows cloning from nanogram quantities of genomic DNA.

Analysis of PCR-amplified DNA with Allele-specific Oligonucleotide Probes

Allelic sequence variation has been analyzed by oligonucleotide hybridization probes capable of detecting single-base substitutions in human genomic DNA that has been digested by restriction enzymes and resolved by gel electrophoresis (Conner et al. 1983). The basis of this specificity is that, under appropriate hybridization conditions, an allele-specific oligonucleotide (ASO) will anneal only to those sequences to which it is perfectly matched, a single-base-pair mismatch being

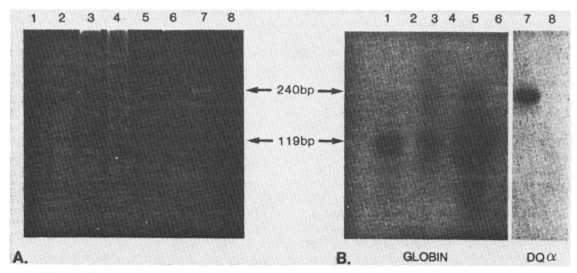

Figure 6. PCR amplification with oligonucleotide linker primers. (*A*) Ethidium-bromide-stained gel showing total amplified products. (*1*) Primers PC03 and PC04 on Molt-4 DNA; (*2*) PC03 and PC04 on GM2064 DNA; (*3*) GH18 and GH19 on Molt-4; (*4*) GH18 and GH19 on GM2064; (*5*) GH18 and GH19 on pBR328::β^A DNA; (*6*) PC03 and PC04 on Molt-4, no enzyme; (*7*) primers GH26 and GH27 on LG-2 DNA; (*8*) GH26 and GH27 on LCL721.180 DNA. (*B*) Southern blots showing specific amplified products. Lanes are numbered as in *A*. Lanes *1* through *6* were hybridized to the labeled RS06 oligonucleotide probe. Lanes *7* and *8* were hybridized to a cloned DBα cDNA probe labeled by nick translation.

sufficiently destabilizing to prevent hybridization. To improve the sensitivity, specificity, and simplicity of this approach, we have used the PCR procedure to amplify enzymatically a specific β-globin sequence in human genomic DNA prior to hybridization with ASOs. The PCR amplification, which produces a greater than 10^5-fold increase in the amount of target sequence, permits the analysis of allelic variation with as little as 1 ng of genomic DNA and the use of a simple "dot-blot" for probe hybridization. As a further simplification, PCR amplification has been performed directly on crude cell lysates, eliminating the need for DNA purification.

To develop a simple and sensitive method, we have chosen the sickle cell anemia and hemoglobin C mutations in the sixth codon of the β-globin gene as a model system for genetic diagnosis. We have used ASOs specific for the normal ($β^A$), sickle cell ($β^S$), and hemoglobin C ($β^C$) sequences as probes to detect these alleles in PCR-amplified genomic samples. The sequences of the 19-base ASO probes used here are identical to those described previously (Studenski et al. 1985).

DNA was extracted from six blood samples from individuals whose β-globin genotypes comprise each possible diploid combination of the $β^A$, $β^S$, and $β^C$ alleles and from the cell line GM2064, which has a homozygous deletion of the β-globin gene. Aliquots (1 μg) of each sample were subjected to 25 cycles of PCR amplification, and one thirtieth of the reaction product (33 ng) was applied to a nylon filter as a dot-blot. Three replicate filters were prepared and each was hybridized with one of the three ^{32}P-labeled ASOs under stringent conditions. The resulting autoradiogram (Fig. 7) clearly indicates that each ASO annealed only to those DNA samples containing at least one copy of the β-globin allele to which the probe was perfectly matched and not at all to the GM2064 deletion mutant. The frequency of the specific β-globin target to PCR-amplified DNA has been estimated by the analysis of cloned amplification products to be approximately 1% (Table 2), an enrichment of greater than 10^5 over unamplified genomic DNA. It is this substantial reduction of complexity that allows the application of 19-base probes in a dot-blot format and the use of shorter oligonucleotide probes, capable of allelic discrimination using less stringent conditions. In addition, this approach has been applied to the analysis of genetic polymorphism in the HLA-DQα locus by using four different ASO probes and is being extended to the HLA-DQβ and DRβ loci.

PCR amplification has been used to detect β-globin genotypes in as little as 0.5 ng of genomic DNA and, recently, in the crude lysate of 75 cells. This ability to rapidly analyze genetic variation of minute amounts of purified DNA or in cell lysates has important implications for a wide variety of clinical genetic analyses. With the use of nonisotopic probes and PCR automation, this procedure combining in vitro target amplification and ASO probes in a dot-blot format promises to be a general and simple method for the detection of allelic variation.

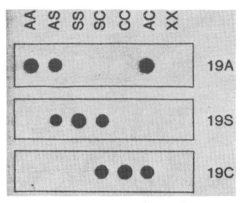

Figure 7. Genotype analysis of PCR-amplified genomic DNA using ASO probes. DNA extracted from the blood of individuals of known β-globin genotype and from the β-globin deletion mutant cell line GM2064 were subjected to PCR amplification. Aliquots of PCR reactions equivalent to 33 ng starting DNA were applied to replicate filters in a dot-blot format and probed with allele-specific oligonucleotides. (AA) $β^Aβ^A$; (AS) $β^Aβ^S$; (SS) $β^Sβ^S$; (SC) $β^Sβ^C$; (CC) $β^Cβ^C$; (AC) $β^Aβ^C$; (XX) GM2064.

ACKNOWLEDGMENTS

We thank Corey Levenson, Lauri Goda, and Dragan Spasic for the synthesis of oligonucleotides, Eric Ladner for graphic support, and Kathy Levenson for preparation of the manuscript.

REFERENCES

Cohen, S., A. Chang, H. Boyer, and R. Helling. 1973. Construction of biologically functional bacterial plasmids *in vitro*. *Proc. Natl. Acad. Sci.* **70:** 3240.

Conner, B.J., A. Reyes, C. Morin, K. Itakura, R. Teplitz, and R.B. Wallace. 1983. Detection of sickle cell β-S-globin allele by hybridization with synthetic oligonucleotides. *Proc. Natl. Acad. Sci.* **80:** 272.

Davies, J.E. and H.G. Gaessen. 1983. Synthetic gene fragments in genetic engineering—The renaissance of chemistry in molecular biology. *Angew Chem. Int. Chem. Ed. Engl.* **22:** 13.

Mullenbach, G.T., A. Tabrizi, R.W. Blacher, and K.S. Steimer. 1986. Chemical synthesis and expression in yeast of a gene encoding connective tissue activating peptide-III. *J. Biol. Chem.* **261:** 719.

Mullis, K.B. and F. Faloona. 1986. Specific synthesis of DNA in vitro via a polymerase catalyzed chain reaction. *Methods Enzymol.* (in press).

Roberts, R.J. 1985. Restriction and modification enzymes and their recognition sequences. *Nucleic Acids Res.* **9:** 75.

Rossi, J.J., R. Kierzek, T. Huang, P.A. Walker, and K. Itakura. 1982. An alternate method for synthesis of double-stranded DNA segments. *J. Biol. Chem.* **257:** 9226.

Saiki, R.K., S. Scharf, F. Faloona, K.B. Mullis, G. Horn, H.A. Erlich, and N. Arnheim. 1985. Enzymatic amplification of β-globin genomic sequences and restriction site analysis for diagnosis of sickle cell anemia. *Science* **230:** 1350.

Smith, H.O. and K.W. Wilcox. 1970. A restriction enzyme from *Hemophilus influenzae*. *J. Mol. Biol.* **51:** 379.

Studencki, A.B., B. Conner, C. Imprain, R. Teplitz, and R.B. Wallace. 1985. Discrimination among the human β-a, β-s, and β-c-globin genes using allele-specific oligonucleotide hybridization probes. *Am. J. Hum. Genet.* **37:** 42.

Urdea, M.S., L. Ku, T. Horn, Y.G. Gee, and B.D. Warner. 1985. Base modification and cloning efficiency of oligodeoxynucleotides synthesized by phosphoramidite method; methyl versus cyanoethyl phosphorous protection. *Nucleic Acids Res.* **16:** 257.

Watt, V.M., C.J. Ingles, M.S. Urdea, and W.J. Rutter. 1985. Homology requirements for recombination in *Escherichia coli. Proc. Natl. Acad. Sci.* **82:** 4768.

A Further Study of the T4 Ligase-catalyzed Joining of DNA at Base-paired Ends

V. Sgaramella and H.G. Khorana

T4 polynucleotide ligase catalyzes the end-to-end joining of DNA duplexes at the base-paired ends. The reaction has been shown to be bi-molecular in character by using a mixture of two different duplexes, when covalent joining of the two duplexes was observed. The reaction showed a temperature optimum of 25°C.

1. Introduction

It has recently been reported that the T4 polynucleotide ligase brings about joining of DNA duplexes at their completely base-paired ends when the required 5′-phosphate and 3′-hydroxyl groups are present (Sgaramella, van de Sande & Khorana, 1970). Because of its novelty and potential importance, the reaction has been further studied. In particular, the mode of the joining reaction has been examined. As illustrated in Figure 1, with DNA duplexes containing one base-paired end, the reaction could occur intermolecularly or intramolecularly. In the latter case, the joining reaction itself

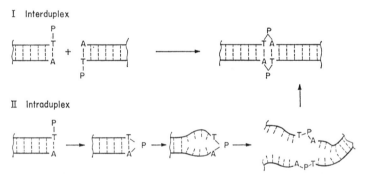

FIG. 1. Alternative modes in the T4 ligase-catalyzed joining of duplexes at the base-paired ends. In the intermolecular mechanism (I), two molecules of the duplex(es) are brought together at the terminal base pairs and the covalent joining then occurs. In the intramolecular mechanism (II), the frayed end of a duplex may be covalently closed by the ligase. The resulting hairpin structure could be extended by denaturation and then could form a duplex from two molecules.

†Paper CXV in this series is Caruthers et al., 1972.

‡Present address: Department of Genetics, Stanford University Medical School, Palo Alto, Calif. 94305, U.S.A.

§Present address: Departments of Biology and Chemistry, Massachusetts Institute of Technology, Cambridge, Mass. 02139, U.S.A.

would not lead to a doubling of the molecular weight, but fraying of the duplex followed by coming together of two such extended molecules would lead to a longer duplex indistinguishable from that formed by the intermolecular mechanism. The results now described show that the intermolecular mode is correct. A preliminary report of these findings has been made (Sgaramella, 1971).

2. Materials and Methods

The methods used have all been described in a recent paper (Sgaramella & Khorana, 1972). The enzymes, T4 polynucleotide kinase and ligase were also prepared as described earlier while the exonuclease III was prepared by the method of Jovin, Englund & Bertsch (1969).

All of the DNA duplexes used have been described in recent papers (Sgaramella & Khorana, 1972; van de Sande, Caruthers, Sgaramella, Yamada & Khorana, 1972; Caruthers *et al.*, 1972).

The duplexes, such as duplex II and duplex III, which contained "cold" 5'-phosphate groups at the unpaired ends, were phosphorylated at the base-paired 5'-ends by using $[\gamma\text{-}^{32}P]$ (or ^{33}P)-ATP and polynucleotide kinase. In the experiment with duplex II, after phosphorylation with $[\gamma\text{-}^{32}P]$ATP the duplex was characterized by degradation to 5'-nucleotides. Radioactivity found in the different mononucleotides was as follows: d-pT, 2358 cts/min; d-pC, 556 cts/min; d-pA, 55 cts/min; and d-pG, 65 cts/min. The radioactivity associated with d-pC probably arose from the phosphorylation of 5'-deoxycytidine unit at the opposite end. The latter, although already phosphorylated, presumably underwent some dephosphorylation by a phosphatase activity contaminating the polynucleotide ligase. Therefore, the phosphorylation of duplex III using $[\gamma\text{-}^{33}P]$ATP was carried out in the presence of 2 mм-sodium phosphate. On degradation of the product to 5'-nucleotides, the radioactivity distribution was as follows: d-pG, 1159 cts/min; d-pA, 74 cts/min; d-pC, 19 cts/min; and d-pT, 23 cts/min.

3. Results

(a) *Further study of joining at completely base-paired ends*

(i) *Oligomerization of duplex I and characterization of oligomers*

Duplex I was prepared as shown in Figure 2. Thus, the hexanucleotide, d-C-C-A-C-C-A, and the decanucleotide, d-C-C-G-G-A-C-T-C-G-T, were labeled at the 5'-end with $[^{32}P]$phosphoryl group. When joined in the presence of the dodecanucleotide, d-T-G-G-T-G-G-A-C-G-A-G-T, a symmetrical duplex (duplex I, Fig. 2) consisting of 28 base pairs was obtained. This was again phosphorylated at the 5'-thymidine ends using polynucleotide kinase and $[\gamma\text{-}^{32}P]$ATP. Again the duplex was submitted to the ligase and the kinetics of joining are shown in Figure 3. Phosphatase resistance at the start of the reaction was 63% because the same sample of $[\gamma\text{-}^{32}P]$-ATP had been used in all the kinase reactions. Phosphatase resistance increased during reaction with the ligase as shown in the inset to Figure 3. The products were fractionated on an agarose (Biogel-A (5 m)) column and the pattern obtained is shown in Figure 3. The various fractions were pooled as shown and analyzed to give the results shown in Table 1. An approximate estimate of size of the various peaks would indicate that material labeled as peak I was not larger than a decamer, the material in peak II had the average size of a pentamer and peak III contained dimers and trimers. On the basis of its elution volume and phosphatase sensitivity, peak IV probably contained the starting duplex.

The evidence given above, and especially the position of elution of the peaks I–III (Fig. 3) clearly showed that products were oligomers with increasing molecular

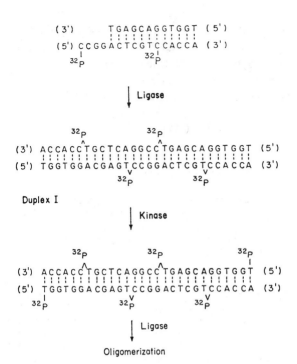

FIG. 2. Scheme for the preparation and oligomerization of duplex 1.

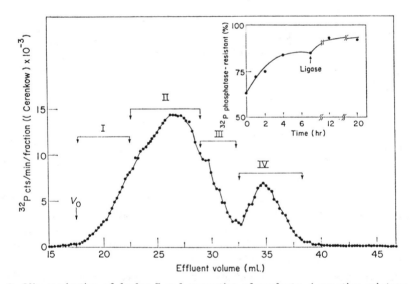

FIG. 3. Oligomerization of duplex I and separation of products. A reaction mixture (21 μl.) containing duplex I (1·2 μM) was set up as usual; about 2 units of ligase were added and the incubation performed at 15°C. The mixture was passed through a Biogel-A (5 m) (0·9 cm × 50 cm) column; fractions of 0·25 ml. were collected in minivials every 30 min and counted directly without scintillation medium. The material was pooled as shown.

TABLE 1

Phosphomonoesterase susceptibility of the fractions in Figure 3

| | Radioactivity (cts/min) | |
	Origin	Inorganic phosphate
Peak I	1563 (93%)	111
Peak II	5969 (91%)	594
Peak III	3115 (84%)	588
Peak IV	183 (9%)	1895

weights. A further experiment which supported this conclusion is shown in Figure 4. This consisted of degradation of the material in peak I by exonuclease III (Richardson, Lehman & Kornberg, 1964), which is known to cause stepwise degradation of a double-stranded DNA. Figure 4 shows the acid solubilization of the "oligomer" and of the "monomer" as a function of time. This oligomer was degraded to about 50%, whereas the monomer, under identical conditions, was degraded almost completely. The survival of a larger amount of acid-insoluble material, presumably single-stranded, at plateau value in the ligase product would be consistent with the latter possessing much higher molecular weight than duplex I as well as a continuous bi-helical structure.

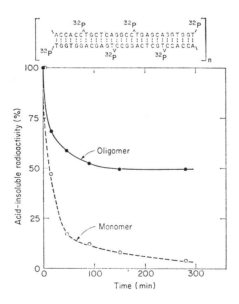

FIG. 4. Exonuclease III degradation of the oligomers of duplex I. The material in peak I of Fig. 3 was digested with exonuclease III in parallel with the isolated monomer. The mixtures contained in 100 μl.: the polynucleotide (corresponding to 3 pmoles of the monomer), about 1 o.d. unit of salmon sperm DNA as carrier, 10 mM-Tris·HCl (pH 8), 2 mM-MgCl$_2$, and 13 units of exonuclease III. Samples of 5 μl. were taken out at various times and spotted on quarters of 3MM Whatman filters, pre-soaked with 50 μl. of 10^{-3} M-EDTA, 0·1 M-P$_i$. The filters were immersed in 5% trichloroacetic acid plus 10^{-4} M-inorganic pyrophosphate, washed at 0°C three times with this solution and twice with alcohol/ether (1:1), dried and counted for radioactivity.

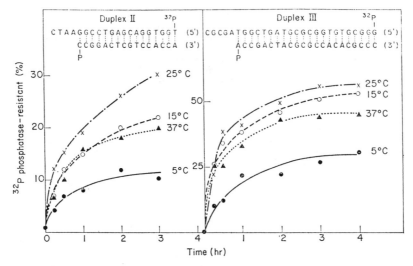

FIG. 5. Temperature effect on the self-joining of duplexes II and III. Four 15-µl. reaction mixtures were set up for each of the duplexes, at a concentration of 1·5 µM of the duplexes. The reactions were run as usual and the time-course was monitored by the DEAE-cellulose paper assay.

(ii) Influence of temperature on joining at the base-paired ends

Because of the unusual nature of the reaction, it was of interest to study the influence of temperature on the reaction. Duplexes II and III were used and the self-joining of each was studied at four different temperatures. The results obtained are shown in Figure 5. As is seen, in both cases, optimal rates were obtained at 25°C, 15 and 37°C being inferior.

(b) Joining at base-paired ends: inter- versus intramolecular reaction

(i) Joining reactions using 5'-³²P-labeled duplex II and 5'-³³P-labeled duplex III

Distinction between inter- and intramolecular modes of joining at the base-paired ends can be made by the use of two different duplexes which carry distinctive labels,

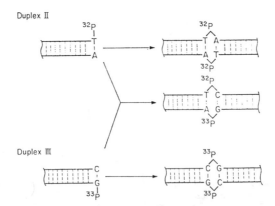

FIG. 6. Test for interduplex joining in T4 ligase-catalyzed reaction. One duplex contains A·T as the terminal base pair while the second contains G·C as the terminal base pair. The first duplex was labeled at the 5'-T end with ³²P$_i$ and the second was labeled at the 5'-G end with ³³P. Only the formation of the heteroduplex (middle) should give, on degradation to 3'-nucleotides, ³²P radioactivity in d-Cp and ³³P radioactivity in d-Ap.

^{32}P and ^{33}P, at their base-paired ends. Thus, when a mixture of duplex II (detailed structure in Fig. 7), which carries an A·T base pair at the matched end, and duplex III, which carries a G·C base pair at the matched end, is used with the 5'-labels as shown in Figure 6, then the following results would be expected from the two mechanisms. If the joining occurs intramolecularly, then only the two homodimers would be formed. On degradation to 3'-nucleotides, ^{32}P label would be only in d-Ap and ^{33}P label only in d-Cp. On the other hand, if the joining occurs intermolecularly, then, in addition to the two homo-dimers, the heteroduplex consisting of duplex II and duplex III would also result. On degradation of the total joined product to 3'-nucleotides, ^{32}P would be found not only in d-Ap but also in d-Cp. Similarly, ^{33}P would be

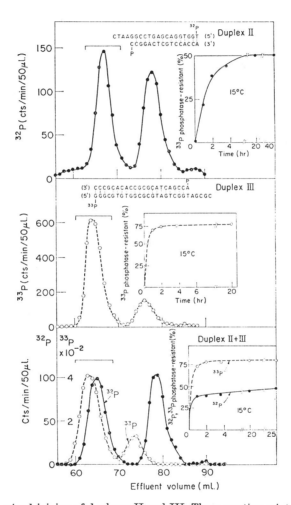

FIG. 7. Self and mixed joining of duplexes II and III. Three reaction mixtures were prepared, containing 2 nmoles of duplex molecules/ml., 10 mM-Tris·HCl (pH 7·6), 10 mM-MgCl$_2$, 10 mM-dithiothreitol, 60 μM-ATP and 280 units of T4 ligase per ml. The first tube (25 μl.) contained duplex II only, the second (25 μl.) duplex III, and the third (50 μl.) an equimolar mixture of the two. The joining was performed at 15°C, and was monitored with the phosphatase DEAE-cellulose paper assay. The time-courses of the reactions are given in the insets. The mixtures were fractionated by gel filtration through a Biogel-A (0·5 m) column (1·2 cm × 140 cm) equilibrated at 5°C with 0·05 M-triethylammonium bicarbonate.

found not only in d-Cp but also in d-Ap. Indeed, the proportion of ^{32}P present in d-Cp should match the proportion of ^{33}P present in d-Ap. This was found to be the case.

In Figure 7 are shown the results of three joining experiments. The top panel shows the kinetics (inset) of self-joining of duplex II labeled with [^{32}P]phosphate at the 5'-end (the label at the 5'-C end had completely decayed by the time of the present experiment) and the fractionation of the reaction product. The middle panel shows the self-joining of duplex III, in which the 5'-G group at the base-paired end was labeled with [^{33}P]phosphate, and the fractionation of the products. The bottom panel shows the kinetics of joining observed when a mixture of ^{32}P-labeled duplex II and ^{33}P-labeled duplex III was used, as well as the fractionation of the products formed.

The joining in the case of duplex II occurred to more than 50% while with duplex III, joining occurred to the extent of more than 75%. A similar difference in the extent of joining of the two duplexes was observed in the mixed experiment shown in the bottom panel of Figure 7.

The first peaks in the three fractionations were characterized for resistance to the phosphatase and by degradation to 3'-nucleotides. The results are given in Table 2. Thus, they were all essentially resistant to the phosphatase. (The second peaks corresponding to the unreacted starting duplexes showed 10 to 20% phosphatase resistance. This was most probably due to the presence of the AMP-pyrophosphate derivative of the 5'-^{32}P (or ^{33}P) end-group.) When peak I of the experiment in the top panel (Fig. 7) was degraded to 3'-nucleotides, most of the radioactivity was found in d-Ap, as expected. Similarly, degradation of the self-joined product of duplex III (middle panel) to 3'-nucleotides gave practically all of the radioactivity in the expected nucleotide d-Cp. In the mixed experiment (bottom panel), degradation of the pooled peak to 3'-nucleotides showed that both labels (^{32}P and ^{33}P) were associated

TABLE 2

Characterization of the joined products obtained by using duplex II and duplex III (Fig. 7)

(1) Resistance to phosphomonoesterase (DEAE-cellulose paper assay)

	Radioactivity (cts/min)	
	Origin	Inorganic phosphate
Peak I in top panel of Fig. 7 (^{32}P)	986 (83%)	211
Peak I in middle panel of Fig. 7 (^{33}P)	1682 (96%)	70
Peak I in bottom panel of Fig. 7 (^{32}P)	460 (87%)	70
(^{33}P)	2010 (97%)	80

(2) 3'-Mononucleotide analysis†

	Radioactivity (cts/min)			
	d-Ap	d-Gp	d-Tp	d-Cp
Peak I in top panel of Fig. 7 (^{32}P)	689	10	189	2
Peak I in middle panel of Fig. 7 (^{33}P)	25	40	10	1844
Peak I in bottom panel of Fig. 7 (^{32}P)	106	0	47	356 (1·1 pmole)
(^{33}P)	769 (1·1 pmole)	25	20	681

†The mixtures were fractionated by electrophoresis at pH 3·5. Similar results were obtained using solvent I.

with d-Ap as well as d-Cp (Table 2) and furthermore as predicted for the hetero-duplex, the molar fraction (pmoles) of ^{32}P label found in d-Cp was equal to the molar fraction (pmoles) of ^{33}P label found in d-Ap (Table 2). Thus, since the specific activity of ^{32}P was 322 cts/min/pmole, the radioactivity found in d-Cp was 1·1 pmoles and, since the specific activity of ^{33}P was 720 cts/min/pmole, the radioactivity found in d-Ap was also equal to 1·1 pmole.

(ii) *Mixed joining using duplex III and duplex IV*

It was hoped to confirm the above conclusion further by using two duplexes of different lengths so that the heteroduplex formed in the joining reaction could be separated from the two homoduplexes expected. Duplex III and duplex IV, which was also available from work on the DNA corresponding to the yeast alanine transfer-RNA, were used. Their structures, the labels used at the 5'-ends (base-paired ends) and the kinetics of the joining reactions (self and mixed) are shown in Figure 8. The reaction mixture obtained in the mixed joining experiment was separated and gave the pattern shown in Figure 9. Although not well resolved, three peaks, which contained both isotopes, were present. It seems highly likely that peak II, the first peak containing both labels, consisted of the mixed duplex. The material after this, peak

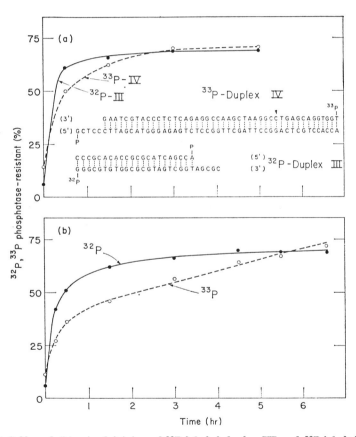

FIG. 8. (a) Self- and (b) mixed joining of ^{32}P-labeled duplex III and ^{33}P-labeled duplex IV. Three reaction mixtures of 30 μl. each were set up as described in the legend to Fig. 7. The concentration of the duplexes was 0·3 μM. The joining was performed at 25°C, and was preceded by an incubation bringing the three mixtures in 5 mM-MgCl$_2$ to 80°C and slowly cooling to 25°C overnight.

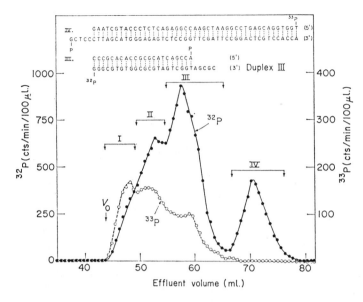

FIG. 9. Fractionation of the reaction mixture containing ^{32}P-labeled duplex III and ^{33}P-labeled duplex IV (Fig. 8). Gel filtration was performed through the same column and under the same conditions as described in the legend to Fig. 7.

III, would be expected to be a mixture of the dimer of duplex III and unreacted duplex IV. The last peak (peak IV) probably represented unreacted duplex III. The various peaks were analyzed for sensitivity to phosphomonoesterase and the results are given in Table 3. Although the radioactivity was low, the results were as expected.

4. Comments

The experiments reported show that T4 ligase brings about the joining of duplexes at their base-paired ends by a bimolecular type of reaction and that an intramolecular mechanism (Fig. 1) can be ruled out. This conclusion was shown not only by the formation of heteroduplexes in the experiments in which two different duplexes were used together in joining reactions but also by the oligomerization of duplex I (Fig. 2).

TABLE 3

Characterization of the different peaks of Figure 9

Resistance to phosphomonoesterase†

		Radioactivity (cts/min)	
		Origin	Inorganic phosphate
Peak I	^{32}P	278 (89%)	39
	^{33}P	166 (75%)	56
Peak II	^{32}P	437 (75%)	147
	^{33}P	150 (78%)	41
Peak III	^{32}P	1048 (84%)	210
	^{33}P	101 (40%)	147
Peak IV	^{32}P	143 (28%)	361

†DEAE-cellulose paper assay.

It would be very difficult to imagine the formation of duplexes as long as or longer than 100 nucleotides by the steps postulated in the intramolecular mechanism. The possibility may further be considered that in the duplexes some of the bases in one of the strands may loop out and, thus, there may be single-stranded protruding ends instead of the base-paired ends. The duplexes would then come together in the conventional way by base pairing at the single-stranded ends. This can be ruled out because in the duplexes with defined sequences used in the present work no possibility exists for generating complementary sequences near the base-paired ends.

As pointed out earlier (Sgaramella *et al.*, 1970) the ability of the T4 ligase to join two different DNA duplexes may be biologically important. Recombination of DNA molecules, at least in some instances, could take place in this way and non-site-specific integration of episomes could be understood without postulating regions of weak homologies between the interacting duplexes. The same mechanism could be postulated for translocations, inversion and repetitions of specific DNA sequences. In this connection, the presence or otherwise of this activity in the other known ligases (bacterial and mammalian) is of great interest.

The present findings make available a simple approach to the laboratory joining of DNA duplexes of different origin, natural as well as synthetic. It may also be useful for some studies to prepare DNA duplexes containing repeats of nucleotide sequence as was demonstrated by the oligomerization of duplex I.

This work was supported by grants from the National Cancer Institute of the National Institutes of Health, U.S. Public Health Service (grants nos. 72576 and CA05178), The National Science Foundation, Washington, D.C., (grants nos. 73078 and GB-7484X), and the Life Insurance Medical Research Fund.

REFERENCES

Caruthers, M. H., Kleppe, K., van de Sande, J. H., Sgaramella, V., Agarwal, K. L., Büchi, H., Gupta, N., Kumar, A., Ohtsuka, E., Raj Bhandary, U. L., Terao, T., Weber, H., Yamada, T. & Khorana, H. G. (1972). *J. Mol. Biol.* **72**, 475.

Jovin, T. M., Englund, P. T. & Bertsch, L. L. (1969). *J. Biol. Chem.* **244**, 2996.

Richardson, C. C., Lehman, I. R. & Kornberg, A. (1964). *J. Biol. Chem.* **239**, 251.

Sgaramella, V. (1971). *Fed. Proc.* **30**, 1524

Sgaramella, V. & Khorana, H. G. (1972). *J. Mol. Biol.* **72**, 445.

Sgaramella, V., van de Sande, J. H. & Khorana, H. G. (1970). *Proc. Nat. Acad. Sci., Wash.* **67**, 1468.

van de Sande, J. H., Caruthers, M. H., Sgaramella, V., Yamada, T. & Khorana, H. G. (1972). *J. Mol. Biol.* **72**, 457.

A Restriction Enzyme from *Hemophilus Influenzae*

I. Purification and General Properties

H.O. Smith and K.W. Wilcox

Extracts of *Hemophilus influenzae* strain Rd contain an endonuclease activity which produces a rapid decrease in the specific viscosity of a variety of foreign native DNA's; the specific viscosity of *H. influenzae* DNA is not altered under the same conditions. This "restriction" endonuclease activity has been purified approximately 200-fold. The purified enzyme contains no detectable exo- or endonucleolytic activity against *H. influenzae* DNA. However, with native phage T7 DNA as substrate, it produces about 40 double-strand 5'-phosphoryl, 3'-hydroxyl cleavages. The limit product has an average length of about 1000 nucleotide pairs and contains no single-strand breaks. The enzyme is inactive on denatured DNA and it requires no special co-factors other than magnesium ions.

1. Introduction

A number of bacteria are capable of recognizing and degrading ("restricting") foreign DNA, such as the DNA of a virus grown on another bacterial strain. The DNA of the host is protected by a "host-controlled modification" (Arber, 1965). Recently, Meselson & Yuan (1968) have purified a restriction endonuclease from *Escherichia coli* K12. The enzyme has the interesting properties: (1) that it is site-specific in action, producing only a limited number of double-strand breaks in unmodified DNA, and (2) that it requires adenosine triphosphate and *S*-adenosyl methionine in addition to magnesium ions.

We have made the chance discovery of what appears to be a similar type of enzyme in *Hemophilus influenzae*, strain Rd. In the course of some experiments in which competent *H. influenzae* cells were incubated with radioactively labeled DNA from the *Salmonella* phage P22, we found that this DNA was apparently degraded since it could not be recovered in cesium chloride density gradients. It seemed likely that the effect was one of restriction. We were able to show the presence in crude extracts of an endonuclease activity which produced a rapid decrease in viscosity of foreign DNA preparations and which was without effect on the *H. influenzae* DNA. We describe in this report the purification and properties of the endonuclease. As with the *E. coli* restriction enzyme, our enzyme produces double-strand breaks in a limited number of specific sites. The enzyme requires only magnesium ions as a co-factor, unlike the *E. coli* enzyme. A preliminary report has been published (Smith & Wilcox, 1969).

SMITH, H.O. and WILCOX, K.W.
A restriction enzyme from *Hemophilus influenzae*. I. Purification and general properties. Reprinted by permission from *J. Mol. Biol.* v. 51: pp. 379-391 (1970).

2. Materials and Methods

(a) Bacterial and phage strains

H. influenzae strain Rd was obtained from Dr Roger Herriott as a frozen culture. The phage P22 c_2 clear plaque mutant (Levine, 1957) grown on *S. typhimurium* LT2 was used as a source of phage P22 DNA. Phage T7 and its host *E. coli* B were obtained from Dr Bernard Weiss.

(b) Enzymes and standards

Bacterial alkaline phosphatase was obtained from Worthington Biochemical Corp. Polynucleotide kinase, 5000 units/ml. (Richardson, 1965) and rechromatographed bacterial alkaline phosphatase, 20 units/ml. (Weiss, Live & Richardson, 1968) for use in the ^{32}P terminal labeling procedure were kindly supplied by Dr Bernard Weiss.

Bovine serum albumin and sperm whale myoglobin with molecular weights of 67,000 and 17,800, respectively, were obtained in a molecular weight marker kit from Mann Research Laboratories.

(c) Nucleic acids and nucleotides

Phage P22 was purified from L broth lysates (Levine, 1957) by differential centrifugation and banding in a CsCl step gradient (Thomas & Abelson, 1966). The DNA was extracted with 1 vol. of cold, redistilled phenol and then precipitated with 2 vol. of cold ethanol and redissolved in 1 vol. of NaCl–Tris buffer (0·05 M-NaCl, 0·01 M-Tris–HCl, pH 7·4, formerly ST buffer). The precipitation was repeated and the DNA was finally redissolved in the above buffer at 1·30 mg/ml. This DNA stock was used for the viscometry assays (see below).

Unlabeled phage T7 DNA was similarly prepared from phage grown on *E. coli* B. Phage T7 labeled with ^{32}P was prepared from phage grown in synthetic medium containing 2 μg phosphorus/ml. (Smith, 1968) and 5 μc of carrier-free [^{32}P]orthophosphate/ml. (New England Nuclear Corp.). The labeled phage were purified as described above and then extracted once with cold phenol. The DNA-containing aqueous phase was removed and dialysed for 20 hr against three 500-ml. changes of NaCl–Tris buffer.

Unlabeled *H. influenzae* DNA was extracted by the procedure of Marmur (1961) from a culture grown to saturation in Difco brain–heart infusion supplemented with NAD, 2 μg/ml., and hemin (Eastman), 10 μg/ml. *H. influenzae* DNA labeled with [^{3}H]thymidine was prepared from cells grown in a synthetic medium as described by Carmody & Herriott (1970).

[γ-^{32}P]ATP was obtained from Dr Bernard Weiss. The method of preparation has been described (Weiss *et al.*, 1968).

(d) Zone sedimentation of DNA in sucrose density-gradients

Native DNA was sedimented in linear gradients of 5 to 20% (w/v) sucrose in NaCl–Tris buffer. Denatured DNA was centrifuged in similar gradients containing 0·1 M-NaOH. The DNA was denatured by addition of 0·1 vol. of 1 N-NaOH for 5 min at room temperature followed by neutralization with 0·1 vol. of 1·1 N-HCl, 0·2 M-Tris as described by Studier (1965). Centrifugation was carried out in an SW50 rotor at 4°C in a Spinco model L centrifuge. Approximately 30 ten-drop fractions were collected from the bottom of the tube directly into scintillation vials containing 15 ml. of scintillation medium (2-methoxy-ethanol, 373 ml.; PPO, 9·3 g; dimethyl-POPOP, 30 mg; toluene to 1 liter) and counted in a Packard scintillation spectrometer.

(e) Assay of the enzymic activity by DNA viscometry

DNA viscosity measurements were performed at 30°C in an Ostwald viscometer having a flow-time for water of approximately 60 sec. The viscometer was filled with 3·5 ml. of phage P22 DNA solution, 40 μg/ml. in Tris–Mg–mercaptoethanol buffer (6·6 mM each of Tris buffer, pH 7·4, MgCl$_2$, and mercaptoethanol). Several flow-time measurements were taken after the DNA solution had reached thermal equilibrium and these were generally reproducible to within 0·1 sec. Five to 50 μl. of extract or purified enzyme was then introduced into the reservoir bulb of the viscometer and rapidly mixed by blowing air retrograde into the bulb. Flow-time measurements were taken as rapidly as was practical. The DNA viscosity was expressed as specific viscosity, $\eta_{sp} = (t/t_D) - 1$, where t_D represents the flow time at the end of the experiment, 5 min after addition of 50 μg of pancreatic DNase (this corresponded to the flow-time for pure solvent within the accuracy of the measurements). Specific viscosity measurements were plotted against time on semi-logarithm paper as fractional values of the zero-time value. One unit of enzyme activity is defined as that amount which produces a decrease in the DNA specific viscosity of 25% in 1 min under the conditions described above. It should be pointed out that the viscometric assay is valid even on crude extracts since, as will be shown in the Results section, no activity is found in crude extracts against homologous DNA.

(f) Purification of endonuclease R

H. influenzae cells were grown in 12 l. of brain–heart infusion, supplemented with 10 μg hemin/ml. and 2 μg NAD/ml. to O.D.$_{650}$ = 0·7, harvested by centrifugation, washed once, and resuspended in 20 ml. of 0·05 M-Tris (pH 7·4), 0·001 M-glutathione. The cells were disrupted by sonication for 4 min at 8 A output on a Bronson sonicator while being cooled in an ice–salt water bath. All subsequent operations were carried out at 0 to 4°C. Cellular debris was removed by centrifugation for 30 min at 100,000 g. The supernatant (27 ml.) was brought to 1 M-NaCl by addition of 1·58 g of NaCl and layered onto a 2·5 cm × 49 cm Bio-Gel A 0·5 M (200 to 400 mesh) column prewashed with 10 vol. of 1 M-NaCl, 0·02 M-Tris–HCl, pH 7·4, 0·01 M-mercaptoethanol. Elution was carried out at 1·2 ml./min using the same buffer solution and 6-ml. fractions were collected. Fractions 18 to 28, containing nearly all the activity and only 10% of the O.D.$_{260}$ absorbing material, were pooled. The pooled fractions (65 ml.) were diluted with 140 ml. of 0·02 M-Tris, pH 7·4, and stirred in an ice bath. Ammonium sulfate, 64·3 g, was added slowly over a 30-min period. The precipitate (0 to 50%) was removed by centrifugation. The supernatant solution was precipitated with ammonium sulfate, 15·8 g, to obtain a 50 to 60% precipitate. The supernatant was again reprecipitated with 16·8 g of ammonium sulfate and this precipitate (60 to 70%) was combined with the 50 to 60% precipitate and dissolved in 20 ml. of 0·05 M-NaCl, 0·02 M-Tris (pH 7·4), 0·001 M-mercaptoethanol.

Part of the 50 to 70% ammonium sulfate fraction was further purified by column chromatography. A phosphocellulose (Whatman, P11) column, 0·5 cm × 9·5 cm, was equilibrated with 100 ml. of 0·01 M-potassium phosphate buffer, pH 7·4. 6 ml. of the ammonium sulfate fraction, containing 60 mg. of protein, was diluted with 54 ml. of 0·01 M-phosphate buffer, pH 7·4, and loaded onto the column at a flow rate of 5 ml./hr at 4°C. Protein was then eluted stepwise with 5-ml. portions of 0·01 M-potassium phosphate buffer, pH 7·4, containing increasing molarities of KCl as follows: 0·0, 0·1, 0·2, 0·3 and 0·4 M. Fractions of 1 ml. were collected. The bulk of the activity was eluted at 0·2 M-KCl. The first two fractions at 0·3 M-KCl contained a small amount of activity and were combined with the 0·2 M-KCl fractions.

The enzyme was finally concentrated by precipitation of the combined phosphocellulose fractions (7 ml.) with 3·3 g of ammonium sulfate. The precipitate was redissolved in 1·5 ml. of 0·20 M-NaCl, 0·02 M-Tris (pH 7·4) containing bovine serum albumin, 0·3%, at a final activity of 16 units/ml. The activity is stable at 4°C in this solution for at least 8 months. Table 1 summarizes the purification procedure. The significant increase in total enzyme units observed following ammonium sulfate precipitation is not explained, but could be due to removal of interfering activities.

(g) ^{32}P-labeling of the 5'-end of the DNA

The 5'-phosphoryl ends of DNA were first dephosphorylated by incubation with alkaline phosphatase and then rephosphorylated with [γ-^{32}P]ATP in the polynucleotide kinase reaction as described by Weiss et al. (1968). The reactions were carried out as follows: DNA, 20 mμmoles, in 0·26 ml. 0·1 M-Tris–HCl, pH 8·0, was incubated with 5 μl. of bacterial alkaline phosphatase (20 units/ml.) at 37°C for 30 min to remove the terminal 5'-phosphoryl groups. A 0·035-ml. volume of a kinase reaction mixture (containing 0·3 M-MgCl$_2$; 0·03 M-potassium phosphate buffer, pH 7·4; [γ-^{32}P]ATP, 1 μmole/ml., 2·2 × 10^8 cts/min/ μmole; and 1·0 M-dithiothrietol, in a ratio of 2 : 2 : 2 : 1 by vol.), and 5 μl. of polynucleotide kinase, 5000 units/ml., were then added and the reaction mixture was incubated for 30 min at 37°C. The terminally labeled DNA was precipitated at 0°C by adding 0·3 ml. of 0·1 M-sodium pyrophosphate followed by 2·5 ml. of cold 6% trichloroacetic acid containing 0·01 M-sodium pyrophosphate. The precipitate was collected on a glass filter, washed with nine 2-ml. portions of 6% trichloroacetic acid containing pyrophosphate, and two 2-ml. portions of ethanol. After drying, the filters were counted in a scintillation counter. Under these conditions labeling takes place only at the ends of native DNA. Internal single-strand breaks (nicks) are not labeled because at 37°C the phosphatase is inactive at the nicks (Weiss *et al.*, 1968). If the DNA is denatured with alkali before labeling, then the nicks are exposed and can be labeled by the above procedure.

(h) *Molecular weight estimation by gel filtration*

The molecular weight of purified enzyme was estimated by filtration through a 0·8 cm × 17·2 cm column of superfine Sephadex G200 (Pharmacia Fine Chemicals). The column was washed with several volumes of NaCl–Tris buffer at a flow rate of 0·5 ml./hr under 10 cm of hydrostatic pressure at 4°C. A mixture containing 250 μg of bovine serum albumin, 250 μg of myoglobin, 0·5 optical density unit (at a wavelength of 650 nm) of dextran blue 2000 (Pharmacia Fine Chemicals), and 25 μl. of purified *H. influenzae* enzyme (16 units/ml.) in a volume of 0·1 ml. was layered onto the column and eluted with NaCl–Tris buffer under the above conditions. Forty 10-drop fractions (0·23 ml.) were collected. Dextran blue was measured by absorbance at 650 nm and the bovine serum albumin and myoglobin by absorbance at 230 nm. The enzyme activity was measured in the following way. Ten μl. of each fraction was incubated for 30 min at 37°C with 25 μl, of ^{32}P-labeled T7 DNA (28 μg/ml., 6·5 × 10^6 cts/min/ml.), 1 μl. of bacterial alkaline phosphatase, 57 units/ml., and 0·2 ml. of Tris–Mg–mercaptoethanol buffer containing 0·05 M-NaCl. The reaction tubes were chilled and 0·1 ml. vol. of salmon sperm DNA, 500 μg/ml. was added as carrier. The DNA was precipitated with 0·3 ml. of chilled 10% trichloroacetic acid. After 5 min the tubes were centrifuged at 4000 *g* for 10 min and 0·5 ml. of each supernatant was removed and counted in scintillation medium. This assay procedure gives results similar to the more cumbersome viscometry method but can only be used with the purified enzyme. The phosphatase serves to cleave off ^{32}P groups exposed by the enzyme digestion and these then appear as trichloroacetic acid-soluble radioactivity.

3. Results

(a) *Detection of an* H. influenzae *nuclease specific for foreign DNA*

H. influenzae extracts appear to contain no detectable endonuclease activity against native *H. influenzae* DNA by the viscometric assay (see Materials and Methods). Addition of 20 μl. of an extract from sonicated cells (containing 16 mg protein/ml.) to a viscometer containing *H. influenzae* DNA caused no decrease in η_{sp} during 60 minutes at 30°C (Fig. 1). However, the η_{sp} of phage P22 DNA was significantly decreased by as little as 10 μl. of extract under the same conditions. The extract thus apparently contains a nuclease with specificity toward the foreign DNA. Addition of 50 μl. of extract produced approximately a fivefold greater rate of fall of the η_{sp} of the phage DNA. Fractional decrease in η_{sp} proceeded logarithmically with time until the value was below 0·7, after which the decrease became less rapid. The proportionality between initial rate of decrease in η_{sp} and the amount of extract added provides a quantitative assay. A unit of the enzyme activity can be defined as that amount which produces a 25% decrease in η_{sp} in one minute.

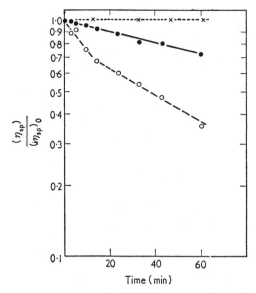

FIG. 1. Effect of a sonicated cell extract from *H. influenzae* on the specific viscosity of phage P22 and *H. influenzae* DNA.

Two viscometers containing 3·5 ml. of phage P22 DNA at a concentration of 40 μg/ml. in Tris–Mg–mercaptoethanol buffer were equilibrated at 30°C. At zero time, 10 μl. (—●—●—) of sonicated cell extract from *H. influenzae*, containing 16 mg protein/ml. was added to the first viscometer and 50 μl. (--○--○--) was added to the second. Measurements of the specific viscosity (η_{sp}) were then taken at intervals and were plotted as fractional values of the zero time specific viscosity ($\eta_{sp})_0$. Control measurements were obtained from a viscometer containing 3·5 ml. of *H. influenzae* DNA at a concentration of 40 μg/ml. in the same buffer to which 20 μl. (--×--×--) of cell extract was added at zero time. (The initial specific viscosity of the *H. influenzae* DNA solution was comparable to that of the phage P22 DNA solution.)

TABLE 1

Purification procedure

	Total (units)	Total (mg)	Spec. act. (units/mg)
100,000 *g* supernatant	135	1350	0·10
Bio-Gel column	185	650	0·28
Ammonium sulfate (50 to 70% ppt)	455	200	2·2
Phosphocellulose column†	144	5·5	26·0

† Entries are calculated assuming that all of the ammonium sulfate fraction was purified through the phosphocellulose step.

(b) *Properties of the purified nuclease*

A preparation of the *H. influenzae* nuclease approximately 200-fold purified from the sonicated cell extract was obtained as described in Materials and Methods.

(i) *Magnesium ion requirements*

Enzymic activity is dependent on the presence of Mg^{2+} ions (Table 2). The purified enzyme was optimally active when assayed at $6·6 \times 10^{-3}$ M-Mg^{2+} in 0·06 M-NaCl. The activity was decreased by a factor of about 50 when assayed at 10^{-4} M-Mg^{2+} and was unmeasureable at 10^{-5} M-Mg^{2+}. About 40% of the maximum activity was obtained at 10^{-3} M-Mg^{2+}.

TABLE 2

Effects of sodium chloride and magnesium ion concentration on the nuclease activity

Mg^{2+} (M)	NaCl (M)	Relative activity
6.6×10^{-3}	0.00	1.0
6.6×10^{-3}	0.02	2.3
6.6×10^{-3}	0.04	2.9
6.6×10^{-3}	0.06	3.4
6.6×10^{-3}	0.08	3.2
6.6×10^{-3}	0.10	2.2
10^{-5}	0.06	< 0.01
10^{-4}	0.06	0.06
10^{-3}	0.06	1.3

DNA viscosity measurements were carried out with phage P22 DNA, 40 μg/ml. in 3.5 ml. of solvent containing 6.6 mM-Tris–HCl, pH 7.4, and 6.6 mM-mercaptoethanol at the various listed NaCl and Mg^{2+} concentrations. Activities are expressed relative to that obtained under standard assay conditions in Tris–Mg–mercaptoethanol solvent.

(ii) *Salt requirements*

Optimum activity was obtained in 0.06 M-NaCl (Table 2). At this molarity an approximately threefold increase was obtained over that found under the standard assay conditions in Tris–Mg–mercaptoethanol buffer.

(iii) *Molecular weight estimate*

In the agarose column purification step the nuclease activity was recovered in fractions corresponding to an approximate molecular weight of 80,000. Gel filtration

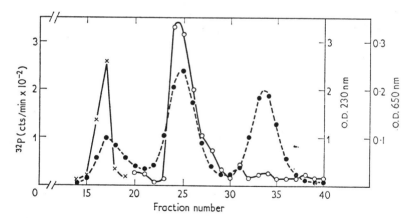

Fig. 2. Gel filtration of the purified nuclease on a Sephadex G200 column.
Dextran blue was measured by absorbancy at 650 nm (—×—×—). Bovine serum albumin and myoglobin were measured by absorbancy at 230 nm (--●--●--). The nuclease activity was measured as released ^{32}P activity as described in Materials and Methods (—○—○—).

of the purified enzyme on Sephadex G200 was carried out with bovine serum albumin and myoglobin as known molecular weight markers. Dextran blue 2000 was used to measure the exclusion volume. The nuclease activity was found in a peak closely associated with the albumin (67,000 molecular weight). The enzyme is thus approximately the same molecular weight (Fig. 2).

(iv) *Foreign DNA versus* H. influenzae *DNA as substrate*

A mixture of native [³H]thymidine-labeled *H. influenzae* DNA and native ³²P-labeled phage T7 DNA was incubated at 37°C for 30 minutes with an excess of enzyme. Samples were removed from the reaction mixture at zero time (before enzyme addition), five minutes and 30 minutes for assay of trichloroacetic acid-soluble radioactivity and additional fractions were removed for sucrose gradient analysis. One-half of each of the latter fractions was zone sedimented on a neutral sucrose density-gradient and the other half was first denatured in alkali and then sedimented on an alkaline sucrose gradient. At zero time both phage and bacterial DNA sedimented approximately to mid-position in the neutral gradient tube (Fig. 3). The bacterial DNA was more heterogeneous in size and produced a broad band in comparison to the phage DNA. In the alkaline gradient the phage DNA showed a trailing shoulder of smaller pieces but appeared greater than 50% intact. The bacterial peak was slightly broader than in the neutral gradient. After five minutes of treatment with the *H. influenzae* nuclease, the bacterial DNA peaks in both neutral and alkaline gradients were unaltered but the phage DNA was degraded to an average molecular weight of 1.45×10^6 in the neutral gradient and a molecular weight of 0.77×10^6 in the alkaline gradient (calculated according to Studier, 1965) as compared to a molecular weight of 26.4×10^6 for the intact molecule (Studier, 1965). No apparent decrease in size over that found at five minutes was obtained after 30 minutes. The ³²P-labeled trichloroacetic acid-soluble radioactivity was $< 0.1\%$, $<0.1\%$ and 0.26% at 0, 5 and 30 minutes, respectively. No trichloroacetic acid-soluble ³H radioactivity was detectable. The nuclease is thus inactive on homologous native DNA, producing neither double- nor single-strand breaks. On heterologous native DNA it appears to act by making a limited number of double-strand breaks. Since the denatured product is approximately one-half the molecular size of the native product, no nicks appear to be produced over and above the double-strand cleavages. The enzyme will be referred to in the remainder of this report as endonuclease R.

(v) *Absence of exonucleolytic activity*

In the above experiment, little, if any, ³²P radioactivity was released as trichloroacetic acid-soluble material during the digestion with endonuclease R. Therefore it appears unlikely that the enzyme itself has an exonuclease activity or that it is contaminated with a significant exonuclease activity. In order to substantiate this further the digest was examined for released nucleotides. A solution of ³²P-labeled phage T7 DNA was extensively digested with an excess of endonuclease R. A sample was then mixed with unlabeled marker nucleotides and chromatographed in two dimensions. As seen in Table 3, essentially no activity was found associated with the nucleotide spots.

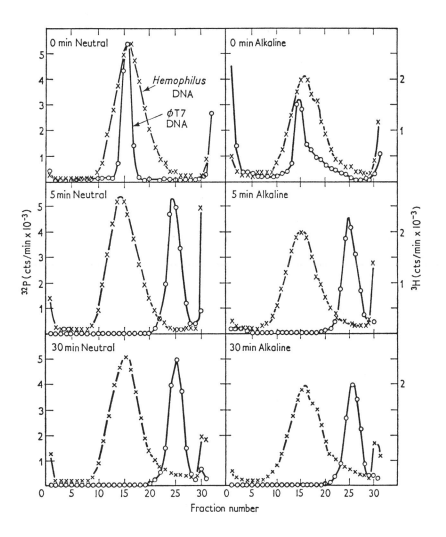

FIG. 3. Zone sedimentation in neutral and alkaline sucrose gradients of the products of the *H. influenzae* nuclease digestion of phage T7 DNA.

The reaction mixture (0·975 ml.) contained ^{3}H-labeled *H. influenzae* DNA, 17·1 μg, 8·7 × 10^4 cts/min; ^{32}P-labeled phage T7 DNA, 0·94 μg, 6·4 × 10^4 cts/min; 50 mM-NaCl; 6·6 mM-Tris–HCl, pH 7·4; 6·6 mM-MgCl$_2$; 6·6 mM-mercaptoethanol. At zero time 0·1 ml. was removed for trichloro-acetic acid precipitation and 0·15 ml. was pipetted into 0·15 ml. of 0·1 M-EDTA, pH 8·0, on ice for sucrose gradient analysis. Purified *H. influenzae* nuclease, 5 μl., was then added and the reaction mixture was incubated at 37°C. Additional samples for trichloroacetic acid precipitation (0·1 ml.) and sucrose analysis (0·15 ml.) were removed at 5 and 30 min. A 0·1-ml. portion of each of the samples that had been taken for sucrose gradient analysis was layered on a neutral gradient and centrifuged at 50,000 rev./min for 2 hr at 4°C. The remaining 0·2 ml., was alkali-denatured by addition of 10 μl. of 4 M-NaOH and 0·1 ml. was then centrifuged at 50,000 rev./min for 2·5 hr. Liquid fractions collected drop wise from the bottom of the centrifuge tubes were counted directly in scintillation medium. (— × — × —), ^{3}H radioactivity; (—○—○—), ^{32}P radioactivity. Trichloro-acetic acid precipitation was carried out by addition of 0·2 ml. of salmon sperm DNA (500 μg/ml.) and 0·3 ml. of 10% trichloroacetic acid on ice. The samples were centrifuged at 4500 g for 10 min and 0·3 ml. of the supernatant was counted in scintillation medium. The results are reported in the text.

TABLE 3

Release of nucleotides by endonuclease R digestion

Nucleotide species	Counts/50 min	%
dAMP	38	0·0004
dTMP	121	0·001
dGMP	0	0
dCMP	0	0
origin	$1·04 \times 10^7$	100

A reaction mixture (0·1 ml.) containing ^{32}P-labeled phage T7 DNA 78 mμmoles, $5·5 \times 10^4$ cts/min/mμmole in Tris–Mg–mercaptoethanol buffer was incubated with 5 μl. of endonuclease R for 120 min at 37°C and for 42 min with an additional 5 μl. of enzyme. Five μl. were then removed, mixed with the 4 standard 5′-nucleotides, spotted on the corner of a thin layer PEI cellulose square and chromatographed in the first dimension with 1 M-formic acid followed by a second dimension with 1 M-LiCl using methods described by Kelly & Smith (1970). The marker spots were located with a shortwave ultraviolet mineralight, cut out and counted in toluene scintillation medium. The origin including the surrounding 1 cm of thin layer material was also counted. The results are presented as 50-min counts corrected for background.

(vi) *Native* versus *denatured DNA as substrate*

Endonuclease R is active only on native DNA. No decrease in the size of denatured T7 DNA was demonstrable by zone sedimentation on alkaline sucrose gradients after incubation with enzyme (Fig. 4(a) and (b)), whereas native DNA was degraded (Fig. 4(c)).

(vii) *3′-Hydroxyl, 5′-phosphoryl cleavage*

Polynucleotide kinase catalyses the transfer of a ^{32}P-labeled phosphoryl group from [γ-^{32}P]ATP to the 5′ terminus of the polynucleotide chain. This reaction is known to depend upon the presence of a free hydroxyl group at the 5′ terminus of the substrate and will not occur if a 5′-phosphoryl group is present (Weiss *et al.*, 1968). It follows that if pretreatment of the substrate with bacterial alkaline phosphatase is required in order to obtain transfer then the 5′ termini of the substrate must be phosphorylated. As seen in Table 4, phage P22 DNA which has been digested to completion with endonuclease R, incubated with phosphatase, and then with kinase and [γ-^{32}P]ATP, incorporated 3740 cts/min/20 mμmole of phage DNA whereas omission of the phosphatase treatment results in only 136 cts/min incorporation. Thus, endonuclease R produces a 3′-hydroxyl, 5′-phosphoryl cleavage. Without endonuclease R treatment only the ends of the intact DNA molecules are labeled. Subtracting the blank of 66 cts/min obtained when kinase is omitted, 114 cts/min are incorporated per 20 mμmole of phage DNA. Using this and the known specific activity of the [γ-^{32}P]ATP, the molecular weight of the phage DNA is calculated as 26×10^6 (see legend of Table 3 for method of calculation). This figure agrees well with the measured molecular weight of $26·3 \times 10^6$ obtained by Rhoades, MacHattie & Thomas (1968). The average calculated molecular weight of the limit product DNA fragments is $0·81 \times 10^6$. Approximately 32 breaks are produced per phage P22 DNA molecule by the enzyme.

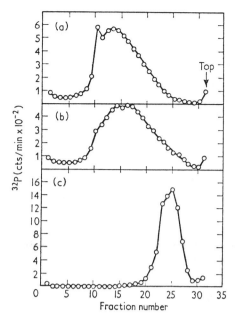

FIG. 4. Nuclease activity on denatured and native DNA.

Three reaction mixtures (a), (b) and (c) were set up in the following way. A control reaction mixture (a) contained 10 μl. of alkali-denatured ^{32}P-labeled phage T7 DNA, $1\cdot15\times10^4$ cts/min/ mμmole, 100 mμmole/ml. and 50 μl. of Tris–Mg–mercaptoethanol buffer containing 0·04 M-NaCl. Reaction mixture (b) contained 5 μl. of purified endonuclease R in addition to the above ingredients. Reaction mixture (c) was the same as in (b) except that 10 μl. of native ^{32}P-labeled T7 DNA was used. The three reaction mixtures were incubated for 15 min at 37°C after which 5 μl. of 0·5 M-EDTA was added and then each was alkali-denatured. Each mixture was then layered onto an alkaline sucrose gradient and centrifuged for 2·5 hr at 50,000 rev./min. Fractions were collected and counted as in Fig. 2. (The ^{32}P-labeled T7 DNA used in this experiment had suffered considerable radiation damage due to storage, and this accounts for the broad peaks obtained.)

TABLE 4

^{32}P-labeling of the 5'-end of endonuclease R-treated phage P22 DNA

DNA treatment	Cts/min	Calc. mol. wt
+ Bacterial alkaline phosphatase + kinase	180	26×10^6
Endonuclease + Bacterial alkaline phosphatase	66	
Endonuclease + kinase	136	
Endonuclease + Bacterial alkaline phosphatase + kinase	3740	$0\cdot81\times10^6$

The reaction mixture for endonuclease R digestion (0·45 ml.) contained 180 mμmoles of phage P22 DNA, 50 mM-NaCl, and 6·6 mM each of Tris–HCl (pH 7·4), MgCl$_2$, and mercaptoethanol. At zero time, endonuclease R (5 μl.) was added and the mixture was incubated at 37°C for 30 min. Samples of 0·05 ml. were removed at zero time (before addition of enzyme) and after the completion of digestion and treated with the various listed combinations of phosphatase or polynucleotide kinase. The [γ-^{32}P]ATP used in the kinase labeling reaction had a specific activity of $2\cdot25\times10^5$ cts/min/mμmole. The amount of ^{32}P incorporated terminally was measured as trichloroacetic acid-precipitable counts.

Molecular weight $= 660\ ma/(c-b)$, where m is the mμmoles of DNA, c is the incorporated terminal ^{32}P radioactivity, b is the blank in which kinase was absent, a is the specific activity of [γ-^{32}P]ATP, and the molecular weight of the base pair is 660.

(viii) *Mechanism of double-strand cleavage*

The sucrose gradient experiments indicate that native foreign DNA possesses a limited number of substrate sites which are subject to double-strand cleavage by endonuclease R. The question arises as to whether the introduction of the second single-strand break within a site is independent of the first. If so, then a limited digestion with endonuclease R should result in a large excess of nicks over duplex breaks. On the other hand, if the introduction of the second nick is coupled to that of the first, then duplex breaks should appear in significant amount early in the reaction.

To determine by which mechanism endonuclease R acts, ^{32}P–5′ end-labeling techniques were again used. It is possible, as described in Materials and Methods, to discriminate nicks from ends by this procedure taking advantage of the fact that nicks will be labeled only if the DNA is denatured before phosphatase treatment. Separate reaction mixtures were set up containing phage T7 DNA and various concentrations of endonuclease R ranging from an amount capable of producing relatively very few breaks during the incubation to an amount which was saturating. After incubation with the enzyme the DNA was labeled at the 5′-termini with ^{32}P either before or after denaturation (Fig. 5). The data show that with incomplete digestion, more label

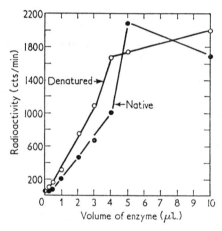

FIG. 5. Production of nicks and double-strand breaks in phage T7 DNA.
The reaction mixtures for endonuclease R digestion each contained unlabeled phage T7 DNA, 66 mμmoles and ^{3}H-labeled *H. influenzae* DNA, 0·23 mμmole, 1·5 × 10⁴ cts/min/mμmole in a 0·15-ml. vol of Tris–Mg–mercaptoethanol buffer containing 0·06 M-NaCl. To each reaction mixture the amount of enzyme shown in the Figure was added and incubation was carried out for 20 min at 37°C. Endonuclease R was then inactivated by treatment at 65°C for 10 min. The DNA in each tube was precipitated by 2 ml. of 70% ethanol and washed with 2 ml. of acetone. The DNA was redissolved in 0·15 ml. of NaCl–Tris buffer. An 0·5 ml. sample from each digest was then terminally labeled by the kinase reaction with or without alkali denaturation preceding the phosphatase step. The ^{3}H-labeled DNA was used to monitor recovery from the precipitation step. Recovery averaged about 60% and corrections were made to 100% recovery.

was incorporated into the DNA after denaturation than before, indicating that some nicks are produced first. However, the early appearance of a significant proportion of breaks clearly suggests an association between the two nicking events which result in a duplex break. This association might be explained by either of two mechanisms. Either the enzyme binds to a site and breaks first one strand and then the other without becoming detached, or else it introduces nicks in separate binding events with the probability for rebinding being much greater following the first nick. In the latter case we might imagine that the presence of the first nick makes the other chain more accessible.

An additional result of the experiments is that the DNA fragments produced as a limit product with high levels of enzyme contained no nicks, within the limits of error of the terminal labeling procedure, since both curves reached a plateau together at an average of about 1875 cts/min incorporated terminally. Since about 45 cts/min were incorporated into the undigested intact DNA, about 40 to 45 breaks were produced per phage T7 molecule.

4. Discussion

Endonuclease R produces double-strand, 5′-phosphoryl, 3′-hydroxyl cleavages at a limited number of sites on foreign native DNA. In addition to phage T7 and phage P22 DNA, we have tested *S. typhimurium* DNA, salmon sperm DNA and *Bacillus subtilis* DNA. All are degraded to about the same extent. The fragments produced in the limit digest of phage T7 DNA contain no excess of nicks, and essentially no trichloroacetic acid-soluble products or nucleotides are released by the digestion. The enzyme has no demonstrable activity on native *H. influenzae* DNA or on denatured foreign DNA. There are no special co-factor requirements for activity other than magnesium ions.

The similarity of endonuclease R to the restriction enzyme purified from *E. coli* by Meselson & Yuan (1968) is considerable. Both enzymes produce a small number of specific double-strand breaks in foreign DNA and are inactive on DNA of the bacterial strain from which they have been purified. An interesting feature of the *E. coli* enzyme is that it requires *S*-adenosyl methionine and adenosine triphosphate as co-factors in addition to magnesium ions. This suggests a connection between the restriction activity and the DNA modification activity which is known to protect the host DNA from cleavage. Arber (1968) has identified the DNA modification as methylation of adenine to form 6-methyl amino purine and recently Kühnlein, Linn & Arber (1969) have demonstrated modification *in vitro*. We have not as yet investigated DNA modification in *H. influenzae*.

We have been particularly interested in the ability of endonuclease R to "recognize" only a few specific sites on rather large foreign DNA molecules. It appears likely that this recognition specificity resides in the base sequence of the sites. An estimate of the size of the base sequence can be made. For phage T7 DNA we observed 40 to 45 breaks per molecule. Since T7 DNA is about 40,000 base pairs in length, the average fragment is about 1000 base pairs in length. For phage P22 DNA, the average fragment is approximately 1300 base pairs in length. To attain this degree of specificity a site would have to be five to six bases in length providing that the enzyme recognizes a completely unique sequence. In the accompanying paper (Kelly & Smith, 1970) the base sequence recognized by endonuclease R is completely identified and provides confirmation of this estimate.

We wish to acknowledge especially the considerable help received from Dr Bernard Weiss. He was very generous in supplying us with labeled ATP, several enzymes and much good advice. We thank Dr Paul Englund who donated enzymes and good advice; Dr Nagaraja Rao who helped during the initial stages of the work and Dr Thomas J. Kelly, Jr. who gave helpful suggestions during preparation of the manuscript.

This work was supported by U.S. Public Health Service grant no. AI-07875.

One of the authors (H.O.S.) holds a U.S. Public Health Service Career Development Award no. AI-17902.

REFERENCES

Arber, W. (1965). *Ann. Rev. Microbiol.* **19**, 365.

Arber, W. (1968). *18th Symp. Soc. Gen. Microbiol.: Molecular Biology of Viruses*, p. 295. London: Cambridge University Press.

Carmody, J. & Herriot, R. M. (1970). *J. Bact.* **101**, 525.

Kelly, T. J. & Smith, H. O. (1970). *J. Mol. Biol.* **51**, 393.

Kühnlein, U., Linn, S. & Arber, W. (1969). *Proc. Nat. Acad. Sci., Wash.* **63**, 556.

Levine, M. (1957). *Virology*, **3**, 203.

Marmur, J. (1961). *J. Mol. Biol.* **3**, 208.

Meselson, M. & Yuan, R. (1968). *Nature*, **217**, 1110.

Rhoades, M., MacHattie, L. A. & Thomas, C. A., Jr. (1968). *J. Mol. Biol.* **37**, 21.

Richardson, C. C. (1965). *Proc. Nat. Acad. Sci., Wash.* **54**, 158.

Smith, H. O. (1968). *Virology*, **34**, 203.

Smith, H. O. & Wilcox, K. W. (1969). *Fed.Proc.* **28**, 465.

Studier, F. W. (1965). *J. Mol. Biol.* **11**, 373.

Thomas, C. A., Jr. & Abelson, J. (1966). *Procedures in Nucleic Acid Research*, ed. by G. Cantoni & D. Davies, p. 553. New York: Harper & Row.

Weiss, B., Live, T. R. & Richardson, C. C. (1968). *J. Biol. Chem.* **243**, 4530.

RNA-dependent DNA Polymerase in Virions of Rous Sarcoma Virus

H.M. Temin and S. Mizutani

INFECTION of sensitive cells by RNA sarcoma viruses requires the synthesis of new DNA different from that synthesized in the S-phase of the cell cycle (refs. 1, 2 and D. Boettiger and H. M. T. (*Nature*, in the press)); production of RNA tumour viruses is sensitive to actinomycin D[3,4]; and cells transformed by RNA tumour viruses have new DNA which hybridizes with viral RNA[5,6]. These are the basic observations essential to the DNA provirus hypothesis—replication of RNA tumour viruses takes place through a DNA intermediate, not through an RNA intermediate as does the replication of other RNA viruses[7].

Formation of the provirus is normal in stationary chicken cells exposed to Rous sarcoma virus (RSV), even in the presence of 0·5 μg/ml. cycloheximide (our unpublished results). This finding, together with the discovery of polymerases in virions of vaccinia virus and of reovirus[8-11], suggested that an enzyme that would synthesize DNA from an RNA template might be present in virions of RSV. We now report data supporting the existence of such an enzyme, and we learn that David Baltimore has independently discovered a similar enzyme in virions of Rauscher leukaemia virus[12].

The sources of virus and methods of concentration have been described[13]. All preparations were carried out in sterile conditions. Concentrated virus was placed on a layer of 15 per cent sucrose and centrifuged at 25,000 r.p.m. for 1 h in the 'SW 25.1' rotor of the Spinco ultracentrifuge on to a cushion of 60 per cent sucrose. The virus band was collected from the interphase and further purified by equilibrium sucrose density gradient centrifugation[14]. Virus further purified by sucrose velocity density gradient centrifugation gave the same results.

The polymerase assay consisted of 0·125 μmoles each of dATP, dCTP, and dGTP (Calbiochem) (in 0·02 M Tris-HCl buffer at pH 8·0, containing 0·33 Mm EDTA and 1·7 mM 2-mercaptoethanol); 1·25 μmoles of $MgCl_2$ and 2·5 μmoles of KCl; 2·5 μg phosphoenolpyruvate (Calbiochem); 10 μg pyruvate kinase (Calbiochem); 2·5 μCi of ^{3}H-TTP (Schwarz) (12 Ci/mmole); and 0·025 ml. of enzyme (10^8 focus forming units of disrupted Schmidt-Ruppin virus, $A_{280\,nm} = 0·30$) in a total volume of 0·125 ml. Incubation was at 40° C for 1 h. 0·025 ml. of the reaction mixture was withdrawn and assayed for acid-insoluble counts by the method of Furlong[15].

Table 1. ACTIVATION OF ENZYME

System	³H-TTP incorporated (d.p.m.)
No virions	0
Non-disrupted virions	255
Virions disrupted with 'Nonidet'	
At 0° + DTT	6,730
At 0° − DTT	4,420
At 40° + DTT	5,000
At 40° − DTT	425

Purified virions untreated or incubated for 5 min at 0° C or 40° C with 0·25 per cent 'Nonidet P–40' (Shell Chemical Co.) with 0 or 1 per cent dithiothreitol (DTT) (Sigma) were assayed in the standard polymerase assay.

To observe full activity of the enzyme, it was necessary to treat the virions with a non-ionic detergent (Tables 1 and 4). If the treatment was at 40° C the presence of dithiothreitol (DTT) was necessary to recover activity. In most preparations of virions, however, there was some activity, 5–20 per cent of the disrupted virions, in the absence of detergent treatment, which probably represents disrupted virions in the preparation. It is known that virions of RNA tumour viruses are easily disrupted[16,17], so that the activity is probably present in the nucleoid of the virion.

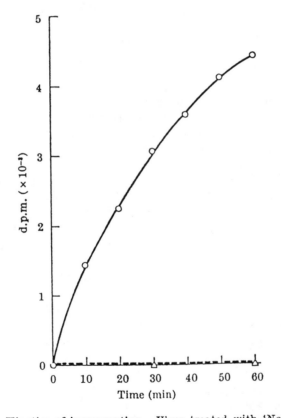

Fig. 1. Kinetics of incorporation. Virus treated with 'Nonidet' and dithiothreitol at 0° C and incubated at 37° C (O—O) or 80° C (△ - - - △) for 10 min was assayed in a standard polymerase assay.

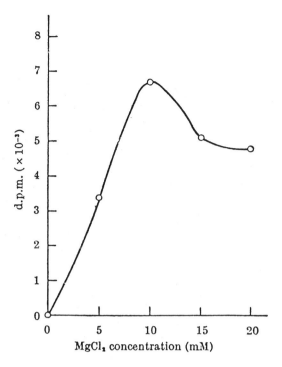

Fig. 2. MgCl₂ requirement. Virus treated with 'Nonidet' and dithio-threitol at 0° C was incubated in the standard polymerase assay with different concentrations of MgCl₂.

The kinetics of incorporation with disrupted virions are shown in Fig. 1. Incorporation is rapid for 1 h. Other experiments show that incorporation continues at about the same rate for the second hour. Preheating disrupted virus at 80° C prevents any incorporation, and so does pretreatment of disrupted virus with crystalline trypsin.

Fig. 2 demonstrates that there is an absolute requirement for $MgCl_2$, 10 mM being the optimum concentration. The data in Table 2 show that $MnCl_2$ can substitute for $MgCl_2$ in the polymerase assay, but $CaCl_2$ cannot. Other experiments show that a monovalent cation is not required for activity, although 20 mM KCl causes a 15 per cent stimulation. Higher concentrations of KCl are inhibitory: 60 per cent inhibition was observed at 80 mM.

Table 2. REQUIREMENTS FOR ENZYME ACTIVITY

System	³H-TTP incor-porated (d.p.m.)
Complete	5,675
Without MgCl₂	186
Without MgCl₂, with MnCl₂	5,570
Without MgCl₂, with CaCl₂	18
Without dATP	897
Without dCTP	1,780
Without dGTP	2,190

Virus treated with 'Nonidet' and dithiothreitol at 0° C was incubated in the standard polymerase assay with the substitutions listed.

Table 3. RNA DEPENDENCE OF POLYMERASE ACTIVITY

Treatment	³H-TTP incorporated (d.p.m.)
Non-treated disrupted virions	9,110
Disrupted virions preincubated with ribonuclease A (50 μg/ml.) at 20° C for 1 h	2,650
Disrupted virions preincubated with ribonuclease A (1 mg/ml.) at 0° C for 1 h	137
Disrupted virions preincubated with lysozyme (50 μg/ml.) at 0° C for 1 h	9,650

Disrupted virions were incubated with ribonuclease A (Worthington) which was heated at 80° C for 10 min, or with lysozyme at the indicated concentration in the specified conditions, and a standard polymerase assay was performed.

When the amount of disrupted virions present in the polymerase assay was varied, the amount of incorporation varied with second-order kinetics. When incubation was carried out at different temperatures, a broad optimum between 40° C and 50° C was found: (The high temperature of this optimum may relate to the fact that the normal host of the virus is the chicken.) When incubation was carried out at different pHs, a broad optimum at pH 8–9.5 was found.

Table 2 demonstrates that all four deoxyribonucleoside triphosphates are required for full activity, but some activity was present when only three deoxyribonucleoside triphosphates were added and 10–20 per cent of full activity was still present with only two deoxyribonucleoside triphosphates. The activity in the presence of three deoxyribonucleoside triphosphates is probably the result of the presence of deoxyribonucleoside triphosphates in the virion. Other host components are known to be incorporated in the virion of RNA tumour viruses[18,19].

The data in Table 3 demonstrate that incorporation of thymidine triphosphate was more than 99 per cent abolished if the virions were pretreated at 0° with 1 mg ribonuclease per ml. Treatment with 50 μg/ml. ribo-

Table 4. SOURCE OF POLYMERASE

Source	³H-TTP incorporated (d.p.m.)
Virions of SRV	1,410
Disrupted virions of SRV	5,675
Virions of AMV	1,875
Disrupted virions of AMV	12,850
Disrupted pellet from supernatant of uninfected cells	0

Virions of Schmidt-Ruppin virus (SRV) were prepared as before (experiment of Table 2). Virions of avian myeloblastosis virus (AMV) and a pellet from uninfected cells were prepared by differential centrifugation. All disrupted preparations were treated with 'Nonidet' and dithiothreitol at 0° C and assayed in a standard polymerase assay. The material used per tube was originally from 45 ml. of culture fluid for SRV, 20 ml. for AMV, and 20 ml. for uninfected cells.

Table 5. NATURE OF PRODUCT

Treatment	Residual acid-insoluble ³H-TTP (d.p.m.)	
	Experiment A	Experiment B
Buffer	10,200	8,350
Deoxyribonuclease	697	1,520
Ribonuclease	10,900	7,200
KOH	—	8,250

A standard polymerase assay was performed with 'Nonidet' treated virions. The product was incubated in buffer or 0·3 M KOH at 37° C for 20 h or with (*A*) 1 mg/ml. or (*B*) 50 μg/ml. of deoxyribonuclease I (Worthington), or with 1 mg/ml. of ribonuclease A (Worthington) for 1 h at 37° C, and portions were removed and tested for acid-insoluble counts.

nuclease at 20° C did not prevent all incorporation of thymidine triphosphate, which suggests that the RNA of the virion may be masked by protein. (Lysozyme was added as a control for non-specific binding of ribonuclease to DNA.) Because the ribonuclease was heated for 10 min at 80° C or 100° C before use to destroy deoxyribonuclease it seems that intact RNA is necessary for incorporation of thymidine triphosphate.

To determine whether the enzyme is present in supernatants of normal cells or in RNA leukaemia viruses, the experiment of Table 4 was performed. Normal cell supernatant did not contain activity even after treatment with 'Nonidet'. Virions of avian myeloblastosis virus (AMV) contained activity that was increased ten-fold by treatment with 'Nonidet'.

The nature of the product of the polymerase assay was investigated by treating portions with deoxyribonuclease, ribonuclease or KOH. About 80 per cent of the product was made acid soluble by treatment with deoxyribonuclease, and the product was resistant to ribonuclease and KOH (Table 5).

To determine if the polymerase might also make RNA, disrupted virions were incubated with the four ribonucleoside triphosphates, including ³H-UTP (Schwarz, 3·2 Ci/mmole). With either $MgCl_2$ or $MnCl_2$ in the incubation mixture, no incorporation was detected. In a parallel incubation with deoxyribonucleoside triphosphates, 12,200 d.p.m. of ³H-TTP was incorporated.

These results demonstrate that there is a new polymerase inside the virions of RNA tumour viruses. It is not present in supernatants of normal cells but is present in virions of avian sarcoma and leukaemia RNA tumour viruses. The polymerase seems to catalyse the incorporation of deoxyribonucleoside triphosphates into DNA from an RNA template. Work is being performed to characterize further the reaction and the product. If the present results and Baltimore's results[12] with Rauscher leukaemia virus are upheld, they will constitute strong evidence that the DNA provirus hypothesis is correct and that RNA tumour viruses have a DNA genome when they are in cells and an RNA genome when they are in virions. This result would have strong implications for theories of viral carcinogenesis and, possibly, for theories of information transfer in other biological systems[20].

This work was supported by a US Public Health Service research grant from the National Cancer Institute. H. M. T. holds a research career development award from the National Cancer Institute.

HOWARD M. TÉMIN
SATOSHI MIZUTANI

McArdle Laboratory for Cancer Research,
University of Wisconsin,
Madison,
Wisconsin 53706.

Received June 15, 1970.

[1] Temin, H. M., *Cancer Res.*, 28, 1835 (1968).

[2] Murray, R. K., and Temin, H. M., *Intern. J. Cancer*, 5, 320 (1970).

[3] Temin, H. M., *Virology*, 20, 577 (1963).

[4] Baluda, M. B., and Nayak, D. P., *J. Virol.*, 4, 554 (1969).

[5] Temin, H. M., *Proc. US Nat. Acad. Sci.*, 52, 323 (1964).

[6] Baluda, M. B., and Nayak, D. P., in *Biology of Large RNA Viruses* (edit. by Barry, R., and Mahy, B.) (Academic Press, London, 1970).

[7] Temin, H. M., *Nat. Cancer Inst. Monog.*, 17, 557 (1964).

[8] Kates, J. R., and McAuslan, B. R., *Proc. US Nat. Acad. Sci.*, 57, 314 (1967).

[9] Munyon, W., Paoletti, E., and Grace, J. T., *Proc. US Nat. Acad. Sci.*, 58, 2280 (1967).

[10] Borsa, J., and Graham, A. F., *Biochem. Biophys. Res. Commun.*, 33, 895 (1968).

[11] Shatkin, A. J., and Sipe, J. D., *Proc. US Nat. Acad. Sci.*, 61, 1462 (1968).

[12] Baltimore, D., *Nature*, 226, 1209 (1970) (preceding article).

[13] Altaner, C., and Temin, H. M., *Virology*, 40, 118 (1970).

[14] Robinson, W. S., Pitkanen, A., and Rubin, H., *Proc. US Nat. Acad. Sci.*, 54, 137 (1965).

[15] Furlong, N. B., *Meth. Cancer Res.*, 3, 27 (1967).

[16] Vogt, P. K., *Adv. Virus. Res.*, 11, 293 (1965).

[17] Bauer, H., and Schafer, W., *Virology*, 29, 494 (1966).

[18] Bauer, H., *Z. Naturforsch.*, 21b, 453 (1966).

[19] Erikson, R. L., *Virology*, 37, 124 (1969).

[20] Temin, H. M., *Persp. Biol. Med.*, 5, 320 (1970).

The Intervening Sequence RNA of *Tetrahymena* Is An Enzyme

A.J. Zaug and T.R. Cech

A shortened form of the self-splicing ribosomal RNA (rRNA) intervening sequence of *Tetrahymena thermophila* acts as an enzyme in vitro. The enzyme catalyzes the cleavage and rejoining of oligonucleotide substrates in a sequence-dependent manner with $K_m = 42$ μM and $k_{cat} = 2$ min^{-1}. The reaction mechanism resembles that of rRNA precursor self-splicing. With pentacytidylic acid as the substrate, successive cleavage and rejoining reactions lead to the synthesis of polycytidylic acid. Thus, the RNA molecule can act as an RNA polymerase, differing from the protein enzyme in that it uses an internal rather than an external template. At pH 9, the same RNA enzyme has activity as a sequence-specific ribonuclease.

IN RNA SELF-SPLICING, THE FOLDED STRUCTURE OF AN RNA molecule mediates specific cleavage-ligation reactions (*1–5*). Self-splicing exemplifies intramolecular catalysis (*6*) in that the reactions are accelerated many orders of magnitude beyond the basal chemical rate (*7, 8*). The reactions are highly specific, as seen in the choice of a free guanosine nucleotide as a substrate in the self-splicing of the *Tetrahymena* ribosomal RNA precursor (pre-rRNA)

Thomas R. Cech is a professor and Arthur J. Zaug is a research associate in the Department of Chemistry and Biochemistry, University of Colorado, Boulder, 80309-0215. Send correspondence to T.R.C.

and other RNA's containing group I intervening sequences (*1–3, 7*). Furthermore, the cleavage-ligation activity mediates a series of splicing, cyclization, and reverse cyclization reactions, suggesting that the active site is preserved in each reaction (*9, 10*). However, the RNA is cleaved and rejoined during self-splicing; because the RNA is not regenerated in its original form at the end of the reaction, it is not an enzyme. The RNA moiety of ribonuclease P, the enzyme responsible for cleaving transfer RNA (tRNA) precursors to generate the mature 5′ end of the tRNA, has been the only example of an RNA molecule that meets all criteria of an enzyme (*11–13*).

Following self-splicing of the *Tetrahymena* rRNA precursor, the excised IVS RNA (*14*) undergoes a series of RNA-mediated cyclization and site-specific hydrolysis reactions. The final product, the L − 19 IVS RNA, is a linear molecule that does not have the first 19 nucleotides of the original excised IVS RNA (*9*). We interpreted the lack of further reaction of the L − 19 species as an indication that all potential reaction sites on the molecule that could reach its active site (that is, intramolecular substrates) had been consumed; and we argued that the activity was probably unperturbed (*9*). We have now tested this by adding oligonucleotide substrates to the L − 19 IVS RNA. We find that each IVS RNA molecule can catalyze the cleavage and rejoining of many oligonucleotides. Thus, the L − 19 IVS RNA is a true enzyme. Although the enzyme can act on RNA molecules of large size and complex sequence, we have found that studies with simple oligoribonucleotides like pC$_5$ (pentacytidylic acid) have been most valuable in revealing the minimum substrate requirements and reaction mechanism of this enzyme. These studies are presented below.

The L − 19 IVS RNA catalyzes the cleavage and rejoining of oligonucleotides.

Unlabeled L − 19 IVS RNA was incubated with $5'$-^{32}P–labeled pC_5 in a solution containing 20 mM MgCl$_2$, 50 mM tris-HCl, pH 7.5. The pC_5 was progressively converted to oligocytidylic acid with both longer and shorter chain length than the starting material (Fig. 1A). The longer products extended to at least pC_{30}, as judged by a longer exposure of an autoradiogram such as that shown in Fig. 1A. The shorter products were exclusively pC_4 and pC_3. Incubation of pC_5 in the absence of the L − 19 IVS RNA gave no reaction (Fig. 1C).

Phosphatase treatment of a 60-minute reaction mixture resulted in the complete conversion of the ^{32}P radioactivity to inorganic phosphate, as judged by polyethyleneimine thin-layer chromatography (TLC) in 1M sodium formate, pH 3.5 (15). Thus, the $5'$-terminal phosphate of the substrate does not become internalized during the reaction, and the substrate is being extended on its $3'$ end to form the larger oligonucleotides. When $C_5\overset{*}{p}C$ was used as the substrate and the products were treated with ribonuclease T$_2$ or ribonuclease A, the ^{32}P radioactivity was totally converted to $C\overset{*}{p}$ (15). Thus, the linkages being formed in the reaction are exclusively $3',5'$-phosphodiester bonds. The products of the $C_5\overset{*}{p}C$ reaction were totally resistant to phosphatase treatment.

The reaction was specific for ribonucleotides, no reaction taking place with d-pC_5 (Fig. 1B) or d-pA_5 (15). Among the oligoribonucleotides, pU_6 was a much poorer substrate than pC_5 or pC_6 (Fig. 1D), and pA_6 gave no reaction (15).

No reaction occurred when magnesium chloride was omitted. The enzyme activity was approximately constant in the range 5 to 40 mM MgCl$_2$ (15). The 20 mM concentration was routinely used to circumvent the potential effect of chelation of Mg^{2+} by high concentrations of oligonucleotide substrates.

The L − 19 IVS RNA is regenerated after each reaction, such that each enzyme molecule can react with many substrate molecules. For example, quantitation of the data shown in Fig. 1G revealed that 16 pmol of enzyme converted 1080 pmol of pC_5 to products in 60 minutes. Such numbers underestimate the turnover number of the enzyme; because the initial products are predominantly C_6 and C_4, it is likely that the production of chains of length greater than six or less than four involves two or more catalytic cycles. Quantitation of the amount of radioactivity in each product also provides some indication of the reaction mechanism. At early reaction times, the amount of radioactivity (a measure of numbers of chains) in products larger than pC_5 is approximately equal to that found in pC_4 plus pC_3, consistent with a mechanism in which the total number of phosphodiester bonds is conserved in each reaction. As the reaction proceeds, however, the radioactivity distribution shifts toward the smaller products. This is most likely due to a competing hydrolysis reaction also catalyzed by the L − 19 IVS RNA, as described below.

The rate of conversion of 30 μM pC_5 to products increases linearly with L − 19 IVS RNA enzyme concentration in the range 0.06 to 1.00 μM (15). At a fixed enzyme concentration (Fig. 1, E to G), there is a hyperbolic relation between the reaction rate and the concentration of pC_5. The data are fit by the Michaelis-Menten rate law in Fig. 2. The resulting kinetic parameters are $K_m = 42$ μM and $k_{cat} = 1.7$ min^{-1}.

The stability of the enzyme was determined by preliminary incubation at 42°C for 1 hour in the presence of Mg^{2+} (standard reaction conditions) or for 18 hours under the same conditions but without Mg^{2+}. In both cases, the incubated enzyme had activity indistinguishable from that of untreated enzyme tested in parallel, and no degradation of the enzyme was observed on polyacrylamide gel electrophoresis (15). Thus, the L − 19 IVS RNA is not a good

substrate. The enzyme is also stable during storage at −20°C for periods of months. The specific activity of the enzyme is consistent between preparations.

Covalent intermediate. When $C_5\overset{*}{p}$ was used as a substrate, radioactivity became covalently attached to the L − 19 IVS RNA (Fig. 3A) (16). This observation, combined with our previous knowledge of the mechanism of IVS RNA cyclization (9, 10, 17,

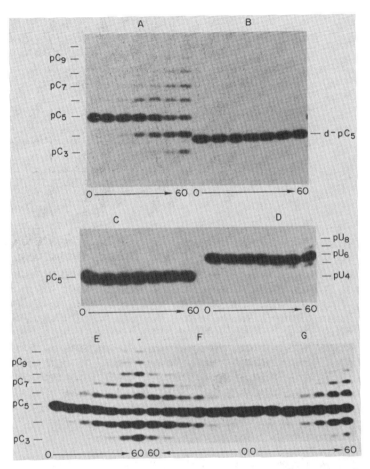

Fig. 1. The L − 19 IVS RNA catalyzes the cleavage and rejoining of oligoribonucleotide substrates; (A) 10 μM pC_5 and (B) 10 μM d-pC_5, both with 1.6 μM L − 19 IVS RNA; (C) 45 μM pC_5 in the absence of L − 19 IVS RNA; (D) 45 μM pU_6 with 1.6 μM L − 19 IVS RNA; (E) 10 μM pC_5, (F) 50 μM pC_5 and (G) 100 μM pC_5, all with 1.6 μM L − 19 IVS RNA. Oligonucleotides were $5'$-end labeled by treatment with [γ-^{32}P]ATP and polynucleotide kinase; they were diluted with unlabeled oligonucleotide of the same sequence to keep the amount of radioactivity per reaction constant. The L − 19 IVS RNA was synthesized by transcription and splicing in vitro. Supercoiled pSPTT1A3 DNA (30) was cut with Eco RI and then transcribed with SP6 RNA polymerase (31) for 2 hours at 37°C in a solution of nucleoside triphosphates (0.5 mM each), 6 mM MgCl$_2$, 4 mM spermidine, 10 mM dithiothreitol, 40 mM tris-HCl, pH 7.5, with 100 units of SP6 RNA polymerase per microgram of plasmid DNA. Then NaCl was added to a final concentration of 240 mM and incubation was continued at 37°C for 30 minutes to promote excision and cyclization of the IVS RNA. Nucleic acids were precipitated with three volumes of ethanol and redissolved in 50 mM CHES, pH 9.0; MgCl$_2$ was added to a final concentration of 20 mM, and the solution was incubated at 42°C for 1 hour to promote site-specific hydrolysis of the circular IVS RNA to give L − 19 IVS RNA (9). The reaction was stopped by the addition of EDTA to 25 mM. The L − 19 IVS RNA was purified by preparative gel electrophoresis and Sephadex G-50 chromatography. Labeled oligonucleotides were incubated with unlabeled L − 19 IVS RNA at 42°C in 20 mM MgCl$_2$, 50 mM tris, pH 7.5, for 0, 1, 2, 5, 10, 30, and 60 minutes. Reactions were stopped by the addition of EDTA to a final concentration of 25 mM. Products were analyzed by electrophoresis in a 20 percent polyacrylamide, 7M urea gel, autoradiograms of which are shown.

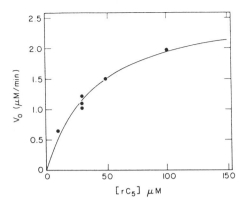

Fig. 2. Kinetics of conversion of pC_5 to larger and smaller oligonucleotides with 1.6 μM L – 19 IVS RNA. Products were separated by polyacrylamide gel electrophoresis. With the autoradiogram as a guide, the gel was cut into strips and the radioactivity in each RNA species was determined by liquid scintillation counting. The amount of reaction at each time was taken as the radioactivity in $pC_3 + pC_4 + pC_6 + pC_7 + \ldots$ divided by the total radioactivity in the lane. The initial velocity of product formation, V_o, was determined from a semilogarithmic plot of the fraction of reaction as a function of time. V_o was then plotted as a function of substrate concentration; the line is a least-squares fit to the Michaelis-Menten equation. The resulting kinetic parameters are $K_m = 42$ μM, $V_{max} = 2.8$ μM min^{-1}, and $k_{cat} = 1.7$ min^{-1}. The kinetic parameters for the first and second steps in the reaction have not yet been determined separately.

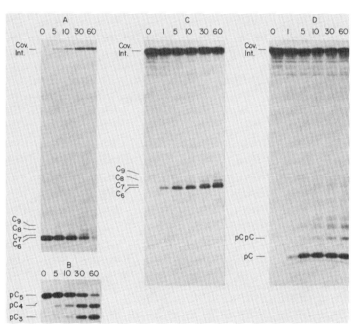

Fig. 3. Formation and resolution of the covalent enzyme-substrate intermediate. (A) To drive the formation of the covalent L – 19 IVS RNA-substrate intermediate, 8.5 nM C_5pC was treated with 0.16 μM L – 19 IVS RNA under standard reaction conditions for 0 to 60 minutes. (B) $\overset{*}{p}C_5$ (0.01 μM) was reacted with 0.16 μM L – 19 IVS RNA. Cleavage occurred normally, but there was very little rejoining. (C) Labeled covalent intermediate was prepared as in (A) (60 minutes) and purified by electrophoresis in a 4 percent polyacrylamide, 8M urea gel. It was then incubated with 10 μM unlabeled C_5 under standard reaction conditions for 0 to 60 minutes. The product designated C_6 comigrated with labeled C_6 marker (not shown). (D) Isolated covalent intermediate as in (C) incubated under site-specific hydrolysis conditions (20 mM MgCl$_2$, 50 mM CHES, pH 9.0) at 42°C for 0 to 60 minutes. Positions of labeled mono- and dinucleotide markers are indicated. In the 10- and 30-minute lanes of (A) and the 10-, 30-, and 60-minute lanes of (C), band compression (reduced difference in electrophoretic mobility) is seen between C_6 and C_7 and to a lesser extent between C_7 and C_8. This is due to the absence of a 5′ phosphate. Thus, the charge-to-mass ratio is increasing with chain length, whereas with 5′-phosphorylated oligonucleotides the charge-to-mass ratio is independent of chain length. When such products were phosphorylated by treatment with polynucleotide kinase and ATP, the distribution was converted to the normal spacing as in Fig. 1 (15).

18), led to a model for the reaction mechanism involving a covalent enzyme-substrate intermediate (Fig. 4).

This reaction pathway is supported by analysis of reactions in which a trace amount of $\overset{*}{p}C_5$ was incubated with a large molar excess of L – 19 IVS RNA. The cleavage reaction occurred with high efficiency, as judged by the production of $\overset{*}{p}C_4$ and $\overset{*}{p}C_3$, but there was very little synthesis of products larger than the starting material (Fig. 3B; compare to Fig. 1A). These data are easily interpreted in terms of the proposed reaction pathway. The first step, formation of the covalent intermediate with release of the 5′-terminal fragment of the oligonucleotide, is occurring normally. The first step consumes all the substrate, leaving insufficient C_5 to drive the second transesterification reaction.

The model shown in Fig. 4 was tested by isolating the covalent enzyme-substrate complex prepared by reaction with $C_5\overset{*}{p}C$ and incubating it with unlabeled C_5. A portion of the radioactivity was converted to oligonucleotides with the electrophoretic mobility of C_6, C_7, C_8, and higher oligomers (Fig. 3C). In a confirmatory experiment, the covalent complex was prepared with unlabeled C_5 and reacted with $\overset{*}{p}C_5$. Radioactivity was again converted to a series of higher molecular weight oligonucleotides (15). In both types of experiments the data are readily explained if the covalent complex is a mixture of L – 19 IVS RNA's terminating in . . .GpC, . . .GpCpC, . . .GpCpCpC, and so on. Because they can react with C_5 to complete the catalytic cycle, these covalent enzyme-substrate complexes are presumptive intermediates in the reaction (Fig. 4). A more detailed analysis of the rate of their formation and resolution is needed to evaluate whether or not they are kinetically competent to be intermediates. We can make no firm conclusion about that portion of the enzyme-substrate complex that did not react with C_5. This unreactive RNA could be a covalent intermediate that was denatured during isolation such that it lost reactivity, or it could represent a small amount of a different enzyme-substrate complex that was nonproductive and therefore accumulated during the reaction.

The G^{414}-A^{16} linkage in the C IVS RNA, the G^{414}-U^{20} linkage in the C′ IVS RNA, and the G^{414}-U^{415} linkage in the pre-rRNA are unusual phosphodiester bonds in that they are extremely labile to alkaline hydrolysis, leaving 5′ phosphate and 3′-hydroxyl termini (9, 19). We therefore tested the lability of the G^{414}-C linkage in the covalent enzyme-substrate intermediate by incubation at pH 9.0 in a Mg^{2+}-containing buffer. This treatment resulted in the release of products that comigrated with pC and pCpC markers and larger products that were presumably higher oligomers of pC (Fig. 3D). Thin-layer chromatography was used to confirm the identity of the major products (15). In those molecules that released pC, the release was essentially complete in 5 minutes. Approximately half of the covalent complex was resistant to the pH 9.0 treatment. Once again, we can make no firm conclusion about the molecules that did not react. The lability of the G^{414}–C bond forms the basis for the L – 19 IVS RNA acting as a ribonuclease (Fig. 4).

A competitive inhibitor. Deoxy C_5, which is not a substrate for L – 19 IVS RNA–catalyzed cleavage, inhibits the cleavage of pC_5 (Fig. 5A). Analysis of the rate of the conversion of pC_5 to pC_4 and pC_3 as a function of d-C_5 concentration is summarized in Fig. 5, B and C. The data indicate that d-C_5 is a true competitive inhibitor with the inhibition constant $K_i = 260$ μM. At 500 μM, d-A_5 inhibits the reaction only 16 percent as much as d-C_5. Thus, inhibition by d-C_5 is not some general effect of introducing a deoxyoligonucleotide into the system but depends on sequence.

The formation of the covalent enzyme-substrate intermediate (EpC) can be represented as

$$E + C_5 \underset{k_{-1}}{\overset{k_1}{\rightleftharpoons}} E \cdot C_5 \overset{k_2}{\longrightarrow} EpC + C_4$$

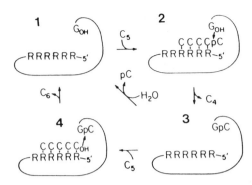

Fig. 4. Model for the enzymatic mechanism of the L − 19 IVS RNA. The RNA catalyzes cleavage and rejoining of oligo(C) by the pathway $1 \rightarrow 2 \rightarrow 3 \rightarrow 4 \rightarrow 1$. The L − 19 IVS RNA enzyme (1) is shown with the oligopyrimidine binding site (RRRRRR, six purines) near its 5′ end and G^{414} with a free 3′-hydroxyl group at its 3′ end. The complex folded core structure of the molecule (23, 24, 32) is simply represented by a curved line. The enzyme binds its substrate (C_5) by Watson-Crick base-pairing to form the noncovalent enzyme-substrate complex (2). Nucleophilic attack by G^{414} leads to formation of the covalent intermediate (3). With the pentanucleotide C_5 as substrate, the covalent intermediate is usually loaded with a single nucleotide, as shown; with substrates of longer chain length, an oligonucleotide can be attached to the 3′ end of G^{414}. If C_5 binds to the intermediate (3) in the manner shown in (4), transesterification can occur to give the new product C_6 and regenerate the enzyme (1). Note that all four reactions in this pathway are reversible. When acting as a ribonuclease, the L − 19 IVS RNA follows the pathway $1 \rightarrow 2 \rightarrow 3 \rightarrow 1$. The covalent intermediate (3) undergoes hydrolysis, releasing the nucleotide or oligonucleotide attached to its 3′ end (in this case pC) and regenerating the enzyme (1).

If $k_{-1} >> k_2$, then $K_m = k_{-1}/k_1$, the dissociation constant for the noncovalent E · C_5 complex. The observation that the K_i for d-C_5 is within an order of magnitude of the K_m for C_5 can then be interpreted in terms of d-C_5 and C_5 having similar binding constants for interaction with the active site on the enzyme. This fits well with the idea that the substrate binds to an oligopurine (R_5) sequence in the active site primarily by Watson-Crick base-pairing, in which case the $C_5 \cdot R_5$ duplex and the d-$C_5 \cdot R_5$ duplex should have similar stability.

Enzyme mechanism and its relation to self-splicing. The stoichiometry of the reaction products (equimolar production of oligonucleotides smaller than and larger than the starting material), the lack of an ATP or GTP (adenosine triphosphate; guanosine triphosphate) energy requirement, the involvement of a covalent intermediate, the specificity for oligoC substrates, and the competi-tive inhibition by d-C_5 lead to a model for the enzyme mechanism (Fig. 4). The L − 19 IVS RNA is proposed to bind the substrate noncovalently by hydrogen-bonded base-pairing interactions. A transesterification reaction between the 3′-terminal guanosine residue of the enzyme and a phosphate ester of the substrate then produces a covalent enzyme-substrate intermediate.

Transesterification is expected to be highly reversible. If the product C_4 rebinds to the enzyme, it can attack the covalent intermediate and reform the starting material, C_5. Early in the reaction, however, the concentration of C_5 is much greater than the concentration of C_4; if C_5 binds and attacks the covalent intermediate, C_6 is produced (Fig. 4). The net reaction is $2 C_5 \rightarrow C_6 + C_4$. The products are substrates for further reaction, for example, $C_6 + C_5 \rightarrow C_7 + C_4$ and $C_4 + C_5 \rightarrow C_3 + C_6$. The absence of products smaller then C_3 is explicable in terms of the loss of binding

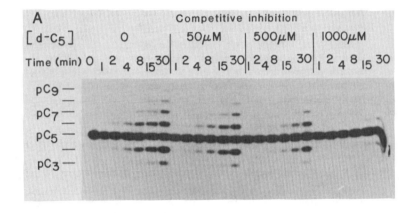

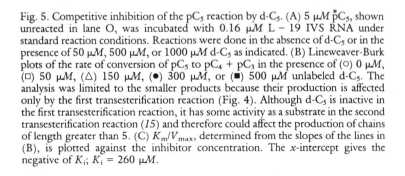

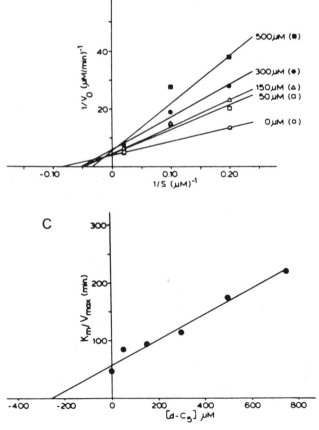

Fig. 5. Competitive inhibition of the pC_5 reaction by d-C_5. (A) 5 μM pC_5, shown unreacted in lane O, was incubated with 0.16 μM L − 19 IVS RNA under standard reaction conditions. Reactions were done in the absence of d-C_5 or in the presence of 50 μM, 500 μM, or 1000 μM d-C_5 as indicated. (B) Lineweaver-Burk plots of the rate of conversion of pC_5 to pC_4 + pC_3 in the presence of (○) 0 μM, (□) 50 μM, (△) 150 μM, (●) 300 μM, or (■) 500 μM unlabeled d-C_5. The analysis was limited to the smaller products because their production is affected only by the first transesterification reaction (Fig. 4). Although d-C_5 is inactive in the first transesterification reaction, it has some activity as a substrate in the second transesterification reaction (15) and therefore could affect the production of chains of length greater than 5. (C) K_m/V_{max}, determined from the slopes of the lines in (B), is plotted against the inhibitor concentration. The x-intercept gives the negative of K_i; K_i = 260 μM.

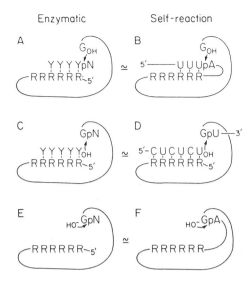

Fig. 6. Relation of reactions catalyzed by the L − 19 IVS RNA to self-splicing and the related IVS RNA-mediated reactions. Formation of the covalent enzyme-substrate intermediate (A) is analogous to IVS RNA autocyclization (B). Resolution of the enzyme-substrate intermediate (C) is analogous to exon ligation (D) or the reversal of cyclization (10). Hydrolysis of the enzyme-substrate intermediate (E) is analogous to site-specific hydrolysis of the circular IVS RNA (F) or of the pre-rRNA (19).

interactions relative to C_4 (C_3 could form only two base pairs in the binding mode that would be productive for cleavage).

The transesterification reactions are conservative with respect to the number of phosphodiester bonds in the system. Thus, RNA ligation can occur without an external energy source as is required by RNA or DNA ligase. Hydrolysis of the covalent intermediate competes with transesterification. The net reaction is $C_5 + H_2O \rightarrow C_4 + pC$, with the L − 19 IVS RNA acting as a ribonuclease.

On the basis of our current understanding of the reaction, the catalytic strategies of the L − 19 IVS RNA enzyme appear to be the same as those used by protein enzymes (20). First, the RNA enzyme, like protein enzymes, forms a specific noncovalent complex with its oligonucleotide substrate. This interaction is proposed to hold the oligonucleotide substrate at a distance and in an orientation such as to facilitate attack by the 3'-hydroxyl of the terminal guanosine of the enzyme. Second, a covalent enzyme-substrate complex is a presumptive intermediate in the L − 19 IVS RNA reaction. Covalent intermediates are prevalent in enzyme-catalyzed group transfer reactions. Third, the phosphodiester bond formed in the covalent intermediate is unusually susceptible to hydrolysis, suggesting that it may be strained or activated to facilitate formation of the pentavalent transition state upon nucleophilic attack (8, 9). Similarly, protein catalysts are thought to facilitate the formation of the transition state, for example, by providing active site groups that bind the transition state better than the unreacted substrate (6, 21). Thus far there is no evidence that another major category of enzyme catalysis, general acid-base catalysis, occurs in the L − 19 IVS RNA reactions, but we think it likely that it will be involved in facilitating the required proton transfers.

Each L − 19 IVS RNA-catalyzed transesterification and hydrolysis reaction is analogous to one of the steps in *Tetrahymena* pre-rRNA self-splicing or one of the related self-reactions (Fig. 6). Thus, the finding of enzymatic activity in a portion of the IVS RNA validates the view that the pre-rRNA carries its own splicing enzyme as an intrinsic part of its polynucleotide chain. It seems likely that the C_5 substrate binding site of the L − 19 IVS RNA is the

oligopyrimidine binding site that directs the choice of the 5' splice site and the various IVS RNA cyclization sites (10, 18, 19, 22). Although the location of this site within the IVS RNA has not been proved, the best candidate is a portion of the "internal guide sequence" proposed by Davies and co-workers (23). Michel and Dujon (24) show a similar pairing interaction in their RNA structure model. The putative binding site, GGAGGG, is located at nucleotides 22 to 27 of the intact *Tetrahymena* IVS RNA and at positions 3 to 8 very near the 5' end of the L − 19 IVS RNA. If this is the substrate binding site, site-specific mutation of the sequence should change the substrate specificity of the enzyme in a predictable manner.

RNA polymerase or RNA restriction endonuclease? With C_5 as a substrate, the L − 19 IVS RNA makes poly(C) with chain lengths of 30 nucleotides and longer. The number of P-O bonds is unchanged in the process. In the synthesis of poly(C) on a poly(dG) template by RNA polymerase, one CTP is cleaved for each residue polymerized. Thus, the RNA polymerase reaction is also conservative with respect to the number of P-O bonds in the system. The L − 19 IVS RNA can therefore be considered to be a poly(C) polymerase that uses C_4pC instead of pppC as a substrate. It incorporates pC units at the 3' end of the growing chain and releases C_4; the C_4 is analogous to the pyrophosphate released by RNA polymerase. Synthesis is directed by a template, but the template is internal to the RNA enzyme. It may be possible to physically separate the template portion from the catalytic portion of the RNA enzyme with retention of activity. If so, the RNA enzyme could conceivably act as a primordial RNA replicase, catalyzing both its own replication and that of other RNA molecules (25).

In its ribonuclease mode, the L − 19 IVS RNA is expected to have specificity similar to that of the IVS RNA cyclization reaction (10, 18). That is, it recognizes three or more nucleotides in choosing a reaction site. Protein ribonucleases that are active on single-stranded RNA substrates have specificity only at the mononucleotide level (for example, ribonuclease T_1 cleaves after guanosine). Thus the L − 19 has more base-sequence specificity for single-stranded RNA than any known protein ribonuclease, and may approach the specificity of some of the DNA restriction endonucleases. An attractive feature of this new RNA ribonuclease is the possibility of completely and predictably changing its substrate specificity by altering the sequence of the internal binding site.

How good an enzyme? The L − 19 IVS RNA catalyzes the cleavage-ligation of pC_5 with $K_m = 42 \ \mu M$, $k_{cat} = 2 \ min^{-1}$, and $k_{cat}/K_m = 1 \times 10^3 \ sec^{-1} \ M^{-1}$. The K_m is typical of that of protein enzymes. The k_{cat} and k_{cat}/K_m are lower than those of many protein enzymes. However, k_{cat} is well within the range of values for proteins that recognize specific nucleic acid sequences and catalyze chain cleavage or initiation of polymerization. For example, Eco RI restriction endonuclease cleaves its recognition sequence in various DNA substrates, including a specific 8-bp DNA fragment, with $k_{cat} = 1 \ min^{-1}$ to $18 \ min^{-1}$ (26). The k_{cat} is also similar to that of the RNA enzyme ribonuclease P, which cleaves the precursor to tRNA with $k_{cat} = 2 \ min^{-1}$ (11, 13).

Another way to gauge the catalytic effectiveness of the L − 19 IVS RNA is to compare the rate of the catalyzed reaction to the basal chemical rate. A transesterification reaction between two free oligonucleotides has never been observed, and hence the uncatalyzed rate is unknown. On the other hand, the rate of hydrolysis of simple phosphate diesters has been studied (27, 28). The second-order rate constant for alkaline hydrolysis of the labile phosphodiester bond in the circular IVS RNA (8) is 12 orders of magnitude higher than that of dimethyl phosphate (27) and ten orders of magnitude higher than that expected for a normal phosphodiester bond in RNA (29). On the basis of the data of Fig. 3D, the covalent enzyme-substrate complex undergoes hydrolysis at approximately the same rate as the equivalent bond in the circular IVS RNA. Thus, we estimate that the $L - 19$ IVS RNA in its ribonuclease mode enhances the rate of hydrolysis of its substrate about 10^{10} times.

REFERENCES AND NOTES

1. T. R. Cech, A. J. Zaug, P. J. Grabowski, *Cell* **27**, 487 (1981); K. Kruger *et al.*, *ibid.* **31**, 147 (1982).
2. G. Garriga and A. M. Lambowitz, *ibid.* **39**, 631 (1984).
3. G. Van der Horst and H. F. Tabak, *ibid.* **40**, 759 (1985).
4. F. K. Chu, G. F. Maley, M. Belfort, F. Maley, *J. Biol. Chem.* **260**, 10680 (1985).
5. C. L. Peebles *et al.*, *Cell*, in press; R. Van der Veen *et al.*, *ibid.*, in press.
6. A. Fersht, *Enzyme Structure and Mechanism* (Freeman, New York, ed. 2, 1985).
7. B. L. Bass and T. R. Cech, *Nature (London)* **308**, 820 (1984).
8. A. J. Zaug, J. R. Kent, T. R. Cech, *Biochemistry* **24**, 6211 (1985).
9. _____, *Science* **224**, 574 (1984).
10. F. X. Sullivan and T. R. Cech, *Cell* **42**, 639 (1985).
11. C. Guerrier-Takada *et al.*, *ibid.* **35**, 849 (1983).
12. C. Guerrier-Takada and S. Altman, *Science* **223**, 285 (1984).
13. T. L. Marsh, B. Pace, C. Reich, K. Gardiner, N. R. Pace, in *Sequence Specificity in Transcription and Translation*, R. Calendar and L. Gold Eds., *UCLA Symposium on Molecular and Cellular Biology* (Plenum, New York, in press); T. L. Marsh and N. R. Pace, *Science* **229**, 79 (1985).
14. Abbreviations: IVS, intervening sequence or intron; $L - 19$ IVS RNA (read "L minus 19"), a 395-nt RNA missing the first 19 nt of the L IVS RNA (the direct product of pre-ribosomal RNA splicing); p, ^{32}P within an oligonucleotide, that is, C_5pC is $CpCpCpCpC^{32}pC$ and pC_5 is $^{32}pCpCpCpCpC$; d-C_5, deoxyC$_5$.
15. A. Zaug and T. R. Cech, unpublished data.
16. The radioactive phosphate was bonded covalently to the $L - 19$ IVS RNA as judged by the following criteria: it remained associated when the complex was isolated and subjected to a second round of denaturing gel electrophoresis; it was released in the form of a mononucleotide upon RNase T_2 treatment; and it was released in the form of a series of unidentified oligonucleotides upon RNase T_1 treatment (*15*). These results are consistent with a series of covalent enzyme-substrate complexes in which various portions of C_5pC are linked to the $L - 19$ IVS RNA via a normal $3',5'$-phosphodiester bond. A more complete structural analysis of the covalent complexes is in progress.
17. A. J. Zaug, P. J. Grabowski, T. R. Cech, *Nature (London)* **301**, 578 (1983).
18. M. Been and T. R. Cech, *Nucleic Acids Res.* **13**, 8389 (1985).
19. T. Inoue, F. X. Sullivan, T. R. Cech, *J. Mol. Biol.*, in press.
20. W. P. Jencks, *Catalysis in Chemistry and Enzymology* (McGraw-Hill, New York, 1969).
21. T. N. C. Wells and A. R. Fersht, *Nature (London)* **316**, 656 (1985).
22. T. Inoue, F. X. Sullivan, T. R. Cech, *Cell* **43**, 431 (1985).
23. R. W. Davies *et al.*, *Nature (London)* **300**, 719 (1982); R. B. Waring, C. Scazzocchio, T. A. Brown, R. W. Davies, *J. Mol. Biol.* **167**, 595 (1983).
24. F. Michel and B. Dujon, *EMBO J.* **2**, 33 (1983).
25. T. R. Cech, in preparation.
26. P. J. Greene *et al.*, *J. Mol. Biol.* **99**, 237 (1975); P. Modrich and D. Zabel, *J. Biol. Chem.* **251**, 5866 (1976); R. D. Wells, R. D. Klein, C. K. Singleton, *Enzymes* **14**, 157 (1981); C. A. Brennan, M. B. Van Cleve, R. I. Gumport, in preparation; B. Terry, W. Jack, P. Modrich, in preparation.
27. J. Kumamoto, J. R. Cox, Jr., F. H. Westheimer, *J. Am. Chem. Soc.* **78**, 4858 (1956); P. C. Haake and F. H. Westheimer, *ibid.* **83**, 1102 (1961).
28. A. J. Kirby and M. Younas, *J. Chem. Soc. Ser. B.* (1970), p. 1165; C. A. Bunton and S. J. Farber, *J. Org. Chem.* **34**, 767 (1969).
29. The rate of nucleophilic attack by hydroxide ion on phosphate esters is sensitive to the pK_a of the conjugate acid of the leaving group. A phosphate in RNA should be more reactive than dimethyl phosphate, because $pK_a = 12.5$ for a nucleoside ribose and $pK_a = 15.5$ for methanol [values at 25°C from P. O. P. Ts'o, *Basic Principles in Nucleic Acid Chemistry* (Academic Press, New York, 1974), vol. 1, pp. 462–463 and P. Ballinger and F. A. Long, *J. Am. Chem. Soc.* **82**, 795 (1960), respectively]. On the basis of the kinetic data available for the alkaline hydrolysis of phosphate diesters (*27, 28*), the slope of a graph of the logarithm of the rate constant for hydrolysis as a function of pK_a can be roughly estimated as 0.6. Thus, RNA is expected to be more reactive than dimethyl phosphate by a factor of $10^{0.6 \ (15.5-12.5)} = 10^{1.8}$. The estimate for RNA pertains to direct attack by OH^- on the phosphate, resulting in $3'$-hydroxyl and $5'$-phosphate termini. Cleavage of RNA by OH^--catalyzed transphosphorylation, producing a $2',3'$-cyclic phosphate, is a much more rapid (intramolecular) reaction but is not relevant to the reactions of the $L - 19$ IVS RNA.
30. J. V. Price and T. R. Cech, *Science* **228**, 719 (1985).
31. E. T. Butler and M. J. Chamberlin, *J. Biol. Chem.* **257**, 5772 (1982); D. A. Melton *et al.*, *Nucleic Acids Res.* **12**, 7035 (1984).
32. T. R. Cech *et al.*, *Proc. Natl. Acad. Sci. U.S.A.* **80**, 3903 (1983); T. Inoue and T. R. Cech, *ibid.* **82**, 648 (1985).
33. We thank O. Uhlenbeck for gifts of oligoribonucleotides; J. Beltman, J.-Y. Tang, and M. Caruthers for oligodeoxyribonucleotides; and A. Sirimarco and M. Gaines for preparation of the manuscript and illustrations. Supported by American Cancer Society grant NP-374B, the National Foundation for Cancer Research, NIH grant GM28039, and an NIH Research Career Development Award (T.R.C.).

25 November 1985; accepted 27 December 1985

Supplementary Readings

Dreyer, G.B. & Dervan, P.B., Sequence-specific cleavage of single-stranded DNA: Oligodeoxynucleotide-EDTA•Fe(II). *Proc. Natl. Acad. Sci. U.S.A* 82:968–972 (1985)

Janda, K.D., Schloeder, D., Benkovic, S.J. & Lerner, R.A., Induction of an antibody that catalyzes the hydrolysis of an amide bond. *Science* 241:1188–1191 (1988)

Kim, S.H. & Cech, T.R., Three dimensional model of the active site of the self-splicing rRNA precursor of *tetrahymena*. *Proc. Natl. Acad. Sci. U.S.A.* 84:8788–8792 (1987)

Pollack, S.J., Jacobs, J.W., & Schultz, P.G., Selective chemical catalysis by an antibody. *Science* 234:1570–1573 (1986)

Schultz, P.G. & Dervan, P.B., Sequence-specific double-strand cleavage of DNA by penta-*N*-methylpyrrole-carboxamide-EDTA•Fe(II). *Proc. Natl. Acad. Sci. U.S.A.* 80:6834–6837 (1983)

Tramontano, A., Janda, K.D. & Lerner, R.A., Catalytic antibodies. *Science* 234:1566–1569 (1986)

Uhlenbeck, O.C., A small catalytic oligoribonucleotide. *Nature* 328:596–600 (1987)

Wilson, G.G. & Murray, N.E., Molecular cloning of the DNA ligase gene from bacteriophage T4. *J. Mol. Biol.* 132:471–491 (1979)

Zaug, A.J., Been, M.D. & Cech, T.R., The *Tetrahymena* ribozyme acts like an RNA restriction endonuclease. *Nature* 324:429–433 (1986)

Sequencing, Synthesis, and Mutation

To be a "genetic engineer" is to be able to manipulate the structure of the target genome. In later sections of this text we shall address this question on an overall genome level. In this section we shall restrict ourselves to individual genes. Of course, there is no real division except one of scale, and as the technology develops one expects (and we are already seeing!) the approaches highlighted in this section expanded to entire genomes in selected cases. However, the particular papers chosen for this section have a much smaller focus—the gene.

Prior to manipulating the structure of a gene we must be able to determine its primary structure. This involves both the determination of its encoded product's primary sequence (typically a peptide sequence) and the gene's DNA sequence. The Hunkapillar and Hood communication, found in the supplementary reading, reviews modern enhanced applications of existing protein sequencing technology. These procedures were based on the pioneering work by Edman in the 1960s, but automated and refined to such a high degree that a "spot" on an electropherogram is sufficient material for sequence analysis. Many of the genes that have been cloned in recent years were obtained by the use of oligonucleotide probes whose design were based upon sequencing such extremely small amounts of protein.

With the gene in hand (achieved through cloning, PCR or synthetic techniques described in this and other Sections [1 and 4], one must know its exact nucleotide sequence. Gilbert's laboratory and Sanger's laboratory solved this problem in 1977 with the publication of two different approaches; nucleotide specific chemical cleavage (Maxam and Gilbert) and nucleotide specific enzymatic blockage (Sanger et al.). Although these approaches generated their nucleotide specific terminated molecules by different procedures, they both relied on similar high resolution polyacrylamide gel electrophoretic fractionation procedures and similar autoradiographic detection mechanisms. These two publications revolutionized molecular biology and established an industry (providing reagents and equipment) and led to the establishment of a major international funding initiative—the human genome sequencing project.

A particularly cogent example of the technical interdisciplinary nature of molecular biology research and biotechnology is the important role of synthetic organic chemistry. The synthetic DNA synthesis techniques first developed by Khorana and his coworkers have had an enormous impact. Starting with approaches which yield defined molecules a few nucleotides in length, the technology has progressed so that one can reliably synthesize defined sequences 100 or more nucleotides long, and the entire process can be performed in an automated format. A review of this methodology is presented in the Matteucci and Caruthers paper.

Synthetic DNA molecules are used routinely to perform three operations: the introduction of defined mutations into cloned genes, the detection of specific DNA sequences within extraordinarily complex mixtures and the synthesis of entire genes. The Gillam and Smith paper describes the site specific mutagenesis procedure for which synthetic oligonucleotides have been so important. Site specific mutagenesis allows us to perform microsurgery on cloned genes and their associated regulatory elements, modifying them to give the precise desired structure. The use of synthetic oligonucleotides to detect specific DNA molecules will be discussed in a subsequent section (3). Finally, the Itakura et al. paper will describe the construction of the *first* synthetic eukaryotic gene which was used to produce protein in bacteria (*E. coli*). This technology has enormous ramifications both in regards to our basic understanding of genetics (it allows us to test in detail our understanding of the genetic code and genetic regulatory signals), and in regards to genetic engineering (allowing us to design, clone, and express a gene corresponding to any polypeptide; even one unrelated to any natural protein).

Although synthetic peptide technologies have not had the same level of impact, they are very important tools in modern biotechnology. These procedures derived from the work described in the supplemental Merrifield paper. The original applications primarily involved the synthesis of simple peptide hormones. More recently rapid microsynthesis methods have been developed which enable entire libraries of synthetic epitopes to be generated corresponding to a given gene or protein sequence. Synthetic peptide vaccines have not yet been a success but the use of peptide antigens in screening and diagnosis has great potential.

Site-specific Mutagenesis Using Synthetic Oligodeoxyribonucleotide Primers:

I. Optimum Conditions and Minimum Oligodeoxyribonucleotide Length

S. Gillam and M. Smith

SUMMARY

A synthetic oligodeoxyribonucleotide mismatched at a single nucleotide to a specific complementary site on wild-type circular ϕX174 DNA can be used to produce a defined point mutation after in vitro incorporation into closed circular duplex DNA by elongation with DNA polymerase and ligation followed by transfection of *Escherichia coli* (Hutchison et al., 1978; Gillam et al., 1979). The present study is an investigation of the optimum conditions required for the oligodeoxyribonucleotide-primed reaction for production of transition and transversion mutations in ϕX174 DNA, using the large (Klenow) fragment of *E. coli* DNA polymerase I. Under optimum conditions up to 39% of the progeny of transfection are the desired mutant and significant mutation is observed using a heptadeoxyribonucleotide.

INTRODUCTION

The only readily discernible method to program any specific type of point mutation at a defined site in a DNA requires a polynucleotide as the specific mutagen. The practical problem is the development of a mutagenic strategy which is efficient and which uses the shortest possible mutagenic polynucleotide, since synthesis of the latter is the limiting factor in the utility of the method. We have shown that quite short oligodeoxyribonucleotides can form stable duplex structures with one mismatched nucleotide pair (Gillam et al., 1975). An oligonucleotide can act as a primer for *E. coli* DNA polymerase I and, with a circular DNA template, the product can be ligated

GILLAM, S. and SMITH, M.
Site-specific mutagenesis using synthetic oligodeoxyribonucleotide primers: I. Optimum conditions and minimum oligodeoxyribonucleotide length. Reprinted with permission from *Gene* v. 8: pp. 81-97.

to produce closed circular duplex DNA (Goulian, 1968a,b; Goulian et al., 1967; 1973) which in turn can be used to transfect cells. This strategy has been used to produce both types of transition mutation, A-T → G-C; G-C → A-T, at position 587 of the ϕX174 (duplex) genome, i.e. A → G and G → A in viral DNA, and two types of transversion mutation, G-C → T-A; T-A → G-C, at position 5276 of ϕX174 duplex DNA, i.e. G → T and T → G in viral DNA (Hutchison et al., 1978; Gillam et al., 1979). The transition mutations were produced using dodecadeoxyribonucleotides mismatched at position 7 from the 5′-end and the transversion mutations by undecadeoxyribonucleotides, mismatched at position 9 from the 5′-end. The DNA polymerase used in these experiments was the large (Klenow) fragment of *E. coli* DNA polymerase I. This enzyme was chosen because it appears to be the most effective DNA polymerase with a longer single-stranded template (Sherman and Gefter, 1976) and the large fragment was used to eliminate the 5′ → 3′ exonuclease activity which would be expected to edit out the desired nucleotide mismatch (Brutlag et al., 1969; Klenow and Henningsen, 1970; Klenow et al., 1971).

In principle, both from the perspective of duplex stability (Gillam et al., 1975), and hence of priming efficiency, and the length of oligodeoxyribonucleotide required to be a site-specific probe in a DNA the size of ϕX174 DNA (Astell and Smith, 1972), it should be possible to obtain specific mutagenesis using shorter oligodeoxyribonucleotides than those described above. The objective of the present study was to establish optimum conditions for oligodeoxyribonucleotide mutagenesis and to define the mutagenicity of shorter oligonucleotides under these conditions.

MATERIALS AND METHODS

(a) Oligodeoxyribonucleotides, enzymes, phage and bacterial strains

Oligodeoxyribonucleotides were synthesized enzymatically using *E. coli* polynucleotide phosphorylase (Gillam et al., 1978) as described previously (Hutchison et al., 1978; Gillam et al., 1979).

The large fragment of *E. coli* DNA polymerase I was from Boehringer, Mannheim. DNA ligase from T4-infected *E. coli* was a gift from Dr. R.C. Miller Jr. One unit of ligase catalysed the conversion of 1 pmol of ^{32}P-labelled 5′-terminus of pT_{10} on a polydeoxyadenylate template to a form resistant to bacterial alkaline phosphatase in 1 min at 20°C (Panet et al., 1973). Endonuclease S1 was purified as described by Vogt (1973), one unit of enzyme solubilized 10 μg of sonicated, denatured calf thymus DNA in 10 min at 45°C. The restriction endonuclease *Fnu*DI ($\equiv$ *Hae*III) was a gift from A.C.P. Lui (Lui et al., 1979).

Bacteriophage ϕX174 were as described previously (Hutchison et al., 1978; Gillam et al., 1979). *E. coli* CQ$_2$ and HF4714 are ϕX174 sensitive and su^+. *E. coli* C is ϕX174 sensitive and su^-. Media were as described by Hutchison and Sinsheimer (1966).

(b) Oligodeoxyribonucleotide-primed synthesis of covalently closed heteroduplex DNA

ϕX174 viral DNA (0.6 pmol) and oligodeoxyribonucleotide (20—40 pmol) in 5 μl of buffer (100 mM NaCl, 40 mM Tris·HCl, pH 7.5, 20 mM MgCl$_2$ 2 mM 2-mercaptoethanol) were heated in a sealed capillary tube for 3 min at 80°C and then incubated at the indicated temperature for 1 h. An aliquot (2 μl) was placed in a polypropylene tube together with 3 μl of buffer containing 22 mM Tris·HCl, pH 7.5, 11 mM MgCl$_2$, 1 mM 2-mercaptoethanol, 0.83 mM each of dATP, dCTP, dGTP and dTTP, 0.4 mM ATP, 0.45 units of *E. coli* DNA polymerase I, large (Klenow) fragment and 4.8 units of T4 DNA ligase. The mixture was incubated at the specified temperature for 30 min and then at 23°C for 5 h.

Single-stranded DNA and gapped duplexes were inactivated by treatment with S1 endonuclease as described previously (Hutchison et al., 1978). The amount of S1 endonuclease used was sufficient to reduce the infectivity of single-stranded ϕX174 viral DNA by 1000-fold.

(c) Transfections and plaque assays

Spheroplasts were prepared and transfected as described previously for mutagenesis at ϕX174 nucleotide 587 (Hutchison et al., 1978) and nucleotide 5276 (Gillam et al., 1979). The resultant phage were assayed on plates of *E. coli* CQ$_2$ (*su*$^+$) at 37°C (nucleotide 587) or 30°C (nucleotide 5276). Wild-type phage were assayed on *E. coli* C (*su*$^-$) at the same temperatures. The mutant phage, which are *am*, were assayed by transferring replicate stabs from plaques grown on *E. coli* CQ$_2$ (*su*$^+$) to lawns of *E. coli* C (*su*$^-$) and CQ$_2$ (*su*$^+$).

(d) Preparation of DNA from mutated phage

Wild-type or *am* phage were isolated as described previously (Hutchison et al., 1978; Gillam et al., 1979) and DNA was released by extraction with phenol followed by dialysis against 1 mM Tris·HCl, pH 7.5 0.05 mM EDTA.

(e) Determination of DNA sequences

The enzymatic method (Sanger et al., 1977; Sanger and Coulson, 1978) was used for all sequence determinations using the appropriate viral DNA template. Sequence in the region of nucleotide 587 was obtained using duplex restriction fragment Z5 as primer followed by *Fnu*DI cleavage and sequence in the region of nucleotide 5276 using fragment H15 (Sanger et al., 1978).

RESULTS

(a) Mutation at nucleotide 587, A → G, of ϕX174 viral DNA (am3 → wild-type)

The oligonucleotide used to induce this change in the earlier studies at

wild-type complementary
oligomer (am+ 12-mer)   3'   A-A-A-C-A-C̲-C-C-T-A-T-Gp   5'

am3 complementary
oligomer (am 12-mer)   3'   A-A-A-C-A-T̲-C-C-T-A-T-Gp   5'
                                           A (am3)
                                           ↑

wild-type
viral DNA   5'   -T-G-C-G-T-T-T-A-T-G-T-A-C-G-C-T-G-G-A-C̲-T-T-T-G-T-G-G̲-G-A-T-A-C-C-C-T-C-G-C-T̲   3'
                           570   580   590   600

gene D   - Cys - Val - Tyr - Gly - Thr - Leu - Asp - Phe - Val - Gly - Tyr - Pro - Arg -

gene E   (Met) - Val - Arg - Trp - Thr - Leu - Trp - Asp - Thr - Leu - Ala -
                                                          Ter (am3)

Fig. 1. Sequence of φX174am3 viral DNA in vicinity of position 587 and the synthetic dodecadeoxyribonucleotides complementary to wild-type and am3 DNA. The changes A → G and G → A at position 587 interconvert am and Trp codons in the reading frame of gene E, but do not change the amino acid, Val, coded in the reading frame of gene D. Ter, termination (= amber).

TABLE I

EFFECT OF LENGTH OF OLIGODEOXYRIBONUCLEOTIDES AND TEMPERATURE ON THE CHANGE A → G AT POSITION 587 OF φX174 VIRAL DNA (am3 → wild-type)[a]

Primer (mismatch is underlined)	Total and mutated progeny plaques ($\cdot 10^{-6}$) produced at different temperatures										
	33°C			25°C			10°C			0°C		
	Total	am+	%	Total	am+	%	Total	am+	%	Total	am+	%
pGTATCCCACAA	12	0.06	0.5	120	2.4	2	200	13	7	210	21	10
pGTATCCCACA	110	0.93	0.9	270	1.1	4	310	49	16	300	48	16
pGTATCCCAC	56	0.11	0.2	210	4.8	2.2	220	20	9	230	36	15
pGTATCCCA	60	0.03	0.05	70	0.6	0.9	140	4.4	3.2	210	7.4	4
pGTATCCC	130	0.01	$7 \cdot 10^{-3}$	160	0.1	0.07	220	0.24	0.11	96	0.22	0.23

[a] For details see text. The template viral DNA was am3cs70 and the progeny plaques listed above were produced by 0.1 µg of viral template DNA. In a control experiment with the homologous primer, pGTATCCTAC, the percentage of revertants obtained at 0°C was $2 \cdot 10^{-2}$%.

TABLE II

EFFECT OF LENGTH OF OLIGODEOXYRIBONUCLEOTIDES AND TEMPERATURE ON THE CHANGE A → G AT POSITION 587 OF φX174 VIRAL DNA (am3 → wild-type) WHEN ENDONUCLEASE S1 WAS USED TO DEGRADE SINGLE STRANDED PHAGE DNA[a]

Primer (mismatch is underlined)	Total and mutated progeny plaques ($\cdot 10^{-4}$) produced at different temperatures											
	33°C			25°C			10°C			0°C		
	Total	am$^+$	%	Total	am$^+$	%	Total	am$^+$	%	Total	am$^+$	%
pGTATCCCACAA	0.5	0.01	2	6	0.9	15	20	7.5	38	27	10	39
pGTATCCCACA	7.7	0.25	3	25	6.5	26	65	24	37	78	27	37
pGTATCCCAC	5.4	$2.5 \cdot 10^{-2}$	0.5	20	2.5	13	52	10	19	67	17	25
pGTATCCCA	3.6	$2 \cdot 10^{-3}$	$5 \cdot 10^{-2}$	9	0.25	3	27	27	10	27	5.6	20
pGTATCCC	3.6	$3.6 \cdot 10^{-5}$	$1 \cdot 10^{-3}$	16	0.03	0.2	18	0.12	0.7	23	0.1	0.5

[a]For details see text. The template viral DNA was am3cs70 and the progeny plaques listed above were produced by 0.1 μg of viral template DNA. In a control experiment with the homologous primer, pGTATCCTAC, the percentage of revertants obtained at 0°C was $4 \cdot 10^{-2}$%.

this site was the dodecadeoxyribonucleotide pGTATCCCACAAA, the underlined nucleotide, seventh from the 5'-end, mismatching with A in φX174 am3 viral DNA (Fig. 1). Because this oligodeoxyribonucleotide was synthesized stepwise from the 5'-end (Hutchison et al., 1978), a set of potential mutagenic oligodeoxyribonucleotides was available. This set of nucleotides was tested for its ability to revert am3 DNA to wild-type using four different temperatures ranging from 0° to 33°C for the initial priming reaction with the large (Klenow) fragment of E. coli DNA polymerase I. The results of these experiments are recorded in Table I. Because it was possible that conversion of the single-stranded template viral DNA to duplex DNA was incomplete, and because single-stranded φX174 viral DNA is particularly infective in the E. coli spheroplast assay (Goulian et al., 1967; Sinsheimer, 1968), a second set of experiments, where the mutagenized DNA was treated with the single-strand specific endonuclease S1, prior to infection to inactive single-stranded DNA, was carried out (Table II). Comparison of the data in Tables I and II leads to several conclusions. Firstly, decrease in the temperature of the initial priming of E. coli DNA polymerase I (large fragment) by the mutagenic oligodeoxyribonucleotide has a marked stimulation of the production of the mutation (>20-fold) with each of the oligonucleotides tested. Secondly, under all conditions, S1 endonuclease treatment resulted in 2—5-fold enrichment of the mutated progeny. Thirdly, under optimum conditions, with priming at 0°C and S1 endonuclease treatment, the octadeoxyribonucleotide pGTATCCCA was a more effective mutagen than was pGTATCCCACAAA under the previously described conditions (Hutchison et al., 1978). It is striking that a significant number of mutants were produced with the heptadeoxyribonucleotide pGTATCCC where the 3'-terminal nucleotide was mismatched.

(b) Mutation at nucleotide 587, G → A, of φX174 viral DNA (wild-type → am3)

The series of oligodeoxyribonucleotides pGTATCCT to pGTATCCTACAAA were used to define optimum conditions for induction of the am3 mutation in wild-type φX174 viral DNA (Fig. 1). Two temperatures, 25° and 0°C, were used for the priming reaction and experiments without and with endonuclease S1 treatment were carried out (Table III). In general, the S1 endonuclease treatment resulted in similar increased yields of the desired mutant, although usually the comparable experiment resulted in lower production of the G → A mutation than in its reversal discussed earlier. The beneficial effect of reducing the temperature of DNA polymerase priming was much less, and under equivalent conditions the longer oligodeoxyribonucleotides did not show a significant increase in yield of mutants at the lower priming temperature. In addition, the heptanucleotide d(pGTATCCT) did not produce any mutants. Apart from the generally lower efficiency of this set of mutagenic oligodeoxyribonucleotides, the complement of the heptanucleotide occurs starting at position

TABLE III

EFFECT OF LENGTH OF OLIGODEOXYRIBONUCLEOTIDES AND TEMPERATURE ON THE CHANGE G → A AT POSITION 587 OF φX174 VIRAL DNA (wild-type → am3)[a]

Primer (mismatch is underlined)	Total and mutated progeny plaques produced at different temperatures without and with endonuclease S_1											
	No S1 treatment						S1 treatment					
	25°C			0°C			25°C			0°C		
	Total $\cdot 10^{-6}$	Mutant Progeny	%	Total $\cdot 10^{-6}$	Mutant Progeny	%	Total $\cdot 10^{-6}$	Mutant Progeny	%	Total $\cdot 10^{-6}$	Mutant Progeny	%
pGTATCCTACAAA	1.9	19/200	9.5	1.3	17/200	8.5	0.3	66/300	22	0.22	64/300	21
pGTATCCTACAA	0.4	13/200	6.5	0.6	14/200	7	0.2	69/300	20	0.3	66/300	22
pGTATCCTACA	1.3	8/200	4	1.2	8/200	4	0.2	53/300	18	0.22	70/300	23
pGTATCCTAC	1.5	9/200	4.5	0.9	6/200	3	0.2	24/300	8	0.22	32/300	11
pGTATCCTA	0.8	0/200	0	1	2/200	1	0.13	2/300	0.7	0.2	13/300	4.3
pGTATCCT	0.2	0/200	0	1.4	0/200	0	0.05	0/300	0	0.2	0/300	0

[a]For details see text. The template viral DNA was am^+sB1 and the progeny listed above were produced from 0.1 μg of template DNA. In a control experiment with the homologous primer pGTATCCCAC, at 0°C 0/300 mutants were obtained.

wild-type complementary
oligomer   3'-A-A-C-T-C-C-G-A-C-C-Cp-5'

am16 complementary
oligomer   3'-A-A-A-T-C-C-G-A-C-C-Cp-5'

                                                    T (am16)
                                                    ↑
wild-type
viral DNA   5'-G-C-A-A-A-A-A-G-A-G-A-G-A-T-G-A-G-A-T-T-G-A-G-G-C-T-G-G-A-A-3'
                            5260   5270   5280

gene B   - Lys - Lys - Arg - Asp - Glu - Ile - Glu - Ala - Gly - Lys -
                                                    Ter (am16)
                                                    ↑
gene A   - Lys - Arg - Glu - Met - Arg - Leu - Arg - Leu - Gly -
                                                    ↓
                                                   Phe

Fig. 2. Sequence of φX174 viral DNA in the vicinity of position 5265 and the synthetic undecadeoxyribonucleotides complementary to am16 and to wild-type DNA. The changes G → T and T → G at position 5265 interconvert Glu and am codons in the reading frame of gene B and Leu and Phe codons in the reading frame of gene A. It is not known whether the change in gene A or in (suppressed) am16 is responsible for the ts phenotype of the suppressed am mutant.

TABLE IV

EFFECT OF LENGTH OF OLIGODEOXYRIBONUCLEOTIDES AT 0°C ON THE CHANGE G → T AT POSITION 5276 OF ϕX174 VIRAL DNA (wild-type → *am*16)[a]

Primer (mismatch is underlined)	Genotype	No S1 treatment: Total progeny plaques	No S1 treatment: *am*16 progeny	%	S1 treatment: Total progeny plaques	S1 treatment: *am*16 progeny	%
PCCCAGCCTAAA	*am*16	$3.63 \cdot 10^5$	22/300	7	$1.2 \cdot 10^5$	52/400	13
pCCCAGCCTAA	*am*16	$3.5 \cdot 10^5$	13/300	4.3	$7.3 \cdot 10^4$	30/400	7.5
pCCCAGCCTA	*am*16	$3.8 \cdot 10^5$	3/300	0.7	$7.5 \cdot 10^4$	2/300	0.7
pCCCAGCCTCAA	*am*$^+$	$9 \cdot 10^4$	0/300	0	$1.8 \cdot 10^4$	0/300	0

[a]For details see text. The template viral DNA was sB1 and the progeny plaques listed above were produced from 0.1 μg of template DNA.

TABLE V

EFFECT OF LENGTH OF OLIGODEOXYRIBONUCLEOTIDES AT 10°C ON THE CHANGE T → G AT POSITION 5276 OF ϕX174 VIRAL DNA (*am*16 → wild-type)[a]

Primer (mismatch is underlined)	Genotype	No S1 treatment: Total progeny plaques	No S1 treatment: Wild-type progeny	%	S1 treatment: Total progeny plaques	S1 treatment: Wild-type progeny	%
pCCCAGCCTCAA	*am*$^+$	$3.6 \cdot 10^6$	$8.1 \cdot 10^5$	22	$9 \cdot 10^5$	$1.5 \cdot 10^5$	19
		$4.5 \cdot 10^6$	$1 \cdot 10^6$	21	$2 \cdot 10^6$	$3.5 \cdot 10^5$	18
pCCCAGCCTCA	*am*$^+$	$2.4 \cdot 10^6$	$3.8 \cdot 10^5$	16	$6.8 \cdot 10^5$	$1.4 \cdot 10^5$	22
pCCCAGCCTC	*am*$^+$	$2.9 \cdot 10^6$	$1.1 \cdot 10^5$	3.5	$1.3 \cdot 10^6$	$4.3 \cdot 10^4$	3.5
pCCCAGCCTAAA	*am*16	$5.4 \cdot 10^6$	$2.7 \cdot 10^2$	$5 \cdot 10^{-3}$	$1.1 \cdot 10^6$	68	$6 \cdot 10^{-3}$

[a]For details see text. The template viral DNA was *am*16 and the progeny plaques listed above were produced from 0.1 μg of template DNA.

3274 of the φX174 viral DNA sequence (Sanger et al., 1978) providing an alternate, perfectly matched, and probably more efficient priming site (Hutchison et al., 1978).

(c) Mutation at nucleotide 5276, G → T, of φX174 viral DNA (wild-type → am16)

The undecadeoxyribonucleotides used to induce these mutations in earlier studies (Gillam et al., 1979), pCCCAGCCTAAA and pCCCAGCCTCAA mismatch with φX174 viral DNA at the underlined ninth nucleotides (Fig. 2). Studies on the effect of oligodeoxyribonucleotide length on mutagenic efficiency were carried out with pCCCAGCCTAAA at 0°C (Table IV) and with pCCCAGCCTCAA at 10°C (Table V). The mutation G → T was studied using a temperature of 0°C for priming E. coli DNA polymerase I (large fragment) because preliminary experiments at 25°C did not produce the desired mutation (results not shown). In addition, production of the mutation involves a purine-purine (G-A) mismatch in the duplex DNA, and would therefore be expected to be sterically less favoured than the transition mutations described above. At 0°C, all three of the oligodeoxyribonucleotides tested produced the desired mutation (Table IV), although at lower efficiency than that of the transition producing oligodeoxyribonucleotides of the same length (Tables I—III). Endonuclease S1 treatment also resulted in significant (two-fold) enrichment of the mutant.

The experiments directed at the mutation T → G at position 5276 (am16 → wild-type) were carried out under different experimental conditions (Gillam et al., 1979) because the complement of the 3′-terminal octadeoxyribonucleotide in pCCCAGCCTCAA occurs twice, starting at positions 533 and 4060, in φX174 viral DNA (Sanger et al., 1978). Therefore, priming in the presence of only three deoxynucleoside-5′ triphosphates, dATP, dCTP and dTTP was carried out at 10°C before shifting to 37°C and addition of dGTP. Only limited extension of a primer is possible at the sites at positions 553 and 4060 in the absence of dGTP whereas at the described priming site more extensive synthesis is possible. This, therefore, results in selective priming at the desired site at 37°C in the presence of dGTP. Preliminary experiments (results not shown) indicated that the standard priming procedure did not produce significant numbers of the desired mutation. The modified conditions resulted in significant production of the desired G → T mutation with all three oligodeoxyribonucleotides tested (Table V). The most striking result in these experiments was the insignificant effect of endonuclease S1 treatment on the production of the desired mutant. The reason for this is not established, but it seems possible that the modified priming conditions described above resulted in more efficient conversion to closed circular duplex DNA.

(d) Characterization of mutants

In experiments directed at site-specific mutagenesis it is essential to characterize the sequence of the mutated DNA. In the case of the change A → G at position 587 of φX174 viral DNA (am3 → wild-type) induced by pGTATCCC, sixteen randomly chosen clones of phage which grew on su⁻ host were combined and used to prepare phage DNA. The result of sequence

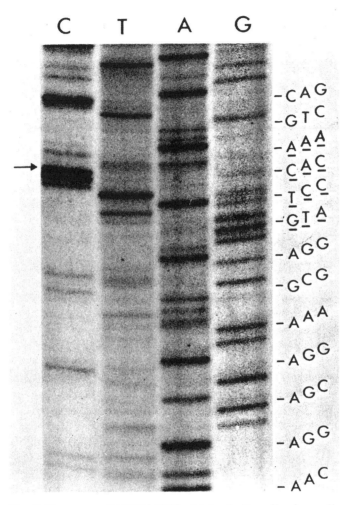

Fig. 3. Sequence of φX174 DNA obtained after oligodeoxyribonucleotide mutation of φX174am3 viral DNA. The sequence shown is that of the complementary strand in the vicinity of position 587 (arrow). The underlined sequence corresponds to the synthetic oligodeoxyribonucleotide pGTATCCCACAAA; the DNA in this specific experiment was produced by pGTATCCC. The sequence was determined by the enzymatic terminator method (Sanger et al., 1977; Sanger and Coulson, 1978) which is particularly suitable for detecting contamination in site-specific mutation experiments (for experimental details see text).

determination on the DNA in the region of position 587 is shown in Fig. 3. The genotype of the DNA used for inducing the mutation was *am*3*cs*70 which grows at 37°C but not at 25°C on *E. coli* C (*su⁻*). All revertants of *am*3 were cold-sensitive, which indicates that no contamination with other wild-type φX174 had occurred.

In the case of the change G → A at position 587 of φX174 viral DNA (wild-type → *am*3) induced by pGTATCCTAC and pGTATCCTACA, five clones of *am* phage induced by each oligonucleotide were combined and

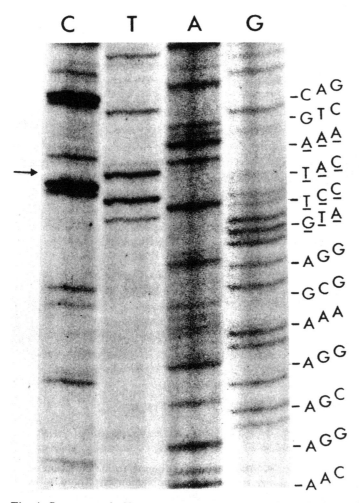

Fig. 4. Sequence of φX174*am*3 DNA obtained after oligodeoxyribonucleotide mutation of φX174 viral DNA. The sequence shown is that of the complementary strand in the vicinity of position 587 (arrow). The underlined sequence corresponds to the synthetic oligodeoxyribonucleotide pGTATCCTACAAA; the DNA in this specific experiment was produced by pGTATCCTAC and pGTATCCTACA (for experimental details see text).

used to prepare phage DNA. The result of sequence determination in the region of position 587 is shown in Fig. 4. Since the *am*3 mutation is defective in lysis because of the absence of the gene E protein, such mutants can be phenotypically suppressed by growth on *su*⁻ host in lysozyme-deoxycholate containing medium (Hutchison et al., 1978). All *am* mutants obtained by mutagenesis directed at position 587 were, in this way, shown to be defective in gene E.

Sequence determination of mutants obtained by the oligodeoxyribonucleotides directed at position 5276, G → T and T → G, has been described previously (Gillam et al., 1979). In addition, in the case of the change G → T (wild-type → *am*16), the phage grown on *su*⁺ host is temperature sensitive, growing poorly at 37°C and well at 30°C. All *am* phage produced by the oligodeoxyribonucleotide directed at position 5276 exhibited this phenotype.

DISCUSSION

The use of synthetic oligodeoxyribonucleotides as site-specific mutagens requires that they be integrated into DNA in vitro (Hutchison et al., 1978). Integration is achieved by using the synthetic oligodeoxyribonucleotide, mismatched at a single specific nucleotide, as a primer for DNA polymerase on a complementary circular DNA which is used to transfect cells. ϕX174 DNA provides a well characterized model system and has been the focus of studies to date (Hutchison et al., 1978; Gillam et al., 1979). However, the method should be applicable to small bacterial plasmids and to small circular DNAs which replicate in eukaryote systems. The objective of the present studies has been to define optimum conditions for the production of two transition and two transversion mutations in ϕX174 and subsequently to use these conditions to investigate the relationship between length of the oligodeoxyribonucleotide and its efficiency as a site-specific mutagen. Clearly, a crucial step is to establish optimum conditions for priming of a DNA polymerase by the synthetic oligodeoxyribonucleotide.

DNA polymerase I of *E. coli* has been shown to be particularly effective using a long single-stranded template (Sherman and Gefter, 1976) and, in fact, has been used in oligodeoxyribonucleotide-primed reactions to convert single-stranded ϕX174 DNA to duplex DNA which can be ligated to form closed circular duplexes (Goulian et al., 1967; Goulian, 1968a). This made *E. coli* DNA polymerase I the enzyme of choice in initial and continuing experiments. One problem with this enzyme is that it has a 5'-exonuclease which excises the primer oligodeoxyribonucleotide from the product (Goulian, 1968b). Therefore, it seemed wise to use the large fragment of this enzyme from which the 5'-exonuclease had been removed proteolytically (Brutlag et al., 1969; Klenow and Henningsen, 1970; Klenow et al., 1971). This assumption has proved to be justified, since an attempt to use the native *E. coli* DNA polymerase I with *short* oligodeoxyribonucleotides to produce the A → G change at position 587 of φX174 viral DNA was not successful (Razin et al., 1978). At 20°C, perfectly matched oligodeoxyribonucleotide primers for *E. coli* DNA polymerase I are less effective if they contain less than nine nucleotides, although there is detectable priming with tri- and tetra-deoxyribonucleotides (Goulian et al., 1973). Physical studies indicate that duplexes containing an oligodeoxyribonucleotide with one mismatched nucleotide pairing have the same temperature stability as those duplexes containing perfectly matched oligodeoxyribonucleotides with one to two fewer nucleotides (Gillam et al., 1975). Hence, it was logical to direct our studies at the effect of reduced temperature for the priming of *E. coli* DNA polymerase I (large fragment) on the efficiency of mutant production. The data in Table I show that for the change A → G at position 587 of φX174 viral DNA, reduced priming temperature increased the efficiency of mutation for all sizes of oligodeoxyribonucleotide from pGTATCCC to pGTATCCCACAA (mismatch nucleotide is underlined). On average, the increase in efficiency on changing the temperature from 33° to 0°C is 20- to 100-fold, the relative improvement being greatest for the shortest oligodeoxyribonucleotide primers, and most of the improvement being obtained once the temperature was decreased to 10°C. The reduced temperatures also resulted in an increased yield of total progeny phage.

φX174 viral DNA is up to twenty times as infective as double-stranded φX174 DNA in spheroplast assays (Goulian et al., 1967; Sinsheimer, 1968). Consequently, residual single-stranded DNA resulting from inefficient priming of DNA polymerase I by the mutagenic oligodeoxyribonucleotide and/or incomplete ligation could strongly bias the results of the above experiments against the production of the desired mutant. Duplex DNA containing a single mismatched nucleotide pair is relatively resistant to the single-strand specific endonuclease S1 (Shenk et al., 1975; Dodgson and Wells, 1977). Hence treatment of the closed circular duplex product obtained after oligonucleotide priming of DNA polymerase and ligation should selectively degrade single-stranded φX174 DNA and result in enrichment of the mutant fraction of the progeny of spheroplast transfection. In earlier experiments using the oligodeoxyribonucleotide pGTATCCCACAAA, as a primer at 25°C, endonuclease S1 produced enrichment of about 20-fold (Hutchison et al., 1978). In the present experiments (Table II) enrichment was 2- to 7-fold suggesting a lower level of contamination by single-stranded DNA. The net result of using low temperature priming and S1 endonuclease was a very significant level of mutant production with the heptadeoxyribonucleotide, pGTATCCC, and close to theoretic yields (50% assuming no biological bias against or toward the mutant) with pGTATCCCACA and pGTATCCCACAA.

The success of these studies encouraged analogous experiments directed at a G → A change at position 587 (wild-type → *am*3, Fig. 1). Analogous experiments were carried out at 25° and 0°C using the series of mutagenic oligodeoxyribonucleotides pGTATCCT to pGTATCCTACAAA (Table III).

In this case the effect of temperature decrease was less dramatic, except for the octadeoxyribonucleotide pGTATCCTA. The enhancement of mutant yield by endonuclease S1 was 2- to 6-fold (Table III). Whilst the yields were not as good as in the *am*3 → wild-type experiments, the successful production of mutants by pGTATCCTA and their highly efficient production by an oligodeoxyribonucleotide as short as pGTATCCTACA was very satisfactory.

The above experiments producing transition mutations involve purine-pyrimidine mismatches at the mutated site. Transversion mutations involve duplexes containing purine-purine or pyrimidine-pyrimidine mismatches which might be less stable or be susceptible to cellular repair mechanisms. Consequently it was of interest to establish the mutagenic efficiency of the two sets of oligodeoxyribonucleotides related to pCCCAGCCTAAA and pCCCAGCCTCAA in producing and reverting the *am*16 mutation at position 5276 of φX174 viral DNA. The results (Tables IV and V) show that the oligodeoxyribonucleotides where the mismatch is at the 3′-end, pCCCAGCCTA and pCCCAGCCTC produced mutants and that mutagenic efficiency is good with an additional nucleotide on the 3′-end. Clearly, all types of mutation involving nucleotide replacements, i.e. transitions and transversions, can be achieved with high efficiency using mutagenic oligodeoxyribonucleotides containing nine or ten nucleotides.

The reduced efficiency of shorter oligodeoxyribonucleotides could be due to reduced priming efficiency, if this was limiting, or due to editing out of the mismatched nucleotide by the intrinsic 3′-exonuclease activity of *E. coli* DNA polymerase I (large fragment). In comparable experiments involving homologous series of oligodeoxyribonucleotides (Tables I to V) the total yields of progeny were similar, indicating that priming efficiency was not a limiting factor. This suggests that the reduced yields of mutants using shorter mutagenic oligodeoxyribonucleotides is a consequence of the greater access of the 3′-exonuclease to the mismatched nucleotide. The data in Tables I to V also clearly indicate that often one additional nucleotide at the 3′-end of a mutagenic oligodeoxyribonucleotide and certainly two additional nucleotides give excellent protection against the editing activity of 3′-exonuclease of *E. coli* DNA polymerase I. It is interesting that this activity is not completely efficient even when a mismatched nucleotide is at the 3′-end of a mutagenic oligodeoxyribonucleotide.

The general conclusion that can be made from these experiments is that with a circular genome of the complexity of φX174 (i.e. 5386 nucleotides), it is possible to use quite short, and readily accessible synthetic oligodeoxyribonucleotides (Gillam et al., 1978; 1979; Hutchison et al., 1978) as efficient specific mutagens. The general requirement of the oligodeoxyribonucleotide is that it be long enough to recognize a unique site in the genome (Astell and Smith, 1972; Hutchison et al., 1978) and to form a primer-template duplex with complementary DNA. This can be achieved with oligodeoxyribonucleotides containing as few as seven nucleotides. In addition, to obtain good protection of the mutagenizing nucleotide in the oligodeoxy-ribonucleotide from editing by the 3'-exonuclease of *E. coli* DNA polymer-ase I (large fragment), there should be one and preferably two nucleotides on the 3'-side of the mismatched nucleotide.

ACKNOWLEDGEMENTS

We are grateful to Dr. R.C. Miller Jr. for a gift of T4 DNA ligase and to Miss A.C.P. Lui for the restriction endonuclease *Fnu*DI and for the φX174 DNA duplex fragment Z5. This research was supported by the Medical Research Council of Canada of which M.S. is a Career Investigator.

REFERENCES

Astell, C.R. and Smith, M., Synthesis and properties of oligonucleotidecellulose columns, Biochemistry, 11 (1972) 4114—4120.

Brutlag, D., Atkinson, M.R., Setlow, P. and Kornberg, A., An active fragment of DNA polymerase produced by proteolytic cleavage, Biochem. Biophys. Res. Commun., 37 (1969) 982—989.

Dodgson, J.B. and Wells, R.D., Action of single-strand specific nucleases on model DNA heteroduplexes of defined size and sequence, Biochemistry, 16 (1977) 2374—2379.

Gillam, S., Waterman, K. and Smith, M., The base-pairing specificity of cellulose-pdT$_9$, Nucl. Acids Res., 2 (1975) 625—634.

Gillam, S., Jahnke, P. and Smith, M., Enzymatic synthesis of oligodeoxyribonucleotides of defined sequence, J. Biol. Chem., 253 (1978) 2532—2539.

Gillam, S., Jahnke, P., Astell, C., Phillips, S., Hutchison, C.A. III and Smith, M., Defined transversion mutations at a specific position in DNA using synthetic oligodeoxyribonucleotides as mutagens, Nucl. Acids Res., 6 (1979) 2973—2985.

Goulian, M., Incorporation of oligodeoxynucleotides into DNA, Proc. Natl. Acad. Sci. USA, 61 (1968a) 284—291.

Goulian, M., Initiation of the replication of single-stranded DNA by *Escherichia coli* DNA polymerase, Cold Spring Harbor Symp. Quant. Biol., 33 (1968b) 11—20.

Goulian, M., Kornberg, A. and Sinsheimer, R.L., Enzymatic synthesis of DNA, XXIV. Synthesis of infectious phage φX174 DNA, Proc. Natl. Acad. Sci. USA, 58 (1967) 2321—2328.

Goulian, M., Goulian, S.H., Codd, E.E. and Blumenfield, A.Z., Properties of oligodeoxynucleotides that determine priming activity with *Escherichia coli* deoxyribonucleic acid polymerase I, Biochemistry, 12 (1973) 2893—2901.

Hutchison, C.A. III and Sinsheimer, R.L., The process of infection with bacteriophage φX174, X. Mutations in a φX lysis gene, J. Mol. Biol., 18 (1966) 429—447.

Hutchison, C.A. III, Phillips, S., Edgell, M.H., Gillam, S., Jahnke, P. and Smith, M., Mutagenesis at a specific position in a DNA sequence, J. Biol. Chem., 253 (1978) 6551—6560.

Klenow, H. and Henningsen, I., Selective elimination of the exonuclease activity of the deoxyribonucleic acid polymerase from *Escherichia coli* B by limited proteolysis, Proc. Natl. Acad. Sci. USA, 65 (1970) 168—175.

Klenow, H., Overgaard-Hansen, K. and Patkar, S.A., Proteolytic cleavage of native DNA polymerase into two different catalytic fragments. Influence of assay conditions on the change of exonuclease activity and polymerase activity accompanying cleavage, Eur. J. Biochem., 22 (1971) 371—381.

Lui, A.C.P., McBride, B.C., Vovis, G.F. and Smith, M., Site specific endonucleases from *Fusobacterium nucleatum*, Nucl. Acids Res., 6 (1979) 1—15.

Panet, A., van de Sande, J.H., Loewen, P.C., Khorana, H.G., Raae, A.J., Lillehang, J.R. and Kleppe, K., Physical characterization and simultaneous purification of bacteriophage T$_4$ induced polynucleotide kinase, polynucleotide ligase, and deoxyribonucleic acid polymerase, Biochemistry, 12 (1973) 5045—5050.

Sanger, F., Nicklen, S. and Coulson, A.R., DNA sequencing with chain-terminating inhibitors, Proc. Natl. Acad. Sci. USA, 74 (1977) 5463—5467.

Sanger, F. and Coulson, A.R., The use of thin acrylamide gels for DNA sequencing, FEBS Lett., 87 (1978) 107—110.

Sanger, F., Coulson, A.R., Friedmann, T., Air, G.M., Barrell, B.G., Brown, N.L., Fiddes, J.C., Hutchison, C.A. III, Slocombe, P.M. and Smith, M., The nucleotide sequence of bacteriophage φX174, J. Mol. Biol., 125 (1978) 225—246.

Shenk, T.E., Rhodes, C., Rigby, P.W.J. and Berg, P., Biochemical method for mapping mutational alterations in DNA with S$_1$ nuclease: The location of deletions and temperature-sensitive mutations in Simian Virus 40, Proc. Natl. Acad. Sci. USA, 72 (1975) 989—993.

Sherman, L.A. and Gefter, M.L., Studies on the mechanism of enzymatic DNA elongation by *Escherichia coli* DNA polymerase II, J. Mol. Biol., 103 (1976) 61—76.

Sinsheimer, R.L., Bacteriophage φX174 and related viruses, in Davidson, J.N. and Cohn, W.E. (Eds.), Progress in Nucleic Acid Research and Molecular Biology, Vol. 8, Academic Press, New York, 1968, pp. 115—167.

Razin, A., Hirose, T., Itakura, K. and Riggs, A.D., Efficient correction of a mutation by use of chemically synthesized DNA, Proc. Natl. Acad. Sci. USA, 75 (1978) 4268—4270.

Vogt, V.M., Purification and further properties of single-strand-specific nuclease from *Aspergillus oryzae*, Eur. J. Biochem., 33 (1973) 192—200.

Communicated by Z. Hradecna.

Expression in *Escherichia Coli* of a Chemically Synthesized Gene for the Hormone Somatostatin

K. Itakura, H. Tadaaki, R. Crea, A.D. Riggs, H.L. Heyneker, F. Bolivar and H.W. Boyer

Abstract. A gene for somatostatin, a mammalian peptide (14 amino acid residues) hormone, was synthesized by chemical methods. This gene was fused to the Escherichia coli β-galactosidase gene on the plasmid pBR322. Transformation of E. coli with the chimeric plasmid DNA led to the synthesis of a polypeptide including the sequence of amino acids corresponding to somatostatin. In vitro, active somatostatin was specifically cleaved from the large chimeric protein by treatment with cyanogen bromide. This represents the first synthesis of a functional polypeptide product from a gene of chemically synthesized origin.

The chemical synthesis of DNA and recombinant DNA methods provide the technology for the design and synthesis of genes that can be fused to plasmid elements for expression in *Escherichia coli* or other bacteria. As a model system we have designed and synthesized a gene for the small polypeptide hormone, somatostatin (Figs. 1 and 2). The major considerations in the choice of this hormone were its small size and known amino acid sequence (*1*), sensitive radioimmune and biological assays (*2*), and its intrinsic biological interest (*3*). Somatostatin is a tetradecapeptide; it was originally discovered in ovine hypothalamic extracts but subsequently was also found in significant quantities in other species and other tissues (*3*). Somatostatin inhibits the secretion of a number of hormones, including growth hormone, insulin, and glucagon. The effect of somatostatin on the secretion of these hormones has attracted attention to its potential therapeutic value in acromegaly, acute pancreatitis, and insulin-dependent diabetes.

The overall construction of the somatostatin gene and plasmid was designed to result in the in vivo synthesis of a precursor form of somatostatin (see Fig. 1). The precursor protein would not be expected to have biological activity, but could be converted to a functional form by cyanogen bromide cleavage (*4*) after cellular extraction. The synthetic somatostatin gene was fused to the lac operon because the controlling sites of this operon are well characterized.

Given the amino acid sequence of somatostatin, one can design from the genetic code a short DNA fragment containing the information for its 14 amino acids (Fig. 2). The degeneracy of the code allows for a large number of possible sequences that could code for the same 14 amino acids. Therefore, the choice of codons was somewhat arbitrary except for the following restrictions. First, amino acid codons known to be favored in *E. coli* for expression of the MS2 genome were used where appropriate (*5*). Second, since the complete sequence would be constructed from a number of overlapping fragments, the fragments were designed to eliminate undesirable inter- and intramolecular pairing. And third, G·C-rich (guanine-cytosine) followed by A·T-rich (adenine-thymine) sequences were avoided since they might terminate transcription (*6*).

Eight oligonucleotides, varying in length from 11 to 16 nucleotides, labeled

ITAKURA, K., TADAAKI, H., CREA, R., RIGGS, A.D., HEYNEKER, H.L., BOLIVAR, F. and BOYER, H.W.
Expression in *Escherichia coli* of a chemically synthesized gene for the hormone somatostatin. Reprinted with permission from *Science* v. 198: pp. 1056-1063. Copyright 1977 by the AAAS.

in Fig. 2 as A through H, were synthesized by the triester method (7). In addition to the 14 codons for the structural information of somatostatin, several other features were built into the nucleotide sequence. First, to facilitate insertion into plasmid DNA, the 5' ends have single-stranded cohesive termini for the Eco RI and Bam HI restriction endonucleases. Second, a methionine codon precedes the normal NH$_2$-terminal amino acid of somatostatin, and the COOH-terminal codon is followed by two nonsense codons.

In the cloning and expression of the synthetic somatostatin gene we used two plasmids. Each plasmid has an Eco RI substrate site at a different region of the β-galactosidase structural gene (see Figs. 3 and 4). The insertion of the synthetic somatostatin DNA fragment into the Eco RI sites of these plasmids brings the expression of the genetic information in that fragment under control of the lac operon controlling elements. After the insertion of the somatostatin fragment into these plasmids, translation should result in a somatostatin polypeptide preceded either by ten amino acids (pSOM1) or by virtually the whole β-ga-

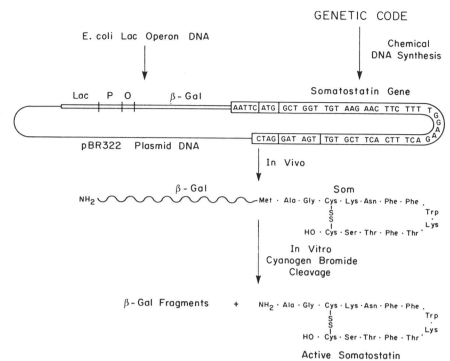

Fig. 1. Schematic outline of the experimental plan. The gene for somatostatin, made by chemical DNA synthesis, was fused to the *E. coli* β-galactosidase gene on the plasmid pBR322. After transformation into *E. coli*, the chimeric plasmid directs the synthesis of a chimeric protein that can be specifically cleaved in vitro at methionine residues by cyanogen bromide to yield active mammalian peptide hormone.

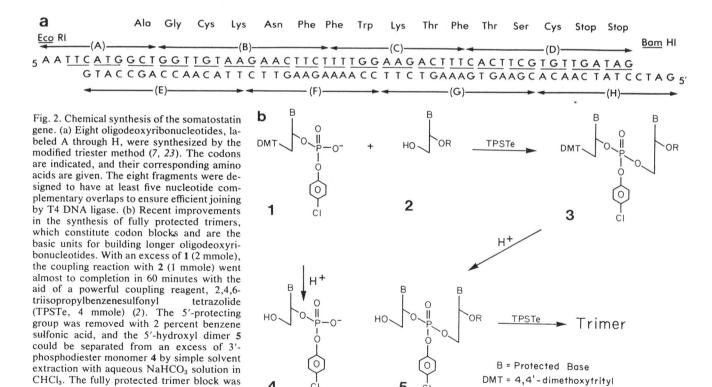

Fig. 2. Chemical synthesis of the somatostatin gene. (a) Eight oligodeoxyribonucleotides, labeled A through H, were synthesized by the modified triester method (7, 23). The codons are indicated, and their corresponding amino acids are given. The eight fragments were designed to have at least five nucleotide complementary overlaps to ensure efficient joining by T4 DNA ligase. (b) Recent improvements in the synthesis of fully protected trimers, which constitute codon blocks and are the basic units for building longer oligodeoxyribonucleotides. With an excess of **1** (2 mmole), the coupling reaction with **2** (1 mmole) went almost to completion in 60 minutes with the aid of a powerful coupling reagent, 2,4,6-triisopropylbenzenesulfonyl tetrazolide (TPSTe, 4 mmole) (2). The 5'-protecting group was removed with 2 percent benzene sulfonic acid, and the 5'-hydroxyl dimer **5** could be separated from an excess of 3'-phosphodiester monomer **4** by simple solvent extraction with aqueous NaHCO$_3$ solution in CHCl$_3$. The fully protected trimer block was prepared successively from the 5'-hydroxyl dimer **5**, **1** (2 mmole), and TPSTe (4 mmole) and isolated by chromatography on silica gel (24). These improvements simplify the purification step and lead to an increase in the overall yields of trimer blocks and to a decrease in the working time by at least a factor of 2 (21). The eight oligodeoxyribonucleotides then were synthesized from the trimers by published procedures (7). The final products, after removal of all protecting groups, were purified by high-pressure liquid chromatography on Permaphase AAX (25). The purity of each oligomer was checked by homochromatography on thin-layer DEAE-cellulose and also by gel electrophoresis in 20 percent acrylamide (slab) after labeling of the oligomers with [γ-^{32}P]ATP in the presence of polynucleotide kinase. One major labeled product was obtained from each DNA fragment.

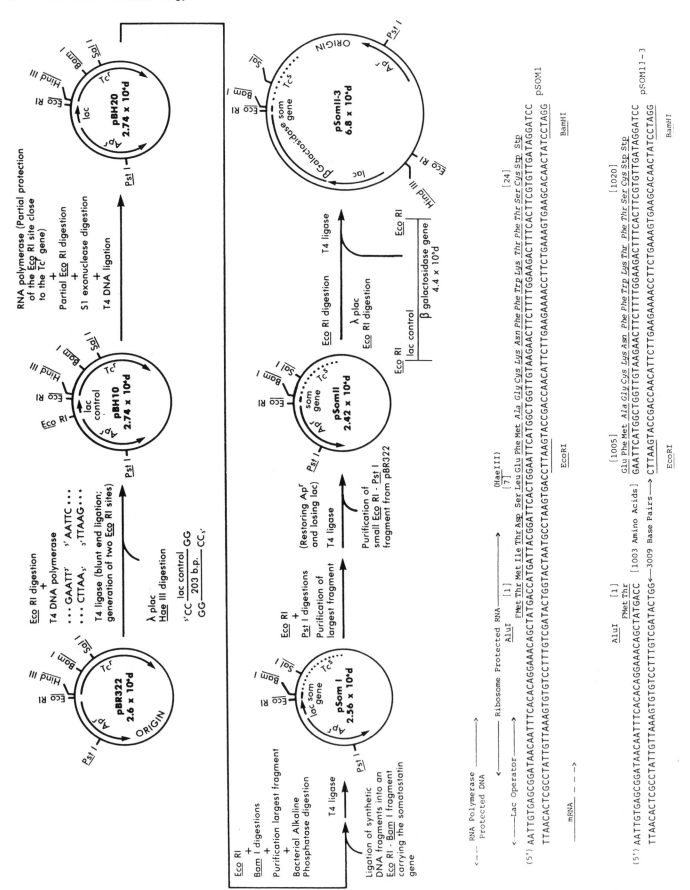

lactosidase subunit structure (pSOM11-3).

The plasmid construction scheme (Fig. 3) begins with plasmid pBR322, a well-characterized cloning vehicle (8). The lac elements were introduced to this plasmid by insertion of an Hae III restriction endonuclease fragment (203 nucleotides) carrying the lac promoter, catabolite-gene-activator-protein binding site, operator, ribosome binding site, and the first seven amino codons of the β-galactosidase structural gene (9) (Figs. 3 and 4). The Hae III fragment was derived from λplac5 DNA. The Eco RI-cleaved pBR322 plasmid, which had its termini repaired with T4 DNA polymerase and deoxyribonucleotide triphosphates, was blunt-end ligated to the Hae III fragment to create Eco RI termini at the insertion points. Joining of these Hae III and repaired Eco RI termini generate the Eco RI restriction site (Figs. 3 and 4) at each terminus. Transformants of *E. coli* RR1 (8) with this DNA were selected for resistance to tetracycline (Tc) and ampicillin (Ap) on 5-bromo-4-chloro-indolylgalactoside (X-gal) medium (10). On this indicator medium, colonies constitutive for the synthesis of β-galactosidase by virtue of the increased number of lac operators titrating repressor, are identified by their blue color. Two orientations of the Hae III fragment are possible, but these were distinguished by the asymmetric location of an Hha restriction site in the fragment. Plasmid pBH10 was further modified to eliminate the Eco RI endonuclease site distal to the lac operator (pBH20).

The eight chemically synthesized oligodeoxyribonucleotides (Fig. 2) were labeled at the 5′ termini with [γ-^{32}P]ATP (adenosine triphophatase) by T4 polynucleotide kinase and joined with T4 DNA ligase. Through hydrogen bonding between the overlapping fragments, the somatostatin gene self-assembles and eventually polymerizes into larger molecules because of the cohesive restriction site termini. The ligated products were treated with Eco RI and Bam HI restriction endonucleases to generate the somatostatin gene (Fig. 2).

The synthetic somatostatin gene fragment with Eco RI and Bam HI termini was ligated to the pBH20 plasmid, previously treated with the Eco RI and Bam HI restriction endonucleases and alkaline phosphatase. The treatment with alkaline phosphatase provides a molecular selection for plasmids carrying the inserted fragment (11). Ampicillin-resistant transformants obtained with this ligated DNA were screened for tetracycline sensitivity, and several were examined for the insertion of an Eco RI-Bam HI fragment of the appropriate size.

Both strands of the Eco RI-Bam HI fragments of plasmids from two clones were analyzed by a nucleotide sequence analysis (12) starting from the Bam HI and Eco RI sites. The sequence analysis was extended into the lac-controlling elements; the lac fragment sequence was in-

Fig. 3 (facing page, left). Construction of recombinant plasmids. Plasmid pBR322 was used as the parental plasmid (8). Plasmid DNA (5 μg) was digested with the restriction endonuclease Eco RI. The reaction was terminated by extraction with a mixture of phenol and chloroform; the DNA was precipitated with ethanol and resuspended in 50 μl of T4 DNA polymerase buffer (26). The reaction was started by the addition of 2 units of T4 DNA polymerase. The reaction (held for 30 minutes at 37°C) was terminated by extraction with phenol and chloroform and precipitation with ethanol. The λplac5 DNA (3 μg) was digested with the endonuclease Hae III (8). The digested pBR322 DNA was blunt-end ligated with the Hae III–digested λplac5 DNA in a final volume of 30 μl with T4 DNA ligase (hydroxylapatite fraction) (27) in 20 mM tris-HCl (pH 7.6), 10 mM MgCl₂, 10 mM dithiothreitol, and 0.5 mM ATP for 12 hours at 12°C. The ligated DNA mixture was dialyzed against 10 mM tris-HCl (pH 7.6) and used to transform *E. coli* strain RR1 (8). Transformants were selected for tetracycline resistance (Tc^r) and ampicillin resistance (Ap^r) on antibiotic (20 μg/ml) minimal X-gal (40 μg/ml) medium (10). Colonies constitutive for the synthesis of β-galactosidase were identified by their blue color. After 45 independently isolated blue colonies were screened, three of them were found to contain plasmids with two Eco RI sites separated by approximately 200 base pairs (28). Plasmid pBH10 was shown to carry the fragment in the desired orientation, that is, lac transcription going into the Tc^r gene of the plasmid. Plasmid pBH10 was further modified to eliminate the Eco RI site distal to the lac operator and plasmid pBH20 was obtained (29). The nucleotide sequence from the Eco RI site into the lac-control region of pBH20 (data not shown), was confirmed. This plasmid was used for cloning the synthetic somatostatin gene. Plasmid pBH20 (10 μg) was digested with endonucleases Eco RI and Bam HI and treated with bacterial alkaline phosphatase (0.1 unit of BAPF, Worthington), and incubation was continued for 10 minutes at 65°C. The reaction mixtures were extracted with a mixture of phenol and chloroform, and the DNA was precipitated with ethanol (30). Somatostatin DNA (50 μl of a solution containing 4 μg/ml) was ligated with the Bam HI–Eco RI, alkaline phosphatase–treated pBH20 DNA in a total volume of 50 μl with the use of 4 units of T4 DNA ligase for 2 hours at 22°C (31). In a control experiment, Bam HI–Eco RI alkaline phosphatase–treated pBH20 DNA was ligated in the absence of somatostatin DNA under similar conditions. Both preparations were used to transform *E. coli* RR1. Transformants were selected on minimal X-gal antibiotic plates. Ten Tc^s transformants were isolated. In the control experiment no transformants were obtained. Four out of the ten transformants contained plasmids with both an Eco RI and a Bam HI site. The size of the small Eco RI-Bam HI fragment of these recombinant plasmids was in all four instances similar to the size of the in vitro prepared somatostatin DNA. Base sequence analysis (12) revealed that the plasmid pSOM1 had the desired somatostatin DNA fragment inserted (data not shown). Because of the failure to detect somatostatin activity from cultures carrying plasmid pSOM1, a plasmid was constructed in which the somatostatin gene could be located at the COOH-terminus of the β-galactosidase gene, keeping the translation in phase. For the construction of such a plasmid, pSOM1 (50 μg) was digested with restriction enzymes Eco RI and Pst I. A preparative 5 percent polyacrylamide gel was used to separate the large Pst I–Eco RI fragment that carries the somatostatin gene from the small fragment carrying the lac control elements (12). In a similar way plasmid pBR322 DNA (50 μg) was digested with Pst I and Eco RI restriction endonucleases, and the two resulting DNA fragments were purified by preparative electrophoresis on a 5 percent polyacrylamide gel. The small Pst I–Eco RI fragment from pBR322 (1 μg) was ligated with the large Pst I–Eco RI DNA fragment (5 μg) from pSOM1. The ligated mixture was used to transform *E. coli* RR1, and transformants were selected for Ap^r on X-gal medium. Almost all the Ap^r transformants (95 percent) gave white colonies (no lac operator) on X-gal indicator plates. The resulting plasmid, pSOM11, was used in the construction of plasmid pSOM11-3. A mixture of 5 μg of pSOM11 DNA and 5 μg of λplac5 DNA was digested with Eco RI. The DNA was extracted with a mixture of phenol and chloroform; the extract was precipitated by ethanol, and the precipitate was resuspended in T4 DNA ligase buffer (50 μl) in the presence of T4 DNA ligase (1 unit). The ligated mixture was used to transform *E. coli* strain RR1. Transformants were selected for Ap^r on X-gal plates containing ampicillin and screened for constitutive β-galactosidase production. Approximately 2 percent of the colonies were blue (such as pSOM11-1 and 11-2). Restriction enzyme analysis of plasmid DNA obtained from these clones revealed that all the plasmids carried a new Eco RI fragment of approximately 4.4 megadaltons, which carries the lac operon control sites and most of the β-galactosidase gene (13, 14). Two orientations of the Eco RI fragment are possible, and the asymmetric location of a Hind III restriction in this fragment can indicate which plasmids had transcription proceeding into the somatostatin gene. The clones carrying plasmids pSOM11-3, pSOM11-5, pSOM11-6, and pSOM11-7 contained the Eco RI fragment in this orientation.

Fig. 4 (facing page, right). Nucleotide sequences of the lac-somatostatin plasmids. The nucleotide sequence of the lac control elements, β-galactosidase structural gene, and the synthetically derived somatostatin DNA, are depicted (9, 14, 27) along with the restriction endonuclease substrate sites. The nucleotide sequence of pSOM1, as depicted, was confirmed (legends to Figs. 3 and 5). The nucleotide sequence of pSOM11-3 was inferred from published data (9, 13, 14, 27). The amino acid sequence of somatostatin is italicized. The amino acid sequence numbers of β-galactosidase are in brackets.

tact, and in one case, pSOM1, the nucleotide sequence of both strands were independently determined, each giving the sequence shown in Fig. 3. In the other case, the sequence was identical except for a base pair deletion (A·T) at a position equivalent to the junction of the B-C oligonucleotides in the original DNA fragment. The basis for the deletion is unclear.

The standard radioimmune assays (RIA) for somatostatin (2) were modified by decreasing the assay volume and by using phosphate buffer (Fig. 6). This modification proved suitable for the detection of somatostatin in *E. coli* extracts. Bacterial cell pellets, extracts, or cultures were treated overnight in 70 percent formic acid containing cyanogen bromide (5 mg/ml). Formic acid and cyanogen bromide were removed under vacuum over KOH before the assay. Initial experiments with extracts of *E. coli* strain RR1 (the recipient strain) (10) indicated that less than 10 pg of somatostatin could easily be detected in the presence of 16 μg or more of cyanogen bromide–treated bacterial protein. More than 2 μg of protein from formic acid–treated bacterial extracts interfered somewhat by increasing the background, but cyanogen bromide cleavage greatly reduced this interference. Reconstruction experiments showed that somatostatin is stable in cyanogen bromide–treated extracts.

The DNA sequence analysis of pSOM1 indicated that the clone carrying this plasmid should produce a peptide containing somatostatin. However, to date all attempts to detect somatostatin radioimmune activity from extracts of cell pellets or culture supernatants have been unsuccessful. Negative results were also obtained when the growing culture was added directly to 70 percent formic acid and cyanogen bromide. We calculate that *E. coli* RR1 (pSOM1) contains less than six molecules of somatostatin per cell. In a reconstruction experiment we have observed that exogenous somatostatin is degraded very rapidly by *E. coli* RR1 extracts. The failure to find somatostatin activity might be accounted for by intracellular degradation by endogenous proteolytic enzymes.

If the failure to detect somatostatin activity from pSOM1 was due to proteolytic degradation of the small protein (Fig. 4), attachment to a large protein might stabilize it. The β-galactosidase structural gene has an Eco RI site near the COOH-terminus (13). The available data on the amino acid sequence of this protein (13, 14) suggested that it would be possible to insert the Eco RI–Bam HI somatostatin gene into the site and maintain the proper reading frame for the correct translation of the somatostatin gene (Fig. 4).

The construction of this plasmid is outlined in Fig. 3. The Eco RI–Pst fragment of the pSOM1 plasmid, with the lac-controlling element, was removed and replaced with the Eco RI–Pst fragment of pBR322 to produce the plasmid pSOM11. The Eco RI fragment of λplac5, carrying the lac operon control region and most of the β-galactosidase structural gene, was inserted into the Eco RI site of pSOM11. Two orientations of the Eco RI lac fragment of λplac5 were expected. One of these orientations would maintain the proper reading frame into the somatostatin gene, the other would not.

A number of independently isolated clones (with plasmid designations pSOM11-2 and pSOM11-3) were analyzed for somatostatin activity, as described above. In contrast to the results of experiments with pSOM1, four clones (pSOM11-3, 11-5, 11-6, and 11-7) were

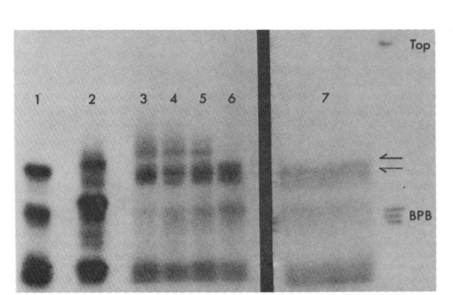

Fig. 5. Ligation and acrylamide gel analysis of somatostatin DNA. The 5′-OH termini of the chemically synthesized fragments A through H (Fig. 2a) were labeled and phosphorylated separately. Just prior to the kinase reaction, 25 μc of [γ-³²P]ATP (~ 1500 c/mmole) (12) was evaporated to dryness in 0.5-ml Eppendorf tubes. The fragment (5 μg) was incubated with 2 units of T4 DNA kinase (hydroxylapatite fraction, 2500 unit/ml) (26), in 70 mM tris-HCl, pH 7.6, 10 mM MgCl₂, and 5 mM dithiothreitol in a total volume of 150 μl for 20 minutes at 37°C. To ensure maximum phosphorylation of the fragments for ligation purposes, 10 μl of a mixture consisting of 70 mM tris-HCl, pH 7.6, 10 mM MgCl₂, 5 mM dithiothreitol, 0.5 mM ATP, and 2 units of DNA kinase were added, and incubation continued for an additional 20 minutes at 37°C. The fragments (250 ng/μl) were stored at −20°C without further treatment. Kinase-treated fragments A, B, E, and F (1.25 μg each) were ligated in a total volume of 50 μl in 20 mM tris-HCl (pH 7.6), 10 mM MgCl₂, 10 mM dithiothreitol, 0.5 mM ATP, and 2 units of T4 DNA ligase (hydroxylapatite fraction, 400 unit/ml) (26), for 16 hours at 4°C. Fragments C, D, G, and H were ligated under similar conditions. Samples (2 μl) were removed for analysis by electrophoresis on a 10 percent polyacrylamide gel and subsequent autoradiography (16) (lanes 1 and 2, respectively). The fast migrating material represents unreacted DNA fragments. Material migrating with the bromophenol blue dye (BPB) is the monomeric form of the ligated fragments. The slowest migrating material represents dimers, which form by virtue of the cohesive ends, of the ligated fragments A, B, E, and F (lane 1) and C, D, G, and H (lane 2). The dimers can be cleaved by restriction endonuclease Eco RI or Bam HI, respectively (data not shown). The two half molecules (ligated A + B + E + F and ligated C + D + G + H) were joined by an additional ligation step carried out in a final volume of 150 μl at 4°C for 16 hours. A sample (1 μl) was removed for analysis (lane 3). The reaction mixture was heated for 15 minutes at 65°C to inactivate the T4 DNA ligase. The heat treatment does not affect the migration pattern of the DNA mixture (lane 4). Enough restriction endonuclease Bam HI was added to the reaction mixture to cleave the multimeric forms of the somatostatin DNA in 30 minutes at 37°C (lane 5). After the addition of NaCl to a concentration of 100 mM, the DNA was digested with Eco RI endonuclease (lane 6). The restriction endonuclease digestions were terminated by phenol-chloroform extraction of the DNA. The somatostatin DNA fragment was purified from unreacted and partially ligated DNA fragments by preparative electrophoresis on a 10 percent polyacrylamide gel. The band indicated with an arrow (lane 7) was excised from the gel, and the DNA was eluted by slicing the gel into small pieces and extracting the DNA with elution buffer (0.5M ammonium acetate, 10 mM MgCl₂, 0.1 mM EDTA, and 0.1 percent sodium dodecyl sulfate) overnight at 65°C (12). The DNA was precipitated with two volumes of ethanol, centrifuged, redissolved in 200 μl of 10 mM tris-HCl (pH 7.6), and dialyzed against the same buffer, resulting in a somatostatin DNA concentration of 4 μg/ml.

found to have easily detectable somatostatin radioimmune activity (Fig. 6, a and b). Restriction fragment analysis revealed that pSOM11-3, pSOM11-5, pSOM11-6, and pSOM11-7 had the desired orientation of the lac operon, whereas pSOM11-2 and 11-4 had the opposite orientation. Thus, there is a perfect correlation between the correct ori-

entation of the lac operon and the production of somatostatin radioimmune activity.

The design of the somatostatin plasmid predicts that the synthesis of somatostatin would be under the control of the lac operon. The lac repressor gene is not included in the plasmid, and the recipient strain (*E. coli* RR1) contains the wild-type chromosomal lac repressor gene, which produces only 10 to 20 repressor molecules per cell (*15*). The plasmid copy number (and therefore the number of lac operators) is approximately 20 to 30 per cell and complete repression is impossible. The specific activity of somatostatin in *E. coli* RR1 (pSOM11-3) was increased by IPTG, an inducer of the lac operon (Table 1). As expected, the level of induction was low, varying from 2.4- to 7-fold. In experiment 7 (Table 1), the α activity (*14*), a measure of the first 92 amino acids of β-galactosidase, also was induced by a factor of 2.

In several experiments (Table 1 and other experiments not shown), no somatostatin radioimmune activity was detected prior to cyanogen bromide cleavage of the total cellular protein. Since the antiserum used in the radioimmune assay, S39, requires a free NH_2-terminal alanine, no activity was expected prior to cyanogen bromide cleavage. After cleavage by cyanogen bromide, cell extracts were chromatographed on Sephadex G-50 in 50 percent acetic acid (Fig. 6c). In this system, somatostatin is well separated from excluded large peptides and fully included small molecules. Only extracts of clones positive for somatostatin exhibited radioimmune activity in the column fractions, and this activity elutes in the same position as chemically synthesized somatostatin.

The strains carrying the Eco RI lac operon fragment (such as pSOM11-2 and pSOM11-3) segregate with respect to the plasmid phenotype. For example, after

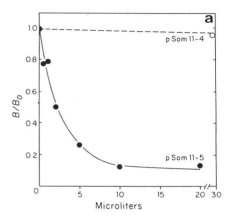

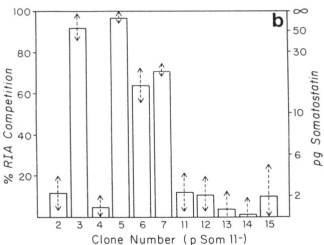

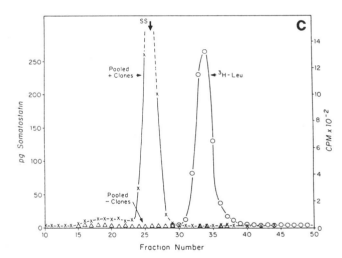

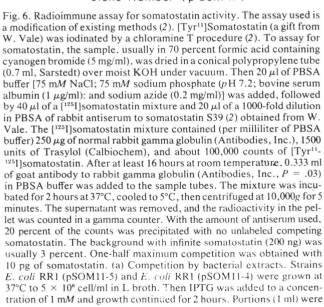

Fig. 6. Radioimmune assay for somatostatin activity. The assay used is a modification of existing methods (*2*). [Tyr[11]]Somatostatin (a gift from W. Vale) was iodinated by a chloramine T procedure (*2*). To assay for somatostatin, the sample, usually in 70 percent formic acid containing cyanogen bromide (5 mg/ml), was dried in a conical polypropylene tube (0.7 ml, Sarstedt) over moist KOH under vacuum. Then 20 μl of PBSA buffer [75 m*M* NaCl; 75 m*M* sodium phosphate (*p*H 7.2; bovine serum albumin (1 μg/ml); and sodium azide (0.2 mg/ml)] was added, followed by 40 μl of a [125I]somatostatin mixture and 20 μl of a 1000-fold dilution in PBSA of rabbit antiserum to somatostatin S39 (*2*) obtained from W. Vale. The [125I]somatostatin mixture contained (per milliliter of PBSA buffer) 250 μg of normal rabbit gamma globulin (Antibodies, Inc.), 1500 units of Trasylol (Calbiochem), and about 100,000 counts of [Tyr[11]-125I]somatostatin. After at least 16 hours at room temperature, 0.333 ml of goat antibody to rabbit gamma globulin (Antibodies, Inc., P = .03) in PBSA buffer was added to the sample tubes. The mixture was incubated for 2 hours at 37°C, cooled to 5°C, then centrifuged at 10,000*g* for 5 minutes. The supernatant was removed, and the radioactivity in the pellet was counted in a gamma counter. With the amount of antiserum used, 20 percent of the counts was precipitated with no unlabeled competing somatostatin. The background with infinite somatostatin (200 ng) was usually 3 percent. One-half maximum competition was obtained with 10 pg of somatostatin. (a) Competition by bacterial extracts. Strains *E. coli* RR1 (pSOM11-5) and *E. coli* RR1 (pSOM11-4) were grown at 37°C to 5 × 10[8] cell/ml in L broth. Then IPTG was added to a concentration of 1 m*M* and growth continued for 2 hours. Portions (1 ml) were

centrifuged for a few seconds in an Eppendorf centrifuge, and the pellets were suspended in 500 μl of 70 percent formic acid containing cyanogen bromide (5 mg/ml). After approximately 24 hours at room temperature, the samples were diluted tenfold in water, and the indicated volumes were assayed in triplicate for somatostatin. B/B_0 is the ratio of [125I]somatostatin bound in the presence of competing somatostatin to that bound in the absence of competing somatostatin. Each point is the average of triplicate tubes. The protein content of the undiluted samples were determined to be 2.2 mg/ml for *E. coli* RR1 (pSOM11-5) and 1.5 mg/ml for *E. coli* RR1 (pSOM-4). (b) The initial screening of clones for somatostatin. Cyanogen bromide-treated extracts of 11 clones (such as pSOM11-2 and pSOM11-3) were made as described above for (a). A sample (30 μl) of each extract was taken in triplicate for radioimmune assay. The range of assay points is indicated. The values for picograms of somatostatin were read from a standard curve obtained as part of the same experiment. (c) Gel filtration of cyanogen bromide-treated extracts. Formic acid and cyanogen-treated extracts of the positive clones (11-3, 11-5, 11-6, and 11-7) were pooled (total volume, 250 μl), dried, and resuspended in 0.1 ml of 50 percent acetic acid. [3H]Leucine was added, and the sample was applied to a column (0.7 by 47 cm) of Sephadex G-50 in 50 percent acetic acid. Portions (50 μl) of the column fractions were assayed for somatostatin. Pooled negative clone extracts (11-2, 11-4, and 11-11) were treated identically. On the same column known somatostatin (Beckman Instruments, Inc.) elutes as indicated (*SS*).

Table 1. Somatostatin radioimmune specific activity. Abbreviations: LB, Luria broth, IPTG, isopropylthiogalactoside; CNBr, cyanogen bromide; SS, somatostatin. Protein was measured by the method of Bradford (32).

Experiment	Strain	Medium	IPTG 1 mM	CNBr 5 mg/ml	SS/protein (pg/mg)
1	11-2	LB	+	+	< 0.1
	11-3	LB	+	+	12
	11-4	LB	+	+	< 0.4
	11-5	LB	+	+	15
2	11-3	LB	+	+	12
	11-3	LB	+	−	< 0.1
3	11-3	LB	+	+	61
	11-3	LB	−	+	8
	11-3	LB	+	−	< 0.1
4	11-3	LB	+	+	71
	11-3	VB + glycerol*	+	+	62
5	11-3	LB + glycerol	+	+	250
6	11-3	LB	+	+	320
	11-2	LB	+	+	< 0.1
7	11-3	LB	+	+	24
	11-3	LB	−	+	10

*Vogel-Bonner minimal medium plus glycerol.

about 15 generations, about one-half of the E. coli RR1 (pSOM11-3) culture was constitutive for β-galactosidase, that is, carried the lac operator, and about half of the nonconstitutive colonies were ampicillin-sensitive. Strains positive (pSOM11-3) and negative (pSOM11-2) for somatostatin are unstable, and, therefore, the growth disadvantage presumably comes from the overproduction of the large but incomplete and inactive galactosidase. The yield of somatostatin has varied from 0.001 to 0.03 percent of the total cellular protein (Table 1) probably as the result of the selection for cells in culture having plasmids with a deleted lac region. The highest yields of somatostatin have been from preparations where growth was started from a single Ap-resistant, constitutive colony. Even in these cases, 30 percent of the cells at harvest had deletions of the lac region.

Several moderate scale (up to 10 liters) attempts have been made to purify somatostatin from E. coli strain RR1 (pSOM11-3). The initial purification scheme was based on known purification properties of β-galactosidase followed by purification of the cyanogen bromide cleavage products of the chimeric protein. However, essentially all of the somatostatin activity found in the crude extract is insoluble and is found in the pellet from the first low speed centrifugation. The activity can be solubilized in 70 percent formic acid, 6M guanidinium hydrochloride, 8M urea, or 2 percent sodium docecyl sulfate. Somatostatin activity has been enriched approximately 100-fold from the cellular debris by cyanogen bromide cleavage, and subsequent alcohol extraction and chromatography on Sephadex G-50 in 50 percent acetic acid.

Recent improvements in the chemical synthesis of DNA provide the opportunity to synthesize quickly DNA with biological interest for genetic manipulation and experimentation. As illustrated earlier (16, 17), in vitro recombinant DNA techniques and molecular cloning enhance the experimental value of chemically synthesized DNA. There are two well-established methods for the synthesis of DNA. The phosphodiester method of Khorana and co-workers (18) and the more recently developed modified phosphotriester method (7). Both methods are capable of producing functional DNA (16, 17, 19, 20); however, the triester method is probably faster. Moreover, a method for rapidly synthesizing trimer blocks (codons) as building units for longer oligodeoxyribonucleotides (21) (Fig. 2b) has increased the speed of the triester method. From the trimer block library, a hexadecadeoxyribonucleotide now can be obtained in a week. We have established here that the DNA made with this improvement is functional.

The data establishing the synthesis of a polypeptide containing the somatostatin amino acid sequence are summarized as follows. (i) Somatostatin radioimmune activity is present in E. coli cells having the plasmid pSOM11-3, which contains a somatostatin gene of proven correct sequence and has the correct orientation of the lac Eco RI DNA fragment. Cells with the related plasmid pSOM11-2, which has the same somatostatin gene but an opposite orientation of the lac Eco RI fragment, produce no detectable somatostatin activity. (ii) As predicted by the design scheme, no detectable somatostatin radioimmune activity is observed until after cyanogen bromide

treatment of the cell extract. (iii) The somatostatin activity is under control of the lac operon as evidenced by induction by IPTG, an inducer of the lac operon. (iv) The somatostatin activity cochromatographs with known somatostatin on Sephadex G-50. (v) The DNA sequence of the cloned somatostatin gene is correct. If translation is out of phase, a peptide will be made which is different from somatostatin at every position. Radioimmune activity is detected indicating that a peptide closely related to somatostatin is made, and translation must be in phase. Since translation occurs in phase, the genetic code dictates that a peptide with the exact sequence of somatostatin is made. (vi) Partially purified samples have been independently assayed by W. Vale (Salk Institute). He has confirmed our radioimmune activity with both antiserum S39, which is directed by the NH₂-terminal, and with antiserum S201 which interacts mainly with somatostatin positions 6 through 14. (vii) Finally, the above samples of E. coli RR1 (pSOM11-3) extract inhibit the release of growth hormone from rat pituitary cells, whereas samples of E. coli RR1 (pSOM11-2) prepared in parallel and with identical protein concentration have no effect on growth hormone release (22).

Our results represent the first success in achieving expression (that is, transcription into RNA and translation of that RNA into a protein of a designed amino acid sequence) of a gene of chemically synthesized origin. The large number of plasmid molecules per cell results in a substantial amount (at least 3 percent) of the cellular protein as the β-galactosidase-somatostatin hybrid. This molecule appears to be relatively resistant to endogenous proteolytic activity. There is evidence that abnormally short β-galactosidase peptides are degraded in E. coli (14) suggesting that the hybrid protein molecule expected from the first somatostatin-lac plasmid (pSOM1) is also rapidly degraded. The synthesis of many gratuitous proteins in E. coli, whether large enzymes or smaller polypeptides, may be undetectable for this reason. In cases where the amino acid composition of the protein is appropriate, the precursor technique described here can be employed. This approach could possibly be extended by taking advantage of proteolytic enzymes with amino acid sequence specificity.

The amount of somatostatin synthesized was variable and about a factor of 10 less than the maximum predicted yield. This variability could be interpreted in several ways. Protein degradation by endogenous proteases, the inability to fully solubilize the chimeric protein, and the selection of altered plasmids could all be contributing factors to the variability in yield. Although recombinant DNA experiments with chemically synthesized DNA are inherently less hazardous than those with DNA from natural sources, consideration should be given to the possible toxicity of the peptide product. A major factor in the choice of somatostatin was its proven low toxicity (3). In addition, the experiment was deliberately designed to have the cells produce not free somatostatin but rather a precursor, which would be expected to be relatively inactive. The cloning and growth of cell cultures were performed in a P-3 containment facility.

KEIICHI ITAKURA
TADAAKI HIROSE
ROBERTO CREA
ARTHUR D. RIGGS

Division of Biology,
City of Hope National Medical Center,
Duarte, California 91010

HERBERT L. HEYNEKER*
FRANCISCO BOLIVAR†
HERBERT W. BOYER

Department of Biochemistry and
Biophysics, University of California,
San Francisco 94143

References and Notes

1. P. Brazeau, W. Vale, R. Burgus, N. Ling, M. Butcher, J. Rivier, R. Guillemin, *Science* **179**, 77 (1973).
2. A. Arimura, H. Sato, D. H. Coy, A. V. Schally, *Proc. Soc. Exp. Biol. Med.* **148**, 784 (1975).
3. W. Vale, N. Ling, J. Rivier, J. Villareal, C. Rivier, C. Douglas, M. Brown, *Metabolism* **25**, 1491 (1976); W. Vale, G. Grant, M. Amoss, R. Blackwell, R. Guillemin, *Endocrinology* **91**, 562 (1972); R. Guillemin and J. E. Gerich, *Annu. Rev. Med.* **27**, 379 (1976); W. Vale, C. Rivier, M. Brown, *Annu. Rev. Physiol.* **39**, 473 (1977).
4. E. Gross and B. Witkop, *J. Am. Chem. Soc.* **83**, 1510 (1961); E. Gross, *Methods Enzymol.* **11**, 238 (1967).
5. W. Fiers *et al.*, *Nature (London)* **260**, 500 (1976).
6. K. Bertrand, L. Korn, F. Lee, T. Platt, C. L. Squires, C. Squires, C. Yanofsky, *Science* **189**, 22 (1975).
7. K. Itakura, N. Katagiri, C. P. Bahl, R. H. Wightman, S. A. Narang, *J. Am. Chem. Soc.* **97**, 7327 (1975); K. Itakura, N. Katagiri, S. A. Narang, C. P. Bahl, R. Wu, *J. Biol. Chem.* **250**, 4592 (1975).
8. F. Bolivar, R. L. Rodriguez, P. J. Greene, M. C. Betlach, H. L. Heyneker, H. W. Boyer, J. H. Crosa, S. Falkow, *Gene* **2**, 95 (1977).
9. W. Gilbert, J. Gralla, J. Majors, A. Maxam, in *Protein-Ligand Interactions*, H. Sund and G. Blauer, Eds. (De Gruyter, Berlin, 1975), pp. 193–210.
10. J. H. Miller, *Experiments in Molecular Genetics* (Cold Spring Harbor Laboratory, Cold Spring Harbor, N.Y., 1972).
11. A. Ullrich, J. Shine, J. Chirgwin, R. Pictet, E. Tischer, W. J. Rutter, H. M. Goodman, *Science* **196**, 1313 (1977).
12. A. M. Maxam and W. Gilbert, *Proc. Natl. Acad. Sci. U.S.A.* **74**, 560 (1977).
13. B. Polisky, R. J. Bishop, D. H. Gelfand, *ibid.* **73**, 3900 (1976); P. H. O'Farrell, B. Polisky, D. H. Gelfand, personal communication.
14. A. V. Fowler and I. Zabin, *Proc. Natl. Acad. Sci. U.S.A.* **74**, 1507 (1977); A. I. Bukhari and D. Zipser, *Nature (London) New Biol.* **243**, 238 (1973); K. E. Langley, A. V. Fowler, I. Zabin, *J. Biol. Chem.* **250**, 2587 (1975).
15. W. Gilbert and B. Muller-Hill, *Proc. Natl. Acad. Sci. U.S.A.* **56**, 1891 (1966).
16. H. L. Heyneker *et al.*, *Nature (London)* **263**, 748 (1976).
17. K. Marians, R. Wu, J. Stawinski, T. Hozumi, S. A. Narang, *ibid.*, p. 744.
18. H. G. Khorana, *Pure Appl. Chem.* **17**, 349 (1968).
19. D. V. Goeddel *et al.*, *Proc. Natl. Acad. Sci. U.S.A.* **74**, 3292 (1977).
20. H. G. Khorana, personal communication.
21. T. Hirose, R. Crea, K. Itakura, in preparation.
22. W. Vale, personal communication; M. M. Bradford, *Anal. Biochem.* **72**, 248 (1976).
23. Y. Stawinski, T. Hozumi, S. A. Narang, C. P. Bahl, R. Wu, *Nucleic Acids Res.* **4**, 353 (1977).
24. B. T. Hunt and W. Rigby, *Chem. Ind.* (1967), p. 1868.
25. R. A. Henry, J. A. Schmidt, R. C. Williams, *J. Chromatogr. Sci.* **11**, 358 (1973).
26. A. Panet, J. H. van de Sande, P. C. Loewen, H. G. Khorana, A. J. Raae, J. R. Lillehaug, K. Kleppe, *Biochemistry* **12**, 5045 (1973).
27. R. C. Dickson, J. Abelson, W. M. Barnes, W. S. Reznikoff, *Science* **187**, 27 (1975).
28. The position of an asymmetrically located Hha I site in the 203 base pair Hae III *lac* control fragment (9) allows for the determination of the orientation of the Hae III fragment, now an Eco RI fragment, in these plasmids.
29. This was accomplished by preferential Eco RI endonuclease cleavage at the distal site by partial protection with RNA polymerase of the other Eco RI site localized between the Tc^r and lac promoters, which are only about 40 base pairs apart. After binding RNA polymerase, the DNA (5 μg) was digested with Eco RI (1 unit) in a final volume of 10 μl for 10 minutes at 37°C. The reaction was stopped by heating at 65°C for 10 minutes. The Eco RI cohesive termini were digested with S1 nuclease in a solution of 25 mM sodium acetate (pH 4.5), 300 mM NaCl, and 1 mM ZnCl$_2$ at 25°C for 5 minutes. The reaction mixture was stopped by the addition of EDTA (10 mM, final) and tris-HCl (pH 8) (50 mM final). The DNA was extracted with phenol-chloroform, precipitated with ethanol, and resuspended in 100 μl of T4 DNA ligation buffer. The T4 DNA ligase (1 μl) was added and the mixture was incubated at 12°C for 12 hours. The ligated DNA was transformed in *E. coli* strain RR1, and Ap^{r}Tc^r transformants were selected on X-gal-antibiotic medium. Restriction enzyme analysis of DNA screened from ten isolated blue colonies revealed that these clones carried plasmid DNA with one Eco RI site. Seven of these colonies had retained the Eco RI site located between the lac and Tc^r promoters.
30. The alkaline phosphatase treatment effectively prevents self-ligation of the Eco RI–Bam HI treated pBH20 DNA, but circular recombinant plasmids containing somatostatin can still be formed upon ligation. Since *E. coli* RR1 is transformed with very low efficiency by linear plasmid DNA, the majority of the transformants will contain recombinant plasmids (11).
31. After 10, 20, and 30 minutes, additional somatostatin DNA (40 ng) was added to the reaction mixture (the gradual addition of somatostatin DNA may favor ligation to the plasmid over self-ligation). Ligation was continued for 1 hour and then the mixture was dialyzed against 10 mM tris-HCl (pH 7.6).
32. M. M. Bradford, *Anal. Biochem.* **72**, 248 (1976).

33. Supported by contracts from Genentech, Inc. to the City of Hope National Medical Center and the University of California. H.W.B. is an investigator of the Howard Hughes Medical Research Institute. We thank L. Shively, Y. Lu, L. Shih, and L. Directo for their assistance in various aspects of the project, and R. A. Swanson for his assistance and encouragement throughout the design and execution of the project. We also thank D. Gelfand and P. O'Farrell for discussing their unpublished data with us.
* Present address: Department of Molecular Genetics, University of Leiden, Wassenaarseweg 64, Leiden, Netherlands.
† Present address: Departamento de Biologia Molecular, Instituto de Investigaciones Biomedicas, Universidad Nacional Autonoma de Mexico, Mexico 20 D.F. Apdo Postal 70228.

2 November 1977

Synthesis of Deoxyoligonucleotides on a Polymer Support

M.D. Matteucci and M.H. Caruthers

Abstract: The development of a new method for synthesizing deoxyoligonucleotides is described. The synthesis begins by derivatizing high-performance liquid chromatography grade silica gel to contain 5'-O-(dimethoxytrityl)deoxynucleosides linked through the 3'-hydroxyl to a carboxylic acid functional group on the support. This matrix is then packed into a column which is attached to a pump and a series of valves. The chemical steps for the addition of one nucleotide to the support are as follows: (1) detritylation using $ZnBr_2$ in nitromethane (30 min); (2) condensation of a 5'-O-(dimethoxytrityl)deoxynucleoside (3'-methoxytetrazoyl)phosphine with the support-bound nucleoside (60 min); (3) blocking unreacted, support-bound nucleoside hydroxyl groups with diethoxytriazolylphosphine (5 min); (4) oxidation of phosphites to phosphates with I_2 (5 min). Completed deoxyoligonucleotides are isolated by sequential treatment with thiophenol and ammonium hydroxide, purification by reverse-phase chromatography, and treatment with 80% acetic acid. The method is extremely fast (less than 2.5 h are needed for each nucleotide addition cycle), yields in excess of 95% per condensation are obtained, and isolation of the final product is a simple one-step column purification. The syntheses of d(C-G-T-C-A-C-A-A-T-T) and d(A-C-G-C-T-C-A-C-A-A-T-T) were carried out as a test of this method. Yields of support-bound deoxyoligonucleotides were 64% and 55%; the isolated yield of deoxydecanucleotide was 30%. Both synthetic products were homogeneous and biologically active by every criteria so far tested.

Synthetic deoxyoligonucleotides of defined sequence have been used to solve important biochemical[2,3] and biophysical[4,5] problems. Moreover, recent advances in chemical methods have led to the synthesis of genes[6-10] and of deoxyoligonucleotides useful for manipulating natural DNA and RNA.[11] Despite these achievements, the synthesis and isolation of deoxyoligonucleotides remains a difficult and time consuming task. Ideally, chemical methods should be simple, rapid, versatile, and completely automatic. In this way the rapid synthesis of genes and gene control regions can be realized and many important biochemical studies which are presently not possible can be initiated.

Our approach to solving this problem has involved developing methods for synthesizing deoxyoligonucleotides on polymer supports. This concept is not new and has been investigated extensively.[12,13] Recently several promising approaches have been

(1) This is paper IV in a series on nucleotide chemistry. Paper III: M. H. Caruthers, S. L. Beaucage, J. W. Efcavitch, E. F. Fisher, M. D. Matteucci, and Y. Stabinsky, *Nucleic Acids Symp. Ser.*, No. 7, 215 (1980). This research was supported by the National Institutes of Health (Grants GM 21120 and GM 25680).

(2) H. G. Khorana, H. Büchi, H. Ghosh, N. Gupta, T. M. Jacob, H. Kössel, R. Morgan, S. A. Narang, E. Ohtsuka, and R. D. Wells, *Cold Spring Harbor Symp. Quant. Biol.*, **31**, 39 (1966).

(3) M. H. Caruthers, *Acc. Chem. Res.*, **13**, 155 (1980).

(4) A. H.-J. Wang, G. J. Quigley, F. J. Kolpak, J. L. Crawford, J. H. van Boom, G. vd Marel, and A. Rich, *Nature (London)*, **282**, 680 (1979).

(5) F. H. Martin and I. Tinoco, Jr., *Nucleic Acids Res.*, **8**, 2295 (1980).

(6) K. L. Agarwal, Büchi, M. H. Caruthers, N. Gupta, H. G. Khorana, K. Kleppe, A. Kumar, E. Ohtsuka, U. L. Rajbhandary, J. H. van de Sande, V. Sgaramella, H. Weber, and T. Yamada, *Nature (London)*, **227**, 27 (1970).

(7) H. G. Khorana, *Science*, **203**, 614 (1979).

(8) K. Itakura, T. Hirose, R. Crea, A. D. Riggs, H. L. Heyneker, F. Bolivar, and H. W. Boyer, *Science*, **198**, 1056 (1977).

(9) D. V. Goeddel, D. G. Kleid, F. Bolivar, H. L. Heyneker, D. Yansura, R. Crea, T. Hirose, A. Kraszewski, K. Itakura, and A. Riggs, *Proc. Natl. Acad. Sci. U.S.A.*, **76**, 106 (1979).

(10) D. V. Goeddel, H. L. Heyneker, T. Hozumi, R. Arentzen, K. Itakura, D. G. Yansura, M. J. Ross, G. Miozzari, R. Crea, and P. Seeburg, *Nature (London)*, **281**, 544 (1979).

(11) R. J. Rothstein, L. F. Lau, C. D. Bahl, S. A. Narang, and R. Wu, *Methods Enzymol.*, **68**, 98 (1979).

(12) C. B. Reese, *Tetrahedron*, **34**, 3143 (1978).

(13) V. Amarnath and A. D. Broom, *Chem. Rev.*, **77**, 183 (1977).

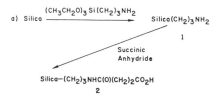

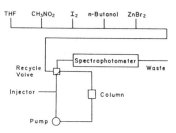

Figure 1. Steps in the synthesis of silica gel containing covalently joined deoxynucleosides.

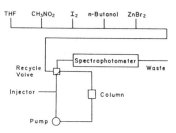

Figure 2. Schematic diagram of the apparatus used for polynucleotide synthesis. Five flasks containing two solvents (tetrahydrofuran and nitromethane) and three reactants are represented in the top part of the diagram. The complete compositions of these reaction solutions (I_2, n-butanol, and $ZnBr_2$) are included in the Experimental Section.

proposed.[14-17] These methods utilize various organic polymers and phosphate-activated di- and trinucleotides as condensing agents. We report in this paper a completely different method. The polymer is high-performance liquid chromatography (HPLC) grade silica gel, and activated nucleotide phosphites are used as condensing agents. A preliminary account of this approach in a form applicable only to deoxyoligopyrimidines has been reported.[18]

Outline of the Procedure

Support. Extensive research has been directed toward developing silica gels containing different functional groups for use in HPLC. These polymers have been designed for efficient mass transfer. We therefore anticipated and have since shown that reactants and reagents can rapidly be removed from these derivatized silica gels after various synthesis steps. Additionally, silica gel is a rigid, nonswellable matrix in common organic solvents. It can be packed into a column and reactants merely pumped through the column. These features make HPLC grade silica gel an attractive support for deoxyoligonucleotide synthesis. For our work, we use a silica gel that has been derivatized to contain a carboxylic acid group (200 μmol/g of silica). The scheme developed for synthesis of this support is outlined in Figure 1, reaction scheme a. The functionalized, insoluble support was prepared from a 20-μm particle size, macroporous (300 Å) silica gel. The initial step involved refluxing (3-aminopropyl)triethoxysilane with silica gel in dry toluene for 3 h. Succinic anhydride was next reacted with **1** in order to generate **2**. Excess silanol groups were eliminated by treatment with $(CH_3)_3SiCl$. For the synthesis of **3** (Figure 1, reaction scheme b), 5'-O-(dimeth-

oxytrityl)thymidine was condensed with **2** by using dicyclohexylcarbodiimide (DCC) in anhydrous pyridine.[19] After 40 h, residual acid groups were converted to an inert amide by addition first of p-nitrophenol and then morpholine. The yield of **3** was 40 μmol/g of reactant. No attempt has been made to maximize the amount of nucleoside covalently joined to the support. This procedure is quite general and has been used for the synthesis of supports containing the 5'-O-dimethoxytrityl derivatives of N-benzoyldeoxycytidine, N-benzoyldeoxyadenosine and N-isobutyryldeoxyguanosine.[20]

Synthesis Machine. A schematic of the apparatus is shown in Figure 2. Machine-assisted synthesis of deoxyoligonucleotides begins by loading the column with 0.20–0.30 g of **3** or one of the other silica gels containing a covalently joined deoxynucleoside. The protocol listed in Table I is then followed for the addition of one nucleotide. Solvents and reactants can be selected through a series of three-way valves and then either cycled once or recycled many times through the column. An injector port is used for the addition of activated nucleotides, and the efficiency of various wash cycles is monitored by a spectrophotometer. The major steps that form a part of this protocol are outlined in the following sections.

Removal of 5'-O-Dimethoxytrityl Ethers. p-Toluenesulfonic acid in acetonitrile rapidly removes the 5'-O-dimethoxytrityl group from either a deoxypyrimidine nucleoside such as **3** or a deoxyoligopyrimidine attached covalently to silica gel.[18] However, when purines were examined, this procedure was unsatisfactory. These results are outlined in Table II. When a dinucleotide attached to silica gel and containing N-benzoyldeoxyadenosine and thy-

Table I. Protocol for Machine-Assisted Polynucleotide Synthesis

reagent or solvent[a,b]	time, min	machine mode
satd $ZnBr_2/CH_3NO_2$	30	flush
$CH_3(CH_2)_2CH_2OH/2,6$-lutidine/THF	5	flush
THF	10	flush
activated nucleotide	60	recycle
$(CH_3CH_2O)_2P$ (triazole)	5	recycle
THF	2	flush
I_2 oxidation	5	flush
THF	5	flush
CH_3NO_2	3	flush

[a] THF = tetrahydrofuran. [b] A nitromethane solution saturated with $ZnBr_2$ is approximately 0.1 M in $ZnBr_2$.

Table II. Investigation of Detritylation and Depurination Conditions[a,b]

conditions[c]	detritylation time, min	depurination %/time
0.1 N toluenesulfonic acid/THF	5	50/20 min
5% BF_3/1% 2,6-di-tert-butylpyridine/acetonitrile	10	50/120 min
2.5% $AlCl_3$/1% 2,6-di-tert-butylpyridine/acetonitrile	10	50/8 h
2.0% $TiCl_4$/3% 2,6-di-tert-butylpyridine/acetonitrile	3	10/4 h
<0.1 M $ZnBr_2$ (saturated)/ nitromethane	15	none/24 h

[a] Analytical reaction: $d[(MeO)_2TrbzA-T\textcircled{P}] \xrightarrow{acid} d(bzA-T\textcircled{P}) + (MeO)_2Tr-OH$. [b] All depuration results were obtained by analyzing reaction mixtures using reverse-phase HPLC. Estimates of the time required for 100% detritylation were by trityl assay (see Experimental Section). [c] 2,6-Di-tert-butylpyridine was used to specifically buffer protic acids [H. C. Brown and B. Kanner, *J. Am. Chem. Soc.*, 88, 986 (1966)] which potentially could be generated via hydrolysis of highly reactive Lewis acids (BF_3, $AlCl_3$, and $TiCl_4$) by trace amounts of water.

(14) M. J. Gait, N. Singh, R. C. Sheppard, M. D. Edge, A. R. Greene, G. R. Heathcliff, T. C. Atkinson, C. R. Newton, and A. F. Markham, *Nucleic Acids Res.*, 8, 1081 (1980).

(15) K. Miyoshi and K. Itakura, *Tetrahedron Lett.*, 38, 3635 (1979).

(16) V. Potapov, V. Veiko, O. Koroleva, and Z. Shabarova, *Nucleic Acids Res.*, 6, 2041 (1979).

(17) R. Crea and T. Horn *Nucleic Acids Res.*, 8, 2331 (1980).

(18) M. D. Matteucci and M. H. Caruthers, *Tetrahedron Lett.*, 21, 719 (1980).

(19) Abbreviations for nucleosides, nucleotides, oligonucleotides, and protected deoxyoligonucleotides are according to the IUPAC-IUB Commission on Biochemical Nomenclature Recommendations [*Biochemistry*, 9, 4022 (1970)]. The symbol ⓟ represents the insoluble, derivatized silica gel.

(20) Unpublished results of S. Beauage and C. Becker.

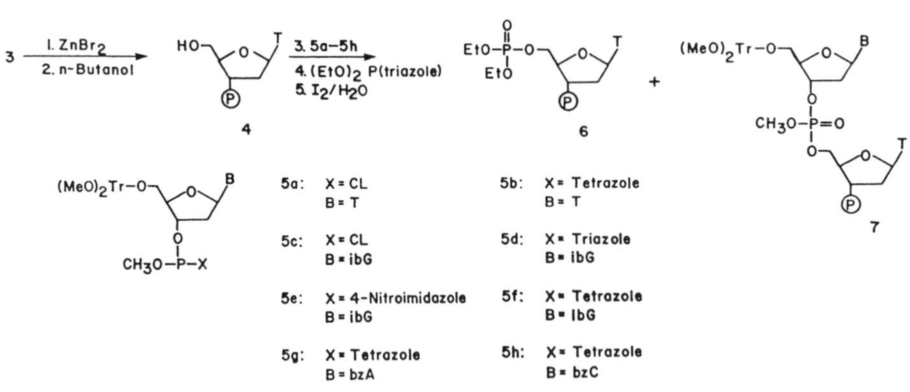

Figure 3. Steps in the synthesis of a dinucleotide. Purine bases (B) are abbreviated as bzA for *N*-benzoyladenine and ibG for *N*-isobutyrlguanine. Pyrimidine bases (B) are abbreviated as bzC for *N*-benzoylcytosine and T for thymine.

midine, d[(MeO)$_2$TrbzA-T℗], was treated with *p*-toluenesulfonic acid, detritylation was complete in 5 min. However, in 20 min, 50% depurination had also occurred. Thus this procedure cannot be used for repetitive detritylation of purine containing deoxyoligonucleotides during their synthesis on a polymer support. Because of this depurination problem, various Lewis acids were investigated as potential detritylating reagents.[21] As can be seen from the results reported in Table II, a saturated solution containing ZnBr$_2$ in nitromethane was superior to the other Lewis acids tested. Detritylation was complete within 15 min and depurination was not detected even after 24 h. Detritylation was also most rapid when the 5'-deoxynucleoside was a purine (10–15 min) rather than a pyrimidine (30 min). On the basis of the results presented in Table II, TiCl$_4$ also appeared to be a potentially useful detritylating reagent. However, since TiCl$_4$ is very easily hydrolyzed, it was not examined further. The reaction of nucleosides with ZnBr$_2$ has been studied in some detail.[22,23] Deprotection appears to proceed via a bidentate chelation mechanism involving the 5' and deoxyribose ring oxygens. Furthermore, in an anhydrous nitromethane solution containing ZnBr$_2$, deprotection generated a compound attached to the support which was not fully reactive toward the phosphite coupling reagent. Presumably this intermediate was a zincate ester. A mildly basic hydrolytic wash with *n*-butanol in tetrahydrofuran and 2,6-lutidine was sufficient to regenerate a free 5'-hydroxyl group. The detritylation reaction as outlined in Figure 3 therefore consisted of two steps. Intially, compound 3 or any other protected nucleoside attached to silica gel was allowed to react for 30 min with a saturated solution of ZnBr$_2$ in nitromethane. The next step (5 min) was a hydrolytic wash with *n*-butanol in tetrahydrofuran and 2,6-lutidine. This procedure was also used for removing a 5'-dimethoxytrityl group from a deoxyoligonucleotide attached to the silica gel.

Condensation of Activated Nucleotides to the Polymer Support. Letsinger and Lunsford have shown that thymidine deoxyoligonucleotides can be synthesized by using phosphite triester intermediates and that protecting groups commonly used in deoxyoligonucleotide synthesis are stable to phosphorodichloridites.[24] Because these reactions proceed rapidly in high yield (95% for synthesis of the dinucleotide), this method appeared very attractive for adaptation to polymer-supported deoxyoligonucleotide synthesis. We have modified this procedure by using methyl phosphorodichloridite in place of trichloroethyl phosphorodichloridite and by using an intermediate activated deoxynucleoside phosphite containing tetrazole in place of the deoxynucleoside phosphorochloridite. As outlined in Figure 3 for the synthesis of a dinucleotide, the formation of the internucleotide bond therefore involves a reaction between a deoxynucleoside attached to the silica

Table III. Comparison of Phosphitylating Reagents[a]

deoxynucleotide[b]	reaction time, min	dinucleotide yield,[c] %
5a	60	90
5b	60	95
5a	5	30
5b	5	85
5c	60	80
5d	60	75
5e	60	75
5f	60	95

[a] The analytical reaction involved step 3 and step 5 of the reaction scheme outlined in Figure 3.   [b] The deoxynucleotides tested are defined in the key presented in Figure 3.   [c] The yield of dinucleotide was obtained by analyzing reaction mixtures using reverse-phase HPLC.

gel and an appropriately protected deoxynucleoside phosphite.

Selection of the methyl triester protecting group was based on several considerations. Our initial experiments with trichloroethyl dichlorophosphite gave satisfactory condensation yields which were comparable to those reported previously. Removal of this blocking group, however, created serious problems. Since a base-labile protecting group anchors the deoxyoligonucleotide to the support, the triester protecting group must be removed prior to release of the deoxyoligonucleotide from the support. Otherwise the basic conditions present during release from the support would cause considerable internucleotide bond cleavage and rearrangement of the deoxyribose skeleton.[25,26] Reductive cleavage procedures involving heterogeneous solutions[27] were not investigated, and those involving homogeneous solutions[28] gave unsatisfactory results. We therefore turned our attention to the methyl group which can be removed by using thiophenol.[29,30] Extensive investigations have not revealed any internucleotide bond cleavage with this reagent. A 5-min exposure of the methyl triester of d(T-T) to the thiophenoxide reagent resulted in complete deprotection to the diester. Further treatment of d(T-T) for 50 h resulted in no detectable degradation. This presumably is because a methyl group is much more reactive than a methylene group toward attack by thiophenoxide, and, once the methyl group has been removed, the phosphodiester is relatively stable toward nucleophilic attack.

Initial studies involving CH$_3$OPCl$_2$ suggested that the yields per deoxynucleotide additon were satisfactory (90%) but not

(21) We thank R. L. Letsinger for suggesting that BF$_3$ might be useful for detritylating nucleosides.

(22) M. D. Matteucci and M. H. Caruthers, *Tetrahedron Lett.*, **21**, 3243 (1980).

(23) V. Kohli, H. Blöcker and H. Köster, *Tetrahedron Lett.*, **21**, 2683 (1980).

(24) R. L. Letsinger and W. B. Lunsford, *J. Am. Chem. Soc.*, **98**, 3655 (1976).

(25) J. H. van Boom, P. M. J. Burgers, P. H. van Deursen, J. F. M. de Rooy, and C. B. Reese, *J. Chem. Soc., Chem. Commun.*, 167 (1976).

(26) A. Myles, W. Hutzenlaub, G. Reitz, and W. Pfleiderer, *Chem. Ber.*, **108**, 2857 (1975).

(27) F. Eckstein and I. Rizk, *Angew. Chem., Int. Ed. Engl.*, **6**, 695 (1967).

(28) R. L. Letsinger and J. L. Finnan, *J. Am. Chem. Soc.*, **97**, 7197 (1975).

(29) G. W. Daub and E. E. van Tamelen, *J. Am. Chem. Soc.*, **99**, 3526 (1977).

(30) Recently amines have also been shown to remove the methyl group without oligonucleotide degradation [D. J. H. Smith, K. K. Ogilvie, and M. F. Gillen, *Tetrahedron Lett.*, **21**, 861 (1980)].

sufficient for a repetitive, multistep synthesis.[18] Furthermore, the production of collidine hydrochloride during the coupling step led to blockage and flow restriction in our machine-assisted synthesis. Investigations of leaving groups other than chloride (triazole, tetrazole, and 4-nitroimidazole) were therefore initiated and these results are reported in Table III. As can be seen by comparing these results, the activated nucleotide containing tetrazole was superior since the yield exceeded 95% and the condensation rate was faster than the parent chloridite.

The activated deoxynucleotides were prepared via a two-step procedure involving the formation of the deoxynucleoside phosphorochloridite followed by conversion to the tetrazolide. An important feature of this procedure is that all key steps are completed under an inert gas atmosphere (nitrogen or argon). 5'-O-Dimethoxytrityl and base-protected deoxynucleosides in tetrahydrofuran were added dropwise to a well-stirred solution of CH_3OPCl_2 and 2,4,6-trimethylpyridine (collidine) in tetrahydrofuran at $-78\ °C$. Collidine hydrochloride was removed by filtration. Initial results with 5a, the phosphorochloridite, in the condensation reaction gave only about a 50% yield of d(T-T) and the synthesis of several side products. These side products were shown to result from trace amounts of CH_3OPCl_2 which were present in the reaction mixture. Since reactions on polymer supports usually require an excess (typically tenfold) of the incoming activated deoxynucleotide, trace amounts of unreacted CH_3OPCl_2 can become significant. Methyl dichlorophosphite would be kinetically more reactive and therefore effectively compete with the activated deoxynucleotide for the 5'-hydroxyl of the deoxynucleoside or deoxyoligonucleotide attached to the support. The unwanted side products persisted even when the molar ratio of deoxynucleoside to CH_3OPCl_2 was 1:0.75. However, these side products were completely eliminated and the condensation yield was improved to 85–90% by repeatedly concentrating reaction mixtures containing the activated nucleotide (the monochloridite) to a gum by using a solution of toluene and THF. This procedure presumably removes the volatile CH_3OPCl_2. By use of this repetitive evaporation step, the molar ratio of 5'-O-(dimethoxytrityl)deoxynucleoside (thymidine, N-benzoyldeoxycytidine, and N-benzoyldeoxyadenosine) to CH_3OPCl_2 could be lowered to 1:1.2. This ratio produces a minimum amount of 3'–3' dimer and a maximum yield of the expected product. By use of polymer-support chemistry, the presence of 3'–3' dimer is not a serious problem but simply reduces the yield of activated nucleotide. It is inert toward further reaction and, unlike conventional solution approaches, can be readily removed simply by washing the silica gel. For N-benzoyl- or N-isobutyryldeoxyguanosine, a molar ratio of deoxynucleoside to CH_3OPCl_2 of 1:0.9 was used because of difficulties encountered in removing methyl dichlorophosphite, even by repetitive evaporation with toluene and tetrahydrofuran. These deoxynucleoside phosphorochloridites were next converted to the tetrazolide by further reaction with 1 molar equiv of tetrazole in tetrahydrofuran. Collidine hydrochloride was removed by filtration, and the activated nucleotides were isolated by precipitation into pentane. After removal of pentane, these tetrazolides are stable for at least 3 months when stored as anhydrous glasses under an inert gas at $-20\ °C$.

Condensations usually were completed by using approximately a tenfold excess of the activated mononucleotides. These activated nucleotides as anhydrous glasses were dissolved in tetrahydrofuran, applied to the column through the injector, and recycled through the column for 1 h at room temperature. Usually each reconstituted aliquot contained enough activated nucleotide for three or four condensations on 0.2–0.3 g of derivatized silica gel.

Capping and Oxidation Steps. On the basis of HPLC analysis of condensation reactions, approximately 1–5% of the deoxynucleoside or deoxyoligonucleotide bound to the support does not react with the activated deoxynucleotide. These unreactive compounds must be blocked or capped in order to prevent the formation of several deoxyoligonucleotides with heterogeneous sequences. This capping step can best be accomplished by using a large excess of a very reactive phosphite such as diethoxytriazoylphosphine which would react with deoxynucleosides and

deoxyoligonucleotides to form a 5'-diethylphosphite, a relatively nonhydrophobic triester (step 4, Figure 3). More traditional reagents such as acetic anhydride and phenyl isocyanate were tried but reacted much more slowly with the unblocked 5'-hydroxyl group. Since purification involves reverse-phase HPLC, this capping reagent assures that all nonhydrophobic failure sequences can be readily separated from the synthetic deoxyoligonucleotide product containing a hydrophobic 5'-O-dimethoxytrityl group.

The final step involves oxidation of these phosphites to the corresponding phosphates with I_2 in water, 2,6-lutidine, and tetrahydrofuran (step 5, Figure 3). As has been reported previously[24] and confirmed by us, this oxidation step is sufficiently mild so that side products are not generated. We have attempted to postpone the oxidation until after all condensation steps, but the results have not been encouraging. Several uncharacterized side products were observed.

Removal of Protecting Groups. Upon completion of the synthesis, silica gel containing the deoxyoligonucleotide can be stored at $-20\ °C$. The initial deprotection step involves removal of the methyl group from phosphotriesters by using triethylammonium thiophenoxide in dioxane. This step is followed by treatment with concentrated ammonium hydroxide at $20\ °C$ for 3 h to hydrolyze the ester joining the deoxyoligonucleotide to the support. After centrifugation and recovery of the supernatant containing the deoxyoligonucleotide, the N-benzoyl groups from deoxycytosine, deoxyadenosine, and the N-isobutyryl group from deoxyguanosine are removed by warming at $50\ °C$ for 12 h. The final purification steps are isolation of the 5'-O-dimethoxytrityl containing deoxyoligonucleotide by reverse-phase HPLC followed by removal of the 5'-dimethoxytrityl group by using 80% acetic acid.[31]

Results

The feasibility of this approach for synthesizing deoxyoligonucleotides was examined by preparing d(A-C-G-C-T-C-A-C-A-A-T-T) and d(C-G-T-C-A-C-A-A-T-T). These syntheses provide a rigorous test of our methodology because all four bases are present in each deoxyoligonucleotide. Moreover, N-benzoyldeoxyadenosine, the purine most susceptible to depurination, is present at multiple positions within the first half of the synthesized molecules. Therefore, N-benzoyldeoxyadenosine must undergo several detritylation cycles and, consequently, multiple exposures to potential depurination conditions before each synthesis is completed. Initially, d(T-C-A-C-A-A-T-T), a deoxyoctanucleotide common to both final products, was synthesized by using 5b,g,h and the apparatus diagrammed in Figure 2. The procedure outlined in Table I was followed. Aliquots of silica gel containing this deoxyoctanucleotide were then separately extended to complete the synthesis of the deoxydecanucleotide and deoxydodecanucleotide. Since the dimethoxytrityl group was alternately joined to the support as part of each activated deoxynucleotide and then removed before the next deoxynucleotide addition, condensation reactions were monitored by measuring the amount of dimethoxytritanol released following each detritylation. A constant amount would be expected if condensations were quantitative. We observed in all cases that condensations did proceed with high yields of approximately 95%. The amount of dimethoxytritanol released after the final condensation steps indicated that the overall yields of deoxydecanucleotide and deoxydodecanucleotide were 64% and 55%, respectively. At the conclusion of each deoxyoligonucleotide synthesis, the triesters were converted to diesters by treatment with triethylammonium thiophenoxide. The expected products and all polymer-bound intermediates were then freed from the support and base-labile protecting groups by treatment with concentrated ammonium hydroxide and analyzed by reverse-phase HPLC. The elution profile of an aliquot of the reaction mixture obtained from the synthesis of d(C-G-T-C-A-C-A-A-T-T) is shown in Figure 4, panel A. Peak I which elutes in the void volume contains deoxyoligo-

---

(31) The preferred detritylating reagent after the removal of amino protecting groups is 80% acetic acid. Depurination is not observed with completely deprotected deoxyoligonucleotides, and, unlike $ZnBr_2$, 80% acetic acid is volatile and easily removed.

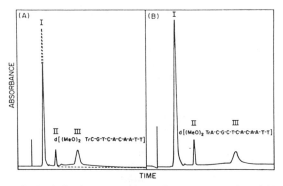

Figure 4. Analysis of reaction mixtures from the preparation of d(C-G-T-C-A-C-A-A-T-T) in panel A and d(A-C-G-C-T-C-A-C-A-A-T-T) in panel B: the solid line is the HPLC elution profile of reaction mixtures containing deoxyoligonucleotides with a 5'-dimethoxytrityl group; the dashed line is the HPLC elution profile of a reaction mixture containing deoxyoligonucleotides in a completely deprotected form. The eluting buffer was 0.1 M triethylammonium acetate in 26% acetonitrile.

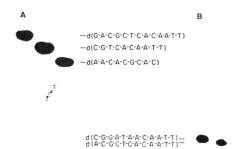

Figure 5. Analysis of ^{32}P-phosphorylated deoxyoligonucleotides by gel electrophoresis: gel A, the results with [5'-^{32}P]d(pC-G-T-C-A-C-A-A-T-T); gel B, the results with [5'-^{32}P]d(pA-C-G-C-T-C-A-C-A-A-T-T). The mobility of characterized markers of known sequence are also shown on the gels. Both gels were 20% acrylamide and 1% N,N-methylenebisacrylamide in 89 mM Tris borate (pH 8.3), 2.2 mM EDTA, and 7 M urea. (The positive electrode was at the bottom of the figure.)

nucleotides corresponding to failure sequences. Peak II was nonnucleotidic. Peak III was shown to be the expected product, d[(MeO)$_2$TrC-G-T-C-A-C-A-A-T-T], isolated in 30% overall yield based on the amount of thymidine attached to the silica gel.[32] As a first step in further characterization of the reaction mixture, a second aliquot was treated with 80% acetic acid and then analyzed by reverse-phase HPLC. The results are also presented in panel A of Figure 4. Peak III has disappeared, and the amount of UV-absorbing material recovered in the column void volume has increased considerably. Both observations are consistent with peak III containing a deoxyoligonucleotide with a hydrophobic dimethoxytrityl group. The deoxyoligonucleotide recovered from peak III was treated with 80% acetic acid and further characterized. One aliquot was shown to be completely degraded with snake venom phosphodiesterase. A second aliquot was analyzed by reverse-phase column chromatography using several acetonitrile concentrations in 0.1 M triethylammonium acetate as the eluting buffer. In all cases the detritylated materials eluted from the column as one peak, suggesting that the sample was homogeneous. A third aliquot was further characterized after phosphorylating the 5'-hydroxyl with [^{32}P]phosphate by using [γ-^{32}P]ATP and T4 kinase. On the basis of the UV absorbance and a calculated extinction coefficient, the phosphorylation was quantitative.[33] After removal of excess [γ-^{32}P]ATP by gel filtration, the phosphorylated material isolated from peak III was analyzed by gel electrophoresis, and these results are reported in Figure 5, panel

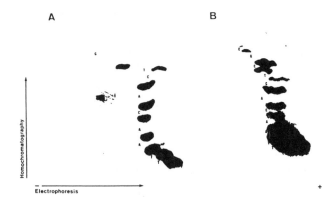

Figure 6. Two-dimension sequence analysis: panel A, [5'-^{32}P]d(pC-G-T-C-A-C-A-A-T-T); panel B, [5'-^{32}P]d(pA-C-G-C-T-C-A-C-A-A-T-T). Electrophoresis was along the longitudinal axis and homochromatography along the vertical axis. Nucleotide losses are recorded between the appropriate spots.

A. As can be seen by inspection of this gel, only one radioactive deoxyoligonucleotide was observed, indicating that the capping step was extremely efficient. Because separation on the reverse-phase column was dependent strictly on the presence of a dimethoxytrityl group and not the size of the deoxyoligonucleotide, incomplete capping would have led to several bands corresponding to deoxyoligonucleotides of shorter length. Furthermore, the mobility of the phosphorylated compound relative to characterized markers on the same gel indicated that the compound was a deoxydecanucleotide. The labeled deoxyoligonucleotide was then analyzed by the two-dimensional procedure,[34] and these results are reported in Figure 6, panel A. As can be seen by these results, the data were consistent with a decanucleotide having the structure [5'-^{32}P]d(pC-G-T-C-A-C-A-A-T-T). Moreover, the analysis indicated that the deoxydecanucleotide was essentially homogeneous since only trace amounts of labeled material migrate at positions other than those expected. This data also indicated that the capping procedure was quantitative and that the chemistry did not cause modifications of various nucleotides. Either circumstance would have led to additional spots as part of the two-dimensional analysis pattern.

A similar purification and analytical procedure was used to characterize d(A-C-G-C-T-C-A-C-A-A-T-T). Fractionation by reverse-phase HPLC of the ammonium hydroxide hydrolysate obtained from silica gel is shown in Figure 4, panel B. The isolated yield of deoxydodecanucleotide was not determined accurately. Peak III containing nucleotidic material and a hydrophobic dimethoxytrityl group was characterized following treatment with acetic acid and phosphorylation with [^{32}P]phosphate. Quantitative phosphorylation was observed. The analysis by gel electrophoresis is shown in Figure 5 (panel B), and the two dimensional analysis is shown in Figure 6, panel B. Once again, all these results are consistent with peak III being composed of an essentially homogeneous sample of the expected deoxyoligonucleotide, d(A-C-G-C-T-C-A-C-A-A-T-T).

Several new chemical reagents and procedures were developed in conjunction with this approach. These include the design of a silica gel matrix, the development of a new phosphitylating reagent, and the discovery of new procedures for removing trityl groups. The approach appears extremely promising as a method for synthesizing biologically active polynucleotides. Nucleotides can be added sequentially every 2 h to a growing deoxyoligonucleotide with yields exceeding 95% for each condensation. Work is in progress to interface this polymer-support procedure into a completely automatic, microprocessor-controlled machine.

Experimental Section

The solvents must be anhydrous and pure. Reagent grade tetrahydrofuran was distilled from sodium benzophenone ketal under inert gas

---

(32) We routinely recover only 40–60% of various deoxyoligonucleotides which are applied to C$_{18}$ HPLC columns. Such low recoveries have been observed irrespective of which synthetic methodology is used.

(33) Because of hypochromic effects, the absolute amount of deoxydecanucleotide cannot be estimated exactly, and therefore the possibility exists that 5–10% of the sample was not phosphorylated.

(34) F. Sanger, J. E. Donelson, A. R. Coulson, H. Kössel, and O. Fischer, *Proc. Natl. Acad. Sci. U.S.A.*, **70**, 1209 (1973).

as needed. Reagent grade toluene and reagent grade pentane were dried by distillation from CaH$_2$. 2,4,6-Trimethylpyridine, pyridine, and 2,6-lutidine were distilled first from p-toluenesulfonyl chloride and then CaH$_2$. Reagent grade nitromethane was distilled from CaH$_2$ and then stored over Linde 4A molecular sieves. 1,2,4-Triazole, 1H-tetrazole, 4-nitroimidazole, and diethyl phosphorochloridite were purchased from Aldrich Chemical Co. and used without further purification. 1,2,4-Triazole, 4-nitroimidazole, and 1H-tetrazole were dried in vacuo over Drierite before use.

The amount of 5'-O-(dimethoxytrityl)deoxynucleoside covalently joined to silica gel was determined quantitatively by measuring the dimethoxytrityl cation released after acid treatment of the silica gel. An aliquot of silica gel was accurately weighed (1–2 mg) and then treated with 5 mL of 0.1 M toluenesulfonic acid in acetonitrile.[35] After centrifugation, the absorbance was measured at 498 nm, and the amount of dimethoxytrityl cation was determined by using an extinction coefficient of 7×10^4. This assay was reproducible within ±5%.

Diethoxytriazolylphosphine was prepared by adding diethyl phosphorochloridite (1.0 mL, 5.8 mmol) dropwise to dry tetrahydrofuran (18 mL) containing 2,6-lutidine (2 mL) and dry triazole (0.50 g, .72 mmol) at 20 °C under nitrogen. After centrifugation to remove 2,6-lutidine hydrochloride, the supernatant containing diethoxytriazolylphosphine can be used directly as a capping reagent. This reagent can also be stored for at least 1 week at 20 °C under nitrogen. The activity of the diethoxytriazoylphosphine solution was assayed by measuring its ability to phosphitylate 5'-O-(dimethoxytrityl)thymidine in 1 min at 20 °C.

Thin-layer chromatography (TLC) was routinely carried out on Merck analytical silica gel plates (No. 5775), and HPLC was completed on a Waters Associates apparatus equipped with a solvent programmer and a Waters Associates, C$_{18}$, reverse-phase, μ-Bondapak analytical column. The organic phase was acetonitrile, and the aqueous phase was 0.1 M triethylammonium acetate (pH 7).

Synthesis of the Support. HPLC grade silica gel (2 g, Vydac TP-20, Separation Group, 100 m^2/g surface area, 300-Å pore size, 20-μm particle size) was exposed to a 15% relative humidity atmosphere (saturated LiCl) for at least 24 h. The silica (2.0 g) was then treated with 3-(triethoxysilyl)propylamine (2.3 g, 0.01 M in toluene) for 12 h at 20 °C and 12 h at reflux under a Drierite drying tube.[36] This reaction was completed on a shaking apparatus because magnetic stir bars pulverize the silica gel and must be avoided. Compound 1 was isolated by centrifugation, washed successively (twice each) with toluene, methanol, and ether, and air-dried.

The carboxylic acid group was introduced by agitating 1 (2 g) and succinic anhydride (2.5 g, 0.025 M) in water. The pH was controlled (pH 2–6) by addition of 2 M NaOH. Completeness of the carboxylation reaction was qualitatively monitored by using a picrate sulfate test.[37] An aliquot of silica (approximately 2 mg) was treated with 0.5 mL of 0.1 M picrate sulfate in saturated sodium borate buffer (pH 10). Compound 1 reacted within 10 min and stained a bright yellow whereas compound 2 remained white. The succinic anhydride reaction was allowed to continue until the silica gel remained white during the picrate sulfate test. Usually the total reaction time was 1 h, and a second addition of succinic anhydride was required. After being washed successively (twice each) with water, 0.1 M trichloroacetic acid, water, methanol, and ether, compound 2 was air-dried, dried in vacuo, and then treated with trimethylsilyl chloride (1.25 mL, 0.01 M) in pyridine (7 mL) for 24 h at 25 °C. Compound 2 was then washed with methanol (four times) and ether. Analysis for extent of carboxylation involved a two-step procedure. An accurately weighed aliquot was treated with DCC and p-nitrophenol in pyridine. After several washings with tetrahydrofuran to remove unreacted p-nitrophenol, 10% piperidine in pyridine was added to the silica gel, and the amount of p-nitrophenol released was measured at 410 nm by using 1.57×10^4 as the extinction coefficient of p-nitrophenoxide. The incorporation of carboxylic acid was 200 μmol/g.

Deoxynucleosides were joined to 2 by using DCC. This is a general procedure, and therefore only one example will be presented. 5'-O-(Dimethoxytrityl)thymidine (1.1 g, 2.16 mmol), DCC (2 g, 0.01 mol), and 2 (4 g, 0.8 mmol, carboxylic acid) were agitated in dry pyridine (21 mL) for 2 days. p-Nitrophenol (1.4 g, 0.01 mol) was added, the mixture was agitated for an additional day, and then the reaction was quenched with morpholine (1 mL, 0.011 mol). After being washed with methanol and ether, the silica gel was analyzed for unreacted carboxylic acid by the procedure outlined previously. Usually a second treatment with DCC (2 g, 0.01 mol) and p-nitrophenol (1.4 g, 0.01 mol) in dry pyridine (20

mL) and finally morpholine (1 mL) was necessary to completely block the trace amount of free carboxylic acid (<10 μmol/g) that remains from the first blocking procedure.

Preparation of Activated Deoxynucleotides. Compounds 5b, 5g, and 5h were synthesized by the same procedure. 5'-O-(Dimethoxytrityl)-thymidine, 5'-O-(dimethoxytrityl)-N-benzoyldeoxycytidine, and 5'-O-(dimethoxytrityl)-N-benzoyldeoxyadenosine were prepared by published procedures.[38] 5'-O-(Dimethoxytrityl)-N-isobutyrldeoxyguanosine was prepared according to a published procedure[39] except that dimethoxytrityl chloride was used instead of monomethoxytrityl chloride. These 5'-O-dimethoxytrityl deoxynucleosides were dried in vacuo over Drierite before use.

Methyl phosphorodichloridite was prepared according to a published procedure[40] and must be of the highest purity. After distillation through a column packed with glass helices, aliquots of methyl phosphorodichloridite were stored in vials under argon at −20 °C. Analysis by phosphorus-31 NMR indicated that the dichloridite was completely stable over several months when prepared in this manner. The phosphitylation procedure was completed in an apparatus designed so that all operations (additions, filtrations, concentrations in vacuo) could be completed under an inert gas atmosphere.

Compound 5b was prepared by the following procedure. 5'-O-(Dimethoxytrityl)thymidine (1.6 g, 2.9 mmole in anhydrous tetrahydrofuran (5 mL) was added dropwise to a well-stirred solution at −78 °C of CH$_3$OPCl$_2$ (0.33 mL, 3.5 mmol) and collidine (1.86 mL, 14.1 mmol) in anhydrous tetrahydrofuran (5 mL). A white precipitate formed during the addition. The mixture was stirred for 15 min at −78 °C and then filtered through a sintered-glass funnel to remove collidine hydrochloride. The collidine hydrochloride was washed with dry tetrahydrofuran (1 mL). The filtrate was then diluted with dry toluene and concentrated to a gum. After dry argon had been bled into the apparatus, a solution (6 mL) containing toluene–tetrahydrofuran (2:1) was added, and the gum was allowed to dissolve completely in this solution. Solvent was removed by concentration in vacuo. This reconcentration cycle using a solution of toluene and tetrahydrofuran was repeated three times. After the final concentration, the gum was dissolved in dry tetrahydrofuran (3 mL) and cooled to −78 °C, and a solution of tetrazole (0.18 g, 2.6 mmol) in dry tetrahydrofuran (3 mL) was added dropwise. A white precipitate of collidine hydrochloride formed during the addition. The mixture was stirred an additional 10 min at −78 °C and then transferred by using positive argon pressure and a cannula to a centrifuge tube filled with argon. The supernatant recovered after centrifugation contained 5b which can be used directly for synthesis of deoxyoligonucleotides. Alternatively, 5b can be stored as a precipitate and reconstituted as needed. Typically aliquots (2 mL) of the supernatant (8 mL) were precipitated into 15 mL of pentane at 20 °C. Each aliquot therefore contained enough activated nucleotide for three or four condensations with 0.2–0.3 g of silica gel. The precipitation was carried out in 15-cm, screw-cap test tubes fitted with Teflon–silicon septums (Pierce Chemical Co.). After centrifugation, cannula decanting, and careful drying in vacuo, the test tubes containing the samples were recovered after reequilibrating the desiccator with argon. Each test tube was cooled in a dry ice/acetone bath under an argon atmosphere, sealed with Parafilm, and stored in a desiccator over Drierite at −20 °C. With these storage conditions, aliquots were stable for at least 3 months.

Compounds 5g and 5h were prepared by the same general procedure. However, for the preparation of 5f, the stoichiometry of deoxynucleoside and CH$_3$OPCl$_2$ was changed. Typically 5'-O-(dimethoxytrityl)-N-isobutyrldeoxyguanosine (1.0 g, 1.56 mmol) and methyl phosphorodichloridite (0.13 mL, 1.4 mmol) were allowed to react in tetrahydrofuran (5 mL). Otherwise the preparation of 5f was the same as for the other activated deoxynucleotides.

The amounts of 5b, 5f, 5g, or 5h per preparation were estimated qualitatively by the following procedure. One aliquot of the activated nucleotide in tetrahydrofuran was quenched with dry methanol and a second with water. A TLC comparison on silica gel plates (ethyl acetate; 10% methanol in CHCl$_3$) between the two quenches identified the bis-(methyl nucleoside) phosphite. Estimated yields were 70% for 5b, 5g, and 5h and 50% for 5f. These estimates were based on the intensity of orange stain due to the dimethoxytrityl group when compounds containing this group were exposed to concentrated hydrochloric acid vapor.

Synthesis of Deoxyoligonucleotides. Syntheses were completed in a machine that is drawn schematically in Figure 2. The apparatus consists of a Milton Roy Minipump, three-way Altex slide valves, a recycle valve (a modified Altex valve), and an injector loop (a three-way Altex valve).

(35) M. H. Caruthers, Ph.D. Dissertation, Northwestern University, Evanston, IL, 1968.
(36) R. Majors and M. Hopper, J. Chromatogr. Sci., 12, 767 (1974).
(37) D. M. Benjamin, J. J. McCormack, and D. W. Gump, Anal. Chem., 45, 1531 (1973).
(38) H. Schaller, G. Weimann, B. Lerch, and H. G. Khorana, J. Am. Chem. Soc., 85, 3821 (1963).
(39) H. Büchi and H. G. Khorana, J. Mol. Biol., 72, 251 (1972).
(40) D. R. Martin and P. J. Pizzolato, J. Am. Chem. Soc., 72, 4584 (1950).

All connections were with Teflon tubing and were designed to minimize the tubing volume in the recycle loop. The column was an 11-mm Ace glass column that had been shortened to approximately 1-mL capacity. Cellulose filters were used to support the silica bed. The filters were acetylated with a solution of acetic anhydride and pyridine (1:1 based on volume) for 4 h at 50 °C before use. the total volume contained within the recycle loop of this apparatus was approximately 2.5 mL. The tetrahydrofuran reservoir was protected from air with a nitrogen bubbler, and the $ZnBr_2$ solution was protected from moisture with a Drierite tube.

The various chemical operations that must be performed for the addition of one nucleotide to the silica are listed in Table I. Typically 0.25 g (10 μmol) of 3 (thymidine) was loaded into the column and the silica washed with nitromethane. The 5'-O-dimethoxytrityl group was removed by flushing the column (30 min) with nitromethane saturated with $ZnBr_2$ (approximately 0.1 M in $ZnBr_2$) at a pump speed of 1 mL/min. The completeness of deprotection was monitored visually or spectrophotometrically by observing the release of a bright orange dimethoxytrityl cation. By measuring the absorbance at 498 nm, the completeness of the previous condensation step was monitored. The step was followed successively by a wash with a solution of n-butanol–2,6-lutidine–tetrahydrofuran (4:1:5) for 5 min at a flow rate of 2 mL/min. The next step was a wash for 5 min (5 mL/min) with dry tetrahydrofuran. During the course of this washing step, the recycle valve and the injector port were also flushed with dry tetrahydrofuran, and the effectiveness of this wash was monitored at 254 nm by using a spectrophotometer. The condensation step was next completed by using activated nucleotide that had been reconstituted with dry tetrahydrofuran. The reconstituted solution was stored in a dry ice/acetone bath over argon, but condensation reactions were carried out at room temperature. When reconstituted, the activated nucleotide stored in this way was stable for several days. Approximately 10 equiv of activated nucleotide (100 μmol for 0.25 g of 4) in 0.5–0.8 mL of tetrahydrofuran was injected into the apparatus and the machine switched to the recycle mode. The activated nucleotide was circulated through the silica gel for 1 h at a pump speed of 2 mL/min. Aliquots of activated nucleotide from the apparatus were then collected directly into dry methanol and water. Analysis as described previously indicated whether activated nucleotide was still present in the system. Usually this is the case. However, occasionally (approximately one time in ten) the dimethyl phosphite of the deoxynucleotide was not observed by this assay. When this occurred, the condensation step was repeated to prevent the possibility of incomplete reaction. The next step involves capping unreacted 4 or deoxyoligonucleotides containing a 5'-hydroxyl by adding diethoxytriazoylphosphine (1 mL of a 0.3 M solution in tetrahydrofuran) directly to the solution of activated nucleotide and continuing the recycle mode for 5 min at a pump speed of 2 mL/min. Residual activated nucleotide and the capping reagent were then flushed from the apparatus with dry tetrahydrofuran (2 min at 5 mL/min). This step was followed by the oxidation of phosphites with a solution of tetrahydrofuran–2,6-lutidine–water (2:1:1) containing 0.2 M I_2. The solution was flushed through the apparatus for 5 min (2 mL/min). Finally the cycle was completed by flushing the system first with dry tetrahydrofuran for 3 min (5 mL/min) and then with nitromethane for 2 min (5 mL/min).

Isolation of Deoxyoligonucleotides. The completely deprotected deoxyoligonucleotides were isolated by the following procedure. An aliquot (10 mg) of the silica gel containing the deoxydecanucleotide triester in protected form was first treated with thiophenol–triethylamine–dioxane (1:1:2 v/v). After 45 min of gentle shaking, the silica gel was recovered by centrifugation and washed with methanol (four times) and ethyl ether. After air drying, the deoxyoligonucleotide was removed from the support by a 3-h treatment with concentrated ammonium hydroxide at 20 °C followed by centrifugation. Base protecting groups were removed by warming the supernatant at 50 °C for 12 h in a sealed tube. The 5'-O-(dimethoxytrityl)deoxyoligonucleotide was isolated by concentrating the hydrolysate in vacuo, dissolving the residue in 0.1 M triethylammonium acetate (pH 7.0), and chromatographing this material on a C_{18} reverse-phase, HPLC column (Waters Associates). The eluting buffer was 0.1 M triethylammonium acetate containing 26% acetonitrile. The peak containing 5'-O-(dimethoxytrityl)deoxyoligonucleotide was concentrated in vacuo, and the residue was treated at 20 °C for 15 min with acetic acid–water (4:1 v/v) to remove the 5'-O-dimethoxytrityl group. The completely deprotected deoxyoligonucleotide was isolated by concentration of the acetic acid solution in vacuo, dissolving the residue in 25 mM triethylammonium bicarbonate (pH 7), and extraction of dimethoxytritanol with water-saturated ether.

Characterization of Deoxyoligonucleotides. The 5'-hydroxyl of each deoxyoligonucleotide was phosphorylated by using [γ-^{32}P]ATP and T4-kinase.[41] The amount of deoxyoligonucleotide used in a phosphorylation reaction was determined by measuring the absorbance and using a calculated extinction coefficient which assumed no hypochromicity for the deoxyoligonucleotide. Phosphorylated deoxyoligonucleotides were separated from excess ATP by desalting on a G-50-40 Sephadex column with 10 mM triethylammonium bicarbonate (pH 7) as eluant. Gel electrophoresis on polyacrylamide[42] and two-dimensional analysis[34] were completed by using standard procedures.

Synthesis of d(C-G-T-C-A-C-A-A-T-T). Compound 3 (0.25 g, 40 μmol/g) was loaded into the column, and the cycle was started by washing the silica gel with nitromethane and removing the 5'-dimethoxytrityl group with $ZnBr_2$. Elongation was performed as previously described by using an approximate tenfold excess of the incoming activated nucleoside phosphite (0.1 mM) at each condensation. Synthesis was continued to the completion of the deoxyoctanucleotide, d(T-C-A-C-A-A-T-T). At this point the silica was divided into two approximately equal portions. One portion was elongated to the deoxydecanucleotide in standard fashion. The overall yield was 64% based on the amount of dimethoxytrityl group bound to the support, and 30% was the yield isolated from a reverse-phase HPLC column.

Synthesis of d(A-C-G-C-T-C-A-C-A-A-T-T). The remaining portion of d(T-C-A-C-A-A-T-T) was elongated in the standard fashion in the machine to the deoxydodecanucleotide. The overall yield was 55% based on the dimethoxytrityl group bound to the support. The isolated yield was not accurately determined.

(41) V. Sgaramella and H. G. Khorana, *J. Mol. Biol.*, **72**, 427 (1972).
(42) T. Maniatis, A. Jeffrey, and H. van de Sande, *Biochemistry*, **14**, 3787 (1975).

A New Method for Sequencing DNA

A.M. Maxam and W. Gilbert

ABSTRACT   DNA can be sequenced by a chemical procedure that breaks a terminally labeled DNA molecule partially at each repetition of a base. The lengths of the labeled fragments then identify the positions of that base. We describe reactions that cleave DNA preferentially at guanines, at adenines, at cytosines and thymines equally, and at cytosines alone. When the products of these four reactions are resolved by size, by electrophoresis on a polyacrylamide gel, the DNA sequence can be read from the pattern of radioactive bands. The technique will permit sequencing of at least 100 bases from the point of labeling.

We have developed a new technique for sequencing DNA molecules. The procedure determines the nucleotide sequence of a terminally labeled DNA molecule by breaking it at adenine, guanine, cytosine, or thymine with chemical agents. Partial cleavage at each base produces a nested set of radioactive fragments extending from the labeled end to each of the positions of that base. Polyacrylamide gel electrophoresis resolves these single-stranded fragments; their sizes reveal *in order* the points of breakage. The autoradiograph of a gel produced from four different chemical cleavages, each specific for a base in a sense we will describe, then shows a pattern of bands from which the sequence can be read directly. The method is limited only by the resolving power of the polyacrylamide gel; in the current state of development we can sequence inward about 100 bases from the end of any terminally labeled DNA fragment.

We attack DNA with reagents that first damage and then remove a base from its sugar. The exposed sugar is then a weak point in the backbone and easily breaks; an alkali- or amine-catalyzed series of β-elimination reactions will cleave the sugar completely from its 3' and 5' phosphates. The reaction with the bases is a limited one, damaging only 1 residue for every 50 to 100 bases along the DNA. The second reaction to cleave the DNA strand must go to completion, so that the molecules finally analyzed do not have hidden damages. The purine-specific reagent is dimethyl sulfate; the pyrimidine-specific reagent is hydrazine.

The sequencing requires DNA molecules, either double-stranded or single-stranded, that are labeled at one end of one strand with ^{32}P. This can be a 5' or a 3' label. A restriction fragment of any length is labeled at both ends—for example, by being first treated with alkaline phosphatase to remove terminal phosphates and then labeled with ^{32}P by transfer from γ-labeled ATP with polynucleotide kinase. There are then two strategies: either (*i*) the double-stranded molecule is cut by a second restriction enzyme and the two ends are resolved on a polyacrylamide gel and isolated for sequencing or (*ii*) the doubly labeled molecule is denatured and the strands are separated on a gel (1), extracted, and sequenced.

THE SPECIFIC CHEMISTRY

A Guanine/Adenine Cleavage (2). Dimethyl sulfate methylates the guanines in DNA at the N7 position and the adenines at the N3 (3). The glycosidic bond of a methylated purine is unstable (3, 4) and breaks easily on heating at neutral pH, leaving the sugar free. Treatment with 0.1 M alkali at 90° then will cleave the sugar from the neighboring phosphate groups. When the resulting end-labeled fragments are resolved on a polyacrylamide gel, the autoradiograph contains a pattern of dark and light bands. The dark bands arise from breakage at guanines, which methylate 5-fold faster than adenines (3).

This strong guanine/weak adenine pattern contains almost half the information necessary for sequencing; however, ambiguities can arise in the interpretation of this pattern because the intensity of isolated bands is not easy to assess. To determine the bases we compare the information contained in this column of the gel with that in a parallel column in which the breakage at the guanines is suppressed, leaving the adenines apparently enhanced.

An Adenine-Enhanced Cleavage. The glycosidic bond of methylated adenosine is less stable than that of methylated guanosine (4); thus, gentle treatment with dilute acid releases adenines preferentially. Subsequent cleavage with alkali then produces a pattern of dark bands corresponding to adenines with light bands at guanines.

Cleavage at Cytosines and Thymines. Hydrazine reacts with thymine and cytosine, cleaving the base and leaving ribosylurea (5–7). Hydrazine then may react further to produce a hydrazone (5). After a partial hydrazinolysis in 15–18 M aqueous hydrazine at 20°, the DNA is cleaved with 0.5 M piperidine. This cyclic secondary amine, as the free base, displaces all the products of the hydrazine reaction from the sugars and catalyzes the β-elimination of the phosphates. The final pattern contains bands of similar intensity from the cleavages at cytosines and thymines.

Cleavage at Cytosine. The presence of 2 M NaCl preferentially supresses the reaction of thymines with hydrazine. Then, the piperidine breakage produces bands only from cytosine.

AN EXAMPLE

Consider a 64-base-pair DNA fragment, cut from *lac* operon DNA by the *Alu* I enzyme from *Arthrobacter luteus*, which cleaves flush at an AGCT sequence between the G and the C (8). After dephosphorylation, the two 5' ends of this fragment were labeled with ^{32}P. The autoradiograph in Fig. 1 shows that the two strands separate during electrophoresis, after denaturation, on a neutral polyacrylamide gel (1); they can be easily excised and extracted. For each strand, aliquots of the four

MAXAM, A.M. and GILBERT. W.
A new method for sequencing DNA. *Proc. Natl. Acad. Sci. U.S.A.* 74:560-564 (1977). Reprinted with the authors' permission.

cleavage reactions (strong G/weak A, strong A, strong C, and C + T) were electrophoresed at 600–1000 V on a 40-cm 20% polyacrylamide/7 M urea gel. Twelve hours later, a second portion of each sample was loaded on the gel and electrophoresis was continued. Fig. 2 displays autoradiographs showing two regions of the sequence of each strand derived from this single gel: one close to the labeled end of the molecule in those samples that had been electrophoresed in a short time, and a region further into the molecule expanded by electrophoresis for a longer time. The sequence can easily be read from the pattern of bands. The spacing between fragments decreases (roughly as an inverse square) from the bottom toward the top of the gel. The slight variations in the spacing are sequence-specific and reflect the last nucleotide added, a T or G decreasing the mobility more than an A or C. The fragments on the gel end with the base just before the one destroyed by the chemical attack; the labels on the bands in the figure represent the attacked bases. In Fig. 2, 62 bases can be read for both strands, the last 2 bases at the two 5′ ends not being determined by this gel. The sequence of each strand is consistent with and confirms that of the other:

stranded DNA. There are sequence-specific effects on the methylation reaction with double-stranded DNA that do not appear with single-stranded DNA: in the sequence GGA the reactivity of the middle G is suppressed; in the sequence AAA the reactivity of the central A is enhanced. Since these effects are absent with single-stranded DNA, they must arise through steric hindrance or stacking interactions; however, they do not interfere with sequencing because they appear equivalently in both displays of the base. Although in single-stranded DNA the N1 of adenine is exposed to methylation and should methylate as readily as the N7 of guanine, methylation at this position does not destabilize adenine on the sugar. Under our conditions the methyl group will migrate to the N^6 position and the extra charge will disappear (3).

The sequencing method is limited only by the resolution attainable in the gel electrophoresis. On 40-cm gels we can, without ambiguities, sequence out to 100 bases from the point of labeling. If there is other information available to support the sequencing, such as an amino acid sequence, one can often read further. The availability of restriction endonucleases is now such that any DNA molecule, obtainable from a phage, virus, or plasmid, can be sequenced.

pXXGGCACGACAGGTTTCCCGACTGGAAAGCGGGCAGTGAGCGCAACGCAATTAATGTGAGTTAG

GACCGTGCTGTCCAAAGGGCTGACCTTTCGCCCGTCACTCGCGTTGCGTTAATTACACTCAAXX$_p$

Fig. 3 is an expansion of one region of the sequencing gel to show the base specificity of the cleavage reactions.

DISCUSSION

The chemical sequencing method has certain specific advantages. First, the chemical treatment is easy to control; the ideal chemical attack, one base hit per strand, produces a rather even distribution of labeled material across the sequence. Second, each base is attacked, so that in a run of any single base all those are displayed. The chemical distinction between the different bases is clear, and, as in our example, the sequence of both strands provides a more-than-adequate check.

We have chosen this specific set of chemical reactions to provide more than enough information for the sequencing. The *Techniques* section describes another reaction that displays the Gs alone as well as an alternative reaction for breaking at As and Cs. However, it is more useful to have a strong G/weak A display, in which there is generally enough information to distinguish both the Gs and the As, than just a pure G pattern alone, because redundant information serves as a check on the identifications. In principle, one could sequence DNA with three chemical reactions, each of single-base specificity, using the absence of a band to identify the fourth position. This would be a nonredundant method in which every bit of information was required. Such an approach is subject to considerable error, and any hesitation in the chemistry would be misinterpreted as a different base. For that reason we have chosen redundant displays, which increase one's confidence in the sequencing.

5-Methylcytosine and N^6-methyladenine are occasionally found in DNA. 5-Methylcytosine can be recognized by our method because the methyl group interferes with the action of hydrazine [thymine reacts far more slowly than does uracil (5)]; thus, a 5-methylcytosine cleavage does not appear in the pattern, producing a gap in the sequence opposite a guanine (observed in this laboratory by J. Tomizawa and H. Ohmori). However, we do not expect to recognize an N^6-methyladenine; the glycosidic bond should not be unstable, and an earlier methylation of adenine at the N^6 position should not prevent the later methylation at N3.

These methods work equally well on double- or single-

TECHNIQUES

[γ-^{32}P]ATP Exchange Synthesis (9). The specific activity routinely attains 1200 Ci/mmol. Dialyze glyceraldehyde-3-phosphate dehydrogenase against 3.2 M ammonium sulfate, pH 8/50 mM Tris·HCl, pH 8/10 mM mercaptoethanol/1 mM EDTA/0.1 mM NAD$^+$; and dialyze 3-phosphoglycerate kinase (ATP:3-phospho-D-glycerate 1-phosphotransferase, EC 2.7.2.3) against the same solution minus NAD$^+$ (enzymes from Calbiochem). Combine 50 µl of the dialyzed dehydrogenase and 25 µl of the dialyzed kinase, sediment at 12,000 × g, and redissolve the pellet in 75 µl of twice-distilled water to remove ammonium sulfate. Dissolve 25 mCi (2.7 nmol) of HCl-free, carrier-free ^{32}P$_i$ in 50 µl of 50 mM Tris·HCl, pH 8.0/7 mM MgCl$_2$/ 0.1 mM EDTA/2 mM reduced glutathione/1 mM sodium 3-phosphoglycerate/0.2 mM ATP (10 nmol); add 2 µl of the dialyzed, desalted enzyme mixture; and allow to react at 25°. Follow the reaction by thin-layer chromatography on PEI cellulose in 0.75 M sodium phosphate, pH 3.5, by autoradiography of the plate. At the plateau, usually 30 min, add 250 µl of twice-distilled water and 5 µl of 0.1 M EDTA, mix, and heat at 90° for 5 min to inactivate the enzymes. Then chill, add 700 µl of 95% ethanol, mix well, and store at −20°. The theoretical limit of conversion is 79%, and if this is achieved the [γ-^{32}P]ATP would have a specific activity near 2000 Ci/mmol.

Labeling 5′ Ends. 5′-Phosphorylation (10, 11) includes a heat-denaturation in spermidine which increases the yield 15-fold with flush-ended restriction fragments. Dissolve dephosphorylated DNA in 75 µl of 10 mM glycine·NaOH, pH 9.5/1 mM spermidine/0.1 mM EDTA; heat at 100° for 3 min and chill in ice water. Then add 10 µl of 500 mM glycine·NaOH, pH 9.5/100 mM MgCl$_2$/50 mM dithiothreitol, 10 µl of [γ-^{32}P]ATP (100 pmol or molar equivalent of DNA 5′ ends, 1000 Ci/mmol), and several units of polynucleotide kinase to a final volume of 100 µl. Heat at 37° for 30 min; add 100 µl of 4 M ammonium acetate, 20 µg of tRNA, and 600 µl of ethanol, mix well, chill at −70°, centrifuge at 12,000 × g, remove the supernatant phase, rinse the pellet with ethanol, and dry under vacuum.

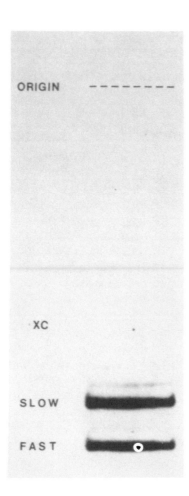

FIG. 1. Strand separation of a restriction fragment: 1.5 μg of a 64-base-pair DNA fragment (75 pmol of 5′ ends) was phosphorylated with [γ-^{32}P]ATP (800 Ci/mmol) and polynucleotide kinase, denatured in alkali, layered onto a 0.3 cm × 3 cm surface of an 8% polyacrylamide slab gel (see under *Techniques*), and electrophoresed at 200 V (regulated) and 20 mA (average), until the xylene cyanol (XC) dye moved 9 cm. The gel on one glass plate was then tightly covered with Saran Wrap and exposed to Kodak XR-5 x-ray film for 10 min.

Labeling 3′ Ends. To adenylate with [α-^{32}P]ATP and terminal transferase (12), dissolve DNA in 70 μl of 10 mM Tris·HCl, pH 7.5/0.1 mM EDTA, heat at 100° for 3 min, and chill at 0°. Then add, in order, 10 μl of 1.0 M sodium cacodylate (pH 6.9), 2 μl of 50 mM CoCl$_2$, mix, 2 μl of 5 mM dithiothreitol, 10 μl of [α-^{32}P]ATP (500 pmol, 100 Ci/mmol), and several units of terminal transferase to a final volume of 100 μl. Heat at 37° for several hours, add 100 μl of 4 M ammonium acetate, 20 μg of tRNA, and 600 μl of ethanol, and precipitate, centrifuge, rinse, and dry the DNA as described above. Dissolve the pellet in 40 μl of 0.3 M NaOH/1 mM EDTA, heat at 37° for 16 hr, and either add glycerol and dyes for strand separation or neutralize, ethanol precipitate, and renature the DNA for secondary restriction cleavage.

Strand Separation. Dissolve the DNA in 50 μl of 0.3 M NaOH/10% glycerol/1 mM EDTA/0.05% xylene cyanol/0.05% bromphenol blue. Load on a 5–10% acrylamide/0.16–0.33% bisacrylamide/50 mM Tris·borate, pH 8.3/1 mM EDTA gel and electrophorese. The concentration of DNA entering the gel is critical and must be minimized to prevent renaturation. Use thick gels with wide slots (0.3- to 1-cm-thick slabs with 3-cm to full width slots), and run cool (at 25°).

Gel Elution. Insert an excised segment of the gel into a 1000 μl (blue) Eppendorf pipette tip, plugged tightly with siliconized glass wool and heat-sealed at the point. Grind the gel to a paste with a siliconized 5-mm glass rod, add 0.6 ml of 0.5 M ammonium acetate/0.01 M magnesium acetate/0.1% sodium dodecyl sulfate/0.1 mM EDTA (and 50 μg of tRNA carrier if the DNA has already been labeled); seal with Parafilm and hold at 37° for 10 hr. Cut off the sealed point, put the tip in a siliconized 10 × 75 mm tube, centrifuge for a few minutes, rinse with 0.2 ml of fresh gel elution solution, and alcohol precipitate twice.

Partial Methylation of Purines. Combine 1 μl of sonicated carrier DNA, 10 mg/ml, with 5 μl of ^{32}P-end-labeled DNA in 200 μl of 50 mM sodium cacodylate, pH 8.0/10 mM MgCl$_2$/0.1 mM EDTA. Mix and chill in ice. Add 1 μl of 99% (10.7 M) dimethyl sulfate, mix, cap, and heat at 20° for 15 min. To stop the reaction, add 50 μl of a stop solution (1.0 M mercaptoethanol/1.0 M Tris·acetate, pH 7.5/1.5 M sodium acetate/0.05 M magnesium acetate/0.001 M EDTA), 1 mg/ml of tRNA, and mix. Add 750 μl (3 volumes) of ethanol, chill, and spin. Reprecipitate from 250 μl of 0.3 M sodium acetate, rinse with alcohol, and dry.

Strong Guanine/Weak Adenine Cleavage. Dissolve methylated DNA in 20 μl of 10 mM sodium phosphate, pH 7.0/1 mM EDTA, and collect the liquid on the bottom of the tube with a quick low-speed spin. Close the tube and heat in a water bath at 90° for 15 min. Chill in ice and collect the condensate with a quick low-speed spin. Add 2 μl of 1.0 M NaOH, mix, and draw the liquid up into the middle of a pointed glass capillary tube, seal with a flame, and hold at 90° for 30 min. Open the capillary and empty into 20 μl of urea-dye mixture, heat, and layer on the gel.

Strong Adenine/Weak Guanine Cleavage. Dissolve methylated DNA in 20 μl of distilled water. Chill to 0°, add 5 μl of 0.5 M HCl, mix, and keep the sample at 0° in ice, mixing occasionally. After 2 hr, add 200 μl of 0.3 M sodium acetate and 750 μl of ethanol, chill, spin, rinse, and dry. Then dissolve in 10 μl of 0.1 M NaOH/1 mM EDTA and heat at 90° for 30 min in a sealed capillary. Add contents to urea-dye mixture, heat, and layer.

An Alternative Guanine Cleavage. Dissolve methylated DNA in 20 μl of freshly diluted 1.0 M piperidine. Heat at 90° for 30 min in a sealed capillary. [This reaction opens 7-MeG adjacent to the glycosidic bond (13), displaces the ring-opened product from the sugar, and eliminates both phosphates to cleave the DNA wherever G was methylated.] Return the contents of the capillary to the reaction tube, lyophilize, wet the residue, and lyophilize again. Finally, dissolve the last residue in 10 μl of 0.1 M NaOH/1 mM EDTA and prepare for the gel.

An Alternative Strong Adenine/Weak Cytosine Cleavage. Combine 20 μl of 1.5 M NaOH/1 mM EDTA with 1 μl of sonicated carrier DNA (10 mg/ml) and 5 μl of ^{32}P end-labeled DNA, and heat at 90° for 30 min in a sealed capillary. [The strong alkali opens the adenine and cytosine rings (13); then, the ring-opened products can be displaced and phosphates eliminated with piperidine.] Rinse the capillary into 100 μl of 1.0 M sodium acetate, add 5 μl of tRNA (10 mg/ml), add 750 μl of ethanol, chill, spin, rinse, and dry. Dissolve the pellet in 20 μl of freshly diluted 1.0 M piperidine, and heat at 90° for 30 min in a sealed capillary. Lyophilize twice, dissolve the last residue in 10 μl of 0.1 M NaOH/1 mM EDTA, and add urea-dye mixture.

Cleavage at Thymine and Cytosine. Combine 20 μl of distilled water, 1 μl of sonicated carrier DNA (10 mg/ml), and 5 μl of ^{32}P end-labeled DNA. Mix and chill at 0°. Add 30 μl of

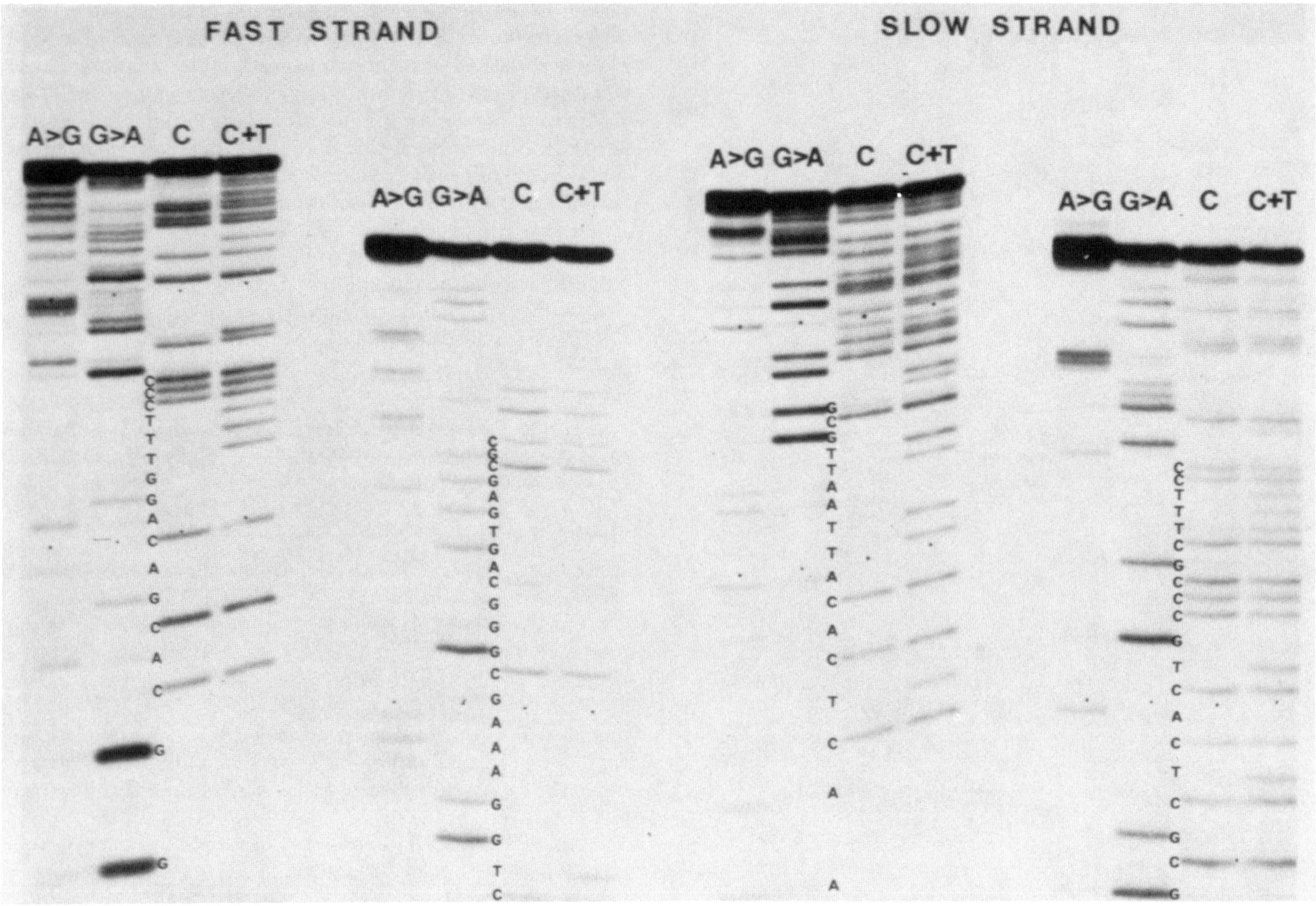

FIG. 2. Autoradiograph of a sequencing gel of the complementary strands of a 64-base-pair DNA fragment. Two panels, each with four reactions, are shown for each strand; cleavages proximal to the 5′ end are at the bottom on the left. A strong band in the first column with a weaker band in the second arises from an A; a strong band in the second column with a weaker band in the first is a G; a band appearing in both the third and fourth columns is a C; and a band only in the fourth column is a T. To derive the sequence of each strand, begin at the bottom of the left panel and read upward until the bands are not resolved; then, pick up the pattern at the bottom of the right panel and continue upward. One-tenth of each strand, isolated from the gel of Fig. 1, was used for each of the base-modification reactions. The dimethyl sulfate treatment was 50 mM for 30 min to react with A and G; hydrazine treatment was 18 M for 30 min to react with C and T and 18 M with 2 M NaCl for 40 min to cleave C. After strand breakage, half of the products from the four reactions were layered on a 1.5 × 330 × 400 mm denaturing 20% polyacrylamide slab gel, pre-electrophoresed at 1000 V for 2 hr. Electrophoresis at 20 W (constant power), 800 V (average), and 25 mA (average) proceeded until the xylene cyanol dye had migrated halfway down the gel. Then the rest of the samples were layered and electrophoresis was continued until the new bromphenol blue dye moved halfway down. Autoradiography of the gel for 8 hr produced the pattern shown.

95% (30 M) hydrazine*, mix well, and keep at 0° for several minutes. Close the tube and heat at 20° for 15 min. Add 200 μl of cold 0.3 M sodium acetate/0.01 M magnesium acetate/0.1 mM EDTA/0.25 mg/ml tRNA, vortex mix, add 750 μl of ethanol, chill, spin, dissolve the pellet in 250 μl of 0.3 M sodium acetate, add 750 μl of ethanol, chill, spin, rinse with ethanol, and dry. Dissolve the pellet and rinse the walls with 20 μl of freshly diluted 0.5 M piperidine. Heat for 30 min at 90° in a sealed capillary. Lyophilize twice, dissolve in 10 μl of 0.1 M NaOH/1 mM EDTA, add urea-dye mixture, heat, and layer on gel.

Cleavage at Cytosine. Replace the water in the hydrazinolysis reaction mixture with 20 μl of 5 M NaCl, and increase the reaction time to 20 min. The freshness and the concentration of the hydrazine are critical for base-specificity.

Reaction Times. The reaction conditions provide a uniformly labeled set of partial products of chain length 1 to 100. To distribute the label over a shorter region, increase the reaction time, and vice versa.

---

* CAUTION: Hydrazine is a volatile neurotoxin. Dispense with care in a fume hood, and inactivate it with concentrated ferric chloride.

Reaction Vessels. We use 1.5-ml Eppendorf conical polypropylene tubes with snap caps, treated with 5% (vol/vol) dimethyldichlorosilane in CCl₄ and rinsed with distilled water.

Alcohol Precipitation, Wash, and Rinse. Unless otherwise specified, the initial ethanol precipitation is from 0.3 M sodium acetate/0.01 M magnesium acetate/0.1 mM EDTA, with 50 μg of tRNA as carrier. Add 3 volumes of ethanol, cap and invert to mix, chill at −70° in a Dry Ice-ethanol bath for 5 min, and spin in the Eppendorf 3200/30 microcentrifuge at 15,000 rpm (12,000 × g) for 5 min. Reprecipitate with 0.3 M sodium acetate and 3 volumes of ethanol, chill, and spin. Rinse the final pellet with 1 ml of cold ethanol, spin, and dry in a vacuum for several minutes.

Gel Samples. All samples for sequencing gels are in 10 or 20 μl of 0.1 M NaOH/1 mM EDTA to which is added an equal volume of 10 M urea/0.05% xylene cyanol/0.05% bromphenol blue. Heat the sample at 90° for 15 sec, then layer on the gel.

Sequencing Gels. These are commonly slabs 1.5 mm × 330 mm × 400 mm with 18 sample wells 10 mm deep and 13 mm

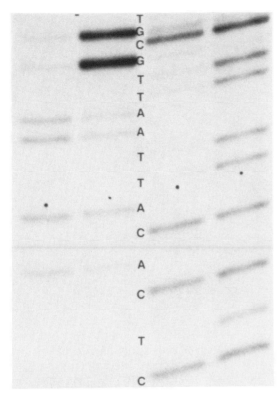

FIG. 3.   Detail of the sequence gel. The four lanes are (from left to right) A > G, G > A, C, C + T; the dots show the position of the bromphenol blue dye marker, between fragments 9 and 10 long.

wide separated by 3 mm (fitting on a 35.5 × 43 cm x-ray film). They are 20% (wt/vol) acrylamide (Bio-Rad)/0.67% (wt/vol) methylene bisacrylamide/7 M urea/50 mM Tris·borate, pH 8.3/1 mM EDTA/3 mM ammonium persulfate; 300 ml of gel solution is polymerized with TEMED within 30 min (generally 50 μl of TEMED). Age the gel at least 10 hr before using it. Electrophorese with some heating (30–40°), to help keep the DNA denatured, between 800 and 1200 V. Load successively whenever the previous xylene cyanol has moved halfway down the gel. Bromphenol blue runs with 10-nucleotide-long frag-

ments, xylene cyanol with 28. With three loadings at 0, 12, and 24 hr, a 1000-V run for 36 hr permits reading more than 100 bases. To sequence the first few bases from the labeled end, use a 25% acrylamide/0.83% bisacrylamide gel in the usual urea buffer and pre-electrophorese this gel for 2 hr at 1000 V.

Autoradiography. Freeze the gel for autoradiography. Remove one glass plate, wrap the gel and supporting plate with Saran Wrap, and mark the positions of the dyes with ^{14}C-containing ink. Place the gel in contact with film in a light-tight x-ray exposure holder (backed with lead and aluminum) at −20° under pressure from lead bricks.

Special Materials. Dimethyl sulfate (99%) was purchased from Aldrich Chemical Co., hydrazine (95%) from Eastman Organic Chemicals, and piperidine (99%) from Fisher Scientific; these were used without further purification.

This work was supported by National Institute of General Medical Sciences, Grant GM 09541. W.G. is an American Cancer Society Professor of Molecular Biology.

1. Hayward, G. S. (1972) *Virology* **49,** 342–344.
2. Gilbert, W., Maxam, A. & Mirzabekov, A. (1976) in *Control of Ribosome Synthesis,* eds. Kjeldgaard, N. O. & Maaløe, O. (Munksgaard, Copenhagen), pp. 139–148.
3. Lawley, P. D. & Brookes, P. (1963) *Biochem. J.* **89,** 127–128
4. Kriek, E. & Emmelot, P. (1964) *Biochim. Biophys. Acta* **91,** 59–66.
5. Temperli, A., Turler, H., Rust, P., Danon, A. & Chargaff, E. (1964) *Biochim. Biophys. Acta* **91,** 462–476.
6. Hayes, D. H. & Hayes-Baron, F. (1967) *J. Chem. Soc.,* 1528–1533.
7. Cashmore, A. R. & Petersen, G. B. (1969) *Biochim. Biophys. Acta* **174,** 591–603.
8. Roberts, R. J., Myers, P. A., Morrison, A. & Murray, K. (1976) *J. Mol. Biol.* **102,** 157–165.
9. Glynn, I. M. & Chappell, J. B. (1964) *Biochem. J.* **90,** 147–149.
10. van de Sande, J. H., Kleppe, K. & Khorana, H. G. (1973) *Biochemistry* **12,** 5050–5055.
11. Lillehaug, J. R. & Kleppe, K. (1975) *Biochemistry* **14,** 1225–1229.
12. Roychoudhury, R., Jay, E. & Wu, R. (1976) *Nucleic Acids Res.* **3,** 863–878.
13. Kochetkov, N. K. & Budovskii, E. I. (1972) *Organic Chemistry of Nucleic Acids* (Plenum, New York), Part B, pp. 381–397.

DNA Sequencing with Chain-terminating Inhibitors

F. Sanger, S. Nicklen and A.R. Coulson

ABSTRACT   A new method for determining nucleotide sequences in DNA is described. It is similar to the "plus and minus" method [Sanger, F. & Coulson, A. R. (1975) *J. Mol. Biol.* 94, 441–448] but makes use of the 2′,3′-dideoxy and arabinonucleoside analogues of the normal deoxynucleoside triphosphates, which act as specific chain-terminating inhibitors of DNA polymerase. The technique has been applied to the DNA of bacteriophage φX174 and is more rapid and more accurate than either the plus or the minus method.

The "plus and minus" method (1) is a relatively rapid and simple technique that has made possible the determination of the sequence of the genome of bacteriophage φX174 (2). It depends on the use of DNA polymerase to transcribe specific regions of the DNA under controlled conditions. Although the method is considerably more rapid and simple than other available techniques, neither the "plus" nor the "minus" method is completely accurate, and in order to establish a sequence both must be used together, and sometimes confirmatory data are necessary. W. M. Barnes (*J. Mol. Biol.*, in press) has recently developed a third method, involving ribo-substitution, which has certain advantages over the plus and minus method, but this has not yet been extensively exploited.

Another rapid and simple method that depends on specific chemical degradation of the DNA has recently been described by Maxam and Gilbert (3), and this has also been used extensively for DNA sequencing. It has the advantage over the plus and minus method that it can be applied to double-stranded DNA, but it requires a strand separation or equivalent fractionation of each restriction enzyme fragment studied, which makes it somewhat more laborious.

This paper describes a further method using DNA polymerase, which makes use of inhibitors that terminate the newly synthesized chains at specific residues.

Principle of the Method. Atkinson *et al.* (4) showed that the inhibitory activity of 2′,3′-dideoxythymidine triphosphate (ddTTP) on DNA polymerase I depends on its being incorporated into the growing oligonucleotide chain in the place of thymidylic acid (dT). Because the ddT contains no 3′-hydroxyl group, the chain cannot be extended further, so that termination occurs specifically at positions where dT should be incorporated. If a primer and template are incubated with DNA polymerase in the presence of a mixture of ddTTP and dTTP, as well as the other three deoxyribonucleoside triphosphates (one of which is labeled with ^{32}P), a mixture of fragments all having the same 5′ and with ddT residues at the 3′ ends is obtained. When this mixture is fractionated by electrophoresis on denaturing acrylamide gels the pattern of bands shows the distribution of dTs in the newly synthesized DNA. By using analogous terminators for the other nucleotides in separate incubations and running the samples in parallel on the gel, a pattern of bands is obtained from which the sequence can be read off as in the other rapid techniques mentioned above.

Two types of terminating triphosphates have been used—the dideoxy derivatives and the arabinonucleosides. Arabinose is a stereoisomer of ribose in which the 3′-hydroxyl group is oriented in *trans* position with respect to the 2′-hydroxyl group. The arabinosyl (ara) nucleotides act as chain terminating inhibitors of *Escherichia coli* DNA polymerase I in a manner comparable to ddT (4), although synthesized chains ending in 3′ araC can be further extended by some mammalian DNA polymerases (5). In order to obtain a suitable pattern of bands from which an extensive sequence can be read it is necessary to have a ratio of terminating triphosphate to normal triphosphate such that only partial incorporation of the terminator occurs. For the dideoxy derivatives this ratio is about 100, and for the arabinosyl derivatives about 5000.

METHODS

Preparation of the Triphosphate Analogues. The preparation of ddTTP has been described (6, 7), and the material is now commercially available. ddA has been prepared by McCarthy *et al.* (8). We essentially followed their procedure and used the methods of Tener (9) and of Hoard and Ott (10) to convert it to the triphosphate, which was then purified on DEAE-Sephadex, using a 0.1–1.0 M gradient of triethylamine carbonate at pH 8.4. The preparation of ddGTP and ddCTP has not been described previously; however we applied the same method as that used for ddATP and obtained solutions having the requisite terminating activities. The yields were very low and this can hardly be regarded as adequate chemical characterization. However, there can be little doubt that the activity was due to the dideoxy derivatives.

The starting material for the ddGTP was *N*-isobutyryl-5′-*O*-monomethoxytrityldeoxyguanosine prepared by F. E. Baralle (11). After tosylation of the 3′-OH group (12) the compound was converted to the 2′,3′-didehydro derivative with sodium methoxide (8). The isobutyryl group was partly removed during this treatment and removal was completed by incubation in NH₃ (specific gravity 0.88) overnight at 45°. The didehydro derivative was reduced to the dideoxy derivative (8) and converted to the triphosphate as for the ddATP. The monophosphate was purified by fractionation on a DEAE-Sephadex column using a triethylamine carbonate gradient (0.025–0.3 M) but the triphosphate was not purified.

ddCTP was prepared from *N*-anisoyl-5′-*O*-monomethoxytrityldeoxycytidine (Collaborative Research Inc., Waltham, MA) by the above method but the final purification on DEAE-Sephadex was omitted because the yield was very low and the solution contained the required activity. The solution was used directly in the experiments described in this paper.

An attempt was made to prepare the triphosphate of the intermediate didehydrodideoxycytidine because Atkinson *et*

Abbreviations: The symbols C, T, A, and G are used for the deoxyribonucleotides in DNA sequences; the prefix dd is used for the 2′,3′-dideoxy derivatives (e.g., ddATP is 2′,3′-dideoxyadenosine 5′-triphosphate); the prefix ara is used for the arabinose analogues.

SANGER, F., NICKLEN, S., and COULSON, A.R.
DNA sequencing with chain-terminating inhibitors. *Proc. Natl. Acad. Sci. U.S.A.* 74:5463-5467 (1977).
Reprinted with the authors' permission.

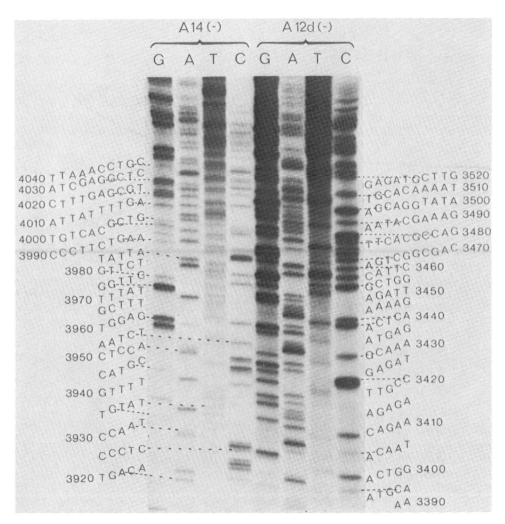

FIG. 1. Autoradiograph of the acrylamide gel from the sequence determination using restriction fragments A12d and A14 as primers on the complementary strand of φX174 DNA. The inhibitors used were (left to right) ddGTP, ddATP, ddTTP, and araCTP. Electrophoresis was on a 12% acrylamide gel at 40 mA for 14 hr. The top 10 cm of the gel is not shown. The DNA sequence is written from left to right and upwards beside the corresponding bands on the radioautograph. The numbering is as given in ref. 2.

al. (4) have shown that the didehydrodideoxy-TTP is also active as a terminator. However, we were unsuccessful in this. These compounds seem much less stable than the dideoxy derivatives.

araATP and araCTP were obtained from P-L Biochemicals Inc., Milwaukee, WI.

Sequencing Procedure. Restriction enzyme fragments were obtained from φX174 replicative form and separated by electrophoresis on acrylamide gels. The material obtained from 5 μg of φX174 replicative form in 5 μl of H_2O was mixed with 1 μl of viral or complementary strand φX174 DNA (0.6 μg) and 1 μl of H × 10 buffer (13) and sealed in a capillary tube, heated to 100° for 3 min, and then incubated at 67° for 30 min. The solution was diluted to 20 μl with H buffer and 2 μl samples were taken for each incubation and mixed with 2 μl of the appropriate "mix" and 1 μl of DNA polymerase (according to Klenow, Boehringer, Mannheim) (0.2 units). Each mix contained 1.5 × H buffer, 1 μCi of $[\alpha\text{-}^{32}P]$dATP (specific activity approximately 100 mCi/μmol) and the following other triphosphates.

ddT: 0.1 mM dGTP, 0.1 mM dCTP, 0.005 mM dTTP, 0.5 mM ddTTP

ddA: 0.1 mM dGTP, 0.1 mM dCTP, 0.1 mM dTTP, 0.5 mM ddATP

ddG: 0.1 mM dCTP, 0.1 mM dTTP, 0.005 mM dGTP, 0.5 mM ddGTP

ddC: 0.1 mM dGTP, 0.1 mM dTTP, 0.005 mM dCTP, approximately 0.25 mM ddCTP

(The concentration of the ddCTP was uncertain because there was insufficient yield to determine it, but the required dilution of the solution was determined experimentally.)

araC: 0.1 mM dGTP, 0.1 mM dTTP, 0.005 mM dCTP, 12.5 mM araCTP

Incubation was at room temperature for 15 min. Then 1 μl of 0.5 mM dATP was added and incubation was continued for a further 15 min. If this step (chase) was omitted some termination at A residues occurred in all samples due to the low concentration of the $[\alpha\text{-}^{32}P]$dATP. With small primers, where it was unnecessary to carry out a subsequent splitting (as in the experiment shown in Fig. 1), the various reaction mixtures were denatured directly and applied to the acrylamide gel for electrophoresis (1). If further splitting was necessary (see Fig. 2), 1 μl of the appropriate restriction enzyme was added shortly after the dATP "chase," and incubation was at 37°.

The single-site ribo-substitution procedure (N. L. Brown, unpublished) was carried out as follows. The annealing of template and primer was carried out as above but in "Mn buffer" (66 mM TrisCl, pH 7.4/1.5 mM 2-mercaptoethanol/

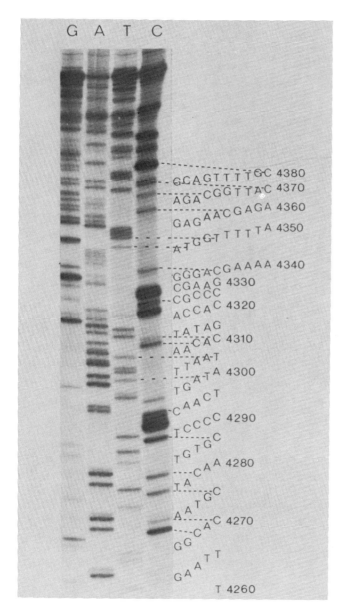

FIG. 2. Autoradiograph from an experiment using fragment R4 as primer on the complementary strand of φX174 DNA. Conditions were as in Fig. 1 with the following exceptions: ddCTP was used as inhibitor instead of araCTP. After incubation of the solutions at room temperature for 15 min, 1 μl of 0.5 mM dATP and 1 μl of restriction enzyme *Hae* III (4 units/μl) were added and the solutions were incubated at 37° for 10 min. The *Hae* III cuts close to the *Hind*II site and it was used because it was more readily available. The electrophoresis was on a 12% acrylamide gel at 40 mA for 14 hr. The top 10.5 cm of the gel is not shown.

0.67 mM MnCl₂) rather than in H buffer. To 7 μl of annealed fragment was added 1 μl of 10 mM rCTP, 2 μl of H₂O, and 1 μl of 10 × Mn buffer. Five microcuries of dried [α-³²P]dTTP (specific activity approximately 1 mCi/μmol) was dissolved in this and 1 unit DNA polymerase (Klenow) was added. Incubation was for 30 min in ice. One microliter of 0.2 M EDTA was added before loading on a 1-ml Sephadex G-100 column. Column buffer was 5 mM Tris, pH 7.5/0.1 mM EDTA. The labeled fragment was followed by monitor, collected in a minimum volume (approximately 200 μl), dried down, and redissolved in 30 μl of 1 × H buffer. Samples (2 μl) of this were taken for treatment as above. Following the chase step, 1 μl of 0.1 M

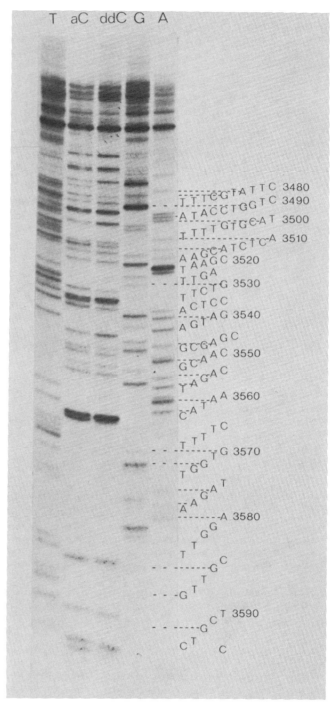

FIG. 3. Autoradiograph of an experiment with fragment A8 as primer on the viral strand of φX174 DNA using the single-site ribosubstitution method. Electrophoresis was on a 12% gel at 40 mA for 6 hr. The top 5.5 cm is not shown. Inhibitors used were (left to right) ddTTP, araCTP, ddCTP, ddGTP, and ddATP.

EDTA and 1 μl of pancreatic ribonuclease A at 10 mg/ml were added and incubated for 60 min at 37°.

RESULTS

Figs. 1–3 show examples of the use of the method for determining sequences in the DNA of φX174. In the experiment shown in Fig. 1 two small restriction enzyme fragments (A12d and A14, ref. 2) were used as primers on the complementary strand and there was no final digestion step to cut between the primer and the newly synthesized DNA. This is the most simple and rapid procedure, requiring only a preliminary annealing of template and primer, incubation of the four separate samples with DNA polymerase and appropriate triphosphates, followed by a chase with unlabeled dATP and application to the gel for electrophoresis. In these experiments the inhibitors used were ddGTP, ddATP, ddTTP, and araCTP. The conditions used for the "T" samples were not entirely optimal, resulting in the faster-moving bands being relatively weak.

The sequences can be read with reasonable accuracy starting at 88 nucleotides from the 5′ end of the primer for about 80 nucleotides (apart from some difficulty at position 3459 with A12d). For the next 50 nucleotides there is some uncertainty in the number of nucleotides in "runs" because bands are not actually resolved.

With longer restriction enzyme fragments as primers it is necessary to split them off from the newly synthesized DNA chains before the electrophoresis. This is normally done by digestion with a restriction enzyme. Fig. 2 shows such an experiment in which fragment R4 was used as primer on the complementary strand of ϕX174 DNA. In this experiment only dideoxynucleoside triphosphates were used as inhibitors because the results with araC were much less satisfactory when a restriction enzyme was used for the subsequent splitting. This may be due to the araC being removed by the 3′-exonuclease activity of the DNA polymerase during the incubation at 37° (which is necessary for the restriction enzyme splitting), resulting in a few C bands being either very faint or missing. Alternatively, the enzyme may be able to extend some chains beyond the araC at the higher temperature while being unable to do so at lower temperatures. araATP, which has been used only under these conditions, shows the same limitations as araCTP. These problems do not arise when ddCTP is used in this reaction.

With one exception (positions 4330–4343, see below), a sequence of 120 nucleotides, starting at a position 61 nucleotides from the restriction enzyme splitting site, could be read off; the sequence agreed with the published one. This region is believed to contain the origin of viral strand replication (2, 14). The bands beyond position 4380 indicated that there was an error in the provisional sequence (2), and further work (to be published later) has shown that the trinucleotide C-G-C should be inserted between positions 4380 and 4381.

When this technique is used the products are cut with a restriction enzyme as above, difficulties arise if there is a second restriction enzyme site close to the first one, because this will give rise to a separate pattern of bands that is superimposed on the normal one, making interpretation impossible. One way in which this can be avoided is by the single-site ribo-substitution method (N. L. Brown, unpublished). After annealing of the template and primer a single ribonucleotide is incorporated by incubation with DNA polymerase in the presence of manganese and the appropriate ribonucleoside triphosphate. Extension of the primer is then carried out with the separate inhibitors as above and the primer is split off at the ribonucleotide by ribonuclease or alkali. The method is particularly suitable for use with fragments obtained with the restriction enzyme *Alu*, which splits at the tetranucleotide sequence A-G-C-T. This enzyme is in fact inhibited by single-stranded DNA and cannot be used for the subsequent splitting of the primer from the newly synthesized DNA chain. The initial incorporation is carried out in the presence of rCTP and [α-^{32}P]dTTP. The incorporation of the ^{32}P facilitates subsequent purification on the Sephadex column.

Fig. 3 shows an example of the use of this method with fragment A8 on the viral strand of ϕX174 DNA. A sequence

of about 110 nucleotides starting 33 residues from the priming site can be read off. In the provisional sequence (2) this region was regarded as very tentative. Most of it is confirmed by this experiment, but there is a clear revision required at positions 3524–3530. The sequence of the viral strand should read A-T-C-A-A-C, replacing A-T-T-C- - -A-C given in the provisional sequence. There is difficulty in reading the sequence at 3543–3550, where there is considerable variation in the distance between bands, suggesting the presence of a looped structure. Further work in which the electrophoresis was carried out at a higher temperature indicates that the sequence here is actually G-C-T-C-G-C-G (viral strand); i.e., an insertion of C between positions 3547 and 3548 in the provisional sequence.

DISCUSSION

The method described here has a number of advantages over the plus and minus methods. First, it is simpler to perform because it requires no preliminary extension, thus avoiding one incubation and purification on a Sephadex column. It requires only the commercially available DNA polymerase I (Klenow fragment). The results appear to be more clear-cut with fewer artefact bands, and can usually be read further than with the plus and minus methods. Intermediate nucleotides in "runs" show up as bands, thus avoiding a source of error in the plus and minus method—estimating the number of nucleotides in a run. Theoretically one would expect the different bands in a run to be of the same strength, but this is not always the case. Frequently, the first nucleotide is the strongest, but in the case of ddCTP the second is the strongest (see Fig. 2). The reasons for these effects are not understood, but they do not usually cause difficulties with deducing the sequences. For the longer sequences in which the separate bands in a run are not resolved, experience has shown that it is frequently possible to estimate the number of nucleotides from the strength and width of the band.

The inhibitor method can also be used on a smaller scale than the plus and minus method because better incorporation from ^{32}P-labeled triphosphates is obtained. This is presumably due to the longer incubation period used, which allows a more quantitative extension of primer chains.

In general, sequences of from 15 to about 200 nucleotides from the priming site can be determined with reasonable accuracy using a single primer. Frequently it is possible to read the gels further and, on occasions, a sequence of about 300 nucleotides from the priming site has been determined. Occasional artefacts are observed, but these can usually be readily identified. It seems likely that these are usually due to contaminants in the fragments. The most serious difficulties are due to "pile-ups" of bands, which are usually caused by the DNA forming base-paired loops under the conditions of the acrylamide gel electrophoresis. These pile-ups are seen as a number of bands in the same position or unusually close to one another on the electrophoresis. They generally occur at different positions when the priming is carried in opposite directions along the DNA over the same sequence. An example of this effect is seen in Fig. 2 at position 4330, where there is a single strong band in the G channel that in fact represents four G residues. They are presumably forming a stable loop by pairing with the four Cs at positions 4323–4326. Another example is in Fig. 3 at positions 3545–3550. This effect is likely to be found in all the rapid techniques that use gel electrophoresis.

It is felt that for an accurate determination of sequence one should not rely completely on single results obtained by this method alone but that confirmation should be obtained by some other technique or by priming on the opposite strand. This consideration probably applies to all other available methods also. The main disadvantage of the present method is the difficulty in obtaining all the inhibitors—particularly ddGTP, which is not commercially available.

We wish to thank Dr. K. Geider for a gift of ddTTP, Dr. F. E. Baralle for a gift of N-isobutyryl-5′-O-monomethoxytrityldeoxyguanosine, and Dr. M. J. Gait for useful advice on the synthetic work.

1. Sanger, F. & Coulson, A. R. (1975) *J. Mol. Biol.* **94**, 441–448.
2. Sanger, F., Air, G. M., Barrell, B. G., Brown, N. L., Coulson, A. R., Fiddes, J. C., Hutchison, C. A., Slocombe, P. M. & Smith, M. (1977) *Nature* **265**, 687–695.
3. Maxam, A. M. & Gilbert, W. (1977) *Proc. Natl. Acad. Sci. USA* **74**, 560–564.
4. Atkinson, M. R., Deutscher, M. P., Kornberg, A., Russell, A. F. & Moffatt, J. G. (1969) *Biochemistry* **8**, 4897–4904.
5. Hunter, T. & Francke, B. (1975) *J. Virol.* **15**, 759–775.
6. Russell, A. F. & Moffatt, J. G. (1969) *Biochemistry* **8**, 4889–4896.
7. Geider, K. (1974) *Eur. J. Biochem.* **27**, 555–563.
8. McCarthy, J. R., Robins, M. J., Townsend, L. B. & Robins, R. K. (1966) *J. Am. Chem. Soc.* **88**, 1549–1553.
9. Tener, G. M. (1961) *J. Am. Chem. Soc.* **83**, 159–168.
10. Hoard, D. E. & Ott, D. G. (1965) *J. Am. Chem. Soc.* **87**, 1785–1788.
11. Büchi, H. & Khorana, H. G. (1972) *J. Mol. Biol.* **72**, 251–288.
12. Robins, M. J., McCarthy, J. R. & Robins, R. K. (1966) *Biochemistry* **5**, 224–231.
13. Air, G. M., Sanger, F. & Coulson, A. R. (1976) *J. Mol. Biol.* **108**, 519–533.
14. Slocombe, P. M. (1976) Ph.D. Dissertation, University of Cambridge.

Supplementary Readings

Beaucage, S.L. & Caruthers, M.H., Deoxynucleoside phosphoramidites—a new class of key intermediates for deoxypolynucleotide synthesis. *Tetrahedron Lett.* 22:1859–1862 (1981)

Caruthers, M.H., Gene synthesis machines: DNA chemistry and its uses. *Science* 230:281–285 (1985)

Crea, R., Kraszewski, A., Hirose, T. & Itakura, K., Chemical synthesis of genes for human insulin. *Proc. Natl. Acad. Sci. U.S.A.* 75:5765–5769 (1978)

Dalbadie-McFarland, G., Cohen, L.W., Riggs, A.D., Morin, C., Itakura, K. & Richards, J.H., Oligonucleotide-directed mutagenesis as a general and powerful method for studies of protein function. *Proc. Natl. Acad. Sci. U.S.A.* 79:6409–6413 (1982)

Edge, M.D., Greene, A.R., Heathcliffe, G.R., Meacock, P.A., Schuch, W., Scanlon, D.B., Atkinson, T.C., Newton, C.R. & Markham, A.F., Total synthesis of a human leukocyte interferon gene. *Nature* 292:756–762 (1981)

Hunkapiller, M.W., & Hood, L.E., Protein sequence analysis: automated microsequencing. *Science* 219:650–659 (1983)

Hunkapiller, M., Kent, S., Caruthers, M., Dreyer, W., Firca, J., Giffin, C., Horvath, S., Hunkapiller, T., Tempst, P. & Hood, L., A microchemical facility for the analysis and synthesis of genes and proteins. *Nature* 310:105–111 (1984)

Hutchison, C.A., Phillips, S., Edgell, M.H., Gillam, S., Jahnke, P. & Smith, M., Mutagenesis at a specific position in a DNA sequence. *J. Am. Chem. Soc.* 253: 6551-6560 (1978)

Letsinger, R.L. & Lunsford, W.B., Synthesis of thymidine oligonucleotides by phosphite triester intermediates. *J. Am. Chem. Soc.* 98:3655–3661 (1976)

Merrifield, R.B., Solid phase peptide synthesis. I. The synthesis of a tetrapeptide. *J. Am. Chem. Soc.* 85:2149-2154 (1963)

Montgomery, D.L., Hall, B.D., Gillam, S. & Smith, M., Identification and isolation of the yeast cytochrome c gene. *Cell* 14:673-680 (1978)

Smith, M., Leung, D.W., Gillam, S., Astell, C.R., Montgomery, D.L. & Hall, B.D., Sequence of the gene for iso-1-cytochrome c in Saccharomyces cerevisiae. *Cell* 16:753–761 (1979)

Tabor, S. & Richardson, C.C., DNA sequence analysis with a modified bacteriophage T7 DNA polymerase. *Proc. Natl. Acad. Sci. U.S.A.* 84:4767–4771 (1987)

Winter, G., Fersht, A.R., Wilkinson, A.J., Zoller, M. & Smith, M., Redesigning enzyme structure by site-directed mutagenesis: tyrosyl-tRNA-synthetase and ATP binding. *Nature* 299:756–758 (1982)

Detection and Separation

Biological macromolecules have two critical properties which challenge any technology; a given species of macromolecule is typically a very rare member of an incredibly complex mixture, and the macromolecule has some very precise distinguishing structural features. In order to study these molecules it was necessary to develop fractionation procedures which resolved the macromolecular mixtures and detection schemes which allowed one to precisely find the desired "needle" within the "haystack" in which it was located. Many fractionation and detection strategies are employed in biotechnology research. The papers which follow touch on only three: gel electrophoretic separation of polynucleotides, nucleic acid hybridization, and immunological detection of specific antigens.

The most widely used technique in fractionating DNA and RNA molecules is electrophoresis through either agarose or polyacrylamide gels. The separation of the different types of molecules is typically a function of their frictional coefficients, which in turn usually is a measure of their lengths. Standard electrophoretic techniques fail to separate large DNA molecules (typically above 20,000 base pairs), but the pulse field techniques described in the supplemental reading by Schwartz and Cantor extend this range to whole chromosomes!

Following fractionation it is necessary to detect the molecule of interest often when in low abundance and/or in the presence of many other contaminating species. This can be accomplished using nucleic acid hybridization technology. Nucleic acid hybridization techniques are critical to modern molecular biology experimentation and are based upon one of the most fundamental of all molecular biology principals: single stranded polynucleotides tend to form precise, stable, double stranded structures with (nearly) exact *complementary* polynucleotide sequences. The Southern paper presents the methodology for fractionating DNA molecules on an agarose gel, denaturing the molecules to their component single strands, transferring these single strands to a solid support membrane and then detecting the location of specific sequences using a radiolabeled oligonucleotide "probe" in a nucleic acid hybridization experiment. The scientific community names this widely used combination of techniques "Southern Blot" after its inventor. When the protocol was modified for the analysis of RNA molecules, the "Northern Blot" procedure was born. "Western Blots" also exist (detecting electrophoretically separated proteins immunologically, see Towbin et al.), but we are anxiously awaiting the development of "Eastern Blots!" The Southern Blot protocol is the basis of a whole field of research, medical and industrial analyses (Restriction Fragment Length Polymorphism [RFLP] tests; see a description of "finger printing" in Section 10), and an entire support industry devoted to equipment specifically designed for the procedure.

Scientists have fractionated proteins using electrophoresis for a number of years. The Towbin et al. paper (which describes the so-called "Western Blot" mentioned above) presents the technology for immunologically detecting specific fractionated proteins. The approach involves making a membrane bound representation of the separated proteins and then allowing specific antibodies to locate the presence of the relevant protein.

The immense power of nucleic acid hybridization technology is well illustrated in the two papers by Grunstein and Hogness, and Suggs et al. (and in a supplementary paper by Benton and Davis), which describe the basic protocol used in almost all recombinant DNA experiments for detecting colonies or viral plaques containing specific clones; clearly the molecular biologist's version of hunting for a "needle in a haystack." Single stranded DNA representations of a large population of possible clone colonies or plaques are immobilized on membranes and then are probed with radiolabeled oligonucleotides complementary to a portion of the desired sequence.

Nucleic acid hybridization detection technology should revolutionize all procedures involving the identification of particular organisms or specific genetic changes in a given organism. These are procedures which are or could be the basis of most medical diagnoses. This technology requires that there be a "tag" or a label on the detecting probe so that its localization on the solid support can be easily determined. In the research laboratory, radiolabels make convenient and extremely sensitive tags; however, in commercial laboratories this is an unsatisfactory solution. The supplemental paper by Langer et al. describes the first of many nonisotopic detection schemes, one of which will hopefully bring the revolutionary sensitivity and precision of nucleic acid hybridization out of the research laboratory and into common usage.

The nucleic acid hybridization technologies are instrumental in facilitating the identification (and isolation) of specific clones. However, other modes of detection have also been important. An important example is described in the paper by Broome and Gilbert. In this case the product encoded by the target cloned gene is sought through an immunological examination of membrane bound extracts of bacterial colonies; therefore, the gene is detected through its expression into protein, the usual "end product" for the biotechnology industry. A number of clever variations have been developed on this theme, e.g., epitope cloning.

Immunological Screening Method to Detect Specific Translation Products

S. Broome and W. Gilbert

ABSTRACT   We describe a very sensitive method to detect as antigens the presence of specific proteins within phage plaques or bacterial colonies. We coat plastic sheets with antibody molecules, expose the sheet to lysed bacteria so that a released antigen can bind, and then label the immobilized antigen with radioiodinated antibodies. Thus, the antigen is sandwiched between the antibodies attached to the plastic sheet and those carrying the radioactive label. Autoradiography then shows the positions of antigen-containing colonies or phage plaques. A few molecules of antigen released from each bacterial cell generate an adequate signal.

How can we detect small amounts of proteins made in a bacterial cell without an enzymatic assay? Immunological methods offer an approach: one should be able to detect not only a complete protein, but even a fragment of a protein sequence by virtue of its antigenic determinants. This problem is posed in recombinant DNA experiments in which one might desire to identify a bacterial cell containing a fragment of a gene from a higher cell. If that gene fragment were inserted within a bacterial protein, in phase, antigenic determinants on the higher cell protein could be synthesized and detected. Two immunological screening techniques have been reported (1, 2): they depended on precipitation of the antigen by antibodies included in the agar of the plate or in an agarose overlay. We have devised an extremely sensitive method for screening plaques or colonies that detects antigen-containing areas by a solid-phase "sandwich" assay.

Antibody molecules adsorb strongly to plastics such as polystyrene or polyvinyl and are not significantly dislodged by washing. This is the basis of very sensitive and simple two-site radioimmune assays (3, 4). Thus, we coat a flat disk of flexible polyvinyl with the IgG fraction from an immune serum and press this disk onto an agar plate so that antigen released from bacterial cells during the formation of a phage plaque or through *in situ* lysis of a colony can bind to the fixed antibody. We then incubate the plastic disk with the same total IgG fraction labeled with radioactive iodine so that other determinants on the bound antigen can in turn bind the iodinated antibody. The radioactive areas on the disk expose x-ray film during autoradiography. Since the polyvinyl disk can be thoroughly washed and treated with carrier serum, the background labeling is low; we have detected as little as 5 pg of antigen distributed over an area somewhat larger than a bacterial colony. Microgram amounts of antibody saturate the plastic disk, so an Ig fraction prepared from 5 ml of immune serum will serve to screen 1000 plates of colonies or plaques.

We describe the application of this technique to the detection of an independently assayable protein whose level of production is well characterized and subject to manipulation, *Escherichia*

coli β-galactosidase (β-galactoside galactohydrolase; EC 3.2.1.23).

MATERIALS AND METHODS

Solid-phase radioimmunodetection

Preparation of Solid-Phase IgG. Press 8.25-cm diameter disks cut from clear, flexible polyvinyl, 8 mil thickness (Dora May Co., New York), between sheets of smooth paper to flatten them. Place 10 ml of 0.2 M $NaHCO_3$ (pH 9.2) containing 60 μg of IgG per ml in a glass petri dish and set a flat polyvinyl disk upon the liquid surface. After 2 min at room temperature, remove the disk and wash it twice with 10 ml of cold wash buffer: phosphate-buffered saline (PBS), 0.5% normal rabbit serum (vol/vol), and 0.1% bovine serum albumin (wt/vol). Wash by gently swirling the buffer over the disk, then pouring and aspirating off the wash solution. We coat 10 polyvinyl disks successively with the same IgG solution and use each disk immediately after it is coated.

Release of Antigen from Cells in Bacterial Colonies. Heat induction of λcI857 prophages conveniently lyses cells in colonies (2). After growth of colonies of lysogens for 24 hr at 32°, incubate the plates for 2 hr at 42°.

Alternatively, apply 2 μl of 10 mM $MgSO_4$ containing 10^5 λvir to each colony and leave the plates for 3 hr at 37°, to effect a direct phage-mediated lysis.

Finally, chloroform vapor will lyse the bacteria *in situ*. Place a small volume of chloroform in the bottom of a tightly coverable container and set open petri dishes on a glass support above the liquid. After 10 min, transfer the plates to a desiccator and apply a vacuum to remove any residual chloroform.

Immunoadsorption of Antigen onto Solid Phase. Gently place the IgG-coated surface of a polyvinyl disk in contact with the agar and lysed colonies or the top agar and phage plaques within a petri dish. Smooth any air bubbles that form between disk and agar to the side, since the plastic is flexible. Leave the plates for 3 hr at 4°; then remove the disks and wash them three times with 10 ml of cold wash buffer, aspirating off any adhering cellular material during the washing.

Reaction of ^{125}I-Labeled Antibodies with Solid-Phase Antigen and Autoradiography. Pipet 1.5 ml of wash buffer containing 5×10^6 cpm (γ emission) of ^{125}I-labeled IgG (^{125}I-IgG) onto the center of an 8.25-cm diameter flat disk of nylon mesh (carried by most fabric stores) placed in the bottom of a petri dish. The mesh serves a necessary spacer function. Set a polyvinyl disk on the mesh and the solution so that the entire lower polyvinyl surface is accessible to the radioactive antibody. By repeating the layering process, generating a stack of alternating nylon mesh and polyvinyl disks, 15–20 polyvinyl disks

Abbreviations: PBS, phosphate-buffered saline; wash buffer, PBS containing 0.5% normal rabbit serum and 0.1% bovine serum albumin; ^{125}I-IgG, ^{125}I-labeled IgG.

BROOME, S. and GILBERT, W.
Immunological screening method to detect specific translation products.
Proc. Natl. Acad. Sci. U.S.A. 75:2746-2749 (1978). Reprinted with the authors' permission.

can be incubated in a single petri dish with 5×10^6 cpm of ^{125}I-IgG each. Incubate overnight at 4°; then wash each disk twice with 10 ml of cold wash buffer and twice with water. Lightly blot the disks to remove water droplets and let them dry at room temperature. Finally, autoradiograph the disks with either Kodak No Screen film or Kodak X-OMAT R film and a Du Pont Cronex Lighting Plus intensifying screen (5).

Procedures specific for β-galactosidase detection

Plating of Bacteria and Phage. Colonies of FMA-10 (W3102 r^- thy^-; from F. Ausubel) (λcI857) or RV(Δlac) (λcI857) were grown for 24 hr at 32° on YT plates or on Minimal A plates containing 0.2% glycerol and 40 μg of 5-bromo-4-chloro-3-indolyl-β-D-galactoside per ml (6). Colonies of nonlysogenic FMA-10 were grown overnight at 37° on YT plates.

Phage strains were λvir (kindly provided by J. G. Sutcliffe) and λplac5 cI857 Sam7. Host cells and phage were plated in 2.5 ml of H soft agar (0.8% agar) over H bottom agar (1% agar) (6), and phage plaques were allowed to form overnight at 37° on lawns of QD5003($suIII^+$) or at 32° on FMA-10(λcI857). Because FMA-10 is thy^-, all plates and media contained 10 μg of thymidine per ml.

Antiserum. New Zealand White rabbits were immunized with 1 mg of electrophoretically pure β-galactosidase (7, 8) in complete Freund's adjuvant (Difco). Booster injections were administered in incomplete Freund's adjuvant (Difco) 2 and 3 weeks after the initial injection, and the rabbits were bled 1 week later. Ten microliters of immune serum precipitated 10 μg of pure β-galactosidase.

The IgG fractions of rabbit pre-immune and rabbit anti-β-galactosidase immune sera were prepared by ammonium sulfate precipitation followed by DEAE-cellulose (Whatman, DE-52) chromatography (9) in 25 mM potassium phosphate, pH 7.3/1% glycerol. Fractions containing the bulk of the flow-through material were pooled, and protein was precipitated by adding ammonium sulfate to 40% saturation. The resulting pellet was resuspended in one-third the original serum volume of 25 mM potassium phosphate, pH 7.3/0.1 M NaCl/1% glycerol, and dialyzed against the same buffer. After dialysis, any residual precipitate was removed by centrifugation. IgG fractions were stored in aliquots at −70°.

Iodination of IgG. IgG fractions were radioiodinated by the method of Hunter and Greenwood (10). The 25-μl reaction mixture contained 0.5 M potassium phosphate (pH 7.5), 2 mCi of carrier-free Na^{125}I, 150 μg of IgG, and 2 μg of chloramine T. After 3 min at room temperature, 8 μg of sodium metabisulfite in 25 μl of PBS was added, followed by 200 μl of PBS containing 2% normal rabbit serum. The ^{125}I-labeled IgG was purified by chromatography on a Sephadex G-50 column equilibrated with PBS containing 2% normal rabbit serum. The ^{125}I-IgG elution fraction was diluted to 5 ml with PBS containing 10% normal rabbit serum, filtered through a sterile Millipore VC filter (0.1 μm pore size), divided into aliquots, and stored at −70°. The specific activities were 1.5×10^7 cpm per μg.

RESULTS

Specificity of solid-phase ^{125}I-antibody binding

Lac^+ colonies growing on a minimal plate containing glycerol as the carbon source and the β-galactosidase indicator substrate 5-bromo-4-chloro-3-indolyl-β-D-galactoside develop a blue coloration while lac^- colonies remain white. Fig. 1A shows such a plate bearing colonies of either FMA-10(λcI857) (lac^+) or RV(λcI857) (lac deletion) cells, after 24 hr of colony growth at 32°. Since no inducer of lac transcription was present in the plate, each cell in the five dark (lac^+) colonies contained approximately 10–20 molecules of β-galactosidase (7).

We lysed the cells within the lac^+ and lac^- colonies by heat induction of the λcI857 prophage and adsorbed any β-galactosidase released onto a polyvinyl disk coated with anti-β-galactosidase antibodies. Fig. 1B shows an autoradiograph of this plastic disk after labeling with I^{125}-anti-β-galactosidase. The regions of ^{125}I-antibody binding clearly show the positions of lac^+ colonies, while the lac^- colonies do not label.

This detection of the basal level of β-galactosidase represents essentially a full signal for this assay. If the cells had been fully induced, the labeled spots would have been larger but not more intense. Since only background labeling is observed when uninduced lac^+ colonies are screened with uncoated polyvinyl or coated with pre-immune serum IgG, the fixed, specific antibody is required. Faintly detectable amounts of antigen can be adsorbed, however, by uncoated polyvinyl exposed to lysed, fully induced lac^+ colonies in which approximately 2% of total protein is β-galactosidase.

Fig. 2A shows that ^{125}I-antibody binding is dependent upon lac^+ cell lysis and demonstrates an application of this solid-phase radioimmunodetection to phage plaques. This disk was applied to a plate bearing plaques of λvir phage on a lawn of

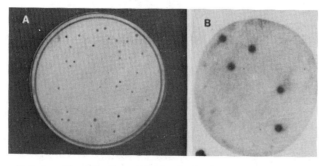

FIG. 1. Identification of lac^+ colonies by solid-phase radioimmunodetection. A mixture of RV (Δlac) (λcI857) and FMA-10 (λcI857) cells was spread on a glycerol Minimal A plate containing 5-bromo-4-chloro-3-indolyl-β-D-galactoside. (A) Colonies formed after 24 hr of growth at 32°. Due to the presence of hydrolyzed indicator substrate, the five lac^+ strain colonies appear darker than the lac^- strain colonies. Cells within each colony on this plate were lysed by prophage induction; released antigen was adsorbed to a polyvinyl disk that had been coated with anti-β-galactosidase IgG. Immobilized antigen was labeled by incubating the polyvinyl disk with radioiodinated anti-β-galactosidase IgG. (B) Autoradiograph of this polyvinyl disk exposed on No Screen film for 48 hr.

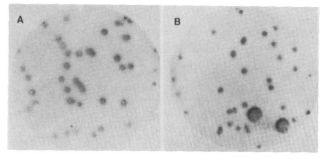

FIG. 2. Solid-phase radioimmunodetection of β-galactosidase present in phage plaques. Antigen released from host cells lysed during phage plaque formation was immobilized on a polyvinyl disk and labeled as described in the legend to Fig. 1. The autoradiographs are of disks imprinted on plates bearing (A) λvir plaques on a lawn of FMA-10(λcI857) cells or (B) λvir and λplac5 Sam7 plaques on QD5003 cells, exposed on No Screen film for 48 hr.

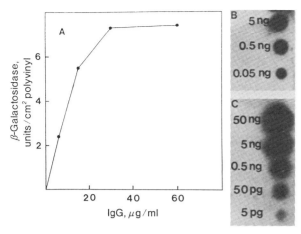

FIG. 3. Assays of IgG adsorption to polyvinyl and solid-phase radioimmunodetection of two different antigens. (A) Polyvinyl disks (8.25-cm) were coated with various dilutions of anti-β-galactosidase IgG in 20 mM NaHCO$_3$ (pH 9.2) for 1 min at room temperature. The coated disks were washed twice with 10 ml of cold wash buffer, and 1-cm^2 pieces cut from these disks were each incubated with 1 ml of wash buffer containing 1 mM MgSO$_4$, 500 mM 2-mercaptoethanol, and 1 μg of β-galactosidase for 4 hr at 4°. Unbound antigen was then removed by washing each 1-cm^2 piece of polyvinyl twice with 5 ml of cold wash buffer. Solid-phase-bound β-galactosidase was measured by o-nitrophenyl-β-D-galactoside hydrolysis (6). One unit of enzyme hydrolyzes 1 nmol of o-nitrophenyl-β-D-galactoside per min at 28°. (B) Aliquots (1-μl) of wash buffer containing the indicated amounts of β-galactosidase were applied to the surface of an agar plate. After the liquid had entered the agar, the antigen was adsorbed in the usual way to a polyvinyl disk that had been coated with anti-β-galactosidase IgG. Immobilized antigen was labeled and detected as described in the legend to Fig. 1, except autoradiography was for 18 hr with an intensifying screen at −70°. (C) Aliquots (1-μl) of wash buffer containing the indicated amounts of E. coli penicillinase (EC 3.5.2.6) were used in a similar experiment. The anti-penicillinase IgG was prepared from an immune serum of a titer comparable to that of our rabbit anti-β-galactosidase immune serum.

uninduced FMA-10(λcI857). The IgG-coated polyvinyl disk detects the 10 β-glactosidase molecules per cell released as the cells lyse during the formation of the plaque but does not pick up any antigen from the unlysed lawn. The control labeling of IgG-coated polyvinyl disks placed upon λvir plaques on the lac^- strain RV(λcI857) is completely uniform and of background intensity (data not shown).

The autoradiographic image is larger than the original colony or phage plaque because the antigen diffuses in the plate. The spread from phage plaques is especially apparent since infection and lysis of host cells (embedded in 0.8% top agar) occurs over

an extensive period of time. Fig. 2B shows the distribution of antigen released during λplac5 Sam7 and λvir growth on a lac^+ $suIII^+$ host, QD5003. The large exposed areas on the autoradiograph correspond to λplac5 Sam7 plaques and reflect the high level of β-galactosidase produced during the I^-, Z^+ lac phage infection. Each λplac5 Sam7 plaque contained on the order of 10–20 ng of β-galactosidase (assuming 10^6 cells lysed per plaque and 10^4 molecules of enzyme per cell). The smaller exposed areas on the film correspond in both size and position to λvir plaques, each of which contained approximately 0.1 ng of β-galactosidase (10^7 cells lysed per plaque, 10 molecules of enzyme per cell). The identity of the phage within each plaque was confirmed since only phage isolated from ostensible λvir plaques produced "macroplaques" on the lysogenic, nonsuppressing host RV(λcI857). Fig. 2B demonstrates that the enlargement of the autoradiographic image is proportional to the amount of β-galactosidase released from lysed cells, and further shows, therefore, that the antigen recognized by ^{125}I-antibody is β-galactosidase.

Selection of solid-phase assay conditions

We determined the appropriate polyvinyl coating conditions by measuring the amounts of β-galactosidase immobilized on 1-cm^2 pieces of polyvinyl cut from disks that had been incubated with various dilutions of anti-β-galactosidase IgG. The bound β-galactosidase retained enzymatic activity and was assayed by o-nitrophenyl-β-D-galactoside hydrolysis (7). Fig. 3A shows that a 1-min incubation with 30 μg of IgG per ml saturates the antibody-adsorbing capacity of an 8.25-cm diameter polyvinyl disk. In addition, the same amount of β-galactosidase is bound by polyvinyl coated for 1 min or for 5 hr with 30 μg of IgG per ml. By the same procedure, we found that 10 disks coated successively with one 10-ml solution of 60 μg of IgG per ml possess identical capacities to bind β-galactosidase. A decrease in binding was observed, however, as disks were coated successively in 10 ml of 30 μg of IgG per ml, from which we estimate that each disk binds about 20 μg of antibody.

In order to estimate the minimum amount of protein detectable using 5 × 10^6 cpm of ^{125}I-IgG, we applied a series of dilutions of antigen directly to the surface of a typical agar plate. The antigen then was adsorbed to an IgG-coated polyvinyl disk and labeled with ^{125}I-antibodies. Fig. 3 B and C shows that 50 pg of β-galactosidase in one experiment and 5 pg of penicillinase (EC 3.5.2.6) in another were detected, spread over an area somewhat larger than a phage plaque or a bacterial colony. Because the antigen had diffused into the agar, these

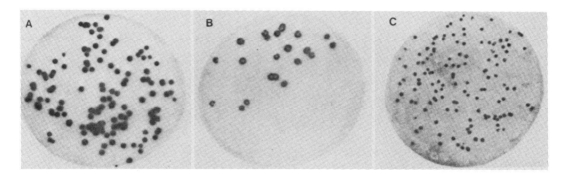

FIG. 4. Autoradiographs illustrating three methods for releasing antigens from bacterial colonies. (A) Colonies of FMA-10(λcI857) were grown for 24 hr at 32° on a YT plate. Cells within these colonies were lysed by prophage induction during a 2-hr incubation at 42°. (B) Cells within FMA-10 colonies, formed overnight at 37° on YT plates, were lysed *in situ* by applying 2 μl of 10 mM MgSO$_4$ containing 10^5 λvir to each colony on one-half of a plate and then incubating this plate for 3 hr at 37°. The other half of this plate serves as a control. (C) Similar colonies were lysed by a 10-min exposure to chloroform vapor in a tightly sealed glass container, as described in *Materials and Methods*.

experiments detected only a fraction of the initial sample.

The specific labeling of the immobilized antigen is maximal after an overnight incubation with ^{125}I-IgG. One-half maximal labeling is reached after 5 hr.

Alternative methods of lysing colonies

As Fig. 4 shows, comparable amounts of β-galactosidase are released from colonies by prophage induction or by direct application of λvir. Chloroform vapor also will lyse cells sufficiently to permit detection of β-galactosidase present in colonies of cells uninduced for *lac* expression (Fig. 4C). Colonies of RV(λcI857) (*lac* deletion) cells lysed by any of these methods did not release any material that reacted with ^{125}I-anti-β-galactosidase IgG.

Viable bacteria exist within a colony of λcI857 lysogens after a 2-hr incubation at 42°. Many of the survivors are lysogens and can be recovered, to confirm a positive response, by picking from the site of a colony. Replica plates must be used if colonies are to be treated with λvir or with chloroform vapor.

DISCUSSION

This solid-phase screening method is simple and sensitive. It detects a few picograms of protein antigen, a few molecules from each bacterial cell, using the IgG fraction from an immune serum of moderate titer. The only clear requirement for this approach is that the antigen bind at least two antibody molecules simultaneously.

Enzymatic assays suggest that about 2×10^8 molecules of β-galactosidase can bind per mm^2 of coated plastic. This is consistent with our estimate of the IgG-adsorbing capacity of the polyvinyl, 2×10^{10} molecules per mm^2, since the specific antibodies constitute only a few percent of the immune IgG fraction. Direct counting of samples of labeled plastic showed that only about 5×10^7 labeled antibodies were bound to each mm^2 of fixed antigen under the conditions described; at least 10-fold more label could bind, but at a price in terms of a higher background. The lower limit of detection presumably could be extended by using affinity-purified antibodies.

Uses for this immunological screening procedure include direct identification of clones containing specific foreign DNA segments, if they express a translation product either fortuitously or after *in vitro* genetic manipulations to that end. Furthermore, this technique provides a simple way to follow the movement of antigen on columns or on slab gels.

This two-site detection is particularly suited for the recognition of certain novel genetic constructions which are much less easily assayed by *in situ* immunoprecipitation approaches. For instance, by coating polyvinyl disks with an IgG fraction prepared from an immune serum directed against one protein and labeling the immobilized antigen with ^{125}I-antibodies directed against another protein, only hybrid polypeptide molecules, synthesized as the result of *in vitro* or *in vivo* DNA sequence rearrangement, would produce an autoradiographic response.

We thank Dr. Abe Fuks for advice about radioiodination, Drs. Jeremy Knowles and Alan Hall for samples of *E. coli* penicillinase and rabbit anti-penicillinase serum, and Roger Brent for suggesting the chloroform release. This work was supported by the National Institutes of Health, Grant GM09541-17. W.G. is an American Cancer Society Professor. S.B. was supported by a National Institutes of Health training grant.

1. Sanzey, B., Mercereau, O., Ternynck, T. & Kourilsky, P. (1976) *Proc. Natl. Acad. Sci. USA* **73**, 3394–3397.
2. Skalka, A. & Shapiro, L. (1976) *Gene* **1**, 65–79.
3. Catt, K. & Tregear, G. W. (1967) *Science* **158**, 1570–1571.
4. Miles L. E. M. (1977) in *Handbook of Radioimmunoassay*, ed. Abraham, G. E. (Dekker, New York), pp. 131–177.
5. Laskey, R. A. & Mills, A. D. (1977) *FEBS Lett.* **82**, 314–316.
6. Miller, J. M. (1972) in *Experiments in Molecular Genetics*, ed. Miller, J. M. (Cold Spring Harbor Laboratory, Cold Spring Harbor, New York), pp. 432–434.
7. Platt, T. (1972) in *Experiments in Molecular Genetics*, ed. Miller, J. M. (Cold Spring Harbor Laboratory, Cold Spring Harbor, New York), pp. 398–404.
8. Fowler, A. V. (1972) *J. Bacteriol.* **112**, 856–860.
9. Livingston, D. M. (1974) in *Methods in Enzymology*, ed. Jakoby, W. B. & Wilchek, M. (Academic, New York), Vol. 34, 723–731.
10. Hunter, W. M. Greenwood, F. C. (1964) *Biochem. J.* **91**, 43–46.

Colony Hybridization: A Method for the Isolation of Cloned DNAs that Contain a Specific Gene

M. Grunstein and D.S. Hogness

ABSTRACT   A method has been developed whereby a very large number of colonies of *Escherichia coli* carrying different hybrid plasmids can be rapidly screened to determine which hybrid plasmids contain a specified DNA sequence or genes. The colonies to be screened are formed on nitrocellulose filters, and, after a reference set of these colonies has been prepared by replica plating, are lysed and their DNA is denatured and fixed to the filter *in situ*. The resulting DNA-prints of the colonies are then hybridized to a radioactive RNA that defines the sequence or gene of interest, and the result of this hybridization is assayed by autoradiography. Colonies whose DNA-prints exhibit hybridization can then be picked from the reference plate. We have used this method to isolate clones of ColE1 hybrid plasmids that contain *Drosophila melanogaster* genes for 18 and 28S rRNAs. In principle, the method can be used to isolate any gene whose base sequence is represented in an available RNA.

Segments of DNA from *Drosophila melanogaster* chromosomes (Dm segments) can be isolated by cloning hybrid DNA molecules that consist of a Dm segment inserted into the circular DNA of an *Escherichia coli* plasmid. We have previously reported on the use of such cloned segments in the analysis of DNA sequence arrangements in the *D. melanogaster* genome (1–3). However, that analysis has been limited by our inability to isolate cloned Dm segments that contain a specified DNA sequence or gene. In this article we describe a procedure that permits the isolation of such specific Dm segments, and which can be extended to DNA segments from any organism.

Experimental Plan. Consider an experiment in which the Dm segments in a random set are individually inserted into a given *E. coli* plasmid. Transformation of *E. coli* by these hybrid plasmids to a phenotype conferred by genes in the parental plasmid will yield colonies that individually contain a single cloned Dm segment (1–3). If these segments are randomly distributed and exhibit a mean length of 10,000 base pairs, or 10 kb, then we expect that about one colony in 16,000 will contain a particular nonrepetitive *D. melanogaster* DNA sequence the length of a typical structural gene, i.e., 1–2 kb. Hence, the goal is to devise a screening procedure whereby one can rapidly determine which colony in thousands contains such a sequence.

The screening procedure that we have developed is designed to detect sequences that can hybridize with a given radioactive RNA. In this procedure the colonies to be screened are first grown on nitrocellulose filters that have been placed on the surface of agar petri plates prior to inoculation. A reference set of these colonies is then obtained by replica plating (4) to additional agar plates that are stored at 2–4°C. The colonies on the filter are lysed and their DNAs are denatured and fixed to the filter *in situ* to form a "DNA-print" of each colony. The defining, labeled RNA is hybridized to this DNA and the result of the hybridization is monitored by autoradiography on x-ray film. The colony whose DNA-print exhibits hybridization with the defining RNA can then be picked from the reference set.

The characteristics of this procedure and its application to the isolation of hybrid plasmids containing the *D. melanogaster* genes for '18' and '28'S rRNAs are described in this paper.

MATERIALS AND METHODS

Bacteria. *E. coli* K12 strains HB101, HB101 [pDm103], and C600 [pSC101] are those used previously (plasmids are indicated in brackets) (3). Strain W3110 has been described (5), and W3110 [ColE1] was obtained from D. R. Helinski.

DNAs, Complementary RNAs (cRNAs), and Enzymes. pDm103 (3) and ColE1 (6) DNAs were generously provided by D. M. Glover and D. J. Finnegan, respectively, and were prepared from HB101 [pDm103] and W3110 [ColE1] according to the indicated references, except that the ColE1 was amplified by overnight incubation of W3110 [ColE1] in the presence of chloramphenicol (7) prior to lysis. ^{32}P- and ^{3}H-labeled cRNAs were transcribed *in vitro* from these DNAs with *E. coli* RNA polymerase (8), as described by Wensink *et al.* (1). The RNA polymerase was prepared according to the indicated reference, and was the generous gift of W. Wickner. Pancreatic ribonuclease and proteinase K were obtained from Worthington Biochemical Corp. and E. Merck Laboratories, respectively.

Colony hybridization

Formation of the Filter and Reference Sets of Colonies. Colonies are formed on Millipore HA filters (0.45 μm pores) that have been washed three times in boiling H$_2$O (1 min per wash), placed between sheets of absorbant paper, autoclaved at 120° for 10 min, and dried for 10 min in the autoclave. The filter is then placed on an L-agar petri plate (1) and the desired bacteria are transferred to the filter surface either by spreading or using sterile toothpicks to obtain ≤7 colonies per cm^2 after incubation of the filter-plate at 37°. The reference set is produced by replica plating of the colonies that develop on the filter to L-agar plates and is stored at 2–4°.

Abbreviations: kb (kilobases), 1000 bases or base pairs in single- or double-stranded nucleic acids, respectively; Dm, a segment of *Drosophila melanogaster* DNA; cDm and pDm, hybrid plasmids consisting of a Dm segment inserted into ColE1 and pSC101 DNAs, respectively; SSC = 0.15 M NaCl, 0.015 M sodium citrate; cRNA, RNA complementary to DNA; rDNA, DNA coding for ribosomal RNA.

* Present address: Molecular Biology Institute and Department of Biology, University of California, Los Angeles, Calif. 90024.

† To whom reprint requests should be sent.

GRUNSTEIN, M. and HOGNESS, D.S.
Colony hybridization: A method for the isolation of cloned DNAs that contain a specific gene.
Proc. Natl. Acad. Sci. U.S.A. 72:3961-3965 (1975). Reprinted with the authors' permission.

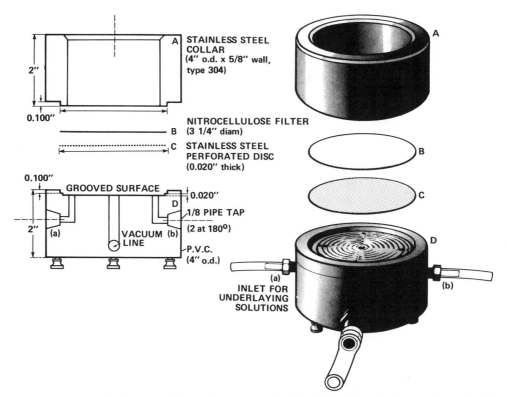

FIG. 1. Apparatus for treatment of colonies on filters. To wet the underside of the filter, solutions are introduced through ports (a) or (b), while the tube connected to the vacuum port is clamped off. Solutions are removed through the vacuum port which is connected to a water aspirator. Other procedures are described in the text. ", inches (2.54 cm); o.d., outside diameter; P.V.C., polyvinyl chloride.

Lysis, DNA Denaturation, and Fixation. To prevent movement of the bacteria or DNA from their colonial sites during lysis, denaturation and fixation, the solutions used to effect these reactions are applied to the underside of the filter and allowed to diffuse into the colony. The apparatus shown in Fig. 1 has been designed for this purpose. The filter is lifted from the agar plate and placed on the perforated disc that is set in a plastic cylinder which has ports cut into it to introduce solutions sequentially to the underside of the filter and to apply vacuum. Unless otherwise indicated, all operations are carried out at room temperature (20–25°).

Lysis and DNA denaturation are effected by introducing 0.5 N NaOH beneath the filter until it barely floats. After 7 min the NaOH is slowly removed with a minimum of vacuum, and replaced by 1.0 M Tris·HCl (pH 7.4) for 1 min. This solution is replaced with the same buffer, after which the pH of the solution in contact with the filter should be approximately neutral. The last wash is replaced by 1.5 M NaCl, 0.5 M Tris·HCl (pH 7.4), which is removed after 5 min. The stainless steel collar is then placed over the filter, and full vacuum is applied for approximately 2 min until the colonial residues assume a dry appearance. At this point there is less danger of movement from the colonial site and the remaining solutions can be layered on the upper side of the filter.

A 2 mg/ml solution of proteinase K in 1 × SSC (0.15 M NaCl, 0.015 M sodium citrate) is added to just cover the filter. After 15 min, it is removed by vacuum filtration, and 95% ethanol (1 ml/cm^2 of filter) is similarly passed through the filter. After five washes effected by passing chloroform through the filter (2 ml/cm^2 per wash), the filter is removed from the apparatus, dipped into 0.3 M NaCl to remove loose cellular debris, and baked at 80° *in vacuo* for 2 hr.

Hybridization and ^{32}P-Autoradiography or ^{3}H-Fluorography. The dry filter is moistened with a 5 × SSC, 50% formamide solution containing the labeled RNA, using 10–15 μl/cm^2 of filter. The filter is covered with mineral oil, incubated for 16 hr at 37° to allow hybridization, and then washed for 10 min in a beaker containing chloroform that is gently agitated on a shaking platform. Two more identical chloroform washes are followed by 10 min washes in 6 × SSC, 2 × SSC, and 2 × SSC containing 20 μg/ml of pancreatic ribonuclease. If the RNA is ^{32}P-labeled, the filter is blotted to remove excess liquid, covered with Saran Wrap, and placed under Kodak RPS/54 x-ray film for autoradiography. If the RNA is ^{3}H-labeled, the filter is dried for 30 min at 80° *in vacuo*, and 40 μl of 7% 2,5-diphenyloxazole (PPO) in ether is applied per cm^2 of filter. The dry filter is then placed under x-ray film for fluorography at −82° (9).

RESULTS

Colony hybridization distinguishes between [ColE1]$^+$ and [ColE1]$^-$ bacteria

We have turned increasingly toward the use of the colicinogenic plasmid, ColE1, as a cloning vector because one can obtain much higher cellular concentrations of its hybrids (7) than is the case for the tetracycline resistance plasmid, pSC101, which we used previously (1–3). The first test system for colony hybridization therefore consisted of ^{32}P-labeled cRNA made by transcription of ColE1 DNA *in vitro* with *E. coli* RNA polymerase, and *E. coli* containing or not containing ColE1, i.e., [ColE1]$^+$ or [ColE1]$^-$ bacteria.

Fig. 2A shows the autoradiographic response obtained after hybridization of [^{32}P]cRNA to the DNA-prints of [ColE1]$^+$ and [ColE1]$^-$ colonies formed on nitrocellulose fil-

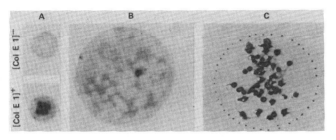

FIG. 2. Hybridization of ColE1 cRNA to [ColE1]⁻ and [ColE1]⁺ colonies. The procedures for colony hybridization, autoradiography, and fluorography are described in *Materials and Methods*, as are the W3310 and W3110 [ColE1] *E. coli* strains used to form the [ColE1]⁻ and [ColE1]⁺ colonies, respectively. (A) 1 × 10⁵ cpm of [³²P]cRNA (5 × 10⁷ cpm/μg) were applied to each 13-mm filter (area = 1.3 cm²) in a 20 μl volume. After hybridization, the DNA-prints of [ColE1]⁺ contained an average of 1.8 × 10² cpm per colony, which is 30-fold greater than the background radiation from an equivalent area on the filter. Exposure time = 45 min. (B) A mixture of [ColE1]⁺ and [ColE1]⁻ bacteria in a 1:100 ratio was spread on a 47-mm filter (area = 17.3 cm²) to obtain a total of 1 to 2 × 10² colonies per filter; 5 × 10⁵ cpm of [³²P]cRNA (3 × 10⁷ cpm/μg) in 250 μl were applied to the filter. Exposure time = 4 hr. (C) A 1:1 mixture of [ColE1]⁺ and [ColE1]⁻ bacteria was spread on a 47-mm filter to obtain a total of 93 colonies, of which 52 gave the A⁺ response seen in the figure; 1 × 10⁶ cpm of [³H]cRNA (2 × 10⁷ cpm/μg) in 200 μl were applied to the filter. Exposure time = 24 hr.

ters. The positive response given by the [ColE1]⁺ colonies is abbreviated by A⁺ and the negative response of [ColE1]⁻ colonies by A⁻. Colonies obtained by spreading mixtures of [ColE1]⁺ and [ColE1]⁻ bacteria in different ratios gave the expected frequencies of A⁺ and A⁻ responses. Fig. 2B shows the result obtained when [ColE1]⁺/[ColE1]⁻ = 1/100.

A more precise measure of the specificity of colony hybridization of mixtures is given by the following experiment in which a 1:1 mixture of [ColE1]⁺ and [ColE1]⁻ bacteria was spread on a filter to yield 31 colonies. Hybridization and autoradiography revealed that 16 were A⁺ and 15 A⁻. Bac-

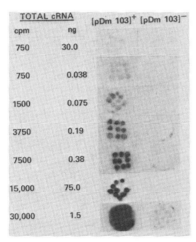

TOTAL cRNA		[pDm 103]⁺	[pDm 103]⁻
cpm	ng		
750	30.0		
750	0.038		
1500	0.075		
3750	0.19		
7500	0.38		
15,000	75.0		
30,000	1.5		

FIG. 3. Hybridization of different amounts of pDm103 [³²P]cRNA to [pDm103]⁺ and [pDm103]⁻ colonies. Colonies were obtained by transferring HB101 [pDm103] or HB101 bacteria, respectively, to 13-mm filters with toothpicks. In the experiments where ≤1.5 ng of cRNA were applied per filter, the specific activity = 2 × 10⁷ cpm/μg. The lower specific activities used for the other two experiments were obtained by mixing this cRNA with unlabeled pDm103 cRNA. The weak response observed for [pDm103]⁻ colonies could result either from *E. coli* DNA impurities in the pDm103 DNA preparations used to prepare the [³²P]cRNA, or from some similarity of sequence in pDm103 and *E. coli* DNAs.

teria from each of the corresponding colonies on the agar replica plate were then tested for colicin production according to an overlay technique described by Finnegan and Willets (10). All 16 A⁺ colonies were colicin-positive (i.e., [ColE1]⁺); all 15 A⁻ colonies were colicin-negative and therefore presumed to be [ColE1]⁻.

Fig. 2A and B show that the position of A⁻ colonies can be detected on the autoradiograph because of the higher background radiation from the filter itself. While this background radiation is convenient for the direct visualization of A⁻ colonies and is not critical to the observation of the A⁺ response obtained with cRNAs, it may become an important factor with other RNAs if they give a weaker A⁺ response. Our observations indicate that the level of this background varies with the preparation of labeled RNA and, possibly, with the batch of filters, but we have not examined such factors in detail.

Fig. 2C shows that the colony hybridization procedure can be adapted to ³H-labeled cRNA by impregnating the filter with 2,5-diphenyloxazole after hybridization and prior to placement on the x-ray film (*Materials and Methods*). Of the 93 colonies obtained by spreading a 1:1 mixture of [ColE1]⁺ and [ColE1]⁻ bacteria, 52 were A⁺ and 41 A⁻. We estimate from the extent of the A⁺ response that this ³H-fluorography is about one-twentieth as efficient as the ³²P-autoradiography.

The autoradiographic response is proportional to the total radioactivity of the applied cRNA and insensitive to its specific activity

We next examined the dependence of the A⁺ response on the total and the specific radioactivity of the applied cRNA. In this case, the ³²P-labeled cRNA was transcribed *in vitro* from a hybrid plasmid called pDm103, and hybridized to DNA-prints of colonies that either contained this hybrid, [pDm103]⁺, or did not, [pDm103]⁻. The pDm103 hybrid was formed between pSC101 plasmid DNA (9 kb) and a segment of *D. melanogaster* DNA (Dm103; 17 kb) that contains the gene for '18' and '28'S rRNAs (3).

Fig. 3 shows that the autoradiographic response obtained when pDm103 [³²P]cDNA was hybridized to 13-mm filters containing [pDm103]⁺ colonies is roughly proportional to the total radioactivity. It is clearly insensitive to the mass of cRNA containing that radioactivity, i.e., to its specific activity. For example, the response to 750 cpm of [³²P]cRNA is approximately the same whether contained in 0.038 ng or in 30 ng. Similarly the response to 15,000 cpm contained in 75 ng is intermediate between that to 7,500 cpm and 30,000 cpm, although the last two samples contained only 0.38 and 1.5 ng, respectively. This would suggest that the RNA·DNA hybridization is occurring under conditions of DNA excess even when 75 ng of pDm103 cRNA are applied per 13 mm filter. However, we have calculated that there is only some 2 ng of pDm103 DNA per colony [i.e., (2 × 10⁷ cells per colony) × (4 pDm103 per cell) 2.9 × 10⁻⁸ ng DNA per pDm103]. This value is based on our observation of 2 × 10⁷ cells per 1 mm colony and the presence of 4 pDm103 per cell in liquid culture (3). Evidently only a small fraction of the applied cRNA can react with the DNA-prints on the filter even though the reaction is occurring ostensibly in DNA excess. A similar result was observed when ColE1 cRNA was hybridized to [ColE1]⁺ colonies (legend, Fig 2A). Of 2 ng cRNA applied to each filter only 0.004 ng (i.e., 0.2%) hybridized per [ColE1]⁺ colony. A 1 mm [ColE1]⁺ colony is estimated to contain 3–4 ng of ColE1 DNA.

A simple explanation of these results is obtained if one assumes that most or all of the cRNA in the small fraction of the RNA solution which wets a DNA-print will hybridize, and that the remainder of the cRNA will not hybridize at a significant rate, due perhaps to its slow diffusion through the nitrocellulose, or because of other barriers. Thus a DNA-print from a 1-mm colony, which occupies 0.6% of the area of a 13-mm filter, would be expected to hybridize ≤0.6% of the applied RNA, an expectation that is compatible with the 0.2% observed. For a given ratio of colony to filter area, the fraction of applied cRNA that hybridizes to a DNA-print, in conditions of local DNA excess, would therefore be constant and independent of the total applied cRNA over a wide range of values.

Colony hybridization with cRNA to pDm103 provides a screen for cDm plasmids containing *D. melanogaster* rDNA

Hybrid plasmids consisting of a Dm segment inserted into ColE1 DNA are called cDm plasmids, as distinguished from pDm plasmids where the Dm segment has been inserted into pSC101. In this section we describe two applications of colony hybridization that result in the isolation of cDm plasmids that contain DNA from the repeating gene-spacer units for '18–28'S rRNAs (i.e., rDNA) in *D. melanogaster* (3). In the first application, [^{32}P]cRNA to pDm103 was used to isolate clones of cDm103 plasmids; i.e., plasmids in which the Dm103 segment is inserted into ColE1 DNA at its single *Eco*RI endonuclease cleavage site (7). In the second application, the same [^{32}P]cRNA was used to screen a large set of random cDm clones for rDNA. cRNA formed by transcription of the entire pDm103 DNA can be used for these purposes since we have demonstrated that pSC101 and ColE1 sequences do not interact to give a significant A$^+$ response (data not shown).

Cleavage of circular pDm103 DNA with the *Eco*RI restriction endonuclease yields intact Dm103 segments and linear pSC101 DNA (3). In cooperation with D. M. Glover, we treated a mixture of *Eco*RI-cleaved pDm103 and ColE1 DNAs with *E. coli* ligase under previously described conditions (3), and then transformed colicin-sensitive *E. coli* to colicin E1 immunity with this mixture of ligated DNAs (11). Since the *Eco*RI termini of the linear ColE1, pSC101, and Dm103 molecules can be randomly joined by the ligase, any of the following circular products of this ligation may be present in the colonies of transformants: (*i*) recycled ColE1 (monomers, dimers, etc), (*ii*) molecules containing one ColE1 and one pSC101 segment [abbreviated by (c)$_1$(p)$_1$], (*iii*) (c)$_1$(Dm103)$_1$ molecules, i.e., the desired cDm103 plasmids, or (*iv*) rarer more complex combinations, such as (c)$_1$(p)$_1$(Dm103)$_1$, which contain one or more copies of ColE1.

Forty-eight of the transformants were screened for the presence of either pSC101 or Dm103 segments by colony hybridization with [^{32}P]cRNA to pDm103 (Fig. 4A), and for the presence of the pSC101 segment by testing for resistance to tetracycline. Of the eight A$^+$ transformants shown in Fig. 4A, six were tetracycline resistant and probably contain (c)$_1$(p)$_1$ plasmids. They were not examined further. The remaining two (indicated by 1 and 2 in Fig. 4A) were tetracycline sensitive, and were assumed to contain cDm103 plasmids; they were designated cDm103/1 and cDm103/2, respectively.

Proof of this assumption was obtained by electron microscopic examination of the plasmids isolated from the two

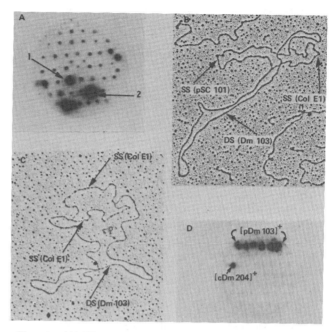

FIG. 4. (A) The screen for cDm103 hybrids. 5 μg of pDm103 DNA and 0.25 μg of ColE1 DNA were cleaved to completion with *Eco*RI endonuclease (in 0.120 ml of 0.1 M Tris·HCl, pH 7.5, 0.01 M MgSO$_4$), heated for 5 min at 65° to inactivate the enzyme and brought to 4°. The DNAs were then incubated at 14° with DNA ligase (14 μg/ml) in 0.1 M Tris·HCl, pH 7.5, as well as a reaction buffer consisting of 0.1 mM DPN, 1 mM EDTA, 10 mM (NH$_4$)$_2$SO$_4$, 10 mM MgSO$_4$ with 100 μg/ml of bovine serum albumin for 120 min in a total volume of 0.140 ml. The solution was then diluted 3-fold with the same reaction buffer and incubated for 36 hr at 14° in the presence of ligase (10 μg/ml). The ligated mixture of *Eco*RI-cleaved pDm103 and ColE1 DNAs (see *text*) was used to transform HB101 to colicin E1 immunity as described previously (11). Each of 48 transformants were transferred by toothpick to a 47-mm filter for colony hybridization (*Materials and Methods*), and to L-agar plates containing 15 μg of tetracycline per ml. 5 × 10^5 cpm of pDm103 [^{32}P]cRNA (2 × 10^7 cpm/μg) were used for the colony hybridization, which after a 6-hr exposure yielded the above autoradiograph. The colonies marked 1 and 2 contain cDm103/1 and cDm103/2 hybrids, respectively. (B) Electron micrograph of a pDm103·cDm103/2 heteroduplex. pDm103 and cDm103/2 circular DNAs were randomly nicked (broken in one strand) by x-rays. The procedures for denaturation and renaturation of these DNAs to form heteroduplexes, for spreading in 40% formamide prior to electron microscopy, and for measuring contour lengths have been described (1). pSC101 (9.2 kb; ref. 1) was used as an internal reference for double-stranded lengths (DS in the figure); no reference was used for single-stranded lengths (SS), as only the ratio of two SS-lengths is used in the analysis (see *text*). (C) Electron micrograph of a cDm103/1·cDm103/2 heteroduplex. The procedures are given in (B) above. See *text* for explanation. (D) The screen for cDm hybrids containing *D. melanogaster* rDNA. Hybrids between *Eco*RI-cut ColE1 and randomly broken Dm segments were formed as indicated in the *text*, and then used to transform HB101 to colicin E1 immunity as in (A) above. 300 independent transformants were transferred to six 47-mm filters, each of which contained six control colonies of HB101 [pDm103] at the top of the pattern. 5 × 10^5 cpm of pDm103 [^{32}P]cRNA (2 × 10^7 cpm/μg) was applied per filter for the colony hybridization. The autoradiograph in the figure resulted from one of the six filters after a 5-hr exposure, and shows one of the five rDNA hybrids (cDm204) identified by this screening procedure.

transformants, and of heteroduplexes formed between pDm103 and cDm103/2, and between cDm103/1 and cDm103/2. The mean lengths ±SD ($n = 18$) of cDm103/1 and cDm103/2 are 23.0 (±1.2) kb and 21.7 (±1.5) kb, respectively. The sum of the lengths of Dm103 (17 kb) and

ColE1 (6 kb; ref. 7) is 23 kb, in reasonable agreement with these values.

A heteroduplex formed between pDm103 and cDm103/2 is shown in Fig. 4B. It consists of a 17 kb double-stranded element whose ends are connected by each of two single-stranded elements that exhibit a length ratio of 1.5. This is the structure expected if cDm103/2 consists of a Dm103 segment inserted at the *Eco*RI cleavage site of ColE1; i.e., the double-stranded element represents the paired Dm103 segments of the two plasmid strands, and the larger and smaller single-stranded elements represent the pSC101 and ColE1 segments respectively (expected length ratio = 9 kb/6 kb = 1.5).

The heteroduplex formed between cDm103/1 and cDm103/2 consists of a 17 kb duplex whose ends are connected by two single-stranded elements of equal length (Fig. 4C). The simplest explanation of this structure is that the Dm103 segments were oppositely inserted into ColE1 during formation of cDm103/1 and cDm103/2. If the Dm103 segments in the single strands of two such oppositely oriented plasmids pair to create a 17 kb duplex element, then the two single-stranded ColE1 segments would contain identical rather than complementary base sequences, and could not pair.

The last experiment consists in screening hundreds of different [cDm]$^+$ colonies for rDNA. The [cDm]$^+$ colonies were obtained by transformation of colicin-sensitive *E. coli* to immunity with a heterogeneous population of cDm molecules constructed from *Eco*RI-cleaved ColE1 and random Dm segments (obtained by shear breakage) by the poly(dA)·poly(dT) joining method (1). These transformants were provided by D. J. Finnegan and G. Rubin. They were individually transferred by toothpick to six 47-mm nitrocellulose filters, each filter containing about 50 independent transformants. Colony hybridization with pDm103 [^{32}P]cRNA indicated no A$^+$ colonies on three filters, 1 A$^+$ colony on two filters, and 3 A$^+$ colonies on one filter. The autoradiograph of one of the two filters containing a single A$^+$ colony, cDm204, is given in Fig. 4D (the top row of A$^+$ colonies on the filter are [pDm103]$^+$ controls). When each of the 5 A$^+$ colonies was retested by repeating this colony hybridization on subclones, such subclones were consistently A$^+$.

Since pSC101 and ColE1 sequences do not interact to give an A$^+$ response, we presume that the cDm plasmids in these 5 A$^+$ colonies contain sequences present in Dm103; i.e., they contain rDNA from *D. melanogaster*. Indeed, D. M. Glover and R. L. White (personal communication) have shown recently that the 28 kb Dm segment in cDm204 contains the same arrangement of '18'-'28'S and spacer sequences as is found in Dm103.

DISCUSSION

In principle, colony hybridization of cloned hybrid plasmids can be used to isolate any gene, or other DNA segment, whose base sequence is represented in an available RNA. We used cRNA to pDm103 for the isolation of cDm plasmids containing rDNA. However, as we have observed that [pDm103]$^+$ colonies give an adequate A$^+$ response with ^{3}H-labeled '18' plus '28'S rRNAs isolated from *D. melanogaster* cell cultures (3), the isolation could have been accomplished with these rRNAs. For rRNA the genes are repeated hundreds of times per genome, and this is the reason that we were able to isolate several hybrids containing rDNA by screening only a few hundred colonies.

By contrast, we calculate that it would be necessary to screen approximately 50,000 hybrid clones to have a 95% chance of finding a hybrid containing a nonrepeated structural gene of typical length from *D. melanogaster*. From the data given in Fig. 3 and assuming 24-hr exposures, we estimate that this would require a total of approximately 4 × 10^6 cpm of [^{32}P]mRNA (specific activity ≥ 4 × 10^5 cpm/μg) applied to about one hundred thirty-five 82-mm filters. Thus a screen of this size is quite feasible. The isolation of nonrepeated genes from larger genomes would, of course, proportionally increase the number of colonies to be screened and hence the total required radioactivity.

An important advantage of colony hybridization is that it facilitates containment of any potentially hazardous hybrid plasmids that may be cloned in such large screening operations. By confining the reproductive state of the hybrid-clones to colonies, the probability of escape is reduced over that for liquid cultures because the number of bacteria per clone is generally smaller and aerosols or accidental spills are less likely. Furthermore the screening operation can be confined to small, controllable areas.

M.G. is grateful to R. T. Schimke for his support and encouragement. We thank D. J. Finnegan, D. M. Glover, G. M. Rubin, and R. L. White for helpful discussions and for providing materials. We are especially thankful to D. M. Glover for his help and advice in isolating and characterizing the cDm103 hybrids. This work was supported by grants from the National Science Foundation (BMS74-21774) and the National Institutes of Health (GM20158, GM14931).

1. Wensink, P. C., Finnegan, D. J., Donelson, J. E. & Hogness, D. S. (1974) *Cell* 3, 315–325.
2. Hogness, D. S., Wensink, P. C., Glover, D. M., White, R. L., Finnegan, D. J. & Donelson, J. E. (1975) in *The Eukaryote Chromosome*, eds. Peacock, W. J. & Brock, R. D. (Australian National University Press, Canberra), in press.
3. Glover, D. M., White, R. L., Finnegan, D. J. & Hogness, D. S. (1975) *Cell* 5, 149–157.
4. Hayes, W. (1965) in *The Genetics of Bacteria and Their Viruses* (John Wiley & Sons, New York), pp. 185–188.
5. Lederberg, E. M. (1960) *Symp. Soc. Gen. Microbiol.* 10, 115–131.
6. Katz, L., Kingsbury, D. K. & Helinski, D. R. (1973) *J. Bacteriol.* 114, 577–591.
7. Hershfield, V., Boyer, H. W., Yanofsky, C., Levett, M. A. & Helinski, D. R. (1974) *Proc. Nat. Acad. Sci. USA* 71, 3455–3459.
8. Berg, D., Barrett, K. & Chamberlin, M. (1971) in *Methods in Enzymology*, eds. Grossman, L. & Moldave, K. (Academic Press, New York), Vol XXI, pp. 506–519.
9. Randerath, k. (1970) *Anal. Biochem.* 34, 188–205.
10. Finnegan, D. J. & Willetts (1972) *Mol. Gen. Genet.* 119, 57–66.
11. Glover, D. (1975) in *New Techniques in Biophysics and Cell Biology*, eds. Pain, R. & Smith, B. (John Wiley & Sons, New York), in press.

Detection of Specific Sequences Among DNA Fragments Separated By Gel Electrophoresis

E.M. Southern

This paper describes a method of transferring fragments of DNA from agarose gels to cellulose nitrate filters. The fragments can then be hybridized to radioactive RNA and hybrids detected by radioautography or fluorography. The method is illustrated by analyses of restriction fragments complementary to ribosomal RNAs from *Escherichia coli* and *Xenopus laevis*, and from several mammals.

1. Introduction

Since Smith and his colleagues (Smith & Wilcox, 1970; Kelly & Smith, 1970) showed that a restriction endonuclease from *Haemophilus influenzae* makes double-stranded breaks at specific sequences in DNA, this enzyme and others with similar properties have been used increasingly for studying the structure of DNA. Fragments produced by the enzymes can be separated with high resolution by electrophoresis in agarose or polyacrylamide gels. For studies of sequences in the DNA that are transcribed into RNA, it would clearly be helpful to have a method of detecting fragments in the gel that are complementary to a given RNA. This can be done by slicing the gel, eluting the DNA and hybridizing to RNA either in solution, or after binding the DNA to filters. The method is time consuming and inevitably leads to some loss in the resolving power of gel electrophoresis. This paper describes a method for transferring fragments of DNA from strips of agarose gel to strips of cellulose nitrate. After hybridization to radioactive RNA, the fragments in the DNA that contain transcribed sequences can be detected as sharp bands by radioautography or fluorography of the cellulose nitrate strip. The method has the advantages that it retains the high resolving power of the gel, it is economical of RNA and cellulose nitrate filters, and several electrophoretograms can be hybridized in one day. The main disadvantage is that fragments of 500 nucleotide pairs or less give low yields of hybrid and such fragments will be under-represented or even missing from the analysis.

2. Materials, Methods and Results

(a) *Restriction endonucleases*

EcoRI prepared according to the method of Yoshimuri (1971) was a gift of K. Murray. HaeIII prepared by a modification of the method of Roberts (unpublished data) was a gift of H. J. Cooke.

SOUTHERN, E.M.
Detection of specific sequences among DNA fragments separated by gel electrophoresis. Reprinted with permission from *J. Mol. Biol.* 98:503-517 (1975).

(b) *Gel electrophoresis*

Gels were cast between glass plates (de Wachter & Fiers, 1971). The plates were separated by Perspex side pieces 3 mm thick and along one edge was placed a "comb" of Perspex, which moulded the sample wells in the gel. The Perspex pieces were sealed to the glass plates with silicone grease and the plates clamped together with Bulldog clips. The assembly was stood with the comb along the lower edge. Agarose solution (Sigma electrophoresis grade agarose) was prepared by dissolving the appropriate weight in boiling electrophoresis buffer (E buffer of Loening, 1969). The solution was cooled to 60 to 70°C and poured into the assembly, where it was allowed to set for at least an hour. The assembly was then inverted, the comb removed and the wells filled with electrophoresis buffer. Samples made 5% with glycerol were loaded from a drawn-out capillary by inserting the tip below the surface and blowing gently. Electrophoresis buffer was layered carefully to fill the remaining space and a filter-paper wick inserted between the glass plates along the top edge. The lower end of the assembly was immersed in a tray of electrophoresis buffer containing the platinum anode, and the paper wick dipped into a similar cathode compartment. Electrophoresis was at 1·0 to 1·5 mA/cm width of gel for a period of about 18 h. Bromophenol blue marker travels about 3/4 the length of the gel under these conditions, but it should be noted that small DNA fragments move ahead of the bromophenol blue, especially in dilute gels. Cylindrical gels were cast in Perspex tubes 9 mm i.d. and either 12 or 24 cm long. These were run at 3 to 5 mA/tube in standard gel electrophoresis equipment.

Dr J. Spiers donated ribosomal DNA that had been purified on actinomycin/caesium chloride gradients from DNA made from the pooled blood of several animals, and also ^{3}H-labelled 18 S and 28 S RNAs prepared from cultured *Xenopus laevis* kidney cells. *Escherichia coli* DNA was prepared by Marmur's (1961) procedure from strain MRE600. ^{32}P-labelled *E. coli* RNA was prepared from cells grown in low phosphate medium with ^{32}Pi at a concentration of 50 μCi/ml and fractionated by electrophoresis on 10% acrylamide gels. ^{32}P-labelled rat DNA was a gift of M. S. Campo. DNA from human placenta was a gift of H. J. Cooke, DNA from rat liver was a gift of A. R. Mitchell, DNA from mouse and rabbit livers were gifts of M. White. Calf thymus DNA was purchased from Sigma Biochemicals. For digestion with restriction endonucleases, the DNAs were dissolved in water to a concentration of approximately 1 mg/ml. One-tenth volume of the appropriate buffer was added and sufficient enzyme to give a complete digestion overnight at 37°C. Enzyme activity was checked on phage λ DNA and digests of this DNA were also used as size markers in gel electrophoresis, using the values given by Thomas & Davis (1975).

(c) *Method of transfer*

This section describes the method finally adopted: preliminary experiments and controls are described in later sections.

After electrophoresis, the gel is immersed for 1 to 2 h in electrophoresis buffer containing ethidium bromide (0·5 μg/ml), and photographed in ultraviolet light (254 nm) with a red filter on the camera. A rule laid alongside the gel aids in matching the photograph of the fluorescence of the DNA to the final radioautograph of the hybrids. Strips to be used for transfer from flat gels are cut from the gel using a flamed blade. The strips should be 0·5 cm to 1 cm wide and normally extend from the origin to the anode end of the gel. The gels used in this laboratory are 3 mm thick, and the length from the origin to the anode end is 18 cm but the method can be adapted to gels with different dimensions and to cylindrical gels. Strips of gel are then transferred to measuring cylinders containing 1·5 M-NaCl, 0·5 M-NaOH for 15 min and this solution is then replaced by 3 M-NaCl, 0·5 M-Tris·HCl (pH 7) and the gel is left for a further 15 min. The depth of liquid in the cylinders should be greater than the length of the gel strips and the cylinders should be inverted from time to time. For cylindrical gels (9 mm diam.), the times required for denaturation and neutralization are 30 and 90 min. Each gel transfer requires:

One piece of thick filter paper 20 cm × 18 cm, soaked in 20 × SSC (SSC is 0·15 M-NaCl, 0·015 M-sodium citrate).

Two pieces of thick filter paper 2 cm × 18 cm soaked in 2 × SSC.

One strip of cellulose nitrate filter (e.g. Millipore 25 HAWP), 2·2 cm × 18 cm, soaked in 2 × SSC. These strips are immersed first by floating them on the surface of the solution; otherwise air is trapped in patches, which leads to uneven transfer.

Three pieces of glass or Perspex, 5 cm × 20 cm and the same thickness as the gel.

Four or five pieces of thick, dry filter paper, 10 cm × 18 cm.

Transfer of the denatured DNA fragments is carried out as follows.

The large filter paper soaked in 20 × SSC is laid on a glass or plastic surface, care being taken to avoid trapping air bubbles below the paper. 20 × SSC is poured on so that the surface is glistening wet. One of the glass or Perspex sheets is laid on top of the wet paper. The gel strip is taken from the neutralizing solution and laid parallel to the glass or Perspex sheet, 2 to 3 mm away from it. The second glass or Perspex sheet is laid 2 to 3 mm away from the other side of the gel (Fig. 1(a)). The cellulose nitrate strip is then laid on top of the gel with its edges resting on the sheets of Perspex or glass, so that it bridges the two air spaces (Fig. 1(b)). The two narrow pieces of filter paper, moistened with 2 × SSC are laid with their edges overlapping the cellulose nitrate strip by about 5 mm (Fig. 1(c)) and the dry filter paper is then placed on top of these (Fig. 1(d)).

For cylindrical gels, the arrangement is similar, but in this case, the Perspex that supports the Millipore filter may be in contact with the gel because an air space is retained over the top of the gel. Several cylindrical gels can be transferred at the same time using the apparatus shown in Fig. 2 and similar arrangements can be used for flat gels.

20 × SSC passes through the gel drawn by the dry filter paper and carries the DNA, which becomes trapped in the cellulose nitrate. The minimum time required for complete transfer has not been measured: it depends on the size of the fragments and probably also depends on the gel concentration. A period of 3 h is enough to transfer completely all HaeIII fragments of *E. coli* DNA from 2% agarose gels 3 mm thick. But even after 20 h, transfer of large EcoRI fragments of mouse DNA from 9 mm diam. cylindrical gels is not complete. DNA remaining in the gel can be seen by the fluorescence of the ethidium bromide, which is not completely removed during treatment of the gel. During the period of the transfer, it is necessary occasionally to add more 20 × SSC to the bottom sheet of filter paper. If the paper dries too much, the gel shrinks against the cellulose nitrate strip and liquid contact is broken. The paper may be flooded, but care must be taken that liquid does not fill the air spaces between the gel and the side-pieces and soak the paper, bypassing the gel. It may be found convenient to leave the cellulose nitrate in position overnight: if the supply of

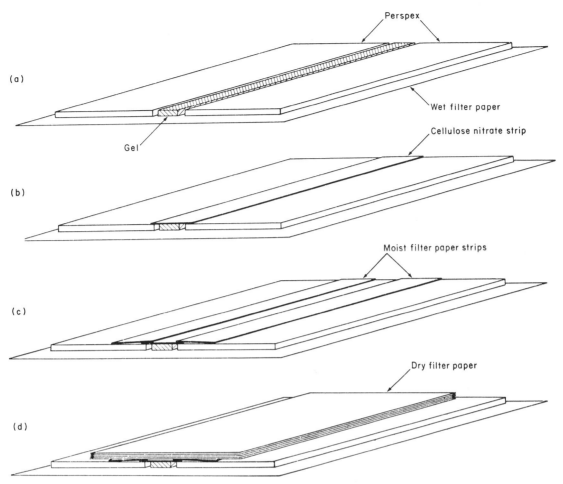

Fig. 1. Steps in the procedure for transferring DNA from agarose gels to cellulose nitrate strips.

$20 \times$ SSC has dried up it will be found that the gel has shrunk against the cellulose nitrate, but this does not impair the transfer. At the end of the transfer period the cellulose nitrate strip is lifted carefully so that the gel remains attached to its underside. It is turned over and the outline of the gel marked in pencil by a series of dots. The gel is peeled off the cellulose nitrate, the area of contact cut out with a flamed blade, and immersed in $2 \times$ SSC for 10 to 20 min. The strip is then baked in a vacuum oven at 80°C for 2 h.

(d) *Hybridization*

Radioactive RNAs are usually available in small quantities only and it is important to keep the volume of the solution used for hybridization as small as possible so that the RNA has a reasonable concentration. Two procedures can be used for hybridizing the cellulose nitrate strips after transferring the restriction fragments.

The procedure that uses the smallest volume is carried out by moistening the strip in hybridization mixture and then immersing it in paraffin oil. A drop of RNA solution (0.3 ml for a strip 1 cm $\times$ 18 cm) is placed on a plastic sheet. One end of the

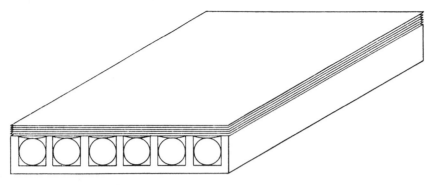

FIG. 2. Apparatus for transferring DNA from a number of cylindrical gels.

The apparatus is constructed of Perspex. The uprights which separate the gels and support the sheet of cellulose nitrate should be about 0·5 mm higher than the diameter of the gels, so that the cellulose nitrate sheet dips down to touch the gel. Thus an air gap is left between the cellulose nitrate sheet and the filter paper, above the line of contact between the gel and cellulose nitrate sheet. The apparatus is laid in a shallow tray containing $20 \times$ SSC and the gels are then inserted into the troughs, care being taken to avoid trapping air bubbles beneath the gel. The cellulose nitrate sheet, wet with $2 \times$ SSC, is laid over the gels and one piece of wet filter paper is laid over this. A stack of dry filter paper is then placed over the whole assembly. If necessary, a glass plate can be used to weigh down the filter papers. The depth of $20 \times$ SSC in the tray should be enough to cover the lower part of the gels, but not so much that the air space between the Perspex and the cellulose nitrate becomes flooded.

cellulose nitrate strip is floated on the drop and when liquid is seen to soak through, the strip is drawn slowly over the surface of the drop. When it is completely wetted from one side, it is turned over and any remaining liquid is used to wet the other side. The strip is then immersed in paraffin oil saturated with the hybridization solution at the hybridization temperature. It should be borne in mind that baking the strip in $2 \times$ SSC introduces salt, which must be taken into account when deciding on a solvent for the RNA if this method of hybridization is used. For example, if hybridization is to be carried out in $6 \times$ SSC the RNA should be dissolved in $4 \times$ SSC. Though this method can give good results (see Plate I) it often leads to high and uneven background. Kourilsky *et al.* (1974) found that this problem is removed if the hybridization is carried out in $2 \times$ SSC, 40% formamide at 40°C. I have not tried this method, because this solvent removed DNA from the filters (see later section). It may well be the best method for hybridization to large fragments. I have found it convenient to carry out the hybridization in a vessel designed to hold the strip in a small volume of liquid.

The vessel (Fig. 3), which is easily made from Perspex, has internal dimensions of 0·8 mm deep by 2 cm high and about 1 cm longer than the strip to be hybridized. The vessel is filled with the solvent to be used for hybridization and the strip is fed in through the narrow opening in the top. The solvent is then drained off and the RNA solution introduced. Around 1 ml of solution is needed for a strip 1 cm × 18 cm. The wide sheets of cellulose nitrate used for transferring several gels (e.g. using the apparatus shown in Fig. 2) are too wide to be hybridized in this type of vessel. They can be hybridized in a small volume by wrapping them around a cylinder of Perspex, which is then inserted into a close-fitting tube. In this way, it is possible to hybridize a sheet 24 cm × 8 cm with about 4 ml of solution. If hybridization is carried out in a water-bath, it is not necessary to seal the top of the vessel provided the water-bath

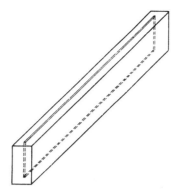

FIG. 3. Vessel used for hybridization of narrow strips.

itself is covered. The liquid in the vessel evaporates very slowly and can be replenished by small additions of water. A further advantage of this method of hybridization is that the RNA can be recovered and used again.

The period allowed for hybridization depends on the RNA concentration, its sequence complexity, its purity, and on the conditions of hybridization (see for example Bishop, 1972). After the appropriate period, strips are removed from the solution or paraffin oil, blotted between sheets of filter paper and washed, with stirring, for 20 to 30 min in a large volume of the hybridization solvent at the hybridization temperature. If the background is high, they may then be treated with a solution of RNAase A (20 μg/ml in 2 × SSC for 30 min at 20°C). After a final rinse in 2 × SSC they are dried in air.

So far the method has been tested with ^{32}P, ^{3}H, ^{35}S and ^{125}I-labelled RNAs. [^{32}P]RNAs have been detected by radioautography. For this the cellulose nitrate strips are laid on X-ray film and flattened against it with light pressure. ^{3}H, ^{125}I, ^{35}S and ^{14}C may be detected by fluorography. The cellulose nitrate strip is dipped through a solution of PPO in toluene (20%, w/v) dried in air, laid against X-ray film (Kodak RP-Royal Xomat) and kept at −70°C.

(e) Completeness of transfer and retention of DNA

Preliminary experiments showed that loading of DNA on to cellulose nitrate filters in 6 × SSC, conditions widely used in hybridization work, did not give complete retention of small fragments and a systematic study was made of the effect of salt concentration on retention. ^{3}H-labelled X. laevis DNA was sonicated to a single-strand molecular weight of 10^{4} and denatured by boiling in 0·1 × SSC. Samples were made up to various salt concentrations and 0·1-ml portions of these solutions were pipetted on to cellulose nitrate filters, previously moistened with 2 × SSC, which were resting on glass-fibre filters. The solution that passed through the cellulose nitrate filter was thus collected in the glass-fibre filter. Both filters were then immersed in 5% trichloroacetic acid for 10 min, dried for 30 min in a vacuum oven at 80°C, and counted. It can be seen (Fig. 4) that the fraction of DNA retained by the cellulose nitrate increases with the salt concentration, and at concentrations above 10 × SSC the DNA is almost completely retained.

Losses of DNA at various stages of the transfer procedure were measured using ^{32}P-labelled E. coli DNA. The DNA was digested with EcoRI to give fragments in

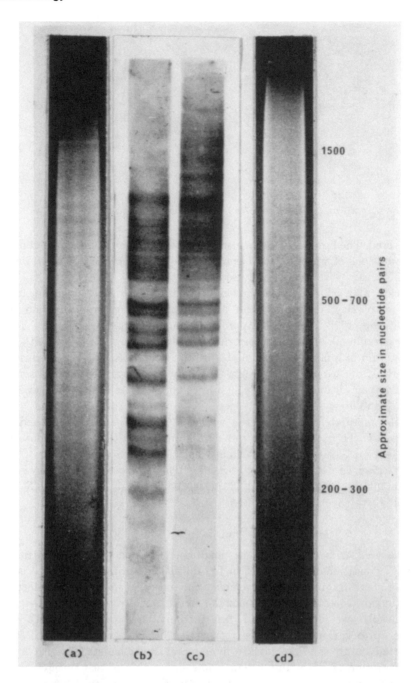

PLATE I. HaeIII digest of *E. coli* MRE600 DNA analyzed by electrophoresis on 2% agarose gel. DNA was then transferred to cellulose nitrate and hybridized with ^{32}P-labelled, high molecular weight RNA. (a) and (d) Photographs of ethidium bromide fluorescence. (b) and (c) Radioautographs of hybrids.

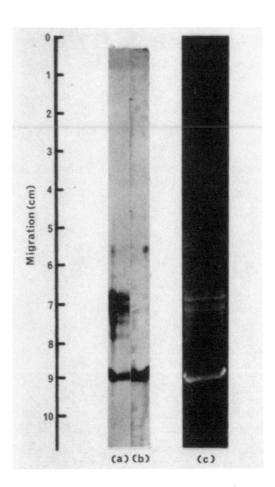

PLATE II. EcoRI digest of purified *X. laevis* ribosomal DNA analyzed by electrophoresis on 1% agarose gel. DNA was transferred to a cellulose nitrate strip, which was then cut longitudinally in two. The left-hand side was hybridized to 18 S RNA and the right-hand side to 28 S RNA (spec. act. of RNAs, $1·5 \times 10^6$ c.p.m. per μg). Hybridization was done in $1 \times$ SSC at 65°C using the vessel shown in Fig. 3. A large excess of cold 28 S RNA was added to the labelled 18 S RNA to compete out any 28 S contamination. After hybridization, the strips were washed in $1 \times$ SSC at 65°C for 1·5 h, and dried. They were then dipped through a solution of PPO in toluene (20%, w/v) dried in air and placed against Kodak RP Royal X-ray film at -70°C for 2 months. Photograph of ethidium bromide fluorescence (c). Fluorograph of 18 S hybrids (a). Fluorograph of 28 S hybrids (b).

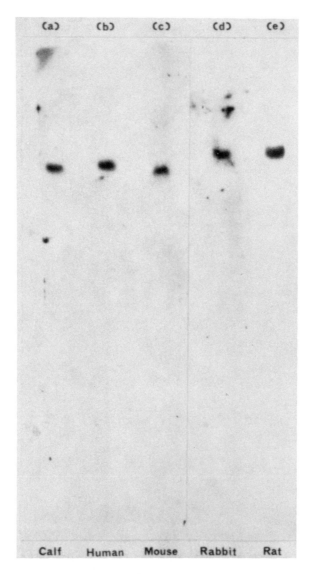

PLATE III. EcoRI digests of five mammalian DNAs, hybridized to 28 S RNA. Calf (a), human (b), mouse (c), rabbit (d) and rat (e) DNAs were digested to completion with EcoRI and separated by electrophoresis on 1% agarose gels (9mm × 12 cm, approx. 40 μg DNA per tube, 3 mA/tube for 16 h). The gels were pretreated as usual and the DNA fragments transferred to a single sheet of cellulose nitrate filter (12 cm × 8 cm) using the apparatus shown in Fig. 2. The top end of each gel was carefully aligned with one edge of the cellulose nitrate sheet. After 20 h, traces of DNA could still be seen, by ethidium bromide fluorescence, in the high molecular weight region of the gel. The filter was hybridized with 28 S RNA and radioautographed as described in the legend to Fig. 8.

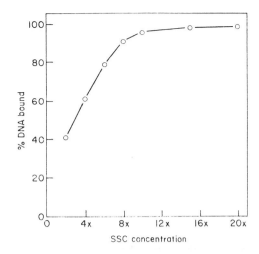

F_IG. 4. Effect of salt concentration on efficiency of binding sonicated DNA to cellulose nitrate filters.

the large size range and with HaeIII to give small fragments. The fragments were then separated on a flat 1% agarose gel and transferred in the usual way. The solutions, the gel and the cellulose nitrate strip were counted. It can be seen (Table 1) that, whereas a small proportion of the DNA is leached out into the solutions during denaturation and neutralization, only traces remain in the gel after transfer.

T_ABLE 1

Losses of DNA at stages of the procedure

	EcoRI fragments	HaeIII fragments
	DNA lost (%)	
Denaturing solution	2·1	4·8
Neutralizing solution	1·3	4·4
Remaining in gel after transfer	0·21	0·31

Two samples of *E. coli* DNA (0·1 μg; spec. act. approx. 10^6 c.p.m. per μg) were digested with EcoRI and HaeIII. The fragments were separated by electrophoresis on 1% gels in 1-cm wide slots, and then transferred to cellulose nitrate strips as described in Materials and Methods. The transfer was left overnight. The radioactivity leached out of the gel by the denaturing and neutralizing solutions, that remaining in the gel, and that which had been trapped on the cellulose nitrate filter were measured in a liquid scintillation counter (Cerenkov radiation).

(f) *Effect of DNA size on yield of hybrid*

Melli & Bishop (1970) have shown that hybridization by the filter method gives low yields with low molecular weight DNA. Their results were obtained using a single set of hybridization conditions and it seemed possible that losses might be reduced by using high salt concentrations. The effect of salt concentration on loss of

DNA from the filters was examined by loading filters with radioactive *X. laevis* DNA, single-strand molecular weight about 10^4, and incubating them in various salt solutions at different temperatures. Increasing the salt concentration does improve the retention of the DNA at any given temperature (Table 2) but the gain does not appear to be useful, because with increasing salt concentration it is necessary to use higher temperatures for hybridization, and this cancels the advantage of the high salt concentration. For example, the loss in $2 \times$ SSC at 65°C is the same as that in $6 \times$ SSC at 80°C and these are both typical hybridization conditions. Further experiments showed that it is disadvantageous to perform hybridization at high salt concentrations, below the optimum temperature. The optimum temperature for rate of hybridization of *X. laevis* 28 S RNA is around 80°C in $6 \times$ SSC but the rate at 70°C is still appreciable (Fig. 5). Below 70°C the rate falls rapidly. 28 S RNA was hybridized

TABLE 2

Effects of temperature and solvent on retention of sonicated DNA on cellulose nitrate filters

Solvent	50°C	Temperature		90°C
		65°C 80°C		
		DNA retained (%)		
$2 \times$ SSC		77	62	48
$6 \times$ SSC		97	76	56
$10 \times$ SSC		95	83	73
$20 \times$ SSC		97	88	81
$6 \times$ SSC in 50% formamide	58	50		

[3]H-labelled *X. laevis* DNA (spec. act. approx. 5×10^5 c.p.m. per μg) was dissolved in ice-cold $0 \cdot 1 \times$ SSC and sonicated in six 15-s bursts. Between each treatment the solution was cooled in ice for 1 min. The solution was boiled for 5 min, made to $20 \times$ SSC and cooled. Samples of this solution were pipetted on to 13-mm circles of cellulose nitrate, which were then washed in $2 \times$ SSC at room temperature. Approximately 650 c.p.m. were loaded on each filter, and there was no loss caused by washing in $2 \times$ SSC. The filters were dried, baked at 80°C for 2 h in a vacuum oven and immersed in 10 ml of the solvent equilibrated at the temperature used for incubation. After 90 min, the filters were removed, washed in $2 \times$ SSC at room temperature, dried under vacuum and counted in a liquid scintillation counter.

to high molecular weight and sonicated DNA in $6 \times$ SSC at 70 and 80°C (Fig. 6). As expected, the rate of hybridization at 70°C was lower than the rate at 80°C, but against expectation, both the rate and the final extent of hybridization were lower at the lower temperature, for the sonicated but not for the high molecular weight DNA. This result was unexpected because Melli & Bishop did not find an effect of DNA size on the rate of hybridization. They suggested that the decrease in yield for low molecular weight DNA is due to a loss of hybrid from the filter and it would be expected that such losses would increase with temperature. The lower yield for low molecular weight DNA at low temperature remains unexplained, but shows that there is no advantage to be gained in using high salt concentrations and low temperatures to retain small fragments of DNA during hybridization reactions. The advantage of using $6 \times$ SSC at optimum temperature is that the rate is greatly increased over the rate with, say, $2 \times$ SSC. A disadvantage is that the background of RNA that sticks to filters that have no DNA, increases with increasing salt concentration.

(g) *Methods of detecting and measuring hybrids: advantages of film detection*

Radioactive RNA may be detected and measured either by radioautography (or fluorography for weak β-emitters) or by cutting the strip into pieces, which can be counted in a scintillation counter. Film detection methods have the advantages over

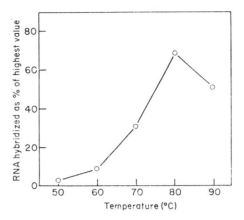

Fɪɢ. 5. Temperature dependence of hybridization of 28 S rRNA to *X. laevis* DNA.

X. laevis DNA was loaded on cellulose nitrate filters (17 μg DNA/13-mm diameter disc), which were pretreated as usual for hybridization. ³H-labelled 28 S RNA from *X. laevis* kidney cells (spec. act. 1.5×10^6 c.p.m./μg) was dissolved in $6 \times$ SSC (0.28 μg/ml) and warmed to the temperature used for hybridization. Two filters loaded with DNA and 2 blank filters were introduced into the solutions and left for 30 min. They were washed in 2 l of $2 \times$ SSC at room temperature, treated with 200 ml of RNAase A (20 μg/ml in $2 \times$ SSC) at room temperature for 20 min, washed in 200 ml of $2 \times$ SSC for 10 min, dried under vacuum and counted. Hybridization is expressed as a percentage of that obtained after 5 h at 80°C.

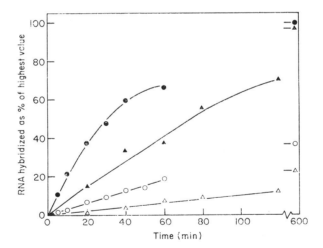

Fɪɢ. 6. Time course of hybridization of 28 S RNA to sonicated and high molecular weight DNA at 70 and 80°C.

Filters were loaded as described in the legend to Fig. 5. Two sets were loaded: one with high molecular weight DNA and one with DNA sonicated as described in the legend to Table 2. Hybridization and subsequent treatment of the filters was carried out as described in the legend to Fig. 6 and filters removed at the times indicated. $6 \times$ SSC at 80°C, high molecular weight DNA (●); $6 \times$ SSC at 70°C, high molecular weight DNA (▲): $6 \times$ SSC, 80°C sonicated DNA (○): $6 \times$ SSC at 70°C, sonicated DNA (△).

34

counting that they are more sensitive, give higher resolution, and can reveal artifacts not seen by counting.

The high sensitivity is illustrated by the analysis of *E. coli* rDNA (Plate I(b)). None of the bands that is clearly visible in the radioautograph contained more than 10 c.p.m. The strip of cellulose nitrate was cut into 150, 1-mm pieces and the pieces counted in a liquid scintillation counter. None of the pieces gave counts more than twice background and none of the features visible in the radioautograph was discernible from the counts. Around 100 c.p.m. of ^{32}P in a single band 1 cm wide can be detected with an overnight exposure. The radioautograph shown in Plate I was exposed for 1 week. Fluorography of ^{3}H is not so sensitive; about 3000 d.p.m. in a 1-cm band are needed to give a visible exposure overnight. The fluorograph shown in Plate II was exposed for 2 months.

The greater resolution of film detection is illustrated by a comparison of Plate II with Figure 7(c). Plate II is a fluorograph of the strip and Figure 7(c) shows the pattern of counts obtained by cutting the strip into 1-mm pieces. Many of the bands seen in the fluorograph are not discernible in the pattern of counts (compare also the tracing of the fluorograph (Fig. 7(b)) with (c)).

For ionizing radiation, blackening of the X-ray film is proportional to the amount of incident radiation, up to the limit where a high proportion of silver grains are exposed. The relative amount of radioactivity in bands can therefore be compared by tracing radioautographs in a densitometer and comparing peak areas. However, like all other photosensitivie materials, X-ray films suffer from "reciprocity failure" at low intensities of illumination by non-ionizing radiation and it is likely that bands which contain only a few counts of ^{3}H will not be detected by fluorography even after long exposures. I have not determined the lower limit of detection. Bonner & Laskey (1974) found that 500 d.p.m. of ^{3}H in a band 1 cm $\times$ 1 mm could be detected in one week and in my own experience, less than 20 d.p.m. can be detected with longer exposure. Reciprocity failure could affect quantitation of fluorographs by densitometry but comparison of Figure 7(b) and (c) suggests that the response of the film is linear within the limits of this experiment. Clearly, quantitation of ^{32}P by densitometry can be accurate and more sensitive than counting, but film response to ^{3}H may not be linear for low amounts.

An additional advantage of film detection is that non-specific binding of RNA to the cellulose nitrate is more easily distinguished from bands of hybrid. Plate III illustrates this point. In this radioautograph, non-specific binding can be seen as dots and streaks with an appearance clearly different from that of a band. Had this. strip been analysed by counting, non-specific binding would not have been distinguishable from the hybrids.

(h) *Analysis of ribosomal DNA in* X. laevis

A total of 0·6 μg of purified *X. laevis* rDNA was digested with EcoRI and the fragments separated by electrophoresis in 1% agarose gels (Plate II(c)). The pattern of fragments is similar to that described by Wellauer *et al.* (1974). They compared the secondary structures of the denatured DNA fragments with those of the ribosomal RNAs and showed that the fastest running fragment (M_r approx. 3×10^6) contained most of the DNA coding for 28 S RNA, all of the transcribed spacer, and a small portion of the DNA coding for 18 S RNA. The larger fragments (M_r 4 to 6×10^6) contained most of the DNA coding for 18 S RNA, all of the non-transcribed

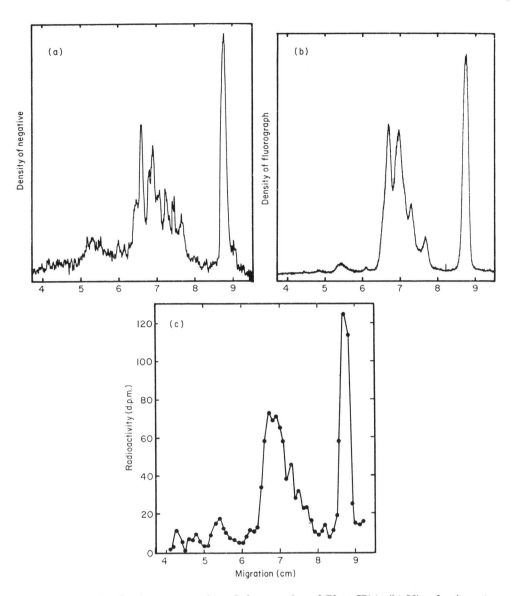

FIG. 7. (a) Microdensitometer tracing of the negative of Plate II(c). (b) Microdensitometer tracing of Plate II(a). (c) Distribution of counts in the Millipore strip which on fluorography gave Plate II(a). The strip was cut into 1 mm pieces, which were counted in a liquid scintillation counter at an efficiency of 40%.

spacer, and a small portion of the DNA coding for 28 S RNA. Different lengths of non-transcribed spacer DNA accounted for the variation in size of the longer fragments. The digest shown in Plate II(c) was transferred to cellulose nitrate as described previously. The strip was cut longitudinally into 2 parts and 1 part was hybridized with 18 S RNA and the other with 28 S RNA. Hybrids were detected by fluorography of the ^{3}H-labelled RNA (Plate II(a) and (b)). Comparison of Plate II(a) and (c) shows that the resolution of the fine bands containing the 18 S coding sequence is not as high in the fluorograph as it is in the photograph of the gel. Whereas 9 bands can be distinguished in the photograph, only 7 can be distinguished with confidence in the fluorograph. From this analysis it is possible to locate the EcoRI site within the DNA coding for 18 S RNA. As Wellauer *et al.* (1974) showed, 1 of the 2 breaks in the rDNA occurs towards one end of the 18 S region and the other is close to the distal end of the 28 S region. The 3×10^6 mol. wt fragment accounts for virtually all of the hybridization to 28 S RNA and for about 30% of the hybridization to the 18 S RNA (27% measured from the tracing of the fluorograph (Fig. 7(b)) and 31% from the counts). Only traces of 28 S RNA hybridize to the heterogeneous collection of fragments with molecular weights between 4 and 6×10^6, whereas about 70% of the 18 S hybridization is accounted for in these fragments. Thus the break in the 28 S region of the DNA is very close to the end of the coding sequence and the break in the 18 S region is about one-third of the way into the coding sequence.

(i) *Analysis of mouse and rabbit ribosomal DNAs: evidence for long, non-transcribed spacer DNA*

An EcoRI digest of total mouse DNA was separated by electrophoresis on cylindrical 1% agarose gels and transferred to strips of cellulose nitrate paper. One strip was hybridized to 18 S RNA and another to 28 S RNA prepared from rat myoblasts labelled with ^{32}P. The 28 S hybrids showed a strong, sharp band at the position of about $5 \cdot 2 \times 10^6$ daltons and a very faint, broad band in the region around 14×10^6 daltons (Fig. 8(b)). The 18 S hybrids showed corresponding bands but in this case the slower moving, broad band was relatively more intense (Fig. 8(a)). From this information, a partial structure can be derived for the ribosomal DNA in mouse. Assuming that the ribosomal genes are tandemly linked, it is clear that EcoRI makes at least 2 breaks in the sequence; one in the 18 S and one in the 28 S region. Transcription of ribosomal genes in mammals produces a precursor RNA corresponding to a DNA mol. wt of about 6×10^6, and it follows that the EcoRI fragment of about $5 \cdot 2 \times 10^6$, which contains both 28 S and 18 S sequences, must also encompass much of the transcribed spacer. The heterogeneous fragments with a mol. wt of 14×10^6 must contain a long stretch of non-transcribed spacer, and may contain some of the transcribed spacer too.

A similar analysis was carried out with rabbit DNA and gave similar results, although the size of the fragments was different from the corresponding fragments from mouse DNA. The band containing most of the 28 S sequence was larger (M_r approx. 6×10^6), whereas that containing most of the 18 S sequence was smaller (M_r approx. 12×10^6) and more homogeneous than the corresponding fragment in the mouse. The structures of mouse and rabbit ribosomal DNAs are thus rather similar to that of *X. laevis* but with longer spacer regions. The overall length of the unit in mouse is at least twice as long as that in *X. laevis*.

(j) *EcoRI sites in the rDNA of five mammals*

The analyses described above, taken with those of Wellauer *et al.* (1974) suggest that the two EcoRI sites in the ribosomal genes have been conserved since the amphibians and mammals diverged. In this case it would be expected that all

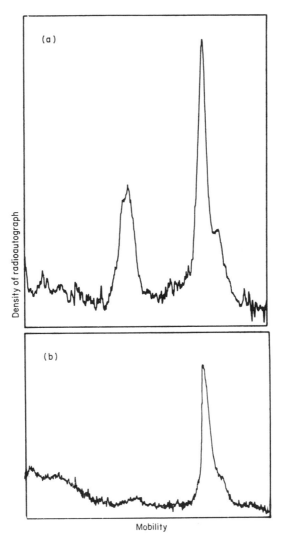

FIG. 8. EcoRI digest of mouse DNA hybridized to 18 S and 28 S RNA.

Total mouse DNA was digested to completion with EcoRI. The digest was separated by electrophoresis on 1% cylindrical agarose gels (9 mm × 24 cm, 5 mA/tube for 20 h, 40 μg of DNA/gel).

The gels were stained, photographed, and the DNA transferred to cellulose nitrate as described in Materials and Methods. One gel was hybridized to ^{32}P-labelled 18 S RNA and another to 28 S RNA. The RNA concentration was 0·1 μg/ml in 6×SSC and hybridization was carried out at 80°C for 4 h. The filters were then washed in 2×SSC (4 l) at 60°C for 30 min, dried and radioautographed using Kodak Blue Brand X-ray film.

(a) Densitometer tracing of the 18 S hybrids. (b) Densitometer tracing of the 28 S hybrids.

mammalian rDNAs would have equivalent EcoRI sites. Total DNAs from calf thymus, human placenta, and from livers of mouse, rabbit and rat were digested with EcoRI and the fragments separated by electrophoresis on cylindrical gels. The fragments were then transferred to a single sheet of cellulose nitrate filter and hybridized with ^{32}P-labelled rat 28 S RNA. All 5 DNAs showed a strong band in the radioautograph of the sheet. Each band was in the mol. wt region of 5 to 6×10^6, but there were small differences in their mobilities (Table 3). This result suggests that the two EcoRI sites have indeed been conserved in the rDNA of the mammals. The different fragment size

TABLE 3

Size of EcoRI fragments that hybridize to ribosomal RNAs

Species	Size of RI fragment bearing 28 S sequences $(\times 10^{-6})$	Size of fragments bearing 18 S sequences $(\times 10^{-6})$
Calf	5·7	
Human	5·7	
Mouse	5·2	5·2 and approx. 14
Rabbit	6·0	6·0 and approx. 12
Rat	6·0	
X. laevis	3·0	3·0 and 4 to 6

Sizes were estimated from mobilities in 1% agarose gels by comparison with EcoRI fragments of λ-phage DNA. The sizes of the large fragments from mouse and rabbit DNAs hybridizing to 18 S RNA are approximate estimates because there was only one marker in this region of the gel and in this region large differences in size result in small mobility differences.

can readily be accounted for by differences in the size of the transcribed spacer between 28 S and 18 S regions. Different sizes for the ribosomal RNA precursor have been reported for HeLa cells and mouse L-cells (Grierson *et al.*, 1970).

3. Conclusion

The method described here provides a simple way of detecting DNA fragments that are complementary to RNAs, after the DNA frqgments have been separated by gel electrophoresis. Transfer of the DNA from the gel to the cellulose nitrate filter is almost complete for a wide range of fragment sizes. However, large fragments ($M_r > 10^7$) diffuse rather slowly and small fragments hybridize inefficiently. These factors should be taken into account when the method is used for quantitative work.

Much of this work was carried out when I was on leave of absence in the Institut fur Molekularbiologie II, Zurich University, supported in part by the Swiss Science Foundation (grant no. 3.8630.725R) and I am grateful to Professor M. L. Birnstiel for hospitality during this period.

REFERENCES

Bishop, J. O. (1972). In *Karolinska Symposia on Research Methods in Reproductive Endo-crinology* (DiczFalnsy, E. & DiczFalnsy, A., eds), pp. 247–273, Karolinska Institute, Stockholm.
Bonner, M. & Laskey, R. A. (1974). *Eur. J. Biochem.* **46**, 83–88.
Grierson, D., Rogers, M. E., Sartirana, M. L. & Loening, U. E. (1970). *Cold Spring Harbor Symp. Quant. Biol.* **35**, 589–598.
Kelly, T. J. & Smith, H. O. (1970). *J. Mol. Biol.* **51**, 393–409.
Kourilsky, Ph., Mercereau, O. & Tremblay, G. (1974). *Biochimie*, **56**, 1215–1221.
Loening, U. E. (1969). *Biochem. J.* **113**, 131–138.

Marmur, J. (1961). *J. Mol. Biol.* **3**, 208–218.

Melli, M. & Bishop, J. O. (1970). *Biochem. J.* **120**, 225–235.

Smith, H. O. & Wilcox, K. (1970). *J. Mol. Biol.* **51**, 379–391.

Thomas, M. & Davis, R. W. (1975). *J. Mol. Biol.* **91**, 315–328.

de Wachter, R. & Fiers, W. (1971). In *Methods in Enzymology* (Grossman, L. & Moldave, K., eds), vol. 21D, pp. 167–178, Academic Press Inc., New York and London.

Wellauer, P. K., Reeder, R. H., Carroll, D., Brown, D. D., Deutch, A., Higashinakagawa, T. & Dawid, I. B. (1974). *Proc. Nat. Acad. Sci., U.S.A.* **71**, 2823–2827.

Yoshimuri, R. N. (1971). Doctoral Thesis, University of California at San Francisco.

Use of Synthetic Oligonucleotides as Hybridization Probes: Isolation of Cloned cDNA Sequences for Human β_2-microglobulin

S.V. Suggs, R.B. Wallace, T. Hirose, E.H. Kawashima and K. Itakura

ABSTRACT We have synthesized two sets of 15-base-long oligodeoxyribonucleotides corresponding to all possible coding sequences for a small portion of human β_2-microglobulin. Labeled oligonucleotides were used as hybridization probes to screen bacterial clones containing cDNA sequences primed with oligo(dT) and inserted into the plasmid vector pBR322. One β_2-microglobulin cDNA clone was detected in the 535 bacterial plasmid clones that were screened. The clone has been characterized by blotting and nucleotide sequence analysis. The cloned β_2-microglobulin sequence contains 217 base pairs of the 3' untranslated region of the mRNA and 328 base pairs (97%) of the coding region.

The hybridization properties of oligodeoxyribonucleotides have been characterized by a number of techniques (1–4). Under appropriate conditions, oligonucleotides hybridize to specific sites in DNA (4, 5). Furthermore, perfectly base paired oligonucleotide duplexes can be discriminated from duplexes containing a single mismatched base pair (4–6). We have taken advantage of the hybridization properties of oligonucleotides in developing a method for the isolation of specific cloned DNA sequences (5). Our general approach is to chemically synthesize a mixture of oligonucleotides that represent all possible codon combinations for a small portion of the amino acid sequence of a given protein. Within this mixture must be one sequence complementary to the DNA coding for that part of the protein. This complementary oligonucleotide will form a perfectly base paired duplex with the DNA from the coding region for the protein, whereas the other oligonucleotides in the mixture will form mismatched duplexes. Under stringent hybridization conditions only the perfectly matched duplex will form, allowing the use of the mixture of oligonucleotides as a specific hybridization probe. Mixed sequence oligonucleotide probes should allow isolation of cloned DNA sequences for any protein for which the amino acid sequence is known.

We have applied this method to the isolation of cloned cDNA sequences for human β_2-microglobulin (β_2m). β_2m is a small protein (molecular weight 11,800) that was isolated from urine (7). Subsequently, β_2m was found associated with cell surface antigens of the major histocompatibility locus (8, 9). The exact function of β_2m is unclear, although recent evidence suggests that the molecule may stabilize the tertiary structure of associated proteins (10). The amino acid sequence has been determined for β_2m from four species, including human (11). We have used the amino acid sequence to design probes for the isolation of a cloned cDNA for human β_2m.

MATERIALS AND METHODS

General Methods. Plasmid DNA was isolated by a cleared lysate procedure (12) and purified by chromatography on Bio-

Table 1. Oligonucleotide probes for the isolation of β_2m

Amino acid sequence		95 Trp	96 Asp	97 Arg	98 Asp	99 Met	
Possible codons	5'	UGG	GA$_C^U$	AG$_G^A$ CGN	GA$_C^U$	AUG	3'
Probe β_2mI	3'	ACC	CT$_G^A$	TC$_C^T$	CT$_G^A$	TAC	5'
Probe β2mII	3'	ACC	CT$_G^A$	GCN	CT$_G^A$	TAC	5'
		75 Lys	76 Asp	77 Glu	78 Tyr		
Amino acid sequence							
Possible codons	5'	AA$_G^A$	GA$_C^U$	GA$_G^A$	UA$_C^U$	3'	
Probe β_2mIII	3'	TT$_C^T$	CT$_G^A$	CT$_C^T$	AT	5'	

N = A, C, G, or T (U).

Gel A-50m (Bio-Rad) followed by ethidium bromide/CsCl equilibrium sedimentation. Restriction enzymes were purchased from Bethesda Research Laboratories (Rockville, MD). Restriction enzyme reactions were performed at 37°C in 60 mM NaCl/10 mM Tris·HCl (pH 7.6)/7 mM MgCl$_2$/6 mM 2-mercaptoethanol. Agarose and polyacrylamide gels (acrylamide to methylenebisacrylamide ratio, 20:1) were run in Tris/borate electrophoresis buffer (13). Electroelution of DNA fragments from polyacrylamide gels has been described (14). Transfer of DNA fragments from agarose (SeaKem; Marine Colloids, Rockland, ME) gels to nitrocellulose was by the standard Southern procedure (15). Prehybridization and hybridization conditions have been described (5). Recombinant DNA was handled in accordance with National Institutes of Health Guidelines.

Simultaneous Synthesis of Oligodeoxyribonucleotides of Mixed Sequence. The oligonucleotide mixtures shown in Table 1 were synthesized by the strategy described previously (5). Probes β2mI and β2mII were synthesized by the triester method in solution (16). β2mIII was synthesized by a solid-phase method (17). All oligonucleotides were purified by high-performance liquid chromatography on an ion-exchange (Du Pont Permaphase AAX) column (5).

Labeling of Oligonucleotides. Oligonucleotides were labeled at the 5' end by transfer of ^{32}P from [γ-^{32}P]ATP, using bacteriophage T4 polynucleotide kinase (New England Nuclear) as described (4). [γ-^{32}P]ATP was synthesized by the method of Walseth and Johnson (18).

Synthesis of Double-Stranded cDNA. The source of RNA used to prepare cDNA was a human lymphoblastoid cell line, Raji [obtained from S. Ohno (19)]. Cytoplasmic RNA was isolated and fractionated on an oligo(dT)-cellulose (Collaborative Research, Waltham, MA) column as described (20). Double-

Abbreviation: β_2m, β_2-microglobulin.
* This paper is no. 3 in a series. Paper no. 2 is ref. 5.
† Present address: Keio University, Tokyo, Japan.
‡ Present address: Biogen S.A., Geneva, Switzerland.
§ To whom reprint requests should be addressed.

stranded cDNA was synthesized by using successive reactions with oligo(dT) (P-L Biochemicals) plus reverse transcriptase (RNA-dependent DNA polymerase; a gift from J. Beard), *Escherichia coli* DNA polymerase I Klenow fragment A (Boehringer Mannheim), and S1 nuclease (Sigma) as described by Goeddel *et al.* (21). The 3′ termini of the double-stranded cDNA were extended with dC residues by using terminal deoxynucleotidyltransferase (Bethesda Research Laboratories) as described (22).

Isolation of cDNA Clones. The C-tailed cDNA was subjected to electrophoresis on a 5% polyacrylamide gel. The fragments 500–800 base pairs in length were recovered by electroelution. Fifteen nanograms of size-selected cDNA was annealed with 100 ng of pBR322 that had been cleaved with restriction endonuclease *Pst* I and tailed with dG residues (23). The mixture was heated to 65°C for 3 min and then allowed to gradually cool from 43°C to 22°C over the course of 20 hr. The annealed mixture was used to transform *E. coli* strain MC1061 (24) by the Kushner procedure (25).

Screening of cDNA Clones. After transformation, recombinant clones were selected on L agar plates containing 25 μg of tetracycline per ml (23). Individual clones were picked and put on fresh tetracycline plates in an ordered array in duplicate. The bacterial clones were allowed to grow overnight, transferred to Whatman 541 filter paper, amplified with chloramphenicol, and prepared for hybridization as described by Gergen *et al.* (26). The filters were prehybridized, hybridized, and washed as described in our previous paper (5).

Nucleotide Sequence Determination. Restriction endonuclease fragments of cloned cDNA were labeled at their 5′ ends by the sequential action of bacterial alkaline phosphatase (P-L

Biochemicals) and polynucleotide kinase plus [γ-^{32}P]ATP (27). *Eco*RI-digested and *Sau*3A-digested DNA was labeled at the 3′ end by repair labeling of the termini: after restriction enzyme digestion, 2 μM [α-^{32}P]dATP (New England Nuclear), polymerase I Klenow fragment A at 2 units/ml, and, in the case of *Sau*3A-digested DNA, 400 μM dGTP (P-L Biochemicals), were added and the reaction mixture was incubated 5 min at 20°C. *Pst* I-digested DNA was labeled at the 3′ termini with ^{32}P-labeled 3′-dATP and terminal transferase as described in the kit supplied by New England Nuclear.

RESULTS

Strategy for Designing β_2m-Specific Oligodeoxyribonucleotide Probes. Our general approach for the isolation of cloned DNA sequences specific for β_2m involved synthesis of a set of oligodeoxyribonucleotides complementary to all the possible coding sequences for a small portion of the protein. In designing the probe sequences, we chose two regions of the protein for which there are relatively few potential coding sequences. The two regions of amino acid sequence we used in designing the β_2m-specific probes are shown in Table 1. Amino acid residues 95–99 of β_2m can be coded for by 24 possible sequences in the mRNA. We synthesized two sets of pentadecanucleotides corresponding to this region; β2mI is a mixture of 8 sequences and β2mII is a mixture of 16 sequences. Amino acid residues 75–78 of β_2m were used to design another probe, β2mIII, which is a set of 8 undecanucleotides. Each of the three sets of probes was synthesized as a mixture of sequences (5).

Isolation of a Bacterial Clone Containing β_2m cDNA Sequences. Using poly(A)-containing cytoplasmic RNA from a human lymphoblastoid cell line, we prepared double-stranded

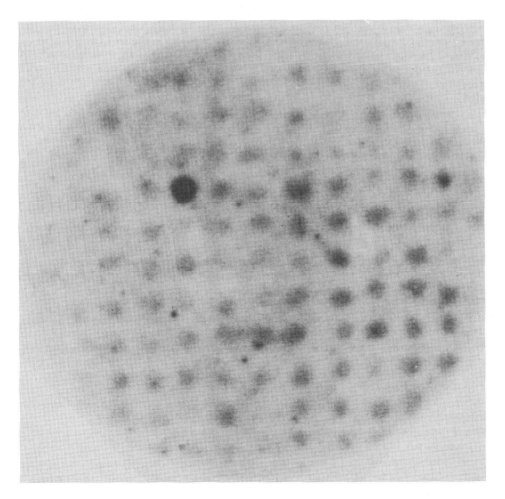

FIG. 1. Hybridization of oligonucleotide probe β2mII to colonies transformed with G-tailed plasmid plus C-tailed cDNA. Bacterial cDNA clones were isolated and screened. ^{32}P-Labeled β2mII probe was hybridized to the filters at 41°C overnight and washed with 0.9 M NaCl/0.09 M sodium citrate at 41°C.

cDNA 500–800 base pairs in length (20) and inserted it into the *Pst* I site of the plasmid vector pBR322 by the standard G·C tailing method (22). The recombinant DNA was used to transform *E. coli* strain MC1061. From the transformants, 535 tetracycline-resistant clones were obtained and placed on fresh plates in an ordered array. The bacterial clones were transferred to Whatman 541 filter paper, amplified with chloramphenicol, and prepared for hybridization as described by Gergen *et al.* (26). The filters were hybridized with ^{32}P-labeled oligonucleotide probes and washed as described in *Materials and Methods*. An autoradiogram of one of the five filters hybridized with the β_2mII probe is shown in Fig. 1. The amount of labeled probe hybridized to one of the clones is clearly greater than that hybridized to any of the other clones. Using the β_2mI and β_2mIII probes to screen the same collection of clones, we could not distinguish specific hybridization to any clone (data not shown).

Characterization of the Cloned cDNA for β_2m by Southern Blot Analysis. Plasmid DNA was prepared from the presumptive bacterial clone for β_2m cDNA observed in Fig. 1. The β_2m plasmid DNA was cleaved with restriction endonucleases *Hind*II + *Hind*III or *Sau*3A. Fig. 2A shows the results of a Southern blot analysis of *Hind*II + *Hind*III-cut DNA. Both the β_2mII and β_2mIII probes hybridize specifically with the smallest *Hind*II + *Hind*III fragment (lanes A2 and A3). The hybridization of β_2mII to the largest *Hind*II + *Hind*III fragment observed in lane A2 represents background hybridization and is not detectable with shorter exposure times. Fig. 2B shows the results of blot analysis of *Sau*3A-digested β_2m plasmid DNA. The β_2mIII probe hybridizes to a restriction fragment 620 base pairs in length (lane B3). There is no detectable hybridization of the β_2mII probe to any of the major DNA bands (lane B2) even though this autoradiogram was exposed for the same amount of time as that in lane A2 showing hybridization of β_2mII to *Hind*II + *Hind*III-digested DNA. The absence of hybridization of β_2mII to *Sau*3A-digested plasmid DNA is due to the presence of a *Sau*3A recognition site within the probe hybridization site. By cleaving the insert DNA with *Sau*3A at a position in the center of the region to which β_2mII hybridizes, hybridization of the probe is eliminated. There is a faint band of hybridization observed in lane B2. This represents hybridization of the probe to a partial digestion product in which the *Sau*3A site within the probe hybridization site is not cleaved. This partial digestion product is not detectable in the photography of the stained gel. Fig 2C is a map of restriction endonuclease sites showing the sites of hybridization of β_2mII and β_2mIII probes.

Nucleotide Sequence Analysis of the Cloned cDNA for β_2m. The nucleotide sequence of the β_2m plasmid DNA was determined by using the base-specific cleavage reactions of Maxam and Gilbert (27). The nucleotide sequence of the cloned β_2m cDNA is shown in Fig. 3. The cloned cDNA contains most of the coding region for the protein, including part of the coding region for the leader peptide, and a large portion of the 3' untranslated region.

The amino acid sequence for β_2m predicted from the nucleotide sequence differs from the published amino acid sequence (11) at two positions: (*i*) position 42 is asparagine and not aspartic acid, and (*ii*) the serine at position 67 of the published amino acid sequence is not present in the sequence deduced from the cloned cDNA. While these differences may reflect polymorphism in human β_2m, the latter difference in the amino acid sequence would require the insertion of three nucleotides (serine codon) in the gene. Accordingly, the amino acid sequence for human β_2m determined in the present study establishes the length of the mature protein as 99 residues, the same

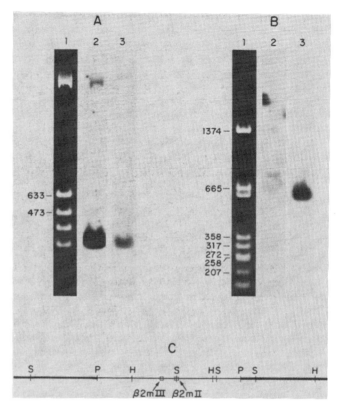

FIG. 2. Southern blot analysis of the cloned cDNA for β_2m. The plasmid DNA containing the β_2m cDNA sequences was digested with restriction enzymes *Hind*II + *Hind*III or *Sau*3A in *A* and *B*, respectively. The DNA fragments were subjected to electrophoresis on a 2% agarose gel, visualized by staining with ethidium bromide, and transferred to nitrocellulose filter paper by the standard Southern blotting technique (18). Duplicate blots were hybridized with ^{32}P-labeled β_2mII at 41°C or ^{32}P-labeled β_2mIII at 14°C. The blots were washed at 37°C and 20°C, respectively. In both *A* and *B*, lane 1 is a photograph of a lane from the stained gel, lane 2 is an autoradiogram of the blot hybridization with ^{32}P-labeled β_2mII, and lane 3 is an autoradiogram of the blot hybridization with ^{32}P-labeled β_2mIII. The sizes (in base pairs) of the restriction fragments derived solely from the vector pBR322 are indicated. (*C*) Restriction maps of the region of the plasmid containing the inserted β_2m sequences. The thick line indicates sequences from the plasmid vector pBR322 and the thin line indicates β_2m sequences. The sites of hybridization of the two probes are indicated. H, *Hind*II site; P, *Pst* I site; and S, *Sau*3A site.

length as β_2m from the three other species for which the sequence has been determined (28, 29).

The sequence of the leader peptide for human β_2m determined from the nucleotide sequence differs from the previously published sequence for mouse β_2m (30) at several positions: (*i*) position −2 is glutamic acid in the human sequence and not tyrosine as in the mouse sequence; (*ii*) position −8 is leucine, not valine; and (*iii*) position −10 is alanine, not valine. Only the first of these changes represents a nonconservative amino acid substitution.

The sequence of the 3' untranslated region of β_2m cDNA is, in general, A+T-rich (65% A·T base pairs). The sequence A-A-T-A-A-A, which is common to many polyadenylylated mRNAs (31), is not observed in the sequence. The absence of this sequence may indicate that a portion of the 3' untranslated region is not present in the clone.

DISCUSSION

Our results demonstrate the usefulness of mixtures of chemically synthesized oligodeoxyribonucleotides as hybridization

	-11	-10	-9	-8	-7	-6	-5	-4	-3	-2	-1	1	2	3	4	5	6	7	8	9	10	11	12	13	14	15	16
	Leu	Ala	Leu	Leu	Ser	Leu	Ser	Gly	Leu	Glu	Ala	Ile	Gln	Arg	Thr	Pro	Lys	Ile	Gln	Val	Tyr	Ser	Arg	His	Pro	Ala	Glu
5'-	CTC	GCG	CTA	CTC	TCT	CTT	TCT	GGC	CTT	GAG	GCT	ATC	CAG	CGT	ACT	CCA	AAG	ATT	CAG	GTT	TAC	TCA	CGT	CAT	CCA	GCA	GAG

	17	18	19	20	21	22	23	24	25	26	27	28	29	30	31	32	33	34	35	36	37	38	39	40	41	42	43
	Asn	Gly	Lys	Ser	Asn	Phe	Leu	Asn	Cys	Tyr	Val	Ser	Gly	Phe	His	Pro	Ser	Asp	Ile	Glu	Val	Asp	Leu	Leu	Lys	Asn	Gly
	AAT	GGA	AAG	TCA	AAT	TTC	CTG	AAT	TGC	TAT	GTG	TCT	GGG	TTT	CAT	CCA	TCC	GAC	ATT	GAA	GTT	GAC	TTA	CTG	AAG	AAT	GGA

	44	45	46	47	48	49	50	51	52	53	54	55	56	57	58	59	60	61	62	63	64	65	66	67	68	69	70
	Glu	Arg	Ile	Glu	Lys	Val	Glu	His	Ser	Asp	Leu	Ser	Phe	Ser	Lys	Asp	Trp	Ser	Phe	Tyr	Leu	Leu	Tyr	Tyr	Thr	Glu	Phe
	GAG	AGA	ATT	GAA	AAA	GTG	GAG	CAT	TCA	GAC	TTG	TCT	TTC	AGC	AAG	GAC	TGG	TCT	TTC	TAT	CTC	TTG	TAT	TAT	ACT	GAA	TTC

	71	72	73	74	75	76	77	78	79	80	81	82	83	84	85	86	87	88	89	90	91	92	93	94	95	96	97
	Thr	Pro	Thr	Glu	Lys	Asp	Glu	Tyr	Ala	Cys	Arg	Val	Asn	His	Val	Thr	Leu	Ser	Gln	Pro	Lys	Ile	Val	Lys	Trp	Asp	Arg
	ACC	CCC	ACT	GAA	AAA	GAT	GAG	TAT	GCC	TGC	CGT	GTG	AAC	CAC	GTG	ACT	TTG	TCA	CAG	CCC	AAG	ATA	GTT	AAG	TGG	GAT	CGA

98	99
Asp	Met	Stop

```
         98  99
         Asp Met Stop         10          20          30          40          50          60          70          80          90
         GAC ATG TAA  GCAGCATCAT GGAGGTTGA AGATGCCGCA TTTGGATTGG ATGAATTCAA AATTCTGCTT GCTTGCTTTT TAATATTGAT ATGCTTATAC

            100         110         120         130         140         150         160         170         180         190
         ACTTACACTT TATGCACAAA ATGTAGGGTT ATAATAATGT TAACATGGAC ATGATCTTCT TTATAATTCT ACTTTGAGTG CTGTCTCCAT GTTTGATGTA

            200         210
         TCTGAGCAGG TTGCTCCACA GGTAGCT-3'
```

FIG. 3. Nucleotide sequence of the cloned cDNA for human β_2m. The nucleotide sequence of the noncoding strand of the cloned cDNA is shown. Above the nucleotide sequence is the amino acid sequence of the protein, with the amino acids of the leader peptide indicated by negative numbers.

probes for the isolation of specific cloned DNA sequences. Our previous results with model systems (4, 5) indicated that the screening procedure should be highly specific. The results with β_2m show that this is indeed the case: (i) The only clone detected with the β_2m probes was the correct one. (ii) The β2mI probe, a mixture of eight 15-base-long oligonucleotides, did not hybridize specifically with the β_2m clone in the colony screening experiment. One of the sequences within the β2mI mixture differs from the corresponding sequence in the cloned β_2m cDNA at only a single nucleotide (G in place of T). This result indicated that under stringent hybridization criteria the oligonucleotide probes do not hybridize with closely related sequences containing a single base substitution.

The screening method described in this study is very rapid. The most tedious step in screening for the β_2m clone was the picking of individual bacterial clones and placing them in an ordered array. This step has been eliminated in subsequent experiments in which large numbers of unordered colonies were successfully screened for mouse H-2K^b antigen cDNA clones (unpublished data).

Apparently contradictory results were obtained with the β2mIII probe, a mixture of eight 11-base-long oligonucleotides. One of the sequences in this mixture is perfectly complementary to the cloned cDNA. In the Southern blot analysis, specific hybridization of the β2mIII probe to the cloned β_2m cDNA was observed (Fig. 2); however, in the colony screening experiment, no specific hybridization of β2mIII to the β_2m clone was observed. This discrepancy in the results of the blot and colony screening experiments may reflect differences in the hybridization properties of oligonucleotides to nitrocellulose filters (used in the blot) versus cellulose filters (used in the colony screening) or may reflect a lower background hybridization of oligonucleotides to purified DNA (in the blot) versus DNA in

lysed colonies (in the colony screening). In our subsequent experiments isolating cloned sequences for other proteins, oligonucleotide probes longer than 11 bases have been used for colony screening.

In previous studies, chemically synthesized oligodeoxyribonucleotides have been employed in the isolation of specific cloned sequences in basically two ways: oligonucleotides have been used directly as hybridization probes or used as primers for the synthesis of radiolabeled cDNA that is used as the probe. Montgomery *et al.* (32) demonstrated the usefulness of synthetic oligonucleotides as hybridization probes in the isolation of the cloned yeast iso-1-cytochrome *c* gene. In this special case, an oligonucleotide probe of unique sequence could be predicted by comparison of the amino acid sequences of cytochrome *c* from several species. Subsequently, Goeddel *et al.* (33) have screened purified plasmid DNAs enriched for human leukocyte interferon sequences by using labeled oligonucleotides as hybridization probes. Several groups have isolated specific cloned sequences by using synthetic oligonucleotides as primers for the synthesis of cDNA probes (21, 34, 35).

We believe the method using oligonucleotides as hybridization probes is superior to the method employing oligonucleotides as primers in the synthesis of cDNA probes for two main reasons. First, greater specificity can be obtained by using the hybridization approach than by using the priming approach. Under appropriate conditions, a mismatched base pair does not allow formation of oligonucleotide·polynucleotides duplexes (4, 5), whereas base mismatches can be tolerated in the priming method (21, 35, 36). Second, with the priming approach, the amount of probe obtained is dependent on the amount of mRNA available for template. In isolating cloned sequences for very low abundance mRNAs, use of the primer approach would entail the isolation of large amounts of mRNA to produce sufficient cDNA probe for screening. In addition, it may be increasingly more difficult to produce a specific cDNA probe as the abundance of a mRNA decreases.

In summary, we believe the screening technique described in this report may be applied to the isolation of cloned DNAs for any protein for which the amino acid sequence is known.

Note Added in Proof. Since the preparation of this manuscript the isolation and partial DNA sequence analysis of three mouse β_2m cDNA clones has been reported (37).

We thank Ting H. Huang for purification and sequence analysis of synthetic oligonucleotides. This work was supported by National Institutes of Health Postdoctoral Fellowship GM07591 (to S.V.S.) and National Institutes of Health Grants GM26391 (to R.B.W.) and GM25658 (to K.I.). K.I. and R.B.W. are members of the Cancer Research Center (CA16434) at the City of Hope Research Institute.

1. Astell, C. R. & Smith, M. (1972) *Biochemistry* **11**, 4114–4120.
2. Astell, C. R., Doel, M. T., Jahnke, P. A. & Smith, M. (1973) *Biochemistry* **12**, 5068–5074.
3. Dodgson, J. B. & Wells, R. D. (1977) *Biochemistry* **16**, 2367–2372.
4. Wallace, R. B., Shaffer, J., Murphy, R. F., Bonner, J., Hirose, T. & Itakura, K. (1979) *Nucleic Acids Res.* **6**, 3543–3557.
5. Wallace, R. B., Johnson, M. J., Hirose, T., Miyake, T., Kawashima, E. H. & Itakura, K. (1981) *Nucleic Acids Res.* **9**, 879–894.
6. Gillam, S., Waterman, K. & Smith, M. (1975) *Nucleic Acids Res.* **2**, 625–633.
7. Berggård, I. & Bearn, A. G. (1968) *J. Biol. Chem.* **243**, 4095–4103.
8. Nakamuro, K., Tanigaki, N. & Pressman, D. (1973) *Proc. Natl. Acad. Sci. USA* **70**, 2863–2865.
9. Peterson, P. A., Rask, L. & Lindblom, J. B. (1974) *Proc. Natl. Acad. Sci. USA* **71**, 35–39.
10. Lancet, D., Parham, P. & Strominger, J. L. (1979) *Proc. Natl. Acad. Sci. USA* **76**, 3844–3848.
11. Cunningham, B. A., Wang, J. L., Berggård, I. & Peterson, P. A. (1973) *Biochemistry* **12**, 4811–4821.
12. Kahn, M., Kolter, R., Thomas, C., Figurski, D., Meyer, R., Remant, E. & Helinski, D. R. (1979) *Methods Enzymol.* **68**, 268–280.
13. Peacock, A. C. & Dingman, C. W. (1968) *Biochemistry* **7**, 668–674.
14. Rossi, J. J., Ross, W., Egan, J., Lipman, D. & Landy, A. (1979) *J. Mol. Biol.* **128**, 21–47.
15. Southern, E. M. (1975) *J. Mol. Biol.* **98**, 503–517.
16. Hirose, T., Crea, R. & Itakura, K. (1978) *Tetrahedron Lett.* **28**, 2449–2452.
17. Broka, C., Hozumi, T., Arentzen, R. & Itakura, K. (1980) *Nucleic Acids Res.* **8**, 5461–5471.
18. Walseth, T. F. & Johnson, R. A. (1979) *Biochim. Biophys. Acta* **526**, 11–31.
19. Epstein, M. A., Achong, B. G., Barr, Y. M., Zajac, B., Henle, G. & Henle, W. (1966) *J. Natl. Cancer Inst.* **37**, 547–555.
20. Singer, R. H. & Penman, S. (1973) *J. Mol. Biol.* **78**, 321–334.
21. Goeddel, D. V., Shepard, H. M., Yelverton, E., Leung, D. & Crea, R. (1980) *Nucleic Acids Res.* **8**, 4057–4074.
22. Roychoudhury, R. & Wu, R. (1980) *Methods Enzymol.* **65**, 43–62.
23. Chang, A. C. Y., Nunberg, J. H., Kaufman, R. J., Ehrlich H. A., Schimke, R. T. & Cohen, S. N. (1978) *Nature (London)* **275**, 617–624.
24. Casadaban, M. & Cohen, S. N. (1980) *J. Mol. Biol.* **138**, 179–207.
25. Kushner, S. R. (1978) in *Genetic Engineering*, eds. Boyer, H. W. & Nicosia, S. (Elsevier/North-Holland, Amsterdam), pp. 17–23.
26. Gergen, J. P., Stern, R. H. & Wensink, P. C. (1979) *Nucleic Acids Res.* **7**, 2115–2136.
27. Maxam, A. & Gilbert, W. (1980) *Methods Enzymol.* **65**, 499–560.
28. Gates, F. T., III, Coligan, J. E. & Kindt, T. J. (1979) *Biochemistry* **18**, 2265–2272.
29. Gates, F. T., III, Coligan, J. E. & Kindt, T. J. (1981) *Proc. Natl. Acad. Sci. USA* **78**, 554–558.
30. Lingappa, V. R., Cunningham, B. A., Jazwinski, S. M., Hopp, T. P., Blobel, G. & Edelman, G. M. (1979) *Proc. Natl. Acad. Sci. USA* **76**, 3651–3655.
31. Proudfoot, N. J. & Brownlee, G. G. (1976) *Nature (London)* **263**, 211–214.
32. Montgomery, D. L., Hall, B. D., Gillam, S. & Smith, M. (1978) *Cell* **14**, 673–680.
33. Goeddel, D. V., Yelverton, E., Ullrich, A., Heyneker, H. L., Miozzari, G., Holmes, W., Seeburg, P. H., Dull, T., May, L., Stebbing, N., Crea, R., Maeda, S., McCandliss, R., Sloma, A., Tabor, J. M., Gross, M., Familletti, P. C. & Pestka, S. (1980) *Nature (London)* **287**, 411–416.
34. Noyes, B. E., Mevarechi, M., Stein, R. & Agarwal, K. L. (1979) *Proc. Natl. Acad. Sci. USA* **76**, 1770–1774.
35. Chan, S. J., Noyes, B. E., Agarwal, K. L. & Steiner, D. F. (1979) *Proc. Natl. Acad. Sci. USA* **76**, 5036–5040.
36. Houghton, M., Stewart, A. G., Doel, S. M., Emtage, J. S., Eaton, M. A. W., Smith, J. C., Patel, T. P., Lewis, H. M., Porter, A. G., Birch, J. R., Cartwright, T. & Carey, N. H. (1980) *Nucleic Acids Res.* **8**, 1913–1931.
37. Parnes, J. R., Velan, B., Felsenfeld, A., Ramanathan, L., Ferrini, U., Appella, E. & Seidman, J. F. (1981) *Proc. Natl. Acad. Sci. USA* **78**, 2253–2257.

Electrophoretic Transfer of Proteins from Polyacrylamide Gels to Nitrocellulose Sheets: Procedure and Some Applications

H. Towbin, T. Staehelin and J. Gordon

ABSTRACT A method has been devised for the electrophoretic transfer of proteins from polyacrylamide gels to nitrocellulose sheets. The method results in quantitative transfer of ribosomal proteins from gels containing urea. For sodium dodecyl sulfate gels, the original band pattern was obtained with no loss of resolution, but the transfer was not quantitative. The method allows detection of proteins by autoradiography and is simpler than conventional procedures. The immobilized proteins were detectable by immunological procedures. All additional binding capacity on the nitrocellulose was blocked with excess protein; then a specific antibody was bound and, finally, a second antibody directed against the first antibody. The second antibody was either radioactively labeled or conjugated to fluorescein or to peroxidase. The specific protein was then detected by either autoradiography, under UV light, or by the peroxidase reaction product, respectively. In the latter case, as little as 100 pg of protein was clearly detectable. It is anticipated that the procedure will be applicable to analysis of a wide variety of proteins with specific reactions or ligands.

Polyacrylamide gel electrophoresis has become a standard tool in every laboratory in which proteins are analyzed and purified. Most frequently, the amount and location of the protein are of interest and staining is then sufficient. However, it may also be important to correlate an activity of a protein with a particular band on the gel. Enzymatic and binding activities can sometimes be detected *in situ* by letting substrates or ligands diffuse into the gel (1, 2). In immunoelectrophoresis, the antigen is allowed to diffuse (3) or electrophoretically move (4) against antibody. A precipitate is then formed where the antigen and antibody interact. Modifications have been described in which the antigen is precipitated by directly soaking the separation matrix in antiserum (5, 6). The range of gel electrophoretic separation systems is limited by the pore size of the gels and diffusion of the antibody. The systems are also dependent on concentration and type of antigen or antibody to give a physically immobile aggregate.

Analysis of cloned DNA has been revolutionized (7) by the ability to fractionate the DNA electrophoretically in polyacrylamide/agarose gels first and then to obtain a faithful replica of the original gel pattern by blotting the DNA onto a sheet of nitrocellulose on which it is immobilized. The immobilized DNA can then be analyzed by *in situ* hybridization. The power of immobilized two-dimensional arrays has been extended to the analysis of proteins by use of antibody-coated plastic sheets to pick up the corresponding antigen from colonies on agar plates (8). Sharon *et al.* (9) have used antigen-coated nitrocellulose sheets to pick up antibodies secreted by hybridoma clones growing in agar.

In this report we describe a procedure for the transfer of proteins from a polyacrylamide gel to a sheet of nitrocellulose in such a way that a faithful replica of the original gel pattern is obtained. A wide variety of analytical procedures can be applied to the immobilized protein. Thus, the extreme versatility of nitrocellulose binding assays can be combined with high-resolution polyacrylamide gel electrophoresis. The procedure brings to the analysis of proteins the power that the Southern (7) technique has brought to the analysis of DNA.

MATERIALS AND METHODS

Immunogens and Immunization Procedures. *Escherichia coli* ribosomal proteins L7 and L12 were extracted (10) from 50S subunits and purified as described (11) by ion-exchange chromatography on carboxymethyl- and DEAE-cellulose. Antibodies were raised in a goat by injecting 250 μg of protein emulsified with complete Freund's adjuvant intracutaneously distributed over several sites. *Bacillus pertussis* vaccine (1.5 ml of Bordet–Gengou vaccine, Schweizerisches Serum- und Impfinstitut, Bern, Switzerland) was given subcutaneously with every antigen injection. Booster injections of the same formulation were given on days 38, 79, and 110. The animal was bled on day 117.

Subunits from chicken liver ribosomes (12) were combined in equimolar amounts, and 200-μg aliquots were emulsified with 125 μl of complete Freund's adjuvant injected at one intraperitoneal and four subcutaneous sites into BALB/c mice. Booster injections of 400 μg of ribosomes in saline were given intraperitoneally on days 33, 57, 58, and 59. The animals were bled on day 71.

Electrophoretic Blotting Procedures. Proteins were first subjected to electrophoresis in the presence of urea either in two dimensions (12) or in one-dimensional slab gels corresponding to the second dimension of the same two-dimensional system. The proteins were then transferred to nitrocellulose sheets as follows. The physical assembly used is shown diagrammatically in Fig. 1. A sheet of nitrocellulose (0.45 μm pore size in roll form, Millipore) was briefly wetted with water and laid on a scouring pad (Scotch-Brite) which was supported by a stiff plastic grid (disposable micropipette tray, Medical Laboratory Automation, Inc., New York). The gel to be blotted was put on the nitrocellulose sheet and care was taken to remove all air bubbles. A second pad and plastic grid were added and rubber bands were strung around all layers. The gel was thus firmly and evenly pressed against the nitrocellulose sheet. The assembly was put into an electrophoretic destaining chamber with the nitrocellulose sheet facing the cathode. The chamber contained 0.7% acetic acid. A voltage gradient of 6 V/cm was applied for 1 hr.

For polyacrylamide electrophoresis in the presence of sodium dodecyl sulfate (13) instead of urea, the procedure was as de-

TOWBIN, H., STAEHELIN, T., and GORDON, J.
Electrophoretic transfer of proteins from polyacrylamide gels to nitrocellulose sheets: Procedure and some applications. *Proc. Natl. Acad. Sci. U.S.A.* 76:4350-4354 (1979). Reprinted with the authors' permission.

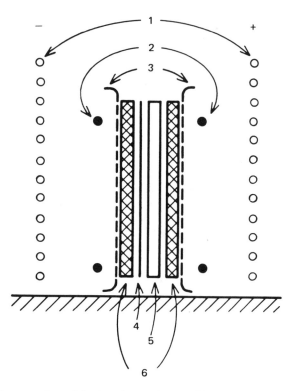

FIG. 1. Assembly for electrophoretic blotting procedure. 1, Electrodes of destainer; 2, elastic bands; 3, disposable pipette-tip tray; 4, nitrocellulose sheets; 5, polyacrylamide gel; 6, Scotch-Brite pads. Assembly parts are shown separated for visualization only.

scribed above except that the polarity of the electrodes was reversed and the electrode buffer was 25 mM Tris·192 mM glycine/20% (vol/vol) methanol at pH 8.3.

Staining for Protein. The blot may be stained with amido black (0.1% in 45% methanol/10% acetic acid) and destained with 90% methanol/2% acetic acid (see ref. 14).

Immunological Detection of Proteins on Nitrocellulose. The electrophoretic blots (usually not stained with amido black) were soaked in 3% bovine serum albumin in saline (0.9% NaCl/10 mM Tris·HCl, pH 7.4) for 1 hr at 40°C to saturate additional protein binding sites. They were rinsed in saline and incubated with antiserum appropriately diluted into 3%·bovine serum albumin in saline also containing carrier serum with concentration and species as indicated in the legends. The sheets were washed in saline (about five changes during 30 min, total) and incubated with the second (indicator) antibody directed against the immunoglobulins of the first antiserum. As indicator antibodies we used ^{125}I-labeled sheep anti-mouse IgG. This had been purified with affinity chromatography on Sepharose-immobilized myeloma proteins and labeled by a modified version of the chloramine T method in 0.5 ml with 0.5 mg of IgG and 1 mCi of Na^{125}I (1 Ci = 3.7 × 10^{10} becquerels) for 60 sec at room temperature. The specific activity was approximately 1.5 μCi/μg of IgG. ^{125}I-Labeled IgG was diluted to 10^6 cpm/ml in saline containing 3% bovine serum albumin and 10% goat serum, and 3 ml of this solution was used for a nitrocellulose sheet of 100 cm^2. Incubation was in the presence of 0.01% NaN$_3$ for 6 hr at room temperature. The electrophoretic blots were washed in saline (five changes during 30 min, total) and thoroughly dried with a hair dryer. The blots were exposed to Kodak X-Omat R film for 6 days.

Fluorescein- and horseradish peroxidase-conjugated rabbit anti-goat IgG (Nordic Laboratories, Tilburg, Netherlands) were reconstituted before use according to the manufacturer's instructions. Fluorescein-conjugated antibodies were used at 1:50

dilution in saline containing 3% bovine serum albumin and 10% rabbit serum. After incubation for 30 min at room temperature, the blots were washed as above and inspected or photographed with a Polaroid camera under long-wave UV light through a yellow filter.

Horseradish peroxidase-conjugated IgG preparations were used at 1:2000 dilution in saline containing 3% bovine serum albumin and 10% rabbit serum. The blots were incubated for 2 hr at room temperature and washed as described above. For the color reaction (15), the blots were soaked in a solution of 25 μg of o-dianisidine per ml/0.01% H$_2$O$_2$/10 mM Tris·HCl, pH 7.4. This was prepared freshly from stock solutions of 1% o-dianisidine (Fluka) in methanol and 0.30% H$_2$O$_2$. The reaction was terminated after 20–30 min by washing with water. The blots were dried between filter paper. Drying considerably reduced the background staining. The blots were stored protected from light.

RESULTS

Electrophoretic Transfer of Ribosomal Proteins from Polyacrylamide Gels to Nitrocellulose Sheets. Most proteins or complexes containing protein adsorb readily to nitrocellulose filters (16), whereas salts, many small molecules, and RNA are usually not retained. These binding properties are widely used for binding assays with nitrocellulose filters. We found that proteins were retained on these filters equally well when carried towards the filter in an electric field. If the electric field was perpendicular to a slab gel containing separated proteins (see Fig. 1), we obtained a replica of the protein pattern on the nitrocellulose sheet. This is demonstrated with ribosomal proteins from E. coli; a conventionally stained gel (Fig. 2A) and a stained electrophoretic blot of an identical gel (Fig. 2B) are shown. All ribosomal proteins from chicken liver and E. coli ribosomes detectable on two-dimensional gels could be seen on the electrophoretic blots produced from them. An example of a blot from a two-dimensional gel is given in Fig. 3. When the original polyacrylamide gel was stained after blotting, no protein could be detected. Thus, the blotting procedure removed all protein from the gel.

To establish whether the proteins removed from the gels were quantitatively deposited on the nitrocellulose sheet, we separated ^{3}H-labeled proteins from chicken liver 60S ribosomal subunit by two-dimensional electrophoresis and compared the radioactivity that could be recovered from the blot with that recovered directly from the gel (Table 1). Single proteins or groups of poorly separated proteins were cut out and radioactivity was measured after combustion of the samples. The results were within the variability inherent to two-dimensional analyses. Variations could be accounted for by variable transfer of proteins into the second dimension gel and the acuity with which spots can be cut out.

At loads exceeding the capacity of nitrocellulose, losses of protein occurred. Titration with radioactive ribosomal proteins under blotting conditions showed that at concentrations below 0.15 μg/mm^2 all protein was adsorbed. Overloading became apparent when a second sheet of nitrocellulose directly underneath the first one took up protein or when protein became visible on the cathodal surface of amido black-stained blots.

The conservation of resolution together with the high recovery of ribosomal proteins simplifies the procedure for autoradiography. The common procedure involving drying of polyacrylamide gels under heat and reduced pressure (19), which is tedious and time consuming, may be eliminated. Because the proteins become concentrated on a very thin layer, autoradiography from ^{14}C- and ^{35}S-labeled proteins should be highly efficient even without 2,5-diphenyloxazole impregnation

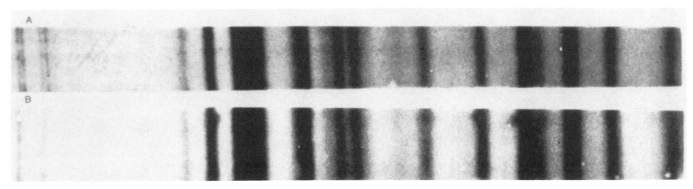

FIG. 2. Electrophoretic blotting of ribosomal proteins from one-dimensional gels. Total ribosomal proteins from *E. coli* were separated on an 18% polyacrylamide slab gel containing 8 M urea. (*A*) A section of the gel was stained with Coomassie blue; (*B*) another section was electrophoretically blotted and the blot was stained as described in *Materials and Methods*. Electrophoresis was from left to right.

(19). We have successfully obtained such autoradiograms from gels of ^{35}S-labeled proteins (not shown). Further, preliminary experiments with tritiated proteins have shown that dried blots may be processed for fluorography by brief soaking in 10% diphenyloxazole in ether (20).

The above experiments were done with ribosomal proteins separated on polyacrylamide gels containing urea. We have electrophoretically blotted proteins from sodium dodecyl sulfate by the modified procedure also described in *Materials and Methods*. Again, there was no loss of resolution. However, differences of staining intensities between proteins on the gel

and the blot were apparent. In spite of the apparently incomplete recovery, blots from polyacrylamide gels containing sodium dodecyl sulfate may be used for detection of antigen in the same way as described below for ribosomal proteins (unpublished experiments).

Detection of Antigen by Antibody Binding on Blots *In Situ*. We found that proteins transferred to nitrocellulose sheets remained there without being exchanged over several days. Because a blot could be saturated with bovine serum albumin to block the residual binding capacity of the sheet, it can be treated as a solid-phase immunoassay. In the following immunological applications, we used indirect techniques throughout. Thus, antibody bound by the immobilized antigen was detected by a second, labeled antibody directed against the first antibody, and in each case excess unbound antibody was washed out.

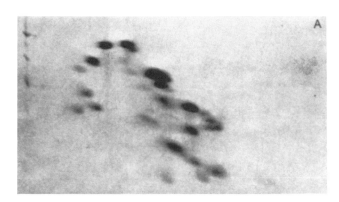

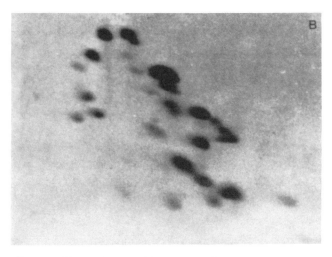

FIG. 3. Electrophoretic blotting of ribosomal proteins from two-dimensional gels. Proteins (35 μg) extracted from the 60S ribosomal subunit of chicken liver (12) were separated by two-dimensional gel electrophoresis. (*A*) Gel stained with Coomassie blue; (*B*) blot of an identical gel. Electrophoresis: 1st dimension, from left to right (towards cathode); 2nd dimension, top to bottom.

Table 1. Efficiency of transfer of ribosomal proteins to nitrocellulose sheets

Protein or group of proteins analyzed	Recovery on blot, %
3	123
4, 4A	104
5	111
6	107
7, 8	86
9	80
10	112
11	79
12, 16	93
13	95
15, 15A, 18	125
17	115
19	139
21, 23	118
26	114
27	143
28, 29	69
31	117
33	131

Ribosomal large-subunit proteins from chicken liver were tritiated by reductive methylation (17) and separated by two-dimensional electrophoresis (12) in the presence of 35 μg of carrier protein. Two identical gels were run. One was stained; the other was electrophoretically blotted on a nitrocellulose sheet. Spots were identified according to our nomenclature for chicken ribosomes (12), which differs only in minor respects from that established for rat ribosomes (18). Corresponding spots or groups of spots were cut from the gel and the blot. Their radioactivity was determined after conversion to tritiated water in a sample oxidizer (Oxymat).

FIG. 4.   Detection of *E. coli* ribosomal proteins L7 and L12 by (*A*) horseradish peroxidase- and (*B*) fluorescein-conjugated antibodies. Total ribosomal proteins from *E. coli* were separated and blotted as in Fig. 2. The anti-L7/L12 serum had a titer of 340 pmol of 70S ribosomes per ml of serum as determined by turbidity formation (20). Incubation was for 2 hr at room temperature in goat antiserum diluted 1:10 in saline containing 3% bovine serum albumin and 10% rabbit carrier serum and then with conjugated anti-goat IgG. In each case the lower strip is a control with preimmune antiserum. Electrophoresis was from left to right.

In Fig. 4 the detection of *E. coli* ribosomal proteins L7 and L12 with a goat serum specific for proteins L7 and L12 is shown. L7 is identical to L12, except for its *N*-acetylated NH$_2$-terminal amino acid (21). L7 and L12 fully crossreact immunologically (22) and are separated on acidic polyacrylamide gels (21). Both peroxidase- (Fig. 4*A*) and fluorescein-conjugated (Fig. 4*B*) antibodies were able to reveal immunoglobulin that was specifically retained by proteins L7 and L12. In each case, the lower gel is a control with preimmune serum. Peroxidase-conjugated antibodies were far more sensitive than fluorescein-conjugated ones. They could therefore be used at much higher dilution. This also permitted the detection of very small amounts of antigen. With a rabbit serum (23) we could detect 100 pg of L7 and L12 with serum and incubation conditions similar to those of the experiment described in Fig. 4 (not shown).

Because we can use the procedure to detect a specific antibody reacting with a specific protein after electrophoresis in polacrylamide, we should also be able to determine which proteins have elicited antibodies in a complex mixture of immunogens. In the experiment of Fig. 5, individual sera of five

mice immunized with chicken liver ribosomes were tested. We used ^{125}I-labeled sheep anti-mouse immunoglobulins to detect the presence of mouse immunoglobulins. In all mice, antibodies were preferentially produced against slowly moving proteins, presumably of high molecular weight. The procedure can thus characterize the antigen population against which specific antibodies have been raised in a mixture of immunogens.

DISCUSSION

The electrophoretic blotting technique described here produces replicas of proteins separated on polyacrylamide gels with high fidelity. We obtained quantitative transfer with proteins from gels containing urea. This was established here with ribosomal proteins. More generally, nitrocellulose membranes have been used to retain proteins from dilute solutions for their subsequent quantitative determination (16). Still, there remains the possibility that certain classes of protein do not bind to nitrocellulose. In this case absorbent sheets other than nitrocellulose or different blotting conditions may be helpful.

We have demonstrated that proteins immobilized on nitrocellulose sheets can be used to detect their respective antibodies.

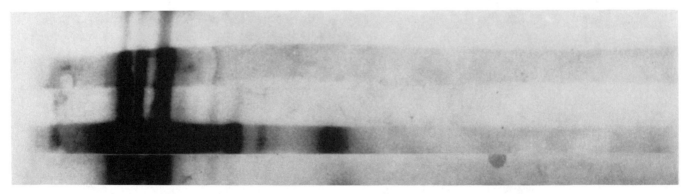

FIG. 5.   Detection of immunoglobulin from individual mice directed against ribosomal proteins from chicken liver. Total protein from chicken liver ribosomes (12) was electrophoretically separated and blotted as in Fig. 2. Sera were obtained from five individual mice immunized against combined 40S and 60S subunits. The antisera were diluted 1:50 in saline containing 3% bovine serum albumin and 10% goat carrier serum. The blots were incubated in 250 µl of the diluted antiserum for 6 hr at room temperature. The blots were combined and treated with ^{125}I-labeled sheep anti-mouse IgG and autoradiographed. Electrophoresis was from left to right.

With radioactively labeled or peroxidase-conjugated antibodies the method is sensitive enough to detect small amounts of electrophoretically separated antigen, and this simple procedure can also be used to show the presence of small amounts of antibody in a serum of low titer. Because the antigen is immobilized on a sheet, the antibody is not required to form a precipitate with the antigen. The blotting technique therefore has the potential for immunoelectrophoretic analysis of proteins by using binding of Fab fragments or binding of antibodies against a single determinant, such as monoclonal antibodies produced by hybridomas (24). This could not be done by current immunoelectrophoretic techniques. If hybridoma clones are obtained from a mouse immunized with impure immunogen, it will be possible to use the technique to screen for clones making antibody directed against a desired antigen. Provided the desired antigen has a characteristic mobility in polyacrylamide gel electrophoresis, the appropriate clone can be selected without ever having pure antigen.

The procedure described here also has potential as a tool for screening pathological sera containing auto-antibodies—e.g., those against ribosomes (25–27). The precise identification of the immunogenic components may be a useful diagnostic tool for various pathological conditions.

A further advantage of immobilization of proteins on nitrocellulose is the ease of processing for autoradiography. Conventional staining, destaining, and drying of polyacrylamide gels takes many hours, and the exact drying conditions are extremely critical, especially for 18% gels as used in the second dimension for ribosomal proteins (12). When the proteins are transferred to a nitrocellulose support, as described here, the electrophoretic blotting takes 1 hr, staining and destaining less than 10 min, and drying an additional 5 min. This is thus both faster and simpler than conventional procedures, and it eliminates the tedious and hazardous procedure of soaking the gels in diphenyloxazole (19).

The technique has been developed to detect specific antisera against ribosomal proteins. However, it is applicable to any analytical procedure depending on formation of a protein–ligand complex. With the blotting technique, the usual procedure of forming a complex in solution and retaining it on a membrane would have to be reversed: the protein, already adsorbed to the membrane, would have to retain the ligand from a solution into which the membrane is immersed. Interactions that can possibly be analyzed in this way include hormone–receptor, cyclic AMP–receptor, and protein–nucleic acid interactions. The ligand may also be a protein. Enzymes separated on polyacrylamide gels could also be conveniently localized on blots by *in situ* assays. A critical requirement for these applications is that the protein is not damaged by the adsorption process and that binding sites remain accessible to ligands and substrates. In this respect, considerations similar to those in affinity chromatography and insoluble enzyme techniques pertain.

The method could also be adapted to the procedure of Cleveland *et al.* (28) for the analysis of proteins eluted from bands in polyacrylamide gels by one-dimensional fingerprints: one could label by iodination *in situ* on the nitrocellulose and then carry out the proteolytic digestion.

In preliminary experiments we have attempted to identify ribosomal RNA binding proteins by binding RNA to ribosomal proteins immobilized on nitrocellulose by the procedure of this paper, followed by staining for RNA (unpublished data), and have found a tendency for nonspecific binding. However, J. Steinberg, H. Weintraub, and U. K. Laemmli (personal communication) have independently developed a similar procedure for identifying DNA binding proteins.

We thank Drs. J. Schmidt and F. Dietrich for advice and help with immunization procedures and Mrs. M. Towbin for advice on setting up the peroxidase assay.

1. Gordon, A. H. (1971) in *Laboratory Techniques in Biochemistry and Molecular Biology*, eds. Work, T. S. & Work, E. (North-Holland, Amsterdam), p. 62.
2. Williamson, A. R. (1971) *Eur. J. Immunol.* **1**, 390–394.
3. Grabar, P. & Williams, C. A. (1955) *Biochim. Biophys. Acta* **17**, 67–74.
4. Laurell, C.-B. (1965) *Anal. Biochem.* **10**, 358–361.
5. Zubke, W., Stadler, H., Ehrlich, R., Stöffler, G., Wittmann, H. G. & Apirion, D. (1977) *Mol. Gen. Genet.* **158**, 129–139.
6. Showe, M. K., Isobe, E. & Onorato, L. (1970) *J. Mol. Biol.* **107**, 55–69.
7. Southern, E. M. (1975) *J. Mol. Biol.* **98**, 503–517.
8. Broome, S. & Gilbert, W. (1978) *Proc. Natl. Acad. Sci. USA* **75**, 2746–2749.
9. Sharon, J., Morrison, S. L. & Kabat, E. A. (1979) *Proc. Natl. Acad. Sci. USA* **76**, 1420–1424.
10. Hamel, E., Koka, M. & Nakamoto, T. (1972) *J. Biol. Chem.* **247**, 805–814.
11. Möller, W., Groene, A., Terhorst, C. & Amons, R. (1972) *Eur. J. Biochem.* **25**, 5–12.
12. Ramjoué, H.-P. R. & Gordon, J. (1977) *J. Biol. Chem.* **252**, 9065–9070.
13. Laemmli, U. K. (1970) *Nature (London)* **227**, 680–685.
14. Schaffner, W. & Weissmann, C. (1973) *Anal. Biochem.* **56**, 502–514.
15. Avrameas, S. & Guilbert, B. (1971) *Eur. J. Immunol.* **1**, 394–396.
16. Kuno, H. & Kihara, H. K. (1967) *Nature (London)* **215**, 974–975.
17. Moore, G. & Crichton, R. R. (1974) *Biochem. J.* **143**, 604–612.
18. McConkey, E. H., Bielka, H., Gordon, J., Lastick, S. M., Lin, A., Ogata, K., Reboud, J.-P., Traugh, J. A., Traut, R. R., Warner, J. R., Welfle, H. & Wool, I. G. (1979) *Mol. Gen. Genet.* **169**, 1–6.
19. Bonner, W. M. & Laskey, R. L. (1974) *Eur. J. Biochem.* **46**, 83–88.
20. Randerath, K. (1970) *Anal. Biochem.* **34**, 188–205.
21. Terhorst, C., Wittmann-Liebold, B. & Möller, W. (1972) *Eur. J. Biochem.* **25**, 13–19.
22. Stöffler, G. & Wittmann, H. G. (1971) *J. Mol. Biol.* **62**, 407–409.
23. Howard, G., Smith, R. L. & Gordon, J. (1976) *J. Mol. Biol.* **106**, 623–637.
24. Köhler, G. & Milstein, C. (1976) *Eur. J. Immunol.* **6**, 511–519.
25. Schur, P. H., Moroz, L. A. & Kunkel, H. G. (1967) *Immunochemistry* **4**, 447–453.
26. Miyachi, K. & Tan, E. M. (1979) *Arthritis Rheum.* **22**, 87–93.
27. Gerber, M. A., Shapiro, J. M., Smith, H., Jr., Lebewohl, O. & Schaffner, F. (1979) *Gastroenterology* **76**, 139–143.
28. Cleveland, D. W., Fischer, S. G., Kirschner, M. W. & Laemmli, U. K. (1977) *J. Biol. Chem.* **252**, 1102–1106.

Supplementary Readings

Benton, W.D. & Davis, R.W., Screening λgt recombinant clones by hybridization to single plaques in situ. *Science* 196:180–182 (1977)

Berk, A.J. & Sharp, P.A., Sizing and mapping of early adenovirus mRNAs by gel electrophoresis of S1 endonuclease-digested hybrids. *Cell* 12:721-732 (1977)

Langer, P.R., Waldrop, A.A. & Ward, D.C., Enzymatic synthesis of biotin-labeled polynucleotides: Novel nucleic acid affinity probes. *Proc. Natl. Acad. Sci. U.S.A.* 78:6633-6637 (1981)

Lathe, R., Synthetic oligonucleotide probes deduced from amino acid sequence data. Theoretical and practical considerations. *J. Mol. Biol.* 183:1–12 (1985)

O'Farrell, P.H., High resolution two-dimensional electrophoresis of proteins. *J. Biol. Chem.* 250:4007–4021 (1975)

Schwartz, D.C. & Cantor, C.R., Separation of yeast chromosome-sized DNAs by pulsed field gradient gel electrophoresis. *Cell* 37:67–75 (1984)

Wallace, R.B., Johnson, M.J., Suggs, S.V., Miyoshi, K., Bhatt, R. & Itakura, K., A set of synthetic oligodeoxyribonucleotide primers for DNA sequencing in the plasmid vector pBR322. *Gene* 16:21–26 (1981)

Winter, G., Fields, S., Gait, M.J. & Brownlee, G.G., The use of synthetic oligodeoxynucleotide primers in cloning and sequencing segment 8 of influenza virus (A/PR/8/34). *Nucleic Acids Res.* 9:237–245 (1981)

Cloning I

In 1973 Stanley Cohen and coworkers published a paper entitled "Construction of biologically functional bacterial plasmids *in vitro*." This paper described of the first recombinant DNA experiments. Many of our readers will set this as the first event in the field of genetic engineering and modern biotechnology. It was also the basis of the well known Cohen and Boyer or Stanford patent (see Section 13). In this first paper will be described the experiments in which these investigators mixed two plasmid DNAs digested with the restriction enzyme EcoR1 and, after ligation, introduced the recombinant or chimeric DNA into *Escherichia coli*. They were then able to propagate the bacteria and isolate greatly amplified recombinant plasmid DNA.

Even though the Cohen et al. paper had a major impact on the field of biotechnology, the reader will notice that the experiments were built on techniques which were previously described; for instance the discovery of type two restriction enzymes (Smith and Wilcox, 1970) and the use of ligase to join DNA molecules (Sgaramella and Khorana, 1972) (see Section 1). Another critical technique used in this original cloning experiment and still used in recombinant DNA laboratories is the introduction of purified DNA into the host cells or bacterial transformation. This procedure had been long known for gram positive bacteria and in fact was used in the original studies by Avery to demonstrate that DNA was the genetic material. However, *E. coli* and other gram negative organisms had been refractory to this procedure. The technical breakthrough came with the work of Mandel and Higa (1970) who found that a Ca++ shock would permeabilize a fraction of *E. coli* cells to DNA.

Cloning, or recombinant DNA generation, involves the joining of a target DNA molecule to a "vector" or "vehicle" DNA molecule which provides, at a minimum, the ability to replicate the recombinant DNA in the desired host. This replication results in the selective amplification of the target DNA, increasing its abundance and purity. The Cohen et al. paper introduced the use of bacterial plasmids as vectors. Plasmids are circular DNA molecules which replicate autonomously in the cell. The further development of plasmid vectors was largely pioneered by the work described by Bolivar et al. Plasmid cloning vectors were developed which were relatively small, existing in multiple copies within the host, encoded an easily selectable marker and contained one or more restriction sites for use in cloning. Since the original plasmid vectors were constructed, still more carefully designed plasmids have been created to facilitate the development of new cloning techniques.

Several important alternative vector systems have been described. The Rimm et al. paper describes one of the most important, the bacteriophage lambda (λ). The development of this system is an excellent case study of how the preexisting extensive information about *E. coli* and its viruses provided tools which were essential for the initial steps in recombinant DNA technology development. For instance, the λ genetic functions which are required (and those which are dispensable) for viral propagation were known in detail, and the mechanism of viral assembly had been studied sufficiently to understand its DNA requirements and to provide an efficient *in vitro* system. The resulting protocol was used to package chimeric DNA molecules into viral capsids, and the viral capsid was used to inject the DNA into host cells with a high efficiency. The Collins and Hohn paper introduced a hybrid vehicle system (cosmids) allowing the λ capsid packaging system to be used in conjunction with plasmid cloning. This powerful approach provided the capability of generating large (~35 kbp) clones.

Messing and Viera describe the adaptation of the single stranded M13 virus into a cloning vector. These were of critical importance because they allowed the isolation of single stranded clones needed for dideoxy (Sanger) sequencing and site specific mutagenesis procedures. In addition, the procedure made use of insertional inactivation of a vector gene to identify successful clones (the blue - white screen). This is a classic genetic protocol.

Burke et al. developed an important yeast cloning system; the generation of yeast artificial chromosomes (YACs). YACs permit the stable maintenance and therefore the immortalization of cloned DNAs considerably larger in size than any other system (several hundred kilobase pairs in length). A YAC is a linear molecule which carries a centromere, a yeast selectable function, the cloned DNA and telomeres which define the ends. The development of YACs is particularly useful for the cloning of large genomic segments, and thus for various aspects of the human genome project.

Most of the cloning strategies described in this collection are designed to pick out one specific sequence. The Clarke and Carbon manuscript introduces an alternative approach; generating a random population of cloned molecules sometimes called a "bank" or a "library." This has become a common approach in which the library once generated can be a source from which hopefully any desired sequence can be isolated.

The early cloning experiments were designed to isolate cloned copies of genomic DNA. However, since many genes whose expression were desired were eukaryotic and the expression host was frequently *E. coli*, this presented significant technical problems. Eukaryotic genes typically contain introns which *E. coli* cannot remove from the mRNA. A solution to this is to clone a DNA copy of the mature mRNA. The discovery of reverse transcriptase (see Temin and Mizutomi and Baltimore in Section 1) provided the basic tool for this process. The earliest attempts yielded partial cDNA clones (see Rougeon et al.), a common problem due to incomplete cDNA synthesis by reverse transcriptase. Within a year, however, nearly full length cDNA clones were achieved (see supplemental readings by Maniatis et al., and Rabbits). The cDNA cloning experiments demonstrate the use of many of the techniques described in other highlights: reverse transcriptase, size fractionation of polynucleotides, terminal transferase tailing, plasmid cloning, colony hybridization, restriction digestion.

cDNA cloning techniques were extended to allow the cloning of still larger genes (see supplemental article by Noda et al.) or ones which led to the production of important pharmacological agents (e.g., insulin, see Ullrich et al.). The Okayama and Berg paper describes an important enhancement in cDNA cloning technology. Through the use of a plasmid vector which serves also as the primer for DNA synthesis, and an interesting adaptation of the tailing strategy favoring full length products, much higher efficiency cloning of full length cDNAs was accomplished. This facilitated the subsequent isolation of full length cDNAs corresponding to rare mRNAs. Subsequently this vector was modified to provide for gene expression as described in the second Okayama and Berg paper (see Section 5).

Construction and Characterization of New Cloning Vehicles

II. A Multipurpose Cloning System

F. Bolivar, R.L. Rodriguez, P.J. Greene, M.C. Betlach,
H.L. Heynker, H.W. Boyer, J.H. Crosa and S. Falkow

SUMMARY

In vitro recombination techniques were used to construct a new cloning vehicle, pBR322. This plasmid, derived from pBR313, is a relaxed replicating plasmid, does not produce and is sensitive to colicin E1, and carries resistance genes to the antibiotics ampicillin (Ap) and tetracycline (Tc). The antibiotic-resistant genes on pBR322 are not transposable. The vector pBR322 was constructed in order to have a plasmid with a single PstI site, located in the ampicillin-resistant gene (Ap^r), in addition to four unique restriction sites, EcoRI, HindIII, BamHI and SalI. Survival of Escherichia coli strain X1776 containing pBR313 and pBR322 as a function of thymine and diaminopimelic acid (DAP) starvation and sensitivity to bile salts was found to be equivalent to the non-plasmid containing strain. Conjugal transfer of these plasmids in bi- and triparental matings were significantly reduced or undetectable relative to the plasmid ColE1.

*Present addresses: (F.B.) Departamento de Biologia Molecular, Instituto de Investigaciones Biomedicas, Universidad Nacional Autonoma de Mexico, Mexico 20, D.F., Apdo Postal 70228; (H.L.H.) Department of Molecular Genetics, University of Leiden, Wassenaarseweg 64, Leiden (The Netherlands); (R.L.R.) Department of Genetics, Briggs Hall, University of California, Davis, CA 95616 (U.S.A.).

Abbreviations: Ap^r, ampicillin-resistant; Cm^r, chloramphenicol-resistant; Col^{imm}, colicin immunity; DAP, diaminopimelic acid; DTT, dithiothreitol; Km^r, kanamycin-resistant; LB, Luria broth; Nx^r, nalidixic-resistant; SDS, sodium dodecyl sulfate; Sm^r, streptomycin-resistant; Su^r, sulfonamide-resistant; Tc^r, tetracycline-resistant.

INTRODUCTION

Bacterial plasmids and bacteriophage have a key role in recombinant DNA technology. Segments of DNA from diverse origins can be excised with the appropriate restriction endonuclease and added to plasmids or bacteriophage (Hershfield et al., 1974; Morrow et al., 1974; Cameron et al., 1975). If these new molecules contain an intact replicon, they can be propagated in a suitable host to yield large quantities of recombinant DNA and in some instances, specific gene products (Hershfield et al., 1974). Several bacterial plasmids have been used as cloning vectors: pSC101 (Cohen et al., 1973), ColE1 (Hershfield et al., 1974) and pCR1 (Covey et al., 1976). However, these plasmids and their derivatives (Hamer et al., 1975; Hershfield et al., 1976; So et al., 1976) have limited versatility in terms of genetic markers for selection of transformants and screening for recombinant plasmids.

We have described the construction of a series of plasmids containing Ap- and Tc-resistant genes derived from pRSF2124 (So et al., 1976) and pSC101 respectively in combination with replication elements of a ColE1-like plasmid (Betlach et al., 1976; Rodriguez et al., 1976). One of these plasmids, pBR313, provides single cleavage sites for the *Hind*III, *Bam*HI, *Eco*RI, *Hpa*I, *Sal*I and *Sma*I restriction endonucleases (Bolivar et al., 1977). In the case of the *Hind*III, *Bam*HI and *Sal*I endonuclease cloning sites, the insertion of DNA fragments inactivates the Tc^r gene. In this paper, we report the construction of another plasmid (pBR322) which is less than half the size of pBR313 and provides additional cloning advantages. The plasmid pBR322 contains a unique *Pst*I cleavage site located in the Ap^r gene as well as two *Hinc*II sites located in the Ap^r and Tc^r genes. The *Pst*I site can be used for molecular cloning of DNA fragments via homodeoxy polymeric extension (Lobban and Kaiser, 1973) and the *Hinc*II site for blunt-end ligation techniques (Sgaramella et al., 1970; Sugino et al., 1977). The properties of pBR313 and pBR322 in the *E. coli* strain X1776 are also presented.

MATERIAL AND METHODS

(a) Bacterial strains

E. coli K12 strain RR1 F⁻*pro leu thi lac*Y Str^r r_k^- m_k^- was used as the recipient cells in the transformation experiments. *E. coli* B strain HB50 *pro leu try his arg met thr gal lac*Y Str^r was used to prepare unmethylated plasmid DNA for *Eco*RII digestions (Yoshimori et al., 1972). *E. coli* K12 strain X1776 F⁻*tau*A53 *dap*D8 *mer*A1 *sup*E42 Δ40(*gal-uvr*B)λ⁻ *min*B2 *mal*A25 *thy*A57 *met*C65 Δ29(*bio*H-*asd*) *cys*B2 *cyc*A1 *Hsd*R2 was kindly provided by R. Curtiss III.

(b) Media and buffers

For transformation RR1 was grown in either LB or M9-glucose minimal media, before CaCl₂ treatment. X1776 was also grown in LB supplemented with DAP 200 μg/ml and thymine (thy) 50 μg/ml. The BSG buffer solution used for washing X1776 in the DAP-less death experiments was 0.85% NaCl, 0.03% KH_2PO_4, 0.06% Na_2HPO_4 100 μg/ml gelatin.

(c) Preparation of plasmid DNA

Plasmid DNA was prepared by first amplifying M9-glucose-grown cultures by the addition of 170 μg/ml of chloramphenicol during logarithmic phase of growth (Clewell et al., 1972). Extraction and purification of plasmid DNA was achieved by a cleared lysate technique previously described (Betlach et al., 1976).

(d) Enzymes

All the restriction enzymes used in this work, except for *Hpa*I (BRL laboratories) were purified according to the procedure by Greene et al. (1977) and are itemized in Table I. Reaction conditions for the various restriction endonucleases have been described previously (Bolivar et al., 1977). T4 DNA ligase was purified from T4 am N82 infected *E. coli* B, according to the procedure described by Panet et al. (1973). The final preparation (500 U/ml) was homogeneous as judged by SDS-polyacrylamide gel electrophoresis.

(e) Ligation of DNA

Ligations were carried out in 66 mM Tris—HCl pH 7.6, 6.6 mM $MgCl_2$ 10 mM DTT and 0.5 mM ATP at 12°C for 2—12 h. The concentration of T4 DNA ligase and of DNA termini varied to promote polymerization or circularization. When blunt-ended DNA fragments were ligated, the concentration of ends was at least 0.2 μM and approximately 50 U of T4 DNA ligase per ml was added to the reaction mixture (Heyneker et al., 1976). When DNA fragments with cohesive ends were ligated, 5 U of T4 DNA ligase per ml was sufficient and the concentration of ends was adjusted in such a way that linear molecules were favored (Dugaiczyk et al., 1975).

(f) Agarose and acrylamide gel electrophoresis

The conditions for agarose and acrylamide electrophoresis have been previously described (Bolivar et al., 1977).

(g) Transformation of E. coli K12

E. coli RRI cells were prepared for transformation by the method described by Cohen et al. (1972). 100 μl of DNA in 30 mM $CaCl_2$ were added to 200 μl of $CaCl_2$-treated cells (5 · 10^9 cells/ml) and the mixture was chilled in ice for 60 min, after which it received a 75-sec, 42°C heat pulse. The pulse was terminated with the addition of 3 ml of LB. The cells were grown for 2 h at 37°C before plating. Transformation of X1776 was achieved using the procedure described by R. Curtiss III (personal communication). An overnight culture of X1776 in LB + DAP + thy was diluted 1/10 with 20 ml of fresh LB + DAP + thy and incubated in a shaker et 37°C for 3 to 4 h until the

TABLE I

RESTRICTION ENDONUCLEASES

Endonucleases	Substrate site	Reference	Endonuclease	Substrate site	Reference
AluI	A G↓C T	Roberts et al., 1976	HaeIII	G G↓C C	Roberts et al., unpublished observations
BamHI	G↓G A T C C	Wilson and Young, 1975	HincII	G T Py↓PuA C	Landy et al., 1974
BglI	---	Wilson and Young, unpublished observations	HindIII	A↓A G C T T	Danna et al., 1973
EcoRI	G↓A A T T C	Greene et al., 1976	HpaI	G T T↓A A C	Gromkova and Goodgal, 1972
EcoRII	↓C C ^A_T G G	Yoshimori et al., 1975	PstI	C T G C A↓G	Smith et al., 1976
HaeII	PuG C G C↓Py	Roberts et al., unpublished observations	SalI	G↓T C G A C	Bolivar et al., 1977

culture reached an absorbance of 0.5 to 0.6 A_{600}. The culture was centrifuged at 7000 rpm for 10 min at 4°C, and the cells washed in 10 ml cold 10 mM NaCl. The suspension was again centrifuged as above and the pellet resuspended in 10 ml of freshly prepared cold 75 mM $CaCl_2$ (pH 8.4) and placed in ice for 25 min. Cells were centrifuged as described above and the pellet resuspended in 2 ml of 75 mM $CaCl_2$ pH 8.4, of which 200 μl was added to 100 μl of plasmid DNA in 10 mM Tris pH 8. The mixture was kept in ice for 60 min, then heated 60 sec at 42°C. Tubes were chilled for 10 min and 3 ml of LB + DAP + thy were added. The cells were incubated at 37°C for 3 h and plated in selective media. The plates were incubated 2 to 3 days at 37°C.

RESULTS

I. Construction of pBR321 and pBR322

We have described the construction of a series of cloning vehicles, one of which, pBR313, a 5.8 · 10^6 dalton Ap^r Tc^r Col^{imm} plasmid (Fig.1), has been extensively mapped using 14 restriction endonucleases (Bolivar et al., 1977). Experiments with pBR313 indicated that one of its PstI sites was located in the Ap^r gene. Therefore molecular cloning into this PstI site would result in recombinant molecules which could be detected by screening for Ap^s phenotypes. In order to construct a molecular cloning vector with one PstI site in the Ap^r gene, it was necessary to construct two derivatives of pBR313. An $Ap^s Tc^r Col^{imm}$ plasmid, pBR318, containing one PstI site was obtained by transforming E. coli RRI with ligated PstI fragments of pBR313 and selecting for Tc^r transformants. Tc^r transformants which were Ap^s were found to carry plasmids that lack the 1.25 and 0.42 · 10^6 dalton PstI fragments present in pBR313 (Fig.1). Another pBR313 derivative, pBR320, an $Ap^r Tc^s Col^s$ plasmid with a molecular weight of 1.95 · 10^6 daltons was obtained by transforming E. coli RRI with unligated EcoRII fragments of pBR313 and selecting for Ap^r transformants. Sixteen $Ap^r Tc^s$ clones were examined, and one was found to carry a plasmid, pBR320, containing only one PstI site. This clone was found to be sensitive to colicin E1. Fig. 1 shows a tentative restriction endonuclease map of pBR320.

An in vitro recombination experiment using pBR318 and pBR320 was designed to restore the $Ap^r Tc^r$ markers in a single low molecular weight relaxed plasmid containing one PstI substrate site. The construction of this plasmid was accomplished by the digestion of pBR318 with PstI and HpaI endonucleases which resulted in two pieces of DNA with molecular weights of 1.95 and 2.2 · 10^6 daltons; the smaller DNA fragment carried the Tc^r gene(s) (Rodriguez et al., 1976; Tait et al., 1976; Bolivar et al., 1977) and part of the Ap^r gene as shown in Fig. 1. The plasmid pBR320 was cleaved with the restriction enzymes PstI and HincII to yield three fragments of DNA. The largest fragment, 1.15 · 10^6 daltons, carries the "origin" of replication and the remaining portion of the Ap^r gene not present in the 1.95 · 10^6 dalton fragment of pBR318.

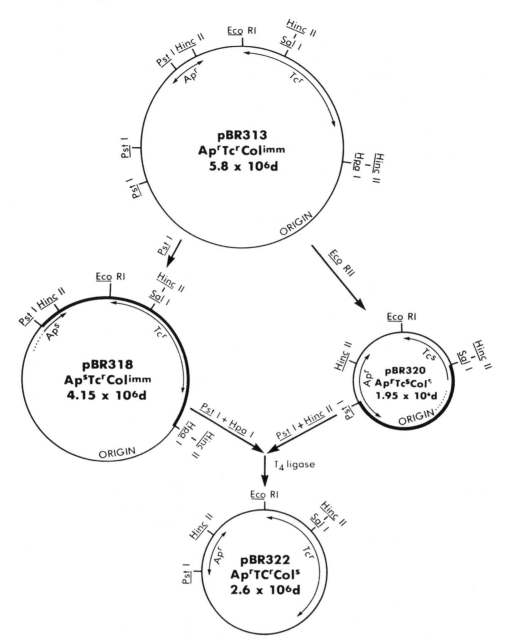

Fig. 1. Diagrammatic representation of the construction of pBR322. The parental plasmid pBR313 was used to construct pBR318 and pBR320 by *Pst*I and *Eco*RII endonuclease digestions respectively. These two plasmids were separately digested with *Pst*I and *Hpa*I endonucleases (pBR318) and with *Pst*I and *Hinc*II endonucleases (pBR320). The heavy lined regions from the *Pst*I to *Hpa*I sites in pBR318 and from *Pst*I to *Hinc*II sites (*Sal*I) in pBR320 represent the two DNA fragments that were ligated to each other to generate pBR321 and pBR322. The origins of replication in these plasmids were determined by restriction endonuclease analysis and electron microscopic examinations (unpublished observations). For detailed explanation, see the text.

The digested DNAs were mixed, ligated in vitro, and transformed into *E. coli* RRI. Since neither pBR320 nor the $1.95 \cdot 10^6$ dalton fragment of pBR318 carry the colicin E1 immunity gene, transformants were selected for $Ap^r Tc^r$ and then screened for sensitivity to colicin E1. This transformation yielded numerous $Ap^r Tc^r Col^s$ clones which carried plasmids (e.g. pBR 321) with a molecular weight of $3.1 \cdot 10^6$ daltons. As expected, this plasmid resulted from the addition of the $1.95 \cdot 10^6$ dalton *Hpa*I-*Pst*I fragment of pBR318 and the $1.15 \cdot 10^6$ d. *Pst*I-*Hinc*II DNA piece from pBR320.

From this transformation we obtained in one instance a smaller, $2.6 \cdot 10^6$ dalton, $Ap^r Tc^r Col^s$ plasmid. This plasmid, possibly the result of in vivo recombination event near unligated termini, was missing $0.5 \cdot 10^6$ daltons of DNA from a region of pBR321, not associated with Ap^r, Tc^r or DNA replication (Fig. 2). Because of its lower molecular weight compared with pBR 321, pBR 322 was chosen for further characterization of number and position of restriction sites.

II. Mapping of pBR322 restriction endonuclease digestions

As determined by agarose and acrylamide gel electrophoresis of DNA digests, pBR322 was found to carry unique substrate sites for the *Bam*HI, *Eco*RI, *Hind*III, *Pst*I and *Sal*I restriction endonucleases. Double and triple endonuclease digests of the plasmid (data not shown) showed that the relative positions of these sites were identical to those mapped in pBR313. As can be seen in Figs. 2 and 3, there are only two *Hinc*II sites in pBR322, one located $0.17 \cdot 10^6$ daltons from the *Pst*I site in the Ap^r gene and the other in Tc^r gene, which is also a *Sal*I site (Bolivar et al., 1977).

The *Eco*RII restriction endonuclease was used to further characterize pBR322. As shown in Fig. 4a (slot 3), pBR322 has five *Eco*RII sites which yield fragments of 1.25, 0.64, 0.53, 0.22 and $0.04 \cdot 10^6$ daltons upon digestion. Slots 3, 4, 5, 6 and 7 (Fig. 4a) show respectively, *Eco*RI, *Hinc*II, *Sal*I, *Bam*HI and *Bgl*I digestions of *Eco*RII-digested pBR322 DNA. In the case of the *Eco*RI endonuclease digest, the largest *Eco*RII fragment when cleaved gives two new fragments of 0.08 and $1.07 \cdot 10^6$ daltons. Slots 5 and 6 show that *Sal*I and *Bam*HI endonucleases cleave the same $0.53 \cdot 10^6$ dalton *Eco*RII fragment as in pBR313 (Bolivar et al., 1977) and generates 0.29 and $0.24 \cdot 10^6$ dalton DNA fragments after the *Sal*I endonuclease digestion and 0.4 and $0.13 \cdot 10^6$ dalton fragments after *Bam*HI endonuclease digestion. Slot 4 shows that the same $0.53 \cdot 10^6$ dalton *Eco*RII piece is cleaved by *Hinc*II endonuclease, generating the same 0.29 and $0.24 \cdot 10^6$ dalton fragments that the *Sal*I endonuclease produces. The $1.25 \cdot 10^6$ dalton *Eco*RII fragment is also cleaved by *Hinc*II endonuclease into 0.82 and $0.43 \cdot 10^6$ dalton fragments. Slot 7 shows the double digestion pattern of pBR322 DNA using *Eco*RII and *Bgl*I endonucleases. Three *Eco*RII fragments are cleaved by *Bgl*I endonuclease into smaller pieces. The largest *Eco*RII fragment ($1.25 \cdot 10^6$ daltons) is cleaved into 0.57 and $0.67 \cdot 10^6$ dalton fragments. *Bgl*I endonuclease also cleaves the $0.53 \cdot 10^6$ dalton fragments into 0.47 and $0.06 \cdot 10^6$ dalton

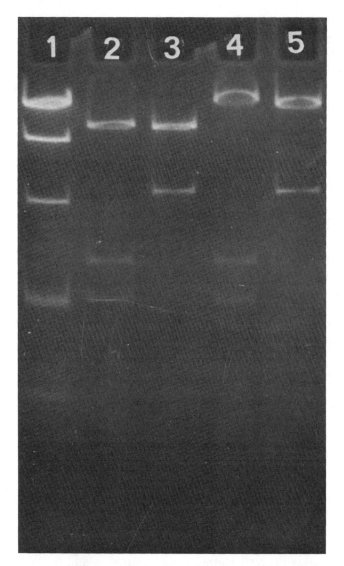

Fig. 2. *Hinc*II-*Eco*RI endonuclease analyses of pBR321 and pBR322. Molecular weight estimates are based in the seven PM2 fragments generated by the *Hind*III endonuclease (31.4, 1.34, 0.6, 0.31, 0.29, 0.14, and 0.06, the last one not seen in the gel, Wes Brown, personal communication) (slot 1). The *Hinc*II endonuclease and *Hinc*II-*Eco*RI endonuclease single and double digestions of pBR322 are shown in slots 3 and 2 respectively while the *Hinc*II endonuclease and *Hinc*II-*Eco*RI endonuclease digestions of pBR321 are shown in slots 5 and 4 respectively. It can be seen that the $0.64 \cdot 10^6$ *Hinc*II band present in pBR322 (slot 3) (see also Fig. 3) is also present in pBR321 (slot 5). This band carries the *Eco*RI site (slots 2 and 4). These data indicate that the spontaneous deletion that generates pBR322 does not extend to the Ap[r] or Tc[r] genes nor the region located in the small *Hinc*II fragment ($0.64 \cdot 10^6$ daltons) in pBR321.

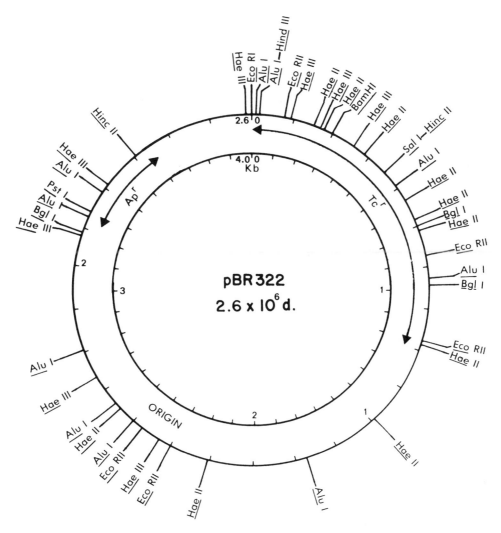

Fig. 3. The circular restriction map of pBR322. The relative position of restriction sites are drawn to scale on a circular map divided into units of $1 \cdot 10^5$ daltons (outer circle) and 0.1 kilobases (inner circle). The estimated size of the Ap^r and Tc^r genes represented in the figure were determined indirectly on the basis of the reported values for the size of the TEM β-lactamase (Datta and Richmond, 1966) and the Tc^r-associated proteins detected in the minicell system (Levy and McMurry, 1974; Tait et al., 1976). Positioning the left-hand boundary of the Tc^r gene was based on our knowledge that cloning into the *Eco*RI site of pBR313 did not affect Tc^r while cloning into the *Hind*III site did affect the Tc^r mechanism. The position and size of the Tc^r region is also consistent with the orientation of the TnA in pBR26 (Bolivar et al., 1977). Only ten out of twelve *Hae*II and *Alu*I and seven out of seventeen *Hae*III substrate sites are represented on the circular map of pBR322. The position of two of the ten *Alu*I sites located at $1.8 \cdot 10^6$ daltons and $1.86 \cdot 10^6$ daltons were mapped on a $0.7 \cdot 10^6$ dalton plasmid which encompasses this region of pBR322 (unpublished observation).

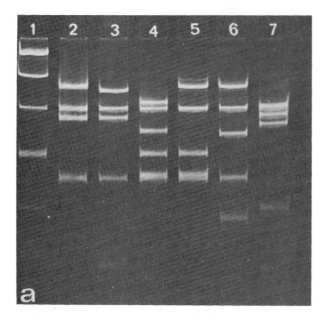

Fig. 4a. Acrylamide gel electrophoresis of *Eco*RII-cleaved pBR322 DNA. Analysis of *Eco*RII endonuclease (slot 2) and double endonuclease *Eco*RII-*Eco*RI digestions (slot 3), *Eco*RII-*Hinc*II (slot 4), *Eco*RII-*Sal*I (slot 5), *Eco*RII-*Bam*HI (slot 6) and *Eco*RII-*Bgl*I (slot 7) of pBR322 DNA. The seven *Hind*III-generated fragments from the DNA of phage PM2 (slot 1) with molecular weights of 3.5, 1.34, 0.6, 0.31, 0.29, 0.15 and 0.06 (Wes Brown, personal communication) were used as molecular weight standards. For explanation see the text.

pieces. The $0.22 \cdot 10^6$ dalton *Eco*RII fragment is cleaved by *Bgl*I endonuclease into 0.16 and $0.06 \cdot 10^6$ dalton DNA fragments. These data allow us to localize four *Eco*RII sites in the pBR322 map (Fig. 3).

The *Alu*I endonuclease cleaves pBR322 into approximately 12 fragments (Fig. 4b, slot 2). Four *Alu*I sites were mapped by analysis of the molecular weights of DNA fragments generated by double endonuclease digestion (Fig. 3). By comparing the *Alu*I fragments of pBR318 and pBR320 (data not shown), we have tentatively localized four additional *Alu*I sites in the map of pBR322. The *Alu*I site located between the *Eco*RI and the *Hind*III sites (Fig. 3) was determined by DNA sequencing (J. Shine et al., unpublished observations).

Using the same strategy of the analysis of double digestion patterns we were able to locate 10 out of 12 *Hae*II sites and the 3 *Bgl*I sites present in the pBR322 map (Fig. 3) (data not shown).

III. Cloning properties of pBR322

(a) *Cloning in the Tc^r gene.* It has been previously shown (Rodriguez et al., 1976; Bolivar et al., 1977) that the *Hind*III, *Bam*HI and *Sal*I sites are present

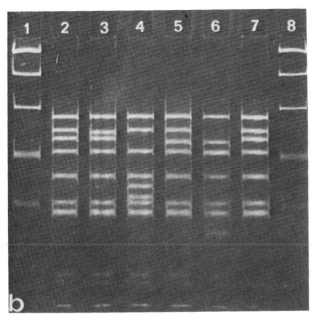

Fig. 4b. Acrylamide gel electrophoresis of *Alu*I endonuclease digested pBR322 DNA. Analysis of *Alu*I endonuclease (slot 2) and double digestions of *Alu*I-*Eco*RI (slot 3), *Alu*I-*Bam*HI (slot 4), *Alu*I-*Sal*I (slot 5), *Alu*I-*Hinc*II (slot 6), and *Alu*I-*Pst*I (slot 7) endonucleases of pBR322 DNA. The seven *Hin*dIII generated fragments from the DNA of phage PM2 (slot 1 and 8) were used as molecular weight standards. It can be seen that the second largest *Alu*I endonuclease generated fragment ($0.43 \cdot 10^6$ daltons) is cleaved by the *Eco*RI endonuclease generating a $0.42 \cdot 10^6$ daltons band and a $0.01 \cdot 10^6$ dalton fragment not seen in the gel. *Bam*HI endonuclease cleaves the $0.38 \cdot 10^6$ dalton fragment into two fragments of 0.2 and $0.18 \cdot 10^6$ daltons. *Sal*I endonuclease cleaves the same $0.38 \cdot 10^6$ dalton *Alu*I fragment generating a piece of $0.34 \cdot 10^6$ daltons and a small one of $0.04 \cdot 10^6$ daltons (not seen in the gel). *Hinc*II endonuclease cleaves two *Alu*I fragments; cleaves in the *Sal*I site generating the same $0.34 \cdot 10^6$ dalton *Alu*I fragment. The relative positions of these fragments can be seen in Fig. 3.

in the Tc^r gene(s) carried by pBR313. Since these sites are in the same relative position in pBR322, we assumed they are associated with the Tc^r gene(s) present in this plasmid. To confirm this point, DNA fragments from *E. coli*, *Drosophila melanogaster*, and *Neurospora crassa* were produced by digestion with *Hin*dIII, *Bam*HI and *Sal*I restriction endonucleases and cloned into their respective sites in pBR322. These recombinants were Ap^{r}Tc^s (Table II). Insertion of DNA fragments into the *Eco*RI site, as in pBR313, does not affect the expression of the Tc^r mechanism.

(b) Cloning in the Ap^r gene. 1. Cloning into the *Pst*I site. To confirm the observation that the *Pst*I site in pBR322 is located in the Ap^r gene, we have cloned several fragments of DNA in *Pst*I site of this plasmid (see Table II). *Pst*I generated fragments of the plasmid pMB1 (Betlach et al., 1976) were

ligated into the *Pst*I site of pBR322 and transformed into RR1. Transformants were selected for Tc^r and screened for Ap^s phenotypes. This experiment resulted in the cloning of five different *Pst*I fragments representing nearly the whole pMB1 genome (Table II).

The unique *Pst*I site in pBR322 provides two advantages for the molecular cloning of DNA by means of the homodeoxypolymeric DNA extention technique (Lobban and Kaiser, 1973). First, the *Pst*I site, which has the sequence C T G C A$^{\downarrow}$G (Smith et al., 1976) provides a protruding 3'OH which

TABLE II

MOLECULAR CLONING OF VARIOUS DNA FRAGMENTS IN pBR322

Sources	Restriction endonuclease substrate site					
	*Eco*RI	*Hind*III	*Bam*HI	*Sal*I	*Hinc*II[a]	*Pst*I
N. crassa	2.8	2.0	4.5	2.6		
	1.1	1.8	3.2	1.1		
			2.1	0.8		
			0.5			
pMB1[b]					2.75[c]	
					2.7[d]	
pMB1						2.82[e]
						1.16
						0.7
						0.5
						0.25
pCB61[f]		1.7				
D. melanogaster						2.3[g]
						4

[a] *Hinc*II site located in the β lactamase (Ap^r) gene.

[b] pMB1 is a clinical isolate plasmid that carries the *Eco*RI restriction and modification genes as well as the colicin E1 production and colicin E1 immunity genes (Betlach et al., 1976).

[c] pMB1 *Hinc*II fragment carrying the *Eco*RI restriction and modification genes (Greene et al., 1976, manuscript in preparation).

[d] pBM1 *Hinc*II fragment carrying colicin E1 production and colicin E1 immunity genes

[e] There are six *Pst*I fragments in pMB1. The sixth has a molecular weight of $0.046 \cdot 10^6$ daltons and was not identified in the agarose gel screening procedure.

[f] pCB61 is a pBR322 derivative plasmid carrying a $1.7 \cdot 10^6$ dalton fragment of DNA cloned in the *Hind*III site. This fragment was isolated from the TnA transposon, and possibly carries a gene(s) involved in the translocation of the ampicillin translocon (Covarrubias et al., unpublished observations).

[g] These fragments were cloned by the homopolymer extension technique in the *Pst*I site of pBR322 and can be recovered after *Pst*I digestion of the recombinant plasmid DNA (A. Dugaiczyk, personal communication).

is a direct substrate for terminal transferase. Secondly, by extending the *Pst*I site in the plasmid with poly dG and the DNA to be cloned with poly dC, it is possible to regenerate two *Pst*I sites after annealing and repairing these two species of DNA in vivo or in vitro.

By fllowing this procedure, dC tailed *Drosophila melanogaster* DNA has been successfully cloned into the dG tailed *Pst*I site of pBR322. Transformants were selected for Tc^r and screened for Ap^s phenotypes. Two out of five *Drosophila melanogaster* DNA fragments cloned in this way had restored *Pst*I sites (Table II).

(b) 2. Cloning in the *Hinc*II sites of pBR322. The *Hinc*II restriction endonuclease which generates blunt-ended DNA fragments, recognizes two sites on pBR322. One of these sites is located in the Tc^r and is also recognized by the *Sal*I endonuclease, while the other site is located in the Ap^r gene, 0.17 · 10^6 daltons from the *Pst*I site. To demonstrate the cloning of blunt-ended DNA fragments into the *Hinc*II site of the Ap^r gene, we used *Hinc*II digested pMB1 DNA which consisted of two fragments, 2.7 · 10^6 daltons and 2.75 · 10^6 daltons (P. Greene, unpublished observations). In order to preferentially cleave the *Hinc*II site in the Ap^r gene, pBR322 DNA was first cleaved with *Sal*I endonuclease followed by a *Hinc*II endonuclease digestion. These digestions generated two DNA fragments each possessing one cohesive end (*Sal*I site) and one blunt end (*Hinc*II site). The pMB1 and pBR322 DNA were ligated under conditions to promote blunt-end ligation (see MATERIALS AND METHODS) and transformed into *E. coli* RRI. After selecting Tc^r transformants and screening for Ap^s, two out of four Tc^{r}Ap^s transformants were found to contain recombinant plasmids carrying each of the two *Hinc*II fragments of pMB1. The two remaining Tc^{r}Ap^s transformants were found to carry pBR322 plasmids with no detectable change in molecular weight but only one *Hinc*II site equivalent to the *Sal*I site. We believe that these plasmids may result from in vivo recombination at the unligated *Hinc*II termini leading to the loss of both the *Hinc*II site and Ap^r.

IV. Properties of pBR322 in the E. coli strain X1776

The plasmids pBR322 and pBR313 could be used in *E. coli* X1776 as an EK2 system if it can be demonstrated that these vectors do not affect the survivability of *E. coli* 1776 (National Institutes of Health, USA, Recombinant DNA Research Guidelines, 1976). Therefore, pBR322 and pBR313 were transformed to the *E. coli* strain X1776 and examined for cell death in medium devoid of DAP, thymine and medium containing bile salts. DAP-less death of *E. coli* X1776 harboring these plasmids is as efficient as in that shown for *E. coli* X1776 alone (Fig. 5). Table III shows the plating efficiencies of the strains X1776 with and without the plasmids on media containing increasing concentrations of bile salts or lacking thymine. As in the DAP-less experiments, neither the pBR322 nor pBR313 affects the plating efficiency of X1776 under these conditions.

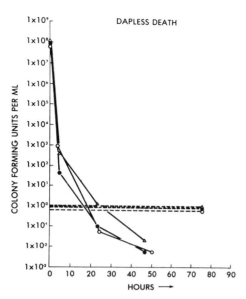

Fig. 5. DAP-less death of X1776, X1776(pBR313) and X1776(pBR322). For each of the strains the following procedure was performed. 500 ml early log cultures (~5 · 10⁷ to 1 · 10⁸ cells/ml) grown in LB + DAP (200 μg/ml) + Thy (50 μg/ml) were centrifuged. The cells were washed with BSG* and resuspended in 5 ml BSG. The resuspended cells were used to inoculate DAP⁻ medium (LB + Thy − NaCl) and DAP⁺ medium (LB + Thy + NaCl) using 0.5 ml of cells to 100 ml medium in each of 10 separate 250 ml erlenmeyer flasks. Cultures were incubated at 37°C with shaking. Samples were removed at various times from a fresh flask each time in order to reduce risk of contamination. The samples were titered on LB agar + DAP + Thy. Time points at 0.2 and 5 h were taken on the control cells grown in medium with DAP to verify their continued growth. At 24 h and later, cells were concentrated by filtering and the Millipore filters placed on LB agar. The dotted line indicates the titer at which the number of colony-forming unit have decreased by 10⁸ from the original titer. △——△, X1776 alone; ○——○, X1776 (pBR313); ●——●, X1776 (pBR322).

Data presented in Table IV show the mobilization of pBR313 and pBR 322 by de-repressed FI or FII R plasmids from "wild-type" E. coli K12 C600. In this case both plasmids are co-resident in the same cell and are mated directly with a wild-type E. coli K12 recipient cell line, SF185. As shown in Table IV, one can demonstrate that if F is coresident with wild-type ColE1, at the time of conjugation, ColE1 is mobilized at a frequency of about 0.5/ recipient cell in 24 h. When F-Km or Rldrd19 are co-resident with pBR313 or pBR322, it is clear that mobilization occurs at a very low frequency. It is also clear that, relative to ColE1, pBR313 and pBR322 are less likely to be directly mobilized from either wild-type E. coli K12 or X1776 by de-repressed F-like plasmids.

Table V shows the transfer of pBR322 and pBR313 from E. coli C600 and E. coli X1776 in triparental matings using conjugative plasmids of various

TABLE III
PLATING EFFICIENCIES OF STRAIN X1776 (pBR313) AND X1776 (pBR322) ON MEDIA CONTAINING BILE SALTS OR WITHOUT THYMINE

Cultures were grown with shaking at 37°C in LB containing DAP (200 μg/ml) and thymine (50 μg/ml) to about 1—2 · 10^8/ml. Cells were pelleted by centrifugation, washed with BSG and concentrated 100-fold in BSG. Dilutions were made in growth medium and plated on the appropriate plates. Plates were incubated at 37°C.

Medium	Plating efficiency[a]		
	X1776	X1776 (pBR313)	X1776 (pBR322)
L agar + DAP + Thy + 0.15% bile salts	$<6.25 \cdot 10^{-6}$	$<7.14 \cdot 10^{-6}$	$2.0 \cdot 10^{-7}$
L agar + DAP + Thy + 0.37% bile salts	$<6.25 \cdot 10^{-8}$	$4.29 \cdot 10^{-9}$	$2.87 \cdot 10^{-8}$
L agar + DAP + Thy + 0.75% bile salts	$<6.25 \cdot 10^{-9}$	$2.6 \cdot 10^{-9}$	$2.4 \cdot 10^{-8}$
M9CAA agar + DAP + biotin (0.5 μg/ml)	$2 \quad \cdot 10^{-9}$	$<7.14 \cdot 10^{-8}$	$6 \quad \cdot 10^{-9}$

[a] Plating efficiency was calculated using the titer of X1776, X1776 (pBR313) or X1776 (pBR322) on L agar + DAP + Thy (without bile salts).

incompatibility groups. Since the FII, I and N compatibility groups comprise over 85% of the conjugative plasmids found in *E. coli* of healthy humans and animals, we focused upon these and two additional conjugative plasmids. Given the results shown in Table IV, it is not surprising that we could not detect ($<10^8$ per final recipient cell in 24 h) any instance for the mobilization of pBR313 and pBR322 from either C600 or X1776 in triparental matings. As a control an intermediate strain carrying the non-conjugative plasmid pRSF2124 was included. pRSF2124, a ColE1 derivative to which the Ap^r gene has been transposed, could be mobilized by F-Km at a frequency of 10^{-4} per final recipient cell within 4 h.

DISCUSSION

A new amplifiable cloning vehicle, the plasmid pBR322, which improves the cloning characteristics of the parental plasmid pBR313 (Bolivar et al., 1977), has been constructed by in vitro recombination techniques.

pBR322 was constructed in order to have a plasmid cloning vector with a single *Pst*I site in addition to those unique restriction sites already present in pBR313. Of the three *Pst*I sites in pBR313, the one located in the Ap^r gene provided the most useful site for cloning purposes. The advantages of having a single *Pst*I site located in the Ap^r gene are the following: (1) molecular cloning of *Pst*I endonuclease digested DNA fragments into the *Pst*I site will lead to the inactivation of the Ap^r gene thus allowing for the detection of cells carrying recombinant plasmids by means of their Ap^s-Tc^r phenotypes; (2) the benefits for molecular cloning by means of homodeoxypoly-

TABLE IV

MOBILIZATION OF pBR313 AND pBR322 by triparental mating[a]

E. coli K12 donor	R plasmid	Inc. group	Drug resistance	Donor cells ml($\cdot 10^8$)	C600 1 h		C600 24 h		C600 control 24 h pRSF2124	X1776 control 4 h pRSF2124	X1776 1 h		X1776 24 h	
					pBR313	pBR322	pBR313	pBR322			pBR313	pBR322	pBR313	pBR322
J5-3	F·Km	FI	Km^r	3	$<10^{-8}$	$<10^{-8}$	$<10^{-8}$	10^{-8}	$2\cdot10^{-4}$	N.D.[b]	$<10^{-8}$	$<10^{-8}$	$<10^{-8}$	$<10^{-8}$
J5-3	Rldrd	FII	Km^r,Su^r, Ap^r, Sm^r, Cm^r	1	$<10^{-8}$	$<10^{-8}$	$<10^{-8}$	10^{-8}	N.D.	$<10^{-8}$	$<10^{-8}$	$<10^{-8}$	$<10^{-8}$	$<10^{-8}$
J5-3	R144	I	Km^{r}Tc^r	5	$<10^{-8}$	$<10^{-8}$	$<10^{-8}$	10^{-8}	$<10^{-8}$	N.D.	$<10^{-8}$	$<10^{-8}$	$<10^{-8}$	$<10^{-8}$
J5-3	N3	N	Tc^r, Sm^r Su^r	5	$<10^{-8}$	$<10^{-8}$	$<10^{-8}$	10^{-8}	$<10^{-8}$	$<10^{-8}$	$<10^{-8}$	$<10^{-8}$	$<10^{-8}$	$<10^{-8}$
J5-3	Sa	W	Sm^r, Su^r, Cm^r, Km^r	2	$<10^{-8}$	$<10^{-8}$	$<10^{-8}$	10^{-8}	$<10^{-8}$	N.D.	$<10^{-8}$	$<10^{-8}$	$<10^{-8}$	$<10^{-8}$
J5-3	RSF1040	X	Ap^r	4.8	$<10^{-8}$	$<10^{-8}$	$<10^{-8}$	10^{-8}	$<10^{-8}$	N.D.	$<10^{-8}$	$<10^{-8}$	$<10^{-8}$	$<10^{-8}$

[a] A triparental mating is one in which the conjugative plasmid resides in one strain (donor), the nonconjugative plasmid to be tested for mobilization in another (intermediate) and a third strain carrying a chromosomal marker serves as the final recipient. Selection is made for the resistance markers present on the nonconjugative plasmid and the chromosome of the final recipient. The intermediate strains were E. coli K12 either C600 or X1776 containing the cloning vehicle pBR313 or pBR322. The final recipient was E. coli K12 W1485-1 Nx^r. Cultures were grown at 37°C in LB plus DAP and thymine. Matings were done by addition to a 250 ml flask of 5 ml each of a culture of the donor strain containing the conjugative R plasmid either C600 or X1776 containing either pBR313 or pBR322 and the final recipient W1485-1. Matings were carried out without shaking for the indicated periods of time. The mating mixtures were plated (0.1 ml) on MacConkey agar containing 20 µg/ml of either tetracycline or ampicillin. C600 (pRSF2124) was included as a control. pRSF2124 is already impaired in its mobilization as compared to ColE1.

Concentrations (cells/ml) of intermediate and final recipient strains:

C600 (pBR313)	C600 (pBR322)	C600 (pRSF2124)	X1776 (pRSF2124)	X1776 (pBR313)	X1776 (pBR322)	W1485-1
Exp. 1 $9.2\cdot10^8$	$4.7\cdot10^8$	$3.7\cdot10^8$	$2\cdot10^8$	$2\cdot10^8$	$2\cdot10^8$	$1\cdot10^8$

[b] N.D., not done.

meric extension are two-fold. Not only is the protruding 3'OH of the cleaved *Pst*I site a direct substrate for *N*-terminal transferase, but the insertion of C-tailed DNA into a G-tailed plasmid generates two *Pst*I sites after polymerization and ligation in vitro or in vivo. This will allow for the recovery of the cloned DNA fragment after digesting the recombinant plasmid with the *Pst*I endonuclease. In addition, the main properties of pBR313 are conserved in pBR322 in that the cloning of *Hind*III, *Sal*I and *Bam*I endonuclease generated fragments in pBR322 continues to inactivate the Tc^r gene. Screening for Tc sensitivity and therefore recombinant plasmids can be imposed on the transformed culture. Cloning of *Pst*I fragments in the appropriate site of pBR322 also inactivates one of the antibiotic-resistance genes. In all of these cases recombinant plasmids in transformed cells possess only one functional antibiotic-resistant gene.

Although the relative positions of the *Eco*RI and *Hind*III sites appear to be the same in both pSC101 and pBR322, cloning into the *Eco*RI site of pBR322 does not affect the level of Tc^r as reported for pSC101 (Tait et al., 1977). Cloning of DNA fragments with one *Eco*RI terminus and the second terminus generated by one of the four endonucleases mentioned above into the appropriately digested plasmid will inactivate the Tc^r or Ap^r function.

Recent experiments of Heyneker et al. (1976) have demonstrated the usefulness of molecular cloning by blunt-end ligation. Therefore, the presence in pBR322 of only two substrate sites recognized by the *Hinc*II endonuclease, which generates blunt-ended DNA fragments, makes this plasmid a potential vector for cloning by blunt-end ligation. Cloning by blunt-end ligation has been achieved in the *Hinc*II site of the Ap^r gene when this site was preferentially cleaved by the *Hinc*II endonuclease by prior digestion of the *Hinc*II site in the Tc^r gene with *Sal*I endonuclease.

A low frequency of plasmid transmissibility has been established by the National Institutes of Health, USA, Recombinant DNA Guidelines as one of the most important safety features of a plasmid cloning vector. On the basis of the data presented in this paper, we feel that pBR313 and pBR322 within *E. coli* X1776 constitutes an improved EK2 host-vector system which can be used for the cloning of a variety of DNAs. Although the mechanism for mobilizing non-conjugative plasmids has received little attention, it is interesting to note that the frequency of mobilization of pBR313, pBR322 and pMB9 (R. Curtiss III, personal communication) has been significantly reduced with respect to wild-type ColE1. It has been recently reported (Gordon Dougan, personal communication) that a gene associated with high mobilization frequency has been mapped on the wild-type ColE1 plasmid. Therefore, ColE1 derivatives which involve deletions (pVH51) and enzymatic rearrangements (pMB9, pBR313 and pBR322) may result in the loss or alteration of mobilization. Although transposition of the Ap^r and Tc^r genes of pBR313 and pBR322 cannot be ruled out, experiments designed to test for this possibility have proven negative (data not shown).

ACKNOWLEDGEMENTS

H.W.B. is an Investigator for the Howard Hughes Medical Institute. This work was supported by grants to H.W.B. from the National Science Foundation (PCM75-10468 A01) and National Institutes of Health (5 R01 CA14026-05). H.L.H., R.L.R. and F.B. were supported by postdoctoral fellowships from the Netherlands Organization for the Advancement of Pure Research (ZWO), the A.P. Giannini Foundation for Medical Research, and CONACYT, Mexico, respectively. We would like to acknowledge Dr. Istvan Fodor, David Russel, Alejandra Covarrubias and Linda K. Luttropp for their discussion and technical assistance. We are also grateful to Patricia L. Clausen for her expert preparation of this manuscript.

REFERENCES

Betlach, M.C., Hershfield, V., Chow, L., Brown, W., Goodman, H.M. and Boyer, H.W., A restriction endonuclease analysis of the bacterial plasmid controlling the *Eco*RI restriction and modification of DNA, Fed. Proc., 35 (1976) 2037—2043.

Bolivar, F., Rodriguez, R., Betlach, M. and Boyer, H.W., Construction and characterization of new vehicles, I. Ampicillin-resistant derivatives of the plasmid pMB9, Gene, 2 (1977) 75—93.

Cameron, J.R., Panasenko, S.M., Lehmon, I.R. and Davis, R.W., In vitro construction of bacteriophage λ carrying segments of *Escherichia coli* chromosome: Selection of hybrids containing the gene for DNA ligase, Proc. Natl. Acad. Sci. USA, 72 (1975) 3416 —3420.

Clewell, D.B., Nature of ColE1 plasmid replication in the presence of chloramphenicol, J. Bacteriol., 110(1972) 667—676.

Cohen, S.N., Chang, A.C.Y., Boyer, H.W. and Helling, R., Construction of biologically functional plasmids in vitro. Proc. Natl. Acad. Sci. USA, 70 (1973) 3240—3244.

Covey, C., Richardson, D. and Carbon, J., A method for the deletion of restriction sites in bacterial plasmid deoxyribonucleic acid, Mol. Gen. Genet., 145 (1976) 155—158.

Danna, K.J., Sack, G.H. and Nathans, D., Studies of simian virus 40 DNA, VII. A cleavage map of the SV40 genome, J. Mol. Biol., 78 (1973) 363—376.

Datta, N. and Richmond, M.D., The purification and properties of a penicillinase whose synthesis is mediated by an R-factor in *Escherichia coli*, Biochem. J., 98 (1966) 204.

Dugaiczyk, A., Boyer, H.W. and Goodman, H.M., Ligation of *Eco*RI endonuclease-generated DNA fragments into linear and circular structures, J. Mol. Biol., 96 (1975) 171.

Greene, P.J., Betlach, M.C., Goodman, H.M. and Boyer, H.W., The *Eco*RI restriction endonuclease, in Wickner, R.B. (Ed.), Methods in Molecular Biology, Marcel Dekker, New York, 1974, pp. 87—111.

Greene, P.J., Heyneker, H., Betlach, M.C., Bolivar, F., Rodriguez, R., Covarrubias, A., Fodor, I. and Boyer, H.W., General method for restriction endonuclease purification, Manuscript in preparation.

Gromkova, R. and Goodgal, S.H., Action of *Haemophilus* endodeoxyribonuclease on biologically active deoxyribonucleic acid, J. Bacteriol., 109 (1972) 987.

Guerry, P., LeBlanc, D.J. and Falkow, S., General method for the isolation of plasmid deoxyribonucleic acid, J. Bacteriol., 116 (1973) 1064-1066.

Hamer, D. and Thomas, C., Molecular cloning of DNA fragments produced by restriction endonucleases *Sal*I and *Bam*I, Proc. Natl. Acad. Sci. USA, 78 (1976) 1537—1541.

Heffron, F., Bedinger, P., Champoux, J.J. and Falkow, S., Proc. Natl. Acad. Sci. USA, 74 (1977) 702—706.

Hershfield, V., Boyer, H.W., Lovett, M., Yanofsky, C. and Helinski, D., Plasmid ColE1 as a molecular vehicle for cloning and amplification of DNA, Proc. Natl. Acad. Sci. USA, 71 (1974) 3455—3461.

Hershfield, V. , Boyer, H.W., Chow, L. and Helinski, D., Characterization of a mini-ColE1 plasmid, J. Bacteriol., 126 (1976) 447—453.

Heyneker, H.L., Shine, J., Goodman, H.M., Boyer, H.W., Rosenberg, J., Dickerson, R.E., Narang, S.A., Itakura, K., Lin, S. and Riggs, A.D., Synthetic *lac* operator DNA is functional in vivo, Nature, 263 (1976) 748—752.

Landy, A., Ruedisueli, E., Robinson, L., Foeller, C. and Ross, W., Digestion of deoxyribonucleic acids from bacteriophage T7, λ and φ80h with site-specific nucleases from *Haemophilus influenzae* strain Rc and strain Rd, Biochemistry, 13(1974) 2134—2141.

Levy, S. and McMurry, L., Detection of an inducible membrane protein associated with R-factor-mediated tetracycline resistance, Biochem. Biophys. Res. Commun., 56 (1974) 1060—1068.

Lobban, P. and Kaiser, A.P., Enzymatic end-to-end joining of DNA molecules, J. Mol. Biol., 78 (1973) 453.

Morrow, J.F., Cohen, S.N., Chang, A.C.Y., Boyer, H.W., Goodman, H.M. and Helling, R.B., Replication and transcription of eukaryotic DNA in *Escherichia coli*, Proc. Natl. Acad. Sci. USA, 71 (1974) 1743—1747.

Panet, A., Van de Sande, J.H., Loewen, P.C., Khorana, H.G., Raae, A.J., Lillehaug, J.R. and Kleppe, K., Physical characterization and simultaneous purification of bacteriophage T4 induced polynucleotide kinase, polynucleotide ligase and deoxyribonucleic acid polymerase, Biochemistry, 12 (1973) 5045—5050.

Recombinant DNA Research Guidelines, National Institutes of Health, USA, Federal Register, 41, 1976, No. 131, 27901—27943.

Roberts, R.J., Myers, P.A., Morrison, A. and Murray, K., A specific endonuclease from *Arthrobacter luteus*, J. Mol Biol., 102 (1976) 157—165.

Rodriguez, R.L., Bolivar, F., Goodman, H.M., Boyer, H.W. and Betlach, M.C., Construction of new cloning vehicles, in Nierlich, D.P., Rutter, W.J. and Fox, C.F. (Eds.), Molecular Mechanisms in the Control of Gene Expression, Academic Press, New York, 1976, pp. 471—477.

Sgaramella, V., Van de Sande, J.H. and Khorana, H.G., A novel joining reaction catalyzed by T4 polynucleotide ligase, Proc. Natl. Acad. Sci. USA, 67 (1970) 1468—1475.

Smith, R., Blattner, R.F. and Davis, A., The isolation and partial characterization of a new restriction endonuclease from *Providencia stuartii*, Nucleic Acids Res. Commun., 3 (1976) 343.

So, M., Gill, R. and Falkow, S., The generation of a ColE1-Ap^r cloning vehicle which allows detection of inserted DNA, Mol. Gen. Genet., 142 (1976) 239—249.

Sugino, A., Cozarelli, N.R., Heyneker, H.L., Shine, J., Boyer, H.W. and Goodman, H.M., Interaction of bacteriophage T4 RNA and DNA ligases in the joining of duplex DNA at base-paired ends, J. Biol. Chem. (1977) in press.

Tait, R.C., Rodriguez, R.L. and Boyer, H.W., Altered tetracycline resistance in pSC101 recombinant plasmids, Mol. Gen. Genet., 151 (1977) 327—331.

Wilson, G.A. and Young, F.F., Isolation of a sequence-specific endonuclease (*Bam*I) from *Bacillus amyloliquefaciens* H, J. Mol. Biol., 97 (1975) 123—126.

Yoshimori, R., Roulland-Dussoix, D., and Boyer, H.W., R-factor controlled restriction and modification of deoxyribonucleic restriction mutants, J. Bacteriol., 112 (1972) 1275—1283.

Communicated by D.R. Helinski.

Cloning of Large Segments of Exogenous DNA Into Yeast by Means of Artificial Chromosome Vectors

D.T. Burke, G.F. Carle and M.V. Olson

Fragments of exogenous DNA that range in size up to several hundred kilobase pairs have been cloned into yeast by ligating them to vector sequences that allow their propagation as linear artificial chromosomes. Individual clones of yeast and human DNA that have been analyzed by pulsed-field gel electrophoresis appear to represent faithful replicas of the source DNA. The efficiency with which clones can be generated is high enough to allow the construction of comprehensive libraries from the genomes of higher organisms. By offering a tenfold increase in the size of the DNA molecules that can be cloned into a microbial host, this system addresses a major gap in existing experimental methods for analyzing complex DNA sources.

S TANDARD RECOMBINANT DNA TECHNIQUES INVOLVE THE in vitro construction of small plasmid and viral chromosomes that can be transformed into host cells and clonally propagated. These cloning systems, whose capacities for exogenous DNA range up to 50 kilobase pairs (kb), are well suited to the analysis and manipulation of genes and small gene clusters from organisms in which the genetic information is tightly packed. It is increasingly apparent, however, that many of the functional genetic units in higher organisms span enormous tracts of DNA. For example, the bithorax locus in *Drosophila*, which participates in the regulation of the development of the fly's segmentation pattern, encompasses approximately 320 kb (1). The factor VIII gene in the human, which encodes the blood-clotting factor deficient in hemophilia A, spans at least 190 kb (2). Recent estimates of the size of the gene that is defective in Duchenne's muscular dystrophy suggest that this

The authors are in the Department of Genetics, Washington University School of Medicine, St. Louis, MO 63110.

Fig. 1. Yeast artificial chromosome (YAC) cloning system. In the diagram of the vector pYAC2, pBR322-derived sequences are shown as a thin line. *SUP4, TRP1, HIS3,* and *URA3* are yeast genes: *SUP4* is an ochre-suppressing allele of a tyrosine transfer RNA gene that is interrupted when exogenous DNA is cloned into the vector; *TRP1* and *URA3* are present in the artificial chromosomes and allow selection for molecules that have acquired both chromosome arms from the vector; *HIS3* is discarded during the cloning process. *ARS1* and *CEN4* are sequences that are naturally adjacent to *TRP1* on yeast's chromosome IV: *ARS1* is an autonomous-replication sequence while *CEN4* provides centromere function. The *TEL* sequences are derived from the termini of the *Tetrahymena* macronuclear ribosomal DNA (rDNA) molecules. The vector was constructed as follows. The Sma I site in the *URA3* gene of YIp5 (*36*) was deleted by digestion with Ava I followed by religation (*37*); the resultant plasmid was cleaved at the single Pvu II and Eco RI sites present in its pBR322-derived sequences (*38*) and the Eco RI end was filled in with the Klenow fragment of DNA polymerase I to produce a blunt-ended 3.2-kb fragment. This fragment was ligated to a 5.5-kb fragment, containing *TRP1, ARS1,* and *CEN4*, which was produced by Pvu II cleavage of YCp19, a plasmid derived by Mann and Davis by cloning a Bam HI–Eco RI fragment from Sc4137 into the pBR322-derived sequences of YRp17 (*39*). The resultant plasmid, pPM662, contains the regenerated Pvu II site of pBR322 and its adjacent replication origin and ampicillin-resistance genes, the *TRP1-ARS1-CEN4* region of YCp19, and the *URA3* gene of YIp5. The *SUP4-o* gene was cloned into the filled-in Bam HI site of pPM662 on a 262-bp Alu I fragment derived from pSU4-A (*40*) by way of synthetic Sfi I–Not I linkers whose sequence was GCGGCCGCXGCGGCCGC (X is a mixture of G and C); the resultant plasmid was named pPM664. A short portion of the *CEN4* region of pPM664, containing an unwanted Xho I site, was deleted by digesting the plasmid at the nearby sites for Eco RI and Kpn I; the Klenow fragment of DNA polymerase I was used to create blunt ends by filling in the 5′ extension left by Eco RI and degrading the 3′ extension left by Kpn I, and then religating. This procedure regenerated an unwanted, new Eco RI site

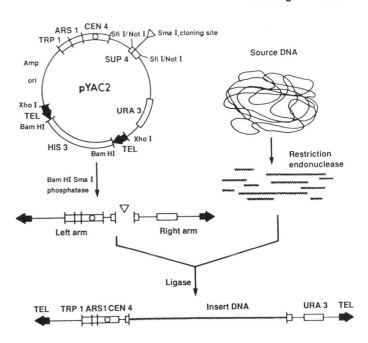

that was destroyed by cleavage with Eco RI, filling in, and religation to create pPM668. Finally, pYAC2 was created by inserting the *TEL-HIS3-TEL*-Xho I fragment of A240 p1 (provided by A. Murray) into the pBR322-derived Pvu II site of pPM668 by way of Xho I linkers; A240p1 contains the same *TEL-HIS3-TEL*-module as the plasmids A252p6 (*15*) and A142p1 (*19*).

single genetic locus, whose protein-coding function could be fulfilled by as little as 15 kb of DNA, actually covers more than a million base pairs (*3*).

Although techniques exist for cloning large genes or gene clusters in many overlapping pieces, this process is laborious, prone to error, and poorly suited to functional studies of the cloned DNA. Furthermore, there are a number of problems in molecular genetics that require the characterization of even more extensive tracts of DNA than those present in the largest known genes. For example, the regulated somatic DNA rearrangements that give rise to functional immunoglobulin genes and T cell receptor genes involve deletions of whole segments of chromosomes, while some of the genetic events that have been implicated in the induction or progression of malignant tumors involve the amplification or deletion of similarly large regions (*4–6*). In other instances, including efforts to define the primary defects in such genetic diseases as Huntington's chorea and cystic fibrosis, only the genetic linkage between the closest identified clones and the disease locus is known (*7, 8*); in typical cases, the search for the locus itself will require the analysis of megabase-pair regions of DNA. Finally, there is increasing interest in the global mapping of the DNA of intensively studied organisms (*9, 10*). Particularly in the case of the human, or other organisms with comparably complex genomes, such projects would require the ordering of hundreds of thousands of conventional clones. A cloning system that allowed the same objective to be achieved with many fewer clones would not only improve mapping efficiency but might also have dramatic effects on the reliability and continuity of the final map.

We report here the development of a high-capacity cloning system that is based on the in vitro construction of linear DNA molecules that can be transformed into yeast, where they are maintained as artificial chromosomes. Several considerations favored this combination of replicon and host. The basic functional units of yeast chromosomes—centromeres, telomeres, and ARS's (autonomous-replication sequences, with properties expected of replication ori-

gins)—have all been defined (*11–13*). In each case, DNA segments that display full functional activity in vivo are confined, at most, to a few hundred base pairs. When these elements are combined on artificial chromosomes that are approximately 50 kb or larger, the chromosomes display enough mitotic and meiotic stabilities to make their genetic manipulation straightforward (*14, 15*). Larger artificial chromosomes, consisting primarily of concatemers of bacteriophage lambda DNA, display increased mitotic stability as they increase in size, a conclusion that is reinforced by comparable studies on natural yeast chromosomes whose structures have been manipulated by homology-directed transformation experiments (*16–19*).

These prior studies of yeast artificial chromosomes have been done on molecules that formed in vivo by recombination between very small linear plasmids and transforming DNA. Typically, the linear plasmid contained all sequences required for replication and segregation, while the transforming DNA contained a marker that can be used to select for the recombinant molecule. Although such systems allow artificial chromosomes to be custom-tailored for genetic studies, they are poorly suited to cloning applications. In contrast, the system described here involves only the standard steps associated with conventional cloning protocols: in vitro ligation of vector and source-DNA fragments followed by transformation of the intact replicons into host cells.

Vector system. The vector (Fig. 1) incorporates all necessary functions into a single plasmid that can replicate in *Escherichia coli*. This plasmid, called a "yeast artificial chromosome" (YAC) vector, supplies a cloning site within a gene whose interruption is phenotypically visible (*SUP4*), an ARS (*ARS1*), a centromere (*CEN4*), selectable markers on both sides of the centromere (*TRP1* and *URA3*), and two sequences that seed telomere formation in vivo (labeled *TEL*). As described by Murray *et al.*, cleavage at the Bam HI sites adjacent to the *TEL* sequences produces termini that heal into functional telomeres in vivo (*17*).

The overall cloning protocol is shown schematically in Fig. 1. Double digestion of the particular YAC vector shown, pYAC2, with

Bam HI and Sma I yields three parts, which can be regarded as a left chromosome arm, including the centromere, a right chromosome arm, and a throwaway region that separates the two *TEL* sequences in the circular plasmid. The two arms are treated with alkaline phosphatase to prevent religation, and then ligated onto large insert molecules derived from the source DNA by partial or complete digestion with an enzyme that leaves Sma I–compatible (that is, blunt) ends. The ligation products are then transformed into yeast spheroplasts by standard methods, which involve embedding the transformed spheroplasts in agar on a selective medium.

Primary transformants are selected for complementation of a *ura3* marker in the host by the *URA3* gene on the vector. The transformants are screened for complementation of a host *trp1* marker, which ensures that the artificial chromosomes have derived both their arms from the vector, and for loss of expression of the ochre suppressor *SUP4*, which is interrupted by insertion of exogenous DNA at the Sma I cloning site, a naturally occurring restriction site in the region coding for *SUP4*'s tRNA^Tyr gene product. *SUP4* is a particularly advantageous interruptible marker since, in an *ade2-ochre* host, cells that are expressing the suppressor form white colonies and those in which the suppressor has been inactivated form red colonies (*20*).

Pilot experiments. An initial test of the vector system involved cloning Sma I limit-digest fragments of yeast and human DNA into pYAC2. The limit-digest fragments produced by cleaving either of these DNA's with Sma I, which recognizes the sequence CCCGGG, are predominantly in the size range 20 to 200 kb. In the case of the human digest, the insert DNA was size-fractionated by velocity sedimentation to eliminate fragments smaller than 40 kb. Because we anticipated a low cloning efficiency, the ligation mixtures were

carried out on a large scale (50 μg of vector plus 25 μg of insert in 200 μl).

The initial yeast-into-yeast experiment produced only a handful of transformants of the desired phenotype (Ura⁺, Trp⁺, Ade⁻, can^R, red), but a high fraction of those obtained contained novel DNA molecules that behaved electrophoretically as though they were linear DNA molecules between 40 and 130 kb in size. Control experiments suggested that the main reason for the low efficiency was that the host strain transformed poorly, even with conventional *Escherichia coli*–yeast shuttle vectors. For the cloning of human sequences into yeast, a new host strain, AB1380, was employed with dramatically improved results. When only half the ligation mixture was transformed into 5×10^7 cells, 1×10^4 Ura⁺ transformants were obtained. In a sample of 48 randomly picked colonies, 28 had all the phenotypes expected for bona fide recombinants, while 16 contained artificial chromosomes large enough to detect on ethidium bromide(EtBr)–stained pulsed-field gels (>40 kb). This sampling suggested that the experiment had an overall yield of 300 usable clones per microgram of insert DNA.

Structure of representative clones. A number of clones from the yeast-yeast and human-yeast pilot experiments were analyzed in more detail to determine whether or not the artificial chromosomes that had been produced had the expected structures (Fig. 2). The artificial chromosomes are visible on the EtBr-stained gel as 50- to 130-kb molecules, migrating ahead of the smallest natural yeast chromosome (Fig. 2A). The DNA was transferred from this gel to nitrocellulose and assayed sequentially by DNA-DNA hybridization with ³²P-labeled probes prepared from plasmid pBR322 (Fig. 2B) and total human genomic DNA (Fig. 2C). All five artificial chromosomes hybridized to pBR322, by way of the pBR322 sequences

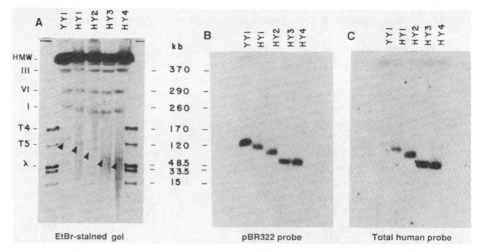

Fig. 2. Characterization of five YAC clones by pulsed-field gel electrophoresis. YY1 (yeast–yeast 1) is a clone containing yeast DNA cloned into yeast; the HY (human-yeast) clones contain human DNA cloned into yeast. (**A**) An EtBr-stained pulsed-field gel of the transformants, in which the artificial chromosomes are visible as faint bands migrating ahead of chromosome 1, which at 260 kb is the smallest natural yeast chromosome (*33, 34, 41*). The separation was made on a modified OFAGE apparatus (*34*) with a pulse time of 20 seconds; samples were prepared in agarose plugs, as described (*33, 41*). (**B**) An autoradiogram showing hybridization of all five clones to ³²P-labeled pBR322 by way of vector-derived sequences. For this experiment, the DNA in the gel shown in (A) was transferred to a nitrocellulose sheet as described by Southern (*42*); the probe was labeled by the hexamer-priming method (*43*). (**C**) An autoradiogram showing hybridization of only the four HY clones to ³²P-labeled total-human DNA. The radioactivity associated with the pBR322 hybridization was stripped off the filter before rehybridization with the total human DNA probe. All the clones were produced by ligating pYAC2-derived "arms" to source DNA that had been digested to completion with Sma I and transforming the ligation mixtures into AB1154 [used for YY1; *MATa* ψ⁺ *ura3 trp1 ade2-1 can1-100 lys2-1 met4-1 tyr1*; *ade2-1, can1-100, lys2-1,* and *met4-1* are ochre alleles, suppressible by *SUP4-o*; ψ⁺ is a cytoplasmic determinant that enhances suppression (*44, 45*)] or AB1380 (used for HY clones; *MATα* ψ⁺ *ura3 trp1 ade2-1 can1-100 lys2-1 his5*). The yeast DNA, which was derived from strain AB972, was prepared as described (*34*); the human DNA, which was derived from the neuroblastoma cell line NLF, was provided by G. Brodeur. Both DNA samples had been prepared as liquid solutions; nonetheless, the average size of the fragments present before cleavage exceeded 500 kb. After Sma I digestion, the human sample was size-fractionated on a sucrose gradient. Fractions larger than 40 kb were pooled. For the Sma I digests, 25 μg of source DNA was digested to completion, gently extracted first with phenol and then with chloroform, dialyzed against

TE8 (10 m*M* tris-HCl, 1 m*M* EDTA, *p*H 8), and concentrated in a collodion bag concentrator (Schleicher & Schuell UH 100/1). Vector DNA was prepared by digesting 50 μg of pYAC2 DNA to completion with Sma I and Bam HI, treating with an excess of calf-intestinal alkaline phosphatase (Boehringer Mannheim, molecular biology grade), extracting with phenol and then chloroform, and concentrating by ethanol precipitation; the throwaway Bam HI fragment containing the *HIS3* gene was not separated from the other two vector fragments. The ligation reaction was carried out for 12 hours at 15°C in a volume of 200 μl with 50 units of T4 ligase (Boehringer Mannheim) in 50 m*M* tris-HCl, 10 m*M* MgCl₂, 1 m*M* adenosine triphosphate, *p*H 7.5; after ligation, the reaction mixture was subjected to sequential extractions with phenol and chloroform and dialyzed against TE8. Half the ligation mixture was transformed into 5×10^7 cells, which had been converted to spheroplasts with lyticase, and plated onto four 100-mm petri plates with the use of a synthetic spheroplast-regeneration medium lacking uracil (*46*); the transformation protocol was as described (*47*).

Fig. 3. The insert in the human-yeast clone HY1 is a single large Sma I fragment. (**A**) An EtBr-stained gel on which uncleaved and Sma I–cleaved DNA from the transformed yeast strain has been fractionated along with size markers (M) in two identical sets of lanes. The high background and absence of intact chromosomes larger than IX (460 kb) in the uncleaved DNA is accounted for by the use of liquid DNA samples (*34*) in this experiment. (**B**) An autoradiogram showing that, in the uncleaved sample, ^{32}P-labeled pBR322 hybridizes to intact HY1; but in the Sma I digest it hybridizes only to the 5.6- and 3.6-kb arms contributed by pYAC2. The left side of the gel shown in (A) was used in this experiment; methods were similar to those described for Fig. 2. (**C**) An autoradiogram showing that in both the uncleaved and Sma I–cleaved samples, total human DNA hybridizes to DNA molecules of similar size; in the cleaved sample, the band corresponds to the single human Sma I fragment present as an insert in HY1. The right side of the gel shown in (A) was used in this experiment.

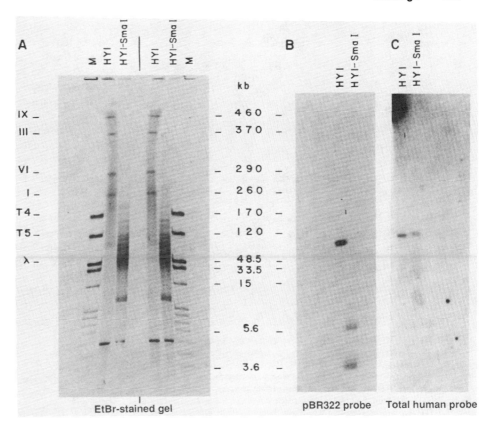

present in pYAC2. As expected, only the four human-yeast chromosomes hybridized to human DNA; under the hybridization conditions employed, the hybridization to total human DNA is expected to involve primarily dispersed repetitive human sequences present both in the probe and in the cloned segments of human DNA. The absence of minor bands in Fig. 2, even when the autoradiograms are overexposed, suggests that the yeast artificial chromosomes are propagated faithfully.

Further analysis of two of the larger artificial chromosomes, HY1 and YY1, demonstrated that both clones have the structures expected for molecules formed by the simple pathway shown in Fig. 1. For example, the data in Fig. 3 demonstrate that the insert in HY1 can be released from vector sequences as a single 120-kb fragment by Sma I digestion. In Fig. 3A, an EtBr-stained gel is shown on which two sets of samples of uncleaved and Sma I–cleaved yeast DNA from the transformant containing HY1 have been fractionated on two identical half-gels. The DNA was transferred to nitrocellulose; the samples on the left were assayed with a pBR322 probe (Fig. 3B) and those on the right were assayed using a total-human probe (Fig. 3C). The pBR322 probe detects the intact artificial chromosome in the uncleaved sample but only the two short vector arms in the Sma I–cleaved sample. In contrast, while the total-human probe again detects the intact chromosome in the uncleaved sample, it detects a single large fragment in the Sma I–cleaved sample, which is not significantly different in size from that of the intact chromosome (*21*). Similar results were obtained for YY1.

Comparison of homologous cloned and genomic Sma I fragments. As a more stringent test of whether or not HY1 and YY1 represent authentic clones, we also showed that the Sma I fragments cloned in these artificial chromosomes are the same size as homologous fragments in the source DNA. This test required the isolation of DNA fragments from the YAC inserts that could be used to probe size-fractionated Sma I digests of genomic human and yeast DNA. We isolated these fragments by a plasmid-rescue technique that takes advantage of the presence of pBR322 sequences in the

original YAC vector. Digestion of insert-containing artificial chromosomes with Xho I would be expected to produce an Xho I fragment starting adjacent to the left *TEL* sequence and extending into the insert to the first Xho I site in the cloned DNA (Fig. 1). Such a fragment would contain the pBR322 origin of replication and ampicillin resistance gene (labeled ori and Amp in Fig. 1), which are the only portions of the plasmid that are essential for replication and selection in *E. coli*. In practice, plasmid rescue is a three-step procedure: total yeast DNA from transformants is digested with Xho I, ligated under conditions that favor formation of monomer circles, and transformed into *E. coli* with selection for ampicillin resistance. Plasmids with the expected structures were readily isolated by this method from both YY1 and HY1; for example, the YY1-derived plasmid was used to demonstrate that the large Sma I fragment cloned in this artificial chromosome has a counterpart in the source DNA (Fig. 4). When DNA was transferred from the gel shown in Fig. 4A and assayed by hybridization with the *E. coli* plasmid that contained sequences rescued from YY1, the results confirmed that the Sma I fragment cloned into YY1 is of the same size in the clone and in genomic yeast DNA. In AB972, the probe hybridizes to a large chromosome in the uncleaved DNA (identified as XII in a separate experiment) and to a 120-kb fragment in the Sma I–cleaved DNA. In the sample containing uncleaved DNA from the transformed strain (lane 3), it hybridizes both to the large natural chromosome and to the small artificial chromosome, while in the sample containing cleaved DNA from the transformed strain (lane 4), it hybridizes only to the 120-kb Sma I fragment. In the latter sample, the band represents a direct superposition of homologous Sma I fragments released from the natural and artificial chromosomes. Similar experiments were carried out with a probe rescued from HY1; the only discrete fragment to which this probe hybridized in Sma I digests of human DNA was of the correct size, but the probe also weakly cross-hybridized to a heterogeneous smear of smaller Sma I fragments.

Indirect end-label mapping of YAC clones. With any primary

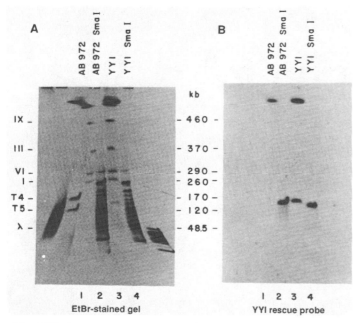

Fig. 4. The large yeast Sma I fragment cloned into the yeast-yeast clone YY1 is also present in the genome of the yeast strain AB972 from which the source DNA was extracted. (**A**) An EtBr-stained OFAGE (orthogonal field alternation gel electrophoresis) (*34*) gel on which both uncleaved and Sma I–cleaved DNA from AB972 and the transformed strain containing YY1 have been fractionated (lanes 1 to 4). The outside lanes on the gel contain size markers. (**B**) An autoradiogram showing hybridization of the samples described in (A) to a plasmid probe "rescued" from the insert of YY1. This probe hybridizes in lane 1 to sequences present at the normal chromosomal site of the cloned Sma I fragment in AB972; the large hybridizing chromosome was identified as XII in a separate experiment. In lane 2, it hybridizes to a single large Sma I fragment released from chromosome XII of AB972. In lane 3, it hybridizes both to the transformation host's chromosome XII and to the YY1 itself. In lane 4, it hybridizes to the comigrating Sma I fragments released from the transformation host's chromosome XII and from YY1.

cloning system, it is essential to have relatively simple methods of surveying the restriction sites present in newly isolated clones. Because of their large sizes, YAC clones are difficult to map by standard techniques. They are, however, particularly well suited to indirect end-label mapping, a method that has been widely employed for genomic sequencing and the mapping of hypersensitive sites in chromatin (*22, 23*). For indirect end-label mapping, a partial digest of the chromosome is size-fractionated and then probed with an end-adjacent sequence, thereby revealing a ladder of bands, the sizes of which correspond to the distances from the end to the various cleavage sites for the restriction enzyme. YAC clones are particularly well suited to indirect end-label mapping because (i) they have natural ends, (ii) the pBR322-derived sequences adjacent to each telomere allow redundant mapping of all clones with just two universal probes; and (iii) the need to detect partial-digest fragments that are present in much less than single-copy amounts is facilitated by the low sequence complexity of yeast DNA (0.5 percent of the mammalian case).

All the sites throughout a 100-kb segment could be mapped with just the left-end probe, on a single high-resolution field-inversion gel (Fig. 5). Complementary data were obtained with a right-end probe, which allowed completion of the map and also provided confirmation of the whole central region of the map. Although separate gels were used in these experiments, the two probes could equally well have been used sequentially on the same filter.

Improved vectors. The most serious limitation on pYAC2 as an all-purpose YAC vector is the inflexibility of the cloning site. Sma I produces blunt ends, which do not ligate as efficiently as "sticky"

ends and also limit the range of ways in which the source DNA can be prepared for cloning. Although various methods exist for overcoming these limitations, they are all more complex than a simple sticky-end ligation and carry the attendant risk of reducing the cloning efficiency. In seeking to adapt YAC vectors to the cloning of fragments generated by a variety of restriction enzymes that leave cohesive ends, we were able to preserve the attractive features of *SUP4* as an interruptible marker—while circumventing the inflexibility of the coding-region Sma I site—by moving the cloning site into the gene's 14-bp intron. We expected that the need to maintain *SUP4* function in the vector would place few constraints on the intron's sequence or precise length, but that the cloning of huge inserts into the intron would still inactivate the gene. Site-directed mutagenesis was used to make a single nucleotide change in the wild-type intron, creating a Sna BI site that occurs only once in the vector. In the process of reconstructing a vector that contained the Sna BI site, the Sfi I and Not I sites flanking the *SUP4* gene in pYAC2 were eliminated in order to allow these enzymes to be used more conveniently either in preparing source DNA for cloning or in the analysis of clones. In this way, a new series of vectors was constructed that offer the following cloning sites: the Sna BI site itself (pYAC3), an Eco RI site created by insertion of an Eco RI linker into the Sna BI site (pYAC4), and a similarly constructed new Not I site (pYAC5). In all cases, the cloning sites occur only once in the vector and these manipulations preserved *SUP4* function.

The most extensively tested of the new vectors is pYAC4, which allows the direct cloning of inserts produced by Eco RI partial digestion. We tested this vector on a population of Eco RI partial-digest fragments prepared from the DNA of circulating human leukocytes. The uncleaved source DNA had an average size of more than 1000 kb, while the partial-digest fragments, which were not size-fractionated, were predominantly in the size range 50 to 700 kb. In this experiment, the cloning efficiency was similar to that reported above for pYAC2 cloning (several hundred clones per microgram of source DNA), but the proportion of the primary transformants that had all the phenotypes expected of bona fide recombinants was much higher (>90 percent); furthermore, nearly all such colonies contained a single artificial chromosome that hybridized to human DNA. For example, when DNA was prepared from ten clones that were picked at random all ten clones proved to contain human DNA in artificial chromosomes ranging in size up to more than 400 kb (Fig. 6).

Future prospects. The above data suggest that the generation of yeast artificial chromosomes with YAC vectors may provide a general method of cloning exogenous DNA fragments of several hundred kilobase pairs. Although only "anonymous" clones have been analyzed, these test cases appear to be propagated as faithful copies of the source DNA. The efficiency with which clones can be generated is ample to allow generation of multi-hit comprehensive libraries of the genomes of higher organisms, particularly in applications in which the availability of source DNA is not limiting. In such situations, the number of clones that can be obtained per petri plate after transformation is the most relevant measure of practicality; the present procedures with YAC vectors allow the recovery of thousands of clones per plate, a number that compares favorably with the need for 2×10^4 clones with 150-kb inserts to obtain single-hit coverage of a mammalian genome.

Further progress toward characterizing the YAC cloning system and applying it to specific biological objectives requires the development of efficient methods of screening libraries for sequences that are present in a single copy in the source DNA. Colony screening methods have been described for yeast (*24, 25*), and the screening of YAC libraries should not pose fundamental difficulties. There are, however, several practical issues that are likely to make the process

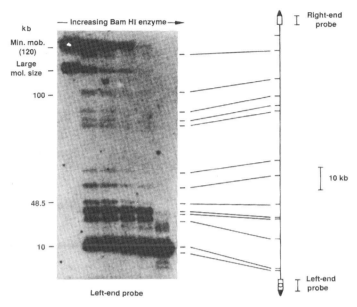

Fig. 5. Indirect end-label mapping of YY1 with Bam HI. DNA prepared from the transformant containing the artificial chromosome YY1 was subjected to partial digestion with increasing concentrations of Bam HI. The samples were fractionated by field-inversion gel electrophoresis under conditions that maximize resolution in the size range 50 to 110 kb. The DNA was transferred to nitrocellulose and assayed by hybridization with a probe consisting of vector sequences present at the left end, as defined in Fig. 1, of all YAC clones. The composite map shows the interpretation of the bands in the autoradiogram in terms of Bam HI sites in YY1. The correlation between bands and sites requires careful size calibration of the gel since gels of this type show a complex relation between size and mobility: there is a substantial compression in the range 15 to 30 kb and a short double-valued region at the top of the gel (molecules of 120 kb have minimal mobility while a heterogeneous population of larger molecules comigrate at a slightly higher mobility; there is some nonspecific hybridization to the large accumulation of yeast DNA migrating in this high molecular weight band). Sites near the right end were mapped in an experiment identical to that shown, except that a probe specific for the right end of YAC clones was employed. The DNA used in these experiments was a liquid sample prepared as previously described (34). The electrophoresis conditions involved a field strength of 10.5 V/cm (measured in the gel), a forward pulse time of 2 seconds, a reverse pulse time of 0.667 second, and a total running time of 12 hours; other conditions were as described (35). The left-end probe was the larger, and the right-end probe the smaller, of the two fragments produced by double-digestion of pBR322 with Pvu II and Bam HI; both fragments were gel-purified before labeling.

Fig. 6. Sizing of the artificial chromosomes present in ten transformants generated by cloning Eco RI partial digest fragments of human DNA into the Eco RI vector pYAC4. DNA from the transformants was prepared in agarose blocks, subjected to electrophoresis at a pulse time of 30 seconds on a pulsed-field gel apparatus that produces uniform, transverse fields intersecting at 120° (48), transferred to nitrocellulose, and assayed by hybridization with ³²P-labeled human DNA. The human DNA for the Eco RI partial digestion was prepared from circulating leukocytes by a liquid sample method whose application to yeast has been described (34). The average size of the DNA before partial digestion with Eco RI was approximately 1000 kb; digestion of 40 μg of this sample was carried out with 0.001 unit of Eco RI for 15 minutes at 25°C. Ligation and transformation conditions were as described for Fig. 1. The transformation host was AB1380.

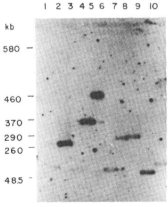

more difficult than conventional colony screening (for example, both the number of the cloned molecules per cell and the number of cells per colony are much lower in YAC cloning than in plasmid or cosmid cloning in *E. coli*); the need to regenerate transformants in agar also precludes direct screening of primary transformants.

Further experience with the YAC cloning system will be required to assess such issues as the stability of clones, the extent to which the source DNA is randomly sampled, and the biological activity of the cloned DNA. Nonetheless, there are grounds for optimism that YAC vectors could even offer important advantages over standard cloning systems in these areas. In particular, there is reason to expect the yeast DNA-replication system to be more compatible with the sequence organization of typical eukaryotic DNA's than is the *E. coli* system. Essentially all sources of eukaryotic DNA that have been tested contain sequences that can function in yeast as ARS's at a spacing that is similar to that found in yeast itself (13). This observation suggests that the existence of a single ARS from the vector will rarely limit the amount of passenger DNA that can be accommodated. Also, the ubiquity of ARS's in eukaryotic DNA and their apparent lack in *E. coli* (26) may hint at a basic functional homology among the replication systems of eukaryotic organisms. Finally, although yeast has relatively little repetitive DNA compared to higher organisms, it has all the qualitative types of repeated sequence that have been described in these systems—for example, dispersed repetitive sequences (27), scrambled clusters of repeats (27), alternating purine-pyrimidine tracts (28), perfect palindromes (27, 29), long tandem arrays (30), and satellite-like simple sequences (28). Consequently, while these sequences are often difficult to clone in *E. coli* (31), they may pose no special problems in yeast. However, present experience suggests that few genomic sequences from higher organisms will be functionally expressed in yeast (32), thereby limiting the likelihood that particular sequences will be selected against because of their genetic content.

Whether or not the YAC system proves to be broadly useful, the demonstration of the basic feasibility of generating large recombinant DNA's in vitro and transforming them into easily manipulated host cells may stimulate experimentation with other combinations of replicons and hosts. There is a strong incentive to develop such systems since they are directed toward the major remaining gap in our ability to dissect the genomes of higher organisms. The resolution of both cytogenetic analysis and linkage mapping is of the order of a few megabase pairs of DNA, while cloning techniques have been limited to a size range of tens of kilobase pairs. These two levels of analysis, with their 100-fold difference in inherent resolution, are now bridged only by pulsed-field gel electrophoresis (33–35). The ability to isolate and amplify large DNA molecules in a simple genetic background would complement this powerful analytical technique and set the stage for expanded structural and functional studies of complex DNA sources.

REFERENCES AND NOTES

1. F. Karch *et al.*, *Cell* **43**, 81 (1985).
2. J. Gitschier *et al.*, *Nature (London)* **312**, 326 (1984).
3. A. P. Monaco *et al.*, *ibid.* **323**, 646 (1986).
4. S. Tonegawa, *ibid.* **302**, 575 (1983).
5. M. Kronenberg, G. Siu, L. E. Hood, N. Shastri, *Annu. Rev. Immunol.* **4**, 529 (1986).
6. J. M. Bishop, *Science* **235**, 305 (1987).
7. A. Beaudet *et al.*, *Am. J. Hum. Genet.* **39**, 681 (1986).
8. J. F. Gusella *et al.*, *Nature (London)* **306**, 234 (1983).
9. M. V. Olson *et al.*, *Proc. Natl. Acad. Sci. U.S.A.* **83**, 7826 (1986).
10. A. Coulson, J. Sulston, S. Brenner, J. Karn, *ibid.*, p. 7821.
11. E. H. Blackburn and J. W. Szostak, *Annu. Rev. Biochem.* **53**, 163 (1984).
12. J. Carbon, *Cell* **37**, 351 (1984).
13. D. H. Williamson, *Yeast* **1**, 1 (1985).
14. A. W. Murray and J. W. Szostak, *Nature (London)* **305**, 189 (1983).
15. D. S. Dawson, A. W. Murray, J. W. Szostak, *Science* **234**, 713 (1986).
16. P. Hieter, C. Mann, M. Snyder, R. W. Davis, *Cell* **40**, 381 (1985).
17. A. W. Murray, N. P. Shultes, J. W. Szostak, *ibid.* **45**, 529 (1986).

18. R. T. Surosky, C. S. Newlon, B.-K. Tye, *Proc. Natl. Acad. Sci. U.S.A.* **83**, 414 (1986).
19. A. W. Murray and J. W. Szostak, *Mol. Cell. Biol.* **6**, 3166 (1986).
20. If necessary, it is also possible to select against nonrecombinants with the *can1-ochre* marker in the host. Cells expressing *SUP4* are canavanine-sensitive, while those lacking suppression are canavanine-resistant [R. J. Rothstein, *Genetics* **85**, 55 (1977)].
21. These data support the hypothesis that HY1 arose by the simple pathway diagrammed in Fig. 1. However, the cloning of a human insert containing internal Sma I sites would not necessarily indicate that the source DNA had been inadequately digested with Sma I or that more complex ligation events had occurred either in vitro or after transformation. Some human Sma I sites are protected from cleavage by CG methylation, and this protection would almost certainly be lost during propagation of the sites in yeast [W. R. A. Brown and A. P. Bird, *Nature (London)* **322**, 477 (1986)].
22. G. M. Church and W. Gilbert, *Proc. Natl. Acad. Sci. U.S.A.* **81**, 1991 (1984).
23. C. Wu, *Nature (London)* **286**, 854 (1980).
24. A. Hinnen, J. B. Hicks, G. R. Fink, *Proc. Natl. Acad. Sci. U.S.A.* **75**, 1929 (1978).
25. W. L. Fangman and B. Dujon, *ibid.* **81**, 7156 (1984).
26. D. T. Stinchcomb, M. Thomas, J. Kelly, E. Selker, R. W. Davis, *ibid.* **77**, 4559 (1980).
27. J. R. Cameron, E. Y. Loh, R. W. Davis, *Cell* **16**, 739 (1979).
28. R. W. Walmsley, C. S. M. Chan, B.-K. Tye, T. D. Petes, *Nature (London)* **310**, 157 (1984).
29. H. L. Klein and S. K. Welch, *Nucleic Acids Res.* **8**, 4651 (1980).
30. J. H. Cramer, F. W. Farrelly, R. H. Rownd, *Mol. Gen. Genet.* **148**, 233 (1976).
31. A. R. Wyman, L. B. Wolfe, D. Botstein, *Proc. Natl. Acad. Sci. U.S.A.* **82**, 2880 (1985).
32. J. D. Beggs, J. van den Berg, A. van Ooyen, C. Weissmann, *Nature (London)* **283**, 835 (1980).
33. D. C. Schwartz *et al.*, *Cold Spring Harbor Symp. Quant. Biol.* **47**, 189 (1982); D. C. Schwartz and C. R. Cantor, *Cell* **37**, 67 (1984).
34. G. F. Carle and M. V. Olson, *Nucleic Acids Res.* **12**, 5647 (1984).
35. G. F. Carle, M. Frank, M. V. Olson, *Science* **232**, 65 (1986).
36. K. Struhl *et al.*, *Proc. Natl. Acad. Sci. U.S.A.* **76**, 1035 (1979).
37. In our isolate of YIp5, the Sma I site is flanked by two closely spaced Ava I sites, only one of which is predicted by the *URA3* gene sequence [M. Rose, P. Grisafi, D. Botstein, *Gene* **29**, 113 (1984)].
38. J. G. Sutcliffe, *Cold Spring Harbor Symp. Quant. Biol.* **43**, 77 (1978).
39. C. Mann and R. W. Davis, *Mol. Cell. Biol.* **6**, 241 (1986).
40. K. J. Shaw and M. V. Olson, *ibid.* **4**, 657 (1984).
41. G. F. Carle and M. V. Olson, *Proc. Natl. Acad. Sci. U.S.A.* **82**, 3756 (1985).
42. E. M. Southern, *Methods Enzymol.* **68**, 152 (1979).
43. A. P. Feinberg and B. Vogelstein, *Anal. Biochem.* **137**, 266 (1984).
44. S. Liebman and F. Sherman, *J. Bacteriol.* **139**, 1068 (1979).
45. M. F. Tuite *et al.*, *Proc. Natl. Acad. Sci. U.S.A.* **80**, 2824 (1983).
46. F. Sherman, G. Fink, C. Lawrence, *Methods in Yeast Genetics* (Cold Spring Harbor Laboratory, Cold Spring Harbor, NY, 1979).
47. P. M. J. Burgers and K. J. Percival, *Anal. Biochem.*, in press.
48. G. Chu, D. Vollrath, R. W. Davis, *Science* **234**, 1582 (1986).

49. We thank A. W. Murray and J. W. Szostak for providing helpful information on artificial chromosome construction and for the telomere-containing plasmid A240p1; R. W. Davis, for the YIp5 and YCp19 plasmids; G. Brodeur for the DNA from the human cell line NLF; D. Schlessinger for the human leukocytes; D. Garza for numerous helpful discussions; and P. Burgers for providing unpublished information and lyticase enzyme for the transformation procedure. Supported, in part, by the Monsanto Company and by NIH training grants 5T32-GM08036 and 5T32-ES07066.

10 March 1987; accepted 21 April 1987

A Colony Bank Containing Synthetic Col EI Hybrid Plasmids Representative of the Entire *E. Coli* Genome

L. Clarke and J. Carbon

Summary

Using the poly(dA·dT) "connector" method (Lobban and Kaiser, 1973), a population of annealed hybrid circular DNAs was constructed in vitro; each hybrid DNA circle contained one molecule of poly(dT)-tailed Col EI-DNA (L$_{RI}$) annealed to any one of a collection of poly(dA)-tailed linear DNA fragments, produced originally by shearing total E. coli DNA to an average size of 8.5×10^6 daltons. This annealed DNA preparation (12 μg) was used to transform an F$^+$ *recA* E. coli strain (JA200), selecting transformants by their resistance to colicin EI. A collection or "bank" of over 2000 colicin EI-resistant clones was thereby obtained, 70% of which were shown to contain hybrid Col EI-DNA (E. coli) plasmids. This colony bank is large enough to include hybrid plasmids representative of the entire E. coli genome. Individual plasmids have been readily identified by replica mating the collection onto plates seeded with cultures of various F$^-$ auxotrophic recipients, selecting for complementation of the auxotrophic markers by F-mediated transfer of hybrid plasmids to the F$^-$ recipients. In this manner, over 80 hybrid Col EI-DNA (E. coli) plasmid-bearing clones have been identified in the colony bank, and about 40 known E. coli genes have been tentatively assigned to these various plasmids. The hybrid plasmids are transferred efficiently from F$^+$ donors to appropriate F$^-$ recipients. The use of this method to establish similar colony banks in E. coli containing hybrid plasmids representative of various simple eucaryotic genomes is discussed.

Introduction

The molecular cloning of specific regions of the Escherichia coli genome on hybrid Col EI plasmids has previously been carried out by transformation of suitable E. coli auxotrophic strains with synthetic recombinant DNA, selecting directly for complementation of a host chromosomal mutation by the hybrid plasmid DNA (Clarke and Carbon, 1975; Hershfield et al., 1974). Using circular Col EI-DNA (E. coli) hybrid molecules constructed in vitro by the poly(dA·dT) "connector" method (Lobban and Kaiser, 1973; Jackson, Symons, and Berg, 1972), this selection procedure has been used to obtain

several E. coli gene systems cloned onto hybrid Col EI plasmids for studies on gene organization and expression, and for the amplification of specific gene products (Clarke and Carbon, 1975; P. Schimmel and J. Carbon, unpublished results).

The direct selection method used in our early studies is somewhat wasteful of the valuable synthetic hybrid DNA preparation, since it requires a separate transformation and selection for each gene system cloned. In this paper, we report that the same amount of hybrid DNA used to select directly for one specific hybrid plasmid can be used to transform an E. coli strain, selecting instead for the vector determinant, colicin EI resistance, thus establishing a collection of transformant clones which carry different hybrid plasmids representative of most of the bacterial genome. In addition, we have prepared our synthetic recombinant DNA from fragments generated by hydrodynamic shearing of total E. coli DNA, rather than by scission with restriction endonuclease *Eco* RI, thus insuring that any desired gene system will be intact in a fraction of the final hybrid DNA.

To make use of hybrid plasmid colony banks, rapid and simple screening methods for the detection of clones bearing specific hybrid plasmids are needed. It is known that E. coli strains harboring the fertility plasmid F along with the nontransmissible plasmid Col EI transfer both plasmids with high efficiency to an F$^-$ recipient (Fredericq and Betz-Bareau, 1953; Clowes, 1964). We find that hybrid Col EI plasmids are also transferred efficiently from F$^+$ donors to auxotrophic F$^-$ recipients, and that this F-mediated transfer provides a convenient screening technique for the identification of specific hybrid plasmid clones in the transformant bank.

These studies were undertaken not only to isolate hybrid Col EI plasmids carrying bacterial genes of particular interest, but also to provide a model system for testing the feasibility of establishing similar transformant colony banks containing hybrid plasmids representative of the entire genome of simple eucaryotic organisms. An application of convenient screening techniques based on F-mediated transfer to these banks should permit a definitive assessment of the ability of segments of DNA derived from simple eucaryotes, such as yeast and Drosophila, to be expressed in the procaryotic host cell.

Results

Construction of Hybrid Col EI-E. coli Annealed Circular DNA

The construction of hybrid Col EI-DNA (E. coli) annealed circles using poly(dA·dT) connectors was carried out as described (Clarke and Carbon, 1975),

CLARKE, L. and CARBON, J.
A colony bank containing synthetic Col EI hybrid plasmids representative of the entire *E. coli* genome.
Reprinted with permission from *Cell* 9:91-99 (1976).

except that the E. coli DNA was fragmented by hydrodynamic shearing (instead of restriction endonuclease cleavage) to an average size of $8.4 \pm 3.0 \times 10^6$ daltons, determined by measuring 47 molecules in the electron microscope. The preparation of annealed hybrid Col El-DNA (E. coli) contained approximately 25% circles, 12% branched circles, 53% linears, 2% branched linears, and 8% unscorable tangles (100 molecules scored by electron microscopy).

Establishment of a Collection of E. coli Clones Containing Hybrid Col El-DNA (E. coli) Plasmids

Using approximately the same amount (12 μg) of annealed hybrid Col El-DNA (E. coli) employed to transform an E. coli auxotroph and to select directly for a specific hybrid plasmid-containing strain (Clarke and Carbon, 1975), we have transformed E. coli strain JA200 (C600 ΔtrpE5 recA/F⁺), selecting instead for the vector determinant, colicin El resistance. In a single transformation and selection, a collection or "bank" of clones was obtained which carry different hybrid Col El-DNA (E. coli) plasmids representative of a large portion of the bacterial genome.

This transformation experiment differed in several ways from the direct selection experiments we have previously described (Clarke and Carbon, 1975). First, the annealed hybrid DNA used for the transformation of JA200 was constructed of Eco RI-cleaved Col El-DNA linked by poly(dA·dT) connectors to randomly sheared E. coli DNA. Sheared DNA was used to permit the cloning of any portion of the E. coli chromosome. Second, after exposure to DNA, the transformed cell culture was not incubated long enough to permit significant cell division before plating. Thus each transformed clone was the consequence of a distinct and separate transformation event. Finally, the recipient for transformation was a strain harboring the sex factor, F, to permit the identification of a clone carrying a particular hybrid plasmid by F-mediated transfer of the hybrid through replica mating of the colony collection to a particular E. coli F⁻ auxotroph (see below).

A recA transformation recipient was chosen to avoid recombination of hybrid plasmids with host chromosomal DNA.

We have determined the transformant colony bank size needed to obtain a plasmid collection which represents 90–99% of the E. coli genome as follows. Given a preparation of cell DNA that has been fragmented to a size such that each fragment represents a fraction (f) of the total genome, then the probability (P) that a given unique DNA sequence is present in a collection of N transformant colonies is given by the expression:

$$P = 1 - (1 - f)^N$$

or

$$N = \frac{\ln(1 - P)}{\ln(1 - f)}$$

Thus using a preparation of E. coli DNA randomly sheared to an average of 8.5×10^6 daltons for the construction of annealed hybrid circular DNA, a colony bank of only 720 transformants would be adequate to give a probability of 90% that any E. coli gene would be on a hybrid plasmid in one of the clones (Table 1; it is assumed that the desired gene is small in comparison with the size of the cloned fragments). At a probability level of 99%, the colony bank size (N) is only about 1400 colonies for E. coli. As the genetic complexity of the organism increases, N at high probability levels increases dramatically (Table 1).

Approximately 2100 colicin El-resistant transformants were picked and transferred to plates in a grid array of 48 colonies per plate as described in Experimental Procedures. The colonies were also individually maintained and stored as 8% DMSO cultures in MicroTest dishes at −80°C. Use of the Triton lysis procedure on ten clones from a pilot experiment identical to the one described (Experimental Procedures) and analysis of supercoiled plasmid DNAs in 1.2% agarose gels revealed that about 70% of the clones in the collection (1400 transformant colonies) contained plasmids appreciably larger than plasmid Col El. The remaining

Table 1. Colony "Bank" Sizes (N) Needed to Contain a Particular Hybrid Plasmid Transformant at Various Probability Levels

DNA Source	Average Size of DNA Fragment Cloned, Daltons	"Bank" Size (N), Number of Colonies		
		P = 0.90	P = 0.95	P = 0.99
E. coli	8.5×10^6	720	940	1,440
Yeast	1×10^7	2,300	3,000	4,600
Drosophila	1×10^7	23,000	30,000	46,000

The above calculations are based on the formula, $P = 1-(1-f)^N$, and assume that each transformant colony in the "bank" arises from an independent transformation event and that each hybrid molecule transforms with the same efficiency. It is also assumed that the length (x) of the desired DNA segment is small in comparison with the length (L) of the DNA fragment actually cloned, to minimize the effect of random breaks occurring within the desired length (x). More accurately, a correct f value (f°) could be obtained from the expression, $f° = (1 - \frac{x}{L}) f$, and substituted for f in the above probability equation.

30% of the colonies were tolerant to colicin EI, but probably contained no plasmid (Davis and Reeves, 1975). Many of the latter colonies were also resistant to colicin E2. In more recent experiments, we have been able to reduce the background of colicin EI-tolerant cells by pregrowing the transformation recipient in 0.05% sodium deoxycholate (C. Ilgen and J. Carbon, unpublished results).

Use of F-Mediated Transfer for the Identification of Specific Hybrid Plasmid-Bearing Clones in the Colony Bank

Once a transformant colony bank is established, it is essential to have a simple rapid way of identifying a desired hybrid plasmid-bearing clone within the collection. It has been shown that strains which har-

bor the fertility plasmid F and which also contain the nontransmissible plasmid Col EI will transfer both plasmids to an F⁻ recipient with high efficiency (Fredericq and Betz-Bareau, 1953; Clowes, 1972; Hardy, 1975). Preliminary reconstruction experiments using strain MV12/pLC19 (Col EI-trp plasmid, Clarke and Carbon, 1975) to which an F factor had been transferred indicated that this Col EI-DNA (E. coli) hybrid transferred readily to an F⁻/trp recipient (Table 4). Furthermore, the transfer was of high enough efficiency to be easily detected by replica mating on plates, using appropriate selections and counter-selections.

A number of replica mating experiments were therefore performed using the entire colony collection and various F⁻ auxotrophs. Some of these ex-

Table 2. Hybrid Plasmids Identified in the Colony Bank

Approximate Map Location (Min)	E. coli Markers Complemented	pLC #	Approximate Map Location (Min)	E. coli Markers Complemented	pLC #	Approximate Map Location (Min)	E. coli Markers Complemented	pLC #
1	araC	24–41	35–38	flaD[c]	7–18	75	ilv[b]	21–35
					13–12			22–3
4	dnaE[a]	26–43						22–31
7	proA[b]	28–33		flaD, hag, flaN[c]	24–16			26–3
		44–11			26–7			27–15
9	lacZY	20–30		flaN, flaBCOE, flaAPQR[c]	41–7			30–15
								30–17
12	dnaZ[a]	5–1						44–7
		5–2	38	his[b]	14–29			
		6–2			26–21		cya[f]	23–3
		10–24	43	glpT[d]	3–46			29–5
		10–26			8–12			36–14
		30–3			8–24			41–4
		30–4			8–29			43–44
22–26	flaKLM[c]	24–46			14–12	79	argH	20–10
		35–44			19–24			41–13
		36–11			42–17			
						81	dnaB[a]	11–9
27	trpE	4–6	51	recA, srl[b]	17–43			44–14
		5–23			18–42			
		29–41			21–33	89	dnaC[a]	4–39
		32–12			22–40			8–9
		32–27			24–32			25–8
								30–24
		41–15		recA[b]	17–38			31–39
35–38	flaGH, cheB[c]	21–2			24–27			
		24–15			30–20	90	serB, trpR[f]	32–33
	flaGH, cheB, mot[c]	1–28	70	glyS, xyl[e]	1–3		trpR, thr[f]	35–1
		1–29			42–22			35–21
	mot, cheA, fla[c]	27–20		xyl	10–15			
		38–14			32–9			
		38–36						

[a] R. McMacken, personal communication.
[b] L. Margossian and A. J. Clark, personal communication. These identifications are tentative and, in some cases, await confirmation.
[c] M. Silverman and M. Simon, personal communication.
[d] J. Weiner, personal communication.
[e] K. Wadey and G. Nagel, personal communication.
[f] J. Schrenk and D. Morse, personal communication.

Genes on hybrid plasmids identified as complementing specific E. coli markers in this table are not necessarily the only E. coli genes carried by the plasmids. In addition, plasmids listed here as complementing a specific marker do not always represent all the plasmids in the total collection which carry that particular E. coli gene. For example, the Col E1-lacZY hybrid plasmid (pLC20–30) was identified by screening as the only plasmid in the collection capable of complementing a lacY mutation (see Table 3). pLC20–30 was later found to also complement lacZ, but other Col E1-lacZ plasmids may occur in the colony bank.

periments were carried out in our laboratory and are discussed in detail below. The remainder were performed by other laboratories using the same colony collection. Table 2 lists the hybrid plasmids tentatively identified to date and some of the markers carried by these plasmids. On the average, three hybrid plasmid-containing clones were identified in the total collection for each marker sought in a replica mating experiment. The list includes about 40 known E. coli genes, some of which were assigned after clones harboring hybrid plasmids which carried neighboring genes were identified. For example, the four clones in the collection which were found by replica mating to contain Col El-*xyl* plasmids were tested for over-production of glycine–tRNA synthetase, the product of the closely neighboring *glyS* gene (Figure 1), and two of the four clones produced elevated levels of this enzyme (G. Nagel, personal communication).

The transformant colony collection appears to contain hybrid plasmids representative of nearly the entire E. coli genome, in that the probability of finding any cloned gene system chosen at random appears to be high (about 80%). For example, in our laboratories, the collection was screened for hybrid plasmids capable of complementing any of six mutations chosen at random (*araC, lacY, trpE, argH, xyl,* and *metE*). Of these, only the *metE* mutation could not be complemented by any hybrid plasmid from the bank (see Table 2). All of the known E. coli motility gene systems were found on hybrid

plasmids in this same collection (M. Silverman and M. Simon, personal communication).

Characterization of Several Col EI-DNA (E. coli) Hybrid Plasmids from the Colony Bank

A semi-confluent to confluent patch on selective plates in the mating experiments described above was scored as representing a clone carrying the desired hybrid plasmid. Phenotypic reversion of a recipient marker could, however, have been a consequence of suppression, if the recipient marker was a point mutation and the hybrid plasmid carried a suppressor transfer RNA gene, or a consequence of suppression possibly resulting from overproduction of an RNA or protein whose gene was carried by a hybrid plasmid. Complementation of a recipient marker in the F-mediated replica matings could also have been the result of Hfr formation (if the recipient was *recA*+ or if the hybrid plasmid carried a *recA*+ region) or F′ formation and subsequent mobilization of donor chromosomal genes. However, chromosomal mobilization generally occurs at a much lower frequency than F-mediated hybrid plasmid transfer (Table 4) and could usually be distinguished from it.

To help establish that clones tentatively identified in the colony bank did carry the desired hybrid plasmids, a number of these strains and their plasmids were further characterized. Table 3 lists the hybrid plasmid strains identified by four replica mating experiments selecting for hybrid plasmids carrying the *ara, trp, arg,* and *xyl* regions. From one to six candidates were obtained from each screening of the bank. Since the parent bank strain was *lacY*, the entire colony collection was also screened for Col El-*lacY*-bearing clones by replica plating the bank onto indicator plates. One Lac+ candidate was found by this method.

Covalently closed, supercoiled hybrid plasmid DNA was purified from at least one representative of each marker class in Table 3. These DNAs were then used to transform either the original recipients or a recipient with a mutation in the same gene as the original. In all cases tested, a high frequency of transformation to the expected phenotype was obtained (Table 3), and all transformants screened were both resistant to colicin El and resistant to fr male-specific phage. These results indicated that the genes responsible for reversion of recipient markers in the original mating experiments were carried by Col El-DNA (E. coli) hybrid plasmids. It is improbable in the case of the *ara, trp,* and *lac* plasmids that reversion was due to suppressor tRNA genes carried by the hybrids, since the mutations complemented by these plasmids were deletions. Suppression mediated by a suppressor tRNA

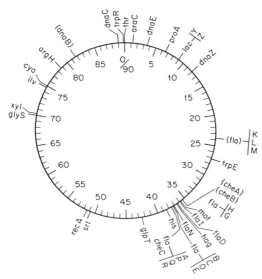

Figure 1. Relative Position on the E. coli Genetic Map of Gene Systems Complemented by Hybrid Col E1-DNA (E. coli) Plasmids For references see Table 2.

is improbable in the case of the *arg* and *xyl* plasmids, because the only known suppressor tRNA gene carried by the DNA from which the hybrid plasmids were constructed is *glyVsu58* (Guest and Yanofsky, 1965; Fleck and Carbon, 1975), and neither the *xyl* nor *arg* plasmids complemented the *trpA58* mutation (suppressible by $tRNA_{suA58}^{Gly\,3}$). Suppression by overproduction of a gene product, such as a tRNA, a protein with weak catalytic activity or a ribosomal component, is difficult to rule out in any case. However, two of the Col El-*xyl* plasmids were shown to carry the closely neighboring *glyS* gene (G. Nagel, personal communication) and the Col El-*lac* plasmid, originally detected in a *lacY* strain, transformed to Lac+ a strain which carries a deletion in at least *lac* Y and Z. Thus these three hybrid plasmids certainly carry the desired portion of E. coli DNA.

The contour lengths of a number of hybrid plasmids were measured by electron microscopy of relaxed circles using Col El-DNA (4.2×10^6 daltons; Bazarol and Helinski, 1968) as a standard on the same grid. Data for four of these measurements are found in Table 3. Other plasmids measured were pLC18-42 ($11.9 \pm 0.1 \times 10^6$ daltons), pLC21-33 ($10.3 \pm 0.1 \times 10^6$ daltons), pLC32-33 ($18.1 \pm 0.1 \times 10^6$ daltons), pLC36-14 ($14.9 \pm 0.2 \times 10^6$ daltons), and pLC50 (Col El-*lac*, $13.6 \pm 0.1 \times 10^6$ daltons; isolated with the same annealed hybrid DNA, but in a separate transformation). The average size of the E. coli inserts based on this relatively small sample is $9.9 \pm 3.4 \times 10^6$ daltons, as compared with the average size of the sheared CS520 DNA ($8.4 \pm 3.0 \times 10^6$ daltons) used originally to construct the plasmids.

Efficiency of Transfer of Several Col El-*trp* Plasmids as Mediated by the F Factor

Conjugal co-transfer of plasmids Col El and F has been shown to occur with high frequency (Fredericq and Betz-Bareau, 1953; Clowes, 1972; Hardy, 1975). After a normal 2 hr mating with F+ Col El+ donors, 80% of the recipients acquired both F and Col El, and only if the mating was interrupted after short periods of time could cells be found which contained one or the other plasmid (Clowes, 1964). Treatment of the cells with acridine orange led to almost complete elimination of the F factor without curing of the Col El plasmid, indicating physical independence of the two (Clowes, Moody, and Pritchard, 1965).

The F-mediated transfer of three Col El-*trp* plasmids was studied (Table 4). Strain MV12/pLC19(F−) has been described (Clarke and Carbon, 1975) and harbors a Col El-*trp* plasmid which carries all or most of the *trp* operon genes. This strain was made F+ by mating and designated JA200/pLC19. Strains JA200/pLC29-41 and JA200/pLC32-12 are F+ clones from the colony bank which were identified as carrying Col El-*trpE* plasmids (Tables 2 and 3). The F− strains, MV12/pLC29-41 and MV12/pLC32-12, were constructed by transformation of strain MV12(F−) with purified pLC29-41 and pLC32-12 DNAs isolated from their respective strains in the colony bank (Table 3). These six strains were used as donors in mating experiments described in the

Table 3. Characterization of Several Hybrid Plasmids

Mating Recipient	Selection	pLC #	Plasmid Size (X 10⁻⁶ Daltons)	Transformation Recipient	Selection	Transformants μg DNA
JA208 (△araC766)	Ara+	24–41	–	JA208 (△araC766)	Ara+	3×10^4
DM493 (△trpE5)	Trp+	4–6	–			–
		5–23	–	MV12 (△trpE5)	Trp+	–
		29–41	–			4×10^4
		32–12	–			4×10^4
		32–27	–			–
		41–15	–			–
CH754 (argH)	Arg+	20–10	11.8 ± 0.2	CH754 (argH)	Arg+	$>2 \times 10^5$
		41–13	10.0 ± 0.1			$>3 \times 10^5$
CH754 (xyl)	Xyl+	1–3	–	CH754 (xyl)	Xyl+	–
		10–15	17.0 ± 0.1			$>2 \times 10^5$
		32–9	–			$>10^4$
		44–22	–			–
MV12 (lacY)/ F+/Col E1-DNA (E. coli) (Screened on MacConkey-lactose plates)	Lac+	20–30	19.4 ± 0.1	JA208 (△lacZ-Y 514)	Lac+	$>3 \times 10^3$

A dash indicates "not measured".

legend to Table 4, with strain DM493 ($\triangle$trpE5/F−) serving as the recipient in all cases. All the Col EI-*trp* plasmids were donated with high efficiency from strains that also contained F. No transfer was detected from strains that contained only the hybrid plasmids. Of the recipients in the matings receiving hybrid plasmids, 88–98% of them also received F and were able to transfer further Col EI-*trp* plasmids. A small fraction of the original recipients, however, picked up only the hybrid plasmids, since they could no longer transfer these plasmids and were resistant to phage fr. Several of the Col EI-*arg* and Col EI-*xyl* plasmids were also very efficiently transferred from F+ strains (D. Richardson and J. Carbon, unpublished results). These data indicate that Col EI-DNA (E. coli) plasmids behave in a very similar manner to plasmid Col EI in an F+ background, and that the hybrid plasmids and the F factor act as physically independent units.

Hybrid Plasmid DNA Preparation Representing a Large Portion of the E. coli Genome

An alternative way to maintain the genetic information of a complete genome as a collection of hybrid plasmids is in the form of a covalently closed hybrid plasmid DNA preparation. Such a preparation was made by scraping a set of master replica plates containing all the clones in the colony bank, allowing the cells to grow nonselectively for approximately two generations, amplifying plasmid DNA in the presence of chloramphenicol, and purifying supercoiled total plasmid DNA by the Triton lytic method described in Experimental Procedures. To identify individual plasmids, this DNA was then used to transform the five recipients listed in Table 5 to the Ara+, Trp+, Arg+, Xyl+, and Lac+ phenotypes. The observed and calculated transformation frequencies are given in Table 5. No Ara+ or Lac+ transformants were found, but the experiment yielded a

disproportionally large number of Arg+ transformants. Thus this purified DNA preparation was not as representative of the bacterial genome as the colony bank and appeared to contain a relatively large amount of Col EI-*arg* DNA but less Col EI-*ara* and Col EI-*lac* DNAs.

Generally, we have observed that strains harboring various hybrid plasmids may grow at different growth rates, and that certain hybrid plasmids segregate at high frequency and others not at all. These observations have indicated to us that the most suitable way of maintaining a hybrid plasmid collection without loss of specific plasmids is as a set of individual clones, each harboring a unique plasmid.

Discussion

Many of the advantages of having particular gene systems isolated and cloned on plasmids present in multiple copies per bacterial cell have been described (Hershfield et al., 1974; Clarke and Carbon, 1975). This paper reports procedures by which most of the genetic information contained in the E. coli genome can be established as a collection of hybrid Col EI-DNA (E. coli) plasmids, each of which contains a unique segment of E. coli DNA of approximate molecular weight of 8.5×10^6 daltons. We have maintained this plasmid collection in two ways: first, as a colony bank of about 2000 clones, each harboring a unique hybrid plasmid; and second, as a covalently closed, supercoiled DNA preparation isolated from the total colony bank.

There are a number of ways of identifying specific plasmids of interest within the collection. In this paper, we have used complementation of various E. coli mutations to isolate plasmids carrying specific genes by F-mediated transfer of hybrid plasmids from the F+ clones in the colony bank to F− auxo-

Table 4. Efficiency of Transfer of Col E1-*trp* Plasmids from F+ and F− Strains

Donors	Recipient	% Donors Transferring Hybrid Plasmid in 1 Hr	% Trp+ Recipients That Were F+
MV12/pLC19 (F−)	DM493 $\triangle$trpE5/F−	<0.004	−
MV12/pLC29-41 (F−)	$\triangle$trpE5/F−	<0.002	−
MV12/pLC32-12 (F−)	$\triangle$trpE5/F−	<0.002	−
JA200/pLC19 (F+)	$\triangle$trpE5/F−	64	88
JA200/pLC29-41 (F+)	$\triangle$trpE5/F−	46	98
JA200/pLC32-12 (F+)	$\triangle$trpE5/F−	40	96

Liquid matings were performed as described by Miller (1972) at a 10:1 ratio of recipients to donors. A 1 hr incubation with gentle shaking at 37°C was followed by Vortex interruption, dilution, and plating onto selective plates. The fraction of Trp− recipients that were also F+ was determined by picking 50 recipients from each mating, replica mating these as donors, using JA198 ($\triangle$trpE5str^r) as a recipient in each case, and selecting for Trp+. Those donors which did not transfer the Trp+ phenotype were confirmed as being F− by their resistance to phage fr.

trophs on selective plates, or by transformation of auxotrophs and direct solution using the mixed plasmid DNA preparation. In several cases, neighboring genes have been found to occur on the same hybrid plasmid.

Maintaining a plasmid collection representative of most of the E. coli genome in the form of a colony bank permits the use of additional methods to identify specific plasmids. The presence of hybrid Col EI-DNA (E. coli) plasmids in multiple copies per cell frequently leads to overproduction of the products of genes carried by these plasmids (Hershfield et al., 1974; Clarke and Carbon, 1975). Overproduction of a protein, a repressor, for example, might result in pleiotropy or auxotrophy in cells harboring specific plasmids, because the gene system(s) under the control of this repressor might not be inducible. Thus screening the clones in the colony bank for auxotrophy may result in further assignment of regulatory genes to plasmids or reveal hitherto unknown regulatory phenomena.

Another method for identifying specific genes on plasmids in the colony bank is the hybridization technique described by Grunstein and Hogness (1975). Many gene systems cannot be identified by F-mediated transfer and selection, either because a suitable auxotrophic recipient is unavailable (for example, most tRNA and ribosomal RNA genes) or because the transformant would not be viable. For example, strains carrying a tRNA-derived suppressor mutation on an amplifying plasmid may not survive because of overproduction of a deleterious su^+ gene product. We are presently using hybridization techniques to detect plasmids in the collection that carry E. coli tRNA genes.

Several difficulties arise in attempting to maintain a colony bank of the type described in this report. In our experience, hybrid plasmids seem to be readily maintained in recA strains if selective pressure, either colicin EI resistance or complementation of a host marker, is kept on the strains. In the absence of selective pressure, however, many plasmids segregate. In $recA^+$ strains, some plasmids appear to be lost, presumably through recombination with the genome or through segregation, and others become smaller, possibly by looping out segments of the E. coli DNA. For these reasons, we have maintained the colony collection as a set of individual clones, have attempted to keep selective pressure on plasmid-bearing strains, and have stored the original transformant collection in the form of separate cultures in 8% DMSO at –80°C. We have gone back to these original cultures to inoculate fresh plates for the experiments described here and have avoided continued growth of clones and replication of plasmids.

The successful establishment and ease of maintenance of a colony bank containing plasmids representative of most of the E. coli genome and the relative ease with which specific plasmids can be identified by F-mediated transfer have prompted us to establish similar colony banks containing hybrid plasmids representative of most of the yeast and Drosophila genomes (B. Ratzkin, C. Ilgen, and J. Carbon, unpublished). Recent experiments in this and other laboratories have indicated that it is possible to complement or suppress E. coli mutations with segments of yeast DNA cloned onto a phage vector (Struhl, Cameron, and Davis, 1976) or on a plasmid Col EI vector (B. Ratzkin and J. Carbon, unpublished results). Col EI-DNA (yeast) and Col EI-DNA (Drosophila) plasmids are readily transferable from F+ strains to F− recipients (B. Ratzkin, C. Ilgen, and J. Carbon, unpublished results). Thus entire collections of F+ clones containing such plasmids may be rapidly tested for complementation of E. coli auxotrophic markers through F-mediated transfer of plasmids with the knowledge that each collection should be representative of most of the genome under investigation.

Finally, F-mediated transfer of Col EI-like plasmids provides a valuable tool for the identification of specific hybrid plasmids constructed of Col EI DNA and the DNAs of procaryotic and simple eu-

Table 5. Efficiency of Transformation by Mixed Plasmid DNA Prepared from the Entire Colony Bank

Transformation Recipient	Selection	Transformants per μg DNA	
		Observed	Expected
JA208(ΔaraC766)	Ara+	<0.5	20
MV12(ΔtrpE5)	Trp+	6.3×10^2	2×10^2
CH754(argH)	Arg+	3.3×10^4	4×10^2
CH754(xyl)	Xyl+	5.9×10^3	8×10^2
JA208(ΔlacZ-Y514)	Lac+	<0.5	>2

Transformations and selections were carried out as described (Clarke and Carbon, 1975) using 2 μg of DNA for each experiment. The expected number of transformants per μg DNA was calculated from known efficiencies of transformation of purified individual plasmid DNAs using these same recipients (Table 3) and from the number of representatives of each plasmid in the colony bank, assuming all plasmids were approximately the same size.

caryotic organisms. However, the procedure converts normally nontransmissible hybrid plasmids, such as those derived from Col EI, into a state which permits their ready transfer to other bacterial hosts. Because of biohazard considerations, we do not recommend the establishment in an F+ background of strains or banks of clones containing hybrid plasmids constructed of Col EI-DNA and random fragments of DNA from higher eucaryotic organisms until the biological properties of such plasmids have been further characterized.

Experimental Procedures

Bacterial Strains

The following strains, all derivatives of E. coli K12, were used as recipients for transformations or in bacterial mating experiments: JA198 ($\Delta trpE$ recA thr leu lacY str^r), JA 200 (F$^+$ $\Delta trpE5$ recA thr leu lacY), JA208 ($\Delta araC766$ $\Delta lacZ$-Y514 recA), MV12 ($\Delta trpE5$ recA thr leu lacY), DM493 (W3110 $\Delta trpE5$), CH754 (argH metE xyl trpA36 recA56). The F plasmid in JA200 was derived from E. coli strain Ymel.

Construction of Hybrid Col EI-DNA (E. coli) Annealed Circles

Linear poly(dT)-tailed plasmid Col EI-DNA [(L_{RI}exo)-(dT)$_{150}$] was prepared as described (Clarke and Carbon, 1975). High molecular weight E. coli DNA was purified from strain CS520 (HfrC trpA58 metB glyVsu58) as described (Clarke and Carbon, 1975), resuspended at 100 μg/ml in 0.01 M Tris–HCl (pH 7.5), 0.01 M NaCl, 0.001 M Na$_2$EDTA (STE), and fragmented by hydrodynamic shearing. DNA was sheared in a stainless steel cup (capacity = 1 ml) at 0°C for 45 min using a setting of 4.5 (approximately 5400 rpm) on a Tri-R Stir-R motor (model S63C) fitted with a Virtis shaft and micro homogenizer blades. The fragmented CS520 DNA (average molecular weight = 8.4 × 10^6 daltons) was treated with λ 5′–exonuclease, and poly(dA)$_{150}$ extensions were added to the 3′ ends of the DNA with the calf thymus deoxynucleotidyl terminal transferase (Clarke and Carbon, 1975). An equimolar mixture of Col EI-DNA [(L_{RI}exo)-(dT)$_{150}$] and CS520 DNA [L_{1h}exo)-(dA)$_{150}$] was annealed, the DNA was concentrated by precipitation in 67% ethanol, dissolved in STE, and examined under the electron microscope (Clarke and Carbon, 1975). The annealed mixture contained approximately 25% hybrid DNA circular molecules.

Establishment of E. coli Hybrid Plasmid Colony Bank

12 μg of annealed hybrid DNA were used to transform strain JA200 (C600 recA/F+) according to a modification of the method of Mandel and Higa (1970) described by Wensink et al. (1974), except that after exposure to DNA, cells were diluted 10 fold into L broth and grown for 30 min. The cells were washed once and resuspended in 1/3 vol L broth. In each of 20 tubes, 0.25 ml cells were mixed with 0.1 ml of crude colicin EI (Spudich, Korn, and Yanofsky, 1970) and 0.65 ml L broth. After incubation for 15 min at room temperature, 4 ml of warm 0.8% L-agar were added to each tube, and the contents were plated onto 1.5% L-agar plates. After incubating the plates for 24 hr at 37°C, approximately 2000 colonies were transferred to 1.5% L-agar plates (48 colonies per plate) freshly prepared with an overlay of colicin EI in 5 ml of 1.5% L-agar. These plates were incubated at 37°C for 24 hr and served as masters in subsequent mating experiments. The colonies were arranged in a grid pattern of 48 per plate such that they could easily be transferred via a wooden block of 48 needles to standard 96-well MicroTest II dishes (Falcon Plastics), with each well containing 0.2 ml L broth, colicin EI, and 8% dimethyl sulfoxide (Roth, 1970). The

colonies in the collection could therefore be individually maintained, permanently stored at –80°C, and used repeatedly by inoculating fresh plates.

A separate control culture of JA200 was simultaneously treated in an identical manner to the transformed culture described above, except that cells were not exposed to DNA. From the number of colicin EI-resistant cells in this mock culture, it was estimated that about 67% of the colicin-resistant cells in the transformed culture contained Col EI plasmids, and the remaining clones were colicin-tolerant but contained no plasmid (Davis and Reeves, 1975). 12 of 15 clones from the transformant culture were resistant to colicin EI and sensitive to colicin E2 (Hershfield et al., 1974), and the remainder were resistant to both colicins. Use of the Triton lysis procedure (described below) on ten clones from a pilot experiment identical to the one described here and electrophoretic analysis of supercoiled plasmid DNAs in 1.2% agarose gels revealed that about 90% of the clones in the transformant culture which were resistant to colicin EI and sensitive to colicin E2 contained plasmids appreciably larger than Col E1, while the remaining 10% contained no plasmid. From these data, we conclude that about 70% of the clones in the colony bank contain hybrid plasmids.

Transformations and Selections

Subsequent transformations with purified hybrid Col E1-DNA (E. coli) plasmid DNAs were carried out exactly as described (Clarke and Carbon, 1975). Standard techniques for bacterial selections, use of indicator plates, and liquid and replica matings were used (Miller, 1972).

Electron Microscopy

DNA was spread by the aqueous method of Davis, Simon, and Davidson (1971), and all contour lengths were measured relative to reference Col E1 relaxed circles or Eco RI-cleaved Col E1 DNA linears on the same grid.

Hybrid Plasmid DNA Isolation

Col E1-DNA (E. coli) hybrid plasmid DNA was isolated from clones in the colony bank according to a modification of a method developed in Dr. Herbert Boyer's laboratory (personal communication). A 10 ml cell culture was grown, and plasmid DNA was amplified by addition of chloramphenicol as described by Clewell (1972). Cells were pelleted and resuspended in 1 ml of cold 25% sucrose in 0.05 M Tris–HCl (pH 8.0) and 0.2 ml of lysozyme [5 mg/ml in 0.25 M Tris–HCl (pH 8.0)] were added. The mixture was incubated at 0°C for 5 min, and 0.4 ml of 0.25 M Na$_2$EDTA (pH 8.0) were added for an additional 5 min at 0°C. 25 μg of pancreatic RNAase [2 mg/ml in 0.01 M Tris–HCl (pH 7.4), 0.001 M Na$_2$EDTA] were added to the cell suspension with 0.5 ml of Triton lytic mix [0.3% Triton X 100, 0.15 M Tris–HCl (pH 8.0), 0.18 M Na$_2$EDTA]. The suspension was swirled gently, incubated at 0°C for 15 min, and spun at 17,000 rpm for 30 min in a Sorvall SS-34 rotor. The supernatant was deproteinized by extraction with 1/3 vol phenol and passed through a 0.5 ml Bio-Beads SM-2 (BioRad) column to remove Triton X 100 (Holloway, 1973). DNA was precipitated in 67% ethanol, and covalently closed supercoiled plasmid DNA was purified by banding to equilibrium in an ethidium bromide–CsCl density gradient (Clewell, 1972). Such a procedure routinely yielded 2–5 μg of purified plasmid DNA. The procedure could be scaled up proportionally for larger amounts of cells.

Acknowledgments

The authors are indebted to Denise Richardson for expert technical assistance. Several laboratories contributed to the data summarized in Table 2; we are grateful to these investigators for permission to mention their results prior to publication. This work was supported by research grants from the National Cancer Institute.

Received April 12, 1976; revised May 17, 1976

References

Bazarol, M., and Helinski, D. R. (1968). J. Mol. Biol. *36*, 185–194.

Clarke, L., and Carbon, J. (1975). Proc. Nat. Acad. Sci. USA *72*, 4361–4365.

Clewell, D. B. (1972). J. Bacteriol. *110*, 667–676.

Clowes, R. C. (1964). Ann. Inst. Pasteur *107*, suppl. 5, 74–92.

Clowes, R. C. (1972). Bacteriol. Rev. *36*, 361–405.

Clowes, R. C., Moody, E. E. M., and Pritchard, R. H. (1965). Genet. Res. *6*, 147–152.

Davis, J. K., and Reeves, P. (1975). J. Bacteriol. *123*, 102–117.

Davis, R., Simon, M., and Davidson, N. (1971). In Methods in Enzymology, *21*, L. Grossman and K. Moldave, eds. (New York: Academic Press), pp. 413–428.

Fleck, E. W., and Carbon, J. (1975). J. Bacteriol. *122*, 492–501.

Fredericq, P., and Betz-Bareau, M. (1953). C.r. Seanc. Soc. Biol., Paris, *147*, 2043.

Grunstein, M., and Hogness, D. (1975). Proc. Nat. Acad. Sci. USA *72*, 3961–3965.

Guest, J. R., and Yanofsky, C. (1965). J. Mol. Biol. *12*, 793–804.

Hardy, K. G. (1975). Bacteriol. Rev. *39*, 464–515.

Hershfield, V., Boyer, H. W., Yanofsky, C., Lovett, M. A., and Helinski, D. R. (1974). Proc. Nat. Acad. Sci. USA *71*, 3455–3459.

Holloway, P. W. (1973). Anal. Biochem. *53*, 304–308.

Jackson, D. A., Symons, R. H., and Berg, P. (1972). Proc. Nat. Acad. Sci. USA *69*, 2904–2909.

Lobban, P., and Kaiser, D. (1973). J. Mol. Biol. *78*, 453–471.

Mandel, M., and Higa, A. (1970). J. Mol. Biol. *53*, 159–162.

Miller, J. (1972). Experiments in Molecular Genetics (New York: Cold Spring Harbor Laboratory).

Roth, J. (1970). In Methods in Enzymology, *17*, H. Tabor and C. W. Tabor, (New York: Academic Press), pp. 3–35.

Spudich, J. A., Korn, V., and Yanofsky, C. (1970). J. Mol. Biol. *53*, 49–67.

Struhl, K., Cameron, J. R., and Davis, R. W. (1976). Proc. Nat. Acad. Sci. USA *73*, 1471–1475.

Wensink, P. C., Finnegan, D. J., Donelson, J. E., and Hogness, D. S. (1974). Cell *3*, 315–325.

Construction of Biologically Functional Bacterial Plasmids *In Vitro*

S.N. Cohen, A.C.Y. Chang, H.W. Boyer and R.B. Helling

ABSTRACT The construction of new plasmid DNA species by *in vitro* joining of restriction endonuclease-generated fragments of separate plasmids is described. Newly constructed plasmids that are inserted into *Escherichia coli* by transformation are shown to be biologically functional replicons that possess genetic properties and nucleotide base sequences from both of the parent DNA molecules. Functional plasmids can be obtained by reassociation of endonuclease-generated fragments of larger replicons, as well as by joining of plasmid DNA molecules of entirely different origins.

Controlled shearing of antibiotic resistance (R) factor DNA leads to formation of plasmid DNA segments that can be taken up by appropriately treated *Escherichia coli* cells and that recircularize to form new, autonomously replicating plasmids (1). One such plasmid that is formed after transformation of *E. coli* by a fragment of sheared R6-5 DNA, pSC101 (previously referred to as Tc6-5), has a molecular weight of 5.8×10^6, which represents about 10% of the genome of the parent R factor. This plasmid carries genetic information necessary for its own replication and for expression of resistance to tetracycline, but lacks the other drug resistance determinants and the fertility functions carried by R6-5 (1).

Two recently described restriction endonucleases, *Eco*RI and *Eco*RII, cleave double-stranded DNA so as to produce short overlapping single-stranded ends. The nucleotide sequences cleaved are unique and self-complementary (2–6) so that DNA fragments produced by one of these enzymes can associate by hydrogen-bonding with other fragments produced by the same enzyme. After hydrogen-bonding, the 3′-hydroxyl and 5′-phosphate ends can be joined by DNA ligase (6). Thus, these restriction endonucleases appeared to have great potential value for the construction of new plasmid species by joining DNA molecules from different sources. The *Eco*RI endonuclease seemed especially useful for this purpose, because on a random basis the sequence cleaved is expected to occur only about once for every 4,000 to 16,000 nucleotide pairs (2); thus, most *Eco*RI-generated DNA fragments should contain one or more intact genes.

We describe here the construction of new plasmid DNA species by *in vitro* association of the *Eco*RI-derived DNA fragments from separate plasmids. In one instance a new plasmid has been constructed from two DNA species of entirely different origin, while in another, a plasmid which has itself been derived from *Eco*RI-generated DNA fragments of a larger parent plasmid genome has been joined to another replicon derived independently from the same parent plasmid. Plasmids that have been constructed by the *in vitro* joining of

*Eco*RI-generated fragments have been inserted into appropriately-treated *E. coli* by transformation (7) and have been shown to form biologically functional replicons that possess genetic properties and nucleotide base sequences of both parent DNA species.

MATERIALS AND METHODS

E. coli strain W1485 containing the RSF1010 plasmid, which carries resistance to streptomycin and sulfonamide, was obtained from S. Falkow. Other bacterial strains and R factors and procedures for DNA isolation, electron microscopy, and transformation of *E. coli* by plasmid DNA have been described (1, 7, 8). Purification and use of the *Eco*RI restriction endonuclease have been described (5). Plasmid heteroduplex studies were performed as previously described (9, 10). *E. coli* DNA ligase was a gift from P. Modrich and R. L. Lehman and was used as described (11). The detailed procedures for gel electrophoresis of DNA will be described elsewhere (Helling, Goodman, and Boyer, in preparation); in brief, duplex DNA was subjected to electrophoresis in a tube-type apparatus (Hoefer Scientific Instrument) (0.6 × 15-cm gel) at about 20° in 0.7% agarose at 22.5 V with 40 mM Tris–acetate buffer (pH 8.05) containing 20 mM sodium acetate, 2 mM EDTA, and 18 mM sodium chloride. The gels were then soaked in ethidium bromide (5 μg/ml) and the DNA was visualized by fluorescence under long wavelength ultra-violet light ("black light"). The molecular weight of each fragment in the range of 1 to 200×10^5 was determined from its mobility relative to the mobilities of DNA standards of known molecular weight included in the same gel (Helling, Goodman, and Boyer, in preparation).

RESULTS

R6-5 and pSC101 plasmid DNA preparations were treated with the *Eco*RI restriction endonuclease, and the resulting DNA products were analyzed by electrophoresis in agarose gels. Photographs of the fluorescing DNA bands derived from these plasmids are presented in Fig. 1b and c. Only one band is observed after *Eco*RI endonucleolytic digestion of pSC101 DNA (Fig. 1c), suggesting that this plasmid has a single site susceptible to cleavage by the enzyme. In addition, endonuclease-treated pSC101 DNA is located at the position in the gel that would be expected if the covalently closed circular plasmid is cleaved once to form noncircular DNA of the same molecular weight. The molecular weight of the linear fragment estimated from its mobility in the gel is 5.8×10^6, in agreement with independent measurements of the size of the intact molecule (1). Because pSC101 has a single *Eco*RI cleavage site and is derived from R6-5, the equivalent DNA sequences of

COHEN, S.N., CHANG, A.C.Y., BOYER, H.W. and HELLING, R.B.
Construction of biologically functional bacterial plasmids *in vitro*.
Proc. Natl. Acad. Sci. U.S.A. 70:3240-3244 (1973). Reprinted with the authors' permission.

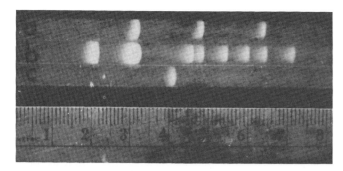

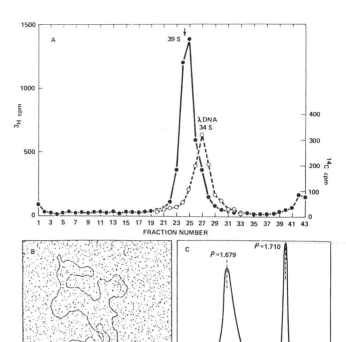

Fig. 1. Agarose-gel electrophoresis of *Eco*RI digests. (*a*) pSC102. The three fragments derived from the plasmid correspond to fragments III, V, and VIII of R6-5 (Fig. 1*b* below) as shown here and as confirmed by electrophoresis in other gels (see *text*). (*b*) R6-5. The molecular weights calculated for the fragments, as indicated in *Methods*, are (from *left* to *right*) I, 17.0; II & III (double band), 9.6 and 9.1; IV, 5.2; V, 4.9; VI, 4.3; VII, 3.8; VIII, 3.4; IX, 2.9. All molecular weight values have been multiplied by 10^{-6}. (*c*) pSC101. The calculated molecular weight of the single fragment is 5.8×10^6. Migration in all gels was from *left* (cathode) to *right*; samples were subjected to electrophoresis for 19 hr and 50 min.

the parent plasmid must be distributed in two separate *Eco*RI fragments.

The *Eco*RI endonuclease products of R6-5 plasmid DNA were separated into 12 distinct bands, eight of which are seen in the gel shown in Fig. 1*b*; the largest fragment has a molecular weight of 17×10^6, while three fragments (not shown in Fig. 1*b*) have molecular weights of less than 1×10^6, as determined by their relative mobilities in agarose gels. As seen in the figure, an increased intensity of fluorescence, of the second band suggests that this band contains two or more DNA fragments of almost equal size; when smaller amounts of *Eco*RI-treated R6-5 DNA are subjected to electrophoresis for a longer period of time, resolution of the two fragments (i.e., II and III) is narrowly attainable. Because 12 different *Eco*RI-generated DNA fragments can be identified after endonuclease treatment of covalently closed circular R6-5, there must be at least 12 substrate sites for *Eco*RI endonuclease present on this plasmid, or an average of one site for every 8000 nucleotide pairs. The molecular weight for each fragment shown is given in the caption to Fig. 1. The sum of the molecular weights of the *Eco*RI fragments of R6-5 DNA is 61.5×10^6, which is in close agreement with independent estimates for the molecular weight of the intact plasmid (7, 10).

The results of separate transformations of *E. coli* C600 by endonuclease-treated pSC101 or R6-5 DNA are shown in Table 1. As seen in the table, cleaved pSC101 DNA transforms *E. coli* C600 with a frequency about 10-fold lower than was observed with covalently closed or nicked circular (1) molecules of the same plasmid. The ability of cleaved pSC-101 DNA to function in transformation suggests that plasmid DNA fragments with short cohesive endonuclease-generated termini can recircularize in *E. coli* and be ligated *in vivo*; since the denaturing temperature (T_m) for the termini generated by the *Eco*RI endonuclease is 5–6° (6) and the transformation procedure includes a 42° incubation step (7), it is unlikely that the plasmid DNA molecules enter bacterial cells with their termini already hydrogen-bonded. A corresponding observation has been made with *Eco*RI endonuclease-cleaved

Fig. 2. Physical properties of the pSC102 plasmid derived from *Eco*RI fragments of R6-5. (*A*) Sucrose gradient centrifugation analysis (1, 8) of covalently closed circular plasmid DNA (●——●) isolated from an *E. coli* transformant clone as described in text. 34 S linear [^{14}C]DNA from λ was used as a standard (O- - -O) pSC102 DNA. (*B*) Electron photomicrograph of nicked (7) pSC102 DNA. The length of this molecule is approximately 8.7 μm. (*C*) Densitometer tracing of analytical ultracentrifugation (8) photograph of pSC102 plasmid DNA. Centrifugation in CsCl ($\rho = 1.710$ g/cm³) was carried out in the presence of d(A-T)$_n$·-d(A-T)$_n$ density marker ($\rho = 1.679$ g/cm³).

SV40 DNA, which forms covalently closed circular DNA molecules in mammalian cells *in vivo* (6).

Transformation for each of the antibiotic resistance markers present on the R6-5 plasmid was also reduced after treatment of this DNA with *Eco*RI endonuclease (Table 1). Since the pSC101 (tetracycline-resistance) plasmid was derived from R6-5 by controlled shearing of R6-5 DNA (1), and no tetracycline-resistant clone was recovered after transformation by the *Eco*RI endonuclease products of R6-5, [whereas tetracycline-resistant clones are recovered after transformation with intact R6-5 DNA (1)], an *Eco*RI restriction site may separate the tetracycline resistance gene of R6-5 from its replicator locus. Our finding that the linear fragment produced by treatment of pSC101 DNA with *Eco*RI endonuclease does not correspond to any of the *Eco*RI-generated fragments of R6-5 (Fig. 1) is consistent with this interpretation.

A single clone that had been selected for resistance to kanamycin and which was found also to carry resistance to neomycin and sulfonamide, but not to tetracycline, chloramphenicol, or streptomycin after transformation of *E. coli* by *Eco*RI-generated DNA fragments of R6-5, was examined further. Closed circular DNA obtained from this isolate (plasmid designation pSC102) by CsCl–ethidium bromide gradient

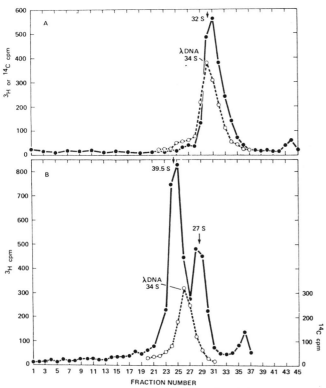

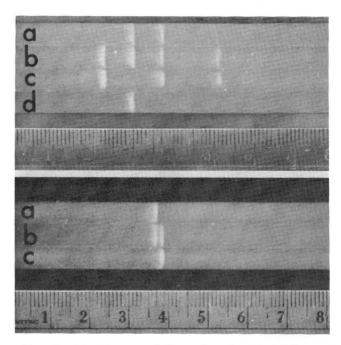

FIG. 3. Sucrose gradient centrifugation of DNA isolated from *E. coli* clones transformed for both tetracycline and kanamycin resistance by a mixture of pSC101 and pSC102 DNA. (*A*) The DNA mixture was treated with *Eco*RI endonuclease and was ligated prior to use in the transformation procedure. Covalently closed circular DNA isolated (7, 8) from a transformant clone carrying resistance to both tetracycline and kanamycin was examined by sedimentation in a neutral 5–20% sucrose gradient (8). (*B*) Sucrose sedimentation pattern of covalently closed circular DNA isolated from a tetracycline and kanamycin resistant clone transformed with an *untreated* mixture of pSC101 and pSC102 plasmid DNA.

FIGS. 4 and 5.   Agarose-gel electrophoresis of *Eco*RI digests of newly constructed plasmid species.   Conditions were as described in *Methods*.

FIG. 4.   (*top*) Gels were subjected to electrophoresis for 19 hr and 10 min. (*a*) pSC105 DNA. (*b*) Mixture of pSC101 and pSC102 DNA. (*c*) pSC102 DNA. (*d*) pSC101 DNA.

FIG. 5.   (*bottom*) Gels were subjected to electrophoresis for 18 hr and 30 min. (*a*) pSC101 DNA. (*b*) pSC109 DNA. (*c*) RSF1010 DNA. Evidence that the single band observed in this gel represents a linear fragment of cleaved RSF1010 DNA was obtained by comparing the relative mobilities of *Eco*RI-treated DNA and untreated (covalently closed circular and nicked circular) RSF1010 DNA in gels. The molecular weight of RSF-1010 calculated from its mobility in gels is 5.5×10^6.

centrifugation has an S value of 39.5 in neutral sucrose gradients (Fig. 2*A*) and a contour length of 8.7 μm when nicked (Fig. 2*B*). These data indicate a molecular weight

TABLE 1.   *Transformation by covalently closed circular and EcoRI-treated plasmid DNA*

Plasmid DNA species	Transformants per μg DNA		
	Tetracycline	Kanamycin (neomycin)	Chloramphenicol
pSC101 covalently closed circle	3×10^5	—	—
*Eco*RI-treated	2.8×10^4	—	—
R6-5 covalently closed circle	—	1.3×10^4	1.3×10^4
*Eco*RI-treated	<5	1×10^2	4×10^1

Transformation of *E. coli* strain C600 by plasmid DNA was carried out as indicated in *Methods*. The kanamycin resistance determinant of R6-5 codes also for resistance to neomycin (15). Antibiotics used for selection were tetracycline (10 μg/ml), kanamycin (25 μg/ml) or chloramphenicol (25 μg/ml).

about 17×10^6. Isopycnic centrifugation in cesium chloride of this non-self-transmissible plasmid indicated it has a buoyant density of 1.710 g/cm³ (Fig. 2*C*). Since the nucleotide base composition of the antibiotic resistance determinant (R-determinant) segment of the parent R factor is 1.718 g/cm³ (8), the various component regions of the resistance unit must have widely different base compositions, and the pSC102 plasmid must lack a part of this unit that is rich in high buoyant density G+C nucleotide pairs. The existence of such a high buoyant density *Eco*RI fragment of R6-5 DNA was confirmed by centrifugation of *Eco*RI-treated R6-5 DNA in neutral cesium chloride gradients (Cohen and Chang, unpublished data).

Treatment of pSC102 plasmid DNA with *Eco*RI restriction endonuclease results in formation of three fragments that are separable by electrophoresis is agarose gels (Fig. 1*a*); the estimated molecular weights of these fragments determined by gel mobility total 17.4×10^6, which is in close agreement with the molecular weight of the intact pSC102 plasmid determined by sucrose gradient centrifugation and electron microscopy (Fig. 2). Comparison with the *Eco*RI-generated fragments of R6-5 indicates that the pSC102 fragments correspond to fragments III (as determined by long-term electrophoresis in gels containing smaller amounts of DNA), V, and VIII of the parent plasmid (Fig. 1*b*). These results suggest that *E. coli* cells transformed with *Eco*RI-generated DNA fragments of R6-5

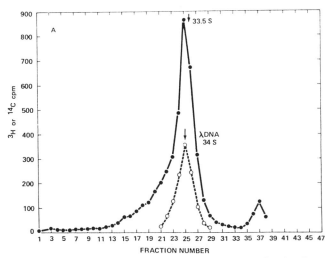

FIG. 6. Sucrose gradient sedimentation of covalently closed circular DNA representing the pSC109 plasmid derived from RSF1010 and pSC101.

TABLE 2. *Transformation of E. coli C600 by a mixture of pSC101 and pSC102 DNA*

Treatment of DNA	Transformation frequency for antibiotic resistance markers		
	Tetracycline	Kanamycin	Tetracycline + kanamycin
None	2×10^5	1×10^6	2×10^2
EcoRI	1×10^4	1.1×10^3	7×10^1
EcoRI + DNA ligase	1.2×10^4	1.3×10^3	5.7×10^2

Transformation frequency is shown in transformants per μg of DNA of each plasmid species in the mixture. Antibiotic concentrations are indicated in legend of Table 1.

can ligate reassociated DNA fragments *in vivo*, and that reassociated molecules carrying antibiotic resistance genes and capable of replication can circularize and can be recovered as functional plasmids by appropriate selection.

A mixture of pSC101 and pSC102 plasmid DNA species, which had been separately purified by dye–buoyant density centrifugation, was treated with the *Eco*RI endonuclease, and then was either used directly to transform *E. coli* or was ligated prior to use in the transformation procedure (Table 2). In a control experiment, a plasmid DNA mixture that had not been subjected to endonuclease digestion was employed for transformation. As seen in this table, transformants carrying resistance to both tetracycline and kanamycin were isolated in all three instances. Cotransformation of tetracycline and kanamycin resistance by the untreated DNA mixture occurred at a 500- to 1000-fold lower frequency than transformation for the individual markers. Examination of three different transformant clones derived from this DNA mixture indicated that each contained two separate covalently closed circular DNA species having the sedimentation characteristics of the pSC101 and pSC102 plasmids (Fig. 3B). The ability of two plasmids derived from the same parental plasmid (i.e., R6-5) to exist stably as separate replicons (12) in a single

bacterial host cell suggests that the parent plasmid may contain at least two distinct replicator sites. This interpretation is consistent with earlier observations which indicate that the R6 plasmid dissociates into two separate compatible replicons in *Proteus mirabilis* (8). Cotransformation of tetracycline and kanamycin resistance by the *Eco*RI treated DNA mixture was 10- to 100-fold lower than transformation of either tetracycline or kanamycin resistance alone, and was increased about 8-fold by treatment of the endonuclease digest with DNA ligase (Table 2). Each of four studied clones derived by transformation with the endonuclease-treated and/or ligated DNA mixture contained only a single 32S covalently closed circular DNA species (Fig. 3A) that carries resistance to both tetracycline and kanamycin, and which can transform *E. coli* for resistance to both antibiotics. One of the clones derived from the ligase-treated mixture was selected for further study, and this plasmid was designated pSC105.

When the plasmid DNA of pSC105 was digested by the *Eco*RI endonuclease and analyzed by electrophoresis in agarose gels, two component fragments were identified (Fig. 4); the larger fragment was indistinguishable from endonuclease-treated pSC101 DNA (Fig. 4d) while the smaller fragment corresponded to the 4.9×10^6 dalton fragment of pSC102 plasmid DNA (Fig. 4c). Two endonuclease fragments of pSC102 were lacking in the pSC105 plasmid; presumably the sulfonamide resistance determinant of pSC102 is located on one of these fragments, since pSC105 does not specify re-

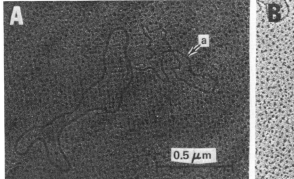

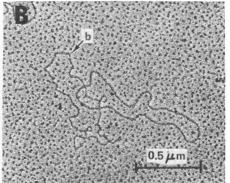

FIG. 7. (A) Heteroduplex of pSC101/pSC109. The single-stranded DNA loop marked by *a* represents the contribution of RSF1010 to the pSC109 plasmid. (B) Heteroduplex of RSF1010/pSC109. The single-stranded DNA loop marked by *b* represents the contribution of pSC101 to the pSC109 plasmid. pSC101 and RSF1010 homoduplexes served as internal standards for DNA length measurements. The scale is indicated by the bar on each electron photomicrograph.

sistance to this antibiotic. Since kanamycin resistance *is* expressed by pSC105, we conclude that this resistance gene resides on the 4.9×10^6 dalton fragment of pSC102 (fragment V of its parent, R6-5). The molecular weight of the pSC105 plasmid is estimated to be 10.5×10^6 by addition of the molecular weights of its two component fragments; this value is consistent with the molecular weight determined for this recombinant plasmid by sucrose gradient centrifugation (Fig. 3A) and electron microscopy. The recovery of a biologically functional plasmid (i.e., pSC105) that was formed by insertion of a fragment of another plasmid fragment into pSC101 indicates that the *Eco*RI restriction site on pSC101 does not interrupt the genetic continuity of either the tetracycline resistance gene or the replicating element of this plasmid.

We also constructed new biologically functional plasmids *in vitro* by joining cohesive-ended plasmid DNA molecules of entirely different origin. RSF1010 is a streptomycin and sulfonamide resistance plasmid which has a 55% G+C nucleotide base composition (13) and which was isolated originally from *Salmonella typhimurium* (14). Like pSC101, this non-self-transmissible plasmid is cleaved at a single site by the *Eco*RI endonuclease (Fig. 5c). A mixture of covalently closed circular DNA containing the RSF1010 and pSC101 plasmids was treated with the *Eco*RI endonuclease, ligated, and used for transformation. A transformant clone resistant to both tetracycline and streptomycin was selected, and covalently closed circular DNA (plasmid designation pSC109) isolated from this clone by dye–buoyant density centrifugation was shown to contain a single molecular species sedimenting at 33.5 S, corresponding to an approximate molecular weight of 11.5×10^6 (Fig. 6). Analysis of this DNA by agarose gel electrophoresis after *Eco*RI digestion (Fig. 5b) indicates that it consists of two separate DNA fragments that are indistinguishable from the *Eco*RI-treated RSF1010 and pSC101 plasmids (Fig. 5a and c).

Heteroduplexes shown in Fig. 7A and B demonstrate the existence of DNA nucleotide sequence homology between pSC109 and each of its component plasmids. As seen in this figure, the heteroduplex pSC101/pSC109 shows a double-stranded region about 3 μm in length and a slightly shorter single-stranded loop, which represents the contribution of RSF1010 to the recombinant plasmid. The heteroduplex formed between RSF1010 and pSC109 shows both a duplex region and a region of nonhomology, which contains the DNA contribution of pSC101 to pSC109.

SUMMARY AND DISCUSSION

These experiments indicate that bacterial antibiotic resistance plasmids that are constructed *in vitro* by the joining of *Eco*RI-treated plasmids or plasmid DNA fragments are bio-

logically functional when inserted into *E. coli* by transformation. The recombinant plasmids possess genetic properties and DNA nucleotide base sequences of both parent molecular species. Although ligation of reassociated *Eco*RI-treated fragments increases the efficiency of new plasmid formation, recombinant plasmids are also formed after transformation by *unligated Eco*RI-treated fragments.

The general procedure described here is potentially useful for insertion of specific sequences from prokaryotic or eukaryotic chromosomes or extrachromosomal DNA into independently replicating bacterial plasmids. The antibiotic resistance plasmid pSC101 constitutes a replicon of considerable potential usefulness for the selection of such constructed molecules, since its replication machinery and its tetracycline resistance gene are left intact after cleavage by the *Eco*RI endonuclease.

We thank P. A. Sharp and J. Sambrooke for suggesting use of ethidium bromide for staining DNA fragments in agarose gels. These studies were supported by Grants AI08619 and GM14378 from the National Institutes of Health and by Grant GB-30581 from the National Science Foundation. S.N.C. is the recipient of a USPHS Career Development Award. R.B.H. is a USPHS Special Fellow of the Institute of General Medical Sciences on leave from the Department of Botany, University of Michigan.

1. Cohen, S. N. & Chang, A. C. Y. (1973) *Proc. Nat. Acad. Sci. USA* **70**, 1293–1297.
2. Hedgepeth, J., Goodman, H. M. & Boyer, H. W. (1972) *Proc. Nat. Acad. Sci. USA* **69**, 3448–3452.
3. Bigger, C. H., Murray, K. & Murray, N. E. (1973) *Nature New Biol.*, **224**, 7–10.
4. Boyer, H. W., Chow, L. T., Dugaiczyk, A., Hedgepeth, J. & Goodman, H. M. (1973) *Nature New Biol.*, **224**, 40–43.
5. Greene, P. J., Betlach, M. C., Goodman, H. M. & Boyer, H. W. (1973) "DNA replication and biosynthesis," in *Methods in Molecular Biology*, ed. Wickner, R. B. Marcel Dekker, Inc. New York), Vol. 9, in press.
6. Mertz, J. E. & Davis, R. W. (1972) *Proc. Nat. Acad. Sci. USA* **69**, 3370–3374.
7. Cohen, S. N., Chang, A. C. Y. & Hsu, L. (1972) *Proc. Nat. Acad. Sci. USA* **69**, 2110–2114.
8. Cohen, S. N. & Miller, C. A. (1970) *J. Mol. Biol.* **50**, 671–687.
9. Sharp, P. A., Hsu, M., Ohtsubo, E. & Davidson, N. (1972) *J. Mol. Biol.* **71**, 471–497.
10. Sharp, P. A., Cohen, S. N. & Davidson, N. (1973) *J. Mol. Biol.* **75**, 235–255.
11. Modrich, P. & Lehman, R. L. (1973) *J. Biol. Chem.*, in press.
12. Jacob, F., Brenner, S. & Cuzin, F. (1963) *Cold Spring Harbor Symp. Quant. Biol.* **23**, 329–484.
13. Guerry, P., van Embden, J., & Falkow, S. (1973) *J. Bacteriol.*, in press.
14. Anderson, E. S. & Lewis, M. J. (1965) *Nature* **208**, 843–849.
15. Davies, J., Benveniste, M. S. & Brzezinka, M. (1971) *Ann. N.Y. Acad. Sci.* **182**, 226–233.

Cosmids: A Type of Plasmid Gene-cloning Vector that is Packageable *In Vitro* in Bacteriophage λ Heads

J. Collins and B. Hohn

ABSTRACT   Evidence is presented that ColE1 hybrid plasmids carrying the cohesive-end site (*cos*) of λ can be used as gene cloning vectors in conjunction with the λ *in vitro* packaging system of Hohn and Murray [(1977) *Proc. Natl. Acad. Sci. USA* 74, 3259–3263]. Due to the requirement for a large DNA molecule for efficient packaging, there is a direct selection for hybrids carrying large sections of foreign DNA. The small vector plasmids do not contribute a large background in the transduced population, which is therefore markedly enriched for large hybrid plasmids (over 90%). The efficiency of the *in vitro* packaging system is on the order of 10^5 *hybrid* clones per microgram of foreign DNA for hybrids in the 20–30 million dalton range.

The mechanism of packaging DNA into the head of *Escherichia coli* bacteriophage λ has been extensively studied through the development of *in vitro* packaging systems (1–4; for review, see ref. 5). These and studies *in vivo* (6) led to the following findings: monomeric circular DNA was not packaged; head-to-tail polymers (concatemers) of the unit-length λ DNA molecules were efficiently packaged if the cohesive-end site (*cos*), substrate for the packaging-dependent cleavage that produces the cohesive ends of mature λ DNA, was 23–33 megadaltons (MDal) apart; and only a small region in the proximity of the cleavage site was required for recognition by the packaging system (7, 8).

This information implies that *cos*-containing plasmids of less than 23 MDal would not be efficiently packaged due to the circular form of their DNA and their size, but that concatemeric derivatives with DNA inserts would be a packaging substrate. The latter DNA structure resembles a ligation mixture between a cleaved *cos*-containing plasmid and DNA to be cloned. It was expected, therefore, that cloning in a *cos*-containing plasmid in conjunction with *in vitro* packaging selects against re-ligated vector molecules but selects for hybrids in the size range of λ DNA, molecules that are recovered only poorly upon transformation.

In our present study, experiments are described in which a *cos*-containing ColE1 *rpo* plasmid (9, 10) was packaged *in vitro* after restriction and re-ligation. The results of this experiment, as well as of RI plasmid and *Pseudomonas* cloning experiments, suggest the use of packageable plasmids as a gene cloning system that is both highly efficient and selective for recovery of large hybrids.

Plasmids containing a *cos* site, which are useful as vectors for gene cloning in conjunction with the packaging system, we refer to as "cosmids."

MATERIALS AND METHODS

Plasmids and Bacteria. Preparation of plasmids pJC720 and pJC703 (Fig. 1) has been described (9, 10). The detailed mapping of these plasmids with restriction endonucleases is unpublished. *E. coli* N205, an *E. coli* K-12 strain ($r_k^+m_k^+$ $recA^-su^-$), was from N. Sternberg; *strain 5K* ($r_k^-m_k^+$ thr^-thi^-) was from S. Glover; strain HB101 ($r_k^-m_k^-$ leu^- pro^-recA^-) was from H. Boyer; and strain GL1 (*pel21*; W3101) was from S. W. Emmons (11).

Packaging System. Exogenous DNA was packaged *in vitro* as described (12), with some slight modifications: single colonies of strains N205 ($\lambda imm_{434}cI_{ts}b2\ red3\ E$am4 Sam7)/λ and N205 ($\lambda imm_{434}cI_{ts}b2red3\ E$am 15 Sam7/λ were streaked out on LA plates (1) and grown overnight at 30°C. Controls were plated to check temperature sensitivity at 42°C. Single colonies were inoculated into warmed LB medium (1) at an OD_{600} of not more than 0.15 and incubated with shaking until an OD_{600} of 0.3 was reached. Prophages were induced by incubation of the cultures at 45°C for 15 min while standing. Thereafter they were transferred to 37°C and incubated for 3 additional hr with vigorous aeration. (A small sample of each culture, which is lysis-inhibited as a result of the mutation in gene S, was checked for induction: upon addition of a drop of chloroform the culture cleared.) The two cultures were then mixed, centrifuged at 5000 rpm for 10 min, and resuspended at 0°C in 1/500th the original culture volume in complementation buffer (40 mM Tris·HCl, pH 8.0/10 mM spermidine hydrochloride/10 mM putrescine hydrochloride/0.1% mercaptoethanol/7% dimethyl sulfoxide) which was made 1.5 mM in ATP. Biological activity of endogenous DNA can be destroyed by UV irradiation prior to concentration (12). This cell suspension was distributed in 20-μl portions in 1.5-ml Eppendorf polyallomer centrifuge tubes, frozen in liquid N_2, and stored at −60°C. When needed, a sample was transferred in liquid N_2 and put on ice. Immediately on thawing (3–4 min on ice), the DNA to be packaged (0.01–0.2 μg) was added in a volume of 1–5 μl. The DNA was usually added in the buffer in which it had just been ligated. The solutions were carefully mixed and bubbles were removed by a few seconds' centrifugation in an Eppendorf desk-top centrifuge. The mix was incubated for 30 min at 37°C. At the end of this incubation period, 20 μl of a frozen and thawed packaging mixture, which had been made 10 mM in $MgCl_2$ and to which a final 10 μg of DNase per ml was added, was mixed to each sample and incubation was continued for 20–60 min. SMC (1) buffer (0.5 ml) and a drop of chloroform were added. After mixing, denatured material was centrifuged off and the solution was used as a phage lysate.

Abbreviations: MDal, megadaltons; cosmid, plasmid containing a *cos* site.
‡ Present address: Friedrich Miescher-Institut, P.O. Box 273, CH 4002 Basel, Switzerland.

COLLINS, J. and HOHN, B.
Cosmids: a type of plasmid gene-cloning vector that is packageable in vitro in bacteriophage λ heads.
Proc. Natl. Acad. Sci. U.S.A. 75:4242-4246 (1978). Reprinted with the authors' permission.

Transduction was carried out by adding 0.4 ml of this phage suspension to 1 ml of N205 or *pel*⁻ cells from a late exponential culture (OD_{600} = 2.0) in L broth–maltose [1% Bacto-Tryptone/0.5% yeast extract (Oxoid)/0.5% NaCl/0.4% maltose]. For the experiment with *Pseudomonas* DNA, HB101 was used as recipient. After a 10-min adsorption at 30°C, the mixture was diluted 1:20 in fresh L broth and incubated for 2 hr at 30°C to allow expression of rifampicin resistance.

Transformation. Strain 5K was used. Cultures grown to an OD_{600} of 0.5 were cooled rapidly on ice, centrifuged, and resuspended in 0.5 vol of 10 mM NaCl on ice. After 30 min on ice, the cells were centrifuged and resuspended in 0.5 vol of 50 mM CaCl_2 and again incubated for 30 min at 0°C. After centrifugation, the cells were resuspended in 0.1 vol of 30 mM CaCl_2 in 20% glycerol. This competent cell preparation, divided into 1-ml aliquots, was kept frozen at −60°C until needed. For transformation the sample was thawed out on ice and 0.5 ml of 40 mM Tris, pH 8.0/40 mM NaCl/1 mM EDTA, containing the DNA for the transformation (0.1–1 μg), was added. After 30 min on ice, the mixture was heated to 42°C for 2 min and rapidly cooled on ice. The cells were diluted 1:30 in L broth and incubated for 2 hr at 37°C to allow expression of rifampicin resistance (13).

Rifampicin resistance was tested on L broth plates containing either 100 μg of rifampicin per ml, when plasmid pJC703 was used, or 30 μg/ml, when plasmid pJC720 was used. The colonies derived after transduction or transformation of pJC720 grow slowly on rifampicin, taking 2 days at 37°C to form large colonies. Since rifampicin is light sensitive, the plates must be kept dark during this prolonged incubation to prevent growth of background colonies.

Restriction and Ligation Reactions. Restriction with *Hin*dIII (Boehringer) was carried out in 30 mM Tris·HCl, pH 7.6/10 mM MgCl_2/10 mM NaCl to completion. Digestion with *Sal* I, *Eco*RI, and *Bgl* II was carried out in the same buffer. *Sal* I and *Eco*RI were generous gifts from H. Mayer and H. Schütte, and *Bgl* II was a generous gift from E. Eichenlaub. Digestion

with *Kpn* I (Bio-Labs) was in 10 mM Tris·HCl, pH 7.9/6 mM MgCl_2/6 mM NaCl/10 mM dithiothreitol containing 100 μg of bovine serum albumin per ml. Gel electrophoresis was in 1% agarose (14). ·

Ligation was carried out after heat inactivation of the restriction endonucleases at 70°C for 10 min in 6 mM MgCl_2, 10 mM Tris·HCl (pH 7.9), 10 mM dithiothreitol, 100 μM ATP, 100 μg of bovine serum albumin per ml, and 5 × 10⁻² unit of T4 DNA ligase (Boehringer) in 100-μl aliquots. The concentration of DNA ends was between 20 and 60 pM or as described in the text. Before the ligase was added the samples were mixed, heated to 70°C for 5 min, and cooled on ice for 30 min. The ligation reaction was continued for 15 hr at 8°C. Completion of the ligation was checked by agarose gel electrophoresis: no sample was used in which linear monomers could still be detected.

Nomenclature of Plasmids. The numbering of the plasmids in the pJC series coming from John Collins will be confined to the first 500 in each thousand, with the exception of those already published, so as to avoid confusion with the collection of Alvin John Clark; pJC703 and pJC720 are two of these exceptions.

Safety Regulations. All experiments described here were carried out under P1 conditions as defined by the National Institutes of Health guidelines for recombinant DNA research.

RESULTS

Packaging of restricted and re-ligated plasmid DNA

Production of Packageable Substrate from Plasmid DNA. Plasmid pJC703 (Fig. 1) yields two *Hin*dIII restriction fragments: the 10-MDal fragment A containing the λ *cos* site and the ColE1 replicon, and the 7-MDal fragment B containing the gene for rifampicin resistance (*rpoB*). It was hoped that cleavage and re-ligation of plasmid pJC703 would produce a population of polymers (some of which are diagrammed in Fig.

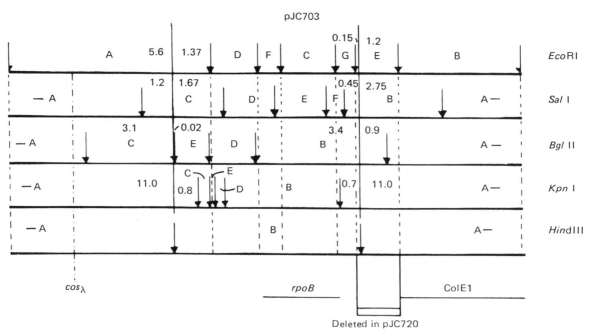

FIG. 1.   Restriction endonuclease map of pJC703 and pJC720. Fragments obtained with each enzyme are alphabetically labeled according to size. Dashed lines indicate the relative positions of the *Eco*RI cleavage site on each map, and the continuous vertical lines the positions of the *Hin*dIII cleavage sites. The distances from the *Hin*dIII sites to the nearest cleavage sites for each restriction enzyme are indicated in MDal. These values are used in the analysis of plasmids containing polymeric *Hin*dIII fragments (Table 2). The ColE1 part of these plasmids is actually derived from a freak isolate (pJC309) which contains a *Sal* I site not present in ColE1.

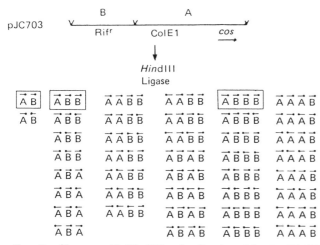

FIG. 2.   Cleavage with *Hin*dIII and re-ligation of the two *Hin*dIII fragments (A and B) of plasmid pJC703 theoretically can yield a series of polymers. The diagram shows all the possible cyclic permutations that contain at least one A *and* one B fragment, up to the tetrameric forms. The relative orientations are indicated by the small "matchsticks" over each fragment designation. Those structures actually obtained are framed.

2) that could mimic the natural substrate for packaging. Cleavage of such molecules at the *cos* site during packaging, and recircularization subsequent to transduction, would lead to the loss of one or more entire A fragments, thus generating plasmids of the form ABB, ABBB, and AAB (or AB).

After the re-ligated *Hin*dIII fragments of pJC703 were packaged, several thousand rifampicin-resistant clones were obtained by transduction into N205. The yield of Rif^r clones is dependent on the concentration of the vector DNA during the ligation (Exp. 1 *a* and *b*, Table 1). This supports the hypothesis that efficient packaging is dependent on the formation of long polymers. In contrast, the formation of such highly polymerized chains is most detrimental to the efficiency of transformation, as has been noted elsewhere (14). The transformation data are included merely as an additional test that the ligation was successful rather than as a direct comparison of packaging and transformation, since the ligation conditions are strongly biased in favor of packaging experiments.

Structure of Packaged Plasmids. Fifty-two colonies were picked at random for further testing. They were all found to be colicin E1 resistant and colicin E2 sensitive, indicative that the plasmid coded E1-immunity carried on the *Hin*dIII A fragment was present. Small cleared lysates were made from each clone. To check for the presence and approximate size of the plasmid DNA, we electrophoresed 5-μl samples (plus 0.1% sodium dodecyl sulfate) on 0.8% agarose gels (14). Supercoiled DNA was prepared from the first 12 samples and, from the remaining 40, from those showing the presence of plasmids larger than pJC703. These DNAs (Exp. 2, Table 1), and in some cases the products of a second packaging step, were analyzed more thoroughly with the restriction enzymes *Bgl* II, *Kpn* I, *Sal* I, *Eco*RI, and *Hin*dIII (Fig. 3).

Three size classes of plasmid were obtained: 17 MDal, corresponding to the starting plasmid, and about 23 MDal and 29 MDal. To test the orientation of the fragments with respect to one another, we cleaved the plasmids with an enzyme other than *Hin*dIII. This would generate fragments overlapping the *Hin*dIII junctions and yield "junctional" fragments diagnostic of the *Hin*dIII fragment arrangement (Table 2 and Fig. 3).

Only a small number of the possible structures (Fig. 2) are found amongst the large number analyzed: $\vec{A}\vec{B}$, $\vec{A}\vec{B}\vec{B}$, and $\vec{A}\vec{B}\vec{B}\vec{B}$ (Table 2 and Fig. 3). Absent are plasmids containing duplicates of the *Hin*dIII A fragment, i.e., that fragment carrying the replicon origin and the *cos* site. Such structures would be eliminated if every *cos* site were cleaved during *in vitro* packaging. The *in vivo* probability, however, of cutting a pair of *cos* sites decreases as the amount of DNA between them decreases (6). The existence of plasmids having tandem ColE1 origins (unpublished observation) would support the argument that there is no *a priori* reason to expect plasmids containing two A fragments to be unstable. The absence of the double A combination is therefore not easily explained.

The absence of the opposed orientation of the AB fragments, namely, $\vec{A}\vec{B}$, may be due to the dependence of expression of the Rif^r *rpoB* gene on readthrough transcription from the A fragment in the correct orientation (9).

The absence of all palindromic structures (perfect inverted repeats extending to the axis of the symmetry) is also remarkable, as is the high number of deletions found (four from 52 isolates). Whether or not the elimination of palindromic

Table 1.   Transformation and packaging efficiencies of plasmids containing *cos* sites, before and after cleavage with *Hin*dIII and ligation with DNA ligase

Exp.	DNA	MDal of vector	DNA, μg/ml	Transformation		Packaging	
				Rif^r colonies/ μg DNA	% hybrids	Rif^r colonies/ μg DNA	% hybrids
1*a*	pJC703 × *Hin*dIII, ligated	17	500	30	NT	1.0×10^5	~50
1*b*	pJC703 × *Hin*dIII, ligated	17	50	1.4×10^3	NT	1.0×10^2	
2	Supercoils of pJC703,						
	configuration AB	17		1.4×10^5		1×10^2	
	ABB	23		1.7×10^4		4×10^3	
	ABBB	29		1.8×10^4		1×10^4	
3	pJC720 × *Hin*dIII	16	146				
	+		+	4.1×10^2	0.15	2.4×10^3	~90
	R*Idrd*19 × *Hin*dIII		17				
4	pJC720 × *Hin*dIII	16	330				
	+		+	NT		5×10^3	~80
	Pseudomonas AM1 × *Hin*dIII		75				

Transformants or (subsequent to packaging) transductants were selected for on media containing rifampicin. Yields are given as Rif^r colonies per μg of input (vector and foreign) DNA. In Exp. 1, the percentage hybrid clones refers to the percent containing more than one copy of the *Hin*dIII B fragment. About 90% of the Rif^r colonies from Exp. 3 also contained the 11.5-MDal *Hin*dIII fragment from R*Idrd*19, which carries ampicillin resistance. The efficiency of packaging λb2 DNA in parallel experiments was about 10^7–10^8 plaque-forming units per μg of input DNA. NT, not tested.

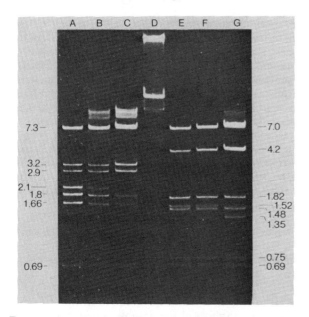

FIG. 3.   Agarose gel (1%) electrophoresis of products of restriction endonuclease digestions. (Gels A, B, and C) Digests of *Sal* I endonuclease; (gels E, F, and G) digests of *Eco*RI; (gel D) uncut supercoiled pJC703. Gels A and E contain digests of a 29-MDal isolate derived from Exp. 2 in Table 1; B and F contain digests of a 23-MDal plasmid from the same experiment; C and G contain digests of pJC703. MDal are indicated.

structures, which are certainly present in the ligation mixture, takes place during the packaging step or after the transduction is not known.

Size Selectivity of Packaging. Of the 52 Rif[r] clones from the experiment of packaging *Hin*dIII cleaved and re-ligated pJC703 DNA, 14 were of the 23-MDal ABB class, 7 were of the 29-MDal ABBB class, 4 were of intermediate sizes showing also aberrant fragments indicating deletions, and 27 were indistinguishable in size from pJC703. Of these latter, five were tested by restriction enzyme analysis and appeared identical to pJC703.

An even stronger size selection was obtained when Exp. 1 (Table 1) was repeated with a *pel*[-] host as recipient. The *pel*[-] mutation increases the DNA size dependency of DNA injection by lambdoid bacteriophages (15). From 29 Rif[r] *pel*[-] clones tested, 19 had plasmids in the 24- to 25-MDal range and 8 in the 29- to 30-MDal range, with a single plasmid of 17 MDal. With the exception of three clones that had small deletions at the junction of the tandemly repeated B fragment and the A

Table 2.   Detection of different possible molecular forms in plasmids containing polymeric regions

Molecular form	Junctional fragments Expected (MDal)		Found
	Sal I	*Eco*RI	
$\overrightarrow{A}\overrightarrow{A}$	3.95	6.8	−
$\overrightarrow{A}\overleftarrow{A}$	5.5	2.4	−
$\overleftarrow{A}\overrightarrow{A}$	2.4	11.2	−
$\overrightarrow{A}\overrightarrow{B}$	3.2	1.35	+
$\overrightarrow{A}\overleftarrow{B}$	4.42	2.6	−
$\overleftarrow{A}\overrightarrow{B}$	1.65	5.7	−
$\overrightarrow{B}\overrightarrow{B}$	2.12	1.52	+
$\overrightarrow{B}\overleftarrow{B}$	3.34	2.74	−
$\overleftarrow{B}\overrightarrow{B}$	0.9	0.3	−

Plasmids containing the molecular form indicated (after the convention adopted for Fig. 2) would produce "junctional" fragments of the indicated MDal.

fragment, all of the larger plasmids were found to be of the form $\overrightarrow{A}\overrightarrow{B}\overrightarrow{B}$ or $\overrightarrow{A}\overrightarrow{B}\overrightarrow{B}\overrightarrow{B}$.

Considering A as a cosmid vector molecule and B as foreign DNA, Exp. 1 (Table 1) can be taken as a model cloning situation. The optimum yield of "hybrid" clones in this experiment would therefore be 3×10^5 per μg of the (foreign) B fragment.

Packaging of Supercoiled DNA. Packaging of supercoiled DNAs of the plasmids (Exp. 2, Table 1) is several orders of magnitude lower than packaging of λ DNA (10^7–10^8 plaque-forming units/μg of λ DNA in parallel experiments), the efficiency for the smallest plasmid being the lowest. Moreover, their structure appears to be unaltered by the packaging-transduction step, as shown by restriction analysis of the supercoiled plasmids isolated from the transductants. Earlier studies on *in vivo* and *in vitro* packaging (summarized in ref. 5) led to the conclusion that supercoiled DNA with a single *cos* site is not packageable without a recombination step, although a low level of *in vivo* packaging of monomeric circular DNA has recently been reported (16). We do not known whether or not the low level of *in vitro* packaging of circular DNA is dependent on a low level of dimer or higher multimers in the supercoiled preparation.

Cloning of RI*drd*19 DNA

The high percentage of larger hybrids in Exp. 1 (Table 1) is evidence that a size selection is occurring in the packaging of cosmid–hybrid DNA. This size dependency was further tested by using the cosmid pJC720 (16 MDal), which contains a single *Hin*dIII site (Fig. 1), to clone fragments from the R factor RI*drd*19 (Exp. 3, Table 1). The *Hin*dIII fragments generated from this plasmid are 42.8, 11.5, 2.9, 2.0, 1.95, 1.8, 0.15, and 0.1 MDal (17). The 11.5-MDal fragment carries the gene for ampicillin resistance.

Of the Rif[r] clones obtained after packaging, 90% were also found to be ampicillin resistant and therefore to be carrying at least the 11.5-MDal *Hin*dIII fragment from RI*drd*19. It would seem, therefore, that a very strong size selection had been imposed by packaging, in which the 27.5-MDal (to 30-MDal?) hybrids were produced in preference to the 16- to 19-MDal plasmids. In Exp. 3, the yield is about 2×10^4 *hybrid* clones per μg of *foreign* DNA, even though 80% of the fragments in this mixture were probably either too small (0.1–2.9 MDal) or too large (42.8 MDal) to be efficiently transduced by this method. The efficiency of packaging hybrid cosmids is therefore on the order of 10^5 per μg of foreign DNA of the correct length.

Transformation with the same DNA yielded few Rif[r] Amp[r] hybrids, the overall transformation efficiency being low due to the high DNA concentration used during ligation.

Cloning of *Pseudomonas* DNA

pJC720 was used to clone fragments from *Pseudomonas* AM1 chromosomal DNA partially digested with *Hin*dIII (Exp. 4, Table 1). By gel electrophoresis it was estimated that more than 80% of the *Pseudomonas* fragments were larger than 16 MDal and probably too large to be clonable by packaging with this vector. In spite of this, the efficiency of *hybrid* formation is about 3×10^4 per μg of foreign DNA. Twenty-seven of the first 32 clones tested carried new DNA fragments. The average size of the DNA insert in these 27 was 10 MDal. On this basis, a few hundred of the clones obtained should constitute a gene bank (18) of *Pseudomonas* AM1 chromosomal DNA in *E. coli*.

DISCUSSION

We have demonstrated that the packaging of plasmid DNA in λ bacteriophage particles can be used as a method for obtaining plasmid hybrids in the 20- to 30-MDal size range when using plasmid DNA that has been linked *in vitro* to foreign DNA fragments. The yield of clones containing these hybrids is of the order of 3×10^5, under optimal conditions, per μg of foreign DNA. Furthermore, by the use of small plasmids (less than 8 MDal) that are themselves very inefficiently packaged (unpublished results), the background of nonhybrid clones is effectively eliminated in a single step without resort to either modification of the DNA (e.g., alkaline phosphatase treatment or polynucleotide tailing) or to elaborate selection or screening procedures which are usually the most time-consuming steps in plasmid cloning experiments. In addition, the use of small cosmids will allow efficient recovery of cloned fragments in the size of up to 25–30 MDal, the selection being imposed by the requirement for packaging of a full or nearly full head.

In vitro packaging of λ cloning vectors can be made independent (12) or dependent (19) of the size of the DNA in the range of 24–30 MDal, but a lower size limit for the vector is set by the requirement for the bacteriophage genes for plaque formation. This requirement is circumvented in the cosmid cloning system, which is independent of phage genes responsible for lytic growth. The space thus provided can be taken up by DNA to be cloned.

Because of the small region required for plasmid replication it is to be expected that new derivatives for use with other restriction enzymes will be rapidly developed. Cosmid derivatives of pJC720 and pJC703 have been produced in which cloning with *Bgl* II or *Bam*HI can be carried out by using rifampicin selection, with *Sal* I, *Eco*RI, *Bgl* II, or *Bam*HI by using ampicillin selection, or with *Xma* I, *Kpn* I, and *Pst* I by using selection for colicin immunity (unpublished results). In addition, a series of cosmid vectors have been developed (unpublished results), including an 8-MDal cosmid (pJC75-58) for use with *Eco*RI, *Bam*HI, and *Bgl* II which is temperature sensitive, ampicillin resistant, and mobilization-minus. It is hoped that in conjunction with incapacitated host strains (20) this latter cosmid will provide an EKII host-vector system that will be most effective in the production of gene banks of eukaryotic DNA.

The packaging of cosmids in λ particles should allow the use of many standard genetic tricks, previously only applicable to λ, for the selection of deletions or insertions in cloned fragments. Such selection methods are based on the instability of full bacteriophage heads in chelating agents, the positive selection for large molecules on infection of *pel*⁻ hosts, or the physical separations possible on the basis of density differences between full and partially filled λ particles.

We thank Hildburg Stephan for her expert technical assistance, Ulla Hartmann for preparation of the partially *Hin*dIII-cleaved *Pseudomonas* AM1 DNA, and Dietmar Blohm and Renate Bonewald for preparation of the RI*drd*19 plasmid DNA. Financial support was provided in part by a grant to T. Hohn from the Swiss National Fonds.

1. Hohn, B. & Hohn, T. (1974) *Proc. Natl. Acad. Sci. USA* **71**, 2372–2376.
2. Kaiser, D., Syvanen, M. & Masuda, T. (1975) *J. Mol. Biol.* **91**, 175–186.
3. Becker, A., Murialdo, H. & Gold, M. (1977) *Virology* **78**, 227–290.
4. Becker, A., Marko, M. & Gold, M. (1977) *Virology* **78**, 291–305.
5. Hohn, T. & Katsura, I. (1977) *Curr. Top. Microbiol. Immunol.* **78**, 69–110.
6. Feiss, M., Fisher, R. A., Crayton, M. A. & Egner, C. (1977) *Virology* **77**, 281–293.
7. Syvanen, M. (1974) *Proc. Natl. Acad. Sci. USA* **71**, 2496–2499.
8. Hohn, B. (1975) *J. Mol. Biol.* **98**, 93–106.
9. Collins, J., Fiil, N. P., Jørgensen, P. & Friesen, J. D. (1976) in *Control of Ribosome Synthesis*, Alfred Benson Symposium 9, eds. Maaløe, O. & Kjeldgaard, V. O. (Munksgaard, Copenhagen), pp. 356–369.
10. Collins, J., Johnsen, M., Jørgensen, P., Valentin-Hansen, P., Karlström, H. O., Gautier, F., Lindenmaier, W., Mayer, H. & Sjöberg, B. M. (1978) in *Microbiology 1978*, eds. David, J. & Novik, R. (American Society of Microbiology, Washington, DC), pp. 150–153.
11. Emmons, S. W., Maccosham, V. & Baldwin, R. L. (1975) *J. Mol. Biol.* **91**, 133–146.
12. Hohn, B. & Murray, K. (1977) *Proc. Natl. Acad. Sci. USA* **74**, 3259–3263.
13. Morrison, D. A. (1977) *J. Bacteriol.* **132**, 349–351.
14. Collins, J. (1977) *Curr. Top. Microbiol. Immunol.* **78**, 121–170.
15. Emmons, S. W. (1974) *J. Mol. Biol.* **93**, 511–525.
16. Umene, K., Shimada, K. & Takagi, Y. (1978) *Mol. Gen. Genet.* **159**, 39–45.
17. Blohm, D. (1978) *Proceedings of the 2nd International Symposium on Microbeal Drug Resistance*, ed. Mitsuhashi, S. (University Press, Tokyo), Vol. 2, in press.
18. Clarke, L. & Carbon, J. (1976) *Cell* **9**, 91–99.
19. Sternberg, N., Tiemeier, D. & Enquist, L. (1977) *Gene* **1**, 255–280.
20. Curtiss, R., III, Inoue, M., Pereira, D., Hsu, J., Alexander, L. & Rock, L. (1977) in *Molecular Cloning of Recombinant DNA*, eds. Scott, W. & Lerner, R. (Academic, New York), pp. 99–114.

Calcium-dependent Bacteriophage DNA Infection

M. Mandel and A. Higa

Escherichia coli cells of strain K12 and C can be made competent to take up temperate phage DNA without the use of "helper phage". This competence is dependent on the presence of calcium ions and is effective for both linear and circular DNA molecules.

It has been known that DNA extracted by phenol treatment from temperate coliphages such as λ, 434, 186 or P2 can infect sensitive *Escherichia coli* cells in the presence of "helper phage" (Kaiser & Hogness, 1960; Mandel, 1967). However, the exact role of the helper phage is still unknown. It seems that injection of the DNA of the helper phage and the presence of the intact helper phage DNA in a cell (Takano & Watanabe, 1967) are required for the cell to become competent in incorporating free DNA.

To be infective the DNA molecule must possess at least one free cohesive end (Strack & Kaiser, 1965; Kaiser & Inman, 1965). Moreover, there is a correlation between the specificity of the cohesive ends of the helper-phage DNA and the infectious DNA and the capacity of the phage to serve as a helper for DNA infectivity (Mandel & Berg, 1968; Kaiser & Wu, 1968). The DNA infection seems to depend on the homology between cohesive ends of the infecting DNA and of the DNA of the helper phage.

Since previous work by one of the authors (Mandel, 1967) had shown that changes in cell wall permeability occurred in *E. coli* (strain C600) when made competent by infection with helper phage, we became interested in the effects of both monovalent and divalent ions on *E. coli* cell wall permeability and its correlation with DNA uptake.

During the course of this investigation we found that the DNA of temperate phages P2 and λ could infect a sensitive host in the absence of helper phage and that DNA uptake depended on the presence of calcium ions.

Bacterial strains used were *E. coli* K12 strain C600 (designated K38) as a host for λi^{434} DNA and *E. coli* C1a as a host for P2 DNA. Phage λi^{434} was obtained by ultraviolet induction of a λi^{434} lysogen and P2 by infection of sensitive cells. Phages were purified by differential centrifugation and phage DNA extracted with buffer-saturated phenol (0·01 M-Tris, pH 8·0). DNA of a streptomycin-resistant mutant of K38 was extracted by the methods suggested by Smith (1968) and Avadhani, Mehta & Rege (1969). Competent cells were prepared by inoculating supplemented P medium (Radding & Kaiser, 1963) with a 1 to 500 dilution of an overnight culture of K38 or C1a and grown with aeration at 37°C until an optical density of 0·6 was reached (1×10^9 cells/ml.). The cells were then quickly chilled, centrifuged and resuspended in 0·5 volume $CaCl_2$, kept cold for 20 minutes, then centrifuged and resuspended in 0·1 volume of cold $CaCl_2$. Chilled DNA samples, 0·1 ml. in volume, in standard saline citrate (0·15 M-NaCl, 0·015 M-sodium citrate, pH 7·0) were added to 0·2 ml. of competent cells, further chilled for 15 minutes and incubated for 20 minutes at 37°C. At the end of the incubation period, the reaction mixture was either chilled or treated with DNase for five minutes at 37°C. Dilutions of the mixture were made and plated

MANDEL, M. and HIGA, A.
Calcium-dependent bacteriophage DNA infection. Reprinted with permission from
J. Mol. Biol. 53:159-162 (1970).

on appropriate indicators. Under those assay conditions we obtained approximately 10^5 to 10^6 plaques per μg of DNA.

Our work shows that *E. coli* K12 and *E. coli* C grown in P medium can take up phage DNA quite readily in the presence of calcium ions. In Figure 1 we see the

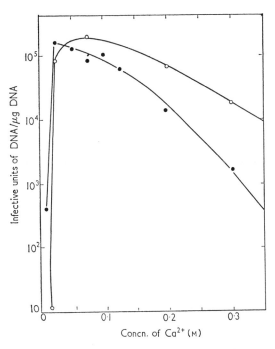

FIG. 1. DNA infectivity as a function of Ca^{2+} concentration. Experimental procedure as described in text. —○—○—, P2 DNA; —●—●—, λi^{434} DNA.

extremely rapid rise in competence of both K38 and C1a in going from 0·01 to 0·025 M-Ca^{2+} and the much slower decline in competence at molarities above 0·1. At 0·5 M-Ca^{2+} concentration the competence of K38 is reduced to practically zero while that of C1a is reduced by a factor of about 100 relative to its peak competence. This may be related to the survival of K38 and C1a cells which when incubated for 20 minutes in 0·5 M-Ca^{2+} give 0·1 and 25% survival, respectively. As a further test of changes in cell wall permeability we incubated competent K38 cells (0·1 M-Ca^{2+}) in the presence of 10 μg actinomycin D/ml. and found 25% survival compared to 100% survival for K38 in 0·01 M-Tris–0·01 M-Mg^{2+} in the presence of actinomycin.

The time-course of the interaction between λi^{434} DNA and K38 at 37°C is shown in Figure 2. The number of plaques increases very rapidly with time, reaching a maximum at about 20 minutes under the conditions of this assay. Since the competent cells and DNA are pre-chilled then mixed at cold room temperatures, and kept at 0°C for 15 minutes before starting the experiment, the plaques obtained at zero time represents DNA taken up by the cells at 0°C and protected from DNase. When the experiment is done with cells and DNA mixed at room temperature and assayed immediately, we find no DNA infectivity at zero time. In both cases the major portion of the interaction is completed in two minutes, which is quite rapid compared to the kinetics of the

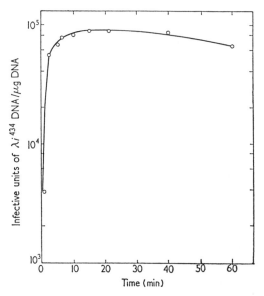

FIG. 2. Time-course of the reaction between λi^{434} DNA at a concentration of $0\cdot1$ μg/ml. and K38 cells at a concentration of 2×10^{10}/ml. in the presence of $0\cdot05$ M-Ca^{2+}. Procedure was the same as that described in the text except for proportionately larger volumes. At the times indicated on the abscissa, samples of $0\cdot3$ ml. each were removed from the incubation mixture and added to $0\cdot1$ ml. of 20 μg pancreatic DNase/ml., incubated 5 min at 37°C and plated for assay.

helper-phage DNA infection under similar conditions. The decrease of infective units after 20 minutes reflects the survival rate of K38 incubated in $0\cdot05$ M-Ca^{2+}.

The number of infectious centers obtained is linearly proportional to the concentration of phage DNA and concentration of cells. The saturation level of DNA is dependent on the cell concentration and under conditions of our assay (cell concentration~2×10^{10}ml.), 10 μg DNA/ml. was not saturating.

In contrast to helper-phage DNA infectivity, cells, competent in the presence of calcium ions, take up both linear and hydrogen-bonded circular DNA (Hershey & Burgi, 1965) with equal efficiency, as shown in Table 1.

Our attempts to transform K38 streptomycin-sensitive cells to streptomycin resistance with DNA extracted from K38 streptomycin-resistant cells met with failure.

TABLE 1

Effect of DNA form on infectivity

| | P2 DNA infectivity per μg DNA | |
	Linear DNA† (heated)	Circular DNA† (not heated)
Ca^{2+} assay	$1\cdot1 \times 10^6$	$1\cdot0 \times 10^6$
Helper-phage assay	$6\cdot0 \times 10^5$	$1\cdot5 \times 10^5$

† Circular DNA prepared as described in Hershey & Burgi (1965). The percentage of circles in DNA preparations depends on salt concentration and duration of storage. Heating at 75°C for 5 min and quick cooling to 0°C converts the closed, circular form to the open linear form.

However, since the two DNA extracts which we carefully prepared showed no hyper-chromic shift on melting, the problem may well be in the technique of extracting undegraded DNA from *E. coli*.

We would like to thank Professor R. Calendar for first pointing out to us that a calcium-rich salt solution improved the helper-phage P2 DNA infectivity assay, and Leslie Jensen for her technical assistance.

This work was supported by research grant no. AI–07919 from the National Institutes of Health to one of us (M.M.).

Department of Biochemistry and Biophysics   M. MANDEL
University of Hawaii   A. HIGA†
Honolulu, Hawaii, U.S.A.

Received 26 May 1970

REFERENCES

Avadhani, N.-G., Mehta, B. M. & Rege, D. V. (1969). *J. Mol. Biol.* **42**, 413.
Hershey, A. D. & Burgi, E. (1965). *Proc. Nat. Acad. Sci., Wash.* **53**, 325.
Kaiser, A. D. & Hogness, D. S. (1960). *J. Mol. Biol.* **2**, 392.
Kaiser, A. D. & Inman, R. B. (1965). *J. Mol. Biol.* **13**, 78.
Kaiser, A. D. & Wu, R. (1968). *Cold Spr. Harb. Symp. Quant. Biol.* **33**, 729.
Mandel, M. (1967). *Molec. Gen. Genetics,* **99**, 88.
Mandel, M. & Berg, A. (1968). *Proc. Nat. Acad. Sci., Wash.* **60**, 265.
Radding, C. M. & Kaiser, A. D. (1963). *J. Mol. Biol.* **7**, 225.
Smith, H. O. (1968). *Virology,* **34**, 203.
Strack, H. B. & Kaiser, A. D. (1965). *J. Mol. Biol.* **12**, 36.
Takano, T. & Watanabe, T. (1967). *Virology,* **31**, 722.

† Present address: Hawaiian Sugar Planters' Association, Honolulu, Hawaii, U.S.A.

A New Pair of M13 Vectors for Selecting Either DNA Strand of Double-digest Restriction Fragments

J. Messing and J. Vieira

SUMMARY

The strategy of shotgun cloning with M13 is based on obtaining random fragments used for the rapid accumulation of sequence data. A strategy, however, is sometimes needed for obtaining subcloned sequences preferentially out of a mixture of fragments. Shotgun sequencing experiments have shown that not all DNA fragments are obtained with the same frequency and that the redundant information increases during the last third of a sequencing project. In addition, experiments have shown that particular fragments are obtained more frequently in one orientation, allowing the use of only one of the two DNA strands as a template for M13 shotgun sequencing. Two new M13 vectors, M13mp8 and M13mp9, have been constructed that permit the cloning of the same restriction fragment in both possible orientations. Consequently, each of the two strands becomes a (+) strand in a pair of vectors. The fragments to be cloned are cleaved with two restriction enzymes to produce a fragment with two different ends. The insertion of such a fragment into the vector can occur only in one orientation. Since M13mp8 and M13mp9 have their array of cloning sites in an antiparallel order, either orientation for inserting a double-digest fragment can be selected by the choice of the vector.

INTRODUCTION

A number of techniques have been used for the preparation of single-stranded DNA. These include poly(UG)-CsCl gradients (see review by Szybalski et al., 1971), alkaline CsCl gradients (Vinograd et al., 1963), polyacrylamide gels (Maxam and Gilbert, 1977; Szalay et al., 1977), and exonuclease treatment (Smith, 1979).

An alternative biological approach has been developed involving the bacteriophage M13. The RF of the phage DNA is a circular double-stranded molecule; it can be isolated from infected cells, used to clone DNA fragments, and reintroduced into *Escherichia coli* cells by transfection. The infected cells extrude the M13 phage particles, each of which contains a circular SS DNA molecule. Large amounts of SS DNA containing a cloned

Abbreviations: kb, kilobase pairs; PEG, polyethylene glycol; RF, double-stranded replicative form; SS, single-stranded viral form.

MESSING, J. and VIEIRA, J.
A new pair of M13 vectors for selecting either DNA strand of double-digest restriction fragments. Reprinted with permission from *Gene* v. 19: pp. 269-276. Copyright 1982, Elsevier Science Publishers B.V.

insert (5–10 µg phage SS DNA/ml bacterial culture) can be easily and rapidly obtained (Messing et al., 1977; Barnes, 1978; Ray and Kook, 1978; Ohsumi et al., 1978; Hermann et al., 1978; Nomura et al., 1978).

The cloning of DNA fragments into the RF of M13 has been facilitated by a series of improvements which produced the M13mp7 cloning vehicle (Gronenborn and Messing, 1978; Messing, 1979; Messing et al., 1981). A fragment of the *E. coli lac* operon (the promoter and N-terminus of the β-galactosidase gene) was inserted into the M13 genome. In addition, a small DNA fragment synthesized in vitro and containing an array of restriction cleavage sites was inserted into the structural region of the β-galactosidase gene fragment. In spite of these insertions the M13mp7 DNA is still infective and the modified *lac* DNA is able to encode the synthesis of a functional β-galactosidase α-peptide (Langley et al., 1975; Messing et al., 1981).

The synthesized DNA fragment contains two sites each for the *Eco*RI, *Bam*HI, *Sal*I, *Acc*I, and *Hin*cII restriction enzymes arranged symmetrically with respect to a centrally located *Pst*I site. Thus, by chance either strand of a cloned restriction fragment can become part of the viral (+) strand. This depends on the fragment's orientation relative to the M13 genome after they have been joined by ligase. The insertion of a DNA fragment into one of these restriction sites is readily monitored because the insertion results in a nonfunctional α-peptide and the loss of β-galactosidase activity. Under appropriate plating conditions, the functional α-peptide results in blue plaques; a nonfunctional α-peptide results in colorless plaques (Messing and Gronenborn, 1978). M13mp7 has found wide application in the dideoxy nucleotide sequencing procedure (Sanger et al., 1977).

This paper reports the construction of two new SS DNA bacteriophage vectors, M13mp8 and M13mp9, and their applications to DNA sequencing and strand-specific hybridization. The nucleotide sequence of M13mp7, containing the multiple restriction sites, has been modified to have only one copy of each restriction site and in addition, single *Hin*dIII, *Sma*I, and *Xma*I sites. Thus, DNA fragments whose ends correspond to two of these restriction sites can be "forced cloned" by ligation to one of these new M13 cloning vehicles that has also been "cut" with the same pair of restriction enzymes. M13mp8 and M13mp9 have this modified multiple restriction site region arranged in opposite orientations relative to the M13 genome. Thus, a given restriction fragment can be directly orientated by forced cloning. This procedure guarantees that each strand of the cloned fragment will become the (+) strand in one or the other of the clones and thus be extruded as SS DNA in phage particles.

MATERIALS AND METHODS

(a) Strains

Two new plasmids, pUC8 and pUC9, have been constructed from the pBR322 plasmid and the M13mp7 vector (Bolivar et al., 1977; Sutcliffe, 1979; Messing et al., 1977, 1981; Ruther, 1980; Vieira and Messing, 1982). The 2297-bp *Eco*RI/*Pvu*II fragment of pBR322 has had its *Pst*I and *Hin*cII sites removed by single base pair changes without loss of β-lactamase activity. The *Acc*I site in this fragment has been removed by cleavage and treatment with BAL31 (Legerski et al., 1978). The 433-bp *lac Hae*II fragment of M13mp7 RF containing the multiple restriction sites was then inserted into this modified pBR322 fragment to yield the plasmid pUC7. The nucleotide sequence containing the multiple restriction sites was then modified to contain the following restriction sites ordered in the same polarity as *lac* transcription: *Eco*RI, *Xma*I, *Sma*I, *Bam*HI, *Sal*I, *Acc*I, *Hin*cII, *Pst*I, and *Hin*dIII. This plasmid is termed pUC8. A second modified form of the multiple restriction site was used to produce the plasmid pUC9. This plasmid contains the multiple restriction sites of pUC8 in a reverse orientation relative to the *lac* promoter. A detailed description of these plasmids, their construction, and their uses are presented in Vieira and Messing (1982).

(b) Construction of M13mp8 and M13mp9

M13mp7 RF DNA was partially cleaved with the *Hae*II enzyme (all enzymes were obtained from the Bethesda Research Laboratory and New England Biolabs) and the linear forms were purified by gel electrophoresis and eluted from the gel (Heidecker et al., 1980). This DNA was then cleaved with the *Eco*RI enzyme to produce phage DNA molecules with *Hae*II and *Eco*RI ends. The modified *lac Hae*II fragment was excised from pUC8 DNA with *Hae*II and also digested with *Eco*RI and *Pvu*I. Cleavage at the *Eco*RI site of the *lac Hae*II fragment produces two fragments with *Eco*RI and *Hae*II ends. *Pvu*I cleaves one of these fragments and blocks its later incorporation into a vector with *Eco*RI and *Hae*II ends.

The linearized M13mp7 DNA was mixed with the triple-enzyme-digested DNA from pUC8 and joined with ligase. This leads to the "forced" insertion of the *Eco*RI-*Hae*II subfragment of the modified *lac Hae*II fragment. The DNA products of the ligation reaction were used to transform competent *E. coli* JM103 cells (Messing et al., 1981). Candidates for properly constructed vectors were identified as blue plaques. These isolates were saved for further characterization and designated according to their plasmid origin as M13mp8.

A similar procedure was used to construct M13mp9. In this case the M13mp7 DNA was prepared as before. The modified *lac Hae*II fragment was excised from pUC9 DNA with *Hae*II and digested with *Eco*RI and *Hin*fI. *Hin*fI cleavage blocks the later incorporation of one of the *Eco*RI-*Hae*II subfragments. The restriction-enzyme-digested phage and the pUC9 DNAs were joined by ligase and used to transform competent JM103 cells. Blue plaques were picked for further characterization.

(c) Preparation of templates

Templates for DNA sequencing reactions with chain terminating inhibitors (Sanger et al., 1977) were prepared as follows. Infected cells from a blue plaque were diluted into 1.5 ml of *2YT* (Miller, 1972), mixed with about 10^7 non-infected log phase JM103 cells (Messing et al., 1981), and grown for 8 h at 37°C. The infected cells were sedimented by a short centrifugation and the phage in the supernatant were concentrated by the addition to 1.3 ml of the supernatant of 200 μl of 27% PEG 6000 (Sigma) in 3.3 M NaCl. The mixture became turbid

after about 30 min at room temperature and was cleared by centrifugation in a Beckman Microfuge B. The supernatant was removed with a Pasteur pipet avoiding any damage to the soft pellet; residual fluid was removed with a Kimwipe. The pellet was resuspended in 800 μl of low Tris buffer (10 mM NaCl, 1 mM EDTA, 10 mM Tris·HCl pH 7.5). Phage were precipitated again by adding 100 μl of a 20% PEG 6000 solution and 100 μl of 5 M NaCl. After another 30 min at room temperature the phage was collected by a short centrifugation in a Beckman Microfuge B. The pellet was then resuspended again in low Tris buffer in a volume of 300 μl and extracted twice with an equal volume of a 1:1 mixture of phenol–chloroform. The DNA was precipitated with ethanol from the aqueous phase, washed once with ethanol, taken up in 10 μl of low Tris buffer, and used directly to provide template for the DNA sequencing reaction with chain terminators (Sanger et al., 1977).

(d) Primer and DNA sequencing

An aliquot from the template solution (1 μl) was annealed with the synthetic master primer and subjected to the chain termination reaction (Sanger et al., 1977) as described previously (Messing et al., 1981; Gardner et al., 1981).

RESULTS AND DISCUSSION

(a) Scheme of forced cloning

Restriction endonuclease cleavage fragments with non-complementing ends are rarely, if at all, joined during a ligation reaction. If DNA is cleaved with two different restriction endonucleases and a fragment with two non-complementing ends is produced, the resulting fragment can be neither circularized nor joined to another fragment in both orientations during a ligation reaction. Consequently, two "double-digest" fragments can be forced to form a chimeric molecule under the appropriate ligation conditions. In addition, a recombinant molecule is formed during the ligation reaction with the defined order of the two fragments. Since the orientation of a cloned DNA

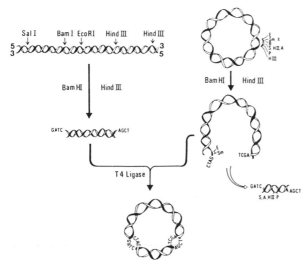

Fig. 1. Scheme of forced cloning. Restriction cleavage sites used as cloning sites are unique and located in a region of the circular vector molecule convenient for the insertion of foreign DNA. The DNA sequence which has been selected for cloning is cut out of its original DNA with the use of two restriction endonucleases. In the described example the enzymes used are *Bam*HI and *Hin*dIII. The DNA fragment produced has two different cohesive ends which cannot join in a DNA ligation experiment. The vector molecule is treated with the same enzymes and the same type of cohesive ends are produced. Both DNA fragments, the vector and the DNA to be cloned cannot form a circular molecule themselves and therefore favor the formation of a circular chimeric molecule, which can be selected for by the transformation of competent cells. The location of the two cloning sites determines how the cloned DNA is oriented within the vector molecule.

fragment in the RF of M13 vectors determines which of the two DNA strands is going to be the viral strand, this procedure allows the direct preparation of one of the two DNA strands by cloning. Fig. 1 illustrates this scheme when the DNA to be cloned is cut out with *Hin*dIII and *Bam*HI. In addition to the strand separation, a "double-digest" restriction fragment with two non-complementing ends is easily selected from DNA fragments in the pool that have been generated by cleavage with only one of the two enzymes.

(b) Modification of the *lacZ* gene

To avoid the random selection of either strand of an insert fragment arising from the use of

cloning into M13mp7, the symmetry around the *Pst*I site of M13mp7 has been replaced by an asymmetric array of sites. The two resulting orders of sites have been produced on two different phage molecules so that both strands of a DNA fragment can serve as the viral strand. In addition, *Hin*dIII, *Xma*I and *Sma*I have been added as new cloning sites. The latter serves as a second blunt end cloning site in the case that the *Hin*cII site cannot be used. A detailed cloning guide which illustrates the combinatorial variety of cohesive and blunt ends produced by a large number of different restriction endonucleases has been described elsewhere (Messing and Seeburg, 1981). The engineering of this segment of *lac* DNA has been conducted with pUC7 which contains the *lac* region of M13mp7 on a plasmid vector as outlined in MATERIALS AND METHODS. The modified cloning sites from the two plasmid vectors pUC8 and pUC9, which contain the cleavage sites in the two possible orientations, were transferred to M13mp7 as described in MATERIALS AND METHODS.

(c) DNA sequence determination for both DNA strands by the same primer

A blue plaque from both transfer experiments has been retained and single-stranded DNA prepared as described in MATERIALS AND METHODS. Using the synthetic master primer (Messing et al., 1981), sequencing reactions with chain terminators (Sanger et al., 1977) have been performed and analyzed by polyacrylamide gel electrophoresis. The pertinent sequence containing the cloning sites can be read from the autoradiogram in both orientations (Fig. 2). The primary structure of the Z gene of M13mp7 is compared with the structures from the two new isolates which are named M13mp8 and M13mp9 (Fig. 3). Both strands of the insert have been sequenced with the same primer because of the reverse polarity of the synthetic *Eco*RI/*Hin*dIII insert in M13mp8 and M13mp9. Therefore, every "double-digest" restriction fragment cloned in both vectors will provide both strands of the restriction fragment directly as template which can be sequenced with the same primer.

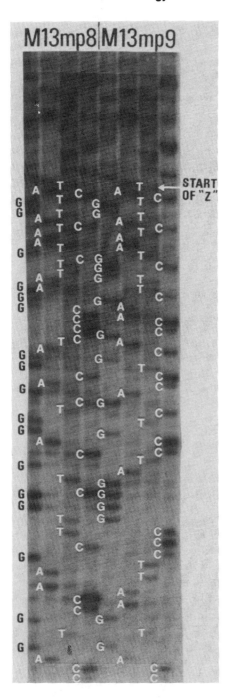

Fig. 2. DNA sequence analysis of M13mp8 and M13mp9. M13mp8 and M13mp9 have been constructed as described in MATERIALS AND METHODS. Starting from a single plaque template from both phage vectors were prepared and a primer extension reaction in presence of chain terminators was carried out as described in MATERIALS AND METHODS. Reaction products were loaded on a 8% polyacrylamide gel in the GATC order and the sequence was read from the HaeIII site (bottom) in the position of the 8th amino acid residue of the Z gene to the ATG start codon. The resulting sequences are compared with M13mp7 in Fig. 3.

(d) Phenotypic comparison of the different lacZ sequences

The addition of the synthetic DNA to the lac region of M13mp2 (Gronenborn and Messing, 1978) leads to the synthesis of a modified α-peptide with the insertion of 11 amino acid residues in M13mp8 and M13mp9 compared to 14 in M13mp7 (Messing et al., 1981) and 19 in M13mp5, a HindIII cloning phage (Messing, 1979). In all cases the polypeptides were functional in α-complementation and gave rise to blue colored plaques under the appropriate plating conditions (Messing et al., 1977). We have noted that the insertion of these small synthetic DNAs affects the complementation test to different degrees, and the various M13mp phage molecules can be differentiated by the intensity of the blue color of their plaques. Indeed, the new phage vectors produce a more intense color than M13mp7. Moreover, the sequencing analysis of this different lacZ sequence causes a different degree of secondary structure affecting compression in the gel electrophoresis (not shown). As expected, M13mp7, with a perfect symmetry, shows the highest degree of compression followed by M13mp5. M13mp8 and M13mp9, however, do not give an indication for compression as shown in Fig. 2.

(e) "Double-digest" cloning in shotgun DNA sequencing

Since all the constructions are based on the use of a master primer (Heidecker et al., 1980) any insert can be conveniently copied by DNA polymerase. Using chain terminators (Sanger et al., 1977) the sequence of the DNA can be deduced from selected templates. The shotgun sequencing method, for instance, leads to a less efficient effort in determining the last 10–20% of the nucleotides of a contiguous sequence (Gardner et al., 1981). A direct way of obtaining missing sequences of both strands is the use of selective cleavage and forced cloning into the RF of M13mp8 and M13mp9. If these resulting clones, and therefore the gaps between two blocks of sequences, are too large to be sequenced with one primer extension reaction these clones in turn can be used as master sequences to identify smaller templates from both strands within

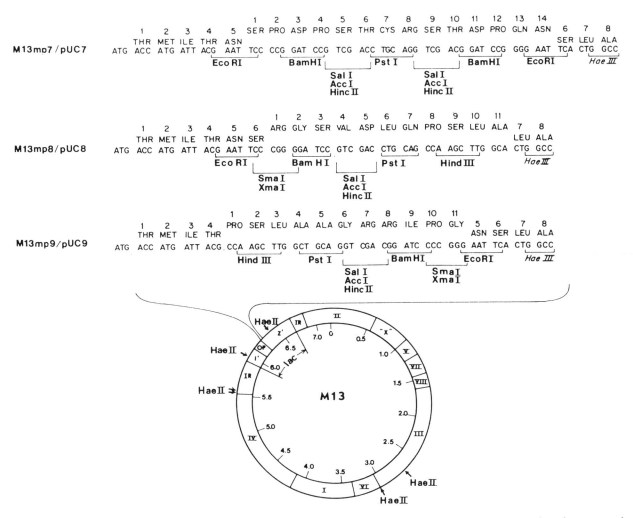

Fig. 3. A genetic map of M13mp7, M13mp8 and M13mp9. The *Hae* II-cleavage sites of the three different vectors have been mapped in respect to the genetic map (Messing et al., 1977). Map units are given in kb. The difference between the three molecules is given in detail by the nucleotide sequence of the pertinent region of the *lacZ* gene. The recognition sites for restriction endonucleases are underlined and labeled. The interruption of the wild-type amino acid sequence in M13mp2 (Gronenborn and Messing, 1978) by the inserted amino acid sequences is set up and numbered. The same *lac* cloning sites exist also on the pBR322 derivative with ampicillin as a selection marker, pUC7, pUC8 and pUC9 as described in MATERIALS AND METHODS. All strains are made available without charge from Bethesda Research Laboratory upon request.

the pool of the different shotgun clones obtained in the first cloning experiment. Using a new procedure of making M13 probes and a dot hybridization procedure, a library of M13 shotgun clones can be rapidly screened for statistically underrepresented templates (Hu and Messing, 1982).

(f) Comparative sequencing, transcriptional mapping, and in vitro mutagenesis

The same template selection can be used to compare mutant sequences. Once the primary structure of a DNA sequence is known its genetic structure has to be determined by comparison of the primary structure with mutant sequences. Since this requires the comparison of a small part of the entire sequence, the appropriate doubly digested fragment of interest can be used to readily obtain mutant sequences in template form in M13mp8 and M13mp9.

Defining the orientation of a cloned fragment in M13 is very useful in at least two other experimental procedures; first, for making single-strand specific hybridization probes (Hu and Messing,

1982), second for in vitro mutagenesis. Single-strand specific hybridization probes are helpful in determining the strand polarity of inserts in other M13 viral DNAs or for physically mapping transcripts. If site-specific changes in a particular nucleotide sequence are planned by using a synthetic primer as a mutagen (Smith and Gillam, 1981), it may be useful to have the extension of the mutagen primer carried out in the absence of the complementary strand. A discussion of an integrative strategy of shotgun DNA sequencing and exploration of gene structure has been described in more detail elsewhere (Messing, 1981; 1982).

ACKNOWLEDGEMENTS

This work was supported by the Minnesota Experiment Station, grant No. MN 15-030, and a grant No. DE-AC02-81 ER 10901 from the Department of Energy. We thank Gisela Heidecker for her stimulating discussions and Jim Fuchs, Perry Hackett, Irwin Rubenstein, Kris Kohn and Bonnie Allen for their aid in preparing this manuscript.

REFERENCES

Barnes, W.M.: DNA sequence from the histidine operon control region: Seven histidine codons in a row. Proc. Natl. Acad. Sci. USA 75 (1978) 4261–4285.

Bolivar, F., Rodriguez, R.L., Greene, P.J., Betlach, M.C., Heynecker, H.L., Boyer, H.W., Crosa, J.H. and Falkow, S.: Construction and characterization of new cloning vehicles, II. A multi-purpose cloning system. Gene 2 (1977) 95–113.

Gardner, R.C., Howarth, A.J., Hahn, P., Brown-Luedi, M., Shepherd, R.J. and Messing, J.: The complete nucleotide sequence of an infectious clone of cauliflower mosaic virus by M13mp7 shotgun sequencing. Nucl. Acids Res. 9 (1981) 2871–2888.

Gronenborn, B. and Messing, J.: Methylation of single-stranded DNA in vitro introduces new restriction endonuclease cleavage sites. Nature 272 (1978) 375–377.

Heidecker, G., Messing, J. and Gronenborn, B.: A versatile primer for DNA sequencing in the M13mp2 cloning system. Gene 10 (1980) 69–73.

Hermann, R., Neugebauer, K., Zentgraf, H. and Schaller, H.: Transposition of DNA sequence determining kanamycin resistance into the single-stranded genome of bacteriophage fd. Mol. Gen. Genet. 153 (1978) 171–178.

Hu, N. and Messing, J.: The making of strand specific M13 probes. Gene 17 (1982) 271–277.

Langley, K.E., Villarejo, M.R., Fowler, A.V., Zamenhof, P.J. and Zabin, I.: Molecular basis of β-galactosidase-α complementation. Proc. Natl. Acad. Sci. USA 72 (1975) 1254–1257.

Legerski, R.H., Hodnett, J.L. and Gray, H.B.: Extracellular nucleases of *Pseudomonas* BAL31, III. Use of the double-stranded deoxyriboexonuclease activity as the basis of a convenient method for mapping of fragments of DNA produced by cleavage with restriction enzymes. Nucl. Acids Res. (1978) 1445–1463.

Maxam, A. and Gilbert, W.: A new method for sequencing DNA. Proc. Natl. Acad. Sci. USA 74 (1977) 560–564.

Messing, J.: A multi-purpose cloning system based on the single-stranded DNA bacteriophage M13. Recombinant DNA Technical Bulletin, NIH Publication No. 79–99, 2, No. 2 (1979) 43–48.

Messing, J.: M13mp2 and derivatives: A molecular cloning system for DNA sequencing, strand specific hybridization, and in vitro mutagenesis, in Walton, A.G. (Ed.), Proceedings of the Third Cleveland Symposium on Macromolecules. Elsevier, Amsterdam, 1981, pp. 143–153.

Messing, J.: An integrative strategy of DNA sequencing and experiments beyond, in Hollaender, A. and Setlow, J.K. (Eds.), Genetic Engineering, Principles and Methods, Vol. 4. Plenum, New York, 1982, pp. 19–36.

Messing, J., Gronenborn, B., Muller-Hill, B. and Hofschneider, P.H.: Filamentous coliphage M13 as a cloning vehicle: Insertion of a *Hin*dII fragment of the *lac* regulatory region in M13 replicative form in vitro. Proc. Natl. Acad. Sci. USA 74 (1977) 3642–3646.

Messing, J. and Gronenborn, B.: The filamentous phage M13 as a carrier DNA for operon fusions in vitro, in Denhardt, D.T., Dressler, D. and Ray, D.S. (Eds.), The Single-stranded DNA Phages. Cold Spring Harbor Laboratory, Cold Spring Harbor, NY, 1978, pp. 449–453.

Messing, J., Crea, R. and Seeburg, P.H.: A system for shotgun DNA sequencing. Nucl. Acids Res. 9 (1981) 309–321.

Messing, J. and Seeburg, P.H.: A strategy for high speed DNA sequencing, in Brown, D. (Ed.), Developmental Biology using Purified Genes, ICN-UCLA Symposia on Molecular and Cellular Biology, Vol. XXIII. Academic Press, New York, 1981, 659–663.

Miller, J.H.: Experiments in Molecular Genetics. Cold Spring Harbor Laboratory, Cold Spring Harbor, NY, 1972.

Nomura, N., Yamagichi, H. and Oka, A.: Isolation and characterization of transducing coliphage fd carrying a kanamycin resistance gene. Gene 3 (1978) 39–51.

Ohsumi, M., Vovis, G.F. and Zinder, N.D.: The isolation and characterization of an in vivo recombinant between the filamentous bacteriophage f1 and the plasmid pSC101. Virology 89 (1978) 438–449.

Ray, D.S. and Kook, Y.: Insertion of the Tn3 transposon into the genome of the single-stranded DNA phage M13. Gene 4 (1978) 103–119.

Ruther, U.: Construction and properties of a new cloning vehicle, allowing direct screening for recombinant plasmids. Mol. Gen. Genet. 1978 (1980) 475–477.

Sanger, F., Nicklen, S. and Coulsen, A.R.: DNA sequencing with chain-terminating inhibitors. Proc. Natl. Acad. Sci. USA 74 (1977) 5463–5467.

Smith, A.J.H.: The use of exonuclease III for preparing single-stranded DNA for use as a template in the chain termination sequencing methods. Nucl. Acids Res. 6 (1979) 831–848.

Sutcliffe, J.G.: Complete nucleotide sequence of the *Escherichia coli* plasmid pBR322. Cold Spring Harbor Symposium 43 (1979) 77–90.

Szalay, A.A., Grohmann, K. and Sinsheimer R.L.: Separation of the complementary strands of DNA fragments on polyacrylamide gels. Nucl. Acids Res. (1977) 1563–1578.

Szybalski, W., Kubinski, H., Hradečná, Z. and Summers, W.C.: Analytical and preparative separation of the complementary DNA strands, in Grossman, L. and Moldave, K. (Eds.), Methods in Enzymology, Vol. 21D. Academic Press, New York, 1971, pp. 383–413.

Vieira, J. and Messing, J.: The pUC plasmids, a M13mp7 derived system for insertion mutagenesis and sequencing with synthetic universal primers. Gene 19 (1982) 259–268.

Vinograd, J., Morris, J., Davidson, N. and Dove, W.F.: The buoyant behavior of viral and bacterial DNA in alkaline CsCl. Proc. Natl. Acad. Sci. USA 49 (1963) 12–17.

Communicated by Z. Hradečná.

Note added in proof

In the meantime, a new M13 vector pair has been constructed. They have two additional cloning sites for *Sst*I and *Xba*I in their polylinker. M13mp10 is derived from M13mp8, and M13mp11 from M13mp9, except that M13mp11 does not contain the extra piece of pBR322 sequence as one of the isolates of M13mp9 does.

High-efficiency Cloning of Full-length cDNA

H. Okayama and P. Berg

A widely recognized difficulty of presently used methods for cDNA cloning is obtaining cDNA segments that contain the entire nucleotide sequence of the corresponding mRNA. The cloning procedure described here mitigates this shortcoming. Of the 10^5 plasmid-cDNA recombinants obtained per μg of rabbit reticulocyte mRNA, about 10% contained a complete α- or β-globin mRNA sequence, and at least 30 to 50%, but very likely more, contained the entire globin coding regions. We attribute the high efficiency of cloning full- or nearly full-length cDNA to (i) the fact that the plasmid DNA vector itself serves as the primer for first- and second-strand cDNA synthesis, (ii) the lack of any nuclease treatment of the products, and (iii) the fact that one of the steps in the procedure results in preferential cloning of recombinants with full-length cDNA's over those with truncated cDNA's.

The availability of complementary DNA (cDNA) copies of mRNA's provides an extremely powerful tool for analyzing the structure, organization, and expression of eukaryote genes (4, 5, 9, 10, 16, 19, 36, 38). Aside from the utility of cDNA's for defining the initiation, coding, and termination sequences of mRNA's, their use as hybridization probes makes it possible to search for, isolate, identify, and characterize the corresponding genes from chromosomal DNA. Indeed, it was comparisons between cloned cDNA's and their genomic counterparts that uncovered the existence of intervening sequences (4, 5, 9, 19, 38) and splicing (37) and the occurrence of genomic rearrangements in the formation of functional immunoglobulin genes (3). More recently, the cloning of cDNA copies of RNA virus genomes (e.g., vesicular stomatitis [32], polio [22], and influenza viruses [35]) has opened the way to a more refined understanding of these viruses' structure, replication, and expression, as well as providing a simpler route for the development of antiviral vaccines (17).

With present techniques for cDNA cloning (12, 31, 33, 42) the yield of recombinant DNAs that have full-length cDNA sequences, i.e., cDNA's containing the entire nucleotide sequence of the mature mRNA, is low. Generally, most cDNA clones contain 3'-untranslated and variable amounts of the protein-coding sequence; obtaining cloned cDNA's with complete 5'-untranslated and protein-coding sequences is rarer, particularly if the mRNA codes for a large protein (32, 35). Although such truncated cDNA's are still useful as hybridization probes (15), they cannot direct the synthesis of complete proteins after their introduction into bacterial or mammalian cells via appropriate expression vectors. Consequently, we have sought to devise a cDNA cloning method that increases the probability of obtaining recombinants with full-length cDNA inserts.

Presently, the initial cDNA copy of a mRNA is synthesized with reverse transcriptase (39) using as primer either oligodeoxythymidylate [oligo(dT)] annealed to the polyadenylate [poly(A)] tail (11, 18) or an oligonucleotide annealed to a complementary sequence in the body of the mRNA (36, 40). The quality of the reverse transcriptase, the integrity and secondary structure of the mRNA, and the reaction conditions influence the length of the primary reverse transcript and, therefore, the completeness of the subsequently cloned cDNA. Second-strand synthesis is at best a poorly controlled step, since it relies on the ability of *Escherichia coli* DNA polymerase I (*Pol*I) to use the initial reverse transcript as both a primer and template, the end result being a hairpin double-stranded DNA with the 5' end of the mRNA sequence in the form of a single-stranded loop of variable size and location (11). S1 nuclease digestion of the single-stranded DNA loop, which must precede the addition of homopolymeric tails to the ends of the cDNA, invariably removes portions of the cDNA corresponding to coding or 5'-proximal portions of the mRNA. Several innovations, such as enrichment for the particular mRNA sequence prior to cDNA synthesis (6), fractionation of single-strand or double-strand cDNA to enrich for particular size classes (6, 10, 36), or even alternative procedures for priming the synthesis of the second strand that eliminate the need for the nuclease digestion of 5'-proximal sequences (24, 33), have improved the yield of full- or nearly full-length cDNA or at least produced cDNA's with intact 5'-proximal nucleotide sequences (24).

Here we describe a modification in the cDNA cloning procedure that permits the recovery, in high yields, of plasmid recombinants with full- or nearly full-length cDNA inserts. The procedure uses a plasmid DNA vector which itself serves as the primer for first- and ultimately second-strand cDNA synthesis; moreover, one of the steps is designed to enrich for recombinants containing full-length cDNA's over those with truncated cDNA's. The procedure has been applied successfully to the cloning of full-length α- and β-globin cDNA's from rabbit reticulocyte mRNA.

MATERIALS AND METHODS

Chemicals, enzymes, and plasmids. Oligo(dT)- and oligodeoxyadenylate [oligo(dA)]-celluloses (both type 6) were purchased from Collaborative Research Inc. The reverse transcriptase was the avian myeloblastosis virus enzyme obtained from J. Beard at the National Institutes of Health. *E. coli* DNA ligase was provided by I. R. Lehman, *E. coli* DNA polymerase I was provided by J. Widom and S. Sherer, and *Eco*RI endonuclease came from J. Carlson (all from Stanford University). Calf thymus terminal deoxynucleotidyl transferase and *E. coli* RNase H were purchased from PL Biochemical Co.; *Hind*III and *Kpn*I endonucleases were products of Bethesda Research Laboratories Inc., *Hpa*I, *Ava*II, and *Pvu*II endonucleases were from New England BioLabs, and *Pst*I endonuclease was obtained from Boehringer-Mannheim. The pBR322-SV40 recombinants that provide the vector-primer DNA and linker DNA fragment (Fig. 1) were constructed by S. Subramani. One contained a simian virus 40 (SV40) DNA segment corresponding to map position 0.71 to 0.86 cloned between the *Pvu*II and *Hind*III sites of pBR322 DNA, and the other had a segment from map position 0.19 to 0.32 inserted between the *Bam*HI and *Hind*III sites of the plasmid DNA.

Preparation of mRNA. Rabbit reticulocyte mRNA, enriched with α-globin mRNA, was prepared from a postpolysomal supernatant of a reticulocyte lysate obtained from phenylhydrazine-treated rabbits (28). The mRNA was recovered after phenol extraction by alcohol precipitation following two cycles of adsorption and elution from an oligo(dT)-cellulose column (1). In recent attempts to prepare cDNA libraries from other cells (H. Okayama and P. Berg, unpublished data) we have found the guanidinium thiocyanate method (7) to be superior for the preparation of mRNA that is readily reverse transcribed.

Preparation of vector primer and oligo dG-tailed linker DNAs. (For a diagram of this procedure, see Fig. 1). A 400-μg sample of pBR322-SV40 (map units 0.71–0.86) DNA was digested at 37°C with 700 U of *Kpn*I endonuclease in a reaction mixture (0.4 ml) containing 6 mM Tris-hydrochloride (pH 7.5), 6 mM MgCl₂, 6 mM NaCl, 6 mM 2-mercaptoethanol, and 0.1 mg of bovine serum albumin (BSA) per ml. After 5 h, the digestion was terminated with 40 μl of 0.25 M EDTA

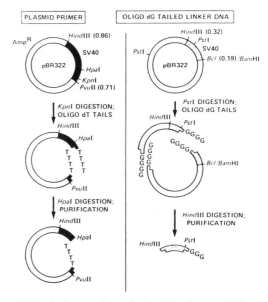

PREPARATION OF PLASMID PRIMER AND OLIGO dG-TAILED LINKER DNA

FIG. 1. Preparation of plasmid primer and linker DNA. The unshaded portion of each ring is pBR322 DNA, and the shaded or stippled segments are from SV40 DNA. The numbers next to the restriction site designations are the corresponding SV40 DNA map coordinates.

(pH 8.0) and 20 μl of 10% sodium dodecyl sulfate (SDS); the DNA was recovered after extraction with water-saturated phenol-CHCl₃ (1:1) (hereafter referred to as phenol-CHCl₃) and ethanol precipitation. Homopolymer tails averaging 60, but not more than 80, deoxythymidylate (dT) residues per end were added to the *Kpn*I endonuclease-generated termini with calf thymus terminal deoxynucleotidyl transferase as follows. The reaction mixture (0.2 ml) contained 140 mM sodium cacodylate–30 mM Tris-hydrochloride (pH 6.8) as buffer, with 1 mM CoCl₂, 0.1 mM dithiothreitol, 0.25 mM dTTP, the *Kpn*I endonuclease-digested DNA, and 400 U of the terminal deoxynucleotidyl transferase. After 30 min at 37°C the reaction was stopped with 20 μl of 0.25 M EDTA (pH 8.0) and 10 μl of 10% SDS, and the DNA was recovered after several extractions with phenol-CHCl₃ by ethanol precipitation. The DNA was then digested with 17 U of *Hpa*I endonuclease in 0.2 ml containing 10 mM Tris-hydrochloride (pH 7.4), 10 mM MgCl₂, 20 mM KCl, 1 mM dithiothreitol, and 0.1 mg of BSA per ml for 5 h at 37°C. The large DNA fragment, which contained the origin of pBR322 DNA replication and the gene conferring ampicillin resistance, was purified by agarose (1%) gel electrophoresis and recovered from the gel by a modification of the glass powder method (41). The dT-tailed DNA was further purified by adsorption and elution from an oligo(dA)-cellulose column as follows. The DNA was dissolved in 1 ml of 10 mM Tris-hydrochloride (pH 7.3) buffer containing 1 mM EDTA and 1 M NaCl, cooled to 0°C, and applied to an oligo (dA)-cellulose column (0.6 by 2.5 cm) equilibrated with the same buffer at 0°C. The column was washed with the same buffer at 0°C and eluted with water at room temperature. The eluted DNA (140 μg) was precipitated with ethanol and dissolved in 100 μl of 10 mM Tris-hydrochloride (pH 7.3) with 1 mM EDTA.

The oligodeoxyguanylate [oligo(dG)]-tailed linker DNA was prepared by digesting 100 μg of pBR322-SV40 (map units 0.19–0.32) with 120 U of PstI endonuclease in 0.2 ml containing 6 mM Tris-hydrochloride (pH 7.4), 6 mM MgCl₂, 6 mM 2-mercaptoethanol, 50 mM NaCl, and 0.1 mg of BSA per ml. After 1.5 h at 37°C the reaction mixture was extracted with phenol-CHCl₃ and the DNA was precipitated with alcohol. Tails of 10 to 15 deoxyguanylate (dG) residues were then added per end with 60 U of terminal deoxynucleotidyl transferase in the same reaction mixture (50 μl) described above, except for 0.1 mM dGTP replacing dTTP. After 20 min at 37°C the mixture was extracted with phenol-CHCl₃, and after the DNA was precipitated with ethanol it was digested with 50 U of HindIII endonuclease in 50 μl containing 20 mM Tris-hydrochloride (pH 7.4), 7 mM MgCl₂, 60 mM NaCl, and 0.1 mg of BSA at 37°C for 1 h. The small oligo (dG)-tailed linker DNA was purified by agarose gel (1.8%) electrophoresis and recovered as described above.

Preparation and cloning of globin cDNA. For a diagram of the preparation and cloning of globin cDNA, see Fig. 2.

Step 1. cDNA synthesis. The reaction mixture (10 μl) contained 50 mM Tris-hydrochloride (pH 8.3), 8 mM MgCl₂, 30 mM KCl, 0.3 mM dithiothreitol, 2 mM each dATP, dTTP, dGTP, and [³²P]dCTP (850 cpm/pmol), 0.2 μg of the reticulocyte mRNA (1 pmol of globin mRNA), 1.4 μg of the vector-primer DNA (0.7 pmol of primer end), and 5 U of reverse transcriptase. [The molar ratio of poly(A)⁺ mRNA to vector-primer DNA should be greater than 1.0, and in our experiments

ranged from 1.5 to 3). cDNA synthesis was initiated by the addition of reverse transcriptase and continued at 37°C for 20 min. By this time the rate of dCTP incorporation had leveled off and more than 60% of the primer was utilized for cDNA synthesis. The reaction was stopped with 1 μl of 0.25 M EDTA (pH 8.0) and 0.5 μl of 10% SDS; 10 μl of phenol-CHCl₃ was added, and the solution was blended vigorously in a Vortex mixer and then centrifuged. After adding 10 μl of 4 M ammonium acetate and 40 μl of ethanol to the aqueous phase, the solution was chilled with dry ice for 15 min, warmed to room temperature with gentle shaking to dissolve unreacted deoxynucleoside triphosphates that had precipitated during chilling, and centrifuged for 10 min in an Eppendorf microfuge. The pellet was dissolved in 10 μl of 10 mM Tris-hydrochloride (pH 7.3) and 1 mM EDTA, mixed with 10 μl of 4 M ammonium acetate, and reprecipitated with 40 μl of ethanol, a procedure which removes more than 99% of unreacted deoxynucleoside triphosphates. The pellet was rinsed with ethanol.

Step 2: Oligodeoxycytidylate [oligo(dC)] addition. The pellet containing the plasmid-cDNA:mRNA was dissolved in 15 μl of 140 mM sodium cacodylate–30 mM Tris-hydrochloride (pH 6.8) buffer containing 1 mM CoCl₂, 0.1 mM dithiothreitol, 0.2 μg of poly(A), 66 μM [³²P]dCTP (6,000 cpm/pmol), and 18 U of terminal deoxynucleotidyl transferase. The reaction was carried out at 37°C for 5 min to permit the addition of 10 to 15 residues of dCMP per end and then terminated with 1.5 μl of 0.25 M EDTA (pH 8.0) and 0.75 μl of 10% SDS. After extraction with 15 μl of phenol-CHCl₃, the aqueous phase was mixed with 15 μl of 4 M ammonium acetate, the DNA was precipitated and reprecipitated with 60 μl of ethanol, and the final pellet was rinsed with ethanol.

Step 3: HindIII endonuclease digestion. The pellet was dissolved in 10 μl of buffer containing 20 mM Tris-

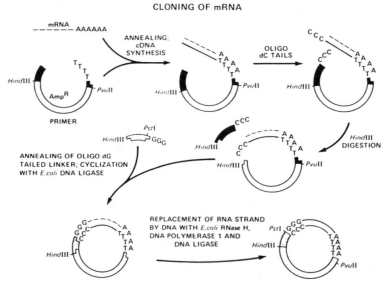

CLONING OF mRNA

FIG. 2. Steps in the construction of plasmid-cDNA recombinants. The designations for the DNA segments are as mentioned in Fig. 1. Experimental details and comments on the procedure are presented in Materials and Methods and Results, respectively.

hydrochloride (pH 7.4), 7 mM MgCl$_2$, 60 mM NaCl, and 0.1 mg of BSA per ml and then digested with 2.5 U of HindIII endonuclease for 1 h at 37°C. The reaction was terminated with 1 μl of 0.25 M EDTA (pH 8.0) and 0.5 μl of 10% SDS, and, after extraction with phenol-CHCl$_3$ followed by the addition of 10 μl of 4 M ammonium acetate, the DNA were precipitated with 40 μl of ethanol. The pellet was rinsed with ethanol and then dissolved in 10 μl of 10 mM Tris-hydrochloride (pH 7.3) and 1 mM EDTA, and 3 μl of ethanol was added to prevent freezing during storage at −20°C.

Step 4: Cyclization mediated by the oligo(dG)-tailed linker DNA. A 1-μl sample of the HindIII endonuclease-digested oligo(dC)-tailed cDNA:mRNA plasmid (0.02 pmol) was incubated in a mixture (10 μl) containing 10 mM Tris-hydrochloride (pH 7.5), 1 mM EDTA, 0.1 M NaCl, and 0.04 pmol of the oligo(dG)-tailed linker DNA (this amount is a twofold molar excess over the quantity of the vector-cDNA:mRNA and of the fragment which remains as a result of the HindIII endonuclease digestion in the previous step) at 65°C for 2 min, shifted to 42°C for 30 min, and then cooled to 0°C. The mixture (10 μl) was adjusted to a volume of 100 μl containing 20 mM Tris-hydrochloride (pH 7.5), 4 mM MgCl$_2$, 10 mM (NH$_4$)$_2$SO$_4$, 0.1 M KCl, 50 μg of BSA per ml, and 0.1 mM β-NAD; 0.6 μg of E. coli DNA ligase was added, and the solution was then incubated overnight at 12°C.

Step 5: Replacement of RNA strand by DNA. To replace the RNA strand of the insert, the ligation mixture was adjusted to contain 40 μM of each of the four deoxynucleoside triphosphates, 0.15 mM β-NAD, 0.4 μg of additional E. coli DNA ligase, 0.3 μg of E. coli DNA polymerase I, and 1 U of E. coli RNase H. This mixture (104 μl) was incubated successively at 12°C and room temperature for 1 h each to promote optimal repair synthesis and nick translation by PolI (23). The reaction was terminated by the addition of 0.9 ml of cold 10 mM Tris-hydrochloride (pH 7.3), and 0.1-ml aliquots were stored at 0°C.

Step 6: Transformation of E. coli. Transformation was carried out using minor modifications of the procedure described by Cohen et al. (8). E. coli K-12 (strain HB101) was grown to 0.5 absorbancy unit at 600 nm at 37°C in 20 ml of L-broth. The cells were collected by centrifugation, suspended in 10 ml of 10 mM Tris-hydrochloride (pH 7.3) containing 50 mM CaCl$_2$, and centrifuged at 0°C for 5 min. The cells were resuspended in 2 ml of the above buffer and incubated again at 0°C for 5 min; then, 0.2 ml of the cell suspensions was mixed with 0.1 ml of the DNA solution (step 5) and incubated at 0°C for 15 min. After the cells were kept at 37° for 2 min and at room temperature for 10 min, 0.5 ml of L-broth was added, and the culture was incubated at 37°C for 30 min, mixed with 2.5 ml of L-broth soft agar at 42°C, and spread over L-broth agar containing 50 μg of ampicillin per ml. After incubation at 37°C for 12 to 24 h, individual colonies were picked with sterile toothpicks.

Characterization of cDNA clones. (i) Colony hybridization and isolation of recombinant plasmids. E. coli transformants were screened for the presence of globin cDNA by in situ colony hybridization (15). One hundred transformants were grown on three replica nitrocellulose filter disks, lysed with alkali, and hybridized with [^{32}P]-cDNA synthesized by oligo(dT)-primed reverse transcription of reticulocyte mRNA. Alternatively, ^{32}P-nick-translated α-globin or β-globin cDNA clones, which had been respectively constructed and identified in the present work or prepared previously (29) from pBG1 (10), were the globin cDNA probes; each of the cDNA's was removed from the vector by restriction enzyme digestions and isolated by agarose gel electrophoresis before use as hybridization probes. Colonies that gave positive hybridization signals were grown in L broth containing 50 μg of ampicillin per ml, and their plasmid DNAs were isolated by standard techniques (21).

(ii) Restriction mapping. DNAs were digested with EcoRI, PstI, and PvuII, or AvaII restriction endonucleases under conditions recommended by the suppliers, and analyzed by agarose gel (1.5%) electrophoresis.

(iii) DNA sequencing. DNAs were digested with PstI endonuclease, incubated with E. coli alkaline phosphatase to remove the terminal phosphates, and terminally labeled with [γ^{32}P]ATP and polynucleotide kinase (26). After digestion with EcoRI endonuclease, the ^{32}P-labeled fragment containing the 5' end of the globin cDNA was purified by agarose gel (1.5%) electrophoresis and sequenced by the method of Maxam and Gilbert (26).

RESULTS

Experimental details of the procedure for preparing and cloning cDNA's are presented above. Here we consider the rationale and several general features of the method and illustrate its application to the cloning of full- and nearly full-length α- and β-globin cDNA's from rabbit reticulocyte mRNA.

Key features of the protocol (outlined in Fig. 2) are that (i) the plasmid vector DNA functions as the primer for the synthesis of the first cDNA strand, an innovation first introduced by Rabbitts (31); (ii) the full- or nearly full-length reverse transcripts of the mRNA are preferentially converted to duplex cDNA's and cloned as recombinants in E. coli in the subsequent steps; and (iii) nick-translation repair of the cDNA:mRNA hybrid, mediated by E. coli RNase H, E. coli PolI, and E. coli DNA ligase, is used to synthesize the second cDNA strand.

The vector-primer, a linear DNA with a poly(dT) tail at one end, was constructed as outlined in Fig. 1. The pBR322-SV40 (map units 0.71–0.86) DNA recombinant was a convenient precursor because it contains both a unique restriction site (KpnI), at which 3' single-strand ends can be generated for the efficient attachment of poly(dT) tails, and a second unique restriction site (HpaI) near one end to permit removal of one of the poly(dT) tails. This particular DNA was useful for the purpose described, but other DNAs with similar arrangements of appropriate restriction sites could be substituted.

Annealing of the poly(A)$^+$ mRNA to the poly(dT)-tailed vector-primer DNA generates the substrate for reverse transcription of the mRNA sequence. A 1.5- to 3-fold molar excess of mRNA over poly(dT)-tailed DNA is advantageous to minimize the possibility that unreacted poly(dT) tails of the vector-primer DNA will be an acceptor for oligo(dC) tails in the next step. Under the conditions used, more than 60% of the vector-primer DNA acquired a covalently linked reverse transcript of the mRNA during the first 10 min of incubation (data not shown); moreover, from pilot experiments with a poly(dT) primer and $[^{32}P]dCTP$ as one of the deoxynucleoside triphosphates, at least 50 to 60% of the cDNA copies attained the length of the globin mRNA.

The aim of the next step is to generate a cohesive tail at the end of the cDNA so that it can be ligated to the other end of the vector DNA and, thereby, provide the template for second-strand cDNA synthesis. This has been achieved by adding oligo(dC) tails to the 3' ends of both the cDNA and vector DNA and removal of the oligo(dC) tail from the vector DNA terminus by cleavage at the unique HindIII restriction site near that end. Since the mRNA:cDNA hybrid is a very poor substrate for HindIII endonuclease (27), it remains intact during digestion with limiting quantities of the restriction enzyme.

The ensuing vector-cDNA:mRNA derivative, with a HindIII cohesive end and an oligo(dC) tail at the respective termini, is cyclized by E. coli DNA ligase using a short linker DNA segment containing a HindIII cohesive end and an oligo (dG) tail. Covalent joining of the linker and vector DNAs occurs via their HindIII cohesive ends, and a noncovalent, base-paired join is made to the cDNA:mRNA duplex via the oligo(dG) and oligo(dC) tails. In practice about 20% of the linear vector-cDNA:mRNA was converted to circular structures as judged by the electrophoretic shift in agarose gel.

Since transformation of E. coli (HB101) with recombinants containing cDNA:mRNA inserts is inefficient and yields cDNA clones with variable but extensive deletions in the inserts (43), we replaced the mRNA strand by the corresponding DNA strand in vitro. This is accomplished by using E. coli RNase H to introduce nicks in the RNA strand (25), E. coli PolI and the four deoxynucleoside triphosphates to replace the RNA segments by nick translation (23), and E. coli DNA ligase to join the newly synthesized DNA fragments into a continuous second cDNA strand. In the repair synthesis, the oligo(dG) tail of the linker DNA serves as the primer for copying any unpaired deoxyribosylcytidine (dC) sequence and extending the strand to the cDNA region. Similarly, if necessary, the E. coli PolI extends the oligo(dC) and poly(A)

tails to produce complete pairing with the oligo(dG) and poly(dT) of their respective opposite strands. E. coli DNA ligase was chosen in place of the T4 enzyme because of its inability to join adjacent RNA and DNA segments (13) that arise during second-strand cDNA synthesis.

Applying this procedure to rabbit reticulocyte mRNA produced about 10^5 ampicillin-resistant HB101 clones per µg of starting mRNA (Table 1). Failure to bridge or ligate the ends of the cDNA and vector by omission of the linker DNA segment or the DNA ligase resulted in drastic reductions in the number of bacterial transformants. If the mRNA strand was not replaced by the second strand of cDNA, the number of bacterial transformants was reduced fivefold.

Characterization of α- and β-globin cDNA clones. To estimate the yield and identify the clones containing α- and β-globin cDNA, colony hybridizations (15) were performed on 100 randomly chosen bacterial transformants (Fig. 3). Figure 3A, B, and C show that about 80% of the transformants contained recombinant plasmids with nucleotide sequences homologous to total reticulocyte cDNA, and 85% of these were accounted for by α-globin (50%) or β-globin (35%) cDNA derivatives. Presumably, the clones with cDNA inserts that failed to hybridize with α- or β-globin cDNA contain nonglobin cDNA's.

A group of clones containing 21 α- or 12 β-globin cDNA's were analyzed further by restriction enzyme digestions to determine the size of the cDNA inserts. The diagram in Fig. 4 summarizes the expected fragments based on the known restriction sites in the recombinant with a complete α-globin cDNA segment. Figure 5 presents the same information for recombinants with a complete β-globin cDNA segment. EcoRI endonuclease digestion of the α- and β-globin cDNA recombinant DNAs revealed that 10 of 21 α-globin and 4 of 12 β-globin cDNA clones gave the expected sized fragment (shown by the position of the arrow) for a complete cDNA copy.

TABLE 1. Yield of ampicillin-resistant transformants obtained by cDNA cloning procedure

Procedure[a]	Transformants (no. per µg of mRNA)
Complete	100,000
− Oligo(dG)-tailed linker DNA	2,400
− DNA ligase	50
− DNA polymerase and RNase H	18,000

[a] "Complete" refers to the procedure as described in the text. The other entries refer to modifications in which the indicated component or step was omitted.

A

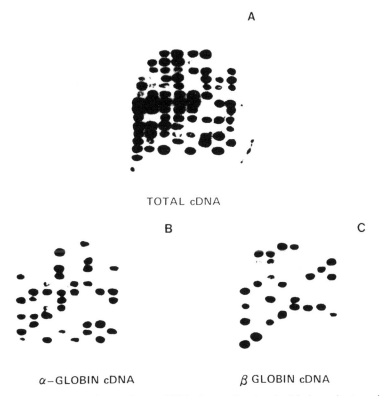

TOTAL cDNA

B   C

α–GLOBIN cDNA   β GLOBIN cDNA

FIG. 3. In situ colony hybridization to detect cDNA clones. One hundred independent ampicillin-resistant transformants were picked and grown on three replica nitrocellulose filters. Detection of cDNA clones was accomplished by colony hybridization (15) using three different probes: (A) total cDNA synthesized by reverse transcription of the rabbit reticulocyte mRNA used for cloning; (B) cloned rabbit α-globin cDNA prepared in this work and identified by restriction analyses; (C) cloned β-globin cDNA made by Efstratiadis et al. (10) and recloned by Mulligan et al. (29).

Each of these plasmid DNAs (10 for α-globin and 4 for β-globin) was digested with PstI endonuclease to cleave at the reconstructed PstI restriction site adjacent to the oligo(dG:dC) join and with PvuII endonuclease to cleave the vector DNA sequence adjacent to the poly(dA:dT) join. All but 2 of the 10 clones, judged to contain nearly full-length α-globin cDNA by the EcoRI endonuclease digestion, contained these two restriction sites and yielded the predicted length fragment from a double digest; similarly, 3 of the 4 β-globin cDNA clones, judged to be nearly full length by EcoRI endonuclease cleavage, produced the expected sized fragments after PstI-PvuII endonuclease digestions. The fuzzy and heterogeneous bands seen in the PstI and PvuII endonuclease digests of α-globin cDNA clone 2 and β-globin cDNA clone 12 are probably due to heterogeneity in the poly(dA:dT) segment that arises during their propagation in E. coli. The eight putative complete α-globin cDNA clones also contain the two known AvaII restriction

sites, spaced to give the expected sized fragment (Fig. 4).

The 5′-proximal nucleotide sequences of the eight α- and three β-globin cDNA segments were determined to establish the completeness of their cDNA's. After ^{32}P labeling at the reconstructed PstI restriction sites adjacent to the oligo(dG:dC) join, about 120 nucleotides in each cDNA were sequenced by the Maxam-Gilbert method (26) (Fig. 6 and 7). Each figure shows the determined cDNA sequence adjacent to the linker oligo(dG) segment and the known 5′-terminal nucleotide sequence of the corresponding globin mRNA. The nucleotide sequences distal to those shown here were identical to the already reported sequence for rabbit α- (18) and β-globin (10) cDNA's.

Two of the eight α-globin cDNA's contained all the nucleotides at the 5′ end of the mRNA sequence, preceded by 14 or 26 dG residues of the oligo(dG:dC) linker. The other six lacked the first nucleotide at the 5′ end of the mRNA

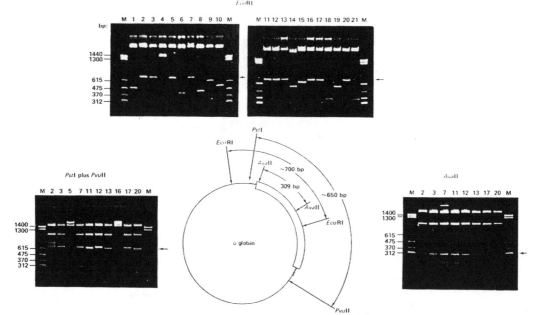

FIG. 4. Restriction endonuclease analyses of α-globin cDNA clones. Plasmid DNA prepared from 21 recombinants with α-globin cDNA inserts were digested with *Eco*RI (upper), *Pst*I and *Pvu*II (left), or *Ava*II (right) endonucleases and electrophoresed on 1.5% agarose gels. The DNA size markers were from *Taq*I endonuclease-digested pBR322 DNA. The presence of a 615-base pair fragment and the relatively low amount of a 475-base pair fragment are due to the presence of a modified and, therefore, resistant *Taq*I restriction site (2). Arrows indicate the position of restriction fragments that would be produced if the cloned cDNA is full length. The diagram in the center summarizes the various enzyme cleavage sites and the sizes predicted for recombinants with complete α-globin cDNA.

sequence and had oligo(dG:dC) linkers ranging in length from 15 to 21 residues. Of the three β-globin cDNA's that were sequenced, one contained the entire 5' end of the mRNA sequence, one lacked one nucleotide, and the other lacked three nucleotides. The linker segments in the β-globin cDNA clones were 14, 16, and 20 nucleotides in length, but the latter two contained, in addition to dG, the unexpected nucleotides dC

and dT. These were probably inadvertently introduced because of incomplete removal of dGTP and dATP after the reverse transcription and before the addition of the oligo(dC) tails to the end of the cDNA strand.

Assuming that the cDNA's analyzed are a representative sample of the cloned cDNA population, we estimate that 30 to 50% of the cDNA segments are full- or nearly full-length copies of

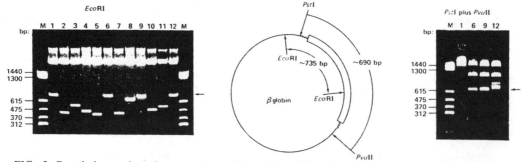

FIG. 5. Restriction endonuclease analyses of β-globin cDNA clones. Plasmid DNA prepared from 12 recombinants with β-globin cDNA inserts were digested with *Eco*RI (left) or *Pst*I and *Pvu*II (right) endonucleases and electrophoresed on 1.5% agarose gels. The markers are the same as in Fig. 4 and the arrows have the same meaning. The diagram summarizes the various enzyme cleavage sites and the sizes predicted for recombinants with complete β-globin cDNA.

FIG. 6. Comparison of 5'-terminal nucleotide sequences of eight putative full-length α-globin cDNA's with the known 5'-terminal sequence of α-globin mRNA (18).

their mRNA's. Quite likely, still more of the globin clones contain an entire coding sequence since the initiator AUGs for α-globin and β-globin are 36 (18) and 53 (10) nucleotides, respectively, from the 5'-capped nucleotide.

DISCUSSION

The cDNA cloning method described here is general, being applicable to purified or mixed mRNA, to poly(A)$^+$ viral RNAs, or to cellular RNAs to which poly(A) tails can be added at the 3' end. The procedure is relatively rapid, taking less than 2 days from reverse transcription to the collection of transformants containing the cloned cDNA's. Moreover, it uses commercially available reagents and enzymes and standard biochemical procedures. Aside from the relative ease and rapidity, the most significant advantage of this procedure, relative to those previously used, is the high efficiency of cloning full- or nearly full-length cDNA's. Applying the procedure to rabbit reticulocyte mRNA yielded approximately 10^5 globin cDNA clones per μg of mRNA, of which 10% had the complete mRNA sequence at the 5' end and 20 to 40% had all but two or three at the 5'-most nucleotides of the globin mRNA sequences. As anticipated, incorporation of one of the complete α-globin cDNA segments into the vector SVGT5 (29) yielded SV40-like late mRNA's containing the α-globin cDNA sequence and immunoprecipatable α-globin protein (Okayama and Berg, unpublished data).

Although each of the factors contributing to the relatively high cloning efficiency of full-length cDNA's has not been fully evaluated, much of the success, we believe, can be ascribed to the following features. (i) Priming by the plasmid DNA for the synthesis of the first cDNA strand achieves the first join of the cDNA to the vector. (ii) Selective cloning of complete or nearly complete cDNA copies of the mRNA's is necessary because the oligo (dC) tails needed to effect ring closure and complete second-strand cDNA synthesis are synthesized more efficiently by terminal transferase with substrates containing base-paired, full-length reverse transcripts than with shortened or truncated cDNA

strands base-paired to longer mRNA's (34); (iii) Replacement of the mRNA by a second cDNA strand is effected by the actions of *E. coli* RNase H (to produce nicks and gaps in the RNA strand), *E. coli* Pol I-mediated nick translation (to replace RNA with DNA sequences), and *E. coli* DNA ligase (to join the DNA fragments into the continuous second cDNA strand).

In this paper the procedure has been used to clone full-length rabbit α- and β-globin cDNA. However, the high cloning efficiency makes it feasible to isolate rare as well as abundant cDNA's from a variety of cellular mRNA populations. For example, a cDNA library containing 2 × 10^5 clones per μg of mRNA, made with mRNA from Chinese hamster cells, contained several clones that code for the 35,000-dalton subunit of adenine phosphoribosyl transferase, one of which contained a cDNA segment of 0.95 kilobases (kb) and is currently being sequenced to determine its structure (R. Axel, private communication). An analogous cDNA library from hamster cells (A29) that synthesize a methotrexate-resistant dihydrofolate reductase (14) contains cDNA clones whose lengths are sufficient to code for that enzyme (M. Dieckmann, H. Okayama, and P. Berg, unpublished data). In both of these cases the relevant mRNA's are estimated to be present in about 5 to 100 copies per cell. M. Ballivet, J. M. Lee, J. Patrick, and S. F. Heinemann have informed us (private communication) that, using this procedure, they have isolated full- or nearly full-sized cDNA clones that code for the 60,000-dalton subunit of the *Torpedo californica* acetyl choline receptor (20); moreover, Shigesada and Stark (private communication) have obtained a cDNA clone of about 6.5 kb from cell RNA containing the 7.9-kb CAD messenger (30). The efficient production and nucleotide sequencing of nearly full-length cloned cDNA's has also facilitated the characterization of normal and aberrant splicing patterns in their corresponding mRNA's (S. C. Clark, M. C. Nguyen-Huu, H. M. Goodman, and P. Berg, unpublished data). Additional studies to establish the efficiency of this cloning

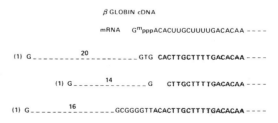

FIG. 7. Comparison of 5'-terminal sequences of three putative full-length β-globin cDNAs with the known 5'-terminal sequence of β-globin mRNA (10).

procedure and the completeness of the cDNA copies of other cellular genes are in progress.

Ours and other approaches for cloning cDNA's (and genomic DNA as well) are clearly limited by the ability to detect particular cDNA's; currently, nucleic acid hybridization (6, 10, 31, 36) or functional assays for the mRNA (16) are the sole means to identify specific cDNA clones. The relatively high yields of cDNA's with a complete protein coding sequence offers another alternative, namely, the identification or even selective propagation of cDNA clones on the basis of the product or function they express. By appropriate choice of the vector-primer and linker DNAs we are exploring the construction of recombinants that can express the cloned cDNA's directly in either bacterial or mammalian cells.

ACKNOWLEDGMENTS

This research was funded by Public Health Service grant GM-13235 from the National Institute of General Medical Sciences, by Public Health Service grant CA-15513-7 from the National Cancer Institute, and by grant MV-35J from the American Cancer Society. H.O. is a fellow of the Jane Coffin Childs Fund for Medical Research.

LITERATURE CITED

1. **Aviv, H., and P. Leder.** 1972. Purification of biological active globin messenger RNA by chromatography on oligothymidylic acid-cellulose. Proc. Natl. Acad. Sci. U.S.A. **69**:1408–1412.
2. **Backman, K.** 1980. A cautionary note on the use of certain restriction endonucleases with methylated substrates. Gene **11**:169–171.
3. **Brack, C., M. Hirama, R. Lenhard-Schuller, and S. Tonegawa.** 1978. Complete immunoglobulin gene is created by somatic recombination. Cell **15**:1–14.
4. **Breathnach, R., J. L. Mandel, and P. Chambon.** 1977. Ovalbumin gene is split in chicken DNA. Nature (London) **270**:314–319.
5. **Catterall, J. F., J. P. Stein, E. C. Lai, S. L. C. Woo, M. L. Mace, A. R. Means, and B. W. O'Malley.** 1979. The ovomucoid gene contains at least six intervening sequences. Nature (London) **278**:323–327.
6. **Chang, A. C. Y., J. H. Nunberg, R. J. Kaufman, H. A. Erlich, R. T. Schimke, and S. N. Cohen.** 1978. Phenotypic expression in E. coli of a DNA sequence coding for mouse dihydofolate reductase. Nature (London) **275**:617–624.
7. **Chirgwin, J. M., A. E. Przybyla, R. J. MacDonald, and W. J. Rutter.** 1979. Isolation of biological active ribonucleic acid from sources enriched in ribonuclease. Biochemistry **18**:5294–5299.
8. **Cohen, S. N., A. C. Y. Chang, and L. Hsu.** 1972. Nonchromosomal antibiotic resistance in bacteria: genetic transformation of Escherichia coli by R-factor DNA. Proc. Natl. Acad. Sci. U.S.A. **69**:2110–2114.
9. **Cordell, B., G. Bell, E. Tischer, F. DeNoto, A. Ullrich, R. Dictet, W. J. Rutter, and H. M. Goodman.** 1979. Isolation and characterization of a cloned rat insulin gene. Cell **18**:533–543.
10. **Efstratiadis, A., F. C. Kafatos, and T. Maniatis.** 1977. The primary structure of rabbit β-globin mRNA as determined from cloned DNA. Cell **10**:571–585.
11. **Efstratiadis, A., F. Kafatos, A. M. Maxam, and T. Maniatis.** 1976. Enzymatic in vitro synthesis of globin genes. Cell **7**:279–288.
12. **Efstratiadis, A., and L. Villa-Komaroff.** 1979. Cloning of double-stranded cDNA, p. 1–14. In J. K. Setlow and A. Hollaender, ed., Genetic Engineering, vol. 1. Plenum Press, New York.
13. **Fareed, E. G., E. M. Wilt, and C. C. Richardson.** 1971. Enzymatic breakage and joining of deoxyribonucleic acid. J. Biol. Chem. **246**:925–932.
14. **Flintoff, W. F., S. V. Davidson, and L. Siminovitch.** 1976. Isolation and partial characterization of three methotrexate-resistant phenotypes from Chinese hamster ovary cells. Somatic Cell Genet. **2**:245–261.
15. **Grunstein, M., and D. S. Hogness.** 1975. Colony hybridization: a method for the isolation of cloned DNAs that contain a specific gene. Proc. Natl. Acad. Sci. U.S.A. **72**:3961–3965.
16. **Harpold, M. M., P. R. Dobner, R. M. Evans, and E. C. Bancroft.** 1978. Construction and identification by positive hybridization-translation of a bacterial plasmid containing a rat growth hormone structural gene sequence. Nucleic Acids Res. **5**:2039–2053.
17. **Heiland, J., and M.-J. Gething.** 1981. Cloned copy of the haemagglutinin gene codes for human influenza antigenic determinants in E. coli. Nature (London) **292**:851–852.
18. **Heindell, H. C., A. Lin, G. V. Paddock, G. M. Studnicka, and W. A. Salser.** 1978. The primary sequence of rabbit α-globin mRNA. Cell **15**:43–54.
19. **Jeffreys, A. J., and R. A. Flavell.** 1979. The rabbit β globin gene contains a large insert in the coding sequence. Cell **12**:1097–1108.
20. **Karlan, A.** 1980. Molecular properties of nicotinic acetylcholine receptors. Cell Surface Rev. **6**:192–246.
21. **Katz, L., D. T. Kingsbury, and D. R. Helinski.** 1973. Stimulation by cyclic adenosine monophosphate of plasmid deoxyribonucleic acid replication and catabolite repression of the plasmid deoxyribonucleic acid-protein relaxation complex. J. Bacteriol. **114**:577–591.
22. **Kitamura, N., B. L. Semler, P. G. Rothberg, G. R. Larsen, C. J. Adler, A. J. Dorner, E. A. Emini, R. Hanecak, J. J. Lee, S. Van der Werf, C. W. Anderson, and E. Wimmer.** 1981. Primary structure, gene organization and polypeptide expression of polio virus RNA. Nature (London) **291**:547–553.
23. **Kornberg, A.** 1980. DNA polymerase 1 of E. coli, p. 101–166. In DNA replication. H. Freeman and Co., San Francisco.
24. **Land, H., M. Grez, H. Hansen, W. Lindermaier, and G. Schuetz.** 1981. 5'-Terminal sequences of eucaryotic mRNA can be cloned with high efficiency. Nucleic Acids Res. **9**:2251–2266.
25. **Leis, J. P., I. Berkower, and J. Hurwitz.** 1973. Mechanism of action of ribonuclease H isolated from avian myeloblastosis virus and Escherichia coli. Proc. Natl. Acad. Sci. U.S.A. **70**:466–470.
26. **Maxam, A., and W. Gilbert.** 1980. Sequencing end-labeled DNA with base-specific chemical cleavages. Methods Enzymol. **65**:499–560.
27. **Molloy, P. L., and R. H. Symons.** 1980. Cleavage of DNA-RNA hybrids by Type II restriction enzymes. Nucleic Acids Res. **8**:2939–2946.
28. **Morrison, M. R., S. A. Brinkley, J. Gorski, and J. R. Lingrel.** 1974. The separation and identification of α- and β-globin messenger ribonucleic acids. J. Biol. Chem. **249**:5290–5295.
29. **Mulligan, R. C., B. H. Howard, and P. Berg.** 1979. Synthesis of rabbit β globin in cultured monkey kidney cells following infection with a SV40-β globin recombinant genome. Nature (London) **277**:108–114.
30. **Padgett, R. A., G. M. Wahl, P. F. Coleman, and G. R. Stark.** 1979. N-(phosphonacetyl)-L-aspartate-resistant hamster cells overaccumulate a single mRNA coding for the multifunctional protein that catalyzes the first step of UMP synthesis. J. Biol. Chem. **245**:974–980.
31. **Rabbitts, T. H.** 1976. Bacterial cloning of plasmids carrying copies of rabbit globin messenger RNA. Nature (London) **260**:221–225.
32. **Rose, J. K.** 1980. Complete intergenic and flanking gene

sequences from the genome of vesicular stomatitis virus. Cell 19:415–421.

33. Rougeon, F., P. Kourilsky, and B. Mach., 1975. Insertion of a rabbit β-globin gene sequence into an *E. coli* plasmid. Nucleic Acids Res. 2:2365–2378.

34. Roychoudhury, R., E. Jay, and R. Wu. 1976. Terminal labeling and addition of homopolymer tracts to duplex DNA fragments by terminal deoxynucleotidyl transferase. Nucleic Acids Res. 3:863–877.

35. Sleigh, M. J., G. W. Both, and G. G. Brownlee. 1979. The influenza virus haemagglutinin gene: cloning and characterization of a double-stranded DNA copy. Nucleic Acids Res. 7:879–893.

36. Sood, A., D. Pereira, and S. M. Weissman. 1981. Isolation and partial nucleotide sequence of a cDNA clone for human histocompatibility antigen HLA-B by use of an oligonucleotide primer. Proc. Natl. Acad. Sci. U.S.A. 78:616–620.

37. Tilghman, S. M., P. J. Curtis, D. C. Tiemeier, P. Leder, and C. Weissman. 1978. The intervening sequence of a mouse β globin gene is transcribed within the 15S β globin mRNA percursor. Proc. Natl. Acad. Sci. U.S.A. 75:725–729.

38. Tilghman, S. M., P. J. Curtis, D. C. Tiemeir, J. G. Seidman, B. M. Peterlin, M. Sullivan, J. V. Maizel, and P. Leder. 1978. Intervening sequence of DNA identified in the structural portion of a mouse β globin gene. Proc. Natl. Acad. Sci. U.S.A. 75:1309–1313.

39. Verma, I. M. 1977. The reverse transcriptase. Biochim. Biophys. Acta 473:1–38.

40. Villa-Komaroff, L., A. Efstratiadis, S. Broome, P. Lomedico, R. Tizard, S. P. Naber, W. L. Chick, and W. Gilbert. 1978. A bacterial clone synthesizing proinsulin. Proc. Natl. Acad. Sci. U.S.A. 75:3727–3731.

41. Vogelstein, B., and D. Gillespie. 1979. Preparation and analytical purification of DNA from agarose. Proc. Natl. Acad. Sci. U.S.A. 76:615–619.

42. Wood, K. O., and J. C. Lee. 1976. Integration of synthetic globin genes into an *E. coli* plasmid. Nucleic Acids Res. 3:1961–1971.

43. Zain, S., J. Sambrook, R. J. Roberts, W. Keller, M. Fried, and A. R. Dunn. 1979. Nucleotide sequence analysis of the leader segment in a cloned copy of adenovirus 2 fiber mRNA. Cell 16:851–861.

Construction of Coliphage Lambda Charon Vectors with *Bam*H1 Cloning Sites

D.L. Rimm, D. Horness, J. Kucera and F.R. Blattner

SUMMARY

Three new λ phage cloning vectors have been constructed, Charons (Ch) 27, 28 and 30. Ch27 and Ch30, are suitable for cloning small and large DNA fragments, respectively, cut with *Bam*HI, *Bgl*II, *Bcl*I, *Mbo*I, *Sau*3A, *Eco*RI, *Hin*dIII, *Sal*I and *Xho*I. Ch30 is similar to Ch28, but since it does not contain extraneous host DNA it should replace Ch28 in all future cloning applications. In vitro selection for the removal of the *Bam*HI site, as used in this study, employs several cycles of in vitro packaging.

INTRODUCTION

The adaptation of bacteriophage λ for use as a DNA cloning vehicle (Murray and Murray, 1974; 1975; Rambach and Tiollais, 1974; Thomas et al., 1974; Blattner et al., 1977; Murray et al., 1977) combined with the in vitro packaging technique (Becker and Gold, 1975; Hohn and Murray 1977; Sternberg et al., 1977) and the plaque hybridization technique (Benton and Davis, 1977) has provided a powerful method for isolation of specific genes and for creation of gene libraries of both eukaryotic and prokaryotic organisms (Blattner et al., 1978; Maniatis et al., 1978). The number of different λ cloning vectors that are available now approaches 100 (for a review see Williams and Blattner, 1979; 1980).

Abbreviations: bp, base pairs; Ch, Charon phage λ vectors; NRE, no-cut right end (Blattner et al., 1977); SDS, sodium dodecyl sulfate.

For technical and historical reasons the first λ cloning vectors were designed for use with the *Eco*RI restriction endonuclease. This restriction system occurs on a plasmid of *E. coli*, the host for λ, and thus mutations could be selected in vivo that eliminate restriction sites from the essential portion of the λ genome. Subsequently λ cloning vectors for other restriction enzymes were constructed from fragments of naturally occurring lambdoid phages or other derivatives of λ that lack or contain specific restriction sites (Murray et al., 1979; Blattner et al., 1977). This dependence on "found objects" is one reason for the great proliferation of useful DNA cloning vectors, with no one of them able to serve all purposes.

The in vitro packaging technique now makes it possible to select mutants of λ that are missing class-II restriction sites without the need to propagate λ in those species that are endowed with a specific restriction endonuclease. Successive application of this approach would in principle permit the development of a single vector suitable for cloning DNA fragmented with practically any class-II restriction endonuclease.

In this communication we describe a step toward that goal, namely, the construction of Ch27, Ch28 and Ch30. Ch27 and Ch30 are vectors suitable for cloning *Bam*HI fragments as well as the products of digestion with *Mbo*I (or *Sau*3A), *Bgl*II or *Bcl*I which produce the same cohesive ends as *Bam*HI. In addition *Hind*III, *Eco*RI, *Sal*I, and *Xho*I fragments as well as combinations of the above enzymes can be used. Ch28 was our first vector for cloning *Bam*HI fragments and it has been used extensively in cloning experiments (Liu et al., 1980). We found subsequently that this vector has a small segment of DNA probably of *E. coli* origin that prevents the use of this vector in cloning large *Eco*RI fragments. It is described here although Ch30 is now recommended for all applications for which Ch28 could be used. While this work was in progress, two other reports have appeared describing *Bam*HI vectors (Klein and Murray, 1979; Loenen and Brammar, 1980).

MATERIALS AND METHODS

(a) Strains, crosses and selection

Genetic crosses except for cross 5 were done by cross-streaking on plates seeded with *E. coli* strains Ymel or K802 according to Blattner et al. (1974). Cross 5 was done in vitro by mixing equal portions of DNA isolated from Ch28-2270 and Ch28-2222 and cutting with *Bam*HI. This preparation was then heated to 70°C for 10 min to disassociate the cohesive ends, ligated, and packaged in vitro (Blattner et al., 1978).

Sources and particulars on bacterial and phage strains and selection methods used in this study are given in Williams et al. (1979) except for those involving *E. coli* K750su3. This *himA⁻* strain, from David Friedman (Miller and Friedman, 1980), was useful because it will plate phages with *QSR* genes of λ but not those with *QSR* genes of φ80 (D. Friedman, personal communication). Lac⁺ phages were scored using 5-bromo-4-chloro-3-indolyl-β-D-galactoside (XG) plates. Selections used in specific genetic crosses were as follows (see Figs. 1–3 for strain designations):

Cross 1: Ch14 × Ch21A: selected for am⁺ lac⁺ plaques on W3350; checked for immunity on Ymel-(φ80) and Ymel(λ).

Cross 2: BVC4 × Ch21A: selected for am⁺ lac⁻ plaques on W3350.

Cross 3: Ch10 × Ch13A: selected for lac⁻ on himA⁻ checked on Ymel(P2) for lack of bio256.

Cross 4: XDR 3.24 × BVC4: selected for am⁺ lac⁻ on W3350.

Cross 5: Ch28-2222 × Ch28-2270 (done in vitro) was screened by gel electrophoretic analysis of individual candidates.

(b) Removal of restriction sites

Selection for removal of the *Bam*HI site in the left arm of λ was done by five successive steps of cutting BP1 DNA with *Bam*HI followed by in vitro packaging. After each packaging step the resulting phage population was grown and purified. At least one microgram of DNA corresponding to 2×10^{10} phage equivalents was used for each cycle. At each step the efficiency of plating of packaged *Bam*HI-cut DNA was compared with uncut DNA. Cycling was stopped when these were nearly equal. At this point six single plaques were purified and DNA isolated. All were devoid of *Bam*HI sites, as judged by agarose gel electrophoresis of the phage DNA.

(c) Phage propagation and analysis of DNA

Bacteriophage preparation, phenol extraction of DNA and in vitro packaging were done as described by Blattner et al. (1978). For initial analysis of phages by gel electrophoresis we used a variant of the "mini-lysate" procedure of Cameron et al. (1977). A confluent-lysis plate stock was done on nutrient plates containing agarose (not agar) and harvested in 4 ml of 10 mM Tris buffer containing 0.1 M NaCl. To 0.6 ml of the cleared lysate we added 1 μl of diethyl pyrocarbonate followed by 10 μl of 10% SDS and 75 μl of 0.2 M EDTA in 2 M Tris buffer pH 8.5. After 5 min at 70°C, SDS was precipitated by addition of 0.75 ml of 5 M potassium acetate and removed by centrifugation in an eppendorf microcentrifuge. The supernatant was extracted with phenol, precipitated twice with 2 vols. of ethanol and the precipitate was resuspended in 40 μl of 10 mM Tris buffer with 1 mM EDTA containing 10 μg/ml RNase A. After digestion, 10 μl samples were subjected to electrophoresis. Restriction endonucleases were purchased from New England Biolabs and digestion was carried out according to the manufacturers recommendation. Gel electrophoresis followed the procedure of Shinnick et al. (1975). After construction each vector was analysed with at least four restriction enzymes to confirm its structure. Then a much more detailed numerical restriction map including many sites not specifically tested was predicted by the subsection approach of deWet et al. (1980).

RESULTS

The steps leading to construction of Ch27, Ch28 and Ch30 are shown in Figs. 1–3. The map of wild-type λ is depicted at the top of Fig. 1. The J to p_R region of DNA, which contains no genes essential for growth of λnin5chi phages, is termed the replaceable region. A region termed nin between genes P and Q was deleted in those phages. The object of vector construction is to remove restriction endonuclease sites from the essential region of the phage and introduce them where needed into the replaceable portion. Wild-type λ has five BamHI sites, one in the irreplaceable part of the genome, three in the replaceable portion and one in the nin area. In addition, two of the EcoRI sites of λ and one of the HindIII sites are located in the irreplaceable portion. We have previously described phages (termed NRE) from which the EcoRI and HindIII sites had been removed from the essential region (Blattner et al., 1977). For construction of the new vectors we wished to use this genetic background so that the final result would be suitable for cloning with all three enzymes. This was done in two steps. First, strain BP1 was constructed by a genetic cross between Ch14 and Ch21A to produce the combination of substitutions and deletions, lac5, b1007, imm80, nin5 and the NRE right end (see Fig. 1). This eliminated all the BamHI sites except the one located about 5200 bp from the left end, and preserved the right end with no HindIII or EcoRI sites.

To enrich for spontaneous point mutations eliminating the single remaining BamHI site, we carried out a series of cycles of cutting with BamHI, followed by in vitro packaging, as described in METHODS, to enrich for mutants resistant to BamHI digestion.

After five cycles (in one of which BamHI digestion was less than complete) we found that virtually all phages in the population were devoid of BamHI sites. One of these, BVC4, was chosen for detailed restriction mapping with BglII and EcoRI as well as BamHI. The patterns confirmed the structure as being identical to BP1 except for the missing BamHI site. To complete construction of the insertion vector Ch27, BVC4 was crossed with Ch21A to introduce a single BamHI site into the replaceable region.

Construction of the replacement vector Ch28 involved production of another precursor, termed XDR3.24, which was obtained by crossing Ch13A with Ch10. Ch28 was then constructed as shown in Fig. 2 by crossing BVC4 with XDR3.24 and selecting for am$^+$ lac$^-$ phages. Two isolates from this cross, stocks 2221 and 2222, which had the proper plating phenotypes, were examined by gel electrophoresis. We found that stock 2222 had an insertion of several thousand base pairs in the right arm. Stock 2221 was examined by gel electrophoresis after digestion with BamHI, EcoRI and BglII, as well as all combinations of these, and the fragments expected for Ch28 were all present. Samples of stock 2221 were distributed as special cloning kits to a number of laboratories, whereas a single plaque isolate from this stock was propagated in our laboratory to produce stock 2270, which we used for a series of cloning experiments (Liu et al., 1980).

The total length of DNA in Ch28 is about 39 390 bp which is very close to the minimum size DNA molecule that is normally assumed to be packageable by λ (approx. 38 000 bp) (Bellett et al., 1971). Phage with small genomes, however, are known to be at a growth disadvantage relative to λ with normal genome, leading to a selection for phages with added DNA.

We found subsequently that the phage DNA of stock 2270 carries a substitution in the left arm (Fig. 3). Other stocks isolated and propagated by the recipients of stock 2221 also showed the same or a similar substitution, although some isolates were reported to be normal Ch28 (2221) (Fig. 2). The net increase of DNA caused by the substitution in Ch28 stock 2270 is estimated to be about 900 bp, and this vector is missing the EcoRI site adjacent to the leftmost BamHI site of the vector (Fig. 2). The insert contains a SmaI site but has not been extensively mapped.

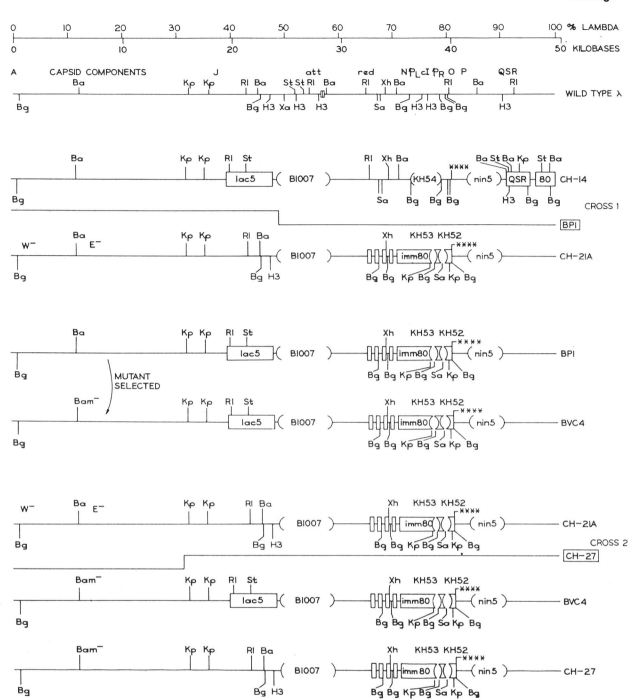

Fig. 1. Construction of Charon 27. The phages and steps used to construct Ch27 are presented beneath the map of bacteriophage λ. Corresponding genes are aligned so that the portions of DNA originating from λ are shown by lines, deletions by parentheses, substitutions by boxes and partially homologous regions by broken boxes. The lengths of substitutions are not to scale. Crosses are described in METHODS. Recombination points are shown by a zigzag line between the two parents of each cross, and the structure of the recombinant selected is presented below the cross. Selection for the *Bam⁻* mutant is described in the text; two-letter abbreviations used for enzymes are: Ba, *Bam*HI; Bg, *Bgl*II; H3, *Hind*III; Kp, *Kpn*I; RI, *Eco*RI; Sal, *Sal*I; St, *Sst*I; Xa, *Xba*I; and Xh, *Xho*I. *att* refers to the phage λ attachment site. **** indicates the position of the *Eco*RI site defining position 40 000 bp on the λ map that was eliminated by point mutation. Symbols p_L and p_R designate λ promoters and B1007 the *B*1007 deletion. For other λ genes see Daniels et al. (1980) and Szybalski and Szybalski (1979).

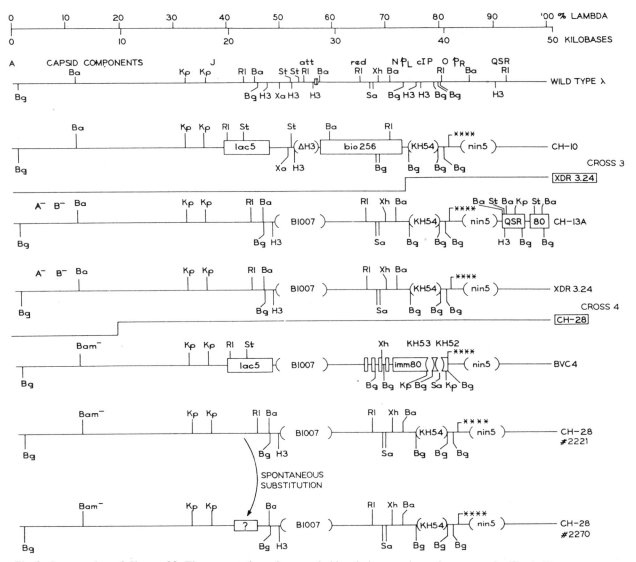

Fig. 2. Construction of Charon 28. The presentation scheme and abbreviations used are the same as for Fig. 1. The spontaneous substitution in stock 2270 is discussed in the text.

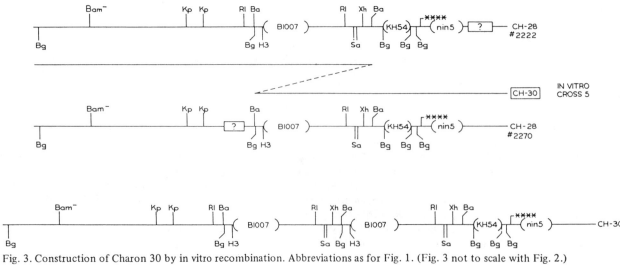

Fig. 3. Construction of Charon 30 by in vitro recombination. Abbreviations as for Fig. 1. (Fig. 3 not to scale with Fig. 2.)

TABLE I

Numerical restriction map

This table shows the numerical restriction maps (number of bp counted from L-end) for Charons 27, 28, and 30. This was predicted from λ maps provided by Donna L. Daniels (personal communication) as explained in the text. L and R indicate left and right termini (ends) of λ DNA.

Ch27		Ch28		Ch30	
Site	bp	Site	bp	Site	bp
L-end	0	L-end	0	L-end	0
*Bgl*II	482	*Bgl*II	472	*Bgl*II	472
*Kpn*I	17 290	*Hpa*I	747	*Hpa*I	747
*Kpn*I	18 812	*Ava*I	4826	*Ava*I	4826
L-vl	19 700	*Hpa*I	5326	*Hpa*I	5326
*Eco*RI	21 569	*Hpa*I	5771	*Hpa*I	5771
*Bam*HI	22 703	*Hpa*I	6015	*Hpa*I	6015
*Bgl*II	22 784	*Hpa*I	8262	*Hpa*I	8262
*Hin*dIII	23 513	*Bcl*I	8894	*Bcl*I	8894
*Bgl*II	28 898	*Bcl*I	9397	*Bcl*I	9397
*Xho*I	29 654	*Hpa*I	11 713	*Hpa*I	11 713
*Bgl*II	29 780	*Pvu*I	12 055	*Pvu*I	12 055
*Kpn*I	32 525	*Bcl*I	13 998	*Bcl*I	13 998
*Bgl*II	33 256	*Kpn*I	17 288	*Kpn*I	17 288
*Sal*I	33 547	*Kpn*I	18 790	*Kpn*I	18 790
R-vl	33 552	L-vl	19 600	L-vl	19 600
*Kpn*I	34 182	*Sma*I	19 660	*Sma*I	19 660
*Bgl*II	35 131	*Sst*II	20 597	*Sst*II	20 597
****	35 146	*Sst*II	20 812	*Sst*II	20 812
R-end	41 783	*Ava*I	21 293	*Ava*I	21 293
		*Sst*II	22 817	*Eco*RI	21 524
		*Hpa*I	23 101	*Sst*II	21 917
		*Bam*HI	23 567	*Hpa*I	22 201
		*Bgl*II	23 656	*Bam*HI	22 667
		*Hin*dIII	24 362	*Bgl*II	22 756
		*Sma*I	28 083	*Hin*dIII	23 462
		*Eco*RI	28 212	*Sma*I	27 183
		*Hpa*I	28 279	*Eco*RI	27 312
		*Hpa*I	26 680	*Hpa*I	27 379
		*Bcl*I	29 198	*Hpa*I	27 780
		*Sal*I	29 235	*Bcl*I	28 298
		*Sal*I	29 739	*Sal*I	28 335
		*Xho*I	29 999	*Sal*I	28 839
		*Bam*HI	30 996	*Xho*I	29 099
		*Hpa*I	31 763	*Bam*HI	30 096
		*Bgl*II	32 215	*Bgl*II	30 184
		*Pvu*I	32 281	*Hin*dIII	30 890
		R-vl	32 296	*Sma*I	34 611
		*Bgl*II	32 511	*Eco*RI	34 740
		*Ava*I	32 622	*Hpa*I	34 807
		*Bgl*II	33 162	*Hpa*I	35 208
		*Bgl*II	33 222	*Bcl*I	35 726
		****	33 576	*Sal*I	35 763
		*Hpa*I	34 016	*Sal*I	36 267
		*Hpa*I	34 243	*Xho*I	36 527

TABLE I (continued)

Ch 27		Ch 28		Ch 30	
Site	bp	Site	bp	Site	bp
		*Sma*I	34 297	*Bam*HI	37 524
		*Sst*II	34 829	*Hpa*I	38 291
		*Bcl*I	35 308	*Bgl*II	38 743
		*Bcl*I	38 077	*Pvu*I	38 809
		*Bcl*I	39 646	R-vl	38 852
		R-end	40 229	*Bgl*II	39 039
				*Ava*I	39 150
				*Bgl*II	39 690
				*Bgl*II	39 750
				****	40 104
				*Hpa*I	40 544
				*Hpa*I	40 771
				*Sma*I	40 825
				*Sst*II	41 357
				*Bcl*I	41 836
				*Bcl*I	44 605
				*Bcl*I	46 174
				R-end	46 757

[a] L-vl and R-vl are the left and right viability limits

[b] **** Location of the *Eco*RI site in gene *O* that has been removed by point mutation

Although the vector Ch28-2270 did not cause any problems in cloning with *Bam*HI, it is unsuitable for cloning large *Eco*RI fragments. Moreover, the uncharacterized insertion makes it difficult to map inserts. We therefore constructed Ch30 by the in vitro cross illustrated in Fig. 3. The result of this cross is a vector whose left arm is from Ch28-2222 and whose right arm is from Ch28-2270. Ch30 carries two copies of the replaceable "stuffer" piece and therefore is much larger than Ch28 (45 840 bp). When Ch30 is propagated no accumulation of aberrant phages is observed. Some excisive recombinations do occur between the two copies of the stuffer piece during propagation leading to the appearance of two bands in CsCl gradients. However, both are equivalent for cloning purposes.

All three of the new vectors, Ch27, Ch28 and Ch30, can be grown to titres of approx. 5×10^{10}/ml. A major advantage of Ch27 and Ch30 is the fact that all the DNA segments in their genomes come from λ. Thus the detailed restriction mapping data that are available for λ (Daniels et al., 1980 and D.L. Daniels and F.R.B., unpublished results) can be used to predict the detailed restriction maps of the vectors. Table I presents such predictions for all three vectors, although the substitution region of Ch28 causes significant uncertainties. The predicted cloning capacities of each vector are shown in Table II. These tables were generated by a computer program described by deWet et al. (1980).

DISCUSSION

One of the most immediate and promising applications of these new vectors is the construction of shotgun libraries of eukaryotic DNA partially digested with *Mbo*I or *Sau*3A. These enzymes recognize the four-base core of the *Bam*HI sequence leaving an identical cohesive end and thus *Sau*3A- or *Mbo*I-cut DNA can be cloned into the *Bam*HI site of the vector. Of course, the four-base recognizing enzymes should cut much more frequently than does *Eco*RI or *Bam*HI. This means that a quite random collection of long fragments can be obtained by partial digestion

TABLE II

Predicted cloning capacities

This table shows predicted cloning capacities assuming the minimum and maximum sizes for λ are 38 000 bp and 51 000 bp, respectively.

Enzymes		Ch27 Maximal size	Minimal size
BamHI	Alone	9217	0
HindIII	Alone	9217	0
EcoRI	Alone	9217	0
SalI	Alone	9217	0
XhoI	Alone	9217	0
BamHI	HindIII	10 027	0
EcoRI	BamHI	10 351	0
BamHI	SalI	20 061	7061
BamHI	XhoI	16 168	3168
EcoRI	HindIII	11 161	0
HindIII	SalI	19 251	6251
HindIII	XhoI	15 358	2358
EcoRI	SalI	21 195	8195
EcoRI	XhoI	17 302	4302
XhoI	SalI	13 110	110

Enzymes		Ch28 Maximal size	Minimal size
BamHI	Alone	18 200	5200
HindIII	Alone	10 771	0
EcoRI	Alone	10 771	0
SalI	Alone	11 275	0
XhoI	Alone	10 771	0
HindIII	EcoRI	14 621	1621
HindIII	SalI	16 148	3148
HindIII	XhoI	16 408	3408
EcoRI	SalI	12 298	0
EcoRI	XhoI	12 558	0
SalI	XhoI	11 535	0

Enzymes		Ch30 Maximal size	Minimal size
BamHI	Alone	19 100	6100
HindIII	Alone	11 671	0
EcoRI	Alone	17 459	4459
SalI	Alone	12 175	0
XhoI	Alone	11 671	0
EcoRI	BamHI	20 243	7243
HindIII	SalI	17 048	4048
HindIII	XhoI	17 308	4308
EcoRI	SalI	18 986	5986
EcoRI	XhoI	19 246	6246
SalI	XhoI	12 435	0

with Sau3A or MboI. Maniatis (1978) has developed a different method for construction of shotgun collections from randomly fragmented DNA using EcoRI linkers to clone HaeIII- or AluI-cut DNA or sheared DNA. However, he had to use several steps, including treatment of the target DNA preparation with EcoRI methylase and trimming with EcoRI after linker attachment, before cloning. With partial Sau3A or MboI digestion (Liu et al., 1980), the procedures are simple and efficient.

Although the goal of developing a universal pair of λ cloning vectors suitable for cloning large and small fragments with all useful enzymes remains in the future, it is clear now such vectors could be derived. The first aim would be to remove all restriction sites from the arms. In Ch30 we have a left and right arm combination devoid of sites for EcoRI, HindIII, BamHI, XbaI, SstI, XhoI and SalI. This DNA has only a single XmaI site which could be removed in the same way as for BamHI. There are two KpnI sites in the Ch30 arms. Calculations show that it should take no more cycles to remove two sites than to remove only one by the method we have used. However, the input phage DNA would have to exceed a milligram to insure the preexistence of a double mutant. Thus, a practical limit may be a removal of no more than two sites in one selecting process.

The second aspect of development of the universal vector set would be the engineering of an adapter DNA segment containing each useful cloning restriction site. This adapter would be placed at each end of the replaceable region in the universal replacement vector and at only one position in the universal insertion vector. The adapter should be in the same orientation at each end of the replaceable region of the replacement vector so that all combinations of restriction enzymes can be used. In addition, the adapter should contain one site in duplicate at each end that could always be used to excise a fragment. Once a vector along these general lines has been constructed we will have a powerful and general tool for DNA cloning, but an endpoint will have been reached that is imposed by λ itself. If it is desired to go beyond this limitation, it will require developments that eliminate, rearrange, or transpose to the host the essential genes of the phage.

ACKNOWLEDGEMENTS

We wish to thank all those recombinant DNA workers who supplied primary support for this line of research through generous monetary donations, in exchange for phage cloning kits. This work has also been supported by grant GM21812 from NIH. This is paper No. 2461 from the Laboratory of Genetics, University of Wisconsin, Madison, WI.

REFERENCES

Becker, A., and Gold, M.: Isolation of the bacteriophage lambda A protein. Proc. Natl. Acad. Sci. USA 72 (1975) 581–585.

Bellett, A.J.D., Busse, H.G. and Baldwin, R.L.: Tandem genetic duplications in a derivative of phage lambda, in A.D. Hershey (Ed.), The Bacteriophage Lambda. Cold Spring Harbor Laboratory, Cold Spring Harbor, NY, 1971, pp. 501–513.

Benton, W.D. and Davis, R.W.: Screening lambda gt recombinant clones by hybridization to single plaques in situ. Science 196 (1977) 180–182.

Blattner, F.R., Williams, B.G., Blechl, A.E., Denniston-Thompson, K., Faber, H.E., Furlong, L., Grunwald, D.O., Kiefer, D.O., Moore, D.D., Schumm, J.W., Sheldon, E.L. and Smithies, O.: Charon phages: safer derivatives of bacteriophage lambda for DNA cloning. Science 196 (1977) 161–169.

Blattner, F.R., Blechl, A.E., Denniston-Thompson, K., Faber, H.E., Richards, J.E., Slightom, J.L., Tucker, P.W. and Smithies, O.: Cloning fetal γ-globin and mouse α-type globin DNA: Preparation and screening of shotgun collections. Science 202 (1978) 1279–1283.

Cameron, J.R., Philippsen, D. and Davis, R.W.: Analysis of chromosomal integration and deletions of yeast plasmids. Nucl. Acids Res. 4 (1977) 1429–1448.

Daniels, D.L., deWet, J.R., Blattner, F.R.: New map of bacteriophage lambda DNA. J. Virol. 33 (1980) 390–400.

deWet, J.R., Daniels, D.L., Shroeder, J.L., Williams, B.G., Denniston-Thompson, K., Moore, D.D. and Blattner, F.R.: Restriction maps for twenty-one Charon vector phages. J. Virol. 33 (1980) 401–410.

Hohn, B. and Murray, K.: Packaging recombinant DNA molecules into bacteriophage particles in vitro. Proc. Natl. Acad. Sci. USA 74 (1977) 3259–3263.

Klein, B. and Murray, K.: Phage lambda receptor chromosomes for DNA fragments made with restriction endonuclease I of *Bacillus amyloliquefaciens* H. J. Mol. Biol. 133 (1979) 289–294.

Liu, C.P., Tucker, P.W., Mushinski, J.F. and Blattner, F.R.: Mapping of heavy chain genes for mouse immunoglobulin M and D. Science 209 (1980) 1348–1353.

Loenen, W.A.M. and Brammar, W.J.: A bacteriophage lambda vector for cloning large DNA fragments made with several restriction enzymes. Gene 10 (1980) 249–259.

Maniatis, T., Hardison, R.L., Lacy, E., Laur, J., O'Connell, C., Quon, D., Sim, G.K. and Efstratiadis, A.: Isolation of structural genes from libraries of eukaryotic DNA. Cell 15 (1978) 687–701.

Miller, H.I. and Friedman, D.I.: An *E. coli* gene product required for λ site specific recombination. Cell 20 (1980) 711–719.

Murray, K. and Murray, N.E.: Phage lambda receptor chromosomes for DNA fragments made with restriction endonuclease I of *Escherichia coli*. J. Mol. Biol. 98 (1975) 551–564.

Murray, N.E. and Murray, K.: Manipulation of restriction targets in phage lambda to form receptor chromosomes for DNA fragments. Nature 251 (1974) 476–481.

Murray, N.E., Brammar, W.J. and Murray, K.: Lambdoid phages that simplify the recovery of in vitro recombinants. Mol. Gen. Genet. 150 (1977) 53–61.

Murray, N.E., Bruce, S. and Murray, K.: Molecular cloning of the DNA ligase gene from T4, II. Expression of the gene and purification of its product. J. Mol. Biol. 132 (1979) 493–505.

Rambach, A. and Tiollais, P.: Bacteriophage λ having *Eco*RI endonuclease sites only in the non-essential region of the genome. Proc. Natl. Acad. Sci. USA 71 (1974) 3927–3930.

Shinnick, T.M., Lund, E., Smithies, O. and Blattner, F.R.: Hybridization of labeled RNA to DNA in agarose gels. Nucl. Acids Res. 2 (1975) 1911–1929.

Sternberg, N., Tiemeier, D. and Enquist, L.: In vitro packaging of a λ*D*am vector containing *Eco*RI DNA fragments of *Escherichia coli* and phage P1. Gene 1 (1977) 255–280.

Szybalski, E.H. and Szybalski, W.: A comprehensive molecular map of bacteriophage lambda. Gene 7 (1979) 217–270.

Thomas, M., Cameron, J.R. and Davis, R.W.: Viable molecular hybrids of bacteriophage λ and eukaryotic DNA. Proc. Natl. Acad. Sci. USA 71 (1974) 4579–4583.

Williams, B.G. and Blattner, F.R.: Construction and characterization of the hybrid bacteriophage lambda Charon vectors for DNA cloning. J. Virol. 29 (1979) 555–575.

Williams, B.G. and Blattner, F.R.: Bacteriophage lambda vectors for DNA cloning. In Setlow, J.K. and Hollander, A. (Eds.), Genetic Engineering, Vol. 2, Plenum, New York, 1980 pp. 201–281.

Communicated by W. Szybalski.

Insertion of a Rabbit β-globin Gene Sequence into an *E. Coli* Plasmid

F. Rougeon, P. Kourilsky and B. Mach

ABSTRACT

        Double stranded DNA has been synthesized in vitro from rabbit globin messenger RNA and elongated with homopolymeric dG tails. An E.coli plasmid was cleaved by EcoRI. The cohesive ends were repaired and dC tails added, to permit reconstitution of the EcoRI sites upon annealing with the dG elongated globin DNA. Transformation of E.coli with the globin-plasmid DNA hybrid has yielded a clone which harbours a recombinant plasmid (pCR1-βG1), as demonstrated by hybridization experiments with radioactive globin cDNA. The sequence carried by the recombinant plasmid corresponds to part of the gene sequence coding for the β chain of rabbit globin. Circular DNA of the purified recombinant plasmid exhibits sensitivity to EcoRI.

INTRODUCTION

        Two possible strategies can be envisaged for the insertion of specific eukaryotic sequences into bacterial or viral DNA :

(1) Fragments of cellular DNA containing whole or part of a given gene can be inserted into a molecular vehicule such as a plasmid or a phage. This manipulation can be accomplished either by pairing the cohesive ends created by certain restriction endonucleases (1-2) or by hybridizing complementary homopolymeric tails synthesized enzymatically (3-4). Random insertion of DNA fragments can be followed by selection or screening for a given sequence among the cloned recombinants. Alternatively, the isolation of a given DNA fragment, or an enrichment, can preceed insertion.

(2) The second possibility is to use a given RNA sequence (cellular or viral) as a template for the in vitro synthesis of a DNA sequence which can then be integrated into a molecular vehicle for the purpose of cloning and amplification.

        We have used double stranded DNA (ds-DNA) synthesized in vitro from a rabbit globin mRNA template for insertion into a bacterial plasmid. In this

ROUGEON, F., KOURILSKY, P. and MACH, B.
Insertion of a rabbit β-globin gene sequence into an *E. coli* plasmid. *Nucleic Acids Res.* 2:2365-2378 (1975).
Reprinted by permission, Oxford University Press.

report, we describe a new E.coli plasmid which carries a sequence specific
of the rabbit gene coding for the β chain of globin.

MATERIALS AND METHODS

Isolation of plasmid DNA. E. coli plasmid pCRI was constructed by
Dr. J. Carbon from plasmid pML2, which is the colEl plasmid carrying the
kanamycine resistant marker (5). As pML2, pCRI can be amplified by the
addition of chloramphenicol,but pCRI has only one site of EcoRI action, while
pML2 has 2. It was kindly provided by Dr. J.D. Rochaix. E. coli K12 harbouring
pCRI was grown in mineral medium supplemented with glucose and casaminoacids
(5). When the density reached 3×10^8 cells per ml, chloramphenicol was added
at a final concentration of 200 µg/ml, and incubation was continued overnight
(5). Cells were then centrifuged and the pellets were eventually stored at -30°C.
Closed circular plasmid DNA was isolated by 2 cycles of density centrifugation
in CsCl-Ethidium bromide gradients (6) followed by centrifugation in a neutral
sucrose gradient (5 to 20 % sucrose in 10 mM Tris-Hcl pH 7.6, 2 mM EDTA,0.1 M NaCl,
4 hours at 37 000 rpm in the SW 41 rotor of a Spinco ultracentrifuge).

Purification of mRNAs. Rabbit 9 S globin mRNA was prepared as
described (7). It was further fractionated into α and β globin mRNA by
electrophoresis in 4.5 % acrylamide gels with 98 % formamide (8-9). The RNAs
were then eluted and rechromatographed on oligodT-cellulose columns as
described elsewhere (10).Cross-contamination is estimated at 10-20%.

Mouse immunoglobulin 14 S L chain mRNA was prepared from MOPC 41
myeloma tumors as previously described (11) with an additional purification in
acrylamide gels (10). The synthesis of MOPC 41 L chain cDNA has also been
described (12).

Enzymes. Avian myeloblastosis virus (AMV) DNA polymerase was a gift
of Dr. J. Beard. Calf thymus deoxynucleotidyl terminal transferase (TdT)
was purified to homogeneity according to Chang and Bollum (13). Single-strand
specific nuclease from Aspergillus oryzae (S 1 nuclease) was purified according
to the procedure of Vogt up to the DEAE cellulose step (14). Eco RI endonuclease
was prepared according to Greene et al (15).E. coli DNA polymerase I (enzyme
A according to Klenow et al (16)) was obtained from Boehringer (Mannheim).

Synthesis of complementary DNA. The reaction mixture contained :
50 mM Tris-HCl pH 8.2, 50 mM KCl, 7.5 mM $MgCl_2$, 1 mM dithiothreitol, 0.2 mM
(each) dATP, dTTP and dGTP, and (^3H)-dCTP (20 Ci/mMole), 50 µg/ml of actinomycin D,
10-50 µg/ml of either 9 S or 14 S mRNA, 1-5 µg/ml of dT_{12-18} (P-L Biochemicals)
and 3 units of AMV DNA polymerase per µg of template. The incubation was for

1 hour at 37°C. After alkaline hydrolysis in 0.3 M NaOH the mixture was filtered on a G 75 Sephadex column (0.5 x 10 cm), in H$_2$O. The excluded material was precipitated with ethanol and fractionated in an alkaline sucrose gradient (12).

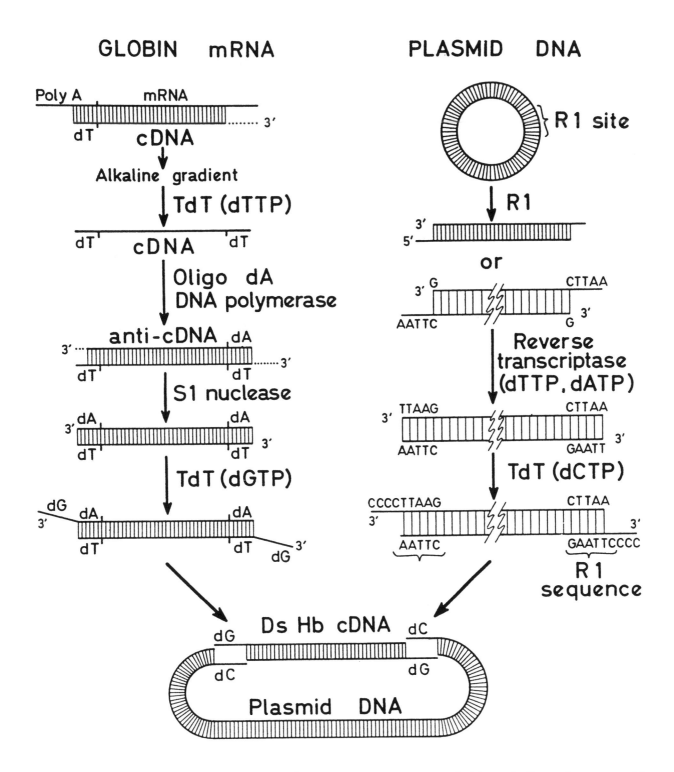

Figure 1. Schematic representation of the approach followed for the annealing of globin DNA to plasmid DNA.

Synthesis of the double stranded DNA. The details of the procedure will be published elsewhere (17). Briefly, unlabelled single-stranded cDNA with a minimum chain length of 200 nucleotides was isolated in a neutral sucrose gradient. About 20-40 dTMP residues on the average were added per molecule by terminal transferase. The dT elongated cDNA was annealed with oligodA$_{10}$ and converted into a duplex structure in a mixture containing 50 mM Tris pH 7.6, 10 mM MgCl2, 20 mM KCl, 1 mM DTT, 0.2 mM (each) dATP, dTTP, dCTP, and dGTP and E. coli DNA polymerase. Incubation was for 30 min at 30 °C. After phenol extraction, the mononucleotides were removed by filtration on a G 75 Sephadex column in H$_2$0 . Control experiments have shown that this procedure results in a well matched, globin mRNA-specific double stranded DNA. It was also observed that the conversion of cDNA into dsDNA is not entirely primer dependent. Both E.coli and AMV DNA polymerases are able to convert cDNA into dsDNA (17).

Elongation of ds globin DNA. Double stranded (ds) globin DNA was incubated for 30 minutes at 45°C with nuclease S 1 in a 0.1 ml mixture containing 0.1 M sodium acetate buffer pH 4.5, 1 mM ZnCl2, 0.2 M NaCl, 0.5 µg of ds-DNA. There was enough S 1 nuclease to digest 1 µg of denatured DNA in 60 minutes at 45°C. The mixture was extracted by chloroform-.isoamylalcohol (24:1) and passed through a G 75 Sephadex column in H$_2$0. Homopolymeric extension of the duplex was achieved by incubation of 0.5 µg of ds- DNA in a 0.05 ml mixture containing : 0.1 M Hepes pH 7.2, 8 mM MgCl$_2$, 1 mM β-mercaptoethanol, 0.2 mM dGTP, and terminal transferase for 1 hour at 37 °C. This resulted in the addition of about 10 residues of dGMP per 3' OH terminus on the average. After S 1 digestion of the ds-DNA, the length of the anti-cDNA strand, determined by electrophoresis in acrylamide gels in 98 % formamide with appropriate molecular weight markers, was heterogeneous, ranging from 100 to 300 nucleotides (17).

Addition of homopolymer to linear pCR1 plasmid DNA molecule. Closed circular pCR1 DNA was converted into linear molecules by incubation in a mixture containing 100 µg per ml of DNA, 50 mM Tris-HCl pH 7.6, 10 mM MgCl$_2$ and Eco RI endonuclease. After 1 hour at 37°C the mixture was made 50 mM in EDTA and extracted with phenol. Linear molecules were isolated by sedimentation at 20 ° in a neutral 5-20 % sucrose gradient (10 mM Tris HCl pH 7.6, 0.1 M NaCl, 1 mM EDTA) for 4 h. at 37.000 rpm in a SW 41 rotor. After ethanol precipitation of the DNA, the 3'OH termini generated by EcoRI were repaired as described (18), using AMV DNA polymerase as the repair enzyme. The incorporation was that expected from complete repair. The mixture was phenol extracted and filtered through a Sephadex G 75 column and the excluded fractions were precipitated with ethanol. The repaired pCR1 (20 µg/ml) in 0.1 M Hepes (pH 7.2), 8 mM

$MgCl_2$, 1 mM β-mercaptoethanol, 0.2 mM dCTP, was incubated in the presence of terminal transferase at 37°C in siliconized glass tubes. The addition of dCMP residues was determined by the measure of the (^{3}H) dCMP incorporated in TCA precipitable material. When 20 monomers per terminus on the average were added, the mixture was phenol extracted and precipitated with ethanol. Control experiments showed that intact circular molecules did not act as primers for the terminal transferase. This suggests that the enzyme does not introduce a significant number of nicks in the duplex plasmid DNA.

Annealing of the pCR1 DNA with globin ds- DNA. The polydC- elongated pCR1 DNA (1 µg/ml) and the poly dG-elongated globin ds-DNA (about 0.01 µg/ml) were annealed in Tris 10 mM (pH 7.6) 0.1 M NaCl, 0.1 mM EDTA for 4 hours at 65°C. Then the mixture was allowed to cool slowly (in about 4 hours) at room temperature.

Transformation. Transformation of C600 (r_k^- m_k^- recBC$^-$) was carried out as described (2). Before plating, the cells (0.3 ml) were incubated with L broth (0.3 ml) for 45 min at 37°C. The plates contained 20 µg/ml of kanamycin. Kanamycin-resistant colonies, visible after overnight incubation at 37°C, were reisolated on kanamycin containing plates, inoculated in broth with 10 µg/ml kanamycin and transferred into mineral medium prior to amplification with chloramphenicol (5). Total DNA was isolated by treatment with proteinase K, phenol extraction and ethanol precipitation. Alternatively, the plasmid was purified from a clear lysate (19).

Hybridization experiments. DNA-DNA hybridization on filters in formamide by a macro or a microtechnique, has been described (20-21). The ^{3}HcDNA had a specific activity of 7.5 x 10^6 cpm/ µg.

For DNA-DNA hybridization in solution, sonicated plasmid or lysate DNA and ^{3}H cDNA were incubated in 0.05 - 0.1 ml 20 mM Tris-HCl (pH 7.6), 0.3 M NaCl, 1 mM EDTA for 4 hours at 65 °C, after which time aliquots were diluted into 2 ml of the buffer used for digestion with nuclease S1 (see above) together with 20 µg of denatured calf thymus DNA. The mixture was divided in two portions for the counting of the total radioactivity and the mesure of the S 1 resistance after incubation with the enzyme for 1 hour at 45°C.

Duplex formation was also determined by chromatography on hydroxylapatite as described (22). The hybridization mixture was diluted in cold 40 mM sodium phosphate buffer and absorbed at 60°C on hydroxylapatite columns (Bio gel HTP, DNA grade, batch 13579). Single strand DNA was eluted with 0.16 M phosphate and double strand DNA with 0.4 M phosphate.(> 95% of input).

RESULTS   1°/ Isolation of a bacterial clone carrying a globin DNA sequence.

Double stranded DNA was synthesized <u>in vitro</u> from purified 9S rabbit globin
mRNA as described in the Materials and Methods section. After treatment with
S1 nuclease (to remove possible single stranded tails in 5') the molecule
was elongated with oligodG by terminal transferase. Purified closed circular
DNA of plasmid pCR1 (derived from <u>col</u> E1 by the adjunction of resistance to
kanamycin and containing only one <u>EcoRI</u> site) was treated with <u>EcoRI</u>. The
cohesive ends were repaired with AMV reverse transcriptase before treatment
with terminal transferase to elongate the molecule with oligodC "tails".

The dG-elongated globin ds-DNA and the dC-elongated plasmid DNA
were annealed in molar ratio estimated to be 1:1 and mixed with $CaCl_2$
treated <u>E. Coli</u> cells for transformation. Kanamycin resistant transformants
were then selected. The dC-elongated plasmid alone should not be able to
transform since it cannot circularize. The addition of the elongated globin
dsDNA should allow circularization and thus restore the transforming ability.
This latter expectation is met with low efficiency since only 5 kanamycin
resistant clones were observed, versus 1 in the absence of globin DNA. These
five clones were analysed - as well as a dozen other clones which originated
from similar experiments and which are presently being studied. In addition,
it is evident from Fig. 1 that the formation of a covalent circular hybrid
molecule should lead to the reconstruction of two <u>EcoRI</u> sites located on
both sides of the globin DNA sequence.

Cultures of these bacterial clones were grown and the amplification
of the plasmid was induced by chloramphenicol. Total DNA was extracted,
denatured, immobilized on nitrocellulose filters, and hybridized with ^{3}H globin
cDNA. The results (Table II) show that one of the 5 clones was positive in
this test. As a control of specificity, the filters were hybridized with ^{3}H
cDNA made from the 14SmRNA coding for the light chain of a mouse immunoglobulin. No hybridization to this heterologous cDNA was observed (Table II). The
plasmid with resulted in a positive hybridization will be referred to as
pCR1-βG1 (see below).

2°) <u>Characterization of plasmid pCR1-βG1</u> : a) increasing amounts of
total DNA extracted from cells carrying either pCR1-βG1 or the parent plasmid
pCR1 were immobilized on nitrocellulose filters and incubated with a constant
amount of ^{3}H cDNA. The results (Table III) show no hybridization to DNA containing the pCR1 plasmid. In contrast, more than 60 % of the ^{3}H cDNA bound
to the filters carrying pCR1-βG1 DNA. Since the radioactive probe was
synthesized from 9S RNA, which is a mixture of the RNAs coding for the α and β
chains of globin, this result suggests extensive hybridization to the
sequences for one of the two chains.

TABLE I. Transformation of E.coli cells by ds globin DNA/pCR1 DNA hybrids

DNA	Input	number of kanamycin-resistant colonies
pCR1 native	0.01 μg	1600
pCR1 cleaved with ECoRI	0.2 μg	0
pCR1-dC	0.2 μg	1
pCR1-dC/ds globin DNA	0.2 μg	5

Transformation was performed as described in Materials and Methods. For each DNA sample, 0.8 ml of $CaCl_2$-treated cells was spread over 4 kanamycin-plates.

TABLE II· Identification of a clone carrying globin DNA by filter hybridization.

source of DNA	cDNA hybridized (cpm)			
	Experiment 1		Experiment 2	
	9S (4hrs)	9S (18hrs)	9S (4hrs)	14S (4hrs)
None	7	11	18	19
pCR1	6	10	18	22
clone ≠ 1	9	8	16	20
2	6	7	22	15
3	480	3000	485	28
4	7	10	20	22 —
5	6	6	18	18

Total DNA was extracted from 10 ml of chloramphenicol treated cells, denatured and immobilized onto 25 mm nitrocellulose filters out of which microfilters (diameter 5 mm) were punched .   Microfilters corresponding to the various isolates were numbered and incubated together, for the times indicated, in 0.5 ml of 50 % formamide, 2xSSc at 37°C with 30 000 cpm of ^{3}H globin cDNA or 20 000 cpm of 14 S cDNA. The backgrounds of the counters were about 5 cpm in experiment 1 and 18 cpm in experiment 2.

b) For additional assays, the pCR1-βG1 plasmid was purified. Aliquots were sonicated, hybridized in liquid with [3]H cDNA and the products of the reaction were analysed by chromatography on hydroxylapatite. Extensive hybridization to the [3]H cDNA is, again, observed (Table IV) which indicates that the globin sequence is indeed carried by the plasmid.

c) It can be predicted that, eventhough the globin ds DNA was made from a mixture of α and β chain mRNA templates, only one of the two sequences has been incorporated into the plasmid. Purified 9S RNA was subjected to electrophoresis through acrylamide gels in the presence of 98 % formamide (Fig.2),

TABLE III. Hybridization of ^{3}H globin cDNA to increasing amounts of DNA
on filters.

Amount of total DNA on filters (micrograms)	cpm hybridized to the total DNA extracted from	
	C600 (pCR1)	C600 (pCR1- βG1)
0	10	10
0.01	9	338
0.025	8	396
0.05	7	435
0.1	8	445
0.2	10	396

DNA-DNA hybridization was carried out on individual microfilters, loaded
with the amounts of DNA indicated, for 18 hours at 37°C in 4 µl of 50 %
formamide, 2xSSC under paraffin oil (21). There was 700 cpm of ^{3}H globin
cDNA per microfilter.

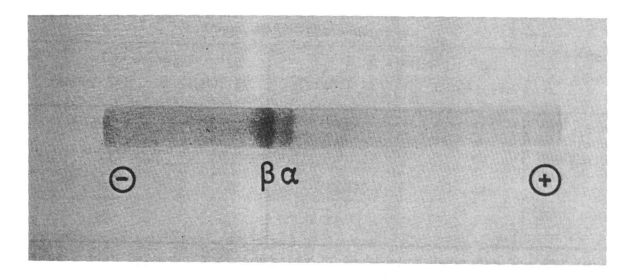

Figure 2.   Acrylamide gel electrophoresis of rabbit 9 S globin mRNA (8 µg)
in 98 % formamide. Staining was with Pyronin G (Serva). The slower and faster
migrating RNA bands have been identified as β and α globin chain mRNA
respectively (9).

and the separated α and β chain mRNAs were recovered. They served as
templates for the synthesis of α and β specific ^{3}H cDNAs, which were used
to analyse pCR1-βG1 further. Hybridization was carried out in liquid, and
the products of the reaction analysed by chromatography on hydroxylapatite.
The results show that pCR1-βG1 carries a sequence specific of the β chain
of rabbit globin mRNA (Table IV). The melting temperature of the hybrids
formed between β globin ^{3}H cDNA and purified pCR1-βG1 DNA was measured
(Fig.3). The observed T_m is high (86°) and compares favourably with

TABLE IV. Analysis of hybrids by chromatography on hydroxylapatite (HAP) and by digestion with nuclease S 1.

| | Unlabelled DNA (or RNA) | | mRNA template for ^{3}H-cDNA | | | |
			globin 9S	α chain	β chain	14 S
Part I HAP	pCR1	0.16M	10131	5450	3860	3998
		0.4 M	600(5.6%)	849(13.9%)	611(13.6%)	197(4.7%)
	pCR1-βG1	0.16M	5086	6183	2952	5820
		0.4 M	3196(38.5%)	1723(21.8%)	5319(64.3%)	1108(15.9%)
Part II Nuclease S1	none	− S1	--	--	879	729
		+ S1	--	--	27(3%)	24(3.2%)
	pCR1	− S1	2355	--	1740	1701
		+ S1	84(3.5%)	--	94(5.4%)	41(2.4%)
	pCR1-βG1	− S1	2448	2346	2334	1561
		+ S1	217(8.8%)	174(7.4%)	471(20.1%)	56(3.5%)
	9 S mRNA	− S1	1680	--	--	--
		+ S1	1330(79%)	--	--	--
	14S mRNA	− S1	--	--	--	1719
		+ S1	--	--	--	896(52%)

Hybridisations (4 hours , 65°C) and assays for hybrid formation were as described in Materials and Methods. In part I, the results are expressed as cpm in the single-strand (0.16M PO_4) and duplex (0.4 M PO_4) fractions and as % of the ^{3}H-cDNA found as duplex. In part II the cpm in controls and in S1 nuclease-treated samples, are presented, with the % of ^{3}H-cDNA resistant to S1.

reported data on the melting of globin cDNA hybridized to homologous cellular DNA (22).

(d) Hybrids formed between pCR1-βG1 and ^{3}H cDNA were also analysed by digestion with nuclease S1. The results of this experiment and the appropriate controls are shown in Table IV. It can be seen that the % of cDNA

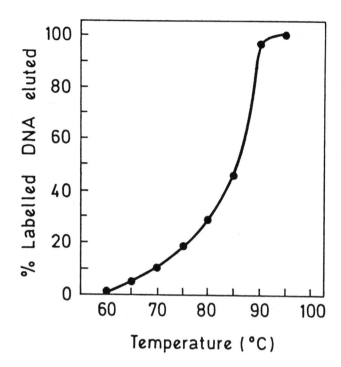

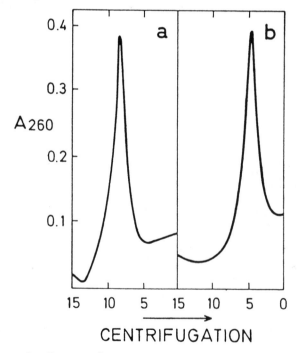

Figure 3. Thermal denaturation of globin β chain ^{3}H-cDNA/plasmid duplexes.
After hybridisation, the sample was loaded onto a hydroxylapatite column in
0.16 M PO$_4$ buffer. After elution of single stranded DNA, the temperature was
raised and elution continued with 0.16 M PO$_4$ buffer. Radioactivity in TCA-
precipitable material was measured in each fraction and expressed as per cent
of the total radioactivity eluted from 65° to 95° (=65% of total ^{3}H-cDNA
applied to the column).

Figure 4. Cleavage of close circular pCR1-βG1 plasmid DNA by EcoR1.
(a). 10 μg of purified pCR1-βG1 plasmid DNA was incubated for 1 hour at 37°
(90 mM Tris HCl, pH 7.6, 10 mM MgSo$_4$) in the presence of EcoR1, made 10 mM
EDTA and centrifuged in a 5-20% sucrose gradient (10 mM Tris-HCl, pH 7.6,
1 mM EDTA, 0.1 M NaCl) at 20° and at 45 000 rpm for 130 min (SW 50.1 Spinco
rotor) (b). Control undigested sample.

found resistant to S1 digestion corresponds to about 1/3 of the radioactivity measured in duplex form with hydroxylapatite. Since the size of the ^{3}H cDNA used corresponds to about 300 to 400 nucleotides (17), it can be estimated that the length of the integrated sequence of globin DNA might be about 100 to 150 nucleotides.

e) Does pCRl-βG1 carry EcoRI sensitive sites ? Purified plasmid DNA was analysed by centrifugation in neutral sucrose gradients with or without prior exposure to the EcoRI nuclease. As shown in Fig.4, treatment with the nuclease causes complete conversion of the rapidly sedimenting closed circular form to the more slowly sedimenting linear form. From this, it can be concluded that pCRl-βG1 carries at least one site sensitive to EcoRI endonuclease.

DISCUSSION

In this report, we describe the integration of a specific mammalian DNA sequence into a bacterial plasmid. Double stranded DNA was synthesized in vitro from purified rabbit globin 9S RNA. The preparation of this globin ds-DNA is described in details elsewhere (17), but it should be mentionned that several steps could not be entirely controlled. The ds-DNA molecules were shorter, on the average, thant the cDNA used as template. This might reflect either uncomplete synthesis of the second DNA strand or endonucleolytic activity in some of the enzymes used (such as the commercial "Klenow" fraction of E.coli DNA polymerase or nuclease S1). A second uncontroled step was the elongation of the globin ds-DNA with terminal transferase. Eventhough the ds-DNA had been treated with S1 nuclease to provide the exposed 3'OH groups necessary for terminal addition, there was no evidence that the latter occured in a symetrical fashion on all molecules.

The strategy for integration was such that the dC-elongated kanamycin resistant plasmid could not form circles unless paired with the dG-elongated globin ds-DNA. Upon transformation of E.coli to kanamycin resistance, this should confer a positive selection in favour of those plasmids in which foreign DNA was inserted. In this experiment, however, the positive transformation events were rare, perhaps because the globin ds-DNA had not been elongated symetrically. We obtained five clones, out of which only one was found to be positive with respect to hybridization with globin ^{3}H cDNA. These clones, as well as other globin-positive clones obtained in other similar experiments are currently under study.

The positive clone was shown to harbour a plasmid (named pCRl-βG1) with the following properties (a).Excess plasmid DNA, in liquid or on filter, hybridizes extensively with ^{3}H globin cDNA. (b). Little hybridization is obtained with ^{3}H cDNA made from globin α chain mRNA, while extensive hybridization is obtained with ^{3}H cDNA made from β chain mRNA, and the Tm of the hybrids formed is high. (c). The DNA of the plasmid can protect about one third of the hybridized ^{3}H cDNA probe from digestion by nuclease S1. (d). The plasmid DNA is susceptible to the EcoRI nuclease.

The specificity of the hybridization reaction was verified in a number of ways. The reaction was carried out under several conditions (in liquid or on filters) and different assays were used. DNA from the parent plasmid pCR1 does not hybridize to ^{3}H globin cDNA. pCR1-βG1 DNA shows little or no hybridization with ^{3}H cDNA made from another eukaryotic mRNA (the 14S RNA of a mouse immunoglobulin light chain) or from globin α chain RNA. In contrast it hybridizes to cDNA made from globin β chain mRNA. The Tm of the hybrids formed (86° in 0.16 M phosphate buffer) is high, showing good matching of the duplex.

The homogeneity of the 9S RNA used as a template for the synthesis of the globin ds-DNA or of the ^{3}H cDNA probe has been estimated by means of RNA-cDNA hybridization kinetics, and found to be higher than 80 % (17). Since over 60 % of total or β globin cDNA could be hybridized to pCR1-βG1, it is unlikely that the sequence incorporated into the plasmid represents the DNA transcript of a contaminant RNA species.

From these observations, we conclude that pCRI-βG1 carries a DNA sequence specific of the β chain of rabbit globin.

The length of this sequence can only be estimated indirectly at this stage. The sequence is longer than that required to ensure a Tm of 86°C in 0.16 M phosphate buffer, as monitored by hydroxylapatite chromatography. The ^{3}H globin cDNA used as a probe has an average size of about 300 to 400 nucleotides. Comparison of hybridization assays with nuclease S1 and hydroxylapatite suggest that pCR1-βG1 carries about one third of the sequence of the probe, or 100 to 140 nucleotides. This figure could be underestimated : the portion of the globin sequence corresponding to the 5' region of the mRNA is not detected in this assay, since the ^{3}H cDNA probe represents less than 2/3 of the length of the 9S mRNA sequence. If the inserted globin ds-DNA has resulted from the replication of longer unlabelled cDNA molecules, part of the sequence carried by the plasmid could remain undetected.

The strategy of integration was devised in such a way as to reconstruct EcoRI sites on both sides of the integrated sequence (Fig.1). Our preliminary data indicate that pCR1-βG1 carries at least one EcoRI site (Fig.4). Work is in progress to determine whether this site lies within the integrated sequence, or whether EcoRI sites have been reconstructed.

The principle of the method described here is general and could be applied to the integration, into plasmid or phage DNA, of other eukaryotic DNA sequences synthesized in vitro from viral or cellular RNAs. An interesting aspect of the procedure is that, starting from a non homogeneous preparation of RNA, one can achieve the purification of a given sequence upon integration and cloning. This is illustrated here by the cloning of a β globin sequence from a mixture of α and β sequences.

An important and obvious limitation of the method which we describe is that only a portion of a given gene, rather than the entire gene and the neighbouring regions, is inserted. However, the fact that milligram amounts of the plasmid carrying a rabbit globin gene sequence can be easily obtained, makes available large amounts of a specific DNA probe. This will increase the sensitivity and feasability of hybridization experiments to detect trace amounts of complementary RNA or DNA sequences, and should be of importance, for instance, in studies of erythroid differenciation or in the analysis of gene expression in native and reconstituted chromatin. Also, it should now be possible to purify large amounts of α and β globin mRNA by preparative hybridization to DNA of the new plasmid. Finally, and, perhaps, most important, the use of the recombinant plasmid might make it possible to purify, by hybridization, the two complementary strands of the rabbit DNA fragments produced by restriction endonucleases and containing globin genes. After reassociation, the specific fragments could be inserted into plasmid or phage DNA for the analysis of the entire gene. It is hoped that the method presented in this paper will help in the analysis of other eukaryotic genes.

ACKNOWLEDGEMENTS

This work was supported, in Geneva, by the Swiss National Science Foundation and, in Paris, by the C.N.R.S. and the D.G.R.S.T.

F.R. is an Eleonor Roosevelt fellow (A.C.S.-U.I.C.C.). We are grateful to Dr. Beard and to Dr. Chirigos for the gift of AMV reverse transcriptase, to Dr. Rochaix for helpful suggestions, to Dr. Chabbert for providing laboratory facilities for some of the experiments, and to Miss A. Bernardin for typing the manuscript. We also thank Dr. P. Vassalli and Dr. F. Gros for stimulating discussions.M. Olgiati, P.A. Briand and D. Gros have provided excellent technical assistance.

* Department of Molecular Biology, Pasteur Institute, Paris, France.

REFERENCES

1 Cohen, S.N., Chang, A.C.Y., Boyer, H.W. and Helling, R.B. (1973) Proc. Nat. Acad. Sci. USA, 70, 3240-3244.

2 Glover, D.M., White, R.L., Finnegar, D.J. and Hogness, D.S. (1975) Cell., 5, 149-155.

3 Jackson, D.A., Symons, R.M. and Berg, P. (1972) Proc. Nat. Acad. Sci. USA, 69, 2904-2909.

4 Wensink, P.C., Finnegan, D.J., Donelson, J.E. and Hogness, D.S. (1974) Cell., 3, 315-325.

5 Herslfield, V., Boyer, H.W., Yanofsky, C., Lovett, M.A. and Helinski, D.R. (1974) Proc. Nat. Acad. Sci. USA, 71, 3455-3459.

6 Sidikaro, J. and Nomura, M. (1975) J. Biol.Chem., 250, 1123-1131.

7 Aviv, H. and Leder, P. (1972) Proc. Nat. Acad. Sci. USA, 69, 1408-1412.

8 Pinder, J.C., Staynov, D.Z., Gratzer, W.B. (1974) Biochemistry 13, 5367-5378.

9 Morrison, M.R., Brinkley, S.A., Gorski, J. and Lingrel, J.B. (1974) J. Biol. Chem., 249, 5290-5295.

10 Farace, M.G., Aellen, M.F., Briand,P.A., Faust, C.H., Vassalli, P. and Mach, B. (1975), submitted.

11 Mach, B., Faust, C.H. and Vassalli, P. (1973) Proc. Nat. Acad. Sci. USA, 70, 451-455.

12 Diggelmann, H., Faust, C.H. and Mach, B. (1973) Proc. Nat. Acad. Sci. USA, 70, 693-696.

13 Chang, L.M.S. and Bollum, F.J. (1971) J. Biol. Chem., 246, 909-916.

14 Vogt, V.M. (1973) Eur. J. Biochem., 33, 192-200.

15 Greene, P.F., Betlach, M., Goodman, H.M. and Boyer, H.W. (1974) in Methods in Molecular Biology, ed. Wickner, R.B. (Publisher, Dekker, M., N.Y.) Vol. 7, 87-111.

16 Klenow, H., Overgaard-Hansen, K. and Patkars, A. (1971) Eur. J. Biochem., 22, 371-381.

17 Rougeon, F. and Mach, B., in preparation.

18 Hedgpeth, J., Goodman, H.M. and Boyer, H.W. (1972) Proc. Nat. Acad. Sci. USA, 69, 3448-3452.

19 Katz, L., Kingsbury, D.T. and Helinski, D.R. (1973) J. Bact., 114, 577-591.

20 Kourilsky, P., Leidner, J. and Tremblay, G.Y. (1971) Biochimie, 53, 1111-1114.

21 Kourilsky, P., Mercereau, O., Gros, D. and Tremblay, G.Y. (1974) Biochimie, 56, 1215-1221.

22 Harrison, P.R., Birnie, G.D., Hell, A., Humphries, S., Young, B.D. and Paul, J. (1974) J. Mol. Biol., 84, 539-554.

Rat Insulin Genes: Construction of Plasmids Containing the Coding Sequences

A. Ullrich, J. Shine, J. Chirgwin, R. Pictet, E. Tischer, W.J. Rutter and H.M. Goodman

Abstract. Recombinant bacterial plasmids have been constructed that contain complementary DNA prepared from rat islets of Langerhans messenger RNA. Three plasmids contain cloned sequences representing the complete coding region of rat proinsulin I, part of the preproinsulin I prepeptide, and the untranslated 3' terminal region of the mRNA. A fourth plasmid contains sequences derived from the A chain region of rat preproinsulin II.

In the 55 years since the first isolation of insulin (1), enormous progress has been made in understanding the role of this hormone in normal glucose homeostasis and in diabetes. Little is known, however, about the control of insulin gene expression in normal and pathological states. Although insulin is composed of two polypeptide chains (A and B), it is the product of a single gene. The immediate precursor of insulin is a single polypeptide, termed proinsulin, that contains the two insulin chains A and B connected by another peptide, C (2). Recently it has been reported that the initial translation product of insulin messenger RNA (mRNA) is not proinsulin itself, but (another precursor preproinsulin) that contains more than 20 additional amino acids on the amino terminus of proinsulin (3). Thus, the structure of the preproinsulin molecule can be represented schematically as NH_2–(prepeptide)–B chain–(C peptide)–A chain–COOH.

To determine the structure of the insulin gene and to study its regulation in normal and pathological states, as well as to investigate the possibility of the synthesis of insulin in an alternate biological system such as bacteria, we have isolated the coding region of the insulin gene by cloning in bacterial plasmids the complementary DNA (cDNA) synthesized in vitro from rat insulin mRNA. We describe here the construction of

these plasmids, and the nucleotide sequence of the cloned DNA. The data show that the plasmids contain the coding region for most of the translated portion of the gene for rat insulin I and the segment coding for the A chain of rat insulin II (4).

The general scheme used for the isolation of bacterial plasmids containing nucleotide sequences coding for rat preproinsulin is illustrated in Fig. 1, and is described below.

Isolation of cDNA complementary to rat islet mRNA. The isolation of insulin mRNA is complicated by the low proportion of the endocrine B cells (which produce insulin) in the pancreas and the high levels of ribonuclease in the dominant acinar cells. These problems have been circumvented by adapting procedures for the relatively large scale isolation of islets of Langerhans (5) (in which the majority of the cells are B cells) from the rat pancreas, and by using a method that allows the extraction of intact, translatable mRNA from sources rich in ribonuclease (6).

Polyadenylated RNA was isolated by chromatography of the total RNA preparation on oligodoxythymidylate-cellulose and transcribed into cDNA with the use of avian myeloblastosis virus (AMV) reverse transcriptase and dT_{12-18} as primer (7), and the RNA removed from the mRNA-cDNA hybrid by alkali treat-

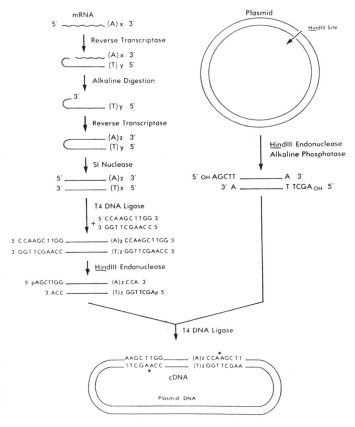

Fig. 1. Schematic diagram for insertion of cDNA into bacterial plasmids with the use of chemically synthesized restriction site linkers. The asterisks in the recombinant plasmid indicate the position where a phosphodiester bond has not formed because of the absence of a 5' terminal phosphate. The example shown here is for the Hind III decanucleotide. In certain cases the Eco RI octanucleotide (5')TGAATTCA(3') was used; in these cases insertion was made into the Eco RI site of the plasmid.

Fig. 2. Analysis of complementary DNA (cDNA) derived from total rat islet of Langerhans mRNA. Avian myeloblastosis virus (AMV) reverse transcriptase (provided by D. J. Beard, Life Sciences, Inc., St. Petersburg, Florida) was used to transcribe total polyadenylated RNA from rat islets of Langerhans into cDNA (7). The reactions were carried out in 50 m*M* tris-HCl, *p*H 8.3, 9 m*M* MgCl$_2$, 30 m*M* NaCl, 20 m*M* β-mercaptoethanol, 1 m*M* unlabeled deoxyribonucleoside triphosphates, 250 μ*M* α-^{32}P-labeled nucleoside triphosphate (specific activity: 50 to 200 c/mole) (each reaction contains three different unlabeled and one labeled triphosphates), 20 μg of oligo(dT$_{12-18}$) (Collaborative Research) per milliliter, 100 μg of polyadenylated RNA per milliliter, and 220 units of reverse transcriptase per milliliter. The mixture was incubated at 45°C for 15 minutes. After addition of the disodium salt of EDTA (EDTA-Na$_2$) to a concentration of 25 mmole/liter, the solution was extracted with an equal volume of water-saturated phenol, followed by chromatography of the aqueous phase on a Sephadex G-100 column (0.3 by 10 cm) in 10 m*M* tris-HCl, *p*H 9, 100 m*M* NaCl, 2 m*M* EDTA. Nucleic acid eluted in the void volume was precipitated with ethanol after addition of ammonium acetate, *p*H 6.0, to a concentration of 0.25*M*. After centrifugation, the pellet was dissolved in 50 μl of 0.1*M* NaOH (freshly prepared) and the RNA was hydrolyzed at 70°C for 20 minutes. Sodium acetate (1*M*, *p*H 4.5) was added for neutralization, and the ^{32}P-labeled cDNA was precipitated with ethanol and dissolved in water. This cDNA was 5 to 8 percent resistant to digestion by S1 nuclease at 45°C (13). Synthesis of the second strand was performed in 50 m*M* tris-HCl, *p*H 8.3, 9 m*M* MgCl$_2$, 10 m*M* dithiothreitol, 500 μ*M* each of the three unlabeled deoxyribonucleoside triphosphates, 100 μ*M* α-^{32}P-labeled deoxynucleoside triphosphate (specific activity: 1 to 10 c/mmole), 50 μg of cDNA per milliliter, and 220 units of reverse transcriptase per milliliter at 45°C for 120 minutes. The reaction was stopped by addition of EDTA-Na$_2$ to a concentration of 25 m*M*; the mixture was extracted with phenol and subjected to chromatography on Sephadex G-100, and the nucleic acids were precipitated by ethanol. Portions (5,000 to 10,000 count/min) of single- or double-stranded cDNA were digested with an excess of restriction endonuclease Hae III (9) and analyzed by electrophoresis on polyacrylamide slab gels (8). The gels were dried, and the ^{32}P-labeled DNA was detected by autoradiography with the use of Kodak No-Screen (NS-2T) film. (A) Lane 1, total single-stranded cDNA transcribed from mRNA isolated from islets of Langerhans from the rat pancreas; lane 2, Hae III cleavage pattern of this cDNA; and lane 3, single-stranded Hae III fragments of M13 DNA. (B) Lane 1, double-stranded cDNA from mRNA isolated from islets of Langerhans from the rat pancreas; and lane 2, Hae III fragments of this cDNA. Sizes were determined with unlabeled Hae III fragments of SV40 DNA as markers (28).

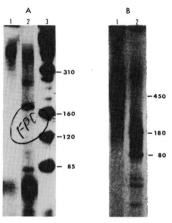

ment. This cDNA was heterodisperse, as judged by polyacrylamide gel electrophoresis (8). It contained at least one predominant cDNA species of about 450 nucleotides (Fig. 2A, lane 1), which gave rise to two prominent bands on cleavage with the restriction endonuclease Hae III (Fig. 2A, lane 2) (9). The single-stranded cDNA was converted to the double-stranded form with AMV reverse transcriptase, labeled deoxynucleoside triphosphates, and the self-priming ability of single-stranded cDNA (see Fig. 1) (7). The product of this reaction is known to consist of a long hairpin loop structure (10). Digestion of this DNA (Fig. 2B, lane 1) by Hae III restriction endonuclease gave fragments (Fig. 2B, lane 2) corresponding to those obtained by digestion of the single-stranded cDNA. It seemed likely that the Hae III fragments were derived from insulin cDNA because of their prominence in the total cDNA preparation. Therefore, these fragments as well as the complete cDNA preparations were used in the cloning experiments.

Molecular cloning of cDNA fragments. We have developed an improved method of cloning cDNA into bacterial plasmids that involves the ligation of chemically synthesized restriction site linkers to cDNA and then cleavage with the appropriate restriction endonuclease to produce cDNA molecules with cohesive termini for ligation to similarly cleaved plasmid DNA. This approach, illustrated in Fig. 1, was first used for the

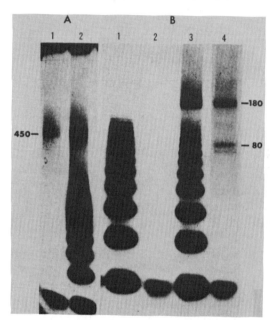

Fig. 3. Blunt-end ligation of Hind III decanucleotide linkers to rat islet cDNA. Double-stranded cDNA (2 to 5 μg/ml) was treated with 30 units of S1 nuclease (1200 units per milliliter; Miles) in 0.03*M* sodium acetate, *p*H 4.6, 0.3*M* NaCl, 4.5 m*M* ZnCl$_2$ at 22°C for 30 minutes followed by 15 minutes at 10°C. Addition of tris base to 0.1*M*, EDTA to 25 m*M*, and *Escherichia coli* transfer RNA to 50 μg/ml was used to stop the digestion. After phenol extraction of the reaction mixture and Sephadex G-100 chromatography, the excluded ^{32}P-labeled cDNA was precipitated with ethanol. This treatment results in a high yield of cDNA molecules with base-paired ends necessary for the blunt-end ligation to chemically synthesized decanucleotides. Double-stranded cDNA that was not treated with S1 nuclease but fragmented by Hae III endonuclease to generate blunt-ended fragments (9) was also ligated as follows. Ligation of Hind III decamers to cDNA was carried out by incubation at 14°C in 66 m*M* tris-HCl, *p*H 7.6, 6.6 m*M* MgCl$_2$, 1 m*M* ATP, 10 m*M* dithiothreitol m*M*, 3 μ*M* 5'-^{32}P-labeled Hind III decamers (10^5 count/min per picomole), and T4 DNA ligase (~ 500 units per milliliter) for 1 hour. The reaction mixture was then heated to 65°C for 5 minutes to inactivate the ligase; KCl (to 50 m*M*), β-mercaptoethanol (to 1 m*M*), and EDTA (to 0.1 m*M*) were then added, and the mixture was digested with 150 units of Hind III endonuclease per milliliter for 2 hours at 37°C. (A) Lanes 1 and 2, Hind III decamer ligation to total cDNA derived from rat islets of Langerhans RNA before (lane 2) and after (lane1) cleavage with Hind III endonuclease. (B) Lanes 1 and 2, Hind III decamer ligation without added cDNA before (lane 1) and after (lane 2) Hind III endonuclease digestion. Lanes 3 and 4, Hind III decamer ligation to Hae III fragments of total cDNA derived from rat islets of Langerhans RNA before (lane 3) and after (lane 4) Hind III endonuclease digestion. Autoradiograph B has been enlarged in order to show ligation of multiple Hind III decamers to the 180 base pair Hae III fragment. SV40 DNA digested by Hae III has been used to determine the sizes of the cDNA Hae III fragments (28).

cloning of synthetic DNA fragments (*11*) and makes use of the ability of T4 DNA ligase to catalyze the joining of blunt-ended DNA molecules (*12*).

Since the double-stranded cDNA molecules generated by self-priming of the single strand are in the form of long hairpin structures, it was first necessary to remove the hairpin and any non-base-paired regions, so that the final product would be a perfectly base-paired duplex DNA. This was achieved by digestion of the double-stranded cDNA with the single-strand specific nuclease Sl (*13*); it was not necessary for the cloning of the cDNA–Hae III fragments since Hae III cleavage itself produces molecules with base-paired ends (*9*). The ligation of blunt-ended cDNA to an excess of the Hind III restriction site decamer (5')CCAAGCTTGG(3') (*14*) is shown in Fig. 3 (C. cytosine; A. adenine; G. guanine; T. thymine). This decamer was chosen since neither the major com-

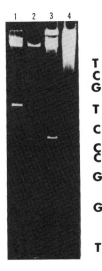

Fig. 4. Restriction endonuclease analysis of the isolated recombinant plasmids. Crude plasmid preparations (2 to 5 μg) were digested with an excess of the appropriate restriction endonuclease (*19a*). EDTA-Na$_2$ (10 mM) and sucrose (10 percent final concentration) were then added, and the mixture was subjected to electrophoresis on an 8 percent polyacrylamide gel. Hind III digest of pAU-1 (lane 1), Eco RI digests of pAU-2 (lane 2), pAU-3 (lane 3), and pAU-4 (lane 4), respectively. The stained material at the bottom of lanes 2 and 3 is RNA that is still present in the crude extract.

Fig. 5. Autoradiogram of a sequence gel for pAU-2. The 180 base pair DNA fragment released by Hind III digestion of pAU-2 was isolated by electrophoresis in an 8 percent polyacrylamide gel. After being stained with ethidium bromide, the cloned DNA fragment was excised and eluted electrophoretically. After elution, the 5' termini of the fragment were labeled with [γ-^{32}P]ATP and polynucleotide kinase (*21*). The terminally labeled DNA was then digested with Alu I endonuclease (*14*) for 4 hours at 37°C, and the two labeled products were separated by electrophoresis in an 8 percent polyacrylamide gel. After elution from the gel, each of the DNA fragments was subjected to the base-specific chemical reactions described by Maxam and Gilbert (*21*). The products of each of the four reactions (G', A', T', C') were separated by electrophoresis in a 20 percent polyacrylamide gel containing 7M urea, and fragments labeled at the 5' terminus were visualized by autoradiography. Identical samples from each reaction were placed on the gels and run for 36, 24, and 12 hours respectively, in the order 3 to 1. The nucleotide sequence of the larger fragment generated by Alu I digestion can be read directly from the autoradiogram as shown on the left figure. The 5' terminal pentanucleotide sequence (5')CTTGG is derived from the 3'-end of the Hind III linker. The 5'-terminal dinucleotide (pAG) has been run off the gel.

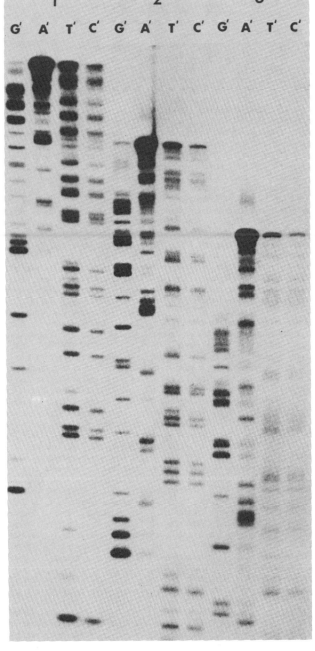

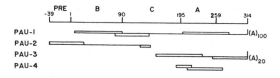

Fig. 6. Schematic diagram of rat preproinsulin mRNA and the corresponding cloned DNA fragments. The boxed areas on each DNA fragment represent the regions directly sequenced and correspond to the italic type in Fig. 7.

ponent of the total islet cDNA nor the two Hae III fragments were themselves cleaved by Hind III endonuclease (data not shown). The self-ligation of the decamers is shown in Fig. 3B, lane 1, and the ligation of multiple decamers to full-length cDNA and to cDNA cleaved by Hae III endonuclease are shown in Fig. 3A, lane 2, and Fig. 3B, lane 3, respectively. Treatment of the ligation mixtures with Hind III endonuclease yields duplex cDNA with Hind III single-stranded cohesive termini (Fig. 3A, lane 1; Fig. 3B, lane 1) as well as the cleaved decanucleotide (Fig. 3B, lane 2). Since the cleaved decamers also contained Hind III termini, and hence would compete with cDNA for ligation to the plasmid, the cDNA was purified by polyacrylamide gel electrophoresis (8) before ligation to the plasmid.

We used the bacterial plasmid pMB9, a 3.5-million-dalton molecule containing single Hind III and Eco RI sites (15). Infection of *Escherichia coli* with pMB9 confers tetracycline resistance and colicin immunity. Since the Hind III site is localized within the promoter for the gene responsible for tetracycline resistance, not all bacteria containing a plasmid formed by insertion of DNA in the Hind III site are tetracycline-sensitive, but some retain variable levels of tetracycline resistance (16).

To ensure ligation of most of the cDNA molecules to plasmid DNA, it is necessary to add a molar excess of the plasmid DNA. However, this results in the majority of the plasmid DNA circularizing without an inserted cDNA fragment, and thus the subsequently transformed cells contain mainly pMB9 and not the recombinant plasmids. In order to reduce the number of colonies to be screened for recombinant plasmids, we first treated the Hind III–cut pMB9 DNA with alkaline phosphatase (see Fig. 1) (17). This removes the 5′ terminal phosphates from the Hind III endonuclease-generated ends of the plasmid and prevents self-ligation of the plasmid DNA, ensuring that circle formation (and hence transformation) is dependent on the insertion of a DNA fragment containing 5′-phosphorylated termini.

The mixtures containing recombinant cDNA–pMB9 were used to transform the EK2 host *E. coli* χ1776 (18). Transformants were selected by growth on medium containing tetracycline (16, 19) and screened for recombinant plasmids (19a). One combinant plasmid (pAU-1) obtained by transformation with the total rat islet cDNA contained an inserted DNA fragment approximately 410 nucleotides in length, which was released from the plasmid by Hind III endonuclease digestion (Fig. 4, lane 1). This cloned DNA fragment, which hybridized to rat islet cDNA (data not shown), was isolated and subjected to DNA sequence analysis (see below). Transformation with the electrophoretically purified Hae III fragments (Fig. 3b, lane 4) did not yield any clones in the Hind III site of pMB9 (16). The Hind III termini were therefore converted to Eco RI cohesive termini. This was achieved by "filling in" the Hind III ends using reverse transcriptase, followed by blunt-end ligation to the Eco RI restriction site octanucleotide (5′)TGAATTCA(3′) (20). Cleavage of this duplex cDNA with Eco RI yields cohesive termini, thus allowing insertion of the fragment into the Eco RI site of pMB9 (20). Transformations by means of the electrophoretically purified Hae III fragments treated in this manner yielded a number of clones, all of which contained insertions of foreign DNA at the Eco RI site (16). The size of these insertions corresponded to that of the original Hae III fragments used for ligation to the plasmid DNA. Two recombinant plasmids, pAU-2 and pAU-3, obtained by transformation with the approximately 80-base-pair Hae III fragments (Fig. 3B, lane 4) and one, pAU-4, from transformation with the smaller Hae III fragment (80 base pairs; Fig. 3B, lane 4) were isolated (Fig. 4, lanes 2, 3, and 4, respectively).

Sequence analysis of cloned DNA. Purified plasmid DNA from pAU-1 was cleaved with endonuclease Hind III and the 410-base-pair insertion isolated by electrophoresis on a 6 percent polyacrylamide gel (similar to those shown in Fig. 4). After elution from the gel, the DNA was labeled at the 5′ termini by incubation with [γ-^{32}P]ATP and polynucleotide kinase (21). The labeled DNA was cleaved with Hae III endonuclease and the two labeled fragments (about 265 and 135 base pairs) separated on a polyacrylamide gel. The isolated fragments were subjected to the base specific cleavage reactions developed by Maxam and Gilbert (21). The cleaved DNA was separated on a 20 percent polyacrylamide gel containing 7M urea and analyzed (21). By these methods, the nucleotide sequence of the DNA can be read directly from the gel (for example, Fig. 5).

The nucleotide sequences of the smaller cloned DNA fragments from plasmids pAU-2 (Fig. 5), pAU-3, and pAU-4 were similarly determined after isolation of the inserted DNA fragments from a polyacrylamide gel.

Rat insulin gene sequences are contained in the DNA clones. The determined nucleotide sequences of cloned DNA's from plasmids pAU-1, -2, -3 are clearly overlapping (Figs. 6 and 7) and thus represent regions of the same molecule. The reconstructed sequence is a continuous stretch of 354 nucleotides, terminated at one end by varying lengths of poly(dA · dT). The 5′ portion of the mRNA is not established by these experiments (22). Since the mRNA contains a terminal poly(A) sequence, the DNA strand containing 3′ terminal poly(dA) is of the same sense. This strand determines an amino acid sequence which exactly corresponds to the entire coding region for rat proinsulin I and 13 out of 23 amino acids of the prepeptide sequence. The nucleotide sequences of the mRNA's therefore confirm the previously determined amino acid sequence for rat proinsulin I (4), and in addition show that the C peptide is connected to the B chain by the sequence -Arg-Arg- (arginine-arginine) and to the A chain by a -Lys-Arg-(lysine-arginine) sequence. Previous data did not provide assignments for the basic arginine or lysine residues connecting the B-C-A peptides, although they had been inferred by analogy with the bovine, porcine, and human proinsulin sequences (4).

The portion of the amino acid sequence of the prepeptide determined from the available mRNA sequence (Fig. 8) is consistent with the partial amino acid sequence reported by Chan *et al.* (3). In agreement with these workers we find leucine at positions −9, −11, and −12, but we find alanine at −13, where they had made a tentative assignment of lysine on the basis of a single sequencer run. The mRNA sequence also estab-

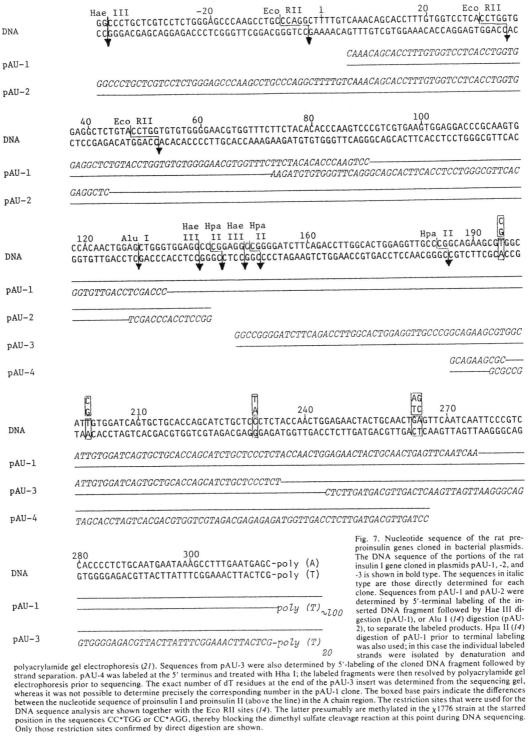

Fig. 7. Nucleotide sequence of the rat pre-proinsulin genes cloned in bacterial plasmids. The DNA sequence of the portions of the rat insulin I gene cloned in plasmids pAU-1, -2, and -3 is shown in bold type. The sequences in italic type are those directly determined for each clone. Sequences from pAU-1 and pAU-2 were determined by 5'-terminal labeling of the inserted DNA fragment followed by Hae III digestion (pAU-1), or Alu I (*14*) digestion (pAU-2), to separate the labeled products. Hpa II (*14*) digestion of pAU-1 prior to terminal labeling was also used; in this case the individual labeled strands were isolated by denaturation and polyacrylamide gel electrophoresis (*21*). Sequences from pAU-3 were also determined by 5'-labeling of the cloned DNA fragment followed by strand separation. pAU-4 was labeled at the 5' terminus and treated with Hha I; the labeled fragments were then resolved by polyacrylamide gel electrophoresis prior to sequencing. The exact number of dT residues at the end of the pAU-3 insert was determined from the sequencing gel, whereas it was not possible to determine precisely the corresponding number in the pAU-1 clone. The boxed base pairs indicate the differences between the nucleotide sequence of proinsulin I and proinsulin II (above the line) in the A chain region. The restriction sites that were used for the DNA sequence analysis are shown together with the Eco RII sites (*14*). The latter presumably are methylated in the χ1776 strain at the starred position in the sequences CC*TGG or CC*AGG, thereby blocking the dimethyl sulfate cleavage reaction at this point during DNA sequencing. Only those restriction sites confirmed by direct digestion are shown.

```
                                                                          1
    -----Ala Leu Leu Val Leu Trp Glu Pro Lys Pro Ala GlN Ala Phe Val Lys GlN His Leu Cys
         GCC CUG CUC GUC CUC UGG GAG CCC AAG CCU GCU CAG GCU UUU GUC AAA CAG CAC CUU UGU


         10                                              20                              30
    Gly Pro His Leu Val Glu Ala Leu Tyr Leu Val Cys Gly Glu Arg Gly Phe Phe Tyr Thr Pro Lys Ser Arg Arg
    GGU CCU CAC CUG GUG GAG GCU CUG UAC CUG GUG UGU GGG GAA CGU GGU UUC UUC UAC ACA CCC AAC UCC CGU CGU


                                    40                              50
    Glu Val Glu Asp Pro GlN Val Pro GlN Leu Glu Leu Gly Gly Gly Pro Glu Ala Gly Asp Leu GlN Thr Leu Ala
    GAA GUG GAG GAC CCG CAA GUG CCA CAA CUG GAG CUG GGU GGA GGC CCG GAG GCC GGG GAU CUU CAG ACC UUG GCA


         60                              70                              80
    Leu Glu Val Ala Arg GlN Lys Arg Gly Ile Val Asp GlN Cys Cys Thr Ser Ile Cys Ser Leu Tyr GlN Leu Glu
    CUG GAG GUU GCC CGG CAG AAG CGU GGC AUU GUG GAU CAG UGC UGC ACC AGC AUC UGC UCC CUC UAC CAA CUG GAG
                                        C       C                           U
```

```
AsN Tyr Cys AsN
AAC UAC UGC AAC UGA GUUCAAUCAAUUCCCGAUCCACCCCUCUGCAAUGAAUAAAGCCUUUGAAUGAGC-poly A
                UAG
```

Fig. 8. Nucleotide sequence of rat preproinsulin I mRNA. The nucleotide sequence of the mRNA was deduced from the composite sequences of the cloned DNA fragments as shown in Fig. 7. The amino acid sequence was predicted from the mRNA sequence and agrees with that previously published for rat proinsulin I (4). The amino acid sequence of the prepeptide (−1 to −13) determined from the nucleotide sequence of pAU-2 was previously unknown except for amino acids −1 to −13, which are identical to previously published data except for −10 [see text and (3)]. The underlined sequence (AAUAAA) is that previously found in the 3' untranslated region of eukaryotic mRNA's (27). The sequences underlined with a broken line have not been fully confirmed. The sequence shown is for preproinsulin I mRNA; nucleotide changes observed in the A chain region of proinsulin II are boxed below the line.

lishes the amino acid sequence at residue −10 and −8 to −1, which have not been previously determined, thereby demonstrating the utility of determination of the amino acid sequence via the sequence of mRNA or its cDNA.

The junction of the prepeptide with rat proinsulin I is at an alanine-phenylalanine (Ala-Phe) bond (Fig. 8). This bond must be enzymatically cleaved to form proinsulin. A protease specific for the Ala-Phe bond has not yet been identified, but it is noteworthy that all of the exocrine secretory proteins (that is, digestive enzymes) whose prezymogens have been partially sequenced from dog pancreas have an Ala-Phe sequence near the end of the prepeptide (23). In addition, 15 percent of all of the dipeptide sequences in the precursor portions are either Ala-Phe or Ala-Tyr; thus a protease specific for these sequences could rapidly remove and degrade the prepeptides from the secretory proteins (24). This processing mechanism is consistent with the proposed common origin of exocrine and endocrine cells (25).

Previous studies have demonstrated the presence of two types of proinsulin (I and II) in the rat (4). Rat proinsulin II has four amino acids (two in the B chain and two in the C chain) that differ from the amino acid sequence of proinsulin I. No clones have thus far been obtained that encompass these regions of proinsulin II.

The cloned cDNA fragment in pAU-4, however, may represent the A chain region of proinsulin II mRNA. The original cDNA fragment which gave rise to this clone was clearly transcribed from the A chain region of insulin mRNA. Since the amino acid sequence of the A chain of insulins I and II are identical but the nucleotide sequence of this fragment is slightly different from the corresponding region of proinsulin I mRNA, we infer that it is derived from proinsulin II mRNA. There are three base changes in the amino acid coding portion of the gene in the proinsulin II clone; all of these involve interconversion of C and T residues in the third position of the codons (Fig. 8). In one case such a change has produced an Hha I restriction site (nucleotide 195, Fig. 7). A further difference is found in the termination codons; for rat proinsulin II it is UAG, whereas for proinsulin I, it is UGA (26) (U, uracil).

An examination of the coding regions of the mRNA's (Fig. 8) suggests there may be some bias in the use of particular codons; for example, for glutamic acid (two possible codons). GAG is used in 8 out of 11 occurrences; for leucine (six codons), CUG is used 9 out of 16 times, and UUA and CUA are not used; for tyrosine (two codons), UAC is used 6 out of 6 times; and for valine (four codons), GUG is used 6 out of 9 times.

The sequence AAUAAA (nucleotides 295 to 300, Fig. 8), which has been reported to occur in many eukaryotic mRNA's in a similar position in the 3' untranslated region (27), is also present in the corresponding region of rat preproinsulin I mRNA.

Our results demonstrate the utility of molecular cloning in analyzing gene structures. These clones containing insulin gene segments should be very useful for studying regulation of insulin mRNA biosynthesis and its expression in bacteria.

AXEL ULLRICH, JOHN SHINE
JOHN CHIRGWIN, RAYMOND PICTET
EDMUND TISCHER, WILLIAM J. RUTTER
HOWARD M. GOODMAN
Department of Biochemistry and Biophysics, University of California, San Francisco, 94143

References and Notes

1. F. G. Banting and C. H. Best, *J. Lab. Clin. Med.* 7, 251 (1922).
2. D. F. Steiner, D. Cunningham, L. Spigelman, B. Aten, *Science* 157, 697 (1967).
3. S. J. Chan, P. Keim, D. F. Steiner, *Proc. Natl. Acad. Sci. U.S.A.* 73, 1964 (1976); P. T. Lomedico and G. F. Saunders, *Nucleic Acids Res.* 3, 381 (1976).
4. There are two nonallelic genes, I and II, for insulin in the rat. L. F. Smith, *Am. J. Med.* 40, 662 (1966); J. L. Clark and D. F. Steiner, *Proc. Natl. Acad. Sci. U.S.A.* 62, 278 (1969); J. Markussen and F. Sundby, *Eur. J. Biochem.* 25, 153 (1972).
5. The procedure of P. E. Lacy and M. Kostianovsky, [*Diabetes* 16, 35 (1967)], was modified by digesting each pancreas with 12 mg of collagenase (Worthington CLS IV) plus 1 mg of soybean trypsin inhibitor. Islets were isolated on Ficoll gradients as described by D. W. Scharp, C.

B. Kemp, M. J. Knight, W. F. Ballinger. P. E. Lacy [*Transplantation* **16**, 686 (1973)].

6. Islets were homogenized in 4*M* guanidinium thiocyanate (Tridom/Fluka). 1*M* mercaptoethanol, buffered to *p*H 5.0 (J. M. Chirgwin, A. Pryzybyla, W. J. Rutter, in preparation). The homogenate was layered over 5.7*M* cesium chloride, and the RNA was sedimented as described by V. Glisin, R. Crkvenjakov, C. Byus. [*Biochemistry* **13**, 2633 (1974)].

7. H. Aviv and P. Leder, *Proc. Natl. Acad. Sci. U.S.A.* **69**, 1408 (1972); A. Efstratiadis, F. C. Kafatos, A. M. Maxam, T. Maniatis, *Cell* **7**, 279 (1976).

8. C. W. Dingman and A. C. Peacock, *Biochemistry* **7**, 659 (1968).

9. K. Horiuchi and N. D. Zinder, *Proc. Natl. Acad. Sci. U.S.A.* **72**, 2555 (1975); R. W. Blakesley and R. D. Wells, *Nature (London)* **257**, 421 (1975).

10. W. A. Salser, *Ann. Rev. Biochem.* **43**, 923 (1974).

11. H. L. Heyneker, J. Shine, H. M. Goodman, H. W. Boyer, J. Rosenberg, R. E. Dickerson, S. A. Narang, K. Itakura, S. Lin, A. D. Riggs, *Nature (London)* **263**, 748 (1976).

12. V. Sgaramella, J. H. Van de Sande, H. G. Khorana, *Proc. Natl. Acad. Sci. U.S.A.* **67**, 1468 (1970); A. Sugina, N. Cozzarelli, H. Heyneker, J. Shine, H. Boyer, H. M. Goodman, *J. Biol. Chem.*, in press.

13. V. M. Vogt, *Eur. J. Biochem.* **33**, 192 (1973).

14. R. H. Scheller, R. E. Dickerson, H. W. Boyer, A. D. Riggs, K. Itakura, *Science* **196**, 177 (1977); Hsu I endonuclease was used interchangeably with Hind III; R. J. Roberts, *Crit. Rev. Biochem.* **3**, 123 (1976).

15. R. L. Rodriguez, F. Bolivar, H. M. Goodman, H. W. Boyer, M. Betlach, in *ICN/UCLA Symposium on Molecular Mechanisms in the Control of Gene Expression*, D. P. Nierlich, W. J. Rutter, C. F. Fox, Eds. (Academic Press. New York, 1976), pp. 471–477.

16. Insertion of DNA fragments into the Hind III site of pMB9 reduces the level of tetracycline resistance of cells carrying such recombinant plasmids to varying degrees dependent on the sequences cloned into this site [R. L. Rodriguez, R. Tait, J. Shine, F. Bolivar, H. Heyneker, M. Betlach, H. W. Boyer in *Tenth Annual Miami Winter Symposium* (Academic Press, New York, in press)]. We have previously observed that the insertion of DNA molecules containing poly dA · dT regions allows the expression of

tetracycline resistance at reduced levels (5 to 10 μg/ml). Screening for recombinant plasmids was therefore carried out at 5 μg of tetracycline per milliliter for transformation into the Hind III site, and 20 μg of tetracycline per milliliter for transformation into Eco R1 site.

17. Plasmid DNA cleaved by Hind III or Eco RI was treated with bacterial alkaline phosphatase (Worthington, BAPF, 0.1 unit per microgram of DNA) at 65°C in 25 m*M* tris-HCl, *p*H 8, for 30 minutes, followed by phenol extraction to remove the phosphatase. After precipitation by ethanol, the phosphatase-treated plasmid DNA was added to cDNA (containing Hind III or Eco RI cohesive terminals) at a molar ratio of 3 : 1 (plasmid : cDNA). The mixture was incubated in 66 m*M* tris, *p*H 7.6, 6.6 m*M* MgCl₂, 10 m*M* dithiothreitol, 1 m*M* adenosine triphosphate (ATP) for 1 hour at 14°C in the presence of 50 units of T4 DNA ligase per milliliter. The ligation mixture was added directly to the χ1776 cells for transformation.

18. A modification (W. Salser, personal communication) of the transformation procedure originally provided to all recipients of χ1776 by R. Curtiss III was used. The work with this strain was done in a P3 physical containment facility in the EK-2 host-vector system *E. coli* χ1776–pMB9 in compliance with the NIH guidelines for recombinant DNA research.

19. F. Bolivar, R. L. Rodriguez, M. C. Betlach, H. W. Boyer, *Gene*, in press.

19a. R. B. Meagher, R. C. Tait, H. Betlock, H. W. Boyer, *Cell* **10**, 521 (1977).

20. Complementary DNA containing Hind III cohesive terminals was incubated at 37°C for 30 minutes with AMV reverse transcriptase under the conditions described in the legend to Fig. 2 for synthesis of single-stranded cDNA. The reaction was followed by using α-³²P-labeled dTTP (deoxythymidine triphosphate). Ligation of the Eco RI site octanucleotide (5′)TGAATTCA(3′) to the "filled in" cDNA was performed as described with the Hind III decamer in the legend to Fig. 3. Digestion with an excess of Eco RI endonuclease produced cDNA with Eco RI cohesive terminals. Since a Hind III site is recreated by this process, this procedure has the additional advantage that the cloned fragment also can be released from the plasmid by Hind III restriction endonuclease.

21. A. Maxam and W. Gilbert, *Proc. Natl. Acad. Sci. U.S.A.* **74**, 560 (1977).

22. J. R. Duguid, D. F. Steiner, W. L. Chick [*ibid.*

73, 3539 (1976)] have reported that insulin mRNA isolated from a rat islet cell tumor is approximately 600 nucleotides in length. Assuming a poly(A) length of 100 residues, the sequence of approximately 150 nucleotide residues remains to be established. Of these, approximately 30 residues are required to define the remaining portion of the known prepeptide sequence, leaving 120 residues of 5′-noncoding sequence. However, Duguid *et al.* used for part of their analyses ribosomal RNA markers in formamide polyacrylamide gels to arrive at their estimate of molecular weight. This method is believed to result in an overestimate of mRNA sizes [S. L. C. Woo, J. M. Rosen, C. D. Liarakos, Y. C. Choi, H. Busch, A. R. Means, B. N. O'Malley, D. L. Robberson, *J. Biol. Chem.* **250**, 7027 (1975)]. Our single-stranded cDNA forms a band on polyacrylamide gels with an estimated size of 450 nucleotides (Fig. 3A, lane 1). If this band comes from molecules initiated by dT₁₂₋₁₈ very close to the poly(A)-nontranslated region junction, then there remain only about 50 nucleotides left to be determined. This latter number is more consistent with the small 5′ noncoding regions that have been observed to date for eukaryotic mRNA's.

23. A. Devillers-Thiery, T. Kindt, G. Scheele, G. Blobel, *Proc. Natl. Acad. Sci. U.S.A.* **72**, 5016 (1975).

24. P. N. Campbell and G. Blobel, *FEBS Lett.* **72**, 215 (1976).

25. R. L. Pictet and W. J. Rutter, in *Handbook of Physiology*, section 7, *Endocrinology*, vol. 1, *Endocrine Pancreas* (American Physiological Society, Washington, D.C., 1972), p. 25; C. de Haen, E. Swanson, D. C. Teller, *J. Mol. Biol.* **106**, 639 (1976).

26. The sequences in this region of proinsulin I are not completely confirmed.

27. N. J. Proudfoot, *J. Mol. Biol.* **107**, 491 (1976).

28. K. N. Subramanian, J. Pan, S. Zain, S. M. Weissman, *Nucleic Acids Res.* **1**, 727 (1974).

29. Supported in part by NIH grants GM 21830 (to W.J. R.), CA 14026 (to H.M.G.), and GM 05385 (to J.C.); by a postdoctoral fellowship from Deutsche Forschungs Gemeinschaft (to A.U.); and by an NIH career development award (to R.P.). We are indebted to W. Imagawa for aid in the isolation of rat islets. The Hind III decamer and T4 ligase were provided by R. H. Scheller, California Institute of Technology, and H. Heyneker, University of California, San Francisco, respectively.

Supplementary Readings

Brenner, S., Cesarini, G. & Karn, J., Phasmids: Hybrids between ColEl plasmids and *E. coli* bacteriophage lambda. *Gene* 17:27–44 (1982)

Hanahan, D., Studies on transformation of *Escherichia coli* with plasmids. *J. Mol. Biol.* 166:557–580 (1983)

Kahn, M., Kolter, R., Thomas, C., Figurski, D. Meyer, R., Ramaut, E. & Helinski, D.R., Plasmid cloning vehicles derived from plasmids ColEl, F, R6K and RK2. *Methods in Enzymol.* 68:268–280 (1979)

Kreft, J., Bernhard, K. & Goebel, W., Recombinant plasmids capable of replication in *B. subtilis* and *E. coli. Mol. Gen. Genet.* 162:59–67 (1978)

Maniatis, T., Kee, S.G., Efstratiadis, A., & Kafatos, F.C., Amplification and characterization of a β-globin gene synthesized *in vitro. Cell* 8:163–182 (1976)

Noda, M., Takahashi, H., Tanabe, T., Toyosato, M., Furutaini, Y., Hirose, T., Asai, M., Inayana S., Miyate, T., and Numa, S., Primary structure of α-subunit precursor of *Torpedo californica* acetylcholine receptor deduced from cDNA sequence. *Nature* 299:793–797 (1982)

Rabbitts, T.H., Bacterial cloning of plasmids carrying copies of rabbit globin messenger RNA. *Nature* 260:221–225 (1976)

Sutcliffe, G., Complete nucleotide sequence of the *Escherichia coli* plasmid pBR322. *CSH Symp. Quant. Biol.* 43:77–90 (1979)

Thompson, C.J., Kieser, T., Ward, J.M. & Hopwood, D.A., Physical analysis of antibiotic-resistance genes from *Streptomyces* and their use in vector construction. *Gene* 20:51–62 (1982)

Gene Expression

Genes have two obvious roles. They are the information content which is inherited, and they are expressed to yield the organism's phenotype. In this section, papers will be presented which deal with two aspects of the expression role. First we will present some studies which deal with strategies for effecting the expression of the desired foreign gene once introduced into the target organism. The second group of papers describes two methodologies, still in the research stage, for turning off specific targeted genes.

Most of the early days of genetic engineering companies were devoted to the race to produce high value added proteins (interferons, etc.) in *E. coli*. This involved much empirical combining of protein coding sequences with translation initiation signals, promoters, etc. within existing plasmid cloning vectors. Rules gradually evolved and it is now possible to obtain more than 30% of total *E. coli* protein as a single commercial product in some cases. The diversion of massive cell resources into product synthesis requires that the expression systems be regulated in order to operationally separate growth of the producing organism from production. In some cases this is absolutely critical since the resulting foreign protein is lethal to the host. Finally, although *E. coli* is still the "workhorse" of the genetic engineering industry, it has been essential to also use other hosts if only to achieve proper protein modification. In this section we will present three articles which will describe regulated expression systems suitable for use in prokaryotic and eukaryotic cloning systems. Several other examples are presented in the papers found in other sections.

The Guarente et al. paper describes the general development of what has become a "classical" system for expressing foreign genes in *E. coli* based upon the lactose operon regulatory system.The necessary components, the promoter required for transcription initiation, the transcription regulatory signals and the translation initiation signals immediately followed by a suitable cloning site for the insertion of the target gene, are described. Although conceptually complete, the actual implementation of this strategy is often problematic. For instance, the precise mRNA sequence corresponding to the translation initiation signal, the cloning site and the N-terminus of the encoded protein sometimes forms a secondary structure which blocks efficient translation initiation.

The strategy described in the Guarente et al. article in essence uses the endogenous *E. coli* gene transcription machinery to provide expression of the cloned gene. Although in many cases this is quite successful, it suffers from two inherent problems; it is typically quite difficult to completely turn off the gene in question when desired, and the same machinery is also expressing many of the other host genes. The Tabor and Richardson paper describes a different strategy in which the target gene is transcribed by an alternate, normally viral encoded, RNA polymerase. In this case the expression of the gene is activated indirectly by turning on the expression of the T7 RNA polymerase. The T7 RNA polymerase has an extremely high specificity for a promoter sequence; a sequence which has been placed immediately upstream of the cloned gene. This typically generates very high levels of selective transcription of the cloned gene, but only at the desired time during the fermentation. It has proven to be very good for the synthesis of lethal gene products.

Prokaryotes cannot perform the secondary modifications (e.g., glycosylation) required for many eukaryotic proteins to achieve their "correct" structure. Thus it has been necessary to extend the general schemes developed for heterologous gene expression in prokaryotes to eukaryotes. The Okayama and Berg paper is an excellent and early example of how this can be accomplished. This paper describes an SV40 based vector in which eukaryotic expression signals are placed adjacent to the cloning sites. A second feature of this paper is that it extends the efficient cDNA cloning technologies described in another Okayama and Berg paper (see Section 4). Finally, the cloning vector was a "shuttle" plasmid which could replicate in either *E. coli* or animal cells, thus allowing *E. coli* to be the host for DNA manipulations and the animal cells to be the host for protein production.

The virulent property of a pathogen or a tumor cell is sometimes the result of the expression of a single gene. It follows that the ideal chemotherapeutic agent, a true "magic bullet," would be one which selectively inhibited the expression of this single critical gene. Recent studies have suggested two modes for accomplishing this selective blocking of gene expression. In both cases the technique has been shown to work in the research laboratory. These techniques act by blocking translation either by forming a duplex RNA structure at the gene's translation initiation signal or by cleaving the gene's mRNA after duplex formation. A major future challenge will be to develop means to target one of the "bullets" to the critical cells.

The Coleman et al. paper describes the practical extension of the same laboratory's previous discovery that intracellular synthesis of a specific antisense RNA can block translation and therefore expression of the target gene. This suggests that any gene can have its expression blocked by introducing into the appropriate cell a construct which expresses an RNA which is complementary (antisense) to the target gene's mRNA, with particular emphasis being at the mRNA's translation initiation signals. This strategy could also be used to block replication or gene expression of RNA viruses.

The discovery of ribozymes (RNA enzymes with endoribonuclease activity) has had a profound impact upon the fields of molecular biology and evolutionary biology (see the article by Zaug and Cech in Section 1). In addition, this discovery may have an important practical contribution; the ability to selectively destroy target mRNAs. Haseloff and Gerlach describe the engineering of such a sequence specific, transacting ribozyme which uses complementary base pairing to target the ribozyme to the desired mRNA. As with the antisense RNA strategy this technique has been shown to work in selected systems in the research laboratory. Several biotechnology companies have staked their futures on the development of these technologies.

The Use of RNAs Complementary to Specific mRNAs to Regulate the Expression of Individual Bacterial Genes

J. Coleman, P.J. Green and M. Inouye

Summary

A naturally occurring small RNA molecule (*micF* RNA), complementary to the region encompassing the Shine–Dalgarno sequence and initiation codon of the *ompF* mRNA, is known to block the expression of that mRNA in E. coli. We have constructed a plasmid that produces a complementary RNA to the E. coli *lpp* mRNA (*mic[lpp]* RNA). Induction of the *mic(lpp)* gene efficiently blocked lipoprotein production and reduced the amount of *lpp* mRNA. Two *mic(ompC)* genes were similarly engineered and their expression was found to inhibit drastically production of OmpC. Analysis of several types of *mic(ompA)* genes suggests that micRNAs complementary to regions of the mRNA likely to come in contact with ribosomes were most effective. The novel capabilities of this artificial mic system provide great potential for application in both procaryotic and eucaryotic cells.

Introduction

It is well documented that the expression of certain genes is regulated at the level of transcription (Rosenberg and Court, 1979; Miller and Reznikoff, 1978). This transcriptional regulation is carried out either negatively (repressors) or positively (activators) by a protein factor. It is also known that specific protein factors regulate translation of specific mRNAs (Gold et al., 1981; Kozak, 1983).

Recently, it has become evident that RNAs are also involved in regulating the expression of specific genes (Mizuno et al., 1983b, 1984; Simons and Kleckner, 1983). We have reported that a small RNA transcript of 174 bases is produced upon growing E. coli cells in a medium of high osmolarity, which inhibits the expression of the gene for an outer-membrane protein (OmpF protein) (Mizuno et al., 1983b, 1984). The inhibition of OmpF protein production by the small RNA transcript (micRNA; mRNA-interfering complementary RNA) is likely due to the formation of the hybrid between the micRNA and the *ompF* mRNA over a region of approximately 80 bases, including the Shine–Dalgarno sequence and the initiation codon (Mizuno et al., 1983b, 1984). A similar regulation by a small complementary RNA has also been described for the Tn10 transposase (Simons and Kleckner, 1983). In this case, however, the gene for the transposase and the gene for the micRNA are transcribed in opposite directions of the same segment of DNA such that the 5′ ends of their transcripts can form a complementary hybrid. The hybrid is thought to

inhibit translation of the transposase mRNA (Simons and Kleckner, 1983). The transposase situation is in contrast to the *ompF* situation, in which the *ompF* gene and the micRNA gene (*micF*) are completely unlinked and map at 21 and 47 min, respectively, on the E. coli chromosome.

These findings prompted us to attempt to construct an artificial mic system designed to regulate the expression of any specific gene in E. coli. One can construct a micRNA system for a gene by inserting a small DNA fragment from the gene, in the opposite orientation, after a promoter. Such a system provides a novel way to regulate specifically the expression of any gene. In particular, by inserting the micDNA fragments under the control of an inducible promoter, the expression of essential E. coli genes can be regulated. The inducible lethality thus created may be a more effective tool in the study of essential genes than conventional temperature-sensitive mutations.

Here we report the construction of an artificial mic system and demonstrate its function using several E. coli genes. We propose that the mic system is an effective way to regulate the expression of specific procaryotic genes. We also feel its principle provides the basis for accomplishing similar regulation of biologically important genes in eucaryotes. For example, the mic system can be used to block the expression of harmful genes such as oncogenes or viral genes.

Results

Construction of an Artificial mic Gene

We have shown that the *micF* gene produces a 174 base RNA that blocks production of the OmpF protein. This small RNA has two stem–loop structures, one at the 3′ end ($\Delta G = -12.5$ kcal) and the other at the 5′ end ($\Delta G = -4.5$ kcal) (Mizuno et al., 1984). Since these structures may play an important role for the function and/or stability of the micRNA, we attempted to use these features in the construction of an artificial mic system using the gene for the major outer membrane lipoprotein (*lpp*) cloned in an inducible expression vector, pIN-II (Nakamura and Inouye, 1982). pIN-II vectors are high-expression vectors that have the *lac*^{po} downstream of the lipoprotein promoter, thus allowing high-level inducible expression of an inserted gene. The pIN-II promoter was fused to the *lpp* gene at a unique Xba I site immediately upstream of the Shine–Dalgarno sequence of the *lpp* mRNA. The resulting plasmid was designated pYM140. When the expression of the *lpp* gene, in pYM140, is induced by isopropyl-β-D-thiogalactoside (IPTG), a *lac* inducer, the RNA transcript derived from the *lpp* gene, has a possible stem–loop structure (at the 5′ end) with a free energy of −5.1 kcal, encoded immediately upstream of the unique Xba I site (Figure 1B). In the 3′ portion of the lpp mRNA another stable stem–loop structure can form (Figure 1B). This latter loop is derived from the ρ-independent transcription termination signal ($\Delta G = -21.4$ kcal; Nakamura et al., 1980) of the *lpp* gene. The construction of a general mic cloning vector,

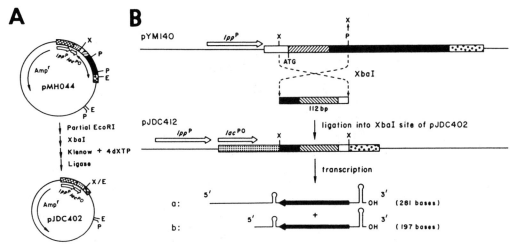

Figure 1. Construction of mic Vector pJDC402 and *mic(lpp)*

(A) Construction of pJDC402. The plasmid, pMH044 was constructed as follows: An Rsa I site immediately upstream of the transcription-termination site of *lpp* was changed to an Eco RI site by partial digestion of pYM140 (see text) followed by insertion of an Eco RI linker. The resulting plasmid, pMH044 was partially digested with Eco RI, followed by a complete digestion with Xba I. The single-stranded portions of the linear DNA fragment were filled in with DNA polymerase I (large fragment), and then treated with T4 DNA ligase, resulting in the formation of the plasmid, pJDC402, that lost the fragment between the Xba I and the Rsa I sites. As a result of this procedure, both an Eco RI and an Xba I site were recreated at the junction. Restriction sites are indicated as follows: X, Xba I; P, Pvu II; E, Eco RI. *lpp^p* and *lac^po* are the lipoprotein promoter and the lactose promoter operator, respectively. Amp^r is the ampicillin-resistance gene. Cross-hatches represent the lipoprotein promoter. Solid dots represent the lactose promoter operator. Slashes indicate the lipoprotein signal sequence, and the solid bar represents the coding region for the mature portion of the lipoprotein. The open dots represent the transcription-termination region derived from the *lpp* gene. The open bar represents the 5′ nontranslated region of the lipoprotein mRNA.
(B) Construction of *mic(lpp)* pJDC412. Open arrows represent promoters. The solid arrow represents the portion of the *mic(lpp)* RNA that is complementary to the *lpp* mRNA. The Pvu II site was converted to an Xba I site by inserting an Xba I linker (CTCTAGAG). This fragment was inserted into the unique Xba I site of pJDC402 in the reverse orientation forming pJDC412. (a) and (b) show the *mic(lpp)* RNAs initiating at the *lpp* and *lac* promoters, respectively.

pJDC402, was achieved by removing the DNA fragment in pMH044 between the two loops as shown in Figure 1A. In pJDC402, the unique Xba I site can serve as the insertion site for any DNA fragment, and the RNA transcript from the artificial *mic* gene has a structure similar to the *micF* RNA; the portion derived from the inserted DNA is sandwiched by two loop structures, one at the 5′ and one at the 3′ end.

In the context of this paper we will use the term "*mic* gene" to define a gene capable of encoding an RNA complementary to all or part of a specific mRNA. The target gene for the micRNA will be shown in parentheses. For example, a *mic(lpp)* gene is capable of encoding an RNA complementary to the *lpp* mRNA.

Construction of the *mic(lpp)* Gene

Using this mic cloning vector, pJDC402, we first attempted to create a mic system for the *lpp* gene of E. coli, in order to block the synthesis of the lipoprotein upon induction of the *mic(lpp)* gene. For this purpose we first isolated the DNA fragment containing the Shine–Dalgarno sequence for ribosome binding, and the coding region for the first few amino acid residues of prolipoprotein. To do this, the Pvu II site immediately after the coding region of the prolipoprotein signal peptide was changed to an Xba I site by inserting an Xba I linker at this position. The resulting plasmid was then digested with Xba I, and the 112 bp Xba

I–Xba I (originally Pvu II–Xba I) fragment was purified. This fragment, encompassing the Shine–Dalgarno sequence and the coding region for the first 29 amino residues from the amino terminus of prolipoprotein (Nakamura and Inouye, 1979), was purified. This fragment was then inserted into the unique Xba I site of pJDC402 in the opposite orientation from the normal *lpp* gene. The resulting plasmid, designated as pJDC412 (see Figure 1B), is able to produce *mic(lpp)* RNA, an RNA transcript complementary to the *lpp* mRNA, upon induction with IPTG.

It should be pointed out that another important feature of the mic expression vector, pJDC402, is that it contains a Hinf I site immediately upstream of the *lpp* promoter, and another one immediately downstream of the transcription-termination site. These two Hinf I sites can be used to remove a DNA fragment containing the entire mic transcription unit, which can then be inserted back into the unique Pvu II site of the vector. In this manner, the entire *mic* gene can be duplicated in a single plasmid. One would expect a plasmid containing two identical *mic* genes to produce twice as much micRNA as a plasmid containing a single *mic* gene. Such a plasmid was constructed, containing two *mic(lpp)* genes and designated pJDC422.

Expression of the *mic(lpp)* Gene

In order to examine the effect of the artificial *mic(lpp)* RNA, we pulse-labeled cells for 1 min with ^35S-methionine 1 hr after induction of the *mic(lpp)* RNA with 2 mM IPTG. The autoradiograph of the isolated outer membranes is shown in Figure 2. The cells harboring the vector pJDC402 produce the same amount of lipoprotein either in the absence (lane 1, Figure 2) or the presence (lane 2) of the inducer, IPTG, as quantitated by densitometric scanning of the autoradiogram and normalizing to an unaffected outer membrane protein, band a (Figure 2). Lipoprotein production was reduced approximately 2-fold in the case of cells carrying a pJDC412 in the absence of IPTG (lane 3) and approximately 16-fold in the presence of IPTG (lane 4). The reduction in lipoprotein synthesis in the absence of IPTG is considered to be due to incomplete repression of the *mic(lpp)* gene. In the case of cells carrying pJDC422, where the *mic(lpp)* gene was duplicated, lipoprotein production is now reduced 4-fold in the absence of IPTG (lane 5), and 31-fold in the presence of IPTG. These results clearly demonstrate that the production of the artificial *mic(lpp)* RNA inhibits lipoprotein production, and that the inhibition appears to be proportional to the amount of the *mic(lpp)* RNA produced. It should be noticed that the *mic(lpp)* RNA is specifically blocking the production of lipoprotein, and that it does not block the production of any other detectable outer membrane proteins except for OmpC protein. The fact that the induction of the *mic(lpp)* gene reduces the production of OmpC plus OmpF proteins (lanes 4 and 6 in Figure 2) was found to be due to unusual homology between the *lpp* gene and the *ompC* gene, as will be discussed below.

We also examined the rate of inhibition of lipoprotein production upon induction of the *mic(lpp)* RNA, by pulse-labeling E. coli JA221/F'*lacI*^{q} harboring pJDC412 with ^{35}S-methionine at various times after induction with IPTG. As can be seen in Figure 3, lipoprotein production was maximally inhibited by 16-fold within 5 min after the addition of IPTG, which is much shorter than the functional half-life of the *lpp* mRNA (12 min; Hirashima and Inouye, 1973).

lpp mRNA Production in the Presence of *mic(lpp)* RNA

It is interesting to examine whether the *mic(lpp)* RNA also affects the level of the *lpp* mRNA, since the expression of the *micF* gene substantially reduced the amount of the *ompF* mRNA (Mizuno et al., 1984). For this purpose, we isolated total cellular RNA 1 hr after the induction of the *mic(lpp)* gene with IPTG. The RNA preparation was analyzed after electrophoresis in a formaldehyde agarose gel and subsequently transferred onto nitrocellulose paper. The paper was then hybridized with a probe specific to the *mic(lpp)* RNA or to the *lpp* mRNA. We also used a probe specific for the *ompA* mRNA as an internal control. Again, pJDC402 shows little difference in the production of the *lpp* mRNA in the absence or presence of IPTG (lanes 3 and 4, Figure 4A); the small apparent difference is possibly due to differing amounts of total RNA in each lane. Because the primer used to make the probe for these experiments (Messing, 1983) contains a portion of the *lac* operon, the probes hybridize to any transcript containing the *lac* promoter, such as the *mic(lpp)* RNA from JDC412 and the short nonsense transcript from pJDC402. Cells harboring pJDC412 contain a reduced amount of the *lpp* mRNA in the absence of IPTG (lane 1, Figure 4A), and a greatly reduced amount of the *lpp* mRNA in the presence of IPTG (lane 2, Figure 4A). Figure 4B shows the production of the *mic(lpp)* RNA in the absence (lane 1) and the presence (lane 2) of IPTG in cells harboring pJDC412. It can be seen that even in the absence of IPTG, a significant amount of the *mic(lpp)* RNA is produced, which is consistent with the results of the lipoprotein production shown earlier (lane 3, Figure 2). However, a much larger amount

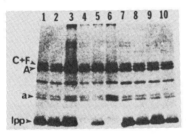

Figure 2. The Effect of *mic(lpp)* Expression on *lpp* Synthesis

Cells were pulse-labeled for 1 min 1 hr after induction (lanes 2, 4, 6, 8, and 10) with 2 mM IPTG. Outer membrane fractions from JA221/F'*lacI*^{q} harboring various plasmids were resolved using SDS-PAGE and autoradiography. Lanes 1 and 2, pJDC402; lanes 3 and 4, pJDC412; lanes 5 and 6, pJDC422; lanes 7 and 8, pJDC413; lanes 9 and 10, pJDC414. The lipoprotein was quantitated by comparing the density of the lipoprotein to the density of the protein migrating to position a on the gel. (lpp, A, C, and F) mark the positions of lipoprotein, OmpA protein, OmpC protein, and OmpF protein, respectively. Lanes 7–10 will be considered later in the text.

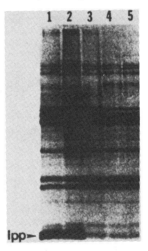

Figure 3. Time Course of the *mic(lpp)* Effect

Membrane fractions from JA221/F'*lacI*^{q} harboring pJDC402, lane 1, or pJDC412, lanes 2–5, were isolated after cells were pulse-labeled for 1 min at various times after induction with IPTG and resolved as described for Figure 2. Lanes 1 and 2, 0 min after induction; lane 3, 5 min after induction; lane 4, 30 min after induction; and lane 5, 60 min after induction.

of *mic(lpp)* RNA (band C, Figure 4B, lane 2) is detected in pJDC412-containing cells induced with IPTG. The fact that the *lpp* mRNA disappears upon induction of the *mic(lpp)* RNA indicates that the mechanism of action of the micRNA is not solely at the level of translation. In Figure 4B (lane 2) there are two *mic(lpp)* RNAs of different sizes. The sizes of these transcripts were estimated to be 281 and 197 bases, which correspond to transcripts initiating at the lipoprotein promoter (the larger RNA) and initiating at the *lac* promoter (the smaller RNA) (see Figure 1B).

Inhibition of OmpC Production with the *mic(ompC)* Gene

We have also been able to achieve an almost complete inhibition of OmpC synthesis by artificially constructing *mic(ompC)* genes. The first construct, pAM320, carrying two *mic(ompC)* genes, gives rise to an RNA molecule complementary to 20 nucleotides of the leader region and 100 nucleotides of the coding region of the *ompC* mRNA. This was achieved by changing the unique Bgl II site in the *ompC* structural gene and the Mnl I site, 20 nucleotides upstream of the ATG initiation codon (Mizuno et al., 1983a) to Xba I sites. The resulting 128 bp Xba I fragment was then inserted into pJDC402 in the opposite orientation from the *ompC* gene and a second copy of the *mic(ompC)* gene was introduced in a manner similar to that described for the pJDC422 construction. In the resulting plasmid, pAM320, the two *mic(ompC)* genes are in the opposite orientations. Reversing the orientation of the second *mic* gene did not change the expression or stability of the plasmid (data not shown). A second construct, pAM321, was designed to extend the complementarity between the micRNA and the *ompC* mRNA to include a longer leader sequence than in the case of pAM320, 72 nucleotides of the leader region instead of 20. This plasmid was constructed as described for pAM320 except that the Mnl I site changed to an Xba I site was located 72 bp upstream of the *ompC* initiation codon. This construct also contains the two *mic(ompC)* genes in opposite orientations.

Figure 5A shows the Coomassie Brilliant Blue-stained gel patterns of the outer membrane proteins isolated from E. coli JA221/F′*lacI*^{q} harboring one of the following plasmids: pJDC402, pAM320, or pAM321. The effect of the addition of IPTG is clearly seen by the appearance of β-galactosidase (band b) in lanes 2, 4, and 6 in Figure 5A. The induction of the *mic(ompC)* RNA from pAM320

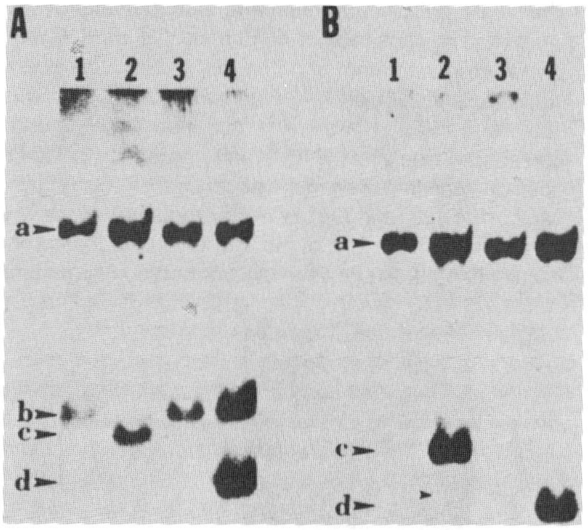

Figure 4. The Effect of *mic(lpp)* Expression on RNA Levels In Vivo

Cells were grown in supplemented M9 medium (see Experimental Procedures) to a Klett reading of 10 (blue filter), at which time 90 μCi of ^{3}H-uridine was added to each culture. At Klett 30, IPTG was added to a final concentration of 2 mM as indicated below. After 1 hr RNA was extracted as described in Experimental Procedures, and quantitated by counting cold TCA-precipitable counts. Equal counts were loaded on a 1.5% agarose-formaldehyde gel. Lanes 1 and 2 contain RNA from JA221/F′*lacI*^{q}) harboring pJDC412 in the absence (lane 1) or presence (lane 2) of IPTG. Lanes 3 and 4 contain RNA from cells harboring pJDC402 in the absence (lane 3) or presence (lane 4) of IPTG. (A) was hybridized with a probe specific for the *lpp* mRNA and one specific for the *ompA* mRNA (as an internal control). (B) was hybridized with a probe specific for the *mic(lpp)* RNA, and a probe specific for the *ompA* mRNA. (a) designates the position of the *ompA* mRNA (approximately 1217 bases); (b) designates the *lpp* mRNA (322 bases); (c) designates the *mic(lpp)* RNA transcript from the *lpp* promoter and the small arrow designates the *mic(lpp)* RNA transcript from the *lac* promoter; (d) designates the transcript from the vector pJDC402 transcribed through the *lac* promoter, described in the text.

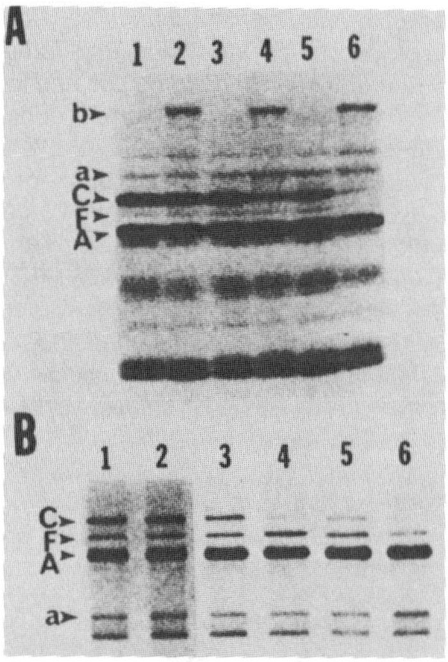

Figure 5. Effects of *mic(ompC)* Expression on OmpC Synthesis

(A) Outer membranes were isolated from cultures of JA221/F′*lacI*^{q} harboring various plasmids, grown overnight in nutrient broth containing 200 mM NaCl and 50 μg/ml ampicillin with (lanes 2, 4, 6) or without (lanes 1, 3, 5) IPTG. Proteins were resolved using urea-SDS-PAGE and stained with Coomassie Brilliant Blue. Lanes 1 and 2, pJDC402; lanes 3 and 4, pAM320; lanes 5 and 6, pAM321. (a, b, C, F, and A) designate the positions of the unaffected band used for normalization, β-galactosidase, OmpC, OmpF, and OmpA, respectively.

(B) Outer membranes were isolated from cells grown and labeled as described for Figure 2 and resolved using urea-SDS-PAGE and autoradiography. Lanes 1 and 2, pJDC402; lanes 3 and 4, pAM320; lanes 5 and 6, pAM321. (a, C, F, and A) designate the positions of the unaffected band used for normalization, OmpC, OmpF, and OmpA, respectively.

caused a substantial decrease (approximately 5-fold) in OmpC production (lane 4, Figure 5A) compared to pJDC402. Induction of the longer *mic(ompC)* RNA from pAM321 (lane 6) decreased OmpC synthesis more dramatically (approximately 20-fold compared to pJDC402). As shown in lane 6, Figure 5B, OmpC production could hardly be detected in the cells harboring pAM321 when they were pulse-labeled for 1 min after a 1 hr induction with IPTG. In the same experiment, OmpC synthesis decreased approximately 7-fold when the *mic(ompC)* gene in cells harboring pAM320 was induced with IPTG. Marked decreases in OmpC expression were also observed when plasmids containing single copies of the *mic(ompC)* genes were induced (data not shown). Again, the longer *mic(ompC)* gene had a greater effect (data not shown). The increased efficiency of mic-mediated inhibition with pAM321 over pAM320 may indicate that the effectiveness of micRNA function is related to the extent of complementarity to the 5′ end of the mRNA. This possibility is explored further in the next section.

It is interesting to note that the synthesis of either of the *mic(ompC)* RNAs described above caused a decrease, not only in OmpC synthesis, but also in lipoprotein synthesis (see Figure 6). This inhibitory effect of the *mic(ompC)*

RNA on lipoprotein production appears to be due to the unexpected homology between the *lpp* mRNA sequence and the *ompC* mRNA, as illustrated in Figure 7. This feature explains why pAM320 and pAM321 are exerting a mic effect on lipoprotein production. Such an explanation would predict that induction of the *mic(lpp)* RNA from pJDC412 and pJDC422 should decrease the synthesis of OmpC protein, and this was found to be the case, as can be seen in lanes 4 and 6 in Figure 2.

Inhibition of OmpA Production with *mic(ompA)* RNA

In an effort to determine which components contribute to the effectiveness of a micRNA, several *mic* genes were constructed from the *ompA* gene. The *ompA* gene was selected for this because the leader and the coding regions of the *ompA* mRNA have been characterized extensively (Movva et al., 1980a, 1980b, 1980c; Cole et al., 1982; Green and Inouye, 1984). Five DNA fragments (I through V, Figure 8) were individually cloned into the Xba I site of pJDC402 in the orientation promoting the production of *mic(ompA)* RNAs. The resulting *mic(ompA)* plasmids containing fragments I–V were designated pAM301, pAM307, pAM313, pAM314, and pAM318, respectively. Each plasmid contains only one copy of the described *mic(ompA)* gene.

E. coli JA221/F′*lacl*^q containing each of the *mic(ompA)*

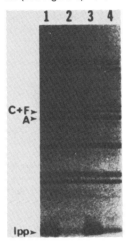

Figure 6. Effects of *mic(ompC)* Expression on Lipoprotein Synthesis

The same outer membrane protein preparations used for lanes 3, 4, 5, and 6 in Figure 5B were separated using SDS-PAGE in lanes 1, 2, 3, and 4, respectively. C + F corresponds to both OmpC and OmpF, which comigrate in this gel system. A marks the position of OmpA and lpp marks the position of lipoprotein.

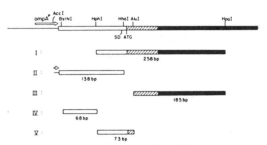

Figure 8. Fragments Used to Construct *mic(ompA)* Genes

The top line shows the structure of the E. coli *ompA* gene. The arrow represents the promoter and the open bar represents the region encoding the 5′ leader region of the *ompA* mRNA. The slashed bar and shaded bar represent the portions of the *ompA* gene encoding the signal sequence and the mature OmpA protein, respectively. Restriction fragment I (Hph I–Hpa I) was inserted into the Xba I site of pJDC402 (Figure 1A) in the orientation opposite that depicted here (as outlined in Figure 1B for *mic(lpp)*) to create the plasmid pAM301. The other *mic(ompA)* plasmids were similarly constructed from: fragment II, pAM307; fragment III, pAM313; fragment IV, pAM314; fragment V, pAm318. The positions of the Shine–Dalgarno sequence (SD), ATG initiation codon (ATG), and relevant restriction sites are shown.

```
lpp : 5'   ---AAUCUA GAGG GUAUUAAUA-- AUC AAACCUACUAAACUGGUACUG---(22 bases)---CUCUGCUGG---CAGGUUGCUCCAGCAA-CGCU---  3'
            II I I IIIII  IIIIII     III IIII I IIII  IIIIII              IIIIIIIII  IIII    IIIIII IIII
ompC : 5'  ---AA-C-A GAGC --UUAAUAAC AUG AAAG-U--UAAA---GUACUG---(16 bases)---CUCUGCUGGUAGCAGG-C--G-CAGCAAACGCU---  3'
                     S.D.          Initiation
                                   Codon
```

Figure 7. Homology between the *ompC* mRNA and the *lpp* mRNA

A region of homology between the *lpp* mRNA (top lines) and the *ompC* mRNA (bottom line) is shown. Bars connect identical bases. Both *mic(ompC)* RNAs described in the text have the potential to hybridize across this homologous region. The Shine–Dalgarno Sequences (S. D.) and AUG initiation codons are boxed.

plasmids was pulse-labeled with ^{35}S-methionine for 1 min with and without a 1 hr prior preincubation with IPTG. Electrophoretic patterns of the outer membrane proteins isolated from these cultures are shown in Figure 9A. The autoradiographs revealed that each of the five *mic(ompA)* genes is capable of inhibiting OmpA synthesis 45%, 51%, 54%, 18%, and 45% for mic constructions using fragments I through V in Figure 8, respectively. The *mic(ompA)* genes appear to be less effective than the *mic(lpp)* and *mic(ompC)* genes described earlier. However, as will be discussed later, we were easily able to circumvent this problem by increasing the *mic(ompA)* gene dosage.

The *mic(ompA)* gene in pAM314, synthesizing a micRNA complementary to fragment IV in Figure 8 is clearly less effective than the other four *mic(ompA)* constructions. Unlike fragments I and V, which encompass the *ompA* translation-initiation site, fragment II, which includes the Shine–Dalgarno Sequence, or fragment III, which includes only structural gene sequences, fragment IV does not include sequences likely to come in contact with ribosomes. This may be the reason why the *mic(ompA)* produced by pAM314 is a weaker inhibitor of OmpA synthesis than are the other *mic(ompA)* plasmids.

The effectiveness of the micRNA encoded by pAM313 that is complementary only to the portion of the *ompA*

structural gene covered by fragment III (Figure 8), which spans the coding region for amino acid residues 4–45 of pro-OmpA, indicates that the micRNA does not need to hybridize to the initiation site for protein synthesis and/or to the 5′ leader region of the target mRNA in order to function. This was also confirmed using mic(*lpp*) genes. Two *mic(lpp)* RNAs that were complementary only to the coding region of the *lpp* mRNA have also been found to inhibit lipoprotein production. Figure 2 shows the effect of the *mic(lpp)* genes in pJDC413 and pJDC414 that were constructed from the *lpp* structural gene fragments coding for amino acid residues 3–29 and 43–63 of prolipoprotein, respectively. Both pJDC413 and pJDC414, however, exhibit only a 2-fold inhibition of lipoprotein synthesis, indicating that a DNA fragment covering the translation initiation site (which caused a 16-fold inhibition) is more effective in the case of the *mic(lpp)* genes.

In order to construct a plasmid capable of inhibiting OmpA synthesis more effectively than those discussed above, we constructed plasmids containing more than one *mic(ompA)* gene. An example of our results is shown in Figure 9B where the expression of pAM307 and its derivatives pAM319 and pAM315 are compared. The latter two plasmids contain two and three copies of the *mic(ompA)* gene in pAM307, respectively. While pAM307 inhibited OmpA synthesis by approximately 47% (lane 4, Figure 9B), pAM315 and pAM319 inhibited OmpA synthesis by 69% and 73% (lanes 6 and 8, Figure 9B, respectively).

Discussion

The present results clearly demonstrate that the artificial mic system can be used for specifically regulating the expression of a gene of interest. In particular, the inducible mic system for a specific gene is a novel and very effective way to study the function of a gene. If the gene is essential, conditional lethality may be achieved upon the induction of the mic system, somewhat similar to a temperature-sensitive mutation. It should be noted, however, that the mic system blocks the synthesis of the specific protein itself while temperature-sensitive mutations block only the function of the protein without blocking its synthesis.

From this study, the following has become evident. The production of an RNA transcript (micRNA) that is complementary to a specific mRNA inhibits the expression of that mRNA. The production of a micRNA specifically blocks the expression of only those genes that share complimentarity to the micRNA. The induction of micRNA production blocks the expression of the specific gene very rapidly in less than the half-life of the mRNA. The micRNA also reduces the amount of the specific mRNA in the cell, as was found when the natural *micF* gene was expressed (Mizuno et al., 1984), as well as when the artificially constructed *mic(lpp)* gene was expressed in the present paper. There is a clear effect of gene dosage; the more a micRNA is produced, the more effectively the expression of the target gene is blocked.

At present, it is not clear what constitutes the best

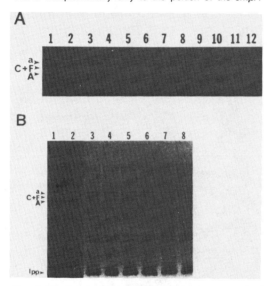

Figure 9. Effects of *mic(ompA)* Expression on OmpA Synthesis

Outer membranes were isolated from JA221/F′*lacI*^{q} harboring various plasmids after the cells were grown and labeled as described for Figure 2. For even-numbered lanes, the cells were induced with IPTG prior to labeling. For odd-numbered lanes, the cells were not induced. (a) indicates the protein used to normalize the densitometric measurements. (C + F) marks the position of OmpC and OmpF, which comigrate on this gel system. (A) and (lpp) mark the position of OmpA and lipoprotein, respectively. (A) Lanes 1 and 2, pJDC402; lanes 3 and 4, pAM301; lanes 5 and 6, pAM307; lanes 7 and 8, pAM313; lanes 9 and 10 pAM314; lanes 11 and 12, pAM318. (B) Lanes 1 and 2, pJDC402; lanes 3 and 4, pAM307; lanes 5 and 6, pAM319; Lanes 7 and 8, pAM315.

micRNA. Our studies indicate that micRNA's complementarity to regions of the mRNA known to interact with ribosomes is the most effective. Using the *lpp* gene as an example, it appears that a *mic(lpp)* RNA that can hybridize to the Shine–Dalgarno sequence and the translation-initiation site of the *lpp* mRNA inhibits lipoprotein synthesis more efficiently than one that cannot. However, for the *ompA* gene, micRNAs complementary to both the Shine–Dalgarno sequence and the translation-initiation site, just the Shine–Dalgarno sequence, or the structural gene alone were equally effective. More detailed studies will be required to determine what role the features of the target gene have in determining the differential efficiencies of structurally different micRNAs. The role of the two stem-loop structures, sandwiching the complementary region of the natural *micF* RNA and our artificial micRNAs, also warrants investigation.

For some genes such as *ompC* and *lpp,* the region of the gene encompassing the translation-initiation site may not contain a unique sequence, and micRNA induction results in the inhibition of the production of more than one protein. In these cases, another region of the gene may be used to construct the *mic* gene. The length of the micRNA is another important variable to be considered. The longer *mic(ompC)* RNA was 4-fold more effective at inhibiting OmpC production than the shorter *mic(ompC)* RNA. It should be noted that the inhibition of lipoprotein expression by the *mic(ompC)* RNA was less effective with the longer *mic(ompC)* RNA, in spite of the fact that the region of the two *mic(ompC)* RNAs complementary to the lipoprotein mRNA is the same. This indicates that higher specificity may be achieved by using longer micRNAs. In contrast to the *mic(ompC)* genes, length did not appear to be a significant factor for the *mic(ompA)* RNA-mediated inhibition of OmpA production. In addition, the secondary structure of the micRNA most likely plays an important role in micRNA function that should be further explored in order to establish the most effective mic system.

There are several mechanisms by which the micRNA may function to inhibit expression of the specific gene. It is most likely that the micRNA acts primarily by binding to the mRNA, thereby preventing the interaction with ribosomes as proposed earlier (Mizuno et al., 1983b, 1984; Simons and Kleckner, 1983). This hypothesis is supported by the fact that the *mic(lpp)* RNA inhibited lipoprotein production much faster than would be expected if only transcription was affected based on the half-life of the *lpp* mRNA (see Figure 3). How does the micRNA cause a reduction in the amount of lipoprotein mRNA? A plausible model to explain this reduction is that the mRNA is less stable when ribosomes are not traversing the entire mRNA. Another possible model to explain this reduction in mRNA level is that complementary hybrid formation between the micRNA and the mRNA causes premature termination of transcription or destabilization of the mRNA. Alternatively, although not mutually exclusive from the hypotheses proposed above, the micRNA may directly inhibit the initiation of transcription, or cause pausing of mRNA elongation in a manner similar to that described for a small complementary RNA species involved in Col E1 replication (Tomizawa and Itoh, 1982).

The mic system appears to have great potential in its application in procaryotic, as well as eucaryotic, cells to block the expression, permanently or upon induction, of various toxic or harmful genes, such as drug-resistance genes, oncogenes, and phage or virus genes.

Experimental Procedures

Strain and Medium

E. coli JA221 (*hsdR leuB6 lacY thi recA ΔtrpE5*)/F′(*lacI^q proAB lacZYA*) (Nakamura and Inouye, 1982) was used in all experiments. This strain was grown in M9 medium (Miller, 1972) supplemented with 0.4% glucose, 2 μg/ml thiamine, 40 μg/ml each of leucine and tryptophan, and 50 μg/ml ampicillin unless otherwise indicated.

Materials

Restriction enzymes were purchased from either Bethesda Research Laboratories or New England BioLabs. T4 DNA ligase and E. coli DNA polymerase I (large fragment) were purchased from Bethesda Research Laboratories. All enzymes were used in accordance with the instructions provided by the manufacturer. Xba I linkers (CTCTAGAG) were purchased from New England BioLabs.

DNA Manipulation

Plasmids pJDC402, pJDC412, and pJDC422 were constructed as described in the text and in Figure 1. Plasmids pJDC413 and pJDC414 were constructed by isolating the 80 bp Alu I fragment from the *lpp* gene encoding amino acid residues 3–29 of prolipoprotein for pJDC413 and the 58 bp Alu I fragment encoding amino acid residues 43–63 of prolipoprotein for pJDC414. The fragments were blunt-end-ligated into pJDC402, which was first digested with Xba I, followed by treatment with DNA polymerase I (large fragment).

The isolation of the appropriate *ompC* fragments for *mic(ompC)* construction involved a subcloning step due to the absence of suitable unique restriction sites between the *ompC* promoter and structural gene. Two derivatives of the *ompC*-containing plasmid, pMY150 (Mizuno et al., 1984), lacking either the 471 bp or the 419 bp Xba I–Mnl I *ompC* promoter-containing fragment (pDR001 and pDR002, respectively), but containing an Xba I site in its place, were isolated. The unique Bgl II sites in each of these plasmids were changed to Xba I sites by treatment with DNA polymerase I (large fragment) and ligation with synthetic Xba I linkers. Following Xba I digestion, the 123 bp Xba I fragment from pDR001 and the 175 bp Xba I fragment from pDR002 were individually isolated and cloned into the Xba I site of pJDC402 to create pAM308 and pAM309, respectively. pAM320 contains the Hinf I fragment covering the *mic(ompC)* gene isolated from pAM308 cloned into the Pvu II site of pAM308. pAM321 was similarly constructed from pAM309, also to contain two *mic(ompC)* genes.

The *mic(ompA)* plasmids pAM301, pAM307, pAM313, pAM314, and pAM318 were constructed as described in the text in a manner similar to the construction of the *mic(lpp)* and the *mic(ompC)* genes. To construct pAM319, the Hinf I fragment containing the *mic(ompA)* gene was isolated from pAM307 and inserted back into the Pvu II site of pAM307. pAM315 was constructed in the same manner as pAM319, except that it contains two Hinf I fragments inserted into the Pvu II site of pAM307.

Analysis of Outer Membrane Protein Production

E. coli JA221/F′*lacI^q* carrying the appropriate plasmid were grown to a Klett–Summerson colorimeter reading of 30, at which time IPTG was added to a final concentration of 2 mM. After 1 additional hr of growth (approximately one doubling), 50 μCi of ³⁵S-methionine (Amersham, 1000 Ci/mMole) was added to 1 ml of the culture. The mixture was then incubated with shaking for 1 min, at which time the labeling was terminated by addition of 1 ml ice-cold stop solution (20 mM sodium phosphate, pH 7.1, containing 1% formaldehyde, and 1 mg/ml methionine). Cells were washed once with 10 mM sodium phosphate, pH 7.1, suspended in 1 ml of the same buffer, and sonicated with a Heat Systems Ultrasonics sonicator model W-220E with a cup horn adapter for 3 min (in 30 sec pulses). Unbroken cells were removed by low-speed centrifugation prior to collecting the outer membrane. Cytoplasmic membranes were solubilized during a 30 min incubation at room temperature in the presence of 0.5% sodium lauroyl sarcosinate (Chemical Additives Co.) (Fillip et al., 1973) and the outer fraction was precipitated by centrifugation at 105,000 ×g for 2 hr.

Lipoprotein and OmpA were analyzed by Tris-SDS polyacrylamide gel electrophoresis as described by Anderson et al. (1973). To analyze OmpC production, urea-SDS polyacrylamide gel electrophoresis was used according to the procedure described by Mizuno and Kageyama (1978). Proteins were dissolved in the sample buffer described by Laemmli (1970) and the solution was incubated in a boiling water bath for 8 min prior to gel application. The autoradiographs of dried gels were directly scanned by a Shimadzu densitometer. To determine relative amounts of the band of interest, the ratio of the area of the peak of interest to the area of an unaffected protein peak was determined for each sample. The unaffected proteins used for normalization are indicated by an "a" Figures 2, 3, 5, and 8. Such determinations were generally reproducible to within 15%.

RNA Analysis

Cells were grown and labeled with ^{3}H-uridine as described in Figure 4, then cell growth was stopped by rapidly chilling the culture on ice for less than 5 min. The cells were collected by centrifugation at 8000 rpm for 5 min. RNA was isolated using the following procedure. The cells were quickly resuspended in hot lysis solution (10 mM Tris-HCl [pH 8.0], 1 mM EDTA, 350 mM NaCl, 2% SDS, and 7 M urea) with vigorous vortexing for 1 min. The mixture was immediately extracted twice with phenol:chloroform (1:1) and twice with chloroform alone. One-tenth volume of 3 M sodium acetate (pH 5.2) was added to the mixture and 3 vol ethanol was added to precipitate the RNA. The precipitate was then dissolved in TE buffer (10 mM Tris-HCl [pH 7.5], 1 mM EDTA). For gel electrophoresis, equal counts were loaded in each lane. The RNA was separated on a 1.5% agarose gel containing 6% formaldehyde essentially as described by Goldberg (1980). The running buffer was 20 mM MOPS (3-[N-morpholino]propanesulfonic acid [Sigma]), 5 mM sodium acetate, and 1 mM EDTA, pH 7.0.

RNA was transferred to nitrocellulose paper as described by Thomas (1983) and filters were hybridized according to Rave et al. (1979). M13 hybridization probes specific for the mic(lpp) RNA and lpp mRNA were individually constructed by cloning the 112 bp Xba I fragment shown in Figure 1B into M13 mp9 in the appropriate orientation. A probe specific for the ompA mRNA was constructed by inserting a 1245 bp Xba I–Eco RI fragment (originally an Eco RV–Pst I (Movva et al., 1980c) fragment into M13 mp10 (Messing 1983). The probes were labeled according to Messing (1983).

Acknowledgments

We would like to thank Masatoshi Inukai for constructing pMH044, Dorothy Comeau-Fuhrman for constructing pDR001 and pDR002, and Paul Bingham and Zuzanna Zacher for their advice concerning the RNA analysis. We are also grateful to Peggy Yazulla and Myra Ward for their secretarial assistance. This work was supported by grant GH19043 from the United States Public Health Service and Grant NP3871 from the American Cancer Society. P. J. G. was supported by National Research Service Award IT32 GM08065 from the National Institutes of Health.

The costs of publication of this article were defrayed in part by the payment of page charges. This article must therefore be hereby marked "advertisement" in accordance with 18 U.S.C. Section 1734 solely to indicate this fact.

Received February 6, 1984; revised March 15, 1984

References

Anderson, W. W., Baum, P. R., and Gesteland, R. F. (1973). Processing of adenovirus 2-induced proteins. J. Virol. 12, 241–252.

Cole, S., Bremer, E., Hindennach, I., and Henning, U. (1982). Characterization of the promoters for the ompA gene which encodes a major outer membrane protein of Escherichia coli. Mol. Gen. Genet. 188, 472–479.

Filip, C., Fletcher, G., Wulff, J. L., and Earhart, C. F. (1973). Solubilization of the cytoplasmic membrane of Escherichia coli by the ionic detergent sodium-lauryl sarcosinate. J. Bacteriol. 115, 717–722.

Gold, L., Pribnow, D., Schneider, T., Shineding, S., Singer, B. S., and Starmo, G. (1981). Translational initiation in prokaryotes. Ann. Rev. Microbiol. 35, 365–403.

Goldberg, D. A. (1980). Isolation and partial characterization of the Drosophila alcohol dehydrogenase gene. Proc. Nat. Acad. Sci. USA 77, 5794–5798.

Green, P. J., and Inouye, M. (1984). Roles of the 5′ leader region of the ompA mRNA. J. Mol. Biol., in press.

Hirashima, A., and Inouye, M. (1973). Specific biosynthesis of an envelope protein of Escherichia coli. Nature 242, 405–409.

Kozak, M. (1983). Comparison of initiation of protein synthesis in prokaryotes, eukaryotes, and organelles. Micro. Biol. Rev. 47, 1–45.

Laemmli, U. K. (1970). Cleavage of structural proteins during the assembly of the head of bacteriophage T4. Nature 227, 680–685.

Messing, J. (1983). New M13 vectors for cloning. Meth. Enzymol. 101, 20–78.

Miller, J. H. (1972). Experiments in Molecular Genetics. (Cold Spring Harbor, New York: Cold Spring Harbor Laboratory).

Miller, J. H., and Reznikoff, W. S. (1978). The Operon. (Cold Spring Harbor, New York: Cold Spring Harbor Laboratory).

Mizuno, T., and Kageyama, M. (1978). Separation and characterization of the outer membrane of Pseudomonas aeruginosa. J. Biochem. 84, 179–191.

Mizuno, T., Chou, M. Y., and Inouye, M. (1983a). A comparative study on the genes for three porins of the Escherichia coli outer membrane. J. Biol. Chem. 258, 6932–6940.

Mizuno, T., Chou, M., and Inouye, M. (1983b). Regulation of gene expression by a small RNA transcript (micRNA) in Escherichia coli K-12. Proc. Japan Acad. 59, 335–338.

Mizuno, T., Chou, M., and Inouye, M. (1984). A novel mechanism regulating gene expression: translational inhibition by a complementary RNA transcript (micRNA). Proc. Nat. Acad. Sci. USA, in press.

Movva, N. R., Nakamura, K., and Inouye, M. (1980a). Amino acid sequence of the signal peptide of OmpA protein, a major outer membrane protein of Escherichia coli. J. Biol. Chem. 255, 27–29.

Movva, N. R., Nakamura, K., and Inouye, M. (1980b). Regulatory region of the gene for the OmpA protein, a major outer membrane protein of Escherichia coli. Proc. Nat. Acad. Sci. USA 77, 3845–3849.

Movva, N. R., Nakamura, K., and Inouye, M. (1980c). Gene structure of the OmpA protein, a major surface protein of Escherichia coli required for cell–cell interaction. J. Mol. Biol. 143, 317–328.

Nakamura, K., and Inouye, M. (1979). DNA sequence of the gene for the outer membrane lipoprotein of E. coli: an extremely AT-rich promoter. Cell 18, 1109–1117.

Nakamura, K., and Inouye, M. (1982). Construction of versatile expression cloning vehicles using the lipoprotein gene of Escherichia coli. EMBO J. 1, 771–775.

Nakamura, K., Pirtle, R., Pirtle, I., Takeishi, K., and Inouye, M. (1980). Messenger ribonucleic acid of the lipoprotein of the Escherichia coli outer membrane II. The complete nucleotide sequence. J. Biol. Chem. 255, 210–216.

Rave, N., Crtvenjakov, R., and Boedtker, H. (1979). Identification of procollagen mRNAs transferred to diazobenzyloxymethyl paper from formaldehyde agarose gels. Nucl. Acids Res. 6, 3559–3567.

Rosenberg, M., and Court, D. (1979). Regulatory sequences involved in promotion and termination of RNA transcription. Ann. Rev. Genet. 13, 319–353.

Simons, R. W., and Kleckner, N. (1983). Translational control of IS10 transposition. Cell 34, 683–691.

Thomas, P. (1983). Hybridization of denatured RNA transferred or dotted to nitrocellulose paper. Meth. Enzymol. 100, 255–266.

Tomizawa, J.-I., and Itoh, T. (1982). The importance of RNA secondary structure in ColE1 primer formation. Cell 31, 575–583.

A Technique for Expressing Eukaryotic Genes in Bacteria

L. Guarante, T.M. Roberts and M. Ptashne

There are two salient differences between the genetic signals required for gene expression in prokaryotes and eukaryotes. The first difference is in those sequences that direct RNA polymerase binding to DNA and initiation of transcription, that is, promoters. Although the exact nature of eukaryotic promoters is not understood, it is apparent that these signals do not, in general, function in bacteria. The second difference is in the signals that allow messenger RNA (mRNA) to be translated into protein.

are not efficiently read by the *Escherichia coli* translational machinery.

The above considerations suggest that the minimal requirements for the expression in *E. coli* of a cloned eukaryotic gene are that it be transcribed from a bacterial promoter and that the resultant mRNA bear an appropriately positioned SD sequence. Our method satisfies these requirements of fusing a fragment of DNA encoding a promoter and leader transcript of an *E. coli* gene is fused to a second fragment of DNA that bears the

relative to the ATG at the start of the eukaryotic gene is critical. The techniques we have used for this purpose have progressed through several stages. The promoter fragment was first used to express the phage λcI gene that encodes the λ repressor (4). The placement of the promoter in that case depended upon the availability of a restriction enzyme site at a specific location near the start of cI (6). The method was then generalized (with the use of λ's *cro* gene as a model system) to include genes that do not have restriction sites located very close to their starts (7, 8).

We begin by cloning the gene of interest on a plasmid (the coding sequences of eukaryotic—but not of prokaryotic—genes are often interrupted by intervening sequences and therefore the eukaryotic gene is usually cloned as a complementary DNA copy of the corresponding mRNA). We then introduce, if one does not already exist, a restriction enzyme site in the 5' flanking region within approximately 100 bp of the gene start. For this purpose, we usually use a synthetic linker fragment (9) that encodes a restriction site not found elsewhere on the plasmid. Next, we open the plasmid at that site and excise varying amounts of DNA with exonuclease. We then insert the promoter fragment, which encodes an SD sequence near one end, and close the plasmid. This produces a set of plasmids bearing the promoter fragment separated by varying distances from the gene. This means, of course, that the DNA encoding the bacterial SD sequence will also be separated by varying distances from the ATG at the start of the gene.

How do we recognize those placements of the *lac* promoter fragment that elicit efficient expression of the desired gene? It is possible in some cases to recognize expression of the gene by functional or immunological assay of bacterial clones [see (10) and (11)]. However, we have designed a strategy that is applicable to a cloned gene even in the absence of an assay for the gene product (12) (Fig. 1). Our method exploits the properties of a specially constructed plasmid (pLG) bearing a portion of the *lac*Z gene of *E. coli*. This *lac*Z DNA encodes a large COOH-terminal fragment of β-galactosidase which is enzymatically active regardless of the sequence of amino acids fused to its NH$_2$-terminus (13). The plasmid does not direct expression of β-galactosidase, however, because the DNA encoding the promoter

Summary. Methods are described that allow efficient expression in *Escherichia coli* of cloned eukaryotic genes. The methods require that the coding sequence of the gene in question be available in a form uninterrupted by intervening sequences (for example, as a complementary DNA clone). The gene products are synthesized unfused to other amino acid sequences. The genetic manipulations are simple, and require the plasmids described and commercially available enzymes.

In eukaryotic mRNA's, the only sequence required for eukaryotic ribosome binding and translation initiation seems to be the AUG (A, adenine; U, uracil; G, guanine) encoding the NH$_2$-terminal methionine which, in general, is the first AUG triplet of the message (1). Also required in at least some cases is a posttranscriptional modification (that is, cap) at the immediate 5' end of the transcript [for reviews, see (1)]. In bacterial mRNA's, there is a sequence in addition to the AUG that is apparently required for efficient bacterial ribosome binding and initiation of translation (2, 3). This 3- to 12-base pair (bp) sequence, known as the Shine-Dalgarno (SD) sequence, occurs in the 5' untranslated leader region of the mRNA; the SD sequence begins 3 to 11 nucleotides upstream from the AUG. This SD sequence is complementary to the 3' end of 16S ribosomal RNA, and the duplexing allowed by this complementarity is thought to play a role in stabilizing the initiation complex formed between the mRNA and the ribosome (3). It is, therefore, not surprising that mRNA's which lack SD sequences

coding sequences of the eukaryotic gene. The mRNA produced from this fusion contains a "hybrid" ribosome binding site (4) consisting of the SD sequence from the *E. coli* gene and the initiating AUG of the eukaryotic gene. As a convenient source of a promoter and leader, we have used a DNA fragment of the *lac* operon encoding the promoter and the *lac*Z leader through the SD sequence. This fragment ends two base pairs before the initiating ATG (T, thymine) codon of *lac*Z. The promoter fragment bears a mutation (UV^5) that renders it functional in the absence of the CAP protein (catabolite gene action protein) and cyclic adenosine monophosphate ordinarily required to stimulate this promoter. Moreover, the fragment bears the *lac* operator, the site at which the *lac* repressor controls the promoter. As a consequence, the levels of protein synthesized under direction of this promoter can be regulated by inducers of the *lac* operon such as isopropyl thiogalactoside (5).

The placement of the promoter fragment and, hence, the *lac*Z SD sequence

The authors are research investigators at the Biological Laboratories, Harvard University, Cambridge, Massachusetts 02138.

and start point of translation have been deleted. As shown in Fig. 1, we first fuse a 5' portion of the eukaryotic (or prokaryotic) gene to be expressed—called gene *X*—to plasmid pLG or a derivative thereof, so that an in-frame fusion protein is encoded (see below). If *X* is a eukaryotic gene, the fused gene *X'-'Z* is usually neither transcribed nor translated. Figure 1B shows how restriction enzymes and nucleases are used as outlined above to position the portable promoter at varying distances from the ATG encoding the NH$_2$-terminal methionine of protein X. A plasmid bearing an optimally positioned promoter directs synthesis of the enzymatically active X'-β-galactosidase fused protein. This protein begins with the NH$_2$-terminal methionine of protein X. These desired plasmids are recognized by transforming Lac$^-$ bacteria with the products of the reaction of Fig. 1B and picking clones that score strongly Lac$^+$ on the appropriate lactose indicator plates. Plasmids from clones producing high levels of the hybrid protein are thus recovered, and the eukaryotic gene is reconstituted. This is done by replacing the *lac*Z part of the hybrid gene with the 3' portion of the eukaryotic gene by means of recombination in vitro (Fig. 1C). We can then monitor expression of the unfused eukaryotic protein by, for example, specific incorporation of radioactive amino acids into plasmid-encoded proteins by the "maxicell" technique (*14*) (Fig. 2). [The *lac*Z plasmid, pLG, actually bears a *lac*I-*lac*Z fusion and comes in three forms (pLG-200, -300, and -400). Each bears a unique restriction cut just 5' to the *lac*Z coding sequence. When cut with the appropriate restriction enzyme, each plasmid is opened in a different reading frame so that if the appropriate plasmid is chosen, any coding sequence can be fused in frame with *lac*Z (see *12*).]

We have produced bacterial clones that express (separately) four eukaryotic proteins: t antigen of SV40, rabbit β-globin, and human fibroblast interferon (F-IF) with (pre-F-IF) and without (F-IF) its NH$_2$-terminal signal peptide (*8, 12, 15*). The sequences around the ribosome binding sites of genes that direct synthesis of the various eukaryotic proteins are shown in Fig. 3. In each case, a hybrid ribosome binding site has been formed in which the number of base pairs separating the SD sequence and the ATG (between 7 and 11 nucleotides) matches that of some known *E. coli* ribosome binding sites (*3*). In each of these cases (except t antigen), the *lac*Z fusion technique was used to recognize optimal promoter placements. In some cases,

these placements were very rare. For example, in the case of F-IF, only 0.01 percent of the colonies bearing plasmids with various promoter-hybrid gene fusions were strongly Lac$^+$. The fact that plasmids recognized in this way bear "hybrid" ribosomal binding sites at the beginning of the eukaryotic gene provides strong support for the hypothesis of Shine and Dalgarno concerning the key role of the SD-AUG sequences in efficient translation of mRNA.

We have estimated the amounts of protein produced by each of our plasmids in two ways. First, we have determined the levels of β-galactosidase syn-

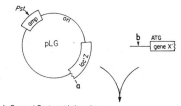

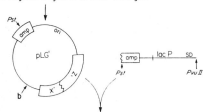

A 1. Open pLG at restriction site *a*
2. Insert amino terminal fragment of gene X to make fused gene

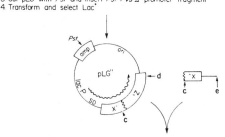

B 1 Open pLG' with restriction enzyme *b*
2 Resect with nucleases for varying distances
3 Cut pLG' with *Pst* and insert *Pst-Pvu II* promoter fragment
4 Transform and select Lac$^+$

C 1 Cut pLG'' with *c* and *d*
2 Replace c-d fragment with c-e fragment to reconstitute X

Fig. 1. A general method to maximize expression in *E. coli* of a eukaryotic gene. (A) A fragment of DNA bearing the NH$_2$-terminal region of gene *X* is inserted into restriction site *a* of a pLG plasmid, thereby fusing gene *X* to *lac*Z. (B) Plasmid pLG' bearing the fused gene is opened at a unique restriction site *b* which precedes the ATG of the fused gene. Resection and insertion of a portable promoter fragment that has a Shine-Dalgarno sequence is performed. Transformed clones that bear plasmids that direct the synthesis of high levels of β-galactosidase are identified as Lac$^+$ colonies on the appropriate indicator plates. (C) Gene *X* is reconstituted from plasmids that direct the synthesis of high levels of β-galactosidase as a fused product. This can be carried out, for example, by cutting the hybrid gene plasmid at any site (c) present in the gene *X* portion of the hybrid and at another site (d) in *lac*Z. Gene *X* is then reconstituted by the insertion of a DNA fragment that contains the COOH-terminal region of gene *X* extending from site c to a site (e) past the end of the gene.

```
......TAACAATTTCACAC AGGA AACAG CT ATG      β-galactosidase
......TAACAATTTCACAC AGGA AACAG AAAG ATG    SV40 t antigen (pTR436)
......TAACAATTTCACAC AGGA AACAG ACAGA ATG   rabbit β-globin (pLG302-3)
......TAACAATTTCACAC AGGA AACAG AC ATG      preFIF (pLG104)
......TAACAATTTCACAC AGGA AACAG CC ATG..    FIF (pLG117)
```

———— ·β-lactamase

Fig. 2 (left). SV40 t antigen synthesized in bacteria. The maxicell technique (see text) was used to specifically label plasmid-encoded proteins with radioisotopes. After the cells were labeled, they were disrupted, and the contents were examined directly by polyacrylamide gel electrophoresis and autoradiography. The positions of the β-lactamase (29 kilodaltons) and SV40 small t antigen (20 kilodaltons) are indicated.   Fig. 3 (right). The DNA sequences (of one strand) around the regions encoding the hybrid ribosome binding sites of several eukaryotic genes expressed in *E. coli*. The top line shows the sequence of the corresponding region of *lacZ*. The *lacZ* SD sequence is boxed as are the protein initiating ATG's. Sequences to the right of the vertical lines are from the indicated eukaryotic gene and the sequence to the left is from the portable promoter fund. The plasmids bearing the eukaryotic gene are pTR436 (*8*), pLG302-3 (*12*), pLG104 (*15*), and pLG117 (*15*).

———◀ · t

thesis directed by the plasmid-borne eukaryotic-*lacZ* fused genes. Second, following reconstitution of the eukaryotic gene, we have compared the amount of the unfused eukaryotic proteins synthesized with that of β-lactamase in so-called maxicell experiments. [As described by Sancar *et al*. (*14*), the maxicell procedure specifically labels with radioisotopes plasmid-encoded proteins. These proteins are then readily visualized by gel electrophoresis and autoradiography.] These methods give comparable results and indicate that our strains usually produce 5,000 to 15,000 molecules of the eukaryotic protein per cell.

In two cases, the identity of the eukaryotic proteins produced in bacteria was confirmed by automated amino acid sequence analysis of radioactively labeled proteins. Both t antigen and β-globin synthesized in bacteria were found to retain their NH₂-terminal methionine. The t antigen bears an NH₂-terminal methionine when produced in animal cells, but β-globin produced in the rabbit does not.

Many proteins are normally synthesized as precursors containing NH₂-terminal signal sequences that are cleaved as the protein is excreted. Our methods will readily produce the precursor forms of these proteins. It has been reported that rat preproinsulin synthesized in bacteria was converted to the mature form by these bacteria (*16*). In the case we

have examined, human pre-F-IF, we failed to detect correct processing of the precursor protein. If the precursor form of a protein is not efficiently processed by bacteria, it might be necessary to add (by DNA synthesis) an appropriately positioned ATG codon so that the mature form can be expressed directly using a bacterial promoter and a hybrid ribosome binding site (*17*). In the case of human F-IF, we were able to express the final form (F-IF) directly because the processed form of the protein begins with methionine.

The four eukaryotic proteins made by our method differ in stability in the bacterium. Pre-F-IF is quite unstable, β-globin is as stable as bacterial β-lactamase, and t antigen and mature F-IF are intermediate in stability. Extracts of bacteria producing F-IF, but not those producing pre-F-IF, display antiviral activity characteristic of authentic human F-IF (*15*). It is possible that differential sensitivity to proteases accounts for this result [see (*15*)].

There are several factors that might influence mRNA translation that we have not systematically investigated. Among these are the following: the identity of bases in the leader, including those between the SD and the AUG, and the identity of bases in the coding sequence that might effect codon usage or mRNA secondary structure. We know, for example, that in the case of the *cro* gene of phage λ, slight changes in the leader in

the region 5′ to the SD had a large influence on the level of gene expression (*7*). We are now in a position to systematically analyze these effects by isolating mutations that increase expression of *lacZ* fused genes.

References and Notes

1. J. A. Steitz, in *The Ribosomes: Structure, Function, and Genetics*, G. Chambliss *et al*., Eds. (University Park Press, Baltimore, 1979), pp. 479–495; F. Sherman, J. W. Stewart, A. M. Shweingruber, *Cell* **20**, 215 (1980); M. Kozak, *ibid*. **15**, 1109 (1978).
2. J. Shine and L. Dalgarno, *Nature (London)* **254**, 34 (1975).
3. J. A. Steitz, in *Biological Regulation and Development*, R. F. Goldberger, Ed. (Plenum, New York, 1979), vol. 1, pp. 349–389.
4. K. Backman and M. Ptashne, *Cell* **13**, 65 (1978).
5. J. H. Miller and W. S. Reznikoff, Eds., *The Operon* (Cold Spring Harbor Laboratory, Cold Spring Harbor, N.Y., 1978).
6. Our original goal in this research was to produce plasmids that would direct the synthesis of large amounts of the λ repressor and cro protein. These proteins play key roles in determining the alternative modes of growth (lytic as compared to lysogenic) of the phage. We were driven to the "hybrid ribosome binding site" solution for expressing the repressor when many alternative strategies failed (*4*). The availability of strains that make large amounts of repressor and cro has greatly facilitated isolation and analysis of the properties of these proteins [see, for example (*18*)]. Moreover, the fact that production of repressor and cro may be varied over a wide range in vivo has greatly facilitated unraveling the complexities of gene control in phage λ (*19, 20*).
7. T. M. Roberts, R. Kacich, M. Ptashne, *Proc. Natl. Acad. Sci. U.S.A.* **76**, 760 (1979).
8. T. M. Roberts, I. Bikel, R. R. Yocum, D. M. Livingston, M. Ptashne, *ibid*., p. 5596.
9. R. H. Sheller, R. E. Dickerson, H. W. Boyer, A. D. Riggs, K. Itakura, *Science* **196**, 177 (1977).
10. K. Struhl and R. W. Davis, *Proc. Natl. Acad. Sci. U.S.A.* **74**, 5255 (1977); A. C. Y. Chang, J. H. Nunberg, R. J. Kaufman, H. A. Erlich, R. T. Schimke, S. N. Cohen, *Nature (London)* **275**, 617 (1978).
11. S. Broome and W. Gilbert, *Proc. Natl. Acad. Sci. U.S.A.* **75**, 2746 (1978).
12. L. Guarente, G. Lauer, T. Roberts, M. Ptashne, *Cell* **20**, 543 (1980).
13. P. Bassford *et al*., in *The Operon*, J. H. Miller and W. S. Reznikoff, Eds. (Cold Spring Harbor Laboratory, Cold Spring Harbor, N.Y., 1978), pp. 245–262.
14. A. Sancar, A. Hack, D. Rupp, *J. Bacteriol.* **137**, 692 (1979).
15. T. Taniguchi, L. Guarente, T. M. Roberts, D. Kimelman, J. Douhan III, M. Ptashne, *Proc. Natl. Acad. Sci. U.S.A.*, in press.
16. K. Talmadge, J. Kaufman, W. Gilbert, *ibid.* **77**, 3988 (1980).
17. D. V. Goeddel *et al*., *Nature (London)* **281**, 544 (1979).
18. A. D. Johnson, C. O. Pabo, R. T. Sauer, *Methods Enzymol.* **65**, 839 (1980).
19. M. Ptashne, A. Jeffrey, A. D. Johnson, R. Maurer, B. J. Meyer. C. O. Pabo, T. M. Roberts, R. T. Sauer, *Cell* **19**, 1 (1980).
20. R. Maurer, B. J. Meyer, M. Ptashne, *J. Mol. Biol.* **139**, 147 (1980); B. J. Meyer, R. Maurer, M. Ptashne, *ibid*., p. 163; B. J. Meyer and M. Ptashne, *ibid*., p. 195.
21. We thank A. Johnson, Carl Pabo, and R. Brent for comments on the manuscript. L.G. is supported by the Jane Coffin Childs Memorial Fund for Medical Research.

25 July 1980

Simple RNA Enzymes with New and Highly Specific Endoribonuclease Activities

J. Haseloff and W.L. Gerlach

In vitro *mutagenesis of sequences required for the self-catalysed cleavage of a plant virus satellite RNA has allowed definition of an RNA segment with endoribonuclease activity. General rules have been deduced for the design of new RNA enzymes capable of highly specific RNA cleavage, and have been successfully tested against a new target sequence.*

CERTAIN naturally occurring RNA molecules possess the property of self-catalysed cleavage[1]. One class of this reaction is shared by a number of small circular RNA molecules which replicate in plants, either alone (viroid RNAs) or dependent on a helper virus (satellite RNAs). Self-cleavage has been demonstrated *in vitro* for avocado sunblotch viroid (ASBV)[2] and the satellite RNAs of tobacco ringspot virus (sTobRV)[3,4] and lucerne transient streak virus (sLTSV)[5] and appears to be an essential and unique part of the life cycle of these RNAs. During replica-

tion, circular forms of the RNAs are thought to act as templates for the production of longer than unit-length RNA transcripts[6-8], and the subsequent self-catalysed cleavage of these concatameric transcripts gives rise to unit-length progeny. In addition to these replicating RNAs, *in vitro* self-catalysed cleavage has now been observed in RNA transcripts of certain tandemly repeated DNA sequences from newt[9].

These self-catalysed RNA cleavage reactions share a requirement for divalent metal ions and neutral or higher *p*H, and

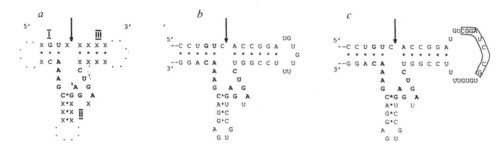

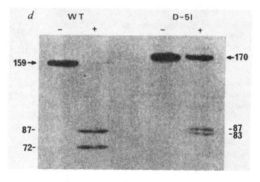

Fig. 1 A mutated self-cleavage domain which is active. *a*, The conserved structures associated with naturally-occurring RNA cleavage sites in ASBV[2,11], newt satellite DNA transcripts[9] and the satellite RNAs of TobRV[4], LTSV[5], velvet tobacco mottle virus[22], Solanum nodiflorum mottle virus[22] and subterranean clover mottle virus[23] are summarized. Nucleotide sequences which are absolutely conserved between these structures are shown, while others are represented as X. The three RNA helices (I, II and III) vary in length from 2 to 7 base pairs. Base-pairing is represented by (*) and the site for RNA cleavage is arrowed. (¹ an extra U is positioned after this residue in LTSV (+) strand.) *b*, The conserved 52-nucleotide structure associated with self-cleavage of (+) strand sTobRV RNA is shown schematically. We refer to this as the wild-type cleavage structure. Conserved residues are highlighted, and the site for cleavage is arrowed. *c*, The D-51 *in vitro* mutant of sTobRV possesses an altered self-cleavage domain, containing an insertion of 8-nucleotides (CGGAUCCG, shown boxed) together with a flanking duplication of three nucleotides (UGU, 7-9). *d*, Subcloned *Hae*III fragments of wild-type sTobRV and the D-51 *in vitro* mutant were each transcribed in both (−) and (+) orientations and radiolabelled transcripts were fractionated by PAGE. The positions of uncleaved 159 and 170 base transcripts from the wild-type (WT) and mutant (D-51) sequences are arrowed; sizes of cleavage products are shown.
Methods. 97 and 108 base-pair *Hae*III fragments containing the sites for self-cleavage were excised from sequenced plasmid clones containing wild-type and D-51 sTobRV sequences, respectively. The fragments were each ligated into the *Sma*I site of pGEM 4 and screened to obtain both orientations of the insert. The plasmids were linearized using *Eco*RI, and (+) and (−) strand RNAs of lengths 159 and 170 bases were transcribed using 200 units ml⁻¹ T7 RNA polymerase in 50 mM Tris-HCl *p*H 7.5, 10 mM NaCl, 6 mM MgCl₂, 2 mM spermidine, 1,000 unit ml⁻¹ RNasin, 500 µM ATP, CTP and GTP with 200 µM [α-³²P]UTP RNAs were fractionated by 10% polyacrylamide, 7 M urea, 25% formamide gel electrophoresis, and autoradiographed.

result in the production of RNA with termini possessing 5', hydroxyl and 2', 3' cyclic phosphate groups[3,5,9,10]. The cleavage reactions presumably result from RNA conformation bringing reactive groups into close proximity. The sites of cleavage are specific and associated with domains of conserved sequence and secondary structure. For sTobRV (ref. 4 and J.H. and W.L.G., unpublished results), ASBV[11] and sLTSV[12], it is known that precisely these conserved regions are required for cleavage, and may thus be directly involved in the reaction. A consensus of the domains associated with known RNA self-cleavage reactions is shown schematically in Fig. 1a. They consist of three branched RNA helices which flank the susceptible phosphodiester bond and two single-strand regions which are highly conserved in sequence. The helices are numbered I, II and III, following the convention of Forster and Symons[5], but the structures are redrawn with the sequences immediately adjacent to the site of cleavage shown across the top of the diagram. In the different self-cleavage domains, the base-paired stems are variously terminated by a small single-strand loop, or connected to the remainder of the viroid or satellite RNA molecule.

The self-catalysed cleavage of these RNAs is normally an intra-molecular reaction, that is a single molecule contains all the RNA-encoded functions required for cleavage. For example, cleavage of sTobRV (J.H. and W.L.G., unpublished results) and sLTSV RNAs[12] requires single contiguous sequences of about 52 nucleotides in length, which contain the conserved structures shown in Fig. 1a. The self-cleavage of ASBV in contrast requires the participation of two sequences which are widely separated on the intact RNA. Similar sequences have been transcribed in vitro as separate 19- and 24-nucleotide RNA fragments[11]. When mixed, the two fragments may base-pair (via stems I and II) to form the conserved structure associated with self-cleavage (Fig. 1a) and efficient cleavage of the 24-nucleotide RNA was observed. The 19-nucleotide fragment remained unaltered and could participate in many cleavage reactions, and therefore possessed the properties of an RNA enzyme. In this case, however, the cleaved RNA also contained conserved sequences and secondary structure, making this system unsuitable as a general model for the design of ribonucleolytic RNA enzymes with wide sequence specificity. Similarly, the more complex RNA component of RNAse P[13] and shortened forms of the self-splicing ribosomal RNA intervening sequence from Tetrahymena thermophila[14,15], which have both been shown to act as endoribonucleases, have so far proved of limited practical use for the design of new endonuclease activities. The term 'ribozyme' has been used to describe such RNAs with catalytic activity[15].

We have taken the single self-cleaving domain from the (+) strand of sTobRV and dissected its RNA substrate and enzyme activities. An RNA substrate was defined which possessed little conserved sequence and no essential secondary structure. Inspection of the separated substrate and ribozyme activities, and comparison with other naturally occurring self-cleaving domains, led to a model for the design of oligoribonucleotides which possess new and highly sequence-specific endoribonuclease activities. This model was successfully tested by the design and construction of ribozymes targeted against three sites within the Tn9 chloramphenicol acetyl-transferase (CAT) messenger RNA sequence.

A mutant that self-cleaves

Encapsidated forms of the satellite RNA of tobacco ringspot virus (+strand sTobRV) have been shown to undergo self-catalysed cleavage at a specific phosphodiester bond[3] between bases 359 and 1. RNA transcripts of sTobRV made in vitro also undergo efficient self-cleavage[4], and in vitro mutagenesis and transcription of cloned sTobRV complementary DNA has led to mapping of the sequences required for the reaction (J.H. and W.L.G., unpublished results). In these experiments, cloned sTobRV cDNAs were mutagenized using an oligonucleotide linker insertion protocol. A library of sTobRV mutants resulted,

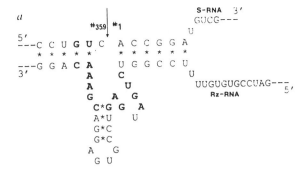

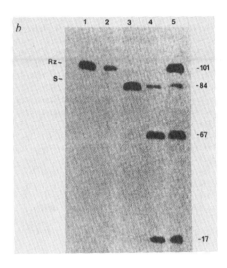

Fig. 2 Separation of substrate and catalytic activities. *a*, The inserted nucleotides in the D-51 mutant (Fig. 1c) contain a BamHI restriction endonuclease site. BamHI was used to split the mutant DNA, and the two sequences were subcloned and transcribed separately in vitro. The RNA transcripts are shown schematically, with potential base-pairings between the RNAs indicated (*). The fragment containing the arrowed site for cleavage was termed S-RNA, and the other Rz-RNA. *b*, [32P]Rz transcript (101 bases) was incubated alone (lane 1), and with unlabelled S-RNA (lane 2). [32P]S RNA was incubated alone (lane 3), and with unlabelled and [32P]Rz RNAs (lanes 4 and 5, respectively). The sizes of the two products are consistent with cleavage of the S-RNA (84 bases) at the normal site between nucleotides 359 and 1, to give 5' and 3' proximal fragments of 67 and 17 nucleotides, respectively. **Methods.** Isolated D-51 HaeIII fragment was digested with BamHI and the Rz and S fragments ligated into SmaI-BamHI digested pGEM3 and 4, respectively. T7 DNA polymerase treated, KpnI-digested Rz-pGEM3 and XhaI digested S-pGEM4 were transcribed using SP6 RNA polymerase under the same conditions used for T7 RNA polymerase (Fig. 1), either without radiolabel or in the presence of 200 μM [α-32P]UTP. The Rz and S RNAs were incubated alone in 50 mM Tris-HCl pH 8.0, 20 mM MgCl₂ at 50 °C for 60 min, fractionated on a 10% polyacrylamide, 7 M urea, 25% formamide gel and autoradiographed.

and nucleotide sequence analysis showed that each mutant contained an inserted BamHI linker sequence (CGGATCCG) together with flanking duplicated or deleted sTobRV sequences. The mutants were transcribed in vitro and the RNAs assayed for their ability to undergo cleavage. The only mutations affecting cleavage of (+) strand sTobRV were those which altered a 52-nucleotide sequence adjacent to the cleaved bond. This sequence contains the domain of conserved sequence and possible secondary structure required for self-cleavage in other RNAs (Fig. 1a,b).

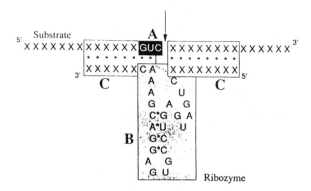

Fig. 3 Model for design of ribozymes. Three structural domains are boxed. (A) contains conserved sequences in the RNA substrate, immediately adjacent to the site of cleavage. (B) comprises the highly conserved sequences maintained in the ribozyme, and the regions (C) consist of flanking helices, with base-pairing between the substrate and ribozyme RNAs.

One mutant, designated D-51, which contained an alteration in this domain proved an exception (Fig. 1c). It contained an 8-nucleotide *Bam*HI linker sequence inserted between three duplicated sTobRV nucleotides numbered 7 to 9, yet still underwent cleavage. The wild-type and D-51 cleavage domains were subcloned as 97 and 108 base-pair *Hae*III fragments, respectively, and RNAs were transcribed in both orientations *in vitro*. As expected, no cleavage of (−) strand RNA transcripts was observed, but (+) strands of both the wild-type and D-51 sequences underwent cleavage (Fig. 1d), with cleavage of the D-51 RNA being somewhat less efficient than that of the wild-type. The extra 11-nucleotides contained in the D-51 mutant were precisely located within a proposed single-strand loop in the conserved structural domain associated with self-cleavage (Fig. 1c). This provided further experimental support for the structural model, and suggested that this single-strand loop region may not be required for cleavage.

Separation of substrate and nuclease activities

Using the *Bam*HI restriction endonuclease site inserted into D-51, flanking *Hae*III–*Bam*HI and *Bam*HI–*Hae*III fragments were obtained and each was subcloned to an *Escherichia coli* plasmid suitable for *in vitro* RNA transcription. This allowed us to effectively eliminate the mutated single-strand loop from the self-cleavage domain, splitting the region into two RNA segments (Fig. 2a). The smaller *Hae*III–*Bam*HI transcript contained sTobRV nucleotides 321 to 9, including the actual site of cleavage, and was termed the substrate or S-RNA. The *Bam*HI–*Hae*III transcript containing sTobRV nucleotides 7 to 48 was termed the ribozyme or Rz-fragment. These were separately transcribed with and without radiolabelled ribonucleotide as tracer. Both the S- and Rz-RNAs showed no significant degradation when incubated alone (Fig. 2b, lanes 1 and 3) under conditions suitable for highly efficient self-cleavage (50 °C, 20 mM MgCl₂, pH 8.0). The Rz-RNA also appeared unaltered after incubation with the S-RNA (Fig. 2b, lanes 2 and 5), but efficient cleavage of the S-RNA occurred (Fig. 2b, lanes 4 and 5) producing two fragments. The product sizes were consistent with cleavage at the normal site. These data show that the S-RNA acted as a substrate for ribonucleolytic cleavage by the Rz-RNA, which apparently acted in a catalytic fashion.

The substrate defined above contains none of the highly conserved secondary structures and few of the conserved sequences thought to be essential for cleavage. Instead, base-pairing between the sTobRV enzyme and substrate RNA segments (via stems I and III) may result in formation of an RNA complex with the required properties for cleavage. Inspection of the separated substrate and enzyme activities derived from

the sTobRV (+) strand, and comparison with other naturally occurring self-cleavage sites has led to a model for the design and construction of novel ribozymes.

Design of new ribozymes

A model showing the proposed minimal structural requirements for ribozyme-catalysed RNA cleavage is presented in Fig. 3, and consists of three elements. First, a region (A) containing the sequence GUC adjacent to the site for cleavage is brought into close proximity to a second region (B) of highly conserved sequence and secondary structure. Third, flanking regions (C) of base-paired RNA helix stabilize this interaction. The model provides the basis for the *de novo* design of ribozymes with new sequence-specific endoribonuclease activities by consideration of three main parameters.

(1) Before a ribozyme can be designed, a particular site for cleavage must be chosen within the target RNA. In naturally occurring self-cleavage, the site in the RNA substrate is 5' flanked by several nucleotides which are highly conserved (A). This sequence, GUC, usually immediately precedes the site of cleavage. In one exception, GUA precedes the active site of cleavage in the (−) strand of sLTSV[5]. Alteration of a self-cleaving sequence similar to that found in transcripts of newt satellite DNA also showed that the residues immediately preceding the cleavage site could be changed[16]. When the normal sequence GUC was changed to GUA or GUU, cleavage activity remained

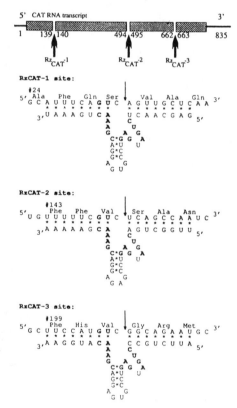

Fig. 4 Design of ribozymes targeted against the CAT gene transcript. Following the design rules described in this paper, ribozymes, termed Rz_CAT1, 2 and 3, were targeted against three sites within an 835 base *in vitro* transcript of the CAT gene. The relative locations of the cleavage sites on the transcript are shown schematically with the flanking bases numbered. The three ribozyme sequences are shown with the actual target sequences. Amino-acid sequences of the CAT gene are numbered and the predicted sites for RNA cleavage arrowed. Rz_CAT1 and 3 contain 24 base sequences derived from (+) strand sTobRV (region B, Fig. 3), while Rz_CAT2 contains a single U to A change in this region.

essentially unaltered, but no cleavage occurred after the sequence GUG. Further, cleavage was shown after the sequences CUC and, to a lesser extent, AUC and UUC, provided that the conserved base-pairing (Fig. 1a) was maintained. In the experiments described below, the sequence GUC was used as the sole requirement in the target RNA sequence for design of a corresponding enzyme. Future experiments will determine the full extent of the sequence requirements for efficient RNA cleavage.

(2) The region (B) contains sequences highly conserved in naturally occurring cleavage domains. In Fig. 1a, those nucleotides which are conserved in all known self-cleavage domains are specified, while only the positions and required base-pairing of the other nucleotides are indicated. The lengths of the base-paired stem II and associated loop do not appear to be conserved and the loop may be dispensable, as for ASBV. In this case, an RNA enzyme might be active as two halves, or subunits, each held to the other by formation of an extended base-paired stem II. But, conserved tertiary folding within this domain might not be evident from sequence comparisons, being maintained by compensating changes in nucleotide sequence, and initially we have used sequences similar to those of sTobRV for this region.

(3) The catalytic region and RNA substrate are held together by flanking regions of base-pairing (C) which must allow accurate positioning of the enzyme relative to the potential cleavage site in the substrate. The extent and type of base-pairing will directly affect the specificity, affinity (K_m) and turnover (k_{cat}) of an RNA enzyme. For the enzymes described below, a size of eight base-pairs has been arbitrarily chosen for each of the flanking regions (C).

Testing of synthetic ribozymes

To test these principles an indicator gene, chloramphenicol acetyl transferase (CAT)[17], was used to provide RNA transcripts as a target for cleavage by three newly designed RNA enzymes. Expression of the CAT gene can provide antibiotic resistance

Fig. 5 Cleavage of the CAT gene transcript. a, The [^{32}P]-labelled CAT RNAs were gel fractionated after incubation alone (−) or with one of the three ribozymes Rz$_{CAT}$1, 2 and 3 (lanes 1, 2 and 3, respectively). The location of the full-length transcript is arrowed. CAT mRNAs incubated with each of the ribozymes underwent efficient cleavage. In each case only two fragments were produced after incubation with a given ribozyme, and their nucleotide sizes were consistent with the predicted sites for cleavage (that is, 139 and 696, 494 and 341, 662 and 173 base fragments were the 5′ and 3′ products from R$_{zCAT}$1, 2, 3-catalysed cleavage respectively). b, 5′ Terminal base analysis. The 3′ fragments produced by ribozyme cleavage of CAT mRNA were [5′-^{32}P]-kinased, gel purified, subjected to complete nuclease digestion and the released terminal residues were fractionated by pH 3.5 PAGE. The 5′ terminal nucleotides, determined by reference to markers (lane M), were A, U and G for the fragments produced by Rz$_{CAT}$1, 2 and 3 (lanes 1, 2 and 3, respectively). c, 5′ terminal sequence analysis. [5′-^{32}P]-labelled CAT mRNA fragments were subjected to base-specific partial ribonucleolytic cleavage. The products were polyacrylamide gel fractionated to allow 5′ terminal sequence determination. Panels 1, 2 and 3 correspond to products of Rz$_{CAT}$1, 2 and 3, respectively. Base specific cleavages are shown above each lane except for lane N which is a size ladder control. Sequences from 4 to 6 to about 25 nucleotides 3′ of the cleavage site are indicated for each fragment, and are consistent with the expected sites for cleavage by each ribozyme.

Methods. The CAT gene was obtained from pCM4[24] and subcloned as a BamHI fragment into pGEM-blue. This plasmid was linearized with HindIII and CAT gene transcripts were obtained using T7 RNA polymerase with 200 μM [α-^{32}P]UTP. Ribozyme sequences were synthesized as oligodeoxynucleotides using automated phosphoamidite chemistry. The 47-, 48- and 44-mer oligonucleotides, Rz$_{CAT}$1, 2 and 3, respectively, were kinased, ligated with phosphatase treated, EcoRI-PstI cut pGEM4 and incubated with the Klenow fragment of DNA polymerase 1 before bacterial transformation. EcoRI-linearized plasmids were transcribed with T7 RNA polymerase to produce ribozyme RNAs. Ribozymes were incubated with CAT transcript in 50 mM Tris-HCl pH 8.0, 20 mM MgCl$_2$ at 50 °C for 60 min, and the products fractionated by 5% polyacrylamide 7 M urea, 25% formamide gel electrophoresis before autoradiography. Ribozyme-cleaved CAT mRNA fragments (2 μg) were incubated in 50 mM Tris-HCl pH 9.0, 10 mM dithiothreitol with 50 μCi [γ-^{32}P]ATP and 5 units T4 polynucleotide kinase for 30 min at 37 °C. Radiolabelled fragments were purified on a 5% polyacrylamide gel and digested with an equal volume of 500 units ml^{-1} RNase T1, 25 units ml^{-1} RNase T2 and 0.125 mg ml^{-1} RNAse A in 50 mM ammonium acetate pH 4.5 for 120 min at 37 °C. Products were fractionated on a 20% polyacrylamide gel containing 25 mM sodium citrate pH 3.5 and 7 M urea[25], and were detected by autoradiography. Terminal sequences were determined directly using the partial enzymatic digestion technique[26].

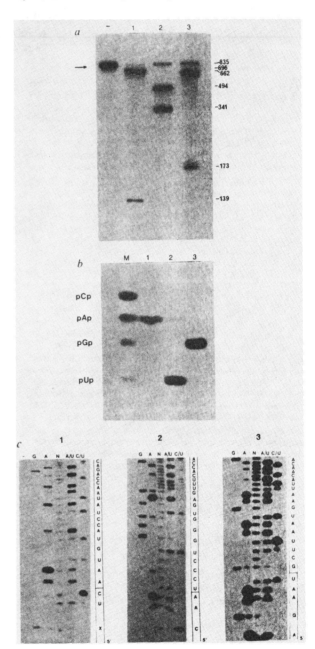

in bacteria, plants and animals and this can be easily assayed. The coding region of the CAT gene transcript was scanned for the occurrence of the sequence GUC as potential sites for RNA cleavage. Three such sites were arbitrarily chosen, and their flanking sequences used to define the eight-nucleotide regions of complementary sequence within the ribozymes, which flanked conserved sequences based on those of the sTobRV (+) strand self-cleavage site (Fig. 4).

The ribozyme sequences (Rz$_{CAT}$ 1, 2 and 3) were synthesized as single-stranded oligodeoxyribonucleotides, with the addition of sequences containing half-sites for EcoRI (AATTC) and PstI (CTGCA) restriction endonucleases at the 5' and 3' termini, respectively. These additional sequences are partly complementary to the overhanging termini left by EcoRI and PstI digestion of double-stranded DNA. The kinased oligomers were incubated with EcoRI–PstI digested, phosphatase-treated plasmid DNA in the presence of DNA ligase and DNA polymerase, and after transformation the resulting clones were screened for presence of the EcoRI–PstI insert. This procedure allowed efficient cloning of the ribozyme sequences using single-strand oligomers, obviating the need for synthesis of a second complementary oligomer.

The ribozyme sequences and the CAT gene were cloned downstream of promoters for T7 RNA polymerase, and this enzyme was used to obtain RNA transcripts in vitro. When the 835-nucleotide CAT transcript was incubated with any one of the three ribozymes, efficient and highly sequence-specific cleavage occurred producing two RNA fragments (Fig. 5a). The conditions required for ribozyme-catalyzed cleavage were similar to those observed for the naturally-occurring cleavage reactions[3,5,9,11] with more efficient cleavage occurring at elevated pH, temperature and divalent cation concentrations (data not shown). When present in molar excess, the three ribozymes catalysed almost complete cleavage of the CAT RNA substrate after 60 minutes in 50 mM Tris-HCl pH 8.0, 20 mM MgCl$_2$ at 50 °C (Fig. 5a). Under similar conditions with 0.1 μM substrate and 3 μM ribozyme, the T$_{1/2}$ of CAT mRNA substrate was 3.5, 3.5 and 2.5 min, in the presence of Rz$_{CAT}$ 1, 2 and 3, respectively. The ribozyme sequences were inactive in the form of oligodeoxyribonucleotides or against the complement of the substrate RNA (data not shown).

The sizes of the CAT RNA fragments produced by each ribozyme were consistent with the predicted sites for cleavage. The 3' terminal cleavage fragments from each ribozyme-catalysed reaction were isolated and 5'^{32}P-kinased. Efficient kinasing of the fragments indicated that they possessed 5' terminal hydroxyl groups, similar to those produced in naturally-occurring cleavage reactions. Terminal nucleotide analysis of these fragments demonstrated that cleavage of CAT sequences by Rz$_{CAT}$ 1, 2 and 3 occurred precisely before nucleotides A, U and G, respectively (Fig. 5b) and RNA sequence determination confirmed that cleavage had occurred at the expected locations within the CAT RNA (Fig. 5c).

Enzymatic catalysis

To demonstrate that these ribozymes cause cleavage of the CAT mRNA substrate in a catalytic manner, each was incubated with a molar excess of substrate, under conditions which should favour both efficient cleavage and product dissociation. Figure 6 shows the results of an experiment where, after 75 min at 50 °C, pH 8.0 in 20 mM MgCl$_2$, 10 pmol of Rz$_{CAT}$1 had catalysed specific cleavage of 163 pmol of a truncated CAT mRNA substrate. On average, each ribozyme had participated in greater than 10 cleavage events. Similar results were obtained for Rz$_{CAT}$ 2 and 3 (data not shown), and thus each acts as an RNA enzyme.

Mechanism of cleavage

All three RNA enzymes which were designed according to the principles outlined above have been demonstrated to catalyse effective and precise cleavage of their target RNA sequences. It is therefore likely that these design principles are generally valid,

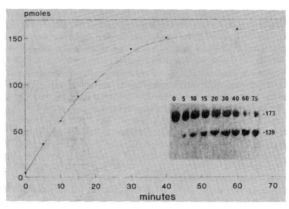

Fig. 6 Enzymatic catalysis. Rz$_{CAT}$1 (15 pmol) was incubated for varying times with a 20-fold molar excess of truncated (173 base) CAT mRNA containing the site for Rz$_{CAT}$1 cleavage. Rz$_{CAT}$1 catalysed cleavage of the CAT RNA to form 5' and 3' fragments, of 139 and 34 bases, respectively. The inset shows the accumulation of the 139 base fragment with time, after PAGE. The amounts of the 139 base fragment were quantified and plotted. After 75 min at 50 °C some non-specific cleavage of RNA was noticed due to the extreme conditions, but 70% of the remaining intact RNAs (163 pmol) had accumulated as the 139 base fragment.
Methods. PvuII-linearized pGEM-blue containing the CAT gene sequence, and EcoRI-linearized pGEM4 containing Rz$_{CAT}$1 were transcribed in vitro with T7 RNA polymerase and 200 μM [α-^{32}P]UTP. 300 pmol of 173 base long, truncated CAT mRNA and 15 pmol of Rz$_{CAT}$1 were mixed in 100 μl of 50 mM Tris-HCl pH 8.0, 20 mM MgCl$_2$. 10 μl aliquots were incubated for varying times at 50 °C. Samples were fractionated by 7 M urea, 5% PAGE, stained with toluidine blue, the bands were excised, and amounts of radiolabelled RNA were determined.

as is the model upon which they are based. The model contains two main assumptions: (1) that catalysis requires that the RNA enzyme and substrate interact through base-pairing (although tertiary interactions presumably exist between enzyme and substrate, they do not need to be considered in the design), and (2) that base-pairing allows precise positioning of a conserved domain (B, Fig. 3) adjacent to a site for cleavage. It may not only be useful, but of functional significance, to consider this conserved domain as the catalytic section or active site of the RNA enzymes.

The RNA enzyme- and related self-catalysed cleavage reactions appear to proceed via concomitant cleavage of the susceptible 3',5'phosphodiester linkage and formation of termini containing 2',3'cyclic phosphodiester and 5' hydroxyl groups[3,5,9,11], a mechanism similar to that occurring during base-catalysed hydrolysis of RNA. There exists a previously described case of similar, highly specific RNA cleavage for which detailed structural information is available. During X-ray crystallographic studies of yeast tRNA^{phe}, heavy-atom derivatives were sought. When the tRNA^{phe} crystals or solutions were incubated with lead (Pb(II)) salts, specific cleavage of the RNA occurred between nucleotides 17 and 18 (refs 18, 19). The cleavage was pH dependent, with increased rates observed around neutral pH, and produced termini containing 2',3' cyclic phosphodiester and 5' hydroxyl groups. Three Pb(II) ions were located within the tRNA^{phe} molecule and one, covalently bound to bases 59 and 60 within the TΨC-loop, was positioned in close proximity to the cleaved bond in the D-loop. Brown et al.[18] suggested that this Pb(II) ion acted as a source of metal-bound hydroxyl ions near neutral pH, as Pb(II) bound water molecules possess a pK$_a$ of ~pH 7.4. The perhaps fortuitously but precisely positioned (Pb-OH)$^+$ moiety could abstract a proton from the 2' hydroxyl group of nucleotide 17, thus initiating nucleophilic attack of the phosphate atom in the adjacent phosphodiester bond between nucleotides 17 and 18, and resulting in formation of a 2′,3′ cyclic phosphate group and 5′ hydroxyl leaving group. A similar mechanism has been proposed for the initial step in cleavage by the protein RNase A, where the precisely positioned imidazole of histidine-12 acts as a general base to deprotonate the 2′ hydroxyl of the RNA substrate[20].

The ribozyme- and self-catalysed cleavage reactions discussed in this paper require divalent metal ions for activity, are pH-dependent, produce similar terminal groups, and thus may involve similar mechanisms to those responsible for (PbII)-catalysed tRNA[phe] cleavage. In addition, the involvement of such mechanisms would be consistent with the simple structural properties of the ribozymes outlined above. For example, the domain of a highly conserved sequence within the ribozyme might constitute a metal-ion binding site. The flanking base-paired regions joining the ribozyme and substrate would serve to both stabilize the structure of this binding site and bring a bound metal ion in close proximity to the 2′ hydroxyl group adjacent to the susceptible phosphodiester bond. The precisely positioned metal-bound hydroxyl group could then abstract the proton from the 2′ hydroxyl group and initiate cleavage of the substrate RNA in a fashion similar to that seen in tRNA[phe] cleavage. Alternatively, cations may play a structural role in catalysis, stabilizing the interaction between substrate and reactive group(s) present in the RNA enzyme. Detailed structural and kinetic studies are required to test such models for ribozyme action, and may provide a further basis for manipulation of the substrate specificity, and catalytic activity of ribozymes.

Potential applications

The RNA enzymes described in this paper have been shown to be capable of efficient and specific cleavage of RNA sequences *in vitro*, and are based on RNA self-cleavage reactions which occur normally *in vivo*. Accordingly, the ability to design and use specific endoribonucleases for cleavage of particular RNA species or target sequences may prove useful both *in vitro* and *in vivo*.

Ribozymes could be used for the *in vitro* manipulation of RNAs either to produce large quantities of particular RNA fragments or, by using ribozymes with lesser specificities, as a means of physically mapping RNAs. In addition, it may be possible to construct ribozymes as two halves, with the halves held together by extended base-pairing. Each half of such a ribozyme would be specific for one side of a particular cleavage site. By combining ribozyme halves it would be possible to generate new substrate specificities. Given that certain self-catalysed cleavage reactions are reversible to some extent[3,21], it may also be possible to isolate and splice together specific RNA fragments using this approach.

A major potential application for these highly sequence-specific endoribonucleases is in cleavage, and thereby inactivation, of gene transcripts *in vivo*. It is now possible to express foreign genes in a number of bacterial, fungal, plant and animal species either through stable chromosomal transformation of a particular genotype, transient expression via episomal or viral vectors, or by microinjection. Such methods could also be used for the introduction and expression of ribozyme sequences. Genes are universally expressed via RNA transcripts, giving rise to structural/functional RNAs or messenger RNAs for polypeptide synthesized. Essentially any RNA is a potential substrate for cleavage by a ribozyme. Provided that the transcribed sequences of the gene are known, it should be possible to target one or more ribozymes against specific RNA transcripts. Expression *in vivo* of such ribozymes and cleavage of the transcripts would in effect inhibit expression of the corresponding gene. This 'anti-gene' activity of the ribozymes could provide a basis for various gene and viral therapies and analyses.

We are grateful to Tim Close for pCM4 plasmid and Kim Newell and Paul Whitfeld for synthesis of oligonucleotides. We thank George Bruening for helpful discussion, and Mark Young, Jim Tokuhisa, John Watson, Bob Symons and Jim Peacock for comments on the manuscript. J.H. was supported by a Queen Elizabeth II Research Fellowship.

Received 13 June; accepted 30 June 1988.

1. Cech, T. R. *Science* **236**, 1532–1539 (1987).
2. Hutchins, C. J., Rathjen, P. D., Forster, A. C. & Symons, R. H. *Nucleic Acids Res.* **14**, 3627–3640 (1986).
3. Prody, G. A., Bakos, J. T., Buzayan, J. M., Schneider, I. R. & Bruening, G. *Science* **231**, 1577–1580 (1986).
4. Buzayan, J. M., Gerlach, W. L. & Bruening, G. B. *Proc. natn Acad. Sci. U.S.A.* **83**, 8859–8862 (1986).
5. Forster, A. C. & Symons, R. H. *Cell* **49**, 211–220 (1987).
6. Kiefer, M. C., Daubert, S. D., Schneider, I. R. & Bruening, G. B. *Virology* **121**, 262–273 (1982).
7. Bruening, G. B., Gould, A. R., Murphy, P. J. & Symons, R. H. *FEBS Lett.* **148**, 71–78 (1982).
8. Hutchins, C. J. *et al. Plant molec. Biol.* **4**, 293–304 (1985).
9. Epstein, L. M. & Gall, J. G. *Cell* **48**, 535–543 (1987).
10. Buzayan, J. M., Gerlach, W. L., Bruening, G. B., Keese, P. & Gould, A. R. *Virology* **151**, 186–199 (1986).
11. Uhlenbeck, O. C. *Nature* **328**, 596–600 (1987).
12. Forster, A. C. & Symons, R. H. *Cell* **50**, 9–16 (1987).
13. Guerrier-Takada, C., Gardiner, K., Marsh, T., Pace, N. & Altman, S. *Cell* **35**, 849–957 (1983).
14. Zaug, A. J., Been, M. D. & Cech, T. *Nature* **324**, 429–433 (1986).
15. Kim, S-H. & Cech, T. R. *Proc. natn Acad. Sci. U.S.A.* **84**, 8788–8792 (1987).
16. Koizumi, M., Iwai, S. & Ohtsuka, E. *FEBS Lett.* **228**, 228–230 (1988).
17. Alton, N. K. & Vapnek, D. *Nature* **282**, 864–869 (1979).
18. Brown, R. S., Hingerty, B. E., Dewan, J. C. & Klug, A. *Nature* **303**, 543–546 (1983).
19. Brown, R. S., Dewan, J. C. & Klug, A. *Biochemistry* **24**, 4785–4801 (1985).
20. Wlodawer, A., Miller, M. & Sjolin, L. *Proc. natn Acad. Sci. U.S.A.* **80**, 3628–3631 (1983).
21. Buzayan, J. M., Gerlach, W. L. & Bruening, G. *Nature* **323**, 349–352 (1986).
22. Haseloff, J. & Symons, R. H. *Nucleic Acids Res.* **10**, 3681–3691 (1982).
23. Haseloff, J. thesis, Univ. Adelaide (1983).
24. Close, T. J. & Rodriguez, R. *Gene* **20**, 305–316 (1982).
25. Dewachter, R. & Fiers, W. *Analyt. Biochem.* **49**, 184–197 (1974).
26. Donis-Keller, H., Maxam, A. M. & Gilbert, W. *Nucleic Acids Res.* **4**, 2527–2538 (1980).

A cDNA Cloning Vector that Permits Expression of cDNA Inserts in Mammalian Cells

H. Okayama and P. Berg

This paper describes a plasmid vector for cloning cDNAs in *Escherichia coli*; the same vector also promotes expression of the cDNA segment in mammalian cells. Simian virus 40 (SV40)-derived DNA segments are arrayed in the pcD vector to permit transcription, splicing, and polyadenylation of the cloned cDNA segment. A DNA fragment containing both the SV40 early region promoter and two introns normally used to splice the virus 16S and 19S late mRNAs is placed upstream of the cDNA cloning site to ensure transcription and splicing of the cDNA transcripts. An SV40 late region polyadenylation sequence occurs downstream of the cDNA cloning site, so that the cDNA transcript acquires a polyadenylated 3′ end. By using pcD-α-globin cDNA as a model, we confirmed that the α-globin transcript produced in transfected cells is initiated correctly, spliced at either of the two introns, and polyadenylated either at the site coded in the cDNA segment or at the distal SV40 polyadenylation signal. A cDNA clone library constructed with mRNA from SV40-transformed human fibroblasts and this vector (about 1.4×10^6 clones) yielded full-length cDNA clones that express hypoxanthine-guanine phosphoribosyltransferase (Jolly et al., Proc. Natl. Acad. Sci. U.S.A., in press).

Cloned cDNA copies of cellular and viral mRNAs have provided invaluable aids for the molecular analysis of eucaryote gene structure, arrangement, and expression (4, 5, 7, 10, 12, 23, 24, 43, 44, 47, 49). Virtually all of the diverse procedures used to construct cDNAs rely on primer-initiated reverse transcription to create the complement of the mRNA sequence (14, 48) and either self (13, 14)- or oligo deoxynucleotide-primed (29, 40) second-strand synthesis to yield double-stranded cDNAs. For molecular cloning, the duplex cDNAs are joined to plasmid (13, 14, 29, 40) or bacteriophage (42; R. A. Young and R. W. Davis, Proc. Natl. Acad. Sci. U.S.A., in press) vectors via complementary homopolymeric tails (12, 14, 40) or cohesive ends created with linker segments containing appropriate restriction sites (43, 44; Young and Davis, in press). The detection and subsequent isolation of specific cloned cDNAs are simplified if pure or enriched samples of the corresponding mRNA are available (12, 23). Although the procedures are tedious and time consuming, rare cDNAs present in complex cDNA libraries can be detected with suitable hybridization probes. However, the search is vexing and diffi-

cult if appropriate hybridization probes are lacking. In the latter instances, cDNA libraries have been screened for the ability of individual cloned segments to hybridize mRNA that can be translated to the corresponding proteins in vitro (22, 38) or in Xenopus oocytes (20) (e.g., by measuring antigenic determinants or enzymatic or other biological activities). In some cases (49), the cloned cDNAs have been identified by their ability to block translation of homologous mRNAs by hybrid arrest (39).

Several attempts have been made to devise vectors that promote expression of the cloned cDNA in *Escherichia coli*, so that the desired cDNAs can be identified by detection of proteins, antigens, or specific phenotypes produced in vivo (8, 49; Young and Davis, in press). Recently, we described a procedure for obtaining cloned cDNAs, many of which are full length or nearly full length (37). In that method, the first cDNA strand is primed by polydeoxythymidylic acid [poly(dT)] covalently joined to one end of a linear plasmid vector DNA; then, the plasmid vector is cyclized with a linker DNA segment that bridges one end of the plasmid to the 5′ end of the cDNA coding sequence. In this work, a DNA fragment that contains the simian virus 40 (SV40) early region promoter and a modified SV40 late region intron was used as the linker fragment. This modification creates a recombinant whose cDNA insert can be tran-

† The plasmids pcDV1 and pL1 described here, as well as the corresponding plasmids described previously (H. Okayama and P. Berg, Mol. Cell. Biol. 2:161–170, 1982), can be obtained from M. Olive at P-L Biochemicals, Inc., 1037 McKinley Ave., Milwaukee, WI 53205.

scribed and processed in mammalian cells and, if the cDNA contains the entire protein coding sequence, can direct the production of the relevant protein.

This paper describes the modified cloning vector and demonstrates that α-globin and dihydrofolate reductase (DHFR) cDNAs, inserted at the cloning site of the vector, are efficiently expressed after transfection into cells. A cDNA clone library prepared in the expression vector with mRNA from SV40-transformed human fibroblasts yielded a clone that encodes the entire amino acid sequence of hypoxanthine-guanine phosphoribosyltransferase (HPRT) and expresses functional HPRT (D. J. Jolly, H. Okayama, P. Berg, A. C. Esty, D. Filpula, P. Bohlen, G. G. Johnson, and T. Friedmann, Proc. Natl. Acad. Sci. U.S.A., in press).

MATERIALS AND METHODS

Cells, enzymes, and chemicals. The care and maintenance of SV40-transformed CV1 monkey cells (COS cells) has already been described (17). T. Friedmann (University of California, San Diego) provided the SV40-transformed human fibroblast cell line (GM 637). *Sal*I endonuclease was obtained from K. Burtis (Standord University, Stanford, Calif.), and other restriction enzymes, bacterial alkaline phosphatase, T4 DNA ligase, T4 polynucleotide kinase, S1 nuclease, and terminal transferase were obtained from commercial sources. Avian myeloblastosis virus reverse transcriptase was provided by J. Beard (Life Sciences, Inc., St. Petersburg, Fla.) and was purified further. The purification procedure was carried out at 0 to 4°C as follows. Reverse transcriptase (about 9,000 U in 0.6 ml) was applied to a Sephacryl S200 column (0.7 by 28 cm) that had been equilibrated with buffer containing 0.2 M potassium phosphate (pH 7.2), 2 mM dithiothreitol, 0.2% Triton X-100, and 20% glycerol. Fractions (0.35 ml each), obtained by elution of the column with the same buffer, were assayed for reverse transcriptase activity with polyadenylic acid [poly(A)] and oligo(dT) as template and primer, respectively, and [α-^{32}P]dTTP. Fractions containing the peak of reverse transcriptase activity were pooled and dialyzed against the same buffer containing 50% glycerol. The dialyzed enzyme solution was frozen in liquid nitrogen and stored in small portions at −70°C.

Oligonucleotides containing *Sal*I, *Xho*I, *Eco*RI, *Bam*HI, and *Pst*I restriction sites were purchased from Collaborative Research, Inc., Waltham, Mass. Sucrose (ultrapure) was from Schwarz/Mann, Orangeburg, N.Y., and oligo(dT)-cellulose was from Collaborative Research, Inc.

Plasmids pcDV1 and pL1. The plasmid pcDV1 was the starting material for preparation of the vector-primer used in the cDNA cloning procedure (37). It was prepared from pBR322SV (map units 0.71 to 0.86) (37) by standard recombinant DNA procedures and has the following salient features (Fig. 1): the SV40 DNA segment contributes *Kpn*I and *Eco*RI restriction sites for the preparation of the vector-primer containing a poly(dT) tail at one end; a *Hin*dIII site at the join of pBR322 and SV40 DNA provides the cohesive end at which the linker fragment is joined to the vector

DNA; and an SV40 DNA segment (stippled in Fig. 1) contains the polyadenylation specification sequence from the late region of the virus (6, 15).

Plasmid pL1 provides the linker segment (shown shaded in Fig. 1) that joins the 5' coding end of the cDNA to the plasmid vector. The linker segment contains the SV40 early region promoter immediately upstream of a sequence encoding two functional introns (Fig. 2). One intron corresponds to the sequence that is spliced in the formation of SV40 19S late mRNA (6, 16, 28). The other intron is a modified form of the sequence that is spliced to form SV40 16S late mRNA; the modification substitutes a *Bam*HI linker for 850 base pairs (bp) within the intron between SV40 map positions 0.77 and 0.93. The linker segment contains a *Hin*dIII cohesive sequence at one end and an oligodeoxyguanylic acid [oligo(dG)] tail at the *Pst*I end for use in the cDNA cloning procedure.

Construction of pcD-*dhfr* and pcD-αG. pcD-*dhfr* was made by ligating three fragments together with T4 DNA ligase: the large *Hin*dIII-*Bam*HI fragment from pcDV1 containing the pBR322 *ori* and Amp^r, the *Hin*dIII-*Pst*I fragment from pL1 containing the SV40 early region promoter and late region introns, and the *Pst*I-*Bgl*II fragment containing the entire DHFR protein coding sequence (*dhfr* cDNA clone 26 [8]). A cloned plasmid having the structure shown in Fig. 3 was isolated after transformation of *Escherichia coli* HB101 with the ligated DNA (data not shown). pcD-α-globin (pcD-αG), having the structure shown in Fig. 3, was generated in a similar way, except that a *Pst*I-*Bam*HI fragment containing a full-length α-globin cDNA segment (37) replaced the *dhfr* cDNA.

Construction of human fibroblast cDNA library with pcD expression vector. (i) Preparation of mRNA. SV40-transformed human fibroblasts (GM637) were grown to half confluency in Dulbecco modified Eagle medium supplemented with 10% fetal calf serum, streptomycin, and penicillin. Total RNA was obtained by lysis of the cells in 4 M guanidinium thiocyanate, sedimentation through a CsCl cushion, and alcohol precipitation (9). Polyadenylated [poly(A)$^+$] RNA was obtained by one cycle of oligo(dT) cellulose chromatography (2).

(ii) Preparation of vector-primer and linker DNAs. The vector-primer DNA, tailed with poly(dT) at the *Kpn*I site closest to the polyadenylation signal, was prepared from pcDV1 DNA as described previously (37), except that cleavage with *Eco*RI, instead of *Hpa*I endonuclease, was used to remove the other poly(dT) tail. The linker fragment containing *Hin*dIII and oligo(dG) cohesive ends was prepared from pL1 by methods described previously (37).

(iii) cDNA cloning. With the exception of the source of the mRNA and the modified vector-primer and linker fragment, the cDNA library was constructed with only minor modifications in the procedure described previously (37). About 8 μg of poly(A)$^+$ RNA in 30 μl of 5 mM Tris-hydrochloride (pH 7.5) was heated at 65°C for 5 min, cooled to 37°C, and immediately adjusted to contain 50 mM Tris-hydrochloride (pH 8.3), 8 mM MgCl$_2$, 30 mM KCl, 0.3 mM dithiothreitol, 2 mM each of dATP, dGTP, dTTP, and [α-^{32}P]dCTP (600 cpm/pmol), and 4.5 μg (2.1 pmol) of the poly(dT) singly tailed vector-primer DNA derived from pcDV1 (total volume, 60 μl). The reaction was initiated by the addition of 60 U of reverse transcriptase and continued for 30 min at 37°C. Tailing of the

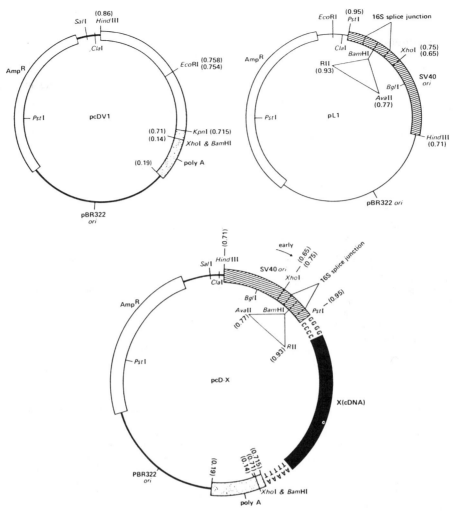

FIG. 1. The structure and component parts of the pcD vector system. pcDV1 is a recombinant of a segment of pBR322 DNA that extends counterclockwise from the *Hind*III restriction site to the position marked 0.19 and two segments of SV40 DNA; one of the SV40 segments, which lies between the same *Hind*III site and the *Kpn*I site, is SV40 DNA corresponding to map positions 0.715 and 0.86, and the other is the SV40 DNA segment flanked by the *Bam*HI and *Bcl*I restriction sites (map positions 0.145 to 0.19). The *Bam*HI to *Bcl*I SV40 DNA segment is retained, but the *Hind*III to *Kpn*I segment is lost from the pcD vector during the construction of the pcD-x recombinants. pL1 is also a recombinant between pBR322 DNA and two segments of SV40 DNA (shown hatched). One portion, which contains the SV40 origin of DNA replication (*ori*) and the early and late promoters (6), corresponds to the SV40 segment between the *Pvu*II and *Hind*III restriction sites at map positions 0.71 to 0.65; the other, which is joined to it on the early region promoter side, derives from the SV40 late region between map positions 0.75 and 0.95, with a *Bam*HI sequence replacing the internal region between map positions 0.77 and 0.93. For use as the linker segment, the hatched fragment contains a *Hind*III cohesive end and an oligo(dG) sequence at the *Pst*I terminus (37). pcD-x depicts the generalized structure of pcD-cDNA recombinants. Shown are the linker segment mentioned above, the dGdC bridge between the linker and the cloned cDNA (shown as the solid stretch), the dAdT stretch derived from the dT primer and the RNA poly(A) tail, and the segment carrying the SV40 late region polyadenylation signal (stippled region).

cDNA with oligodeoxycytidylic acid in the next step and the subsequent cleavage with *Hind*III endonuclease were performed at three times the scale previously described (37). After *Hind*III endonuclease digestion, the material obtained by alcohol precipitation was dissolved in 40 µl of 5 mM Tris-hydrochloride (pH 7.5) containing 0.5 mM EDTA and 50% ethanol and stored at −20°C. The cyclization step, in which the linker fragment bridges the *Hind*III cohesive end of the vector and the oligodeoxycytidylic acid tail of the

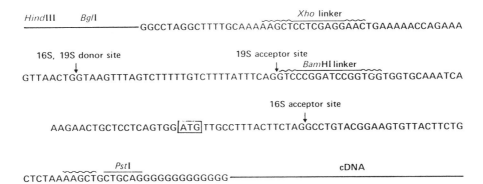

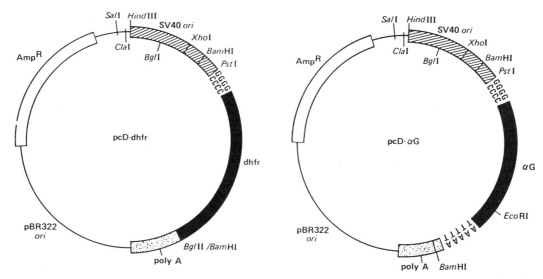

FIG. 2. The nucleotide sequence of the pcD intron segment. The nucleotide sequence from top to bottom and left to right corresponds to the *ori* segment (*BgI*I) across the connecting *Xho*I site, through the two intron sequences and the *Pst*I oligo(dG) sequence, to the cDNA. Arrows indicate the positions of the 5′ splice junction (marked donor site) and 3′ splice junctions (marked acceptor sites) of the 16S and 19S SV40 late mRNAs (6, 16, 28). The *Bam*HI restriction site marks the position of the deletion within the 16S RNA-type intron. The ATG enclosed in a box is a methionine codon in the SV40 capsid protein VP2.

newly formed cDNA, was carried out at 10 times the scale described previously (37) and then terminated by freezing at −20°C.

(iv) Transformation of *E. coli* with vector-cDNA library. *E. coli* strain X1776 (11) and the transformation protocol described by Maniatis et al. (32) were used to obtain maximum transformation efficiency. Transformation-competent cells were prepared as follows. A glycerol stock of X1776 was streaked on an agar plate containing X broth (5 mM MgCl₂, 10 g of tryptone [Difco Laboratories, Detroit, Mich.], 5 g of yeast extract, 5 g of NaCl, 5 g of glucose, 0.1 g of diaminopimelic acid, and 0.04 g of thymidine per liter) and grown at 37°C for 20 h. A single colony was innoculated into 25 ml of XT broth (50 mM Tris-hydrochloride [pH 7.5], 20 mM MgSO₄, 25 g of tryp-

tone, 7.5 g of yeast extract, 0.1 g of diaminopimelic acid, and 0.04 g of thymidine per liter) and incubated at 37°C for 3 h with vigorous shaking; then 10 ml of the culture was added to 1 liter of XT broth and grown at 37°C with vigorous aeration. Cells were harvested at an absorbance at 600 nm of 0.3 by centrifugation (3,000 × *g*) for 10 min at 4°C and suspended in 200 ml of ice-cooled XT buffer (100 mM RbCl₂, 45 mM MnCl₂, 35 mM potassium acetate, 10 mM CaCl₂, 5 mM MgCl₂, 0.5 mM LiCl, and 15% ultrapure sucrose adjusted to pH 5.8 at 20°C with acetic acid). The suspension was cooled at 0°C for 5 min, and the cells were centrifuged at 3,000 × *g* for 10 min at 4°C and suspended in 75 ml of XT buffer at 0°C for 5 min. After the addition of 2.5 ml of dimethyl sulfoxide, the suspension was cooled for 15 min in ice, and then 0.5-ml samples were frozen

FIG. 3. The structure of pcD-*dhfr* and pcD-αG. The symbols and other designations are those explained in the legend to Fig. 1.

in 1.5-ml Eppendorf microfuge tubes in a solid CO_2-ethanol bath and stored at $-70°C$.

Twenty tubes of frozen competent X1776 cells were thawed at 22°C and cooled in ice for 3 min; after receiving 17 μl of dimethyl sulfoxide each, the cultures were cooled for another 3 min in ice. A 13-μl amount of the vector-cDNA preparation [step (iii)] was mixed with each sample of competent cells and incubated at 0°C for 10 min, then at 37°C for 2 min, and again at 0°C. This procedure was repeated four times with the approximately 1.1 ml of vector-cDNA. After pooling all the samples, in 2 liters of X broth, a sample (0.2 ml) was mixed with 2.5 ml of X broth soft agar and spread on an X broth agar plate containing 25 μg of ampicillin per ml to determine the number of transformed cells. The 2-liter culture was incubated at 37°C for 3 h with occasional swirling, ampicillin was added (15 μg/ml), and the incubation was continued for an additional 20 h at 37°C. The stationary culture was adjusted to 7% dimethyl sulfoxide and stored in 1.5-ml portions at $-70°C$.

(v) Preparation of sublibraries based on cDNA insert size. Plasmid DNA was prepared from the transformed X1776 cells by the lysozyme-triton procedure (27) followed by equilibrium sedimentation in CsCl. Approximately 30 μg of plasmid DNA was digested for 1 h at 37°C with 30 U of *Sal*I endonuclease (in buffer containing 6 mM Tris-hydrochloride [pH 7.5], 6 mM $MgCl_2$, 150 mM NaCl, 6 mM 2-merceptoethanol, and 0.1 mg of bovine serum albumin per ml) and electrophoresed in 1% agarose gel in Tris-acetate-EDTA buffer (pH 8.2); DNA fragments whose sizes spanned the range 3 to 8 kilobases (kb) were electrophoresed in adjacent tracks. After staining with ethidium bromide, the gel was sliced into 10 sections corresponding to cDNA insert sizes of 0.3 to 0.6, 0.6 to 1.0, 1.0 to 1.5, 1.5 to 2.0, 2.0 to 2.5, 2.5 to 3.0, 3.0 to 4.0, 4.0 to 5.0, 5.0 to 6.0, and 6.0 to 7.0 kb. DNA was extracted from each slice (50), recycled with T4 DNA ligase, and used to transform X1776 as described above. The individual transformed cultures were stored as mentioned above and constitute approximately sized cDNA sublibraries from SV40-transformed human fibroblasts.

Analysis of mRNAs produced by pcD-αG. COS cells (17) were transfected with pcD-αG DNA in a calcium phosphate precipitate (18), and after 48 h cytoplasmic, poly(A)⁻ RNA was isolated (45). The structure of the α-globin mRNA was determined by using the Weaver-Weissmann modification (51) of the S1 nuclease procedure described by Berk and Sharp (3). The 5' end-labeled DNA probes (see individual experiments in Results) were labeled with [α-^{32}P]ATP and T4 polynucleotide kinase (33) after dephosphorylation of the fragments with bacterial alkaline phosphatase; the 3' end-labeled DNA probes were made by filling in the 5' protruding ends of restriction fragments with *E. coli* DNA polymerase I and α-^{32}P-deoxynucleoside triphosphates (33). To minimize the signal caused by the reannealing of one of the 3' end-labeled probes (see Fig. 6), the fragment was digested with λ-exonuclease to expose the labeled 3' ends. The isolated poly(A)⁺ cell RNA was annealed to the 5' end-labeled DNA probe at 40°C in the hybridization buffer containing 80% formamide and to the exonuclease-digested 3' end-labeled probe at 37°C in the buffer containing 50% formamide. After hybridization, the DNA-RNA hy-

brids were digested with S1 nuclease (1,000 U) for 60 min at 37°C, denatured, and electrophoresed in a urea-polyacrylamide gel (33); the resulting bands were visualized by autoradiography.

RESULTS

Expression of pcD recombinants in mammalian cells. Our objective in developing the pcD vector was to permit the expression of cloned cDNA segments in mammalian cells. The design of the vector ensures that (i) cloned cDNA segments can be transcribed from the SV40 early region promoter (Fig. 1), (ii) the transcript can be polyadenylated within the cDNA segment or at the polyadenylation signal of the vector located beyond the cDNA sequence (Fig. 1), and (iii) the transcript can be spliced at one or both of the SV40 late region introns located between the promoter and the cDNA (Fig. 2).

These suppositions were tested with two model pcD recombinants (Fig. 3). One, pcD-*dhfr*, contains a mouse dihydrofolate reductase (*dhfr*) cDNA with the entire 558-bp DHFR coding sequence, flanked by about 80 bp of 5' and 680 bp of 3' untranslated regions (clone 26) (8); the *dhfr* cDNA segment was truncated at the *Bgl*II restriction site and therefore lacks its own polyadenylation site. The other pcD recombinant, pcD-αG, contains a full-length α-globin cDNA and more closely emulates the structure of the anticipated pcD-cDNA recombinants in that it contains the dG:dC and dA:dT stretches at the 5' and 3' ends, respectively. pcD-*dhfr* and pcD-αG DNAs were transfected into DHFR-negative CHO cells and COS cells, respectively, to test for expression of the cDNA.

pcD-*dhfr*-mediated transformation of DHFR-negative CHO cells. Subramani et al. (45) have shown that pSV2-*dhfr*, a recombinant that contains the same SV40 promoter and dhfr cDNA, but a different intron and polyadenylation signal, transforms DHFR-negative CHO cells to a DHFR-positive phenotype. Comparing the transformation efficiencies of pcD-*dhfr* and pSV2-*dhfr* shows that the two plasmid DNAs are equally effective; mock or transfections with pBR322-*dhfr* DNA, which lacks the eucaryote promoter and processing signals (8), yield no transformants (Table 1). It appears, therefore, that the *dhfr* segment in pcD-*dhfr* is expressed efficiently enough to complement the DHFR deficiency of these cells.

Expression of α-globin after transfection with pcD-αG. Expression of the α-globin cDNA segment was measured by the production of α-globin mRNA after transfection of pcD-αG DNA into COS cells. Cytoplasmic poly(A)⁺ RNA was isolated 48 h after transfection, and the α-globin mRNA was detected and characterized by the S1 nuclease procedure of Berk and

TABLE 1. Transformation of DHFR-negative CHO cells by various vector-*dhfr* cDNA recombinants[a]

DNA	Transformation frequency
Mock	$<10^{-6}$
*dhfr*26	$<10^{-6}$
pSV2-*dhfr*	4×10^{-4}
pcD-*dhfr*	4×10^{-4}

[a] *dhfr* 26 and pSV2-*dhfr* are previously described recombinants containing the clone 26 *dhfr* cDNA segment (8) inserted in pBR322 (8) or pSV2 (45) DNAs, respectively. The test cells were DHFR-negative CHO cells (45), and the transfection and selection conditions for DHFR-positive transformants were as described previously (45). The transformation frequencies are expressed as the fraction of cells converted to a DHFR-positive phenotype after transfection with saturating levels of DNA (10 µg per plate).

Sharp (3) by using 5' or 3' end-labeled DNA probes (51).

One probe (Fig. 4) was ^{32}P labeled at the 5' end of the *Eco*RI site located near the 3' end of the α-globin coding sequence and extends through the intron and promoter segment to the *Hin*dIII site (Fig. 3). S1 nuclease digestion of hybrids formed from the isolated mRNA and this DNA probe yields labeled DNA fragments whose lengths are a measure of the distance from the *Eco*RI site to the 3' splice junctions of the putative α-globin mRNAs. Splicing of the α-globin mRNA at the junction used in the production of SV40 16S late mRNA is indicated by a fragment of 445 nucleotides; splicing analogous to the 19S late mRNA yields a 516-nucleotide fragment. A transcript that initiated at the usual position of SV40 early region RNA and is unspliced would produce a fragment of 638 to 662 nucleotides. Each of these fragments was detected in the gels (Fig. 4); judging from the intensity of the bands, we estimate that 70 to 80% of the α-globin mRNA is spliced and that splicing of the 16S intron is three to four times more frequent than splicing of the 19S intron (Fig. 4).

The 5' splice junction of the α-globin mRNA was mapped with a DNA probe that extends from the *Xho*I restriction site separating the SV40 *ori* and intron segments through the cDNA segment to the *Eco*RI restriction site; the ^{32}P label was at the filled-in 3' end of the *Xho*I site. S1 nuclease digestion of the labeled RNA-DNA hybrids, electrophoresis, and autoradiography revealed, beside the reannealed probe, a single fragment of 32 nucleotides (Fig. 5). This is the expected length of the fragment if the mRNA initiates upstream of the *Xho*I site and is spliced 32 nucleotides downstream at the 5' splice junctions used in the splicing of 16S and 19S SV40

late mRNAs (6, 16, 28). This result and the one described above indicate that transcription of pcD-αG initiates within the SV40 *ori* segment, most probably at the early region promoter sites (15a), and splicing occurs 50 to 60% of the time at the 16S intron junction and 10 to 20% of the time at the 19S intron.

To characterize the 3' end of the globin mRNA, the hybridization probe extended from the *Eco*RI restriction site in the cDNA through the AT bridge and the pBR322 DNA to the *Hin*dIII restriction site (Fig. 3); the fragment was ^{32}P labeled at the filled-in 3' end of the *Eco*RI site. Two types of labeled fragments were found after S1 nuclease digestion of the RNA-DNA hybrids, electrophoresis, and autoradiography (Fig. 6). One, 420 nucleotides long, confirms that some of the α-globin mRNA extends to the SV40 late polyadenylation site (near the *Hpa*I site) in the SV40 DNA segment distal to the cDNA insert (Fig. 1). The other fragments are heterogeneous, ranging in size between 170

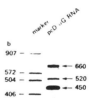

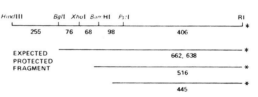

FIG. 4. Analysis of the α-globin mRNA produced by transfection of COS cells with pcD-αG DNA. RNA-DNA hybrids formed from poly(A) RNA (obtained 48 h after transfection) and a DNA probe extending from the ^{32}P-labeled (*) 5' end at the *Eco*RI site through the cDNA to the *Hin*dIII end of the linker segment were digested with S1 nuclease and electrophoresed in a urea-polyacrylamide gel with marker fragments on an adjacent track. Shown at the bottom is a diagram depicting the origins and lengths of the expected protected fragments assuming no splicing (662,638 b), 19S RNA-type splicing (516 b), and 16S RNA-type splicing (445 b). The top-most band in the pcD-αG track corresponds to undegraded probe DNA; the bottom-most band suggests a low level of splicing into the α-globin cDNA segment. b, Base pairs.

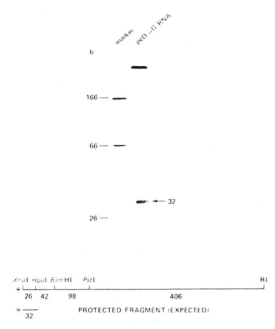

FIG. 5. The position of the 5' splice junction in the α-globin mRNA made from pcD-αG. The analysis was carried out as described in the legend to Fig. 4, except that the DNA probe was a fragment beginning from the *Xho*I site joining the early region promoter and extending through the intron segment to the *Eco*RI site in the α-globin cDNA (see Fig. 3); the fragment was labeled by filling in the 3' end at the *Xho*I site with ^{32}P-labeled deoxynucleoside triphosphates (33). The size of the expected protected fragment assumes a single 5' splice junction for both introns. The band at the top of the pcD-αG track is from undegraded probe DNA. b, Base pairs.

and 240 nucleotides; these fragments signify that the poly(A) termini of some of the α-globin mRNA end within the AT stretch, most probably because polyadenylation occurred at the signal in the 3' untranslated segment of the α-globin cDNA. The smaller fragments are heterogeneous in length because the extent of S1 nuclease digestion at the poly(A-dT) hybrid bp is variable. Judging from the amounts of the 420- and 170- to 240-nucleotide-long fragments, we infer that polyadenylation occurs about equally frequently at the α-globin cDNA and SV40 DNA signals.

Construction of a pcD-human fibroblast cDNA clone library. The pcD vector and cDNA cloning procedure have been tested for the isolation of cDNA clones that are directly expressible in mammalian cells. A cDNA clone library was prepared by using poly (A)⁺ RNA from an SV40-transformed human fibroblast cell line (GM637) (see Materials and Methods for details). The library consisted of about 1.4×10^6

independently transformed *E. coli*, 70 to 80% of which contained plasmids with cDNA inserts ranging in size from several hundred to 5,000 bp (data not shown). To facilitate the search for specific cDNAs, the clone library was subdivided on the basis of the size of the cDNA inserts (see Materials and Methods). Several clones from the sublibrary containing cDNA segments of 1.5 to 2 kb hybridized with a human genomic DNA clone containing sequences encoding part of the amino acid sequence of human HPRT (25). These clones readily transform HPRT-minus mouse L cells to an HPRT-positive phenotype at frequencies comparable to those found with pSV2-*gpt* (10 to 100 hypoxanthine-aminopterin-thymine medium resistant transformants per 10^5 cells) (34, 35; Jolly et al., in press). Experiments are in progress to isolate other functional cDNA segments and to determine whether this clone library or sublibraries can transduce appropriate recipient cells for other specific functions.

DISCUSSION

This paper describes a plasmid vector that can be used to clone cDNAs in *E. coli* and also promote expression of the cDNA segment in mammalian cells. Transcription and processing signals derived from SV40 DNA are arrayed in the pcD plasmid to ensure transcription, splicing, and polyadenylation of the cloned cDNA segment (Fig. 1). The SV40 early region promoter was chosen because, in the absence of repression by SV40 large T antigen, it is a relatively strong RNA polymerase II-specific promoter (21); moreover, no other SV40 functions are needed for transcription from this promoter (34, 35, 45). Other mammalian promoters, e.g., the adenovirus 2 late promoter (46), a retroviral 5' long terminal repeat (31), or appropriate promoters for bacteria, yeast, or other suitable cloning hosts, could readily be substituted for the SV40 early promoter. Indeed, a DNA segment derived from the *E. coli* tryptophan operon that contains the promoter-operator-attenuator and 5' end of *trpE* (30) has been used to express the mouse *dhfr* cDNA in *E. coli* (N. Osheroff, E. R. Shelton, H. Okayama, D. L. Brutlag, and P. Berg, 12th Int. Cong. Biochem., Perth, Australia, 1982, abstract POS-003-079).

The SV40 late region introns used to splice 16S and 19S mRNAs have been employed for splicing of the cDNA transcripts. Both could be obtained on a single DNA segment that was modified to reduce the size of one of the introns. A notable feature of this segment is that it can promote two alternate kinds of splicing (Fig. 2). One splice, occurring at the 16S RNA intron junctions, places the cDNA's initiator AUG codon first in line from the 5' end of the mRNA.

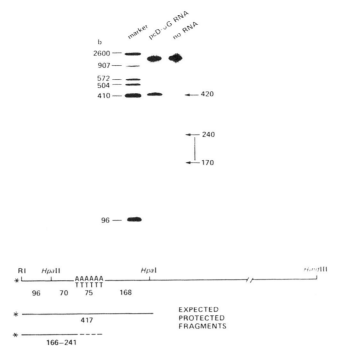

FIG. 6. The location of the 3'-poly(A) tail in the α-globin mRNA made from pcD-αG. The analysis of the S1 nuclease digest of the RNA-DNA hybrids was as described in the legend to Fig. 4. The DNA probe was labeled with ^{32}P at the 3' end of the EcoRI site in the fragment and extended from the EcoRI site in α-globin cDNA through the AT join to the HindIII site (see Fig. 3). The expected fragment sizes, shown at the bottom, assume that polyadenylation occurs at the end of the α-globin cDNA sequence and the poly(A) extends into the AT sequence or at the SV40 polyadenylation signal near the HpaI site. The band at the top of the pcD-αG track derives from undegraded probe. b, Base pairs.

Splicing at the 19S RNA intron retains an AUG upstream of the cDNA in the processed mRNA; therefore, if the clone contains an incomplete cDNA, translation from the upstream AUG would yield a fused protein. The splicing signals were placed 5' proximal, rather than 3' proximal to the cDNA segment, so that splicing of the transcript could occur even if polyadenylation occurred within the cDNA sequence.

The pcD plasmid vector contains several restriction sites that are particularly useful in the manipulation or characterization of the pcD recombinants. For example, the unique SalI and ClaI restriction sites (Fig. 1), which are rare in mammalian DNA, can be used to convert the circular plasmids to linear DNA. Gel electrophoresis of the linear DNA can therefore be used to fractionate the human cDNA clone library on the basis of cDNA size. Two XhoI and two BamHI restriction sites flank the cloned cDNA insert and can be used in the isolation and characterization of the insert DNA. The BamHI restriction site inserted at the point of the deletion in the large intron (Fig. 1) provides a way to introduce additional genetic elements into the

newly created transcription unit. We are presently exploring the utility of introducing a procaryote promoter at the BamHI site to permit expression of the cDNA segment in both procaryote and eucaryote hosts; selectable genetic markers (1, 26) could also be introduced at this site to aid in the detection of transductants and to facilitate recovery of integrated copies of the transducing plasmids.

In the pcD recombinants, the cloned cDNAs are flanked by a short dGdC stretch at their 5' ends and a long dAdT stretch at their 3' ends. Our evidence indicates that these are not serious impediments to transcription, processing, or translation of the cDNA. The nearly equivalent transforming activity of pcD-dhfr and pSV2-dhfr DNA indicates that the dGdC stretch in the former does not significantly impair expression. More specifically, pcD-αG produces the mature transcript, α-globin mRNA, in about the same amounts as the similar transducing plasmids pSV2-β-globin or pSV2-gpt (data not shown). Furthermore, the α-globin mRNA appears to begin at the expected early region transcription initiation site, to terminate at either of the two sequential polyadenylation sites, and to be spliced properly at the two intron junctions. Additional support for the conclusion that the two homopolymer stretches do not impair expression of the cDNA comes from the finding that introduction of the same α-globin cDNA segment with flanking dGdC and dAdT stretches into the virus-transducing vector SVGT5 (36) yields the anticipated amount of spliced and polyadenylated α-globin mRNAs as well as α-globin protein (unpublished data).

By using the pcD vector, we have also constructed several cDNA clone libraries from mRNA obtained from cultured cells. Generally, with *E. coli* X1776 (11) as the cloning host, about 0.5×10^6 to 2×10^6 independent *E. coli* transformants have been recovered per μg of RNA, 50 to 80% of which contain cDNA inserts. Colony hybridization (19) of the cDNA library prepared with mRNA from an SV40-transformed human fibroblast line with a fragment of putative HPRT genomic DNA (25) identified several clones encoding HPRT, one of which contained the entire HPRT coding sequence (Jolly et al., in press; unpublished observations). Moreover, plasmid DNA obtained from one such bacterial clone transformed HPRT-negative mouse L cells to grow in hypoxanthine-aminopterin-thymine medium (Jolly et al., in press). The same cDNA clone library yielded a number of full-length or nearly full-length β (2,050 bp) and γ (2,230 bp) actin cDNAs and α (1,665 bp) and β₂ (1,610 bp) tubulin cDNAs (P. Gunning, P. Ponte, and L. Kedes, personal communication). This and comparable libraries made with other cellular mRNA preparations are being screened for other cDNA clones.

One of the goals of this work was to produce a vector and cloning procedure that could be used to isolate cDNAs on the basis of the function they expressed, rather than by hybridization. Thus, complementation of a mutant cell's defect or alteration of a recipient cell's morphological or growth phenotype could serve to select or detect a particular cDNA for which no hybridization probe exists. To date, our attempts to transduce HPRT-negative cells with plasmid DNA from entire libraries or sublibraries containing the pcD-HPRT recombinant have not succeeded. Relevant to this failure is the observation that if a mixture of cloned pcD-HPRT DNA and increasing amounts of carrier pcD-cDNA recombinants are introduced into cells as a calcium phosphate precipitate (18) or by spheroplast fusion (41), the frequency of HPRT transformants decreases as the ratio of carrier plasmid to pcD-HPRT increases. No transformants were detected among 10^6 transfected cells if pcD-HPRT DNA was present at less than 1 part in 10^3 to 10^4 of the total plasmid DNA. Since the

hybridization screening indicated that HPRT cDNA clones occur at a frequency of about 2×10^{-5} in the cDNA clone library, and only a fraction of these have full-length coding sequences, the failure to detect HPRT transformants can be attributed to the relative rarity of the pcD-HPRT cDNA clones in the library. Quite likely, cloned cDNAs of more abundant mRNAs, e.g., those present at 0.1% or more in the population, could readily be identified and recovered by this method. Nevertheless, even if rare cDNAs cannot yet be identified and isolated by transduction, once they are identified and isolated by hybridization screening procedures, they can be tested for expression without further manipulations. However, we are presently exploring modifications in the cloning and transfection procedures that would increase the efficiency of detecting desired clones directly by transfection of appropriate cells.

ACKNOWLEDGMENTS

This research was funded by Public Health Service grant GM-13235 from the National Institute of General Medical Sciences and by grant CA-31928 from the National Cancer Institutes.

This work was carried out while H.O. was a fellow of the Jane Coffin Childs Fund for Medical Research.

LITERATURE CITED

1. **Alton, N., and D. Vapnek.** 1979. Nucleotide sequence analysis of the chloramphenicol resistance tansposon Tn9. Nature (London) **282**:864–869.
2. **Aviv, H., and P. Leder.** 1972. Purification of biological active globin messenger RNA by chromatography on oligothymidylic acid-cellulose. Proc. Natl. Acad. Sci. U.S.A. **69**:1408–1412.
3. **Berk, J. A., and P. A. Sharp.** 1978. Spliced early mRNAs of simian virus 40. Proc. Natl. Acad. Sci. U.S.A. **75**:1274–1278.
4. **Brack, C., M. Hirama, R. Lenhard-Schuller, and S. Tonegawa.** 1978. Complete immunoglobulin gene is created by somatic recombination. Cell **15**:1–14.
5. **Breathnach, R., J. L. Mandel, and P. Chambon.** 1977. Ovalbumin gene is split in chicken DNA. Nature (London) **270**:314–319.
6. **Buchman, A. R., L. Burnett, and P. Berg.** 1980. The SV40 nucleotide sequence. p. 799–829. *In* J. Tooze (ed.), DNA tumor viruses, part II. Cold Spring Harbor Laboratory, Cold Spring Harbor, N.Y.
7. **Catterall, J. F., J. P. Stein, E. C. Lai, S. L. C. Woo, M. L. Mace, A. R. Means, and B. W. O'Malley.** 1979. The ovomucoid gene contains at least six intervening sequences. Nature (London) **278**:323–327.
8. **Chang, A. C. Y., J. H. Nunberg, R. J. Kaufman, H. A. Erlich, R. T. Schimke, and S. N. Cohen.** 1978. Phenotypic expression in *E. coli* of a DNA sequence coding for mouse dihydofolate reductase. Nature (London) **275**:617–624.
9. **Chirgwin, J. M., A. E. Przybyla, R. J. MacDonald, and W. J. Rutter.** 1979. Isolation of biological active ribonucleic acid from sources enriched in ribonuclease. Biochemistry **18**:5294–5299.
10. **Cordell, B., G. Bell, E. Tischer, F. DeNoto, A. Ulrich, R. Dictet, W. J. Rutter, and H. M. Goodman.** 1979. Isolation and characterization of a cloned rat insulin gene. Cell **18**:533–543.
11. **Curtiss, R., III, M. Inoue, D. Pereira, J. C. Hsu, L. Alexander, and L. Rock.** 1977. Construction and use of safer bacterial host strains for recombinant DNA re-

search, p. 99–111. *In* W. A. Scott and R. Werner (ed.). Molecular cloning of recombinant DNA. Academic Press. Inc., New York.

12. **Efstratiadis, A., F. C. Kafatos, and T. Maniatis.** 1977. The primary structure of rabbit β-globin mRNA as determined from cloned cDNA. Cell **10**:571–585.

13. **Efstratiadis, A., F. Kafatos, A. M. Maxam, and T. Maniatis.** 1976. Enzymatic *in vitro* synthesis of globin genes. Cell **7**:279–288.

14. **Efstratiadis, A., and L. Villa-Komaroff.** 1979. Cloning of double stranded cDNA. p. 1–14. *In* J. K. Setlow and A. Hollaender (ed.), Genetic engineering, vol. 1. Plenum Publishing Corp., New York.

15. **Fiers, W., R. Contreras, G. Haegeman, R. Rogiers, A. Van de Vourde, H. Van Heuverswyn, J. Van Herreweghe, G. Volckaert, and M. Ysebaert.** 1978. Complete nucleotide sequence of SV40 DNA. Nature (London) **273**:113–120.

15a. **Fromm, M., and P. Berg.** 1982. Deletion mapping of DNA regions required for SV40 early region promoter function *in vivo*. J. Mol. Appl. Genet. **1**:457–481.

16. **Ghosh, P. K., V. B. Reddy, J. Swinscoe, P. Lebowitz, and S. M. Weissman.** 1978. Heterogeneity and 5′ terminal structures of the late RNAs of simian virus 40. J. Mol. Biol. **126**:813–846.

17. **Gluzman, Y.** 1981. SV40-transformed simian cells support the replication of early SV40 mutants. Cell **23**:175–182.

18. **Graham, F. L., and A. J. Van der Eb.** 1973. A new technique for the assay of infectivity of human adenovirus 5 DNA. Virology **52**:456–467.

19. **Grunstein, M., and D. S. Hogness.** 1975. Colony hybridization: a method for the isolation of cloned DNAs that contain a specific gene. Proc. Natl. Acad. Sci. U.S.A. **72**:3961–3965.

20. **Gurdon, J. B., C. D. Lane, H. R. Woodland, and G. Marbaix.** 1971. Use of frog eggs in oocytes for the study of mRNA and its translation in living cells. Nature (London) **233**:177–182.

21. **Handa, H., R. J. Kaufman, J. Manley, M. Gefter, and P. Sharp.** 1981. Transcription of simian virus 40 DNA in a Hela whole cell extract. J. Biol. Chem. **256**:478–482.

22. **Harpold, M. M., P. R. Dobner, R. M. Evans, and E. C. Bancroft.** 1978. Construction and identification by positive hybridization-translation of a bacterial plasmid containing a rat growth hormone structural gene sequence. Nucleic Acids Res. **5**:2039–2053.

23. **Heindell, H. C., A. Lin, G. V. Paddock, G. M. Studnicka, and W. A. Salser.** 1978. The primary sequence of rabbit α-globin mRNA. Cell **15**:43–54.

24. **Jeffreys, A. J., and R. A. Flavell.** 1979. The rabbit β-globin gene contains a large insert in the coding sequence. Cell **12**:1097–1108.

25. **Jolly, D. J., A. C. Esty, H. V. Bernard, and T. Friedmann.** 1982. Isolation of a genomic clone partially encoding human hypoxanthine phosphoribosyltransferase. Proc. Natl. Acad. Sci. U.S.A. **79**:5038–5041.

26. **Jorgensen, R. A., S. J. Rothstein, and W. S. Reznikoff.** 1979. A restriction enzyme cleavage map of Tn5 and location of a region encoding neomycin resistance. Mol. Gen. Genet. **177**:65–72.

27. **Katz, L., D. T. Kingsbury, and D. R. Helinski.** 1973. Stimulation by cyclic adenosine monophosphate of plasmid deoxyribonucleic acid replication and catabolite repression of the plasmid deoxyribonucleic acid-protein relaxation complex. J. Bacteriol. **114**:577–591.

28. **Lai, C., R. Dhar, and G. Khoury.** 1978. Mapping of spliced and unspliced late lytic SV40 RNAs. Cell **14**:971–982.

29. **Land, H., M. Grez, H. Hansen, W. Lindermaier, and G. Schuetz.** 1981. 5′-Terminal sequences of eucaryotic mRNA can be cloned with high efficiency. Nucleic Acids Res. **9**:2251–2266.

30. **Lee, F., K. Bertrand, G. Bennett, and C. Yanofsky.** 1978. Comparison of the nucleotide sequences of the initial transcribed regions of the tryptophan operons of *Escherichia coli* and *Salmonella typhimurium*. J. Mol. Biol.

121:193–217.

31. **Lee, F., R. C. Mulligan, P. Berg, and G. Ringold.** 1981. Glucocorticoids regulate expression of dihydrofolate reductase cDNA in mouse mammary tumor virus chimaeric plasmids. Nature (London) **294**:228–232.

32. **Maniatis, T., E. F. Fritsch, and J. Sambrook.** 1982. Transformation of *E. coli* X1776, p. 254–255. *In* Molecular cloning. Cold Spring Harbor Laboratory, Cold Spring Harbor, N.Y.

33. **Maxam, A., and W. Gilbert.** 1980. Sequencing end-labeled DNA with base-specific chemical cleavages. Methods Enzymol. **65**:499–560.

34. **Mulligan, R. C., and P. Berg.** 1980. Expression of a bacterial gene in mammalian cells. Science **209**:1422–1427.

35. **Mulligan, R. C., and P. Berg.** 1981. Selection for animal cells that express the *Escherichia coli* gene coding for xanthine-guanine phosphoribosyl transferase. Proc. Natl. Acad. Sci. U.S.A. **78**:2072–2076.

36. **Mulligan, R. C., B. H. Howard, and P. Berg.** 1979. Synthesis of rabbit β-globin in cultured monkey kidney cells following infection with a SV40-β-globin recombinant genome. Nature (London) **277**:108–114.

37. **Okayama, H., and P. Berg.** 1982. High-efficiency cloning of full-length cDNAs. Mol. Cell. Biol. **2**:161–170.

38. **Parnes, J. R., B. Velan, A. Felsenfeld, L. Ramanathan, U. Ferrini, E. Appella, and J. G. Seidman.** 1981. Mouse β2 microglobulin cDNA clones: a screening procedure for cDNA clones corresponding to rare mRNAs. Proc. Natl. Acad. Sci. U.S.A. **78**:2253–2257.

39. **Paterson, B. M., B. E. Roberts, and E. L. Kuff.** 1977. Structural gene identification and mapping by DNA:mRNA hybrid-arrested cell-free translation. Proc. Natl. Acad. Sci. U.S.A. **74**:4370–4374.

40. **Rougeon, F., P. Kourilsky, and B. Mach.** 1975. Insertion of a rabbit β-globin gene sequence into an *E. coli* plasmid. Nucleic Acids Res. **2**:2365–2378.

41. **Schaffner, W.** 1980. Direct transfer of cloned genes from bacteria to mammalian cells. Proc. Natl. Acad. Sci. U.S.A. **77**:2163–2167.

42. **Scherer, G., J. Telford, C. Baldari, and V. Pirrota.** 1981. Isolation of cloned genes differentially expressed at early and late stages of Drosophila embryonic development. Dev. Biol. **86**:438–447.

43. **Seeburg, P. H., J. Shine, J. A. Martial, J. D. Baxter, and H. M. Goodman.** 1977. Nucleotide sequence and amplification in bacteria of structural gene for rat growth hormone. Nature (London) **270**:486–494.

44. **Shine, J., P. H. Seeburg, J. A. Martial, J. D. Baxter, and H. M. Goodman.** 1977. Construction and analysis of recombinant DNA for human chorionic somatomammotropin. Nature (London) **270**:494–499.

45. **Subramani, S., R. C. Mulligan, and P. Berg.** 1981. Expression of the mouse dihydrofolate reductase cDNA in simian virus 40 vectors. Mol. Cell. Biol. **1**:854–864.

46. **Thummel, C., R. Tjian, and T. Grodzicker.** 1981. Expression of SV40 T antigen under control of adenovirus promoters. Cell **23**:825–836.

47. **Tilghman, S. M., P. J. Curtis, D. C. Tiemeir, J. G. Seidman, B. M. Peterlin, M. Sullivan, J. V. Maizel, and P. Leder.** 1978. Intervening sequence of DNA identified in the structural portion of a mouse β globin gene. Proc. Natl. Acad. Sci. U.S.A. **75**:1309–1313.

48. **Verma, I. M.** 1977. The reverse transcriptase. Biochim. Biophys. Acta **473**:1–38.

49. **Villa-Komaroff, L., A. Efstratiadis, S. Broome, P. Lomedico, R. Tizard, S. P. Naber, W. L. Chick, and W. Gilbert.** 1978. A bacterial clone synthesizing proinsulin. Proc. Natl. Acad. Sci. U.S.A. **75**:3727–3731.

50. **Vogelstein, B., and D. Gillespie.** 1979. Preparation and analytical purification of DNA from agarose. Proc. Natl. Acad. Sci. U.S.A. **76**:615–619.

51. **Weaver, R. F., and C. Weissmann.** 1979. Mapping of RNA by a modification of the Berk-Sharp procedure. Nucleic Acids Res. **75**:1175–1193.

A Bacteriophage T7 RNA Polymerase/Promoter System for Controlled Exclusive Expression of Specific Genes

S. Tabor and C.C. Richardson

ABSTRACT The RNA polymerase gene of bacteriophage T7 has been cloned into the plasmid pBR322 under the inducible control of the λ P_L promoter. After induction, T7 RNA polymerase constitutes 20% of the soluble protein of *Escherichia coli*, a 200-fold increase over levels found in T7-infected cells. The overproduced enzyme has been purified to homogeneity. During extraction the enzyme is sensitive to a specific proteolysis, a reaction that can be prevented by a modification of lysis conditions. The specificity of T7 RNA polymerase for its own promoters, combined with the ability to inhibit selectively the host RNA polymerase with rifampicin, permits the exclusive expression of genes under the control of a T7 RNA polymerase promoter. We describe such a coupled system and its use to express high levels of phage T7 gene 5 protein, a subunit of T7 DNA polymerase.

During bacteriophage T7 infection, the right-most 80% of the genome is transcribed by a phage-encoded RNA polymerase, the product of gene 1 (Fig. 1) (1). In contrast to the multisubunit RNA polymerases of bacteria and eukaryotes, T7 RNA polymerase is a single polypeptide of molecular weight 98,800 (2, 3). The enzyme is specific for its own promoters, a conserved 23-base-pair (bp) sequence (4–6).

T7 RNA polymerase is present in relatively low amounts in T7-infected cells, constituting 0.1% of the cellular protein. To facilitate studies on its role in the initiation of T7 DNA replication (7), we have placed its gene on a plasmid under the control of the λ P_L promoter. When induced, T7 RNA polymerase constitutes 20% of the soluble protein, permitting a simple purification of the enzyme to homogeneity. Davanloo *et al.* (8) have also described the purification of T7 RNA polymerase from cells overexpressing the cloned T7 gene 1.

A logical extension of these studies is to exploit the specificity of T7 RNA polymerase for its promoters to express other cloned genes. Transcription by *Escherichia coli* RNA polymerase can be inhibited selectively by the addition of rifampicin. Here, we use the T7 RNA polymerase/promoter system to overproduce bacteriophage T7 gene 5 protein, a subunit of the T7 DNA polymerase (Fig. 1).

MATERIALS AND METHODS

Strains. *E. coli* HMS262 (*thr⁻ leu⁻ lacY⁻ thi⁻ supE hsdR⁻ tonA⁻ trxA⁻*) is *E. coli* C600 (9) transduced with *hsdR* and *trxA⁻*. *E. coli* B/7004 (10) was the donor of *trxA⁻*. *E. coli* HMS273 [*lacᵃᵐ trpᵃᵐ phoᵃᵐ malᵃᵐ supCᵗˢ rpsL tsx::Tn10 lon(Δ)100 htpRᵃᵐ*] is SG935 from S. Goff (Harvard Medical School). pACYC177 and pACYC184 have been described (11). pJL23, provided by J. Lodge and T. Roberts (Harvard Medical School), is a derivative of pACYC184 that contains

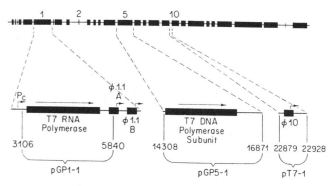

FIG. 1. The genetic map of bacteriophage T7 with inserts showing the cloned DNA fragments used in this study. T7 base pairs are numbered as described (6). Gene sizes reflect molecular weights of protein products. $\phi1.1A$, $\phi1.1B$, and $\phi10$ correspond to the T7 RNA polymerase promoters located before genes *1.1* and *10*, respectively.

the λ repressor gene *cI857*. HMS273/pJL23 is *E. coli* HMS273 containing the plasmid pJL23. pKB280–*cI857*, a pBR322 derivative, contains the λ repressor gene under the control of the lactose promoter (12). pKC30 is a pBR322 derivative that contains the λ P_L promoter (13).

Enzymes and Chemicals. Restriction enzymes were from New England Biolabs. Phage T4 DNA ligase (14), T7 DNA polymerase (15), *E. coli* thioredoxin (16), T7 DNA (17), and plasmid DNAs (14) were prepared as described. pUC12 DNA was from P-L Biochemicals.

Enzyme Assays. The assay (100 μl) for T7 RNA polymerase contained 40 mM Tris·HCl, pH 8.0/20 mM MgCl₂/5 mM dithiothreitol/0.4 mM rNTPs ([³H]rUTP, 30 cpm/pm)/60 μM T7 DNA/50 μg of bovine serum albumin per ml and the indicated amount of T7 RNA polymerase. Enzyme was diluted in 10 mM Tris·HCl, pH 7.5/10 mM 2-mercaptoethanol/1 mg of bovine serum albumin per ml. After incubation at 37°C for 15 min, acid-insoluble radioactivity was determined (2). One unit of activity is the amount that catalyzes the incorporation of 1 nmol of [³H]rUTP into an acid-insoluble form in 60 min.

T7 gene 5 protein was determined by complementation with thioredoxin, the other subunit of T7 DNA polymerase (18), to restore activity (16). One unit of activity catalyzes the incorporation of 10 nmol of total nucleotide into an acid-insoluble form in 30 min.

Purification of T7 RNA Polymerase. Ten liters of cells (*E. coli* HMS273/pJL23/pGP1) were grown with aeration in a New Brunswick fermentor at 30°C in 2% tryptone/1% yeast extract/0.5% NaCl/0.2% glucose, pH 7.4. At OD₅₉₀ = 3.0, the temperature was raised to 42°C. After 30 min, the temperature was lowered to 40°C for 120 min. The cells (95 g) were harvested, washed with 2 liters of 10% (wt/vol) sucrose/20 mM Tris·HCl, pH 8.0/25 mM EDTA, pH 8.0, at 0°C, resuspended in 250 ml of 10% sucrose/20 mM Tris·HCl,

Abbreviation: bp, base pair(s).

TABOR, S. and RICHARDSON, C.C.
A bacteriophage T7 RNA polymerase/promoter system for controlled exclusive expression of specific genes. *Proc. Natl. Acad. Sci. U.S.A.* 82:1074-1078 (1985). Reprinted with the authors' permission.

pH 8.0/1 mM EDTA, and frozen in liquid N_2.

Cells (280 ml) were thawed at 0°C. Lysozyme was added to 0.2 mg/ml. After 45 min at 0°C, the cells were twice frozen in liquid N_2 and thawed at 0°C. The lysate was centrifuged and the supernatant was collected (fraction I, 265 ml).

Ammonium sulfate (93 g) was added to fraction I at 0°C over 60 min. The precipitate was collected and redissolved in 10 mM Tris·HCl, pH 7.5/0.1 mM EDTA/0.5 mM dithiothreitol/10% (vol/vol) glycerol (buffer A) to a conductivity equal to that of buffer A containing 50 mM NH_4Cl (fraction II, 480 ml).

A column of Whatman DE52 DEAE-cellulose (12.6 cm^2 × 23 cm) was equilibrated with buffer A containing 50 mM NH_4Cl. Fraction II was applied to the column and the column was washed with 600 ml of buffer A containing 50 mM NH_4Cl. Proteins were eluted with a 2.5-liter linear gradient from 50 to 300 mM NH_4Cl in buffer A. RNA polymerase activity eluted at 130 mM NH_4Cl (fraction III, 230 ml).

A column of Whatman P11 phosphocellulose (4.9 cm^2 × 20 cm) was equilibrated with 10 mM potassium phosphate buffer, pH 7.5/0.1 mM dithiothreitol/0.1 mM EDTA (buffer B). Fraction III was diluted with buffer B to 100 mM NH_4Cl and applied to the column. The column was washed with 250 ml of buffer B containing 200 mM KCl. Proteins were eluted with a 1-liter linear gradient from 200 to 400 mM KCl in buffer B. RNA polymerase, eluting at 300 mM KCl, was dialyzed against 20 mM potassium phosphate buffer, pH 7.5/50% (vol/vol) glycerol/0.1 mM EDTA/0.1 mM dithiothreitol (fraction IV, 130 ml). The enzyme was stored at −18°C.

Other Methods. Polyacrylamide gel electrophoresis in the presence of 0.1% NaDodSO$_4$ was as described (6). Gels were either stained with Coomassie blue (19) or analyzed by autoradiography. Protein was determined by the method of Lowry *et al.* (20).

RESULTS

Cloning of Gene _1_ of Phage T7. Cloned DNA fragments that contain gene _1_ and adjacent T7 sequences are lethal to _E. coli_ (ref. 8; unpublished results). The nucleotide sequence of this region (3, 6) reveals the presence of several potentially deleterious elements (Fig. 1). Immediately 5' proximal to gene _1_ is a weak _E. coli_ RNA polymerase promoter, promoter C (21), while 3' proximal are two T7 RNA polymerase promoters, φ1.1A and φ1.1B (6). If a T7 RNA polymerase promoter is located in the same orientation as gene _1_ on a plasmid, then even a single T7 RNA polymerase molecule will result in lethal runaway amplification of the gene _1_ product and plasmid transcription.

To circumvent this problem, we isolated a DNA fragment containing all of gene _1_ but lacking the _E. coli_ RNA polymerase promoter C and the two T7 RNA polymerase promoters. Since these elements are extremely close to gene _1_ in T7, we constructed the T7 mutant ST9. This mutant has a deletion of the sequences 5840–5916, which includes the T7 RNA polymerase promoters φ1.1A and φ1.1B, and inserted at that site is the linker C-C-G-G-A-T-C-C-G-G-G-G-A-A-T, which creates a unique *Bam*HI restriction site. The _E. coli_ promoter C was removed by limited digestion (3150 bp) with BAL-31 nuclease. The DNA was digested with *Bam*HI, and fragments 2700 bp long were isolated by gel electrophoresis. These fragments, having a flush end 5' to gene _1_ and a *Bam*HI-generated end 3' to gene _1_, were ligated to the *Hpa* I and *Bam*HI ends of pKC30. The orientation of gene _1_ is such that it is under the control of the λ P_L promoter (Fig. 2). The ligation mixture was used to transform HMS273/pJL23, a strain containing the temperature-sensitive λ repressor, _c_I857. After selecting recombinant plasmids that contain inserts of a size compatible with gene _1_, extracts were pre-

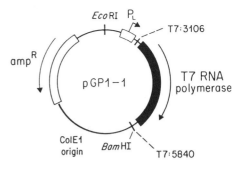

FIG. 2. Structure of pGP1-1. The T7 mutant ST9 was used to isolate the gene _1_ fragment. It has a deletion of sequences 5840–5916 in T7, and inserted into that site is the *Bam*HI linker C-C-G-G-A-T-C-C-G-G-G-G-A-A-T. The right end of the cloned fragment is this *Bam*HI restriction site. The left end, at nucleotide 3106, was generated by digestion with BAL-31 nuclease. The gene _1_ fragment was inserted between the *Hpa* I and *Bam*HI site of pKC30 (13); only the *Bam*HI site is regenerated.

pared from induced cells and assayed for T7 RNA polymerase activity.

One recombinant plasmid, pGP1-1, gives rise to high levels of T7 RNA polymerase activity after induction. DNA sequence analysis reveals that the insert begins at nucleotide 3106 in T7, 65 bp before the start of gene _1_. The insert ends at T7 sequence 5840, 18 bp past the termination codon of gene _1_.

Expression of T7 RNA Polymerase. When cultures of _E. coli_ HMS273/pJL23/pGP1-1 are induced, T7 RNA polymerase accumulates over a 3-hr period (Fig. 3, lane B). The overproduced T7 RNA polymerase is soluble in extracts, repre-

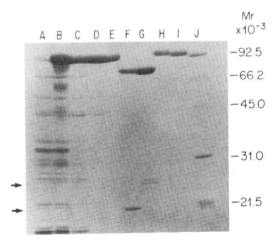

FIG. 3. Expression and purification of T7 RNA polymerase. Conditions for cell growth and enzyme purification are described in the text. A 9%–23% gradient polyacrylamide gel was stained with Coomassie blue after electrophoresis in the presence of NaDodSO$_4$. Lane A contains uninduced _E. coli_ HMS273/pJL23/pGP1-1. Lane B contains induced cells. Lanes C, D, and E show fractions I, III, and IV of purified T7 RNA polymerase. Lane F contains the proteolytically cleaved T7 RNA polymerase (fraction IV). Quantitative cleavage was obtained by incubating extracts (_E. coli_ HMS262/pJL23/pGP1-1) at 37°C in 10% sucrose/50 mM Tris·HCl, pH 8.0/100 mM NaCl for 60 min. Cleaved gene _1_ fusion protein (fraction IV) and intact gene _1_ fusion protein (fraction IV) are shown in lanes G and H. The fusion protein is produced by a derivative of pGP1-1, in which the first 6 codons of T7 gene _1_ are replaced by the first 33 codons of the λ _N_ gene (see text). Lane I contains T7 RNA polymerase purified from T7 phage-infected cells. Lane J contains protein markers (phosphorylase B, ovalbumin, bovine serum albumin, carbonic anhydrase, and trypsin inhibitor). Arrows indicate position of the smaller peptide resulting from proteolytic cleavage of T7 RNA polymerase (lane F) and gene _1_ fusion protein (lane G).

senting 20% of the cellular protein (Fig. 3, lane C). Transcription from the P_L promoter also results in the synthesis of a M_r 6000 polypeptide containing the first 33 amino acids of the N protein of λ fused to 22 amino acids encoded by the region before gene 1, and terminating 2 bp before the start codon for gene 1. This polypeptide has run off the gel shown in Fig. 3.

Purification of T7 RNA Polymerase from Induced Cells. Induced *E. coli* HMS273/pJL23/pGP1-1 provides a source for the isolation of T7 RNA polymerase. We have designed a rapid purification procedure (Table 1) consisting of fractionation by ammonium sulfate, followed by chromatography on DEAE and phosphocellulose. The purity of each fraction is shown in Fig. 3 (lanes C, D, and E). Approximately 570 mg of homogeneous T7 RNA polymerase was obtained from 95 g of cells. The purified enzyme has 3 times the specific activity (190,000 units/mg) of homogeneous T7 RNA polymerase previously purified by us from phage-infected cells. Presumably, the high concentration of T7 RNA polymerase at the outset, as well as the rapid purification procedure, leads to greater stability of the enzyme.

Proteolytic Cleavage of T7 RNA Polymerase. During initial attempts to purify the overproduced T7 RNA polymerase, the enzyme was proteolytically cleaved at a specific region, a phenomenon also observed by Davanloo *et al.* (8). Under optimal conditions for proteolysis (a *lon*$^+$ *E. coli* host and incubation of cell extract in 100 mM NaCl at 37°C for 60 min) all of the gene 1 protein is cleaved at a site 25% of the distance from one end of the molecule. The two peptides (M_r, 75,000 and 23,000) remain associated throughout purification (Fig. 3, lane F). The cleaved form of T7 RNA polymerase has one-eighth the activity (23,000 units/mg) of the intact enzyme.

T7 RNA polymerase is not cleaved in intact cells. *In vivo*, induced cells containing T7 RNA polymerase radioactively labeled with [^{35}S]methionine can be chased for 2 hr without detectable proteolysis (data not shown). In extracts, the protease activity fractionates together with the outer membrane of *E. coli*. The protease is insensitive to EDTA, *o*-phenanthroline, and diisopropyl fluorophosphate; it is inhibited by *N*-ethylmaleimide, low salt concentration, low temperature (0°C), the presence of DNA, and a *lon*$^-$ genetic background. The decreased proteolysis of T7 RNA polymerase in *lon*$^-$ mutants is an indirect effect of the *lon* mutation (22), because the cellular location of the protease and its sensitivity to inhibitors differ from those of the *lon* protease (23).

To determine the location of the cleavage, we purified a derivative of T7 RNA polymerase that is larger in size by M_r 3000. This protein is produced by a plasmid analogous to pGP1-1, except that the T7 insert begins at nucleotide 3189 (Fig. 1). As a consequence, the first six amino acids of T7 RNA polymerase are replaced by the first 33 amino acids of the λ N protein. This M_r 102,000 fusion protein is also susceptible to proteolysis, although the cleavage is less specific, occurring at four sites over a region of 20 amino acids (Fig. 3, lane G). The large proteolytic fragment is the same size as that obtained with gene 1 protein, while the small fragment is larger than the small polypeptide produced by cleavage of gene 1 protein by M_r 3000. The region cleaved therefore lies

25% of the distance from the amino end of T7 RNA polymerase. The specific activity of fraction IV of cleaved fusion protein (19,000 units/mg) is one-sixth that of intact fusion protein (120,000 units/mg).

Expression of Specific Genes with the Use of Cloned T7 Gene 1 and a T7 RNA Polymerase Promoter. The expression system shown in Fig. 4 consists of two compatible plasmids, pGP1-2 and pT7-1. pGP1-2, a derivative of pACYC177, provides for expression of T7 RNA polymerase. It consists of gene 1 of phage T7 under the control of the inducible λ P_L promoter, and the gene for the heat-sensitive λ repressor, *cI857*.

To express a given gene, the gene is inserted into the second plasmid, pT7-1. pT7-1 contains a T7 RNA polymerase promoter, φ10, isolated from a 40-bp T7 fragment (Fig. 1). A polylinker containing eight different restriction sites lies adjacent to the promoter to facilitate the insertion of DNA fragments. Transcription from the Ø10 promoter results in expression of the cloned gene and the β-lactamase gene. Exclusive expression of these genes is achieved, after heat induction of T7 RNA polymerase, by the addition of rifampicin to shut off *E. coli* RNA polymerase transcription; the addition of [^{35}S]methionine results in specific labeling of these plasmid-encoded proteins.

Expression and Overproduction of T7 Gene 5 Protein. With this expression system, we have directed the synthesis of T7 gene 5 protein. T7 DNA polymerase consists of two subunits, a M_r 84,000 protein encoded by T7 gene 5 and the M_r 12,000 thioredoxin of the host (16, 18). A fragment of T7 DNA containing gene 5 was inserted into the polylinker of pT7-1 to create pGP5-1 (Fig. 4). After heat induction and addition of rifampicin, the plasmid proteins were labeled with [^{35}S]methionine. The autoradiogram of the gel electrophoresis in the presence of NaDodSO$_4$ (Fig. 5, lanes G–I) shows the profile of labeled proteins. After induction, the M_r 82,000 gene 5 protein is the predominant labeled protein synthesized (lane I), even in the absence of rifampicin (lane H). β-Lactamase and the M_r 14,000 T7 gene 5.3 protein are also synthesized. Approximately 5×10^6 cpm can be incorporated into these proteins from 10 µCi of [^{35}S]methionine (1 Ci = 37 GBq). Lanes G–I are shown stained with Coomassie blue in lanes G'–I', demonstrating the amount of gene 5 protein present prior to and after induction of *E. coli* HMS262/ pGP1-2/pGP5-1. To maximize the amount of expressed protein, induced cells are incubated in enriched medium with rifampicin for several hours; under these conditions the gene 5 protein is the predominant protein in the cell (lane J). Although up to 30% of the cellular protein is gene 5 protein, only approximately one-third is soluble in the absence of NaDodSO$_4$.

As controls, the comparable experiment was carried out using *E. coli* HMS262/pGP1-2 (cells containing T7 gene 1, but no T7 promoter), and *E. coli* HMS262/pGP1-2/pT7-1 (cells containing T7 gene 1 and T7 φ10 expressing only β-lactamase) (Fig. 4). T7 RNA polymerase is present in all three strains after (Fig. 5, lanes B, E, and H) but not prior to (lanes A, D, and G) induction. If rifampicin is added after induction, no proteins are labeled when the cells lack a T7 RNA polymerase promoter (lane C). When pT7-1 is present,

Table 1.   Purification of T7 RNA polymerase

Fraction	Step	Protein, mg	Total units, $\times 10^6$	Specific activity, units/mg	% recovery
I	Extract	5200	220	42,000	100
II	Ammonium sulfate	2700	210	78,000	95
III	DEAE-cellulose	770	120	160,000	55
IV	Phosphocellulose	570	110	190,000	50

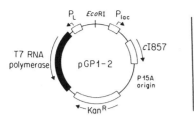
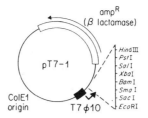
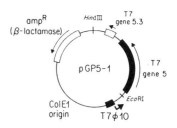

FIG. 4. Coupled T7 RNA polymerase/promoter system. All cells contain pGP1-2 (*Left*), which expresses T7 RNA polymerase. This plasmid, 7200 bp in size, contains the 3200-bp *Bam*HI/*Bgl* II fragment of pGP1-1 (T7 gene *1*), the 1100-bp *Pst* I/*Eco*RI fragment of pKB280-c*I857* (λ repressor gene), and the 2900-bp *Bam*HI/*Pst* I fragment of pACYC177 [the kanamycin-resistance gene (Kan^R) and the P15A origin]. Cells also contain either pT7-1 or pGP5-1. pT7-1, 2400 bp in size, contains the T7 *Taq* I/*Xba* I fragment from nucleotides 22879–22928 (Ø10), the 70-bp polylinker region of pUC12, and nucleotides 2065–4360 from pBR322 (the β-lactamase gene and the ColE1 origin; see ref. 24). pGP5-1, 5900 bp in size, contains the T7 *Nde* I/*Aha* III fragment from nucleotides 14308–16871 (T7 gene *5* and *5.3*), inserted with *Bam*HI linkers into the *Bam*HI site of pT7-1.

only β-lactamase is radioactively labeled (lane F). β-Lactamase appears as three bands: the top is the M_r 29,000 precursor (24), the middle band is the M_r 27,000 processed enzyme, while the lower band is a translated segment reading counterclockwise on pT7-1.

The gene *5* protein produced after induction has full activity as measured by its ability to complement thioredoxin to produce an active T7 DNA polymerase (Fig. 6). DNA synthesis in extracts prepared from induced *E. coli* HMS262/pGP1-2/pGP5-1 is stimulated 100-fold by purified thioredoxin. The specific activity of gene *5* protein in extracts is 1050 units per mg of protein, compared to 10,300 units/mg for homogeneous T7 DNA polymerase (15).

In the absence of thioredoxin (*E. coli trx*A$^-$), cells can tolerate high levels of gene *5* protein. When extracts are prepared from uninduced cells, the gene *5* protein represents 1% of the soluble protein (Fig. 6). This activity results from residual expression of gene *1* from the P_L promoter at 30°C. In *trx*$^+$ cells containing thioredoxin, the expression of gene *5* protein is lethal to some *E. coli* strains (e.g., HMS262 *trx*$^+$); pGP1-2 and pGP5-1 are not compatible in these strains. To express T7 gene *5* in these cells, T7 RNA polymerase must be more tightly repressed (see *Discussion*).

DISCUSSION

We have described the purification of T7 RNA polymerase from *E. coli* cells harboring gene *1* of phage T7 and overproducing the enzyme some 200-fold. Our own incentive for overproducing the enzyme was to have sufficient quantities of homogeneous T7 RNA polymerase to use in studies on its role in T7 DNA replication (7). In addition, the fact that T7 RNA polymerase is a monomeric protein and recognizes a delineated promoter sequence makes it an attractive enzyme to use as a model system for studying protein–DNA interactions.

E. coli DNA does not have promoters for T7 RNA polymerase (2). When >2% of the cellular protein is T7 RNA polymerase, we find no change in the growth rate of *E. coli*. However, when a single T7 RNA polymerase promoter is present, then even low levels of T7 RNA polymerase are lethal to the cell, presumably because its efficient transcription serves as a sink for ribonucleoside triphosphates. The induced T7 RNA polymerase complements T7 phage defective in gene *1*. We, as well as Davanloo *et al.* (8), have constructed T7 phage with deletions in gene *1* that propagate only in *E. coli* cells expressing T7 RNA polymerase.

To analyze the products of cloned genes using the T7 RNA polymerase/promoter system, it is necessary to clone a T7 promoter in conjunction with the gene to be expressed. This can be accomplished by inserting a DNA fragment containing the promoter into the plasmid, or alternatively, by recloning the gene into a parent vector that contains a T7 RNA polymerase promoter. The T7 RNA polymerase/promoter system provides an attractive alternative to the mini- (26) or maxicell (27) procedures for labeling plasmid-encoded proteins. All transcription by T7 RNA polymerase is directed from the unique T7 promoters. The host RNA polymerase is

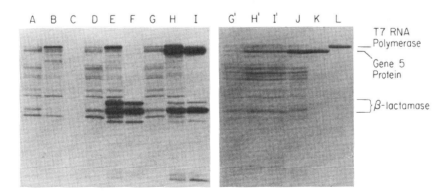

FIG. 5. Expression of T7 gene *5* by T7 RNA polymerase. *E. coli* HMS262/pGP1-2 alone (lanes A–C), containing pT7-1 (lanes D–F), or containing pGP5-1 (lanes G–J) were grown in M9 medium supplemented with thiamine (20 µg/ml) and 18 amino acids (0.1%, minus cysteine and methionine). Cells (1 ml) were grown at 30°C to OD$_{590}$ = 0.4 (uninduced: lanes A, D, and G). Temperature was shifted to 42°C for 20 min (induced: lanes B, E, and H). Rifampicin (200 µg/ml) was added, and after 10 additional min at 42°C, the cells were grown for 20 min at 30°C (induced plus rifampicin: lanes C, F, and I). After pulse labeling with 10 µCi of [^{35}S]methionine, the cells were harvested, resuspended in 60 mM Tris·HCl, pH 6.8/1% NaDodSO$_4$/1% 2-mercaptoethanol/10% glycerol/0.01% bromophenol blue, heated to 95°C for 3 min and loaded onto a 14% polyacrylamide gel containing 0.1% NaDodSO$_4$. Lanes G', H', and I' are identical to lanes G, H, and I except they were stained with Coomassie blue. In lane J, *E. coli* HMS262/pGP1-2/pGP5-1, after induction, was incubated with rifampicin for 2 hr. Lanes K and L contain purified T7 gene *5* and T7 RNA polymerase, respectively.

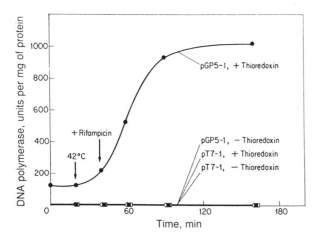

FIG. 6. Induction of gene 5 protein. Cells (*E. coli* HMS262/pGP1-2/pT7-1 and HMS262/pGP1-2/pGP5-1) were grown at 30°C in 2% tryptone/1% yeast extract/0.5% NaCl/ and 0.2% glucose (pH 7.4). At 20 min (A_{590} = 1.5), the temperature was shifted to 42°C. At 40 min, rifampicin was added (200 μg/ml), and the cultures were incubated with aeration at 30°C. At the intervals shown, 50-ml aliquots were removed, and fraction II was prepared from extracts as described (25), with the modification that the lysis buffer contained 1.2 M NaCl, which was necessary to solubilize the overproduced gene 5 protein. Fractions were assayed for DNA polymerase activity in the presence or absence of purified thioredoxin as described (16). □, HMS262/pGP1-2/pT7-1, minus thioredoxin; ■, HMS262/pGP1-2/pT7-1, plus thioredoxin; ○, HMS262/pGP1-2/pGP5-1, minus thioredoxin; ●, HMS262/pGP1-2/pGP5-1, plus thioredoxin.

inhibited specifically by the addition of rifampicin (28). Since *E. coli* mRNA decays rapidly, all mRNA in the cell is produced from the T7 RNA polymerase promoter. T7 RNA polymerase synthesizes RNA by a rapid and processive mechanism. In fact, *in vitro*, transcription by T7 RNA polymerase from a T7 promoter on a plasmid results in transcripts several times the plasmid length (29).

We have presented two examples of the production of high levels of specifically labeled proteins by the T7 RNA polymerase/promoter system. β-Lactamase and the phage T7 gene 5 protein are overproduced and represent the only radioactively labeled proteins. Since T7 RNA polymerase will circumvent a plasmid several times without terminating, the location of the promoter on the plasmid is unimportant; comparable levels of T7 gene 5 protein are obtained if the promoter is placed immediately before or after gene 5. Under optimal conditions, 30% of the cellular protein is gene 5 protein, compared to only 0.01% in T7 phage-infected cells (25).

A problem can arise if the T7 RNA polymerase promoter is directing the expression of a gene product toxic to *E. coli*. For example, we have used this system to express T7 gene 2 protein, which, because it inhibits *E. coli* RNA polymerase, is lethal when expressed in the cell. Under uninduced conditions, the cI857 λ repressor does not repress the P_L promoter tightly enough to completely inhibit synthesis of T7 RNA polymerase. To express T7 gene 2 protein, we have reduced the uninduced level of T7 RNA polymerase by placing a terminator for *E. coli* RNA polymerase into pGP1-2, such that expression of gene 1 from the P_L promoter is dependent on transcriptional readthrough.

In addition to the studies described here, the T7 RNA polymerase/promoter system should be useful *in vivo* to synthesize anti-sense RNA to probe specific gene functions and *in vitro* to generate specific RNA transcripts. Such transcripts are useful as RNA substrates and as single-stranded probes. The T7 RNA polymerase/promoter system should also be useful for directing gene expression in organisms other than *E. coli*.

Note Added in Proof. Amino acid sequence determination of the NH$_2$ terminus of the large proteolytic fragment of T7 RNA polymerase (Fig. 3, lane F) reveals that the cleavage site lies between amino acid 172 (lysine) and amino acid 173 (arginine) (Rodney M. Hewick, Genetics Institute, Cambridge, MA, unpublished results).

We thank Richard Ikeda for his helpful discussions and assistance in the purification of T7 RNA polymerase. This investigation was supported by U.S. Public Health Service Grant AI-06045 and Grant NP-1M from the American Cancer Society.

1. Chamberlin, M., McGrath, J & Waskell, L. (1970) *Nature (London)* **228**, 227–231.
2. Chamberlin, M. & Ring, J. (1973) *J. Biol. Chem.* **248**, 2235–2244.
3. Moffatt, B. A., Dunn, J. J. & Studier, F. W. (1984) *J. Mol. Biol.* **173**, 265–269.
4. Rosa, M. D. (1979) *Cell* **16**, 815–825.
5. Panayotatos, N. & Wells, R. D. (1979) *Nature (London)* **280**, 35–39.
6. Dunn, J. J. & Studier, F. W. (1983) *J. Mol. Biol.* **166**, 477–535.
7. Fuller, C. W., Beauchamp, B. B., Engler, M. J., Lechner, R. L., Matson, S. W., Tabor, S., White, J. H. & Richardson, C. C. (1983) *Cold Spring Harbor Symp. Quant. Biol.* **48**, 669–679.
8. Davanloo, P., Rosenberg, A., Dunn, J. J. & Studier, F. W. (1984) *Proc. Natl. Acad. Sci. USA* **81**, 2035–2039.
9. Appleyard, R. K. (1954) *Genetics* **39**, 440–452.
10. Chamberlin, M. (1974) *J. Virol.* **14**, 509–516.
11. Chang, A. C. Y. & Cohen, S. N. (1978) *J. Bacteriol.* **134**, 1141–1156.
12. Hecht, M. H., Nelson, H. C. M. & Sauer, R. T. (1983) *Proc. Natl. Acad. Sci. USA* **80**, 2676–2680.
13. Shimatake, H. & Rosenberg, M. (1981) *Nature (London)* **292**, 128–132.
14. Davis, R. W., Botstein, D. & Roth. J. R. (1980) *Advanced Bacterial Genetics* (Cold Spring Harbor Laboratory, Cold Spring Harbor, NY).
15. Engler, M. J., Lechner, R. L. & Richardson, C. C. (1983) *J. Biol. Chem.* **258**, 11165–11173.
16. Modrich, P. & Richardson, C. C. (1975) *J. Biol. Chem.* **250**, 5508–5514.
17. Richardson, C. C. (1966) *J. Mol. Biol.* **15**, 49–61.
18. Mark, D. F. & Richardson, C. C. (1976) *Proc. Natl. Acad. Sci. USA* **73**, 780–784.
19. Fairbanks, G., Steck, T. L. & Wallach, D. F. H. (1971) *Biochemistry* **10**, 2606–2617.
20. Lowry, O. H., Rosebrough, N. J., Farr, A. L. & Randall, R. J. (1951) *J. Biol. Chem.* **193**, 265–275.
21. McConnell, D. J. (1979) *Nucleic Acids Res.* **6**, 525–544.
22. Gottesman, S. & Zipser, D. (1978) *J. Bacteriol.* **133**, 844–851.
23. Swamy, K. H. S. & Goldberg, A. L. (1982) *J. Bacteriol.* **149**, 1027–1033.
24. Sutcliffe, J. G. (1979) *Cold Spring Harbor Symp. Quant. Biol.* **43**, 77–90.
25. Hori, K., Mark, D. F. & Richardson, C. C. (1979) *J. Biol. Chem.* **254**, 11591–11597.
26. Dougan, G. & Sherratt, D. (1977) *Mol. Gen. Genet.* **151**, 151–160.
27. Sancar, A., Wharton, R. P., Seltzer, S., Kacinsky, B. M., Clark, N. D. & Rupp, W. D. (1981) *J. Mol. Biol.* **148**, 45–62.
28. Chamberlin, M. & Ring, J. (1973) *J. Biol. Chem.* **248**, 2245–2250.
29. McAllister, W. T., Morris, C., Rosenberg, A. H. & Studier, F. W. (1981) *J. Mol. Biol.* **153**, 527–544.

Supplementary Readings

Amann, E., Brosius, J. & Ptashne, M., Vectors bearing a hybrid *trp-lac* promoter useful for regulated expression of cloned genes in *Escherichia coli*. *Gene* 25:167–178 (1983)

Coleman, J., Hirashima, A., Inokuchi, Y., Green, P.J. & Inouye, M., A novel immune system against bacteriophage infection using complementary RNA (micRNA). *Nature* 315:601–603 (1985)

D'Andrea, A.D., Lodish, H.F. & Wong, G.G., Expression cloning of the murine erythropoietin receptor. *Cell* 57:277–285 (1989)

De Boer, H.A., Comstock, L.J. & Vasser, M., The *tac* promoter: A functional hybrid derived from the *trp* and *lac* promoters. *Proc. Natl. Acad. Sci. U.S.A.* 80:21–25 (1983)

Fuerst, T.R. & Moss, B., Structure and stability of mRNA synthesized by vaccinia virus-encoded bacteriophage T7 RNA polymerase in mammalian cells. Importance of the 5' untranslated leader. *J. Mol. Biol.* 206:333–348 (1989)

Gearing, D.P., King, J.A., Gough, N.M. & Nicola, N.A., Expression cloning of a receptor for human granulocyte-macrophage colony-stimulating factor. *EMBO J.* 8:3667–3676 (1989)

Jay, E., Seth, A.K., Rommens, J., Sood, A. & Jay, G., Gene expression: chemical synthesis of *E. coli* ribosome binding sites and their use in directing the expression of mammalian proteins in bacteria. *Nucleic Acids Res.* 10:6319-6329 (1982)

Kozak, M., Compilation and analysis of sequences upstream from the translational start site in eukaryotic mRNAs. *Nucleic Acids Res.* 12:857–872 (1984)

Nakamura, K. & Inouye, M., Construction of versatile expression cloning vehicles using the lipoprotein gene of *Escherichia coli*. *EMBO J.* 1:771–775 (1982)

Remaut, E., Stanssens, P. & Fiers, W., Plasmid vectors for high-efficiency expression controlled by the p_L promoter of coliphage lambda. *Gene* 15:81–93 (1981)

Rosenberg, M. & Court, D., Regulatory sequences involved in the promotion and termination of RNA transcription. *Annu. Rev. Genet.* 13:319–353 (1979)

Shine, J. & Dalgarno, L., Determinant of cistron specificity in bacterial ribosomes. *Nature* 254:34–38 (1975)

Stormo, G.D., Schneider, T.D. & Gold, L.M., Characterization of translational initiation sites in *E. coli. Nucleic Acids Res.* 10:2971–2996 (1982)

Studier, F.W. & Moffatt, B.A., Use of bacteriophage T7 RNA polymerase to direct selective high-level expression of cloned genes. *J. Mol. Biol.* 189:113–130 (1986)

Cloning II

The practical culmination of the scientific and technical developments described in the previous and following sections is the ability to isolate sequences encoding desired proteins, and to express the synthesis of those proteins in sufficient quantities for purification, analysis and eventual marketing. In this section we present several papers which represent important early examples of this type of accomplishment. These papers also demonstrate elements of a shift from academic-scientific to corporate-scientific competition; such are the fruits of future promise.

The Villa-Komaroff work builds on the work of Ullrich et al. presented previously (Section 4) to develop clones synthesizing human proinsulin. The market for human insulin is obviously immense, and the success of this project and other competitive projects is evidenced by the fact that human recombinant insulin is sold at neighborhood pharmacies the world over.

Another success story involves the cloning and expression of sequences encoding human growth hormone (see Martial et al.). Human growth hormone was available in extremely limited quantities from human cadavers. Related peptides from bovine, etc., potentially available in large quantities from slaughterhouses, had proven totally ineffective for human therapy. Thus the production of the recombinant product was the only practical route to take. This is an example of a desired product for which there was no other acceptable source.

The Gitschier et al. and Toole et al. papers describe the cloning of the gene for factor VIII. This was such a tour de force that the editor of *Nature* wondered whether the cloning of a whole chromosome would be next. The factor VIII gene is 180 kilobase pairs long, is divided into 26 exons and encodes a protein 2332 amino acids in length. Not only was this work technically very important, but it will also hopefully be of medical importance since lack of factor VIII is the causative factor in one of the most widely distributed and famous inherited diseases, hemophilia; and, because of their dependence on blood transfusions, hemophiliacs are a major group at risk in regards to AIDS.

Finally, Nagata et al. and Taniguchi et al. describe examples of projects targeted at producing pharmaceutical "magic bullets" — the interferon/interleukin type proteins which the popular press and Wall Street hoped would cure cancer (amongst other accomplishments). It is worth noting that these and related projects described in the supplemental papers were built on team efforts which were successful not only because of the recombinant DNA technology used, but also because of wise choices in regards to starting materials and sensitive bioassays. In other words, more "classical" cell culture and other biological and biochemical techniques are still essential.

It was the studies such as those described in this section that made *E. coli* into an industrial microorganism; something that most people would have not believed possible at the beginning of the 1970s!

Characterization of the Human Factor VIII Gene

J.Gitschier, W.I. Wood, T.M. Goralka, K.L. Wion, E.Y. Chen, D.H. Eaton, G.A. Vehar, D.J. Capon and R.M. Lawn

The complete 186,000 base-pair (bp) human factor VIII gene has been isolated and consists of 26 exons ranging in size from 69 to 3,106 bp and introns as large as 32.4 kilobases (kb). Nine kb of mRNA and protein-coding DNA has been sequenced and the mRNA termini have been mapped. The relationship between internal duplications in factor VIII and evolution of the gene is discussed.

HAEMOPHILIA has been known for millennia to be a male-specific, inherited disease[1]. Haemophilia A afflicts 10–20 per 100,000 males; a relatively high proportion due to novel mutations of the factor VIII gene[2,3]. The high frequency of haemophilia A compared with other autosomal clotting disorders results from the location of this gene on the X chromosome; one affected gene will thus produce the disease state in males. We report here the isolation and characterization of the 186,000 base pair (bp) region of the human X chromosome containing the complete factor VIII gene. A set of overlapping bacteriophage and cosmid clones was used to map and sequence the 26 exons containing the approximately 9,000 bp of coding sequence. Accompanying articles describe the expression of active human factor VIII from recombinant DNA clones[4] and present the detailed characterization of the protein[5]. Our results provide a basis for studies of the molecular defects associated with haemophilia.

Factor VIII genomic clones

Initial factor VIII clones were recovered from a bacteriophage λ genomic library[4] ('λ4X') containing DNA derived from an individual with 4X chromosomes (karyotyped 49,XXXXY). This library was screened with a unique 36-base oligonucleotide probe ('8.3'), synthesized to represent one codon choice of a sequenced tryptic peptide recovered from purified human factor VIII[5]. These initial clones contained overlapping segments spanning 28 kilobases (kb) of the human X chromosome. We subsequently expanded the set of genomic clones to contain 200 kb of the human genome encompassing the entire factor VIII gene. Both bacteriophage λ and cosmid clones were used in the process of 'genomic walking' (see legends to Figs. 1 and 2).

The restriction endonuclease map of the human factor VIII gene is shown in Fig. 1; the gene map was confirmed by the characterization of at least two independent clones for most of the 210-kb region and by numerous Southern blots of 49,XXXXY and normal DNA. No noticeable discrepancies between genomic DNA and the map derived from recombinant clones were found. In addition, low stringency blot experiments

have so far detected no closely related factor VIII genes or pseudogenes.

DNA sequence analysis

The DNA sequence of the entire mRNA coding portion (~9 kb) of the factor VIII gene has been determined. Exon-containing restriction fragments of genomic clones were identified by hybridization to cDNA and sequenced by dideoxy-chain termination procedures[6]. This DNA sequence analysis is distinct from that of cDNA clones derived from AL-7 cell RNA (see ref. 4). The genomic DNA sequence of exon boundaries (Fig. 3) and the promoter and polyadenylation regions (Fig. 4) are reported here. We found only two nucleotide differences between the genomic and cDNA sequences in the 8,860 nucleotides compared. Nucleotide 3,780 in cDNA is the G of a glutamic acid codon GAG, whereas in the cloned genomic DNA it is C, creating the conservative substitution of an aspartic acid residue. Nucleotide 8,728 in the 3' untranslated region is G in cDNA and A in genomic DNA. The surprising degree of similarity between genomic sequence derived from the 49,XXXXY cell DNA and cDNA sequence from AL-7 cell RNA indicates that the sequence data and the reverse transcription procless are reliable and suggests also a lower level of nucleotide sequence polymorphism of this X-linked gene than has been found for some autosomal genes, an effect noted for other X chromosome loci[7].

The boundaries of exons shown in Fig. 3 were established by restriction mapping and DNA sequence comparison of cDNA with genomic clones. In cases of nucleotide redundancy at either side of an intron, boundaries were chosen to be consistent with established consensus splice sites[8,9]. All of the splice donor and acceptor sites conform to the GT...AG rule for nucleotides immediately flanking exon borders (Fig. 3). Further flanking sequences are in general agreement with favoured nucleotide frequencies noted in other compendia[8,9]. The dinucleotide AG never occurs in the pyrimidine-rich region encompassing the final 15 nucleotides of introns; exons frequently begin with the sequence GT.

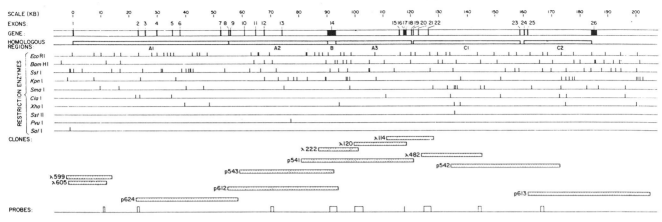

Fig. 1 Map of human factor VIII gene. The structure of the gene is schematically represented by an open bar; the 26 exons are filled-in areas drawn to scale. The direction of transcription is from left to right. The size scale in kb is drawn above the gene; the initiating ATG is positioned at nucleotide 1. Immediately below the gene are the extent of the triplicated 'A' domain, the unique 'B' domain, and the duplicated 'C' domain. The location of the recognition sites for the 10 restricton enzymes used to map the factor VIII gene are given in the next series of lines. Cross-hatched boxes represent the extent of human genomic DNA contained in each λ phage and cosmid clone. Other clones which encompass the same regions are not shown. No genomic clones span the 8.5 kb of intron located between clones λ599 and p624. Other EcoRI, BamHI or SstI sites may lie in this region. The extent of this gap was determined by probing genomic Southern blots with single copy fragments isolated from the 3' end of λ599 and the 5' end of p624. The bottom line shows the location of synthetic oligonucleotide or restriction fragment probes used in the genomic cloning, from left to right: the 2 probes used to map the uncloned region, three 5' walk probes, the original synthetic 36-mer probe '8.3'; three 3' walk probes.

Methods: λ and cosmid libraries were screened with random calf-thymus primed ^{32}P-labelled single copy genomic or cDNA probes[27]. Nitrocellulose filters were hybridized overnight at 42 °C in 50% formamide, 0.1 g l^{-1} salmon sperm DNA, 5×SSC, 0.05 M sodium phosphate (pH 6.8), 5× Denhardt's solution and 10% dextran sulphate[28]. Filters were washed in 0.2×SSC, 0.1 g l^{-1} SDS at 60 °C for 1 h. Library filters were repeatedly rescreened with newly isolated walk probes so that extending clones were easily identified. To map restriction endonuclease cleavage sites, DNA from the clones was digested with restriction enzymes singly or in combinations and characterized by gel electrophoresis (followed by Southern blot hybridization in some cases). DNA fragments generated by EcoRI and BamHI digestion were subcloned into pUC plasmid vectors[29] for further propagation and analysis. Restriction mapping, DNA sequence analysis and blot hybridizations with the 8.3 probe determined the orientation of transcription of the gene. The genome walk was initiated by identifying single copy probe fragments near the ends of the initial 28 kb region. To find suitable fragments, digests of cloned DNA were blot hybridized with total ^{32}P-labelled human DNA under conditions where only DNA fragments containing sequences repeated more than about 50 times in the genome will hybridize[30,31]. Candidate fragments were retested for repeated sequences by hybridization to 50,000 phage from the λ/4X library. To begin extension in the 5' direction, clone λ222 was identified by a triplet of 1 kb NdeI/BamHI probe fragments isolated from λ120 DNA. In the 3' direction, a single copy probe fragment of λ114 failed to yield extending clones in screens of several λ libraries. Based on genomic blotting results, we then constructed a size selected[32] BclI library of human 49,XXXXY DNA in the vector λ1059[33]. λ482 is one of the extending clones so derived. Subsequent overlapping clones (p541, p542, p543, p612, p613 and p624) were derived by screening a recombinant cosmid library containing human DNA (see Fig. 2). The most 5' genomic clones, λ599 and λ605 were obtained by screening with cDNA-derived probes.

Fig. 2 Cosmid vector pGcos4. To facilitate genomic walking, we developed a new cosmid vector with the following features: (1) It can accommodate 45 kb inserts. (2) It confers resistance to tetracycline, rather than ampicillin. (3) The 641 bp AvaI/PvuII fragment of pBR322 has been removed to increase plasmid copy number and to delete sequences interfering with transformation of eukaryotic cells[34]. (4) A methotrexate-resistant dihydfolate reductase gene driven by a simian virus 40 (SV40) early promoter has been incorporated for the selection and propagation of clones in eukaryotic cells[35]. (5) A synthetic DNA fragment containing a unique BamHI site flanked by pairs of PvuI and EcoRI sites has been inserted as the cloning site. The flanking EcoRI sites are useful for subcloning fragments and the PvuI sites allow excision of the entire insert in most cases, since this enzyme cuts eukaryotic DNA only about once per 135,000 bp. Construction: The scale is represented by dots within the circle numbered in kb from the zero point. The 403 base (b) annealed HincII fragment of λcI857S7 (Bethesda Research Lab) containing the cos site was cloned in pBR322 from AvaI to PvuII to generate the plasmid pGcos1. Separately, the 1,624 b PvuII to NaeI fragment of pFR400[35] containing an SV40

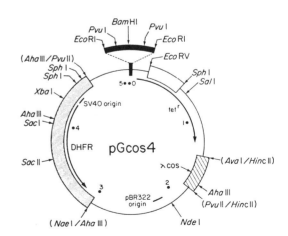

origin and promoter, a methotrexate-resistant dihydropholate reductase gene, and hepatitis B surface antigen polyadenylation sequences, was cloned into the pBR322 AhaIII site to generate the plasmid mp33dhfr. The following three fragments were ligated together to generate the cosmid vector pGcos3: 1,497 base (b) SphI-NdeI fragment of pGcos1, 2,163 b NdeI-EcoRV fragment of mp33dhfr and 376 b EcoRV-SphI fragment of pKT19. pKT19 is a derivative of pBR322 in which the BamHI site in the tetracycline resistance gene has been removed (provided by Herb Boyer). pGcos4 was generated by cloning the synthetic 20mer, 5' AATTCGATCGGATCCGATCG, in the EcoRI of pGcos3. Left and right arms of this vector were prepared by cleavage of two aliquots with SstI or SalI, followed by treatment with alkaline phosphatase, cleavage with BamHI and isolation of the large fragments[28]. 49,XXXXY DNA isolated from the GM1202A cell line (NIGMS Human Genetic Mutant Cell Repository, Camden, New Jersey) was partially cleaved with five levels of Sau3A1 and the 45 kb fragments isolated by sucrose density centrifugation[28]. These fragments were ligated to the BamHI arms of pGcos4, packaged in vitro and used to infect E. coli HB101.

a

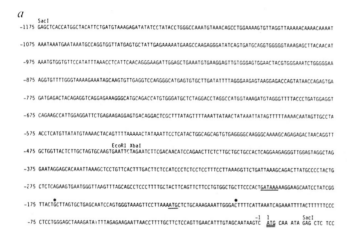

Exon	Location in cDNA	Length	Splice Acceptor	Splice Donor	Acceptor Match (of 16)	Donor Match (of 9)	Distance to previous Acceptor AG	Intervening Sequence Length
1	-170 – 143	313 b		GCAAG|GTAAAGGC		7		22.9 kb
2	144 – 265	122	TTCCTTTCTTCACGCAG|ATT . . . GATGG|GTAATGAA		14	5	38 b	2.6
3	266 – 388	123	ACATCTCATTCTTACAG|GTC . . . TGAGG|GTGAGTAA		14	8	28	3.9
4	389 – 601	213	ATTTCTTCCTGCTATAG|GAG . . . AGAAG|GTAAGTGT		15	9	24	5.4
5	602 – 670	69	TCTCACTTCTTTTTCAG|GGA . . . TGAAG|GTTAGTGA		15	8	23	2.4
6	671 – 787	117	GCCTTCTCTCTCATCAG|GGA . . . GCCAG|GTATGTAC		15	8	30	14.2
7	788 – 1009	222	TTATTCCTACTTTACAG|GTC . . . ACATG|GTAATATC		14	6	20	2.6
8	1010 – 1271	262	GGTTTGTCTGACTCCAG|ATG . . . GACAG|GTAAGCAC		12	8	32	0.3
9	1272 – 1443	172	TTTTTTCTCTTATACAG|AAG . . . TGTTG|GTAAGTTG		14	7	40	4.8
10	1444 – 1537	94	CTCGCTTATACTTTCAG|ATT . . . AAAAG|GTAAATAT		12	8	17	3.8
11	1538 – 1752	215	GTTTTGCTTGTGGGTAG|GTG . . . ACCAG|GTGAGTTC		12	9	24	2.8
12	1753 – 1903	151	ATATATATGTAATTAACAG|ATA . . . GCACA|GTGAGTAA		9	7	37	6.3
13	1904 – 2113	210	CCCCATTGTTTTTGCAG|GCA . . . CCCAG|GTTAGTTA		14	8	104	16.0
14	2114 – 5219	3106	TTCTTCCTCATCTCCAG|GTC . . . AACAG|GTATGAAT		15	7	20	22.7
15	5220 – 5373	154	CTGCTTTTTTTCTCCAG|GGC . . . TCATG|GTGAGTTA		15	8	56	1.3
16	5374 – 5586	213	TCGTTATTGTTCTACAG|GTA . . . ACCTG|GTAAGCAG		13	7	39	0.3
17	5587 – 5815	229	ATGTCTTCCCTCCCTAG|GAA . . . CCATG|GTAATATA		15	6	20	0.2
18	5816 – 5998	183	TCTGTGTCCTTCTCCAG|CAA . . . TCCAG|GTATGAGC		13	7	20	1.8
19	5999 – 6115	117	CTGTTGGTTTTTATAAG|GTG . . . CAATA|GTGATAG		11	7	30	0.6
20	6116 – 6187	72	CACCCGTTTCATTTCAG|AGT . . . ATATG|GTAAATAC		13	7	47	1.6
21	6188 – 6273	86	TACTTACTTGGGCAAAG|GAC . . . TCAAG|GTTAGAA		11	7	>35	3.4
22	6274 – 6429	156	GGTTATTTTAATTGGTAG|GTG . . . TAATG|GTATGTAA		12	7	16	32.4
23	6430 – 6574	145	CTAATCTCTCCATACAG|GTC . . . AAATA|GTAAGTGC		13	7	56	1.4
24	6575 – 6723	149	TTTTTCTTTTTCTTTGAG|GTT . . . CTCAG|GTAAGAGG		15	8	37	1.0
25	6724 – 6900	177	AACTCTATTGCCCTCAG|GTG . . . TAAAG|GTAAGCTG		14	8	18	22.4
26	6901 – 8858	1958	TATCTTTCCTCTTTCAG|GTT		16		30	

b

Acceptor Frequencies   Donor Frequencies

	-17	-16	-15	-14	-13	-12	-11	-10	-9	-8	-7	-6	-5	-4	-3	-2	-1	1	2	3		-5	-4	-3	-2	-1	1	2	3	4	5	6	7	8
G	16%	4	16	8	0	20	8	4	4	16	8	4	8	16	4	0	100	72	16	28		20%	16	0	8	88	100	0	20	0	76	8	24	20
A	20	16	16	8	12	8	4	8	12	12	12	32	8	100	0	24	20	28	8	100		32	12	60	52	12	0	68	84	12	24	48	36
T	40	56	36	60	64	48	64	56	56	48	56	44	64	32	16	0	0	0	60	20		32	16	8	36	0	0	100	12	16	12	56	28	12
C	24	24	32	24	24	24	24	32	28	20	28	36	16	20	72	0	0	4	24			16	44	32	4	0	0	0	0	0	0	12	0	32
Consensus	N	T	T	T	T	T	T	T	T	T	T	T	T	N	C	A	G	G	T	N		N	N	A	A	G	G	T	A	A	G	T	A	N
			C	C			C	C	C	C	C														C										

a

SacI

-1175 GAGCTCACCATGGCTACATTCTGATGTAAAGAGATATATCCTATACCTGGGCCAAATGTAAACAGCCTGGAAAAGTGTTAGGTTAAAAACAAAACAAAAT

-1075 AAATAAATGAATAAATGCCAGGTGGTTATGAGTGCTATTGAGAAAAATGAAGCCAAGAGGGATATCAGTGATGCAGGTGGGGGTAAAGAGCTTACAACAT

-975 AAATGTGGTGTTCCATATTTAAACCTCATTCAACAGGGAAGATTGGAGCTGAAATGTGAAGGAGTTGTGGGAGTGGAACTACGTGGGAAATCTGGGGGAA

-875 AGGTGTTTTGGGTAAAAGAAATAGCAAGTGTTGAGGTCCARGGGCATGAGTGTGCTTGATATTTTAGGGAAGAGTAAGGAGACCAGTATAACCAGAGTGA

-775 GATGAGACTACAGAGGTCAGGAGAAAGGGCATGCAGACCATGTGGGATGCTCTAGGACCTAGGCCATGGTAAAGATGTAGGGTTTTACCCTGATGGAGGT

-675 CAGAAGCCATTGGAGGATTCTGAGAAGAGGAGTGACAGGACTCGCTTTATAGTTTTAAATTATAACTATAAATTATAGTTTTTAAAACAATAGTTGCCTA

-575 ACCTCATGTTATATGTAAAACTACAGTTTTAAAAACTATAAATTCCTCATACTGGCAGCAGTGTGAGGGGCAAGGGCAAAAGCAGAGAGACTAACAGGTT

EcoRI XbaI

-475 GCTGGTTACTCTTGCTAGTGCAAGTGAATTCTAGAATCTTCGACAACATCCAGAACTTCTCTTGCTGCTGCCACTCAGGAAGAGGGTTGGAGTAGGCTAG

-375 GAATAGGAGCACAAATTAAAGCTCCTGTTCACTTTGACTTCTCCATCCCTCTCCTCCTTTCCTTAAAGGTTCTGATTAAAGCAGACTTATGCCCCTACTG

-275 CTCTCAGAAGTGAATGGGTTAAGTTTAGCAGCCTCCCTTTTGCTACTTCAGTTCTTCCTGTGGCTGCTTCCCACT**GATAAAA**GGAAGCAATCCTATCGG

-175 TTACTGCTTAGTGCTGAGCAATCCAGTGGGTAAAGTTCCTTAAA**ATG**CTGCTGCCAAAGAAATTGGGACTTTTCATTAAATCAGAAATTTTACTTTTTTCCC

-1 SacI

-75 CTCCTGGGAGCTAAAGATA͡TTTAGAGAAGAATTAACCTTTTTGCTTCTCCAGTTGAACATTTGTAGCAATAAGTC AŢG CAA ATA GAG CTC TCC

b

8364 GAGCAGTTGGAGGAAGCATCCAAAGATTGCAACCCAGGGCAAATGGAAAACAGGAGATCCTAATATGAAAGAAAAATGGATCCCAATCTGAGAAAAGGCA

BamHI

8464 AAAGAATGGCTACTTTTTTCTATGCTGGAGTATTTTCTAATAATCCTGCTTGACCCTTATCTGACCTCTTTGGAAACTATAACATAGCTGTCACAGTATA

8564 GTCACAATCCACAAATGATGCAGGTGCAAATGGTTTATAGCCCTGTGAAGTTCTTAAAGTTTAGAGGCTAACTTACAGAAATGAATAAGTTGTTTTGTTT

HpaII HpaII

8664 TATAGCCCGGTAGGAGGTTAACCCCAAAGGTGATATGGTTTTATTTCCTGTTATGTTTAACTTAATAATCTTATTTTGGCATTCTTTTCCCATTGACTA

8764 TATACATCTCTATTTCTCAAATGTTCATGGAACTAGCTCTTTTATTTTCCTGCTGGTTTCTTCAGTAATGAGTT**AAATAAAA**C**ATTG**ACACATACAAACA

8864 AATGCCTTTGAGAATTGTGTTTTTACACTGGAAATAAAAATGTGAACACTGATTTTTAAAACAAATAGGGGCACTGAATAGCAAGCATGGACACTCTAGAA

8964 AACCCAAAATTAGTGAGTTAGAAAACCAGATTAAATTGAACTCAGAGTAAAAATGATATAATTCATGAGAGTCTGAATAAAATAAATCAGAAATGGAGCCTC

9064 AATCCAGGAGAACAGCTTATATGGAGAGAGAGAGACTGAGAGAGAAATGGGAGTTTCTGTTCAATGGGCATAAAGTTTCAGCTATGCTGCTGGGCACAGT

9164 GGCTCATGCCTGTAATCCTGGCACTTTGGGAGGCCAAGGCGGGCGGATCACCTGAGGTCAGGAGTTCAAGACCAGCCTAGCCAACATGGCGAAATCCCGT

SmaI

9264 CGCTAATAAAAATATAAAAATTAGTTGGGCATGGTGGCACATGCCTGTAGTCCCAGCTACTTGGGAGGTTGAGGCACAGAATCGCTTGAACCCGGG

Fig. 4 Sequence of the 5' and 3' termini. ***a***, DNA sequence of the genomic *Sst*I (*Sac*I) fragment containing exon 1. The double underline at nucleotide position +1 denotes the ATG translation initiation codon which precedes the signal peptide and mature factor VIII coding sequences. A single underline indicates an ATG triplet located 5' to this site, which is closely followed by stop codons in all three possible reading frames. Closed circle, terminus of the 5'-most cDNA clone recovered; *, the probable mRNA 5' start site. The promoter-like 'ATA' box sequence is also underlined. Selected restriction enzyme sites are indicated. ***b***, DNA sequence of part of the genomic region containing 3' untranslated (exon 26) and flanking regions. *, Beginning of the poly(A) tail found in cDNA clones[4]. The polyadenylation signal sequence AATAAA[22] and the sequence CATTG[23] are underlined. Nucleotides are numbered in accordance with the continuous cDNA sequence beginning at the initiator ATG[4].

Fig. 3 Summary of factor VIII exons, introns and splice junctions. ***a***, Each row lists the exons of the human factor VIII gene, numbered from 5' to 3' in the direction of transcription; the nucleotide numbers of mRNA (cDNA) encompassed by each exon (the 5' end of exon 1 is considered to be the 5' mRNA start site and the 3' end of exon 26 the nucleotide preceding the poly(A) tail). The total exon length, the DNA sequence surrounding the 5' border (acceptor) and the 3' border (donor) of each exon (hence intron sequences are to the left of the vertical line in the acceptor column and to the right of the vertical line for the donor column) are also listed. Further shown are the match of acceptor and donor nucleotides to consensus splice site sequences[9], the distance from the splice acceptor site to the nearest upstream AG dinucleotide[9] and the length of the intron immediately following the numbered exon. ***b***, The % frequency of each nucleotide at the splice site borders of human factor VIII gene and the consensus sequence drawn from these data.

The 26 exons of the factor VIII gene range in size from 69 to 3,106 bp (median size, 164 bp), consistent with published distributions of exon sizes for 20 proteins[10]. However, two factor VIII exons greatly exceed the norm; one may be the largest exon yet reported. The largest exon (14) is 3,106 bp and codes for all of the relative molecular mass (M_r) ~ 100,000 region of the protein connecting the amino-terminal M_r 90,000 and carboxy-terminal M_r 80,000 fragments[5]. Proteolytic removal of this connecting region may be associated with activation of the protein[11]. The other large exon (26) is 1,958 bp, 1,805 bp of which is 3' untranslated sequence. Introns range in size from 207 bp to 32.4 kb. There are six large introns, each containing over 14 kb of DNA. The complete gene consists of 9 kb exon and 177 kb intron. Apparently, there has been no effective selective pressure for small introns to reduce the overall size of this large gene.

Characterization of gene termini

The 5' end of the factor VIII gene is contained in a 1.2 kb *Sst*I fragment whose DNA sequence is shown in Fig. 4*a*. The ATG at position +1 serves as the translation initiation codon and is followed in mRNA by a continuous open reading frame coding for 2,351 amino acids, comprising the signal peptide and the mature factor VIII protein. The genomic DNA sequence exactly matches that of the cDNA clone with the greatest 5' extension (to nucleotide position −109), suggesting that the transcription initiation site may be contained also in the same exon. This 1.2 kb *Sst*I genomic fragment directs transcription of factor VIII mRNA in transfected mammalian cells (data not shown), indicating that the promoter is probably contained in this region. We therefore undertook RNAse mapping[12,13] experiments to determine the mRNA 5' start site.

Labelled probe fragments of 185 bp and, to a lesser extent, 187 bp, are protected from digestion by poly(A)$^+$ RNA obtained from AL-7 and human liver cells, whereas the probe was not protected by RNA obtained from human peripheral blood lymphocyte, HepG2, Alexander cells or mouse T lymphoma cells (Fig. 5). The size of the protected fragments implies that the factor VIII mRNA from these two cell sources starts most frequently at position −170 and less frequently at position −172 (Fig. 4*a*). The sequence at this location does not contain a splice acceptor recognition sequence (although one is located six nucleotides downstream), rendering unlikely the possibility that

Fig. 5 Determination of the 5' mRNA start site by RNAse protection. Autoradiograph of a 6% polyacrylamide, 7 M urea gel. Lanes 1 and 2, a known DNA sequencing ladder employed as size standards; lanes 3–11, results of RNAse protection reactions.

Methods: Cell and tissue RNA was prepared[36]. The 5' genomic *Sst*I fragment was ligated into pSP64 (a gift of P. Mellon and T. Maniatis) and linearized with *Pvu*II. The uniformly labelled ^{32}P-RNA probe was synthesized with 400 Ci mmole[-1] α-[32P]CTP (Amersham) and SP6 RNA polymerase (NEN) followed by DNAse treatment (DPRF, Worthington) as described[12,13]. Sample RNAs were hybridized with 5×10^5 c.p.m. of probe at 37 °C and digested with 80 μg ml[-1] RNAse A and 4 μg ml[-1] RNAse T1 (Worthington) followed by extraction, electrophoresis and autoradiography. ^{32}P-labelled, anti-sense strand RNA probe was prepared by ligating the 1.2 kb *Sst*I genomic fragment containing exon 1 and 5' flanking sequences into pSP64[12] and transcribing the linearized plasmid with SP6 RNA polymerase[12,13]. This labelled probe was hybridized with poly(A)[+] RNA derived from various sources and digested with RNAse A and RNAse T1 before electrophoresis. Protected RNA was derived from: lane 3, AL-7 human hybridoma cells (10 μg); 4, human liver sample 1 (10 μg); 5, human liver sample 2 (10 μg); 6, human Alexander cell line (10 μg); 7, human Hep G2 cell line (10 μg); 8, peripheral blood lymphocyte sample 1 (3 μg); 9, peripheral blood lymphocyte sample 2 (3 μg); 10, mouse T lymphoma cell line S49 (10 μg); 11, no RNA control. An identical pair of protected fragments (arrow), measuring 185 and 187 bases from the *Sst*I site, appear only in RNA obtained from the two human liver samples and the AL-7 hybridoma cells. This positions the probable 5' mRNA start site at ~170 bases 5' of the initiation ATG as shown in Fig. 4. (As RNAses A and T1 cannot cleave after an adenosine residue and the complementary strand is radiolabelled, no protected band was seen corresponding to a terminus at position −171 of the gene (Fig. 4), which is a T residue in the genomic sequence.)

the protected RNA is bounded by an intron instead of the true mRNA 5' start site.

The RNAse protection experiments also show that liver is a site of synthesis of factor VIII. No factor VIII RNA was detected in the other tissues of the protection experiment shown in Fig. 5 nor by Northern blot analysis of RNA from various tissues and cell lines[4]. Liver is only one of several candidates for the site of factor VIII production[14–16].

Inspection of the genomic sequence of the 5' flanking region (Fig. 4a) revealed the sequence GATAAA beginning 30 bp 5' of the presumed 5' mRNA start site, similar to the canonical 'ATA' or 'Goldberg–Hogness' sequence, located about 25–35 bp before eukaryotic cap sites and apparently required for precise initiation of transcription by RNA polymerase II[17,18]. We found no convincing candidate for the less frequently conserved 'CAT' sequence[19] in the −70 bp region.

One additional ATG to nucleotide 5' to +1 is found in the 5' untranslated region at position −131 (Fig. 5). This region cannot serve as the translation start site for factor VIII because seven stop codons, distributed in all three possible reading frames, occur in the 128 nucleotides between this upstream ATG and position +1. Although only 5% of the previously compiled eukaryotic mRNAs contain an ATG 5' of the translation initiation site[20], the presence of an upstream ATG does not necessarily repress translational initiation at the proper start site when it is followed by an in-frame termination codon[21].

The 3' untranslated region of the factor VIII gene contains 1,805 nucleotides (counting the TGA stop triplet and terminating before the first A of the poly(A) tail; Fig. 4b). The site of polyadenylation was inferred from cDNA clones and is preceded 19 bases by the common polyadenylation signal AATAAA[22] as well as a second, less conserved recognition element CATTG[23].

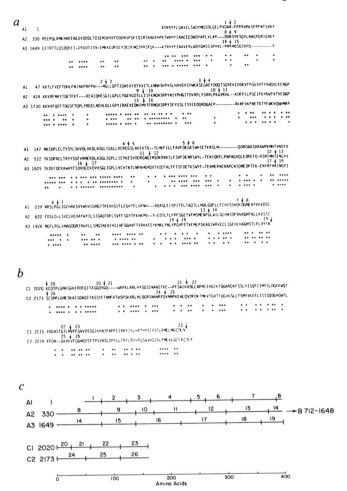

Fig. 6 Homologous factor VIII domains and intron boundaries. *a*, 'A' unit triplication showing the alignment of three amino acid sequences. The amino acid numbers precede each line. Arrows indicate the location of introns and are flanked by the appropriate exon numbers. *, Conserved amino acids. *b*, 'C' unit duplication. *c*, Representation of the A and C repeat units with the location of introns indicated by vertical lines. The exons are numbered.

Introns and protein domains

Searches for internal homologies in the factor VIII DNA sequence and the derived amino acid sequence revealed three types of domains: a triplicated 'A' unit of 330–380 amino acids, a unique 'B' unit of about 925 amino acids and a duplicated 'C' unit of about 160 amino acids. The order of the domains in the gene and protein is A1–A2–B–A3–C1–C2 (refs 4, 5; Fig. 1) (note that the exons constituting each domain are not clustered). The amino acid alignment of the A and C repeats (Fig. 6) demonstrates that the A units have ~30% pairwise amino acid homology (20% of the residues being conserved in all three domains), whereas the two C units have 37% homology[4,5].

If segmental DNA duplication has played a role in the evolution of the large factor VIII gene, one might expect a conservation of the intron boundaries within the A and C repeats of factor VIII. The alignment of most intron boundaries relative to the protein coding sequence is well fixed by the high degree of homology in the A and C repeats (Fig. 6). For the C duplication, intron boundaries occur precisely at the borders of the C1 and C2 repeats, as would be expected for a DNA duplication mechanism where the boundaries fall within the introns. Furthermore, the junction between A3 and C1 also contains an intron precisely at the boundary, again as expected for an intron joining mechanism. On the other hand, the A1–A2 and A2–A3 junctions are each contained on one exon (Fig. 6).

Within the A and C repeats only some of the intron boundaries are conserved. In fact, each of the repeats contains a different number of exons. The precise alignment of some intron boundaries within the A and C repeats suggests that these introns were present in an ancestral precursor of the repeat units and duplicated in the course of the event that generated the repeat. The non-aligned introns appear to have arisen since the duplication, either by loss of one member of a duplicated intron or by the insertion of new introns after the duplication. In an analogous situation, α-feto protein and albumin each have a triplicated gene structure, but unlike factor VIII, all of the nine intron boundaries within the repeats for these two genes are conserved precisely[24-25].

It is also interesting to speculate as to the origin of domain B in factor VIII. The 925 amino acids of this domain encompass nearly all of the 3.1 kb of exon 14, where the end of the A2 repeat (at the 5' end) and the start of the A3 repeat (at the 3' end) are also found. This suggests that during the course of evolution a processed gene may have been inserted into a short exon that contained the A2–A3 boundary. Insertion of a processed gene (derived from an mRNA) would also account for the anomalously large size of exon 14. The B domain connects the M_r 90,000 (made up of A1 and A2) and the M_r 80,000 (A3, C1, C2) fragments of the factor VIII protein. Proteolytic removal of the domain is associated with activation of the protein[5,11];

thus exon 14 is related to a physiological unit of the protein. Whether any other functional domains or subdomains of factor VIII correspond to any of the exons awaits further characterization of the molecule.

In conclusion, we report here the complete isolation and characterization of the human factor VIII gene. It is the largest gene yet characterized, encompassing nearly 0.1% (186 kb) of the human X chromosome. Transcription of such a large gene could take over three hours (based on a transcription rate of 15 nucleotides per second[26]). Probes derived from the factor VIII gene will allow identification of restriction fragment length polymorphisms which can be used further to localize the gene on the X chromosome and lead to improved prenatal diagnosis of haemophilia. We anticipate that the knowledge of the gene structure and DNA sequence will be invaluable in the analysis of the molecular basis of haemophilia.

We thank John McGrath who characterized the λ222 clone and Philip Hollingshead for laboratory assistance; Edward Tuddenham, David Martin, Ingrid Caras and Glenn Nedwin for providing tissue samples; Jeanne Arch and Alane Gray for preparation of the manuscript and figures. We thank David Goeddel, Christian Simonsen, Laurence Lasky and Dennis Kleid for helpful discussions and David Martin and Robert Swanson for their support.

Received 9 August; accepted 27 September 1984.

1. Rosner, F. *Medicine in the Bible and the Talmud* 43-49 (Yeshiva University Press, New York, 1977).
2. Galjaard, H. *Genetic Metabolic Disease* (Elsevier, New York, 1980).
3. Haldane, J. B. S. *J. Genet.* **31**, 317-326 (1935).
4. Wood, W. 1. *et al. Nature* **312**, 330-337 (1984).
5. Vehar, G. A. *et al. Nature* **312**, 337-342 (1984).
6. Sanger, F., Nicklen, S. & Coulson, A. R. *Proc. natn. Acad. Sci. U.S.A.* **74**, 5463-5467 (1977).
7. Cooper, D. N. & Schmidke, J. *Hum. Genet.* **66**, 1-66 (1984).
8. Breathnach, R. & Chambon, P. *A. Rev. Biochem.* **50**, 349-383 (1981).
9. Mount, S. M. *Nucleic Acids Res.* **10**, 459-472 (1982).
10. Blake, C. *Nature* **308**, 535-537 (1983).
11. Fulcher, C. A., Roberts, J. R. & Zimmerman, T. S. *Blood* **61**, 807-811 (1983).
12. Melton, D. A. *et al. Nucleic Acids Res.* (in the press).
13. Zinn, K., Di Maio, D. & Maniatis, T. *Cell* **34**, 865-879 (1983).
14. Zimmerman, T. S. & Meyer, D. in *Haemostasis and Thrombosis* (eds Bloom, A. L. & Duncan, P. T.) 111-123 (Churchill Livingstone, New York, 1981).
15. Stel, H. V., van der Kwast, T. H. & Veerman, E. C. I. *Nature* **303**, 530-532 (1983).
16. Kelly, D. A., Summerfield, J. A. & Tuddenham, E. G. D. *Br. J. Haem.* **56**, 535-543 (1984).
17. Corden, J. *et al. Science* **209**, 1406-1414 (1980).
18. Grosschedl, R. & Birnstiel, M. L. *Proc. natn. Acad. Sci. U.S.A.* **77**, 1432-1436 (1980).
19. Benoist, C. *et al. Nucleic Acids Res.* **8**, 127-142 (1980).
20. Kozak, M. *Nucleic Acids Res.* **8**, 127-142 (1980).
21. Liu, C.-C., Simonsen, C. C. & Levinson, A. D. *Nature* **309**, 82-85 (1984).
22. Proudfoot, N. & Brownlee, G. *Nature* **252**, 359-362 (1981).
23. Berget, S. M. *Nature* **309**, 179-181 (1984).
24. Eiferman, F. A. *et al. Nature* **294**, 713-718 (1981).
25. Sargent, T. D. *et al. Molec. Cell Biol.* **1**, 871-883 (1981).
26. Davidson, E. H. *Gene Activity in Early Development*, 2nd edn 363 (Academic, New York, 1976).
27. Taylor, J. M., Illmensee, R. & Summer, S. *Biochim. biophys. Acta* **442**, 324-330 (1976).
28. Maniatis, T., Fritsch, E. F. & Sambrook, J. *Molecular Cloning* (Cold Spring Harbor Laboratory, New York, 1982).
29. Vieira, J. & Messing, J. *Gene* **19**, 259-268 (1982).
30. Wood, W. I., Nichol, J. & Felsenfeld, G. *J. biol. Chem.* **256**, 1502-1505 (1981).
31. Shen, C-K. J. & Maniatis, T. *Cell* **19**, 379-391 (1981).
32. Lawn, R. M. *et al. Nucleic Acids Res.* **9**, 6103-6114 (1981).
33. Korn, J. *et al. Proc. natn. Acad. Sci. U.S.A.* **77**, 5172-5176 (1980).
34. Lusky, M. & Botchan, M. *Nature* **293**, 79-81 (1981).
35. Simonsen, C. C. & Levinson, A. D. *Proc. natn. Acad. Sci. U.S.A.* **80**, 2495-2499 (1983).
36. Ullrich, A. *et al. Science* **196**, 1313-1315 (1977).

Human Growth Hormone: Complementary DNA Cloning and Expression in Bacteria

J.A. Martial, R.A. Hallewell, J.D. Baxter and H.M. Goodman

Abstract. *The nucleotide sequence of a DNA complementary to human growth hormone messenger RNA was cloned; it contains 29 nucleotides in its 5' untranslated region, the 651 nucleotides coding for the prehormone, and the entire 3' untranslated region (108 nucleotides). The data reported predict the previously unknown sequence of the signal peptide of human growth hormone and, by comparison with the previously determined sequences of rat growth hormone and human chorionic somatomammotropin, strengthens the hypothesis that these genes evolved by gene duplication from a common ancestral sequence. The human growth hormone gene sequences have been linked in phase to a fragment of the* trp D *gene of Escherichia coli in a plasmid vehicle, and a fusion protein is synthesized at high level (approximately 3 percent of bacterial protein) under the control of the regulatory region of the* trp *operon. This fusion protein (70 percent of whose amino acids are coded for by the human growth hormone gene) reacts specifically with antibodies to human growth hormone and is stable in* E. coli.

Growth hormone, along with at least two other polypeptide hormones, chorionic somatomammotropin (placental lactogen) and prolactin, forms a set of proteins with amino acid sequence homology and to some extent overlapping biological activities (*1, 2*). Since the genes of this set of proteins probably have a common ancestral origin (*1*), they constitute an excellent model to study the evolution, structure, and differential regulation of related genes. In addition, since human growth hormone is of considerable medical importance and its supply is limited, the synthesis of growth hormone in bacteria might provide the required alternate source of this critical hormone.

We have previously isolated and analyzed bacterial clones containing copies of complementary DNA (cDNA) transcripts of messenger RNA's (mRNA's) for these hormones. The complete sequence of rat pregrowth hormone mRNA (*3*) has been reported; in addition, sequence data have been presented for fragments of about 550 bases complementary to part of the coding (amino acid residues 24 to 191) and 3' untranslated portions of human chorionic somatomammotropin (hCS) (*3, 4*) and human growth hormone (hGH) mRNA's (*5*). A partial sequence of rat prolactin has been determined by Gubbins *et al.* (*6*). These sequence data showed that, whereas the growth hormone genes of the rat and man had significant homology, they also had diverged substantially, such that they differed more than the genes for the functionally distinct human hormones hCS and hGH.

We now report the synthesis, cloning, and sequence analysis of cDNA containing the entire coding and most of the noncoding portions of hGH mRNA. We also describe the insertion of these sequences into an "expression plasmid" containing part of the *Escherichia coli* tryptophan (*trp*) operon whose construction has been realized by Hallewell and Emtage (*7*). We describe the use of this plasmid to promote the inducible bacterial synthesis of high levels of a hybrid protein, 70 percent of which is composed of amino acids coded for by the hGH gene.

Human growth hormone mRNA isolation. Polyadenylated RNA was isolated (*8*) from human pituitary tumors removed by transphenoidal hypophysectomy. To obtain an indication of the integrity and the relative abundance of growth hormone mRNA in each sample, the individual mRNA preparations were translated in the wheat germ cell-free system, and the products were analyzed by electrophoresis on sodium dodecyl sulfate–polyacrylamide gels (Fig. 1). Among the translation products of the five acromegalic tumor RNA's (Fig. 1, lanes 1 to 5), the most prominent band corresponds to a protein of approximately 24,000 daltons. This protein is assumed to be human pregrowth hormone since it is similar in size to rat pregrowth hormone (Fig. 1, "rat") and is precipitated by antiserum to hGH (data not shown). This assumption is further justified by comparison with the translation products of polyadenylated RNA isolated from bovine pituitary (Fig. 1, "cow") and from a human prolactin-producing tumor (Fig. 1, lane 6). Both of these RNA's directed the synthesis of a protein similar in size to human and rat pregrowth hormone, but also directed the synthesis of a larger quantity of a protein of higher molecular weight, presumably preprolactin. The tumors show variation in the extent to which hGH mRNA is present, as measured by their translational activities. Nevertheless, hGH mRNA appears to be the most abundant mRNA species in the acromegalic tumors (Fig. 1, lanes 1 to 5). These results are consistent with and verify the clinical diagnoses made prior to surgery.

Molecular cloning of hGH cDNA. The polyadenylated RNA from the tumors

that appeared to have the greatest abundance of hGH mRNA by the translational assay were pooled for synthesis of double-stranded cDNA (Fig. 2). Portions of the double-stranded cDNA were analyzed by restriction endonuclease digestion before and after treatment with S1 nuclease. A high proportion of the cDNA was about 1000 nucleotides long (Fig. 2, lanes a and b), the length expected for hGH mRNA, assuming analogy with rat pregrowth hormone mRNA (3). Endonuclease Hae III digestion of the DNA generated a 550–base pair (bp) fragment (Fig. 2, lanes h and j) previously reported to occur in hGH and hCS cDNA's (4, 5). The prominence of this band supports the idea that the cDNA is highly enriched in hGH gene sequences. This is further suggested by the finding of a fragment of about 400 bp generated by digestion with Hinf I and Sma I (Fig. 2, lane g). The fragment of about 500 bp generated by Pvu II (Fig. 2, lanes e and i) extends beyond the previously cloned 550 bp fragment, which contains only one Pvu II site. However,

its presence is predictable from the sequence of rat growth hormone cDNA (3), and by the conservation between species of the amino acid sequence in this region (9). The fragments of about 350 bp and 150 bp generated by combined digestion with Pvu II and Bgl II (Fig. 2, lane f) would also be anticipated from the previously determined structure of the 550 bp hGH fragment and knowledge of the existence of the additional Pvu II site. Therefore, this cDNA preparation appears to be highly enriched in full-length copies of hGH mRNA.

The uncleaved cDNA was cloned in the plasmid pBR322 and *E. coli* χ1776 in a P3 physical containment facility (10) by methods similar to those previously described (3). Briefly, the cDNA was first treated with S1 nuclease and subsequently with DNA polymerase I in the presence of the four deoxynucleoside triphosphates to generate blunt-ended cDNA molecules. Synthetic DNA containing the site for the restriction endonuclease Hind III was then added to

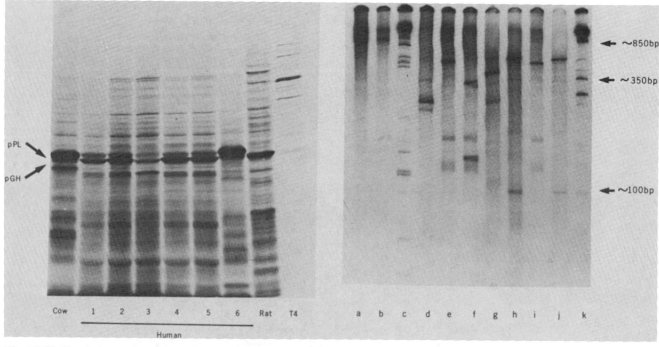

Fig. 1 (left). Translation products of mRNA isolated from growth hormone and prolactin-producing tumors and from bovine pituitary. These tissues were stored in liquid nitrogen shortly after their removal until preparation of the RNA. A portion of polyadenylated RNA isolated from each human tumor, the bovine pituitary, and the cultured rat pituitary tumor (GC) cells was used as a messenger in the wheat germ cell-free protein synthesis system (20). The ^{35}S-labeled proteins were analyzed by electrophoresis and autoradiography on sodium dodecyl sulfate–polyacrylamide gels (12.5 percent) (20). The translation products from five acromegalic and one prolactin-secreting tumor are shown in lanes 1 to 5 and lane 6, respectively. The lanes labeled "cow" and "rat" show the translation products from bovine pituitary and rat GC cell RNA. Lane 7 shows the bacteriophage T4 proteins (21) used as molecular weight markers. The arrows indicate the bands corresponding to pregrowth hormone (pGH) and preprolactin (pPL).   Fig. 2 (right). Analysis of cDNA synthesized from mRNA extracted from the growth hormone–producing pituitary tumors. The polyadenylated RNA from tumors two, four, and five were pooled (135 μg) and used as a template for synthesis of ^{32}P-labeled double-stranded cDNA as described (3). Samples of this cDNA were cleaved with various restriction endonucleases before and after S1 nuclease digestion. The figure shows an autoradiogram of the resulting DNA fragments after electrophoresis on a 4.5 percent polyacrylamide gel (3). (Lane a) Uncleaved cDNA before S1 digestion; (lane b) uncleaved cDNA after S1 digestion; (lane c) bacteriophage fd DNA, Hpa II (molecular weight markers); (lane d) cDNA, Pst I + Bgl II; (lane e) cDNA, Pvu II; (lane f) cDNA, Bgl II + Pvu II; (lane g) cDNA, Hinf I + Sma I; (lane h) cDNA, Hae III; (lane i) cDNA, Pvu II; (lane j) cDNA, Hae III; (lane k) bacteriophage fd DNA, Hae III (molecular weight markers).

each end of the cDNA, and cohesive ends were generated by digestion with endonuclease Hind III. The resulting cDNA was purified on gel and ligated to Hind III-cut and bacterial alkaline phosphatase-treated pBR322 plasmid DNA (8). Bacteria were then transformed with this recombinant plasmid. Colonies with recombinant DNA containing plasmids, selected by antibiotic resistance (tet^{S}amp^R) were grown, and the plasmid DNA's were isolated. The DNA was digested with Hind III, treated with various other restriction endonucleases, and analyzed by gel electrophoresis. One clone contained an insert of about 800 bp whose digestion by Hae III, Pvu II, and Bgl II generated fragments similar in size

```
                         -26                    -20                             -10
                         met ala thr gly ser arg thr ser leu leu leu ala phe gly leu leu cys leu pro trp
GG AUC CUG UGG ACA GCU CAC CUA GCU CCA AUG GCU ACA CGC UCC CCG ACG UCC CUG CUC CUG GCU UUU GGC CUG CUC UGC CUG CCC UGC

      1                      10                     20
leu gln glu gly ser ala phe pro thr ile pro leu ser arg leu phe asp asn ala met leu arg ala his arg leu his gln leu ala
CUU CAA GAG GGC AGU GCC UUC CCA ACC AUU CCC UUA UCC AGG CUU UUU CAC AAC GCU AUG CUC CGC GCC CAU CGU CUG CAC CAG CUG GCC

         30                     40                     50
phe asp thr tyr gln glu phe glu glu ala tyr ile pro lys glu gln lys tyr ser phe leu gln asn pro gln thr ser leu cys phe
UUU GAC ACC UAC CAG GAG UUU GAA GAA GCC UAU AUC CCA AAG GAA CAG AAG UAU UCA UUC CUG CAG AAC CCC CAG ACC UCC CUC UGU UUC

      60                     70                     80
ser glu ser ile pro thr pro ser asn arg glu glu thr gln gln lys ser asn leu glu leu leu arg ile ser leu leu leu ile gln
UCA GAG UCU AUU CCG ACA CCC UCC AAC AGG GAG CAA ACA CAA CAG AAA UCC AAC CUA GAG CUG CUC CGC AUC UCC CUG CUG CUC AUC CAG

      90                     100                    110
ser trp leu glu pro val gln phe leu arg ser val phe ala asn ser leu val tyr gly ala ser asp ser asn val tyr asp leu leu
UCG UCG CUG GAG CCC GUG CAG UUC CUC AGG AGU GUC UUC CCC AAC AGC CUG GUG UAC GGC CCC UCU GAC AGC AAC GUC UAU GAC CUC CUA

      120                    130                    140
lys asp leu glu glu gly ile gln thr leu met gly arg leu glu asp gly ser pro arg thr gly gln ile phe lys gln thr tyr ser
AAC GAC CUA GAG GAA GGC AUC CAA ACG CUG AUG GGG AGG CUG GAA GAU GGC AGC CCC CGG ACU GGG CAG AUC UUC AAG CAG ACC UAC AGC

      150                    160                    170
lys phe asp thr asn ser his asn asp asp ala leu leu lys asn tyr gly leu leu tyr cys phe arg lys asp met asp lys val glu
AAG UUC GAC ACA AAC UCA CAC AAC GAU GAC GCA CUA CUC AAG AAC UAC GGG CUG CUC UAC UGC UUC AGG AAG GAC AUG GAC AAG GUC GAG

      180                    190 191
thr phe leu arg ile val gln cys arg ser val glu gly ser cys gly phe AM
ACA UUC CUG CGC AUC GUG CAG UGC CGC UCU GUG GAG GCC AGC UGU CGC UUC UAG CUG CCC GGG UGG CAU CCU GUG ACC CCU CCC CAG UGC

CUC UCC UGG CCC UCG AAC UUG CCA CUC CAG UGC CCA CCA GCC UUG UCC UAA UAA AAU UAA GUU GCA UCA AAA AAA AAA
```

Fig. 3. Nucleotide sequence of hGH mRNA and the amino acid sequence of human pregrowth hormone. The sequence was determined according to the procedure of Maxam and Gilbert (11) from 5'- or 3'-end-labeled restriction fragments of chGH800/pBR322. The sequence between the two internal Hae III sites was taken from our previous work (5). Most of it, and all of the sequence outside these two internal Hae III sites was resequenced by the chain-termination technique (12) as described in detail elsewhere (22) using as single-stranded templates the chGH800 Hind III fragment recloned in the vector M13 mp5 and as primers restriction fragments of chGH800/pBR322. The RNA sequence has been taken from the DNA sequence. The amino acid sequence has been deduced from the RNA sequence using the genetic code. The termination codon, UAG, is designated by the symbol AM for "amber."

4/UUU/phe	3/UCU/ser	3/UAU/tyr	2/UGU/cys
10/UUC/phe	7/UCC/ser	5/UAC/tyr	3/UGC/cys
1/UUA/leu	3/UCA/ser	0/UAA/OC	0/UGA/OP
0/UUG/leu	1/UCG/ser	1/UAG/AM	2/UGG/trp
2/CUU/leu	0/CCU/pro	1/CAU/his	1/CGU/arg
10/CUC/leu	6/CCC/pro	2/CAC/his	4/CGC/arg
4/CUA/leu	2/CCA/pro	3/CAA/gln	0/CGA/arg
16/CUG/leu	1/CCG/pro	11/CAG/gln	2/CCG/arg
2/AUU/ile	1/ACU/thr	0/AAU/asn	2/AGU/ser
6/AUC/ile	4/ACC/thr	9/AAC/asn	5/AGC/ser
0/AUA/ile	5/ACA/thr	1/AAA/lys	0/AGA/arg
4/AUG/met	2/ACG/thr	8/AAG/lys	5/AGG/arg
0/GUU/val	3/GCU/ala	2/GAU/asp	0/GGU/gly
3/GUC/val	6/GCC/ala	9/GAC/asp	8/GGC/gly
0/GUA/val	1/GCA/ala	6/GAA/glu	0/GGA/gly
4/GUG/val	0/GCG/ala	9/GAG/glu	3/GGG/gly

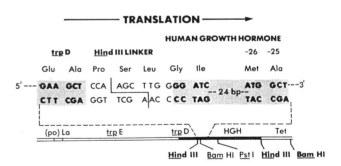

Fig. 4. Codon usage in hGH mRNA. The numbers indicate how many times the codons are used in the region of hGH mRNA coding for the prehormone; OC, OP, and AM designate the stop codons ochre, opal, and amber, respectively. Fig. 5. Postulated nucleotide sequence around the Hind III site in the hybrid gene of expression plasmid ptrpED50-chGH800. Plasmid ptrpED50 was constructed from ptrpED5-1 (7) by linearizing the plasmid with Hind III, filling in the protruding 5' ends with the use of DNA polymerase I (Klenow fragment from Boehringer) and ligating synthetic decamers containing a Hind III site (collaborative research) to the blunt-ended material (3, 8). After digestion with Hind III the plasmid was separated from residual linker molecules by chromatography on Sephadex G-200, recircularized with T4 DNA ligase, and used to transform E. coli W3110 trpoE∇1(23) by a standard procedure (24). Plasmid DNA isolated from one of these colonies was digested with Hind III and treated with alkaline phosphatase (4, 8). A portion (5 μg) of this DNA was end-labeled with [γ-^{32}P]-ATP with the use of T4 DNA kinase (Boehringer) and cut by Hae III. The DNA sequence of the labeled fragments was determined by chemical cleavage (11), after they were isolated by polyacrylamide electrophoresis. The cloned chGH800 DNA was cleaved from pBR322 with Hind III and isolated by polyacrylamide gel electrophoresis. This DNA was ligated to similarly cleaved and alkaline phosphatase-treated ptrpED50. The ligation mixture was used to transform E. coli strains W3110 trpoE∇1 and RR1 (25) in a P3 facility, and transformants resistant to ampicillin were selected. Resistant colonies were examined for the presence of inserted chGH800 sequences by gel analysis of such plasmids after digestion with restriction endonucleases Bam HI and Pst I. Plasmids with the growth hormone initiator codon proximal to the trpD gene sequence showed bands of 250 and 900 bp, whereas plasmids with the inserted cDNA in the opposite orientation showed bands of 250 and 350 bp.

to those from the 550 bp hGH cDNA clone (data not shown) and to the digested uncloned cDNA (Fig. 2). This suggested that this clone did contain cDNA complementary to full-length or nearly full-length hGH mRNA. (This clone is designated chGH800/pBR322.)

Sequence analysis of cloned DNA. The nucleotide sequence of the cloned DNA was determined by the chemical cleavage method of Maxam and Gilbert (*11*) and the chain-termination technique of Sanger, Nicklen, and Coulson (*12*). The hGH mRNA sequence and the corresponding amino acid sequence of human pregrowth hormone can be derived from the DNA sequence (Fig. 3). The amino acid sequence determined from the DNA sequence is consistent with the known amino acid sequence of hGH (*13*) with the following exceptions: the DNA sequence predicts glutamine, asparagine, glutamine, glutamic acid, glutamine, aspartic acid, asparagine, and glutamine at amino acid positions 29, 47, 49, 74, 91, 107, 109, and 122, respectively, while the protein sequence indicates glutamic acid, aspartic acid, glutamic acid, glutamine, glutamic acid, asparagine, aspartic acid, and glutamic acid. It is likely that the DNA sequence is correct in this regard since it is sometimes difficult in protein sequence analysis to differentiate aspartic acid from asparagine and glutamic acid from glutamine. The amino acid sequence of the signal peptide portion of human pregrowth hormone had not been previously determined and is deduced from the mRNA sequence. If translation begins with the methionine codon "in phase," 26 codons proximal to the first amino acid of growth hormone (Fig. 3), then the primary translation product of hGH mRNA would be a protein of 24,851 daltons, a value in agreement with the cell-free translation data shown in Fig. 1.

A comparison of the amino acid and nucleic acid sequence homologies between rat growth hormone, hGH, and hCS and their respective mRNA's is shown in Table 1. In the coding regions, there is higher homology between the nucleic acid sequences than between the amino acid sequences. This difference is consistent with the already mentioned view that the genes of these related hormones evolved from a common evolutionary precursor gene, and is further supported by the marked homology in the 5'-noncoding portions of the mRNA's for rat and human growth hormone. (Data for the 5'-noncoding region of hCS are not yet available.) Human growth hormone has more homology with hCS than with rat growth hormone,

Table 1. Amino acid and nucleic acid sequence homology of growth hormone, chorionic somatomammotropin, and their mRNA's. Data for rat growth hormone (rGH) and human chorionic somatomammotropin (hCS) and their mRNA's are from (*3, 4*). For hCS, only data for amino acids residues 24 to 191 (and the corresponding portion of the mRNA) and a portion of the noncoding 3'-region corresponding to the cloned 550 bp fragment are used for comparisons, since data for the other portions are not available. The amino acid sequence of the prepeptide portion of hCS was determined by Sherwood *et al.* (*27*).

Source	Homology (percent)	
	hGH versus rGH	hGH versus hCS
Nucleic acid:		
5'-Noncoding	73	
Presequence	76	
Coding	76	92
3'-Noncoding	38*	94
Amino acid:		
Presequence	58	84
Coding	67	86

*The homology in this region can be increased to 55 percent by adding appropriate gaps in the sequences. This procedure also reveals that there is a homology of 27 out of 30 bases in the region of the AAUAAA (*28*). Similar conservation between species but with different sequences are found when human and rabbit β-globin (*29*) and human and rat insulin (*30*) are compared.

especially for the 3'-noncoding portions. This finding supports the hypothesis developed earlier that the chorionic somatomammotropin and growth hormone genes probably evolved by a gene duplication mechanism (*1*) at some time after the separation of the human and rat species. In addition, the fact that both hormones exist in both species implies that the same hormones may have evolved independently more than once.

Figure 4 shows the codons used for hGH mRNA. As is the case with rat growth hormone (*3*) and hCS (*6*) mRNA's, there is a nonrandom selection of codons. This appears to be mostly due to the preference for G (guanine) or C (cytosine) over A (adenine) or U (uracil) for the third position of the triplet codon. This is also the case with most (*3, 14*) but not all (*14*) eukaryotic mRNA's whose structures are known.

Construction of a plasmid for growth hormone expression. To see whether hGH gene sequences can be expressed in bacteria, we used the plasmid ptrpED5-1 (*7*), which contains the regulatory region [(po)La], the first gene (trpE), and 15 percent of the second gene (trpD) of the *E. coli* trp operon. Cells containing ptrpED5-1 normally synthesize small amounts of trp gene products. However, if trp operon transcription is derepressed

by addition of 3β-indolylacrylic acid, synthesis of trp gene products increases, so that within 3 hours trp proteins account for about 30 percent of the total cellular proteins (*7*). We hoped that by placing the hGH gene sequence under control of the trp operon, not only would it be expressed, but a higher level of hGH production could be obtained than was previously achieved with rat growth hormone gene sequences under control of the β-lactamase gene (*15*).

The hGH sequences from chGH800 were inserted at the Hind III site of the trpD gene sequence as described in the legend to Fig. 5. In order to insert the hGH codons in phase with those of the trpD sequence, the Hind III site in the trpD gene was manipulated in such a way as to shift the reading frame of any DNA inserted through the Hind III site of the plasmid by one base. To do this, ptrpED5-1 was cleaved with Hind III, the protruding 5'-ends "filled in" with the use of DNA polymerase I (Klenow fragment) and synthetic DNA decamers containing the Hind III site ligated to the blunt-ended material. This DNA was then digested with Hind III to produce new cohesive Hind III ends, and plasmid molecules were recircularized with the use of DNA ligase after the residual Hind III linker molecules were removed on a Sephadex (G-200) column. This material was used to transform *E. coli* and, after selection for ampicillin-resistant colonies, the plasmid DNA was isolated from one of the transformants. The DNA sequence at the Hind III site of the newly constructed plasmid (designated ptrpED50) was determined and showed that the enzymatic reactions had altered the reading frame as predicted. In this way, when chGH800/pBR322 was cleaved with Hind III and the hGH sequences ligated to the Hind III site of the newly constructed plasmid, the codons of hGH would be in phase with those of the trpD gene, provided that the hGH gene sequences were inserted in the proper orientation. This was achieved by obtaining several clones, isolating plasmid DNA, and determining the orientation of the cloned segment by restriction endonuclease analysis (Fig. 5, legend).

As is indicated in Fig. 5, the construction of the hGH "expression" plasmid was such that the anticipated product would be a fusion protein containing the NH₂-terminal region of the trpD protein, amino acids coded by the 5'-untranslated portion of hGH mRNA, the 26 amino acids of the signal peptide, and all of the amino acids of hGH.

Synthesis of growth hormone in bacteria. To determine whether the newly constructed gene can direct the synthesis of large amounts of a new fused polypeptide, and whether its expression is regulated by the *trp* promoter, cells containing the expression plasmid were derepressed for *trp* transcription, and proteins were labeled for 5 minutes with ^{14}C-labeled amino acids. Figure 6 shows an autoradiogram of sodium dodecyl sulfate–polyacrylamide gel of such proteins labeled at various times from 0 to 4 hours after induction of the *trp* operon. Two proteins (53,000 and 32,000 daltons) seem to be specifically derepressed by the inducer. The higher molecular weight protein is the *trp*E gene product (7). The 32,000-dalton protein has approximately the anticipated size for the *trp*D–hGH fusion protein (34,000 daltons). It is immunoprecipitated by antiserum to hGH (Fig. 6, lane a) but not by the control antiserum (Fig. 6, lane b), and precipitation can be blocked by a large excess of hGH (Fig. 6, lanes c to h). Some of the *trp*E protein is immunoprecipitated by antiserum to hGH, but the amount is less when an excess of competitor hGH is added or control antiserum is used. This result might be expected for two reasons. Precipitation of *trp*E protein by control antiserum may be due to the high abundance of this protein. More interestingly, specific precipitation of *trp*E by hGH antiserum (Fig. 6, lane c) and blockage of precipitation by an excess of competitor hGH (Fig. 6, lane e) may be the result of association of the *trp*E protein and the *trp*D–hGH fused polypeptide. The *trp*E and *trp*D proteins are normally associated in *E. coli* as a tetramer containing two subunits of each protein (16); the resulting enzymatic activity (anthranilate synthetase) requires the *trp*E protein and the NH$_2$-terminal 30 percent of the *trp*D protein (17). Thus, the fused *trp*D–hGH protein may contain those *trp*D residues required for binding *trp*E. All of these lines of evidence suggest that the 32,000-dalton protein is a fused *trp*D–hGH polypeptide.

On the basis of the relative quantity of radioactivity incorporated into the *trp*D–hGH gene product, the fusion product appears to be a major protein made by the bacteria, constituting 3 percent of the total bacterial protein synthesis. Thus, the natural hGH gene sequence can be expressed at a high level in bacteria.

The *trp*E and *trp*D protein molecules ordinarily accumulate at a similar rate (molar ratio 1:1) when the *trp* operon is induced (7). However, the molar ratio of *trp*E to *trp*D–hGH is about 6:1, indicating that synthesis of *trp*D–hGH is only 17 percent of the expected level. The reduced level of synthesis is not due to instability of the fused polypeptide since a "pulse-chase" experiment has shown that no significant degradation of the fused polypeptide occurs during the 60-minute period after incorporation of the label (data not shown). There is some evidence that the chick ovalbumin protein is also synthesized in *E. coli* at lower levels than expected (18).

Hypopituitary dwarfism is a fairly common disease treatable only by replacement with hGH (19). Growth hormone may also be useful in the treatment of other disorders. However, the potential uses of this hormone have not been adequately investigated because its only source is pituitaries from human cadavers. In order to have an adequate supply of this hormone, it is necessary to find alternative means of producing it; the synthesis of hGH in bacteria may provide such a means.

Joseph A. Martial
Robert A. Hallewell
John D. Baxter
Howard M. Goodman
Howard Hughes Medical Institute Laboratories, Department of Medicine, Metabolic Research Unit and *Department of Biochemistry and Biophysics, University of California, San Francisco 94143*

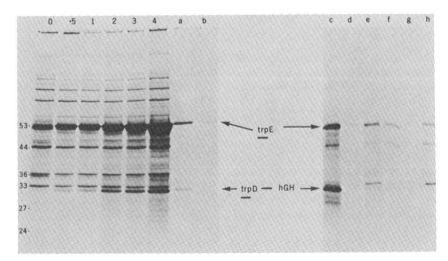

Fig. 6. Autoradiograms of sodium dodecyl sulfate–polyacrylamide gels (10 percent) of ^{14}C- and ^{35}S-labeled proteins from bacteria harboring the p*trp*ED50-chGH800 expression plasmid. Cultures of strains W3110 *trp*oE∇1 and RRI harboring this plasmid were induced with 3β-indolylacrylic acid and 3-ml samples labeled for 5 minutes with 2 μCi of ^{14}C-labeled amino acids (W3110 *trp*oE∇1) or 10 μCi of ^{35}S-labeled methionine (RRI) as described (7). Samples labeled at zero, 0.5, 1, 2, 3, and 4 hours were centrifuged and resuspended (50 μl) by sonication (26) prior to loading (5 μl per gel slot) in sodium dodecyl sulfate sample buffer (20). Samples were immunoprecipitated by means of the SAC technique (20) in order to collect antigen-antibody complexes. Immunoprecipitations contained 10 μl of sonicated cells, 390 μl of 0.5 percent NP40 (Particle Data Laboratory, Elmhurst, Ill.) in phosphate saline buffer (0.025M potassium phosphate, pH 7.4, 0.1M NaCl), 20 μl of rabbit antiserum to hGH (Antibodies Inc., 1000 unit/ml) or 20 μl of nonimmune rabbit antiserum each diluted 50-fold in phosphate saline buffer containing bovine serum albumin (2 mg/ml). Competitor hGH was added at 40 μg per reaction mixture. The *E. coli* proteins in lane 0 were used as molecular weight markers (8). (a and b) Immunoprecipitates of ^{14}C-labeled proteins from the 4-hour time point with (a) antiserum against hGH and with (b) nonimmune serum. (c to e) Immunoprecipitates of ^{35}S-labeled proteins from the 4-hour time point, with (c) antiserum against hGH, (d) nonimmune serum, and (e) antiserum against hGH together with an excess of competitor hGH. (f to h) Immunoprecipitates of ^{35}S-labeled proteins from the zero time point with the use of (f) antiserum against hGH, (g) nonimmune serum, and (h) antiserum against hGH together with an excess of competitor hGH.

References and Notes

1. T. A. Bewley, J. S. Dixon, C. H. Li, *Int. J. Pept. Protein Res.* **4**, 281 (1972); H. D. Niall, M. L. Hogan, R. Sayer, I. Y. Rosenblum, F. C. Greenwood, *Proc. Natl. Acad. Sci. U.S.A.* **68**, 866 (1971).
2. A. G. Frantz, in *Peptide Hormone*, J. A. Parson, Ed. (University Park Press, Baltimore, 1976), p. 199.
3. P. H. Seeburg, J. Shine, J. A. Martial, J. D. Baxter, H. M. Goodman, *Nature (London)* **270**, 5637 (1977).
4. P. H. Seeburg, J. Shine, J. A. Martial, A. Ullrich, J. D. Baxter, H. M. Goodman, *Cell* **12**, 157 (1977).
5. P. H. Seeburg, J. Shine, J. A. Martial, J. D. Baxter, H. M. Goodman, in preparation; H. M. Goodman, P. H. Seeburg, J. Shine, J. A. Martial, J. D. Baxter, *Alfred Benson Symposium 1978* (Munsgaard, Copenhagen, in press).
6. E. J. Gubbins, R. A. Maurer, J. L. Hartley, J. E. Donelson, *Nucleic Acids Res.* **6**, 915 (1979); N. E. Cooke and J. A. Martial, unpublished results.
7. R. A. Hallewell and S. Emtage, in preparation.
8. A. Ullrich, J. Shine, J. Chirgwin, R. Pictet, E. Tischer, W. J. Rutter, H. M. Goodman, *Science* **196**, 1313 (1977).
9. M. O. Dayhoff, *Atlas of Protein Sequence and Structure* (National Biomedical Research Foundation, Washington, D.C., 1972), vol. 5, pp. D-201-D-204.
10. "Recombinant DNA research, revised guidelines," *Fed. Reg.* **43**, 60080 (1978); "Recombi-

nant DNA research: actions under guidelines,'' *ibid.* **44**, 21730 (1979).

11. A. Maxam and W. Gilbert, *Proc. Natl. Acad. Sci. U.S.A.* **74**, 560 (1977).

12. F. Sanger, S. Nicklen, A. R. Coulson, *ibid.*, p. 5463.

13. Reference (*9*), p. D-202, was used as the "standard" hGH sequence, except that an additional Gln residue was placed at position 69 as indicated in (*1*) and (*2*). References (*1*) and (*2*) present slightly different versions of the hGH sequence, which match the sequence predicted from the cDNA in several residues, which do not match the standard sequence.

14. L. McReynolds, B. W. O'Malley, A. D. Nisbet, J. E. Fothergill, D. Givol, S. Fields, M. Robertson, G. G. Brownlee, *Nature (London)* **273**, 723 (1978).

15. P. H. Seeburg, J. Shine, J. A. Martial, R. D. Ivarie, J. A. Morris, A. Ullrich, J. D. Baxter, H. M. Goodman, *ibid.* **276**, 5690 (1978).

16. J. Ito and C. Yanofsky, *J. Bacteriol.* **97**, 734 (1969).

17. C. Yanofsky, V. Horn, M. Bonner, S. Stasiowski, *Genetics* **69**, 409 (1971).

18. T. H. Fraser and B. J. Bruce, *Proc. Natl. Acad. Sci. U.S.A.* **75**, 5936 (1978).

19. J. M. Tanner, *Nature (London)* **237**, 433 (1972).

20. J. A. Martial, J. D. Baxter, H. M. Goodman, P. H. Seeburg, *Proc. Natl. Acad. Sci. U.S.A.* **74**, 1816 (1977).

21. P. A. O'Farrell and L. H. Gold, *J. Biol. Chem.* **248**, 7066 (1973).

22. B. Cordell, G. Bell, E. Tischer, F. M. DeNoto, A. Ullrich, R. Pictet, W. J. Rutter, H. M. Goodman, in preparation.

23. N. E. Murray and W. J. Brammar, *J. Mol. Biol.* **77**, 615 (1973).

24. D. M. Glover, in *New Techniques in Biophysics and Cell Biology*, R. H. Pain and B. J. Smith, Eds. (Wiley, New York, 1976), vol. 8, pp. 125–145.

25. F. Bolivar, R. L. Rodriguez, M. C. Betlach, H. W. Boyer, *Gene* **2**, 75 (1977).

26. P. H. O'Farrell, *J. Biol. Chem.* **250**, 4007 (1975).

27. L. M. Sherwood, Y. Burstein, I. Schechter, *Conference on Precursor Processing in The Biosynthesis of Proteins* (New York Academy of Sciences, 2 to 4 May 1979), Abstr.; S. Birken, D. L. Smith, R. E. Canfield, I. Boime, *Biochem. Biophys. Res. Commun.* **74**, 106 (1977).

28. N. J. Proudfoot and G. G. Brownlee, *Nature (London)* **263**, 211 (1976).

29. A. Efstratiadis, F. C. Kafatos, T. Maniatis, *Cell* **10**, 571 (1977).

30. G. I. Bell, W. F. Swain, R. Pictet, B. Cordell, H. M. Goodman, W. J. Rutter, *Nature (London)*, in press.

31. Supported by a grant from Eli Lilly Co. and a postdoctoral fellowship from the British Science Research Council (to R.A.H.). We thank Dr. C. Yanofsky for communicating information prior to publication and for helpful suggestions on the manuscript; J. Messing for M13mp5 and advice on its use; Drs. W. Swain and P. O'Farrell for their respective gifts of plasmid pBR322 and radioactive T4 protein; Drs. P. Seeburg and J. Shine for helpful discussion and participation in some experiments; and D. Coit and E. Tischer for technical assistance. Avian myeloblastosis virus reverse transcriptase was provided by the Office of Program Resources and Logistics, NCI. J.D.B. and H.M.G. are investigators of the Howard Hughes Medical Institute.

8 June 1979

Synthesis in *E. Coli* of a Polypeptide with Human Leukocyte Interferon Activity

S. Nagata, H. Taira, A. Hall, L. Johnsrud, M. Streuli, J. Ecsödi, W. Boll, K. Cantell and C. Weissmann

Double-stranded cDNA prepared from the 12S fraction of poly(A) RNA from interferon (IF)-producing human leukocytes was cloned in Escherichia coli using the pBR322 vector. One of the resulting clones had a 910-base pair insert which could hybridise to IF mRNA and was responsible for the production of a polypeptide with biological IF activity. Up to 10,000 units IF activity per g of cells was obtained from some clones.

CELLS of almost all vertebrates, when exposed to certain viruses or inducers, produce one or more (glyco)proteins, known as interferons[1,2]. Interferons (IFs) are characterised biologically by their ability to induce in target cells a virus-resistant state which is associated with the *de novo* synthesis of several proteins, in particular a protein kinase[3], an oligoisoadenylate synthetase[4,5] and a phosphodiesterase[6]. In addition, IFs have a regulatory effect on the immune response[7,8] and their enhancement of

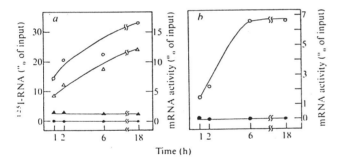

Fig. 1 Hybridisation of IF mRNA and ^{125}I-globin mRNA to filter-bound DNAs. DNA was linked to DPT paper as described elsewhere[36]. *a*, For each time point, one 0.25-cm^2 piece of DPT paper with 500 ng of *Hind*III-excised insert of rabbit β-globin cDNA plasmid and one piece with 700 ng of *Hind*III-digested pBR322 were hybridised as a sandwich with 10 μl of hybridisation medium[22] containing 200 ng ^{125}I-labelled globin mRNA. *b*, As above except that DPT papers contained 250 ng of *Hind*III-excised insert of rabbit β-globin cDNA plasmid and 250 ng of *Pst*I-excised insert of Hif-2h, respectively, and hybridisation was with 5 μg of Le poly(A) RNA in 10 μl. In all cases hybridisation, washing and elution were as described in ref. 22. The RNA was recovered, its ^{125}I radioactivity determined, and injected into 40–50 oocytes. In experiment *b*, oocytes were incubated with 50 μCi ^{3}H-histidine. Oocyte supernatants were assayed for IF activity by the cytopathic effect reduction assay (see Table 3 legend). Le poly(A) RNA (1 μg) injected directly into 20 oocytes gave 2,700 units IF. For the determination of ^{3}H-labelled globin formation, oocytes were homogenised, centrifuged and an aliquot of the supernatant electrophoresed through a 20% polyacrylamide gel[37]. The globin band was cut out and the radioactivity determined in toluene-based scintillator solution. 100 ng globin mRNA injected directly gave 100,000 c.p.m. ^{3}H-globin. *a*, ^{125}I-RNA hybridised to β-globin cDNA (○) or to pBR322 (●). ^{3}H-labelled β-globin formed in oocytes after injection of RNA hybridised to β-globin cDNA or (△) pBR322 (▲). *b*, IF activity formed in oocytes after injection of RNA hybridised to Hif-2h fragment (○) or β-globin cDNA (●).

killer lymphocyte activity[9] may be the basis of their inhibitory effect on tumour growth[5].

Two major classes of acid-stable (type I) IFs have been recognised in man—leukocyte interferon (Le-IF), released by stimulated leukocytes, and fibroblast interferon (F-IF), produced by stimulated fibroblasts. Le-IF and F-IF differ not only immunologically but also in their target cell specificity: whereas both IFs induce a virus-resistant state in human cells, Le-IF is also very active on bovine, porcine and feline cells, whereas F-IF is not[2]. The two IFs are encoded by separate mRNAs[10].

Human Le-IF has been purified more than 80,000-fold, to a specific activity of 4×10^8 units per mg (ref. 11) or 2.5×10^8 units per mg (ref. 12). Two components have been characterised by polyacrylamide gel electrophoresis, with apparent molecular weights (MWs) of 21–22,000 and 15–18,000, respectively[13,14]; they are believed to differ in their degree of glycosylation[14]. Enzymatic[15] or chemical[16] removal of most or all of the carbohydrate moiety seems to have little effect on the biological activity of IF.

We sought to clone human Le-IF cDNA in order to construct bacterial strains producing polypeptide(s) with human IF activity, and to generate the tools required for the analysis of Le-IF gene structure and function. The particular difficulties of this undertaking were the lack of a purified Le-IF mRNA and our ignorance of the structure of Le-IF, which precluded the preparation of pure or highly enriched IF cDNA, or of a probe for the identification of the desired clones.

We describe here the isolation of a hybrid plasmid containing a 872-base pair Le-IF cDNA, which elicits the formation in *Escherichia coli*, of a polypeptide with the immunological and biological properties of human Le-IF.

Isolation of hybrid plasmids containing IF cDNA sequences

Hybrid DNA, consisting of leukocyte cDNA sequences joined to pBR322 at the *Pst*I site by means of dG:dC sequences, was prepared by conventional means, using as starting material a 12S fraction of poly(A) RNA from IF-producing leukocytes, purified about 10-fold for IF mRNA. cDNA cloned in this fashion is usually flanked by *Pst*I sites and, as it is located in the

NAGATA, S., TAIRA, H., HALL, A., JOHNSRUD, L., STREULI, M., ECSÖDI, J., BOLL, W., CANTELL, K. AND WEISSMANN, C.
Synthesis in *E. coli* of a polypeptide with human leukocyte interferon activity. Reprinted by permission from *Nature* v. 284: pp. 316-320. Copyright 1980 Macmillan Magazines Ltd.

Table 1   mRNA hybridisation translation assay for the detection of IF cDNA in hybrid DNA from pools of transformed *E. coli*

DNA sample	Interferon activity
Expt 1: Pools of 512 clones	
I	<60 (<60); <u>110</u> (<20); <110 (<110); <110 (<110); <35 (<35)
δ	<u>20</u> (<20)
N	<u>35</u> (<20); <110 (<110); <u>200</u> (<110)
λ	<60 (<60); <u>60</u> (<20); <110 (<110); <110 (<110)
8 other groups negative	
Expt 2: Pools of 64 clones from sample λ	
λ-I	<35 (<35); <35 (<35)
λ-II	<u>130</u> (<30); <45 (<45)
λ-III	<u>225</u> (<35); <u>35</u>(<30); <u>35</u> (<30); <u>600</u> (<30); <20 (<20)
λ-IV	<u>85</u> (<35); <25 (<25)
λ-V to VIII	negative
Expt 3: Pools of 8 clones from sample λ-III	
λ-III-1	<20 (<20); <20 (60); <u>35</u> (<30)
λ-III-2	<35 (<35); <30 (<30); <u>150</u> (<20); <u>600</u> (<35); <u>110</u> (60)
λ-III-3	<25 (<25); <30 (<30)
λ-III-4	<u>30</u> (<30); <20 (<20); <20 (60)
λ-III-5 to 8	negative
Expt 4: Single clones from sample λ-III-4	
λ-III-4B	<35* (<35); <20 (60)
λ-III-4C	35 (60); <u>60</u>* (<35); <u>111</u>* (<11); <11 (<11); <u>20</u> (<20)

Hybrid DNA containing leukocyte cDNA was prepared as follows. To obtain poly(A) RNA from IF-producing leukocytes, 10^{11} human leukocytes were primed with Le-IF and induced with Sendai virus as described elsewhere[21]. After 5 h at 37°C the cells were collected, suspended in 1 l PBS and added to 17 l 20 mM Tris-HCl (pH 7.5), 1 mM EDTA and 2% SDS. The lysate was digested with Pronase (200 μg ml^{-1}) for 1 h at 20°C. 2 M Tris-HCl buffer (pH 9, 5% vol) was added and the solution extracted with 15 l phenol for 30 min. Chloroform (3 l) was added to aid phase separation, the aqueous phase adjusted to 0.3 M NaOAc buffer (pH 5.5) and the nucleic acid precipitated with ethanol. The precipitate (about 1 g) was dissolved in 900 ml TNE [Tris-HCl (pH 7.5), 100 mM NaCl, 5 mM EDTA] containing 0.5% SDS, extracted three times with phenol and exhaustively with ether, and the poly(A) RNA recovered by three batch adsorptions to 3 × 5 g oligo (dT) cellulose (type 7, P-L Biochemicals) followed by elution with water. The yield was 1.6 mg; 1 μg gave rise to 300 units IF when injected into oocytes. For further purification, 860 μg RNA in 5 mM EDTA were passed through a Chelex-100 column, heated for 90 s at 100°C and centrifuged through a 5–23% sucrose gradient in 50 mM Tris-HCl (pH 7.5), 1 mM EDTA and 0.2 M NaCl. The fractions containing the IF-mRNA activity, sedimenting around 12S, were pooled and the poly(A) RNA recovered by oligo(dT) cellulose chromatography. The yield was 40 μg; 1 μg gave rise to 3,600 units IF when injected into oocytes. pBR322-linked Le-cDNA was prepared essentially as described previously[22]. Sucrose-gradient purified poly(A) RNA from two preparations (48 μg) was used as template for reverse transcriptase to generate 10 μg cDNA 600–1,000 nucleotides long; this was converted into double-stranded DNA by DNA polymerase I and treated with S$_1$ endonuclease (yield, 8 μg preparation A). Of this DNA, 5 μg were centrifuged through a sucrose-density gradient, and material sedimenting faster than a 600-base pair ^{32}P-DNA marker was pooled and precipitated with ethanol (preparation B, 3 μg). cDNA was elongated with dCMP residues, annealed to dGMP-elongated, *Pst*I-cleaved pBR322 (ref. 17) and used to transform *E. coli* χ1776 (ref. 23) (preparation A; 3.3 × 10^4 tetracycline-resistant transformants per μg DNA), or *E. coli* HB101 (preparation B; 4 × 10^4 transformants per μg DNA). Ten thousand colonies of transformed *E. coli* χ1776 were inoculated individually into wells of microtitre plates and stored with 20% glycerol at −20°C. Five thousand colonies of transformed *E. coli* HB101 (from preparation A) were raised on Millipore filters and stored frozen as described by Hanahan and Meselson[22,24]. To carry out the hybridisation translation assay, the number of bacterial clones (from preparation A) indicated in the table were inoculated individually on agar plates, incubated for 24 h and washed off with medium. This suspension was used to inoculate 1-l cultures from which plasmid DNA was purified (method B in ref. 25). The hybrid Le cDNA (20 μg) was cleaved with *Hind*III, mixed with 12 μg Le poly(A) RNA, 5 ng ^{125}I-labelled rabbit β-globin mRNA (5,000 c.p.m.) and 0.1 μg *Pst*I-cleaved rabbit globin cDNA plasmid (Z-pBR322(H3)/RcβG-4.13)[26] in 40 μl 80% formamide, 0.4 M NaCl, 10 mM PIPES buffer (pH 6.4) and 5 mM EDTA, and heated for 4 h at 56 °C. After diluting to 1 ml with 0.9 M NaCl and 0.09 M trisodium citrate (pH7) and adjusting to 4% formamide, the solution was filtered through a Millipore filter (13 mm diameter, 0.4 μm pore size). The filter was washed for 10 min at 37 °C in 0.15 M NaCl, 0.015 M trisodium citrate and 0.5% SDS, and the RNA recovered by heating the filter for 5 min in 0.5 ml 1 mM EDTA, 0.5% SDS and 3 μg ml^{-1} yeast RNA at 75 °C. The RNA was purified by oligo(dT) cellulose chromatography, precipitated with ethanol and assayed for IF-mRNA activity. To determine IF-mRNA activity, the RNA sample (up to 3 μg) was dissolved in 1–3 μl 15 mM Tris-HCl (pH 7.5) and 88 mM NaCl, and injected into 20–60 *Xenopus laevis* oocytes[27] (50 nl per oocyte). Oocytes were incubated for 12–16 h in Barth's medium[28], homogenised in 0.5 ml 50 mM Tris-glycine buffer (pH 8.9) and the supernatant was assayed for IF. In later experiments (Table 2), incubation was for 24–48 h and the incubation medium was assayed for excreted IF[29]. IF was determined by the vesicular stomatitis virus (VSV) plaque reduction assay[21]. All values are expressed in international units. Values marked with an asterisk were obtained by hybridisation to diazobenzyloxymethyl (DBM)-bound DNA, as described in ref. 22. Underlined values are considered positive. The control values (in parentheses) were obtained by hybridisation to pBR322 DNA. All manipulations involving live *E. coli* HB101 or *E. coli* X1776 containing Le cDNA-pBR322 hybrids were carried out in P3 containment conditions as described in the NIH Recombinant DNA Research Guidelines.

β-lactamase gene, can be expressed as a fused protein[17] or, in certain circumstances, as an independent polypeptide[18].

We identified an IF cDNA clone by a mRNA hybridisation translation assay[19]. Hybrid plasmid was prepared from pools of 512 bacterial clones, and 20 μg of each plasmid pool were cleaved with *Pst*I, denatured and annealed with 12 μg crude Le poly(A) RNA. ^{125}I-Labelled globin mRNA and rabbit β-globin cDNA plasmid were added to monitor hybridisation and all subsequent steps. Hybridised RNA was recovered from the filters, purified and injected into oocytes to determine its IF-mRNA activity. Control hybridisations were carried out with pBR322. The overall recovery of β-globin mRNA activity was only about 5% of the input.

Four out of 12 groups of 512 clones (δ, λ, I and N) gave positive results by this assay, albeit erratically; controls were consistently negative in these groups (Table 1). However, in later experiments, controls occasionally gave a positive result, perhaps due to insufficient washing of the filters. A group of clones was scored as being positive if the value was higher than in the parallel control. The bacterial clones of group λ were arranged in 8 subgroups of 64 each, and assayed as above. Three of these subgroups, λ-II, λ-III and λ-IV, gave positive responses; the clones of λ-III were regrouped into eight sets of eight.

The set λ-III-4 was the first to yield a positive result; λ-III-2 subsequently also gave positive results. DNA was prepared from the single λ-III-4 clones and that from λ-III-4C gave positive responses both by liquid and filter-bound hybridisation (Table 1). After recloning in *E. coli* HB101 the hybrid plasmid of clone λ-III-4C, designated Z-pBR322(*Pst*)/HcIF-4c (abbreviated to Hif-4c), was purified and cleaved with *Pst*I; it released a 320-base pair insert, that is, a fragment about one-third of the expected length of complete IF cDNA. The fragment bound IF mRNA efficiently (Table 2).

A set of colonies containing hybrid DNAs related to Hif-4c was identified by *in situ* hybridisation with ^{32}P-labelled Hif-4c *Pst*I fragment. Among the 64 clones of λ-III, three gave a strong hybridisation response, namely 4C, 2H and 7D, and two (1E and 3D) a weak one. λ-III-2H had the largest insert, about 900 base pairs; it was recloned in *E. coli* HB101 and designated Z-pBR322(*Pst*)/HcIF-2h (abbreviated to Hif-2h). In addition, 5,000 clones prepared as described, but using double-stranded Le-IF cDNA selected for length above 600 base pairs (preparation B, cloned in *E. coli* HB101), were screened by *in situ* hybridisation, using the same probe. Of 185 positive clones identified, 95 gave a strong and 90 a weak hybridisation response in the Grunstein–Hogness assay[20]. The former were designated *E. coli* HB101(Z-pBR322(*Pst*)/HcIF-SN1 to -SN95) (abbreviated to SN1 to SN95).

Properties of plasmid Hif-2h

The insert of plasmid Hif-2h, released by *Pst*I cleavage, was attached to diazophenylthioether (DPT) paper and the kinetics of hybridisation to IF mRNA in conditions of DNA excess determined (Fig. 1). In optimal conditions, about 7% of the IF-mRNA activity and 12% of the β-globin-mRNA activity were recovered, as measured in the oocyte system. Thus, the insert of Hif-2h hybridises to IF mRNA with about the same efficiency as does β-globin cDNA to β-globin mRNA.

Restriction and sequence analysis of Hif-2h (M. Schwarzstein, N. Mantei and M.S., unpublished results) showed that the insert has 910 base pairs of which 23 are 5'-terminal and 15 are 3'-terminal GC pairs; there is one site each for *Bsp*I (85), *Bgl*II (335) and *Eco*RI (710) endonucleases, two sites for *Pvu*II (125, 425), and three sites for *Ava*II (190, 385, 655) and none for *Hha*I, *Taq*I, *Hind*III, *Hpa*II, *Pst*I and *Bam*HI. (The values in parentheses indicate the distance in base pairs from the *Pst* terminus corresponding to the 5' end of the mRNA.) The orientation of the cDNA insert, as ascertained by nucleotide sequence analysis, was such that the reading direction of the IF cDNA coincided with that of the β-lactamase gene.

Detection of IF activity in *E. coli* strains transformed with Hif-4c-related hybrid plasmids

The isolated Hif-2h *Pst*I fragment was joined to *Pst*I-cleaved pBR322 and to three plasmids, pKT279, pKT280 and pKT287, derived from pBR322 by deletions in the β-lactamase gene; DNA ligated into the *Pst* site of this set can be translated in the three possible reading frames by readthrough from the β-lactamase sequence (K. Talmadge, personal communication). *E. coli* HB101 strains transformed with these hybrid DNAs were *E. coli* HB101 (Z-pBR322(*Pst*)/HcIF-2h-AH1 to -AH4), *E. coli* HB101 (Z-pKT279(*Pst*)/HcIF-2h-AH1 to -AH8) and so forth, or in abbreviated form, 322-AH1 to -AH4, 279-AH1 to -AH8, and so on. S-30 or S-100 extracts, from 24 of the AH and 49 of the SN strains grown to stationary phase, were tested for IF activity. The original Hif-2h-containing strain and many of the AH strains showed IF activity; three of them, 279-AH8, 280-AH3 and 287-AH6, were selected for further testing. Of

the 49 SN strains, 16 had IF activity; two of the highest producers, SN35 and SN42, and a negative control, SN32, were further examined. Table 3 shows the results obtained with S-100 extracts of log phase bacteria. IF activities ranged from 100 to 1,000 units per ml of S-100 extract derived from a 20-ml resuspension of the 2.0 g (approximately) of bacterial cells contained in 1 l of culture.

Characterisation of the IF activity produced in transformed *E. coli*

We tested the sensitivity of the IF activity to a protease by incubating S-100 extracts of 287-AH6 and SN35 for 30 min at 37 °C with increasing amounts of trypsin. As a control, authentic human Le-IF was mixed with the (inactive) S-100 extract of SN32 (to give a similar protein concentration, 6 mg ml^{-1}) and digested in parallel. In all cases, the activity was partially abolished at 200 μg ml^{-1} and completely abolished at 1 mg ml^{-1} trypsin.

Table 2 Characterisation of the insert of hybrid plasmid Hif-4c by the mRNA hybridisation translation assay

DNA fragment	Amount of leukocyte poly (A) RNA (μg)	Time of hybridisation (h)	IF activity (units ml^{-1})
Hif-4c	2.5	16	250; 100
β-globin cDNA	2.5	16	4; 1
Hif-4c	7.5	16	3,000; 1,000
β-globin cDNA	7.5	16	4; 30
Hif-4c	7.5	5	1,000; 1,000
β-globin cDNA	7.5	5	10; 1

The insert of plasmid Hif-4c was excised with *Pst*I, purified by electrophoresis through a 2% agarose gel and recovered by successive adsorption to and elution from hydroxyapatite and DEAE cellulose. 120-ng fragments were linked to each 0.25 cm^2 DPT paper (B. Seed, personal communication). Pre-hybridisation, hybridisation and elution of RNA were as described elsewhere[22]. The RNA was injected into oocytes and IF activity was determined after 48 h by the cytopathic effect reduction assay[30] (see Table 3 legend).

Table 3 IF activity in extracts of transformed *E. coli*

S-100 extracts of *E. coli* HB101 transformed by:	IF activity (units per ml extract)
a Z-pBR322(*Pst*)/HcIF-2h	100; 100
b Z-pKT279(*Pst*)/HcIF-2h-AH8	100; 300
c Z-pKT280(*Pst*)/HcIF-2h-AH3	1,000; 1,000
d Z-pKT287(*Pst*)/HcIF-2h-AH6	200; 200
e Z-pBR322(*Pst*)/HcIF-SN35	1,000; 1,000
f Z-pBR322(*Pst*)/HcIF-SN42	300; 100
g Z-pBR322(*Pst*)/HcIF-SN32	0; 0

The IF-cDNA insert of Hif-4c, excised with *Pst*I and purified as described in Table 2 legend, was joined to *Pst*I-cleaved pKT279, pKT280 and pKT287, respectively. *E. coli* HB101 was transformed with these products and tetracycline-resistant colonies were screened by *in situ* hybridisation[20] as described by Hanahan and Meselson[24], using the Hif-4c *Pst*I fragment nick-translated with [α-^{32}P]dATP (1,100 Ci mmol^{-1}, NEN) and [α-^{32}P]dCTP (470 Ci mmol^{-1}, NEN) as labelled substrates[31]. Three clones (*b-d*) were selected in preliminary assays for IF activity. Clones *c-g* were from a set of IF-cDNA-containing clones identified among 5,000 *E. coli* HB101 transformed with Le cDNA (preparation B, see Table 1 legend) by *in situ* hybridisation as above. *e* And *f* were shown to produce IF in a preliminary screening, and *g* was chosen as negative control. One-litre cultures of transformed *E. coli* were grown to an A_{650} of about 0.8. The cells (about 2.0 g) were collected, washed with 50 mM Tris-HCl (pH 8), 30 mM NaCl and resuspended in 20 ml of the same buffer. Lysozyme was added to 1 mg ml^{-1}; after 30 min at 0 °C, the suspension was frozen and thawed five times and centrifuged at 10,000 r.p.m. for 20 min. The supernatant was centrifuged at 40,000 r.p.m. for 1 h in a Spinco 60 rotor. The S-100 supernatants (about 6 mg ml^{-1} protein in all cases) were assayed in duplicate by the cytopathic effect reduction assay and their IF content estimated relative to a standard IF preparation. IF activity was determined by the cytopathic effect reduction assay as follows. The IF samples, serially diluted 1:3 were mixed with 10^5 CCL23 cells in the wells of a microtitre plate (Cooke) in MEM-10% newborn calf serum. After 24 h the medium was replaced by an appropriate dilution of Mengo virus in the same medium. 24 h later the medium was replaced with 0.5% crystal violet, 3% formaldehyde, 30% ethanol and 0.17% NaCl for 15 min; the wells were then washed exhaustively with water.

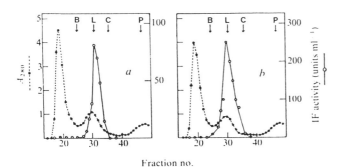

Fraction no.

Fig. 2 Chromatography of Le-IF and *E. coli* IF on Sephadex G-100. (*a*) 0.9 ml S-100 extract of 280-AH3 (500 units, 6 mg protein ml^{-1}) and (*b*) 0.9 ml of a dilution of human Le-IF (1,000 units of preparation P-IF (ref. 34)) in S-100 extract of SN32 (no IF activity, 6 mg ml^{-1}) were mixed with cytochrome *c* (0.2 mg) (C), ^{32}P-phosphate (10^5 c.p.m.) (P) and ^{125}I-labelled β-lactoglobulin (2.5 × 10^5 c.p.m.) (L) and chromatographed on a 0.9 × 49-cm column of Sephadex G-100 in PBS at 4 °C. Fractions of 0.7 ml were collected at 2.3 ml h^{-1}. IF activity was measured by the cytopathic effect reduction assay, and radioactivity, A_{280} and A_{410} (cytochrome *c*) were determined for each fraction. The position of bovine serum albumin (B) was determined in a separate run, relative to C and P.

Table 4 Antibody titres of anti-Le-IF and anti-F-IF measured against different IF preparations

	Le-IF	F-IF	S-100 E. coli extracts	
			SN35	280-AH3
Sheep anti-Le-IF	100,000	3,000	30,000	30,000
Goat anti-F-IF	<10	1,000	<10	<10

About 10 units of each IF preparation were incubated for 1 h at 20 °C with different antiserum dilutions, and the IF activity was determined by the VSV plaque reduction assay[13]. The titres are given as the reciprocal of the highest dilution which raises the plaque counts by a factor of 2. The sheep anti-Le-IF was prepared as described elsewhere[33]. Human Le-IF was purified to the P-IF stage[34].

Human Le-IF is stable at pH 2 (ref. 21). S-100 extracts of 280-AH3 and SN35, as well as 250 units human Le-IF mixed with (inactive) S-100 extract of SN32, were dialysed overnight against 0.1 M NaCl and 50 mM glycine-HCl (pH 2) buffer and then 5 h against phosphate-buffered saline (PBS). A precipitate was removed by centrifugation and the supernatant assayed by both cytopathic effect reduction and plaque reduction. In all cases the initial IF activity was recovered in full.

To compare the MWs of authentic Le-IF and the IF activity in transformed E. coli (E. coli IF), S-100 extracts were chromatographed on Sephadex G-100 columns. IF activity moved with a K_{av} of 0.46 in the case of 280-AH3 (Fig. 2a), 281-AH6 and SN35 (data not shown), which was slightly slower than authentic Le-IF (mixed with S-100 extract of SN32, Fig. 2b); the difference may, however, not be significant.

We compared the serological properties of authentic IFs and E. coli IFs. As shown in Table 4, sheep anti-human Le-IF had a similar titre against Le-IF and E. coli IF of SN35 and 280-AH3, and was $\frac{1}{30}$th as active on fibroblast IF (F-IF); goat anti-human F-IF was active only against F-IF. Thus, E. coli IF is immunologically similar to Le-IF, and quite distinct from F-IF.

Both authentic Le-IF and E. coli IFs show specificity in regard to the cells on which they will act: they are most active on human cells, less active on monkey and mouse cells and inactive on chick cells (Table 5). It is not clear whether the relatively high activity of E. coli IF on monkey cells is significant; further experiments with the purified material are necessary.

As shown by Kerr[4,5] and others, treatment of cells with IF increases 10- to 15-fold their level of oligoisoadenylate synthetase, an enzyme that condenses ATP to $ppp(A2'p)_n5'A$ ($n = 1-4$). Cells were treated with various IF preparations and the cell extracts assayed by measuring the ^3H radioactivity transferred from ^3H-ATP to the dephosphorylation products of $pppA2'p5'A$ and $ppp(A2'p)_25'A$, namely $(2'-5')ApA$ and $(2'-5')ApApA$. As shown in Table 6, $(2'-5')ApA$ radioactivity was six- to ninefold higher in cells treated with Le-IF or E. coli IF than in controls treated with an inactive E. coli extract, and $(2'-5')ApApA$ radioactivity was more than 14–33 times higher than in controls; there was no significant difference between the activity of Le-IF and S-100 extracts of 280-AH3 and SN35.

Table 5 IF activities measured on different cell types

Cells	Interferon activity			
Expt 1	Le-IF	F-IF	E. coli S-100 extracts	
			280-AH3	SN35
Human U amnion	6,000	2,000	600	600
Monkey Vero	600	600	350	350
Monkey GMK	350	200	350	110
Primary chick embryo fibroblasts	<20	<20	<20	<20
Expt 2	Le-IF	mouse-IF	E. coli S-100 extracts	
			287-AH6	SN35
Human CCL23	1,000	ND	300	1,000
Mouse L929	40	120	40	120

Human Le-IF was preparation P-IF (ref. 34), mouse IF was the NIH standard. U cells were maintained by K.C. All cells were challenged with VSV, except for CCL23 cells, where Mengo virus was used. Experiment 1 was assayed by plaque reduction, experiment 2 by the cytopathic effect reduction assay. ND, Not done.

Thus, by all criteria tested, E. coli IF is very similar to authentic Le-IF, although, of course, the molecular structure may well differ in various respects.

Discussion

A strain of E. coli containing IF-cDNA was identified by an IF-mRNA hybridisation translation assay in which DNA from successively smaller pools of strains was screened. Because only 4 of 12 groups of 512 clones had originally given a positive response, we were surprised to find 5 IF clones in a selected group of 64. It is probable that, when used on large pools, the assay was at borderline sensitivity and only detected groups and subgroups particularly rich in IF-cDNA clones. The subsequent screening of 5,000 colonies using an IF-cDNA probe revealed 185 positive clones, a frequency of about 1:27. Taking into account the fact that the poly(A) RNA used to generate the clones had been enriched about 10-fold with respect to IF mRNA, the proportion of IF mRNA in poly(A) RNA from induced leukocytes was not less than 1:270.

The identification of the 910-base pair insert in Hif-2h as a cDNA copy of human Le-IF rests on two lines of evidence: (1) its capacity to hybridise selectively to IF mRNA, and (2) its ability to direct the synthesis, in E. coli, of a polypeptide with the

Table 6 Levels of oligoisoadenylate synthetase in human cells treated with Le-IF or E. coli IF

Cells treated with:	Cell protein (μg)	^3H-A in oligoisoadenylate (% of recovered radioactivity)	
		ApA	ApApA
1. S-100 extract of SN35 (200 units IF ml⁻¹)	7.6	1.4	<0.1
	38	5.2	1.4
2. S-100 extract of 280-AH3 (200 units IF ml⁻¹)	7.6	1.5	0.1
	38	7.8	3.3
3. S-100 extract of SN32 (no IF)	7.6	<0.1	<0.1
	38	0.9	<0.1
4. Le-IF (P-IF) (200 units IF ml⁻¹)	7.6	1.3	0.25
	38	8.0	2.1

Confluent CCL23 cell monolayers in 50-mm dishes were treated with a mixture of 1 ml E. coli S-100 extract and 4 ml minimal essential medium (MEM)–10% newborn calf serum or a dilution of Le-IF (P-IF) in 5 ml medium. After 20 h the cells were lysed and supernatants prepared as described elsewhere[35]. Varying amounts of lysate were adsorbed to poly(rI).(rC)–Sepharose and incubated as described elsewhere[35], except that ^3H-ATP (specific activity 40 Ci mmol⁻¹) was used instead of ^{32}P-ATP. After treatment with bacterial alkaline phosphatase, the products were separated by electrophoresis using ApA and ApApA as markers[35]. The paper was cut into strips and the radioactivity determined by scintillation counting. Most radioactivity was recovered in adenosine. 100% radioactivity was $3-5 \times 10^4$ c.p.m.

biological activity of IF. The polypeptide has properties of human Le-IF in that it induces a virus-resistant state in human cells, to a lesser extent in monkey and mouse cells and not in chick cells, and is neutralised by antibody to human Le-IF but not to human F-IF. Moreover, E. coli IF stimulates the activity of isoadenylate synthetase in human cells to the same extent as does authentic Le-IF.

The IF-cDNA plasmids were constructed to allow synthesis of an IF molecule fused to part of β-lactamase. It seems likely, however, that the biologically active material is a non-fused polypeptide, because its formation is directed by hybrids derived from each of the three pKT plasmids and is therefore independent of the reading frame resulting from the construction. Moreover, a fused β-lactamase fragment should contribute 180 amino acids when the IF cDNA is inserted in the PstI site of pBR322, but not more than 26 or 29 amino acids when it is linked to pKT280 or pKT287 (K. Talmadge, personal communication); in fact, there is no detectable difference in the size of the biologically active IF polypeptides made by the three strains. At the structural level, E. coli IF probably differs from authentic Le-IF by the absence of appropriate glycosylation. Also, it is possible that E. coli IF consists of the Le-IF sequence preceded by a signal sequence, as nucleotide sequence analysis of the cloned IF cDNA revealed a region coding for 22 amino acids which follows the first AUG and precedes the stretch coding for mature IF (M. Schwarzstein, N. Mantei and M.S., unpublished results).

We do not know whether *E. coli* IF has the same specific activity as authentic Le-IF. If this were the case, the amount of active IF produced in transformed *E. coli*, about 20,000 units per 1 of culture, would correspond to one to two fully active molecules per cell. This would be consistent with the occurrence of rare translational events at the physiological initiation site of the IF sequence, and appropriate modifications of the hybrid plasmid should allow a considerable increase in the yield of active IF. If, however, lack of appropriate glycosylation diminishes the activity of the molecule, we shall have a problem on our hands.

Received 24 January; accepted 15 February 1980.

We thank Drs P. Curtis and T. Taniguchi for preliminary work in connection with this project, Ms S. Hirvonen for technical assistance, Dr V. Edy for his advice and for two batches of human leukocytes, and Dr J. Beard, for reverse transcriptase. pKT287, pKT280 and pKT279 were obtained from Karen Talmadge, goat anti-F-IF was from Dr E. de Clercq, F.-IF from Dr K. E. Mogensen and mouse L929 cells from Dr P. Lengyel. The work was supported by Biogen SA.

Note added in proof: Taniguchi and his colleagues have recently identified by nucleotide sequence analysis (T. Taniguchi, personal communication) a fibroblast IF cDNA hybrid prepared from induced fibroblast poly(A) RNA and selected by hybridisation procedures (T. Taniguchi *et al., Proc. Jap. Acad.* **55**, Ser. B. 464–469 (1979)).

1. Isaacs, A. & Lindenmann, J. *Proc. R. Soc.* B147, 258–267 (1957).
2. Stewart, W. E. II *The Interferon System* (Springer, Berlin, 1979).
3. Lebleu, B., Sen, G. C., Shaila, S., Cabrer, B. & Lengyel, P. *Proc. natn. Acad. Sci. U.S.A.* 73, 3107–3111 (1976).
4. Hovanessian, A. G., Brown, R. E. & Kerr, I. M. *Nature* 268, 537–540 (1977).
5. Hovanessian, A. G. & Kerr, I. M. *Eur. J. Biochem.* 93, 515–526 (1979).
6. Schmidt, A. *et al. Proc. natn. Acad. Sci. U.S.A.* 76, 4788–4792 (1979).
7. Johnson, H. M. *Texas Rep. Biol. Med.* 35, 357–369 (1978).
8. De Maeyer, E. & De Maeyer-Guignard, J. *Texas Rep. Biol. Med.* 35, 370–374 (1978).
9. Herberman, R. R., Ortaldo, J. R. & Bonnard, G. D. *Nature* 277, 221–223 (1979).
10. Cavalieri, R. L., Havell, E. A., Vilcek, J. & Pestka, S. *Proc. natn. Acad. Sci. U.S.A.* 74, 3287–3291 (1977).
11. Rubinstein, M. *et al. Proc. natn. Acad. Sci. U.S.A.* 76, 640–644 (1979).
12. Zoon, K. C., Smith, M. E., Bridgen, P. J., zur Nedden, D. & Anfinsen, C. B. *Proc. natn. Acad. Sci. U.S.A.* 76, 5601–5605 (1979).
13. Bridgen, P. J. *et al. J. biol. Chem.* 252, 6585–6587 (1977).
14. Stewart, W. E. II, Wiranowska-Stewart, M., Koistinen, V. & Cantell, K. *Virology* 97, 473–476 (1979).
15. Bose, S., Gurari-Rotman, D., Ruegg, V. T., Corley, L. & Anfinsen, C. B. *J. biol. Chem.* 251, 1659–1662 (1976).
16. Stewart, W. E. II, Lin, L. S. Wiranowska-Stewart, M. & Cantell, K. *Proc. natn. Acad. Sci. U.S.A.* 74, 4200–4204 (1977).
17. Villa-Komaroff, L. *et al. Proc. natn. Acad. Sci. U.S.A.* 75, 3727–3731 (1978).
18. Chang, A. C. Y. *et al. Nature* 275, 617–624 (1978).
19. Harpold, M. M., Dobner, P. R., Evans, R. M. & Bancroft, F. C. *Nucleic Acids Res.* 5, 2039–2053 (1978).
20. Grunstein, M. & Hogness, D. S. *Proc. natn. Acad. Sci. U.S.A.* 70, 2330–2334 (1975).
21. Cantell, K., Hirvonen, S., Mogensen, K. E. & Pyhälä, L. *In Vitro Monogr.* 3, 35–38 (1974).
22. Hoeijmakers, J. H. J., Borst, P., van den Burg, J., Weissmann, C. & Cross, G. A. M. *Gene* 8, 391–417 (1980).
23. Curtiss, R. III *et al. Miami Winter Symp.* 13, 99–114 (1977).
24. Hanahan, D. & Meselson, M. *Gene* (in the press).
25. Wilkie, N. M. *et al. Nucleic Acids Res.* 7, 859–877 (1979).
26. Mantei, N., Boll, W. & Weissmann, C. *Nature* 281, 40–46 (1979).
27. Gurdon, J. B., Lingrel, J. B. & Marbaix, G. *J. molec. Biol.* 80, 539–551 (1975).
28. Barth, L. G. & Barth, L. J. *J. Embryol. exp. Morph.* 7, 210–222 (1959).
29. Colman, A. & Morser, J. *Cell* 17, 517–526 (1979).
30. Stewart, W. E. II & Sulkin, S. E. *Proc. Soc. exp. Biol. Med.* 123, 650–653 (1966).
31. Jeffreys, A. J. & Flavell, R. A. *Cell* 12, 1097–1108 (1977).
32. Strander, H. & Cantell, K. *Ann. Med. exp. Fenn.* 44, 265–293 (1966).
33. Mogensen, K. E. Pyhälä, L. & Cantell, K. *Acta path. microbiol. scand.* 83B, 443–450 (1975).
34. Cantell, K. & Hirvonen, S. *J. gen. Virol.* 39, 541–545 (1978).
35. Kimchi, A. *et al. Proc. natn. Acad. Sci. U.S.A.* 76, 3208–3212 (1979).
36. Stark, G. R. & Williams, J. G. *Nucleic Acids Res.* 6, 195–203 (1979).
37. Laemmli, U. K. *Nature* 227, 680–685 (1970).

Structure and Expression of a Cloned cDNA for Human Interleukin-2

T. Taniguchi, H. Matsui, T. Fujita, C. Takaoka, N. Kashima,
R. Yoshimoto and J. Hamuro

A cDNA coding for human interleukin-2 (IL-2) has been cloned from a cDNA library prepared from partially purified IL-2 mRNA. The DNA sequence codes for a polypeptide which consists of 153 amino acids including a putative signal sequence. A biologically active polypeptide, characteristic of human IL-2, was produced when the cDNA was fused to a simian virus 40 promoter sequence and used to transfect cultured monkey COS cells.

INTERLEUKIN-2 (IL-2), formerly referred to as T-cell growth factor, is a lymphokine which is produced by lectin- or antigen-activated T cells[1,2]. The reported biological activities of IL-2 include stimulation of the long-term *in vitro* growth of activated T-cell clones[1,3], enhancement of thymocyte mitogenesis[4,5] and induction of cytotoxic T-cell reactivity and plaque-forming cell responses in cultures of nude mouse spleen cells[6,7]. As this lymphocyte regulatory molecule can be used to maintain culture of functional monoclonal T cells it has a key role in the study of the molecular nature of T-cell differentiation, the mechanism of differentiated T-cell functions as well as T-cell antigen receptors. In addition, like interferons (IFNs), IL-2 has been shown to augment natural killer cell activity[8], suggesting a potential use in the treatment of neoplastic diseases.

The IL-2 molecule has been characterized physicochemically in several laboratories. Human IL-2 has a molecular weight (MW) of 15,000 (ref. 9). Molecular heterogeneity of human IL-2 described previously is at least partly due to the different extent of sialylation of the molecules[10]. mRNA encoding IL-2 has also been extracted from various cell sources and translated in *Xenopus laevis* oocytes and in rabbit reticulocyte cell-free systems[9,11,12] but much less data are available with respect to the structure of IL-2.

Here we describe the first cloning and sequence analysis of a cDNA coding for human IL-2. Expression of the cDNA in cultured monkey COS cells gave rise to a protein product characteristic of authentic human IL-2.

Hybrid plasmids containing IL-2 cDNA

We prepared IL-2 mRNA from a human leukaemic T-cell line, designated Jurkat-111, cloned from Jurkat-FHCRC (ref. 13). The cells (3×10^6 cells per ml) were stimulated by concanavalin A (Con A, 25 μg ml^{-1}) for IL-2 production and, 6 h later, polyadenylated RNA (poly(A) RNA) was prepared[14] and fractionated by centrifugation on a sucrose density gradient[15]. IL-2 mRNA activity was monitored by injecting aliquots of each fraction into *X. laevis* oocytes[16] and assaying the incubation medium for the ability to stimulate the incorporation of tritiated thymidine (^{3}H-TdR) by an IL-2-dependent T-cell line (CTLL-2) according to the methods described by Gillis *et al.*[2].

One peak of IL-2 mRNA activity was constantly observed which sedimented at ~11.5S (Fig. 1). As has been reported by Efrat *et al.*[12], we also occasionally found the mRNA activity which corresponds to 13S. We believe that this 13S mRNA in our RNA preparation is an aggregated form of the 11.5S mRNA (see discussion). RNA from the most active fraction was used as template to synthesize double-stranded cDNA by the standard procedure[17,18]. The cDNA was size-fractionated and

material longer than 600 base pairs (bp) was inserted into pBR322 by the standard G-C tailing method[19]. The resulting hybrid plasmid was used to transform *Escherichia coli* K-12

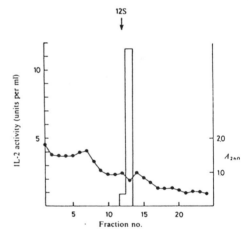

Fig. 1 Sedimentation analysis of human IL-2 mRNA. Human leukaemic T cells (Jurkat-111) were grown in RPMI 1640 medium containing 10% FCS in roller bottles (volume 1:1) to the final cell density of 1.5×10^6 cells per ml. Cells, concentrated to 3×10^6 cells per ml in the same medium containing 1% FCS, 100 μg ml^{-1} each of penicillin and streptomycin, were then induced for IL-2 production with Con A (25 μg ml^{-1}). In these conditions, IL-2 production in culture medium increased to 2,000 units per ml. Total RNA was extracted from the induced cells and poly(A)-containing RNA was purified by oligo(dT)-cellulose chromatography[14]. The mRNA (2.4 mg) was heated in 0.5 mM EDTA at 70 °C for 2 min and fractionated on a 5–25% linear sucrose gradient in 50 mM Tris-HCl, pH 7.5, 0.2 M NaCl and 1 mM EDTA by centrifugation at 26,000 r.p.m. for 26 h at 4 °C in a Beckman SW28 rotor. ^{32}P-labelled *Pst*I-*Hind*III fragments of pBR322 which sediment at 12S and 9S were included as size markers. RNA from individual fractions was ethanol precipitated, dissolved in water and each aliquot was assayed for IL-2 mRNA by injection in *X. laevis* oocytes[16] at a concentration of 0.3 μg ml^{-1}. After 24 h incubation at 24 °C, the media were assayed for IL-2 activity by measuring ^{3}H-TdR incorporation of cloned cytotoxic T cells (CTLL-2) essentially as described previously[2]. Briefly, 4,000 CTLL-2 cells were seeded in 100 μl of RPMI 1640 medium containing 2% FCS in 96-well flat-bottom microplates together with 100 μl of the serially diluted translation products. After 20 h at 37 °C, cells were pulsed for 4 h with 0.5 μCi of ^{3}H-TdR, collected onto glass fibre strips with the aid of automated cell harvester and then the incorporated radioactivity was determined. One unit per ml of IL-2 stimulated 50% of the maximum ^{3}H-TdR incorporation which was observed with standard rat IL-2 preparation.

strain χ1776 (ref. 20) to prepare a cDNA library as described previously[18].

To identify an IL-2 cDNA clone, we screened the cDNA library by a mRNA hybridization translation assay[18,21–23]. Hybrid plasmids were prepared from groups of 24 bacterial clones, and about 50 μg of each plasmid group were cleaved with *Hin*dIII, denatured and bound to nitrocellulose filters. The filters were hybridized with 400 μg of poly(A) RNA from induced Jurkat-111 cells. Hybridized RNA was recovered from the filters, purified and injected into *Xenopus* oocytes to determine the IL-2 mRNA activity. Control hybridization was carried out with pBR322.

One out of 18 groups, each consisting of 24 clones, gave a positive result (group 3) by this assay while others were clearly negative (Table 1). We then prepared plasmid DNAs separately from single clones from group 3 and assayed them for the presence of IL-2 mRNA sequences as above. As shown in Table 1, only plasmid DNA from clone 16, designated p3-16, gave a positive response.

Plasmid p3-16 contained a cDNA insert consisting of about 650 bp, which is apparently shorter than the corresponding 11.5S mRNA. We therefore prepared another cDNA library according to the procedure of Land *et al.*[24] and screened it for cDNA clones containing larger inserts by *in situ* hybridization[25] with ^{32}P-labelled p3-16 cDNA insert. By this screening, we identified a clone containing a plasmid, pIL2-50A, whose cDNA insert consists of about 880 bp.

Jurkat cells do not produce IL-2 without a stimulus such as Con A (ref. 13). Accordingly, we never detected IL-2 mRNA activity in the poly(A) RNA from mock-induced Jurkat-111 cells (unpublished observation). We examined the presence and the size of the mRNA which correspond to pIL2-50A cDNA insert by means of RNA blot analysis[26] in poly(A) RNA isolated from induced or mock-induced Jurkat cells. As shown in Fig. 2, the cloned cDNA hybridized only to the mRNA from induced cells that was of very similar size to IL-2 mRNA.

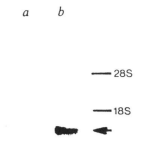

a   b

— 28S

— 18S

Fig. 2 Analysis of the mRNA which corresponds to pIL2-50A cDNA insert. Poly(A) RNA from Con A-induced or mock-induced Jurkat-111 cells was prepared as described in Fig. 1 legend. 10 μg of the RNA was denatured and subjected to RNA blotting analysis as described by Thomas[26] using nick-translated pIL2-50A cDNA as the probe. *a*, Poly(A) RNA from mock-induced Jurkat cells; *b*, poly(A) RNA from Con A-induced Jurkat cells. Ribosomal RNA (28S and 18S) was run separately as size markers. The arrow indicates the position of the positive band which appeared when poly(A) RNA containing 12S human IFN-β₁ mRNA was analysed in parallel using nick-translated IFN-β₁ cDNA[18,35] as the probe.

Nucleotide sequence of the cDNA insert

The complete nucleotide sequence of the *Pst*I insert from pIL2-50A was determined by the procedure of Maxam and Gilbert[27]. Restriction endonuclease cleavage map of the cDNA insert and the sequencing strategy are illustrated in Fig. 3*a*. The DNA sequence of the insert contains a single large open reading frame (Fig. 3*b*). The first ATG, which usually serves as the initiation codon in eukaryotes[28], is found at nucleotides

Table 1   mRNA hybridization translation assay for the detection of IL-2 cDNA from pools of transformed *E. coli*

	DNA samples	IL-2 activity (units per ml)	DNA samples	IL-2 activity (units per ml)	DNA samples	IL-2 activity (units per ml)
First screening: groups of 24 clones	1	<0.1	7	<0.1	13	<0.1
	2	<0.1	8	<0.1	14	<0.1
	3	4.2	9	<0.1	15	<0.1
	4	<0.1	10	<0.1	16	<0.1
	5	<0.1	11	<0.1	17	<0.1
	6	<0.1	12	<0.1	18	<0.1
Second screening: single clones from group 3	1	<0.1	9	<0.1	17	<0.1
	2	<0.1	10	<0.1	18	<0.1
	3	<0.1	11	<0.1	19	<0.1
	4	<0.1	12	<0.1	20	<0.1
	5	<0.1	13	<0.1	21	<0.1
	6	<0.1	14	<0.1	22	<0.1
	7	<0.1	15	<0.1	23	<0.1
	8	<0.1	16	16.0	24	<0.1

Hybrid plasmid DNA containing human IL-2 cDNA was constructed as follows. Double-stranded cDNA was prepared by standard procedures[17,18] using 18 μg of mRNA from fraction 13 (Fig. 1). The cDNA of over 600 bp (2.4 μg) was obtained after fractionation on a sucrose density gradient. The cDNA was then extended with dCMP residues using terminal deoxynucleotidyl transferase[41] and an aliquot (50 ng) was annealed with 250 ng of dGMP-elongated, *Pst*I-cleaved pBR322 (ref. 41). The resulting hybrid plasmids were used to transform *E. coli* χ1776 (ref. 19). From this library (total of 2,000 clones), 432 clones were picked up arbitrarily and individually inoculated into wells of microtitre plates containing 200 μl of L broth[42] and 10 μg ml⁻¹ tetracycline. The plates were incubated at 37 °C for 16 h. To carry out the hybridization translation assay, 100 μl of each bacterial culture from pools of 24 clones was mixed and the resulting mixture was used to inoculate 250-ml cultures from which plasmid DNA was prepared as described previously[42]. Thus, a total of 18 plasmid DNA corresponding to 432 clones were prepared. Each plasmid (50 μg) was cleaved with *Hin*dIII and bound individually to a nitrocellulose filter (Schleicher & Schüll, 2.5 cm diameter, 0.45-μm pores) essentially as described by Harpold *et al.*[20]. Filters were then hybridized with 400 μg poly(A) RNA from induced Jurkat cells in 600 μl the hybridization solution containing 50% (v/v) formamide, 20 mM PIPES at *p*H 6.4, 0.4 M NaCl, 2 mM EDTA and 0.1% SDS. After 18 h at 37 °C, filters were washed three times with 200 ml of 10 mM PIPES, *p*H 6.4, 0.15 M NaCl, 1 mM EDTA and 0.2% SDS at 65 °C and three times with 50 ml of 1 mM PIPES, *p*H 6.4, and 10 mM NaCl at 25 °C. The filter-hybridized RNA was recovered by heating the filter three times in 0.6 ml of 0.5 mM EDTA and 0.1% SDS at 95 °C for 1 min and the RNA was purified by oligo(dT)-cellulose chromatography. The RNA sample was finally dissolved in 2.5 μl of 10 mM Tris-HCl (*p*H 7.5) and 88 mM NaCl and injected into 10–15 *Xenopus laevis* oocytes (50 nl per oocyte). For the second screening, the assay was carried out essentially as above except that 25 μg of plasmid DNA from individual clones was bound to the nitrocellulose filter of 1 cm diameter (binding efficiency was about 75%). IL-2 activity in oocyte medium was determined as described in Fig. 1 legend. No IL-2 mRNA activity was detectable (0.1 units ml⁻¹) when pBR322 DNA was used for the assay.

48–50 from the 5' end. This ATG is followed by 152 codons before the termination triplet TGA is found at nucleotides 507–509. A stretch of A residues corresponding to the 3' poly(A) terminus of the mRNA is found at the end of the cDNA and this is preceded by the hexanucleotide AATAAA (position 771–776) which is usually found in most eukaryotic mRNAs[29]. To examine whether the 5'-terminal region of the mRNA is copied efficiently in the cDNA of pIL2-50A, a single-stranded *Rsa*I–*Dde*I fragment (position 52–84), ^{32}P-labelled at the 5' *Dde*I site, was prepared and used to prime the reverse transcription of 11.5S IL-2 mRNA into single-stranded ^{32}P-cDNA. The size of this ^{32}P-cDNA product was 85–87 nucleotides (unpublished results), indicating that the cDNA insert of pIL2-50A covers the length of the IL-2 mRNA except perhaps for a few nucleotides.

A primary structure of the human IL-2 polypeptide consisting of 153 amino acids could be deduced as shown in Fig. 3*b* and its MW was calculated to be 17,631.7. As has been reported as a common feature in most of the secretion proteins known to date[30], the N-terminal region of the deduced IL-2 polypeptide sequence is also quite hydrophobic and this region probably serves as a signal peptide which is cleaved off in the secretion process of mature IL-2. Such cleavage could occur between Ser and Ala at positions 20 and 21, respectively, as similar cleavage sites have often been found in other secretion proteins[30]. If so, mature human IL-2 would contain 133 amino acids with the calculated MW of 15,420.5; this is comparable with the reported MW of human IL-2 protein from Jurkat cells (15,000)[9], but cleavage could take place at an alternative position somewhere around this site. Human IL-2 from Jurkat cells is reportedly non-glycosylated *in vivo*[10]. Consistent with this observation is the absence of potential *N*-glycosylation sites[31] (AsN-X-Ser or AsN-X-Thr) in the deduced sequence of human IL-2. However, the presence of modifications other than *N*-glycosylation, such as *O*-glycosylation and sialylation, cannot be ruled out[10,32]. The cDNA sequence also predicts that the IL-2 molecule has a neutral isoelectric point, having an equal number of basic (Arg + Lys) and acidic (Asp + Glu) amino acids and this is consistent with the value obtained with partially purified IL-2 by Gillis *et al.*[9] and by Ruscetti and Gallo[33]

($pI = 6.8$–7.1). When the structure of IL-2 polypeptide was compared with those of other soluble factors such as interferons and growth hormones to which IL-2 could have some functional relationships, little homology was observed. With respect to human IFN-γ, another lymphokine, homology covering three consecutive amino acids occurs at amino acids 61–63 (Lys-Asn-Phe)[34] which are also found in human IL-2 at positions 96–98. Homologous amino acid sequences can also be found between human IL-2 (Leu-Lys-Pro-Leu-Glu-Glu-Val-Leu at positions 83–90) and human IFN-β₁ (Leu-Lys-Thr-Val-Leu-Glu-Glu-Lys-Leu at positions 98–106)[35]. Further detailed analysis may be required to search for other homologies with respect to their secondary structure.

Expression of IL-2 cDNA in monkey cells

Despite the extensive characterization of human IL-2 molecules in several laboratories, nothing is known about the primary structure of this protein. It is therefore essential to prove that the cDNA sequence of pIL2-50A actually codes for IL-2 by obtaining expression of this cloned gene.

We used a mammalian cell system to express the cDNA because, unlike bacteria such as *E. coli*, it should be possible to obtain the protein product being properly processed and secreted in the culture medium[34].

A plasmid which should direct the synthesis of human IL-2 in mammalian cells was constructed as follows. A plasmid pCE-1 was constructed from pKCR (ref. 36) and pBR328 (ref. 37) by a series of modification procedures as illustrated in Fig. 4. This plasmid contains a single *Pst*I site just downstream of the SV40 early promoter and upstream of the part of the rabbit β-globin chromosomal gene containing one intron. The plasmid also contains the replication origin of SV40 as well as the polyadenylation site for the early gene[36]. The cDNA insert of pIL2-50A was isolated by *Pst*I cleavage and inserted into the *Pst*I site of pCE-1. Thus a plasmid pCEIL-2, in which the IL-2 structural gene should be transcribed from the early promoter of SV40 in appropriate host cells, was obtained (Fig. 4).

This plasmid was digested by *Hha*I and then introduced by DNA transfection into the transformed monkey cell line COS-7 which allows replication of DNA containing SV40 origin sequences[38]. It seems important to digest the plasmid with *Hha*I before transfection for the efficient expression of cDNA because sequences which could hamper replication of the transfected DNA in COS cells[39] can be removed from the essential part of the plasmid for cDNA expression by this procedure. In our previous experiments, the expression level went down approximately 5–10-fold without this procedure when human IFN-β₁ and -γ cDNA were expressed similarly (unpublished results).

Two to three days after transfection, the cultured cell medium were assayed for human IL-2 activity. As shown in Table 2, media from pCEIL-2 transfected cultures contained IL-2 activity. No IL-2 activity was detectable in the media from cells transfected with pCE-1 DNA.

IL-2 activity from monkey cells

Properties of the IL-2 activity secreted from pCEIL-2 transfected COS cell were examined. As shown in Table 2, it stimulates the ^{3}H-TdR incorporation by the IL-2-dependent CTLL-2 cell line and also stimulates *in vitro* T-cell growth. This activity was neutralized by a monoclonal antibody raised against purified human IL-2 in a BALB/c mouse (unpublished results). The dose–response curve of this IL-2 activity was comparable to that of native IL-2 as shown in Fig. 5. Sensitivity of this activity to trypsin treatment indicated that a protein is responsible for the activity (results not shown). In addition, Sephadex G-100 column chromatography showed that the MW of the IL-2 produced in COS cells was ~15,000 and was indistinguishable from authentic human IL-2 generated by Con A-stimulated Jurkat cells (result not shown). The protein encoded by the cloned cDNA sequence of pIL2-50A therefore shows many of the known properties of human IL-2.

Table 2 Detection of the IL-2 activity in the culture medium of COS-7 cells transfected by pCEIL-2

Transfected DNA	Medium collected after transfection (h)	IL-2 activity measured by	
		^{3}H-TdR incorporation (units per ml)	T-cell growth
pCEIL-2	48	1.5	++
		1.5	++
	72	14.0	++++
		6.0	++++
pCE-1	48	<0.1	—
		<0.1	—
	72	<0.1	—
		<0.1	—

Fresh monolayers of COS-7 cells in wells of 1.5 cm diameter with 1 ml of Dulbecco's minimal essential medium (DMEM, Gibco) were transfected with 1 µg of *Hha*I-digested pCEIL-2 or pCE-1 DNA by using the calcium phosphate co-precipitation technique essentially as described previously[43]. After 15 h at 37 °C, medium were replaced with 1 ml of fresh DMEM containing 8% fetal calf serum (FCS). Each medium, collected at appropriate intervals (indicated in the table) was kept at 4 °C until use. The IL-2 assay by ^{3}H-TdR incorporation was performed as described in Fig. 1 legend. To measure the ability to stimulate *in vitro* T-cell growth[44], five CTLL-2 cells were mixed with the undiluted culture medium to be assayed for IL-2 (0.1 ml) and 0.1 ml of DMEM containing 2% FCS. Cells were cultured for 72 h in 96-well flat bottom microplate and cell number was counted by microscopy: —, <1 viable cell; ++, 5–20 cells; ++++, 50–100 cells.

Fig. 3 *a*, Restriction endonuclease cleavage maps of the cDNA insert of pIL2-50A. To obtain pIL2-50A, another cDNA library was prepared essentially according to the procedure of Land *et al.*[24]. Briefly, 1.6 μg of single-stranded cDNA was synthesized by using 4 μg of mRNA from fraction 13 (Fig. 1), elongated by dCMP residues, and double-stranded cDNA was synthesized by using oligo(dG)$_{12-18}$ as the primer for DNA polymerase I (Klenow fragment). The cDNA (0.6 μg) longer than the 680-bp DNA size marker was obtained by sucrose gradient centrifugation and inserted into the *Pst*I site of pBR322 by the standard G-C tailing method[41]. After transformation of *E. coli* χ1776 by the hybrid plasmids, ~2,500 colonies were screened by *in situ* hybridization[25] with nick-translated p3-16 cDNA insert as the probe and the colony containing pIL2-50A was identified. The rectangle represents the cDNA insert, and the dotted area the protein coding region. Wavy lines indicate G-C junctions. The strategy of the DNA sequence analysis is presented in the lower part of the figure. Arrows indicate the extent of sequencing of each segment analysed without ambiguity. *b*, Nucleotide sequence and deduced amino acid sequence of the plasmid pIL2-50A cDNA insert. The entire sequence was determined by the chemical method of Maxam and Gilbert[27]. The arrow indicates the conjectured cleavage site by a signal peptidase[30]. Numbers above each line refer to amino acid position and numbers below each line to nucleotide position.

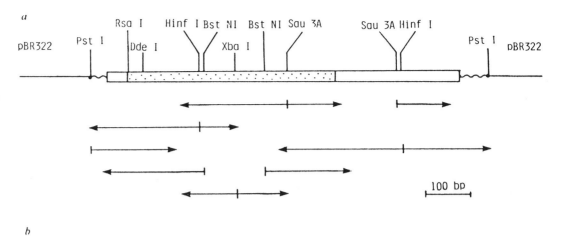

```
a
```

```
                                            1
                              Met Tyr Arg Met Gln Leu Leu Ser Cys Ile Ala
ATCACTCTCTTTAATCACTACTCACAGTAACCTCAACTCCTGCCACA ATG TAC AGG ATG CAA CTC CTG TCT TGC ATT GCA
                                                  50
                              20↓
Leu Ser Leu Ala Leu Val Thr AsN Ser Ala Pro Thr Ser Ser Ser Thr Lys Lys Thr Gln Leu Gln Leu
CTA AGT CTT GCA CTT GTC ACA AAC AGT GCA CCT ACT TCA AGT TCT ACA AAG AAA ACA CAG CTA CAA CTG
                 100
          40
Glu His Leu Leu Leu Asp Leu Gln Met Ile Leu AsN Gly Ile AsN AsN Tyr Lys AsN Pro Lys Leu Thr
GAG CAT TTA CTG CTG GAT TTA CAG ATG ATT TTG AAT GGA ATT AAT AAT TAC AAG AAT CCC AAA CTC ACC
150                                            200
          60                                                                        80
Arg Met Leu Thr Phe Lys Phe Tyr Met Pro Lys Lys Ala Thr Glu Leu Lys His Leu Gln Cys Leu Glu
AGG ATG CTC ACA TTT AAG TTT TAC ATG CCC AAG AAG GCC ACA GAA CTG AAA CAT CTT CAG TGT CTA GAA
                              250
Glu Glu Leu Lys Pro Leu Glu Glu Val Leu AsN Leu Ala Gln Ser Lys AsN Phe His Leu Arg Pro Arg
GAA GAA CTC AAA CCT CTG GAG GAA GTG CTA AAT TTA GCT CAA AGC AAA AAC TTT CAC TTA AGA CCC AGG
                  300                                                        350
                                                       120
Asp Leu Ile Ser AsN Ile AsN Val Ile Val Leu Glu Leu Lys Gly Ser Glu Thr Thr Phe Met Cys Glu
GAC TTA ATC AGC AAT ATC AAC GTA ATA GTT CTG GAA CTA AAG GGA TCT GAA ACA ACA TTC ATG TGT GAA
                                        400
                                   140
Tyr Ala Asp Glu Thr Ala Thr Ile Val Glu Phe Leu AsN Arg Trp Ile Thr Phe Cys Gln Ser Ile Ile
TAT GCT GAT GAG ACA GCA ACC ATT GTA GAA TTT CTG AAC AGA TGG ATT ACC TTT TGT CAA AGC ATC ATC
                              450
          153
Ser Thr Leu Thr
TCA ACA CTA ACT TGA TAATTAAGTGCTTCCCACTTAAAACATATCAGGCCTTCTATTTATTTAAATATTTAAATTTTATATTTATT
     500                                                550

GTTGAATGTATGGTTTGCTACCTATTGTAACTATTATTCTTAATCTTAAAACTATAAATATGGATCTTTTATGATTCTTTTTGTAAGCCCT
          600                                                650

AGGGGCTCTAAAATGGTTTCACTTATTTATCCCAAAATATTTATTATTATGTTGAATGTTAAAATATAGTATCTATGTAGATTGGTTAGTAA
          700                                                750

AACTATTT AATAAA TTTGATAAATATAAAAAAAAAAAAACAAAAAAAAAAA
          800
```

Discussion

A cDNA library was prepared from the fractionated poly(A) mRNA which sedimented at around 11.5S and showed the highest IL-2 mRNA activity. From this library, a hybrid plasmid, p3-16, containing the cDNA copy of human IL-2 was identified by an IL-2 mRNA hybridization translation assay. Subsequent rescreening of the library with the ^{32}P-cDNA probe indicated that the frequency of the IL-2-specific colonies in the cDNA library was only 1:2,000. This means that the frequency of the IL-2-specific mRNA in our poly(A) mRNA fraction is approximately 1:20,000 as about 10-fold enrichment of the IL-2 mRNA is achieved by the sucrose density gradient fractionation[18]. As has been reported by Efrat *et al.*[12], we also occasionally found the IL-2 mRNA peak at 13S. However, when the 13S mRNA was subjected to RNA blot analysis with the same probe as above, only one positive band which co-migrates with 11.5S IL-2 mRNA was detected (data not shown). We conclude that the 13S IL-2 mRNA in our mRNA preparation was an aggregated form of 11.5S mRNA. A similar observation has been made by Gray *et al.* with human IFN-γ mRNA[34].

Fig. 4 Plasmid DNA constructed for the expression of IL-2 cDNA in monkey COS-7 cells. The plasmid pKCR (ref. 36) consists of: (1) Segments of SV40 DNA (shown as hatched blocks) containing the early gene promoter and an origin of replication (0.725–0.648 map units) and a polyadenylation site from the early gene (0.169–0.144 map units). (2) A part of the rabbit β-globin gene (shown as open blocks) (*Bam*HI–*Pvu*II fragment). (3) A segment from pBR (*Eco*RI–*Bam*HI fragment) containing an origin of replication and ampicillin resistance gene. This plasmid was cleaved by *Bam*HI and, after filling both ends by DNA polymerase I (Klenow fragment), a synthetic *Pst*I linker DNA was introduced to construct pKCR(*Pst*I). Plasmid pKCR(*Pst*I) was cleaved by *Sal*I, treated by the Klenow fragment to fill the ends and then partially cleaved by *Eco*RI to obtain *Eco*RI–*Sal*I fragment which contains the whole DNA derived from

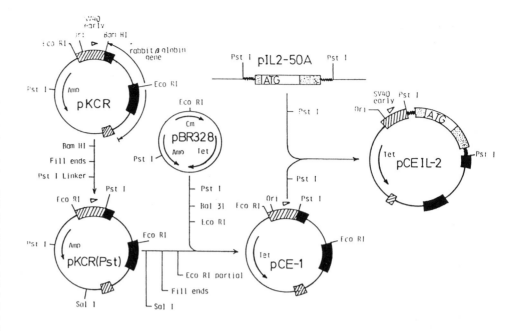

SV40 and the globin gene. This fragment was then ligated to a piece of pBR328 DNA which contains the tetracycline resistance gene and an origin of replication[37] as outlined. The resulting plasmid pCE-1 contains a single *Pst*I site just downstream of the SV40 early promoter. cDNA insert of pIL2-50A was excised by *Pst*I cleavage and ligated to *Pst*I-cleaved pCE-1 to construct pCEIL-2 in which expression of the IL-2 structural gene should be under control of the SV40 early promoter. Plasmid pCE-1 was originally constructed for the cDNA cloning by the G-C tailing method[41] in bacteria and direct expression in mammalian cells.

Fig. 5 Dose–response curve for the stimulation of ^{3}H-TdR incorporation by an authentic and cDNA-directed IL-2 preparation. The IL-2-dependent ^{3}H-TdR incorporation by the CTLL-2 cells has been measured as described in Fig. 1 legend. *a*, Medium alone which was used for COS cell expression. *b*, Cell medium collected 72 h after transfection of pCEIL-2 DNA to COS-7 cells

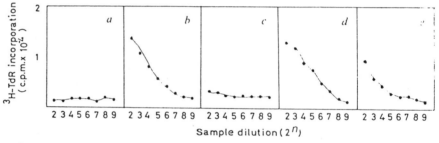

(determined to be 16 units per ml). *c*, Cell media collected 72 h after transfection of pCE-1 DNA to COS-7 cells. *d*, IL-2 produced by Jurkat-111 cells (25 units per ml). *e*, IL-2 produced by Jurkat-111 cells (5 units per ml).

Sequence analysis of the cloned cDNA showed that the cDNA encodes a polypeptide of 153 amino acids, 20 of which could possibly be cleaved off during the secretion of mature IL-2. The identity of the deduced amino acid sequence has mostly been confirmed with a separate cDNA clone (unpublished observation). The deduced amino acid sequence contained no potential *N*-glycosylation site. Robb and Smith[10] reported that IL-2 prepared from human tonsil cells were variably modified, in contrast to the preparation from Jurkat cells. It remains to be seen whether this is due to the difference in the producer cell type or in the gene structure.

To prove ultimately that the cloned cDNA corresponds to that of IL-2, we have expressed it in monkey COS cells. The expression product manifested biological activities characteristic of human IL-2. The activity found in COS cell media (the maximum value we have obtained so far is 70 units per ml) could become higher by removing the G-C stretch which is present between the SV40 promoter and the cDNA as reported

by Simonsen *et al.* in the expression of human IFN-γ cDNA[40]. Expression of the IL-2 cDNA in *E. coli* is in progress and it will soon become possible by the use of cloned cDNA to produce this immunoregulatory molecule in large quantity for various purposes.

The cloning and analysis of the human IL-2 cDNA raise many questions. How many copies of IL-2 genes are present in the human genome? How is the gene organized in the chromosome? Is the gene linked to other lymphokine genes? Our preliminary analysis of the human genomic DNA indicates that the IL-2 gene exists in a single copy. Further studies will provide answers to these questions.

We thank Drs R. Breathnach and Y. Gulzman for providing us with pKCR and COS-7 cells, respectively; Dr H. Sugano for his continuous support and advice; Drs M. Muramatsu, T. Akashi, T. Miyata, K. Sano and S. Nagata for their support and valuable discussions and Dr Y. Fujii-Kuriyama for critical reading of the manuscript.

Received 13 January; accepted 10 February 1983.

1. Morgan, D. A., Ruscetti, F. W. & Gallo, R. *Science* **193**, 1003–2027 (1976).
2. Gillis, S., Ferm, M. M., Ou, W. & Smith, K. J. *Immunology* **120**, 2027–2023 (1978).
3. Gillis, S. & Smith, K. A. *Nature* **268**, 154–156 (1977).
4. Chen, B. M. & DiSabato, G. *Cell. Immun.* **22**, 211–224 (1967).
5. Shaw, J., Monticone, V. & Paetkau, V. J. *Immunology* **120**, 1967–1973 (1978).
6. Gillis, S., Baker, P. E., Union, N. A. & Smith, K. A. *J. exp. Med.* **149**, 1960–1968 (1979).
7. Wagner, H., Hardt, C., Heeg, K., Rollinghoff, M. & Pfizenmaier, K. *Nature* **284**, 278–280 (1980).
8. Henney, C. S., Kuribayashi, K., Kern, D. E. & Gillis, S. *Nature* **291**, 335–338 (1981).
9. Gillis, S. *et al. Immun. Rev.* **63**, 167–209 (1982).
10. Robb, R. J. & Smith, K. A. *Molec. Immun.* **18**, 1087–1094 (1981).
11. Bleackley, R. C. *et al. J. Immun.* **127**, 2432–2435 (1981).
12. Efrat, S., Pilo, S. & Kaempfer, R. *Nature* **297**, 236–239 (1982).
13. Hansen, J. A., Martin, P. J. & Nowinski, R. C. *Immunogenetics* **10**, 247–252 (1980).
14. Aviv, H. & Leder, P. *Proc. natn. Acad. Sci. U.S.A.* **69**, 1408–1412 (1972).
15. Taniguchi, T., Pang, R. H. L., Yip, Y. K., Henriksen, D. & Vilcek, J. *Proc. natn. Acad. Sci. U.S.A.* **78**, 3469–3472 (1981).
16. Gurdon, J. B., Lane, C. D., Woodland, H. R. & Marbaix, G. *Nature* **233**, 177–182 (1971).

17. Wickens, M. P., Buell, G. N. & Shimke, R. T. *J. biol. Chem.* **253**, 2483–2495 (1978).
18. Taniguchi, T. *et al. Proc. Japan Acad.* **553**, 464–469 (1979).
19. Villa-Komaroff *et al. Proc. natn. Acad. Sci. U.S.A.* **75**, 3727–3731 (1978).
20. Curtiss, R. III *et al. Miami Winter Symp.* **13**, 99–114 (1977).
21. Harpold, M. M., Dobner, P. R., Evans, R. M. & Bancroft, F. C. *Nucleic Acids Res.* **5**, 2039–2053 (1978).
22. Nagata, S. *et al. Nature* **284**, 316–320 (1980).
23. Parnes *et al. Proc. natn. Acad. Sci. U.S.A.* **78**, 2253–2257 (1981).
24. Land, H., Grez, N., Hansen, H., Lindermaier, W. & Schuetz, G. *Nucleic Acids Res.* **9**, 2251–2266 (1981).
25. Grunstein, M. & Hogness, D. S. *Proc. natn. Acad. Sci. U.S.A.* **70**, 2330–2334 (1975).
26. Thomas, P. S. *Proc. natn. Acad. Sci. U.S.A.* **77**, 5201–5205 (1980).
27. Maxam, A. W. & Gilbert, W. *Meth. Enzym.* **65**, 499–560 (1980).
28. Kozak, M. *Cell* **15**, 1109–1123 (1978).
29. Proudfoot, N. J. & Brownlee, G. G. *Nature* **263**, 211–214 (1976).
30. Blobel, G. *et al. Symp. Soc. exp. Med.* **33**, 9–36 (1979).
31. Winzler, R. J. in *Hormonal Proteins and Peptides* (ed. Li, C. H.) 1–15 (Academic, New York, 1973).
32. Tomita, M. & Marchesi, V. *Proc. natn. Acad. Sci. U.S.A.* **72**, 2964–2968 (1975).
33. Ruscetti, R. W. & Gallo, R. C. *Blood* **57**, 379–385 (1981).
34. Gray, P. W. *et al. Nature* **295**, 503–508 (1982).
35. Taniguchi, T., Ohno, S., Fujii-Kuriyama, Y. & Muramatsu, M. *Gene* **10**, 11–15 (1980).
36. O'Hare, K., Benoist, C. & Breathnach, R. *Proc. natn. Acad. Sci. U.S.A.* **78**, 1527–1531 (1981).
37. Soberon, X., Covarrubias, L. & Bolivar, F. *Gene* **9**, 287–305 (1980).
38. Gluzman, Y. *Cell* **23**, 175–182 (1981).
39. Lusky, M. & Botchan, M. *Nature* **293**, 79–81 (1981).
40. Simonsen, C. C. *et al. UCLA Symp.* **25**, 1–14 (1982).
41. Chang, A. C. Y. *et al. Nature* **275**, 617–624 (1978).
42. Bolivar, F. *et al. Gene* **2**, 95–113 (1977).
43. Frost, E. & Williams, J. *Virology* **91**, 39–50 (1978).
44. Stull, D. & Gillis, S. *J. Immun.* **126**, 1680–1683 (1981).

Molecular Cloning of a cDNA Encoding Human Antihaemophilic Factor

J.J. Toole, J.L. Knopf, J.M. Wozney, L.A. Sultzman, J.L. Buecker,
D.D. Pittman, R.J. Kaufman, E. Brown, C. Showemaker,
E.C. Orr, G.W. Amphlett, W.B. Foster, M.L. Coe, G.J. Knutson,
D.N. Fass and R.M. Hewick

A complete copy of the mRNA sequences encoding human coagulation factor VIII:C has been cloned and expressed. The DNA sequence predicts a single chain precursor of 2,351 amino acids with a relative molecular mass (M_r) 267,039. The protein has an obvious domain structure, contains sequence repeats and is structurally related to factor V and ceruloplasmin.

HAEMOPHILIA A is a bleeding disorder caused by deficiency or abnormality of a particular clotting protein, factor VIII:C[1] occurring in about 10–20 males in every 100,000. Afflicted individuals suffer episodes of uncontrolled bleeding and are treated currently with concentrates rich in factor VIII:C derived from human plasma. The available therapy, although reasonably effective, is very costly and is associated with a finite risk of infections. We report here significant progress in the use of recombinant DNA technology to provide pure human factor VIII:C as an alternative treatment for haemophiliacs.

Blood clotting begins with injury to a blood vessel. The damaged vessel wall causes adherence and accumulation of platelets activating the plasma proteins which initiate the coagulation process. Sequential activation, via specific proteolytic cleavages and conformational changes, of a series of proteins comprising the coagulation cascade eventually leads to deposi-tion of insoluble fibrin which, together with aggregated platelets, curtails the escape of blood through the damaged vessel wall. Factor VIII:C is a large plasma glycoprotein that functions in the blood coagulation cascade as the cofactor for the factor IXa-dependent activation of factor X. It can be activated proteolytically by a variety of coagulation enzymes including thrombin[2].

In order to provide factor VIII:C for treatment of haemophiliacs we cloned a full-length cDNA. A major obstacle to the cloning effort was the large size of the protein, estimated to be at least M_r 250,000. Purification of factor VIII:C from plasma[3] is made difficult by its low abundance, its extreme sensitivity to degradation by serum proteases and its tight association with polymeric forms of the more abundant protein, von Willebrand factor. Fass *et al.*[4] have described a purification procedure for porcine factor VIII:C using monoclonal antibody

TOOLE, J.J., KNOPF, J.L., WOZNEY, J.M., SULTZMAN, L.A., BUECKER, J.L., PITTMAN, D.D., KAUFMAN, R.J., BROWN, E., SHOWEMAKER, C., ORR, E.C., AMPHLETT, G.W., FOSTER, W.B., COE, M.L., KNUTSON, G.J., FASS, D.N. and HEWICK, R.M.
Molecular cloning of a cDNA encoding human antihaemophilic factor. Reprinted by permission from *Nature* v. 312: pp. 342–347. Copyright 1984 Macmillan Magazines Ltd.

Fig. 1 *a*, *b*; N-terminal amino acid sequence analysis of bovine thrombin cleavage and natural cleavage fragments of porcine factor VIII:C. *c*, The positions of fragments in the factor VIII:C precursor in diagrammatic form. The molecular masses are the apparent porcine values deduced from SDS gel electrophoresis rather than those calculated from the human amino acid sequence. Map lengths are indicated in amino acid residues to correlate with Fig. 4. In *a*, N-terminal sequences were determined for the indicated fragments of porcine factor VIII:C. The first residue in each sequence could not be assigned and is indicated as? In *b*, the porcine amino acid sequences are compared with the corresponding sequences in human factor VIII:C predicted from the nucleotide sequence of the cDNA clone (Fig. 4). Boxed areas, indentity between sequences. Thrombin cleavage fragments of porcine factor VIII:C were purified by SDS polyacrylamide gel electrophoresis and located in the unfixed gel using a small portion of ^{125}I-labelled porcine factor VIII:C as a tracer. After electrophoretic elution from the gel matrix as described previously[28], the polypeptide sample was subjected to N-terminal amino acid sequence analysis using an Applied Biosystems gas-phase sequenator[11].

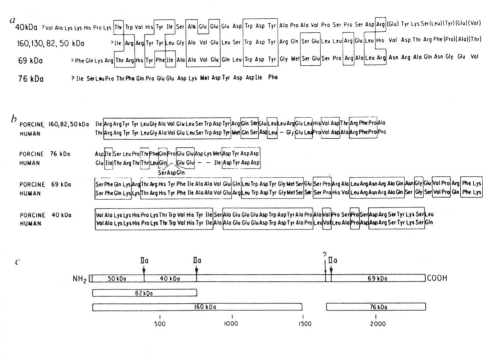

immunoaffinity chromatography, obtaining highly purified polypeptides of M_r 160,000, 130,000 and 76,000, similar in size to human factor VIII:C polypeptides[5]. Although it has been generally supposed that factor VIII:C, like its homologue factor V, is synthesized as a single chain macromolecular precursor which is then cleaved to yield the fragments found in purified preparations, the results presented here are the first conclusive demonstration that this supposition is correct[6].

Additional obstacles to the isolation of a factor VIII:C cDNA clone were the lack of a cell line which made factor VIII:C and uncertainty about its site of synthesis *in vivo*[7-9]. Microsequencing of highly purified porcine factor VIII:C yielded sufficient amino acid sequence to construct oligonucleotide probes. We used these probes to identify a partial clone by direct screening of porcine genomic DNA libraries. The clone provided a probe for identification by cross-species homology of a corresponding human genomic clone. Using this probe, we determined that human liver is a source of factor VIII:C mRNA and obtained subsequently from such tissue the entire 9 kilobase (kb) cDNA sequence as a set of overlapping clones. Reconstruction of these clones into one continuous sequence within an expression vector provided a functional full-length cDNA able to express active protein after transfection into mammalian cells.

Porcine factor VIII:C

Treatment of highly purified porcine factor VIII:C with bovine thrombin (factor IIa) causes an initial activation of up to 60-fold[4] concomitant with loss of the $M_r \sim 160,000$ polypeptide, transient appearance of a M_r 130,000 chain and the appearance of a new chain of M_r 82,000 (Fig. 1*c*). As the digestion proceeds, the M_r 76,000 polypeptide generates a fragment of M_r 69,000 and the M_r 82,000 polypeptide is cleaved, apparently to yield polypeptides of $M_r \sim 50,000$ and 40,000.

Our sequence analysis was performed using factor VIII:C purified from ~10 l of porcine plasma collected from about ten hogs with precautions to minimize proteolysis during collection and subsequent processing[10]. We prepared 0.1–1.5 nmol of each of the M_r 160,000, 130,000 and 76,000 chains originally present in purified porcine factor VIII:C and the 82,000, 69,000,

50,000 and 40,000 thrombin cleavage fragments. N-terminal amino acid sequence analysis using the microsequencing procedures described previously[11] yielded the data presented in Fig. 1*a*, *b*. The M_r 160,000, 130,000, 82,000 and 50,000 chains have the same N-terminal amino acid sequence for at least thirty residues, whereas the 69,000 and 40,000 fragments have distinct N-terminal sequences. There is, however, striking homology between the 69,000, 40,000 and the 160,000 polypeptides and their fragments. The M_r 76,000 fragment shows no N-terminal sequence homology with the N-terminal sequences of any of the other fragments. Comparison of these sequences with partial sequence data for bovine factor V has led to a hypothetical model for the structural relationship between the putative single chain factor VIII:C and its cleavage products[6] similar to that shown diagrammatically in Fig. 1*c*.

Genomic cloning

As the tissue source of mRNA for factor VIII:C was unknown, we tried the novel approach of directly screening genomic libraries with oligonucleotide mixtures to obtain a unique length of porcine factor VIII:C DNA. We used two probes, one a pool of long oligomers containing few sequences and the other a pool of shorter oligomers containing all possible sequences. We identified residues His8–Met22 of the M_r 69,000 chain (see Fig. 1) as containing the least codon ambiguity. The set of 45-mers described in the legend to Fig. 2 was selected on the basis of eukaryotic codon usage[12], the relative stability of G·T versus A·C mismatches[13] and the infrequency of the dinucleotide CpG in eukaryotic genes. Jaye *et al.*[14] used a similar strategy to generate long oligonucleotides for screening a cDNA library, but they did not attempt genomic screening. A set of 15-mers containing all possible sequences that could encode the Trp18–Met22 region of the M_r 69,000 polypeptide served as the secondary probe.

We constructed a genomic library from *Bam*HI digested female porcine DNA and screened it with the radiolabelled 45-mers under relatively nonstringent conditions (see Fig. 2 legend). Rescreening the library with the 15-mers yielded four clones out of 4×10^5 recombinants which hybridized strongly to both the 45-mers and the 15-mers. The four bacteriophages were isolated and, on hybridization to additional oligonucleotides derived from the amino acid sequence of the M_r 69,000 polypeptide, one contained a 6.6 kilobase (kb) BamHI fragment that hybridized to all probes (Fig. 2). Sequence analysis of DNA fragments subcloned into M13 confirmed that we had isolated a recombinant containing DNA, encoding a portion of the M_r 69,000 polypeptide. The 45-mer had 11% mismatch (5 of 45 nucleotide positions) but still hybridized strongly to the porcine genomic clone.

To obtain a human factor VIII:C genomic clone, the 6.6 kb BamHI fragment from the porcine genomic clone was used to screen a human HaeIII/AluI library[15]. From 8×10^5 screened, a single recombinant phage containing an insert of about 16 kb was isolated (Fig. 2). The region of this clone containing sequence homologous to the porcine M_r 69,000 fragment, identified by hybridization to the 45-mers, was subcloned into M13 and its sequence determined. The sequence shows strong (>80%) homology to the porcine factor VIII:C sequence at both the nucleotide and derived amino acid level (see Fig. 1). Additionally, an exon/intron boundary occurs at precisely the same position in both porcine and human genomic DNAs.

To confirm further that we had cloned a segment of the human factor VIII:C gene, we demonstrated linkage to the X chromosome by comparing the Southern blot band intensities between human male diploid DNA and XXXXY DNA (human lymphoblastoid line GM 1202A from the NIGMS mutant cell repository). Several unique DNA fragments from the human genomic clone were used as probes. Each hybridized to a single human fragment which was appropriately overrepresented in the 4X DNA when compared with human albumin DNA internal controls.

The entire human factor VIII:C gene was isolated subsequently as overlapping clones spanning ~200 kb. We thus constructed a genomic library from the human XXXXY cell line and screened sequentially with unique probes from the ends of previously isolated recombinants. The factor VIII:C gene spans more than 180 kb; most of its exons are small (150–250 base pairs) with the exception of the 3.1 kb exon (Fig. 2) and the 1.9 kb 3'-most exon, which includes 1.8 kb of non-coding sequence.

Table 1 Expression of factor VIII:C cDNA in COS-7 cells

	Chromogenic activity (mU ml^{-1})	1 stage APTT assay (mU ml^{-1})
No DNA	<0.05	≪
pCVSVL	<0.05	≪
pCVSVL-VIII	7	10.0
+thrombin	ND	50
+α VIII IgG	<0.05	ND

The full-length cDNA clone was excised from pSP64-VIII with SalI and inserted into the PstI site in pCVSVL[18] by addition to the vector of synthetic oligonucleotide adapters containing SalI cohesive ends. The resulting plasmid containing the factor VIII:C cDNA in the correct orientation for transcription from the adenovirus major late promoter was designated pCVSVL-VIII. Plasmid DNA (8 μg) was transfected into COS-1 monkey cells[23] in a 10 cm dish by the DEAE-dextran transfection protocol[24] with the addition of chloroquin treatment[25]. The transfected cells were incubated for 36 h, then rinsed and fed with 4 ml serum-free media. Samples were taken for assay 24 h later. Factor VIII:C activity was determined by the Kabi Coatest factor VIII:C method modified to afford a sensitivity better than 0.05 mU ml^{-1} and by the one-stage activated partial thromboplastin time (APTT) coagulation assay[26] using factor VIII:C-deficient plasma. For thrombin activation, samples were pretreated 1–10 min with 0.2 units ml^{-1} thrombin at room temperature. The peak of activity (value shown) was reached after 4 min and then gradually declined. Inhibition with anti-factor VIII:C immunoglobulin was measured by the standard Bethesda protocol[27]. Isolated immunoglobulin from serum of a patient with high titre factor VIII:C inhibitor was prepared by ion exchange and protein A Sepharose chromatography[10]. ND, not determined; ≪, not detectable.

cDNA cloning

Our original human factor VIII:C genomic clone provided a probe with which to identify a useful tissue source for human factor VIII:C mRNA. A cell line secreting factor VIII:C is not available, but it had been reported that the protein was synthesized in liver and spleen[7-9]. RNA blot analysis, initially with porcine material, confirmed the presence of very low mRNA levels in these tissues. Northern blot analysis of human liver mRNA, using a highly radioactive single-stranded probe of ~800 nucleotides of human factor VIII:C exon DNA (see Fig. 3 legend), demonstrated a single mRNA of ~9 kb (Fig. 3A, lanes a, b). A messenger of this size could encode a single-chain factor VIII:C precursor of M_r 250,000–300,000. In other experiments, the same size mRNA was detected in placenta but was absent from peripheral blood lymphocytes. All of these observations are consistent with the suggestion that factor VIII:C mRNA is synthesized by endothelial cells[16].

We demonstrate in Fig. 3A (lane c) that the level of factor VIII:C mRNA in human liver is far lower than that of coagulation factor IX mRNA. Quantitative RNA dot blot hybridizations indicated that the amount of factor VIII:C mRNA was 20–40 times lower than factor IX mRNA. We had found previously that 1 in 5,000 recombinant phage in a liver cDNA library contains factor IX sequences. Thus we estimate that libraries of more than 200,000 recombinants would be required to isolate a single factor VIII:C clone.

Several large cDNA libraries were constructed in λ Charon 21A and GT10 (see Fig. 3 legend), using as a first strand primer either oligo(dT) or a unique 38-mer deduced from the sequence of the large human exon in Fig. 2. Ten factor VIII:C cDNA clones were isolated after screening ~2,000,000 recombinants

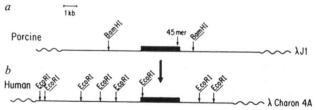

Fig. 2 a, Porcine and b, human factor VIII:C genomic clones. Homologous 3.1 kb exons are indicated by shaded boxes.
Methods: A porcine genomic library was constructed from BamHI digested female porcine liver DNA in the bacteriophage vector λJ1 (a derivative of L47.1) using established procedures. Recombinants (4×10^5) were screened using a modification of the procedure of Woo[29]. A set of 45-mers containing the sequences 5'd (CATGCCATA^A_GTCCCACAGCTG^T_CTCCACAGCAGCAAT^A_GA-AGTAGTG) was synthesized on the Applied Biosystems Model 380A instrument. Filters were hybridized with 5×10^6 c.p.m. ml^{-1} (0.5 pmole ml^{-1}) of ^{32}P-labelled 45-mers for 12 h at 45 °C in 5X SSC, 5X Denhardts, 0.1% SDS, 100 μg ml^{-1} denatured salmon sperm DNA. Filters were washed at 50 °C in 5X SSC, 0.1% SDS, air dried and subjected to autoradiography using Kodak XAR film. After autoradiography, filters were denatured in 0.1M NaOH, neutralized and rehybridized with ^{32}P-labelled 15-mers, sequence 5'd (CATNCC^A_GTA^A_GTCCCA). Hybridization was performed as described above at 37 °C. Recombinants hybridizing to both oligomers were plaque purified and phage DNA prepared. Phage DNA prepared from one recombinant (PB34) contained a 6.6 kb BamHI fragment which hybridized strongly to both sets of oligonucleotides in Southern blot analysis. The 6.6 kb BamHI insert was purified from an agarose gel and ^{32}P-labelled by nick translation. Recombinants (8×10^5) from the human HaeIII/AluI genomic library of Lawn et al.[15] were hybridized with 10^6 c.p.m. ml^{-1} of the nick-translated porcine probe using the same conditions described for the 45-mer hybridization. A single recombinant phage, HH25, was thus isolated. Sau3A fragments of the insert from HH25 were subcloned into M13 and a phage which hybridized to the radiolabelled 45-mer was identified. The sequence of the Sau3A fragment in this subclone proved that HH25 contained DNA from the human factor VIII:C gene. The entire exon (shaded box) was subsequently sequenced. A 0.8 kb BamHI fragment spanning the 3' exon/intron junction was subcloned as the probe to screen for a tissue source of factor VIII:C mRNA (see Fig. 3).

from the oligo(dT) libraries; 55 additional cDNA clones were obtained from among 1,000,000 recombinants from the 38-mer primed library. Four of the several clones characterized in detail are illustrated in Fig. 3*B* together spanning 9,000 base pairs (bp), sufficient to encode the entire mRNA detected by Northern

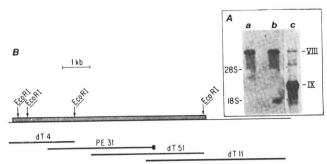

Fig. 3 *A*, Northern blot analysis to locate a tissue source for human factor VIII : C mRNA. Polyadenylated RNA (5 μg) extracted[30] from adult liver (lane *b*) or fetal liver (lanes *a*, *c*) was fractionated on an 0.8% agarose formaldehyde gel and transferred to nitrocellulose as described[31]. The filter in lanes *a*, *b* was hybridized with 5×10^6 c.p.m. of a ^{32}P-labelled factor VIII : C single-stranded probe prepared by primed synthesis from an M13 template which contained an 0.8 kb base *Bam*HI fragment of HH25 (see Fig. 2 legend)[32,33]. The filter in lane *c* was probed in the same manner except 5×10^6 c.p.m. of a human factor IX probe of identical specific activity was used in addition. The exposure time for lanes *a*, *b* was longer than that for lane *c* and the RNA was from a different preparation. *B*, restriction map (*Eco*RI) of the human factor VIII : C cDNA sequence showing the single long open reading frame as a shaded box. An illustrative set of four overlapping bacteriophage λ clones are shown. Those designated dT4, dT51 and dT11 were from oligo(dT) primed λCharon 21A libraries. Clone PE31 was from a library in λGT10 primed with a unique 38-mer (solid rectangle). The cDNA sequence comprises 150 bases of 5'-untranslated region, an open reading frame of 7,053 bases and 1,806 bases of 3'-untranslated region.
Methods: Oligo(dT) primed double-stranded cDNA was synthesized from 10 μg polyadenylated fetal liver RNA as described[34]. 200 pmol of a unique complementary 38-mer (nucleotides 5,308–5,345) and 10 μg polyadenylated fetal liver mRNA were denatured in 10 mM CH₃HgOH. The reaction was made 140 mM in 2-mercaptoethanol, 700 mM KCl, 1 mM EDTA, 20 mM Tris-HCl (pH 8.3 at 42 °C), 1 unit μl^{-1} of RNasin (Promega-Biotec) and incubated at 50 °C for 2 min and at 42 °C for 2 min. First and second strand synthesis was then performed as for the oligo(dT) primed cDNA[34]. The second strand reaction was terminated by the addition of EDTA to 20 mM. The *Eco*RI sites were methylated, after the addition of S-adenosylmethionine to 50 μM and 40 units of *Eco*RI methylase, by incubation at 37 °C for 1 h. These reactions were terminated by phenol-chloroform extraction and chromatographed using Sephadex G-50 equilibrated in 10 mM Tris-HCl (pH 8.0), 1 mM EDTA, 100 mM NaCl. DNA in the excluded volume was pooled and precipitated with ethanol. The double-stranded cDNA was blunted in 50 μl containing 50 mM Tris pH 8.3, 10 mM MgCl₂, 10 mM 2-mercaptoethanol, 50 mM NaCl, 50 μM of deoxynucleotide triphosphate, 100 μg ml⁻¹ ovalbumin and 5 units of T4 polymerase (PL Biochemicals). The reaction was incubated at 37 °C for 30 min and then terminated by phenol-chloroform extraction. Nucleic acids were recovered by ethanol precipitation with 2 M ammonium acetate. Kinased *Eco*RI linkers (NE. Biolabs; 300 ng) were then added to the blunted double standed cDNA by ligation overnight at 16 °C in a final volume of 30 μl. The reaction was diluted to 200 μl in *Eco*RI digestion buffer and digested with 300 units of *Eco*RI (NE. Biolabs) for 2 h at 37 °C. EDTA to 15 mM was added and the reaction extracted with phenol-chloroform and chromatographed over Sepharose CL-4B (Pharmacia) equilibrated with 10 mM Tris-HCl (pH 8.0), 1 mM EDTA, 100 mM NaCl. cDNA in the void volume was collected and precipitated by ethanol in the presence of 10 μg yeast tRNA carrier. cDNA was redissolved in 10 mM Tris-HCl (pH 8.0), 1 mM EDTA and ligated to *Eco*RI cleaved, phosphatased, λCharon 21A DNA for the oligo(dT) primed cDNA and to λGT10 DNA for the primer extended cDNA. The libraries were packaged, plated and initially screened using a unique ^{32}P-labelled 53-mer (nucleotides 5,158–5,210) permitting the isolation of both PE31 and dT51. Recombinant dT4 was identified with a ^{32}P-labelled pool of 64 56-mers containing the sequences; 5'd(TCCCAATCCTCTTCTTC^T_CGC^A_GGA^G_AA-TATAGTG^C_CACCCA^A_GGTCTT^A_TGGGTGCTTCTT) (using the same criteria as was used for the construction of the 45-mer pools) made to residues (Lys3–Asp21) of the M_r 40,000 polypeptide fragment of porcine VIII : C. Recombinant dT11 was isolated using a nick translated 3' end *Pst*I/*Eco*RI fragment (nucleotides 5,902–7,127) of dT51. Numerous other cDNA clones were isolated by similar methods; some were also characterised in detail, sequenced and used in the construction of the full-length cDNA clone. Selected cDNA and genomic exon fragments were ligated to one another two at a time, subcloned and ultimately inserted in the *Sal*I site of pSP64 after the addition of a synthetic *Sal*I site at nucleotide 145. The resulting recombinant, PSP64-VIII, contains the factor VIII : C cDNA from just before the initiator ATG (at nucleotide 150-152) to nucleotide 7,422 (some 200 bp downstream from the terminator at nucleotide 7,201).

blot analysis. (The unique 9,009 bp cDNA sequence we obtained after sequencing numerous overlapping fragments of cDNA and genomic DNA has been deposited in the NIH US Nucleic Acid Sequence Databank and is available through GENBANK.) The sequence comprises a 5'-noncoding region of 150 nucleotides, a single long open reading frame of 7,053 nucleotides (2,351 codons) and 1,806 nucleotides of the 3'-noncoding region. The translated sequence, in a single-letter amino acid code, is presented as Fig. 4.

A full-length cDNA clone was assembled (see Fig. 3 legend) and inserted into the *Sal*I site of the polylinker in plasmid pSP64 (Promega Biotec). We thus obtained a recombinant, pSP64-VIII, with a single continuous factor VIII : C coding sequence that could be readily excised by *Sal*I digestion for insertion into a variety of expression vectors.

Expression and structure

To confirm that the factor VIII : C cDNA indeed encodes active factor VIII, we introduced the full-length cDNA into the mammalian expression vector pCVSVL[17,18]. Plasmid pCVSVL-VIII contains the cDNA in the proper orientation to be expressed from the adenovirus major late promoter. Medium collected 60 h after transfection of COS-1 cells with this plasmid contained readily detectable levels of factor VIII : C activity (Table 1). No activity was detected in controls. The activity was inhibited by antibody to human factor VIII : C and was activated fivefold by thrombin.

The sequence of the cDNA clone comprises 2,351 amino acids from the initiator Met to the terminating Tyr residue and begins with a secretory leader peptide sequence of 19 amino acids. Regions of homology with the previously determined amino acid sequences of porcine factor VIII : C fragments can be located (Fig. 1), confirming the model for the structure of porcine factor VIII : C of Fass *et al.*[6] and predicting the mature N-terminus and thrombin cleavage sites of the human molecule.

Fig. 4 Amino acid sequence of the human factor VIII : C precursor, predicted from the sequences of cDNA and genomic clones. Arrows indicate (1) a presumed signal peptidase cleavage site at residues 19-20 (Ser-Ala); (2) thrombin cleavage sites at residues 391-392, 759-760 and 1,708-1,709 (all Arg-Ser); (3) a possible thrombin cleavage site at residues 1,667-1,668 (Arg-Glu). Before each cleavage site the apparent M_r of the corresponding porcine factor VIII : C fragment C-terminal to the cleavage site[6] is given. *, Amino acid residue identity with porcine factor VIII : C amino acid sequence determined from a variety of proteolytic and chemical cleavage fragments. Potential N-glycosylation sequences (Asn-X-Ser/Thr) are underlined.

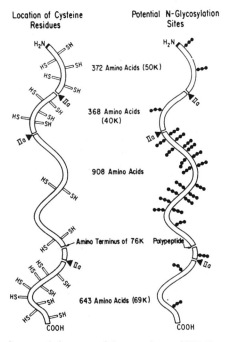

Location of Cysteine Residues   Potential N-Glycosylation Sites

Fig. 5 Structural features of human factor VIII:C, showing schematically the location of cysteine residues and potential N-glycosylation sites (Asn-X-Ser/Thr) with respect to thrombin cleavage domains. The number of amino acid residues in each domain is given as well as the apparent molecular weights of the corresponding regions of porcine VIII:C.

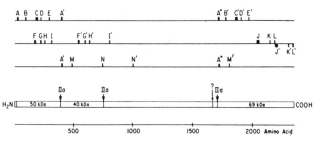

Fig. 6 Amino acid sequence homology within human factor VIII:C. A computer search for internal amino acid sequence homology was performed using the PEP program in the IntelliGenetics software system. Blocks of homology are labelled alphabetically on the linear maps, with their location and relative size indicated. Matching blocks of homology bear the same alphabetical symbol plus a prime or double prime. The actual amino acid sequences within these blocks and their numerical position with the entire precursor molecule can be located in Fig. 4. A(residues 21–35); A'(400–414); A"(1,714–1,728); B(89–101); B'(1,776–1,788); C(177–198); C'(1,856–1,877); D(211–227); D'(1,902–1,908); E(292–298); E'(1,979–1,985); F(164–179); F'(539–554); G(217–227); G'(594–604); H(250–261); H'(633–644); I(318–324); I'(809–815); J(2,038–2,059); J'(2,191–2,212); K(2,144–2,151); K'(2,301–2,308); L(2,181–2,188); L'(2,338–2,345); M(486–492); M'(1,796–1,802); N(749–755); N'(1,002–1,008).

The location of cysteine residues and potential N-glycosylation sites (Asn-X-Ser/Thr) is shown schematically in Fig. 5. Their exact numerical position within the human factor VIII:C precursor molecule is shown in Fig. 4. The cysteine residues appear to be clustered in the M_r 50,000, 40,000 and 69,000 regions. There are no disulphide bridges between these regions in porcine factor VIII:C, where the same pattern of polypeptides is observed on SDS gels in either reducing or non-reducing conditions.

In contrast to the location of cysteine residues, many of the potential N-glycosylation sites appear between the M_r 40,000 and 69,000 regions (Fig. 5). Distinct bands corresponding to this domain of 908 amino acids, or cleavage fragments derived from it, cannot be readily identified on silver stained SDS gels. We have obtained amino acid sequence data from minor components in purified porcine factor VIII:C preparations which are present in the M_r 80,000 and 50,000 regions of SDS gels. The N-terminal sequences of these polypeptides are identical; furthermore, a similar amino acid sequence is found in the predicted human factor VIII:C precursor amino acid sequence shown in Fig. 4 (residues 760–766).

A striking feature of human factor VIII:C is the presence of repeated blocks of amino acid homology (Fig. 6); the three most extensively repeating patterns of homology are within the M_r 50,000 and 69,000 fragments (blocks A, B, C, D, E and A", B', C', D', E'); within the M_r 50,000 and 40,000 fragments (blocks A, F, G, H, I and A', F', G', H', I') and a M_r 69,000 fragment internal repetition (blocks J, K, L and J', K', L'). Another set of repeat blocks (A', M and A", M') is suggestive of a vestige of a M_r 40,000/69,000 duplication. Not only is there a strong M to M' homology, but also the A to M spacing is very similar in both regions. A block of amino acids (N) in the 40,000 region is repeated (N') in the region between the M_r 40,000 and 76,000 regions.

We analysed the precursor human factor VIII:C amino acid sequence by the method of Hopp and Woods[19] for the determination of relative hydrophobicity. As observed for other secreted proteins, the factor VIII:C precursor contains a hydrophobic

leader sequence (residues 1–19). Two other extensive regions of hydrophobicity are also predicted by this analysis; first, 27 amino acids from Ser308 to His334 and second, from Ser630 to Phe677. The hydrophobic nature of these regions is substantiated by the method of Kyte and Doolittle[20]. The role of these hydrophobic regions in factor VIII:C secretion and/or activity is unknown.

Homology with other plasma proteins

We have reported previously[21] amino acid sequence homologies among certain regions in factor VIII:C, factor V, and ceruloplasmin. These homologies were identified by comparison of the N-terminal amino acid sequences of thrombin cleavage fragments of bovine factor V and porcine factor VIII:C with amino acid sequence deduced from a genomic clone of porcine factor VIII:C and with the published amino acid sequence of human ceruloplasmin[22]. The deduced amino acid sequence of human factor VIII:C described here (Fig. 4) contains regions of homology with the known amino acid sequences of bovine factor V and extensive homology with the entire human ceruloplasmin molecule, supporting the suggestion that these three proteins are related evolutionarily. Note the homology between blocks A, A' and A" of human factor VIII:C (Fig. 6) and residues 2–16, 353–366 and 712–726 respectively of human ceruloplasmin[22], as these homology blocks apparently mark the boundary of a primordial gene encoding approximately 350 amino acids. Similarly, block A is homologous with residues 3–17 of the M_r 94,000 amino terminal thrombin cleavage fragment of bovine factor V and block A" is homologous with residues 6–20 of the C-terminal M_r 74,000 thrombin cleavage fragment of factor V[21]. At present, the significance of repeating homologous units in factor VIII:C, factor V and ceruloplasmin is unclear, but they presumably reflect duplications and triplications of some smaller ancestral gene.

We thank Louise Wasley and John Brown for technical assistance; Ed Fritsch and Steve Clark for support and helpful discussions; Joyce Lauer and Katherine Smith for help with the manuscript and Martha Richardson, Marybeth Erker and Cathy Beechinor for word processing. Finally, we thank both Robert Kamen and Gabriel Schmergel for their encouragement throughout the project. This work was supported by Baxter Travenol Inc., the Mayo Foundation and grant HL-17430 to D.N.F.

Received 31 August; accepted 2 October 1984.

1. Hoyer, L. W. *Blood* **58**, 1-12 (1981).
2. Rapaport, S. I., Schiffman, S. & Patch, M. J. *Blood* **21**, 221-235 (1963).
3. Vehar, G. A. & Davie, E. W. *Biochemistry* **19**, 410-416 (1980).
4. Fass, D. N., Knutson, G. J. & Katzman, J. A. *Blood* **59**, 594-600 (1982).
5. Fulcher, C. A., Gardiner, J. E., Griffin, J. H. & Zimmerman, T. A. *Blood* **63**, 486-489 (1984).
6. Fass, D. N., Hewick, R. M., Knutson, G. J., Nesheim, M. E. & Mann, K. G. *Proc. natn. Acad. Sci. U.S.A.* (in the press).
7. Owen, C. A. Jr, Bowie, W. E. J. & Fass, D. N. *Br. J. Haemat.* **43**, 307-315 (1979).
8. Libre, E. P., Cowen, D. H., Watkins, S. P. & Shulman, N. R. *Blood* **31**, 358-368 (1968).
9. Marchioro, T. L., Houghie, C., Ragde, H., Epstein, R. B. & Thomas, E. D. *Science* **163**, 188-190 (1969).
10. Knutson, G. J. & Fass, D. N. *Blood* **59**, 615-624 (1982).
11. Hewick, R. M., Hunkapiller, M. E., Hood, L. E. & Dreyer, W. J. *J. biol. Chem.* **256**, 7990-7997 (1982).
12. Strehler, B. & North, D. *Mech. Ageing Dev.* **18**, 285-313 (1982).
13. Agarwal, K. L., Brunstedt, J. & Noyes, B. E. *J. biol. Chem.* **256**, 1023-1028 (1981).
14. Jaye, M. *et al. Nucleic Acids Res.* **11**, 2325-2335 (1983).
15. Lawn, R. M., Fritsch, E. F., Parker, R. C., Blake, G. & Maniatis, T. *Cell* **15**, 1157-1174 (1978).
16. Stel, H. V., vanderKwast, Th. H. & Veerman, E. C. I. *Nature* **303**, 530-532 (1983).
17. Kaufman, R. J. & Sharp, P. A. *Molec. Cell Biol.* **2**, 1304-1319 (1982).
18. Clark, S. C. *et al. Proc. natn. Acad. Sci. U.S.A.* **81**, 2541-2547 (1984).
19. Hopp, T. P. & Woods, K. R. *Proc. natn. Acad. Sci. U.S.A.* **78**, 3824-3828 (1981).
20. Kyte, J. & Doolittle, R. F. *J. molec. Biol.* **157**, 105-132 (1982).
21. Church, W. R. *et al. Proc. natn. Acad. Sci. U.S.A.* (in the press).
22. Takahashi, N., Ortel, T. L. & Putnam, F. W. *Proc. natn. Acad. Sci. U.S.A.* **81**, 390-394 (1984).
23. Gluzman, Y. *Cell* **23**, 175-182 (1981).
24. Sompayrac, L. M. & Dana, K. J. *Proc. natn. Acad. Sci. U.S.A.* **78**, 7575-7578 (1981).
25. Luthman, H. & Magnusson, G. *Nucleic Acids Res.* **11**, 1295-1308 (1983).
26. Lee, M. L., Maglalang, E. A. & Kingdon, H. S. *Thromb. Res.* **30**, 511-519 (1983).
27. Kasper, C. K. *Thromb. Diath. Haemorrh.* **34**, 869-872 (1975).
28. Hunkapillar, M. W., Lujan, E., Ostrander, F. & Hood, L. E. *Meth. Enzym.* **91**, 227-236 (1983).
29. Woo, S. L. C. *Meth. Enzym.* **68**, 389-395 (1979).
30. Hastie, N. D., Held, W. & Toole, J. J. *Cell* **17**, 449-457 (1979).
31. Derman, E. *et al. Cell* **23**, 731-739 (1981).
32. O'Hare, K., Levis, R. & Rubin, G. M. *Proc. natn. Acad. Sci. U.S.A.* **80**, 6917-6921 (1983).
33. Shaw, P. H., Held, W. A. & Hastie, N. D. *Cell* **32**, 755-761 (1983).
34. Gubler, U. & Hoffman, B. J. *Gene* **25**, 163-269 (1983).

A Bacterial Clone Synthesizing Proinsulin

L. Villa-Komaroff, A. Efstratiadis, S. Broome, P. Lomedico,
R. Tizard, S.P. Naber, W.L. Chick and W. Gilbert

ABSTRACT　　We have cloned double-stranded cDNA copies of a rat preproinsulin messenger RNA in *Escherichia coli* χ1776, using the unique *Pst* endonuclease site of plasmid pBR322 that lies in the region encoding amino acids 181–182 of penicillinase. This site was reconstructed by inserting the cDNA with an oligo(dG)·oligo(dC) joining procedure. One of the clones expresses a fused protein bearing both insulin and penicillinase antigenic determinants. The DNA sequence of this plasmid shows that the insulin region is read in phase; a stretch of six glycine residues connects the alanine at position 182 of penicillinase to the fourth amino acid, glutamine, of rat proinsulin.

Can the structural information for the production of a higher cell protein be inserted into a plasmid in such a way as to be expressed in a transformed bacterium? To attack this problem, we used as a model rat insulin, an interesting protein that can be identified by immunological and biological means.

Although mature insulin contains two chains, A and B, it is the product of a single longer polypeptide chain. The hormone is initially synthesized as a preproinsulin structure (1, 2). A hydrophobic leader sequence of 23 amino acids at the amino terminus of the nascent chain is cleaved off, presumably as the polypeptide chain moves through the endoplasmic reticulum (2–4), producing a proinsulin molecule. The proinsulin chain folds up and then the C peptide is cleaved from its middle (5). Thus each of the two (nonallelic) insulin genes in the rat (6–8) encodes a polypeptide 109 amino acids long, whose initial structure is NH_2—leader sequence—B chain—C peptide—A chain.

Ullrich *et al.* (9) have cloned double-stranded cDNA copies of rat preproinsulin mRNA isolated from pancreatic islets and determined sequences covering much of those two genes. We have made double-stranded cDNA copies of mRNA from a rat insulinoma (10) and cloned these in the *Pst* (*Providencia stuartii* endonuclease) site of pBR322 (11), which lies within the penicillinase gene.

The *Escherichia coli* penicillinase is a periplasmic protein, the gene for which was recently sequenced (12). Penicillinase is synthesized as a preprotein with a 23 amino acid leader sequence (12, 13), which presumably serves as a signal to direct the secretion of the protein to the periplasmic space, and is removed as the protein traverses the membrane. Insertion of the structural information for insulin into the penicillinase gene should cause expression of the insulin sequence as a fusion product transported outside the cell.

MATERIALS AND METHODS

Bacterial Strains. *E. coli* K-12, strain HB101 [*hsm⁻*, *hrs⁻*, *recA⁻*, *gal⁻*, *pro⁻*, *strʳ* (14)] was initially obtained from H. Boyer. *E. coli* K-12 strain χ1776 (15) (F⁻, *tonA53*, *dapD8*, *minA1*, *supE42*, Δ40[*gal–uvrB*], λ⁻, *minB2*, *rfb-2*, *nalA25*, *oms-2*, *thyA57*, *metC65*, *oms-1*, Δ29[*bioH–asd*], *cycB2*, *cycA1*, *hsdR2*) was provided by R. Curtiss.

DNA and Enzymes. pBR322 DNA, a gift from A. Poteete, was used to transform *E. coli* HB101. Plasmid DNA was purified according to the procedure of Clewell (16). Avian myeloblastosis virus reverse transcriptase (RNA-dependent DNA polymerase), *E. coli* DNA polymerase I, and terminal transferase were gifts from T. Papas, M. Goldberg, and J. Wilson, respectively. Restriction enzymes were purchased from Bethesda Research Labs and New England BioLabs.

RNA Purification. An x-ray-induced, transplantable rat beta cell tumor (10) was used as source of preproinsulin mRNA. Tumor slices (20 g per preparation) were homogenized, and a cytoplasmic RNA (about 2 mg/g of tissue) was purified from a postnuclear supernatant by Mg^{2+} precipitation (17), followed by extraction with phenol and chloroform, and enriched for poly(A)-containing RNA by oligo(dT)-cellulose chromatography (18). About 4% of the material binds to the column (data from eight preparations). Further purification of the oligo(dT)-cellulose-bound material by sucrose gradient centrifugation and/or polyacrylamide gel electrophoresis showed that the preproinsulin mRNA was a minor component of the preparation.

Double-Stranded cDNA Synthesis. Oligo(dT)-cellulose-bound RNA was used directly as template for double-stranded cDNA synthesis (19), except that a specific p(dT)₈dG-dC primer (Collaborative Research) was utilized for reverse transcription. The concentrations of RNA and primer were 7 mg/ml and 1 mg/ml, respectively. All four [α-³²P]dNTPs were at 1.25 mM (final specific activity 0.85 Ci/mmol). The reverse transcript was 2% of the input RNA, and 25% of it was finally recovered in the double-stranded DNA product.

Construction of Hybrid DNA Molecules. pBR322 DNA (5.0 μg) was linearized with *Pst*, and approximately 15 dG residues were added per 3′ end by terminal transferase at 15° in the presence of 1 mM Co^{2+} (20) and autoclaved gelatin at 100 μg/ml. Similarly, dC residues were added to 2.0 μg of double-stranded cDNA, which was then electrophoresed in a 6% polyacrylamide gel. Following autoradiography, molecules in the size range of 300 to 600 base pairs (0.5 μg) were eluted from the gel (21). Size selection was done after tailing rather than before because previous experience had indicated that occasionally impurities contaminating DNA extracted from gels inhibits terminal transferase. The eluted double-stranded cDNA was concentrated by ethanol precipitation, redissolved in 10 mM Tris·HCl at pH 8, mixed with 4 μg of dG-tailed pBR322, and dialyzed versus 0.1 M NaCl/10 mM EDTA/10 mM Tris, pH 8. The mixture (4 ml) was then heated at 56° for 2 min, and annealing was performed at 42° for 2 hr. The hybrid DNA was used to transform *E. coli* χ1776.

Transformation and Identification of Clones. Transformation of *E. coli* χ1776 (an EK2 host) with pBR322 (an EK2

VILLA-KOMAROFF, L., EFSTRATIADIS, A., BROOME, S., LOMEDICO, P., TIZARD, R., NABER, S.P., CHICK, W.L. and GILBERT, W.
A bacterial clone synthesizing proinsulin. *Proc. Natl. Acad. Sci. U.S.A.* 75:3727-3731 (1978). Reprinted with the authors' permission.

vector) was performed in a biological safety cabinet in a P3 physical containment facility in compliance with NIH guidelines for recombinant DNA research published in the *Federal Register* [(1976) **41**, 27902–27943].

χ1776 was transformed by a transfection procedure (22) adapted to χ1776 by A. Bothwell (personal communication) and slightly modified as follows: χ1776 was grown in L broth (23) supplemented with diaminopimelic acid at 10 μg/ml and thymidine (Sigma) at 40 μg/ml to OD_{590} of 0.5. Cells (200 ml) were sedimented at 500 × g and resuspended by swirling in 1/10th vol of cold buffer containing 70 mM $MnCl_2$, 40 mM Na acetate at pH 5.6, 30 mM $CaCl_2$, and kept on ice for 20 min. The cells were repelleted and resuspended in $\frac{1}{30}$th of the original volume in the same buffer. Two milliliters of the annealed DNA preparation was added to the cells. Aliquots of this mixture (0.3 ml) were placed in sterile tubes and incubated on ice for 60 min. The cells were then placed at 37° for 2 min. Broth was added to each tube (0.7 ml) and the tubes were incubated at 37° for 15 min; 200 μl of the cells was spread on sterile nitrocellulose filters (Millipore) overlaying agar plates containing tetracycline at 15 μg/ml. (The filters were boiled to remove detergents before use.) The plates were incubated at 37° for 48 hr. Replicas of the filters were made by a procedure developed by D. Hanahan (personal communication): The nitrocellulose filters containing the transformants were removed from the agar and placed on a layer of sterile Whatman filter paper. A new sterile filter was placed on top of the filter containing the colonies and pressure was applied with a sterile velvet cloth and a replica block. A sterile needle was used to key the filters. The second filter was placed on a new agar plate and incubated at 37° for 48 hr. The colonies on the first filter were screened by the Grunstein–Hogness technique (24), using as probe an 80-nucleotide-long fragment produced by *Hae* III digestion of high specific activity cDNA (9). Positive colonies were rescreened by hybrid-arrested translation (25) as described in the legend of Table 1.

Radioimmunoassays. Two-site solid-phase radioimmunoassays were performed (28). Cells from colonies to be tested were transferred with an applicator stick onto 1.5% agarose containing 30 mM Tris·HCl, pH 8, lysozyme at 0.5 mg/ml, and 10 mM EDTA; released antigen was adsorbed to an IgG-coated polyvinyl disk during a 1-hr incubation at 4°. The wash buffer contained streptomycin sulfate at 300 μg/ml and normal guinea pig serum (Grand Island Biological Co.) instead of normal rabbit serum. Guinea pig antiserum to bovine insulin was purchased from Miles Laboratories.

Standard (liquid) radioimmunoassays were performed using the back titration procedure employing alcohol precipitation of insulin–antibody complexes (29).

DNA Sequencing. DNA sequencing was performed as described by Maxam and Gilbert (30).

RESULTS

Construction and Identification of cDNA Clones. We isolated poly(A)-containing RNA from a transplantable rat insulinoma. This preparation contained preproinsulin mRNA, because it directed the synthesis in a cell-free system of a product precipitable with anti-insulin antibody (data not shown). However, the mRNA yield after further purification was not sufficient for cloning, and therefore we decided to clone cDNA synthesized from the total preparation. In an attempt to enrich the reverse transcript for insulin sequences, we utilized the DNA sequence reported by Ullrich *et al.* (9) to choose a specific primer, $(dT)_8dG\text{-}dC$. The product of double-stranded cDNA synthesis (19) was extended by a short oligo(dC) tail about 15 nucleotides in length, and sized on a polyacrylamide

Table 1. Hybrid-arrested translation and immunoprecipitation of the cell-free products

| Source of arresting DNA | Radioactivity, cpm/20 μl | | | % Immuno-precipitable* |
| | Acid insoluble | Immuno-precipitable | | |
		− Insulin	+ Insulin	
Control I (−DNA, −RNA)†	2,570			
Control II (−DNA, +RNA)‡	35,700	12,300	310	36.2
pBR322	28,800	7,850	245	29.0
Clone 3	15,100	3,630	264	26.9
Clone 13	19,600	5,190	350	28.4
Clone 15	18,600	4,850	252	28.7
Clone 16	29,200	8,830	247	32.2
Clone 17	24,000	6,700	316	30.0
Clone 18	15,900	3,690	251	25.8
Clone 19	8,650	587	277	5.0
Clone 20	15,100	4,070	231	30.6
Clone 21	21,100	5,170	223	26.7

Plasmid DNA (about 3 μg) was digested with *Pst*, precipitated with ethanol, and dissolved directly in 20 μl of deionized formamide. After heating for one minute at 95° each sample was placed on ice. Following the addition of 1.5 μg of oligo(dT)-cellulose-bound RNA, piperazine-*N,N′*-bis(2-ethanesulfonic acid) (Pipes) at pH 6.4 to 10 mM, and NaCl to 0.4 M, the mixtures were incubated for 2 hr at 50°. They were then diluted by the addition of 75 μl of H_2O and ethanol precipitated in the presence of 10 μg of wheat germ tRNA, washed with 70% (vol/vol) ethanol, dissolved in H_2O, and added to a wheat germ cell-free translation mixture (26) containing 10 μCi of [³H]leucine (60 Ci/mmol). Fifty-microliter reaction mixtures were incubated at 23° for 3 hr and then duplicate 2-μl aliquots were removed for trichloroacetic acid precipitation. From the remainder two 20-μl aliquots were treated with ribonuclease, diluted with immunoassay buffer, and analyzed for the synthesis of immunoreactive preproinsulin by means of a double antibody immunoprecipitation (27) in the absence or presence of 10 μg of bovine insulin. The washed immunoprecipitates were dissolved in 1 ml of NCS (Amersham) and assayed in 10 μl of Omnifluor (New England Nuclear) by liquid scintillation counting.
* Calculated using the formula [(immunoprecipitable radioactivity in the absence of insulin) − (immunoprecipitable radioactivity in the presence of insulin)]/[(acid-insoluble radioactivity) − (acid-insoluble radioactivity of control I)].
† Reaction mixture incubated in the absence of added RNA.
‡ Cell-free translation by the direct addition of oligo(dT)-cellulose-bound RNA into the reaction mixture.

gel. A broad size cut averaging 500 base pairs was selected in order to enrich for full-length sequences. We inserted these molecules into the *Pst* site of pBR322 after elongating the 3′-terminal extension of the cleavage site with oligo(dG). We used this oligo(dG)·oligo(dC) joining procedure in order to reconstruct the *Pst* recognition sequence (ref. 31; W. Rowenkamp and R. Firtel, personal communication); approximately 40% of the inserts were excisable with *Pst* after cloning. From about 0.25 μg of tailed cDNA we obtained 2355 transformants in *E. coli* strain χ1776. To identify clones containing insulin sequences, we first screened one-third of the transformants, using as a probe an 80-nucleotide-long *Hae* III fragment of cDNA synthesized from oligo(dT)-bound RNA because the results of Ullrich *et al.* (9) suggested that such a fragment should be insulin specific. About 20% of the clones were positive, but restriction analysis of plasmid DNA from a few candidates showed that the inserts were not insulin sequences. We concluded that our probe was not pure and rescreened some of the positive clones, using hybrid-arrested translation (25). This method is based on the principle that mRNA in the form of an RNA·DNA hybrid does not direct cell-free protein synthesis. We incubated aliquots of oligo(dT)-bound RNA with linearized

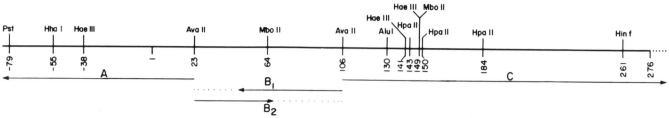

FIG. 1. Restriction map of the insertion in clone pI19. Each restriction site is identified by a number indicating the 5'-terminal nucleotide generated by cleavage at the message strand. Nucleotides are numbered beginning with the first base of the sequence encoding proinsulin. Nucleotides in the 5' direction from position 1 in the message strand are identified by negative numbers, beginning with −1. Arrows indicate the sequenced fragments; those pointing to the left indicate sequences derived from the antimessage strand, and those pointing to the right indicate sequences derived from the message strand. The uniquely labeled restriction fragments were generated as follows: Following excision with *Pst*, DNA of the insertion was digested with *Ava* II and end labeled. Fragments A and C purified from a polyacrylamide gel were sequenced directly because the *Pst* ends do not label significantly. Fragment B was strand separated on a polyacrylamide gel and sequenced in both directions. The exact number of C·G pairs in the right-hand tail before the *Pst* site could not be counted.

DNA from nine clones under conditions favoring DNA·RNA hybridization (32), added them to cell-free translation systems, and assayed for a specific inhibition of insulin synthesis. Table 1 shows that one of the plasmids, pI19, inhibited the synthesis of immunoprecipitable material. Restriction endonuclease digestions of the *Pst*-excised insert of pI19 with several enzymes generated fragments whose sizes were consistent with the sequence of Ullrich *et al.* (9). We confirmed the presence of insulin DNA in pI19 by direct DNA sequence analysis and screened the rest of the clones with purified pI19 insert labeled by nick translation. About 2.5% (48/1745) of the clones hybridized strongly to this probe. There must have been enrichment for insulin sequence at some step of our procedure, because hybridization analysis using cloned insulin DNA as probe showed the presence of only 0.3% insulin mRNA in the original oligo(dT)-bound RNA.

Sequence Information. Fig. 1 shows the restriction map of the insertion in clone pI19 and Fig. 2 shows the sequence of the insert. It corresponds to rat insulin I (5, 33) and encodes the entire preproinsulin chain with the exception of the first two amino acid residues of the reported preregion (1). It therefore extends the sequence determined by Ullrich *et al.* (9) by twenty-five 5'-terminal nucleotides. It also verifies the reported amino acid residues for positions −14, −17, −18, and −20; it identifies the previously uncertain residue −15; and it identifies the unknown residue −19. However, the residues at positions −16 and −21 differ from those reported (1).

The sequence deviates from that determined by Ullrich *et al.* (9) at the region immediately after the UGA terminator, where a GAGTC sequence occurs, predicting a *Hinf* cleavage

site that we have experimentally verified. Furthermore, only moderate agreement exists between the two sequences for the next 15 nucleotides of the 3' untranslated region.

Expression. Almost two-thirds of the clones carrying inserts were ampicillin resistant; thus the active site of penicillinase must lie between amino acid residues 23 and 182 (12). The degree of resistance was variable, suggesting the expression of different sequences from the inserts in the form of fused translation products, probably differing in length and stability.

We therefore screened colonies of the 48 clones containing insulin sequence for the presence of insulin antigenic determinants, using a solid-phase radioimmunoassay (28). Polyvinyl sheets coated with antibody molecules will bind specific antigens released from bacteria. The immobilized antigen can then be detected by autoradiography following exposure of the sheets to ^{125}I-labeled antibody. This method permits detection of as little as 10 pg of insulin in a colony. We coated plastic disks with anti-insulin antibody and used ^{125}I-labeled anti-insulin to detect solely insulin antigenic determinants. Disks coated with anti-penicillinase antibody and exposed to ^{125}I-anti-insulin detect the presence of a fused protein, as do disks coated with anti-insulin and exposed to radioiodinated anti-penicillinase.

One clone, pI47, gave positive responses with all of the combinations described above; this indicates the presence of a penicillinase–insulin hybrid polypeptide. Fig. 3 shows some of the results. To determine whether this fused protein is secreted, we grew clone pI47 in liquid culture and extracted the proteins in the periplasmic space by osmotic shock, a method that does not lyse bacteria (34). Fig. 4 shows that the insulin

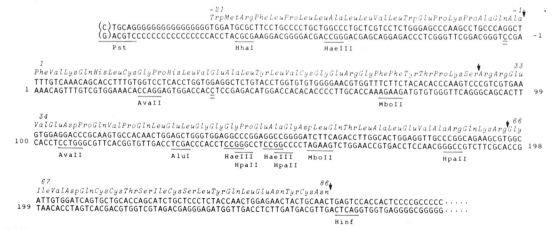

FIG. 2. DNA sequence of the insertion in clone pI19. Nucleotides are numbered using the convention described in Fig. 1. Accordingly, amino acids are numbered beginning with the first amino acid of proinsulin, while the last amino acid of the leader sequence (pre region) is numbered as −1. Restriction endonuclease cleavage sites experimentally verified are underlined and identified. The arrows indicate, in order, the ends of the leader sequence and the peptides B, C, and A. Two nucleotides indicated by double underlining are uncertain.

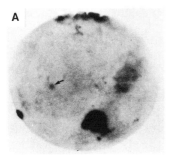 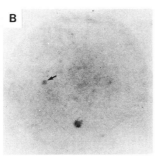

FIG. 3. Initial detection of penicillinase-insulin hybrid polypeptides in an insulin cDNA clone. Cells from colonies of the 48 insulin cDNA clones and from control colonies, χ1776 and χ1776-pBR322, were applied to an agarose/lysozyme/EDTA plate. Positive controls, 5 ng of insulin and 5 ng of penicillinase, each in 1 µl of wash buffer, also were spotted on plate. Antigen was adsorbed to an IgG-coated polyvinyl disk during a 1-hr incubation at 4°. Immobilized antigen was labeled by setting the plastic disk on a solution containing radioiodinated anti-insulin IgG. The autoradiographs are of disks precoated with anti-insulin IgG (A) or anti-pencillinase IgG (B), exposed on Kodak X-Omat R film using a Du Pont Cronex Lightning Plus intensifying screen for 12 hr at −70°. The arrows indicate the signal generated by clone pI47. The large exposed area in the lower right of (A) is the positive control for insulin detection.

antigen was recovered in the distilled water wash of the shock procedure. Table 2 shows that the insulin antigen in the wash is also detectable and quantifiable by a standard radioimmunoassay. The yield of antigen depended on the growth medium; antigen was released by cells grown in M9/glucose/amino acids medium but not by cells grown in brain/heart infusion. We estimate a recovery of about 100 molecules per cell.

Structure of the Fused Protein. We sequenced pI47 to determine the sequence around the junctions. Fig. 5 shows that a proinsulin I cDNA lies in the *Pst* site in the correct orientation and in phase, so that a fused protein can be synthesized. In pI19, the insert is in the correct orientation, but not in phase. In pI47 the oligo(dG)·oligo(dC) region encodes six glycines that connect the penicillinase sequence, ending at amino acid 182 (alanine), to the fourth amino acid (glutamine), of the proinsulin sequence. The cDNA sequence in pI47 extends 26 base pairs past the UGA terminator. Thus, we infer the structure of the fused protein to be penicillinase(24–182)-(Gly)$_6$-proinsulin(4–86).

DISCUSSION

The coding regions of eukaryotic structural genes are often interrupted by introns (35–38), whose transcripts are spliced out of the mature mRNA. Because prokaryotes do not appear to process their messengers, double–stranded cDNA made from a mature messenger is the material of choice to carry eukaryotic structural information into bacteria.

By using cDNA cloning technology and an extremely sensitive method to assay expression, we were able to construct a derivative of *E. coli* strain χ1776 carrying an insulin gene sequence and to detect the synthesis and secretion into the periplasmic space of a fused protein carrying antigenic determinants of both insulin and penicillinase. This was accomplished simply by inserting double-stranded cDNA carrying the structural information for insulin into a restriction site within the structural gene for penicillinase. Not only is the fused DNA sequence expressed as a chain of amino acids, but also the polypeptide folds so as to reveal insulin antigenic shapes. Thus we expect soon to be able to demonstrate biological function for this, or for a similar, fused protein.

We anticipate that the joining of cDNA sequences to nucleotides that lie ahead of the *Pst* site in the penicillinase gene

Table 2. Immunoreactive insulin concentration in distilled water wash of osmotic shock procedure

Exp.	Insulin, µunits/ml	Cells/ml
1	318	1.5×10^{10}
2	166	6.0×10^{9}
3	386	4.2×10^{10}

Duplicate 0.1-ml aliquots of each sample prepared as described in the legend to Fig. 4 were assayed (29) in a final volume of 0.4 ml using rat insulin standard, a gift from J. Schlichtkrull. One unit = 48 µg. The NaCl/Tris wash, the 20% sucrose wash, and the media of χ1776-pI47 as well as the water wash from osmotic shock of χ1776-pBR322 gave values below the sensitivity of the assay (25 µunits/ml).

will also produce fused and secreted molecules. Moreover, if the fusion replaces the preproinsulin leader with that of penicillinase it is likely that the new protein will also be secreted by the *E. coli* cell and may even be correctly matured by cleavage of the leader sequence.

Clearly, we have exploited a general method that should lead to the expression and secretion of any eukaryotic protein provided another protein, such as penicillinase, will serve as a carrier, by virtue of its leader sequence. Moreover, the secretion of the eukaryotic protein sequence to the periplasm or extracellular space will both permit its harvest in a purified form and probably eliminate intracellular sources of instability.

Often just an expression of antigens is the goal. In a "shotgun" screening, the existence of a fused protein antigen could be used to identify transformants carrying desired eukaryotic gene fragments. On the other hand, the insertion of a DNA fragment coding for surface antigenic determinants of a virus into a carrier protein should lead to the secretion of a fused protein that could serve as a vaccine, even though no entirely correct virus product is ever produced.

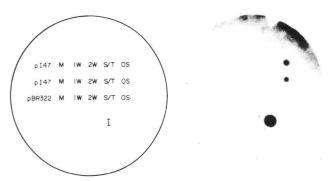

FIG. 4. Release of insulin antigen from χ1776-pI47 cells by osmotic shock. One liter of χ1776-pI47 cells growing at 37° in M9 medium supplemented with 1 g of tryptone, 0.5 g of yeast extract, and 0.5% glucose was harvested at a density of 5 × 10⁷ cells per ml and washed two times in 10 ml of cold 10 mM Tris·HCl, pH 8/30 mM NaCl. The cells were then osmotically shocked (34) in the following manner: The final wash pellet was resuspended in 10 ml of 20% sucrose per 30 mM Tris·HCl, pH 8, at room temperature, made 1 mM in EDTA, shaken at room temperature for 10 min, centrifuged out, resuspended in 10 ml of cold distilled water, shaken in an ice bath for 10 min, and again pelleted. The resulting supernatant was termed the "water wash." As a control, 1 liter of χ1776-pBR322 was grown and treated in a similar manner. Aliquots (1 µl) of each fraction to be assayed for the presence of insulin antigen were applied to the surface of a 1.5% agar plate. (A) Positions of each fraction on the plate. M, medium; 1W, first wash supernatant; 2W, second wash supernatant; S/T, sucrose/Tris supernatant; OS, distilled water wash; I, insulin. (B) Autoradiograph showing results of a two-site radioimmunoassay of these fractions. Antigen was adsorbed to a polyvinyl disk and labeled by using anti-insulin IgG. The labeled areas correspond to the water washes and the positive control (1 ng insulin). A spectrophotometric assay for β-galactosidase (23) indicated that no more than 4% of cells lyse during this procedure.

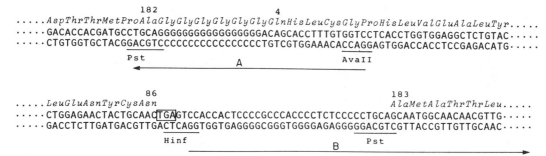

FIG. 5.   Partial DNA sequence of the insertion in clone pI47. Clone pI47 DNA was digested with *Hinf* and two fragments, H1 and H2 (ca. 1700 and 280 base pairs long, respectively) were isolated. H1 contains the amino-terminal portion of the penicillinase gene and the bulk of the cDNA insert. H1 was digested with *Ava* II, end labeled, and digested again with *Pst*. A fragment 39 nucleotides long (fragment A, arrow) was isolated and sequenced. Fragment H2 was end labeled and digested with *Alu* I (which cuts at the region corresponding to amino acid 200 of penicillinase). A fragment 88 base pairs long (fragment B, arrow) was isolated and sequenced. The termination sequence TGA is boxed.

We thank David Baltimore, Philip Sharp, and Salvador Luria for the use of the Massachusetts Institute of Technology P3 laboratory. We thank Macy Koehler for help with the figures; Fotis Kafatos for use of facilities; Philip Sharp, Al Bothwell, Shirley Tilghman, Doug Hanahan, and Richard Firtel for discussions. W.G. is an American Cancer Society Professor of Molecular Biology. W.L.C. is an Established Investigator of the American Diabetes Association. This work was supported by National Institutes of Health Grants AM 21240 and GM 09541-17 to W.G. and AM 15398 to W.L.C.

1. Chan, S. J., Keim, P. & Steiner, D. F. (1976) *Proc. Natl. Acad. Sci. USA* **73**, 1964–1968.
2. Chan, S. J. & Steiner, D. F. (1977) *Trends Biochem. Sci.* **2**, 254–256.
3. Blobel, G. & Dobberstein, B. (1975) *J. Cell Biol.* **67**, 835–851.
4. Blobel, G. & Dobberstein, B. (1975) *J. Cell Biol.* **67**, 852–862.
5. Steiner, D. F., Kemmler, W., Tager, H. S. & Peterson, J. D. (1974) *Fed. Proc. Fed. Am. Soc. Exp. Biol.* **33**, 2105–2115.
6. Smith, L. F. (1966) *Am. J. Med.* **40**, 662–666.
7. Clark, J. L. & Steiner, D. F. (1969) *Proc. Natl. Acad. Sci. USA* **62**, 278–285.
8. Markussen, J. & Sundby, F. (1972) *Eur. J. Biochem.* **25**, 153–162.
9. Ullrich, A., Shine, J., Chirgwin, J., Pictet, R., Tischer, E., Rutter, W. J. & Goodman, H. M. (1977) *Science* **196**, 1313–1319.
10. Chick, W. L., Warren, S., Chute, R. N., Like, A. A., Lauris, V. & Kitchen, K. C. (1977) *Proc. Natl. Acad. Sci. USA* **74**, 628–632.
11. Bolivar, F., Rodriguez, R. L., Greene, P. J., Betlach, M. C., Heyneker, H. L., Boyer, H. W., Crossa, J. H. & Falkow, S. (1977) *Gene* **2**, 95–113.
12. Sutcliffe, J. G. (1978) *Proc. Natl. Acad. Sci. USA* **75**, 3737–3741.
13. Ambler, R. P. & Scott, G. K. (1978) *Proc. Natl. Acad. Sci. USA* **75**, 3732–3736.
14. Boyer, H. W. & Rouland-Dussoix, D. (1969) *J. Mol. Biol.* **41**, 459–472.
15. Curtiss, R., III, Pereira, D. A., Hsu, J. C., Hull, S. C., Clarke, J. E., Maturin, L. J., Sr., Goldschmidt, R., Moody, R., Inoue, M. & Alexander, L. (1977) in *Recombinant Molecules: Impact on Science and Society. Proceedings of the 10th Miles International Symposium*, eds. Beers, R. F., Jr., & Bassett, E. G. (Raven, New York), pp. 45–56.
16. Clewell, D. B. (1972) *J. Bacteriol.* **110**, 667–676.
17. Palmiter, R. (1974) *Biochemistry* **13**, 3603–3615.
18. Aviv, H. & Leder, P. (1972) *Proc. Natl. Acad. Sci. USA* **69**, 1408–1412.
19. Efstratiadis, A., Kafatos, F. C., Maxam, A. M. & Maniatis, T. (1976) *Cell* **7**, 279–288.
20. Roychoudhury, R., Jay, E. & Wu, R. (1976) *Nucleic Acid Res.* **3**, 101–116.
21. Gilbert, W. & Maxam, A. M. (1973) *Proc. Natl. Acad. Sci. USA* **70**, 3581–3584.
22. Enea, V., Vovis, G. F. & Zinder, N. D. (1975) *J. Mol. Biol.* **96**, 495–509.
23. Miller, J. M. (1972) *Experiments in Molecular Genetics* (Cold Spring Harbor Laboratory, Cold Spring Harbor, New York), pp. 431–435.
24. Grunstein, M. & Hogness, D. S. (1975) *Proc. Natl. Acad. Sci. USA* **72**, 3961–3965.
25. Paterson, B. M., Roberts, B. E. & Kuff, E. L. (1977) *Proc. Natl. Acad. Sci. USA* **74**, 4370–4374.
26. Roberts, B. E. & Paterson, B. M. (1973) *Proc. Natl. Acad. Sci. USA* **70**, 2330–2334.
27. Lomedico, P. T. & Saunders, G. F. (1976) *Nucleic Acids Res.* **3**, 381–391.
28. Broome, S. & Gilbert, W. (1978) *Proc. Natl. Acad. Sci. USA* **75**, 2746–2749.
29. Makula, D. R., Vichnuk, D., Wright, P. H., Sussman, K. E. & Yu, P. L. (1969) *Diabetes* **18**, 660–689.
30. Maxam, A. M. & Gilbert, W. (1977) *Proc. Natl. Acad. Sci. USA* **74**, 560–564.
31. Boyer, H. W., Betlach, M., Bolivar, F., Rodriguez, R. L., Heyneker, H. L., Shine, J. & Goodman, H. M. (1977) in *Recombinant Molecules: Impact on Science and Society. Proceedings of the 10th Miles International Symposium*, eds. Beers, R. F., Jr. & Bassett, E. G. (Raven, New York), pp. 9–20.
32. Casey, J. & Davidson, N. (1977) *Nucleic Acids Res.* **4**, 1539–1552.
33. Humbel, R. E., Bosshard, H. R. & Zahn, H. (1972) in *Handbook of Physiology, Section 7 (Endocrinology)*, eds. Steiner, D. F. & Freinkel, N., (American Physiological Society, Washington, DC), Vol. 1, pp. 111–132.
34. Neu, H. C. & Heppel, L. A. (1965) *J. Biol. Chem.* **240**, 3685–3692.
35. Tilghman, S. M., Tiemeier, D. C., Seidman, J. G., Peterlin, B. M., Sullivan, M., Maizel, J. V. & Leder, P. (1978) *Proc. Natl. Acad. Sci. USA* **75**, 725–729.
36. Jeffreys, A. J. & Flavell, R. A. (1977) *Cell* **12**, 1097–1108.
37. Breathnach, R., Mandel, J. L. & Chambon, P. (1977) *Nature* **270**, 314–319.
38. Gilbert, W. (1978) *Nature* **271**, 501.

Supplementary Readings

Derynck, R., Content, J., De Clercq, E., Volckaert, G., Tavernier, J., Devos, R. & Fiers, W., Isolation and structure of a human fibroblast interferon gene. *Nature* 285:542–547 (1980)

Gray, P.W., Leung, D.W., Pennica, D., Yelverton, E., Najarian, R., Simonsen, C.C., Derynck, R., Sherwood, P.J., Wallace, D.M., Berger, S.L., Levinson, A.D. & Goeddel, D.V., Expression of human immune interferon cDNA in *E. coli* and monkey cells. *Nature* 295:503–508 (1982)

Pennica, D. Holmes, W.E., Kohr, W.J., Harkins, R.N., Vehar, G.A., Ward, C.A., Bennett, W.F., Yelverton, E., Seeburg, P.H., Heyneker, H.L., Goeddel, D.V. & Collen, D., Cloning and expression of human tissue-type plasminogen activator cDNA in *E. coli*. *Nature* 301:214–221 (1983)

Pennica, D., Nedwin, G.E., Hayflick, J.S., Seeburg, P.H., Derynck, R., Palladino, M.A., Kohr, W.J., Aggarwal, B.B. & Goeddel, D.V., Human tumour necrosis factor: precursor structure, expression and homology to lymphotoxin. *Nature* 312:724–729 (1984)

Rüther, U., Koenen, M., Sippel, A.E. & Müller-Hill, B., Exon-cloning: Immunoenzymatic identification of exons of the chicken lysozyme gene. *Proc. Natl. Acad. Sci. U.S.A.* 79:6852–6855 (1982)

Tuite, M.F., Dobson, M.J., Roberts, N.A., King, R.M., Burke, D.C., Kingsman, S.M. & Kingsman, A.J., Regulated high efficiency expression of human interferon-alpha in *Saccharomyces cerevisiae. EMBO J.* 1:603–608 (1982)

Vehar, G.A., Keyt, B., Eton, D., Rodriguez, H., O'Brien, D.P., Rotblat, F., Oppermann, H., Keck, R., Wood, W.I., Harkins, R.N., Tuddenham, E.G.D., Lawn, R.M. & Capon, D.J., Structure of human factor VIII. *Nature* 312:337–342 (1984)

Wood, W.I., Capon, D.J., Simonsen, C.C., Eaton, D.L., Gitschier, J., Keyt, B., Seeburg, P.H., Smith, D.H., Hollingshead, P., Wion, K.L., Delwart, E., Tuddenham, E.G.D., Vehar, G.A. & Lawn, R.M., Expression of active human factor VIII from recombinant DNA clones. *Nature* 312:330–337 (1984)

Microbial Engineering

The root of modern biotechnology lies in the use, over the centuries, of microbes to provide all manners of products for the benefit of mankind. In more recent times, the metabolic pathways of microbes have provided antibiotics, enzymes for food and industrial processes, and fine chemicals. These products were normal metabolites, often produced in quantities that were not sufficient for commercial use. The discovery of mutagenesis permitted the process of "strain improvement," an uncharacterized form of microbial alchemy which enabled the antibiotic industry to coax appropriate yields of therapeutically-active products and could provide the wherewithal to control infectious disease throughout the world; it also led to production of industrial enzymes on a large scale for commercial applications.

The advent of recombinant DNA technology provided a new dimension to the use of microbes for the production of a range of metabolites for the biotechnology industries, as illustrated in the papers presented in this section. In addition, the introduction of foreign genes into microbes and their expression to make heterologous proteins became a reality, permitting the use of microbes as "factories " for the making of high-value-added products such as insulin, interferons, cytokines, etc. In a similar fashion, the potential for the metabolic conversion of biomass (waste cellulose products) into fuels and other valuable chemicals was evoked with the expectation that gene cloning would allow the construction of engineered microbes capable of converting wood pulp or sugar stalks into ethanol. Unfortunately, in spite of significant scientific advances, the commercial realization of the latter, highly laudable goals has been retarded by insufficient political and economic pressure. Will we learn from the experience of the Gulf War?

The genetic manipulation of microbes for the purposes described above required the application of many years of knowledge of microbial genetics and biochemistry, coupled with the development of methods for introducing DNA and expressing heterologous genes under a variety of conditions (see the sections on cloning and expression) and also the means of obtaining the products in convenient form. As usual, there were surprises— no one expected that heterologous proteins would form inclusion bodies in the producing organism. The attempts at solving the latter problem has led us into new concepts of the way in which a messenger RNA sequence is converted into a fully functional protein inside or outside of the cell; there is more than one "code"! The steps of protein folding necessary during synthesis to obtain the native and active conformation have yet to be unravelled, as have the full details of the components and functions of secretion pathways.

The papers in this section are divided into topics, not necessarily presented in chronological order. The first group describes the initial demonstration of expression of a yeast protein gene (Struhl *et al.*) and a mammalian gene (Chang *et al.*) in *E. coli*; different methods are used and both demonstrate expression without consideration of amounts—that was to come later!

The work done by Ehrlich and Hinnen *et al.* provided the molecular tools for gene cloning in *Bacillus subtilis* and in *Sacchararomyces cerevisiae* (see supplemental readings). Both were thought likely to offer advantages over *E. coli* as safe and well-understood industrial micro-organisms, but years of experience in biotechnology companies have resulted in obtaining high level expression of a large number of different heterologous proteins in *E. coli*, and this organism is still the paradigm for the production of non-modified heterologous proteins on the large scale.

The papers of Talmadge *et al.* and Bitter *et al.* illustrate the first efforts to establish useful secretion systems for heterologous proteins in *E. coli* and yeast respectively. Even with relatively scant knowledge of the detailed process of secretion, these experiments showed that there was no inherent impediment to being heterologous, and the procedures have since been refined by extensive studies which have unravelled many of the steps involved to provide proteins for the pharmaceutical and food industries. Palva *et al.* were successful in devising heterologous protein secretion processes for *B. subtilis*, which has always been considered a prime candidate for an industrial microorganism. However this microbe is still in need of improvement if it is to be used for the secretion of appropriate quantities of properly processed heterologous molecules.

As mentioned above, the finding of insoluble granules of heterologous proteins in high-producing *E. coli* strains (see the paper by Schoner *et al.*) proved to be a stumbling block in many cases and posed intriguing problems for protein chemists who devised ways of solubilizing the desired proteins only to find that they were biologically inactive. However, these difficulties have been overcome in many instances. The review by Schein (see Supplementary Readings) provides an excellent summary of the problems of microbial protein inclusion bodies and ways of avoiding them, as described by someone with much experience.

Extension of the commercial applications of biotechnology demand, in addition, that we improve the natural metabolic and catabolic characteristics of microbes to suit industrial needs. For example, how to get more of the product? How to construct the correct enzyme combination? This has been achieved by classical genetics and by recombinant DNA methods, but most frequently by a combination of the two. Both the vector system (expression) and the host may require "improvement" to obtain maximum efficacity. The work of Ramos *et al.* illustrates these approaches en route to the goal of obtaining microbes capable of degrading toxic molecules in the environment, an aspect of the important field of bioremediation (see supplemental readings).

By contrast, the work of the Hopwood laboratory at John Innes has concentrated on studies of streptomycetes with the objective of understanding and manipulating the pathways of antibiotic biosynthesis. The Malpartida and Hopwood contribution demonstrated the clustering of biosynthesis genes by the successful cloning of a complete antibiotic biosynthetic pathway. This was the forerunner of the work described in the supplemental reading by Hopwood *et al.* which permitted the mixing of antibiotic biosynthesis genes to produce hybrid molecules. The potential of this approach is high and a number of hybrid or modified antibiotics have been obtained in the search for more active molecules.

Finally, the engineering of micro-organisms for the efficient utilisation of cellulose biomass requires the identification and characterization of the multi-enzyme involved in degrading such complex macromolecules. The work of Shoemaker *et al.* and Joliff *et al.* (see supplemental readings) provide early demonstrations of cloning and expression of genes for cellulose degradation in *E. coli* and the crystallization of an enzyme obtained from this heterologous host. It is believed that a detailed three-dimensional structure of this enzyme will guide future efforts at protein engineering of cellulases so that they will perform more efficiently under industrial fermentation conditions. The ultimate goal is the complete conversion of a given biomass source into a product such as ethanol.

Secretion of Foreign Proteins from *Saccharomyces Cerevisiae* Directed by α-factor Gene Fusions

G.A. Bitter, K.K. Chen, A.R. Banks and P.-H. Lai

ABSTRACT Fusions between the cloned yeast α-factor structural gene and chemically synthesized DNA segments encoding human protein analogs have been constructed. The gene fusions encode hybrid proteins that include the first 89 amino acids of the native α-factor precursor fused to either a small (β-endorphin, 31 amino acids) or large (α-interferon, 166 amino acids) foreign protein. Proteolytic cleavage sites involved in α-factor maturation from the native precursor immediately precede the foreign peptide in the hybrid protein. The α-factor promoter was utilized to express the gene fusions in *Saccharomyces cerevisiae* and resulted in the efficient secretion of the foreign proteins into the culture medium. The processing of the hybrid proteins has been characterized by amino acid sequence analysis of the secreted proteins. The proteolytic cleavages involved in the maturation of α-factor peptides from the native precursor also occur accurately in the hybrid protein. In addition, cleavages occurred on the carboxyl side of two lysines within the β-endorphin peptide. Internal cleavages in the interferon protein were also detected. However, in this case, the cleavages occurred at a very low frequency such that >95% of the secreted interferon remained intact.

The yeast mating pheromone α-factor is a 13-amino acid peptide that is secreted into the culture medium by *MAT*α cells. α-Factor arrests *MAT*a cells in G_1 phase and induces specific biochemical and morphological changes as a prerequisite to mating of the a cell to an α haploid. On a molar basis, α-factor is produced as efficiently as the highly expressed yeast glycolytic enzymes (1). It seems likely, therefore, that the components responsible for the efficient secretion of α-factor might be utilized to direct secretion of foreign proteins into the culture medium.

Kurjan and Herskowitz (2) cloned the *Saccharomyces cerevisiae* α-factor structural gene by using a bioassay to screen for yeast clones that overproduce the pheromone due to the presence of the α-factor gene on a multicopy plasmid. From the DNA sequence of the cloned gene these authors deduced that the 13-amino acid α-factor peptide is synthesized as a 165-amino acid prepropolyprotein precursor containing four copies of the α-factor peptide. A probable processing pathway in α-factor maturation was proposed (2), which has, in large part, been experimentally substantiated (3–5). The precursor contains a hydrophobic amino-terminal 22-residue segment that presumably initiates translocation into the endoplasmic reticulum. The function of the next 61 amino acids (pro-segment) is not known, but it does contain three glycosylation sites and may be involved in directing the precursor into the correct secretory pathway. The reiterated α-factor peptides are separated by spacer peptides and are cleaved from the precursor and processed by three enzymatic activities: trypsin-like or cathepsin B-like cleavage at Lys-Arg; carboxypeptidase B-like cleavage of basic residues

from the excised peptides; and removal of the Glu-Ala or Asp-Ala dipeptides by dipeptidyl aminopeptidase A.

Subsequently, Singh *et al.* (6) reported cloning two α-factor structural genes. One gene (MFα1) appears to be the same as the original Kurjan and Herskowitz isolate, whereas the other clone (MFα2) encodes a precursor containing only two α-factor peptides. We used a synthetic oligonucleotide as a hybridization probe to clone a segment of the yeast genome with complementarity to the α-factor peptide coding region (7). The DNA sequence of our clone, at least in the coding region, is identical to the MFα1 gene of Kurjan and Herskowitz (data not shown). We have constructed fusions between the α-factor structural gene and chemically synthesized DNA segments encoding foreign proteins. Utilizing the endogenous α-factor promoter on the cloned segment to express the gene fusions in *S. cerevisiae*, we have obtained efficient secretion into the culture medium of a number of foreign proteins. Recently, Emr *et al.* (8) reported an α-factor gene fusion that directed secretion of yeast invertase to the periplasmic space. In the present report, we describe the processing events associated with secretion into the medium of β-endorphin and an analog of human α-interferon, IFN-αCon$_1$.

MATERIALS AND METHODS

Vector Construction, Cell Transformation, and Culture Conditions. Plasmid pαF consists of a 2.1-kilobase yeast *Eco*RI DNA fragment containing an α-factor structural gene cloned in pBR322 (7). A 1.7-kilobase *Xba* I–*Eco*RI fragment was subcloned by using *Bam*HI linkers into pBRΔH (pBR322 derivative in which *Hind*III site has been deleted) or pBRΔHS (pBR322 derivative in which *Hind*III and *Sal* I sites have been deleted; constructed by D. Hare) to generate pαC1 and pαC2, respectively, which were used for construction of the gene fusions. The *Xba* I site is 930 base pairs (bp) 5′ to the ATG translation initiation codon and the *Eco*RI site is 330 bp 3′ to the termination codon of the α-factor structural gene. The β-endorphin gene was cloned between the *Hind*III sites at the alanine codon of the first and fourth spacer peptide. The IFN-αCon$_1$ gene was cloned between the *Hind*III site of the first spacer peptide and the *Sal* I site 35 bp beyond the termination codon of the α-factor structural gene. Details of vector construction and DNA sequences of the synthetic genes will be published elsewhere.

After construction of the gene fusion and confirmation of the DNA sequence, the hybrid gene was cloned as a *Bam*HI fragment into a yeast–*Escherichia coli* shuttle vector. pGT41 contains a *LEU2* selectable marker and has been described (9). pYE was constructed from pGT41 by replacing the *Bam*HI–*Sal* I 2.5-kilobase *LEU2* gene segment with the 275-bp *Bam*HI–*Sal* I segment of pBR322 and subsequently cloning the yeast *TRP1* gene (*Eco*RI–*Bgl* II 852-bp fragment; ref. 10) into the *Sal* I site by blunt-end ligation. pYα-E and pYE/α-E contain an α-factor/β-endorphin gene fusion

Abbreviation: bp, base pair(s).

BITTER, G.A., CHEN, K.K., BANKS, A.R. and LAI, P.-H.
Secretion of foreign proteins from *Saccharomyces cerevisiae* directed by α-factor gene fusions.
Proc. Natl. Acad. Sci. U.S.A. 81:5330-5334 (1984). Reprinted with the authors' permission.

cloned into pGT41 and pYE, respectively. pYE/αF-Con₁ contains an α-factor/IFN-αCon₁ gene fusion cloned into pYE. *S. cerevisiae* was transformed according to the method of Hinnen *et al.* (11). Transformants of strain GM3C-2 (*MATα leu2-3 leu2-112 trp1-1 his4-519 cyc1-1 cyp3-1*) were selected by leucine prototrophy in SD medium (0.67% yeast nitrogen base without amino acids/2% glucose) supplemented with 0.01% histidine and 0.01% tryptophan, whereas 20B-12 (*MATα trp1 pep4-3*) transformants were selected in SD medium containing 0.5% Casamino acids.

DNA Sequencing. DNA sequence determinations were by the method of Maxam and Gilbert (12) or by the dideoxy chain-termination method (13) on fragments subcloned into M13.

Chemical Synthesis of DNA. Oligonucleotides were synthesized essentially according to the chemistry outlined by Tanaka and Letsinger (14). The oligonucleotide segments of the β-endorphin gene were designed to allow only one ligation pathway. All purified oligonucleotides except the 5′-end segments were phosphorylated, mixed in equimolar ratios, and ligated and the full-length synthetic β-endorphin gene was purified by polyacrylamide gel electrophoresis. The 5′ termini of the synthetic gene were phosphorylated prior to cloning into M13mp9 as a *Hin*dIII–*Bam*HI segment. After confirming the DNA sequence of the synthetic gene, which utilized optimal yeast codons (ref. 15; unpublished data), it was excised from the phage RF as a *Hin*dIII fragment for subsequent vector construction. The chemical synthesis and cloning of the IFN-αCon₁ gene has been described (ref. 16; unpublished data).

Amino Acid Sequence Determinations. Automated sequence analyses (17, 18) of peptide fragments isolated by HPLC were performed with a gas-phase sequenator (Applied Biosystems, Foster City, CA). Phenylthiohydantoin-amino acids obtained from each sequenator cycle were identified by reverse-phase HPLC (19).

RESULTS

We constructed gene fusions between the α-factor leader region and two chemically synthesized genes. The gene fusions encode a hybrid protein (Fig. 1A), which includes the first 89 amino acids of the α-factor precursor: signal peptide, pro-segment containing three glycosylation sites, and the first spacer peptide. The first amino acid of the foreign protein occupies the same position in the hybrid protein as does the first amino acid (tryptophan) of the first α-factor peptide in the native precursor. One fusion contains a human β-endorphin analog (Leu5) as the last 31 amino acids of the hybrid. The second gene fusion encodes a hybrid of the α-factor leader and a 166-amino acid α-interferon analog (IFN-αCon₁). IFN-αCon₁ is an average amino acid sequence of all the known naturally occurring leukocyte interferons and expression of this gene in *E. coli* results in a protein with a 10-fold higher antiviral activity than any of the known naturally occurring α-interferons (16). The gene fusions include the 930 bp of yeast DNA upstream from the ATG translation initiation codon of the α-factor structural gene (*Materials and Methods*). Thus, we expected this fragment to include all sequences necessary for α-factor promoter function *in vivo* in addition to including 300 bp of 3′-untranslated yeast DNA.

The gene fusions were cloned into various yeast–*E. coli* shuttle vectors (Fig. 1B) and reintroduced into *S. cerevisiae*. Transformants were cultured in liquid medium, the cells were removed by centrifugation, and the conditioned medium was tested for the presence of the foreign protein. β-Endorphin was quantitated by competitive RIA and IFN-αCon₁ antiviral activity was determined by an end-point cytopathic effect assay as described (16). Strain GM3C-2 containing

A

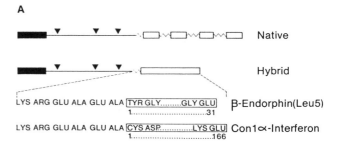

B

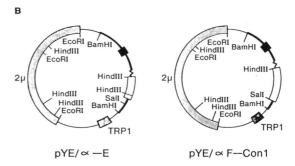

pYE/α —E   pYE/α F—Con1

Fig. 1. (*A*) Native and hybrid precursor proteins. The α-factor precursor signal peptide (closed box), pro-segment (straight line), *N*-linked glycosylation sites (inverted triangles), and spacer peptides (squiggle) are indicated. The α-factor peptides and foreign protein segments are represented by open boxes. The amino acid sequence at the junction between the α-factor and foreign protein segments of the precursors is detailed. (*B*) Vectors for expressing α-factor gene fusions in yeast. pBR322 DNA is represented by the thin line and coding regions of the gene fusions are indicated as above. pYα-E is identical to pYE/α-E, with the exception that the yeast *LEU2* gene is incorporated instead of the *TRP1* gene. The α-factor/β-endorphin gene fusion includes the coding region for the last α-factor peptide, but this is not expressed since a termination codon was included in the synthetic gene.

pYα-E and cultured to an OD₆₀₀ of 1 secreted 150–200 μg of β-endorphin immunoreactive material into the medium per liter of culture. No β-endorphin was detected in conditioned medium from cells containing a gene fusion in which the β-endorphin segment was in the incorrect orientation relative to the α-factor gene. Strain 20B-12 transformed with pYE/αF-Con₁ secreted 2 × 10⁸ units of IFN-αCon₁ into the medium per liter of culture at an OD₆₀₀ of 1. These results demonstrate that the expression system results in the efficient secretion of both small (31 amino acids) and large (166 amino acids) proteins into the culture medium.

The hybrid proteins encoded by pYα-E and pYE/αF-Con₁ contain proteolytic processing sites involved in the maturation of α-factor peptides from the native precursor (Fig. 1A). The gene fusions in our constructs were designed with the anticipation that these sites might also be recognized in the hybrid proteins and thus generate secreted proteins with the native amino termini. We have characterized the processing of the hybrid protein by determining the complete amino acid sequence of the secreted β-endorphin immunoreactive peptides. When conditioned medium is subjected to HPLC, three β-endorphin immunoreactive species are resolved (Fig. 2). Each peptide was sequenced by the automated Edman degradation procedure and the results are included in Table 1. Peak I is the carboxyl 12-amino acid fragment of β-endorphin that was generated by cleavage on the carboxyl side of lysine at position 19 of the β-endorphin peptide. Peak II is the amino-terminal 19-amino acid fragment of β-endorphin and thus represents the cleavage complement of peak I. Since peak II has the authentic β-endorphin amino-terminal sequence, this result demonstrates that the proteolytic cleav-

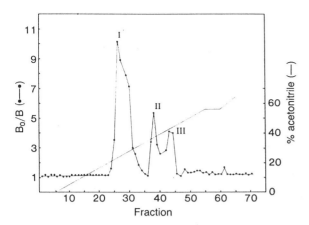

FIG. 2. HPLC fractionation of secreted β-endorphin immunoreactive peptides. *S. cerevisiae* GM3C-2 transformed with pYα-E was cultured to an OD_{600} of 1 and the cells were removed by centrifugation. The conditioned medium was made 1 mM phenylmethylsulfonyl fluoride and concentrated 20-fold by ultrafiltration through an Amicon YM2 membrane. The concentrated medium was adjusted to pH 2.1 and applied to a C_{18} column (0.39 × 30 cm, 10-μm particle size, Waters Associates) and a gradient was developed from 0–56% acetonitrile in 0.1% trifluoroacetic acid. Individual 1-min fractions were assayed by competitive RIA (New England Nuclear) and the results are presented as B_0 (cpm in reaction lacking unlabeled β-endorphin) divided by B (cpm in reaction containing sample). No β-endorphin immunoreactive material was detected in the column flow-through.

ages involved in α-factor biogenesis also occur accurately in the α-factor/β-endorphin hybrid protein. Peak III of Fig. 2, consists of the peptide IIIc, which has also been accurately processed on the amino terminus but cleaved on the carboxyl side of lysine at position 9 in the β-endorphin peptide. Other peptides resulting from cleavage within the β-endorphin peptide (e.g., the fragment corresponding to residues 10–19) would escape detection in our analysis if they are not reactive in the RIA.

The proteolytic cleavages within the β-endorphin peptide do not correspond to cleavage sites in the native α-factor precursor. These cleavages could result from exposure of the hybrid precursor to vacuolar proteases during the secretory process. Alternatively, the cleavages could occur by proteases in the extracellular medium, perhaps released by cell lysis. To test this possibility, we subcloned the α-factor/β-endorphin gene fusion into another shuttle vector and transformed the new plasmid (pYE/α-E) into *S. cerevisiae*

20B-12. This strain lacks 95% of the vacuolar protease activity of wild-type strains (20) due to the presence of the *pep4-3* mutation. For as yet uncharacterized reasons, these transformants secrete more β-endorphin immunoreactive material (up to 450 μg/liter of culture at an OD_{600} of 1). However, when conditioned medium from cultures of these cells was fractionated by HPLC, the three β-endorphin immunoreactive peaks noted above were observed in the same ratios (data not shown). These results suggest that the cleavages within the β-endorphin peptide occur during passage of the hybrid precursor through the normal secretory pathway. HPLC peak III from strain 20B-12 transformed with pYE/α-E was also subjected to amino acid sequence analysis. In this case, multiple phenylthiohydantoin-amino acids were identified in each sequencing cycle. Thus, it did not consist of a single peptide, despite the fact it chromatographed as a single immunoreactive peak. From the sequencing data, we deduced that this peak contained, in addition to peptide IIIc (Table 1), peptides IIIb and IIIa, which contain Glu-Ala and Glu-Ala-Glu-Ala extensions, respectively, on the amino terminus of peptide IIIc.

The processing of the α-factor/IFN-αCon$_1$ hybrid precursor was also investigated. Strain 20B-12 transformed with pYE/αF-Con$_1$ was cultured in liquid medium, the cells were removed by centrifugation, and the conditioned medium was concentrated by ultrafiltration. Denaturing polyacrylamide gel electrophoresis (NaDodSO$_4$/PAGE) of the secreted proteins (Fig. 3A) reveals a 20,000-dalton polypeptide (lane 2) that is not present in medium proteins from a negative control strain (lane 3). The IFN-αCon$_1$ secreted from yeast migrates somewhat slower than IFN-αCon$_1$ purified from *E. coli* (lane 1). From the sequencing results of the secreted β-endorphin immunoreactive peptides (Table 1), we infer that the retarded electrophoretic migration of IFN-αCon$_1$ may be due to incomplete processing of the amino-terminal Glu-Ala dipeptides. Fig. 3B depicts an electrophoretic transfer immunoblot of the same protein samples electrophoresed in the gel shown in Fig. 3A. There is some nonspecific background, as evidenced by reaction with proteins from the negative control yeast strain (lane 3). However, IFN-αCon$_1$ specific reactivity is clear (lanes 1 and 2). In addition to the intact 20,000-dalton polypeptide, secreted yeast proteins include subfragments of IFN-αCon$_1$. A major subfragment of 14,500 daltons and minor species with masses of 13,500, 12,500, and 11,000 daltons are observed. Thus, internal cleavages also occur in IFN-αCon$_1$ during α-factor-directed secretion from yeast. The autoradiograph of the immunoblot was purposely overexposed to detect low-abundance subfragments. However, inspection of the stained gel (Fig. 3A) reveals that these

Table 1. Amino acid sequence of secreted β-endorphin immunoreactive peptides

Peptide	Amino acid sequence
IIIa	Glu-Ala-Glu-Ala-Tyr-Gly-Gly-Phe-Leu-Thr-Ser-Glu-Lys
IIIb	Glu-Ala-Tyr-Gly-Gly-Phe-Leu-Thr-Ser-Glu-Lys
IIIc	Tyr-Gly-Gly-Phe-Leu-Thr-Ser-Glu-Lys 1 9
II	Tyr-Gly-Gly-Phe-Leu-Thr-Ser-Glu-Lys-Ser-Gln-Thr-Pro-Leu-Val-Thr-Leu-Phe-Lys 19
I	Asn-Ala-Ile-Ile-Lys-Asn-Ala-Tyr-Lys-Lys-Gly-Glu 20 31

Purified β-endorphin immunoreactive peptides (*Results*) were subjected to automatic amino acid sequence determination. Numbers refer to residue positions in authentic β-endorphin.

cleavages occur at a low frequency such that >95% of the secreted IFN-αCon₁ remains intact.

DISCUSSION

We have demonstrated that the α-factor leader region directs secretion of foreign proteins from *S. cerevisiae*. Fusions have been constructed between the yeast α-factor leader region (signal peptide, pro-segment, and first spacer peptide) and chemically synthesized DNA segments encoding human protein analogs. Utilizing the α-factor promoter to express the gene fusions results in the efficient secretion of both small (e.g., β-endorphin, 31 amino acids) and large (e.g., IFN-αCon₁, 166 amino acids) foreign proteins into the culture medium. It should be noted that we have not yet determined the intracellular levels of the foreign polypeptides nor the amount secreted but associated with the cell wall. The present report documents the processing events associated with secretion of the foreign protein segment of the hybrid precursors into the culture medium.

All secreted β-endorphin immunoreactive peptides contain only β-endorphin sequences, with the exception of peptides IIIa and IIIb, which contain amino-terminal Glu-Ala-Glu-Ala and Glu-Ala extensions, respectively, derived from

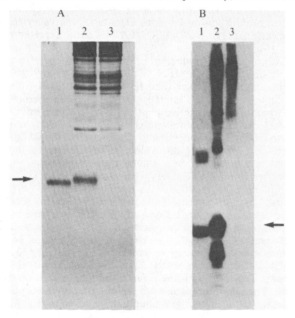

FIG. 3. NaDodSO₄/PAGE analysis of secreted yeast proteins. *S. cerevisiae* 20B-12 transformed with either pYE/αF-Con₁ or pYE/αF-calc was cultured to an OD₆₀₀ of 2 and the cells were removed by centrifugation. The strain containing pYE/αF-calc secretes human calcitonin (M_r 3500) directed by an α-factor gene fusion and will be described elsewhere. The conditioned medium was made 1 mM phenylmethylsulfonyl fluoride and concentrated 50-fold by ultrafiltration through an Amicon YM10 membrane (M_r cutoff, 10,000). Lanes 1, IFN-αCon₁ purified from *E. coli*; lanes 2, pYE/αF-Con₁ conditioned medium; lanes 3, pYE/αF-calc conditioned medium. The 20,000-dalton secreted IFN-αCon₁ is indicated by the arrow. Molecular sizes were determined by comparison to the migration of standards (ovalbumin, M_r 43,000; α-chymotrypsinogen, M_r 25,700; β-lactoglobulin, M_r 18,400; lysozyme, M_r 14,300; cytochrome *c*, M_r 12,300; bovine trypsin inhibitor, M_r 6200). (*A*) A 15% polyacrylamide gel was run as per Laemmli (21) and stained with Coomassie brilliant blue R250. (*B*) Proteins were electrophoresed in a 10–20% polyacrylamide gradient gel and transferred to nitrocellulose, and the immobilized proteins were incubated with rabbit antiserum (23) raised against IFN-αCon₁ produced in *E. coli*. Immune complexes were allowed to react with radioiodinated staphylococcal A protein and the nitrocellulose was exposed to Kodak XAR-2 film.

the α-factor portion of the hybrid protein precursor. These results indicate that processing of the precursor at the trypsin-like Lys-Arg cleavage site is not a rate-limiting step in processing, even when the hybrid protein is overproduced from multicopy vectors. Alternatively, cleavage of the precursor at Lys-Arg may be a prerequisite of secretion of carboxyl peptide segments. Since we have examined only peptides secreted through the cell wall into the extracellular medium, there may, in actual fact, be an accumulation of the immediate polypeptide precursor to the trypsin-like cleavage within the cell. Whether the cleavage at Lys-Arg is absolutely required for secretion may be determined by expressing precursors in which the cleavage site is altered or deleted. The amino-terminal Glu-Ala-Glu-Ala and Glu-Ala extensions on peptides IIIa and IIIb, respectively, indicate that production of the precursor from multicopy vectors results in incomplete processing by dipeptidyl aminopeptidase A. These results are consistent with those of Julius *et al.* (3), which indicated that this protease is rate-limiting in the processing of α-factor peptides from the native precursor. In our analysis, we have determined the molar ratio of peptides IIIa/IIIb/IIIc to be 2.4:1.0:1.5 based on phenylthiohydantoin-amino acid recovery during sequencing. The accumulation of peptide IIIb with only one Glu-Ala dipeptide indicates that dipeptidyl aminopeptidase A is not a processive enzyme. It should be possible to secrete peptides with homogeneous, completely processed amino termini by overproducing dipeptidyl aminopeptidase A, a product of the cloned *STE13* gene (3). Alternatively, hybrid genes may be constructed that encode precursors containing the Lys-Arg cleavage site but lacking the Glu-Ala codons. The processing of such precursors would thus be independent of the rate-limiting dipeptidyl aminopeptidase A.

In addition to the processing events characteristic of the native α-factor precursor, two additional cleavage sites were mapped in the α-factor/β-endorphin hybrid precursor. These cleavages occurred on the carboxyl side of lysines in the β-endorphin portion of the protein at Glu-Lys-Ser [8,9,10] and Phe-Lys-Asn [18,19,20]. Peptides arising from cleavage at the lysines in position 24, 28, or 29 of the β-endorphin peptide were not observed. The cleavage at position 19 occurred in all peptides since no secreted peptides were observed that spanned this region. In contrast, the cleavage at Lys-9 of the β-endorphin peptide occurred in ≈50% of the secreted peptides. It is not clear whether these cleavages are effected by the same protease that cleaves at Lys-Arg in the native precursor or by another protease to which the hybrid precursor is exposed. The mature α-factor peptide contains a lysine (Leu-Lys-Pro [6,7,8]) that has not been observed to be cleaved during maturation. However, it should be noted that Lys-Pro bonds in peptides are not cleaved by trypsin. The fact that different lysine-containing sequences in the β-endorphin peptide were cleaved quantitatively, partially or not at all, indicates that these internal cleavages may be dependent on the conformation of the β-endorphin peptide. Consistent with this hypothesis is the observation that the most susceptible bond in native β-endorphin to mild trypsin digestion is Lys-Asn [19,20] (22) and this bond was also the most susceptible to proteolysis during secretion from yeast.

The processing of the α-factor/IFN-αCon₁ hybrid protein was characterized by NaDodSO₄/PAGE. The electrophoretic mobility of the secreted IFN-αCon₁ suggests that this product may also contain amino-terminal Glu-Ala extensions. This possibility, which is consistent with the results of secreted β-endorphin, must be tested by direct amino acid sequencing. Although internal cleavages could be detected by electrophoretic transfer immunoblot analysis, >95% of the secreted IFN-αCon$_1$ protein remained intact. The widely disparate degrees of internal cleavage, both between different lysines within β-endorphin and between β-endorphin and IFN-αCon$_1$, indicate that the extent to which internal cleavages occur will be determined by the conformation-dependent accessibility of susceptible bonds in the foreign protein.

These findings also have implications for the secretion of native yeast proteins. Thus, for proteins that follow the same secretory pathway as the α-factor precursor, it is likely that susceptible bonds (involving basic amino acids) are protected from proteases to which the precursor is exposed. This may be effected via conformation-determined inaccessibility or protection of the susceptible bond by glycosylation. In either case, this hypothesis implies a co-evolution of the secretory apparatus and secreted protein structure with selection for efficient processing and secretion of biologically active (intact) proteins.

We thank Dr. K. Alton for the cloned INF-αCon$_1$ gene, Dr. L. Goldstein for providing purified IFN-αCon$_1$, Dr. B. Altrock and H. Hockman for performing the electrophoretic transfer blots, and Cheryl Bradley for performing interferon bioassays.

1. Thorner, J. (1981) in *The Molecular Biology of the Yeast Saccharomyces: Life Cycle and Inheritance*, eds. Strathern, J. N., Jones, E. W. & Broach, J. R. (Cold Spring Harbor Laboratory, Cold Spring Harbor, NY), Vol. 1, pp. 143–180.
2. Kurjan, J. & Herskowitz, I. (1982) *Cell* **30**, 933–943.
3. Julius, D., Blair, L., Brake, A., Sprague, G. & Thorner, J. (1983) *Cell* **32**, 839–852.
4. Brake, A. J., Julius, D. J. & Thorner, J. (1983) *Mol. Cell. Biol.* **3**, 1440–1450.
5. Emer, O., Meihler, B., Achstetten, T., Mullen, H. & Wolf, D. H. (1983) *Biochem. Biophys. Res. Commun.* **116**, 822–829.
6. Singh, A., Chen, E. Y., Lugovoy, J. M., Chang, C. N., Hitzeman, R. A. & Seeburg, P. H. (1983) *Nucleic Acids Res.* **11**, 4049–4063.
7. Bitter, G. A. & Chen, K. K. (1983) *J. Cell. Biochem.* Suppl. B, **7**, 1496.
8. Emr, S. D., Schekman, R., Flessel, M. & Thorner, J. (1983) *Proc. Natl. Acad. Sci. USA* **80**, 7080–7084.
9. Tschumper, G. & Carbon, J. (1983) *Gene* **23**, 221–232.
10. Tschumper, G. & Carbon, J. (1980) *Gene* **10**, 157–166.
11. Hinnen, A., Hicks, J. B. & Fink, G. R. (1978) *Proc. Natl. Acad. Sci. USA* **75**, 1929–1933.
12. Maxam, A. M. & Gilbert, W. (1980) *Methods Enzymol.* **65**, 499–560.
13. Sanger, F., Nicklen, S. & Coulson, A. R. (1977) *Proc. Natl. Acad. Sci. USA* **74**, 5463–5467.
14. Tanaka, T. & Letsinger, R. L. (1982) *Nucleic Acids Res.* **10**, 3249–3260.
15. Bennetzen, J. L. & Hall, B. D. (1982) *J. Biol. Chem.* **257**, 3026–3031.
16. Alton, K., Stabinsky, Y., Richards, R., Ferguson, B., Goldstein, L., Altrock, B., Miller, L. & Stebbing, N. (1983) in *The Biology of the Interferon System*, eds. de Maeyer, E. & Schellekens, H. (Elsevier, Amsterdam), pp. 119–128.
17. Edman, P. & Begg, G. (1967) *Eur. J. Biochem.* **1**, 80–91.
18. Hewick, R. M., Hunkapiller, M. W., Hood, L. E. & Dreyer, W. J. (1981) *J. Biol. Chem.* **256**, 7990–7997.
19. Hunkapiller, M. W. & Hood, L. E. (1983) *Science* **219**, 650–659.
20. Jones, E. (1976) *Genetics* **85**, 23–30.
21. Laemmli, U. K. (1970) *Nature (London)* **227**, 680–685.
22. Austen, B. M. & Smyth, D. G. (1977) *Biochem. Biophys. Res. Commun.* **77**, 86–94.
23. Burnette, W. N. (1981) *Anal. Biochem.* **112**, 195–203.

Phenotypic Expression in *E. Coli* of a DNA Sequence Coding for Mouse Dihydrofolate Reductase

A.C.Y. Chang, J.N. Nunberg, R.J. Kaufman, H.A. Erlich, R.T. Schimke and S.N. Cohen

The construction and analysis of bacterial plasmids that contain and phenotypically express a mammalian genetic sequence are described. Such plasmids specify a protein that has enzymatic properties, immunological reactivity and molecular size characteristic of the mouse dihydrofolate reductase, and render host cells resistant to the antimetabolic drug trimethoprim.

SINCE the initial propagation of eukaryotic DNA in bacteria[1], several systems have been used to study the expression in *Escherichia coli* of DNA derived from higher organisms. Biological activity of genes from the lower eukaryotes, *Saccharomyces cerevisiae*[2,3] and *Neurospora crassa*[4], has been demonstrated using phenotypic selection for functions that complement mutationally inactivated homologous bacterial genes. Immunological reactivity with antibody made against human somatostatin was shown for a peptide fragment cleaved *in vitro* from a hybrid protein encoded in part by bacterial DNA and in part by a chemically synthesised somatostatin DNA sequence[5]. Very recently, a protein containing amino acids of rat proinsulin was shown to be made by bacteria that carry a double-stranded complementary DNA (cDNA) transcript of pre-proinsulin mRNA[6]; in that instance, antigenic determinants for both insulin and the bacterial enzyme β-lactamase were detected on a fused peptide transported outside the cell. It is not known, however, whether the mammalian peptide components of such immunologically reactive hybrid proteins have functional biological activity.

Our approach to the study of mammalian gene expression in bacteria has been to generate a heterogeneous population of clones carrying a DNA sequence that codes for a selectable mammalian gene product, and then to select directly those bacteria in the population that phenotypically express the genetic sequence. The mammalian enzyme dihydrofolate reductase (DHFR), which catalyses the conversion of dihydrofolic acid to tetrahydrofolic acid, is especially suitable for this purpose. The mammalian DHFR has a much lower affinity for the antimetabolic drug, trimethoprim (Tp), than does the corresponding bacterial enzyme[7]. Thus, bacteria which biologically express mammalian DHFR activity are resistant to levels of trimethoprim that ordinarily inhibit growth.

When these studies were initiated, the only bacterial host approved for EK2 recombinant DNA experiments[8] was *E. coli* K12 strain χ1776 (ref. 9). As this strain is already resistant to high concentrations of trimethoprim because of its *thy*[-] mutation[10], direct selection of bacteria that synthesise the mammalian DHFR could not be carried out. Therefore, our initial studies used an *in situ* hybridisation procedure[11] to identify χ1776 clones which carried a mouse DHFR cDNA sequence,

¶ To whom correspondence should be addressed at Room S337, Stanford University Medical Center, Stanford, California 94305.

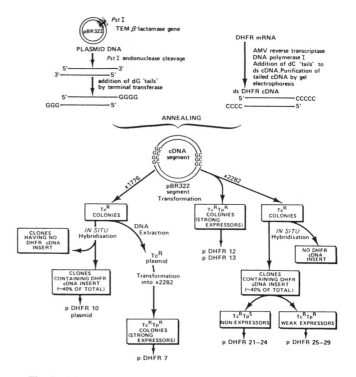

Fig. 1 Scheme used for cloning and expression of mouse DNA sequences that code for DHFR. Additional experimental details are given in the legend to Table 1 and in the text.

with the expectation that a highly sensitive indirect radioimmunoassay[12] would be used to detect any clones that expressed antigenic determinants of the protein product encoded by the gene.

Construction and cloning of chimaeric plasmids containing a DHFR ds cDNA

Figure 1 summarises the experimental scheme used in this study. Partially purified mRNA containing DHFR sequences from methotrexate-resistant AT-3000 mouse cells[13] was used in the preparation of double-strand (ds) cDNA by RNA-dependent DNA polymerase (reverse transcriptase) and DNA polymerase I (ref. 14 and Fig. 2). Homopolymeric deoxy-C 'tails' were added to the unfractionated cDNA by terminal deoxynucleotidyl transferase (Fig. 2) and the section of the gel containing the predominant (1,500 base pairs) tailed cDNA was eluted. The material recovered from the gel was annealed with an equimolar concentration of pBR322 plasmid DNA that had been cleaved in the β-lactamase gene by the *Pst*I endonuclease[15] and treated with terminal transferase to add homopolymeric dG tails at the cleavage sites. The extent of annealing of DHFR ds cDNA with the plasmid vector was monitored by electron microscopy[16] using circle formation as an indicator. *Pst*I sites are regenerated at both ends of the insert as a result of such re-circularisation[17,18].

Constructed chimaeric plasmids were introduced into $MnCl_2$-treated[19] *E. coli* K12 strain χ1776 and tetracycline (Tc)-resistant transformants (yield ~30 colonies per ng DNA) were selected and tested separately for the presence of a DNA species complementary to a highly purified DHFR cDNA probe. About 40% of the Tc-resistant clones gave a positive reaction by an *in situ* hybridisation test[11] (Table 1 and Fig. 3). Plasmids from 14 of the reacting clones were examined by gel electrophoresis and all were found to contain a single DNA insert approximately 1,500 base pairs in length. As both ends of the *Pst*I-cleaved pBR322 DNA receive homopolymeric dG tails, only those molecules that have acquired a dE-tailed cDNA insert or had escaped cleavage and/or dG tailing would be expected to re-circularise and transform; thus, we presume that most of the nonreacting Tc-resistant colonies contain inserts of contaminating non-DHFR cDNA.

When χ2282, a *thy*[+] variant of χ1776, was approved for EK2 use, direct selection of bacteria that phenotypically expressed the eukaryotic DNA sequence became feasible (see Fig. 1). Two populations (25 colonies each) of previously identified Tc-resistant colonies of χ1776 were pooled, and plasmid DNA extracted from these populations was introduced by transformation into χ2282. In other experiments, we transformed χ2282 directly with annealed pBR322–DHFR cDNA (Table 1). Colonies that expressed resistance to both Tc and Tp were obtained in both types of experiments; three independent transformants were found to be capable of immediate growth in media containing at least 1,000 μg ml^{-1} of Tp, and were termed 'strong expressors'. Plasmid DNA isolated from these colonies was shown by repeat transformation to encode both the Tc and Tp resistance phenotypes. However, $Tp^R Tc^R$ transformants occurred in different experiments at only 20–60% of the frequency observed when selection of transformants was carried out for Tc resistance alone. Transfer of Tc-selected colonies on to minimal medium agar plates containing Tp showed that all such clones express resistance to at least 1,000 μg ml^{-1}. Together, these findings suggest that phenotypic expression of Tp resistance may be delayed in transformants until a sufficient quantity of plasmid-specified DHFR has accumulated. Analogous results have been obtained with other antimicrobial drug-resistance determinants encoded by plasmids introduced into *E. coli* by transformation[20].

Structure of the DHFR cDNA

Gel electrophoresis of endonuclease-cleaved plasmid DNA from three separately derived bacterial clones that expressed high levels of trimethoprim resistance showed similar overall patterns. Using such data, a cleavage map (Fig. 4) of the cDNA insert of one of these (pDHFR7) was constructed.

Examination of the amino acid sequence of the mouse DHFR enzyme[21] allowed us to assign one of the *Hae*III cleavage sites to the *trp-pro* (TGG-CCX) present at amino acid positions 24 and 25, thus localising the DHFR structural sequence to a position near the 5' end of the mRNA template used in synthesising the cloned cDNA. Other cleavage sites within the DHFR structural sequence were consistent with positions predicted by computer analysis of the amino acid sequence. DNA sequence analysis of two separate regions of the cDNA provided direct verification that the nucleotide sequence of the insert corresponds to the amino acid sequence reported for the mouse DHFR enzyme, and also confirmed that the coding sequence for DHFR is located at the ds cDNA equivalent of the 5' end of the mRNA.

The nucleotide sequence at the pBR322–cDNA junction nearest the 5' end of the mRNA used as template for DHFR cDNA (that is, at the *Pst*I$_a$ site) is of special interest. The complement of the 'sense' strand of β-lactamase gene of the vector (J. G. Sutcliffe, personal communication) is interrupted at the *Pst*I site by a series of 11 dG residues added by the terminal transferase, and these are followed immediately by (1) an ATG (AUG) protein start codon, and (2) the codon for the first amino acid of the mouse DHFR structural gene. The sequence that codes for the mouse DHFR is in the same orientation as at that encoding the β-lactamase on the vector plasmid; however, the number of incremental G residues (that is, 10) at the vector-insert junction ensures that the DHFR cDNA sequence is not in the same translational reading frame as the β-lactamase gene.

If the phenotypic expression we have observed for the mouse DHFR sequences in bacteria is the result of translational readthrough from signals that initiate protein chains within the β-lactamase gene, then the host bacterial cell must be able to circumvent the observed frame shift by a 'slippage' mechanism of translation. Such an event might potentially be aided by the long run of dG residues introduced at the pBR322–cDNA junction and could account for our observation (Table 1) that plasmids containing a DHFR cDNA insert yield more than twice as many expressors as would be expected from considerations of reading frame and orientation. However, the high level of

---

Table 1 Transformation experiments using χ1776 and χ2282

			Transformants per ng DNA			
Host	Transforming DNA Plasmid vector	Insert	Tc (5 μg ml^{-1})	Tc (10 μg ml^{-1})	Tc/Tp (5 μg ml^{-1} of each)	*In situ* hybridisation with DHFR cDNA probe (% positive)
χ1776	pBR322	None	4×10^2	2×10^2	$<2 \times 10^{-3}$	—
χ1776	pBR322	cDNA(1°)	60	32	—	40
χ2282	pBR322	cDNA(1°)	70	—	2	44
χ2282	pBR322	cDNA(2°)	60	—	1.3×10^{-1}	—
χ2282	pDHFR7		75	—	25	—

pBR322 plasmid DNA that had been annealed *in vitro* with dC-tailed DHFR cDNA (designated 1°) was introduced into χ1776 or χ2282, using a modification of a previously described transfection procedure[19]. 1 ml of an overnight bacterial culture was inoculated into 100 ml of L broth supplemented with diaminopimelic acid (DAP, 50 μg ml^{-1}) and (for χ1776 only) thymidine (4 μg ml^{-1}). Bacterial cultures were grown until exponential phase at 35°C and then collected by centrifugation at 4°C. Cells were washed in 0.3 volume 10 mM NaCl, resuspended in 30 ml freshly prepared MCN buffer (70 mM $MgCl_2$; 40 mM sodium acetate, pH 5.6, and 30 mM calcium chloride) and chilled on ice for 20 min. Cells were collected, resuspended in 1 ml MCN and added in 200-μl aliquots to 50 μl DNA in TEN (10 mM Tris-HCl, pH 7.5, 0.1 mM EDTA, 50 mM NaCl) or MCN buffer. After chilling at 0°C for 30 min, reactions were incubated at 27°C for 5 min, chilled again for 30 min, and 50-μl samples were plated on to Penassay broth agar supplemented with DAP, thymidine (for χ1776), and antibiotics as indicated. When χ2282 was used, the selective medium was M9 minimal agar supplemented with 0.5% casamino acids, biotin (2 μg ml^{-1}), DAP (50 μg ml^{-1}) and Tp (2.5–10 μg ml^{-1}) plus tetracycline (Yc) or kanamycin (Km) as indicated. Plates containing transformants were incubated at 32 °C and colonies were scored 2–3 d after plating. pBR322 plasmid DNA lacking the cDNA insert was used as a control. cDNA preparations labelled as 2° consisted of plasmid DNA isolated from a nonfractionated population of clones that had previously been transformed with chimaeric molecules carrying a cDNA insert. In the experiment shown in the last line the transforming DNA was isolated from a clone (pDHFR7) that expresses resistance to Tp as well as Tc.

Fig. 2 Preparation and characterisation of DHFR cDNA insert. DNA complementary to DHFR mRNA was synthesised essentially as described elsewhere[14], using avian myeloblastosis virus (AMV) reverse transcriptase and polysomal RNA obtained by indirect immunoprecipitation of DHFR-synthesising polysomes from methotrexate-resistant AT-3000 S-180 mouse cells[13]. The RNA had been estimated to contain DHFR mRNA as 20% of its mRNA[13]. The reaction was carried out in 100 μl using 340 μg polysomal RNA (estimated to contain 5 μg polyA–RNA), 45 units AMV reverse transcriptase and dCTP labelled to 4 Ci mmol^{-1} with ^{32}P-dCTP (Amersham). Approximately 1.4 μg cDNA was synthesised in 30 min. The reaction was stopped by the addition of EDTA (to 10 mM) and extracted with phenol, followed by ether, before being passed over a Sephadex G-50 fine column in 10 mM Tris, pH 7.4, 2 mM EDTA and, 10 mM NaCl (TEN). The void volume was collected and precipitated with ethanol. After centrifugation, the RNA was hydrolysed with NaOH (ref. 31), neutralised and precipitated with ethanol. The cDNA was then used as template for the synthesis of the second strand by *E. coli* DNA polymerase I essentially as described[31]. The reaction took place at 42 °C for 10 min in 100 μl using 1.1 μg cDNA, 10 units DNA polymerase I, and 200 μM of each deoxynucleotide triphosphate with dCTP adjusted to 30 Ci mmol^{-1} as above. Approximately 0.85 μg of the second strand was synthesised. The reaction was stopped and extracted as above before being passed over a Sephadex G-50 fine column in TEN containing only 0.1 mM EDTA. Column fractions containing the ds cDNA were then treated with *Aspergillus oryzae* S$_1$ nuclease as described elsewhere[31]. After extraction and precipitation with ethanol, approximately 1.0 μg ds cDNA was obtained. Aliquots of *a*, first strand product; *b*, first strand product after base treatment; *c*, second

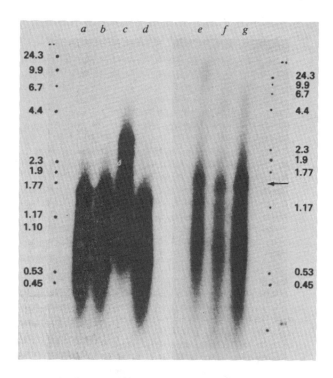

strand product and *d*, second strand product after S$_1$ nuclease treatment were examined on a 1.5% agarose gel in alkaline conditions as described[32]. Terminal addition of dCTP to the ds cDNA by terminal deoxynucleotidyl transferase (TdT), prepared as described elsewhere[33] was carried out by a modification[34] of the Co^{2+} procedure[35]. The reaction was carried out in 500 μl containing 140 mM cacodylic acid, 30 mM Tris base, 110 mM KOH (final pH 7.6), 0.1 mM dithiothreitol, 150 μM dCTP (adjusted to 8 Ci mmol^{-1} with ^{3}H-dCTP (Amersham)), 1 mM CoCl$_2$ (added to prewarmed reaction mix before enzyme addition), approximately 1.0 μg ds cDNA (assuming an average MW giving approximately 600 base pairs, this provides 10 pM 3' termini per ml) and 0.5 μl TdT (2.3 × 10^5 units ml^{-1}). The reaction was allowed to proceed at 37 °C for 10 min before being cooled and sampled to determine incorporation. Approximately 30 dC residues were added per 3' terminus. The reaction was stopped, extracted, desalted and precipitated with ethanol as above. Aliquots of *e*, second strand product; *f*, second strand product after S$_1$ nuclease treatment, and *g*, dC-tailed ds cDNA were analysed on a 1.7% agarose gel in Tris-acetate-NaCl (ref. 36). The dC-tailed ds cDNA was then preparatively electrophoresed on a similar gel and the '1,500-base pair' region (arrow) cut out of the gel and electrophoretically eluted into a dialysis bag as described elsewhere[37]. The eluted material was extracted as above, concentrated by lyophilisation and precipitated with ethanol. After centrifugation, the 1,500-base pair dC-tailed ds cDNA (approximately 80 ng) was redissolved in 10 mM Tris HCl, pH 7.4, 0.25 mM EDTA and 100 mM NaCl (annealing buffer). pBR322 plasmid DNA, isolated as described elsewhere[38] was digested with a 1.5-fold excess of *Pst*I endonuclease in conditions suggested by the vendor (New England Biolabs) and the linear plasmid DNA was cut out and eluted from a 0.7% agarose gel in TBE[39] as described above. The plasmid DNA was 'tailed' with dG residues using procedures similar to those described above. Approximately 15–20 dG residues were added per 3' terminus. Following extraction, the dG-tailed vector was passed over a Sephadex G-50 fine column in annealing buffer and the void volume was collected. Equimolar amounts of dC-tailed ds cDNA and dG-tailed vector DNA were allowed to anneal essentially as described by W. Rowe and R. A. Fratel (personal communication) except that the vector concentration was kept at 75 ng ml^{-1} in the annealing reaction. Circularisation was monitored by electron microscopy[16] and was typically about 20–40%. This annealed DNA was used directly for transformation into χ1776 or χ2282.

---

functional expression observed for both primary and secondarily transformed clones of pDHFR7 does not seem to be readily explained by slippage of tRNA molecules during translation.

Perhaps a more likely explanation is that the *Pst*I–polyG–ATG sequence that has been constructed preceding the coding sequence for DHFR serves as a binding and protein initiation site for the bacterial ribosome. Recent studies[22,23] have identified sequences on mRNA in the 5' direction from the initiator codon that are complementary to the CCUCC sequence at the 3' end of the 16S ribosomal RNA species, proposed by Shine and Dalgarno[24] to be involved in the binding of mRNA to ribosomes. It is tempting to speculate that the mRNA transcript from the sequence at the pBR322–cDNA junction has sufficient complementarity to the CCUCC sequence to allow ribosomal binding when a translational start signal is located an appropriate distance away. In such an event, the ATG protein start signal that immediately precedes the coding sequence for the mouse DHFR might initiate a peptide chain having a size characteristic of the mammalian enzyme. Immunological analysis of extracts derived from pDHFR7 and other expressing clones has yielded results consistent with this interpretation (see below).

Analysis of enzyme activity encoded by pDHFR7 plasmid

Mammalian dihydrofolate reductases can be distinguished from their bacterial counterparts by the ability of mammalian enzymes to use folate as a substrate and by their differential sensitivity to competitive inhibitors[25,26]. In initial experiments, the reduction of folate to tetrahydrofolate was measured using extracts from the pDHFR7 clone, from a Tp-sensitive clone containing a DHFR cDNA insert (the pDHFR10 plasmid), and from cells that contain only the pBR322 vector. Although all three clones are capable of synthesising a chromosomally produced bacterial enzyme, only the enzyme present in extracts from cells containing the pDHFR7 plasmid gave reduction of folate (4 × background). Additional evidence that the reductase encoded by the pDHFR7 plasmid is of mammalian origin was obtained by inhibitor analysis (Fig. 5). The DHFR isolated directly from mouse cells and the activity encoded by the pDHFR7 plasmid showed identical sensitivities to methotrexate, trimethoprim and a triazine derivative (2,4,-diamino-1-(4'-butylphenol)-6,6 dimethyl-1, 6-dihydro-1,3,5–triazine); both enzyme activities were 200 times more sensitive to the triazine than to trimethoprim. In contrast, bacterial dihydrofolate

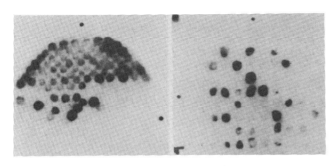

Fig. 3 Detection of colonies containing DHFR cDNA inserts by *in situ* hybridisation. Colonies were screened for DHFR sequences using a modification (G. N. Duell, unpublished) of an *in situ* hybridisation procedure[11]. Tc-resistant colonies were transferred to nitrocellulose filters (Millipore, HAWG) that had been placed on Penassay broth agar plates containing Tc (10 μg per ml). (Filters had been washed twice by boiling in H_2O and autoclaved before being placed on plates.) After 2–3 d of bacterial growth at 32°C, the filter was removed from the plate and placed on a Whatman no. 3 pad saturated with 0.5 M NaOH. After 7 min, the filter was sequentially transferred to a series of similar pads saturated with 1 M Tris, *p*H 7.4 (twice, 7 min each); 1.5 M NaCl, 0.5 M Tris, *p*H 7.5 (once, 7 min); and 0.30 M NaCl, 0.03 M Na citrate (2 × SSC) (once, 7 min). After the excess liquid had been removed by suction, the filter was placed on a pad containing 90% ethanol, dried by suction and baked *in vacuo* at 80°C for 2 h. Before hybridisation, filters were pretreated for 3–6 h at 65°C in hybridisation buffer that contained 5 × SSC, *p*H 6.1 0.2% SDS, 0.02% Ficoll 400 (Pharmacia) and 8 μg ml⁻¹ *E. coli* tRNA. Hybridisations were carried out with individual filters in 1.5 ml hybridisation buffer containing 2×10^4 c.p.m. ^{32}P-labelled purified DHFR cDNA[13] in a sealed plastic bag at 65 °C for 24 h. The filters were then washed in hybridisation buffer (once, 60 min at 65 °C); in 5 × SSC, *p*H 6.1 (three times, 60 min each at 65 °C); and in 2 × SSC, *p*H 7.4 (twice, 10 min at room temperature), air dried, and prepared for autoradiography. Left: top, a collection of χ1776 colonies which contain a DHFR cDNA insert; middle, Tc-resistant χ2282 colonies derived from transformation with annealed pBR322 ds cDNA—both reacting and non-reacting colonies are seen; bottom, colonies containing pBR322 and pACYC101 plasmids which show no visible hybridisation. Right: several positive colonies on a representative filter analysed in screening χ1176 transformants. Negative colonies represent clones containing pBR322 or pACYC101 plasmids which show no visible hybridisation.

reductase is inhibited more effectively by trimethoprim ($K_i = 5 \times 10^{-9}$) than it is by triazine ($K_i = 6.5 \times 10^{-4}$)[25].

As methotrexate binds stoichiometrically to dihydrofolate reductase[27], we can estimate the number of molecules of enzyme in the pDHFR7 plasmid extracts from the methotrexate inhibition date of Fig. 5. We calculate from the specific activity of the extract (3 units per mg of soluble protein) and the specific activity and methotrexate binding parameters of the mouse enzyme[28] that 0.01% of the soluble bacterial protein is active mammalian DHFR.

Immunological characterisation of bacterial cell extracts containing mouse DHFR

Immunological evidence confirming the nature of the DHFR encoded by pDHFR7 and other plasmids that contain a mouse DHFR cDNA insert was obtained using a solid-phase sandwich radioimmunoassay[12]. Tp^R clones of χ2282 containing the independently derived plasmids pDHFR 7, 12 and 13 showed a strong reaction with rabbit antibody directed against mouse DHFR in an *in situ* immunoassay[12] (data not shown); protein that reacted with the antibody was also made by bacteria which showed low levels of phenotypic expression and by some clones that did not make a biologically functional DHFR (that is, were Tp sensitive). The nature of the antigen synthesised by Tp^s and Tp^R clonnes was examined more fully using a newly developed

method (filter affinity transfer, or FAT procedure) for the *in situ* immunological characterisation of proteins in gels[29]. This procedure depends on the covalent coupling of F(ab)'₂ antibody fragments to a chemically derivatised and activated cellulose filter; antigen transferred on to the filter from an SDS–polyacrylamide gel is detected by subsequent incubations with antiserum and ^{125}I-labelled *Staphylococcus aureus* protein A (ref. 12).

Filter affinity transfer analysis of the pDHFR7 extract (Fig. 6, lanes *b*, *c*) shows the presence of protein that reacts immunologically with the antibody to mouse DHFR and further shows that most of the immunologically reactive material has the same electrophoretic mobility as enzyme obtained directly from mouse cells (molecular weight 22,000) (Fig. 6, lane *a*). An immunologically reactive band that migrates at this position was also seen in the material eluted with folic acid from a methotrexate affinity column that contained an extract from χ2282 (pDHFR7) cells (Fig. 6, lane *d*), suggesting that the 22,000 MW protein made by these bacterial cells has binding sites for both methotrexate and folate. Additional immunologically reactive bands which have mobilities consistent with a MW of 30,000–90,000 were seen in varying amounts in different extracts of bacterial cells that contained pDHFR7 (Fig. 6, lanes *b*, *c*); as the 1,500-base pair insert in this plasmid is capable of coding for a polypeptide no larger than 50,000 MW, we conclude that the most slowly moving bands are likely to be hybrid proteins that include antigenic sites of the mouse DHFR. Immunologically reactive high MW proteins are also made by a Tp^s clone (pDHFR21, Fig. 6, lane *e*) which contains a DHFR cDNA insert; however, this clone fails to synthesise an immunologically reactive 22,000 MW band. No immunological reactivity with antibody made to mouse DHFR was detected in extracts of cells carrying only the pBR322 vector (Fig. 6, lane *f*).

Variation in level of expression of mouse DHFR cDNA sequences in bacteria

Thirty-two separately derived clones of χ2282 transformants that had been selected on plates containing only Tc were replated on medium containing both Tc and Tp, and also were tested for the presence of a DHFR cDNA insert by *in situ* hybridisation. Fourteen colonies contained sequences homologous with the purified DHFR cDNA probe, and five of these (termed 'weak expressors') grew on plates containing 5 μg ml⁻¹ or more Tp. The remaining nine clones that contained DHFR sequences failed to show any growth on concentrations of Tp above 2.5 μg ml⁻¹, and were termed 'non-expressors'. Plasmid DNA isolated from four of the nine non-expressors and from all five weak expressors was analysed by gel electrophoresis, and the mean inhibitory concentration (MIC) of Tp was determined for each of the clones. The structural relationship of the cDNA insert to the β-lactamase gene sequence of the vector, the length of each insert, and the minimal inhibitory concentration (MIC) determined for the clone are shown in Fig. 7. As can be seen, plasmids pDHFR7, 12, 13 and 26–29 all contain a complete DHFR structural sequence inserted in the same orientation (that is, orientation a) as the gene encoding the bacterial β-lactamase. The clone carrying each of these plasmids expresses Tp resistance, although the MIC varies from 150 μg ml⁻¹ for pDHFR28 to >1,000 μg ml⁻¹ for pDHFR7, 12 and 13; the greatest reactivity with antibody to mouse DHFR occurs with pDHFR12 (unpublished data).

It is unlikely that translational reading frame is the determining factor in the different levels of expression observed in these clones, as our DNA sequence analysis indicates that a correct reading frame is not essential for efficient expression in pDHFR 7, 12 or 13. However, the positioning of the putative ribosomal binding site in relation to the ATG start codon may potentially influence the strength of expression by affecting the formation of the translational initiation complex (compare with ref. 23) (Figs 4, 7).

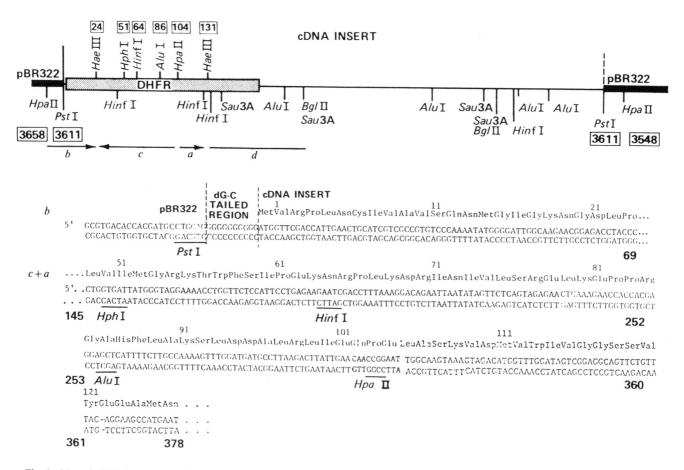

Fig. 4 Map of cDNA insert and adjacent regions of pDHFR 7 plasmid. Endonuclease cleavage map and partial DNA sequence of cDNA insert of the pDHFR 7 plasmid. The shaded area indicates the structural sequence for mouse DHFR. The locations of endonuclease cleavage sites were determined from polyacrylamide gel electrophoresis patterns following simultaneous or sequential digestions of either intact pDHFR or fragments isolated after earlier digestions. Restriction endonucleases (New England Biolabs or Bethesda Research Laboratories) were used according to the vendor's recommendations. Cleavage sites listed above the shaded area were assigned to specific amino acid positions within the mouse enzyme by nucleotide sequence. The boxed numbers of the HpaII and PstI sites indicate the locations of the sites in the pBR322 plasmid as determined by Sutcliffe (personal communication), and were used in orientating the amino acid sequence of the DHFR gene with respect to the β-lactamase sequence. The nucleotide sequence in the vicinity of the pBR322–cDNA junction corresponding to the 5′ end of the mRNA used as template for the cDNA, and the sequence in the region of the HpaII site at amino acid 104 of the structural sequence are shown. Nucleotides and amino acids are numbered from the start of the DHFR coding sequence; nucleotides in the 5′ direction on the mRNA from position 1 have negative numbers. The DNA sequence shown were determined by the method of Maxam and Gilbert[40]. Fragments (d + c) and (a + b) were 5′ end-labelled at the HpaII and BglII-generated ends, treated with HaeIII endonuclease, and subjected to electrophoresis in 6% acrylamide gel and TBE buffer[39]. Fragments a, b and c were eluted from the gel[32], precipitated with ethanol and used directly for base sequence determination using the gel system described in refs 40 and 41.

The end of DHFR structural sequence that corresponds to the 5′ end of the mRNA is not present in plasmids pDHFR 21 and 24, thus explaining the observed lack of functional expression of the cDNA in these clones. However, pDHFR21, which has lost less than 15% of the structural sequence, nevertheless encodes a (probably hybrid) peptide that contains antigenic sites which react with antibody to mouse DHFR (Fig. 6 and unpublished data). It is particularly interesting that the clone carrying pDHFR25 expresses a low level of Tp resistance, although the coding sequence for the DHFR enzyme is inserted in an orientation opposite to that of the β-lactamase gene. This finding, and our detection in extracts of the pDHFR25 clone of protein that reacts immunologically with antibody to the mouse enzyme (unpublished data), suggest that readthrough transcription into the DHFR coding sequence from a promoter sequence located on or near the distal segment of the β-lactamase gene may occur. Consistent with this interpretation are preliminary data suggesting that DHFR antigenic sites are also synthesised by cells carrying pDHFR23, which is a non-expressor of Tp resistance and contains a cDNA insert in orientation b (Fig. 7). Further study is required to determine whether sequences in the distal segment of the β-lactamase gene are capable of serving as weak promoters for the initiation of mRNA chains that extend into the DHFR cDNA.

The findings reported here indicate that the bacterial clones we have constructed are synthesising and phenotypically expressing DHFR encoded by mouse cDNA sequences: (1) the cDNA insert cloned in bacteria has been shown by in situ hybridisation to be homologous with the mouse gene and by direct DNA sequence analysis to encode the amino acid sequence of mouse DHFR, (2) DHFR enzymatic activity and resistance to Tp are specified by nucleotide sequences present on chimaeric plasmids but not on the vector, (3) the DHFR encoded by the constructed plasmids shows differential sensitivity to competitive inhibitors of DHFR characteristic of the mammalian gene product, and (4) the enzyme synthesised by bacteria containing DHFR cDNA is immunologically reactive with antibody made against mouse DHFR.

Note that the clones which express the highest levels of Tc resistance contain an immunologically reactive peptide having a size characteristic of the mammalian DHFR. A peptide of this size could potentially result from proteolytic cleavage of a fused β-lactamase–DHFR protein that was initiated at the β-lactamase ribosomal binding site. This proposal is consistent with the finding of large-sized protein species containing DHFR antigenic sites in extracts from cells that are either Tp^R or Tp^S

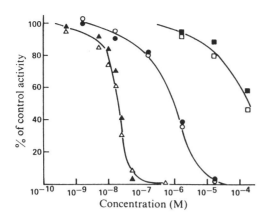

Fig. 5 Inhibitor analysis of DHFR from bacterial cells. Stationary phase cultures of χ2282 expressing trimethoprim resistance were grown in the presence of Tp (1 μg ml^{-1}) in minimal medium, washed with isotonic saline, and suspended in 50 mM potassium phosphate buffer, pH 7.0, containing 10 mM benzamidine and 10 mM phenyl methyl sulphonyl fluoride (3 volumes buffer to 1 volume cells). The suspension was sonicated and centrifuged at 10,000 r.p.m. for 15 min. The supernatant was centrifuged for 1 h at 100,000g before being studied. An R$_2$ methotrexate-resistant mouse cell extract was prepared as described elsewhere[42]. Enzyme activity was measured by the radioactive folic acid assay previously described[28]. Protein was determined by the method of Lowry[43]. Approximately 3 units of activity from the χ2282 extract or 5 units from the methotrexate-resistant mouse cell extract were incubated with inhibitor for 10 min at 24°C before assaying for folate reductase activity; the concentrations shown represent the final concentration of inhibitor in the reaction mixture. Background values, determined by measuring enzyme activity in the presence of 10 mM methotrexate, have been subtracted from all points. The results presented are the average of duplicate samples which generally varied by less than 10% and are expressed as a percentage of the value obtained in the absence of inhibitor. One unit of activity is the amount of enzyme needed to reduce 1 nmol of folate in 15 min at 37°C. △ and ▲ indicate addition of methotrexate, ○ and ● indicate addition of the triazine derivative, and □ and ■ indicate trimethoprim addition for χ2282 and the mouse cell extracts, respectively.

Fig. 6 Filter affinity transfer analysis[29] of bacterial cell extracts. 20 μl of extracts in SDS sample buffer were run at constant current for 3 h in an 11.25% SDS–polyacrylamide slab gel. The gel was incubated in PBS (50 mM phosphate buffer containing 0.15 M NaCl) for 30 min and placed on a blotter wet with PBS. Peptides were specifically transferred from the gel to strips of a dry cellulose filter that had been covalently coupled to anti-DHFR F(ab)$'_2$ fragments[12]. Filters were washed, incubated with antibody to DHFR, washed again and treated with ^{125}I-labelled protein A. After additional washing and drying steps, the filters were analysed by autoradiography. The eluate fraction (lane d) was obtained by passage of an extract of χ2282 cells containing pDHFR7 over a 0.5 ml methotrexate–Sepharose affinity column. The extract was acidified to pH 5.8, passed over the column and the bound fraction was eluted with 2 mM folic acid in a 5 mM NaHCO$_3$ buffer at pH 8.5 as described elsewhere[28]. Lane a, extracts of mouse cell line; lanes b and c, pDHFR 7; lane d, eluate from methotrexate column; lane e, pDHFR21; lane f, pBR322.

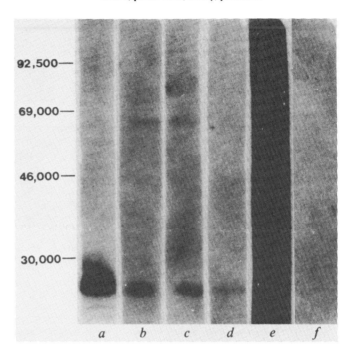

(Fig. 6). More intriguing, however, is the possibility noted above that the PstI–poly dG sequence constructed at the vector–cDNA junction can act together with the nearby ATG (AUG) translational start codon to bind mRNA to the bacterial ribosome and initiate DHFR peptide chains within the cDNA insert. If this interpretation is correct, initiation of peptide chains within other eukaryotic cDNA inserts may be obtainable in bacteria by use of the same structural relationships that have resulted in expression of the mouse DHFR coding sequence. Additional DNA sequence analysis and investigation of the protein products encoded by chimaeric plasmids should provide definitive information on this point and should help elucidate further the structural basis for the different levels of Tp resistance expressed by various clones.

Some of the bacterial clones we have isolated produce proteins that react immunologically with antibody to mouse DHFR but which are not biologically active. Our results suggest that an important obstacle to functional expression of mammalian DNA sequences in bacteria has been the development of an assay capable of detecting those clones that possess both a complete coding sequence and the correct nucleotide relationships to allow such expression. The strong phenotypic selection possible in the present experiments has provided an effective means of identifying and isolating expressing clones.

As the cloned coding sequence for mouse DHFR is selectable in higher organisms as well as in bacteria, it constitutes a powerful tool for the construction of eukaryotic cloning

vectors, for the isolation of replication regions of eukaryotic chromosomal and extrachromosomal genomes (compare ref. 30), and for the isolation and characterisation of signals that control genetic transcription and translation in variety of species.

This work was supported by grants from the NIH, American Cancer Society and NSF to S.N.C. and R.T.S. H.A.E. is a Fellow of the American Cancer Society, California Division and acknowledges the hospitality and support of H. O. McDevitt. AMV reverse transcriptase was provided by Dr J. Beard, RNA by Drs F. Alt and R. Kellems, triazine derivative by Dr J. J. Burchall, and E. coli DNA polymerase by Dr H. Schaller. TdT was prepared by Dr R. L. Ratliff.

Received 4 September; accepted 15 September 1978.

1. Morrow, J. F. et al. Proc. natn. Acad. Sci. U.S.A. **71**, 1743–1747 (1974).
2. Struhl, K., Cameron, J. R. & Davis, R. W. Proc. natn. Acad. Sci. U.S.A. **73**, 1471–1475 (1976).
3. Ratzkin, G. & Carbon, J. Proc. natn. Acad. Sci. U.S.A. **74**; 487–491 (1977).
4. Vapnek, D., Hautala, J. A., Jacobson, J. W., Giles, N. H. & Kushner, S. Proc. natn. Acad. Sci. U.S.A. **74**, 3508–3512 (1977).
5. Itakura, K. et al. Science **198**, 1056–1063 (1977).
6. Villa-Kamaroff, L. et al. Proc. natn. Acad. Sci. U.S.A. **75**, 3727–3731 (1978).
7. Burchall, J. J. & Hitching, G. H. Molec. Pharmacl. **1**, 126–136 (1965).
8. Guidelines for Research Involving Recombinant DNA Molecules, 23 June, (NIH, US Department of Health, Education and Welfare, 1976).
9. Curtiss, R., III et al. in Molecular Cloning of Recombinant DNA, (eds Scott, W. A. & Werner, R.) 99–111 (Academic, New York, 1977).
10. Miller, J. H. in Experiments in Molecular Genetics, 218–220 (Cold Spring Harbor Laboratories, New York, 1972).
11. Grunstein, M. & Hogness, D. S. Proc. natn. Acad. Sci. U.S.A. **72**, 3961–3965 (1975).

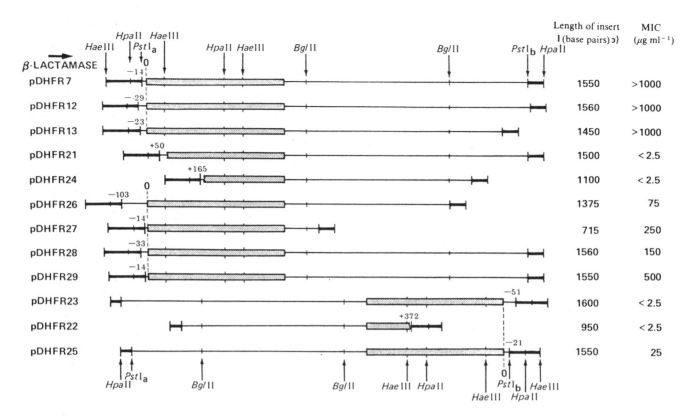

Fig. 7. Some structural and functional properties of chimaeric plasmids and χ2282 clones containing DHFR cDNA insert. The MIC of Tp for each of the χ clones was tested on M9 minimal agar plates containing biotin (2 μg ml^{-1}), casamino acids (0.5%), DAP (50 μg ml^{-1}) and Tp concentrations ranging from 0 to 1,000 μg ml^{-1}. χ2282 (pBR322) was used as a control and is sensitive to Tp at 2.5 μg ml^{-1}, as determined by incubation at 32 °C for 3 d. Plasmid DNA isolated from each clone was digested with *Pst*I endonuclease[15] at 37 °C for 3 h and extracted sequentially with phenol and ether. DNA was precipitated with ethanol, resuspended in 8 μl TE buffer and electrophoresed in 1.2% agarose gels in TBE buffer to determine the length of the inserted fragment. ColEI plasmid DNA digested with *Hae*II endonuclease[42] was added to each sample as an internal molecular weight standard. Additionally, *Hind*III-generated fragments of SV40 DNA were used as an external standard[45]. The standard error is ±100 base pairs with the method used. The orientation of the cDNA inserts in the vector plasmid was determined by gel analysis of plasmid DNA digested with *Bgl*II and *Hinc*II endonucleases; in addition, pDHFR25 was treated with the *Eco*RI and *Bgl*II enzymes to confirm the orientation of its insert. The direction of transcription of the β-lactamase gene of pBR322 is indicated by an arrow. The shaded area in each plasmid map indicates the structural sequence for DHFR. The numbers shown above each *Pst*I$_a$ site indicate the distance (in base pairs) between the cleavage point and the first nucleotide of the DHFR structural sequence as described in Fig. 4. This distance was determined exactly by DNA sequence analysis for plasmids pDHFR 7, 12 and 13 and was estimated for the other plasmids by gel electrophoresis of fragments produced by either the *Hae*III or *Hpa*II endonucleases. For the chimaeric plasmids that we subsequently sequenced, this estimate proved to be accurate within 5 base pairs. *Pst*I–*Bgl*II distances were determined for the chimaeric plasmids by gel electrophoresis using *Hpa*II-cleaved pBR322 DNA as an internal standard. Nucleotides in the 5' direction on the mRNA from position 1 have negative numbers and are estimates obtained from gel analysis for all plasmids except pDHFR7 and 12.

12. Erlich, H. A., Cohen, S. N. & McDevitt, H. O. *Cell* **13**, 681–689 (1978).
13. Alt, F. W., Kellems, R. E., Bertino, J. R. & Schimke, R. T. *J. biol. Chem.* **253**, 1357–1370 (1978).
14. Buell, G. N., Wickens, M. P., Payvar, F. & Schimke, R. T. *J. biol. Chem.* **253**, 2471–2482 (1978).
15. Smith, H., Blattner, N. & Davies, J. *Nucleic Acids Res.* **3**, 343–353 (1976).
16. Kleinschmidt, A. K., Lang, D., Jacherts, D. & Zahn, R. K. *Biochim. biophys. Acta* **61**, 856–864 (1962).
17. Bolivar, F. *et al. Gene* **2**, 95–113 (1977).
18. Chang, A. C. Y. & Cohen, S. N. *J. Bact.* **134**, 1141–1156 (1978).
19. Enea, V., Vovis, G. F. & Zinder, N. D. *J. molec. Biol.* **96**, 495–509 (1975).
20. Cohen, S. N., Chang, A. C. Y. & Hsu, L. *Proc. natn. Acad. Sci. U.S.A.* **69**, 2110–2114 (1973).
21. Stone, D. & Phillips, A. W. *FEBS Lett.* **74**, 85–87 (1977).
22. Steitz, J. A. & Steege, D. A. *J. molec. Biol.* **114**, 545–558 (1977).
23. Steege, D. A. *Proc. natn. Acad. Sci. U.S.A.* **74**, 4163–4167 (1977).
24. Shine, J. & Dalgarno, L. *Proc. natn. Acad. Sci. U.S.A.* **71**, 1342–1346 (1974).
25. Burchall, J. J. *Ann. N. Y. Acad. Sci.* **186**, 143–152 (1971).
26. Baccanari, D., Phillips, A., Smits, S., Sinsky, D. & Burchall, J. J. *Biochemistry* **14**, 5267–5263 (1975).
27. Werkheiser, W. J. *biol. Chem.* **236**, 888–893 (1961).
28. Alt, F. W., Kellems, R. E. & Schimke, R. T. *J. biol. Chem.* **251**, 3063–3074 (1976).
29. Erlich, H. A., Levinson, J., Cohen, S. N. & McDevitt, H. O., in preparation.
30. Timmis, K., Cabello, F. & Cohen, S. N. *Proc. natn. Acad. Sci. U.S.A.* **72**, 2242–2246 (1975).
31. Seeburg, P. H., Shine, J., Martial, J. A., Baxter, J. D. & Goodman, H. M. *Nature* **270**, 486–494 (1977).
32. Wickens, M. P., Buell, G. N. & Schimke, R. T. *J. biol. Chem.* **253**, 2483–2495 (1978).
33. Chang, L. M. S. & Bollum, F. J. *J. biol. Chem.* **246**, 909–916 (1971).
34. Wahl, G., Padgett, R. A. & Stark, G. R., in preparation.
35. Roychoudhury, R., Jay, E. & Wu, R. *Nucleic Acids Res.* **3**, 101–116 (1976).
36. Dugaiczyk, A., Boyer, H. W. & Goodman, H. M. *J. molec. Biol.* **96**, 171–184 (1975).
37. McDonell, M. W., Simon, M. N. & Studier, F. W. *J. molec. Biol.* **110**, 119–146 (1977).
38. Kuperstock, Y. M. & Helinski, D. R. *Biochem. biophys. Res. Commun.* **54**, 1451–1459 (1973).
39. Sharp, P. A., Sugden, B. & Sambrook, J. *Biochemistry* **12**, 3055–3063 (1974).
40. Maxam, A. M. & Gilbert, W. *Proc. natn. Acad. Sci. U.S.A.* **74**, 560–564 (1977).
41. Sanger, F. & Coulson, A. R. *FEBS Lett.* **87**, 107–110 (1978).
42. Kaufman, R. J., Dana, S. L. & Schimke, R. T., in preparation.
43. Lowry, O. H., Rosebrough, N. J., Farr, A. L. & Randall, R. J. *J. biol. Chem.* **193**, 265–275 (1951).
44. Oka, A. & Takanami, M. *Nature* **264**, 193–196 (1976).
45. Fiers, W. *et al. Nature* **273**, 113–120 (1978).

Transformation of Yeast

A. Hinnen, J.B. Hicks and G.R. Fink

ABSTRACT A stable *leu2⁻* yeast strain has been transformed to *LEU2⁺* by using a chimeric ColE1 plasmid carrying the yeast *leu2* gene. We have used recently developed hybridization and restriction endonuclease mapping techniques to demonstrate directly the presence of the transforming DNA in the yeast genome and also to determine the arrangement of the sequences that were introduced. These studies show that ColE1 DNA together with the yeast sequences can integrate into the yeast chromosomes. This integration may be additive or substitutive. The bacterial plasmid sequences, once integrated, behave as a simple Mendelian element. In addition, we have determined the genetic linkage relationships for each newly introduced *LEU2⁺* allele with the original *leu2⁻* allele. These studies show that the transforming sequences integrate not only in the *leu2* region but also in several other chromosomal locations.

Transformation is the process by which naked DNA is introduced into a cell, resulting in a heritable change. Transformation provides the link between the *in vitro* analysis of DNA and its *in vivo* function. Many recent advances in the analysis of eukaryotic DNA sequences have resulted from the facility with which these DNA segments can be attached to bacterial plasmids and subsequently introduced into bacterial cells by transformation. Such cloned sequences can be easily obtained in large amounts and can be altered *in vivo* by bacterial genetic techniques and *in vitro* by specific enzymatic modifications. To determine the effects of these experimentally induced changes on the function and expression of eukaryotic genes, the rearranged sequences must be taken out of the bacteria in which they are cloned and studied and introduced back into the eukaryotic organism from which they were originally obtained. Transformation back into eukaryotic organisms is the missing step in this sequence. There have been many attempts to transform fungi (refs. 1 and 2; reviewed in ref. 3). However, all failed to prove at the molecular level that exogenous DNA had become incorporated as a heritable component of the genome. In the final analysis, only by identification of the exogenous sequences in the recipient DNA can trivial explanations such as reversion be ruled out. Recent advances in recombinant DNA technology have provided the tools for a direct proof of transformation in yeast.

In the studies described here, a hybrid bacterial plasmid containing yeast DNA was used as a source of highly enriched genes as well as a molecular probe for the sequences introduced into the recipient yeast cells. The plasmid pYeleu10 is a hybrid composed of the *Escherichia coli* plasmid ColE1 and a segment of yeast DNA from chromosome III. This plasmid complements *leuB* mutants of *E. coli* and contains the yeast *LEU2⁺* gene which encodes the leucine biosynthetic enzyme β-isopropylmalate dehydrogenase (4, 5). Using the pYeleu10 plasmid, we

can transform a stable *leu2⁻* mutant of yeast to *LEU2⁺*. We have also determined that during the transformation event the ColE1 sequences together with the yeast DNA on the plasmid can integrate into the yeast chromosomes in several different places. Thus, through transformation, bacterial genes become a heritable component of the yeast genome and obey the rules of Mendelian inheritance characteristic of eukaryotic organisms.

MATERIALS AND METHODS

Strains. All yeast strains used as recipients for transformation were derived from wild-type strain S288C. AH22 (*a leu2-3 leu2-112 his4-519 can1*) has a *leu2⁻* double mutation constructed by recombining the *leu2* alleles from strains 5463-8A (*a leu2-3 his4-519 can1*) and ICR112 (*α leu2-112*). MC333 (*α met8-1 trp1-1 leu2-2*) was obtained from M. Culbertson.

DNA Preparation. pYeleu10, a hybrid plasmid isolated by B. Ratzkin and J. Carbon, contains the *LEU2⁺* gene of yeast cloned on the *E. coli* plasmid vector ColE1 (4, 5). Plasmid DNA was isolated by a modification (5) of procedures described by Clewell (6) and Guerry *et al.* (7). Yeast DNA was isolated by the method of Cryer *et al.* (8).

Transformation of *Saccharomyces cerevisiae*. Spheroplasts were prepared as described by Hutchinson and Hartwell (9). A fresh logarithmic phase culture (80 ml; 2×10^7 cells per ml) was concentrated to $1/10$ volume by centrifugation and treated with 1% Glusulase (Endo Laboratories) in 1 M sorbitol for 1 hr at 30°. Spheroplasts were washed three times with 1 M sorbitol and resuspended in 0.5 ml of 1 M sorbitol/10 mM Tris·HCl/10 mM CaCl₂, pH 7.5. Plasmid DNA was added to a final concentration of 10–20 μg/ml and incubated for 5 min at room temperature. Then, 5 ml of 40% polyethylene glycol 4000 (Baker Chemical Co., Phillipsburg, NJ)/10 mM Tris·HCl/10 mM CaCl₂, pH 7.5, was added as recently described by van Solingen and van der Plaat (10). After 10 min the spheroplasts were sedimented by centrifugation and resuspended in 5 ml of the sorbitol/Tris/CaCl₂ mixture; 0.2-ml aliquots were added to 10 ml of regeneration agar and poured on minimal agar plates [regeneration agar is Difco yeast nitrogen base without amino acids, supplemented with 1 M sorbitol, 2% glucose, 2% YEPD, and 3% agar (10)].

Hybridization Analysis of Restriction Digests. Total yeast DNA was digested with restriction endonuclease *Hind*III (New England Biolaboratories, Beverly, MA). Restriction digests were separated on 0.8% agarose gels and transferred to nitrocellulose filters (Millipore HAWP) according to the method of Southern (11). Details of the blotting procedures and hybridization conditions have been described (5). ³²P-Labeled plasmid DNA was prepared by nick translation with *E. coli* DNA polymerase I (Worthington Biochemical Co.) (12).

Yeast Colony Hybridization. Colony hybridization was performed as described by Grunstein and Hogness (13) with

The costs of publication of this article were defrayed in part by the payment of page charges. This article must therefore be hereby marked "*advertisement*" in accordance with 18 U. S. C. §1734 solely to indicate this fact.

HINNEN, A., HICKS, J.B. and FINK, G.R.
Transformation of yeast. *Proc. Natl. Acad. Sci. U.S.A.* 75: 1929-1933 (1978). Reprinted with the authors' permission.

the following modifications developed by J. Walsh in our laboratory. Colonies were grown on Millipore HAWP filters and lysed *in situ* by placing the filter sequentially on blotting paper saturated with the following solutions: 50 mM EDTA/2.5% 2-mercaptoethanol, pH 9.0 for 15 min; 1 mg/1 ml of Zymolyase 60000 (Kirin Brewery) for 2–3 hr at 37°; 0.1 M NaOH/1.5 M NaCl for 1 min; 0.2 M Tris·HCl, pH 7.5/0.30 M NaCl/0.03 M Na citrate for two changes, 2 min each. All treatments were at room temperature unless specified. Filters were air-dried briefly and baked under vacuum at 80° for 1 hr.

Genetic Techniques and Media. Genetic techniques and media, with the exception of regeneration agar (above), have been described (14).

Biohazard Consideration. This work was carried out under P2 laboratory conditions as approved by the local biohazards committee and the National Science Foundation. All yeast and bacterial strains harboring recombinant DNA were destroyed by autoclaving or exposure to Clorox solution before disposal as specified by the *NIH Guidelines for Recombinant DNA Research*, July 1976.

RESULTS

Transformation to *LEU2*⁺ Depends on the pYe*leu*10 DNA. A sensitive detection system for transformation requires that the recipient strain not give rise to colonies in the absence of DNA. For this reason, strains carrying deletions are ideal recipients. However, none exists for the *leu2* region of yeast. Instead, we constructed a stable *leu2*⁻ strain, functionally equivalent to a deletion, by combining within the *leu2* gene two different mutations, each of which has a low background rate of reversion to Leu⁺. This stable double-mutant strain, AH22, fails to revert either spontaneously or by mutagenesis with ICR170 or UV irradiation (<10⁻¹⁰).

Spheroplasts of strain AH22 were mixed with pYe*leu*10 DNA, Ca²⁺, and polyethylene glycol, and the treated spheroplasts were regenerated in 3% agar (10). Ten percent of the spheroplasts plated under these conditions gave rise to colonies. When spheroplasts were plated on medium lacking leucine, colonies appeared after 3–5 days at 30° at a frequency of 1/10⁷ regenerated spheroplasts. No colonies were obtained in controls where spheroplasts were plated without DNA. Putative transformants were purified and subjected to further genetic and biochemical analysis.

Hybridization Tests. In the DNA hybridization experiments, two probes were used: hybrid plasmid pYe*leu*10 and its parent bacterial plasmid ColE1. There are no ColE1 sequences in the original recipient strain, AH22, or in any untransformed yeast strain that we have examined (see AH22 in Fig. 1 *right*). Therefore, the presence of ColE1 sequences in yeast DNA is unambiguous evidence of transformation. We tested each of 42 transformants for the presence of ColE1 sequences by using a modification of the Grunstein–Hogness colony hybridization technique (13). In this case, yeast colonies rather than bacterial colonies were lysed on nitrocellulose filters and subjected to hybridization with ³²P-labeled ColE1 DNA. Of 42 putative transformants tested, 35 were shown to contain ColE1 sequences by this method.

The arrangement of ColE1 and pYe*leu*10 sequences in the transformed strains was demonstrated by hybridization of the native plasmids to specific restriction fragments from the total DNA restriction digest of these strains. There are no *Hin*dIII restriction sites in either the yeast or the bacterial DNA carried by pYe*leu*10; therefore, a *Hin*dIII digest of total yeast DNA contains a single restriction fragment of ~10⁷ daltons capable

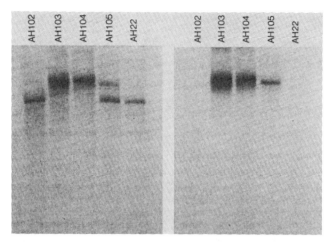

FIG. 1. Radioautographs of ³²P-labeled DNA hybridized to a "Southern blot" containing *Hin*dIII restriction digests of DNA from the recipient strain AH22 and four transformants, AH102–AH105. (*Left*) Hybrid plasmid pYe*leu*10 as the labeled probe. (*Right*) Plasmid ColE1 DNA as probe.

of hybridizing to this plasmid (ref. 5; see also AH22 in Fig. 1 *left*). Furthermore, as shown in Fig. 1 *right*, no sequences in the recipient strain hybridized with ColE1 DNA by itself. Thus, any fragment that hybridizes to ColE1 must contain sequences introduced by transformation. The sequences present in the restriction digests of yeast DNA that were complementary to the transforming DNA were visualized by hybridization with ³²P-labeled pYe*leu*10 or ColE1 plasmid DNA prepared by nick translation.

Approximately 5 μg of total DNA extracted from four representative transformants and the recipient strain AH22 were digested with *Hin*dIII and the restriction fragments were displayed on a 0.8% agarose gel. The restriction digest in the gel was then transferred to a nitrocellulose filter by the Southern blotting procedure (11). Radioautographs of two such hybridization experiments are shown in Fig. 1. The differences in the hybridization patterns of these strains indicate that at least three distinct types of transformation events had occurred. Each of these events is described further below.

Genetic Identification of "New" leu2 Regions. To determine the linkage relationships of the *LEU2*⁺ regions in the transformants two sets of crosses were performed. (*i*) The transformants were crossed by a *leu2*⁻ strain (MC333) carrying the centromere-linked *trp1* allele. The diploids were heterozygous for *his4*, *leu2*, and *trp1*, permitting the detection of linkage between *his4* and *LEU2*⁺ as well as between *LEU2*⁺ and its centromere. (*ii*) The transformants were crossed by a *LEU2*⁺ strain (S288C). This cross allowed us to determine whether the *leu2*⁻ region of the recipient was still present in the transformed strains. The results of these crosses (Tables 1 and 2) again define three classes of transformants congruent with the classes defined by hybridization studies. Each of these three types is described below along with hybridization and genetic mapping data obtained from representative transformants.

Classification of Transformants. Of the more than 100 transformants obtained, 42 have been subjected to both genetic analysis and colony hybridization with ColE1. Representatives of each of the three types were also subjected to restriction enzyme analysis (Fig. 1) and more detailed genetic mapping.

Recipient strain. The recipient strain AH22 showed a single

Table 1. Linkage of *LEU2*$^+$ to *his4* and *trp1* *

	leu2-his4			leu2-trp1		
	PD	NPD	TT	PD	NPD	TT
Expected	50	<1	50	45	45	10
Observed:						
AH-102 × MC-333	18	0	22	17	22	10
AH-103 × MC-333	17	1	15	17	16	6
AH-104 × MC-333	31	0	22	27	25	6
AH-105 × MC-333	2	6	31	10	7	27

These data represent the results of crosses between the transformants (*LEU2*$^+$ *his4*$^-$ *TRP1*$^+$ *MET8*$^+$) by strain MC333 (*leu2*$^-$ *HIS4*$^+$ *trp1*$^-$ *met8*$^-$). In nontransformed strains, *his4* and *leu2* give a ratio parental ditype (PD)/nonparental ditype (NPD)/tetratype (TT) of 1:0.01:0.7, indicating 20% linkage. Thus, in the first three crosses above, *LEU2*$^+$ is linked to *his4*. If PD:NPD:TT is approximately 1:1:4, then two markers are considered unlinked. In the last cross, *his4* and the *LEU2*$^+$ region are unlinked. If the ratio PD:NPD:TT is 1:1:<4, two markers are considered linked to their centromeres. *leu2* and *trp1* in untransformed strains show centromere linkage. In the first three crosses the new *LEU2*$^+$ region shows centromere linkage, whereas in the AH-105 × MC333 cross it does not.
* Numbers represent both complete asci and three-spored asci that could be scored unambiguously.

*Hin*dIII fragment capable of hybridization with pY*eleu*10 (Fig. 1 *left*). No hybridization with ColE1 could be found (Fig. 1 *right*). Crosses with strain AH22 showed the expected linkage relationships for *his4* and *leu2*.

Type I (duplication of leu2 region of chromosome III). The most frequent class of transformation event (30/42) appeared to involve the integration of the complete pY*eleu*10 plasmid into the region of chromosome III homologous with the yeast-derived portion of the plasmid and adjacent to the *leu2* gene. AH103 and AH104 are representatives of such a type I event. In type I transformants, ColE1 DNA was present and the *LEU2*$^+$ allele was closely linked to the original *leu2*$^-$ allele (Tables 1 and 2). Restriction analysis of the DNA from type 1 transformants AH103 and AH104 showed a single *Hin*dIII fragment that hybridized to both pY*eleu*10 and ColE1 DNA. The size of this fragment is consistent with that expected if the pY*eleu*10 plasmid had integrated into the original *Hin*dIII fragment. Fig. 2 illustrates the integration of pY*eleu*10 into the yeast genome near the *leu2* locus by a Campbell-like recombination event (15). Fig. 2 also contains a schematic representation of the hybridization pattern predicted by such a sequence arrangement.

Genetic tests confirmed our interpretation of the hybridization data. Type I transformants gave meiotic segregation patterns indicating that the newly introduced *LEU2*$^+$ region was adjacent to, but did not replace, the resident *leu2*$^-$ region. In crosses of the *LEU2*$^+$ transformants by standard *LEU2*$^+$ strains (Table 2) 1 in 20 tetrads showed a *leu2*$^-$ spore. Each Leu$^-$ spore was tested and shown to carry the *leu2-3leu2-112* double mutation present in the original recipient. We interpret the *leu2-3leu2-112* meiotic segregants of these strains to result from recombination events that separate the *LEU2*$^+$ region from the *leu2-3leu2-112* region. Thus, both a *LEU2*$^+$ and a *leu2*$^-$ region are present in chromosome III in type I transformants (Fig. 2).

The ColE1 sequences in transformant AH103 have been shown to be tightly linked to the *LEU2*$^+$ allele by colony hybridization analysis of the meiotic tetrads (unpublished data).

Type II (pYeleu10 integrated into other chromosomes). AH105 is representative of another class of transformants which

Table 2. Segregation of *LEU2*$^+$ in transformants × wild type (S288C) crosses

	Leu$^+$:Leu$^-$		
	4:0	3:1	2:2
Expected:			
New Leu$^+$ at *leu2*	100	0	0
New Leu$^+$ unlinked to *leu2*	17	66	17
Observed:			
AH-102 × S288C	41	0	0
AH-103 × S288C	38	2	0
AH-104 × S288C	42	2	0
AH-105 × S288C	4	19	6

All strains used in these crosses are phenotypically Leu$^+$. The appearance of Leu$^-$ segregants therefore means that the original *leu2* region of the recipient is still present in the transformed strain. The frequency of Leu$^-$ segregants (3:1 and 2:2 asci) is a measure of the linkage between the old *leu2*$^-$ region and the new *LEU2*$^+$ region introduced by transformation. In the last three crosses, the *leu2*$^-$ region was clearly present. In the second and third crosses, *LEU2*$^+$ and *leu2*$^-$ were closely linked, whereas in the last cross they segregated independently.

contains the bacterial ColE1 sequences; 5 of 42 transformants were of this type. Fig. 1 *left* shows that AH105 has two *Hin*dIII fragments that hybridize with pY*eleu*10. We identified the lower band as the "old" *leu2*$^-$ region of the host genome and the higher band as the "new" region resulting from transformation. This conclusion is based on several facts. First, the lower band showed the same molecular weight as the single band in AH22. Second, the upper band hybridized with ColE1 as well as with pY*eleu*10. Finally, crosses of AH105 indicate that the *LEU2*$^+$ region in this strain is unlinked to the *leu2* region on chromosome III (Table 2). Moreover, this *LEU2*$^+$ region is unlinked to *his4* and the centromere of this chromosome (Table 1). These results can be explained by insertion of pY*eleu*10 into other yeast chromosomes as depicted in Fig. 2.

Type III (no bacterial sequences). Transformant AH102 showed the same gel hybridization pattern with ColE1 and pY*eleu*10 as did the original recipient AH22 (Fig. 1). This transformant, along with six others that failed to show colony hybridization with ColE1, acts genetically as if only a single *leu2* region were present. Transformants of this type showed the standard linkage relationships for *leu2* on chromosome III and, so far, have yielded no Leu$^-$ segregants in crosses with wild type (Table 2).

Several explanations are compatible with these facts. First, AH102 could result from a transformation event in which a double crossover occurred as shown in Fig. 2. Second, this type could arise by excision of the *leu2*$^-$ portion of a type I tandem duplication. Of course, there is no way to distinguish this type from a revertant to *LEU2*$^+$. However, we have never observed revertants in our controls without DNA.

Stability of the Leu$^+$ Transformants. Transformants containing the duplicated *leu2* region (AH103, AH104) were extremely unstable during mitotic growth; 1–2% of the vegetative haploid cells segregated Leu$^-$ clones. These are stable Leu$^-$ strains that have lost the ColE1 sequences but retained both *leu2*$^-$ alleles present in the original recipient. The mechanism responsible for the production of Leu$^-$ segregants is unknown. However, it is possible that the *LEU2*$^+$ region is excised by a reversal of the mechanism by which it became inserted (Fig. 2). No Leu$^-$ segregants have been observed in type II and III transformants.

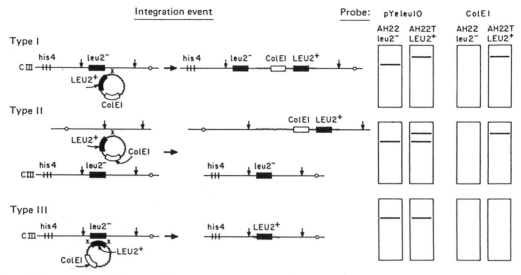

FIG. 2.   Schematic interpretation of the integration events proposed for transformant types I, II, and III. Each type of integration event (*Left*) gives rise to a unique chromosome structure (*Center*) that can be visualized by hybridization of pYeleu 10 and ColE1 DNA to *Hin*dIII restriction digests (*Right*). The arrows (↓) represent *Hin*dIII restriction sites. Type I: integration of plasmid pYeleu 10 into chromosome III at a sequence complementary to a yeast sequence carried by the plasmid. Type II: integration of plasmid pYeleu 10 into a chromosomal location genetically unlinked to the *leu2* region of chromosome III. Type III: integration of yeast DNA sequences of plasmid pYeleu 10 into the *leu2* region by a double crossover event. These different integration events lead to predictable patterns when *Hin*dIII restriction digests of these strains are hybridized with pYeleu 10 or ColE1 DNA. These hypothetical hybridization patterns are in agreement with the actual patterns shown in Fig. 1.

DISCUSSION

Direct proof that transformation has occurred in yeast comes from our hybridization studies that show that the bacterial ColE1 sequences from the pYeleu 10 plasmid are present in the DNA of the transformed yeast strains. In a parallel fashion, genetic analysis confirms that the *LEU2*+ gene from the DNA of the plasmid has integrated in several different places in the yeast genome. The hybridization results predict a genetic behavior of the newly inherited *LEU2*+ sequences that is fully realized when the transformants are analyzed by standard genetic analysis. For example, the presence of a single large *Hin*dIII fragment in type I transformants capable of hybridization with pYeleu 10 predicts insertion of pYeleu 10 into the original *leu2*-containing *Hin*dIII fragment. In agreement with this prediction, genetic analysis shows that the newly obtained *LEU2*+ region is closely linked to the old *leu2*− region of the recipient. Furthermore, the two *leu2* regions in the transformants can be separated by a rare meiotic recombination event. In the type II transformants the presence of two *Hin*dIII fragments capable of hybridization with pYeleu 10 predicts two separable *leu2* regions. In agreement with this prediction, genetic analysis shows that the *LEU2*+ region of these transformants is unlinked to the normal *leu2* region on chromosome III. In fact, the *leu2*− region of the recipient and the *LEU2*+ region show independent assortment. The genetic results reinforce the hybridization data and lend an additional dimension to our studies: The pYeleu 10 plasmid not only transforms yeast to Leu+ but also can integrate into the yeast chromosomes at several locations.

Many questions remain concerning transformation in yeast. The mechanism of insertion of the pYeleu 10 plasmid into yeast DNA is currently unknown. We have illustrated three types of integration events in Fig. 2. Type I and type II probably arise from Campbell-like recombination events which involve a circular transforming plasmid. This type of insertion would result in the integration of the ColE1 sequences together with

the *LEU2*+ sequences as was found for types I and II. If pYeleu 10 can integrate by this mechanism, then yeast must contain enzymes analogous to the λ *int* system (16). The type III transformants have no sequences that hybridize to ColE1 and could result from the event shown in Fig. 2. Alternatively, the integration could result from an intrachromosomal crossover subsequent to a type I event. Because the *leu2* region is duplicated in type I transformants, an intrachromosomal crossover could "pop out" the *leu2*− region together with the ColE1 sequences.

The fact that most of the transformation events occur at or near *leu2* suggests that homology plays a role in the insertion event. The segment of yeast DNA in pYeleu 10 is likely to carry several genes in addition to *LEU2*+ and our evidence suggests that the insertion takes place within one of these genes adjacent to *leu2*− on chromosome III. Genetic analysis of the type II transformants shows that the pYeleu 10 can also insert in at least three other locations. Thus, if the recombination event involves homology, pYeleu 10 must carry a sequence that is repeated at several places on the yeast genome. Integration is not a unique property of the pYeleu 10 plasmid, however. We have also obtained transformation using a plasmid carrying the yeast *his3* gene (pYehis1) (4). These results will be presented elsewhere.

Transformation of yeast makes possible the cloning of eukaryotic genes in a eukaryotic host with a sophisticated genetic system. Baker's yeast has several advantages over enteric bacteria as a host for pharmacologically important genes such as insulin. It is not a pathogen under any known circumstances and, because it is a eukaryote, it probably will allow more efficient expression of such eukaryotic genes. Maximal expression would be especially desirable for commercial applications. Furthermore, transformation permits several novel approaches to the cloning of yeast genes. The association of bacterial sequences, such as ColE1, with yeast genes *in vitro* provides a specific probe for those genes. This bacterial sequence, juxtaposed with a gene of interest, would permit the identification and isolation of that gene. This property could be exploited by transformation back and forth between yeast and *E. coli* as described below.

Many yeast genes of interest fail to function when cloned in *E. coli*. However, it is likely that these genes could be expressed upon transformation back into yeast. The existing banks of yeast–*E. coli* hybrid plasmids could be used to transform yeast auxotrophs. Presumably, the bacterial sequences could often integrate along with the yeast genes of interest (as in type I and type II transformants, Fig. 2) and the position of these sequences could be verified by tetrad analysis. The bacterial plasmid DNA along with the adjacent yeast gene could then be isolated by standard cloning techniques using the original bacterial vector as a hybridization probe. Alternatively, the yeast sequences could be obtained by transforming *E. coli* with a restriction digest of DNA from the yeast transformant. In such a digest, one fragment would contain the original bacterial vector along with the desired yeast sequences. Selection for the bacterial functions carried by the vector (drug or colicin resistance) after transformation into *E. coli* would permit the isolation of this fragment from all other yeast DNA. Thus, the bacterial sequences would provide the vehicle for cloning those yeast sequences that had been adjacent to it on the yeast chromosome.

The successful transformation of yeast by bacterial plasmid DNA raises the possibility that genetic exchange can take place between widely divergent species (17). The behavior of the bacterial plasmid DNA sequences once integrated into yeast chromosomes demonstrates that prokaryotic DNA can be maintained and transmitted by eukaryotes. Genetic exchange by interspecific transformation could lead to the acquisition of blocks of new genes that could contribute to the genetic diversity of the species.

We are deeply grateful to Dr. John Carbon and Dr. Barry Ratzkin who supplied the recombinant plasmids pYeleu10 and pYehis1 and who have shared unpublished results and ideas with us since the inception of this work. We also thank Jean Walsh for her contribution to the development of the yeast colony hybridization technique and to Richard Hallberg for a critical reading of the manuscript. This work was supported by National Science Foundation Grant PCM76-11667 to G.R.F. and a grant from Schweizerischer Nationalfonds to A.H.; J.B.H. was a postdoctoral fellow of the National Institutes of Health.

1. Khan, N. C. & Sen, S. P. (1974) *J. Gen. Microbiol.* **83**, 237–250.
2. Mishra, N. C. & Tatum, E. L. (1973) *Proc. Natl. Acad. Sci. USA* **70**, 3875–3879.
3. Fowell, R. R. (1969) in *The Yeasts*, eds. Rose, A. H. & Harrison, J. S. (Academic, London), Vol. 1, pp. 303–383.
4. Ratzkin, B. & Carbon, J. (1977) *Proc. Natl. Acad. Sci. USA* **74**, 487–491.
5. Hicks, J. & Fink, G. R. (1977) *Nature* **269**, 265–267.
6. Clewell, D. B. (1972) *J. Bacteriol.* **110**, 667–676.
7. Guerry, P., LeBlanc, D. J. & Falkow, S. (1973) *J. Bacteriol.* **116**, 1064–1066.
8. Cryer, D. R., Eccleshall, R. & Marmur, J. (1975) in *Methods in Cell Biology*, ed. Prescott, D. M. (Academic, New York), Vol 12, pp. 39–44.
9. Hutchinson, H. T. & Hartwell, L. H. (1967) *J. Bacteriol* **94**, 1697–1705.
10. van Solingen, P. & van der Plaat, J. B. (1977) *J. Bacteriol.* **130**, 946–947.
11. Southern, E. M. (1975) *J. Mol. Biol.* **98**, 503–517.
12. Maniatis, T., Jeffrey, A. & Kleid, D. (1975) *Proc. Natl. Acad. Sci. USA* **72**, 1184–1188.
13. Grunstein, M. & Hogness, D. S. (1975) *Proc. Natl. Acad. Sci. USA* **72**, 3961–3965.
14. Sherman, F., Fink, G. R. & Lawrence, C. W. (1974) *Methods in Yeast Genetics* (Cold Spring Harbor Laboratory, Cold Spring Harbor, NY).
15. Campbell, A. (1971) in *The Bacteriophage Lambda*, ed. Hershey, A. D. (Cold Spring Harbor Laboratory, Cold Spring Harbor, NY), pp. 13–44.
16. Kellenberger-Gujer, G. & Weisberg, R. A. (1971) in *The Bacteriophage Lambda*, ed. Hershey, A. D. (Cold Spring Harbor Laboratory, Cold Spring Harbor, NY), pp. 407–415.
17. Reanney, D. (1976) *Bacteriol. Rev.* **40**, 552–590.

Molecular Cloning of the Whole Biosynthetic Pathway of a *Streptomyces* Antibiotic and its Expression in a Heterologous Host

F. Malpartida and D.A. Hopwood

The application of molecular cloning to antibiotic-producing microorganisms should lead to enhanced antibiotic productivity and to the biosynthesis of novel antibiotics by *in vitro* interspecific recombination[1,2]. To allow such approaches, the genes for anti-biotic synthesis must be isolated, analysed and perhaps modified. Certain *Streptomyces* species produce nearly two-thirds of the known natural antibiotics[3]; the recent development of cloning systems in the genus[4-7] makes it possible to isolate and analyse *Streptomyces* genes. However, antibiotics are metabolites which require sets of several enzymes for their synthesis and attempts to isolate the corresponding genes have so far yielded clones carrying either individual genes of the set, or only incomplete gene sets[8-11]. We describe here the isolation of a large continuous segment of *Streptomyces coelicolor* DNA which apparently carries the complete genetic information required for synthesis of an antibiotic, actinorhodin, from simple primary metabolites. Not only can the cloned DNA 'complement' all available classes of actinorhodin non-producing mutants of *S. coelicolor* but, on introduction into a different host, *Streptomyces parvulus*, it directs the synthesis of the antibiotic. The tendency for the genes for antibiotic synthesis to be clustered together on the chromosomes of *Streptomyces* species[12] and the availability of plasmid vectors which can carry stable inserts of DNA larger than 30 kilobase pairs (kb) and which can be introduced efficiently into *Streptomyces* protoplasts, suggest that the experiments described have general significance for this area of biotechnology.

Actinorhodin[13] belongs to a class of antibiotics named iso-chromanequinones which also includes granaticin[14], kala-fungin[15] and the nanaomycins[16]. It is derived biosynthetically from acetate units (acetyl coenzyme A) through the polyketide pathway[17]. Rudd and Hopwood[18] isolated 76 actinorhodin non-producing mutants (*act*) and classified them into seven phenotypic classes, where each class represented lesions in a different biosynthetic step. Mutants in six of the classes co-synthesized actinorhodin in combination with other mutants, allowing a tentative biosynthetic sequence to be deduced, as follows:

$$\xrightarrow{\text{I,III}} \xrightarrow{\text{VII}} \xrightarrow{\text{IV}} \xrightarrow{\text{VI}} \xrightarrow{\text{V}} \text{actinorhodin.}$$

Mutants of class II failed to co-synthesize actinorhodin in any combination and perhaps included polar and regulatory muta-tions. At least one mutation of each class was mapped to the same segment of the *S. coelicolor* chromosome[18]. The actino-rhodin system therefore offered the possibility of cloning a DNA fragment encoding a complete biosynthetic pathway provided that a suitable vector was available, with the eventual aim of analysing its physical organization and regulation.

Reconstruction experiments indicated that Act⁺ colonies could be detected easily among a large number of *act* colonies by the formation of the blue diffusible actinorhodin when grown on R2YE medium[6]. 'Complementation' of an *act* mutant, detec-ted by restoration of the pigmented phenotype, was therefore used as the screen for desired clones. A mutant of *act* class V, blocked in a late step of the biosynthesis (and also carrying a *red* mutation to abolish production of the other pigment, undecylprodigiosin[19]) was used as the recipient for cloned DNA from the wild-type *act*⁺ strain.

The plasmid SCP2*, a 31-kb sex factor of *S. coelicolor*[20,21], has recently been developed into a series of cloning vectors by deletion of dispensable DNA and by addition of drug resistance markers[22]. One such vector is pIJ922 (25 kb) which contains a gene for thiostrepton resistance (*tsr*) from *Streptomyces azureus*[6] and has single sites for several restriction enzymes, including *Bam*HI, within a non-essential region. DNA of pIJ922, purified by CsCl–ethidium bromide density gradient centrifugation[21], after alkaline lysis[23], was digested with *Bam*HI. After phenol extraction, 0.5 μg of the linearized plasmid was mixed with 5 μg of chromosomal DNA from an *act*⁺ *S. coelicolor* strain which had been partially digested with *Mbo*I and size-fractionated in a sucrose density gradient[24] to give fragments in the size range 15–30 kb. The DNA mixture, after ethanol precipitation, was dissolved in T4–DNA ligase buffer and ligated[25] at a final DNA concentration of 50 μg ml⁻¹. The ligated DNA was re-precipi-tated and dissolved in 25 μl TE (10 mM Tris-HCl, l mM EDTA, pH 8). An aliquot containing 4.4 μg of ligated DNA was used

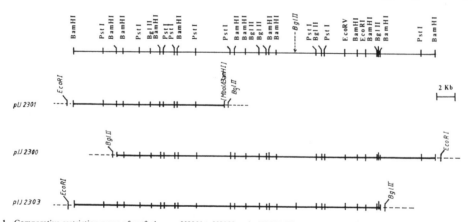

Fig. 1 Comparative restriction maps of *act*⁺ clones pIJ2301, pIJ2300 and pIJ2303. The maps of pIJ2300 and pIJ2303 were derived from the maps of overlapping fragments subcloned into SCP2*-derived vectors. In addition to loss of a 6-kb *Bam*HI fragment from the right-hand end of pIJ2300, pIJ2303 carries a new *Bgl*II site (arrow). From the results of subcloning experiments, this discrepancy does not affect the ability of pIJ2303 to 'complement' all of the *act* mutants tested, and may be explicable by homologous recombination between pIJ2300 (with wild-type DNA) and the chromosome of the *act* mutant in *S. coelicolor* JF4, giving rise to a heterogeneous plasmid population. Dashed lines indicate pIJ922 vector DNA, with the *Eco*RI and *Bgl*II sites shown for orientation.

Table 1 Complementation test of *act*$^+$ recombinant plasmids with *act* *S. coelicolor* mutants and expression of the clones in *S. parvulus*

Recipient strain		Blue pigment produced by strains carrying		
John Innes stock no.	*act* class	pIJ2300	pIJ2301	pIJ2303
S. coelicolor*				
JF1	II	+	−	+
TK17	I	+	−	+
TK18	III	+	−	+
B257	VII	+	+	+
TK16	IV	+	−	+
JF3	VI	−	+	+
JF4	V	+	+	+
S. parvulus 2283†	−	−	−	+
S. parvulus 2618‡	−	−	−	+

* All the strains are mutational and recombinational derivatives of *S. coelicolor* A3(2); their complete genotypes are as previously reported[8].
† Chromosomal marker: *str-1*.
‡ Chromosomal markers: *cys-1 his-5*.

to transform[25] protoplasts of *S. coelicolor* SCP2$^-$ strain JF4 (*proA1 argA1 cysD18 strA1 act109* (V) *redE60*)[8]. After allowing 20 h for phenotypic expression of thiostrepton resistance, the R2YE regeneration medium plates were overlaid with soft nutrient agar (Difco) containing thiostrepton to give a final concentration in the entire plates of 50 µg ml^{-1}.

Two blue colonies (presumptive clones) were detected among a total of ~8,000 transformants growing on 20 plates. Isolation of CCC DNA from the colonies revealed recombinant plasmids (named pIJ2300 and pIJ2301); each was introduced by transformation into *act* mutants of each of the seven classes, with the results shown in Table 1. pIJ2300 complemented all classes except VI, while pIJ2301 complemented classes V, VI and VII. Restriction maps of the plasmids revealed inserts of 34 and 16 kb in pIJ2300 and pIJ2301 respectively, with an apparently overlapping region of 12 kb (Fig. 1).

The entire insert of pIJ2300 was bounded by *Bam*HI sites. Moreover, preliminary subcloning experiments showed that class VI mutations could be complemented by a 4.5-kb *Bam*HI fragment uniquely present at the left-hand end of pIJ2301 (Fig. 1) and presumably the most distal region carrying biosynthetic genes. This suggested a strategy for constructing a single clone able to complement all available *act* mutants, and perhaps containing the entire chromosomal region responsible for actinorhodin biosynthesis. A mixture of pIJ2300 partially digested with *Bam*HI and pIJ2301 completely digested with *Bam*HI (plus *Eco*RI to prevent re-formation of pIJ2301) was ligated and used to transform protoplasts of strain JF3 (*act* class VI). Of 20 blue transformants, one carried a plasmid (pIJ2303) that complemented strongly *act* mutants of all seven classes. The restriction map of pIJ2303 (Fig. 1) demonstrated insertion of the required 4.5-kb *Bam*HI fragment from pIJ2301 at the left end of the pIJ2300 insert and in the original orientation of pIJ2301. At the opposite end, pIJ2303 showed deletion of a 6-kb *Bam*HI fragment present in pIJ2300, but evidently this was not required for complementation of any of the available *act* mutants.

To determine whether pIJ2303 carried the entire set of *act* biosynthesis genes, pIJ2300, pIJ2301 and pIJ2303 were introduced by transformation into genetically marked strains of *S. parvulus* ATCC 12434. This strain does not produce actinorhodin and is not a close relative of *S. coelicolor* A3(2) as judged either by conventional taxonomy or by DNA–DNA hybridization[26]. It is, however, a host for SCP2 plasmids. Only those transformants that grew on R2YE plates containing thiostrepton and which carried pIJ2303 were blue (Table 1). Thin-layer chromatography of methanol extracts[19] of the *S. parvulus* host cultures and the same strains containing each of the three recombinant plasmids showed the presence of a blue pigment only in clones carrying pIJ2303. This blue pigment co-migrated with a corresponding pigment in extracts from *act*$^+$ but not *act*

S. coelicolor cultures and had acid–base indicator properties characteristic of actinorhodin. The amounts of the blue pigment in extracts of *S. parvulus* carrying pIJ2303 were similar to those of wild-type *S. coelicolor*. This result suggests that the entire DNA sequence corresponding to the structural genes for actinorhodin biosynthesis is carried by pIJ2303.

The success of the cloning experiments reported here may well have depended on the use of a low-copy number vector (the estimated copy number of the parent SCP2 is one per chromosome of *S. coelicolor*[21]) so that the recipient strain is not likely to be placed at a disadvantage by possible high-level expression of physiologically active gene products. This view is supported, though not proved, by our failure to isolate any Act$^+$ clones in extensive trials with the high-copy number vector pIJ702 (ref. 27) despite the fact that this vector has proved convenient for the cloning of comparatively short segments of DNA carrying other genes[8,28]. Vectors such as pIJ922, and other SCP2* derivatives recently developed by Lydiate[22], appear to be particularly suitable for the stable maintenance of large fragments of cloned DNA. Moreover, the large recombinant plasmids are readily isolated from *Streptomyces* cultures[23], and transform *Streptomyces* protoplasts efficiently. Thus pIJ2300, of 59 kb, yielded 10^6 transformants per µg of DNA in optimal conditions, compared with 10^7 for the pIJ922 vector.

As the entire carbon skeleton of actinorhodin is derived from acetate[17], no specialized primary metabolites are required for its biosynthesis, a factor which may have contributed to the efficient expression of pIJ2303 in *S. parvulus*, leading to production of actinorhodin. Many important antibiotics are also polyketides—including the macrolides, tetracyclines and rifamycins. It is generally accepted that the primary carbon skeletons of polyketides are assembled on multienzyme complexes (or multifunctional enzymes) analogous to fatty acid synthetases[29], and the genes or gene clusters that encode polyketide synthetases may have had a common evolutionary origin, perhaps shared with fatty acid synthetases. Thus, the availability of clones carrying the presumptive actinorhodin 'synthetase' may be useful, not only as a model system, but possibly as DNA probes for the isolation of other antibiotic synthetases.

We thank Derek Lydiate for making available his unpublished results and for vector plasmids, Jerry Feitelson for helpful discussions, and Mervyn Bibb, Keith Chater and Tobias Kieser for useful comments. F.M. acknowledges a post-doctoral stipend from the Juan March Foundation of Madrid.

Received 23 February; accepted 29 March 1984.

1. Hopwood, D. A. & Chater, K. F. *Phil. Trans. R. Soc.* B290, 313–328 (1980).
2. Hopwood, D. A. in β-Lactam Antibiotics (eds Salton, M. R. J. & Shockman, G. D.) 585–598 (Academic, New York, 1979).
3. Bérdy, J. *Process. Biochem.* Oct./Nov., 28–35 (1980).
4. Bibb, M. J., Schottel, J. L. & Cohen, S. N. *Nature* 284, 526–531 (1980).
5. Suarez, J. E. & Chater, K. F. *Nature* 286, 527–529 (1980).
6. Thompson, C. J., Ward, J. M. & Hopwood, D. A. *Nature* 286, 525–527 (1980).
7. Bibb, M. J., Chater, K. F. & Hopwood, D. A. in *Experimental Manipulation of Gene Expression* (ed. Inouye, M.) 53–82 (Academic, New York, 1983).
8. Feitelson, J. S. & Hopwood, D. A. *Molec. gen. Genet.* 190, 394–398 (1983).
9. Gil, J. A. & Hopwood, D. A. *Gene* 25, 119–132 (1983).
10. Chater, K. F. & Bruton, C. J. *Gene* 26, 67–78 (1983).
11. Hopwood, D. A., Bibb, M. J., Bruton, C. J., Feitelson, J. S. & Gil, J. A. *Trends Biotechnol.* 1, 42–48 (1983).
12. Hopwood, D. A. in *Biochemistry and Genetic Regulation of Commercially Important Antibiotics* (ed. Vining, L. C.) 1–23 (Addison-Wesley, Reading, Massachusetts, 1983).
13. Brockman, H., Zeeck, A., van der Merve, K. & Müller, W. *Justus Liebigs Annln Chem.* 698, 3575–3579 (1966).
14. Carbaz, R. *et al. Helv. chim. Acta* 40, 1262–1269 (1957).
15. Hoeksema, H. & Krueger, W. C. *J. Antibiot., Tokyo* 29, 704–709 (1976).
16. Tanaka, H., Koyama, V., Nagai, T., Marumo, H. & Ōmura, S. *J. Antibiot., Tokyo* 28, 868–875 (1975).
17. Gorst-Allman, C. P., Rudd, B. A. M., Chang, C.-J. & Floss, H. G. *J. org. Chem.* 46, 455–456 (1981).
18. Rudd, B. A. M. & Hopwood, D. A. *J. gen. Microbiol.* 114, 35–43 (1979).
19. Rudd, B. A. M. & Hopwood, D. A. *J. gen. Microbiol.* 119, 333–340 (1980).
20. Schrempf, H., Bujard, H., Hopwood, D. A. & Goebel, W. J. *Bact.* 121, 416–421 (1975).
21. Bibb, M. J., Freeman, R. F. & Hopwood, D. A. *Molec. gen. Genet.* 154, 155–166 (1977).
22. Lydiate, D. J. thesis, Univ. East Anglia, Norwich (1984).
23. Kieser, T. *Plasmid* (in the press).
24. Maniatis, T., Hardison, R. C., Lacy, E., Lauer, J. & O'Connell, C. *Cell* 15, 687–701 (1978).
25. Thompson, C. J., Ward, J. M. & Hopwood, D. A. *J. Bact.* 151, 668–677 (1982).
26. Westpheling, J. thesis, Univ. East Anglia, Norwich (1980).
27. Katz, E., Thompson, C. J. & Hopwood, D. A. *J. gen. Microbiol.* 129, 2703–2714 (1983).
28. Seno, E. T., Bruton, C. J. & Chater, K. F. *Molec. gen. Genet.* 193, 119–128 (1984).
29. Packter, N. M. in *The Biochemistry of Plants* Vol. 4 (ed. Strumpf, P. K.) 535–570 (Academic, London, 1983).

Secretion of *Escherichia Coli* β-lactamase from *Bacillus Subtilis* by the Aid of α-amylase Signal Sequence

I. Palva, M. Sarvas, P. Lehtovaara, M. Sibakov and L. Kääriäinen

ABSTRACT   We describe a secretion vector system for introducing foreign genes into *Bacillus subtilis*. We constructed secretion vectors from the plasmid pUB110 and the promoter and signal sequence region of the α-amylase gene from *Bacillus amyloliquefaciens*. Foreign structural genes can be inserted into the various vectors after the signal sequence region of the α-amylase gene. Demonstrating secretion of a foreign gene product from *Bacillus*, we here report that the *Escherichia coli* β-lactamase gene, devoid of its own signal sequence coding region, can be expressed in *B. subtilis* by the aid of the secretion vectors so that >95% of the enzyme activity is secreted to the growth medium. Efficient secretion of β-lactamase (penicillin amido-β-lactamhydrolase, EC 3.5.2.6) is observed if the complete signal sequence coding region of the α-amylase gene precedes the β-lactamase structural gene. However, an incomplete α-amylase signal peptide lacking the six carboxy-terminal amino acid residues does not promote secretion of the fused β-lactamase, which remains unprocessed and cell-associated.

Bacilli are potential hosts for the biosynthesis of foreign gene products in microorganisms by recombinant DNA technology. Compared to *Escherichia coli*, *B. subtilis* has the following advantages: (*i*) it is nonpathogenic, (*ii*) it is well known as an industrial organism, and (*iii*) it can secrete proteins to the culture medium. However, efficient expression of foreign genes and secretion of the gene products in *Bacilli* is still hampered by several problems. For instance, the secretory mechanisms of gram-positive *Bacilli* are not sufficiently well characterized, and particularly the control mechanisms remain unresolved. There also has been a lack of suitable vectors to promote expression and secretion of the product of an inserted gene. Expression of hepatitis B core antigen and of the major antigen of foot-and-mouth disease virus in *B. subtilis* has been obtained (1), but there have been no reports on secretion of foreign gene products by *Bacilli*.

The enzyme α-amylase (1,4-α-D-glucon gluconohydrolase, EC 3.2.1.1) is secreted in large amounts by some *Bacilli* to the culture medium. We have chosen the α-amylase gene as a model system to study the regulatory gene regions involved in efficient expression and secretion. The chromosomal α-amylase gene from *Bacillus amyloliquefaciens*, including the promoter region, has been isolated and cloned in the plasmid pUB110, with *B. subtilis* as host organism (2). The nucleotide sequence of the promoter and signal sequence region of this gene has been determined (3). Here we have constructed a series of secretion vectors for cloning in *B. subtilis* in hopes of obtaining both efficient expression and secretion of foreign gene products. The vectors are derivatives of the plasmid pUB110, and they contain

the promoter, ribosome binding site, and signal sequence region of the α-amylase gene in front of the site where a foreign gene can be inserted. It has been shown that expression of foreign genes in *B. subtilis* apparently requires transcriptional and translational initiation signals from *Bacillus* or other gram-positive bacteria (4–6). Here we show that *B. subtilis* can secrete enzymatically active β-lactamase (penicillin amido-β-lactamhydrolase, EC 3.5.2.6) to the culture medium after transformation with hybrid plasmids containing the α-amylase promoter and complete signal sequence region fused to the pBR322 structural gene of β-lactamase.

MATERIALS AND METHODS

Bacteria, Plasmids, and Media. *Bacillus subtilis* strain IH6140, a prototrophic derivative of *B. subtilis* Marburg with reduced level of exoprotease activity, was from our collection. Plasmid pHV33, a tetracycline-resistant derivative of pHV14 (7), was obtained from S. D. Ehrlich. Plasmids pKTH38 and pKTH39 were derived from pKTH10 (2, 3) by exonuclease treatment as described below. Plasmid pKN410 carrying the β-lactamase gene (8) was a kind gift from K. Nordström through T. Palva. *B. subtilis* IH6140 transformants were selected on L-broth plates supplemented with 10 μg of kanamycin per ml. To assay the β-lactamase activity in liquid cultures, *B. subtilis* were grown in minimal medium (9) containing 0.5% glycerol, 1% soluble starch, and 10 μg of kanamycin per ml.

Cloning and Plasmid Isolation. *E. coli* K-12 strain HB101 was transformed (10), and plasmids were isolated (11) under standard conditions. Transformation of *B. subtilis* was as described by Gryczan *et al.* (12). To obtain plasmid multimers necessary for the physiological transformation of *B. subtilis*, the ligation was performed at 23°C for 3 hr by using an insert/plasmid ratio of 2:1 and a plasmid concentration of 75 μg/ml. Plasmid isolation in preparative scale was as described (3) and rapid screening was done by the method of Birnboim and Doly (13).

Exonuclease Treatments. Partial digestion of 9 pmol of linearized plasmid DNA with 75 units (18 pmol) of *E. coli* exonuclease III (New England BioLabs) was in 100 μl of 20 mM Tris·HCl, pH 7.6/0.66 mM MgCl$_2$/1 mM 2-mercaptoethanol at 22°C. Aliquots were removed at several time points (10–20 min) to an equal volume of 60 mM sodium acetate, pH 4.5/0.6 M NaCl/6 mM ZnSO$_4$ to stop the reaction. To obtain blunt ends, the pooled DNA was digested with S1 nuclease (0.1 units, P-L Biochemicals) first for 30 min at 37°C and then for 5 min at 4°C, followed by phenol extraction and a fill-in reaction with 10 units of reverse transcriptase (gift of J. Beard, Life Sciences) in 50 mM Tris·HCl, pH 8.3/60 mM KCl/12 mM MgCl$_2$/1 mM dithiothreitol/150 μM dNTPs for 30 min at 37°C.

Partial digestion of 10 pmol of linearized plasmid DNA with 1 unit of BAL-31 (Bethesda Research Laboratories) was in 250

PALVA, I., SARVAS, M., LEHTOVAARA, P., SIBAKOV, M. and KÄÄRIÄINEN, L.
Secretion of *Escherichia coli* β-lactamase from *Bacillus subtilis* by the aid of α-amylase signal sequence.
Proc. Natl. Acad. Sci. U.S.A. 79:5582-5586 (1982). Reprinted with the authors' permission.

μl of 20 mM Tris·HCl, pH 8.1/0.6 M NaCl/12 mM CaCl₂/ 12 mM MgCl₂/1 mM EDTA at 30°C. An excess of EDTA was added to aliquots removed at time points between 5 and 7.5 min, and the pooled DNA was extracted with phenol and filled in with T4 DNA polymerase (P-L Biochemicals) as follows: 3.3 pmol of DNA termini, 10 units of T4 polymerase, 50 μM of each dNTP in 6.3 mM MgCl₂/63 mM Tris·HCl, pH 8.1, in 20 μl for 80 min at 11°C. Linkers were ligated to linearized plasmids at 15°C for 16 hr by using a linker-to-plasmid molar ratio of 40:1.

Other Methods. β-lactamase was assayed from the supernatants of cultured cells by the method of Callaghan *et al.* (14). After electrophoresis in 12.5% NaDodSO₄/polyacrylamide gels (15), β-lactamase was visualized by an immunoblotting method (16) by using anti-β-lactamase anti-serum and ¹²⁵I-labeled protein A. DNA sequences were determined by the method of Maxam and Gilbert (17) as described (3).

RESULTS

Construction of Secretion Vectors for Cloning in *Bacillus*. The parental plasmid, pKTH38, is a derivative of the plasmid pKTH10 (2, 3), which contains the cloned chromosomal α-amylase gene from *B. amyloliquefaciens*. In pKTH38, the promoter and signal sequence for the secreted α-amylase is followed by 90 nucleotides from the coding region and an *Eco*RI linker. The plasmid pKTH38 was linearized with *Eco*RI (Fig. 1A), partially digested with exonuclease BAL-31, followed by ligation of *Hind*III linkers to the digested plasmid population. Linear plasmid monomers with *Hind*III linkers at the termini were purified, ligated, and transformed into *B. subtilis*.

The position of *Hind*III linkers in the plasmid DNA of 30 individual transformants was determined by sequence determination of the purified DNAs labeled at the *Hind*III sites and cleaved with *Cla* I. The sequence ladders A+G and C+T (17), covering nucleotides 1–40 for each sample, unequivocally determined the position of the *Hind*III linkers within the previously determined nucleotide sequence region of the α-amylase gene (3). By this screening, we obtained 12 different potential secretion vectors, with the *Hind*III linkers attached to the codon at the signal-sequence-processing site (pKTH50) or at various amino-terminal codons of the α-amylase structural

gene. The gene coding for mature β-lactamase was fused to the plasmid pKTH50 to construct pKTH78.

We also prepared a pool from 120 isolated transformants containing *Hind*III linkers at various sites and characterized the pool by electrophoresis of the *Cla* I/*Hind*III double digests (Fig. 2). The size distribution of the *Cla* I/*Hind*III fragments showed that the *Hind*III linkers are attached within a 70-nucleotide region that should cover also the signal sequence cleavage site of α-amylase. When *B. subtilis* was transformed with plasmids that were recombinants of this pool and the β-lactamase gene, and when the transformants were screened for expression of β-lactamase, the plasmids pKTH83, -84, and -86 were obtained (Fig. 3).

Isolation of the β-Lactamase Gene Devoid of the Signal Sequence. Expression of an inserted foreign gene in *B. subtilis* and secretion of the product was tested by using β-lactamase gene from pBR322. For isolation of the β-lactamase gene devoid of its signal sequence, a fusion plasmid pHV33 (Fig. 1B) was linearized with *Eco*RI and partially digested with *E. coli* exonuclease III. The use of this fusion plasmid instead of pBR322 enabled the preservation of the tetracycline-resistance marker of pBR322. Exonuclease III/S1 nuclease-digested plasmids were attached to *Eco*RI linkers, sealed, and transformed into *E. coli*. Plasmids from ampicillin-sensitive transformants were isolated, and the *Eco*RI linkers were localized by nucleotide sequence determination from the *Eco*RI sites. Thus, the plasmid pKTH33 obtained has an *Eco*RI linker attached to the β-lactamase gene that is devoid of its signal sequence.

To obtain a construction with an *Eco*RI linker also near the 3′ end of β-lactamase gene, the *Bst*NI fragment containing this gene was isolated from pKTH33, *Eco*RI linkers were attached to it, and, after *Eco*RI digestion, the fragment was cloned in pBR325. For insertion of this β-lactamase gene into the *Bacillus* secretion vectors, the *Eco*RI linkers were replaced by *Hind*III linkers, and the fragment was recloned in pBR322.

Fusion of β-Lactamase Gene with *Bacillus* Secretion Vectors. The β-lactamase gene from pBR322, devoid of its signal sequence and containing *Hind*III termini, was ligated with the secretion vector pool. It also was ligated with one of the secretion vectors of known sequence, pKTH50, that had the linker attached to position −1 in the α-amylase signal sequence. The

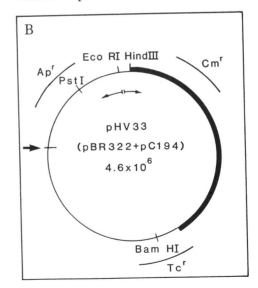

Fig. 1. (A) Plasmid pKTH38. The thin line represents pUB110 DNA, and the heavy black line represents a 0.7-kilobase chromosomal DNA insert from *B. amyloliquefaciens*. The *Eco*RI site is formed by a DNA linker at a position coding for amino acid 32 in the mature α-amylase. The arrows ←→ indicate the BAL-31 digestion starting at the *Eco*RI site. P and ss, promoter–signal sequence area of the α-amylase gene. (B) Plasmid pHV33. The thin line represents pBR322 DNA, and the heavy black line represents the pC194 DNA. The arrows ←→ indicate the *E. coli* exonuclease III/S1 digestion starting at the *Eco*RI site of pBR322. The position of the relevant *Bst*NI site is shown by the thick arrow (➤).

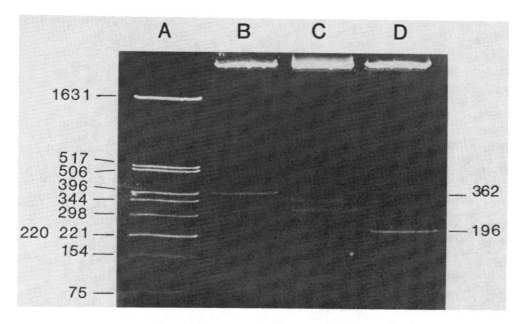

FIG. 2.   Polyacrylamide gradient gel electrophoresis (18) of the BAL-31 digest of pKTH38. Lanes: A, *Hinf*I digest of pBR322; B, *Cla* I/*Eco*RI double digest of plasmid pKTH38, which has an *Eco*RI linker 362 base pairs from the *Cla* I site; C, *Cla* I/*Hind*III double digest of a plasmid pool isolated from 120 independent BAL-31-digested pKTH38 derivatives; D, *Cla* I/*Eco*RI double digest of plasmid pKTH39, a pKTH10 derivative, which has an *Eco*RI linker 196 base pairs from the *Cla* I site. Sizes are shown in base pairs.

hybrid plasmids were transformed into *B. subtilis*, the kanamycin-resistant transformants were screened for β-lactamase activity, and the sequences of some hybrid plasmids isolated from the β-lactamase-positive transformants were partially determined from the *Hind*III sites as described above.

Fig. 3 shows nucleotide sequences at the α-amylase gene–β-lactamase gene junctions of the plasmids pKTH78, pKTH83, pKTH84, and pKTH86. In plasmid pKTH78, only one nucleotide is missing from the full signal sequence of α-amylase, and the linker is attached to the codon for amino acid residue −1 (−1 construction). The plasmids pKTH83 and -84 contain in addition to the complete signal sequence, 11 nucleotides (+4 construction) and 41 nucleotides (+14 construction), respectively, from the α-amylase gene. In plasmid pKTH86, the β-lactamase gene is joined to an incomplete α-amylase signal sequence, which is devoid of 19 nucleotides (−7 construction). The deduced amino acid sequences of these constructions are shown in Fig. 4 together with those of the signal sequence and

amino terminus of α-amylase and β-lactamase. In all four hybrid plasmids, an additional tetrapeptide sequence, Gln-Ala-Cys-Pro, coded by the linker construction, precedes the mature β-lactamase that starts from the proline residue at position 2.

Expression of the β-Lactamase Gene Fused to the Vectors. *B. subtilis* carrying the hybrid plasmids were grown in glycerol minimal medium supplemented with starch, and the β-lactamase activity was assayed from the culture supernatants and from sonicated cells (Fig. 5). The activity curves for strains carrying the plasmids pKTH78 and pKTH83 (construction −1 and +4) were almost identical (Fig. 5 A and B). The maximal activity, about 2,000 units/ml, was obtained 6–8 hr after the logarithmic growth phase.

B. subtilis harboring the plasmids with −1 or +4 vector constructions were able to secrete over 95% of the β-lactamase activity to the growth medium, and those carrying the plasmid pKTH84 (+14 construction) secreted over 90%. The total activity obtained by using the +14 construction was only 20% of

```
                   −31  −30                         −25                          −20                          −15                          −10
pKTH10           ATG  ATT  CAA  AAA  CGA  AAG  CGG  ACA  GTT  TCG  TTC  AGA  CTT  GTG  CTT  ATG  TGC  ACG  CTG  TTA  TTT  GTC  AGT  TTG

                   −5                          −1  +1                          +5                          +10                          +15
                 CCG  ATT  ACA  AAA  ACA  TCA  GCC  GTA  AAT  GGC  ACG  CTG  ATG  CAG  TAT  TTT  GAA  TGG  TAT  ACG  CCG  AAC  GAC  GGC


                                        −4   −3   −2   −1                           +2   +3   +4
pKTH78                       ...  AAA  ACA  TCA  GC G  CAA  GCT  TGC  CC C  CCA  GGA  ACG  ...

                                        +1   +2   +3   +4
pKTH83                       ...  GTA  AAT  GGC  AC G  CAA  GCT  TGC  CC C  CCA  GGA  ACG  ...

                                       +11  +12  +13 +14
pKTH84                       ...  TGG  TAT  ACG  CC G  CAA  GCT  TGC  CC C  CCA  GGA  ACG  ...

                                       −10   −9   −8   −7
pKTH86                       ...  GTC  AGT  TTG  CC G  CAA  GCT  TGC  CC C  CCA  GGA  ACG  ...


                             α-amylase gene                      β-lactamase gene
```

FIG. 3.   The nucleotide sequence of the α-amylase gene–β-lactamase gene junction in the hybrid plasmids pKTH78, pKTH83, pKTH84, and pKTH86. The nucleotides from the linkers are encased (G-C-A-A-G-C-T-T-G-C is the *Hind*III linker and C-C is from an *Eco*RI linker), and those from the β-lactamase structural gene are given in italics. The complete signal sequence (−31 to −1), and the codons for 17 amino-terminal amino acids in the α-amylase gene contained in pKTH10 (3) are shown.

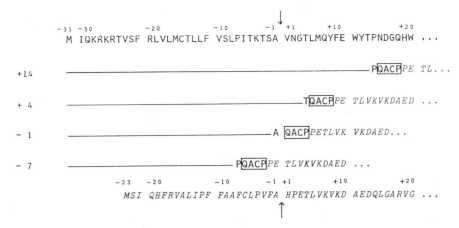

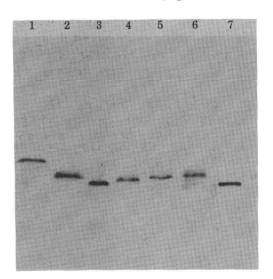

FIG. 4. The deduced amino acid sequences coded by the α-amylase and β-lactamase hybrid constructions +14 (pKTH84), +4 (pKTH83), −1 (pKTH78), and −7 (pKTH86). The sequences coded by the linker are encased, and the β-lactamase sequences are shown in italics. Identity is denoted by a line. The β-lactamase structural gene in the hybrids starts with a codon for proline at position +2. For comparison, the signal peptide and amino terminus of α-amylase (3) is given at the top and those of β-lactamase from pBR322 (19) are given at the bottom of the figure. The arrows indicate the site for processing for α-amylase and β-lactamase. A, Ala; C, Cys; D, Asp; E, Glu; F, Phe; G, Gly; H, His; I, Ile; K, Lys; L, Leu; M, Met; N, Asn; P, Pro; Q, Gln; R, Arg; S, Ser; T, Thr; V, Val; W, Trp; Y, Tyr.

that from the −1 or +4 constructions. The −7 construction, in contrast to the three others, did not contain the full signal sequence of α-amylase, and 90% of the β-lactamase activity was found to be cell bound. Also the total activity was low—only 10% of that obtained with the −1 and +4 vector constructions.

The size of the β-lactamase produced was estimated by NaDodSO₄/polyacrylamide gel electrophoresis in combination with an immunoblotting technique (Fig. 6). The β-lactamase secreted by all of the constructions −1, +4, and +14 (Fig. 6, lanes 4, 5, and 6, respectively) was somewhat larger than the authentic enzyme purified from *E. coli* carrying plasmid pKN410 (Fig. 6, lanes 3 and 7). This can be understood on the basis of the DNA sequence data, which indicates that the β-lactamase is preceded by 3, 7, or 18 extra amino acids in the constructions −1, +4, and +14, respectively. Apparently the signal sequence of α-amylase is processed at or near the correct position in these constructions. However, the cell-bound β-lactamase produced by construction −7 migrated more slowly than the other products (Fig. 6, lane 1), suggesting that the incomplete signal sequence of α-amylase was not cleaved. Somewhat unexpected was the relatively slow mobility of the "secreted" product from the same construction (Fig. 6, lane 2).

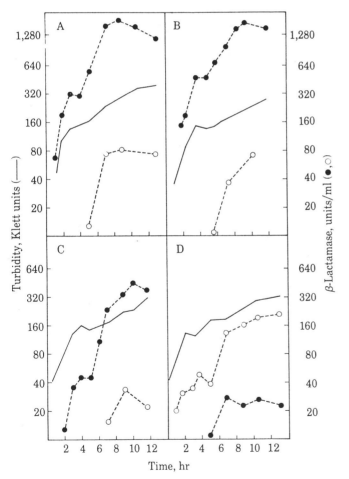

FIG. 5. Activity curves of β-lactamase assayed from the culture supernatants and sonicated cells of the different α-amylase gene–β-lactamase gene hybrid constructions. These constructions were: −1 (*A*); +4 (*B*); +14 (*C*); and −7 (*D*). The cells were grown in Spizizen minimal medium containing 0.5% glycerol, 1% soluble starch, and 10 μg of kanamycin per ml in a rotatory shaker at 37°C. ●----●, β-Lactamase activity in supernatant; ○----○, β-lactamase activity in sonicated cells; ——, turbidity.

FIG. 6. Immunoblotting analysis of the β-lactamase synthesized by α-amylase gene–β-lactamase gene hybrid constructions. Lanes: 1, cells of −7 construction; 2, supernatant of −7 construction; 3 and 7, β-lactamase isolated from *E. coli* carrying the plasmid pKH410; 4, supernatant of −1 construction; 5, supernatant of +4 construction; and 6, supernatant of +14 construction.

DISCUSSION

We report here the construction of hybrid plasmids to be used as expression and secretion vectors for cloning of foreign genes in *Bacillus subtilis*. The ability of the plasmid constructions to promote secretion was first tested by introducing the β-lactamase gene, devoid of its own signal sequence, from pBR322 into the vectors after the promoter and signal sequence region of α-amylase. The transformed *B. subtilis* strain secreted β-lactamase efficiently to the culture medium, thus demonstrating the potential of this system.

The size estimations of the secreted β-lactamase show that it was processed correctly or at least within a few amino acid residues from the correct position in the constructions -1 and $+4$. The production curves for these constructions were nearly indistinguishable. This suggests that the amino acid sequence after the signal peptide cleavage site may not be critical for the specificity of the signal peptidase(s) in *B. subtilis* and that the signal sequence of α-amylase is sufficient for the processing and secretion of β-lactamase. Most of the β-lactamase coded by the -7 construction was membrane bound and apparently unprocessed, showing that the carboxyl-terminal six amino acids of the α-amylase signal peptide are necessary for secretion and processing.

The yield of β-lactamase by the $+14$ construction was surprisingly only 20% of that produced by the -1 and $+4$ constructions. This could be due to (*i*) a lower specific activity of the enzyme with an amino-terminal extension, (*ii*) a lower rate of synthesis, or (*iii*) greater susceptibility to proteolytic degradation. According to the quantitations from immunoblotting analysis (Fig. 6), all of the different constructions produced β-lactamase with approximately the same specific activity (data not shown), which would exclude the first possibility.

Secretion of β-lactamase begins at the end of the logarithmic growth phase, like the secretion of α-amylase. The maximal amount of extracellular β-lactamase in L broth with starch media was obtained 4–5 hr after the logarithmic phase and was about 6,200 units/ml, corresponding to 20 μg/ml. It seems probable that the proteases emerging at the beginning of the stationary growth phase of *B. subtilis* still prevent us from using the full capacity of this system, even though our host has a reduced level of proteases (unpublished data). *B. subtilis* harboring the plasmid pKTH10 continues to accumulate α-amylase in the supernatant for 45 hr (2).

Our approach to construction of the vector resembled that used by Gilbert and co-workers (20, 21), who fused rat proinsulin to the β-lactamase signal sequence and obtained transport of the proinsulin into the periplasmic space of *E. coli*. The complete signal peptide of α-amylase was in our system sufficient to direct secretion and processing of *E. coli* β-lactamase in *B. subtilis*, but a shortened signal peptide lacking the six carboxyl-terminal amino acid residues was not. The gene for β-lactamase was chosen to test our secretion vector because this enzyme is normally secreted into the periplasmic space of *E. coli*. We also have found that human interferon can be secreted from *B. subtilis* to the growth medium (unpublished data). It remains to be clarified whether this system also can be applied to promote secretion of proteins which normally are not secreted.

We thank Mses. Eila Kujamäki, Hannele Lehtonen, and AnnaLiisa Ruuska for their excellent technical help. This work was funded by the Finnish National Fund for Research and Development (SITRA).

1. Hardy, K., Stahl, S. & Küpper, H. (1981) *Nature (London)* **293**, 481–483.
2. Palva, I. (1981) *Gene*, in press.
3. Palva, I., Pettersson, R. F., Kalkkinen, N., Lehtovaara, P., Sarvas, M. Söderlund, H., Takkinen, K. & Kääriäinen, L. (1981) *Gene* **15**, 43–51.
4. Williams, D. M., Schoner, R. G., Duvall, E. J., Preis, L. H. & Lovett, P. S. (1981) *Gene* **16**, 199–206.
5. Goldfarb, D. S., Doi, R. H. & Rodriguez, R. L. (1981) *Nature (London)* **293**, 309–311.
6. McLaughlin, J. R., Murray, C. L. & Rabinowitz, J. C. (1981) *J. Biol. Chem.* **256**, 11283–11291.
7. Ehrlich, S. D. (1978) *Proc. Natl. Acad. Sci. USA* **75**, 1433–1436.
8. Uhlin, B. E., Molin, S., Gustafsson, P. & Nordström, K. (1979) *Gene* **6**, 91–106.
9. Anagnostopoulos, C. & Spizizen, J. (1961) *J. Bacteriol.* **81**, 741–746.
10. Mandell, M. & Higa, A. (1970) *J. Mol. Biol.* **53**, 159–162.
11. Clewell, D. B. & Helinski, D. R. (1969) *Proc. Natl. Acad. Sci. USA* **62**, 1159–1166.
12. Gryczan, T., Contente, S. & Dubnau, D. (1978) *J. Bacteriol.* **134**, 318–329.
13. Birnboim, H. C. & Doly, J. (1979) *Nucleic Acids Res.* **7**, 1513–1523.
14. Callaghan, C. H., Morris, A., Kirby, S. & Schnigler, A. H. (1972) *J. Antimicrob. Chemother.* **1**, 283–288.
15. Laemmli, U. K. (1970) *Nature (London)* **227**, 680–685.
16. Towbin, H., Staehelin, T. & Gordon, J. (1979) *Proc. Natl. Acad. Sci. USA* **76**, 4350–4354.
17. Maxam, A. M. & Gilbert, W. (1980) *Methods Enzymol.* **65**, 499–560.
18. Jeppesen, P. G. N. (1980) *Methods Enzymol.* **65**, 305–319.
19. Sutcliffe, J. G. (1978) *Proc. Natl. Acad. Sci. USA* **75**, 3737–3741.
20. Talmadge, K., Stahl, S. & Gilbert, W. (1980) *Proc. Natl. Acad. Sci. USA* **77**, 3369–3373.
21. Talmadge, K., Kaufman, J. & Gilbert, W. (1980) *Proc. Natl. Acad. Sci. USA* **77**, 3988–3992.

Isolation and Purification of Protein Granules from *Escherichia Coli* Cells Overproducing Bovine Growth Hormone

R.G. Schoner, L.F. Ellis and B.E. Schoner

High-level expression of bovine growth hormone in *Escherichia coli* results in the formation of distinct cytoplasmic granules that are visible with the phase-contrast microscope. These granules were examined by transmission and scanning electron microscopy. Intact granules were isolated from crude cell lysates by differential centrifugation and were further purified by a simple washing procedure that yields nearly homogeneous bovine growth hormone.

The overproduction of many procaryotic and eucaryotic proteins in *Escherichia coli* has been made possible through the use of recombinant DNA technology. High-level expression of these proteins has been achieved by cloning their coding sequences into multicopy plasmids downstream from strong promoters and ribosome binding sites to create recombinant expression plasmids. Using this approach, several investigators have reported that products of cloned genes can accumulate in certain cases up to 50% of total *E. coli* cell protein[1-10]. In many of these cases it was observed that the overproduced proteins aggregate and pellet with the cell debris during low-speed centrifugation of crude lysates. Examples of such aggregated proteins are the *E. coli* sigma subunit of RNA polymerase[1], the product of the *envZ* gene of *E. coli*[2], the cII repressor protein of bacteriophage λ[3], human gamma-interferon[4], interleukin-2[5], a nonsense fragment of the *E. coli* β-galactosidase[6] and several hybrid fusion proteins consisting of procaryotic and eucaryotic sequences[7-10].

In a previous paper[11] we reported the high-level expression of a derivative of bovine growth hormone (bGH) that contains 8 additional N-terminal amino acids. In this report, we show that this bGH derivative forms aggregates in the *E. coli* cytoplasm. These aggregates form distinct granules that are visible with a phase-contrast microscope. Cytoplasmic and isolated granules were examined by electron microscopy, and purified by a simple procedure that yields nearly homogeneous bGH from these granules.

RESULTS

Phase-contrast microscopy of *E. coli* cells overproducing bovine growth hormone. In an earlier paper[11], we described the construction of a thermoinducible runaway-replication plasmid, pCZ101, into which we cloned the coding sequence for bGH. A temperature shift from 25°C to 37°C leads to amplification of the plasmid copy number from about 10 copies per cell in cultures grown at 25°C to about 1000 copies per cell in cultures grown at 37°C and a simultaneous increase in bGH production. In Figure 1, we compare the appearance of *E. coli* RV308 cells harboring

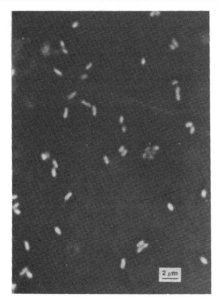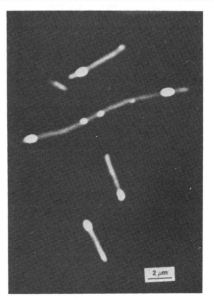

FIGURE 1 Phase-contrast microscopy of RV308/pCZ101 cells grown at 25°C (A) and at 37°C (B). The bar equals 2 μm.

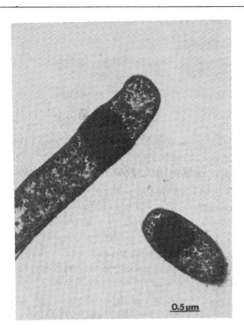

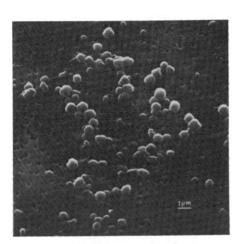

FIGURE 2 A Transmission electron micrograph of RV308/pCZ101 cells grown at 37°C for 6 hrs. The bar equals 0.5 μm. **B** Scanning electron micrograph of bGH granules from RV308/pCZ101. The bar equals 1 μm.

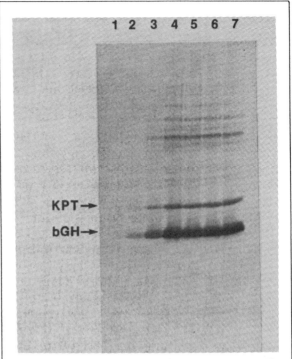

FIGURE 3 bGH synthesis and stability in RV308/pCZ101. [^{35}S] Methionine pulse label analysis: lane 1, 1 min; lane 2, 3 min; lane 3, 10 min; lane 4, 30 min; pulse-chase analysis: lane 5, 3 min; lane 6, 10 min; lane 7, 30 min. The arrows indicate the positions of bGH and kanamycin phosphotransferase (KPT).

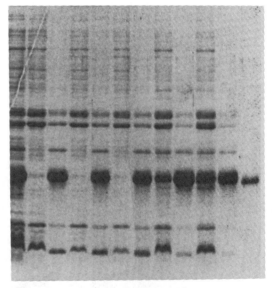

FIGURE 4 Analysis of bGH granules by SDS polyacrylamide gel electrophoresis. The crude granule preparation (lane 1) was washed with increasing concentrations of urea and the granular pellets (lanes 3, 5, 7, 9 and 11) and supernatants (lanes 2, 4, 6, 8, 10) were examined. The urea concentrations were: 0.5 M (lanes 2 and 3); 1.0 M (lanes 4 and 5); 2.0 M (lanes 6 and 7); 5.0 M (lanes 8 and 9); 5.0 M urea and 1% Triton X-100 (lanes 10 and 11). Lane 12 contains the bGH size standard.

pCZ101 when grown for 6 hrs at 25°C (Fig. 1A) and 37°C (Fig. 1B) with a phase-contrast microscope. The culture grown at 25°C contains typical, short and rod-shaped *E. coli* cells whereas the culture grown at 37°C contains abnormally elongated cells with multiple, highly refractile cytoplasmic bodies (granules).

In the 37°C cultures, small granules first become visible 2–3 hours after the temperature shift; they are usually located at one pole of the cell. During the next 4 hours, these "early" granules enlarge and additional granules appear, first at the opposite pole of the cell, and then between poles (Fig. 1B).

In cultures kept at 25°C, no detectable bGH is synthesized, whereas in cultures shifted to 37°C, over 30% of the total cell protein is bGH[11]. To demonstrate that granule formation and cell elongation are the result of bGH synthesis, we analyzed RV308 cells carrying plasmids with a thermoinducible runaway-replicon, but without the bGH gene. When grown at 37°C for 6 hrs, these cells do not form granules and appear identical under the microscope to cells shown in Figure 1A. This observation led us to conclude that granule formation and cell elongation during the 6 hours after the temperature shift are not due to plasmid amplification *per se,* but result from bGH synthesis and accumulation.

Electron microscopy of bovine growth hormone granules. The morphology of RV308/pCZ101 cells grown at 37°C for 6 hours was examined by transmission electron microscopy (Fig. 2A). The granules appear as darkly staining areas and are not enclosed by a distinct membrane. Scanning electron microscopy of granules isolated from cell lysates by sucrose gradient centrifugation (see Experimental Protocol) showed that the granules are roughly spherical in shape and are up to about one μm in diameter (Fig. 2B).

Isolation and characterization of bovine growth hormone granules. To demonstrate that the observed granules contain bGH, we first determined what proteins were synthesized following a temperature shift to 37°C. The result of a pulse-label experiment with [^{35}S] methionine is shown in Figure 3. RV308/pCZ101 cells growing at 25°C in early exponential phase were shifted to 37°C and grown for 2 hrs. Methionine [^{35}S] was added and samples were withdrawn 1, 3, 10 and 30 min later (Fig. 3, lanes 1, 2, 3 and 4). The majority of [^{35}S] methionine was incorporated into a protein that is identical in size to the bGH derivative. This protein is absent from RV308 cells harboring plasmids that lack the bGH gene. The second major protein synthesized after the temperature shift is kanamycin phosphotransferase, also a plasmid encoded protein that confers resistance to the antibiotic kanamycin. The stability of these proteins was determined by a pulse-chase analysis. After the cells were labeled with [^{35}S] methionine for 30 min, a 50,000-fold excess of unlabeled methionine over the [^{35}S] methionine was added and samples were withdrawn at the indicated times. As shown in Figure 3 (lanes 5, 6 and 7), the incorporated [^{35}S] methionine was not lost after addition of excess unlabeled methionine. We therefore concluded from these experiments that RV308 cells harboring pCZ101 synthesize primarily the plasmid encoded proteins and that these proteins are stable for at least 30 min.

Next, we examined the protein content of isolated granules. The granules were separated from the cell lysate by differential centrifugation as described in the experimental protocol and were analyzed on SDS-polyacrylamide gels stained with Coomassie blue (Fig. 4). The major protein band in the crude granule preparation (Fig. 4, lane 1) migrates identically with the bGH standard (Fig. 4, lane 12). Aliquots of the crude granule preparations were washed with increasing concentrations of urea and centrifuged. The pellets and supernatants were then analyzed separately. Successive washing with 0.5 M, 1.0 M or 2.0 M urea releases the majority of the proteins associated with the bGH granule without solubilizing bGH itself. Although the granular bGH begins to solubilize in 5.0 M urea (Fig. 4, lane 8) over 90% remains in the pellet that is free of most contaminating proteins. Washing with 5.0 M urea and 1% Triton X-100 partially solubilizes the granular bGH without further removing the remaining contaminating proteins. The most prominent of these is kanamycin phosphotransferase which can be solubilized only under conditions that also solubilize bGH, suggesting that it is tightly associated with the bGH granule.

The protein isolated by this procedure was shown to be bGH by high pressure liquid chromatography and radioimmune assay (data not shown).

DISCUSSION

We have described a simple procedure for the purification of bGH granules that accumulate in *E. coli* cells as a result of high-level expression of this protein. We have shown that these granules occur only in cells overproducing bGH and are clearly visible with a phase-contrast microscope. While all of the experiments presented in this paper were carried out with a derivative of bGH that contains 8 additional N-terminal amino acids, we have observed granule formation with other bGH derivatives[11], when they are expressed at a level of at least 5–10% of the total cell protein. Cells containing these granules are still capable of carrying out protein synthesis as shown by their ability to incorporate [^{35}S] methionine. The ability of these proteins to form distinct granules has served as a very useful method for quickly identifying clones that overproduce bGH. Further, we demonstrated that the intact granules can be separated from the cell lysates by differential centrifugation and that they can be substantially enriched for bGH by a simple washing procedure.

Bovine growth hormone is not a unique protein in its ability to form intracellular granules. The formation of distinct granules was also observed with human growth hormone (N. Mayne, personal communication) and with human insulin chimeric proteins[9,10]. In addition, there are numerous reports [see introduction and references cited in Simons et al. (4)] of native and fusion proteins that aggregate when expressed at a high level, and are found in the cell pellet fraction following centrifugation of cell lysates. Although not stated explicitly by the authors, it is likely that these proteins form granules similar to those that we have described for bGH, suggesting that aggregation and granule formation is a common property of even native *E. coli* proteins when they are overproduced. We assume that these aggregates form when the intracellular concentration of the overproduced protein exceeds its solubility. In many examples reported to date, the protein constituents of the granules appear to be resistant to proteolytic degradation, and in at least one case, granule formation can confer stability to an otherwise labile protein[6].

A second type of protein granule has been observed in *E. coli* composed of abnormal proteins resulting from the incorporation of amino acid analogs or puromycin[12,13]. These granules also lack an enclosing membrane and cells containing them are viable. However, these abnormal protein granules differ from the bGH granules in that they resemble amorphous aggregates of proteins and are degraded by *E. coli* proteolytic enzymes[13], whereas the bGH granules are stable. It would be of interest to compare these two types of granules further to assess if structural or compositional differences between them can account for differences in their stabilities.

EXPERIMENTAL PROTOCOL

Culture conditions. *E. coli* K-12 RV308 [su⁻, Δ*lac*X74, *gal* ISII::OP308, *str*A] containing pCZ101[11] was propagated in TY medium (Difco) with 50 μg/ml kanamycin (Sigma). Cultures grown to early exponential phase at 25°C were· shifted to 37°C for 2–6 hours as indicated below and in the text.

Light microscopy. Cultures in early logarithmic phase at 25°C were shifted to 37°C for 6 hours. Control cultures were kept at 25°C. Cells from these cultures were immobilized in agar and examined with a Reichert microscope (using a 97X objective and a 10X eyepiece) equipped with a Pentax camera.

Electron microscopy. For transmission electron microscopy, cultures grown at 37°C for 6 hrs were fixed in glutaraldehyde (3% final concentration), centrifuged and washed with 0.1 M sodium cacodylate buffer (Sigma). The cells were resuspended in 1% osmium tetroxide in Millonigs buffer[14] at 4°C, dehydrated in an acetone series and embedded in Spurrs resin[15]. Sections were stained with lead and uranyl acetate and examined with a Philips 400 electron microscope. For scanning electron microscopy, the cells were lysed by sonication and granules were purified on a sucrose density gradient (40–60% w/w sucrose), centrifuged at 80,000 × g for 16 hours. An aliquot from this granule preparation was filtered through a 0.2 µm Nuclepore filter (General Electric), rinsed with ethanol and dried. After coating with gold in a vacuum evaporator, the filter specimen were examined with a scanning electron microscope (Etec Autoscan).

Labeling of cellular proteins. Cultures growing in early exponential phase at 25°C were shifted to 37°C and growth was continued for 2 hours. 1.0 µCi/ml of [^{35}S] methionine (1240 Ci/mmole; Amersham) was added. For the pulse-label experiment, 1 ml culture samples were removed at the indicated times and placed in tubes kept at 4°C, containing a 50,000-fold excess of unlabeled methionine over the [^{35}S] methionine. After centrifugation, the cell pellets were lysed in sample buffer (0.125 M Tris-HCl, pH 6.8, 2% SDS, 30% glycerol, 1 M 2-mercaptoethanol, 6 M urea) and boiled before loading onto a 12.5% SDS-polyacrylamide gel[16]. For the pulse-chase, the cells were labeled for 30 min with [^{35}S] methionine; a 50,000-fold excess of unlabeled methionine over the [^{35}S] methionine was added, 1 ml samples were removed at the indicated times and treated as above. The sample volumes were adjusted to contain extracts from approximately the same number of cells in the sample buffer. Ten µl of the sample was loaded on the gel which was run at 50 mA for 3 hours. The gel was soaked in Enlightening (New England Nuclear) for 15 min, dried and exposed for 4 hours to XAR-5 (Kodak) x-ray film with an intensifying screen at −70°C.

Isolation of bGH granules. One gram of frozen cell pellet was resuspended in 5 ml of 0.1 M Tris-HCl, pH 7.3 and treated with lysozyme (1 mg/ml) for 15 min at 25°C. The cells were disrupted on ice with a sonicator (Heat Systems, Model W225R, Ultrasonics, Inc.) by 2, 15 sec. bursts using a microprobe. The lysate was centrifuged at 1,000 × g for 5 min at 4°C. The granules which are contained in the supernatant were pelleted at 27,000 × g for 15 min at 4°C. The pellet was resuspended in 1 ml H$_2$O and 100 µl aliquots of this suspension were pelleted again in a microfuge. These pellets were washed in 100 µl of 0.1 M Tris-HCl, pH 8.5, containing urea and Triton X-100 at concentrations indicated in the text. The washed suspensions were centrifuged again in a microfuge and 10 µl samples of the supernatants and pellets (resuspended in 100 µl H$_2$O) were analyzed on SDS-polyacrylamide gels, stained with Coomassie Brilliant Blue R (Sigma).

Acknowledgments

We thank J. Paul Burnett for helpful discussions, John Wood for critically reading the manuscript, Richard Schlegel for excellent technical assistance in photography and Cheryl Alexander for an outstanding effort in manuscript preparation.

Received 11 September 1984; accepted 10 October 1984.

References

1. Gribskov, M. and Burgess, R. R. 1983. Overexpression and purification of the sigma subunit of *Escherichia coli* RNA polymerase. Gene **26**:109–118.
2. Masui, Y., Mizuno, T., and Inouye, M. 1984. Novel high-level expression cloning vehicles: 10⁴-fold amplification of *Escherichia coli* minor protein. Bio/Technology **2**:81–85.
3. Ho, Y.-S., Lewis, M., and Rosenberg, M. 1982. Purification and properties of a transcriptional activator. J. Biol. Chem. **257**:9128–9134.
4. Simons, G., Remaut, E., Allet, B., Devos, R., and Fiers, W. 1984. High-level expression of human interferon gamma in *Escherichia coli* under control of the P$_L$ promoter of bacteriophage lambda. Gene **28**:55–64.
5. Devos, R., Plaetinck, G., Cheroutre, H., Simons, G., Degrave, W., Tavernier, J., Remaut, E., and Fiers, W. 1983. Molecular cloning of human interleukin 2 cDNA and its expression in *E. coli*. Nucleic Acids Res. **11**:4307–4323.
6. Cheng, Y.-S. E., Kwoh, D. Y., Kwoh, T. J., Soltvedt, B. C., and Zipser, D. 1981. Stabilization of a degradable protein by its overexpression in *Escherichia coli*. Gene **14**:121–130.
7. Itakura, K., Hirose, T., Crea, R., Riggs, A. D., Heyneker, H. L., Bolivar, F., and Boyer, H. W. 1977. Expression in *Escherichia coli* of a chemically synthesized gene for the hormone somatostatin. Science **198**:1056–1063.
8. Kleid, D. G., Yansura, D., Small, B., Dowbenko, D., Moore, D. M., Grubman, M. J., McKercher, P. D., Morgan, D. O., Robertson, B. H., and Bachrach, H. L. 1981. Cloned viral protein vaccine for foot-and-mouth disease: responses in cattle and swine. Science **214**:1125–1129.
9. Williams, D. C., Van Frank, R. M., Muth, W. L., and Burnett, J. P. 1982. Cytoplasmic inclusion bodies in *Escherichia coli* producing biosynthetic human insulin proteins. Science **215**:687–689.
10. Paul, D. C., Van Frank, R. M., Muth, W. L., Ross, J. W., and Williams, D. C. 1983. Immunocytochemical demonstration of human proinsulin chimeric polypeptide within cytoplasmic inclusion bodies of *Escherichia coli*. European J. Cell Biol. **31**:171–174.
11. Schoner, B. E., Hsiung, H. M., Belagaje, R. M., Mayne, N. G., and Schoner, R. G. 1984. Role of mRNA translational efficiency in bovine growth hormone expression in *Escherichia coli*. Proc. Natl. Acad. Sci. (USA) **81**:5403–5407.
12. Prouty, W. F. and Goldberg, A. L. 1972. Fate of abnormal proteins in *E. coli* accumulation in intracellular granules before catabolism. Nature New Biol. **240**:147–150.
13. Prouty, W. F., Karnovsky, M. J., and Goldberg, A. L. 1975. Degradation of abnormal proteins in *Escherichia coli*. J. Biol. Chem. **250**:1112–1123.
14. Millonig, G. 1961. Advantages of a phosphate buffer for osmium tetraoxide solutions in fixation. J. Appl. Phys. **32**:1637–1639.
15. Spurr, A. R. 1969. A low viscosity epoxy resin embedding medium for electron microscopy. J. Ultrastructure Research **26**:31–36.
16. Laemmli, U. K. 1970. Cleavage of structural proteins during the assembly of the head of bacteriophage T4. Nature **227**:680–685.

Functional Genetic Expression of Eukaryotic DNA in *Escherichia Coli*

K. Struhl, J.R. Cameron and R.W. Davis

ABSTRACT   We have isolated a segment of DNA from the eukaryote *Saccharomyces cerevisiae* (baker's yeast) as a viable molecular hybrid of bacteriophage λ DNA which, when integrated into the chromosome of an *E. coli* histidine auxotroph, allows this bacterium to grow in the absence of histidine. The nonrevertable, histidine auxotroph lacks the enzymatic activity of imidazole glycerol phosphate (IGP) dehydratase (EC 4.2.1.19). From genetic experiments, we conclude that expression of the segment of yeast DNA results in the production of a diffusible substance and that transcription necessary for the complementation is most likely initiated from the segment of eukaryotic DNA.

Techniques have been recently developed for the construction and cloning of viable molecular hybrids between bacteriophage λ and any foreign DNA (1, 2). Pools of molecular hybrids containing many different fragments of DNA from any given genome can be generated (3). Genetic selection systems which depend on the functional expression of the foreign DNA sequence have been used to isolate specific hybrids of interest. Hybrid phage containing specific bacterial genes were isolated by lytic growth in hosts having lesions in those genes. The gene for DNA ligase was isolated from *Escherichia coli* (3), and the genes for DNA polymerase I were isolated from *Klebsiella aerogenes* and *Klebsiella pneumoniae* (Struhl and Davis, in preparation). Since bacteriophage λ is capable of integrating into the bacterial chromosome to form stable lysogens, selections can also be performed without killing the host. Such selections were used to isolate hybrids containing the gene for 3-enolpyruvylshikimate 5-P synthetase (*aroA*) from *K. aerogenes* and *K. pneumoniae* and the gene for chorismate synthetase (*aroC*) from *E. coli* (Struhl and Davis, unpublished results).

In this paper, we have used a selection employing the lysogenic state of λ to obtain a specific hybrid which contains DNA from the yeast *Saccharomyces cerevisiae*. This selection is based upon complementation of a nonreverting bacterial mutant which has a single, defined, enzymatic lesion. Such a selection requires either the synthesis of a functional gene product from eukaryotic genetic information by *E. coli* or suppression of the specific lesion.

MATERIALS AND METHODS

Chemicals, Phage, Bacteria, DNA, and Enzymes. Bacterial strain his463 was obtained from Dr. P. Hartmann. Originally, this strain had been characterized as a *hisC* mutant (4). It was recharacterized as a *hisB* mutant which lacks the activity for imidazole glycerol phosphate (IGP) dehydratase (EC 4.2.1.19) but not the activity for histidinol phosphatase [L-histidinol-phosphate phosphohydrolase (EC 3.1.3.15)]. We constructed

Abbreviations: IGP dehydratase, imidazole glycerol phosphate dehydratase; SSC, standard saline–citrate solution (0.15 M sodium chloride–0.015 M sodium citrate, pH 7); ×SSC means that the concentration of the solution used is times that of the standard saline–citrate solution.

another *hisB* mutant of *E. coli* by mating of the *S. typhimurium* strain TR74 (F′ *hisB2404/ser821 arg501 his*Δ*712*) which harbors an episome containing the histidine operon from *E. coli* with his461 (Δ*his*) and selecting for growth on minimal plates supplemented with histidinol. The resulting strain produces an *E. coli his* B protein which lacks the activity for IGP dehydratase only (B. Cooper and J. Roth, personal communication). Other histidine auxotrophs were constructed in the same manner.

The λgt vector used in this paper contains all the essential genes necessary for phage growth but is too small to be packaged. An insertion of an *Eco*RI endonuclease generated fragment of yeast DNA between the λgt ends results in a viable molecular hybrid (1). λgti is a derivative of λgt-λC (1) which has a mutation in a recognition site for *Eco*RI endonuclease (RI-3). Other reagents have been described previously (1, 3, 5, 6).

Generation of λgt Hybrid Pools with Yeast DNA. DNA from the *Saccharomyces cerevisiae* strain A364a × H79-20.3α was supplied by Dr. B. Hall. We made hybrid DNA molecules using λgt-λB DNA (1) as the vector by the *Eco*RI endonuclease-DNA ligase method described previously (3, 6). The hybrid molecules were transfected into calcium treated *E. coli* SF8 cells (C600 rK⁻ mK⁻ recBC⁻ lop 11 lig⁺). Plaques representing individual viable hybrid phage were pooled by scraping the plates, and the phage pool was grown on C600 rK⁻ mK⁺ to make a high titer stock. The viable hybrids are thereby K modified and can therefore infect any K12 strain of *E. coli* which is sensitive to infection by bacteriophage λ. These pools were more fully characterized by Cameron and Davis (in preparation).

Cultivation of Bacteria. The minimal medium used was standard M9 to which vitamin B1 was added at a concentration of 1 μg/ml, histidine at 100 μg/ml, and histidinol at 1 mM when required as supplements. Carbon sources were added to a final concentration of 0.4%.

Formation of Double Lysogens of HisB463. Since the vector used for hybrid pools was λgt-λB, all hybrid phage were deleted for the gene necessary for integration (*int*) and the phage attachment site (*att*) contained in the *Eco*RI-λC fragment. Hybrids were integrated into the bacterial chromosome as double lysogens by using an integration helper phage (λgti) which is *int*⁺*att*⁺. Double lysogens are formed by linkage of the prophages at the bacterial attachment site (7, 8). Because the hybrids have no phage attachment site, the double lysogens are not formed in the usual tandem linkage. Instead, the hybrids can be integrated into the chromosome by general bacterial recombination by homology with λgti. Recombination mediated formation of double lysogens has been shown to occur at a frequency of 1% (9). When induced, double lysogens formed in this manner yield equal numbers of the two prophages. HisB463 was coinfected with the integration helper and phage from the yeast hybrid pool (each at a multiplicity of in-

STRUHL, K., CAMERON, J.R. and DAVIS, R.W.
Functional genetic expression of eukaryotic DNA in *Escherichia coli. Proc. Natl. Acad. Sci. U.S.A.*
73:1471-1475 (1976). Reprinted with the authors' permission.

fection of 1) in order to obtain a phage which complements this strain.

Curing of λ Prophages from Lysogenic Strains. All phage used have the *cI857* mutation, thus permitting temperature induction of the prophages. From 10^4 to 10^6 lysogens were inoculated into 1 ml of L broth and grown at 30° for 1–2 hr. The culture was shifted to 42° for 5 min to induce prophages and then returned to 30° for 2–4 hr. An aliquot was streaked on L plates at 42° and colonies were subsequently tested.

Induction of Double Lysogens and Purification of Phage. A culture of the double lysogen, growing exponentially at 30°, was induced for 15 min at 42° and then shifted to 37°. After lysis, a loopful of phage was streaked for single plaques. Since the λgti integration helper is *red+* and the hybrid phage is *red−*, it is easy to distinguish between them by their plaque size. A small (*red−*) plaque was picked and tested for the *red−* character by its inability to plate on either lig ts7 or H560 (*polA−*).

Preparation of Complementary RNA. RNA was synthesized (10) with labeled [^{32}P]ATP and [^{32}P]GTP at 3 Ci/mmol and the appropriate phage DNA as a template. Incubation was for 2 hr at 37° with *E. coli* RNA polymerase holoenzyme. RNA was extracted twice with phenol and chromatographed on a Sephadex G-50 column.

Agarose Gel Electrophoresis, Hybridization, and Autoradiography. Agarose gel electrophoresis was performed as in Thomas and Davis (6). DNA was transferred from an 0.7% agarose cylindrical (0.6 × 14 cm) gel to strips of cellulose nitrate as in Southern (11). Complementary RNA (5×10^5 dpm in 0.6 ml) was hybridized to the strips for 18 hr at 43° in 50% formamide and 6 × SSC. Strips were washed with 1 M NaCl, 10 mM EDTA, and 100 mM sodium phosphate at pH 7.0 for 2 hr, soaked for 15 min in 2 × SSC, dried, and autoradiographed.

RESULTS

Complementation of the hisB463 lesion by a phage from the yeast hybrid pool

Two pools of viable molecular hybrids were constructed as described in *Materials and Methods* using the same preparation of yeast DNA. The pools, which were made 3 months apart, contained phage from about 10,000 and 14,000 independent plaques. These hybrids were stably integrated into the chromosomes of cells of the histidine auxotroph hisB463 by the integration helper method. After incubation for 60 hr on glucose minimal plates at 30°, colonies growing without histidine (*his+*) were found at a frequency of approximately 10^{-8}. It was possible to detect complementation at this frequency by using either of the yeast hybrid pools. The colonies isolated in this manner are not revertants of hisB463. The strain reverts to *his+* with a frequency of less than 10^{-11}; in fact, we and previous workers have never seen a revertant. It is also impossible (less than 10^{-10}) to induce reversion of this strain by ultraviolet light, ethyl methane sulfonate, nitrosoguanidine, or ICR 191. Attempts to isolate this gene by this method from *E. coli, K. aerogenes, K. pneumoniae,* and *Dictyostelium discoideum* were unsuccessful, a result probably due to EcoRI endonuclease cleavage within the histidine operon.

Complementation of the hisB lesion depends on a λ prophage

The *his+* colonies presumably containing a prophage capable of complementing the *hisB* mutation, were purified on minimal glucose plates at 30°. Since all phage used in these experiments have a temperature sensitive repressor (*cI857*), stable lysogeny and cell viability should occur only at temperatures less than

35°. All 200 *His+* colonies were unable to grow on rich medium at 42°. If the complementation is in fact dependent upon a λ prophage, removal of the prophage (by curing) from the chromosome of the *his+* lysogen should result in the original hisB463 strain. Indeed, of 50 cured lysogens, all required histidine for growth at both 30 and 42°.

The *his+* colonies should contain two different prophages; the complementing prophage and the integration helper. Temperature induction of the *his+* colonies resulted in the production of approximately 20 phage per bacterium. Phage were isolated by equilibrium cesium chloride density gradient centrifugation. Two bands were clearly seen, confirming the fact that the *his+* colonies contained two different prophages. As expected equal numbers of the two phage were produced (see *Materials and Methods*).

Isolation and characterization of the phage capable of complementing hisB463

The prophage necessary for complementation of hisB463 was induced and a single plaque isolate was obtained as described in the *Materials and Methods*. Phage descendants from this plaque are designated by the strain name λgt-Sc2601. In this paper, they will be referred to by the more descriptive name λgt-Sc his. When hisB463 cells are coinfected with λgt-Sc his and the integration helper, *his+* double lysogens are found at the relatively low frequency of 10^{-4} after 3 days. Double lysogens produced by our methods are typically found at a frequency of 10^{-2}. Nevertheless, the frequency of 10^{-4} represents an enrichment of 10^4 from the original hybrid pool. The *his+* colonies obtained by using purified λgt-Sc his are indistinguishable from the *his+* colonies originally isolated with the yeast hybrid pool. They are temperature sensitive for growth, and curing of the prophages renders the strain *his−*.

Single infection of hisB463 by λgt-Sc his, at 30°, resulted in stable *his+* colonies at the low frequency of 10^{-8}. These single lysogens are temperature sensitive and upon curing are *his−* at all temperatures. Phage can be induced from such a single lysogen although only 0.2 phage per bacterium are produced, an efficiency of only 1% compared to induction of the *his+* double lysogens. This poor efficiency is not due to a lowered burst, but rather to poor excision, since only 1% of the single lysogens induce any phage at all.

λgt-Sc his is also capable of complementing another hisB⁻ strain (B2404) which synthesizes a *hisB* protein which lacks only the IGP dehydratase activity (see *Materials and Methods* for construction of this strain). When complementation was done in an analogous manner as with hisB463, *his+* double lysogens were also found at a frequency of 10^{-4}. λgt-Sc his does not complement any of the histidine auxotrophs tested which belong to other complementation groups (*hisA,C,D,E,F,G,I*) or a single site *hisB* mutant lacking both IGP dehydratase and histidinol phosphate phosphatase activities.

λgt-Sc his is a hybrid containing yeast DNA

DNA from λgt-Sc his was characterized by cleavage with EcoRI endonuclease and gel electrophoresis in 0.5% agarose (Fig. 1). The expected 4-band pattern for a λgt hybrid is seen. The band nearest the top of the gel represents a hydrogen bonded association of the two λgt end fragments. The two bands of intermediate mobility correspond to the individual λgt end fragments. The lowest band corresponds to the inserted segment of DNA. The length of the insert is about 10.3 kilobase pairs as determined by mobility in the agarose gel and length measurements in the electron microscope.

The facts that (1) λgt-Sc his could be obtained from two in-

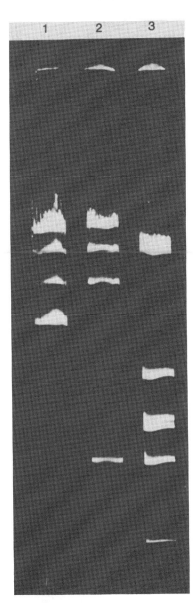

FIG. 1.   Agarose gel electrophoresis. DNA was cleaved with *Eco*RI endonuclease and electrophoretically separated on a 0.5% agarose gel as in Thomas and Davis (1). Samples are (1) λgt-Sc his, (2) λgt-λB, and (3) λcI857 S_7. The upper band in gels 1 and 2 represents the right and left ends of λgt which are hydrogen bonded via the λ cohesive ends.

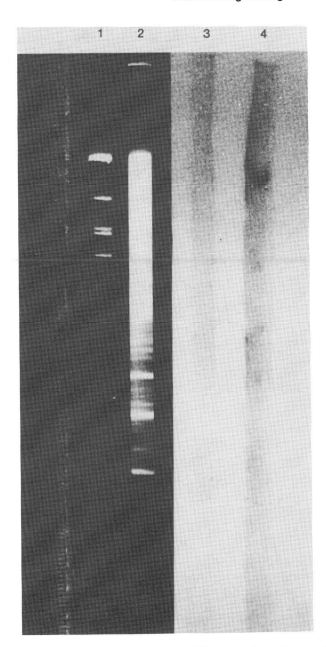

FIG. 2.   Nucleic acid hybridization. This was performed as stated in the *text*. All DNA was cleaved with *Eco*RI endonuclease, and all gels were 0.7% agarose. (1) Gel of λcI857 S_7 DNA, (2) gel of total yeast DNA, (3) autoradiograph of cRNA to λgt-λB DNA hybridized to total *Eco*RI cleaved yeast DNA, (4) autoradiograph of cRNA to λgt-Sc his DNA hybridized to total *Eco*RI cleaved yeast DNA.

dependently prepared pools of yeast hybrids, (2) selections from all other hybrid pools including *E. coli* were unsuccessful, and (3) λgt-Sc his could not complement a *hisB* mutant lacking both activities made it unlikely that the inserted segment of DNA was a non-yeast contaminant. To avoid possible contamination in the preparation of yeast DNA, we made a new preparation of yeast DNA from another strain (D585-11C) as in Cameron and Davis (in preparation). This DNA was cleaved with *Eco*RI endonuclease, electrophoretically separated on a 0.7% agarose gel, denatured in alkali, and then transferred to a strip of cellulose nitrate paper by a technique developed by E. M. Southern which maintains the gel pattern on the strip (11). ³²P-labeled RNA complementary to λgt-Sc his DNA was synthesized *in vitro* with *E. coli* RNA polymerase and hybridized to the cellulose nitrate strip containing the yeast DNA. An autoradiograph from such an experiment is shown in Fig. 2, gel 4. One

band corresponding to a length of DNA of 10-11 kilobases is evident, which is the same size as the inserted segment of DNA in λgt-Sc his. cRNA made to DNA from λgt-λB does not detectably hybridize to yeast DNA.

Complementation of hisB463 by λgt-Sc his does not depend upon orientation of the fragment of yeast DNA

λgt-Sc his DNA was cleaved with *Eco*RI endonuclease and then resealed with DNA ligase. Following transfection of these molecules into calcium treated cells, three plaques were picked and grown. The DNAs from these three phage were characterized by heteroduplex analysis with DNA from the original λgt-Sc his phage (12). Two of the three clearly contained the

Table 1.   Growth rates of bacterial strains

Strain	Supplement*		
	None	Histidinol	Histidine
W3110	2.1	2.0	2.2
hisB463	No growth	2.1	2.1
His+ double lysogen	2.7	2.2	2.2
Cured double lysogen	No growth	2.1	1.9

All cultures were grown at 29° in glucose minimal medium. Supplements were added as in *Materials and Methods*. Growth was monitored with a Klett spectrophotometer. Table entries represent doubling times in hours.

same segment of yeast DNA, but in the inverted orientation as that of λgt-Sc his DNA and are called λgt-Sc his'. DNA from the third phage was identical to λgt-Sc his DNA. When the inversions, λgt-Sc his', were integrated into the chromosome of hisB463, *his+* colonies were obtained with the same frequency as with λgt-Sc his.

Growth of cells dependent upon complementation by the fragment of yeast DNA

The *his+* cells containing a λgt-Sc his prophage grow in glucose minimal medium at 29° with a doubling time of 2.7 hr. When histidine or histidinol was added as a supplement, the doubling time was only 2.1 hr, a rate comparable to the wild strain (W3110) at these conditions. Cells which were cured of their prophages had growth characteristics indistinguishable from hisB463. A summary of growth rates is presented in Table 1.

DISCUSSION

We have isolated a hybrid phage containing a segment of yeast DNA which complements an *E. coli* mutant lacking imidazole glycerol phosphate (IGP) dehydratase activity. The intergeneric complementation is dependent upon the presence of λgt-Sc his since curing of this prophage is accompanied by loss of the his+ phenotype. The hybrid indeed contains yeast DNA. It is important when working with such potent selective pressures (capable of isolating 1 colony out of 10^{10}) to rule out possible artifacts. In addition to the trivial case of bacterial revertants, his+ colonies could result from hybrids with inserted fragments which are DNA contaminants, recombinants, or rearrangements. The possibility of such an artifact is quite real considering that the hybrid phage are passaged on *E. coli* strains which are essentially wild type.

The nature of the complementation indicates that the segment of yeast DNA in hybrid λgt-Sc his codes for the synthesis of a diffusible yeast product which complements hisB463 *in vivo*. The complementation is not a site-specific effect on the *E. coli* histidine operon since the integration site for λ prophages does not map near this operon, inversion of the yeast DNA fragment with respect to λ (and the bacterial chromosome) has no effect, and because λgt-Sc his specifically complements hisB mutants lacking IGP dehydratase. The transcription for such a product is almost certainly initiated in the fragment of yeast DNA for two reasons. First, the expression occurs in a λ lysogen in which the λ promoters are either strongly repressed or have been deleted in λgt-Sc his. Second, since inversion of the yeast fragment has no effect, two as of yet unknown λ promoters would be required if the promoter did not reside in the yeast fragment itself. However, we have no evidence as to whether

transcription begins from the correct yeast promoter or merely from a sequence fortuitously recognized by *E. coli* RNA polymerase.

The frequency of his+ colonies formed after 3 days when purified λgt-Sc his is used is only 1% of the expected value. Typically, 1% of the lysogens formed by our methods are double lysogens. The actual frequency of phenotypically his+ double lysogens (10^{-4}) may be explained by inefficient expression of the yeast DNA, inefficient functioning of the yeast product, instability of the yeast product, or by mutational changes in the sequence of the yeast fragment during lytic growth of the hybrid. The variability in his+ colony size on selective medium and the observation that more his+ colonies become visible after 3 days could be explained by any or a combination of the above. The single his+ lysogens formed by the $int^- att^-$ hybrid alone are almost certainly the result of abnormal integration. The frequency of such his+ single lysogens is extremely low (10^{-8}) and the poor induction of viable phage indicates that abnormal excision is required.

The frequency of isolating a specific phage from the yeast hybrid pool capable of complementing hisB463 appears to be quite low. This is explained by three factors: the actual frequency of the specific yeast hybrid in the pool (about 10^{-4}), the frequency of forming double lysogens (10^{-2}), and the effectiveness of the complementation (10^{-2}). It should be possible to isolate other genes from yeast by similar selections in which different *E. coli* auxotrophs are used. However, selection of yeast and other eukaryotic genes by virtue of complementation is unlikely to be a completely general technique. Such complementation not only requires adequate genetic expression of the eukaryotic DNA in *E. coli*, but in addition requires activity of the protein product. For example, genes coding for eukaryotic enzymes which have significantly different enzymatic behavior from the analogous enzymes in *E. coli*, or enzymes which are part of an enzyme complex are less likely to be isolated by *in vivo* complementation.

Complementation of a single, defined, enzymatic lesion requires either the synthesis and functioning of an equivalent gene product or suppression of the original mutation. Although no direct evidence is presented in this paper, it seems unlikely that the complementation of hisB463 is the result of suppression of this specific lesion. The effect of the integrated yeast DNA is very specific, since it complements strains with a revertable or a nonrevertable mutation in hisB lacking IGP dehydratase activity, but not a strain with a revertable mutation in hisB lacking both IGP dehydratase and histidinol phosphate phosphatase activities. Complementation is not observed for any of the other genes in the histidine operon. A postulated suppressor would have to suppress two different specific mutations in hisB lacking IGP dehydratase activity and not suppress mutations in other his genes.

The mutant hisB463 used in this work is extremely stable. No growth is detected in the absence of histidine for at least 10 days. No revertants have ever been found spontaneously or after many different mutagenic treatments. The hisB463 lesion is most probably a deletion which affects the IGP dehydratase activity but not the histidinol phosphatase activity of the hisB protein. It is difficult to conceive of how the product of a segment of yeast DNA could suppress a mutation which is unrevertable and unsuppressible by some *E. coli* mechanism even after a variety of mutagenic treatments. Nonsense and missense suppression of single base changes can occur via tRNAs but the fragment of yeast DNA does not detectably hybridize to yeast tRNA under conditions in which other λgt-yeast hybrids do. A frameshift suppressor is unlikely since hisB463 does not revert after treatment with ICR 191. It is also unlikely that the yeast DNA codes for a product which stabilizes the mutant hisB463 protein. Besides the unusual nature of such a specific interaction which would be required for two different mutant hisB proteins, bacterial strains with mutations resulting in the production of such proteins would not be expected to have such a stringent requirement for histidine or histidinol.

 Therefore, although we have not yet excluded the possibility of suppression, we believe it more likely that the yeast DNA in the hybrid phage codes for the structural gene for imidazole glycerol phosphate dehydratase (*his* 3) or some other gene with a similar activity, and that this DNA is transcribed and subsequently translated with enough fidelity by *E. coli* to produce a functional protein. The genes for IGP dehydratase and histidinol phosphatase are unlinked in *S. cerevisiae*. Therefore, a single 10 kilobase segment of its DNA should code for only one of these activities. This is consistent with λgt-Sc his complementing the *E. coli* mutants lacking only the IGP dehydratase activity. In any event, functional genetic expression of the fragment of yeast DNA in λgt-Sc his results in the specific complementation of *E. coli* hisB mutants which lack IGP dehydratase.

 We thank Dr. J. Roth for strains of *Salmonella typhimurium* used for the complementation analysis; Elizabeth Zimmer for imidazole glycerol phosphate, aid in the enzyme assays, and for use of laboratory facilities. We thank Dr. Philip Hartmann, Dr. Jonathan Hodgkin, Dr.

A. Dale Kaiser, John A. Ridge, Dr. John Roth, John Scott, and Mariana Wolfner for fruitful discussions. This work was supported in part by Public Health Service Grant GM 21891-1 from the National Institutes of General Medical Sciences.

1. Thomas, M., Cameron, J. R. & Davis, R. W. (1974) *Proc. Natl. Acad. Sci. USA* **71**, 4579–4583.
2. Murray, N. W. & Murray, K. (1974) *Nature* **251**, 476–481.
3. Cameron, J. R., Panasenko, S. M., Lehman, I. R. & Davis, R. W. (1974) *Proc. Natl. Acad. Sci. USA* **72**, 3416–3420.
4. Garrick-Silversmith, L. & Hartmann, P. E. (1970) *J. Bacteriol.* **66**, 231–244.
5. Martin, R. G., Berberich, M. A., Ames, B. N., Davis, W. W., Goldberger, R. F. & Yourno, J. D. (1971) in *Methods in Enzymology*, eds. Tabor, H. & Tabor, C. W. (Academic Press, New York), Vol. 17, Part. B, pp. 3–44
6. Thomas, M. & Davis, R. W. (1975) *J. Mol. Biol.* **91**, 315–328.
7. Campbell, A. (1965) *Virology* **27**, 340–348.
8. Gottesman, M. E. & Yarmolinsky, M. B. (1968) *Cold Spring Harbor Symp. Quant. Biol.* **33**, 735–747.
9. Gottesman, M. E. & Yarmolinsky, M. B. (1968) *J. Mol. Biol.* **31**, 487–505.
10. Berg, D., Barrett, K. & Chamberlin, M. (1971) in *Methods in Enzymology*, eds. Grossman, L. & Moldave, K. (Academic Press, New York), Vol. 21, part D, pp. 506–519.
11. Southern, E. (1975) *J. Mol. Biol.* **98**, 503–517.
12. Davis, R. W., Simon, M. N. & Davidson, N. (1971) in *Methods in Enzymology*, eds. Grossman, L. & Moldave, K. (Academic Press, New York), Vol. 21, pp. 413–438.

Bacteria Mature Preproinsulin to Proinsulin

K. Talmadge, J. Kaufman and W. Gilbert

ABSTRACT   By inserting the rat preproinsulin gene into the bacterial prepenicillinase gene, we formed a variety of hybrid bacterial-eukaryotic signal sequences attached to proinsulin. Among these were the four following constructions: rat proinsulin attached to the entire penicillinase signal sequence and rat preproinsulin fused to all of, to half of, or only to the first four amino acids of the bacterial signal sequence. In all four cases, more than 90% of the rat insulin antigen appeared in the periplasmic space. By immunoprecipitation and determination of the amino acid sequences of the radiolabeled products, we show that the bacteria correctly process both the bacterial and the eukaryotic signal sequences of these hybrid proteins. The cleavage of the eukaryotic signal by bacterial peptidase, in this case, generates proinsulin.

Secretion is an essential feature of cells. The precursors of almost all secreted proteins, both eukaryotic and prokaryotic, contain an amino-terminal extension (ref. 1; see ref. 2 for review). The signal hypothesis (1, 3) proposes that this peptide, the signal sequence, serves to bind the protein to the membrane and then to lead it across. Sometime during transport, the signal sequence is removed and the preprotein is thereby processed to the mature form.

In bacteria, direct evidence establishes that the signal sequence is essential for transport. Mutations have been described for two proteins (4, 5) that lead to the accumulation of the mutant product in the cytoplasm as the preprotein. In each case, the mutation results in an amino acid replacement in the signal sequence. Furthermore, rat proinsulin, attached to a complete bacterial signal sequence, is efficiently transported (6, 7); lacking a signal, it is not (7).

The mechanism of secretion is quite general. Shields and Blobel (8) have used dog pancreas microsomes to segregate and process fish preproinsulin. Moreover, Fraser and Bruce (9) showed that chicken ovalbumin is secreted (50%) from bacterial cells when that gene is cloned in bacteria. Ovalbumin is unique among secreted proteins studied so far: it does not have an amino-terminal extension (10), although it may contain an internal signal sequence (11). We have recently shown (7) that a normal eukaryotic signal sequence, the rat preproinsulin signal sequence, directs the efficient secretion of rat insulin antigen in bacteria. Is this eukaryotic presequence processed?

MATERIALS AND METHODS

Materials. Chicken lysozyme, chicken ovalbumin, sperm whale myoglobin, and iodoacetamide were from Sigma; bovine proinsulin was a gift of Donald Steiner; human β_2-microglobulin was a gift of Cox Terhorst; $H_2{}^{35}SO_4$ (carrier-free) and L-[4,5-^{3}H(N)]leucine (50 Ci/mmol; 1 Ci = 3.7 × 10^{10} becquerels) was purchased from New England Nuclear. An IgG fraction of anti-insulin antiserum (from Miles) was prepared as described by Broome and Gilbert (12). *Staphylococcus aureus* strain Cowan I was heat-killed and formalin-treated by the method of Kessler (13) and resuspended (10% volume/volume) in NET buffer (50 mM Tris·HCl, pH 7.5/5 mM EDTA/0.15 M NaCl) (13).

Radiolabeling of Proteins. *Escherichia coli* K-12 strains PR13 bearing insulin plasmids p287.47 (which produces protein i27/+4), p241.1947 (protein i12/−21), p218.CB6 (protein i4/−21), or pKT41 (a control plasmid with no insulin insert), and FMA10/λcI857 bearing insulin plasmid p280.1947 (protein i25/−21) [all described by Talmadge *et al.* (7)] were grown overnight in 2YT medium (14) supplemented with thymidine at 40 μg/ml for FMA10. Fifty microliters was inoculated into 10 ml of S medium (15) supplemented either with thiamine at 10 μg/ml and thymidine at 40 μg/ml for FMA10 or with L-leucine and L-threonine at 40 μg/ml each for PR13, and then grown to OD$_{550}$ of 0.3. Five millicuries of $H_2{}^{35}SO_4$ was added to all cells (except PR13/p287.47) and incubation was continued 1 hr with shaking at 37°C (PR13) or 34°C (FMA10). PR13 bearing p287.47 was harvested, resuspended in 10 ml of S medium supplemented with L-threonine at 40 μg/ml and incubated for 1 hr at 37°C with shaking with 5 mCi of $H_2{}^{35}SO_4$ and 2.5 mCi of [^{3}H]leucine.

Immunoprecipitations. Labeled cells were harvested, resuspended in 100 μl of Tris·HCl, pH 8/20% sucrose and incubated 15 min with 100 μl of lysozyme at 20 mg/ml in 20 mM EDTA, pH 8. The cells were pelleted by centrifugation for 5 min at 10,000 rpm in a Sorvall SS-34 rotor, and the supernatant was diluted with 800 μl of 150 mM Tris·HCl, pH 8/2% Triton X-100/0.2 M EDTA. Alternately, labeled cells were harvested, resuspended in 100 μl of the Tris/sucrose buffer as above, incubated with 100 μl of lysozyme in EDTA as above, and lysed with 800 μl of Triton buffer as above. The cell debris was pelleted at 16,500 rpm for 1 hr in a Sorvall SA600 rotor. A 200- to 1000-fold excess [as determined by radioimmunoassay (7)] of an IgG fraction of guinea pig anti-insulin serum was added to each supernatant, and the mixture was held for 1 hr at 37°C and then 1 hr on ice. One hundred microliters of heat-killed, formalin-treated *S. aureus* (10% vol/vol) was added, and the mixture was incubated for 30 min on ice and washed by the method of Kessler (13).

Polyacrylamide Gel Electrophoresis. *S. aureus* bacteria complexed to proteins to be analyzed by polyacrylamide gel electrophoresis were resuspended in 100 μl of sample buffer [200 mM Tris·HCl, pH 6.8/10% (vol/vol) glycerol/0.01% (wt/vol) bromophenol blue/5 mM EDTA/2% NaDodSO$_4$/dithiothreitol (freshly added to 10 mM)], boiled 3 min, allowed to cool to room temperature, and incubated for 20 min with 20 μl of 0.5 M iodoacetamide. Thirty microliters of sample buffer made 200 mM in dithiothreitol was added, the room temperature incubation was continued another 10 min, and the bacteria were removed by centrifugation. Aliquots (10–50 μl) were loaded onto a 15% Laemmli NaDodSO$_4$/polyacrylamide gel

---

Abbreviation: HPLC, high-performance liquid chromatography.

TALMADGE, K., KAUFMAN, J. and GILBERT, W.
Bacteria mature preproinsulin to proinsulin. *Proc. Natl. Acad. Sci. U.S.A.* 77:3988-3992 (1980). Reprinted with the authors' permission.

(16) with 7 M urea in the bottom gel and 2 mM EDTA added to all buffers. One microgram each of sperm whale myoglobin, chicken lysozyme, human β_2-microglobulin, and bovine proinsulin were run as molecular weight markers. The stained dried gel was autoradiographed on Kodak XR-5 film. *S. aureus* bacteria complexed to proteins whose sequences were to be determined were resuspended in 100 μl of Maizel gel buffer (17), boiled 3 min, and centrifuged 5 min in a Sorvall SS-34 rotor at 10,000 rpm. The supernatant was run on a 15% Maizel gel (17), the wet gel was autoradiographed on Kodak XR-5 film for 1 hr at 4°C, and the protein was eluted from a crushed gel slice for 8 hr at room temperature with shaking in 1–2 ml of 50 mM ammonium bicarbonate, pH 7.5/0.2 mg of ovalbumin per ml/0.2 mM dithiothreitol/0.1% NaDodSO$_4$. The crushed gel was removed by filtration through silicone-treated glass wool and the protein was lyophilized.

Amino Acid Sequence Analysis. The protein isolated from a Maizel gel was resuspended in 100 μl of distilled water and 3 mg of ovalbumin was added. The proteins were precipitated in 5 vol of acetone and resuspended in 200 μl of 70% (wt/vol) formic acid, and then 3 mg of Polybrene (Aldrich) in 200 μl of 70% formic acid was added. Between 20,000 and 300,000 cpm was loaded onto a Beckman sequenator, updated model 890B, and successive steps of Edman degradation were performed, using a 0.1 M Quadrol program (18). The amino acid derivatives were collected after each cycle, dried under streaming nitrogen, and converted in 200 μl of 0.1 M HCl at 80°C for 10 min. Then 20–100 μl was dried in a vacuum oven, resuspended in 100 μl of distilled water, and mixed with 2 ml of Aquasol, and radioactivity was measured by liquid scintillation counting. Fractions with radioactivity were extracted with ethyl acetate and the aqueous phase of the ^{35}S-labeled fractions and the ethyl acetate phase of all the fractions were analyzed on a Waters high-performance liquid chromatography (HPLC) system, using an RCSS Radial Pak A (C$_{18}$) column.

RESULTS

Description of Hybrid Proteins. In a set of plasmid constructions designed to create a series of hybrid proteins, each containing a fusion of some portion of a bacterial signal sequence (derived from penicillinase) to some part of a eukaryotic signal sequence (derived from rat preproinsulin), four constructions transport more than 90% of the rat insulin antigen into the periplasmic space of *E. coli* (7). Fig. 1 shows the sequences of these four hybrid proteins, named by a lower case "i" and a pair of numbers: the first number referring to the last

prepenicillinase wild-type amino acid before the amino acids encoded by the insertion of the *Pst* restriction site, the second referring to the first amino acid of preproinsulin (negative numbers) or proinsulin (positive numbers). Either the complete bacterial signal sequence or the major part of the eukaryotic signal served to transport efficiently to the periplasm.

Immunoprecipitation and Polyacrylamide Gel Electrophoresis. We grew cells containing the insulin gene plasmids in a low-sulfate medium and labeled the proteins with H$_2^{35}$SO$_4$ or with both H$_2^{35}$SO$_4$ and [^{3}H]leucine. We isolated the labeled protein products from the periplasmic fraction by adding an excess of anti-insulin IgG and immunoprecipitating by incubating with formalin-treated, heat-killed *S. aureus* (13). Fig. 2 shows an autoradiogram of the electrophoresis of the immunoprecipitated proteins on a Laemmli NaDodSO$_4$/polyacrylamide gel (16) containing urea and EDTA. A dark new band appears in the four samples from insulin-antigen-producing cells (Fig. 2, lanes a–d) that is absent in the control precipitation (Fig. 2, lane e). Without processing, i25/−21 would have 142 amino acids, i12/−21 would have 130, i27/+4 would have 121, and i4/−21 would have 118. Instead, i27/+4 (Fig. 2, lane a) is larger than the other three, which are all the same size and run close to, but slower than, the bovine proinsulin standard; bovine proinsulin is 5 amino acids shorter than rat proinsulin [the deletion is in the C peptide (22)]. The gel mobilities in comparison with those of molecular weight standards suggest that i27/+4, which has a complete bacterial signal and no eukaryotic signal, has also been processed. A similar pattern is obtained if the proteins are immunoprecipitated from a Triton lysate of whole cells (data not shown).

Amino-Terminal Sequences of Radiolabeled Proteins. To verify that these proteins had been processed and to determine exactly where they had been clipped, we labeled i12/−21, i25/−21, and i4/−21 with H$_2^{35}$SO$_4$; i27/+4, with both H$_2^{35}$SO$_4$ and [^{3}H]leucine. After electrophoresis of the immunoprecipitates through a 15% Maizel gel (17), we autoradiographed the wet gel for an hour in the cold and then cut the samples directly out of the gel, using the autoradiograph as a template. Automated, successive Edman degradations of the radioactive protein on a Beckman sequenator, using ovalbumin and Polybrene as carriers, determined the positions of the sulfur-containing amino acids, methionine and cysteine, in the sequence. Fig. 1 shows the amino terminus of each protein. If the three candidates with most of the eukaryotic signal sequence (i25/−21, i12/−21, and i4/−21) are matured at the correct preproinsulin clipping site, radioactive cysteine should appear

```
Prepenicillinase                 ↓
         MSIQHFRVALIPFFAAFCLPVFA   HPETLVK...


i27/+4   MSIQHFRVALIPFFAAFCLPVFA   HPET   AAGGGGGG                              QHLCGPHLVEALYLVCGE...

i25/-21  MSIQHFRVALIPFFAAFCLPVFA   HP     LQGGGGG   WRMFLPLLALLVLWEPKPAQA   FVKQHLCGPHLVEALYLVCGE...

i12/-21  MSIQHFRVALIP                     LQGGGGG   WRMFLPLLALLVLWEPKPAQA   FVKQHLCGPHLVEALYLVCGE...

i4/-21   MSIQ                            AAAG      WRMFLPLLALLVLWEPKPAQA   FVKQHLCGPHLVEALYLVCGF...


                                ↓
         MALWRMFLPLLALLVLWEPKPAQA   FVKQHLCGPHLVEALYLVCGE...
         Preproinsulin
```

FIG. 1. Amino acid sequences of hybrid proteins made as fusions between the prepenicillinase signal sequence and the preproinsulin signal sequence, constructed around *Pst* linkers, as described in ref. 7. Each sequence begins at the prepenicillinase fMet and ends at amino acid 21 of proinsulin. Each line represents one continous sequence, which has been grouped from left to right to emphasize similarities and differences as follows: first group, prepenicillinase signal sequence amino acids; second group, matured penicillinase amino acids; third group, amino acids created by the inserted *Pst* linker (underlined) or by poly(G·C) tailing (glycines); fourth group, preproinsulin signal sequence amino acids; fifth group, matured proinsulin amino acids through amino acid 21. The sequence of prepenicillinase is from ref. 19, the sequence of preproinsulin is from refs. 6 and 20. The arrows indicate the sites of prepenicillinase and preproinsulin cleavage maturation. A = Ala, R = Arg, C = Cys, Q = Gln, E = Glu, G = Gly, H = His, I = Ile, L = Leu, K = Lys, M = Met, F = Phe, P = Pro, S = Ser, T = Thr, W = Trp, Y = Tyr, V = Val.

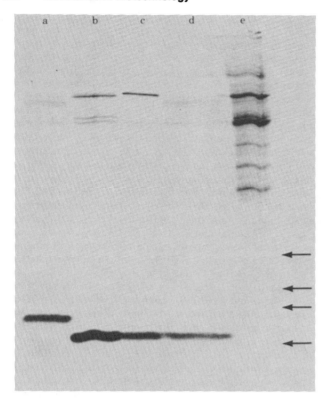

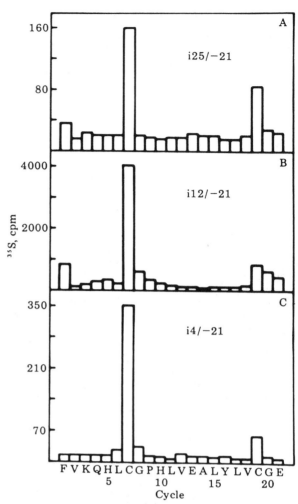

FIG. 2. Immunoprecipitated rat insulin antigen from *E. coli* strains bearing insulin plasmids, isolated and electrophoresed as described in the text. Lane a, i27/+4; b, i25/−21; c, i12/−21; d, i4/−21; e, PR13 bearing pKT41, a control plasmid without an insulin insert (7). The molecular weight markers, indicated by arrows, are, from top to bottom: sperm whale myoglobin (17,200), chicken lysozyme (14,400), human β_2-microglobulin (11,600), and bovine proinsulin (8700). The molecular weight of authentic rat proinsulin is 9100 (21). The dye front is indicated, below the arrows, by a dot. The amount of material in each lane corresponds to an input of 0.5 mCi in the labeling. The dry gel was exposed for 12 hr.

FIG. 3. Location of ^{35}S-containing residues in the amino-terminal region of the insulin products of three constructions containing the DNA encoding the preproinsulin signal sequence. The antigen was purified from $H_2{}^{35}SO_4$-labeled cells by immunoprecipitation and NaDodSO$_4$/polyacrylamide gel electrophoresis and then subjected to automated Edman degradation. The amount of radioactivity released by each cycle of degradation was determined by liquid scintillation counting. The amino-terminal sequence of authentic rat proinsulin is presented for comparison. (*A*) i25/−21: 20,000 cpm loaded, double-coupled at steps 1, 2, and 10, double-cleaved at step 9, 10% of each cycle analyzed. (*B*) i12/−21: 150,000 cpm loaded, double-coupled at step 1, 50% each cycle analyzed. (*C*) i4/−21: 50,000 cpm loaded, double-coupled at step 1, double-cleaved at step 9, 50% of each fraction analyzed.

at the 7th and 19th residues. Fig. 3 shows this unique pattern for all three proteins. HPLC of the radioactive fractions proved that most of the ^{35}S radioactivity was originally in cysteine, except for the first fractions of i12/−21 (Fig. 3*B*) and i25/−21 (Fig. 3*C*), where the radioactivity was not in any amino acid and was probably the result of protein washing out of the sequenator cup (data not shown). To test whether i27/+4 was matured at the end of the penicillinase signal sequence, we examined the positions of ^{35}S- and ^{3}H-labeled leucine. If the bacterial signal has been correctly removed, ^{35}S should appear at residue 16, while ^{3}H should appear at residues 15 and 20. Fig. 4 shows this unique pattern, demonstrating correct maturation of the bacterial signal when it is fused, four amino acids away from the clipping site, to rat proinsulin. Again, HPLC proved that the ^{35}S was originally in cysteine and the ^{3}H was in leucine (data not shown).

DISCUSSION

These experiments test the ability of bacteria to mature a variety of signal sequences fused to rat proinsulin (Fig. 1). Three of these hybrid proteins have most of the eukaryotic signal sequence attached to all of, half of, and only four amino acids of, the bacterial signal sequence. The sequencing data (Fig. 3) demonstrate that, in every case, the bacterial signal peptidase recognizes the eukaryotic clipping site and correctly matures

the hybrid preprotein to proinsulin. Furthermore, when the whole bacterial signal sequence is fused to rat proinsulin only four amino acids from the bacterial clipping site, this hybrid preprotein is also correctly matured (Fig. 4).

The bacterial signal peptidase correctly matures the hybrid preproteins to proinsulin whether the site for clipping is 29 (i4/−21), 40 (i12/−21), or 52 (i25/−21) amino acids from the amino terminus. This clearly demonstrates that the information to determine the site of cleavage is local and not dependent on the distance from the start of the signal peptide.

A simple model to account for the site of clipping of any preprotein would be that the signal sequence extends outside the native folded protein and that the signal peptidase clips back to the protein surface. Our results do not support this model; it is unlikely that proinsulin protects the bacterial signal exactly as does penicillinase, yet we see precise, correct, clipping of

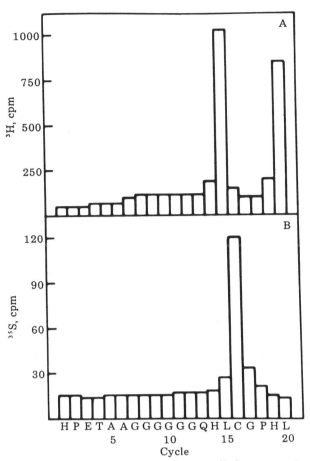

FIG. 4. Location of the ^{35}S-containing and [^{3}H]leucine residues in the amino-terminal sequence of i27/+4. The insulin antigen was purified from cells labeled with both H$_2$^{35}SO$_4$ and [^{3}H]leucine by immunoprecipitation and NaDodSO$_4$/polyacrylamide gel electrophoresis and then subjected to automated Edman degradation. (A) 300,000 cpm loaded; (B) 85,000 cpm loaded. Double-coupling was done at step 1, double-cleaving at steps 2 and 18. The amount of ^{35}S and ^{3}H radioactivity released at each cycle of degradation was determined by liquid scintillation counting, with the crossover into the ^{3}H channel subtracted. Ten percent of each fraction was analyzed. The amino-terminal sequence of i27/+4 matured at the correct bacterial clipping site (see Fig. 1) is presented for comparison.

i27/+4. The information for this clipping must therefore be contained in the signal sequence plus the first four amino acids of the mature protein. Lin *et al.* (23) characterized a mutant prelipoprotein in which a glycine, seven amino acids from the clipping site, was replaced with aspartic acid; the mutant prelipoprotein was transported but not cleaved. Thus, the carboxyl-terminal portion of the signal must participate in the processing.

Does the processing of the eukaryotic signal sequence in bacteria indicate a general phenomenon, or is it a special case of fortuitous signal sequence similarities? If we align the prepenicillinase (bacterial) and preproinsulin (eukaryotic) signal sequences, there are four amino acids which are the same distance from the site of clipping, underlined below (see Fig. 1 for the one-letter amino acid code):

Prepenicillinase

MSIQHFRVALIPFFAAFCLPVFA HPETLVK...

Preproinsulin

MALWMRFLPLLALLVLWEPKPAQA FVKQHLC...

The first two, phenylalanine and leucine, do not stand out

among the generally hydrophobic amino acids found in signal sequences. The last, alanine, is a frequent delineator; about half of the known signal sequences (2) end with alanine. The third is proline, four amino acids from the clipping site. Schechter *et al.* (24) have isolated an immunoglobulin light chain variant that is matured by a cut three amino acids from an invariant glycine (at amino acid −4) despite replacement of the three intervening amino acids, and they propose that helix-breaking amino acids in this region create part of the structure that the signal peptidase recognizes in order to cut three amino acids toward the carboxyl-terminus. Although it is possible that the clipping of the eukaryotic signal by the bacterial signal peptidase is an artifact of this proline at −4, the ability of dog pancreas microsomes to segregate and to process preproteins from species as unrelated as dogs and fish (8), as well as the ability of bacteria to transport a eukaryotic preprotein (7), suggests that the mechanism of secretion is both general and ancient. Thus, we expect that all eukaryotic signals will be recognized by bacteria, that the preproteins will be secreted with some efficiency, and that the secreted protein will be correctly matured.

Inserting a eukaryotic gene into a bacterial gene normally produces a hybrid bacterial-eukaryotic protein. One method to eliminate the extraneous bacterial protein is to insert the codon for an unusual amino acid between the two genes and to subject the fused protein product to chemical cleavage (25). This only works for proteins that lack the unusual amino acid. An alternate strategy involves direct expression of the gene within the bacterium, where the entire eukaryotic gene (including its ATG initiation codon) is inserted downstream from a bacterial promotor. This produces a product with an extra formyl methionine at the amino terminus, arising from the bacterial initiator. The formyl group may be removed [reported for simian virus 40 tumor antigen (26) and rabbit β-globin (27)], leaving an extra methionine in the case of rabbit β-globin. Our results suggest a simple method for the production of the extract, native protein, applicable where the eukaryotic protein is normally secreted, in cases such as insulin, interferon, human growth hormone, and many other medically important proteins. If the gene for the preprotein is inserted downstream from a bacterial promoter, the mature protein, without extraneous bacterial amino acids, can be isolated from the bacterial periplasmic space.

We gratefully acknowledge Peter Lomedico, Harry T. Orr, and Jeremy Knowles for helpful discussions and advice. This work was supported in part by National Institutes of Health Grant AM 21240 and in part by Biogen.

1. Milstein, C., Brownlee, G. G., Harrison, T. M. & Matthews, M. B. (1972) *Nature (London) New Biol.* **239**, 117–120.
2. Blobel, G., Walter, P., Change, C. N., Goldman, B. M., Erikson, A. H. & Lingappa, V. R. (1979) *Symp. Soc. Exp. Biol.* **33**, 9–36.
3. Blobel, G. & Dobberstein, B. (1975) *J. Cell Biol.* **67**, 835–851.
4. Emr, S. D., Hedgpeth, J., Clement, J. M., Silhavy, T. J. & Hofnung, M. (1980) *Nature (London)* **285**, 82–85.
5. Bedouelle, H., Bassford, P., Fowler, A., Zabin, I., Beckwith, J. & Hofnung, M. (1980) *Nature (London)* **285**, 78–81.
6. Villa-Komaroff, L., Efstratiatis, A., Broome, S., Lomedico, P., Tizard, R., Naber, S. P., Chick, W. L. & Gilbert, W. (1978) *Proc. Natl. Acad. Sci. USA* **75**, 3727–3731.
7. Talmadge, K., Stahl, S. & Gilbert, W. (1980) *Proc. Natl. Acad. Sci. USA* **77**, 3369–3373.
8. Shields, D. & Blobel, G. (1977) *Proc. Natl. Acad. Sci. USA* **74**, 2059–2063.
9. Fraser, T. & Bruce, B. J. (1978) *Proc. Natl. Acad. Sci. USA* **75**, 5936–5940.

10. Palmiter, R. D., Gagnon, J. & Walsh, K. (1978) *Proc. Natl. Acad. Sci. USA* **75**, 94–98.
11. Lingappa, V. R., Lingappa, J. R. & Blobel, G. (1979) *Nature (London)* **281**, 117–121.
12. Broome, S. & Gilbert, W. (1978) *Proc. Natl. Acad. Sci. USA* **75**, 2746–2749.
13. Kessler, S. W. (1975) *J. Immunol.* **115**, 1617–1624.
14. Miller, J. H. (1972) *Experiments in Molecular Genetics* (Cold Spring Harbor Laboratory, Cold Spring Harbor, NY).
15. Roberts, R. B., Abelson, P. A., Cowie, D. B., Bolton, E. T. & Britten, R. J. (1957) *Studies of Biosynthesis in Escherichia coli,* (Carnegie Institution of Washington, Washington DC), Publ. No. 607.
16. Laemmli, U. K. (1970) *Nature (London)* **227**, 680–685.
17. Maizel, J. V., Jr. (1970) in *Methods in Virology,* eds. Maramorosch, K. & Koprowski, H. (Academic, New York), Vol. 5, pp. 179–246.
18. Bauer, A. W., Margolies, M. N. & Haber, E. (1975) *Biochemistry* **14**, 3029–3035.
19. Sutcliffe, J. G. (1978) *Proc. Natl. Acad. Sci. USA* **75**, 3732–3736.
20. Lomedico, P., Rosenthal, N., Efstratiadis, A., Gilbert, W., Kolodner, R. & Tizard, R. (1979) *Cell* **18**, 545–558.
21. Dayhoff, M. O. (1972) *Atlas of Protein Sequences and Structure* (National Biomedical Research Foundation, Washington, DC), Vol. 5.
22. Nolan, C., Margoliash, E., Peterson, J. D. & Steiner, D. F. (1971) *J. Biol. Chem.* **246**, 2780–2795.
23. Lin, J. J. C., Kanazawa, H., Ozols, J. & Wu, H. C. (1978) *Proc. Natl. Acad. Sci. USA* **75**, 4891–4895.
24. Schechter, I., Wolf, O., Zemell, R. & Burstein, Y. (1979) *Fed. Proc. Fed. Am. Soc. Exp. Biol.* **38**, 1839–1845.
25. Itakura, K., Hirose, T., Crea, R., Riggs, A. D., Heyneker, H., Bolivar, F. & Boyer, H. F. (1977) *Science* **198**, 1056–1063.
26. Roberts, T., Bikel, I., Yocum, R., Livingston, D. M. & Ptashne, M. (1979) *Proc. Natl. Acad. Sci. USA* **76**, 5596–5600.
27. Guarante, L., Lauer, G., Roberts, T. M. & Ptashne, M. (1980) *Cell* **20**, 543–553.

Supplementary Readings

Beggs, J.D., Transformation of yeast by a replicating hybrid plasmid. *Nature* 275:104–109 (1978)

Bibb, M.J., Ward, J.M. & Hopwood, D.A., Transformation of plasmid DNA into *Streptomyces* at high frequency. *Nature* 274:398–400 (1978)

Chang, S. & Cohen, S.N., High frequency transformation of *Bacillus subtilis* protoplasts by plasmid DNA. *Mol. Gen. Genet.* 168:111–115 (1979)

Chater, K.F. & Bruton, C.J., Mutational cloning in *Streptomyces* and the isolation of antibiotic production genes. *Gene* 26:67–78 (1983)

Ehrlich, S.D., DNA cloning in *Bacillus subtilis. Proc. Natl. Acad. Sci. U.S.A. 75:1433–1436 (1978)*

Gray, G., Selzer, G., Buell, G., Shaw, P., Escanez, S., Hofer, S., Voegeli, P. & Thompson, C.J., Synthesis of bovine growth hormone by *Streptomyces lividans. Gene* 32:21–30 (1984)

Hopwood, D.A., Bibb, M.J., Chater, K.F., Kieser, T., Bruton, C.J., Kieser, H.M., Lydiate, D.J., Smith, C.P., Ward, J.M. & Schrempf, H., *Genetic Manipulation of Streptomyces: a laboratory manual.* Norwich, UK: John Innes Foundation, 1985

Hopwood, D.A., Malpartida, F., Kieser, H.M., Ikeda, H., Duncan, J., Fujii, I., Rudd, B.A.M., Floss, H.G. & Omura, S., Production of 'hybrid' antibiotics by genetic engineering. *Nature* 314:642–644 (1985)

Joliff, G., Béguin, P., Juy, M., Millet, J., Ryter, A., Poljak, R. & Aubert, J.-P., Isolation, crystallization and properties of a new cellulase of *Clostridium thermocellum* overproduced in *Escherichia coli. Bio/Technology* 4:896–900 (1986)

Lynd, L.R., Cushman, J.H., Nichols, R.J. & Wyman, C.E., Fuel ethanol from cellulosic biomass. *Science* 251:1318–1323 (1991)

Palva, I., Lehtovaara, P., Käärääinen, L., Sibakov, M., Cantell, K., Schein, C.H., Kashiwagi, K. & Weissmann, C., Secretion of interferon by *Bacillus subtilis. Gene* 22:229–235 (1983)

Ramos, J.L., Stolz, A., Reineke, W. & Timmis, K.N., Altered effector specificities in regulators of gene expression: TOL plasmid *xylS* mutants and their use to engineer expansion of the range of aromatics degraded by bacteria. *Proc. Natl. Acad. Sci. U.S.A.* 83:8467–8471 (1986)

Ramos, J.L., Wasserfallen, A., Rose, K. & Timmis, K.N., Redesigning metabolic routes: manipulation of TOL plasmid pathway for catabolism of alkylbenzoates. *Science* 235:593–596 (1987)

Rojo, F., Pieper, D.H., Engesser, K.-H., Knackmuss, H.-J. & Timmis, K.N., Assemblage of ortho cleavage route for simultaneous degradation of chloro- and methylaromatics. *Science* 238:1395–1398 (1987)

Rouvinen, J., Bergfors, T., Teeri, T., Knowles, J.K.C. & Jones, T.A., Three-dimensional structure of cellobiohydrolase II from *Trichoderma reesei. Science* 249:380–386 (1990)

Schein, C.H., Production of soluble recombinant proteins in bacteria. *Bio/Technology* 7:1141–1149 (1989)

Shoemaker, S., Schweickart, V., Ladner, M., Gelfand, D., Kwok, S., Myambo, K. & Innis, M., Molecular cloning of exo-cellobiohydrolase I derived from *Trichoderma reesei* strain L27. *Bio/Technology* 691–696 (1983)

Struhl, K., Stinchcomb, D.T., Scherer, S. & Davis, R.W., High-frequency transformation of yeast: Autonomous replication of hybrid DNA molecules. *Proc. Natl. Acad. Sci. U.S.A.* 76:1035–1039 (1979)

Plants

The genetic manipulation of plants has been practised for centuries: disease or pest-resistant races producing higher yields derived by traditional genetic techniques have had a tremendous impact on all aspects of commercial agriculture. The possibility of making stable hybrids has benefitted all aspects of plant production, from ornamental plants such as roses to key crop plants such as corn, cotton, potatoes, and tomatoes. As with all traditional genetic approaches, the methodology was successful but limited to exchanges between closely related species, and the experimental methods employed were often painstakingly difficult.

The potential for using recombinant DNA technology to manipulate the properties of plants in a variety of different ways not accessible to existing methods was apparent once suitable techniques for the introduction of DNA into plants had been developed. The application of genetic engineering methods to plant research promises much in terms of the characterization of functions necessary for nodulation, nitrogen fixation, protein storage and degradation, colour, etc. In the early days of plant genetic engineering (the hype period) there were many over-publicized claims for the prospect of manipulating plants to fix their own nitrogen, obviating the need for fertilizers or bacterial symbionts. Happily, these exaggerated claims (still within the realm of possibility, nonetheless!) were soon replaced by more immediately feasible and equally useful goals employing the powerful tools of molecular and cellular biology. One of the difficulties inherent in plant biotechnology has been the development of appropriate selection markers, vectors, and efficient transformation systems. This last in particular has been a major stumbling block, and only recently have important food stocks such as rice and other monocotyledenous plants become amenable to recombinant DNA manipulations. The papers presented describe the development of the techniques used to obtain transgenic plants. Basic food stocks such as potatoes have been improved, and herbicide resistance genetically engineered into plants, permitting more efficient control of weeds and rapid farming techniques, is being field tested. For example, the resistance gene for the herbicide bialaphos (phosphinothricin) produced by a bacterium has been introduced into several different dicotyledons and found to provide complete protection from the herbicide. (See DeBlock *et al.* in the Supplementary References.) In a similar manner, insecticide-resistant plants (Fischoff *et al.*, Supplementary References) have been obtained by the incorporation of the gene for a bacterial insecticidal protein; the gene is expressed under the control of a plant promoter and has been found to prevent insect infections effectively.

The most significant advances in plant biotechnology have come from the study and use of the *Agrobacterium tumefaciens* Ti plasmids. By genetic engineering techniques the infection process and resulting integration of heterologous DNA induced by infection with this bacterium have been analyzed in great detail. The delivery and integration system of the Ti plasmids allows the manipulation of a large number or dicotyledenous plant species with a variety of different genetic determinants, which have, to a large extent, demonstrated the efficacy of the system in crop manipulation. Unfortunately, since the *Agrobacterium* host range does not include monocotyledons, other transformation procedures had to be developed to extend the range of plant genetic engineering. The "ballistic" approach or "particle gun" has been very successful and heterologous genes controlled by plant promoters have been introduced into new hosts. This is described in the Klein paper.

As with many applications of modern biotechnology, genetic manipulation of plants has not met with universal acclaim. This is understandable, given that traditional farming might be threatened because large scale production methods could be expected to become even more practical and economically advantageous. In addition, plant biotechnology would represent the first extensive release into the environment of organisms genetically engineered using recombinant DNA methods. As mentioned earlier, genetically manipulated plants have been used for many years, but the use of recombinant DNA methods is of recent vintage. There is no fundamental difference between the new and old techniques with respect to "risk," but the recent introduction of genetically-modified commercial crops and the treatment of plants with recombinant organisms such as ice-minus bacteria have been subject to extensive evaluation by both responsible authorities and "biofundamentalists." Much effort is being expended by responsible scientists and industrialists worldwide, to ensure that the public is well informed and that appropriate experimentation and testing is being carried out. Regrettably, over-exposure of the views of the more alarmist factions all too often delays the correct processes of testing and evaluation under field conditions.

We should not think of plants only as components of our cuisine; many plant products are used pharmaceutically (alkaloids, anti-tumour drugs, steroids) and have applications in other domains such as flavourings, etc. The explosion of research in plant biology promises transgenic methodology to modify such natural production processes; accompanying advances in plant cell tissue culture and bioreactors will mean that field processes may not, in all cases, be necessary. One interesting application of recombinant DNA technology in the plant field is the production of mouse antibodies in the transgenic tobacco plants. (See the Supplementary References.) The potential for agricultural biotechnology is considerable, and improvements and extension of the basic discoveries described in these papers will be critical to future developments.

A Chimaeric Antibiotic Resistance Gene as a Selectable Marker for Plant Cell Transformation

M.W. Bevan, R.B. Flavell and M.-D. Chilton

The T-DNA region of *Agrobacterium tumefaciens* tumour-inducing plasmids of the nopaline type[1] contains a gene coding for the enzyme nopaline synthase. This gene is expressed constitutively in host plant cells to which it is transferred during tumour induction[2]. We have exploited the regulatory elements of this gene to construct a chimaeric gene that confers antibiotic resistance on transformed plant cells. The chimaeric gene encodes the expected chimaeric transcripts in plant cells, and confers on transformed cells the ability to grow in the presence of normally lethal levels of the antibiotic G418 (ref. 3). Experiments using *in vitro* transformation techniques on single plant cells indicate that this antibiotic resistance can be used as a selectable marker, and can therefore be used in selecting cells transformed by T-DNA vectors that have had the genes for hormone autotrophy deleted[4]. Plant cells transformed by such 'disarmed' T-DNA vectors can be regenerated into entire plants, whose sexual progeny contain unaltered copies of the inciting T-DNA[5]. The availability of this dominant selectable marker should allow a wider range of experiments to be undertaken using different host plants.

The antibiotic resistance gene used in our experiments (*neo*) is derived from Tn*5* and codes for neomycin phosphotransferase II. The *neo* coding region, when spliced to viral gene regulatory sequences, confers on animal cells resistance to the aminoglycoside antibiotic, G418 (ref. 3). We have placed the coding region of *neo* under the transcriptional control of the nopaline synthase promoter, which has been identified by DNA sequencing and S_1 nuclease mapping experiments[6,7]. We included in our construct the 5'-untranslated region of the nopaline synthase gene, but nothing of the 5' end of the coding region. At a position 3' to the *neo* coding region our construct contains the last portion of the nopaline synthase-coding region, a 3'-untranslated region, and termination and polyadenylation signals.

The 5' end of the chimaeric *neo* gene was constructed as outlined in Fig. 1. A 370-base pair (bp) *Sau*3a fragment containing the nopaline synthase promoter region and the first 16 codons of the coding region was treated with *Bal*31 to remove the coding region including the initiator methionine codon. The resulting population of molecules was ligated to *Bam*HI linkers and inserted into the *Bam*HI site of pBR322. A 270-bp *Bam*HI fragment with all 48 nucleotides of the nopaline synthase-coding region deleted was identified by DNA sequencing[8] after subcloning into M13mp7 (ref. 9). The plasmid containing this insert was designated pJFΔ166.

To construct a nopaline synthase gene containing the appropriate deletion and a unique *Bam*HI site for insertion of the *neo* coding region, the 270-bp promoter fragment was digested with *Bam*HI and *Sst*II and cloned into *Bam*HI/*Sst*II-digested pJF28 (see Fig. 1 legend). This produced a deletion of 814 bp in the nopaline synthase gene in the resulting plasmid, pJF28Δ166 (Fig. 1, step *C*). The resected gene in this plasmid has a unique *Bam*HI site adjacent to the promoter, followed by 423 bp of coding region and the polyadenylation site.

In order to be able to transfer the reconstructed gene to an *Agrobacterium* Ti plasmid by double recombination, it is necessary to have extensive regions of homology on both sides of the gene, and a tightly linked prokaryotic selectable marker[10]. We therefore subcloned the deleted nopaline synthase gene, together with the closely linked kanamycin resistance (*Km*^{r}) marker from pACYC177 (ref. 11), into pJFXhoX to create pJFXhoXΔ166 (Fig. 1, step *D*).

Construction of the chimaeric *neo* gene started with the plasmid pAM1 (ref. 12), which contains a *Hind*III/*Sal*I fragment of Tn*5* (ref. 13). The coding region of the *neo* gene is found on a 1-kilobase (kb) *Bgl*II/*Sma*I fragment[14]. To insert this into the deleted nopaline synthase gene, the *Sma*I site adjacent to the 3' end of the *neo* gene was first converted to a *Bam*HI site using linkers to create pAM1 (B). The coding region was isolated from pAM1 (B) by digestion with *Bam*HI and *Bgl*II, and ligated into *Bam*HI-digested pJFXhoXΔ166 (Fig. 1, step *E*). The orientation of the *neo* coding region in relation to the nopaline synthase promoter was verified in several plasmids by digestion with *Bam*HI and *Sst*II, *Hind*III and *Sst*II, and *Pst*I. Clones with the AUG initiation codon adjacent to the promoter were designated pJFneo$^+$, and those in the reverse orientation pJFneo$^-$.

To transfer the relevant segments of pJFneo$^+$ and pJFneo$^-$ to *A. tumefaciens*, the plasmids were digested with *Eco*RI and ligated into *Eco*RI-digested pRK290 (ref. 28). The resulting DNAs were cloned in *E. coli* and subsequently transformed into *A. tumefaciens* strain A208 (ref. 15) using kanamycin selection. Double recombinants (homogenotes) that incorporated the engineered fragment onto pTiT37 were isolated as described previously[10]. The structure of the homogenotes was confirmed by Southern blot analysis of total bacterial DNA. Clones harbouring the desired recombinant pTiT37 plasmids were designated A208neo$^+$ and A208neo$^-$. These bacteria were used to incite tumours on sterile stem explants of *Nicotiana tabacum* var. Turkish. The resultant tumorous overgrowths were excised, rendered axenic by growth in MS medium[16] containing 500 μg ml^{-1} carbenicillin, and cloned as described elsewhere[17]. Normal teratomatous tumour growth occurred 1 month after cloning. To look for the presence of T-DNA, and to determine its structure in the tumours, DNA was isolated from representative tobacco clones[18], digested with *Eco*RI or *Bam*HI and Southern-blotted[19]. We used nick-translated[20] MINI-Ti plasmid[12] containing the entire T-DNA of pTiT37 (ref. 21) as a hybridization probe for analysis of T-DNA structure. An autoradiogram showing the pattern of hybridization is presented in Fig. 2. Hybridization with MINI-Ti revealed that our tumour lines contained 1–2 copies of T-DNA identical to that of natural HT37 tumour cells except for the size of *Bam*HI fragment 14a and the diverse border fragments. In HT37 tumours *Bam*HI fragment 14a is 4.5 kb (arrowed in Fig. 2). As expected, in neo$^+$ tumours, the corresponding fragment is nearly 6 kb due to the presence of the 1.5-kb pACYC177 kanamycin resistance gene (arrowed in Fig. 2). In neo$^-$ tumours, the counterpart of *Bam*HI fragment 14a is 7 kb due to the insertion of the *Bgl*II/*Bam*HI coding region of *neo* (arrowed).

The transcriptional activity of the chimaeric gene in the cloned tobacco tumours was studied by northern blot analysis[22,23]. Polyadenylated RNA isolated from HT37, neo$^+$ and neo$^-$ tumours[24] was denatured, subjected to electrophoresis and transferred to nitrocellulose. The RNA blots were hybridized with the *neo* coding region fragment, which hybridized to a 1,800-base transcript in neo$^+$ tumours (Fig. 3*B*, lanes 3 and 4) and to 1,700- and 1,800-base transcripts in neo$^-$ tumours (Fig. 3*B*, lanes 8 and 9). As expected, this probe did not hybridize significantly to HT37 RNA (Fig. 3*B*, lanes 1, 2, 6 and 7). A 1.1-kb *Bam*HI/*Hind*III fragment homologous to the 3' end of the nopaline synthase gene[6,7] also hybridized to a 1,800-base transcript in neo$^-$ tumours (Fig. 3*A*, lanes 3 and 4) and to 1,700- and 1,800-base transcripts in neo$^-$ tumours (Fig. 3*A*, lanes 7 and 8). This probe, as expected, hybridized to an unaltered 1,600-base nopaline synthase transcript in HT37 RNA (Fig. 3*A*, lanes 1, 2, 5 and 6). The 1,800-base transcript observed in neo$^-$ and neo$^-$ tumours that hybridized to both probes is expected of a transcript that originates at the normal nopaline synthase initiation (cap) site, traverses the 1,010-bp BglII/BamHI neo coding region (which replaces 814 bp of the nopaline synthase gene), and terminates at the nopaline synthase polyadenylation site. As the 1,700-base transcript observed in neo⁻ tumours is polyadenylated, it seems likely that it arises from either aberrant initiation of transcription 100 bp

from the normal cap site or polyadenylation at the unused poly(A) signal observed 100 bp upstream of the 3′ end[7].

The abundance of neo⁻ and neo⁺ transcripts relative to the natural nopaline synthase transcript can be estimated roughly from these hybridization data. When the different levels of contamination with rRNA are taken into account (visible in ethidium bromide-stained gels), the neo⁺ transcript and both

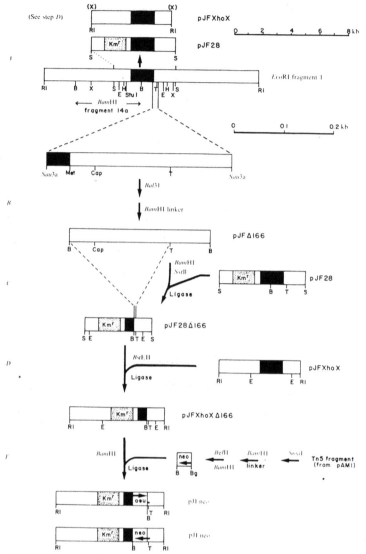

Fig. 1 Construction of the chimaeric *neo* gene. *A.* Construction of pJF28, a plasmid containing *Km'* closely linked to the nopaline synthase gene. A recombinant plasmid containing *Sal*I fragment 10 of pTiT37 was partially digested with *Stu*I, and to this was ligated a 1.5-kb *Hae*II fragment of pACYC177 (ref. 11) (encoding the *Km'* gene). Plasmid pJF28 contained the *Km'* insert in the *Stu*I site adjacent to the 3′ end of the nopaline synthase gene, as determined by analysis of minipreps. *B.* Isolation of the nopaline synthase promoter. A 370-bp *Sau*3a fragment that contains the first 16 codons of the nopaline synthase gene and extends nearly to the right border of T-DNA was isolated from *Hind*III fragment 23 by preparative gel electrophoresis in 4% polyacrylamide; 5 µg of this fragment was treated with limiting *Bal*31 exonuclease, and the resulting fragments were ligated with 100-fold molar excess of *Bam*HI linkers²⁶ (25 µl, 18 h, 4 °C, 10 U T₄ DNA ligase). After heating at 70 °C for 10 min the product was digested with *Bam*HI (100 U, 4 h, 37 °C). After preparative electrophoresis (5% polyacrylamide gel), molecules of ~270 bp were excised and recovered by electroelution, DEAE-Sephadex chromatography and ethanol precipitation. This population of molecules was cloned into *Bam*HI-linearized pBR322 (ref. 27). Minipreps of recombinant plasmids (Ap'Tc') were digested with *Bam*HI and *Sst*II, and analysed for the presence of the desired 192-bp *Bam*HI–*Sst*II fragment. Twelve candidates were subcloned into M13mp7 (ref. 9) and their inserts sequenced using the dideoxy method⁹. One plasmid, pJFΔ166, was found to have been deleted to the nucleotide just 5′ to the ATG initiation codon of the nopaline synthase gene. *C.* Deletion of the 5′ end of the coding region of the nopaline synthase gene. Plasmid pJF28 (1 µg) and plasmid pJFΔ166 (5 µg) were digested with *Bam*HI and *Sst*II, and ligated together (25 µl, 10 U T₄ DNA ligase). *Km'* transformants were screened for the presence of the 192-bp *Bam*HI–*Sst*II fragment. The desired plasmid was designated pJF28Δ166. *D.* Substitution of the truncated nopaline synthase gene into a plasmid containing flanking homology with the Ti plasmid. Plasmid pJFXhoX was made by isolating a 6-kb *Xho*I fragment containing the nopaline synthase gene from *Eco*RI fragment 1, and converting the *Xho*I sites to *Eco*RI sites using molecular linkers²⁶. The desired *Eco*RI fragment of 6 kb was cloned into pBR325-B (ref. 27; pBR325 that has had the *Bam*HI site destroyed by blunting and recircularization). The resulting plasmid, pJFXhoX (1 µg), was digested with *Bst*EII to completion and ligated with 1 µg of *Bst*EII-digested pJF28Δ166 that had been treated with alkaline phosphatase. Substitution in the correct orientation of the kanamycin resistance-conferring *Bst*EII fragment for the *Bst*EII fragment of pJFXhoX was verified by digestion of minipreps with *Bst*EII, *Sma*I and *Eco*RI plus *Bam*HI. The desired recombinant molecule was designated pJFXhoXΔ166. *E.* Construction of chimaeric nopaline synthase/*neo* genes. The coding region of the *neo* gene lies within a *Bgl*II/*Sma*I fragment of Tn5; this fragment includes 30 bp upstream and 180 bp downstream from the coding region¹¹. Plasmid pAM1 contains a *Hind*III–*Sal*I fragment of Tn5¹²

that includes the *neo* gene. The *Sma*I site adjacent to the 3′ end of the *neo* gene was converted to a *Bam*HI site using linkers as described²⁶, producing plasmid pAM1(B). The *neo* coding region was excised from pAM1(B) by digestion with *Bam*HI and *Bgl*II and purified. Plasmid pJFXhoXΔ166 (1 µg) was digested to completion with *Bam*HI, and treated with alkaline phosphatase as described above. To this was ligated 0.1 µg of the gel-purified *Bam*HI–*Bgl*II fragment containing the *neo* coding region. Recombinant plasmids with the *Bgl*II–*Bam*HI fusion adjacent to the *Sst*II site were designated pJFneo⁺ and those with the reverse orientation were called pJFneo⁻. The *Escherichia coli* host for DNA constructions was LE392. Drug concentrations used for selective media were: for *E. coli*, ampicillin (Ap), 100 µg ml⁻¹; kanamycin (Km), 25 µg ml⁻¹; tetracycline (Tc), 25 µg ml⁻¹; chloramphenicol (Cm), 25 µg ml⁻¹; for *A. tumefaciens* A208, Km, 100 µg ml⁻¹; gentamycin, 50 µg ml⁻¹. The scale bar at the top of the figure refers to all constructions except those in step *B*, which are enlarged (see separate scale bar). Cap is the transcription initiation site, and met the initiation of translation codon. Restriction endonuclease sites used in constructions are designated as follows: B, *Bam*HI; E, *Bst*EII; H, *Hind*III; R1, *Eco*RI; S, *Sal*I; T, *Sst*II; X, *Xho*I; (X), destroyed *Xho*I site. The black box represents the coding region of the nopaline synthase gene. The stippled box represents the prokaryotic *Km'* gene from pACYC177. The coding region of neomycin phosphotransferase II from Tn5 is designated *neo*, and the arrow on this DNA fragment indicates the orientation of the gene.

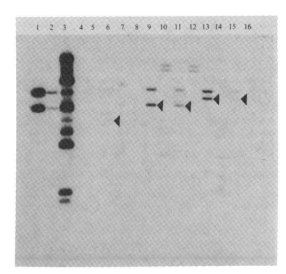

Fig. 2 Southern blot analysis of DNA from cloned tumours. The hybridization probe used is a clone of the entire T-DNA region of pTiT37 cloned in pBR322. Lanes contained the following DNA samples: 1, 10 copy reconstruction: 2, 1 copy reconstruction; 3, molecular weight marker; 4, normal, *Bam*HI; 5, normal, *Eco*RI; 6, HT37, *Bam*HI; 7, HT37, *Eco*RI; 8, no DNA; 9, neo⁺ clone 14, *Bam*HI; 10, neo⁺ clone 14, *Eco*RI; 11, neo⁺ clone 8, *Bam*HI; 12, neo⁺ clone 8, *Eco*RI; 13, neo⁻ clone 2, *Bam*HI; 14, neo⁻ clone 2, *Eco*RI; 15, neo⁻ clone 1, *Bam*HI; 16, neo⁻ clone 1, *Eco*RI. The arrows in the autoradiogram indicate the position of the modified *Bam*HI fragment 14a adjacent to the *neo* gene.
Methods: DNA was isolated[18] from representative cloned tobacco tumour lines incited by A208neo⁺ and A208neo⁻. Positive and negative control DNAs were isolated from HT37-8-70 tumour tissue which contains an unaltered T-DNA insert (G. Jen, unpublished data), and from *N. plumbaginifolia* leaves. Ten µg of each DNA sample were digested to completion with either *Eco*RI or *Bam*HI and electrophoresed through a 0.7% agarose gel. Southern blotting, hybridization and washing were performed as described elsewhere[29]. Labelled probe DNAs ($1.5-3 \times 10^8$ c.p.m. µg⁻¹) were prepared by nick translation[20] using [α³²P]dATP (NEN). The molecular weight marker was λ DNA digested with *Eco*RI and *Hind*III, and the reconstruction lanes contained 15 pg and 150 pg of pJFneo⁺ DNA digested with *Eco*RI. The Southern blots were exposed to X-ray film with two intensification screens for 2 days at −70 °C.

Fig. 3 Northern blot analysis of transcripts. **A**, Hybridization probe specific for 3' end of nopaline synthase gene. Lanes contained the following RNA samples: 1, 20 µg HT37; 2, 10 µg HT37; 3, 5 µg neo⁻ clone 8; 5, 5 µg HT37; 6, 1 µg HT37; 7, 20 µg neo⁻ clone 2; 8, 20 µg neo⁻ clone 2. **B**, Hybridization probe specific for *neo* coding region. Lanes contained the following RNA samples: 1, 20 µg HT37; 2, 10 µg HT37; 3, 5 µg neo⁻ clone 14; 4, 5 µg neo⁻ clone 8; 5, molecular weight marker; 6, 5 µg HT37; 7, 1 µg HT37; 8, 20 µg neo⁻ clone 2; 9, 20 µg neo⁻ clone 1. **C**, Structure and transcription of chimaeric genes. (F) is the *Bam*HI–*Bgl*II fusion. Lanes 1–4 of *A* and 1–5 of *B* are from the same gel, and lanes 5–8 of *A* and 6–9 of *B* are from another gel. Direct size comparisons between these different gels was not possible. Transcript sizes were calculated from standards run in each gel. The diagram showing a truncated chimaeric transcript in neo⁻ tumours shows only one possible RNA molecule. Another possible transcript may be truncated at the 3' end.
Methods: RNA was isolated from the four tumour clones whose DNA was analysed by Southern blotting (Fig. 2). Control RNA was isolated from HT37-8-70 tumour tissue.

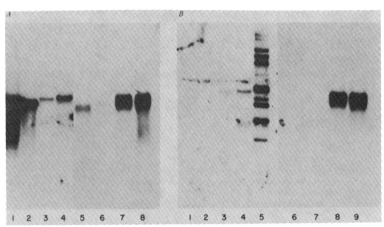

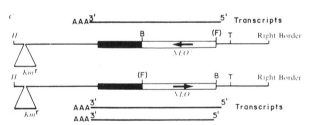

After purification by a guanidine thiocyanate procedure[24], poly(A)-containing RNA was selected by olido-dT cellulose chromatography[30]. RNA was denatured and electrophoresed in a 1.5% agarose–formaldehyde gel[22], and transferred to nitrocellulose[23]. Hybridization probes were nick-translated *Bam*HI–*Bgl*II fragment containing the *neo* coding region excised from pAM1(B) and a 1.1-kb *Bam*HI–*Hind*III fragment complementary to the 3' end of the nopaline synthase gene (see Fig. 1 for location). Molecular weight markers were *Eco*RI and *Hind*III fragments of λ DNA that had been end-labelled with [α³²P]dATP using Klenow polymerase. The blot was hybridized for 16 h at 41 °C, washed as described[29], and exposed to X-ray film at −70 °C for 2 days with two intensification screens.

neo⁻ transcripts are approximately as abundant as the nopaline synthase transcript. A diagram of the transcription of neo⁺ and neo⁻ genes is presented in Fig. 3C.

Having established that the chimaeric *neo* gene is transcribed in the tumours, we investigated its effect on the plant cell

phenotype. The growth rates of neo⁺ and neo⁻ cloned tumours were determined on MS medium without phytohormones, but containing various concentrations of G418. Tumour explants known to contain T-DNA were collected and weighed after 30 days. Figure 4 shows the results obtained for two neo⁺ and

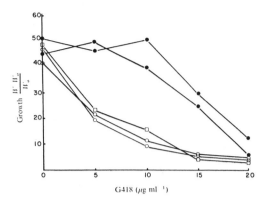

Fig. 4 Growth of tumours on G418-containing medium. Open circles, neo⁻ clone 1 and neo⁻ clone 2 tumours; closed circles, neo⁺ clone 8 and neo⁺ clone 14 tumours; open squares, HT37-8-70 tumour. W_0, initial weight (30 mg); W, final weight after 25 days.

Methods: Approximately 20 clones of neo⁻ and neo⁺ tumours were analysed for growth on MS medium[16] containing 3% sucrose and various concentrations of G418. Ten pieces of tissue from each clone, each weighing ~30 mg, were transferred to the indicated concentrations of G418 and incubated at 25 °C in a 12 h light/dark cycle for 30 days. Tissues were collected and weighed, and the mean weight increase calculated. The data presented here are for the cloned tumour lines whose T-DNA and transcripts are described above (Figs 2 and 3). The remaining 18 clones of each type (neo⁺ and neo⁻) gave similar data.

two neo⁻ clones. Tumours containing the neo⁻ construction exhibited almost complete cessation of growth on medium containing more than 5 μg ml⁻¹ G418, as did natural HT37 tumours. In contrast, tumours harbouring the neo⁺ gene showed no decrease in growth rate until 15–20 μg ml⁻¹ G418. We tested approximately 20 additional clones of neo⁻ and neo⁺ tumours, and these displayed almost identical growth curves to their counterparts in Fig. 4. As the growth rate of tumours is strictly dependent on the orientation of the coding region relative to the nopaline synthase promoter, we conclude that the observed differences in growth rate are due to differences in the activity of the *neo* gene in these tumours.

The results presented here demonstrate that a chimaeric gene composed of T-DNA regulatory, bacterial coding and T-DNA termination elements is transcribed in plant cells, and that the expected enzyme activity associated with the translation of the chimaeric transcript is manifest by an enhanced resistance of those plant cells to the antibiotic G418. Preliminary cocultivation experiments[25] using *Nicotiana plumbaginifolia* and A208neo⁺ and A208neo⁻ indicate that transformants can be selected using 10 μg ml⁻¹ G418 in the growth medium of the regenerating cells (data not shown). Thus, this G418 resistance will be useful in the selection of plant cells transformed by 'disarmed' T-DNA vectors containing no natural selectable marker[4].

This research was supported by DOE contract AC02-81ER 10889. M.W.B. gratefully acknowledges postdoctoral fellowship funds from Monsanto during his stay at Washington University, and NRDC funds during his work at the PBI.
Note added in proof: Recently Herrera-Estrella *et al.*[31] have demonstrated the expression of chloramphenicol transacetylase from nopaline synthase gene constructions nearly identical to those reported here.

Received 4 May; accepted 8 June 1983.

1. Leemans, J. *et al. J. molec. appl. Genet.* **1**, 149–164 (1981).
2. Wullems, G. J., Molendijk, L., Ooms, G. & Schilperoort, R. A. *Cell* **24**, 719–727 (1981).
3. Colbere-Garapin, F., Horodniceanu, F., Kourilsky, P. & Garapin, A.-C. *J. molec. Biol.* **150**, 1–14 (1981).
4. Leemans, J. *et al. J. molec. appl. Genet.* **1**, 149–164 (1981).
5. Barton, K. A., Binns, A. N., Matzke, A. J. M. & Chilton, M.-D. *Cell* **32**, 1033–1043 (1983).
6. Depicker, A., Stachel, S., Dhaese, P., Zambryski, P. & Goodman, H. M. *J. molec. appl. Genet.* **1**, 561–573 (1982).
7. Bevan, M., Barnes, W. M. & Chilton, M.-D. *Nucleic Acids Res.* **11**, 369–385 (1983).
8. Sanger, F., Nicklen, S. & Coulson, A. R. *Proc. natn. Acad. Sci. U.S.A.* **74**, 5463–5467 (1977).
9. Messing, J., Crea, R. & Seeburg, P. H. *Nucleic Acids Res.* **9**, 309–323 (1981).
10. Matzke, A. J. M. & Chilton, M.-D. *J. molec. appl. Genet.* **1**, 39–49 (1981).
11. Oka, A., Sugisaki, H. & Takanami, M. *J. molec. Biol.* **147**, 217–226 (1981).
12. de Framond, A. J., Barton, K. A. & Chilton, M.-D. *Bio/Technology* **1**, 262–269 (1983).
13. Rothstein, S. J., Jorgenson, R. A., Postle, K. & Reznikoff, W. S. *Cell* **19**, 795–805 (1980).
14. Auerswald, E. A., Ludwig, G. & Schaller, H. *Cold Spring Harb. Symp. quant. Biol.* **45**, 107–113 (1980).
15. Holsters, M. *et al. Molec. gen. Genet.* **163**, 181–187 (1978).
16. Murashige, T. & Skoog, F. *Pl. Physiol.* **15**, 473–497 (1962).
17. Meins, F. & Binns, A. *Proc. natn. Acad. Sci. U.S.A.* **74**, 2928–2932 (1977).
18. Chilton, M.-D. *et al. Nature* **295**, 432–434 (1982).
19. Southern, E. M. *J. molec. Biol.* **98**, 503–517 (1975).
20. Maniatis, T., Jeffrey, A. & Kleid, D. G. *Proc. natn. Acad. Sci. U.S.A.* **72**, 1184–1188 (1975).
21. Depicker, A. *et al. Plasmid* **3**, 193–211 (1980).
22. Rave, N., Ckvenjakov, R. & Boedtker, H. *Nucleic Acids Res.* **6**, 3559–3567 (1979).
23. Thomas, P. S. *Proc. natn. Acad. Sci. U.S.A.* **77**, 5201–5205 (1981).
24. Bevan, M. W. & Chilton, M.-D. *J. molec. appl. Genet.* **1**, 539–546 (1982).
25. Wullems, G. J., Molendijk, L., Ooms, G. & Schilperoort, R. A. *Proc. natn. Acad. Sci. U.S.A.* **78**, 4344–4348 (1981).
26. Green, M. R. & Roeder, R. G. *Cell* **22**, 231–242 (1980).
27. Bolivar, F. *Gene* **4**, 121–136 (1978).
28. Ditta, G., Stansfield, S., Corbin, D. & Helinski, D. *Proc. natn. Acad. Sci. U.S.A.* **77**, 7347–7351 (1980).
29. Thomashow, M. F. *et al. Proc. natn. Acad. Sci. U.S.A.* **77**, 6448–6452 (1980).
30. Aviv, H. & Leder, P. *Proc. natn. Acad. Sci. U.S.A.* **69**, 1408–1412 (1972).
31. Herrera-Estrella, L., Depicker, A., Van Montagu, M. & Schell, J. *Nature* **303**, 209–213 (1983).

Propagation of Foreign DNA in Plants Using Cauliflower Mosaic Virus as Vector

B. Gronenborn, R.C. Gardner, S. Schaefer and R.J. Shepherd

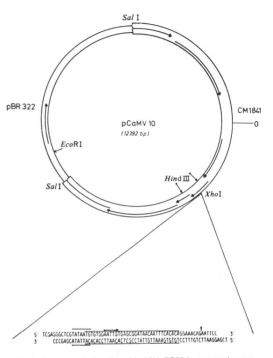

Cauliflower mosaic virus (CaMV) is the best analysed member of the caulimoviruses, a group of small isometric plant viruses that contain a circular double-stranded genome (for review see ref. 1). As most plant viruses have an RNA rather than a DNA genome, caulimoviruses have attracted considerable interest due to their potential use as genetic vectors for plants. DNA purified from CaMV particles can infect plants and cause virus production when rubbed over the surface of susceptible leaves[2]. Moreover, the entire chromosome of several CaMV strains can be propagated through bacterial hosts by plasmid and phage λ vectors and has been used successfully to infect plants[3,4]. Recently, Howell et al.[5] reported the insertion of an 8-base pair (bp) EcoRI linker molecule into the large 'intergenic' region of cloned CaMV strain CM 4-184 DNA without impairment of infectivity. Here we report the successful propagation of foreign DNA in plants using cauliflower mosaic virus as vector. We find that the size of foreign DNA that can be successfully propagated through virus particles has an upper limit of ~250 base pairs.

The nucleotide sequence of CaMV[6,7] suggests that there are extensive regions of the genome which encode protein. To insert new DNA, however, it is necessary to identify non-coding or dispensable regions where insertions will not interfere with essential functions. To identify such regions, we selected a segment of DNA which specifies a phenotype in *Escherichia coli* that is easily detected and selected, and which is sufficiently small so as not to exceed a possible limitation on packaging capacity of the virion. This marker DNA was then inserted into the CaMV genome cloned in an *E. coli* plasmid vector.

The *E. coli* marker chosen for these trial insertions was a 65-bp fragment derived from the *lac* promoter–operator region[8]. In a multicopy plasmid, this *lac* operator fragment leads to a *lac* constitutive phenotype in a suitable host due to titration of *lac* repressor[9]. We provided the fragment with an asymmetrically positioned *EcoRI* restriction site and added synthetic *XhoI* linker molecules to its ends[10]. *XhoI* linkers were chosen because the position of a naturally occurring *XhoI* site on the chromosome of CaMV strain 1841 is covered by a deletion in the variant strain CM 4-184 (refs 11, 12), indicating a non-essential region of the genome. The *lac* operator with *XhoI* ends could be inserted easily to test this hypothesis. As an acceptor molecule we used DNA of plasmid pCaMV10, a derivative of pBR322 carrying the complete genome of CaMV strain 1841 (refs 4, 12). The structure and orientation of two different *lac* operator inserts are given in Fig. 1.

To test whether a CaMV genome containing such inserts is infectious, the viral DNA was released from the plasmid vector by cleavage with *SalI* restriction endonuclease and used to infect leaves of 'Just Right' turnip plants (*Brassica campestris* L.). About 200 ng of DNA per leaf were sufficient to cause 2–30 local lesions on mechanically inoculated leaves and led in all cases to symptoms typical of systemic infection[13]. The appearance of symptoms was slightly delayed compared with infection by the parent molecules and the disease symptoms were milder. Viral DNA prepared from infected plants was analysed[14]. Figure 2a shows that the *lac* operator fragment can be recovered after *XhoI* digestion of the viral DNA. To determine whether only a fraction of the viral population contained the insert, four individual local lesions induced by DNA

5′ TCGASGGCTCGTATAATGTGTGGAATTGTGCAGCGGATAACAATTTCACACAGGAAACAGAATTCC 3′
3′ CCCGAGCATATTACACACTTAACACTCGCCTATTGTTAAAGTGTGTCCTTTGTCTTAAGGAGCT 5′

Fig. 1 Structure and orientation of the 1841–EOP2 *lac* operator insert at the unique *XhoI* site of the CaMV 1841 hybrid plasmid pCaMV10. A 55-bp *HpaII–EcoRI* restriction fragment from the *lac* UV5 promoter–operator region of *E. coli* in plasmid pUR108-1 (ref. 8 and U. Rüther, unpublished results) was given flush ends by incubation with *E. coli* Pol 1 Klenow fragment (Boehringer) in the presence of dATP and dTTP[17]. After purification by electrophoresis through an 8% polyacrylamide gel, 1 pmol of fragment was ligated to 50 pmol of *XhoI* linker molecules (Collaborative Research)[10]. This step restores the *EcoRI* restriction site at the promoter distal end of the fragment. *XhoI* restriction endonuclease digestion (restriction enzymes from New England Biolabs, Boehringer or BRL) was used to cleave off superfluous linkers and a second 8% polyacrylamide gel was used to prepare the now *XhoI*-ended fragment. The purified fragment was ligated to pCaMV10 DNA linearized by *XhoI*. Hybrid molecules were identified after transformation of *E. coli* CSH 51 (ref. 18) and selection of ampicillin resistant bacteria on plates containing ampicillin (50 μg ml⁻¹) and 5-bromo-4-chloro-indolyl-β-D-galactoside (40 μg ml⁻¹). *lac* constitutive bacteria produce blue colonies in these conditions[18]. Two different isolates were analysed further; these carry the *lac* operator in both orientations in the CaMV genome. In addition, pCaMV10-EOP1 has lost one terminal nucleotide pair at the *HpaII* end of the *lac* operator. The six open reading frames of CaMV are indicated by arrows, starting in a clockwise direction from the zero point of the map. The orientation of pBR322 relative to CaMV DNA is indicated by its unique *EcoRI* site and the reading frame of the β-lactamase gene. The *lac* operator insert of pCaMV10-EOP2 is enlarged at the bottom. The Pribnow box homology and the operator sequence are underlined. The arrow pointing to the right indicates the transcription start point in *E. coli*. The additional *EcoRI* site is indicated by ▽. pCaMV10-EOP1 carries the *lac* operator in the opposite orientation and lacks one GC pair at the former *HpaII* end of the fragment.

containing both types of inserts were transferred to several new plants. After systemic symptoms developed, viral DNA from each of the individual transfers was analysed. Figure 2b shows that the offspring of all eight independently transferred lesions carried the *lac* operator insert.

To determine whether the inserts were copied correctly during viral DNA replication we subcloned in plasmid pBR322 a *HindIII* restriction fragment containing the *lac* operator inserts at the *XhoI* site (Fig. 1) from either passaged CaMV DNA or from the parent plasmid used to infect the plants. DNA of these subclones was sequenced according to the Maxam–Gilbert procedure[15]. Figure 3 shows an example of the sequence

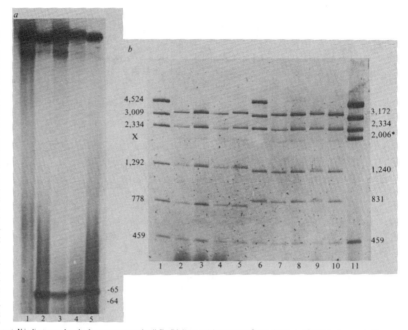

Fig. 2 *a*, Viral DNA was prepared from plants that had been infected with *Sal*I-cut DNA of pCaMV10 (lane 1), pCaMV10-EOP1 (lane 3) and pCaMV10-EOP2 (lane 4). After cleavage by *Xho*I restriction endonuclease and labelling with [α-³²P]dATP by polymerase I (Boehringer), the DNA was electrophoresed on a 10% polyacrylamide gel. Plasmid DNA of pCaMV10-EOP1 (lane 2) and pCaMV10-EOP2 (lane 5) was treated in the same way. The 64-bp operator fragment of -EOP1 can be distinguished from the 65-bp fragment of -EOP2, as indicated at the right-hand side. *b*, *Eco*RI restriction pattern of viral DNA derived from eight independent local lesion isolates of CaMV-1841-EOP1 and EOP2 that had been transferred to a set of new plants. Lanes 1, 6 and 11 contain plasmid DNA from pCaMV10-EOP1, pCaMV10-EOP2 and pCaMV10, respectively. Lanes 2–5 contain DNA from four different isolates of CaMV1841-EOP1; lanes 7–10, DNA of four independent isolates of CaMV1841-EOP2. The *Eco*RI fragment that has been modified by the insertions is indicated by *. The size of the fragments is given in base pairs. The introduction of the *lac* operator insert into the 2,006-bp *Eco*RI fragment also introduces an additional *Eco*RI restriction site and gives rise to two new fragments, the sizes of which (1,292 and 778 bp as opposed to 1,240 and 831 bp) vary depending on the orientation of the insert. X indicates a submolar fragment present in all *Eco*RI digests of CaMV DNA'. The 4,524- and 3,009-bp fragments are 'hybrid' fragments containing CaMV and pBR322 DNA, their CaMV moiety is part of the 3,172- and 2,334-bp fragments of the viral DNA.

comparison. No changes in the sequence of the two *lac* operator regions were detected. Regions adjacent to the site of insertion showed no sequence alteration compared with the pCaMV10 parent DNA sequence.

The stability of the inserts was checked after three successive transfers of virus derived from systemically infected tissue. All eight independent insert lines still contained the *lac* operator fragment. Some revertants occurred in the CaMV-EOP1 progeny, as indicated by the appearance of additional minor bands in the *Eco*RI restriction pattern. With CaMV-EOP2 (opposite orientation) no minor bands due to revertants could be detected. However, after five successive transfers and extended growth of the plants (2 months) the *lac* insert in both orientations was lost. In these conditions rare revertants seem to have a selective advantage over the insertion mutants which have slightly delayed growth.

To determine the maximum insertion capacity at the *Xho*I site, we inserted fragments of 256, 531 and 1,200 bp, respectively. The 256- and 531-bp fragments were derived from phage λ DNA cleaved with *Sal*I and *Xho*I (sites 44, 45 and 46 of a 1981 λ map update, F. R. Blattner *et al.*, personal communication). The 1,200-bp fragment, a *Sal*I segment of pUC5 containing the kanamycin resistance gene of Tn*903* (Vieria and J. Messing, unpublished results), carries *Pst*I sites adjacent to its *Sal*I ends, with the kanamycin resistance gene inserted between the *Pst*I sites by G·C homopolymer tails. Infectivity assays using CaMV DNA containing these inserts were performed as described above. After development of systemic symptoms, viral DNA was prepared from infected leaves and analysed. Figure 4 shows the *Eco*RI digestion pattern of CaMV DNA derived from the 256- and 531-bp insertions; only the former was propagated stably. Insertion of 531 bp gave rise to three different deletion mutants in the region of insertion. These infections were apparent in plants only after a prolonged period: 5–6 weeks after inoculation as opposed to 2 weeks normally required for systemic development. In addition, attempts to recover infectious CaMV DNA molecules containing the 1,200-bp fragment were unsuccessful. Analysis of the viral DNA from

plants that developed symptoms showed that the inserted fragment was lost, perhaps by recombination in the homopolymer tails as most of the isolates gained at least one additional *Pst*I site at the insertion point.

Thus, we have provided evidence for a non-essential region in the CaMV genome. Insertion of foreign DNA from two different sources into an *Xho*I restriction site whithin this region does not interfere with the infectivity of the viral DNA nor with virus production. However, the stability of the 256-bp λ DNA

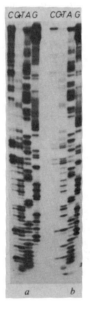

Fig. 3 Sequence comparison of the *lac* operator insert before and after passage through the plants. The respective *Hind*III restriction fragments (see Fig. 1) were subcloned in pBR322 (ref. 19). DNA of the hybrid plasmids was cleaved by *Eco*RI, labelled either with [γ-³²P]ATP (Amersham) and T4 polynucleotide kinase (Boehringer) or by Pol I Klenow fragment plus [α-³²P]dATP to label the opposite strand. After cleavage with *Hind*III and preparation of the fragments on 6% polyacrylamide gels, the DNA sequence of the inserts was determined as described elsewhere[15]. The figure shows an 8% sequencing gel of CaMV-EOP1 DNA (after passage) labelled with [α-³²P]dATP (*a*) and pCaMV10-EOP1 DNA before passage through the plants (*b*). The readable sequence starts with 3' AACAAT 5' in the middle of the *lac* operator (see Fig. 1, upper strand).

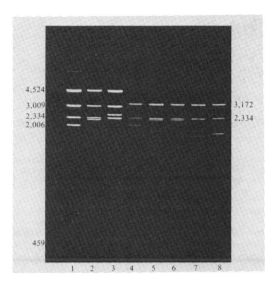

Fig. 4 *Eco*RI restriction pattern of λ DNA insertions in CaMV1841. Lane 1, *Eco*RI-cleaved plasmid DNA of pCaMV10; lanes 2 and 3, pCaMV10 containing a 256- and a 531-bp insert of λ DNA, respectively. The insertions at the *Xho*I site cause differences in mobility of the former 2,006-bp fragment. Lanes 4–8 contain viral DNA isolated from individual plants that showed symptoms after inoculation with the three plasmid DNAs cut by *Sal*I. Lane 4 shows the wild-type CaMV1841 *Eco*RI restriction pattern. The mobility of the *Eco*RI fragment that contains the 256-bp insert seems identical to that of the parent plasmid (lanes 5 and 6). In contrast, the 531-bp insert produced three different deletions, each of which rendered the fragment carrying the insert smaller than the wild-type 2,006-bp fragment (lanes 7 and 8). Only the largest of these new fragments (lane 7) still contains sequences that hybridize to λ DNA (data not shown).

insert in a systematic survey has not been compared with that of the *lac* EOP inserts. The instability of the 64/65-bp *lac* promoter–operator inserts could be due to some peculiarity of the DNA sequence itself. Note for example the similarity of the *lac* UV5 Pribnow box to the 'TATAA' initiation signal sequence for RNA polymerase II in eukaryotes[16]. Whether the transcription patterns of CaMV 1841-EOP1 or 1841-EOP2 differ from that of CaMV 1841 has not been determined.

A packaging limitation of CaMV particles for additional DNA seems to restrict larger inserts—an additional 531 bp could not be successfully propagated in similar experiments. Thus there seems to be an upper limit for additional DNA of 256–531 bp. Recombinant molecules having large insertions in the CaMV genome may replicate, as a whole, only in those cells which become initially infected after inoculation. Perhaps encapsidation occurs only after sufficiently small deletions have arisen and may be a prerequisite for cell-to-cell movement and systemic development in the plant. Further experiments with a 363-bp insertion at the *Xho*I site of pCaMV10 support this assumption as there is an 'eclipse period' of >2 months before the plants start to develop symptoms of virus infection. Analysis of the viral DNA reveals that deletions occur which include part of the DNA inserted and part of reading frame II on the CaMV chromosome (B.G. *et al.*, in preparation).

As insertion of foreign DNA in the *Xho*I site interferes with expression of the assumed reading frame II[3,4], its product might be non-essential. Alternatively, this region may not code for any protein at all. The finding that the deletion in CM4-184 eliminates almost the entire coding region II agrees with these results[12].

Insertion of bacterial DNA has been useful in probing the genome of CaMV. We are now investigating whether there are other regions suitable for insertions and whether inserts of different kinds can be used to express new proteins in plants using CaMV as vector.

We thank U. Rüther for pUR108-1 DNA, J. Schell for providing greenhouse space at Cologne, and E. Czerny and J. Duffus for assistance with the plants. This work was supported by the Deutsche Forschungsgemeinschaft through SFB 74 and by the NSF (PCM-7904960) and the US Department of Agriculture (SEA5901-0273).

Received 1 September; accepted 3 November 1981.

1. Shepherd, R. J. A. *Rev. Pl. Physiol.* **30**, 405–423 (1979).
2. Shepherd, R. J., Bruening, G. E. & Wakeman, R. J. *Virology* **41**, 339–347 (1970).
3. Howell, S. H., Walker, L. L. & Dudley, R. K. *Science* **208**, 1265–1267 (1980).
4. Lebeurier, G., Hirth, L., Hohn, T. & Hohn, B. *Gene* **12**, 139–146 (1980).
5. Howell, S. H., Walker, L. L. & Walden, R. M. *Nature* **293**, 483–486 (1981).
6. Franck, A., Guilley, H., Jonard, G., Richards, K. & Hirth, L. *Cell* **21**, 285–294 (1980).
7. Gardner, R. C. *et al. Nucleic Acids Res.* **9**, 2871–2888 (1981).
8. Gilbert, W. in *RNA Polymerase* (eds Losick, L. & Chamberlin, M.) 193–203 (Cold Spring Harbor Laboratory, New York, 1976).
9. Heynecker, H. L. *et al. Nature* **263**, 743–752 (1976).
10. Scheller, R. H., Dickerson, R. E., Boyer, H. W., Riggs, A. D. & Itakura, K. *Science* **196**, 177–180 (1977).
11. Hull, R. *Virology* **100**, 76–90 (1980).
12. Howarth, A. J., Gardner, R. C., Messing, J. & Shepherd, R. J. *Virology* **112**, 678–685 (1981).
13. Shepherd, R. J. & Lawson, R. W. in *Handbook of Plant Virus Infections* (ed. Kurstak, E.) 848–878 (Elsevier, Amsterdam, 198?).
14. Gardner, R. C. & Shepherd, R. J. *Virology* **106**, 159–161 (1980).
15. Maxam, A. M. & Gilbert, W. *Proc. natn. Acad. Sci. U.S.A.* **74**, 560–564 (1977).
16. Goldberg, M. thesis, Stanford Univ. (1979).
17. Backman, K., Ptashne, M. & Gilbert, W. *Proc. natn. Acad. Sci. U.S.A.* **73**, 4174–4178 (1976).
18. Miller, J. *Experiments in Molecular Genetics* (Cold Spring Harbor Laboratory, New York, 1972).
19. Bolivar, F. *et al. Gene* **2**, 95–113 (1977).

The *Agrobacterium Tumefaciens* Ti Plasmid as a Host Vector System for Introducing Foreign DNA in Plant Cells

J.-P. Hernalsteens, F. Van Vliet, M. De Beuckeleer, A. Depicker,
G. Engler, M. Lemmers, M. Holsters, M. Van Montagu and
J. Schell

The molecular basis of the neoplasmic transformation of plant cells by the crown gall bacterium *Agrobacterium tumefaciens* is the transfer to and stable maintenance of T DNA—a well defined segment of the bacterial Ti plasmid—in plant cells[1-4]. Transformed cells express new biosynthetic capacities, such as the synthesis of opines[5], of which octopine[6] and nopaline[7] are well known examples. Ti plasmids harbour genes enabling the bacteria to degrade opines and use them as carbon and nitrogen sources[8]. pTi-linked genes also determine the specificity of synthesis of opines in the plant tumour cells[8]. On this basis the Ti plasmid can be considered an unusual type of catabolic plasmid, which induces the synthesis of its substrate in transformed plant cells. This relationship, called genetic colonization[2], has opened prospects for the use of Ti plasmids as vectors for genetic manipulation of plants. To evaluate this possibility we inserted a well defined DNA segment, the bacterial transposon Tn7 (ref. 9), into the Ti-plasmid DNA sequence that determines nopaline

* Max Planck Institut für Züchtungsforschung, 30 Köln 5000, FRG.

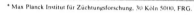

synthesis in *A. tumefaciens* strain T37 Noc[c]1 (ref. 10). As we report here, the inserted Tn7 DNA segment became part of the T DNA because the 9.6×10^6 molecular weight (MW) Tn7 DNA sequence was transferred to, and maintained in, the DNA of tumour tissue cultures induced by this mutant strain.

Tn7, which encodes streptomycin, spectinomycin and trimethoprim resistance, was introduced into the chromosome of strain T37 Noc[c]1. An RP4::Tn7 plasmid[9], pGV1 (our isolate) was transferred from the auxotrophic *Escherichia coli* strain GV1000 (ref. 11) to strain T37 Noc[c]1 by plate conjugation[12]. A transconjugant subsequently cured of its pGV1 plasmid was obtained by selection with the pilus-specific bacteriophage GU5 (J. P. H., in preparation). A culture of a transconjugant colony, exponentially growing in LB medium, was infected with a multiplicity of 5 plaque-forming units of GU5 per bacterium and plated after overnight incubation at 32 °C. About 15% of surviving cells were shown to have lost the carbenicillin, kanamycin and tetracycline resistance markers of RP4 but retained the spectinomycin ($100 \mu g \, ml^{-1}$) and trimethoprim ($1,000 \mu g \, ml^{-1}$) resistance markers of the Tn7 transposon.

One such T37 Noc[c]1 chr::Tn7 strain, GV15, was used for the isolation of Tn7 insertions in the transfer-derepressed Ti plasmid of strain T37 Noc[c]1 by selecting for the Ti-plasmid mediated conjugative transfer of the transposon resistance markers. GV15 was incubated for 24 h on minimal A[13] agar with the rifampicin- and erythromycin-resistant C58C1 derivative GV3102. Tn7-containing transconjugants were found at a frequency of 1.5×10^{-5} per recipient cell after plating on LB[13] medium supplemented with rifampicin ($100 \mu g \, ml^{-1}$), erythromycin ($100 \mu g \, ml^{-1}$) and spectinomycin ($100 \mu g \, ml^{-1}$). In 10% of these strains Ti plasmid could be demonstrated by its ability to degrade nopaline[8] and sensitivity to agrocin 84 (ref. 14). The remaining transconjugants expressed none of the Ti-plasmid markers and therefore probably harboured Tn7 insertions in the cryptic conjugative plasmids of strain

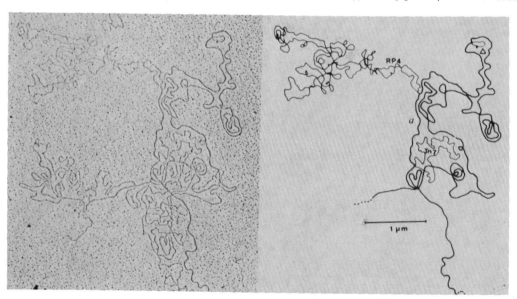

Fig. 1 Physical mapping of the Tn7 insertion in a nopaline-synthesis deficient (Nos⁻) mutant of pTiT37. pGV3106 DNA was prepared from strain GV3171 as described previously[12]. The Tn7 insertion in pGV3106 was localized by electron microscopic analysis of heteroduplex molecules[20] formed by renaturation of single-stranded DNA molecules of pGV3106 and of pGV4000 (ref. 18), a co-integrate plasmid consisting of pTiC58 and the P plasmid RP4. The distance (a) between the Tn7 transposon in pGV3106 and the RP4 insertion (at map position 130.4)[18] in pGV4000 is $1.3 \pm 0.1 \mu m$. The 0.5-μm deletion loop (Δ) in pTiT37 (ref. 4) and present in pGV3106 (at map position 124.3–125.3) (G. E., in preparation) was used as a reference point for the Tn7 integration site. This mapping result agrees with the restriction endonuclease analysis (data not shown) of pGV3106.

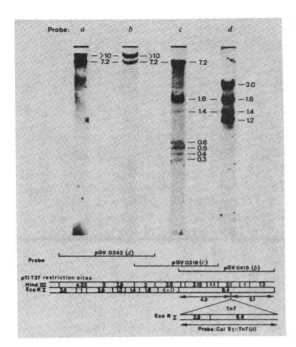

Fig. 2 Analysis of Tn7 and Ti-plasmid related DNA fragments present in restriction endonuclease digests of tobacco tumour DNA. W38T37::Tn7-1, a bacterium-free tumour tissue culture, was started from a tumour induced by strain GV3171 on *N. tabacum* cv Wisconsin 38, essentially as described previously[16]. After 5 months of *in vitro* culture on hormone-free medium, samples of the tissue were extracted with 25 ml buffer (Tris 50mM pH 8.2, EDTA 100 mM, diethylammonium diethyl-dithiocarbamate (DIECA) 100mM, triisopropyl naphthalene sulphonic acid 2%, diethylpyrocarbonate 0.1%) per g lyophilized tissue. The mixture was brought to 0.3 M NaCl and extracted with an equal volume of phenol and chloroform. The aqueous phase was brought to 0.5 M NaCl and re-extracted with the same organic phase. After precipitation with two volumes of ethanol the crude DNA was further purified by CsCl gradient banding (starting density 1.55). After dialysis it was concentrated to about 1 mg ml^{-1} by ethanol precipitation. The DNA was digested to completion with restriction endonuclease *Eco*RI. Digested DNA (20 μg) was separated by electrophoresis through a 0.8% agarose slab gel and transferred to a nitrocellulose filter[19]. The filter was hybridized by Southern's[19] method to cloned Ti-plasmid fragments derived from the T region of pTiC58 (ref. 17) and to ColE1::Tn7 plasmid DNA, labelled *in vitro* with ^{32}P by nick-translation[21]. ColE1::Tn7 plasmid DNA was used as a probe in lane *a*. The extent of the cloned pTi fragments used as probes for the hybridizations of lanes *b*, *c* and *d* is indicated on top of the physical map of the T region of pTiT37. Lanes *a* and *b* demonstrate the presence in the tumour DNA of the complete Tn7 sequence integrated in a sequence homologous to the 9.4×10^6 MW *Eco*RI fragment of pTiT37. The localization of the integration site of Tn7 in the T-DNA segment based on these data and confirmed by similar hybridization experiments with *Hin*dIII digests of the tumour DNA (data not shown), corresponds to the integration site of Tn7 in pGV3106 (Fig. 1). Lanes *b*, *c* and *d* demonstrate that the T-DNA segment in this tumour line is identical, except for the *Eco*RI fragment of MW 9.4×10^6, to the T-DNA segment independently shown to be present in the BT37 tobacco tumour line induced by wild-type pTiT37 (ref. 4). Numbers adjacent to bands and numbers in each segment indicate MW($\times 10^{-6}$).

T37 Noc^{c}1 (G. E., unpublished). Twelve transconjugants expressing both pTi and Tn7 markers were inoculated on sunflower hypocotyl segments[15]. Nine were found to be pathogenic. Nopaline synthesis was demonstrated reproducibly by electrophoretic analysis[16] of the tumours induced by eight of these strains, but could not be detected in tumours induced by strain GV3171. The inability of strain GV3171 to induce nopaline synthesis was confirmed by analysis of tumours induced on potato, *Kalanchoë daigremontiana* and tobacco.

The Tn7 insertion in strain GV3171 was localized on the mutant Ti plasmid (pGV3106) by restriction endonuclease analysis, and mapped in a DNA fragment homologous to the *Hin*dIII-23 fragment[17] of the standard nopaline Ti plasmid pTiC58 (data not shown). This localization was confirmed independently by electron microscopic heteroduplex analysis, which also showed that the Tn7 inserted without detectable

deletion (Fig. 1). This *Hin*dIII-23 fragment was subsequently shown to be homologous with the right-end border fragment of the T DNA[2,4] and thus to be part of the Ti-plasmid DNA segment which is normally transferred to plant cells. To confirm this result, several hundred Tn1 and Tn7 insertions in pTiC58 were screened for nopaline synthesis after induction of tumours. A single nopaline synthesis-deficient mutant was found and shown to be the result of a Tn1 insertion in fragment *Hin*dIII-23 (ref. 18).

A bacterium-free tissue culture, W38T37::Tn7-1, was started from a tumour induced by strain GV3171 on *Nicotiana tabacum* cv Wisconsin 38. Analysis by Southern blot hybridizations[19] of a DNA sample isolated from this culture, using ColE1::Tn7 (our isolate) DNA and various cloned segments of the nopaline plasmid pTiC58 as radioactive probes showed that the complete Tn7 was present in the plant tumour genome,

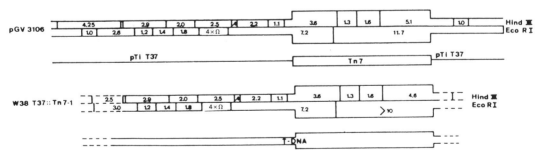

Fig. 3 The colinearity between the T region of the pTiT37::Tn7 plasmid pGV3106 and the T-DNA segment in the tumour line W38T37::Tn7-1 induced by the *A. tumefaciens* strain GV3171 harbouring pGV3106. The figure is based on results obtained by Southern type DNA hybridizations as described in Fig. 2 legend.

integrated at its original site in a normal-size T DNA (Figs. 2, 3).

Our results show that the Ti plasmid can be used as a vector for the experimental introduction and stable maintenance of foreign genes in higher plants. Furthermore, the demonstration that the Ti-plasmid DNA segment specifying nopaline synthesis in transformed plant cells is part of the T DNA, but not functionally required for the induction or maintenance of the neoplasmic state, supports the genetic colonization model of crown gall[2].

We thank Ruth Trimarchi for help. This work was supported by grants from the Kankerfonds van de ASLK, Instituut tot Aanmoediging van het Wetenschappelijk Onderzoek in Nijverheid en Landbouw (IWONL) (2841A), the Fonds Wetenschappelijk Geneeskundig Onderzoek (3.0052.78) to J. S. and M. V. M., and Fonds voor Kollektief Fundamenteel Onderzoek (2.0007.77) to M. Jacobs and M. V. M. M. H. thanks the Belgian NFWO for a fellowship and M. L. and M. D. B. thank the IWONL for financial support.

Received 5 June; accepted 8 August 1980.

1. Chilton, M. D. *et al. Cell* **11**, 263–271 (1977).
2. Schell, J. *et al. Proc. R. Soc.* B**204**, 251–266 (1979).
3. Thomashow, M. F., Nutter, R., Montoya, A. L., Gordon, M. P. & Nester, E. W. *Cell* **19**, 729–739 (1980).
4. Lemmers, M. *et al. J. molec. Biol.* (in the press).
5. Petit, A., Dessaux, Y. & Tempé, J. *Proc. 4th int. Conf. Pl. Path. Bact.*, Angers, 143–152 (1978).
6. Ménagé, A. & Morel, G. *C. r. hebd. Séanc. Acad. Sci., Paris* **259**, 4795–4796 (1964).
7. Goldmann, A., Thomas, D. W. & Morel, G. *C. r. hebd. Séanc. Acad. Sci., Paris* **268**, 852–854 (1969).
8. Bomhoff, G. *et al. Molec. gen. Genet.* **145**, 177–181 (1976).
9. Barth, P. T., Datta, N., Hedges, R. W. & Grinter, N. J. *J. Bact.* **125**, 800–810 (1976).
10. Petit, A. *et al. Nature* **271**, 570–572 (1978).
11. Hernalsteens, J. P., De Greve, H., Van Montagu, M. & Schell, J. *Plasmid* **1**, 218–225 (1978).
12. Van Larebeke, N. *et al. Molec. gen. Genet.* **152**, 119–124 (1977).
13. Miller, J. H. *Experiments in Molecular Genetics*, 466 (Cold Spring Harbor Laboratory, New York, 1972).
14. Engler, G. *et al. Molec. gen. Genet.* **138**, 345–349 (1975).
15. Petit, A. & Tempé, J. *Molec. gen. Genet.* **167**, 147–155 (1978).
16. Aerts, M., Jacobs, M., Hernalsteens, J.-P., Van Montagu, M. & Schell, J. *Pl. Sci. Lett.* **17**, 43–50 (1979).
17. Depicker, A. *et al. Plasmid* **3**, 193–211 (1980).
18. Holsters, M. *et al. Plasmid* **3**, 212–230 (1980).
19. Southern, E. M. *J. molec. Biol.* **98**, 503–518 (1975).
20. Davis, R. W., Simon, M. & Davidson, N. *Meth. Enzym.* **21**D, 413–428 (1971).
21. Rigby, P. W. J., Dieckmann, M., Rhodes, C. & Berg, P. *J. molec. Biol.* **113**, 237–252 (1977).

Expression of Chimaeric Genes Transferred Into Plant Cells Using a Ti-plasmid-derived Vector

L. Herrera-Estrella, A. Depicker, M. Van Montagu and
J. Schell

Foreign genes introduced into plant cells with Ti-plasmid vectors are not expressed. We have constructed an expression vector derived from the promoter sequence of nopaline synthase, and have inserted the coding sequences of the octopine synthase gene and a chloramphenicol acetyltransferase gene into this vector. These chimaeric genes are functionally expressed in plant cells after their transfer via a Ti-plasmid of Agrobacterium tumefaciens.

CROWN gall formation on dicotyledonous plants by *Agrobacterium tumefaciens* is the result of the transfer and covalent integration of a segment (called T-region) of the Ti-plasmid into the chromosomal DNA of plant cells (for reviews see refs 1–4). Insertion of foreign DNA sequences within the T-region of Ti-plasmids leads to their co-transfer and integration into the plant genome[5]. To date, inserts of up to 50 kilobases (kb) have been successfully co-transferred with the T-DNA (unpublished results of this laboratory). The upper limit of the size of the DNA segments that can be transferred to plant genomes with this system has not been determined. The transferred T-region (or T-DNA) is expressed in plant cells[6] and codes for several host polymerase II-dependent polyadenylated transcripts[7-10]. Some of these transcripts were shown to be responsible for the tumorous mode of growth of the transformed plant cells, while others code for enzymes that synthesize novel compounds (termed opines), which are amino acid or sugar derivatives. None of these transcripts was found to be essential for T-DNA transfer[11-13].

To understand the mechanism by which T-DNA-linked genes are expressed in plants, the nucleotide sequence and the precise location of 3′ and 5′ ends of the transcripts of two different opine synthase genes were determined, the octopine synthase gene carried by pTiB6S3 (ref. 14) and the nopaline synthase from pTiT37 (refs 15, 16). Although both genes are encoded by plasmids of bacterial origin, they share more characteristics with eukaryotic genes than with prokaryotic genes. Both octopine and nopaline synthase genes, designated *ocs* and *nos* respectively, have a sequence similar to the so-called 'TATA' or 'Goldberg–Hogness' box[17] in the 5′ region upstream of the start of transcription, and a sequence 'AATAA', similar to the polyadenylation signal of eukaryotic genes, near the 3′ end[18,19].

The opine synthase genes provide useful tools for studying gene expression in plant cells, because they are constitutively expressed in callus cells as well as in all tissues of plants regenerated from T-DNA-transformed cells[20,21]. Of practical importance is the fact that very rapid and sensitive assays for the detection of both nopaline and octopine have been developed[22,23]. Various attempts to express a number of bacterial, animal and some unrelated plant genes (for example, leghaemoglobin) introduced and integrated into the tobacco genome with the use of Ti-plasmid gene vectors[24], have thus far failed (unpublished results of this laboratory), because these foreign genes were not transcribed in the transformed tobacco cells. This lack of expression was presumably due to the fact that the plant transcription machinery could not recognize the transcription signals (promoter sequences) carried by these foreign genes.

HERRERA-ESTRELLA, L., DEPICKER, A., VAN MONTAGU, M. and SCHELL, J.
Expression of chimaeric genes transferred into plant cells using a Ti-plasmid-derived vector. Reprinted by permission from Nature v. 303: pp. 209-213. Copyright 1983 Macmillan Magazines Ltd.

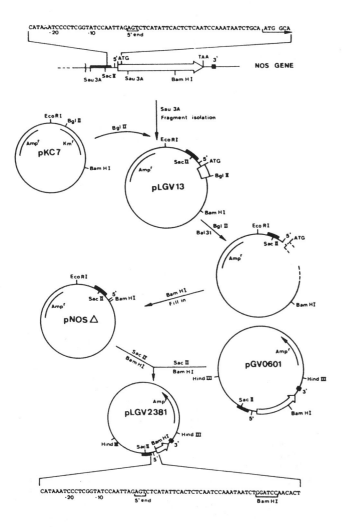

CATAAATCCCTCGGTATCCAATTAGAGTCTCATATTCACTCTCAATCCAAATAATCTGCA ATG GCA

CATAAATCCCTCGGTATCCAATTAGAGTCTCATATTCACTCTCAATCCAAATAATCTGGATCCAACACT

Fig. 1 Construction of a plasmid for expression of genes behind the *nos* promoter. The DNA sequence of the beginning of the *nos* gene is shown at the top of the figure. 10 μg of pGV0601 (a pBR322 derivative lacking the *Bam*HI site and carrying the *Hin*-dIII-23 fragment which contains the complete *nos* gene)²⁴ were digested with *Sau*3A and the 350-bp fragment carrying the *nos* promoter was isolated from a preparative 5% polyacrylamide gel. The promoter fragment was ligated to *Bgl*II-cut pKC7 (ref. 25), previously treated with bacterial alkaline phosphatase (BAP). 20 μg of the resulting plasmid (pLGV13) were digested with *Bgl*II and treated with 7 units of the *Bal*31 exonuclease (Biolabs) for 4–10 min in 400 μl of 12 mM MgCl₂, 12 mM CaCl₂, 0.6 M NaCl, 1 mM EDTA and 20 mM Tris-HCl, pH 8.0, at 30 °C. During this time ~20–50 bp of DNA were removed. The *Bal*31-treated molecules were digested with *Bam*HI and incubated with DNA polymerase (Klenow fragment) and the 4-deoxynucleotide triphosphates (at 10 mM each) to fill in the ends. Plasmids were screened for a regenerated *Bam*HI site derived from the ligation of a filled-in *Bam*HI end and the end of the *Bal*31 deletion. The sizes of the *Bam*HI–*Sac*II fragment of several candidates were estimated in a 6% urea–polyacrylamide gel, and the nucleotide sequences of the candidates with sizes ranging over 160–190 nucleotides were determined by the method of Maxam and Gilbert⁴². A pNOSΔ clone lacking the *nos*-coding sequence and two nucleotides of the leader sequence was used to substitute the *Sac*II–*Bam*HI fragment in the *nos* gene in pGV0601; the final promoter vector is called pLGV2381. All the recombinant plasmids were selected by transformation of the *E. coli* strain HB101. The bases marked 5′ and 3′ constitute the start and stop of transcription, and the AUG and TAA refer to the codons used to start and stop translation. The heavy black line indicates the *nos* promoter region, and the white area indicates the *nos*-coding region. Amp', ampicillin resistance; Km', kanamycin resistance.

To overcome the potential barrier to the expression of foreign genes in plants, we have constructed a number of chimaeric genes consisting of promoter sequences derived from the *nos* gene and of coding sequences derived from foreign genes. These chimaeric genes were introduced in the genome of tobacco with Ti-plasmid vectors and the resulting transformed tobacco cells were shown to express these chimaeric genes and to produce functional proteins.

Expression vector

Figure 1 outlines the construction of an expression vector using the *nos* promoter region. Fragment *Hin*dIII-23 from plasmid pTiC58 carries the complete coding sequence and the sequences responsible for faithful expression of the *nos* gene in plant cells (ref. 15 and C. Konsz *et al.*, in preparation). DNA from pGV0601 (ref. 24) (a pBR322 derivative lacking the *Bam*HI site and carrying the fragment *Hin*dIII-23) was digested with *Sau*3A and a fragment, carrying the transcription signals and the sequences coding for the first 15 amino acids of the *nos* gene, was subcloned into the *Bgl*II site of the cloning vector pKC7 (ref. 25), resulting—because of the specific sequences involved—in the regeneration of a *Bgl*II site near the initiation codon of *nos* (pLGV13).

Eukaryotic ribosomes seem to use preferentially the AUG most proximal to the 5′ end of a messenger RNA to initiate translation²⁶,²⁷. Therefore, pLGV13 was digested with *Bgl*II and treated with the exonuclease *Bal*31 to delete the *nos*-coding sequence remaining in the plasmid. To place a *Bam*HI site at the end of the *Bal*31 deletions, the treated plasmid was digested with *Bam*HI, and the protruding ends were filled in with DNA polymerase. Self-ligation of these molecules followed by screening for the presence of a *Bam*HI site, allowed us to identify a number of pNOSΔ plasmids with *Bam*HI-ended *nos* promoter sequences. Three of these, containing *Sac*II–*Bam*HI fragments of sizes 150–180 base pairs (bp), were chosen and subjected to nucleotide sequence analysis, and shown to have 3′ ends in different positions of the 5′-untranslated sequence of the *nos* gene. A candidate with a deletion removing the *nos*-coding sequence and two nucleotides of the untranslated 5′ leader was chosen for further work (Fig. 1).

The *Sac*II–*Bam*HI fragment from this pNOSΔ plasmid was subsequently cloned back into the plasmid pGV0601 (ref. 24), resulting in a vector which has a *Bam*HI restriction site directly behind the *nos* promoter. This vector contains a deletion of 800 bp of the *nos*-coding sequence, removing 300 of the 400 amino acids of the *nos* protein while retaining the putative signal for polyadenylation and termination of transcription.

An *ocs* 'cassette' fragment

To determine whether the constructed expression vector (pGV2381) would promote expression of coding sequences of other genes, we chose to combine it first with the coding sequence of the octopine synthase gene. Indeed, previous work had established that this enzyme is stable in cells of different plants, and that its activity can readily be assayed; moreover *ocs* activity has never been found to be present in extracts of any of the untransformed plants tested so far²². We therefore constructed a 'cassette' fragment consisting primarily of the coding sequence of the octopine synthase, in such a way that this fragment would be convenient for insertion between this and other potential promoter sequences.

The construction of the octopine cassette is outlined in Fig. 2. A *Bam*HI–*Sma*I fragment from pTiB6S3 carrying the complete coding sequence and part of the promoter of the *ocs* gene¹⁴ was cloned into pBR322 cut with *Bam*HI and *Pvu*II. The resulting plasmid, pAGV828, was digested with *Bam*HI and treated with the exonuclease *Bal*31 to remove the 5′-upstream flanking sequences from the structural part of the *ocs* gene, then digested with *Hin*dIII, filled in and self-ligated. A set of plasmids were obtained carrying the complete coding sequences of the *ocs* gene with different deletion end points in the 5′-nontranslated leader sequence. To bracket the *ocs*-

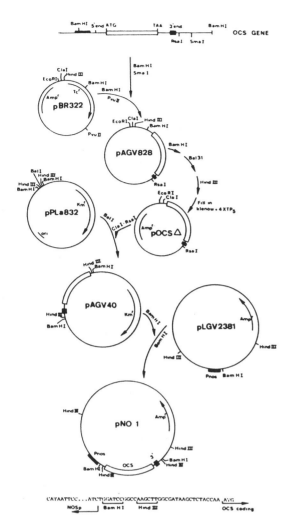

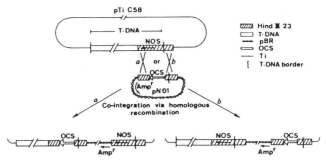

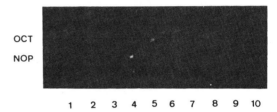

Fig. 3 Possible structure of the co-integrates between pTiC58 and pNO1 in *Agrobacterium*. The helper plasmids that provide the transfer and mobilization functions (R64*drd*11 and pJG28, respectively)[29] were transferred to *E. coli* HB101 (pNO1) by conjugation. The resulting strain carrying all three plasmids were conjugated to *Agrobacterium* (pTiC58) for 16 h on solid Luria broth (LB) medium. Exconjugants carrying the co-integrate pNO1::pTiC58 were selected by plating on LB medium containing 100 μg ml⁻¹ of ampicillin. The structure of the co-integrates was determined by Southern blotting analysis (data not shown), and the two possibilities for co-integration via homologous recombination between the Ti-plasmid and pNO1 to the 1-kb regions of *Hind*III-23 present in pNO1 is shown.

Fig. 4 Detection of octopine in plant tissue transformed with the chimaeric *nos–ocs* gene. Axenic tissue derived from tumours induced with the chimaeric genes were obtained by cutting the tumour into small pieces and subsequently growing *in vitro* on hormone-free Murashige and Skoog (MS) medium in the presence of the antibiotic HR-756 (Hoechst) for several weeks, and then transferring to medium without antibiotic[24]. 10 mg of the bacterium-free tissue was incubated for 12 h in liquid MS medium containing 0.1 M arginine, ground with a glass rod and centrifuged for 3 min at 12,000 r.p.m. 3 μl of the supernatant were spotted onto 3 MM Whatman paper and subjected to electrophoresis[22]. Lane 1 shows the control octopine and nopaline (3 μl from a solution of 1 mg ml⁻¹ each), and extracts obtained from tissues infected with *Agrobacterium* carrying pTiB6S3 (lane 2), pTiC58 (lane 3), co-integrates pNO1::pTiC58 with structure-type *a* (lanes 4, 5), co-integrates pNO1::pTiC58 structure-type *b* (lanes 6, 7), and co-integrates pNO2::pTiC58 (lanes 8–10). Lanes 1 and 3–10 show the presence of nopaline, and lanes 2 and 4–7 show the presence of octopine.

A chimaeric *nos–ocs* gene

A chimaeric gene linking the *nos* promoter sequence to the coding sequence of the octopine synthase gene was constructed as shown in Fig. 2. The *Bam*HI fragment of pAGV40 was cloned into the *Bam*HI site of pLGV2381 in both orientations pNO1 and pNO2, respectively (Fig. 2).

Plasmids pNO1 and pNO2 were transferred by conjugation to *Agrobacterium* and recombined with an acceptor pTiC58 plasmid by a method that allows direct mobilization of pBR322 derivatives or any other ColE1-like plasmid to *Agrobacterium*[29]. As the origin of replication of pBR322 is not functional in *Agrobacterium*[30], all the exconjugants carrying the Amp[r] marker are derived from homologous recombination between pNO1 and pTiC58. The structure of such a co-integrate is illustrated in Fig. 3. The co-integration of pNO1 and the Ti-plasmid results in a plasmid having two right T-DNA borders[31] in tandem, due to a duplication of the part of *Hind*III-23 carried by pNO1. Southern blotting, and DNA hybridization analysis (data not shown) of eight independent co-integrates

Fig. 2 Construction of a cassette containing the complete *ocs*-coding sequence, and its insertion in plasmid pLGV2381 behind the *nos* promoter. 10 μg of *Bam*HI fragment 17a of the octopine Ti-plasmid B6S3 were digested with *Bam*HI and *Sma*I, the fragment containing the *ocs*-coding sequence was isolated from a 1% agarose gel, and ligated to the large *Bam*HI–*Pvu*II fragment of pBR322; 20 μg of the resulting plasmid, pAGV828, were digested with *Bam*HI, treated with the exonuclease *Bal*31 as described in Fig. 1 legend, subsequently digested with *Hind*III, and the ends were filled-in and self-ligated. The sizes of the *Bal*31 deletions were estimated in a 6% polyacrylamide gel. The nucleotide sequences of several candidates were determined, and a candidate having only 7 bp remaining of the 5′-untranslated leader sequence was chosen for further work. To bracket the *ocs* sequence with *Bam*HI sites, the *Cla*I–*Rsa*I fragment was filled in and subcloned into the *Bal*I site of pPLa832. The resulting plasmid pAGV40 was digested with *Bam*HI, the fragment carrying the *ocs* sequence isolated by electroelution from a preparative 1% agarose gel, and ligated to pLGV2381 previously cut with *Bam*HI, and treated with bacterial alkaline phosphatase (BAP). The insertion of the *ocs* sequence in pLGV2381 was obtained in both orientations (pNO1 and pNO2). The junction sequence between the *ocs*-coding sequence and the *nos* promoter in pNO1 is shown. The heavy black line refers to the *ocs* promoter region and the white area to the *ocs*-coding region. Other notations are as for Fig. 1.

coding sequence with *Bam*HI sites, the *Cla*I–*Rsa*I fragment of the resulting pOCSΔ plasmids was isolated, the *Cla*I site was filled in with DNA polymerase and the fragment was cloned into the *Bal*I site of pPLa832 (ref. 28). One of these plasmids (pAGV40) containing only 7 nucleotides of the 5′-untranslated leader sequence and the complete coding sequence, as well as the 3′-untranslated sequence, including the AATAAA signal from the *ocs* gene, was chosen for further work.

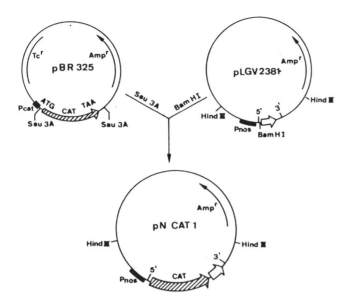

Fig. 5 Construction of a chimaeric gene that directs the expression of the chloramphenicol acetyltransferase in plant cells. 20 μg of pBR325 were digested with *Sau*3A and the fragment containing the complete coding sequence of the *cat* gene was isolated by electroelution from a 2% agarose gel. The isolated fragment was cloned into the *Bam*HI site of pLGV2381, previously treated with BAP. The orientation of the insert was determined by *Eco*RI and *Hind*III digestion. pNCAT1 carries the *cat* gene in the proper orientation to be transcribed from the *nos* promoter, and pNCAT2 carries the gene in opposite orientation. The small blackened box represents the *cat* promoter, the long blackened box the *nos* promoter, and the cross-hatched box represents the structural part of the *cat* gene.

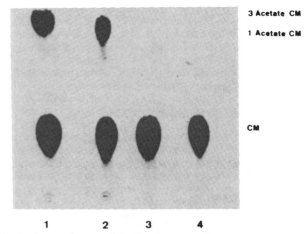

Fig. 6 Assay of *cat* activity in plant cells. 100 mg of tumour tissue was extracted by grinding manually with a glass rod in the presence of an equal volume of buffer containing 0.5 M sucrose, 0.25 M Tris-HCl, 1% ascorbic acid, 0.5 mM leupeptine (Sigma), 10 mM EDTA, 1% cysteine-HCl, pH 7.5, 0.5 mM acetyl CoA and 1 μCi ^{14}C-chloramphenicol (50 μCi mmol^{-1}, NEN), and the mix was incubated at 37 °C for 30 min (ref. 34), followed by centrifugation (5 min at 12,000 r.p.m.) and removal of the supernatant for analysis. A sonicated extract of *E. coli* containing pBR325 (ref. 32) was used as a positive control. Plant and bacterial reaction mixes were extracted with ethyl acetate, concentrated by evaporation, and subjected to ascendent chromatography in a silica gel thin-layer plate with chloroform/methanol (95:5) as eluant. The autoradiogram shown was exposed for 48 h at room temperature. Lane 1 shows the bacterial enzyme control, lane 2 shows the *cat* activity found in cells transformed with *Agrobacterium* containing pNCAT1::pTiC58. No *cat* activity was found in tissue transformed with the strain carrying pNCAT2::pTiC58 (lane 3), even after much longer exposure (data not shown). Lane 4 shows the reconstruction experiment where 10^9 agrobacteria carrying pTiC58::pNCAT1 were added to nopaline tumour tissue (induced by agrobacteria carrying wild-type pTiC58) before extraction.

revealed that five had the orientation *ocs*–border–*nos*–border (*a*), and three the orientation *nos*–border–*ocs*–border (*b*).

Tumours were induced by infecting decapitated tobacco seedlings with eight independently isolated *Agrobacterium* strains carrying the co-integrates pNO1::pTiC58, and four carrying co-integrates pNO2::pTiC58. Tumours were subsequently cultivated *in vitro* in hormone-free medium[24] and 10 mg of bacterium-free tissue were used to test for the presence of octopine and nopaline. As shown in Fig. 4, octopine was present in all the tumours induced with strains carrying the co-integrate pNO1::pTiC58 in amounts slightly higher than those found in tumours induced by the octopine wild-type Ti-plasmid pTiB6S3. Furthermore, the presence of both nopaline and octopine in tumours transformed with the pNO1::pTiC58 co-integrates of either type *a* or type *b* (see Fig. 3) suggests that both genes are transferred in spite of their position on either side of the internal right border of the T-DNA. This might also suggest that either the most external borders are preferentially used to transfer the DNA positioned between them, or that the presence of two borders flanking any DNA sequence may create a new 'T-DNA' that is independently transferred to the plant genome. As the tumour tissue (used for these experiments) was not cloned, the results may reflect several independent integration events, and further experiments are needed to clarify the structure of the T-DNA present in tumour tissue transformed with the co-integrates used in this study. On the other hand, tumours induced with pNO2::pTiC58, which contains the octopine synthase in opposite orientation to the transcription direction of the *nos* promoter, while positive for nopaline, did not contain octopine.

These results indicate that the *Hind*III–*Bam*HI fragment of pLGV2381 contains a functional *nos* promoter able to promote transcription of the octopine synthase coding sequence, and, therefore, presumably of any other coding sequence properly oriented downstream from this region.

A chimaeric *nos–cat* gene

To test further the applicability of the constructed expression vector, the *nos* promoter was combined with the coding sequence of chloramphenicol acetyltransferase (EC 2.3.1.28) (*cat*) from pBR325 (ref. 32). This enzyme of bacterial origin provides a very sensitive and simple assay system, and no similar endogenous activity is present in plant cells. Furthermore, this enzyme has been shown to be expressed in yeast and mammalian cells[33,34].

A *Sau*3A fragment from pBR325, containing the complete coding sequence of the *cat* gene, and 56 bp of the untranslated leader sequence where no other ATG than the initiation codon is present[35,36], was cloned in two orientations into the *Bam*HI site of pLGV2381 as outlined in Fig. 5. Both pNCAT1 and pNCAT2 (opposite orientation) were mobilized from *Escherichia coli* to *Agrobacterium* as described for pNO1 (Fig. 2), and exconjugants containing pTiC58::pNCAT1 and pTiC58::pNCAT2 co-integrates were selected on ampicillin-containing media. Southern blotting DNA hybridizations (data not shown) demonstrated that strain GVCAT1 contained a pTiC58::pNCAT1 co-integrate, and strain GVCAT2 a pTiC58::pNCAT2 co-integrate, which both resulted from a cross-over similar to situation *a* illustrated in Fig. 3.

Strains GVCAT1 and GVCAT2 were used to induce tumours on tobacco seedlings and the resulting tumour tissues were grown in axenic cultures[24]. Extracts from such axenic tissue cultures were tested for *cat* activity. As shown in Fig. 6, the tissue derived from tumours induced by the strain GVCAT1, whose T-DNA contains the *cat* gene-coding sequence properly oriented relative to the *nos* promoter, did indeed contain *cat* activity, as seen by the presence of the same chloramphenicol derivatives as in the control incubated with an extract of *E. coli* (pBR325). The tumour line induced with GVCAT2, whose *cat* gene-coding sequence is in the wrong orientation, was negative. The fact that tumours induced by GVCAT1 were shown to contain nopaline (data not shown) as well as the observed *cat* activity, suggests that the external border was involved in the integration of the T-region of the co-integrate pTiC58::pNCAT1 in the plant DNA, as discussed above for the *nos–ocs* chimaeric gene.

Agrobacterium strains have been found to contain some endogenous *cat* activity[37]. To ensure that bacterial *cat* activity cannot account for the observed activity in tobacco tumours induced by GVCAT1, we took particular care to demonstrate that the tumour lines were completely devoid of any surviving bacteria before extraction. Furthermore, wild-type C58 tobacco crown gall tissue was mixed with 10^9 cells from a fresh culture of GVCAT1, and the extraction procedure described in Fig. 6 legend was repeated. No *cat* activity was observed in these circumstances, demonstrating that the very gentle mechanical extraction procedure used to open the plant cells could not lead to a contamination of the plant cell extract with bacterial *cat* activity.

Discussion

The successful expression of the octopine synthase and chloramphenicol acetyltransferase coding sequence as chimaeric genes linked to the transcription-promoting sequence of the nopaline synthase gene demonstrates that this approach can be used to transfer and express the coding sequences of foreign genes in plants. The *nos* promoter is particularly useful because of its wide host range: it is functional in calli derived from all dicotyledonous plants so far tested, and it appears to be expressed in most tissues of regenerated *nos*-containing plants[38].

The easy and sensitive assay for the presence of octopine in plant cell extracts makes the *ocs* cassette a very versatile tool for studying the structure–function relationships of various promoter sequences in plants, especially with regard to their role in the regulation of gene expression and specificity.

Another important application is the construction of general selectable marker genes for plants. Using the *nos* promoter expression vector (pLGV2381), we constructed two such chimaeric genes, one containing the coding sequence of the neomycin phosphotransferase of Tn5 (ref. 39) and the other the methotrexate-resistant dihydrofolate reductase of plasmid R67 (ref. 40). Both these genes were shown to be expressed in tobacco, and able to be used as selectable marker genes[41], thus confirming the general applicability of the constructions reported here. On the basis of these results it can be expected that gene-splicing techniques, so far applied to bacterial and mammalian cells, can be used for plant cells also. Moreover, the availability of simple and rapid techniques for the regeneration of plants from single cells, makes plants an especially interesting system for the study of development and differentiation, as well as for genetic engineering.

We thank Paul Tenning, Ms June Simpson and Jef Seurinck for technical assistance, and Albert Verstraete and Ms Martine De Cock for help in preparing this manuscript. This research was supported by grants from the 'Kankerfonds van de Algemene Spaar- en Lijfrentekas', from the 'Instituut tot Aanmoediging van het Wetenschappelijk Onderzoek in Nijverheid en Landbouw' (IWONL3849A), from the Services of the Prime Minister (OOA/12052179), and from the 'Fonds voor Geneeskundig Wetenschappelijk Onderzoek' (FGWO3.0001.82) to J.S. and M.V.M. L.H.-E. is indebted to CONACYT México for a PhD fellowship.

Received 8 February; accepted 30 March 1983.

1. Bevan, M. W. & Chilton, M.-D. *A. Rev. Genet.* **16**, 357–384 (1982).
2. Ream, L. & Gordon, M. P. *Science* **218**, 854–858 (1982).
3. Van Montagu, M. & Schell, J. *Curr. Topics Microbiol. Immun.* **96**, 237–245 (1981).
4. Schell, J. *et al. Cell Fusion* (Raven, New York, in the press).
5. Hernalsteens, J. P. *et al. Nature* **287**, 654–656 (1980).
6. Drummond, M. H., Gordon, M. P., Nester, E. W. & Chilton, M.-D. *Nature* **269**, 535–536 (1977).
7. Willmitzer, L., Simons, G. & Schell, J. *EMBO J.* **1**, 139–146 (1982).
8. Willmitzer, L. *et al. Cell* **32**, 1045–1056 (1983).
9. Bevan, M. W. & Chilton, M.-D. *J. molec. appl. Genet.* **1**, 539–546 (1982).
10. Gelvin, S. B., Gordon, M. P., Nester, E. W. & Aronson, A. I. *Plasmid* **6**, 17–29 (1981).
11. Garfinkel, D. J. *et al. Cell* **27**, 143–153 (1981).
12. Leemans, J. *et al. EMBO J.* **1**, 147–152 (1982).
13. Joos, H. *et al. Cell* **32**, 1057–1067 (1983).
14. De Greve, H. *et al. J. molec. appl. Genet.* **1**, 499–512 (1982).
15. Depicker, A., Stachel, S., Dhaese, P., Zambryski, P. & Goodman, H. M. *J. molec. appl. Genet.* **1**, 561–574 (1982).
16. Bevan, M., Barnes, W. M. & Chilton, M.-D. *Nucleic Acids Res.* **11**, 369–385 (1983).
17. Breathnach, R. & Chambon, P. A. *Rev. Biochem.* **50**, 349–383 (1981).
18. Fitzgerald, M. & Shenk, T. *Cell* **24**, 251–260 (1981).
19. Dhaese, P. *et al. EMBO J.* **2**, 419–426 (1983).
20. De Greve, H. *et al. Nature* **300**, 752–755 (1982).
21. Otten, L. *et al. Molec. gen. Genet.* **183**, 209–213 (1981).
22. Otten, L. *Pl. Sci. Lett.* **25**, 15–27 (1982).
23. Aerts, M., Jacobs, M., Hernalsteens, J. P., Van Montagu, M. & Schell, J. *Pl. Sci. Lett.* **17**, 43–50 (1979).
24. Leemans, J. *et al. J. molec. appl. Genet.* **1**, 149–164 (1981).
25. Rao, R. N. & Rogers, S. G. *Gene* **7**, 79–82 (1979).
26. Kozak, M. *Nucleic Acids Res.* **9**, 5233–5252 (1981).
27. Lomedico, P. T. & McAndrew, S. J. *Nature* **299**, 221–226 (1982).
28. Remaut, E., Stanssens, P. & Fiers, W. *Gene* **15**, 81–93 (1981).
29. Van Haute, E. *et al. EMBO J.* **2**, 411–418 (1983).
30. Leemans, J. *et al. Gene* **19**, 361–364 (1982).
31. Zambryski, P., Depicker, A., Kruger, K. Goodman, H. M. *J. molec. appl. Genet.* **1**, 361–370 (1982).
32. Bolivar, F. *Gene* **4**, 121–136 (1978).
33. Cohen, J. *et al. Proc. natn. Acad. Sci. U.S.A.* **77**, 1078–1082 (1980).
34. Gorman, C. M., Moffat, L. F. & Howard, B. H. *Molec. cell. Biol.* **2**, 1044–1051 (1982).
35. Marcoli, R., Iida, S. & Bickle, T. A. *FEBS Lett.* **110**, 11–14 (1980).
36. Prentki, P., Karch, F., Iida, S. & Meyer, J. *Gene* **14**, 289–299 (1981).
37. Zaidenzaig, Y., Fitton, J. E., Packman, L. C. & Shaw, W. V. *Eur. J. Biochem.* **100**, 609–618 (1979).
38. Wöstemeyer, A. *et al. Genetic Engineering in Eukaryotes* (ed. Lurquin, P.) (Plenum, New York, in the press).
39. Beck, E., Ludwing, G., Averswald, E. A., Reiss, B. & Schaller, H. *Gene* **19**, 329–336 (1982).
40. Fling, M. E. & Elwell, L. P. *J. Bact.* **141**, 779–785 (1980).
41. Herrera-Estrella, L. *et al. EMBO J.* (in the press).
42. Maxam, A. M. & Gilbert, W. *Meth. Enzym.* **65**, 499–559 (1980).

Expression of Ti Plasmid Genes in Monocotyledonous Plants Infected with *Agrobacterium Tumefaciens*

G.M.S. Hooykaas-Van Slogteren, P.J.J. Hooykaas and R.A. Schilperoort

When the bacterium *Agrobacterium tumefaciens* infects a fresh wound site on a dicotyledonous plant, it attaches to the plant cell wall and introduces a piece of its Ti plasmid DNA into the plant cell via an unknown mechanism[1-4]. This piece of Ti plasmid DNA (the T-DNA) becomes integrated in the nuclear genome of the plant cell and is transcribed into a specific number of different transcripts. The T-DNA renders the plant cell tumorous, and also codes for enzymes involved in the synthesis of certain tumour-specific compounds called opines. Any DNA segment inserted into the T-region of the Ti plasmid by genetic manipulation seems to be co-transferred to the plant cell by *A. tumefaciens*. Thus, the Ti plasmid offers great potential as a vector for the genetic engineering of plant cells. However, as monocots are not susceptible to tumour formation by *A. tumefaciens*, it is generally believed that the Ti plasmid can be used as a vector for dicotyledonous plants only[1-5]. This would severely limit the applicability of the Ti plasmid as a vector, as most commercially important crops are monocots. However, we report here data which indicate that the Ti plasmid may be a useful vector for transforming monocotyledonous plant species.

Infection of dicotyledonous plants with *A. tumefaciens* usually results in crown gall tumour formation, due to the expression of *onc* genes present in the T-DNA[6,7]. These *onc* genes enable crown gall cells to grow in the absence of the phytohormones auxin and cytokinin[6-8]. Recent evidence suggests that the *onc* genes in the T-DNA possibly code for enzymes that are directly involved in the biosynthesis of these hormones[9].

Tumorous plants can be derived from certain tobacco crown gall lines which spontaneously regenerate shoots. These shoots do not usually form roots, but they can develop into mature plants after grafting onto normal tobacco understems[10]. Superinfection of such plants with *A. tumefaciens* does not lead to tumour induction whereas infection with *A. rhizogenes*, which contains an Ri plasmid, results in the formation of an extensive swelling of the internodes having an infected wound rather than in the formation of a local tumour[10]. Similarly, infection of monocotyledonous plants with *A. tumefaciens* does not result in crown gall formation[11]. It is believed there is no T-DNA transfer to the tumour plant cells[12,13] or monocot cells[14], either

* To whom correspondence should be addressed.

because the bacterium cannot attach to the plant cell walls[12,14] or because of an abnormal phytohormone balance in such cells[14]. An alternative explanation is that T-DNA transfer did occur in these cases, but did not lead to tumour formation.

The opines octopine and nopaline are synthesized via enzymes that are encoded by T-DNA genes[15,16] which only have eukaryotic regulatory sequences for expression[17]. The detection of opines in plant cells is therefore direct evidence for the presence of T-DNA. To establish whether T-DNA is still transferred to and expressed in tumorous plants and monocots after infection with *A. tumefaciens*, we assayed for octopine and nopaline in plant material from infected wound sites via the method described by Otten and Schilperoort[18].

Grafted octopine tumour plants (from line TSO136)[19] were infected with *A. tumefaciens* strain LBA2318 (containing a nopaline Ti plasmid), strain LBA1020 (an Ri plasmid), and strain LBA2347 (carrying a nopaline Ti plasmid and an Ri plasmid). Infection with LBA2318 did not result in any swelling or tumour formation, but after infection with LBA1020 or LBA2347 small swellings were apparent (Fig. 1). Functions determined by the Ri plasmid thus stimulate the occurrence of swellings. Nopaline was detected in plant material isolated from the swellings induced by LBA2347, which shows that the nopaline T-DNA from LBA2347 must have been introduced into the TSO136 crown gall cells. Therefore, we conclude that neither existing T-DNA, nor an altered phytohormone balance nor an altered cell wall are barriers to the transfer of T-DNA into the cells of the tumorous plant. The finding that introduction of a second copy of T-DNA into the tumorous shoots does not lead to tumour formation may be related to the abnormal phytohormone balance in the tumorous plants.

To determine whether monocots, which also do not show tumour formation after exposure to agrobacteria, can take up and express T-DNA after infection with *A. tumefaciens* we tested for production of opines in infected tissues. Two monocot species were selected for these studies, *Chlorophytum capense* (Liliaceae) and *Narcissus* cv. 'Paperwhite' (Amaryllidaceae). Plants of both species were infected with strain LBA2347. Fourteen days after infection small swellings were observed at the wound sites. After 21 days plant material from the wound sites always contained nopaline. In non-infected plant tissue neither

Fig. 2 Infection of *Narcissus* cv. 'Paperwhite' with avirulent *A. tumefaciens* strain LBA288 (*a*) and with virulent strain LBA2318 (*b*). Opine assays were performed as described elsewhere[18]. A small piece of tissue (~60 mm³) was extracted with 30 µl cold buffer, and the supernatant was mixed 1:1 with buffer and incubated for 60 min. At the start ($t=0$) and end ($t=60$) of the incubation period, 2-µl samples from the mix were spotted onto electrophoresis paper. After electrophoresis the paper was dried then stained with phenanthrene quinone reagent. Yellowish spots of guanidine compounds (such as arginine, octopine and nopaline) were visualized under long-wave UV. Tissues from both monocots and dicots (used as a control) infected with the nopaline strains LBA2318 or LBA2347 normally displayed a spot corresponding to nopaline at $t=0$ and $t=60$. In monocot or dicot tissues from plants infected with the octopine strains LBA1010 or LBA1023, a clear spot corresponding to octopine was visible only in the $t=60$ sample. This shows that in both infected dicot and monocot cells sufficient nopaline, but not enough octopine, accumulates for direct detection at $t=0$. Although octopine is not detected at $t=0$, octopine synthase activity is present which converts *in vitro* pyruvate and arginine into octopine so that at $t=60$ an octopine spot becomes visible. Neither octopine nor nopaline synthase activity was present in uninfected tissues from monocots or dicots, nor in tissues from plants infected with the avirulent strains LBA288 or LBA1516.

Fig. 1 Infection of a grafted crown gall shoot from line TSO136 with *A. tumefaciens* strains LBA1020, LBA2347 and LBA2318.

HOOYKAAS-VAN SLOGTEREN, G.M.S., HOOYKAAS, P.J.J. and SCHILPEROORT, R.A.
Expression of Ti plasmid genes in monocotyledonous plants infected with *Agrobacterium tumefaciens*.
Reprinted by permission from *Nature* v. 311: pp. 763-764. Copyright 1984 Macmillan Magazines Ltd.

octopine nor nopaline was detected. Following these positive results, infection tests were done on *Narcissus* cv. 'Paperwhite' with many different *A. tumefaciens* strains, LBA288 (Ti plasmid-cured; avirulent), LBA1516 (octopine Ti plasmid with a *vir*B mutation; avirulent), LBA1010 (octopine Ti plasmid), LBA2318 (nopaline Ti plasmid), LBA1023 (Ri plasmid plus octopine Ti plasmid), LBA2347 (Ri plasmid + nopaline Ti plasmid). After infection with the avirulent strains LBA288 and LBA1516 no swellings were formed, and in plant cells from wound sites infected with these strains, neither octopine nor nopaline was detected. After infection with any of the other strains small swellings were invariably formed (Fig. 2). In the case of the infection with LBA1010 or LBA1023, octopine was detected in the plant cells from the wound sites; in the case of infection with LBA2318 or LBA2347, nopaline was detected.

These results show for the first time both the occurrence of T-DNA transfer to cells of monocotyledonous plants, and the expression of the T-DNA genes for octopine and nopaline synthase in such cells. This latter finding suggests that the regulatory sequences which are essential for transcription and translation of the octopine and nopaline synthase genes in cells of dicots also function in monocots. This result is of great importance in the development of vectors with selectable markers for monocotyledonous plants.

Our results show that T-DNA transfer to monocots occurs from strains with a wild-type Ti plasmid, but not from strain LBA1516, which has a defect in one of its essential *vir* genes. Therefore it is likely that *A. tumefaciens* introduces T-DNA into cells of monocots and dicots via the same mechanism. Some years ago it was found that the attachment of *A. tumefaciens* to the plant cell wall is necessary for the bacterium to accomplish T-DNA transfer[20]. Our results suggest that *Agrobacterium* is able to attach to the cell walls of monocots. In support of this it has been shown recently that *A. tumefaciens* can attach to cells of *Asparagus officinalis* (a monocot)[21] which had not previously been found for other monocots[14]. The apparent discrepancy between these results might be explained by assuming that wounding can trigger transient changes in the cell walls of monocots which eventually enable *A. tumefaciens* to attach for a limited period of time.

T-DNA transfer to monocots is not accompanied by the induction of large crown gall tumours. The finding that the T-DNA genes for octopine and nopaline synthase are expressed in monocotyledonous plant cells suggests that the other T-DNA genes are also expressed in the monocot cells. Therefore the absence of proper tumour formation in monocots may be because monocot cells (like the octopine crown gall cells) do not respond to the products derived from the T-DNA encoded *onc* genes, perhaps because of peculiarities of their phytohormone metabolism.

We suggest that the Ti plasmid may emerge as a useful vector for monocotyledonous plants of the families Liliaceae and Amaryllidaceae. Further research is needed to establish whether crops from other monocot families (for example, Gramineae) can be transformed by the Ti plasmid. Nevertheless, our results have considerable implications for the biotechnology of a vast group of plant species of commercial interest.

We thank Drs A. Hoekema, L. Melchers and R. J. M. van Veen for critical reading of the manuscript.

Received 18 June; accepted 30 July 1984.

1. Kahl, G. & Schell, J. (eds) *Molecular Biology of Plant Tumors* (Academic, London, 1982).
2. Hooykaas, P. J. J. & Schilperoort, R. A. *Adv. Genet.* 22, 209–283 (1984).
3. Ream, L. W. & Gordon, M. P. *Science* 218, 854–859 (1982).
4. Schell, J. & Van Montagu, M. *Bio/Technology* 1, 175–180 (1983).
5. Flavell, R. & Mathias, R. *Nature* 307, 108–109 (1984).
6. Ooms, G., Hooykaas, P. J. J., Moolenaar, G. & Schilperoort, R. A. *Gene* 14, 33–50 (1981).
7. Garfinkel, D. J. *et al. Cell* 27, 143–153 (1981).
8. Braun, A. C. *Proc. natn. Acad. Sci. U.S.A.* 44, 344–349 (1958).
9. Schröder, G., Waffenschmidt, S., Weiler, E. W. & Schröder, J. *Eur. J. Biochem.* 138, 387–391 (1984).
10. Wullems, G. J., Molendijk, L., Ooms, G. & Schilperoort, R. A. *Cell* 24, 719–727 (1981).
11. De Cleene, M. & De Ley, J. *Bot. Rev.* 42, 389–466 (1976).
12. Lippincott, J. A. & Lippincott, B. B. *Science* 199, 1075–1078 (1978).
13. Krens, F. A., Wullems, G. J. & Schilperoort, R. A. in *Structure and Function of Plant Genomes* (eds Ciferri, O. & Dure, L.) 387–408 (Plenum, New York, 1983).
14. Rao, S. S., Lippincott, B. B. & Lippincott, J. A. *Physiologia Pl.* 56, 374–380 (1982).
15. Schröder, J. *et al. FEBS Lett.* 129, 166–168 (1981).
16. Murai, N. & Kemp, J. D. *Proc. natn. Acad. Sci. U.S.A.* 79, 86–90 (1982).
17. De Greve, H. *et al. J. molec. appl. Genet.* 1, 499–512 (1983).
18. Otten, L. A. B. M. & Schilperoort, R. A. *Biochim. biophys. Acta* 527, 497–500 (1978).
19. Van Slogteren, G. M. S., Hoge, J. H. C., Hooykaas, P. J. J. & Schilperoort, R. A. *Plant molec. Biol.* 2, 321–333 (1983).
20. Lippincott, B. B. & Lippincott, J. A. *J. Bact.* 97, 620–628 (1969).
21. Draper, J., MacKenzie, I. A., Davey, M. R. & Freeman, J. P. *Plant. Sci. Lett.* 29, 227–236 (1983).

High-velocity Microprojectiles for Delivering Nucleic Acids Into Living Cells

R.M. Klein, E.D. Wolf, R. Wu and J.C. Sanford

We report here a novel phenomenon, namely that nucleic acids can be delivered into plant cells using high-velocity microprojectiles. This research was conducted in the hope of circumventing some of the inherent limitations of existing methods for delivering DNA into plant cells[1-6]. After being accelerated, small tungsten particles (microprojectiles) pierce cell walls and membranes and enter intact plant cells without killing them. Microprojectiles were used to carry RNA or DNA into epidermal tissue of onion and these molecules were subsequently expressed genetically. This approach can therefore be used to study the transient expression of foreign genes in an intact tissue. It remains to be shown that smaller cell types, as are found in regenerable plant tissues, can be stably transformed by this method. If this proves possible, it would appear to provide a broadly applicable transformation mechanism capable of circumventing the host-range restrictions of *Agrobacterium tumefaciens*[1], and the regeneration problems of protoplast transformation[2-5].

Several devices for accelerating small particles to high velocities have been designed and tested[7]. The device (particle gun) illustrated in Fig. 1*a* was used to accelerate tungsten microprojectiles (spherical particles 4 μm in diameter) into intact epidermal cells of *Allium cepa* (onion) (Fig. 1*b* and *c*). Many cells can be bombarded simultaneously and about 90% of the

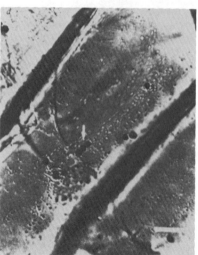

Fig. 1 *a*, Schematic diagram of the particle gun. About 0.05 mg of tungsten particles (average diameter 4 μm, General Electric Corp., Cleveland, Ohio) is placed on the front surface of a cylindrical nylon projectile (diameter, 5 mm; length, 8 mm) as a suspension in 1-2 μl of water. A gun powder charge (GY22AC, gray extra light, No. 1, Speed Fasteners Inc., St Louis, Missouri), detonated with a firing pin, is used to accelerate the nylon projectile down the barrel of the device. The tungsten particles continue toward the target cells through a small 1-mm aperture in a steel plate designed to stop the nylon projectile. The tungsten microprojectiles leave the particle gun with an initial velocity of about 430 m s⁻¹. This value was determined by allowing the nylon macroprojectile to leave the barrel of the device and estimating its velocity in-flight with a chronograph (Ohler Research, Austin, Texas). The target cells are placed ~10-15 cm from the end of the device. *b*-*d*, Nomarski micrographs of *A. cepa* epidermal cells following bombardment with tungsten microprojectiles 4 μm in diameter. The epidermal layer was stripped from the underlying bulb tissue before bombardment. *b*, Cell with 3 microprojectiles on its surface (one arrowed); *c*, microprojectile (arrowed) within the same cell as revealed by focusing 40 μm below the surface of the cell; *d*, Eight microprojectiles (one arrowed) in the interior of a living cell. Scale bar, 20 μm.

cells in a 1-cm² of *A. cepa* epidermal tissue (~2,000 cells) typically contain microprojectiles following bombardment. Cells from *A. cepa* can survive penetration by many microprojectiles (Fig. 1*d*). However, the viability of cells (as determined by the maintenance of cytoplasmic streaming for at least 24 h after bombardment) is adversely affected by penetration by a large number of microprojectiles (Fig. 2).

To demonstrate that nucleic acid can be delivered into cells by this method, RNA isolated from tobacco mosaic virus (TMV) strain U₁ was adsorbed to the surface of 4-μm tungsten particles before their acceleration into *A. cepa* cells. Expression of the viral RNA was monitored by examining the bombarded cells microscopically for the presence of viral inclusion bodies (crystallized virus particles[8]) 3 days after treatment. Crystalline material, in the form of hexagonal plates, round plates and needles, was observed in the cytoplasm and vacuole of 30–40% of the cells that contained microprojectiles (Table 1). These distinctive crystalline inclusions were never observed in unbombarded tissue or in tissue bombarded with microprojectiles containing no nucleic acid. The observed percentage of cells

Table 1 The delivery of TMV RNA to *A. cepa* cells by accelerated microprojectiles

Trial	Cells containing microprojectiles: without inclusions	Cells containing microprojectiles: with inclusions	Proportion penetrated cells with inclusion bodies (%)
1	44	34	43.6
2	85	35	29.2
3	168	98	36.8
4	257	76	22.8
5	147	52	26.1

A. cepa epidermal cells were scored for presence of TMV inclusion bodies following bombardment with microprojectiles to which TMV RNA has previously been adsorbed. About 50 microscope fields (×500) were observed within a 1-cm² area of tissue for each trial.
Methods. For adsorption of RNA to the tungsten microprojectiles, 2 μl of a TMV RNA[21] solution (2 μg of RNA per μl of distilled water) was added to 18 μl of a suspension of tungsten microprojectiles (10 mg of tungsten per ml of distilled water). The RNA was precipitated by the addition of 7.5 μl of 0.25 M CaCl₂ solution and the resulting suspension was centrifuged for 30 min at 13,000g. The particles were resuspended and 2 μl of the suspension was placed on the front surface of the nylon projectile. Centrifugation brings the RNA precipitate into close association with the tungsten surfaces and results in uniform adsorption which can be visualized by staining with the fluorescent dye DAPI[22]. Following bombardment the tissue was incubated for 3 days at 21 °C.

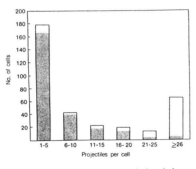

Fig. 2 Relationship between cell survival and the number of microprojectiles within a cell. Bars (solid lines), total cells; shaded part of bar, living cells. Three 1-cm² sections of *A. cepa* tissue were bombarded as described in Fig. 1. About 100 cells from each section were analysed microscopically for viability (based on the maintenance of cytoplasmic streaming) and the number of microprojectiles in these cells were counted. Data from the three sections of tissue were pooled.

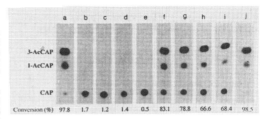

Fig. 3 CAT activity from *A. cepa* epidermal tissue bombarded with microprojectiles coated with plasmid harbouring the *CAT* gene. Lane a, positive control: a colony of *E. coli* carrying pBR325 was sonicated in 1 ml of buffer and centrifuged at 10,000g for 5 min. The supernatant (10 μl) was used for the CAT assay. Lane b, untreated epidermal tissue (10 pooled sections); lane c, epidermal tissue (10 pooled sections) bombarded with naked microprojectiles; lane d, epidermal tissue (10 pooled sections) bombarded with microprojectiles coated with pUC13. Lane e, microprojectiles (30 μl of a suspension prepared as described in the text) were rubbed over the surface of each of 10 sections of epidermal tissue which were pooled for the assay. Lane f–i, epidermal tissue (10 sections pooled for each assay) bombarded with microprojectiles coated with p35S-CAT; lane j: epidermal tissue (one section for the assay) bombarded with p35S-CAT. CAP, chloramphenicol; 3-AcCAP and 1-AcCAP, monoacetylated forms of chloramphenicol. The plasmid (p35S-CAT), which has been previously described[12], includes the following components: a 941-basepair (bp) region including the 35S promoter from cauliflower mosaic virus[23]; a 900-bp region encoding the CAT activity[24]; a 700-bp region carrying the 3′ noncoding sequences and the transcription termination region of the ribulose bisphosphate carboxylase gene and a derivative of the pUC13 plasmid in which the *Hinc*II site was changed to a *Cla*I site. The resulting construct is 5.3 kilobases (kb) long. The plasmid p35S-CAT was grown in *Escherichia coli*, isolated by phenol extraction, and purified by CsCl/ethidium bromide density-gradient centrifugation[25]. The microprojectiles were coated by adding 10 μl of the plasmid DNA (1.0 μg of DNA per μl of TE buffer, pH 7.7) to 100 μl of a suspension of tungsten microprojectiles (0.1 g of tungsten per ml of distilled water). The DNA was precipitated by the addition of 100 μl of a CaCl₂ solution (2.5 M) and 40 μl of a spermidine solution (free base, 0.1 M). The resulting suspension was centrifuged for 30 min at 13,000g. *A. cepa* epidermal tissue (1-cm² sections) was then bombarded with the DNA-coated microprojectiles (2 μl of the suspension placed in the middle of the front surface of the nylon macroprojectile). Assays of CAT activity were performed on extracts prepared from tissue that had been incubating for 3 days following bombardment with the microprojectiles. Tissue extracts were prepared by placing the epidermal sections in an Eppendorf tube (1.5 ml) with 100 μl Tris-HCl (0.25 M, pH 7.8) and grinding the tissue with a pellet pestle mixer (Kontes). CAT activity in the extracts was determined in 30-min assays as previously reported[26], except that the amount of ¹⁴C-chloramphenicol was decreased to 0.1 μCi per assay[18]. All the radioactivity in the Eppendorf tube (~170,000 c.p.m. in each assay) was spotted on the thin-layer chromatography (TLC) plate (Chromagram, Kodak). After resolution of chloramphenicol and its acetylated derivatives by chromatography, an autoradiogram (12-h exposure) was made. Quantitative results were obtained by scintillation counting of separated spots of chloramphenicol and its acetylated derivatives, and the conversion calculated.

expressing TMV following bombardment was comparable to that reported for the delivery of TMV RNA into protoplasts by liposome uptake[9,10] or electroporation[11].

Using the particle gun, DNA can be delivered into intact plant cells, and this can result in transient expression of a foreign gene. Tungsten microprojectiles were coated with plasmid (p35S-CAT)[12] containing a gene that encodes chloramphenicol acetyltransferase (CAT). These microprojectiles were used to bombard 1-cm² sections of *A. cepa* epidermal tissue. Extracts from epidermal tissue bombarded with microprojectiles coated with p35S-CAT showed very high levels of CAT activity with most of the chloramphenicol being converted to its acetylated derivatives during the assay (Fig. 3). Strong CAT activity was detected in extracts from as little as a single 1-cm^2 piece of bombarded epidermal tissue (~2,000 cells) whereas control tissue showed negligible CAT activity.

These findings indicate that particle bombardment can be used to deliver RNA or DNA into large numbers of intact plant cells simultaneously and that the foreign nucleic acids introduced by this process can subsequently be expressed. The introduction of DNA into *A. cepa* epidermal tissue by the particle bombardment process should be a useful tool for the study of transient expression of foreign nucleic acids in plant cells. The process does not require cell culture or the pretreatment of the recipient tissue in any way and only small quantities of DNA are needed (about 0.1 μg per treatment). Because the preparation of particles and the bombardment of tissue can be rapid, many nucleic acid samples can be tested in a short period of time.

Particle bombardment may eventually prove useful for the transformation of other plant species and may be of particular value for those species that cannot be genetically engineered successfully with existing techniques. Although several delivery systems have been used to transfer genes into various dicotyledonous plant species[1,2,13,14], these techniques have not yet been useful for the production of whole, transformed plants of graminaceous species such as rice, wheat, or corn. Although there is evidence that *Agrobacterium* can transfer DNA to *Zea mays*[15], *Asparagus*[16], *Chlorophytum* and *Narcissus*[17], the efficient use of this agent for the transfer of genes to monocots may be hampered by the restricted host range of the bacterium[1]. Genes have been transferred to protoplasts of the monocots *Triticum monococcum* and *Lolium multiflorum* by incubation with plasmid DNA[3,4] and to *Zea mays*[5] and *Oryza sativa*[18] by electroporation, but these techniques are restricted by the difficulties of regenerating whole plants from protoplasts of most monocot species[19,20]. The particle bombardment process should not suffer from the host-range limitations of infectious agents such as *Agrobacterium*, and the problems associated with systems that require regeneration of plants from protoplasts might be circumvented by bombarding regenerable tissues such as meristems or embryogenic callus with DNA-bearing microprojectiles. To accomplish this, the particle bombardment process must be refined so that cells much smaller (10–20 μm) than those of *A. cepa* epidermal tissue can be penetrated by microprojectiles.

This work was supported by grants from the Cornell Biotechnology Program, the US Department of Agriculture, and the Rockefeller Foundation. We thank N. Allen for input on the design of the particle gun and its construction, M. Zaitlin and M. Nishiguchi for TMV RNA, and M. McCann for technical assistance with the CAT assays.

Received 7 January; accepted 2 March 1987.

1. Fraley, R. T., Rogers, S. G. & Horsch, R. B. *CRC crit. Rev. Pl. Sci.* **4**, 1–46 (1986).
2. Steinbiss, H. H. & Broughton, W. *Int. Rev. Cytol. (Suppl.)* **16**, 191–207 (1983).
3. Lörz, H., Baker, B. & Schell, J. *Molec. gen. Genet.* **199**, 178–182 (1985).
4. Potrykus, I., Saul, M. V., Petruska, J., Paszkowski, J. & Shillito, J. R. D. *Molec. gen. Genet.* **199**, 183–188 (1985).
5. Fromm, M., Taylor, L. P. & Walbot, V. *Proc. natn. Acad. Sci. U.S.A.* **82**, 5824–5828 (1985).
6. De la Peña, A., Lörz, H. & Schell, J. *Nature* **325**, 274–276 (1987).
7. Sanford. J. C., Klein, T. M., Wolf, E. D. & Allen, N. *Particle Sci. Technol.* (in the press).
8. Christie, R. G. & Edwardson, J. R. *Light and Electron Microscopy of Plant Virus Inclusions, Florida Agric. Exp. Station Monograph Ser. 9* (University of Florida, Gainesville, 1977).
9. Fukunaga, Y., Nagata, T. & Takebe, I. *Virology* **113**, 752–760 (1981).
10. Fraley, R. T., Dellaporta, S. L. & Papahadjopoulos, D. *Proc. natn. Acad. Sci. U.S.A.* **79**, 1859–1863 (1982).
11. Nishiguchi, M., Langridge, W. H. R., Szalay, A. A. & Zaitlin, M. *Pl. Cell Rep.* **5**, 57–60 (1986).
12. Morelli, G., Nagy, F., Fraley, R. T., Rogers, S. G. & Chua, N.-H. *Nature* **315**, 200–204 (1985).
13. Crossway, A. *et al. Molec. gen. Genet.* **202**, 179–185 (1986).
14. Brisson, N. & Hohn, T. *Meth. Enzym.* **118**, 659–668 (1986).
15. Grimsley, N., Hohn, T., Davies, J. W. & Hohn, B. *Nature* **325**, 177–179 (1987).
16. Hernalsteens, J.-P., Thia-Toong, L., Schell, J. & Van Montagu, M. *EMBO J.* **3**, 3039–3041 (1984).
17. Hooykaas-Van Slogteren, G. M. S., Hooykaas, P. J. J. & Schilperoort, R. A. *Nature* **311**, 763–764 (1984).
18. Ou-Lee, T. M., Turgeon, R. & Wu, R. *Proc. natn. Acad. Sci. U.S.A.* **83**, 6815–6819 (1986).
19. Vasil, I. K. *Int. Rev. Cytol. (Suppl.)* **16**, 79–88 (1983).
20. Potrykus, I. & Shillito, R. F. *Meth. Enzym.* **118**, 549–578 (1986).
21. Zaitlin, M. in *Nucleic Acids in Plants* Vol. 2 (eds Hall, T. C. & Davis, J. W.) 31–64 (CRC Press, Boca Raton, Florida, 1979).
22. Coleman, A. W. & Goff, J. L. *Stain Technol.* **60**, 145–154 (1985).
23. Hohn, T., Richards, K. & Lebeurier, G. *Current Topics Microbiol. Immun.* **96**, 193–236 (1982).
24. Alton, N. K. & Vapnek, D. *Nature* **282**, 864–869 (1979).
25. Maniatis, T., Fritsch, E. F. & Sambrook, J. *Molecular Cloning, a Laboratory Manual* (Cold Spring Harbor Laboratory, New York, 1982).
26. Gorman, C. M., Moffat, L. F. & Howard, B. H. *Molec. cell. Biol.* **2**, 1044–1051 (1982).

Direct Gene Transfer to Plants

J. Paszkowski, R.D. Shillito, M. Saul, V. Vandak, T. Hohn, B. Hohn
and I. Potrykus

Evidence for direct, gene-mediated stable genetic transformation of plant cells of *Nicotiana tabacum* is presented. A selectable hybrid gene comprising the protein coding region of the Tn5 aminoglycoside phosphotransferase type II gene under control of cauliflower mosaic virus gene VI expression signals was introduced into plant protoplasts as part of an *Escherichia coli* plasmid. The gene was stably integrated into plant genomic DNA and constitutively expressed in selected, drug-resistant, protoplast-derived cell clones. The mode of integration of the foreign gene into the plant genome resembled that observed for DNA transfection of mammalian cells. Plants regenerated from transformed cell lines were phenotypically normal and fertile, and they maintained and expressed the foreign gene throughout the development of vegetative and generative organs. Microspores, grown in anther culture, developed into resistant and sensitive haploid plantlets. Genetic crossing analysis of one of the transformed plants revealed the presence of one dominant trait for kanamycin resistance segregating in a Mendelian fashion in the F_1 generation.

Key words: selectable marker genes/plant protoplast transformation/recombinant DNA/plant tissue culture

Introduction

Direct gene transfer has been instrumental in the study of gene regulation and function in bacteria, fungi and animal cells. In plants the only available method for genetic transformation is based on the ability of the soil bacterium *Agrobacterium tumefaciens* to transfer and to integrate part of its genetic material into genomic DNA (e.g., Bevan and Chilton, 1982). This complex natural transformation process has also been exploited for integrating functional foreign genes into the plant genome (Herrera-Estrella *et al.*, 1983a, 1983b; Bevan *et al.*, 1983; Fraley *et al.*, 1983; Horsch *et al.*, 1984; Murai *et al.*, 1983). Further, direct transformation of isolated Ti plasmid into protoplasts (Davey *et al.*, 1980; Krens *et al.*, 1982) has demonstrated that foreign DNA could be taken up by plant cells and integrated into the genome, but did not exclude the possibility that Ti plasmid-specific functions are essential for this transformation process. Here we present proof for *A. tumefaciens*-independent direct gene transfer into plants and transmission of the foreign gene to the sexual offspring in a Mendelian pattern.

Results and Discussion

Construction and properties of the selectable hybrid gene

Part of the bacterial transposon Tn5 codes for the amino-glycoside phosphotransferase II [(APH3')II] gene (Rothstein and Reznikoff, 1981). The protein encoded inactivates a related group of aminoglycoside antibiotics (neomycin, kanamycin, G-418) by phosphorylation and thereby confers resistance to them in bacteria (Rao and Rogers, 1979), fungi (Hirth *et al.*, 1982), mammalian cells (Colbère-Garapin *et al.*, 1981) and plant cells (Herrera-Estrella *et al.*, 1983a, 1983b; Bevan *et al.*, 1983; Fraley *et al.*, 1983). In order to obtain expression of the APH(3')II gene in higher eukaryotic cells, corresponding expression signals have to be provided. We have constructed a functional hybrid marker gene by placing the protein coding sequence of the APH(3')II gene under the control of the cauliflower mosaic virus (CaMV) gene VI expression signals because this gene is expressed in plant cells at very high levels during viral infection (Xiong *et al.*, 1982). The sequences flanking the gene VI mRNA coding region are typical eukaryotic promoter and termination signals (Hohn *et al.*, 1982).

We have constructed a pair of plasmids: pABDI with the correct orientation of the APH(3')II gene with respect to the gene VI promoter region and polyadenylation site; and pABDII with the reverse orientation (Figure 1). Details of the construction are described in the legend of Figure 1. The protein product predicted from the hybrid gene (pABDI) will be an APH(3')II amino-terminal fusion protein, 23 amino acid residues longer than the original APH(3')II. Modifications of the APH(3')II gene product (Beck *et al.*, 1982) have shown that this fusion is likely to have little effect on the phosphorylation activity. We have demonstrated biological activity of this particular hybrid gene and its use in selection by introducing it into tobacco cells *via* integration into T-DNA and standard co-cultivation treatments with *A. tumefaciens* (Paszkowski *et al.*, in preparation). The system thus seemed suitable to approach direct gene transfer into the plant genome by transformation of isolated protoplasts.

Protoplast transformation and selection of kanamycin-resistant cell lines

Nicotiana tabacum c.v. Petit Havana clone SR1 (Maliga *et al.*, 1973) protoplasts were chosen as the recipient system for gene transfer because (i) they are totipotent; (ii) their plating efficiency is high and reproducible; and (iii) they have good survival of treatments favouring DNA uptake. Freshly isolated mesophyll protoplasts were treated with the hybrid plasmids pABDI and pABDII using a DNA uptake procedure slightly modified from that developed by Krens *et al.* (1982). In three independent experiments we recovered resistant clones exclusively from the treatment involving pABDI (Table I). In all control treatments involving >10^8 protoplasts (including experiments verifying the selection procedure) no resistant colonies were recovered. A novel culture system (Shillito *et al.*, 1983) based on the use of agarose beads proved to be clearly superior to other selection regimes tested (i.e., liquid cultures, agar solidified cultures) (Figure 2a).

PASZKOWSKI, J., SHILLITO, R.D., SAUL, M., VANDAK, V., HOHN, T., HOHN, B. and POTRYKUS, I.
Direct gene transfer to plants. *EMBO J.* 3:2717-2722 (1984). Reprinted by permission, Oxford University Press.

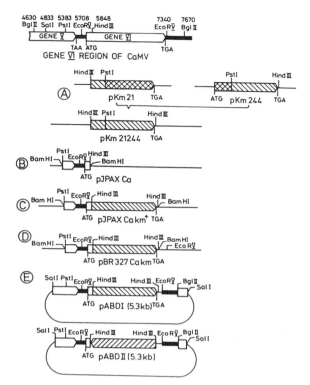

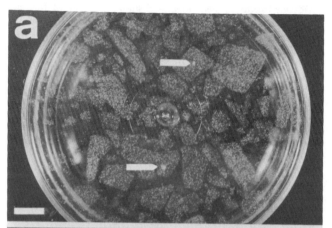

Table I. Recovery of drug-resistant clones from transformation experiments

Experiment	DNA treatment	Protoplasts treated	Resistant clones recovered
1	pABDI	~2 x 10⁶	2 (T_{2-1}, T_{2-2})
	Controls[a]	~10 x 10⁶	0
2	pABDI	~2 x 10⁶	1
	Controls[a]	~10 x 10⁶	0
3	pABDI	~2 x 10⁶	2
	Controls[a]	~10 x 10⁶	0

[a]Every transformation experiment included the following control treatments: (i) pABDII + calf thymus DNA + PEG; (ii) pUC8 + calf thymus DNA + PEG; (iii) calf thymus DNA + PEG; (iv) PEG; (v) untreated protoplasts.

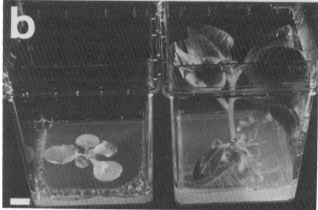

Fig. 1. Construction of the selectable hybrid gene. **(A)** Recombination of *Bal*31-deleted versions of the APH(3′)II gene. The plasmids pKm 21 and pKm 244 (Beck *et al.*, 1982) were digested by *Pst*I endonuclease, the derived fragments were purified by electrophoresis and ligated. The resulting plasmid pKm 21244 contains a combination of the 5′ and 3′ *Bal*31 deletions of the APH(3′)II gene. **(B)** and **(C)** Fusion of the CaMV gene VI promoter sequence to the APH(3′)II gene on the linker plasmid (pJPAX). pJPAX was derived from plasmids pUC8 and pUC9 (Messing and Vieira, 1982). The linker sequence of pUC9 was deleted (of 10 bp) by restriction at the *Hind*III and *Sal*I sites and the resulting cohesive ends filled in by treatment with DNA polymerase I (Klenow fragment) followed by ligation, thus restoring the *Hind*III restriction site. An 8-bp synthetic *Xho*I linker was inserted into the *Sma*I site of this deleted linker sequence. Recombination of the appropriate *Xor*II-*Hind*III fragments of pUC8 and the modified pUC9 plasmid yielded pJPAX with a partially asymmetric linker sequence containing the following sequence of restriction sites: *Eco*RI, *Sma*I, *Bam*HI, *Sal*I, *Pst*I, *Hind*III, *Bam*HI, *Xho*I, *Eco*RI. The joining of the 5′ expression signals of the CaMV gene VI and the *Hind*III fragment of APH(3′)II was carried out on pJPAX by inserting the *Pst*I-*Hind*III fragment of the CaMV gene VI promoter region between its *Pst*I and *Hind*III sites. The resulting plasmid (pJPAXCa) was restricted at its single *Hind*III site and the *Hind*III fragment of pKm 21244 was cloned into it in both orientations, yielding pJPAXCaKm⁺ (restoring the reading frame of the fused protein) and pJPAXCaKm⁻. **(D)** To provide an *Eco*RV site near the 3′-terminal region of the hybrid APH(3′)II gene, *Bam*HI fragments of pJPAXCaKm⁺ and of pJPAXCaKm⁻ were recloned into the *Bam*HI site of pBR327 (giving pBR327CaKm⁺ or ⁻). **(E)** The *Eco*RV fragments of pBR327CaKm⁺ and ⁻ were used to replace an *Eco*RV region of the CaMV gene VI, recloned as a *Sal*I fragment in pUC8, thereby placing the APH(3′)II protein coding region under control of both the 5′ and 3′ gene VI expresson signals. Two analogous plasmids were created: pABDI and pABDII, with correct and inverted orientation respectively of the APH(3′)II gene with respect to the gene VI promoter region. Open boxes represent open reading frames of CaMV, and hatched boxes the protein coding sequences of APH(3′)II. Thick lines represent intragenic regions of CaMV and thin lines represent bacterial vector sequences. Numbers above the restriction sites of the CaMV gene VI region indicate their positions on the CaMV map (Gardner *et al.*, 1981). Distances are not drawn to scale.

Fig. 2. (a) Resistant cell colonies 25 days after transformation of isolated protoplasts growing in bead type culture in 50 mg/l (0.086 mM) kanamycin sulphate. Arrows show proliferating resistant calli growing in the background of dead cell colonies. (b) Wild-type (kanamycin sensitive - left) and transformed (kanamycin resistant - right) protoplast-derived shoots after 4 weeks on a medium containing 150 mg/l kanamycin sulphate. The white bars represent 1 cm.

This method permits the selective medium to be replaced easily and repeatedly without disturbing the development of the cell clones in the agarose beads, thus guaranteeing a continuous and controlled selection pressure during the important early stages of protoplast culture. Resistant calli were transferred 4 weeks after exposure to DNA onto agar-solidified LS culture medium (Linsmaier and Skoog, 1965) containing 75 mg/l kanamycin and were subcultured to fresh selective medium every 4 weeks thereafter. Resistant calli proliferated under selective conditions (50 mg/l kanamycin in bead cultures and 75 mg/l kanamycin on plates) at the same rates as control cultures under non-selective conditions. Both kanamycin concentrations were toxic for untransformed callus colonies.

Plant regeneration from kanamycin-resistant cell lines

Two transformed clones (T_{2-1} and T_{2-2}) (Table I) were first subjected to treatments inducing plant regeneration and then to further analysis at the molecular level. Both clones regenerated numerous shoots. From these shoots ~50 plantlets were regenerated and transferred to potting compost where they continued to develop under greenhouse conditions. These plants were phenotypically identical to untransformed control plants. Plants have been regenerated in both the absence and the presence of kanamycin (150 mg/l) during the entire regeneration process. The presence of kanamycin in the regeneration medium influenced neither shoot nor root formation from the resistant clones. In contrast, nontransformed SR1 shoots, regenerated under non-selective conditions, failed to root on kanamycin-containing medium, ceased development and showed kanamycin-induced chlorophyll bleaching leading finally to complete chlorosis (Figure 2b).

Recovery of resistant cell cultures from explants and protoplasts

In order to follow the introduced trait throughout the development to the mature plant so as to record a possible loss of the kanamycin resistance at any developmental stage, explants from all parts of one selected plant from the base up to the influorescence were taken into culture, under selective (50 mg/l kanamycin) and non-selective conditions. Resistant cell cultures developed from leaves, petioles, stems, floral sepals, petals and carpels.

Protoplasts isolated from leaves proliferated under selective conditions (50 mg/l kanamycin) with the same plating efficiencies as wild-type protoplasts under non-selective conditions. Mixing experiments of protoplasts from the leaves of kanamycin-resistant and wild-type plants in ratios down to three kanamycin-resistant to 150 000 untransformed protoplasts (Figure 3), demonstrated that our selection system recovers putative transformants from large populations of sensitive protoplasts. Recovery of resistant colonies, after correction for the division frequency of the total population, varies between 50 and 100% and is lower (~50%) when the ratio of sensitive to resistant protoplasts is extremely high (>10 000).

Microspore analysis via anther culture

In tobacco, as in numerous other plants, the direct products of meiosis, the microspores, can be induced to develop into haploid plants. This provides an opportunity to follow directly transmission of the introduced trait through the male gametes. Anthers of the correct developmental stage from kanamycin-resistant and from wild-type plants were taken for anther culture (Sunderland and Dunwell, 1977) under nonselective conditions. Microspore-derived green plantlets were then transferred to selective conditions (200 mg/l kanamycin sulphate) and scored for undisturbed development or bleaching. The wild-type plantlets bleached without exception whereas ~50% (106 from 248) of microspore plantlets from T_{2-1} did not bleach and continued to develop to green haploid plants.

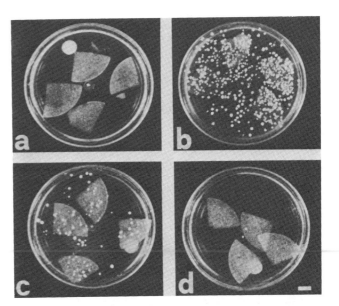

Fig. 3. Reconstruction experiment. Mesophyll protoplasts isolated from the wild-type SR1 and from the kanamycin-resistant plant T_{2-2} were mixed at different ratios and cultured under selective conditions (50 mg/l kanamycin and bead type culture technique). The figures show the plastic containers (9 cm in diameter) with the four quadrants of an agarose-protoplast gel (6 cm in diameter) after 7 weeks in culture. (**a**) 150 000 wild-type protoplasts: no resistant colony. (**b**) 148 500 wild-type protoplasts plus 1500 protoplasts from T_{2-2}: 739 resistant colonies. (**c**) 150 000 wild-type protoplasts plus 150 protoplasts from T_{2-2}: 101 resistant colonies. (**d**) 150 000 wild-type protoplasts plus 15 protoplasts from T_{2-2}: 3 resistant colonies. The white bar represents 1 cm.

Table II. Genetic crossing analysis

	T_{2-1} x self	SR1 x T_{2-1}	T_{2-2} x self	SR1 x SR1
No. of seeds tested	746	647	1731	1450
No. of seeds germinated	732	641	1684	1412
No. of kanamycin-resistant seedlings	536	308	1428	0
No. of kanamycin-sensitive seedlings	196	333	256	1412
Ratio of resistant to sensitive	2.73:1	0.92:1	5.58:1	0:1412

Genetic crossing analysis

The ultimate biological test for the fate of a foreign gene introduced into the plant genome should be provided by an analysis of the acquired trait in sexual offspring. Of the 58 plants regenerated from subclones of the two putative transformed cell lines T_{2-1} and T_{2-2} 57 were fertile. This enabled us to undertake crossing analysis of the kanamycin resistance. Data for two random plants are given (Table II). Fertility, seed set and germination were comparable with the wild-type SR1, so that segregation data could be interpreted without any additional assumptions. Controlled pollinations of emasculated flowers were done in an insectfree growth cabinet. Seed populations of individual seed capsules were sterilized and germinated on 1/10 inorganic NN culture medium (Nitsch and Nitsch, 1969) solidified with

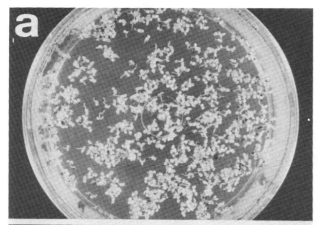

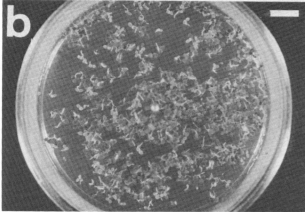

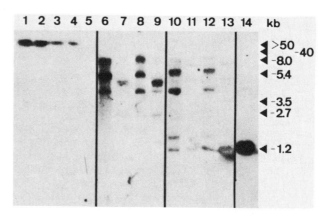

Fig. 4. Genetic crossing analysis of the transformed plant T_{2-1}. Seeds were surface sterilized and germinated on 200 mg/l kanamycin. The photographs were taken 5 weeks after germination (see also Table II). (**a**) Wild-type SR1 x self: seedlings germinate, bleach and die. (**b**) Transformant T_{2-1} x self: seedlings germinate and segregate ~3:1 viable green to bleached, dying seedlings. The white bar represents 1 cm.

Fig. 5. Detection of APH(3')II gene sequence in the DNA of transformed callus lines and leaves of regenerated plants. After restriction and electrophoresis of ~5 μg DNA/slot DNA was transferred onto nitrocellulose filters and hybridized with the nick translated *Hind*III fragment (5 − 10 x 10^8 c.p.m./μg) of pKm21244 (Figure 1). **Lanes 1 − 5**, unrestricted DNA; **6 − 9**, DNA restricted with *Bst*EII; **10 − 14**, DNA restricted with *Eco*RV; **lanes 1,6,10**, DNA from callus of T_{2-1}; **2,7,11**, DNA from callus T_{2-2}; **3,8,12**, DNA from leaf of T_{2-1}; **4,9,13**, DNA from leaf of T_{2-2}; **5**, DNA from wild-type SR1 callus; **14**, pABDI (2 ng, shorter exposure). The leaf DNA was isolated from plants grown in soil, regenerated from callus under non-selective conditions. Plants grown *in vitro* under continuous selective pressure (150 mg/l kanamycin) also show the cell line specific hybridization profile (data not shown).

The hybrid gene is present in selected kanamycin-resistant cell lines and leaves of regenerated plants

In order to correlate the observed kanamycin-resistant phenotype with the presence of the hybrid APH(3')II gene, DNA of the selected kanamycin-resistant cell lines and regenerated plants was analysed. Southern blot analysis (Southern, 1975) (Figure 5) revealed in both of the lines analysed (T_{2-1}, T_{2-2}) the presence of 3 − 5 copies of APH(3')II gene sequences per haploid genome, estimated from reconstruction experiments (data not shown). The APH(3')II specific probes hybridized only to high mol. wt. (>50 kb) DNA suggesting that the APH(3')II sequences had been integrated into the plant genome. There was no hybridization to chloroplast DNA prepared from transformed plants (data not shown). When nuclear DNA was restricted with *Bst*EII, which does not cut pABDI, we observed a transformant-specific hybridization pattern to the APH(3')II probe. This suggests random rather than directed integration of pABDI sequences into the tobacco nuclear genome. The size of the various *Bst*EII fragments (both smaller and larger than the 5.3-kb pABDI) indicates that pieces of the pABDI rather than full length plasmids were integrated. *Eco*RV restriction of DNA from both transformed lines yielded the expected 1.2-kb *Eco*RV fragment containing the intact APH(3')II gene. In addition, fragments of different sizes were visualised which could be interpreted as representing integration of parts of the APH(3')II coding sequence leading to the deletion of one of the flanking *Eco*RV restriction sites. The mode of integration of a non-homologous supercoiled plasmid into plant genomes will be studied in more detail, but clearly resembles non-homologous integration of foreign DNA as reported for animal cells (Colbère-Garapin *et al.*, 1981; Wigler *et al.*, 1979).

The introduced genes were not eliminated during plant development even in cases where differentiation was induced

0.8% agar. Approximately half of each population was germinated under selective conditions (200 mg/l kanamycin), the other half under non-selective conditions. After 2 − 3 weeks incubation in the light (~5000 lux, cool fluorescent tubes, 16 h per day, 27°/20°C) sensitive seedlings germinated, bleached, and died without developing a primary leaf, whereas resistant seedlings developed to green plantlets with primary and secondary leaves. The F_1 population from the self-cross of wild-type SR1 bleached without exception. The F_1 population from the self-cross of T_{2-1} segregated 3:1 resistant to sensitive and the F_1 population of the backcross of T_{2-1} with the wild-type SR1 segregated 1:1 resistant to sensitive, thus providing evidence for the presence of one dominant Mendelian factor for kanamycin resistance in T_{2-1} (Figure 4). Both microspore analysis and genetic crossing analysis of T_{2-1} demonstrate the transmission of one dominant and functional locus for kanamycin resistance.

The pattern of segregation of the kanamycin-resistant trait in plant T_{2-2} is more complex. This will be studied in the next generation. In this case there may be more than one functional copy of the gene integrated into the genomic DNA in such a way that these are genetically linked (i.e., distantly placed on the same chromosome).

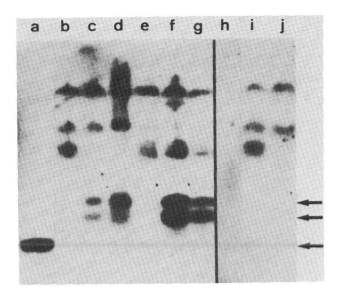

Fig. 6. Assay of APH(3′)II enzyme activity in selected kanamycin-resistant cell lines and in tissues of plants derived from them. **Lanes a,h** bacterial APH(3′)II enzyme from an osmotic shock extract of *E. coli* DH1 carrying plasmid pKC7 (Rao and Rogers, 1979); **b,i** extracts of wild-type leaf tissue; **c,j** extracts of leaf tissue from plants regenerated from transformant T_{2-1}; **d** extract of leaf tissue from plants regenerated from transformant T_{2-2}; **e** extract of wild type SR1 callus; **f,g** extracts of callus of the transformed lines T_{2-1} and T_{2-2}, respectively. **Lanes a−g** reacted with ^{32}P-labelled ATP in the presence of kanamycin; **h−j** reacted in the absence of kanamycin (showing background of non-specific phosphorylation). The arrows show the position of the APH(3′)II activity from the bacterial enzyme and that produced in plant tissues.

on kanamycin-free medium. The pattern of integration in plants was identical to that of the corresponding undifferentiated cell lines (Figure 5). The hybrid gene is also present in the kanamycin-resistant F_1 seedlings (data not shown).

The introduced APH(3′)II gene is expressed

To demonstrate that the drug-resistant phenotype was actually based on the expression of the APH(3′)II gene and the synthesis of the expected protein, the APH(3′)II enzyme activity was assayed according to the method of Reiss *et al.* (1984). The activity was absent in wild-type SR1 tissue and present in both transformed callus clones and in leaves of plants regenerated from them (Figure 6). These data indicate the apparently constitutive expression of the introduced gene independent both of the developmental stage of the cell and of the presence of the antibiotic in the cellular environment. The APH(3′)II activity migrated in non-denaturing polyacrylamide gels as two bands. The reason for this has not been defined, but it could obviously be due to post-translational modifications, proteolysis or association with other cellular protein(s) as previously found in animal cells (Colbère-Garapin *et al.*, 1981). The two bands probably do not arise by changes in the protein coding region due to recombination within the gene, since the same protein pattern was found in the two transformed lines in spite of different patterns of integrated DNA. The introduced APH(3′)II hybrid gene is also expressed in kanamycin-resistant F_1 plantlets (data not shown).

DNA-mediated transformation frequencies in our experiments were in the range of 10^{-6} if related to protoplasts treated or 10^{-5} if calculated on the basis of colonies which could have been recovered without selection, and were thus

low compared with those obtainable when using the *A. tumefaciens* co-cultivation system (Herrera-Estrella *et al.*, 1983a, 1983b; Bevan *et al.*, 1983; Fraley *et al.*, 1983; Horsch *et al.*, 1984; Murai *et al.*, 1983). Experiments are now in progress to optimize the procedure for gene-mediated transformation of protoplasts.

Stable genetic transformation of plant cells by the introduction of a hybrid marker gene into protoplasts has been established on the basis of the following criteria: (i) the kanamycin-resistant phenotype of selected cell lines and the plants derived from them; (ii) the presence of DNA sequences of the introduced gene integrated into plant genomic DNA; (iii) enzymatic activity of the protein product of the introduced gene in transformed lines, plants and their progeny. These results therefore also show that neither *A. tumefaciens* nor Ti plasmid-specific functions are a prerequisite for transfer and stable integration of functional genes into the plant genome.

Further genetic analysis of the progeny obtained from several independently transformed cell lines should give an indication of the site(s) of integration of foreign gene(s) into plant genome in the process of DNA-mediated transformation. These experiments are in progress.

Direct gene transfer in plants should enlarge the scope of biological problems accessible to investigation by molecular genetics; these might include, for example, studies of DNA replication in plants, by the search for and study of autonomously replicating plant vectors, or possibly directing the integration process by providing targets on the vectors for promoting homologous recombination.

Materials and methods

Construction of hybrid gene

The steps of the construction are described in Figure 1. All enzymes used during the construction were used as recommended by suppliers (Boehringer, Bio-Labs, B.R.L.). Other DNA and *E. coli* (strain DHI) manipulation procedures were as described by Maniatis *et al.* (1982).

Protoplast isolation

Leaf mesophyll protoplasts of *N. tabacum* cv. Petit Havana line SR1 (Maliga *et al.*, 1973) were isolated from sterile shoot cultures and purified as described by Nagy and Maliga (1976) modified by addition of one volume of 0.6 M sucrose to the leaf digest before the first centrifugation.

DNA transformation of protoplasts

The transformation method was essentially that of Krens *et al.* (1982): 1 ml aliquots of 2 x 10⁶ protoplasts each in K_3 medium (Nagy and Maliga, 1976) (0.1 mg/l 2,4-D, 1 mg/l NAA, 0.2 mg/l BAP) were incubated for 30 min at room temperature with 13% w/v PEG 6000, 15 μg of intact plasmid pABDI DNA (from 1 mg/ml stock) and 50 μg calf thymus DNA (from 1 mg/ml stock). The pH of the F medium was readjusted to 5.3 after autoclaving. After stepwise dilution with F medium the protoplasts were collected by sedimentation (5 min at 100 g) and resuspended in 30 ml of K_3 medium.

Protoplast culture, selection and culture of transformed cell lines

DNA-treated protoplasts were resuspended in 30 ml of K_3 medium and incubated in 10 cm diameter Corning Petri dishes in 10 ml aliquots (6.6 x 10⁴ protoplasts/ml) in the dark at 24°C. After 3 days the culture medium was diluted with 0.3 volumes of fresh K_3 culture medium and the cultures incubated under continuous light (2000 lux, cool fluorescence SYLVANIA 'daylight') at 24°C. After a total of 7 days the protoplast-derived population of developing cell clones was embedded in medium solidified with 1% Sea Plaque LMT agarose (Marine Colloids) and further cultured under selective conditions (50 mg/l kanamycin sulphate) in the agarose bead type culture system (Shillito *et al.*, 1983) at 24°C in the dark. The selective culture medium surrounding the beads was replaced every 5 days. Following 4 weeks of proliferation under selective conditions, resistant calli (2−3 mm in diameter) were transferred onto agar-solidified LS culture medium (Linsmaier and Skoog, 1965) (0.5 mg/l 2,4-D, 2 mg/l NAA, 0.1 mg/l BAP, 0.1 mg/l kinetin) containing 75 mg/l kanamycin sulphate and subcultured every 4 weeks.

Plant regeneration

Plants were regenerated under selective (150 mg/l kanamycin sulphate) or non-selective conditions by inducing shoots on LS medium containing 0.2 mg/l BAP and rooting these shoots on T medium (Nitsch and Nitsch, 1969). Wild-type SR1 shoots were regenerated in the same way on kanamycin-free media.

DNA isolation from plant material

Plant DNA was isolated by a method modified after Thanh Huyuh (Department of Biochemistry, Stanford University; personal communication). Samples of 0.5 g of callus or leaf tissue were homogenized at 0°C in a Dounce homogenizer in 15% sucrose, 50 mM EDTA, 0.25 M NaCl, 50 mM Tris-HCl pH 8.0. Centrifugation of the homogenate for 5 min at 1000 g resulted in a crude nuclear pellet which was resuspended in 15% sucrose, 50 mM EDTA, 50 mM Tris-HCl pH 8.0. SDS was added to a final concentration of 0.2% w/v. Samples were heated for 10 min at 70°C. After cooling to room temperature, potassium acetate was added to a final concentration of 0.5 M. After incubation for 1 h at 0°C the precipitate formed was sedimented at 4°C by a 15 min spin in an Eppendorf microcentrifuge. The DNA from the supernatant was precipitated with 2.5 volumes of ethanol at room temperature and redissolved in 10 mM Tris-HCl pH 7.5, 5 mM EDTA containing RNase A (10 μg/ml). Afer 10 min incubation at 37°C, proteinase K was added to a final concentration of 250 μg/ml and incubation was continued for 1 h at 37°C. The proteinase K was removed by phenol and chloroform-isoamyl alcohol extractions. DNA from the aqueous phase was precipitated with 60% isopropanol, 0.6 M Na acetate and dissolved in 50 μl of 5 mM EDTA, 10 mM Tris-HCl pH 7.5. DNA concentrations were estimated spectrophotometrically. The preparations yielded high mol. wt. DNA (predominantly >50 kb) susceptible to various restriction endonucleases.

Southern blot analysis

DNA electrophoresed in 1% agarose gel was transferred to nitrocellulose membrane (Southern, 1975) and hybridized with nick-translated (Rigby et al., 1977) DNA (5–10 x 10⁸ c.p.m./μg). Filters were washed three times for 1 h in 2 x SSC at 65°C. Blots were exposed to X-ray film with intensifying screen for 24–48 h.

APH(3')II activity test

This test was carried out essentially after the method of Reiss et al. (1984). Callus or leaf pieces (100–200 mg) were crushed in an Eppendorf centrifuge tube with 20 μl extraction buffer. This buffer was modified from that of Herrera-Estrella et al. (1983a) by omitting the bovine serum albumin and adding 0.1 M sucrose. Extracts were centrifuged for 5 min at 12 000 g and bromophenol blue was added to the supernatant to a final concentration of 0.004%. Proteins in 35 μl of the supernatant were separated by electrophoresis in a 10% non-denaturing polyacrylamide gel. The gel was incubated with kanamycin and γ-³²P-labelled ATP and then blotted onto Whatman p81 phosphocellulose paper. The paper was washed six times with deionized water at 90°C before autoradiography.

Acknowledgements

The authors thank B. Reiss and H. Schaller (University of Heidelberg, FRG) for supplying plasmids pKm 21 and pKm 244. We are also grateful to B. Reiss for releasing pre-publication details of the APH(3')II assay.

References

Beck,E., Ludwig,G., Auerwald,E.A., Reiss,B. and Schaller,H. (1982) Gene, 19, 327-336.

Bevan,M.W. and Chilton,M.D. (1982) Annu. Rev. Genet., 16, 357-384.

Bevan,M.W., Flavell,R.B. and Chilton,M.D. (1983) Nature, 304, 184-187.

Colbère-Garapin,F., Horodniceanu,F., Kourilsky,P. and Garapin,A.C. (1981) J. Mol. Biol., 150, 1-14.

Davey,M.R., Cocking,E.C., Freeman,J., Pearce,N. and Tudor,I. (1980) Plant Sci. Lett., 18, 307-313.

Fraley,R.T., Rogers,S.G., Horsch,R.B., Sanders,P.R., Flick,J.S., Adams, S.P., Bittner,M.L., Brand,L.A., Fink,C.L., Fry,J.S., Galluppi,G.R., Goldberg,S.B., Hoffmann,N.L. and Woo,S.C. (1983) Proc. Natl. Acad. Sci. USA, 80, 4803-4807.

Gardner,R.C., Howarth,A.J., Hahn,P., Brown-Luedi,M., Shepherd,R.J. and Messing,J. (1981) Nucleic Acids Res., 9, 2871-2888.

Herrera-Estrella,L., DeBlock,M., Messens,E., Hernalsteens,J.-P., Van Montagu,M. and Schell,J. (1983a) EMBO J., 2, 987-995.

Herrera-Estrella,L., Depider,A., Van Montagu,M. and Schell,J. (1983b) Nature, 303, 209-213.

Hirth,K.P., Edwards,C.A. and Firtel,R.A. (1982) Proc. Natl. Acad. Sci. USA, 79, 7356-7360.

Hohn,T., Richards,K. and Lebeurier,G. (1982) Curr. Top. Microbiol. Immunol., 96, Springer-Verlag, Berlin-Heidelberg, pp. 193-220.

Horsch,R.B., Fraley,R.T., Rogers,S.G., Sanders,P.R., Lloyd,A. and Hoffmann,N. (1984) Science (Wash.),, 223, 496-498.

Krens,F.A., Molendijk,L., Wullems,G.J. and Schilperoort,R.A. (1982) Nature, 296, 72-74.

Linsmaier,E.M. and Skoog,F. (1965) Physiol. Plant, 18, 100-127.

Maliga,P., Breznovitz,A. and Marton,L. (1973) Nature New Biol., 244, 29-30.

Maniatis,T., Fritsch,E.F. and Sambrook,J. (1982) Molecular Cloning. A Laboratory Manual, published by Cold Spring Harbor Laboratory Press, NY.

Messing,J. and Vieira,J. (1982) Gene, 19, 269-276.

Murai,N., Sutton,D.W., Murray,M.G., Slightom,J.L., Merlo,D.J., Reichert,N.A., Sengupta-Gopalan,C., Stock,C.A., Barker,R.F., Kemp,J.D. and Hall,T.C. (1983) Science (Wash.), 222, 476-482.

Nagy,J.I. and Maliga,P. (1976) Z. Pflanzenphysiol., 78, 453-455.

Nitsch,J.P. and Nitsch,C. (1969) Science (Wash.), 163, 85-87.

Rao,R.N. and Rogers,S.G. (1979) Gene, 7, 79-82.

Reiss,B., Sprengel,R., Willi,M. and Schaller,H. (1984) Gene, in press.

Rigby,W.J., Dieckmann,M., Rhodes,C. and Berg,P. (1977) J. Mol. Biol., 113, 237-251.

Rothstein,S.J. and Reznikoff,W.S. (1981) Cell, 23, 191-199.

Shillito,R.D., Paszkowski,J. and Potrykus,I. (1983) Plant Cell Rep., 2, 244-247.

Southern,E.M. (1975) J. Mol. Biol., 98, 503-517.

Sunderland,N. and Dunwell,J.M. (1977) in Street,H.E. (ed.), Plant Tissue and Cell Culture, University of California Press, Berkeley, pp. 223-265.

Wigler,M., Sweet,R., Gek,K.S., Wold,B., Pellicer,A., Lacy,E., Maniatis, T., Silverstein,S. and Axel,R. (1979) Cell, 16, 777-785.

Xiong,C., Muller,S., Lebeurier,G. and Hirth,L. (1982) EMBO J., 1, 971-976.

Received on 30 July 1984; revised on 24 August 1984

Supplementary Readings

An, G., Watson, B.D., Stachel, S., Gordon, M.P. & Nester, E.W., New cloning vehicles for transformation of higher plants. *EMBO J.* 4:277–284 (1985)

De Block, M., Herrera-Estrella, L., Van Montagu, M., Schell, J. & Zambryski, P., Expression of foreign genes in regenerated plants and their progeny. *EMBO J.* 3:1681–1689 (1984)

De Block, M., Botterman, J., Vandewiele, M., Dockx, J., Thoen, C., Gosselé, V., Rao Movva, N., Thompson, C., Van Montagu, M. & Leemans, J., Engineering herbicide resistance in plants by expression of a detoxifying enzyme. *EMBO J.* 6:2513–2518 (1987)

De Greef, W., Delon, R., De Block, M., Leemans, J. & Botterman, J., Evaluation of herbicide resistance in transgenic crops under field conditions. *Bio/Technology* 7:61–64 (1989)

Fischhoff, D.A., Bowdish, K.S., Perlak, F.J., Marrone, P.G., McCormick, S.M., Niedermeyer, J.G., Dean, D.A., Kusano-Kretzmer, K., Mayer, E.J., Rochester, D.E., Rogers, S.G. & Fraley, R.T., Insect tolerant transgenic tomato plants. *Bio/Technology* 5:807–813 (1987)

Hain, R., Stabel, P., Czernilofsky, A.P., Steinbiss, H.H., Herrera-Estrella, L., & Schell, J., Uptake, integration, expression and genetic transmission of a selectable chimaeric gene by plant protoplasts. *Mol. Gen. Genet.* 199:161–168 (1985)

Herrera-Estrella, L., Van den Broeck, G., Maenhaut, R., Van Montagu, M., Schell, J., Timko, M. & Cashmore, A., Light-inducible and chloroplast-associated expression of a chimaeric gene introduced into *Nicotiana tabacum* using a Ti plasmid vector. *Nature* 310:115–120 (1984)

Hiatt, A., Cafferkey, R. & Bowish, K., Production of antibodies in transgenic plants. *Nature* 342:76–78 (1989)

Hoekema, A., Hirsch, P.R., Hooykaas, P.J.J. & Schilperoort, R.A., A binary plant vector strategy based on separation of *vir-* and T-region of the *Agrobacterium tumefaciens* Ti-plasmid. *Nature* 303:179–180 (1983)

Horsch, R.B., Fraley, R.T., Rogers, S.G., Sanders, P.R., Lloyd, A. & Hoffmann, N., Inheritance of functional foreign genes in plants. *Science* 223:496–498 (1984)

Krens, F.A., Molendijk, L., Wullems, G.J. & Schilperoort, R.A., *In vitro* transformation of plant protoplasts with Ti-plasmid DNA. *Nature* 296:72–74 (1982)

Shimamoto, K., Terada, R., Izawa, T. & Fujimoto, H., Fertile transgenic rice plants regenerated from transformed protoplasts. *Nature* 338:274–276 (1989)

Vaeck, M., Reynaerts, A., Höfte, H., Jansens, S., De Beuckeleer, M., Dean, C., Zabeau, M., Van Montagu, M. & Leemans, J., Transgenic plants protected from insect attack. *Nature* 328:33–37 (1987)

Animals, Insects and Their Cells

Modern genetic engineering has entered the animal kingdom in two important ways. The simplest way, in terms of technology and public acceptance, has been the genetic modification of animal cells which are to be grown in culture. This technology has been used to produce proteins encoded by the genes introduced into the cells. Animal cells are particularly useful if the desired protein product has elaborate secondary modifications (such as glycosylation of factor VIII, tissue plasminogen activator, etc.), in which case bacteria can not be used as a living "factory."

Genetic modifications of domesticated animals is the alternative entree of genetic engineering into the animal kingdom. The domestication of animals including their selective breeding is one of man's earliest efforts in the field of biotechnology. Despite this long history, modern biotechnology applications in this field have been extremely slow in coming. In part this delay has been a consequence of technical challenges. Some papers presented in this section address the technological advances which have allowed the genetic engineering of the germ line of animals. Two goals for this endeavor will be presented; the introduction of new desirable traits into domesticated animals (e.g., leaner meat, faster growth, etc.) or the use of animals to produce specialized proteins.

It has been known for some time that purified DNA can be introduced into tissue culture cells using a calcium phosphate coprecipitation technique. There are some important characteristics of this protocol: it is very inefficient, and the introduced DNA is randomly incorporated into the host genome. The first of these features presented a barrier towards the introduction of most foreign genes into tissue culture cells since the gene of choice would provide no selective advantage to the transformed cell, and thus the transfected cell would be extremely hard to find. The Wigler et al. paper describes a simple solution, co-transformation with a second DNA molecule which encodes a selectable marker. The assumption, which proved to be correct, was that the inefficiency of DNA uptake was due to the limitation in the number of "competent" cells, and that each competent cell was likely to take up several DNAs.

An alternate solution to the inefficiency problem was to ligate the gene to be introduced into a "vector" which itself provided the selectable marker. Two examples of this approach are described by Cone and Mulligan paper (in which a retrovirus vector is used), and the supplemental Saver et al. paper (in which a bovine papilloma virus vector is used). In both cases one achieves the added advantage of having the gene incorporated into a viral replicon which can be reisolated from the cells and remanipulated. The Alt et al. paper introduces the methodology which allows the isolation of cells with the selective amplification of genes linked to the dihydrofolate reductase gene. This provided a methodology which facilitated the overproduction of desired proteins.

An entirely different approach toward using tissue culture cells for the production of foreign proteins is described by Smith et al. These experiments make use of the unique properties of the insect baculovirus. These viruses normally encode a polyhedron coat protein, which during the infectious process becomes approximately 25% of the synthesized protein. The investigators substituted the human β-interferon coding sequence for the coat protein gene, and were successful at achieving high levels of foreign protein synthesis. A subsequent use of baculovirus technology involved the use of infected silkworms instead of insect cell cultures (see the supplementary reading by Maeda et al.).

The ability to introduce defined genes into the germline of animals would allow carefully planned augmentation of the genome of the target organism. Microinjection of purified DNA into fertilized eggs followed by implantation of the egg into a foster mother has proven to be a successful approach towards this goal, although it has some serious limitations. The procedure is of very low efficiency and, since the DNA is randomly inserted into the genome, the DNA can cause mutations in the host genome, and its expression may be modulated by the character of the chromosome region into which it was inserted. Nonetheless, the Brinster et al., Palmiter et al., and Gordon et al. papers report experiments which demonstrate that this procedure is feasible. The Palmiter et al. paper points towards this procedure being used to modify some overall property of the organism (increasing the rate of growth in mice transgenic for the rat growth hormone gene). The Gordon et al. contribution is an early report which describes the coupling of mouse whey acidic protein controlling elements to the gene whose protein expression is desired and then making transgenic mice with this construct. The desired goal, to have the protein selectively made in the mammary glands and excreted into the milk as is the case normally for the whey acidic protein, was realized for those rare cases in which successful germ line inheritance was achieved. This approach holds considerable promise since mammary gland specific controlling elements are well known, and the milk from larger mammals should serve as a convenient source (e.g., see the studies with transgenic sheep described in supplemental article by Simons et al.).

Selective Multiplication of Dihydrofolate Reductase Genes in Methotrexate-resistant Variants of Cultured Murine Cells

F.W. Alt, R.E. Kellems, J.R. Bertino and R.T. Schimke

The rate of dihydrofolate reductase synthesis in the AT-3000 line of methotrexate-resistant murine Sarcoma 180 cells is approximately 200- to 250-fold greater than that of the sensitive, parental line. We have purified cDNA sequences complementary to dihydrofolate reductase mRNA and subsequently used this probe to quantitate dihydrofolate reductase mRNA and gene copies in each of these lines. Analysis of the association kinetics of the purified cDNA with DNA from sensitive and resistant cells indicated that the dihydrofolate reductase gene is selectively multiplied approximately 200-fold in the resistant line. A similar analysis of a partially revertant line of resistant cells indicated that the loss of resistance observed when the AT-3000 line is grown in the absence of methotrexate is associated with a corresponding decrease in the dihydrofolate reductase gene copy number. In each of these lines the relative number of dihydrofolate reductase gene copies is proportional to the cellular level of dihydrofolate reductase and dihydrofolate reductase mRNA sequences.

We have also studied parental and methotrexate-resistant lines of L1210 murine lymphoma cells. Both resistance and an associated 35-fold increase in the level of dihydrofolate reductase appear to be stable properties of the resistant L1210 line since we find no decrease in either parameter in over 100 generations of growth in the absence of methotrexate. Once again, we find that the increased levels of dihydrofolate reductase in the methotrexate-resistant L1210 line are associated with a proportional increase in the number of dihydrofolate reductase gene copies. In this case the dihydrofolate reductase gene copy number appears to be relatively stable in the resistant line. Therefore, we conclude that selective multiplication of the dihydrofolate reductase gene can account for the overproduction of dihydrofolate reductase in both stable and unstable lines of methotrexate-resistant cells.

The resistance of both human neoplasms (1) and various lines of cultured cells (2–11) to the 4-amino analogs of folic acid is often associated with an increase in the cellular content of dihydrofolate reductase. We have been studying the overproduction of this enzyme in variant lines of murine Sarcoma 180 cells that were selected by a step-wise procedure for growth in the presence of high concentrations of methotrexate (a folic acid analogue). Dihydrofolate reductase comprises as much as 6% of the soluble protein in the methotrexate-resistant AT-3000 line, representing an increase of more than 200-fold over the level in the sensitive, parental cells (12). Purified dihydrofolate reductase from resistant cells appears to be identical to that from sensitive cells; and, in addition, the relative half life of the enzyme is similar to these lines (12). We have demonstrated that the increased level of dihydrofolate reductase in resistant cells is due to an increased rate of enzyme synthesis (12), and that, in turn, this increase is correlated with increased cellular levels of translatable dihydrofolate reductase mRNA (13).

One of the most interesting characteristics of the AT-3000 line is that high levels of resistance are lost when these cells are grown in the absence of methotrexate (3). Loss of resistance is associated with a decrease in the level of dihydrofolate reductase (3, 12), and a corresponding decrease in both the rate of dihydrofolate reductase synthesis (12) and the level of the specific mRNA activity (13). Several lines of evidence suggest that these decreases are due to the instability of the variation (mutation?) which leads to increased enzyme synthesis (12). Instability is also a characteristic of methotrexate resistance in a number of other cell lines (9, 14). In contrast, in certain lines of methotrexate-resistant baby hamster kidney (BHK) cells (15) as well as in other resistant lines (2), resistance and increased levels of dihydrofolate reductase appear to be stable characteristics and do not decline when the cells are grown in the absence of the drug. However, other properties of methotrexate resistance in BHK cells appear to be similar to those of Sarcoma 180 cells, including high rates of dihydrofolate reductase synthesis (16) and high levels of translatable dihydrofolate reductase mRNA (17). Various considerations of the possible mechanisms that could lead to stable or unstable changes in the phenotypic expression

* This research was supported by Grant GM 14931 from the National Institute of General Medical Sciences, National Institutes of Health and Grant CA 16318 from the National Cancer Institute, National Institutes of Health. The costs of publication of this article were defrayed in part by the payment of page charges. This article must therefore be hereby marked "advertisement" in accordance with 18 U.S.C. Section 1734 solely to indicate this fact.

‡ Present address, Center for Cancer Research, Massachusetts Institute of Technology, 77 Massachusetts Ave., Cambridge, Mass. 02139.

§ Permanent address, Department of Pharmacology, Sterling Hall of Medicine, Yale University, 333 Cedar Street, New Haven, Conn. 06510.

ALT, F.W., KELLEMS, R.E., BERTINO, J.R. and SCHIMKE, R.T.
Selective multiplication of dihydrofolate reductase genes in methotrexate-resistant variants of cultured murine cells. *J. Biol. Chem.* 253:1357-1370 (1978). Reprinted with the authors' permission.

of cultured cells have been discussed (18, 19).

We report here the purification of cDNA sequences complementary to dihydrofolate reductase mRNA of murine origin, and subsequent use of this probe to quantitate dihydrofolate reductase mRNA and gene copies in a number of different cell lines. We have examined both sensitive and methotrexate-resistant lines of Sarcoma 180 cells, as well as a partially revertant line that was derived by growing resistant cells in the absence of methotrexate for 400 cell doublings (12). In addition, we have also studied parental and methotrexate-resistant lines of L1210 murine lymphoma cells. Both resistance and associated high levels of dihydrofolate reductase appear to be stable properties of the L1210 lines since we find no decrease in either parameter over several hundred generations of growth in the absence of methotrexate. We find that in both the stable (L1210) and unstable (S-180) lines of resistant cells, increased levels of dihydrofolate reductase and dihydrofolate reductase mRNA are associated with a proportional increase in the number of dihydrofolate reductase gene copies. When unstable lines are grown in the absence of selection, loss of resistance is associated with a decrease in the dihydrofolate reductase gene copy number.

EXPERIMENTAL PROCEDURES

Materials — Sources of most of the reagents have been given previously (12, 13). Oligo(dT)-cellulose and oligo(dT)$_{12-18}$ were purchased from Collaborative Research; micrococcal nuclease, salmon sperm DNA, and calf thymus DNA from Sigma; S1 nuclease from Miles; [³H]leucine (5 Ci/mmol) from New England Nuclear; [³H]deoxycytidine triphosphate (20 Ci/mmol) from Amersham/Searle; Chelex 100 and Bio-Gel hydroxylapatite from Bio-Rad. Purified reverse transcriptase (Lot no. G-1176, 39,216 units/mg) was supplied by Dr. J. W. Beard (Life Sciences Inc., St. Petersburg, Florida) and methotrexate by Dr. Paul Davignon, Pharmaceutical Resources Branch, National Cancer Institute.

Generously provided as gifts were purified ovalbumin mRNA and *E. coli* tRNA from Dr. Gray Crouse (Stanford University), and purified chicken oviduct DNA from Dr. Henry Burr (Stanford University).

Cell Culture — The Sarcoma 180 cell line and the 3000-fold methotrexate-resistant AT-3000 subline were grown as described previously (12) except that thymidine and glycine were omitted from the medium of the resistant cells. A partially phenotypic revertant line, Rev-400, was obtained by growing the AT-3000 line for 400 cell doublings in methotrexate-free medium. Some characteristics of the "revision" phenomenon have been described previously (12), and further details are described under "Results."

Suspensions of L1210 murine lymphoma cells (L1210S) were grown in Fischer's Medium for Leukemic Cells of Mice (GIBCO) containing 10% horse serum. A 5000-fold methothrexate-resistant subline (L1210RR) and a 25,000-fold resistant subline (L1210 RR500) were grown in the same medium supplemented with 100 μM and 500 μM methotrexate, respectively. For some of the experiments described the L1210RR line was grown for approximately 100 cell doublings (over 10 months) in methotrexate-free medium. Further characteristics of these lines will be described under "Results" and elsewhere.[1]

Determination of the Relative Rate of Dihydrofolate Reductase Synthesis — The relative rate of dihydrofolate reductase synthesis was determined as described previously by direct immunoprecipitation of the enzyme from extracts of pulse-labeled cells (12).

RNA Preparation — Total cytoplasmic RNA was prepared from each of the cell lines as described previously (13). These preparations were used immediately or stored in liquid nitrogen.

Poly(A)-containing RNA was prepared by oligo(dT)-cellulose chromatography of total cytoplasmic RNA. RNA was dissolved in 10 mM Tris/Cl (pH 7.4) and 0.5% sodium dodecyl sulfate, heated at 68° for 5 min and rapidly cooled in an ethanol-ice bath. This solution was then adjusted to 0.4 M NaCl and oligo(dT)-cellulose chromatography

---

[1] C. Lindquist and J. Bertino, manuscript in preparation.

was carried out essentially as described by Aviv and Leder (20). The bound RNA fraction was eluted with 10 mM Tris/Cl (pH 7.4) and 0.5% sodium dodecyl sulfate, adjusted to 400 mM NaCl, and precipitated overnight at −20° by the addition of 2 volumes of ethanol. The precipitates were dissolved in a minimal volume of H$_2$O and stored in liquid N$_2$.

Polysome Preparation — The various lines were grown in roller bottles and were fed with fresh medium 4 h prior to harvest. Cells were rinsed once with ice cold Hanks' balanced salts solution plus 50 μg/ml cycloheximide, scraped from the bottles with rubber policemen, and washed three times by centrifugation through the same salt solution. Homogenization (13) and preparation of polysomes by the "cushion" method was as described previously by Palacios *et al.* (21), except that the homogenization buffer contained 10 mM MgCl$_2$. Polysomes were dialyzed for 12 h against 25 mM Tris/Cl (pH 7.1), 25 mM NaCl, 5 mM MgCl$_2$, and 1 mg/ml sodium heparin (Buffer A) and then stored in liquid nitrogen for subsequent use.

Antibody Purification — Rabbit anti-dihydrofolate reductase γ-globulin, prepared against purified dihydrofolate reductase protein as described previously (12), was purified by affinity chromatography on dihydrofolate reductase-Sepharose. Conditions for preparation of the resin and affinity chromatography were essentially as described by Shapiro *et al.* (22). Bound γ-globulin, eluted with 4.5 M MgCl$_2$, was enriched approximately 100-fold for anti-dihydrofolate reductase activity. The purified antibody preparation was made ribonuclease-free by passage through a column of DEAE-cellulose overlaid with CM-cellulose (21).

Iodination of Anti-dihydrofolate Reductase Globulin — Anti-dihydrofolate reductase globulin was iodinated by the lactoperoxidase method essentially as described by Taylor and Schimke (23). Iodinated antibody was made ribonuclease-free as described above.

Binding of Iodinated Anti-dihydrofolate Reductase Antibody to Polysomes — Prior to incubation, polysomes prepared as described above were thawed at 4° and centrifuged for 10 min at 5000 × g to remove particulate material. Reaction mixtures containing 30 A_{260} units of polysomes and 1.3 μg of iodinated anti-dihydrofolate reductase (specific radioactivity 77,000 cpm/μg) in 2 ml of Buffer A were incubated for 50 min at 0°. Polysomes were then reisolated from the reaction mixture by the "cushion" method as described above and sedimented through a linear sucrose gradient (0.5 M to 1.5 M in 11 ml of Buffer A) for 1.8 h at 4°. Gradients were fractionated and monitored for A_{260} with an Isco model 640 density gradient fractionator equipped with an ultraviolet flow monitor. For scintillation counting, 0.5-ml fractions were dissolved in 10 ml of Instagel (Packard).

Isolation of Dihydrofolate Reductase-synthesizing Polysomes — Indirect immunoprecipitation of polysomes was carried out essentially as described by Shapiro *et al.* (22). Resistant cell (AT-3000) polysomes at a final concentration of 10 to 15 A_{260} units/ml in 25 mM Tris/Cl (pH 7.1), 4 mM MgCl$_2$, 150 mM NaCl$_4$, 750 μg/ml sodium heparin, and 0.5% w/v Triton X-100 and sodium deoxycholate were incubated with optimal concentrations of purified rabbit anti-dihydrofolate reductase γ-globulin (20 μg/A_{260} unit of polysomes) for 60 min at 0°. The antibody·nascent chain complex was then precipitated by incubation with goat anti-rabbit γ-globulin (80 μg/μg of rabbit γ-globulin) for an additional 90 min at 0°. The precipitated complex was pelleted and washed as described by Shapiro *et al.* (22). Pellets were resuspended in 25 mM Tris/Cl (pH 7.1), 5 mM EDTA, 6 mM MgCl$_2$, 25 mM NaCl, 1 mg/ml sodium heparin, and 1% sodium dodecyl sulfate, and RNA was extracted by the phenol/chloroform procedure described previously (13).

RNA-dependent Rabbit Reticulocyte Lysates — Micrococcal nuclease-treated rabbit reticulocyte lysates were prepared by a modification of the procedure described by Pelham and Jackson (24) as follows: standard rabbit reticulocyte lysate reaction mixtures were prepared as described previously (13) except that [³H]leucine and RNA were omitted. Aliquots (325 μl) of this mixture were combined with 3.3 μl of 100 mM CaCl$_2$ (final concentration, 1 mM) and 3.3 μl of a 1 mg/ml solution of micrococcal nuclease (final concentration 10 μg/ml), and incubated for 15 min at 25°, at which time nuclease action was inhibited by the addition of 7 μl of 100 mM ethylene glycol bis(β-aminoethyl ether)*N,N'*-tetraacetic acid (final concentration, 2 mM). The nuclease-treated reticulocyte lysate mix prepared in this fashion was either used immediately or stored for up to 2 weeks in liquid nitrogen with no significant loss of activity.

Typical *in vitro* protein synthesis assays consisted of 60 μl of nuclease-treated lysate reaction mix, 4.6 μl of 200 μM [³H]leucine (specific radioactivity, 5 Ci/mmol), and 25.4 μl of an aqueous solution of RNA. Following incubation for 1 h at 25°, the reaction was terminated by the addition of 36 μl of 0.1 M leucine and 14 μl of a mixture of 10% (w/v) sodium deoxycholate and 10% (w/v) Triton X-100. Stimulation of total protein synthesis was determined as the difference between trichloroacetic acid-precipitable radioactivity appearing in reactions that had received RNA and that in reactions to which no RNA had been added. Incorporation into dihydrofolate reductase was measured by specific immunoprecipitation as described previously (13) and expressed as a percentage of the total stimulated trichloroacetic acid-precipitable radioactivity in the lysate reaction. Under these conditions, incorporation into total trichloroacetic acid-precipitable radioactivity and dihydrofolate reductase was linear with time for up to 90 min with added poly(A)-containing RNA to 15 μg/ml, and with added total RNA to at least 50 μg/ml. Typical stimulation for a standard translation assay was approximately 100,000 cpm/μg of poly(A)-containing RNA, a level 20- to 30-fold greater than background.

Sodium Dodecyl Sulfate-Polyacrylamide Gel Electrophoresis of Lysate Products — After termination of lysate reactions, aliquots were removed and mixed with an equal volume of dissolving buffer, boiled for 3 min, and subjected to sodium dodecyl sulfate-polyacrylamide gel electrophoresis as described previously (12). Subsequent to electrophoresis, gels were soaked for 16 h in a liter of 7.5% acetic acid and 5% methanol (with one change of solution) in order to remove soluble radioactivity. Gels were then sliced and prepared for scintillation counting as described previously (12).

cDNA Preparation — cDNA was prepared essentially as described by Buell *et al.* (25). The reactions were carried out in 20-μl volumes and contained: 50 mM Tris/Cl (pH 8.3), 140 mM KCl, 30 mM β-mercaptoethanol, 10 mM MgCl$_2$, 100 μg/ml oligo(dT), 0.5 mM dGTP, dATP, and dTTP, 0.5 mM [^{3}H]dCTP (20 Ci/mmol), 16 units of avian myeloblastosis virus reverse transcriptase, and 3 μg of poly(A)-containing RNA prepared from either total cytoplasmic-resistant cell RNA, or RNA extracted from immunoprecipitated dihydrofolate reductase synthesizing resistant cell polysomes.

Reactions were incubated at 42° for 1 h and stopped by the addition of 120 μl of 0.3 M NaOH. After a further incubation at 37° for 20 h, samples were neutralized with 1 N HCl and sodium dodecyl sulfate was added to a final concentration of 0.1%. The reaction mixtures were then extracted with 2 volumes of CHCl$_3$, and the aqueous phase passed over a small (8-ml) column of G-100 Sephadex which was previously equilibrated with H$_2$O. The void volume was pooled and concentrated by ethanol precipitation.

In all cases the yield was approximately 10^6 cpm of cDNA per μg of added RNA. Based on the specific radioactivity of the [^{3}H]dCTP and assuming equal representation of all four bases, this corresponds to approximately 0.1 μg of cDNA synthesized per μg of added RNA. Approximately 7 to 8% of the trichloroacetic acid-precipitable radioactivity in the cDNA preparation was resistant to treatment with S1 nuclease.

RNA/cDNA Hybridizations — All analytical RNA/cDNA hybridizations were done in 20 mM Tris/Cl (pH 7.7), 600 mM NaCl, 2 mM EDTA, and 0.2% sodium dodecyl sulfate except where noted otherwise. Reaction mixtures of 2 to 40 μl were overlaid with mineral oil in plastic tubes and incubated at 68°. The quantities of [^{3}H]cDNA and RNA used in these reactions are described in appropriate figure legends. In all cases, final R_0t values were corrected to standard salt conditions (26).

At the end of the incubation, reaction mixtures were diluted into 1 ml of buffer containing 30 mM Na (C$_2$H$_3$O$_2$) (pH 4.5), 3 mM ZnSO$_4$, 300 mM NaCl, and 10 μg/ml denatured salmon sperm DNA. Each sample was divided into two aliquots: one was digested for 30 min at 45° with 8 μg/ml of S1 nuclease, and the other incubated identically, but without S1 nuclease. After digestion, 100 μg/ml of carrier calf thymus DNA was added to both S1-treated and control samples and nucleic acids precipitated with an equal volume of 10% trichloroacetic acid containing 1% sodium pyrophosphate at 4° for 15 min. Precipitates were collected on Millipore filters, washed three times with 5% trichloroacetic acid, dried, and counted in 10 ml of ScintiLene (Fisher).

Hybrid formation was scored as the amount of trichloroacetic acid-precipitable radioactivity remaining after S1 treatment and expressed as a percentage of the untreated control value. Depending on the cDNA preparation from 1.5 to 8% of the trichloroacetic acid-precipitable counts were resistant to S1 treatment in the absence of added RNA. In all experiments, the appropriate percentage of endogenous S1 resistance was subtracted from treated and control

values before calculation of the per cent hybridization. In calculating R_0t values, we assumed an average value of 346 g of RNA nucleotides per mol.

DNA Preparation — The 27,000 $\times$ g pellets (containing nuclei) resulting from standard RNA preparations (13) were stored at $-20°$. Approximately 5 ml of frozen nuclear pellet was thawed and gently homogenized in 50 ml of 0.15 M NaCl, 0.1 M EDTA (pH 8.0), 0.6 M sodium perchlorate, and 1.0% sodium dodecyl sulfate by five strokes in a dounce homogenizer (loose pestle). The homogenate was slowly stirred at 25° for 30 min, extracted with 2 volumes of chloroform, and DNA was spooled from the aqueous phase after the addition of 2 volumes of ice cold ethanol.

Spooled DNA was dissolved in 10 mM Tris/Cl (pH 7.4) and then treated with 60 μg/ml pancreatic ribonuclease (boiled for 10 min in 20 mM NaCl prior to use) for 2 h at 37°. Sodium dodecyl sulfate and proteinase K were then added to a final concentration of 0.2% and 60 μg/ml, respectively, and the incubation continued for another 5 h at 37°. The solution was then extracted with 2 volumes of CHCl$_3$ and the aqueous phase precipitated overnight at $-20°$ with 2 volumes of ethanol. Precipitated DNA was pelleted by centrifugation for 5 min at 2000 $\times$ g, lyophilized, and dissolved in 100 mM sodium acetate (pH 7.8). DNA was then sheared by passage through the needle valve of a French pressure cell at a pressure of 20,000 p.s.i. Divalent cations were removed by passing the sheared DNA preparations over a small (10 ml) volume of Chelex (equilibrated with 100 mM sodium acetate, pH 7.8), and the DNA was subsequently ethanol-precipitated as described above and redissolved in 20 mM Tris/Cl (pH 7.4) and 1 mM EDTA. 1 M NaOH was then added to a final concentration of 0.3 M and the solution was incubated for 22 h at 37°, at which time the base was neutralized by the addition of an equivalent amount of 1 N HCl.

These preparations were then stored at 4° until subsequent use as described below. All of the DNA samples prepared in this fashion sedimented as symmetrical peaks on isokinetic alkaline sucrose gradients (see below) with a calculated size of approximately 450 base pairs.

Sedimentation Analysis of DNA — DNA was analyzed by sedimentation through isokinetic alkaline sucrose gradients prepared as described by McCarty *et al.* (27) using 5% and 29.4% sucrose containing 0.1 N NaOH and 0.9 M NaCl. The molecular size of the DNA was calculated from S value as described by Studier (28).

cDNA/DNA Association Reactions — DNA/DNA associations were done in reaction mixtures containing 25 mM Tris/Cl (pH 7.4), 1 mM EDTA, 300 mM NaCl, 50 pg of [^{3}H]cDNA (500 cpm) and 500 μg of cellular DNA (prepared as described above) in a final volume of from 0.05 ml to 1.1 ml. Reaction mixtures were overlaid with mineral oil in plastic tubes, heated to 102° for 10 min in an H$_2$O/ethylene glycol bath, cooled, and incubated at 68° for various times in order to achieve the desired C_0t values.

Single- and double-stranded DNA were then fractionated by chromatography on hydroxylapatite. Reaction mixtures were diluted into 5 ml of 0.12 M NaPO$_4$ (pH 6.8) and passed over a column containing 1 g of hydroxylapatite (boiled for 5 min in 5 ml of 0.12 M NaPO$_4$ prior to use and equilibrated in the same buffer) which was maintained at 60° with a recirculating water bath. Single-stranded DNA was eluted with 0.12 M NaPO$_4$ (pH 6.8) and double-stranded material subsequently eluted with 0.5 M NaPO$_4$ (pH 6.8). The single- and double-stranded fractions were monitored for A_{260}, and the DNA then was precipitated by the addition of carrier calf thymus DNA to 25 μg/ml and 0.1 vol of 100% trichloroacetic acid. Trichloroacetic acid-precipitated material was collected and counted as described above. In order to calculate DNA concentration, an A_{260} absorbance of 1 was assumed to correspond to DNA concentrations of 43 μg/ml and 50 μg/ml, respectively, for single- and double-stranded DNA fractions. The per cent double-stranded in each sample was determined by dividing the amount of DNA or [^{3}H]cDNA recovered in the double-stranded fraction by the total amount recovered in the double- and single-stranded fractions. In calculating C_0t values we assumed an average value of 332 g of DNA nucleotides per mol.

RESULTS

Purification of Dihydrofolate Reductase-specific cDNA

Immunoprecipitation of Dihydrofolate Reductase-synthesizing Polysomes — In order to further study the factors responsible for the accumulation of high levels of translatable dihydrofolate reductase mRNA in methotrexate-resistant cells, we needed a cDNA probe complementary to dihydrofolate reductase mRNA. The usual method for the preparation of such a reagent has involved purification of a specific mRNA and subsequent synthesis of a complementary cDNA. Dihydrofolate reductase mRNA contains poly(A), allowing easy separation from rRNA, but its sedimentation rate on sodium dodecyl sulfate or denaturing sucrose gradients is not sufficiently distinct from that of total poly(A)-containing RNA to permit a significant additional purification by size fractionation (13). Therefore, we have employed the specific polysome immunoprecipitation procedure described by Shapiro *et al.* (22) to enrich for dihydrofolate reductase-synthesizing polysomes. The initial step in this procedure involved incubation of purified (100-fold) anti-dihydrofolate reductase antibody with resistant cell polysomes. The data in Fig. 1a demonstrate that this procedure results in the specific binding of the antibody to a size class of resistant cell polysomes (5 to 7 ribosomes) expected for those engaged in the synthesis of dihydrofolate reductase (M_r = 20,000). However, only a low level of apparently nonspecific binding is observed with polysomes from sensitive cells (Fig. 1b), where the rate of dihydrofolate reductase synthesis is below the resolution level of this technique. These results suggest that the incubation procedure results in the binding of purified antibody specifically to dihydrofolate reductase nascent chains. Subsequent to the initial binding reaction, the resulting antibody·nascent chain·polysome complexes were precipitated with a second antibody directed against the first antibody (see "Experimental Procedures" for details). We estimated the purification achieved by this procedure by translating the poly(A)-containing RNA extracted from the immunoprecipitated polysomes in the mRNA-dependent rabbit reticulocyte lysate (25). At the end of the incubation, samples of the total lysate reaction mix were analyzed by electrophoresis on sodium dodecyl sulfate-polyacrylamide gels. Fig. 2a indicates the very low background observed in this system in the absence of added RNA. In contrast, addition of purified ovalbumin mRNA resulted in the stimulation of a single peak of incorporated radioactivity with a mobility characteristic of authentic ovalbumin (Fig. 2b). This result indicates that the generation of incomplete or fragmented polypeptide chains is not a problem with this system. Furthermore, the specificity of the assay is evidenced by the fact that in this experiment more than 95% of the stimulated incorporation was precipitable with anti-ovalbumin antibody (data not shown). The addition of polysomal poly(A)-containing RNA from resistant cells resulted in the synthesis of a broad size distribution of proteins of which approximately 1.9% were precipitable by anti-dihydrofolate reductase antibody (data not shown). This value corresponds well to our estimate of the relative rate of dihydrofolate reductase synthesis as a per cent of total protein synthesis in this line.[2] Poly(A)-containing RNA extracted from the immunoprecipitated polysomes stimulated incorporation into a single major peak of radioactivity which co-migrated with added dihydrofolate reductase marker (Fig. 2d). In this experiment, 25% of the stimulated incorporation was precipitable by anti-dihydrofolate reductase antibody (data not shown). Comparison of the relative incorporation into dihydrofolate reductase

[2] We have previously described dihydrofolate reductase synthesis as a per cent of soluble protein synthesis. In these lines, soluble protein accounts for approximately 20 to 30% of the total protein synthesis (data not shown).

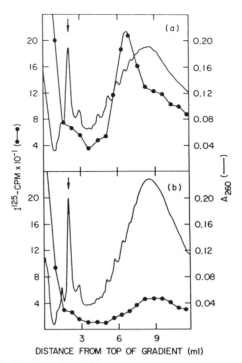

FIG. 1. Binding of anti-dihydrofolate reductase antibody to polysomes. The binding of ^{125}I-labeled anti-dihydrofolate reductase antibody to polysomes from AT-3000 (*a*) and S-3 (*b*) cells was examined as described under "Experimental Procedures." ^{125}I-radioactivity, ●——●; A_{260}, ——.

in the experiments presented in Fig. 2, *c* and *d* indicates an approximately 10-fold purification of dihydrofolate reductase mRNA by the polysome precipitation procedure.

cDNA Synthesis from the Partially Purified Dihydrofolate Reductase mRNA – cDNA was prepared from the partially purified dihydrofolate reductase mRNA resulting from the immunoprecipitation procedure and then analyzed by hybridization to excess poly(A)-containing RNA from sensitive or resistant cells (Fig. 3a). Comparison of the kinetics of these reactions indicates that approximately 15 to 20% of the cDNA sequences hybridize to mRNA sequences that are considerably more abundant in resistant cells than in sensitive cells. This percentage roughly corresponds with our estimate of the proportion of dihydrofolate reductase mRNA sequences in the partially purified RNA preparation from which the cDNA was synthesized (see above). However, these data indicate that this cDNA preparation is not pure enough for use as an analytical reagent. In order to further enrich for cDNA sequences complementary to the dihydrofolate reductase mRNA, we devised the purification scheme described below.

Purification of cDNA Sequences Complementary to Dihydrofolate Reductase mRNA – As a further means of purification of dihydrofolate reductase-specific sequences in the cDNA preparation, we exploited the large and apparently specific increase in the abundance of dihydrofolate reductase mRNA sequences in the RNA population of resistant as compared to sensitive cells. Analysis of the soluble proteins produced by sensitive and resistant cells suggested that the only major difference between the two is the overproduction of dihydrofolate reductase (12). Furthermore, we have used a reticulocyte lysate *in vitro* translation assay to demonstrate that most, if not all, of the several hundred-fold increase in the level of dihydrofolate reductase synthesis in resistant cells can be

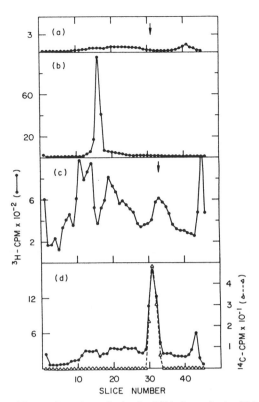

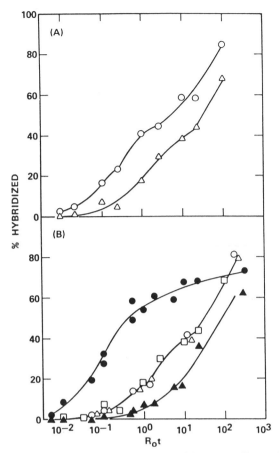

FIG. 2. Electrophoretic analysis of mRNA-dependent rabbit reticulocyte reaction products. Aliquots from mRNA-dependent lysate reactions stimulated with (a) no added RNA, (b) 1.5 μg of purified ovalbumin mRNA, (c) 1.25 μg of poly(A)-containing RNA prepared from AT-3000 cell polysomes, (d) 1 μg of poly(A)-containing RNA prepared from immunoprecipitated dihydrofolate reductase-synthesizing polysomes (see "Experimental Procedures" for details) were examined by sodium dodecyl sulfate-polyacrylamide gel electrophoresis as described under "Experimental Procedures." Other aliquots of the stimulated lysate reactions were used for the specific immunoprecipitation of ovalbumin (reaction b) as described by Rhoads et al. (29) and dihydrofolate reductase (reactions c and d) as described previously (12). The incorporation into each of these proteins relative to total stimulation was measured as described under "Experimental Procedures." ^{3}H-labeled lysate reaction products, ●——●; authentic ^{14}C-labeled dihydrofolate reductase, △- - -△. Arrows mark the migration of added ^{14}C-labeled dihydrofolate reductase in panels a and c.

attributed to a similar increase in the level of dihydrofolate reductase mRNA activity (13). Therefore, we estimate that dihydrofolate reductase mRNA sequences are as much as 200 to 300 times more abundant in resistant cells than in sensitive cells, whereas other mRNA sequences are probably present in similar abundance in the two cell types. Furthermore, since the cDNA was prepared from resistant cell RNA that was enriched an additional 10-fold for dihydrofolate reductase mRNA sequences, the dihydrofolate reductase-specific sequences in the cDNA preparation could be as much as 2000- to 3000-fold more abundant than the complementary sequences in sensitive cell poly(A)-containing RNA. These estimates provide the rationale for Step A of the dihydrofolate reductase-specific cDNA purification procedure that is outlined in Fig. 4. The cDNA preparation was incubated with a 30-fold mass excess of sensitive cell poly(A)-containing RNA to a R_0t value sufficiently high to ensure completion of the reaction (see legend to Fig. 4). Under these conditions, cDNA complementary to mRNA sequences that are present in similar abundance in resistant and sensitive cells (or in greater

FIG. 3. A, hybridization of cDNA prepared from partially purified dihydrofolate reductase mRNA to poly(A)-containing RNA from resistant and sensitive S-180 cells. Poly(A)-containing RNA extracted from the S-3 (△——△) and AT-3000 (○——○) cell lines was reacted with 60 pg (600 cpm) of [^{3}H]cDNA prepared from partially purified dihydrofolate reductase mRNA (see text). Similar quantities of RNA from sensitive and resistant lines, ranging from 0.1 μg to 1 μg per sample were used to drive hybridization reactions that were stopped at corresponding R_0t values. Other reaction conditions and measurement of the extent of hybridization by S1 nuclease hydrolysis are described under "Experimental Procedures." Endogenous S1 resistance of this cDNA preparation was approximately 8%. B, hybridization of partially purified cDNA to RNA from sensitive and resistant cells. The [^{3}H]cDNA recovered in the single- and double-stranded fractions resulting from Step A of the purification procedure outlined in Fig. 4 was hybridized to excess poly(A)-containing RNA from S-3 and AT-3000 cells. Reaction conditions were essentially as described in A. Hybridization to AT-3000 RNA: [^{3}H]cDNA from single- (●——●) and double-stranded (○——○) fractions; hybridization to S-3 RNA: [^{3}H]cDNA from single- (▲——▲) and double-stranded (△——△) fractions. Hybridization of unfractionated [^{3}H]cDNA to S-3 RNA is reproduced from A (□——□).

abundance in sensitive cells) should be driven into hybrids by the excess sensitive cell RNA. However, the majority of the cDNA sequences that are complementary to mRNA sequences present in far greater abundance in resistant cells than in sensitive cells (relative to the 30-fold mass excess of sensitive cell RNA) will remain single-stranded at the end of the reaction. Therefore, based on the estimates described above, these unhybridized cDNA sequences should be greatly enriched for sequences complementary to dihydrofolate reductase mRNA.

Subsequent to the hybridization reaction, single- and double-stranded material was separated by chromatography on hydroxylapatite, RNA removed by alkaline hydrolysis, and the cDNA from both fractions analyzed by hybridization to excess RNA from sensitive and resistant cells. The cDNA

Step A

1. Poly(A)-containing RNA from resistant cells (enriched for dihydrofolate reductase sequences)

↓

[^{3}H]cDNA

↓

2. Hybridized to 30-fold mass excess of poly(A)-containing RNA from sensitive cells (final R_0t = 1600 mol-s/liter)

↓

3. Nonhybridized [^{3}H]cDNA recovered by hydroxylapatite chromatography

Step B

1. [^{3}H]cDNA selected by Step A hybridized to large excess of resistant cell poly(A)-containing RNA (final R_0t = 0.8 mol-s/liter)

↓

2. Hybridized [^{3}H]cDNA recovered by hydroxylapatite chromatography

Fig. 4. Purification of cDNA sequences complementary to dihydrofolate reductase mRNA. *Step A*, hybridization to a limited excess of sensitive cell poly(A)-containing RNA. Approximately 200 ng of [^{3}H]cDNA that was prepared from resistant cell (AT-3000) poly(A)-containing RNA extracted from partially purified dihydrofolate reductase-synthesizing polysomes (see "Experimental Procedures" and text for details) was hybridized to 6 μg of sensitive cell (S-3) poly(A)-containing RNA (30-fold mass excess of RNA). The final reaction volume was 20 microliters, and the other conditions were as described under "Experimental Procedures." The extent to which the preparative reaction approached maximum hybridization was estimated at various times (R_0t values) by measuring the S1 nuclease resistance of control samples identical to the preparative reaction just described except that only 400 pg of [^{3}H]cDNA was used (15,000-fold mass excess of RNA). At an R_0t of 1600 mol-s/liter, a value where the cDNA in the control samples was essentially 100% S1 nuclease-resistant, the preparative sample was diluted with 68 μl of H_2O containing 15 μg each of native and denatured salmon sperm DNA (sheared to 400 base pairs) and 12 μl of 1 M NaPO₄ (final concentration, 0.12 M). Single- and double-stranded material was then fractionated by chromatography on hydroxylapatite essentially as described under "Experimental Procedures." Approximately 23% of the radioactivity failed to bind to the column in 0.12 M NaPO₄. This represented single-stranded material since greater than 93% was sensitive to S1 nuclease digestion. The remainder of the cDNA was eluted in the double-stranded fraction with 0.4 M NaPO₄ and was essentially 100% resistant to S1 nuclease. RNA was removed from the double-stranded fraction by base hydrolysis (see "Experimental Procedures" for details) and the cDNA from each of these fractions was tested by hybridization to excess poly(A)-containing RNA from resistant and sensitive cells. (See text and legend to Fig. 3B for details.) *Step B*, low R_0t fractionation of cDNA-resistant cell poly(A)-containing RNA hybrids. Approximately 30 ng of [^{3}H]cDNA, prepared as described in *Step A*, were hybridized to 165 μg of poly(A)-containing RNA from resistant (AT-3000) cells (approximately 100-fold excess of dihydrofolate reductase-specific RNA sequences) in a 600-μl reaction mixture containing 0.12 M NaPO₄, 1 mM EDTA, and 0.1% sodium dodecyl sulfate. After incubation at 68° to a R_0t of 0.8, the reaction mix was diluted to 3.7 ml with 0.12 M NaPO₄ plus 25 μg of native and denatured salmon sperm DNA. At this point, 43% of the cDNA was resistant to S1 nuclease. Single- and double-stranded material was fractionated by hydroxylapatite chromatography as described above, and approximately 35% of the radioactivity was recovered in the 0.5 M NaPO₄ (double-stranded) fraction. RNA was removed from this fraction by alkaline hydrolysis as described under "Experimental Procedures." Following neutralization, 25 μg of *E. coli* tRNA carrier was added and NaPO₄ was removed by chromatography on Sephadex G-100. The void volume was pooled, concentrated by ethanol precipitation, and dissolved in H_2O.

recovered in the double-stranded fraction should contain sequences present at a similar abundance in both cell types. As expected, this cDNA fraction hybridized to RNA from sensitive (Fig. 3B, △——△) and resistant (Fig. 3B, ○——○) cells with kinetics that were essentially identical to each other and to those with which the unfractionated cDNA preparation hybridized to RNA from sensitive cells (Fig. 3B, □——□). However, most of the cDNA recovered in the single-stranded fraction hybridized to excess RNA from resistant cells (Fig. 3B, ●——●) at a rate approximately 200-fold greater than to that of sensitive cells (Fig. 3B, ▲——▲). This difference is consistent with our estimate of the relative level of dihydrofolate reductase mRNA sequences in these cell types. We recovered approximately 23% of the unfractionated cDNA in the single-stranded fraction, and of this about 65 to 70% had highly accelerated kinetics when hybridized to RNA from resistant cells as opposed to that of sensitive cells. Assuming that the relative abundance of sequences in the cDNA preparation is representative of the mRNA population from which it was derived, this recovery roughly corresponds to that expected for dihydrofolate reductase-specific sequences. The maximum hybridization observed with this cDNA fraction was never above 80%. Presumably, the explanation for this result is that in selecting for single-stranded material after the hybridization described above, we also enrich for any nonhybridizable material present in the unfractionated cDNA preparation.

As a final purification step (*Step B*, Fig. 4), the cDNA selected in Step A was hybridized to a 140-fold mass excess of resistant cell poly(A)-containing RNA, to a final R_0t of 0.8 mol-s/liter. Hybridized sequences were then isolated by chromatography on hydroxylapatite. As can be seen in Fig. 3B, cDNA complementary to RNA sequences that are highly abundant in resistant cells (putative dihydrofolate reductase-specific sequences) are hybridized at this R_0t and are therefore selected. However, both the low level of cDNA sequences that appear to hybridize to less abundant RNA sequences and the nonhybridizable material selected by the previous step are excluded. Approximately 35 to 40% of the cDNA was recovered in the double-stranded fraction in this step. When analyzed by alkaline, isokinetic sucrose gradient centrifugation (see "Experimental Procedures" for details) this material sedimented as a symmetrical peak at 5.4 S, with a calculated size of approximately 350 bases. The specificity of this purified cDNA fraction was then analyzed as described below.

Specificity of the Purified cDNA—The cDNA selected by the final step of the purification procedure (Step B) should represent the portion of the cDNA resulting from the previous step that had highly accelerated kinetics when hybridized to RNA from resistant cells as opposed to that of sensitive cells. As expected, this material still hybridizes to excess poly(A)-containing RNA from resistant cells at a rate approximately 200-fold greater than to that of sensitive cells (Fig. 5). However, these hybridization reactions now approach 100% with kinetics suggestive of a single, pseudo-first order reaction. This result suggests, but does not prove, that the purified cDNA preparation consists mainly of sequences complementary to a single species of mRNA. (See below for further discussion of this point.) Since the purification procedure would enrich for any cDNA sequence complementary to mRNA present at high abundance in the resistant but not the sensitive cells employed in the procedure, we further defined the specificity of the purified cDNA by analyzing the hybridization of this material to poly(A)-containing RNA

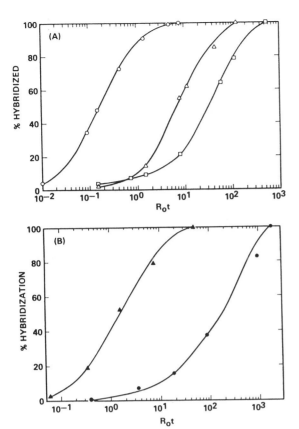

TABLE I

Relative level of dihydrofolate reductase activity, mRNA and gene copies in S-180 and L1210 lines

The origin of each of the lines noted above is described under "Experimental Procedures" and "Results." In each column, values are normalized to those of the sensitive line which was taken as 1. Dihydrofolate reductase-specific activity was assayed as described previously (12). The relative abundance of dihydrofolate reductase mRNA sequences in the S-180 lines was determined from the inverse of $R_0t_{1/2}$ values for the reactions shown in Fig. 5A. The relative number of dihydrofolate reductase gene copies in the S-180 and L1210 lines was determined from the inverse of the $C_0t_{1/2}$ values of the reactions shown in Figs. 8 and 9, respectively. In order to estimate the $C_0t_{1/2}$ of the reaction with L1210S DNA (Fig. 9), we assumed this reaction would proceed to the same extent as the others.

Line	Relative dihydrofolate reductase		
	Specific activity	mRNA sequences	Gene copies
S-180			
S-3	1	1	1
AT-3000	250	220	180
Rev-400	10	7	10
L1210			
S	1		1
RR(+mtx)	35		45
RR(−mtx)	35		35

FIG. 5. *A*, hybridization of purified cDNA to RNA from sensitive, resistant, and partially revertant lines of S-180 cells. Poly(A)-containing RNA isolated from S-3 (0.17 μg to 12 μg/sample, □——□), AT-3000 (0.17 μg to 2.2 μg/sample, ○——○), and Rev-400 (0.17 μg to 5 μg/sample, △——△) were reacted with 30 pg (300 cpm) of the purified [³H]cDNA (selected as described in the legend to Fig. 4) and the extent of hybridization at the indicated R_0t values measured by hydrolysis with S1 nuclease. (See "Experimental Procedures" for details.) Endogenous resistance of the purified cDNA to S1 nuclease hydrolysis was approximately 1.5%. *B*, hybridization of purified cDNA to RNA from sensitive and methotrexate-resistant lines of mouse L1210 lymphoma cells. Poly(A)-containing RNA from the L1210S (0.055 μg to 5.5 μg/sample, ●——●), and L1210 RR500 (0.004 μg ot 4 μg/sample, ▲——▲) cell lines were reacted with 40 pg (400 cpm) of purified [³H]cDNA as described in the legend to Fig. 4.

extracted from several other cell types in which dihydrofolate reductase levels vary widely as a function of methotrexate resistance.

By growing resistant sarcoma 180 cells in the absence of methotrexate for 400 cell doublings (12), we have established a partially revertant line (Rev-400) in which the level of dihydrofolate reductase has declined to an apparently stable[3] value approximately 10-fold greater than that of sensitive cells (Table I). This decrease is also accompanied by a decrease in the relative synthesis of dihydrofolate reductase (12) and the level of translatable dihydrofolate reductase mRNA (13). Comparison of the $R_0t_{1/2}$ value for the reaction of the purified cDNA with RNA from partially revertant cells (Fig. 5a) to those observed in the reactions with sensitive and resistant cell RNA indicates that in each of these lines the abundance of mRNA sequences complementary to the purified cDNA is proportional to the relative level of dihydrofolate reductase (Table I). In addition, we have also examined hybridization of

³ R. Kaufman, unpublished observation.

the purified cDNA to excess RNA from both murine L1210 lymphoma cells, as well as a 25,000-fold methotrexate-resistant subline, L1210 RR500 (Fig. 5b). Relative to the parental line, the RR500 subline has an approximately 80-fold greater level of dihydrofolate reductase activity that is associated with an increase in both the level of dihydrofolate reductase synthesis and translatable dihydrofolate reductase mRNA (data not shown). The data in Fig. 5b indicate that the purified cDNA hybridizes to excess RNA from the L1210 RR500 line at a rate approximately 100-fold greater than that observed with RNA from the sensitive parental line. Therefore, sequences complementary to the purified cDNA are again present at a level proportional to the relative dihydrofolate reductase content of these two cell types. Thus these results, which link the abundance of RNA sequences complementary to the purified cDNA to dihydrofolate reductase levels in a variety of different cell lines, strongly suggest that the purified cDNA preparation consists specifically of sequences complementary to dihydrofolate reductase mRNA. This conclusion is substantiated by two independent lines of evidence which are described below.

We did not size-fractionate either the RNA or the cDNA in the purification procedure. Therefore, another criterion of the specificity of the purified cDNA would be to show that it is specifically complementary to RNA the size of dihydrofolate reductase mRNA. Thus, total RNA from resistant cells was fractionated on isokinetic sucrose gradients, and an equal portion of each fraction was hybridized to purified cDNA under conditions where the per cent hybridization is roughly proportional to the concentration of complementary RNA sequences (30). Other aliquots were used to assay both total and dihydrofolate reductase mRNA activity. As shown by the data in Fig. 6, the purified cDNA hybridizes specifically to a size class of RNA that is distinct from that of total mRNA activity and identical to that of translationally active dihydrofolate reductase mRNA.

Finally, we analyzed the hybridization of the purified cDNA

to both total polysomal RNA from resistant cells, as well as RNA from the same preparation in which dihydrofolate reductase sequences were either enriched or depleted by immunoprecipitation of dihydrofolate reductase-synthesizing polysomes. As shown in Fig. 7, the immunoprecipitation procedure specifically enriched for RNA sequences complementary to the purified cDNA. This result links the specificity of the purified cDNA to the previously demonstrated specificity of the anti-dihydrofolate reductase antibody (12). More importantly, however, the abundance of RNA sequences complementary to the purified probe was directly proportional to the level of dihydrofolate reductase mRNA in each of these fractions (Table II). Therefore, RNA sequences complementary to the purified cDNA are enriched identically to dihydrofolate reductase mRNA by immunoprecipitation of dihydrofolate reductase-synthesizing polysomes.

In summary, in all of the kinetic analyses described above the purified cDNA hybridized with excess RNA to essentially 100% with kinetics suggestive of a single, pseudo-first order reaction. Although this observation suggests that the cDNA is complementary to a single species of mRNA, indistinguishable reaction kinetics would be observed if the cDNA preparation consisted of several different sequences, all of which had complements present at identical abundance in the driver RNA. However, in all RNA preparations that we have examined, the rate at which the purified cDNA hybridized was proportional to the level of dihydrofolate reductase mRNA activity. This was true both in experiments where the abun-

dance of dihydrofolate reductase mRNA varied due to biological factors (*i.e.* in resistant, sensitive, and revertant cells) as well as in experiments where the abundance of these sequences was experimentally manipulated (*i.e.* gradient fractionation or immunoprecipitation). We feel that it is extremely unlikely that any other mRNA would respond identically to translationally active dihydrofolate reductase mRNA with respect to all of these criteria. Therefore, we conclude that our purified cDNA preparation is comprised specifically of sequences complementary to dihydrofolate reductase mRNA.

We have also used the procedure described in Fig. 4 to purify cDNA sequences that were prepared from total poly(A)-containing RNA of resistant (AT-3000) cells that was not further enriched for dihydrofolate reductase mRNA. The cDNA purified in this way was again dihydrofolate reductase-

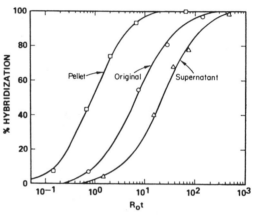

FIG. 7. Hybridization of purified cDNA to RNA extracted from immunoprecipitated dihydrofolate reductase-synthesizing polysomes. Dihydrofolate reductase-synthesizing polysomes were immunoprecipitated from 200 A_{260} units of AT-3000 cell polysomes as described under "Experimental Procedures." Total RNA was extracted from the supernatant and pellet fractions resulting from this procedure, as well as from a reserved sample of the original unfractionated polysomes. RNA from each of these fractions was then reacted with 30 pg (300 cpm) of purified cDNA and the extent of hybridization at the indicated R_0t values determined by hydrolysis with S1 nuclease (see "Experimental Procedures" for details). Hybridization of cDNA to RNA extracted from: pellet (0.115 to 11.5 μg/sample, □——□); supernatant (1.6 to 16 μg/sample, △——△); original polysomes (2.6 μg to 26 μg/sample, ○——○).

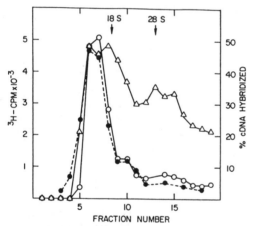

FIG. 6. Hybridization of purified cDNA to size-fractionated resistant cell RNA. Total cytoplasmic RNA from AT-3000 cells (200 μg) was fractionated on isokinetic sucrose gradients as previously described (13). Each fraction was adjusted to contain 0.3 M NaCl and 15 μg of *E. coli* tRNA carrier, and nucleic acids were subsequently precipitated overnight at $-20°$ by the addition of 2 volumes of ethanol. Precipitates were washed twice with 70% ethanol plus 0.1 M NaCl, lyophilized, and dissolved in 100 μl of H_2O. Equal (25 μl) aliquots from each fraction were assayed in the mRNA-dependent reticulocyte lysate system for stimulation of incorporation into total trichloroacetic acid-precipitable material (△——△) and dihydrofolate reductase (○——○) as described under "Experimental Procedures." Other equal aliquots (3 μl) of each fraction were reacted with 25 pg (250 cpm) of purified [³H]cDNA for 45 min in a final reaction volume of 30 μl. Other reaction conditions and measurement of the extent hybridization in each sample (●- - -●) by S1 nuclease hydrolysis are described under "Experimental Procedures." These conditions were devised so that the maximum extent of hybridization in any sample was less than 50%. Therefore, the per cent hybridization of the [³H]cDNA is roughly proportional to the concentration of complementary RNA sequences in the corresponding gradient fraction (30).

TABLE II

Hybridization to partially purified dihydrofolate reductase mRNA

Samples from each of the fractions described in Fig. 7 were also assayed for stimulation of incorporation into dihydrofolate reductase in the mRNA-dependent rabbit reticulocyte lysate system which was then expressed as a per cent of the total stimulated trichloroacetic acid-precipitable radioactivity as described under "Experimental Procedures." Also shown is the inverse of the $R_0t_{1/2}$ values of each of the corresponding hybridization reactions which is proportional to the abundance of complementary sequences in the RNA sample used to drive the reaction. In order to facilitate comparison of the inverse of $R_0t_{1/2}$ and the per cent dihydrofolate reductase synthesis, the values in each column were normalized to the value for the original fraction which was set equal to 1. Normalized values are shown in parentheses.

Sample	$1/R_0t_{1/2}$	% dihydrofolate reductase synthesis
Original	0.14 (1)	3.8 (1)
Supernatant	0.045 (0.32)	1.2 (0.32)
Pellet	1.1 (7.7)	22.5 (5.9)

specific as judged by the criteria described above. This result confirms our assumption that only the level of dihydrofolate reductase mRNA sequences are greatly increased in resistant cells. Furthermore, this result also indicates that the approximately 200-fold increase in the abundance of dihydrofolate reductase mRNA sequences in resistant cells (Fig. 5a) is sufficient to allow purification of dihydrofolate reductase-specific cDNA by this method.

This general approach to specific cDNA purification has been used to purify cDNA sequences complementary to RNA sequences that were absent in mutant cells (31) or viruses (32). Our results indicate that this approach can be extended to situations where it is possible to obtain RNA preparations that have been enriched for specific sequences (for example, by induction or partial purification).

Selective Multiplication of Dihydrofolate Reductase Genes in Resistant Lines

Selective Gene Multiplication in Unstable Lines of Methotrexate-resistant Cells — One possible mechanism consistent with the marked instability of the overproduction of dihydrofolate reductase in methotrexate-resistant lines of Sarcoma 180 cells (12) is selective multiplication of the dihydrofolate reductase structural gene (33). In order to test this possibility, DNA prepared from the nuclei of sensitive, resistant, and revertant cells was denatured and allowed to reanneal in the presence of a trace amount of dihydrofolate reductase specific cDNA. The per cent association at various C_0t values was then determined by fractionation of the double- and single-stranded material on hydroxylapatite. We detected no significant differences in the renaturation of the driver DNA from each of these cell types (Fig. 8, - - -) and, in all of these reactions, association of the dihydrofolate reductase-specific cDNA went to approximately 80 to 85% completion with kinetics characteristic of a unique, second order reaction. Association of the purified cDNA with sensitive cell DNA (Fig. 8, O——O) occurred over roughly the same C_0t range as observed for renaturation of the unique sequence fraction of the genomic DNA suggesting that dihydrofolate reductase genes are present, on the average, at no more than a few copies per cell in this line. However, the dihydrofolate reductase-specific cDNA associated with DNA from resistant cells (Fig. 8, △——△) at a rate approximately 200-fold greater than that with which it associated with sensitive cell DNA. In addition, similar relative rates were obtained when these reactions were assayed by S1 nuclease hydrolysis (data not shown). Thus, the dihydrofolate reductase structural gene is selectively multiplied approximately 200-fold in resistant cells, a level roughly in proportion to the relative increase in the content of dihydrofolate reductase and dihydrofolate reductase mRNA in this variant line (Table I).

Analysis of the association kinetics of the specific cDNA to DNA from partially revertant cells (Fig. 8, □——□) indicates that the dihydrofolate reductase gene copy number is unstable in resistant cells. Comparison of the kinetics of this reaction to those observed for the reaction of the cDNA to DNA from resistant and sensitive cells (Fig. 8, Table I) demonstrates that the number of dihydrofolate reductase gene copies in the partially revertant line has declined to a value approximately 10-fold greater than that of the sensitive cells. Once again, this value is proportional to the level of dihydrofolate reductase in revertant cells relative to sensitive and resistant cells (Table I).

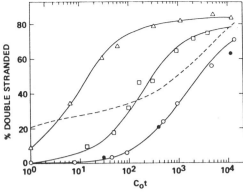

FIG. 8. Association kinetics of purified cDNA with DNA from sensitive, resistant, and partially revertant lines of S-180 cells. DNA was prepared from S-3, AT-3000, and Rev-400 as described under "Experimental Procedures." An aliquot was removed from the S-3 preparation at a point in the preparation immediately preceding NaOH treatment and adjusted to contain 20 μg of poly(A)-containing RNA from resistant cells per mg of DNA. This sample was then processed through the final DNA preparation steps identically to the others. Based on the yield of RNA and DNA from these lines, this ratio of added RNA to DNA approximately represents the relative level of RNA to DNA in these cells. DNA from each of these preparations was melted and allowed to reanneal in the presence of a trace amount of purified [³H]cDNA and the extent of association at various C_0t values measured by chromatography on hydroxylapatite (see "Experimental Procedures" for details). When incubated in the absence of driver DNA, approximately 2% of the [³H]cDNA was retained by hydroxylapatite. Total and double-stranded recoveries of [³H]cDNA from each sample were corrected for this value before calculation of the present double-stranded. The reassociation of the driver DNA from each of these preparations occurred with essentially identical kinetics which are summarized by the *dashed line*. Association of purified [³H]cDNA with DNA from S-3, O——O; S-3 processed with added poly(A)-containing RNA from AT-3000, ●——●; AT-3000, △——△; and Rev-400, □——□.

In order to show that contamination of the DNA preparations by cellular RNA could not have artifactually led to these results, we demonstrated that the rate with which the dihydrofolate reductase-specific probe associates with sensitive cell DNA was not affected by the addition of a large excess of resistant cell poly(A)-containing RNA to the sensitive cell DNA at a point in the DNA preparation procedure immediately preceding base treatment (Fig. 8, ●——●). Since the amount of added RNA represented considerably more than the maximum possible level of RNA contamination at this point (see legend to Fig. 8), this experiment demonstrates that the base hydrolysis step is sufficient to remove any contaminating RNA.

Selective Gene Multiplication in Stable Lines of Methotrexate-resistant Cells — The 5000-fold methotrexate-resistant L1210RR line of murine L1210 lymphoma cells contains an approximately 35-fold increase in the level of dihydrofolate reductase relative to the sensitive, parental line (Table I). We have also shown that this increase is associated with an increase in the rate of dihydrofolate reductase synthesis and the level of dihydrofolate reductase mRNA activity (data not shown). Increased dihydrofolate reductase levels appear to be a stable property of this resistant line, since we find no significant decrease in this parameter after more than 100 cell doublings in the absence of methotrexate (Table I). The stability of increased dihydrofolate reductase levels observed when these and other lines of methotrexate-resistant cells were grown in the absence of methotrexate (15) suggested that the alteration leading to increased enzyme synthesis might be a regulatory mutation. In order to test the generality of the selective gene multiplication phenomenon, we quantitated the relative number of dihydrofolate reductase genes in

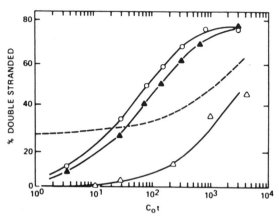

FIG. 9. Association kinetics of purified cDNA with DNA prepared from sensitive and methotrexate-resistant lines of L1210 cells. DNA prepared from various lines of L1210 cells was denatured and allowed to reanneal in the presence of a trace amount of purified [³H]cDNA, and the extent of association at indicated C_0t values determined by chromatography on hydroxylapatite (see "Experimental Procedures" for details). Reassociation of the driver DNA summarized for all three reactions, - - -; association of the purified [³H]cDNA with DNA from L1210S, △——△; L1210RR, ○——○; and L1210RR grown for approximately 100 cell doublings in the absence of methotrexate, ▲——▲.

the L1210 lines just described. The data in Fig. 9 indicate that the dihydrofolate reductase-specific cDNA associates with DNA from the L1210RR line grown in the presence of methotrexate (○——○) at a rate approximately 45-fold more rapid than that observed with DNA from the sensitive parental line (△——△), indicating that the relative number of dihydrofolate reductase genes is approximately 45-fold greater in the resistant line. Again, the relative number of dihydrofolate reductase gene copies is roughly proportional to the relative level of dihydrofolate reductase in these two lines (Table I). We observed only a slight, and probably not significant, decrease (20 to 25%) in the rate with which the probe associated to DNA from the L1210RR line that had been grown in the absence of methotrexate. Therefore, the dihydrofolate reductase gene copy number appears to be relatively stable in this line of methotrexate-resistant cells (Table I).

Thermal Stability of Duplexes between Dihydrofolate Reductase-specific cDNA and DNA from Different Cell Types — The thermal denaturation characteristics of duplexes formed between dihydrofolate reductase-specific cDNA and DNA from either sensitive cells, resistant cells, or mouse liver are essentially indistinguishable (Fig. 10a). The T_m values for these reactions range between 81.5° and 82.5°, and in each, the melting profile occurs as a single transition over a relatively narrow temperature range. The T_m of the driver DNA was similar for DNA from S-180 cells ($T_m = 84$) and human placenta ($T_m = 82$) (Fig. 10b). These reactions also proceeded to the same extent (see legend to Fig. 10). However, although the human DNA contains sequences which can anneal with dihydrofolate reductase-specific cDNA prepared from a murine cell line, the extent of duplex formation (see legend to

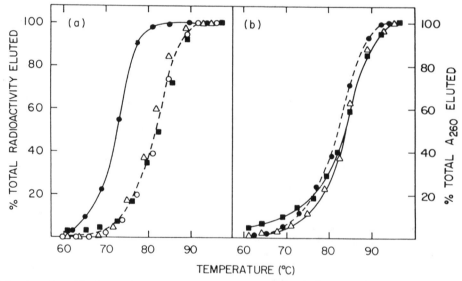

FIG. 10. Thermal denaturation of duplexes formed between dihydrofolate reductase-specific DNA and DNA from various sources. 100 pg of dihydrofolate reductase-specific [³H]cDNA (1000 cpm) was annealed with approximately 1 mg of DNA from AT-3000 (resistant) cells (final $C_0t = 1000$), S-3 (sensitive) cells (final $C_0t = 20,000$), mouse liver (final $C_0t = 14,000$), aborted human placenta (final $C_0t = 10,000$), and chicken oviduct (final $C_0t = 10,000$). The final reaction volume was 200 μl, the temperature 68°, and other conditions were as described under "Experimental Procedures." At the indicated C_0t values, reactions were diluted into 5 ml of 0.12 M NaPO₄ and adsorbed to 1-g columns of hydroxylapatite which were maintained at 60° with a recirculating water bath. The column was washed with 5 ml of 0.12 M NaPO₄ at 60° and subsequently the

temperature of the column and wash buffer was raised to 97° in increments of approximately 3°. At each step, the washing procedure was repeated. The resulting fractions were monitored for A_{260} and trichloroacetic acid-precipitable radioactivity as described under "Experimental Procedures." The final percentage of the driver DNA and [³H]cDNA, respectively, that were recovered as double-stranded in each reaction are listed in parentheses below following the appropriate reaction symbols. *Panel a* shows elution of [³H]cDNA and panel *b* show elution of driver DNA from reactions driven with DNA from: AT-3000, ○——○ (50, 82); S-3, △——△ (77, 60); mouse liver, ■——■ (77, 56); human placenta, ●——● (80, 26), and chicken oviduct (85, 0). Data are presented as the cumulative elution of DNA with increasing temperature.

Fig. 10) and the stability of the duplexes (T_m = 72°) was considerably less than those formed with DNA from murine sources (Fig. 10a). Under the relatively stringent conditions used for these reactions, we observed no duplex formation between dihydrofolate reductase-specific cDNA and chicken oviduct DNA (see legend to Fig. 10).

These results suggest that there is little difference in the nucleotide sequence of individual dihydrofolate reductase genes in resistant cells and, furthermore, that these sequences have diverged little from homologous sequences in sensitive cells or mouse liver (confirming the murine origin of the multiplied dihydrofolate reductase genes in resistant cells). Although there is significant divergence between the nucleotide sequences of the human and murine genes, there appears to be sufficient homology to allow use of the murine probe to analyze methotrexate resistance in human tumors.

DISCUSSION

We have shown that in methotrexate-resistant lines of Sarcoma 180 and L1210 murine lymphoma cells increased synthesis of dihydrofolate reductase is associated with increased copies of the dihydrofolate reductase structural gene. More recently we have also found that increased dihydrofolate reductase synthesis in methotrexate-resistant 3T6 cells is also accompanied by a corresponding increase in the number of dihydrofolate reductase gene copies.[4] These methotrexate-resistant lines represent the first reported examples of mammalian cells in which a structural gene that codes for a protein is selectively multiplied.

In order to understand the processes involved with the selective multiplication of dihydrofolate reductase genes in the resistant lines, we should first examine the role of methotrexate in this phenomenon. This folic acid analogue strongly and specifically inhibits dihydrofolate reductase in a competitive manner (34), and therefore indirectly inhibits the *de novo* synthesis of purines, thymidylate, and glycine (35). Hence, exposure to sufficiently high concentrations of methotrexate kills dividing cells. Resistance to this analogue has been found to result from any of several mechanisms (36), but the most frequently reported for mammalian cells is an increased cellular content of dihydrofolate reductase (2–11). In this case resistant cells accumulate sufficiently high dihydrofolate reductase levels to maintain some free enzyme activity in the presence of the drug (37). Highly resistant lines with greatly increased levels of dihydrofolate reductase (such as those described in this report) have never been selected in a single step. The common method for obtaining such lines involves either gradually increasing the concentration of methotrexate in the medium (3) or progressing in several steps (6), each step using a 10- to 20-fold greater concentration of the drug than is required to inhibit growth by 50%. In the latter case it is possible to estimate the frequency of resistant variants in the cell population. It has been our observation, as well as those of other laboratories (6, 38), that in the initial step this frequency is low (less than 1 in 10^6).

Several lines of evidence suggest that methotrexate does not act directly to induce or maintain the increased synthesis of dihydrofolate reductase in resistant lines. 1. Simple exposure of cells to methotrexate (without selection) has no effect on dihydrofolate reductase synthesis (12). 2. In the methotrexate-resistant lines of L1210 cells that we have studied, as well as a number of other resistant lines (2, 15), increased dihydrofolate reductase synthesis is a stable property and does not decline when cells are grown in the absence of the drug. 3. The kinetics of the decrease in dihydrofolate reductase synthesis observed when unstable lines of resistant cells are grown in the absence of methotrexate do not correspond to dilution of the drug from the cells (12). 4. The decrease in dihydrofolate reductase synthesis observed when unstable lines are grown in the absence of methotrexate is also observed when cells are grown in the continued presence of the drug, but supplemented with a purine source, thymidine, and glycine.[3] Biedler *et al.* (39) have suggested that methotrexate may have mutagenic properties, possibly due to the inhibition of the synthesis of nucleic acid precursors. Such properties could influence the rate (or mechanism) with which resistant variants arise. However, fluctuation analyses done with L1210 cells indicated that in this line methotrexate-resistant variants are generated spontaneously during growth in the absence of the drug (40). In addition, this drug was found to have no mutagenic properties as judged by the *Salmonella* microsome test (41). Therefore, in summary, all available evidence indicates that methotrexate acts only as a selective agent and has no direct role in the resistance (gene multiplication) process.

We propose that exposure of sensitive cells to methotrexate selects for those cells in the population harboring spontaneous multiplications (duplications) of the dihydrofolate reductase structural gene and as a result, increased levels of dihydrofolate reductase. Of course, there are many other conceivable genetic alterations that could lead to increased dihydrofolate reductase levels, including those generating an absolute increase in the transcription rate of the gene or an increased stability of the specific mRNA. However, in all of the lines that we have studied (including 3T6 lines), we observe a proportionality between the relative level of dihydrofolate reductase activity and the relative number of dihydrofolate reductase gene copies (Table I). This result suggests that there is little difference between the activities of individual dihydrofolate reductase genes in sensitive and resistant cells, and that selective gene multiplication is the most important, if not the only mechanism leading to increased dihydrofolate reductase accumulation by these highly resistant lines. A possible explanation for the predominance of this mechanism is that the dihydrofolate reductase gene is expressed at or near the maximum possible activity in sensitive cells, and therefore no type of genetic alteration could greatly increase this activity. Alternatively, the events which lead to duplication or multiplication of dihydrofolate reductase genes may occur at a higher frequency than other types of genetic alterations that would lead to increased expression of a limited number of gene copies.

Clearly, in order to understand the mechanism by which these genes are multiplied, as well as why their number is relatively stable in some lines and not in others, it will be necessary to determine the location and molecular arrangement of the multiple gene copies in the various cell lines. Are they chromosomal or extrachromosomal, and do they exist in tandem arrays or at many locations in the genome? An interesting observation that may reflect on these questions was made by Biedler and her colleagues who consistently detected the appearance of a large homogeneously staining

---

[4] R. E. Kellems, F. W. Alt, and V. Morhen, unpublished observation.

region associated with specific chromosomes of highly metho-trexate-resistant lines of Chinese hamster lung cells (42).[5] In addition, resistance and corresponding high dihydrofolate reductase levels were unstable when these lines were grown in the absence of methotrexate; and, significantly, the size of the chromosomal alteration decreased in parallel to the decrease in enzyme activity (14). It is tempting to speculate that such a chromosomal alteration might correspond to a tandem array of dihydrofolate reductase genes. In some of their resistant lines the specifically altered region represented as much as 6% of the chromosomal complement (14), considerably more than would be necessary to account for an increase in dihydrofolate reductase gene copy number corresponding to the increased enzyme content of the line (approximately 200-fold). However, in bacteria, selected duplication of a specific gene can extend far beyond the vicinity of that gene and involve as much as 20% of the bacterial chromosome (43). If genes other than those coding for dihydrofolate reductase are multiplied in the resistant Sarcoma 180 lines that we have studied, they apparently are not expressed. As judged by both comparison of proteins synthesized by sensitive and resistant cells (12), as well as by the specificity of the cDNA purification procedure (see above), the large increase in dihydrofolate reductase synthesis appears to be unique.

De novo duplication of specific genes in bacteria and phages occurs with relatively high frequencies (43–47); and in these cases duplications appear to be in tandem. A well known example of tandem duplications in eukaryotic cells occurs as a result of unequal crossing over at the bar locus in *Drosophila* (48, 49). If the initial event selected in methotrexate resistance were a tandem duplication of the dihydrofolate reductase gene or alternatively if the genes already existed in multiple, tandem copies in sensitive cells, expansion of the tandem array might occur by homologous but unequal crossover events between dihydrofolate reductase genes on homologous chromosomes (50). However, a more likely mechanism would involve unequal exchanges between sister chromatids. Sister chromatid exchange has been demonstrated to occur in a variety of organisms (51–53) and in *Drosophila* unequal sister chromatid exchanges presumably lead to changes in the number of tandem repeats at the bar locus (54) as well as the number of ribosomal genes at the bobbed locus (55). This process has been discussed in detail as a mechanism for the evolution of repeated DNA sequences (56), the coincidental evolution of members of multi-gene families (57, 58), and the magnification-reduction of the ribosomal gene copy number in *Drosophila* (58, 59). One attractive feature of such a mechanism for the selective multiplication of dihydrofolate reductase genes in methotrexate-resistant cells is that it would be consistent with the multi-step selection procedure necessary to generate these lines.

Alternatively, selective multiplication of dihydrofolate reductase genes may occur by a mechanism which at least initially generates extrachromosomal copies of the multiplied genes. The classic example of such a process in eukaryotic cells is the amplification of ribosomal genes in amphibian oocytes (60). In this case, amplification is specifically regulated as part of a developmental sequence and occurs extrachromosomally, apparently by a rolling circle replication mechanism (61). Other possible amplification mechanisms include reverse transcription of the specific mRNA (62, 63) or disproportionate replication of specific genes (64). The former mechanism may be involved in the production of extrachromosomal copies of mouse mammary tumor virus genes (65) while the latter has recently been implicated in the production of large numbers of extrachromosomal copies of SV40 DNA from the integrated viral genome.[6] One common feature of these mechanisms is that large increases in the number of specific genes might be obtained in a single selective step. In this regard, it will be interesting to measure the absolute number of dihydrofolate reductase gene copies in sensitive cells, and the maximum increase in that number obtainable in a single step.

A selective increase in the number of dihydrofolate reductase genes in resistant cells might also be achieved by the retention of specific chromosomal fragments. Although there are apparently no specific differences between the karyotypes of sensitive and resistant lines (38),[7] chromosome transfer experiments indicate that chromosomal fragments retained by host cells are frequently so small that they may be cytologically undetectable (66, 67). Similarly to dihydrofolate reductase genes in unstable lines, such transferred genetic elements are usually lost rapidly from host cells (1 to 10% loss per generation) (68–71), but can be maintained indefinitely by growth under appropriate selective conditions (70). In addition, prolonged growth of host cells under selective conditions leads to the emergence of lines which stably express the transferred characteristic (66, 70).

Finally, by analogy to bacterial systems, duplication or subsequent multiplication of specific genes (by many of the mechanisms considered above) may also be promoted by flanking sequences (*e.g.* translocatable elements or viral sequences) (72, 73). Such sequences (insertion elements) may be involved in the accumulation of R-factors containing multiple r-determinant segments in chloramphenicol-resistant lines of *Proteus mirabilis*. This phenomenon also shares many features with methotrexate-resistance in Sarcoma 180 cells. High levels of chloramphenicol resistance are unstable in the absence of selection and result from increased production of chloramphenicol transacetylase in association with the selective multiplication of the r-determinant carrying the gene for this enzyme (74, 75).

An intriguing question is why the multiple gene copies are stable in some lines and unstable in others. One possibility is that the multiplication process occurs by a different mechanism in these lines, but recent results indicate that this need not be the case. After growth in the presence of methotrexate for an additional 2 years, the highly unstable lines of metho-trexate-resistant S-180 cells described previously (12) appear to have become much more stable.[3] This phenomenon can be explained as follows: whatever the mechanism of gene loss (see discussion below), unstable lines of resistant cells are presumably constantly generating cells with decreased numbers of dihydrofolate reductase genes. Growth of such lines in the presence of methotrexate would maintain a certain average level of dihydrofolate reductase gene copies per cell by eliminating those cells in which the gene copy number (and correspondingly dihydrofolate reductase levels) had decreased below that necessary for survival. Under these conditions,

---

[5] A similar chromosomal alteration was recently observed in methotrexate-resistant Chinese hamster ovary cells. L. Chasin, personal communication.

[6] M. Botchan, personal communication.

[7] J. Nunberg and R. Kaufman, unpublished observation.

cells in which the gene copy number had become more stable would have an obvious selective advantage (more of their progeny would survive) and eventually outgrow the population.

Loss of chromosomal genes might be associated with specific chromosomal deletions or fragmentations (70). In addition, if the multiple gene copies exist in clusters of tandem repeats in resistant lines, instability in their numbers could be due to the same general types of processes which were considered above for their multiplication. Thus, unequal crossover events would generate as reciprocal products both a cell with increased numbers of dihydrofolate reductase gene copies and one in which the number was reduced; a decrease in the average number of dihydrofolate reductase gene copies per cell would result if in the absence of methotrexate, cells which devoted less of their energy to the production of unnecessarily high levels of dihydrofolate reductase had a selective growth advantage. Tandem duplications in bacterial cells are usually quite unstable (43, 44, 46), presumably due to crossover events between repeats on the same chromosome. Loss of dihydrofolate reductase genes might occur by a similar process, and by analogy to bacterial systems in which such repeats are much more stable in Rec A$^-$ lines (43, 44), stabilization could result from the loss of an enzymatic function that was involved in their excision or exchange. Stabilization might also occur by the inactivation of flanking sequences (by excision or mutation) involved in the multiplication process or by translocation of clustered genes to multiple sites in the genome. Extrachromosomal genes might be lost by a number of different mechanisms. Unstable genetic elements resulting from chromosome transfer experiments (see above) are thought to be extrachromosomal, and recent evidence suggests that stability results from integration of the transferred fragment into the genome of the host cell (67). Stability of extrachromosomal genes, whatever their origin, might be achieved through such a mechanism.

All of our studies were done with murine cell lines; therefore in order to assess the generality of the selective gene multiplication phenomenon it will be necessary to know the mechanism of increased dihydrofolate reductase accumulation in cell lines derived from other organisms (5, 15). The selection of cell lines resistant to highly specific inhibitors of other key enzymes should allow extension of this approach to many different genes. For example, Kempe *et al.* have shown that in certain hamster cell lines, resistance to a specific inhibitor of aspartate transcarbamylase is associated with increased cellular content of that enzyme (76). More recently, this group has found that resistant lines synthesize the enzyme at a greater rate and contain increased levels of the specific mRNA.[8] It will be interesting to know if these lines also contain increased numbers of aspartate transcarbamylase gene copies.

We do not know if the processes leading to selective multiplication of dihydrofolate reductase genes in the permanent cell lines that we have studied have a role in normal cells. Certainly, a mechanism for generating spontaneous and random duplications of genetic material might be important for evolutionary flexibility (77), as well as for the generation of multigene families (33). The unstable lines of resistant Sarcoma 180 cells may, in fact, provide a good model system for studying the evolution and maintenance of multi-gene families; since under appropriate growth conditions it is possible to select lines in which the dihydrofolate reductase gene copy number is increased, decreased, or fixed. In systems where it has been studied, selective gene multiplication as a mechanism for the synthesis of large amounts of differentiated cell proteins has not been observed (78–80). However, other lines of evidence suggest that various types of genomic alterations including duplications, deletions, and translocations may underlie a number of controls of differentiation (see Ref. 81 for further discussion of this point). Our results add further evidence to support the concept that the genome of higher organisms is not constant, but can undergo a variety of changes.

Acknowledgments — We are indebted to Henry Burr and Marvin Wickens for their advice and assistance, to Andi Justice for growing the cells used for many of these experiments, and to Keiko Nakanishi Alt for assistance in the preparation of this manuscript. We are especially grateful to Gray Crouse and Randy Kaufman for advice and thoughtful criticism throughout the course of this work.

[8] R. Padgett, G. Wahl, and G. Stark, personal communication.

REFERENCES

1. Bertino, J. R., Donohue, D. M., Simmons, B., Gabrio, B. W., Silber, R., and Huennekens, F. M. (1963) *J. Clin. Invest.* 42, 466–475
2. Fischer, G. A. (1961) *Biochem. Pharmacol.* 7, 75–80
3. Hakala, M. T., Zakrzewski, S. F., and Nichol, C. A. (1961) *J. Biol. Chem.* 236, 952–958
4. Kashet, E. R., Crawford, E. J., Friedkin, M., Humphreys, S. R., and Golding, A. (1964) *Biochemistry* 3, 1928–1931
5. Littlefield, J. W. (1969) *Proc. Natl. Acad. Sci. U. S. A.* 62, 88–95
6. Friedkin, M., Crawford, E. S., Humphreys, S. R., and Golding, H. (1962) *Cancer Res.* 22, 600–606
7. Perkins, J. P., Hillcoat, B. L., and Bertino, J. R. (1967) *J. Biol. Chem.* 242, 4771–4776
8. Sartorelli, A. C., Both, B. A., and Bertino, J. R. (1964) *Arch. Biochem. Biophys.* 108, 53–59
9. Jackson, R. C., and Huennekens, F. M. (1973) *Arch. Biochem. Biophys.* 154, 192–198
10. Courtenay, V. D., and Robins, A. B. (1972) *J. Natl. Cancer Inst.* 49, 45–53
11. Biedler, J. L., Albrecht, A. M., Hutchison, D. J., and Spengler, B. A. (1972) *Cancer Res.* 32, 153–161
12. Alt, F. W., Kellems, R. E., and Schimke, R. T. (1976) *J. Biol. Chem.* 251 3063–3074
13. Kellems, R. E., Alt, F. W., and Schimke, R. T. (1976) *J. Biol. Chem.* 251, 6987–6993
14. Biedler, J. L., and Spengler, B. A. (1976) *J. Cell Biol.* 70, 117a
15. Nakamura, H., and Littlefield, J. W. (1972) *J. Biol. Chem.* 247, 179–187
16. Hanggi, U. J., and Littlefield, J. W. (1976) *J. Biol. Chem.* 251, 3075–3080
17. Chang, S. E., and Littlefield, J. W. (1976) *Cell* 7, 391–396
18. Thompson, L. H., and Baker, R. M. (1973) *Methods Cell Physiol.* 6, 209–281
19. Demars, R. (1974) *Mutat. Res.* 24, 335–364
20. Aviv, H., and Leder, P. (1972) *Proc. Natl. Acad. Sci. U. S. A.* 69, 1408–1412
21. Palacios, R., Palmiter, R. D., and Schimke, R. T. (1972) *J. Biol. Chem.* 247, 2316–2321
22. Shapiro, D. J., Taylor, J. M., McKnight, G. S., Palacios, R., Gonzalez, C., Kiely, M. L., and Schimke, R. T. (1974) *J. Biol. Chem.* 249, 3665–3671
23. Taylor, J. M., and Schimke, R. T. (1974) *J. Biol. Chem.* 249, 3597–3601
24. Pelham, H. R. B., and Jackson, R. J. (1976) *Eur. J. Biochem.* 67, 247–256
25. Buell, G., Wickens, M., Payvar, F., and Schimke, R. T. (1978) *J. Biol. Chem.*, in press
26. Britten, R. J., Graham, D. E., and Neufeld, B. R. (1974) *Methods Enzymol.* 29, 363–418
27. McCarty, K. S., Jr., Volmer, R. T., and McCarty, K. S. (1974)

Anal. Biochem. **61**, 165–183

28. Studier, W. F. (1965) *J. Mol. Biol.* **11**, 373–390
29. Rhoads, R. E., McKnight, G. S., and Schimke, R. T. (1973) *J. Biol. Chem.* **248**, 2031–2039
30. Fan, H., and Baltimore, D. (1973) *J. Mol. Biol.* **80**, 93–117
31. Ramirez, F., Nutter, C., O'Donnel, J. V., Canale, V., Bailey, G., Sangvensermsvi, T., Maniatis, G., Marks, P., and Bank, A. (1975) *Proc. Natl. Acad. Sci. U. S. A.* **72**, 1550–1554
32. Stehelin, D., Guntaka, R., Varmus, H., and Bishop, J. M. (1976) *J. Mol. Biol.* **101**, 349–365
33. Hood, L., Campbell, J. H., and Elgin, S. C. R. (1975) *Ann. Rev. Genet.* **9**, 306–353
34. Werkeiser, W. C. (1961) *J. Biol. Chem.* **236**, 888–893
35. Huennekens, F. M. (1963) *Biochemistry* **2**, 151–159
36. Blakeley, R. L. (1969) *The Biochemistry of Folic Acid and Related Pteridines,* pp. 139–187, North Holland, Amsterdam
37. Hakala, M. T. (1965) *Biochim. Biophys. Acta* **102**, 198–209
38. Hakala, M. T., and Ishihara, T. (1962) *Cancer Res.* **22**, 987–996
39. Biedler, J. L., Albrecht, A. M., and Hutchinson, D. J. (1965) *Cancer Res.* **25**, 246–257
40. Law, L. W. (1952) *Nature* **169**, 628–629
41. Benedict, W. F., Baker, M. S., Haroun, L., Choi, E., and Ames, B. N. (1977) *Cancer Res.* **37**, 2209–2213
42. Biedler, J. L., and Spengler, B. A. (1976) *Science* **191**, 186–187
43. Anderson, R. P., Miller, C. G., and Roth, J. R. (1976) *J. Mol. Biol.* **105**, 201–218
44. Folk, W. R., and Berg, P. (1971) *J. Mol. Biol.* **58**, 595–610
45. Hill, C. W., and Combriato, G. (1973) *Mol. & Gen. Genet.* **127**, 197–214
46. Hariuchi, T., Hariuchi, S., and Novick, A. (1963) *Genetics* **48**, 157–169
47. Emmons, S. W., MacCosham, U., and Baldwin, R. L. (1975) *J. Mol. Biol.* **91**, 133–146
48. Sturtevant, A. H. (1925) *Genetics* **10**, 117–147
49. Bridges, C. B. (1936) *Science* **83**, 210–211
50. Stern, C. (1936) *Genetics* **21**, 625–730
51. Taylor, J. H., Woods, P. S., and Hughs, W. L. (1957) *Proc. Natl. Acad. Sci. U. S. A.* **43**, 122–128
52. Marin, G., and Prescott, D. M. (1964) *J. Cell Biol.* **21**, 159–167
53. McClintock, B. (1941) *Cold Spring Harbor Symp. Quant. Biol.* **9**, 72–80
54. Peterson, H. M., and Laughnan, J. R. (1963) *Proc. Natl. Acad. Sci. U. S. A.* **50**, 126–133
55. Schalet, A. (1969) *Genetics* **63**, 133–153
56. Smith, G. P. (1976) *Science,* **191**, 528–535
57. Smith, G. P. (1973) *Cold Spring Harbor Symp. Quant. Biol.* **38**, 507–513
58. Tartof, K. D. (1973) *Cold Spring Harbor Symp. Quant. Biol.* **38**, 491–500
59. Tartof, K. D. (1974) *Proc. Natl. Acad. Sci. U. S. A.* **71**, 1272–1276
60. Brown, D. D., and Dawid, I. B. (1968) *Science* **160**, 272–280
61. Hourcade, D., Dressler, D., and Wolfson, J. (1973) *Cold Spring Harbor Symp. Quant. Biol.* **38**, 537–550
62. Baltimore, D. (1970) *Nature* **226**, 1209–1211
63. Temin, H., and Mizutani, S. (1970) *Nature* **226**, 1211–1213
64. Tartof, K. D. (1975) *Ann. Rev. Genet.* **9**, 370
65. Ringold, G. M., Yamamoto, K. R., Shaulo, P. R., and Varmus, H. E. (1977) *Cell* **10**, 19–26
66. Willecke, K., Lange, R., Kruger, A., and Reber, T. (1976) *Proc. Natl. Acad. Sci. U. S. A.* **73**, 1274–1278
67. Fournier, R. E. K., and Ruddle, F. H. (1977) *Proc. Natl. Acad. Sci. U. S. A.* **74**, 3937–3941
68. McBride, O. W., and Ozer, H. L. (1973) *Proc. Natl. Acad. Sci. U. S. A.* **70**, 1258–1262
69. Willecke, K., and Ruddle, F. H. (1975) *Proc. Natl. Acad. Sci. U. S. A.* **72**, 1792–1796
70. Degnen, G. E., Miller, I. L., Eisenstadt, J. M., and Adelberg, E. A. (1976) *Proc. Natl. Acad. Sci. U. S. A.* **73**, 2838–2842
71. Spandidos, D. A., and Siminovitch, L. (1977) *Proc. Natl. Acad. Sci. U. S. A.* **74**, 3480–3484
72. Cohen, S. N. (1976) *Nature* **263**, 731–738
73. Kleckner, N. (1977) *Cell* **11**, 11–23
74. Rownd, R., Kasamatu, H., and Michel, S. (1971) *Ann. N. Y. Acad. Sci.* **182**, 188–206
75. Perlman, D., and Rownd, R. H. (1975) *J. Bacteriol.* **123**, 1013–1034
76. Kempe, T. D., Swyrd, E. A., Bruist, M., and Stark, G. R. (1976) *Cell* **9**, 541–550
77. Ohno, S. (1970) *Evolution of Gene Duplication,* Springer-Verlag, New York
78. Packman, S., Aviv, H., Ross, J., and Leder, P. (1972) *Biochem. Biophys. Res. Commun.* **49**, 813–819
79. Suzuki, Y., Gage, L. P., and Brown, D. D. (1972) *J. Mol. Biol.* **70**, 637–649
80. Sullivan, D., Palacios, R., Stavezer, J., Taylor, J. M., Faras, A. J., Kiely, M. L., Summers, N. M., Bishop, J. M., and Schimke, R. T. (1973) *J. Biol. Chem.* **248**, 7530–7539
81. Schimke, R. T., Alt, F. W., Kellems, R. E., Kaufman, R., and Bertino, J. R. (1977) *Cold Spring Harbor Symp. Quant. Biol.* **42**, in press

Somatic Expression of Herpes Thymidine Kinase in Mice Following Injection of a Fusion Gene into Eggs

R.L. Brinster, H.Y. Chen, M. Trumbauer, A.W. Senear, R. Warren and R.D. Palmiter

Summary

A plasmid, designated pMK, containing the structural gene for thymidine kinase from herpes simplex virus (HSV) fused to the promoter/regulatory region of the mouse metallothionein-I gene, was injected into the pronucleus of fertilized one-cell mouse eggs; the eggs were subsequently reimplanted into the oviducts of pseudopregnant mice. The first experiment produced 19 offspring, one of which expressed high levels of HSV thymidine kinase activity in the liver and kidney. pMK DNA sequences were detected in equal amounts in several tissues of the expressing mouse as well as in three mice that did not express HSV thymidine kinase activity. In all cases, several copies of the pMK plasmid were tandemly duplicated and integrated into mouse DNA. It appears as though multiple copies of the intact plasmid were fused by homologous recombination either before or after integration at a single site in the mouse genome. The overall efficiency of obtaining somatic expression of thymidine kinase in experiments performed to date is about 10% (4/41), and twice this number have integrated pMK DNA. This procedure not only provides a means of introducing new genes into mice, but it will also be a valuable system for studying tissue-specific regulation of gene expression.

Introduction

The recently developed techniques for introducing purified genes into cells via transfection or injection provide powerful tools for examining the DNA sequences required for normal gene expression and regulation. With appropriate vectors, these techniques are limited only by the availability of cell lines that can be propagated or by the sensitivity of the assays for gene expression. Using these techniques, investigators have introduced several eucaryotic genes into foreign cells, and their expression has been demonstrated at the nucleic acid level, protein level, or both (Mulligan et al., 1979; Wigler et al., 1979; DeRobertis

and Olson, 1979; Capecchi, 1980; Grosschedl and Birnstiel, 1980; Lai et al., 1980). Furthermore, there are recent reports of regulation of gene expression in these systems (Kurtz, 1981; Buetti and Diggelmann, 1981; Hynes et al., 1981). For other genes, we anticipate that expression or regulation may be difficult to demonstrate because the appropriate cell type may not be amenable to analysis in culture, or developmental programming may be essential. To overcome these potential problems, it would be desirable to introduce genes into embryos and then to analyze gene expression in differentiated adult cells. Several approaches have been explored to achieve this goal.

One approach for introducing genes into animals has been to inject foreign cells, for example, teratocarcinoma cells, into mouse preimplantation blastocysts and then to reimplant them into pseudopregnant mice. In several cases this has resulted in phenotypic expression of the genes derived from the foreign cell in several tissues of the adult mouse, including the germ line (Brinster, 1974; Mintz and Illmensee, 1975; Papaioannou et al., 1975). Since these foreign cells can participate in the development of the adult, and because genes can be transfected or injected into cultured cells, the stage is set for introducing specific genes into animals. An important extension of this approach is to inject nuclei into enucleated mouse eggs, thereby assuring that the genetic traits of interest will be transmitted to the embryo (Illmensee and Hoppe, 1981).

We have microinjected plasmids directly into germinal vesicles of mouse oocytes or pronuclei of fertilized mouse ova (Brinster et al., 1981) and then implanted them into pseudopregnant mice. Gordon and coworkers (1980) have also used this technique to introduce plasmids containing the herpes thymidine kinase gene and SV40 sequences into mice. Two of 78 mice that they screened by Southern blot analysis were shown to have plasmid sequences; however, in both cases the plasmid DNA was not intact, and viral thymidine kinase activity was not detected. Here we report the successful application of this technique to the integration of a functional fusion gene into mice.

Results

Preparation of the Fusion Plasmid pMK

Preliminary experiments established that fusion of MT-I promoter/regulatory regions to the herpes thymidine kinase structural gene (HSV-TK) as shown in Figure 1 results in a gene, which we call MK, that can be expressed in mouse L cells and in mouse eggs. Furthermore, this gene is subject to regulation by heavy metals in a manner similar to the normal MT-I gene (K. Mayo, R. Warren and R. Palmiter; R. Brinster, H. Chen, R. Warren and R. Palmiter, submitted). This vector thus combines the advantages of a regulatable

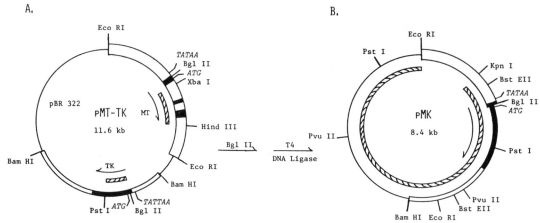

Figure 1. Structure of the Plasmids and DNA Fragments Used in this Study

(A) Plasmid pMT-TK was constructed from plasmid m_1pEE_{38} (Durnam et al., 1980), which contains a 3.8 kb genomic Eco RI fragment that includes the MT-I gene inserted into the Eco RI site of pBR322, by insertion of the 3.5 kb Bam HI fragment of Herpes Simplex Virus Type I containing the thymidine kinase gene (McKnight, 1980) into the Bam HI site. The two genes are present in the same transcriptional orientation, as shown by the arrows.

(B) The fusion plasmid, pMK, was created by digestion of pMT-TK with Bgl II restriction endonuclease followed by ligation with T4 DNA ligase to directly join the 5' region of the MT-I gene to the TK structural gene. pBR322 sequences are shown by a single line, TK gene sequences by a narrow box and MT-I gene sequences by a wide box. mRNA coding regions are represented by closed boxes; nontranscribed and intron regions are shown by open boxes. Hatched boxes inside the circles represent regions of these genes that were used as hybridization probes; the MT-I specific probe, MT-XH, extends from Xba I to Hind III and TK-specific probe, TK-BP, extends from Bgl II to Pst I. The fusion plasmid probe, pMK(-EK), includes the entire plasmid except the Eco RI to Kpn I region because Southern blots revealed that this is a sequence present many times in the mouse genome. Restriction sites relevant to the construction of the plasmids and the gene-specific probes are shown in panel A. All restriction sites used in mapping integrated copies of pMK are shown in panel B. pMK is not cut by Hind III, Xba I or Xho I. Also shown are the locations of the TATAA promoter sequences and ATG translation start codons for the two genes.

promoter and a simple enzyme assay to detect expression.

Expression of Herpes Thymidine Kinase Activity in Mouse Liver

Approximately 200 copies of the plasmid, pMK, were injected into the male pronucleus of fertilized one-cell eggs; an average of 16 eggs then were transferred into the oviducts of 15 pseudopregnant mice. Six of these mice had litters, for a total of 12 male and 7 female offspring. At the age of 4 weeks each of the males was mated with a normal female. Before assaying for gene expression, the mice were injected with $CdSO_4$ (2 mg/kg). This was done in the hope of inducing HSV-TK activity since this dose is known to induce MT-I mRNA in liver and kidney (Durnam and Palmiter, 1981). Eighteen hours later the mice were killed, liver samples were prepared for TK assay and the remainder of each animal was frozen for subsequent nucleic acid analysis.

For initial TK assay, 5 μl of a 20% liver homogenate was tested. One animal (23-2) showed about 40-fold more activity than the others did; however, this activity was so high that it was in the nonlinear range of the assay. After appropriate dilution we measured about 200-fold more TK activity in this mouse compared with litter mates and other mice of similar age (Table

1). To ascertain whether the TK activity was derived from the HSV-TK gene or from the endogenous mouse gene, an antibody specific for HSV-TK was mixed with the liver extracts prior to enzyme assay. Table 1 shows that the TK activity of mouse 23-2 was inhibited 97% with this antisera, whereas the TK activity of the other mice was essentially unaffected.

Additional assays confirmed that the majority of TK activity of mouse 23-2 was due to the HSV gene product. The endogenous mouse TK enzyme cannot phosphorylate iododeoxycytidine (IdC) whereas the HSV enzyme can (Summers and Summers, 1977). Thus, IdC will inhibit the conversion of ^{3}H-thymidine to ^{3}H-thymidylic acid if the enzyme is of viral origin. Table 2 shows that this is observed with the enzyme preparation from mouse 23-2, but not from the litter mates. The substrate specificity of the mouse and HSV-TK enzymes also can be demonstrated using ^{125}IdC and tetrahydrouridine, an inhibitor of cytidine deaminase. In crude liver extracts from normal mice, ^{125}IdC is converted into phosphorylated derivatives due to the action of deaminases that convert ^{125}IdC to iododeoxyuridine which can be phosphorylated by TK. However, when an inhibitor of deaminase (tetrahydrouridine, THU) is included in the assay, labeled substrates for the endogenous TK enzyme are not formed and the apparent activity is inhibited 30-fold.

Table 1. Herpes Thymidine Kinase and Mouse Metallothionein-I Gene Expression in Mouse Liver

Mouse	pMK DNA[a]	Thymidine Kinase Activity[b] (cpm of ^{3}H-TMP formed/mg wet wt/min)		HSV-Thymidine Kinase mRNA (molecules/cell)[c]	Mouse Metallothionein-I mRNA (molecules/cell)
		−Ab	+Ab		
14-1	—	355	261	<2	2650
14-2	—	552	425	<2	1330
19-1	—	593	723	<2	1040
19-2	+ +	920	915	7.5	660
19-3	—	492	368	<2	2760
21-1	—	656	525	<2	1090
21-2	—	716	550	<2	1480
21-3	+	468	579	<2	610
21-4	—	770	654	<2	770
21-5	—	633	571	<2	1780
23-1	+	560	546	<2	2420
23-2	+	23,475	9,416	28	1500
		(121,830)	(4,141)[d]		

[a] The presence of pMK DNA in mouse tissue was determined by Southern blotting analysis as shown in Figure 2.
[b] TK activity was measured as described under Experimental Procedures except that the samples were incubated with either 10 μl of anti–HSV-TK IgG fraction (+Ab) or normal IgG fraction (−Ab) for 15 min at 4° prior to the addition of the reagent mixture.
[c] HSV-TK mRNA and mouse MT-I mRNA were determined as described in Experimental Procedures.
[d] Values in parentheses were obtained when the homogenate from mouse 23-2 was diluted 10-fold prior to assay.

Table 2. Controls to Distinguish HSV-Thymidine Kinase Activity from Endogenous Thymidine Kinase

Mouse	pMK DNA	Thymidine Kinase Activity with ^{3}H-thymidine as Substrate[a,b]		Thymidine Kinase Activity with ^{125}IdC as Substrate[b,c]	
		−IdC	+IdC	−THU	+THU
23-2	+	497,000	187,000	71,300	56,400
23-1	—	14,500	14,700	—	—
14-1	—	7,520	9,640	—	—
C57 × SJL	—	—	—	150,400	4,800

[a] Thymidine kinase was measured as described under Experimental Procedures, except that all of the samples contained tetrahydrouridine at 40 μM and half of the samples included 100 μM iododeoxycytidine (IdC) as indicated.
[b] Data are the means of three determinations and they are expressed as cpm of product formed per assay.
[c] Thymidine kinase activity was measured as described under Experimental Procedures except that 0.5 μCi of ^{125}I-iododeoxycytidine at >700 Ci/mmole was substituted for ^{3}H-thymidine; half of the samples included the cytidine deaminase inhibitor, tetrahydrouridine (THU), at a final concentration of 40 μM. The enzyme from mouse 23-2 was diluted to give an activity comparable to the control.

In contrast, the TK activity in mouse 23-2 is inhibited only 20%, as would be expected with a viral enzyme that can utilize ^{125}IdC directly (Summers and Summers, 1977).

Detection of MK Genes in Several Mice

To assay for the presence of the MK fusion gene in the mice, kidney DNA was digested with restriction enzyme, Bst EII, subjected to electrophoresis on an agarose slab gel and blotted according to the method of Southern (1975). Nick-translated probes were used that would detect both the endogenous MT-I gene and any fusion gene. The endogenous MT-I gene falls within a 6 kb Bst EII fragment, whereas the MK fusion gene would be cut into a 2.3 kb fragment by this enzyme (see Figure 1B). Figure 2 shows that mouse 23-2 and three additional mice have the 2.3 kb band expected of the MK gene. The MK gene band has approximately half the intensity, as measured by densitometry, as the MT-I gene band in all of the mice except 19-2, in which the MK band is about six times more intense. To estimate the number of MK genes per cell, a control experiment was performed in which the same combination of probes was hybridized to equal molar amounts of the MT-I and MK genes. This was done by digesting pMT-TK (Figure 1A) with Eco RI or Pvu II and separating the MT-I gene and MK gene-containing fragments by agarose gel electrophoresis. Different amounts of the digests (40–160 pg of plasmid DNA) were analyzed by electrophoresis to facilitate quantitation. We observed that the autoradiographic band representing the MT-I gene was consistently 4-fold more intense than was the band representing the MK gene (data not shown). Since in our previous experiments the MT-I gene band was only twice as intense as was the MK gene band, we conclude that there must be twice as many MK genes per cell as MT-I genes. Thus, knowing that there are two MT-I genes per cell, we infer that there are four MK genes per cell in these mice. By use of the same

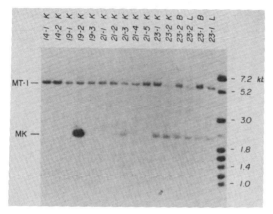

Figure 2. pMK Sequences Are Present in Four Mice

DNA (6.5 μg) from the kidney (K) of each of the twelve mice, as well as DNA (6.5 μg) from liver (L) and brain (B) of mice 23-1 and 23-2, was digested with restriction endonuclease Bst EII (see Figure 1B), subjected to electrophoresis on a 1.2% agarose gel, transferred to nitrocellulose, and hybridized to a mixture of MT-XH and TK-BP probes (see Figure 1A). The 6 kb band present in all samples represents the endogenous MT-I gene; the 2.3 kb band comigrates with the band generated by Bst EII digestion of pMK. We have noted a decreased intensity of the MT-I band in liver and kidney of mouse 23-2 in several experiments with these DNA preparations, but we cannot provide a satisfying explanation for this result.

calculation, we estimate that mouse 19-2 has about 48 copies of the MK gene per cell.

Since HSV-TK enzyme activity was not detected in mice 19-2, 21-3 or 23-1 even though intact MK genes were present (Table 1), we checked whether the mice were actually induced with Cd, by measuring the amount of MT-I mRNA by solution hybridization with ^{32}P-labeled MT-I cDNA. Table 1 shows that all of the mice had between 600 and 2700 molecules of MT-I mRNA per liver cell. The basal level of MT-I mRNA in mouse liver is variable but averages about 150 molecules per cell, whereas after optimal induction, levels of about 2300 molecules per cell are generally obtained (Durnam and Palmiter, 1981). This control indicates that at the time the mice were killed the MT-I gene was still induced and suggests that the lack of thymidine kinase activity was not due to the failure of Cd delivery to the tissues.

We also measured HSV-TK mRNA levels by solution hybridization with a ^{32}P-labeled HSV-TK cDNA. Although TK mRNA was detectable in the liver of mouse 23-2, the level was only 28 molecules per cell. A low amount of HSV-TK mRNA was also detected in mouse 19-2, the mouse with nearly 50 copies of the MK gene (Table 1). All other mice had less than two molecules of TK mRNA per cell.

Somatic Distribution of MK Gene and Thymidine Kinase Activity

Figures 2 and 3B show that the MK gene was present in several different tissues of mouse 23-2, including

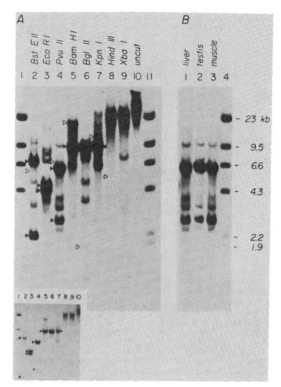

Figure 3. Junction Fragments and Tandem Repeats of pMK

(A) Liver DNA (12 μg) from mouse 23-2 was digested with restriction enzymes that cut pMK twice (Bst EII, Eco RI, Pvu II), once (Bam HI, Bgl II, Kpn I), or not at all (Hind III, Xba I), subjected to electrophoresis on a 0.8% agarose gel, transferred to nitrocellulose, and hybridized with a pMK(-EK) nick-translated probe (see Figure 1B). The inset at the bottom of A reveals the major bands more clearly; it was obtained after a 14 hr exposure compared to 60 hr for the main figure. Restriction fragments predicted from the endogenous MT-I gene (▷) and from cleavage within the repeats of pMK (▶) are indicated; the other bands are presumably junction fragments containing both pMK and mouse DNA.

(B) Comparison of Pvu II digests of liver, testis and muscle DNA (12 μg each) from mouse 23-2. Markers are a Hind III digest of phage λ.

liver, kidney, brain, muscle and testis, and the intensity of the hybridizing band was similar in each tissue, suggesting that the gene copy number is constant in each tissue. HSV-TK activity and mRNA levels were lower in kidney than in liver and were undetectable in brain (Table 3). Thus, MK gene expression closely paralleled the MT-I gene expression in those tissues.

Integration of pMK into the Mouse Genome

To ascertain whether the pMK plasmid was integrated into the mouse genome, DNA from each of the four mice positive for the MK gene was digested with several enzymes that cut twice, once, or not at all within the pMK plasmid. After electrophoresis and blotting, the nitrocellulose was hybridized with a nick-

Table 3. Herpes Thymidine Kinase and Mouse Metallothionein-I Gene Expression in Several Mouse Tissues

Mouse	pMK DNA[a]	Thymidine Kinase Activity[b] (pmole TMP/min/mg protein)			Thymidine Kinase mRNA[c] (molecules/cell)			Metallothionein-I mRNA[c] (molecules/cell)		
		Liver	Kidney	Brain	Liver	Kidney	Brain	Liver	Kidney	Brain
19-2	+	1.37	1.44	ND[d]	7.5	3.4	ND	660	33.4	ND
21-1	−	1.98	1.74	ND	<2	<2	ND	1090	113	ND
21-2	−	1.89	2.26	ND	<2	<2	ND	1480	81.7	ND
21-3	−	1.12	1.78	ND	<2	<2	ND	610	183	ND
23-1	+	1.56	2.05	1.50	<2	<2	<2	2420	81.2	7.2
23-2	+	66.8	4.44	1.23	28	<2	<2	1500	74.5	9.4

[a] The presence of pMK DNA in mouse tissue was determined as in Figure 2.
[b] TK enzyme activity was determined as described in Experimental Procedures. The assay buffer did not contain 5-iododeoxycytidine and therefore represents total enzyme activity.
[c] TK mRNA and MT-I mRNA were determined as described in Experimental Procedures.
[d] ND = not determined.

translated probe that included all of the plasmid except the 1150 bp between Eco RI and Kpn I; this region was omitted because it contains a repeat sequence. Predictions of what size bands would be produced are quite different depending on whether the pMK plasmid is integrated into the mouse genome or not. For example, with enzymes that cut once within a single integrated plasmid we would expect to generate only junction fragments, that is, fragments that combine both plasmid and genomic sequences, and they would be of a size different from that predicted from the plasmid (see Figure 4A).

Restriction of liver DNA from each of the mice positive for MK genes with enzymes Bam HI, Bgl II or Kpn I, which cut only once within pMK, reveals a prominent 8.4 kb fragment that is the same size as that predicted from an unintegrated plasmid (Figure 3A and Figure 5). Likewise, enzymes that cut twice within pMK, such as Bst EII, Eco RI and Pvu II, give two prominent bands that add up to 8.4 kb, the size of pMK (Figure 3A). However, when enzymes were used that do not cut within pMK, such as Hind III and Xba I, the hybridizing DNA was nearly as large as uncut genomic DNA; no bands corresponding to unintegrated single plasmids were observed. A possible resolution of this paradox is that several copies of pMK are duplicated tandemly n times and integrated at a single site as indicated in Figures 4B and 4C. Restriction of DNA with this configuration would generate fragments corresponding to the original plasmid plus two junction fragments that would be less than $1/n$ as intense. Indeed, Figure 3 shows that in addition to the intense bands there are typically several additional fainter bands. One of these faint bands (marked with an open arrow) corresponds to the MT-I gene, which would be expected because about 650 bp of the probe are homologous to sequences 5' of the MT-I gene (see Figure 1). The other faint bands (unmarked) are good candidates for the predicted junc-

tion fragments. The average intensity of the junction fragments from mouse 23-2 relative to the main band(s) is about 1/5, suggesting that the intact pMK plasmid is repeated about five times in this mouse.

To test this idea of tandem duplication, we isolated high molecular weight DNA from mouse 23-2 and cut it with a battery of enzymes that do not cut within plasmid pMK; this procedure was an effort to cut the pMK repeat unit to a minimal size. Figure 6 shows that the size of the hybridizing band is greater than 45 kb (our largest marker) in uncut DNA and between 23 and 45 kb after restriction with Hind III, Xba I and Xho I, whereas total mouse DNA is cut to a weight average size of about 2 kb as shown by ethidium-bromide staining. When Bam HI was added along with the other enzymes, the 8.4 kb Bam HI linear fragment was obtained along with several fainter fragments that probably represent the predicted junction fragments. Thus, we conclude that there are four or five direct repeats of the pMK plasmid in mouse 23-2, a result that is consistent with the relative intensity of the 8.4 kb band and the junction fragments as well as gene dosage.

Inspection of Figure 3A reveals four junction fragments with most enzymes that cut within pMK (lanes 2 through 7) and one or two junction fragments with enzymes that do not cut within pMK (Figure 3A, lanes 8 and 9; Figure 6, lane 2). This suggests that in addition to the insertion of the major repeat of pMK, there may also be one or more integrated fragments of pMK in mouse 23-2, because integration at only a single site would have generated only two junction fragments with enzymes that cut within the plasmid and none with enzymes that do not.

Discussion

The data presented here demonstrate that we have achieved expression of HSV-TK in at least two tissues

A.

B.

C.

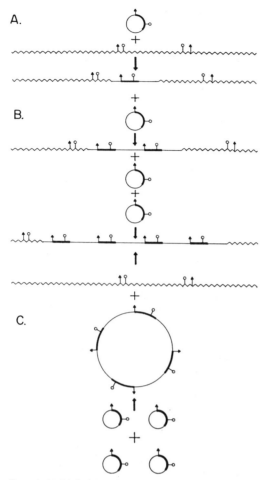

Figure 4. Models for Integration of pMK DNA into Mouse DNA

The plasmid is illustrated as a circular DNA molecule with two unique restriction sites shown symbolically (♀ ♦). (A) A single plasmid DNA molecule recombines (either by homologous or nonhomologous recombination) with mouse DNA to give a single integrated plasmid. The junction between mouse and plasmid sequences might occur at any site in both plasmid and mouse DNAs. Digestion with either of the two enzymes whose sites are shown would be expected to generate two products, neither of which would be likely to be the same size as linear pMK DNA.

(B) After the initial integration event, a number of subsequent events could occur involving homologous recombination of additional copies of pMK into those already integrated, giving rise to a tandem repetition of the integrated pMK sequences. Digestion with either enzyme would generate several copies (three as drawn) of full-length linear pMK molecules, plus single copies of two new junction fragments.

(C) Several copies of the plasmid could homologously recombine with one another to generate a tandemly repetitive plasmid, which would then recombine with mouse DNA, again generating a tandemly repeated integrated plasmid. Restriction enzyme analysis would generate the same products as model B.

of a mouse derived from an injected egg. The HSV-TK enzyme activity, mRNA and gene were detected in both liver and kidney of mouse 23-2, whereas only

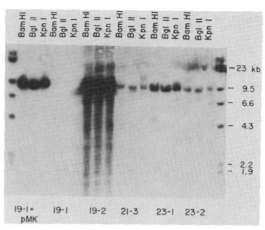

Figure 5. Junction Fragments and Tandem Repeats of pMK Are Present in All MK-Positive Mice

Liver DNA (10 μg) of all four mice that contain pMK sequences (19-2, 21-3, 23-1, 23-2) and from a control mouse (19-1) was digested with Bam HI, Bgl II and Kpn I (each of these enzymes cuts pMK at a single site; see Figure 1B), subjected to electrophoresis on a 0.8°o agarose gel, transferred to nitrocellulose and hybridized to pMK(-EK). Bands common to the 19-1 control and to the other mice come from the endogenous MT-I gene. The major band comigrates with linearized pMK. Additional bands that are presumed to be junction fragments are also observed.

the DNA was detected in the brain of this mouse. We used a fusion plasmid, pMK, that fuses the promoter/regulatory region of the mouse MT-I gene to the structural gene of the HSV-TK gene at the unique Bgl II site which lies a few nucleotides upstream of the initiation codons of both MT-I and HSV-TK genes (see Figure 1). This fusion plasmid can be regulated positively by heavy metals, such as Cd, when injected into mouse eggs or transfected into mouse L cells (R. Brinster, H. Chen, R. Warren and R. Palmiter; K. Mayo, R. Warren and R. Palmiter, submitted). In mouse eggs, the amount of TK activity obtained after exposure of injected eggs to Cd increases 10–20-fold depending on the dosage of pMK injected. Thus, we expected that the MT-I promoter would allow a high level of expression of HSV-TK. We assume that the MK gene is regulated by heavy metals in mouse 23-2 because the amount of TK activity in the liver of this mouse was about 100-fold higher than endogenous mouse TK activity. The tissue specificity of expression of the MK gene in mouse 23-2 is similar to that observed for the endogenous MT-I gene (Table 3) suggesting that the gene has been subject to some of the same developmental influences as the MT-I gene. In subsequent experiments the protocol has been changed so that the mouse is not killed for bioassay. Thus, we hope to be able to demonstrate regulation directly in future experiments.

The efficiency of achieving plasmid integration and expression in our first experiments is remarkably good. Nineteen offspring (12 males and 7 females) of

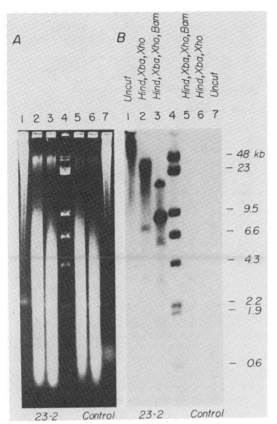

Figure 6. pMK Sequences Are Integrated into Cellular DNA

DNA (16 μg) from the livers of mice 19-1 (control), 23-2, and 19-2 were subjected to electrophoresis on a 0.6% agarose gel without restriction nuclease digestion, after digestion with a mixture of Hind III, Xba I and Xho I (none of these enzymes cuts within pMK), or after digestion with the above enzymes plus Bam HI (which cuts pMK at a single site). (A) shows the ethidium bromide stained gel; (B) shows an autoradiogram after transfer to nitrocellulose and hybridization to a pMK(-EK) probe. The size markers are intact λ DNA and λ DNA digested with Hind III. The strong band seen after Bam HI digestion comigrates with pMK that has been similarly digested.

the first experiment were assayed for TK expression and one, 23-2, was clearly positive. The DNA from the 12 males was assayed and four were positive. In a second experiment, the pMK plasmid was augmented by inserting the Bam HI fragment from Mulligan and Berg's (1980) vector, pSV3-gpt, which contains the SV40 origin and T-antigen gene. Twelve offspring were analyzed for TK expression and pMK DNA and all were negative. In a third experiment, still in progress, the original plasmid was linearized and ligated to mouse DNA prior to injection. In this experiment, five offspring have now been analyzed, and two express HSV-TK and have the MK gene (see cover). In a fourth experiment, a linear 2.3 kb Bst E2 fragment that includes the MK gene was injected, and one out of five mice is positive for HSV-TK expression. The number of offspring is too low for good statistics and hence we do not know whether the variations in protocol are significant. Overall, out of 41 offspring that have been analyzed, we have obtained four mice that express HSV-TK and an additional three that have intact MK genes but do not express TK. We have decided to name the male mice that express the MK fusion gene MaK and the female mice MyK followed by a number denoting their foster mother. Thus mouse 23-2 will be designated MaK-23 in the future.

Analysis of the plasmid DNA from the four MK-positive mice described here suggests that the pMK plasmid is tandemly duplicated several times (Figure 5). If each plasmid were integrated separately, we would expect many junction fragments (fragments containing both pMK and mouse DNA sequences) when the DNA is restricted with an enzyme that cuts once within the pMK sequence. But instead, we observe predominantly the 8.4 kb fragment that represents the original plasmid (Figure 3). The best argument against freely replicating single plasmids is that the hybridizing band from uncut genomic DNA is much larger than the original plasmid (Figures 3 and 6). Thus, we conclude that in all cases the DNA must be duplicated tandemly as a direct repeat. The number of direct repeats is difficult to measure with certainty but it appears to be four or five in MaK-23 (Figure 6). These plasmid sequences must be linked physically to non-pMK sequences, since digestion with enzymes that do not cleave pMK nonetheless results in a significant reduction in the size of DNA band containing pMK (Figure 6). Further evidence that the set of direct repeats is integrated into genomic DNA is the existence of minor bands that may represent the junction fragments that are postulated if the DNA is integrated. The tandem duplication could occur by homologous recombination among the numerous pMK molecules injected, and it could occur either before or after the integration event as depicted in Figures 4B and 4C. Considering the fact that the egg is just completing meiosis and beginning the first mitotic division at the time of plasmid injection, the requisite enzymes for recombination may be abundant. Alternatively, oligomers of pMK that were present in the original plasmid preparation may be integrated preferentially. The presence of the pMK sequences in nearly equal abundance in all tissues that have been examined indicates stable inheritance of the pMK sequences during development, which is most easily explained by integration into the genome.

If the tandemly duplicated pMK plasmids are integrated, then an important parameter for determining the success of achieving expression is when the integration event occurs. The current view of mouse embryogenesis is that after the first four to six cleavages only the internal three to six cells are destined to become the embryo while the other outer cells develop into extraembryonic tissues. Thus, assuming that integration occurs only once, there is a considerable decrease in the probability of the plasmid DNA being represented in the embryo once cleavage begins. If the integration event occurs after the cells that are destined to become the embryo start to cleave, then the result would be a mosaic mouse.

We are exploring the question of whether the pMK sequences are in the germ line by analyzing offspring for pMK DNA or HSV-TK activity. Unfortunately the female to which MaK-23 was mated did not become pregnant. However one offspring (out of six) of MaK-67, the male on the cover, and one offspring (out of three) of MyK-84, a female from the fourth experiment described above, clearly express high levels of HSV-TK in the liver. Thus, in the two cases that have been analyzed to date, an expressible MK gene has been passed on to the second generation. In contrast, six offspring of mice 19-2 and 23-1, mice that contain pMK sequences but do not express HSV-TK, have no detectable pMK DNA sequences in their liver. Analysis of more mice will be necessary to determine whether this correlation between expression and transmission in the germ line is significant.

The basic approach described here provides a means of achieving expression of a gene in many, if not all, tissues of an animal. Although the average efficiency of achieving expression is only 10% so far, further refinements of the methodology may improve these statistics. There are several long-range benefits of this experimental approach for gene implantation. One is that a foreign gene can be introduced into any cell type. Another is that the influence of developmental history on gene commitment and gene expression can be assessed. Finally, this approach may provide a way to correct certain genetic defects.

Experimental Procedures

Plasmid Preparation

Plasmid pMK was constructed as shown in Figure 1. pMK DNA was isolated from a cleared lysate of bacterial cells by SDS-proteinase K treatment followed by phenol:chloroform extraction and ethanol precipitation. The nucleic acids were digested with RNAase A and passed through Bio-Gel A50m column in 0.1 × SET (1 × SET = 1% SDS, 5 mM EDTA, 10 mM Tris [pH 7.5]) to separate DNA from RNA fragments. The preparation used for these experiments contained about 1/3 supercoiled plasmids, 1/3 nicked circles, and about 1/3 larger oligomers of the plasmid, as revealed by agarose gel electrophoresis and ethidium-bromide staining.

Injection and Manipulation of Eggs

Fertilized one-cell ova of C57 × SJL hybrids were flushed from the oviduct using Brinster's (1972) medium on the morning of day one of pregnancy. Cumulus cells were removed from ova with hyaluronidase (300 U/ml) and the ova were washed free of debris and enzyme before manipulation. For injection, the ova were transferred to a depression slide in Brinster's medium containing 5 μg/ml cytochalasin B and were held in place by a blunt pipette while the tip of the injector pipette was inserted through the zona pellucida and vitellus and into the male pronucleus (Brinster et al., 1981). The DNA solution in the injector pipette was discharged slowly into the nucleus using a syringe connected to a micrometer. The larger pronucleus (male) of the fertilized ovum was injected with approximately 2 pl of plasmid solution. Following injection, the ova were washed free of cytochalasin and were returned to the same medium used for collection. When injections were completed, the ova were transferred to the oviducts of pseudopregnant, random-bred Swiss mice (Rafferty, 1970).

Thymidine Kinase Assay

Mouse tissues were homogenized in 4 vol of buffer containing 10 mM KCl, 2 mM MgCl$_2$, 10 mM Tris–Cl, 1 mM ATP, 10 mM NaF and 50 mM ε-aminocaproic acid (pH 7.4), and then centrifuged for 10 min at 15,000 × g (Kit et al., 1974). An aliquot (5 μl) was added to 25 μl of reaction mixture containing 150 mM Tris–Cl (pH 7.5), 10 mM ATP, 10 mM MgCl$_2$, 25 mM NaF and 10 mM β-mercaptoethanol and 5 μCi of ^{3}H-thymidine (80 Ci/mmole; New England Nuclear). The mixture was incubated for 30 min or 2 hr at 37° and the ^{3}H-TMP produced was measured by adsorption to DE-81 paper (Whatman) and subsequent scintillation counting. For specific activity measurements, protein was determined by the method of Bradford (1976).

Isolation of Nucleic Acid

Mouse tissues (liver, kidney, brain) were homogenized in 1 × SET buffer containing 50 μg/ml proteinase K (Beckman), incubated for 60 min at 45°, phenol:chloroform-extracted, ethanol-precipitated, and redissolved in 10 mM Tris–Cl, 0.25 mM EDTA (pH 7.5).

Metallothionein-I and Herpes Thymidine Kinase mRNA Determinations

MT-I mRNA levels were determined as described by Beach and Palmiter (1981). HSV-TK mRNA levels were determined similarly. The 840 bp Pst I–Pst I fragment that includes most of the structural HSV-TK gene (McKnight, 1980) was cloned into the Pst I site of fd 103 (Herrmann et al., 1980) and single-strand phage DNA containing the mRNA strand was prepared. ^{32}P-cDNA was then prepared by nick translation of the 460 bp fragment extending from Bgl II site of HSV-TK to the Sst I site, and isolation of the cDNA strand by hybridization to the fd message strand and subsequent elution as described by Beach and Palmiter (1981).

Restriction Digestion, Gel Electrophoresis and Southern Blot Hybridization

Restriction enzymes (from Bethesda Research Labs and New England Biolabs) were used under standard conditions; RNAase A (Sigma; further treated by preincubation at 80° for 10 min) was added to all restriction digests at 50 μg/ml. Agarose gel electrophoresis was performed on horizontal slabs in 80 mM sodium acetate, 40 mM Tris, 4 mM EDTA (pH 8.0). DNA was transferred to nitrocellulose essentially as described by Southern (1975). DNA probes were labeled by nick translation (Rigby et al., 1977) using α-^{32}P-dNTPs (400–2000 Ci/mmole) purchased from Amersham and New England Nuclear. Hybridizations were at 45° in 50% formamide, 3.2 × SSC, 10% dextran sulfate for 12–16 hr (Wahl et al., 1979). After hybridization, filters were washed in 2 × SSC plus 0.5 × SET at 68° followed by 0.5 × SET at 45°. Size markers are either phage λ DNA digested with Hind III and endlabeled, or a series of restriction fragments derived from the MT-1 gene plasmid, m$_1$pEE$_{3.8}$ (Durnam et al., 1980).

Acknowledgments

We thank Dr. S. Kit for antiserum against herpes simplex virus thymidine kinase. We are grateful to Ms. A. Dudley for preparing this manuscript. This work was supported by grants from the National Institutes of Health and the National Science Foundation.

The costs of publication of this article were defrayed in part by the payment of page charges. This article must therefore be hereby marked "advertisement" in accordance with 18 U.S.C. Section 1734 solely to indicate this fact.

Received September 4, 1981; revised September 25, 1981

References

Beach, L. R. and Palmiter, R. D. (1981). Amplification of the metal-lothionein-I gene in cadmium-resistant mouse cells. Proc. Nat. Acad. Sci. USA *78*, 2110–2114.

Bradford, M. (1976). A rapid and sensitive method for quantitation of microgram quantities of protein utilizing the principle of protein-dye binding. Analyt. Biochem. *72*, 248–254.

Brinster, R. L. (1972). Cultivation of the mammalian embryo. In Growth, Nutrition and Metabolism of Cells in Culture, 2, G. Rothblat and V. Cristofala, eds. (New York: Academic Press), pp. 251–286.

Brinster, R. L. (1974). The effect of cells transferred into the mouse blastocyst on subsequent development. J. Exp. Med. *140*, 1049–1056.

Brinster, R. L., Chen, H. Y. and Trumbauer, M. E. (1981). Mouse oocytes transcribe injected Xenopus 5S RNA gene. Science *211*, 396–398.

Buetti, E. and Diggelmann, H. (1981). Cloned mouse mammary tumor virus DNA is biologically active in transfected mouse cells and its expression is stimulated by glucocorticoid hormones. Cell *23*, 335–345.

Capecchi, M. R. (1980). High efficiency transformation by direct microinjection of DNA into cultured mammalian cells. Cell *22*, 479–488.

DeRobertis, E. M. and Olson, M. V. (1979). Transcription and processing of cloned yeast tyrosine tRNA genes microinjected into frog oocytes. Nature *278*, 137–143.

Durnam, D. M. and Palmiter, R. D. (1981). Transcriptional regulation of the mouse metallothionein-I gene by heavy metals. J. Biol. Chem. *256*, 5712–5716.

Durnam, D. M., Perrin, F., Gannon, F. and Palmiter, R. D. (1980). Isolation and characterization of the mouse metallothionein-I gene. Proc. Nat. Acad. Sci. USA *77*, 6511–6515.

Gordon, J. W., Scangos, G. A., Plotkin, D. J., Barbosa, J. A. and Ruddle, F. H. (1980). Genetic transformation of mouse embryos by microinjection of purified DNA. Proc. Nat. Acad. Sci. USA *77*, 7380–7384.

Grosschedl, R. and Birnstiel, M. L. (1980). Identification of regulatory sequences in the prelude sequences of an H2A histone gene by the study of specific deletion mutants *in vivo*. Proc. Nat. Acad. Sci. USA *77*, 1432–1436.

Herrmann, R., Neugebauer, K., Pirkl, E., Zentgraf, H. and Schaller, H. (1980). Conversion of bacteriophage fd into an efficient single-stranded DNA vector system. Mol. Gen. Genet. *177*, 231–242.

Hynes, N. E., Kennedy, N., Rahmsdorf, V. and Groner, B. (1981). Hormone-responsive expression of an endogenous proviral gene of mouse mammary tumor virus after molecular cloning and gene transfer into cultured cells. Proc. Nat. Acad. Sci. USA *78*, 2038–2042.

Illmensee, K. and Hoppe, P. C. (1981). Nuclear transplantation in Mus musculus: developmental potential of nuclei from preimplantation embryos. Cell *23*, 9–18.

Kit, S., Leung, W.-C., Trkula, D. and Jorgensen, G. (1974). Gel electrophoresis and isoelectric focusing of mitochondrial and viral-induced thymidine kinases. Int. J. Cancer *13*, 203–218.

Kurtz, D. T. (1981). Hormonal inducibility of rat α_{2u} globulin genes in transfected mouse cells. Nature *291*, 629–631.

Lai, E. C., Woo, S. L. C., Bordelon-Riser, M. E., Fraser, T. H. and O'Malley, B. W. (1980). Ovalbumin is synthesized in mouse cells transformed with the natural chicken ovalbumin gene. Proc. Nat. Acad. Sci. USA *77*, 244–248.

McKnight, S. L. (1980). The nucleotide sequence and transcript map of the herpes simplex virus thymidine kinase gene. Nucl. Acids Res. *8*, 5949–5964.

Mintz, B. and Illmensee, K. (1975). Normal genetically mosaic mice produced from malignant teratocarcinoma cells. Proc. Nat. Acad. Sci. USA *72*, 3585–3589.

Mulligan, R. C. and Berg, P. (1980). Expression of a bacterial gene in mammalian cells. Science *209*, 1422–1427.

Mulligan, R. C., Howard, B. H. and Berg, P. (1979). Synthesis of rabbit β-globin in cultured monkey kidney cells following infection with a SV40 β-globin recombinant genome. Nature *277*, 108–114.

Papaioannou, V. E., McBurney, M. W., Gardner, R. L. and Evans, M. J. (1975). Fate of teratocarcinoma cells injected into early mouse embryos. Nature *258*, 70–73.

Rafferty, K. A., Jr. (1970). Methods in Experimental Embryology of the Mouse. (Baltimore: Johns Hopkins Press).

Rigby, P. W. J., Dieckmann, M., Rhodes, C., and Berg, P. (1977). Labeling deoxyribonucleic acid to high specific-activity *in vitro* by nick translation with DNA polymerase I. J. Mol. Biol. *113*, 237–251.

Southern, E. M. (1975). Detection of specific sequences among DNA fragments separated by gel electrophoresis. J. Mol. Biol. *98*, 503–517.

Summers, W. C. and Summers, W. P. (1977). [^{125}I]deoxycytidine used in a rapid, sensitive and specific assay for herpes simplex virus type 1 thymidine kinase. J. Virol. *24*, 314–318.

Wahl, G. M., Stern, M. and Stark, G. R. (1979). Efficient transfer of large DNA fragments from agarose gels to diazobenzyloxymethyl-paper and rapid hybridization by using dextran sulfate. Proc. Nat. Acad. Sci. USA *76*, 3683–3687.

Wigler, M., Sweet, R., Sim, G. K., Wold, B., Pellicer, A., Lacy, E., Maniatis, T., Silverstein, S. and Axel, R. (1979). Transformation of mammalian cells with genes from procaryotes and eucaryotes. Cell *16*, 777–785.

High-efficiency Gene Transfer into Mammalian Cells: Generation of Helper-free Recombinant Retrovirus with Broad Mammalian Host Range

R.D. Cone and R.C. Mulligan

ABSTRACT We have constructed a chimeric retrovirus genome containing ecotropic *gag–pol* sequences from Moloney murine leukemia virus and envelope sequences derived from the amphotropic virus 4070A. This reconstructed genome, termed *pMAV-ψ⁻*, lacks the ψ site required for encapsidation of the viral genome. NIH 3T3 cells transfected with *pMAV-ψ⁻*, called ψ-AM lines, are capable of producing high titer stocks of helper-free recombinant retrovirus with amphotropic host range after transfection with recombinant retroviral vectors carrying the neomycin phosphotransferase gene. Most transfected ψ-AM cells remain helper-free, even after months in culture. ψ-AM virus stocks infect nearly all human and murine cell lines tested thus far, as assayed by resistance to the neomycin analogue G418. Southern and RNA blot analyses of ψ-AM-infected human cells show that recombinant murine retroviruses integrate randomly into genomic DNA as normal proviruses and express high levels of the subgenomic and genomic viral messages in the expected stoichiometry of 1:1.

Recombinant retroviruses, whether created by *in vivo* recombination events or engineered by DNA cloning, typically are defective for replication and require helper virus to be passed from cell to cell. Recently, we constructed a helper cell line, called ψ2, which, upon transfection with recombinant murine retroviral vectors, produces helper-free stocks of recombinant virus (1). Similar helper lines have been created by Watanabe and Temin for use with avian spleen necrosis virus vectors (2). The use of helper-free virus stocks in gene transfer experiments avoids unwanted biological effects of wild-type virus and facilitates efficient gene transfer, because the rapid spread of wild-type virus in tissue culture cells (3) or in whole animals (4) renders the infected cells resistant to superinfection by replication-defective recombinant virus. In addition, the availability of helper-free stocks of transforming retroviruses provides a means of determining the contribution of the helper virus component in retrovirus-mediated tumorigenesis.

ψ2 was created by transfecting NIH 3T3 fibroblasts with pMOV-ψ⁻, an ecotropic Moloney murine leukemia virus (Mo-MuLV) clone. pMOV-ψ⁻ expresses all the viral gene products, yet lacks a sequence known as ψ, required for encapsidation of the viral genome. The ecotropic viral envelope glycoprotein expressed by pMOV-ψ⁻ recognizes a receptor present only on mouse and closely related rodent cells. Hence, the use of ψ2 cells for retrovirus-mediated gene transfer is limited to a small number of species. However, MuLV strains do exist that possess a broad host range (for review, see ref. 5) as a consequence of differences in receptor specificity of the viral envelope gp70. For example, xenotropic, mink cell focus-forming, and amphotropic MuLV strains each recognize a receptor present on many mammalian cell types, including human cells. Moreover, ecotropic MuLV genomes (such as those we have previously used for the generation of recombinant genomes) can form viral pseudotypes that contain the envelope proteins from such strains and thereby acquire widened host range.

Here, we describe the construction and properties of ψ2-like packaging cell lines, termed ψ-AM, useful for the production of recombinant virus with amphotropic host range. These cell lines contain a modified pMOV-ψ⁻ genome in which the ecotropic envelope glycoprotein gene has been replaced with envelope sequences derived from the amphotropic virus 4070A. In addition, we have examined the integration and subsequent expression of a retroviral genome containing the neomycin phosphotransferase gene (23) in a number of human cell lines infected with virus stocks generated from the ψ-AM lines.

MATERIALS AND METHODS

NIH 3T3 cells were grown in Dulbecco's modified Eagle's medium (DME medium) containing 10% calf serum. The adherent HeLa cell line JW36 (provided by P. Sharp) and early passage human foreskin fibroblasts (provided by B. Weinberg) were grown in DME medium containing 10% fetal calf serum. The human histiocytic lymphoma line U937 (6) was grown in RPMI medium containing 10% fetal calf serum. All cell cultures producing recombinant amphotropic viruses were maintained under P3 containment conditions.

The plasmid pL1 was kindly supplied by A. Oliff. Plasmid constructions (7), transfections (8, 9), and reverse transcriptase assays (10) were done using standard techniques. Viral stocks were prepared by adding 10 ml of fresh DME medium containing 10% calf serum to a near confluent monolayer of cells and, after 18 hr, filtering the supernatant through a 0.45-μm filter (Milex). Infections were done by incubating 10^5 cells with 1 ml of virus plus 8 μg of Polybrene (Aldrich) for 2.5 hr.

Cell lines were selected for resistance to the drug G418 (GIBCO) at a concentration of 1 mg/ml, except for the U937 line, which required 2 mg/ml. Genomic DNA was prepared as described (7). Total cellular RNA was prepared using a guanidine thiocyanate procedure (11), denatured for 15 min at 55°C in sample buffer (50% formamide/2.2 M formaldehyde/1× running buffer/3% Ficoll/0.001% bromphenol blue), and electrophoresed on a 1% agarose gel containing 2.2 M formaldehyde/1× running buffer [5 mM Na acetate/1 mM EDTA/20 mM 3-(N-morpholino)propanesulfonic acid, pH 7.0]. Nick-translations and RNA blot and Southern transfers were done as described (7).

RESULTS

MuLVs encode two polyproteins, gag–pol and env (Fig. 1). To insert an amphotropic envelope gene into pMOV-ψ⁻ we

Abbreviations: MuLV, murine leukemia virus; cfu, colony-forming units.

CONE, R.D. and MULLIGAN, R.C.
High-efficiency gene transfer into mammalian cells: Generation of helper-free recombinant retrovirus with broad mammalian host range. *Proc. Natl. Acad. Sci. U.S.A.* 81:6349-6353 (1984). Reprinted with the authors' permission.

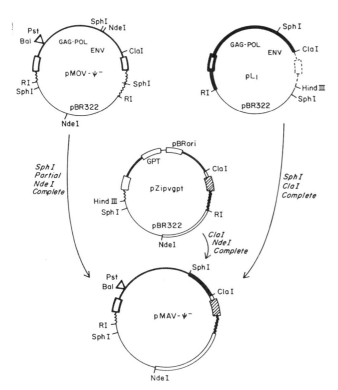

FIG. 1. Construction of pMAV-ψ⁻. pMAV-ψ⁻ was constructed by ligating together fragments from three retroviral clones, pMOV-ψ⁻, pL₁, and pZipvgpt. pBR322 sequences are represented by thin lines, Mo-MuLV sequences by medium thick lines, Ampho 4070A sequences by thick lines, F-MuLV sequences by dashed lines, and genomic sequences flanking proviruses by wavy lines. The three-part construction was necessitated by one or more *Sph* I sites in the genomic sequences flanking the 3′ long terminal repeat of pMOV-ψ⁻. Half of the pBR322 backbone and the 5′ portion of the retrovirus was contributed by pMOV-ψ⁻ (*Nde* I to *Sph* I), the *env* region was contributed by the amphotropic virus plasmid pL₁ (*Sph* I to *Cla* I), and the 3′ long terminal repeat through pBR322 to the *Nde* I site was contributed by pZipvgpt.

relied on the observation that the proviruses of amphotropic MuLVs share many common restriction sites with those of ecotropic strains (12). Very little sequence data exist for the amphotropic viruses, yet the accumulated restriction data suggest that these viruses, as a whole, differ from ecotropic strains primarily in the viral envelope glycoprotein, gp70, which is a cleavage product of the env polyprotein (12). Based on these observations, we chose to insert amphotropic gp70 sequences derived from the 4070A genome into pMOV-ψ⁻ between a *Sph* I site ≈200 amino acids before the carboxyl terminus of gag–pol and a *Cla* I site 15 amino acids before the carboxyl terminus of p15E. Both of these sites are found in similar positions in Mo-MuLV and 4070A.

The presence of one or more unmapped *Sph* I sites in the large mouse DNA sequence flanking the 3′ long terminal repeat of pMOV-ψ⁻ necessitated the three-part construction diagrammed in Fig. 1. The recombinant clone, called pMAV-ψ⁻, was cotransfected with the plasmid pSV2gpt (13) at a ratio of 10 μg/1 μg into NIH 3T3 fibroblasts, and 3 days later the cells were split into gpt selective medium (14). Thirty-one gpt⁺ clones were isolated by using cloning cylinders, and tested for reverse transcriptase activity. Ten clones were found to be positive, and the levels of reverse transcriptase activity varied widely, from levels just above background to levels greater than those found in wild-type virus-infected cells. Six clones with the highest reverse transcriptase activity were expanded for further analysis.

Our characterization of these ψ-AM cell lines was aimed toward determining which clones produced the highest viral titers after transfection with recombinant retroviral vectors, and also whether any of these clones produced detectable levels of wild-type amphotropic virus. After more than a month in culture, ψ-AM clones 22b, 224, and 227 were found to be negative for wild-type virus production by reverse transcriptase assay of NIH 3T3 cells infected with supernatants from these lines (data not shown). A more important test, however, would be to assay for wild-type virus production after stable transfection of these lines with retroviral transducing vectors that contain a ψ site for viral encapsidation. The cells would contain high levels of two viral transcripts, which, if copackaged into virions, might be expected to recombine at high frequency upon subsequent infection of cells, to regenerate fully wild-type amphotropic virus.

Virus Titers of Transfected ψ-AM Lines. The method used for characterizing ψ-AM clones is outlined in Fig. 2. ψ-AM lines were transfected with 10 μg of either pZipNeoSV(X)1 (15) or pMSVneo. Three days later, the cells were split into G418 selection. After 10 days to 2 weeks, G418ʳ clones were isolated by using cloning cylinders, and plates containing 100–1000 clones were pooled (henceforth referred to as ψ-AM populations). After continuous culture of clones and populations for >4 weeks post-transfection, virus harvests were taken as described. NIH 3T3 cells and JW36 HeLa cells were infected with various dilutions of these virus stocks and then split into G418 medium 72 hr later. The titers of ψ-AM virus stocks are comparable whether assayed on the murine NIH 3T3 line or the human HeLa line (Fig. 3).

Generation of Wild-Type Virus in ψ-AM Lines. The G418ʳ titers produced by individual ψ-AM clones varies widely. Titers are commonly as high as 10⁴ colony-forming units (cfu)/ml, and occasionally we have isolated clones such as ψ-AM2275, which produces 2 × 10⁵ G418ʳ cfu/ml. To determine whether any ψ-AM clones or ψ-AM populations were producing wild-type amphotropic virus, supernatants from NIH 3T3 cells infected with 1 ml of undiluted virus from each clone or population were assayed for reverse transcriptase activity (Fig. 2, first passage). The values shown are in arbitrary units, with a corresponding ψ-AM clone set at a value of 100. In addition, 1 ml of these same supernatants was assayed for G418ʳ cfu by reinfection of new NIH 3T3 and HeLa cells (Fig. 2, second passage). From >20 experiments of this type, we have shown that in most cases it is possible to transfect ψ-AM lines and generate G418ʳ cell populations or clones that continuously produce high titer stocks of helper-free recombinant virus, even after many months in culture. Nevertheless, we have also observed two different types of apparent recombinational events. For instance, NIH 3T3 cells infected with virus from the pZipNeoSV(X)1-transfected ψ-AM22b population produce a very low titer of amphotropic G418ʳ cfu (Fig. 2, row 1). However, these cells have no detectable reverse transcriptase activity, even after long-term culture. As expected, 3T3 and HeLa cells infected with supernatants from first passage 3T3 cells show no reverse transcriptase activity (Fig. 2, second passage) and do not produce any G418ʳ cfu (data not shown). Furthermore, the reverse transcriptase activity and titer produced by these first passage 3T3 cells remain stable over time.

A second recombinational event yielded wild-type virus with amphotropic host range. After transfection of ψ-AM224 and ψ-AM222 with the vector pMSVneo, both G418ʳ clones (Fig. 2, rows 17–19, 21) and populations (Fig. 2, rows 14 and 15) produced wild-type virus, as shown by reverse transcriptase assay of infected NIH 3T3 cells and the subsequent ability of first passage 3T3 cells to produce a titer of G418ʳ cfu comparable to that of the parent ψ-AM cells. Second passage 3T3 cells were also reverse transcriptase positive.

Kinetics of Virus Spread Suggests That the Generation of Wild-Type Virus Is Rapidly Detectable. In general, when testing ψ-AM cells for wild-type virus production, we have

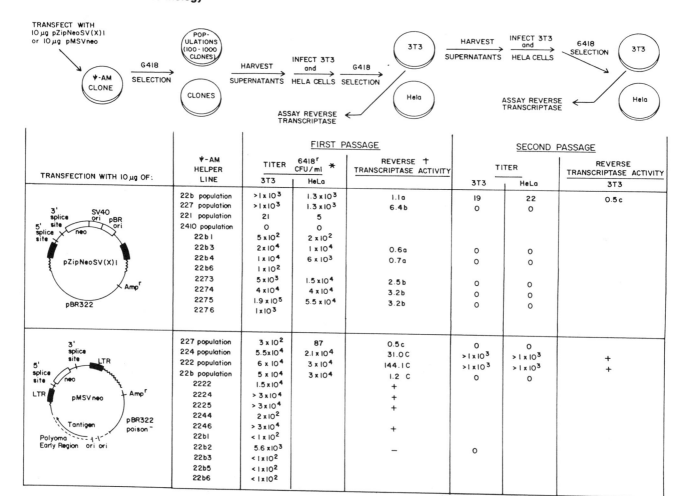

FIG. 2. Summary of ψ-AM transfection experiments. ψ-AM lines were transfected with two different retroviral vectors containing G418^r markers as shown. G418^r clones and populations (100–1000 clones) were isolated, and viral harvests were used to infect NIH 3T3 and HeLa cells (first passage). After several weeks in culture, supernatants from these G418^r 3T3 cells, previously infected with 1 ml of undiluted virus, were assayed for reverse transcriptase activity and used to infect new NIH 3T3 and HeLa cells (second passage). *, Titers were determined by counting the G418^r colonies found after infection with the greatest dilution of virus still yielding colonies. †, Reverse transcriptase activity is in arbitrary units with (a) ψ-AM22b, (b) ψ-AM227, or (c) ψ-AM222 set at a value of 100. +/− determinations were made by autoradiography after spotting the reverse transcriptase assay mix onto DE52 paper.

TRANSFECTION WITH 10 μg OF:	ψ-AM HELPER LINE	FIRST PASSAGE			SECOND PASSAGE		
		TITER 6418^r CFU/ml *		REVERSE † TRANSCRIPTASE ACTIVITY	TITER		REVERSE TRANSCRIPTASE ACTIVITY
		3T3	HeLa		3T3	HeLa	3T3
pZipNeoSV(X)1	22b population	>1x10^3	1.3x10^3	1.1a	19	22	0.5c
	227 population	>1x10^3	1.3x10^3	6.4b	0	0	
	221 population	21	5				
	2410 population	0	0				
	22b1	5x10^2	2x10^2				
	22b3	2x10^4	1x10^4	0.6a	0	0	
	22b4	1x10^4	6x10^3	0.7a	0	0	
	22b6	1x10^2					
	2273	5x10^3	1.5x10^4	2.5b	0	0	
	2274	4x10^4	4x10^4	3.2b	0	0	
	2275	1.9x10^5	5.5x10^4	3.2b	0	0	
	2276	1x10^3					
pMSVneo	227 population	3x10^2	87	0.5c	0	0	
	224 population	5.5x10^4	2.1x10^4	31.0C	>1x10^3	>1x10^3	+
	222 population	6x10^4	3x10^4	144.1C	>1x10^3	>1x10^3	+
	22b population	5x10^4	3x10^4	1.2 C	0	0	
	2222	1.5x10^4		+			
	2224	>3x10^4		+			
	2225	>3x10^4		+			
	2244	2x10^2					
	2246	>3x10^4		+			
	22b1	<1x10^2					
	22b2	5.6x10^3		−	0		
	22b3	<1x10^2					
	22b5	<1x10^2					
	22b6	<1x10^2					

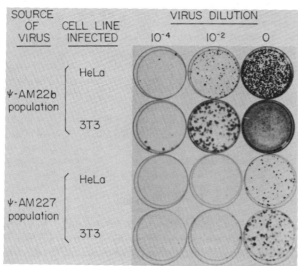

FIG. 3. Viral titering of pMSVneo-transfected ψ-AM cell lines. ψ-AM lines were transfected with a retroviral vector, pMSVneo, and G418^r populations were expanded. Approximately 5 × 10^5 HeLa or NIH 3T3 cells were infected with 1 ml of virus diluted as shown. After 3 days, ≈5 × 10^5 cells were put into medium containing G418. Selective medium was changed every 3 days, and after 10 days to 2 weeks the cells were stained with crystal violet.

maintained the cells in culture 2–4 weeks post-transfection before taking a viral harvest, and then maintained the subsequently infected 3T3 cells in culture, again for 2–4 weeks, before assaying for reverse transcriptase or G418^r cfu. To show that the spread of wild-type virus through a culture could be detected in this period of time, we have conducted a number of reconstitution assays, an example of which is shown in Fig. 4. NIH 3T3 cells infected with 1 ml of undiluted virus from pZipNeoSV(X)1-transfected ψ-AM227 and ψ-AM22b populations were cultured for 17 days. Then, 10^5 of each of these cells were split into 10 ml of normal medium on a 100-mm Petri dish with 10, 1000, or 100,000 wild-type ecotropic Mo-MuLV, or amphotropic 1504A virus-producing cells. The cultures were split 1:20 every 2–3 days, and 18-hr harvests from confluent plates were assayed for reverse transcriptase activity. Reverse transcriptase activity plateaus at wild-type levels in 7–11 days and is easily detectable at 3–4 days, even if only 10 wild-type virus-producing cells are added.

Expression of Recombinant Murine Retrovirus in ψ-AM-Infected Human Cell Lines. A number of investigators have reported on the variety of non-murine cells infectable with the amphotropic viruses (16, 17). In general, the immunofluorescence focus assays used to detect infection indicate that at least some viral antigens must be expressed in the non-murine cell types infected. However, little is known about transcription of murine retroviruses in human cells. To ensure

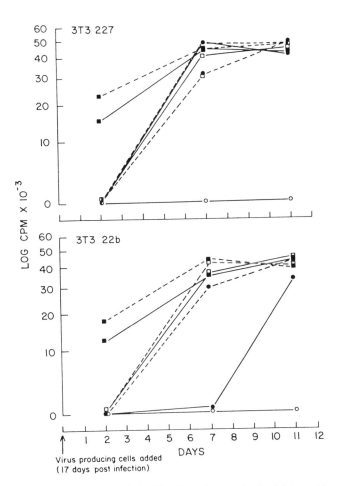

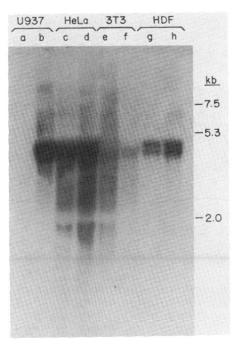

FIG. 4. Analysis of the kinetics of spread of wild-type Mo-MuLV and 1504A virus in NIH 3T3 cultures. NIH 3T3 cells were infected with 1 ml of undiluted virus from populations of pZip-NeoSV(X)1-transfected ψ-AM227 or ψ-AM22b lines. Seventeen days after infection 0 (○), 10 (●), 1000 (□), or 100,000 (■) Mo-MuLV (broken lines) or 1504A (solid lines) producer cells were added to 10⁵ NIH 3T3 cells. Cultures were maintained for 12 days, and the cells were split 1:20 every 2–3 days. Just before confluence, 18-hr viral harvests were collected for reverse transcriptase assays. The logarithm of the total counts incorporated into high-molecular-weight material during a standard reverse transcriptase reaction is plotted against number of days after addition of virus-producing cells.

FIG. 5. RNA blot analysis of ψ-AM infected cells. Cells were infected with virus from pZipNeoSV(X)1-transfected ψ-AM clones or populations. After selection in G418 medium, and cloning in the case of the human diploid fibroblasts (HDF), total cellular RNA was electrophoresed on a formaldehyde gel, transferred to a nylon-type membrane (Zetabind), and probed with a nick-translated pBR322 derivative containing the neomycin phosphotransferase gene (pXLneo). Thirty micrograms was loaded per lane except for lanes a (6 μg), g (5 μg), and h (3 μg). Size markers were 18S rRNA, 2 kilobases (kb); 28S rRNA, 5.3 kb; and polio RNA, 7.5 kb. Cells were infected with virus from mock control (lane a), ψ-AM22b (lanes b, d, and f), ψ-AM227 (lanes c and e), ψ-AM22bl (lane g), and ψ-AM2274 (lane h).

that ψ-AM-packaged retroviral vectors will be useful for gene transduction into human cells, we examined the integration and expression of ψ-AM pZipNeoSV(X)1 virus in three human cell types, JW36 HeLa cells, U937 (a histiocytic lymphoma line), and early passage human diploid fibroblasts. The Mo-MuLV promoter is a moderately strong one in murine cells, because the level of viral mRNA in infected cells is estimated to be 5–10% of the total cellular mRNA (18, 19). We analyzed the levels of viral RNA present in infected human cells by RNA blot analysis of total cellular RNA, using the neo gene as a probe (Fig. 5). Surprisingly, we found a higher level of viral RNA in all three human lines than in NIH 3T3 cells. The slight difference in migration between the human and murine RNAs appears to be an artifact, because the bands comigrated on some gels and in mixing experiments. The vector pZipNeoSV(X)1 constructed by Cepko et al. (15) contains ≈400 nucleotides between the splice donor and splice acceptor. The two bands in Fig. 5 correspond to the 4.7-kilobase full-length viral message and the 4.3-kilobase spliced subgenomic message. Since the two messages always appear to be present in equimolar amounts, this suggests that the unusual partial splice of retrovirus is also regulated properly in human cells. This fulfills an impor-

tant requirement for many of the vectors constructed in our laboratory, which depend on the production of a spliced transcript for expression of a selectable gene such as neo, and production of a genomic RNA for transmission of the recombinant genome and expression of nonselectable DNA sequences. As expected from our RNA blot data, the proviral genome in infected human cells appears to reside in the high-molecular-weight DNA in the form of full-length proviruses (Fig. 6). In addition, there do not appear to be any preferred integration sites in the human genome, at least none that occur more than ≈10% of the time, because no discrete flanking fragments are detectable when the DNA is cut with the endonuclease Bgl II (which cleaves the recombinant proviral genome once) and probed with the neo gene.

DISCUSSION

The construction of ψ-AM packaging lines extends the versatility of murine retroviral vectors in allowing the infection of a broad range of mammalian cell types. With the exception of some lymphoid cell lines, nearly all human cell lines tested in our laboratory have been infectable as assayed by G418ʳ. Some examples include the HeLa, U937, HepG2, and K562 cell lines, several secondary diploid fibroblast cultures, and secondary keratinocyte cultures (unpublished work). The monkey cell lines CV1 and Cos are also infectable as are a variety of murine and rat cell lines. Notably, we have been unable to infect Chinese hamster ovary cell lines at high efficiency with ψ2 or ψ-AM virus, although G418ʳ colonies do appear at low frequency with some lines.

The ψ-AM line is also useful for introducing retroviral vectors into murine cells refractory to ecotropic virus as a result of previous infection (e.g., murine erythroleukemia cells).

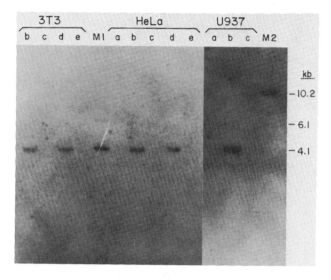

pZipNeoSV(X)1

FIG. 6.   Southern analysis of ψ-AM infected cells. Genomic DNA was isolated from aliquots of the same cells used in Fig. 5 that were infected with virus from mock supernatants (lanes a), ψ-AM22b (lanes b and c), and ψ-AM227 (lanes d and e). DNA (10 μg) was cut with *Xba* I (lanes b and d) or *Bgl* II (lanes c and e), run on a 1% agarose gel, transferred to a nylon-type filter, and probed with pXLneo. Markers are 10 pg of pZipNeoSV(X)1 plasmid cut with *Xba* I (lane M1) or *Bgl* II (lane M2). kb, Kilobases.

Since ψ-AM- and ψ2-generated virus each recognize unique receptors, both of which are present on NIH 3T3 cells, we can also rapidly shuttle retroviral genomes between ψ2 and ψ-AM to change the host range of the virus or to produce virus stocks containing two different genomes.

The titer produced by ψ-AM lines appears, on average, to be ≈1/10th that produced by ψ2 cells transfected with the same retroviral vectors. Since pMAV-ψ⁻ contains primarily ecotropic sequences, except for gp70 and p15E, the amphotropic envelope proteins may be less efficient in mediating infection than are their ecotropic counterparts. Nevertheless, one can readily isolate lines such as ψ-AM2275 that produce >10⁵ recombinant virus per ml. These titers are high enough to facilitate the nonselective introduction of genes into 100% of a population of cells at high enough cell numbers to allow rapid analysis of DNA, RNA, or protein.

We have observed two different events that lead to the ability of NIH 3T3 cells infected with virus from transfected ψ-AM cells to produce G418ʳ cfu. In the first instance, the pZipNeoSV(X)1-transfected ψ-AM22b population appears to pass the *pMAV-ψ⁻* genome without subsequent recombination to generate wild-type amphotropic virus. Since retroviral recombination of copackaged genomes occurs at frequencies estimated as high as 10% (20, 21), our data suggest that the frequency of passage of the *pMAV-ψ⁻* genome and/or the frequency of its copackaging with pZipNeoSV(X)1 must be extremely low. It is noteworthy that retroviruses are known to package nonviral RNAs at low frequency (22).

The second type of event is demonstrated by the pMSVneo-transfected ψ-AM222 and ψ-AM224 populations, which both produce wild-type amphotropic virus. Individual

clones from these transfections also produce wild-type virus, suggesting that a recombinational event occurred soon after transfection, allowing virus to spread throughout the culture. Ordinarily, wild-type ecotropic virus-infected cells are refractory to further infection by a factor of 10⁴ relative to uninfected cells. However, ψ2 cells are only 1/100th as infectable as NIH 3T3 cells (unpublished observations), and this is likely to be true for ψ-AM cells as well. Therefore, any wild-type virus, whether generated by DNA–DNA recombination or viral recombination, should spread rapidly, even through a culture of ψ-AM cells, during the 2–4 weeks required to isolate stably transfected clones and populations. The occasional generation of wild-type virus after transfection of ψ-AM cell lines suggests that transfected ψ-AM clones or populations should be tested for such activity prior to use. Interestingly, the recombinational events that lead to regeneration of fully wild-type amphotropic virus have only been observed soon after transfection, while the majority of ψ-AM producers are helper-free after transfection and remain helper-free even after long-term culture.

We thank Alan Oliff for supplying us with cloned 4070A proviral DNA. We also thank members of the laboratory for helpful discussion, Devon Young for preparing the manuscript, and Claire Pritchard for technical assistance. This work was supported by grants from the National Cancer Institute and the MacArthur Foundation. R.D.C. was supported by National Institutes of Health Training Grant GM07287.

1. Mann, R., Mulligan, R. C. & Baltimore, D. B. (1983) *Cell* **33**, 153–159.
2. Watanabe, S. & Temin, H. M. (1983) *Mol. Cell. Biol.* **3**, 2244–2249.
3. Rubin, H. (1960) *Proc. Natl. Acad. Sci. USA* **46**, 1105–1119.
4. Stuhlmann, H., Cone, R. D., Mulligan, R. C. & Jaenisch, R. (1984) *Proc. Natl. Acad. Sci. USA* **81**, in press.
5. Weiss, R. (1982) in *RNA Tumor Viruses*, eds. Weiss, R., Teich, N., Varmus, H. & Coffin, J. (Cold Spring Harbor Laboratory, Cold Spring Harbor, NY), pp. 232–243.
6. Sundstrom, C. & Nilsson, K. (1976) *Int. J. Cancer* **17**, 565–577.
7. Maniatis, T., Fritsch, E. F. & Sambrook, J. (1982) *Molecular Cloning: A Laboratory Manual* (Cold Spring Harbor Laboratory, Cold Spring Harbor, NY).
8. Graham, R. & Van der Eb, A. (1973) *Virology* **52**, 456–467.
9. Parker, B. A. & Stark, G. R. (1979) *J. Virol.* **31**, 360–369.
10. Goff, S., Traktman, P. & Baltimore, D. (1981) *J. Virol.* **38**, 239–248.
11. Chirgwin, J. M., Przybyla, A. E., MacDonald, R. J. & Rutter, N. J. (1980) *Biochemistry* **18**, 5294–5299.
12. Chattopadhyay, S. K., Oliff, A. I., Linemeyer, D. L., Lander, M. R. & Lowy, D. R. (1981) *J. Virol.* **39**, 777–791.
13. Mulligan, R. C. & Berg, P. (1980) *Science* **209**, 1422–1427.
14. Mulligan, R. C. & Berg, P. (1981) *Proc. Natl. Acad. Sci. USA* **78**, 2072–2076.
15. Cepko, C. L., Roberts, B. E. & Mulligan, R. C. (1984) *Cell* **37**, 1053–1062.
16. Hartley, J. N. & Rowe, W. P. (1976) *J. Virol.* **19**, 19–25.
17. Rasheed, S., Gardner, M. B. & Chan, E. (1976) *J. Virol.* **19**, 13–18.
18. Fan, H. (1977) *Cell* **11**, 297–305.
19. Bishop, J. M. (1978) *Annu. Rev. Biochem.* **47**, 35–88.
20. Vogt, P. K. (1971) *Virology* **46**, 947–952.
21. Wong, P. K. Y. & McCarter, J. A. (1973) *Virology* **53**, 319–326.
22. Linial, M., Medeiros, E. & Hayward, W. S. (1978) *Cell* **15**, 1371–1381.
23. Southern, P. J. & Berg, P. (1982) *J. Mol. Appl. Genet.* **1**, 327–341.

Production of Human Tissue Plasminogen Activator in Transgenic Mouse Milk

K. Gordon, E. Lee, J.A. Vitale, A.E. Smith, H. Westphal and L. Hennighausen

We set out to express an exogenous gene in the mammary epithelium of transgenic mice in the hope that the encoded protein would be secreted into milk. The promoter and upstream regulatory sequences from the murine whey acid protein (WAP) gene were fused to cDNA encoding human tissue plasminogen activator (t-PA) with its endogenous secretion signal sequence. This hybrid gene was injected into mouse embryos, resultant transgenic mice were mated, and milk obtained from lactating females was shown to contain biologically active t-PA. This result establishes the feasibility of secretion into the milk of transgenic animals for production of biologically active heterologous proteins, and may provide a powerful method to produce such proteins on a large scale.

Genes injected into mouse embryos may be incorporated into the germ line and be expressed in patterns that mimic those of their endogenous counterparts[1,2]. The pattern of spatial and temporal expression of foreign genes in transgenic animals can be controlled by prior manipulation of the signals regulating gene expression. We introduced into mice a construct designed to express a foreign protein in the lactating mammary epithelium in which 5' sequences from the whey acid protein gene were fused with a cDNA coding for tissue plasminogen activator. We demonstrate here that such an approach is a feasible means of expression of foreign proteins into secreted milk.

Whey acid protein (WAP) is the most abundant whey protein in mouse milk[3]. During lactation, the level of WAP RNA in the mammary gland increases approximately 340-fold from the barely detectable levels present in the mammary gland of virgin mice[4], and accumulates in lactating tissue at levels of about 15% of the total mRNA[4–6]. Expression of the WAP gene and the stabilization of its mRNA are subject to complex regulation by both steroid and peptide hormones[4], and putative regulatory protein binding sites within the WAP promoter have been described[7]. Since WAP is found in mouse milk at high levels and the gene had been previously cloned and

characterized[8,9], we chose to utilize WAP upstream DNA as a promoter in our expression vector. By demonstrating secretion of a foreign protein into milk, the results reported here extend earlier observations showing upstream sequences from the WAP gene were able to target gene expression to the lactating mammary gland in transgenic mice[10].

RESULTS AND DISCUSSION

Construction of t-PA expression vector. A mammary expression vector was constructed in which 5' sequences from the whey acid protein gene were fused with a cDNA coding for tissue plasminogen activator. t-PA has great potential clinical utility as an agent to dissolve fibrin clots and thus treat victims of myocardial infarction and other life threatening conditions. Its advantage relative to other pharmacological agents such as streptokinase and urokinase lies in its specificity for fibrin. Moreover, assay of its biological activity is both sensitive and convenient and an antibody kit is available for routine screening.

The t-PA gene utilized here was a cDNA clone from a human uterus cDNA library. The t-PA DNA sequence was determined previously and the protein expressed in C127 cells using bovine papilloma vectors[11]. The construct shown in Figure 1 (designated WAP-tPA) is a tripartite fusion consisting of 2.6 kb of upstream DNA from the WAP gene through the endogenous CAP site, t-PA cDNA beginning in the untranslated 5' region, and the polyadenylation/termination signals from SV40. This tPA/SV40 polyadenylation cassette was characterized previously[11]. The secretion signal sequence in this construct derives from the native t-PA gene; the analogous signal encoding region from the WAP gene was removed in the construction. We did not know a priori whether the t-PA secretion signal would function efficiently in native mammary epithelial cells. However, since many proteins with different signal peptides are secreted efficiently by mammary cells and since milk proteins can be efficiently transported by membrane systems from other cells[12], it seemed likely that no specific signal sequence is required for secretion in mammary tissue.

Transient expression in tissue culture. To test WAP-tPA for its ability to specify production of secretable t-PA in mammary epithelial cells, the fusion gene was transfected into the mammary cell line, MCF7. Tissue culture supernatants collected 48 hours after transfection were loaded into wells of an assay plate as shown in Figure 2A. The assay consisted of lysis (clearing) of an artificial fibrin clot laid down as a matrix in agarose poured into the wells of a tissue culture plate. The degree of clearing, determined by estimating the diameter of the cleared ring

GORDON, K., LEE, E., VITALE, J.A., SMITH, A.E., WESTPHAL, H. and HENNIGHAUSEN, L.
Production of human tissue plasminogen activator in transgenic mouse milk. *Bio/Technology* 5:1183-1187.

emanating from the sample loading well, indicates the amount of active t-PA in the sample. In five repetitions from two separate transfections, the level of t-PA secreted into the culture medium was found to be 2.5, 1.5, 10, 5, and 5 ng/ml. The same five samples were also assayed by ELISA using a polyclonal anti-human t-PA antibody. By this assay, the expression levels were either below the detection limit (approximately 2 ng/ml) or 11, 10, and 10 ng/ml, respectively. Thus, whether assayed by biological activity or immunologically, MCF7 cells transfected with WAP-tPA were able to secrete t-PA.

Generation of transgenic animals. The plasmid WAP-tPA was injected into one-cell pronuclear mouse embryos as a purified Hind-3/BamHI fragment containing no procaryotic sequences. The injected embryos were implanted into pseudopregnant females and 29 mice were born. Of these, seven were identified as being transgenic by diagnostic Southern blot hybridization with a human cDNA t-PA probe. Under conditions of high stringency, this probe does not hybridize with the endogenous mouse t-PA gene. The blot patterns of three positive mice, #wt1-26, wt1-25 and wt1-7 are shown in Figure 1. By comparison to the hybridization intensity obtained with positive controls, the number of copies of the injected fragment present in the genomes of these transgenic mice was estimated to be between 20 and 50. Digestion with SacI (lanes b-d) yielded a diagnostic band of 1.75 kb that spans the WAP and t-PA junction and hybridizes to the probe (Fig. 1). The intact plasmid digested with SacI was used as a positive control for this digest (lane a). Exogenous DNA injected into embryos tends to form concatomers even when introduced as a fragment with non-cohesive ends. The 2.3 kb band seen in lanes b-d corresponds to the 3' end of the t-PA gene (which does not contain SacI sites), apparently ligated to the 5' end of the WAP promoter, and through to the first Sac 1 site in the WAP DNA. The presence and size of this fragment is diagnostic of head-to-tail concatomers.

The EcoRI digest (control, lane h; experimentals, lane i-k) showed the expected 472 bp band internal to the t-PA gene. In addition, a 3.3 kb band can be seen that represents the 5' region of the t-PA gene and extends through the WAP gene to 5' boundary EcoRI site. Thus, despite the fact that the WAP EcoRI site was near the end of the injected fragment, it appeared to be intact in the genomic DNA of this transgenic animal. The 1.2 kb band represents the 3'-most region of the t-PA gene, which must have ligated head-to-tail to the 5' end of the WAP gene, leaving the t-PA gene bounded on its 3' end by an EcoRI site. Interestingly, the weak 2.3 kb band indicates that some of the copies of the fragments formed concatomers in a head-to-head configuration. KpnI digestion (lanes e-g) produced a single band of 4.9 kb, as expected. It is impossible to determine from this Southern blot whether all copies of the concatomer integrated at a single or at multiple sites.

Expression of biologically active t-PA in milk. Mice #wt1-26 and wt1-25 were mated to wild type males and had no apparent difficulty in conception or maintenance of pregnancy. Several days after parturition, milk was obtained from the females and was assayed for t-PA activity. Since wt1-11 was a male, it was necessary to obtain transgenic female progeny, mate them, and obtain milk from the second generation females after parturition. We have characterized expression from one progeny animal of this lineage, wt2-102. Since it was not known whether the mouse milk itself would interfere in the fibrin clot assay, we used standards consisting of recombinant t-PA added to milk from nontransgenic mice. As shown in Figure 2b, dilution of standards in milk did not affect the

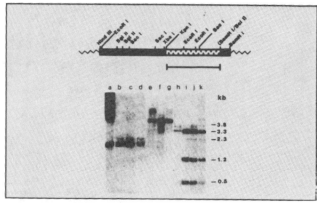

FIGURE 1 Generation of transgenic animals. Top portion: Restriction map of WAP-t-PA; Bottom portion: Southern blot of DNA from tails. Lanes a and h show 500 pg of WAP-tPA DNA digested with Sac I or EcoRI, respectively. Lanes b, c, and d contain 5 µg of DNA from mouse wt1-26, wt1-25 and wt1-7, respectively, digested with Sac I. Lanes e-g are from mouse tails of wt1-26, wt1-25 and wt1-7, respectively, cut with Kpn I, and lanes i-k are these DNAs cut with Eco RI. Lanes a-d were run on a separate gel than the rest of the lanes. Negative control DNAs did not show any hybridization to this probe under these conditions (not shown).

apparent concentration (in comparison to standards diluted in PBS), nor was there background clearing in the negative control sample wells. In this figure it can be seen that milk from wt1-26 cleared the fibrin clot to a significant extent. By comparison with lysis catalyzed by known amounts of added t-PA, the concentration was calculated to be about 200 ng/ml. In parallel assays, milk obtained from wt1-25 and wt2-102 was shown to contain 200 ng/ml and 400 ng/ml of t-PA (data not shown). When plates were incubated longer than 24 hours, minor clearing was seen in control wells containing milk from untransfected mice, but this was always significantly less than clearing seen from milk of any of the transgenic lineages. The origin of the residual fibrinolytic activity in non-transgenic mouse milk is not known. However, the presence of low levels of plasminogen activator (PA) in the lactating mammary gland of rodents[13] raises the possibility that some fibrinolytic protein is present naturally in milk.

Milk from mice wt1-26, wt1-25 and 2-102 was assayed by ELISA using an anti-human t-PA polyclonal antibody (Fig. 3). A standard curve was generated by addition of known amounts of human t-PA to mouse milk. Identical curves were generated by dilution of t-PA in cow milk and aqueous buffer (not shown). The inset of Figure 3 shows results of an assay of serial dilutions of milk from wt1-26 confirming that about 300 ng/ml of t-PA was present in this sample. Milk from wild type mice showed no signal in the ELISA. Milk from wt1-25 and wt2-102 contained t-PA at concentrations of 114 and 460 ng/ml, respectively (data not shown). The measurements by fibrin clot lysis and ELISA were not sufficiently accurate to determine precisely the specific activity of the t-PA produced in milk. Further studies of the purified protein (now in progress) will establish whether the specific activity of the protein is identical to that produced by melanoma cells and by recombinant DNA methods. Pilot studies indicate that t-PA remains stable and bioactive in whole milk for at least 48 hours at 37°C, and can be stored at -80°C (data not shown).

Since WAP RNA constitutes as much as 15% of the poly A (+) mRNA in the lactating mammary gland, it is probable that the level of t-PA in the milk of these mice is far below the level of endogenous whey acid protein. This

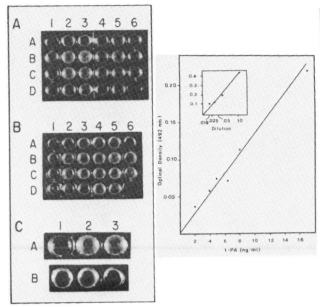

FIGURE 2 Clot lysis bioassay of secreted t-PA. (A) Transient expression of WAP t-PA in MCF7 cells. Two transfections were done, #1 (two repetitions, a and b) and #2 (3 repetitions, a, b, and c). In row A, columns 1 through 6 are recombinant DNA t-PA standards diluted in culture medium at concentrations of 20, 10, 5, 2.5, 1.25, and 0 ng/ml. Row B contains samples from transfection #1. In columns 1, 2, and 3 are three dilutions of sample from one transfection: 1×, .5×, and .25×, respectively. In columns 4, 5, and 6 or row B are similar dilutions from the repeat transfection. Row C, columns 1–3, row C, columns 4–6, and row D, columns 1–3 contain similar dilutions from the three repetitions of transfection #2. (B) Milk from transgenic mouse #wt1–26. Following identification of mouse #wt1–26 as a positive transgenic, the mouse was mated to a wild type male. Seventeen days after the first litter was born, milk was removed from the lactating female following stimulation with oxytocin. Milk was diluted in PBS by 50% and stored frozen. Milk was diluted further in PBS as indicated below just prior to assay and added to the wells of a fibrin clot lysis plate. The positive controls were generated by addition of recombinant t-PA to media composed of either 10% negative mouse milk (row A), 10% negative cow milk (row B), or PBS (row C). Concentrations of t-PA in the milk dilution curves, from columns 1 through 5 are: 40, 20, 10, 5, and 0 ng/ml. The concentrations in the PBS dilution curve, from column 1 through 6 are: 40, 20, 10, 5, 2.5, and 0 ng/ml. In row A, column 6 is the milk from mouse #wt1–26 at a final concentration of 10% and in row B, column 6 is the milk at a concentration of 5%. This photograph was taken after approximately 8 hours of assay incubation time. The negative mouse milk used for these controls was pooled from outbred CD-1 mice in different stages of lactation. In other negative controls (not shown) milk was used from inbred females of the same strain used for microinjection and was obtained at the same stage of lactation as the positive sample. The specificity of recombinant t-PA secreted into mouse milk was shown to be plasminogen in other experiments (not shown) in which plasminogen was omitted from the agarose matrix in similar fibrin clot lysis assays. (C) Enlargement of key data of sections A and B. Row A shows an enlargement of the data of section A (above), row C, columns 1–3; Row B shows one of the data of section B (above), row A, columns 4–6.

FIGURE 3 Quantitation by ELISA of recombinant t-PA secreted into milk of mouse wt1–26. The standard curve was performed in negative mouse milk diluted to a final concentration of 10% with PBS, to which was added t-PA supplied with the kit as indicated. The inset shows milk from mouse #wt1–26 in dilutions as indicated. The dilution of .1 refers to a final concentration of 10% milk. In each dilution of transgenic milk, samples were supplemented with negative mouse milk in order to keep the final concentration at 10%. All points of the control curve and the experimental (inset) curve have the background value (the value determined for negative mouse milk) subtracted.

could be due to many factors. Preliminary data indicate that variation in t-PA expression levels among transgenic mice containing WAP-tPA may be considerable, suggesting that the chromosomal integration site may play a key role in establishing levels of expression from this construction. In fact, one transgenic mouse (not shown) appeared to express virtually no t-PA in the milk. Thus, analysis of additional animals may identify those which produce more t-PA. In addition, intragenic and/or noncoding 5′ and 3′ sequences from the WAP gene, missing from the construction introduced into mice in these experiments, may play important roles in RNA stability. Considerable work remains to be done to configure the t-PA expression vector for maximal expression.

We demonstrate here that a foreign protein, human tissue plasminogen activator, can be secreted into the milk of transgenic mice under the control of a mammary-specific promoter. Thus, concerns that foreign proteins produced in the mammary gland might not be secreted, accurately processed, or be sufficiently stable in milk appear to be reduced by these results. The advantages of producing foreign proteins in this manner include the fact that milk is well characterized biochemically and that many of the genes encoding key milk proteins have been cloned. In addition, many milk-specific genes are expressed in the lactating mammary gland at high levels under hormonal control and in a tissue-specific manner. Thus, with expression cassettes similar to the one described here, it should be possible to target precisely foreign gene expression to the lactating mammary epithelium. Factor IX and t-PA have been produced in the blood of transgenic mice[14,15]; the ability to produce these proteins in milk would facilitate their collection. The ultimate goal of our experiments is to express foreign proteins in the milk of farm animals. Since production of transgenic farm animals has been achieved[16], this presents a reasonable possibility. Although many technical hurdles remain, the data presented here demonstrate that transgenic animals may become an attractive alternative for future production of genetically engineered biologically active proteins.

EXPERIMENTAL PROTOCOL

Construction of expression vector. A Hind-III site was added at the 5′ end of the 2.6 kb WAP promoter sequence ([■■] -Ref. 5) by digestion at the single EcoRI site in the WAP promoter, filling in with Klenow and dNTPs, and ligation of Hind-3 linkers, all by standard protocols. The t-PA cDNA ([M])-SV40 termination/polyadenylation (■) cassette (see Ref. 11) was inserted into the polylinker region of the WAP promoter vector as a KpnI-Bam HI fragment.

Generation and analysis of transgenic animals. To purify the eukaryotic sequences for microinjection, WAP-tPA was digested with Hind-3 and BamHI, the fragments separated by gel electrophoresis, and the 4.9 kb fragment purified by binding to glass filter fiber papers[17]. After elution and concentration by ethanol precipitation, the DNA was suspended for microinjection in 10mM Tris, .05 mM EDTA, pH 7.5 at a concentration of .5 ng/microliter. The regulatory/coding restriction fragment from pWAP-tPA was microinjected into one cell fertilized embryos as described previously[18]. At four weeks of age, tail sections were taken from mice born from these injections, digested with proteinase K, phenol-chloroform extracted, then digested with various restriction enzymes. DNA was electrophoresed on a Tris-borate gel, blotted to nitrocellulose, and hybridized with a probe consisting of the entire coding region of t-PA cDNA (see bold line under the restriction map of Fig. 1). Labeling was done by extension of random hexamers to a specific activity of 1×10^9 cpm/µg.

Calcium phosphate transformation. MCF7 cells were plated in 100 mm dishes at densities of 5×10^6 or more per dish at least one day prior to transfection. Transfections were performed as described previously[19] and transient supernatants collected 48 hours after transfection were assayed for t-PA.

Fibrin clot lysis assay. The fibrin clot assay measures the ability of t-PA to digest fibrinogen matrices which are laid down in a background of agarose, thrombin and plasminogen within the wells of a plate [20]. A small hole is bored through the agarose mixture upon hardening and 25 microliters of the samples are loaded into each of the holes. As t-PA diffuses into the agarose, clearing of the fibrinogen is evident visually and the amount of clearing is directly proportional to the amount of active t-PA. These assays are extremely sensitive and reproducible.

ELISA assay. Assays were performed with the IMUBIND ELISA kit produced by American Diagnostica Inc. The assay is a double antibody sandwich in which the primary antibody is a goat antiserum raised against t-PA from human uterus and the second antibody is a peroxidase conjugated anti-t-PA IgG. The standard curves were performed in negative mouse milk diluted to a final concentration of 10% with PBS, to which was added t-PA supplied with the kit.

Acknowledgments

The authors would like to thank N. Capalucci for help in assembling the figures, and S. Chappel, N. Cole, S. Groet, and G. Moore for discussions and critical review of the manuscript. We are also grateful to J. Khillan for his contribution during the early phases of this work and to G. Parsons for encouragement and stimulating discussions.

Received 16 June 1987; accepted 17 August 1987.

References
1. Gordon, K., and Ruddle, F. H. 1986. Gene transfer into mouse embryos. *In:* Development Biology. Gwatkin, R. B. L. (ed.) Plenum Publishing Corp., New York. **4**:1–36.
2. Palmiter, R. D. and Brinster, R. L. 1986. Germ-line transformation of mice. Ann. Rev. Genet. **20**:465–499.
3. Hennighausen, L. G. and Sippel, A. E. 1982. Characterization and cloning of the mRNAs specific for the lactating mouse mammary gland. Eur. J. Biochem. **125**:131–141.
4. Hobbs, A. A., Richards, D. A., Kessler, D. J., and Rosen, J. M. 1982. Complex hormonal regulation of rat casein gene expression. J. Biol. Chem. **257**:3598–3605.
5. Richards, D. A., Rodgers, J. R., Supowit, S. C., and Rosen, J. M. 1981. Construction and preliminary characterization of the rat casein and lactalbumin cDNA clones. J. Biol. Chem. **256**:526–532.
6. Hennighausen, L. G., Sippel, A. E., Hobbs, A. A., and Rosen, J. M. 1982. Comparative sequence analysis of mRNAs coding for mouse and rat whey acid protein. Nucleic Acids Res. **10**:3732–3744.
7. Lubon, H. and Hennighausen, L. 1987. Nuclear proteins from lactating mammary glands bind to the promoter of a milk protein gene. Nuc. Acids Res. **15**:2103–2121.
8. Hennighausen, L. G. and Sippel, A. E. 1982. Mouse whey acidic protein is a mouse member of the family of four "disulfide core" proteins. Nucleic Acids Res. **10**:2677–2684.
9. Campbell, S. M., Rosen, J. M., Hennighausen, L., Strech-Jurk, U., and Sippel, A. E. 1984. Comparison of the whey acidic protein genes of the rat and mouse. Nucleic Acids Res. **12**:8685–8696.
10. Andres, A.-C., Schonenberger, C.-A., Groner, B., Hennighausen, L., LeMeur, M., and Gerlinger, P. 1987. Ha-*ras* oncogene expression directed by a milk protein gene promoter: tissue specificity, hormonal regulation, and tumor induction in transgenic mice. Proc. Natl. Acad. Sci. USA **84**:1299–1303.
11. Reddy, V. B., Garramone, A. J., Sasak, H., Wei, C.-M., Watkins, P., Galli, J., and Hsuing, N. 1987. Expression of human tissue-type plasminogen activator in mouse cells using BPV vectors. DNA, *in press.*
12. Craig, R. K., Perera, P. A. J., Mellor, A., and Smith, A. 1979. Initiation and processing in vitro of the primary translation products of guinea-pig caseins. Biochem. J. **184**:261–267.
13. Ossowski, L., Biegel, D., and Reich, E. 1979. Mammary plasminogen activator: correlation with involution, hormonal modulation and comparison between normal and neoplastic tissue. Cell **16**:929–940.
14. Busby, S. J., Bailey, M. C., Mulvihill, E. R., Joseph, M. L., and Kumar, A. A. 1986. Expression of human tissue plasminogen activator in transgenic mice. American Heart Assn. **74**:II-247a.
15. Choo, K. H., Raphael, K., McAdam, W., and Peterson, M. G. 1987. Expression of active human blood clotting factor IX in transgenic mice: use of cDNA with complete mRNA sequence. Nuc. Acids Res. **15**:871–884.
16. Hammer, R. E., Pursel, V. G., Rexroad, C. E., Wall, R. J., Bolt, D. J. Ebert, K. M., Palmiter, R. D., and Brinster, R. L. 1985. Production of transgenic rabbits, sheep and pigs by microinjection. Nature **315**:680–683.
17. Yang, R. C.-A., Lis, J., and Wu, R. 1979. Elution of DNA from agarose gels after electrophoresis. Meth. in Enz. **68**:176–182.
18. Overbeek, P. A., Chepelinsky, A. B., Khillan, J. S., Piatigorsky, J., and Westphal, H. 1985. Lens-specific expression and developmental regulation of the bacterial chloramphenicol acetyltransferase gene driven by the murine α-A crystallin promoter in transgenic mice. Proc. Natl. Acad. Sci. USA **82**:7815–7819.
19. Hsiung, M., Fitts, R., Wilson, S., Milne, A., and Hamer, D. 1984. Efficient production of hepatitis B surface antigen using a bovine papilloma virus-metallothionein vector. Jour. Mol. and Appl. Genet. **2**:497–506.
20. Granelli-Piperno, A. and Reich, E. J. 1978. A study of proteases and protease inhibitor complexes in biological fluids. J. Exp. Med. **148**:223–234.

Dramatic Growth of Mice that Develop from Eggs Microinjected with Metallothionein-growth Hormone Fusion Genes

R.D.Palmiter, R.L. Brinster, R.E. Hammer, M.E. Trumbauer,
M.G. Rosenfeld, N.C. Birnberg and R.M. Evans

A DNA fragment containing the promoter of the mouse metallothionein-I gene fused to the structural gene of rat growth hormone was microinjected into the pronuclei of fertilized mouse eggs. Of 21 mice that developed from these eggs, seven carried the fusion gene and six of these grew significantly larger than their littermates. Several of these transgenic mice had extraordinarily high levels of the fusion mRNA in their liver and growth hormone in their serum. This approach has implications for studying the biological effects of growth hormone, as a way to accelerate animal growth, as a model for gigantism, as a means of correcting genetic disease, and as a method of farming valuable gene products.

THE introduction of foreign genes into mammalian embryos could form the basis of a powerful approach for studying gene regulation and the genetic basis of development. Recent studies have clearly established the feasibility of introducing foreign DNA into the mammalian genome by microinjection of DNA molecules of interest into pronuclei of fertilized eggs, followed by insertion of the eggs into the reproductive tracts of foster mothers[1-7]. The data suggest that the so-called transgenic animals that develop from this procedure integrate the foreign DNA into one of the host chromosomes at an early stage of embryo development. As a result, the foreign DNA is generally transmitted through the germ line[3,8,9]. In some cases, expression of these foreign genes has also been detected; low levels of rabbit β-globin and viral thymidine kinase (TK) activity were measured in transgenic mice after introducing the respective genes[6,7]. It is possible to obtain regulated expression of viral TK by fusing the structural gene to the mouse metallothionein-I (MT-I) gene promoter and inducing expression with heavy metals[2,9].

The possibility of introducing genes encoding secreted regulatory peptides into animals via this technology has several potential applications. The growth hormone (GH) gene whose protein product has profound developmental effects is particularly convenient for testing this possibility. Based on our success in achieving regulatable expression of herpes TK using the MT-I promoter[9], we fused this promoter to the rat GH structural gene which has been shown to direct GH synthesis in mouse cells[10]. We show here that when this gene construct (MGH) is introduced into eggs it is expressed efficiently, giving rise to greater than normal levels of growth hormone in transgenic mice. The most obvious consequence of this high level of growth hormone expression is gigantism. This success raises several issues of genetic and practical importance.

Generation of animals carrying MGH fusion genes

A fragment of the cloned rat GH gene, from which the 5′ regulatory region had been deleted, was fused to the MT-I promoter region, generating the plasmid pMGH (see Fig. 1a). In this construction, the fusion occurs in exon 1, retaining the initiation codon of the GH gene and the 5′ regulatory region

of the MT-I gene. Thus, the fusion gene is predicted to direct transcription of a mRNA containing 68 bases contributed by MT-I, followed by 1 base contributed by an XhoI linker, then the rat GH mRNA sequence starting at nucleotide 7. This construction preserves the AUG initiation codon for GH synthesis located at position 124–126 (Fig. 1c), the four intervening sequences and the poly(A) site of the rat GH gene (Fig. 1b), and 3 kilobases (kb) of downstream chromosomal sequences.

A 5.0-kb fragment extending from the BglI site of MT-I (−185) to BamHI (see Fig. 1b) was restricted from pMGH, separated from the other fragments on an agarose gel and used for injection into eggs. This linear fragment with heterologous ends was chosen because our recent experience suggests that such fragments integrate into host DNA more efficiently than supercoiled plasmids. Although this fragment contains only 253 base pairs (bp) of MT-I sequence, we have shown previously that this includes sequences essential for heavy metal regulation of the MT-I promoter[11,12]. The male pronuclei of fertilized eggs were microinjected with 2 pl containing ~600 copies of this fragment and 170 eggs were inserted into the reproductive tracts of foster mothers; 21 animals developed from these eggs[2].

When they were weaned, total nucleic acids were extracted from a piece of tail and used for DNA dot hybridization to determine which animals carried MGH sequences. Seven of the animals gave hybridization signals above background (Fig. 2a) and their DNA was analysed further by restriction with SstI and Southern blotting. This analysis showed that all seven had an intact 1.7-kb SstI fragment predicted from the restriction map shown in Fig. 1. All the animals having a prominent 1.7-kb hybridizing band also revealed a 3.3-kb band. This band is predicted from a circularized version of the 5.0-kb BglI–BamHI fragment. Two of the mice having multiple copies (MGH-10 and -19) gave a 5.0-kb hybridizing fragment when digested with HindIII, an enzyme that cuts only once within the injected fragment. Surprisingly, BamHI also generates a 5.0-kb hybridizing fragment in all the DNAs analysed (MGH-10, -14, -16 and -19), suggesting that this restriction site was restored during circularization, whereas BglI and EcoRI (which does not cut within the fragment) gave larger fragments (data not shown). We conclude that all the MGH-positive animals have

PALMITER, R.D., BRINSTER, R.L., HAMMER, R.E., TRUMBAUER, M.E., ROSENFELD, M.G., BIRNBERG, N.C., and EVANS, R.M.
Dramatic growth of mice that develop from eggs microinjected with metallothionein-growth hormone fusion genes. Reprinted by permission from *Nature* v. 300, pp. 611-615. Copyright 1982 Macmillan Magazines Ltd.

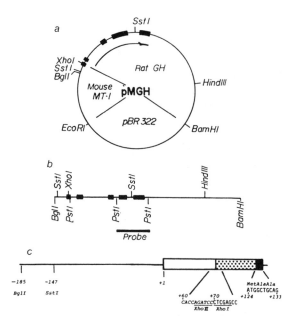

Fig. 1 Construction of the fusion plasmid pMGH. *a*, The unique *Bgl*II site of the MT-I genomic clone[28] was converted to a *Xho*I site by digesting with *Bgl*II, followed by filling in the sticky ends with Klenow fragment of DNA polymerase in the presence of all four dNTPs; then *Xho*I linkers (CCTCGAGG) were ligated to the blunt ends and bacterial strain RR1 was transformed with this DNA. The *Pvu*II site of pBR322 was converted to a *Bam*HI site by a similar procedure, then the 4.1-kb *Xho*I–*Bam*HI fragment containing the MT-I promoter and pBR was ligated to a 4.8-kb *Xho*I–*Bam*HI fragment containing the rat GH structural gene (see refs 10, 30) to give pMGH (8.9 kb). *b*, The fragment used for injection was a 5.0-kb *Bgl*I–*Bam*HI fragment which was isolated from an agarose gel by the NaClO₄ method[29]. For genomic Southern blots, a 1.0-kb *Pst*I fragment spanning exons 4 and 5 was isolated and nick-translated[9] and used as a hybridization probe. *c*, The predicted structure of exon 1 of the fusion gene. The line and open box represent MT-1 untranslated sequences, the stippled box represents GH untranslated sequences and the solid box represents the beginning of the GH coding region (see refs 28 and 30 for complete sequence information).

at least one intact *Bgl*I–*Bam*HI insert and four of the mice have a tandem head-to-tail duplication of this 5.0-kb fragment; however, the details of how and when this fragment circularized and integrated are not discernible. One of the mice, MGH-10, transmitted the MGH genes to half (10 of 19) of its offspring, suggesting that these genes are stably integrated into one of its chromosomes.

Rapid growth of mice carrying MGH genes

At 33 days post-parturition, all of the mice were weaned and maintained on a solid diet supplemented with water containing 5,000 p.p.m. ZnSO₄ (76 mM); their growth rate was recorded periodically. The dose of Zn was chosen on the basis of experiments which indicated that 5,000 p.p.m. induced almost maximal levels of MT-I mRNA in the liver without imparing the breeding potential of mice maintained on this diet for several months. Figure 3 shows the growth of the seven mice carrying MGH sequences. The littermates without MGH sequences served as convenient controls; the weights for each sex were averaged separately. Six of the mice with MGH sequences showed substantially more weight gain (up to 1.8-fold) than the controls. With the exception of MGH-16, there was a correlation between weight gain and MGH gene dosage (Table 1). The animal carrying the most copies of MGH sequences, MGH-21, died after 7 weeks of rapid growth; other mice have, however, grown larger. Most of the animals were already larger

than normal before the Zn diet was instituted. Furthermore, one of the animals (MGH-19) was removed from the Zn diet on day 56 and it continued to grow at an accelerated rate. Thus, the effect of Zn on growth rate is uncertain. Perhaps mice carrying several copies of the MGH gene produce excess GH constitutively without a requirement for heavy metal induction. Thus, the question of Zn-dependent growth will be most easily answered by analysis of offspring of mice such as MGH-16 that contain only a few copies of the MGH gene.

Expression of the MGH gene in the liver

Based on previous studies with mice carrying metallothionein–TK fusion genes, maximal expression of MGH was expected in the liver, intestine and kidney. Evaluation of the tissue specificity of MGH gene expression was initiated by analysis of its activity in liver. A partial hepatectomy on day 56 allowed isolation and quantification of RNA for MGH-specific sequences. Figure 4*b* shows the results of RNA slot hybridization. The level of MGH mRNA expression in the liver correlated with the growth of the transgenic mice. The largest animals, MGH-2, -19 and -21, contained large amounts of MGH mRNA in the liver, whereas MGH-10, -14 and -16, which showed slower growth rates, contained ~50-fold less MGH RNA, not visible on the exposure shown in Fig. 4. Analysis of samples exposed to base hydrolysis before spotting the nucleic acids on to nitrocellulose (Fig. 4*a*) or treated with RNases A plus T₁ after hybridization (not shown) established that RNA rather than potentially contaminating DNA was hybridizing to the probe. From these data, we estimate that 30 μg of liver RNA from MGH-2, -19 and -21 contains the same amount of GH mRNA as 13, 25 and 50 ng of poly(A)⁺ rat pituitary RNA, respectively. The RNA standard used contains about 10% GH mRNA on the basis of Rot analysis; therefore, we estimate that there are ~800–3,000 MGH mRNA molecules per liver cell in these transgenic mice (Table 1).

If all processing signals in the fusion gene are correctly recognized, a fusion mRNA 63 nucleotides larger than *bona fide* rat GH mRNA (~1 kb) would be generated[10,13]. Denaturing gel electrophoresis and RNA Northern blot analysis of the liver RNA from MGH-21 show that its size is similar, as expected, to that of authentic mouse and rat pituitary GH mRNA (Fig. 5*a*). Liver RNA from a control mouse shows no GH-reactive sequences (Fig. 5*a*, lane 1). Because the GH cDNA probe used for the RNA blot analysis recognizes both

Table 1 MGH genes, mRNA, GH levels, and growth of transgenic mice

Mouse	Sex	No. of MGH genes per cell	No. of MGH mRNA molecules per cell	Growth hormone (μg ml^{-1})	Growth* (g)	Ratio
MGH-2	♀	20	800	57.0	41.2	1.87
MGH-3	♀	1	<50	0.87	22.5	1.02
MGH-10	♂	8	<50	0.28	34.4	1.32
MGH-14	♂	2	<50	0.31	30.6	1.17
MGH-16	♂	2	<50	17.9	36.4	1.40
MGH-19	♂	10	1,500	32.0	44.0	1.69
MGH-21	♀	35	3,000	112.0	39.3	1.78
Female littermates (n = 3)	0	0	0.16 ±0.1	22.0 ±0.8	
Male littermates (n = 11)	0	0	0.15 ±0.08	26.0 ±2	

The number of MGH genes per cell was estimated by DNA dot hybridization and scintillation counting (Fig. 2*a*). MGH mRNA molecules per cell were estimated by densitometric scanning of the autoradiogram from RNA slot hybridization (Fig. 4) and assuming 2.5×10^5 mRNA molecules per liver cell. GH was measured by RIA as described elsewhere[10].

* Animal weights when 74 days old. The ratio of body weight compared with littermates of the same sex is indicated (see Fig. 3).

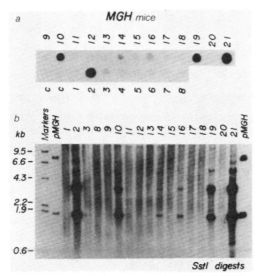

Fig. 2 Analysis of DNA from transgenic mice for MGH sequences. *a*, DNA dot hybridization[9] was used to detect mice with MGH sequences and quantitate their abundance. A nick-translated *Pst*I probe (see Fig. 1*b*) was used (6 h exposure). Numbers correspond to the MGH mice examined; c, controls. *b*, DNA (5 μg) from MGH mice was restricted with *Sst*I electrophoresed on a 1% agarose gel, transferred to nitrocellulose and hybridized with the nick-translated *Pst*I probe shown in Fig. 1*b*. pMGH (13 pg, left and 130 pg, right) was included as a hybridization standard. Markers are *Hind*III-cut λ DNA end-labelled with [32]P.

rat and mouse GH mRNAs, it is necessary to establish that the hybridizing species in the liver is actually the product of the fusion gene and not mouse GH mRNA due to an unexpected activation of the endogenous mouse GH gene. The use of a *Xho*I linker for the construction of the fusion gene generates a sequence that will be uniquely present in MGH mRNA. Thus, the *Xho*I site of pMGH was labelled using [γ-^{32}P]ATP and polynucleotide kinase, followed by cleavage with *Sst*I (Fig. 1*c*). This 221-nucleotide fragment was gel-purified, denatured, and used as hybridization probe in a single-strand-specific nuclease protection assay. Hybridization to MGH mRNA should generate a 74-base nuclease-resistant fragment while mouse GH mRNA or metallothionein mRNA will be unable to protect the kinased end and should therefore be negative. Figure 5*b*, lane 6 shows that the predicted 74-base fragment is present in liver RNA of mouse MGH-21, but not in normal mouse pituitary RNA (lane 5), control mouse liver RNA (lane 4) or in the liver RNA of MGH-3 (lane 3), an animal showing no accelerated growth (Fig. 3). Thus, it seems that transcription is initiating properly at the MT-I promoter and continuing through the putative termination site of the GH gene, the four GH intervening sequences are being properly spliced and the MGH mRNA is polyadenylated.

Increased growth hormone in sera of MGH mice

The correlation between MGH mRNA levels and growth of the mice sugggests that expression of the MGH gene accounts for the observed biological consequences; this was supported by the three independent RNA analyses above. These results suggest that the circulating levels of GH should be increased. Blood was taken from the transgenic mice and from littermates and assayed for GH by radioimmunoassay (RIA). The GH levels in four of the transgenic mice were 100–800 times greater than levels in control littermates; one mouse had 112 μg ml[-1] of GH in its serum (Table 1). Two of the transgenic mice showing a slow but significantly elevated growth rate had serum GH levels at the high end of the normal range. The lack of

physiological regulation and the ectopic production of GH in these animals probably account for their accelerated growth.

Discussion

These data strongly suggest that the altered phenotype of the mice is a direct result of the integration and expression of the metallothionein–growth hormone fusion gene. The elevated level of GH present in some of these mice corresponds to a high level of MGH mRNA in the liver (up to 3,000 molecules per cell). The accumulation of MGH mRNA in the liver is comparable to the endogenous level of MT-I mRNA[14], but is ~100-fold higher than that obtained from metallothionein–TK fusion genes studied previously[2]. This difference is probably the consequence of the intrinsic stability of the GH mRNA relative to TK mRNA; however, differences in transcription rates or processing efficiency due to variations in fusion gene construction are also possible. The high level of MGH gene expression in transgenic mice will greatly facilitate direct comparison of MGH and MT-I mRNA production in different tissues, thus allowing us to determine whether chromosomal location has an important effect on tissue-specific expression of the MT-I promoter.

Growth hormone levels in some of the transgenic mice were as much as 800-fold higher than in normal mice, resulting in animals nearly twice the weight of their unaffected littermates. This greater than normal accumulation of GH undoubtedly reflects both the lack of normal feedback mechanisms and expression of this gene in many large organs including liver, kidney and intestine. The effect of chronic exposure to high levels of GH is well documented[15], resulting in the clinical condition referred to as gigantism. This condition in humans is usually associated with pituitary adenomas[15] and more rarely with ectopic expression of GH by lung carcinomas[16,17].

Some of the diverse effects of GH are mediated directly by the hormone[18-20]. However, it is generally believed that the major effect of GH is stimulation of somatomedin production in the liver[18,19]. Somatomedins are insulin-like growth factors that promote proliferation of mesodermal tissues such as muscle, cartilage and bone[18,19]. The involvement of somatomedins in the GH response provides an explanation for

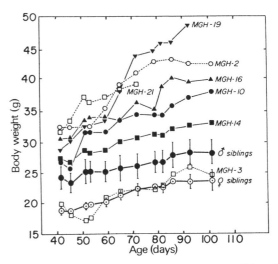

Fig. 3 Growth of MGH mice. Microinjected eggs (SJL×C57) were transferred to oviducts of foster mothers on 14 April; 21 animals were born 3 weeks later. At 33 days old they were weaned and the drinking water was supplemented with 76 mM ZnSO₄. The body weights of the males are shown as solid symbols; the mean weight (±s.d.) of 11 siblings not containing MGH sequences is also shown. The female weights are represented by open symbols; means (±s.d.) of three siblings are indicated also. MGH-21 died on day 72. A partial hepatectomy was performed on MGH-19 on day 56 and it was taken off Zn thereafter; it was killed on day 100. All mice were taken off Zn after day 83.

the growth of animals such as MGH-10, 14 and 16 in which the circulating level of GH was only slightly higher than normal. In these animals, GH produced in the liver may be sufficient to stimulate somatomedin production because the local GH concentration is relatively high. Thus, in these animals, GH may mimic the local paracrine function of some hormones[21]. However, the reason for lack of growth of MGH-3 is unclear.

The implicit possibility is to use this technology to stimulate rapid growth of commercially valuable animals. Benefit would presumably accrue from a shorter production time and possibly from increased efficiency of food utilization. The effects of continuously elevated GH levels on the quality of meat will need to be evaluated. By choosing the appropriate GH gene, milk production may be disproportionately increased considering that GH is homologous to prolactin, that GH of some species binds to prolactin receptors, and that exogenously administered GH has been shown to increase milk yield[22,23]. Having a regulatable promoter may be particularly advantageous for timely expression of GH. Applying these techniques to large animals will be more difficult. Nevertheless, when genes for desired traits can be isolated, this approach should provide a valuable adjunct to traditional breeding methods. Optimizing the conditions for integration and expression of foreign genes in mice should facilitate the eventual application of these techniques to other animals.

Another possibility is the use of this technology either to correct or to mimic certain genetic diseases. There are several inbred dwarf strains of mice, one of which, *little*, lacks GH and is about half normal size when homozygous[24]. Thus, introduction of the fusion gene described here or a natural GH gene might restore normal growth to these animals. On the other

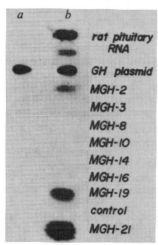

Fig. 4 Slot-blot analysis of liver RNAs from transgenic mice. Mouse liver RNA was purified by homogenizing a liver slice in 1×SET (1% SDS, 5 mM EDTA, 10 mM Tris pH 7.5), 200 μg ml^{-1} proteinase K (Boehringer) followed by phenol/chloroform extraction, ethanol precipitation, DNase I treatment (Worthington; DPFF), a second phenol/chloroform extraction, then re-precipitation with ethanol and finally resuspension for A_{260} determination. Liver RNA (30 μg) was diluted to 100 μl in 2X SSC. Samples were brought to 0.5 M NaOH or 0.5 M NaCl and incubated at 65 °C for 30 min. 140 μl of 20× SSC and 160 μl of 12.3 M formaldehyde were added, incubated at 65 °C for another 15 min and applied to a nitrocellulose sheet resoaked in 20× SSC through a slot-blot device[31]. The nitrocellulose was baked at 80 °C for 2 h, prehybridized for 5 h and then hybridized with nick-translated ^{12}P-rGH cDNA[32] (5× 10^6 c.p.m. in 4 ml at 42 °C overnight), washed in 0.1×SSC at 42 °C, and exposed to X-ray film at −70 °C for 24 h. *a*, Samples were treated with 0.5 M NaOH and neutralized with HCl. *b*, Samples were incubated with an equivalent amount of NaCl. 40 and 20 ng of rat pituitary poly(A)$^+$ RNA in the presence of 30 μg of control liver RNA were applied to slots as GH mRNA controls. 1 ng of rGH gene plasmid, also in the presence of 30 μg of control liver RNA, was applied as a base-resistant control.

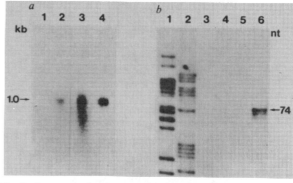

Fig. 5 Structure of MGH mRNAs in liver of transgenic mice. *a*, Northern blot analysis. RNAs were resuspended 10 mM NaH$_2$PO$_4$, pH 7.4/50% formamide/2.2 M formaldehyde, heated at 65 °C for 5 min, and subjected to electrophoresis on a slab gel composed of 1.5% agarose in 10 mM NaH$_2$PO$_4$, pH 7.4/0.55 mM EDTA/1.1 M formaldehyde. The running buffer was 10 mM NaH$_2$PO$_4$, pH 7.4/0.5 M formaldehyde. The gel was stained with acridine orange to identify rRNA markers, photographed, incubated for 90 min with 50 mM NaOH, and then neutralized with two washes of 0.2 M NaOAc, pH 4.3. The RNA was then transferred to diazotized paper and prehybridized as described elsewhere[32]. Lane 1, total liver RNA (15 μg) from a control littermate; lane 2, total liver RNA (15 μg) from MGH-21; 3, total mouse pituitary RNA; 4, 40 ng of rat pituitary mRNA, poly(A)$^+$. Lanes 1 and 2 exposed for 48 h; lanes 3 and 4 exposed for 5 h. *b*, Single-strand-specific nuclease protection assay. 20 μg of RNA were hybridized at 47 °C for 5 h in 40 μl of a solution containing 40 mM PIPES pH 6.4, 0.4 M NaCl, 80% formamide and 50,000 c.p.m. gel-purified 221-bp SstI–XhoI fragment of pMGH end-labelled at the XhoI site with ^{32}P (see Fig. 1c). The samples were then diluted with 0.3 ml of 280 mM NaCl, 30 mM NaOAc pH 4.4, 4.5 mM ZnSO$_4$, 20 μg ml^{-1} salmon sperm DNA and 150 units of mung bean single-strand-specific nuclease (Collaborative Research) and incubated at 47 °C for 1 h. Samples were ethanol-precipitated, resuspended in 90% formamide containing bromophenol blue and Xylene Cyanol FF, loaded on to an 8% acrylamide–urea sequencing gel, electrophoresed for 1.5 h at 2,000 V, dried and autoradiographed for 7 days at −70 °C with an intensifying screen. Lanes 1 and 2, sequencing ladder used for size standards; 3, MGH-3 RNA; 4, control liver RNA; 5, mouse pituitary RNA; 6, MGH-21 RNA.

hand, the experiments described above show that gigantism can be created. Once these mice are inbred to give homozygous stocks, they will provide a valuable model system for biochemical studies on the consequences of excess GH production.

Finally, the exceptionally high levels of GH found in the sera of some of these mice raises the possibility of extending this technology to the production of other important polypeptides in farm animals. The concentrations of GH in MGH-21 serum was 10–100-fold higher than that reported for bacterial or mammalian cell cultures that were genetically engineered for GH production[10,25-27]. This genetic farming concept is comparable to the practice of raising valuable antisera in animals, except that a single injection of a gene into a fertilized egg would substitute for multiple somatic injections; moreover, the expression of that gene is likely to be heritable[9]. This approach may be particularly applicable in those cases where the protein of interest requires special covalent modifications (for example, proteolytic cleavage, glycosylation, or γ-carboxylation) for activity or stability.

Clearly the ability to introduce into mice, and by extrapolation into other animals, functional genes of selected construction offers wide ranging experimental as well as practical opportunities.

We thank Joan Vaughan for performing GH radioimmunoassays; Wylie Vale and Marcia Barinaga for valuable discussions; our colleagues for helpful comments; and Abby Dudley for secretarial assistance in the preparation of the manuscript. N.C.B. was supported by a NCI training grant (CA09370 HHS). The work was supported by research grants from the NIH (HD-09172, HD-15477, GM-26444, AM-21567) and NSF (PCM-8107172).

Received 27 September; accepted 19 October 1982.

1. Gordon, J. W., Scangos, G. A., Plotkin, D. J., Barbosa, J. A. & Ruddle, F. H. *Proc. natn. Acad. Sci. U.S.A.* **77**, 7380–7384 (1980).
2. Brinster, R. L. *et al. Cell* **27**, 223–231 (1981).
3. Constantini, F. & Lacy, E. *Nature* **294**, 92–94 (1981).
4. Harbers, K., Jähner, D. & Jaenisch, R. *Nature* **293**, 540–542 (1981).
5. Rusconi, S. & Schaffner, W. *Proc. natn. Acad. Sci. U.S.A.* **78**, 5051–5055 (1981).
6. Wagner, E., Stewart, T. & Mintz, B. *Proc. natn. Acad. Sci. U.S.A.* **78**, 5016–5020 (1981).
7. Wagner, T. E. *et al. Proc. natn. Acad. Sci. U.S.A.* **78**, 6376–6380 (1981).
8. Gordon, J. W. & Ruddle, F. H. *Science* **214**, 1244–1246 (1981).
9. Palmiter, R. D., Chen, H. Y. & Brinster, R. L. *Cell* **29**, 701–710 (1982).
10. Doehmer, J. *et al. Proc. natn. Acad. Sci. U.S.A.* **79**, 2268–2272 (1982).
11. Brinster, R. L., Chen, H. Y., Warren, R., Sarthy, A. & Palmiter, R. D. *Nature* **296**, 39–42 (1982).
12. Mayo, K. E., Warren, R. & Palmiter, R. D. *Cell* **29**, 99–108 (1982).
13. Evans, R. M., Birnberg, N. C. & Rosenfeld, M. G. *Proc. natn. Acad. Sci. U.S.A.* (in the press).
14. Durnam, D. M. & Palmiter, R. D. *J. biol. Chem.* **256**, 5712–5716 (1981).
15. Richmond, I. L. & Wilson, C. B. *J. Neurosurg.* **49**, 163–167 (1978).
16. Steiner, H., Dahlback, O. & Waldenstarm, J. *Lancet* **i**, 783–785 (1968).
17. Greenberg, P. B., Beck, C., Martin, T. J. & Burger, H. G. *Lancet* **i**, 350–352 (1972).
18. Daughaday, W. H., Herington, A. C. & Phillips, L. S. *A. Rev. Physiol.* **37**, 211–244 (1975).
19. Kostyo, J. L. & Isaksson, O. *Int. Rev. Physiol.* **13**, 255–274 (1977).
20. Morikawa, M., Nixon, T. & Green, H. *Cell* **29**, 783–789 (1982).
21. Krieger, D. T. & Martin, J. B. *New Engl. J. Med.* **304**, 876–885 (1981).
22. Shiu, R. P. C. & Friesen, H. G. *J. biol. Chem.* **249**, 7902–7911 (1974).
23. Peel, C. J., Bauman, D. E., Gorewit, R. C. & Sniffen, C. J. *J. Nutr.* **111**, 1662–1671 (1981).
24. Beamer, W. G. & Eicher, E. M. *J. Endocr.* **71**, 37–45 (1976).
25. Goeddel, D. V. *et al. Nature* **281**, 544–548 (1979).
26. Pavlakis, G. N., Hizuka, N., Gorden, P., Seeburg, P. & Hamer, D. H. *Proc. natn. Acad. Sci. U.S.A.* **78**, 7398–7402 (1981).
27. Robins, D. M., Paek, I., Seeburg, P. H. & Axel, R. *Cell* **29**, 623–631 (1982).
28. Glanville, N., Durnam, D. M. & Palmiter, R. D. *Nature* **292**, 267–269 (1981).
29. Chen, C. W. & Thomas, C. A. Jr *Analyt. Biochem.* **101**, 339–341 (1980).
30. Barta, A., Richards, R. I., Baxter, J. D. & Shine, J. *Proc. natn. Acad. Sci. U.S.A.* **78**, 4867–4871 (1981).
31. Tlsty, T., Brown, P. C., Johnston, R. & Schimke, R. T. in *Gene Amplification* (ed. Schimke, R. T.) 231–238 (Cold Spring Harbor Laboratory, New York, 1982).
32. Potter, E., Nicolaisen, K., Ong, E. O., Evans, R. M. & Rosenfeld, M. G. *Proc. natn. Acad. Sci. U.S.A.* **78**, 6662–6666 (1981).

Production of Human Beta Interferon in Insect Cells Infected with a Baculovirus Expression Vector

G.E. Smith, M.D. Summers and M.J. Fraser

Autographa californica nuclear polyhedrosis virus (AcNPV) was used as an expression vector for human beta interferon. By using specially constructed plasmids, the protein-coding sequences for interferon were linked to the AcNPV promoter for the gene encoding for polyhedrin, the major occlusion protein. The interferon gene was inserted at various locations relative to the AcNPV polyhedrin transcriptional and translational signals, and the interferon-polyhedrin hybrid genes were transferred to infectious AcNPV expression vectors. Biologically active interferon was produced, and greater than 95% was secreted from infected insect cells. A maximum of ca. 5×10^6 U of interferon activity was produced by 10^6 infected cells. These results demonstrate that AcNPV should be suitable for use as a eucaryotic expression vector for the production of products from cloned genes.

Autographa californica nuclear polyhedrosis virus (AcNPV) has a genome of ca. 130 kilobases of double-stranded, circular DNA and is the most extensively studied baculovirus (for recent reviews see references 13 and 20). Its host range is limited to lepidoptera species and cultured lepidoptera cells, and it is currently being considered for agricultural use as a viral insecticide. AcNPV has a biphasic replication cycle and produces a different form of infectious virus during each phase. Between 10 and 24 h postinfection (p.i.), extracellular virus is produced by the budding of nucleocapsids through the cytoplasmic membrane. By 15 to 18 h p.i., nucleocapsids are enveloped within the nucleus and viral occlusions begin to form. Each occlusion contains many virus particles embedded in a paracrystalline protein matrix, which is formed from a single major protein called polyhedrin. In infected *Spodoptera frugiperda* (fall armyworm, Lepidoptera, Noctuidae) cells, AcNPV polyhedrin accumulates to high levels and constitutes 25% or more of the total protein mass in the cell (34); it is probably synthesized in greater abundance than any other protein in a virus-infected eucaryotic cell.

Polyhedrin is encoded by the virus (40), and the gene has been mapped (34) and sequenced (B. J. L. Hooft van Iddekinge, G. E. Smith, and M. D. Summers, Virology, in press). From an analysis of viral deletion mutants, we demonstrated that the polyhedrin gene is not required for the production of infectious extracellular virus (30). Inactivation of the polyhedrin gene by

deletion (30) or, as will be demonstrated in this report, by insertion results in mutants that do not produce occlusions in infected cells. These occlusion-negative viruses form plaques that are different from plaques produced by wild-type viruses, and this distinctive plaque morphology is used as a means to screen for recombinant viruses (30).

Several properties of AcNPV that may make this virus ideally suited as an expression vector for cloned eucaryotic genes have been described (20). These include the potential of the rod-shaped virus to encapsidate viral genomes with large pieces of additional, foreign DNA and the inherent safety of a recombinant viral vector that is not pathogenic to vertebrates. In addition, the polyhedrin gene provides: (i) a nonessential region of the AcNPV genome in which to insert foreign DNA, (ii) a very strong promoter which directs transcription late in infection after extracellular virus is produced and after host genes and most viral genes are turned off, and (iii) a genetic marker to select for recombinant viruses. To examine the utility of AcNPV as an expression vector, we fused the gene for human beta interferon (IFN-β) to the AcNPV polyhedrin promoter and measured its production in infected cells.

IFN-β is known to have antiviral, antiproliferative, and antitumor properties and should prove important in a variety of medical applications. Although cultured human cells normally do not produce interferon, treatment with inducers such as viruses or double-stranded RNA

SMITH, G.E., SUMMERS, M.D. and FRASER, M.J.
Production of human beta interferon in insect cells infected with a baculovirus expression vector.
Mol. Cell Biol. 3:2156-2165 (1983). Reprinted with the authors' permission.

results in a transient and relatively low-level biosynthesis of IFN-β (35). The IFN-β gene has been expressed in *Escherichia coli* cells by fusing interferon protein-coding sequences to bacterial promoter and ribosomal binding sites in plasmid (8, 37) or lambda phage (22) vectors. Nonhuman mammalian cells have also been employed to synthesize IFN-β from the cloned gene. For example, the IFN-β gene has been cloned with its protein-coding sequences under the transcriptional control of a simian virus 40 promoter (7), a herpes simplex virus thymidine kinase promoter (25), or its own genetic signals (2, 11, 21, 24, 43). The highest reported levels of beta interferon activity produced in bacterial and animal host-vector systems are 1×10^4 U/ml in *E. coli* cells (8) and 5×10^4 U/ml in transformed mouse cells (11). By using AcNPV as an expression vector, biologically active IFN-β was efficiently secreted in the media of infected cells and reached a titer of 5×10^6 U/ml. In addition, the IFN-β protein was glycosylated, and our data indicate that the signal peptide at the amino-terminal end of the primary translation product was removed.

MATERIALS AND METHODS

Cells and virus. AcNPV stocks were grown and titers were determined in *S. frugiperda* cells (IPLB-SF21-AE), using Hink medium (12) plus 10% fetal bovine serum (FBS) as described previously (31).

Human amnionic (WISH) cells were maintained in Earle minimum essential medium supplemented with penicillin (100 U/ml), mycostatin (50 U/ml), and 15% FBS.

Interferon. Interferon activity was determined by the virus plaque-reduction assay (18) in WISH cells challenged with vesicular stomatitis virus. Serial dilutions (0.5 log) of culture media from recombinant AcNPV-infected cells were made over 2.5×10^4 WISH cells in 96-well microtiter plates. After 12 h at 37°C, each well was infected with ca. 60 PFU of vesicular stomatitis virus, and the cells were overlaid with 0.5% methyl cellulose in Earle minimum essential medium. At 24 h p.i., the cell monolayers were stained with crystal violet and the plaques were counted. An international reference standard of human interferon was included in all assays, and the activity was expressed in National Institutes of Health reference units (IU) per milliliter.

Transfection and selection for recombinant viruses. The procedure used to transfect *S. frugiperda* cells and the marker-transfer method used to select for recombinant AcNPV expression vectors were as described previously (30).

Protein blot radioimmunoassay. Total cellular proteins were electrophoretically separated and then reacted with antibody as described previously (32). At 30 h p.i., cells were washed with 1× phosphate-buffered saline and ca. 100-μg samples of cell protein were electrophoresed on a sodium dodecyl sulfate–12% polyacrylamide gel (12) and then transferred by diffusion to two nitrocellulose filters. The filters were incubated with either a 1:1,000 dilution of rabbit

antiserum prepared against AcNPV polyhedrin (32) or a 10-μg/ml concentration of monoclonal antibody to recombinant IFN-β (Hoffmann-La Roche Inc., Nutley, N.J.). Specifically bound antibody was detected by using radioiodinated *Staphylococcus aureus* protein A.

Radiolabeling of infected cells and analysis of polypeptides. At various times after infection, cells were labeled for 30 min in leucine-free Grace medium (9) supplemented with 2% dialyzed FBS and L-[³H]leucine at 100 μCi/ml or for 60 min in Grace medium with 10% the normal glucose supplemented with 2% dialyzed FBS and D-[2-³H]mannose at 200 μCi/ml. To label cells in the presence of tunicamycin, infected cells were pretreated for 20 to 30 h p.i. with 5 μg of tunicamycin per ml and then labeled as described above with L-[³H]leucine in the presence of 5 μg of tunicamycin per ml. The labeled cells were washed with Grace medium, disrupted and electrophoresed on sodium dodecyl sulfate–12% polyacrylamide gels, and then processed for fluorography as described previously (33). Molecular weights were calculated by comparing electrophoretic mobility with that of protein markers (Pharmacia Fine Chemicals, Piscataway, N.J.).

Construction of AcNPV transfer vectors. AcNPV *Eco*RI-I, which contains the polyhedrin gene (1, 26, 33), was cloned into the *Eco*RI site of pUC8 (41). The resulting plasmid (p18) has three *Bam*HI recognition sites: one in the polyhedrin gene, one in *Eco*RI-I downstream from the polyhedrin gene, and one in the cloning region of pUC8. The *Bam*HI site in pUC8 was removed by digesting p18 successively with *Pst*I, *Sma*I, and S1 nuclease. *E. coli* JM83 cells were transformed with the digested DNA, and a plasmid was isolated that was missing *Pst*I, *Bam*HI, *Sma*I, and *Eco*RI restriction sites in the cloning region of pUC8. This plasmid was partially digested with *Bam*HI, and full-length linear DNA was isolated from an agarose gel, treated with S1 nuclease (500 U/ml) at 37°C for 15 min, circularized with T4 DNA ligase, and used to transform JM83 cells. A plasmid (pAc101) was identified with a single *Bam*HI site located ca. 220 bases from the polyhedrin transcriptional start site (Fig. 1; 34).

The following procedure was used to modify pAc101 such that some or all of the DNA sequences between the transcriptional start site and the *Bam*HI recognition sequence in the polyhedrin gene were deleted and replaced by a synthetic *Bam*HI linker sequence, 5'-dCGGATCCG-3' (P-L Biochemicals, Inc., Milwaukee, Wis.). The *Eco*RI-to-*Bam*HI fragment in AcNPV *Eco*RI-I (located 0 to 4,210 bases in *Eco*RI-I; see reference 34) was cloned into pUC8. Forty micrograms of this plasmid (pB') was digested with *Bam*HI and then incubated for 40 min at 30°C with 0.5 U of *Bal* 31 exonuclease. At 4-min intervals, 10-μl samples were removed, and the reaction was stopped by adding 10 μl of 0.025 M EDTA. The samples were pooled, phenol extracted, and precipitated with 2 volumes of ethanol. The ends of the DNA were repaired by incubating the DNA in 100 μl for 30 min at 23°C with 4 U of *E. coli* DNA polymerase (Klenow fragment). Phosphorylated *Bam*HI linkers were added to the ends with T4 DNA ligase, the fragments were digested with *Bam*HI and *Eco*RI, and truncated pB' fragments with ca. 100- to 400-base-pair deletions were isolated from

```
GGTCTGCGAGCAGTTGTTTGTTGTTAAAAATAACAGCCATTGTAATGAGACGCACAAACTAATAATCACAAACTGGAAAATGTCTATCAATATATAGTT
-198                                                   -142         -128            -110
```

```
                    pAc380
GCTGATATCATGGAGATAATTAAAATGATAACCATCTCGCAAATAAATAAGTATTTTACTGTTTTCGTAACAGTTTTGTAATAAAAAAACCTATAAATA
-99EcoRV              -78              -58
```

```
       pro asp tyr ser tyr arg pro thr ile gly arg thr tyr val tyr asp asn lys tyr tyr lys asn leu gly
ATG CCG GAT TAT TCA TAC CGT CCC ACC ATC GGG CGT ACC TAC GTG TAC GAC AAC AAG TAC TAC AAA AAT TTA GGT
+1  MspI                                pAc360   pAc311
```

```
   ala val ile lys asn ala lys arg lys lys his phe ala glu his glu ile glu glu ala thr leu asp pro leu
GCC GTT ATC AAG AAC GCT AAG CGC AAG AAG CAC TTC GCC GAA CAT GAG ATC GAA GAG GCT ACC CTC GAC CCC CTA
76                                                      Sau3A                   TaqI
                                                         TaqI
```

```
        pAc101
   asp asn tyr leu val ala glu asp pro phe leu gly pro gly lys asn gln lys leu thr leu phe lys glu ile
GAC AAC TAC CTA GTG GCT GAG GAT CCT TTC CTG GGA CCC GGC AAG AAC CAA AAA CTC ACT CTC TTC AAG GAA ATC
151                   BamHI            MspI
                      Sau3A
```

```
   arg asn val lys pro asp thr met lys leu val val gly trp lys gly lys glu phe tyr arg glu thr trp thr
CGT AAT GTT AAA CCC GAC ACG ATG AAG CTT GTC GTT GGA TGG AAA GGA AAA GAG TTC TAC AGG GAA ACT TGG ACC
226                               HindIII
                                  AluI
```

FIG. 1. Nucleotide and amino sequence of the 5' end of the polyhedrin gene. The DNA sequence of a segment of the polyhedrin gene and 5' flanking region is presented as the sense (+) strand and was determined using the dideoxy termination procedure (29). The sequence around the 5' end of the polyhedrin gene was determined from the following overlapping templates: the minus strand beginning at the HindIII or BamHI sites in pAc101, the minus strand beginning at the synthetic BamHI sites in pAc311, pAc360, and pAc380, and the plus strand beginning at the BamHI or HindIII sites in pAc101 (Fig. 2). The location of the natural BamHI site in the transfer vector pAc101 and positions of the BamHI linkers in pAc311, pAc360, and pAc380 are indicated by arrows. The box at −58 marks the bases that code for the 5' end of polyhedrin mRNA (34). The rectangles centered at −78 and −110 mark the nucleotides that resemble the canonical TATA and CAAT box sequences, respectively. The repeated CACAAACT sequences upstream from the CAAT box are underlined. The translational start signal for polyhedrin is assigned position +1.

an agarose gel. The purified fragments were ligated to EcoRI- plus-BamHI-digested pUC8 and then used to transform JM83 cells. The sizes of the deletions in the resulting clones were estimated by restriction enzyme analysis and in some cases by DNA sequencing. The XhoI-to-BamHI fragment in pAC101 (located at 1,900 to 4,210 bases in EcoRI-I; 34) was removed and replaced with XhoI-to-BamHI linker fragments from several different pB′ mutant plasmids, using standard cloning procedures, to produce the modified transfer vectors pAc311, pAc360, and pAc380. These plasmids have single BamHI cloning sites in the polyhedrin gene at the locations indicated in Fig. 1.

DNA sequencing. AcNPV restriction fragments were cloned into M13mp8 or M13mp9 (19), and the single-stranded recombinant phage DNAs were purified and used as templates in the chain-termination sequencing method (29). The sequence was determined from overlapping DNA templates, and all autoradiographs were read independently at least twice. The sequence across all restriction enzyme sites in the polyhedrin gene was also determined.

RESULTS

Nucleotide sequence of the polyhedrin promoter and the construction of transfer vectors. The nucleotide sequence of the 5' end of the polyhedrin-coding region and 200 bases upstream from

the start of translation is shown in Fig. 1. The site that specifies the 5' end of polyhedrin mRNA has been mapped in a high-resolution S1 nuclease experiment to 220 bases in the 5' direction from the single BamHI site that maps within the gene (Fig. 1; 34). An ATG translational start signal (the A residue was assigned position +1) was located ca. 58 bases from the transcription start site, followed by an open reading frame for 244 amino acids. Proline is known to be the amino-terminal residue for AcNPV polyhedrin (J. Maruniak, Ph.D. thesis, University of Texas, Austin, 1980) and was the penultimate N-terminal amino acid predicted from the DNA sequence (Fig. 1). The predicted sequence for AcNPV polyhedrin (Fig. 1) was 80 to 90% homologous with several other NPV polyhedrins (28). The complete sequence of the AcNPV polyhedrin gene will be presented in another report (Hooft van Iddekinge et al., in press). Located about 78 and 110 bases upstream from the translational start site were sequences that resembled, respectively, the canonical "TATA" and "CAAT" boxes found in similar positions in many eucaryotic structural genes (see reference 5 and references therein). Centered at −128 and −142 were the direct

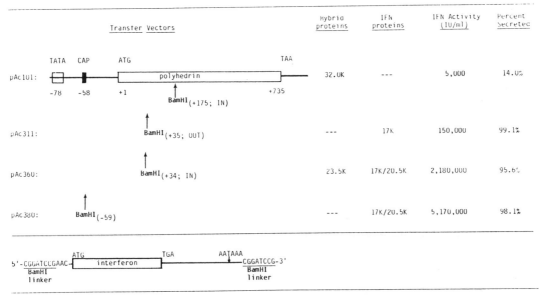

FIG. 2. Location of *Bam*HI cloning sites in the transfer vectors pAc101, pAc311, pAc360, and pAc380 illustrated relative to the coding sequences (open rectangles), the putative cap site, and the TATA box sequences of the polyhedrin gene. The interferon fragment cloned into these *Bam*HI sites is shown in the lower panel. When cloned into pAc101 and pAc360, the interferon coding sequences (open rectangle) are in the same reading frame as polyhedrin, and when cloned into pAc311 they are not. The polyhedrin-interferon hybrid proteins and 17K non-glycosylated and 20.5K glycosylated IFN-β proteins produced by AcNPV expression vectors are indicated. Interferon was produced by AcNPV expression vectors, and the activity was measured as follows. *S. frugiperda* cells were infected at 20 PFU per cell with AcNPV or AcNPV expression vectors. For infection, 4 × 10⁶ *S. frugiperda* cells were seeded in 4 ml of medium in a 25-cm² culture flask and then infected for 1 h at 23°C with 1 ml of virus inoculum. Infected cells were washed, 4 ml of fresh medium plus 10% FBS were added, and the flasks were incubated at 27°C for 48 h. The media were removed and centrifuged for 10 min at 3,000 × *g*. The supernatant from each was centrifuged at 100,000 × *g* for 30 min at 4°C and then stored at −80°C until assayed. Infected cells were washed in phosphate-buffered saline, suspended at 2 × 10⁶ cells per ml, and then disrupted by mixing with an equal volume of 7 M guanidine hydrochloride. Media and disrupted cells were assayed for interferon by the plaque-reduction assay and compared with an international interferon reference standard (18).

repeated nucleotides CACAAACT. These sequences may be analogous to the tandemly repeated promoter element (−100 region) in, for example, the promoter region of rabbit beta globin (5). The AcNPV polyhedrin gene has a 58-base nontranslated leader sequence preceding the translational start codon, and, as suggested from S1 nuclease experiments of AcNPV polyhedrin mRNA (34) and R-loop mapping of another NPV polyhedrin mRNA (27), these were no intervening sequences.

The AcNPV genome has no known unique restriction sites into which genes can be conveniently introduced in a site-specific manner. To incorporate the gene for IFN-β into the viral genome adjacent to the polyhedrin promoter, we first constructed chimeric plasmid vectors (transfer vectors). Each of these has both a cloning site near the polyhedrin promoter and flanking viral DNA linked to the *E. coli* plasmid pUC8. *Eco*RI fragment I was cloned into the *Eco*RI site in pUC8. There are three *Bam*HI sites on this plasmid, one of which is located at

position +175 in the polyhedrin gene (Fig. 1). The other two *Bam*HI sites were removed, resulting in the transfer vector pAc101. From the nucleotide sequence of polyhedrin and interferon, it can be determined that the IFN-β protein-coding sequences, when inserted into the unique *Bam*HI cloning site in pAc101, would be in the same translational reading frame as polyhedrin.

To investigate the importance of various polyhedrin gene sequences in the expression of IFN-β, the transfer vector pAc101 was modified to have *Bam*HI cloning sites at other positions near the 5′ end of the polyhedrin gene. Deletions were produced between the natural *Bam*HI site at +175 and the polyhedrin promoter in pAc101, and a synthetic oligonucleotide containing a *Bam*HI recognition site was added at the points of deletion. The transfer vectors pAc311 and pAc360 have *Bam*HI sites at +35 and +34 bases, respectively, downstream from the start site of polyhedrin translation (Fig. 1). The vector pAc380 has a *Bam*HI cloning site at −59, which is the putative cap site for polyhedrin mRNA (Fig. 1).

The nucleotide sequence for IFN-β and the location of various transcription and translation signals are known (4, 10, 23). A 767-base-pair *Hinc*II fragment from a genomic clone of human IFN-β (pBR13; 11) (containing the entire protein coding sequences for IFN-β, three additional bases before the ATG translation start signal, and all of the nontranslated 3′ sequences including the signal for polyadenylation [23]) was cloned into the *Hinc*II site of pUC8. This fragment was joined to synthetic octanucleotide *Bam*HI linkers and then cloned into the unique *Bam*HI sites in pAc101, pAc311, pAc360, and pAc380 (Fig. 2). Plasmids were examined, using various restriction enzymes to obtain the transfer vectors in which IFN-β was oriented in the same 5′ to 3′ direction as the polyhedrin gene, hereafter referred to as pAc101-IFN-β, pAc311-IFN-β, pAc360-IFN-β, and pAc380-IFN-β. As an example, pAc380 and the location and orientation of the interferon fragment in this transfer vector are shown in Fig. 3.

Construction of AcNPV expression vectors. Next, we transferred these hybrid polyhedrin-interferon genes to the AcNPV genome. AcNPV DNA was mixed with each of the transfer vectors containing the hybrid genes, and *S. frugiperda* cells were transfected as described previously (30). Recombinant AcNPV with insertions of IFN-β DNA was obtained from the resulting viral progeny by screening for plaques formed by viruses that did not produce occlusions (O⁻ mutants). Of the viral plaques resulting from the virus produced in cells cotransfected with viral and plasmid DNAs, an average of 0.5% were formed from putative recombinant O⁻ viruses. DNA restriction analysis demonstrated that each had the addition of IFN-β DNA at the appropriate location in *Eco*RI fragment I and that no other changes in the viral DNAs were detected (data not shown). The recombinant viruses were, therefore, defective in the production of occlusions due to insertional inactivation of the polyhedrin gene. Referred to here as the viral expression vectors Ac101-IFN-β, Ac311-IFN-β, Ac360-IFN-β, and Ac380-IFN-β they resulted from recombination between AcNPV and the corresponding IFN-β transfer vectors. The details of the relatively simple screening procedure used to select for AcNPV recombinants are described in a recent report (30).

Biologically active interferon is produced and secreted in cells infected with recombinant AcNPV. Cells infected with AcNPV-interferon expression vectors underwent morphological changes that are typical of O⁻ viral infections (30). The entire infection process, including viral protein synthesis, viral assembly, and partial

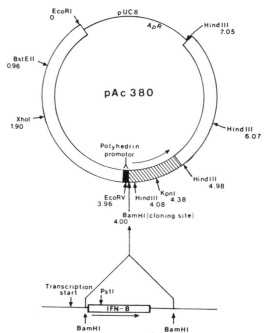

FIG. 3. AcNPV transfer vector pAc380 and IFN-β. Arrows mark the approximate location of the polyhedrin mRNA coding sequences, and the positions of certain restriction sites are indicated in kilobase pairs. The *Bam*HI cloning site at 4.00 is the position at which a deletion from +175 to −59 (Fig. 2) was introduced and replaced by a *Bam*HI octanucleotide linker. *Bam*HI linkers were added to a 767-base-pair *Hinc*II IFN-β fragment and then cloned into the *Bam*HI site in pAc380, producing the recombinant transfer vector pAc380-IFN-β. IFN-β protein-coding sequences are indicated by the rectangle, and the long arrow marks the direction of transcription.

cell lysis, was complete by ca. 72 h p.i. There was a marked reduction in protein synthesis between 50 and 60 h p.i., and cell lysis was first detected at about 50 h p.i. The titer of recombinant AcNPV-infected cells reached a maximum of ca. 5×10^8 PFU/ml of medium, which is typical of AcNPV-infected cells. Thus, the insertion of interferon DNA into the polyhedrin gene had no major effect on the replication of these recombinant viruses.

To determine whether the AcNPV polyhedrin gene sequences could promote expression of IFN-β, the level of interferon activity was measured in *S. frugiperda* cells infected with each of the AcNPV expression vectors. Because interferon is normally secreted, media from infected cells were also tested for interferon activity. We used the virus, plaque-reduction assay (18) in WISH cells challenged with vesicular stomatitis virus to determine interferon activity. If virus particles were removed by centrifugation, no interferon activity was measured in medium

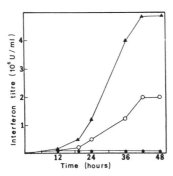

FIG. 4. Production of interferon activity in infected cells S. *frugiperda* cells were infected with AcNPV (●), Ac360-IFN-β (○), or Ac380-IFN-β (▲). The interferon activity in the medium was measured at 0, 12, 20, 24, 36, 42, and 48 h p.i. as described for Fig. 2.

from wild-type AcNPV-infected cells. However, if AcNPV-containing media were used during interferon assays 1,000 to 3,000 U of interferon per ml was produced, indicating that AcNPV virions apparently can induce the expression of endogenous interferon in WISH cells. Because many species of enveloped viruses are known to induce interferon production in human cells (39), these results were expected. To avoid this effect, all subsequent samples were centrifuged before testing.

Significant levels of interferon were detected by 48 h p.i. in cells infected with all of the expression vectors. Ac380-IFN-β, in which interferon was inserted about 19 bases downstream from the polyhedrin TATA box sequences, produced the highest levels (Fig. 2). Very high levels were also produced by Ac360-IFN-β, in which interferon was located at +34 and in the same translational reading frame as polyhedrin. In contrast, very low levels of activity were produced in cells infected with Ac101-IFN-β (Fig. 2), in which interferon was in the same reading frame as polyhedrin but located much further downstream at +175. Somewhat unexpected were the moderate levels of interferon activity produced by cells infected with Ac311-IFN-β (Fig. 2), in which interferon was inserted at +35 and consequently not in the correct reading frame for polyhedrin. Described later is an examination of proteins produced in infected cells, which helps to explain these results.

The kinetics of interferon induction were examined in S. *frugiperda* cells infected with the expression vectors Ac360-IFN-β and Ac380-IFN-β. Samples of the medium were taken from 0 to 48 h p.i. and stored at $-80°C$ until assayed for interferon. The medium from Ac380-IFN-β infected cells had about 10,000 IU of interferon per ml at 12 h p.i., and this level increased to a

maximum of nearly 5×10^6 IU/ml by 42 h p.i. (Fig. 4). A similar pattern of induction was observed in Ac360-IFNβ-infected cells (Fig. 4). The synthesis of polyhedrin in AcNPV-infected cells is known to follow a similar temporal pattern of expression (see reference 33 and references therein).

Effects of IFN-β on S. *frugiperda* cells. A potential limitation in using animal virus expression vectors for interferons (7) is that interferon itself can impose restraints upon the level of expression by interfering with the replication of the virus vector. The high levels of IFN-β produced in S. *frugiperda* cells, using AcNPV as a vector, and the normal titer of virus that was produced suggested that interferon does not affect AcNPV replication. To examine whether human IFN-β induced an antiviral state in S. *frugiperda* cells, the following experiment was conducted. Cells (2×10^6) were treated for 12 h with, per ml, up to 5×10^6 IU of interferon produced in Ac380-IFN-β-infected cells or 5×10^3 IU of an international standard of IFN-β, and the treated cells were infected with 100 PFU of AcNPV or Ac380-IFN-β. Exposure of the cells to human interferon had no measurable effect on the number of virus plaques that developed.

Protein synthesis in cells infected with AcNPV expression vectors. In the expression vector Ac380-IFN-β the nucleotides between -59 and $+175$ were deleted and replaced by an IFN-β gene fragment containing the interferon ATG translational start signal preceded by three bases and a synthetic *Bam*HI linker (Fig. 2 and 3). This placed the interferon gene ATG protein start signal approximately the same number of bases downstream from the polyhedrin gene TATA box as the polyhedrin transcriptional start site, leaving little, if any, nontranslated leader at the 5′ end of the predicted mRNA for this hybrid gene. High levels of interferon activity were measured in the medium of Ac380-IFN-β-infected cells, indicating that the polyhedrin promoter efficiently directed the transcription of this gene. At 30 and 48 h p.i., two polypeptides of 17,000 (17K) and 20.5K molecular weight were being made in Ac380-IFN-β but not in AcNPV-infected cells (Fig. 5A and B). The sizes of nonglycosylated and glycosylated human IFN-β proteins are reported to be comparable to the 17K and 20.5K polypeptides, respectively (3, 15). At 30 h p.i., the 17K polypeptide was being made in Ac360-IFN-β-infected cells (Fig. 5A; see Fig. 7), and by 48 h p.i., both 17K and 20.5K polypeptides were detected (Fig. 5B). Only the 17K polypeptide could be detected in Ac311-IFN-β-infected cells at 30 and 48 h p.i. (data not shown). An abundantly produced 23.5K protein was observed in Ac360-IFN-β-infected cells

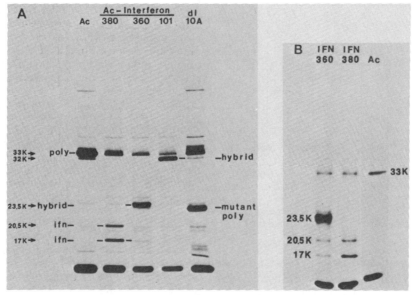

FIG. 5. Recombinant interferon produced in cells at (A) 30 or (B) 48 h p.i. (A) *S. frugiperda* cells were infected with AcNPV (Ac), Ac380-IFN-β, Ac360-IFN-β, Ac101-IFN-β, and *dl*10A (a deletion mutant of AcNPV that produces a truncated polyhedrin protein) and then labeled for 30 min with [³H]leucine at 30 h p.i. (B) *S. frugiperda* cells were infected with AcNPV (Ac), Ac380-IFN-β, and Ac360-IFN-β and then labeled for 30 min with [³H]leucine at 48 h p.i. Total cell lysates were electrophoresed on a sodium dodecyl sulfate–12% polyacrylamide gel; the gel was impregnated with En³Hance (New England Nuclear Corp., Boston, Mass.) and exposed to X-ray film for 2 days at −80°C. The molecular weights (×10³) of certain polypeptides are indicated.

(Fig. 5A and B) and has the size expected for a hybrid protein consisting of the entire interferon protein, including the 21 amino acid signal peptide plus an additional 14 amino acids derived from the first 10 codons of the polyhedrin gene and the *Bam*HI linker sequences (Fig. 1 and 2).

In the expression vector Ac101-IFN-β, the interferon fragment was inserted into the natural AcNPV *Bam*HI site at +175 in the same translational reading frame as polyhedrin. A putative hybrid 32K polypeptide was made in cells infected with this expression vector (Fig. 5). The size of the protein was as predicted from the DNA sequence.

Immunological identification of recombinant products. The 17K and 20.5K proteins being made in Ac380-IFN-β- and, to a lesser extent, Ac360-IFN-β-infected cells reacted with human IFN-β monoclonal antibody (Fig. 6C). The autoradiogram shown in Fig. 6 is a 6-h exposure, and the reaction of IFN-β antibody to the 20.5K protein was more obvious with longer exposure times. The reduced reaction of this antibody to the 20.5K protein as compared with that to the 17K protein was in part due to the fact that 17K accumulates to higher levels in cells than does 20.5K (data not shown). In addition, the antibody may be reacting to an epitope on the 17K polypeptide that is partially masked by, for example, glycosylation of the 20.5K polypep-

tide. The putative hybrid 23.5K and 32K proteins also reacted with IFN-β antibody (Fig. 6C). Polyclonal antibody to polyhedrin recognized the 10 amino acids of the 23.5K protein and the 57 amino acids of the 32K protein that would be predicted from the DNA sequence to be present at the N-terminal ends of the hybrid proteins (Fig. 6B).

Glycosylation of IFN-β in infected cells. Tunicamycin is a specific inhibitor of glycosylation which prevents the assembly of glycoproteins (38). The inhibitor has been used to produce nonglycosylated IFN-β (42) and is effective in AcNPV-infected *S. frugiperda* cells (36). In the presence of tunicamycin, a 64K AcNPV glycoprotein (6) was missing and a new protein of 58K was present (Fig. 7), indicating that the concentration of inhibitor used in this experiment (5 μg/ml) was sufficient to block glycosylation of this protein. Tunicamycin had little effect on the synthesis of the 17K IFN-β or the 23K and 32K hybrid proteins in Ac360-IFN-β- and Ac101-IFN-β-infected cells (Fig. 7). However, no 20.5K IFN-β was produced in the presence of tunicamycin in cells infected with Ac360-IFN-β (Fig. 7) or Ac380-IFN-β (data not shown).

To further demonstrate that 20.5K IFN-β was gylcosylated, Ac380-IFN-β-infected cells were labeled late in infection with [³H]mannose. The 20.5K IFN-β and three additional proteins were

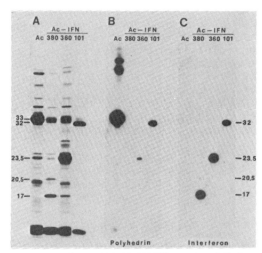

FIG. 6. Immunological identification of interferon and polyhedrin-interferon hybrid proteins. At 40 h p.i., total-cell lysates from AcNPV (Ac)-, Ac380-IFN-β-, Ac360-IFN-β-, and Ac101-IFN-β- infected cells were electrophoresed on a sodium dodecyl sulfate–12% polyacrylamide gel, transferred by diffusion to two nitrocellulose filters, and incubated with either polyhedrin antiserum or IFN-β monoclonal antibody as described previously (32). Specifically bound antibody was detected with ^{125}I-labeled protein A, and the filters were exposed to X-ray film at −80°C for 6 h with the aid of an intensifying screen. (A) Autoradiogram of infected cells labeled with [^{3}H]leucine at 40 h p.i. as described for Fig. 5. The molecular weights (×10³) of certain proteins are marked.

the major mannose-containing glycoproteins labeled in Ac380-IFN-β-infected cells (Fig. 8). A summary of the 17K non-glycosylated and 20.5K glycosylated IFN-β and the polyhedrin-IFN-β hybrid proteins produced by the various expression vectors is given in Fig. 2.

DISCUSSION

We have demonstrated that exceptionally high levels of interferon were produced in cells infected with a recombinant AcNPV expression vector in which the gene for IFN-β was placed under the transcriptional control of the polyhedrin promoter. Based on the specific activity of purified interferon (14), we estimate that ca. 10 μg of interferon was secreted by 10⁶ *S. frugiperda* cells infected with Ac380-IFN-β. In Ac380-IFN-β, the coding sequences for interferon were linked to the polyhedrin promoter at −59 (ca. 19 bases downstream from the canonical TATA sequences), suggesting that (i) polyhedrin gene sequences upstream from −59 were primarily responsible for the regulation of transcription and (ii) the 58-base nontranslated 5′-

leader sequence of polyhedrin was not absolutely required for efficient expression.

Cleavage of the signal peptide of pre-IFN-β is thought to be essential for producing biologically active, mature IFN-β (3, 37). Therefore, the polyhedrin-IFN-β hybrid proteins produced by the vectors Ac360-IFN-β and Ac101-IFN-β, which almost certainly still have the pre-IFN-β signal peptide as well as additional polyhedrin sequences, were most likely not biologically active. Much of the mature IFN-β produced by Ac360-IFN-β was probably a result of the removal of the signal peptide from some of the 23.5K hybrid gene product, in spite of the fact that the 21-amino acid signal sequence was an additional 14 amino acids from the N-terminal end of the protein. Sequence analysis of IFN-β and the hybrid proteins will be needed to confirm these speculations. Mature IFN-β may have also been produced by Ac360-IFN-β by another mechanism. It can be predicted from the scanning model for eucaryotic translation initiation (16) that the polyhedrin protein start sequence AUAAUGC is less favorable for initia-

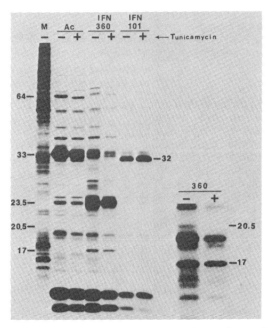

FIG. 7. Effects of tunicamycin on the synthesis of interferon At 30 h p.i. mock (M)-, AcNPV (Ac)-, Ac360-IFN-β-, and Ac101-IFN-β- infected cells were labeled with [^{3}H]leucine for 30 min in the presence or absence of 5 μg of tunicamycin per ml. Infected cells were pretreated for 10 h in medium plus 5 μg of tunicamycin per ml before being labeled in the presence of the inhibitor. A longer exposure of Ac360-IFN-β-labeled proteins is shown in the right panel. The molecular weights (×10³) of certain proteins are marked.

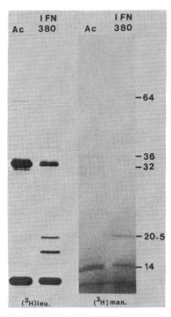

FIG. 8. Incorporation of [³H]mannose into recombinant interferon. AcNPV (Ac)- and Ac380-IFN-β-infected cells were labeled for 1 h at 40 h p.i. with [³H]leucine or [³H]mannose. Total cell lysates were electrophoresed on sodium dodecyl sulfate–12% polyacrylamide gel. The molecular weights ($\times 10^3$) of the polypeptides that were labeled with [³H]mannose are indicated.

tion than is the optimal sequence PuNNAUGG. Accordingly, this suboptimal sequence would be expected to allow some ribosomes to advance to the next potential initiator codon. Therefore, translation of 23.5K hybrid mRNA in Ac360-IFN-β infected cells may have occasionally initiated not at the first (polyhedrin) AUG but at the second (interferon) AUG. The fact that IFN-β was produced by Ac311-IFN-β is good evidence that this occurs, as interferon in this vector was in a different reading frame from polyhedrin and translation initiation from the second AUG is the most likely explanation for these results.

Linking interferon to an efficient promoter explains only in part why high levels of interferon were produced. Another important factor was the lack of antiviral activity on the AcNPV vector itself. The millions of interferon units produced per milliliter in our system had no measurable effect on virus yields, whereas only 100 U of interferon per ml reduces the yield of simian virus 40 by up to 80% (7), limiting the use of this and similarly affected vertebrate animal viruses as vectors for interferon.

AcNPV and other baculoviruses should prove to be important vectors for the production of cloned gene products in insect cells or organisms. The invertebrate cell will provide a unique biochemical environment for the production of foreign products and will complement vertebrate and procaryotic host-vector systems. Potentially, any gene could be linked to the polyhedrin promoter, incorporated into the AcNPV genome, and efficiently expressed in infected cells. We are currently investigating whether genes with intervening sequences will be correctly processed in this system, whether the promoter for the highly expressed AcNPV 10K gene (34) can be used in place of, or in the addition to, the polyhedrin promoter, and whether there are limits on the size of the DNA that can be incorporated into AcNPV expression vectors.

ACKNOWLEDGMENTS

We thank John Collins and Gerhard Gross for generously providing us with the cloned gene for IFN-β. P. W. Trown and Hoffmann-LaRoche Inc. for providing monoclonal antibody to IFN-β. Dennis Brown for the vesicular stomatitis virus. Ann Sorensen and Carol Rix for reviewing the manuscript. Bart Hooft van Iddekinge for assistance in DNA sequencing. and Carol Rix for technical assistance in interferon assays.

This work was supported in part by Public Health Service grant AI 14755 from the National Institutes of Health and by Texas Agricultural Experiment Station project TEXO 6316.

ADDENDUM IN PROOF

After this report was submitted for review a study by Remaut et al. (E. Remaut. P. Stanssens. and W. Fiers, Nucleic Acids Res. 14:4677–4688. 1983) was published which described the synthesis of mature IFN-β under the transcriptional control of a phage lambda promoter in *E. coli* cells at a level similar to that which was reported here.

LITERATURE CITED

1. **Adang, M. J., and L. K. Miller.** 1982. Molecular cloning of DNA complementary to RNA of the baculovirus *Autographa californica* nuclear polyhedrosis virus: location and gene products of RNA transcripts found late in infection. J. Virol. **44**:782–793.

2. **Canaani, D., and P. Berg.** 1982. Regulated expression of human interferon β₁ gene after transduction into cultured mouse and rabbit cells. Proc. Natl. Acad. Sci. U.S.A. **79**:5166–5170.

3. **Content, J., L. De Wit, R. Derynck, E. De Clercq, and W. Fiers.** 1982. *In vitro* cotranslational procession of human pre-interferon β₁ enhances its biological activity. Virology **122**:466–470.

4. **Derynck, R., J. E. Content, G. De Clercq, J. Volckaert, R. Tavernier, R. Devos, and W. Fiers.** 1980. Isolation and structure of a human fibroblast interferon gene. Nature (London) **285**:542–547.

5. **Dierks, P., A. van Ooyen, M. D. Cochran, C. Dobkin, J. Reiser, and C. Weissman.** 1983. Three regions upstream from the cap site are required for efficient and accurate transcription of the rabbit β-globin gene in mouse 3T6 cells. Cell **32**:695–706.

6. **Dobos, P., and M. A. Cochran.** 1980. Proteins in cells infected by *Autographa californica* nuclear polyhedrosis virus (Ac-NPV): the effect of cytosine arabinoside. Virology **103**:446–464.

7. **Gheysen, D., and W. Fiers.** 1982. Expression and excretion of human fibroblast β₁ interferon in monkey cells after transfection with a recombinant SV40 plasmid vector. J. Mol. Appl. Genet. **1**:385–394.

8. **Goeddel, D. V., H. M. Shepard, E. Yelverton, D. Leung, R. Crea, A. Sloma, and S. Pestka.** 1980. Synthesis of

human fibroblast interferon by *E. coli*. Nucleic Acids Res. 8:4057–4074.

9. Grace, T. D. C. 1962. Establishment of four strains of cells from insect tissue grown *in vitro*. Nature (London) 195:788–789.

10. Gross, G., U. Mayr, W. Bruns, F. Grosveld, H.-H. M. Dahl, and J. Collins. 1981. The structure of a thirty-six kilobase region of the human chromosome including the fibroblast interferon gene IFN-β. Nucleic Acids Res. 9:2495–2507.

11. Hauser, H., G. Gross, W. Bruns, H.-K. Hochkeppel, U. Mayr, and J. Collins. 1982. Inducibility of human β-interferon gene in mouse L-cell clones. Nature (London) 297:650–654.

12. Hink, W. F. 1970. Established insect cell line from the cabbage looper, *Trichoplusia ni*. Nature (London) 226:466–467.

13. Kelly, D. C. 1982. Baculovirus replication. J. Gen. Virol. 63:1–13.

14. Knight, E., Jr. 1976. Interferon: purification and initial characterization from diploid cells. Proc. Natl. Acad. Sci. U.S.A. 73:520–523.

15. Knight, E., Jr., and D. Fahey. 1982. Human interferon-beta: effects of deglycosylation. J. Interferon Res. 2:421–429.

16. Kozak, M. 1983. Comparison of initiation of protein synthesis in procaryotes, eucaryotes, and organelles. Microbiol. Rev. 47:1–45.

17. Laemmli, U. K. 1970. Cleavage of structural proteins during the assembly of the head of bacteriophage T4. Nature (London) 227:680–685.

18. Langford, M. P., D. A. Weigent, F. J. Stanton, and S. Baron. 1981. Virus plaque-reduction assay for interferon: microplaque and regular macroplaque reduction assays. Methods Enzymol. 78:339–346.

19. Messing, J., and J. Vieira. 1982. A new pair of M13 vectors for selecting either DNA strand of double-digest restriction fragments. Gene 19:269–276.

20. Miller, L. K., A. J. Lingg, and L. A. Bulla, Jr. 1983. Bacterial, viral, and fungal insecticides. Science 219:715–721.

21. Mitrani-Rosenbaum, S., L. Maroteaux, Y. Mory, M. Revel, and P. M. Howley. 1983. Inducible expression of the human interferon β₁ gene linked to a bovine papilloma virus DNA vector and maintained extrachromosomally in mouse cells. Mol. Cell. Biol. 3:233–240.

22. Mory, Y., Y. Chernajovsky, S. Feinstein, L. Chen, U. Nir, J. Weissenbach, Y. Malpiece, P. Tiollais, D. Marks, M. Ladner, C. Colby, and M. Revel. 1981. Synthesis of human interferon β₁ in *Escherichia coli* infected by a lambda phage recombinant containing a human genomic fragment. Eur. J. Biochem. 120:197–202.

23. Ohno, S., and T. Taniguchi. 1981. Structure of a chromosomal gene for human interferon β. Proc. Natl. Acad. Sci. U.S.A. 78:5305–5309.

24. Ohno, S., and T. Taniguchi. 1982. Inducer-responsive expression of the cloned human interferon β₁ gene introduced into cultured mouse cells. Nucleic Acids Res. 10:967–977.

25. Reyes, G. R., E. R. Gavis, A. Buchan, N. B. K. Raj, G. S. Hayward, and P. M. Pitha. 1982. Expression of human β-interferon cDNA under the control of a thymidine kinase promoter from herpes simplex virus. Nature (London) 297:598–601.

26. Rohel, D. Z., M. A. Cochran, and P. Faulkner. 1983. Characterization of late mRNAs of *Autographa californica* nuclear polyhedrosis virus. Virology 124:357–365.

27. Rohrmann, G. F., D. J. Leisy, K.-C. Chow, G. D. Pearson, and G. S. Beaudreau. 1982. Identification, cloning, and R-loop mapping of the polyhedrin gene from the multicapsid nuclear polyhedrosis virus of *Orgyia pseudotsugata*. Virology 121:51–60.

28. Rohrmann, G. F., M. N. Pearson, T. J. Bailey, R. R. Becker, and G. S. Beaudreau. 1981. N-terminal polyhedrin sequences and occluded *Baculovirus* evolution. J. Mol. Evol. 17:329–333.

29. Sanger, F., S. Nicklen, and A. R. Coulsen. 1977. DNA sequencing with chain-termination inhibitors. Proc. Natl. Acad. Sci. U.S.A. 74:5463–5467.

30. Smith, G. E., M. J. Fraser, M. D. Summers. 1983. Molecular engineering of the *Autographa californica* nuclear polyhedrosis virus genome: deletion mutations within the polyhedrin gene. J. Virol. 46:584–593.

31. Smith, G. E., and M. D. Summers. 1978. Analysis of baculovirus genomes with restriction endonucleases. Virology 89:517–527.

32. Smith, G. E., and M. D. Summers. 1981. Application of a novel radioimmunoassay to identify baculovirus structural proteins that share interspecies antigenic determinants. J. Virol. 39:125–137.

33. Smith, G. E., J. M. Vlak, and M. D. Summers. 1982. In vitro translation of *Autographa californica* nuclear polyhedrosis virus early and late mRNAs. J. Virol. 44:199–208.

34. Smith, G. E., J. M. Vlak, and M. D. Summers. 1983. Physical analysis of *Autographa californica* nuclear polyhedrosis virus transcripts for polyhedrin and 10,000-molecular-weight protein. J. Virol. 45:215–225.

35. Stewart, W. E., II. 1979. The interferon system. Springer, Vienna, Austria.

36. Stiles, B., H. A. Wood, and P. R. Hughes. 1983. Effect of tunicamycin on the infectivity of *Autographa californica* nuclear polyhedrosis virus. J. Invertebr. Pathol. 41:405–408.

37. Taniguchi, T., L. Guarente, T. M. Roberts, D. Kimelman, J. Douhan III, and M. Ptashne. 1980. Expression of the (human) fibroblast interferon gene in *E. coli*. Proc. Natl. Acad. Sci. U.S.A. 77:5230–5233.

38. Tkacz, J. S., and J. O. Lampen. 1975. Tunicamycin inhibition of polyisoprenyl N-acetylglucosaminyl pyrophosphate formation in calf-liver microsomes. Biochem. Biophys. Res. Commun. 65:489–257.

39. Torrence, P. F., and E. De Clercq. 1981. Interferon inducers: general survey and classification. Methods Enzymol. 78:291–299.

40. van der Beek, C. P., J. D. Saaijer-Riep, and J. M. Vlak. 1980. On the origin of the polyhedral protein of *Autographa californica* nuclear polyhedrosis virus. Virology 100:326–333.

41. Vieira, J., and Messing, J. 1982. The pUC plasmids, a M13mp7 derived system for insertion mutagenesis and sequencing with synthetic universal primers. Gene 19:259–268.

42. Yip, Y. K., and J. Vilcek. 1981. Production of human fibroblast interferon in the presence of the glycosylation inhibitor tunicamycin. Methods Enzymol. 78:212–219.

43. Zinn, K., P. Mellon, M. Ptashne, and T. Maniatis. 1982. Regulated expression of an extrachromososmal human β-interferon gene in mouse cells. Proc. Natl. Acad. Sci. U.S.A. 79:4897–4901.

Transformation of Mammalian Cells with Genes from Prokaryotes and Eukaryotes

M. Wigler, R. Sweet, G.K. Sim, B. Wold, A. Pellicer, E. Lacy,
T. Maniatis, S. Silverstein and R. Axel

Summary

We have stably transformed mammalian cells with precisely defined procaryotic and eucaryotic genes for which no selective criteria exist. The addition of a purified viral thymidine kinase (tk) gene to mouse cells lacking this enzyme results in the appearance of stable transformants which can be selected by their ability to grow in HAT. These biochemical transformants may represent a subpopulation of competent cells which are likely to integrate other unlinked genes at frequencies higher than the general population. Co-transformation experiments were therefore performed with the viral tk gene and bacteriophage ΦX174, plasmid pBR322 or the cloned chromosomal rabbit β–globin gene sequences. Tk+ transformants were cloned and analyzed for co-transfer of additional DNA sequences by blot hybridization. In this manner, we have identified mouse cell lines which contain multiple copies of ΦX, pBR322 and the rabbit β–globin gene sequences. The ΦX co-transformants were studied in greatest detail. The frequency of co-transformation is high: 15 of 16 tk+ transformants contain the ΦX sequences. Selective pressure was required to identify co-transformants. From one to more than fifty ΦX sequences are integrated into high molecular weight nuclear DNA isolated from independent clones. Analysis of subclones demonstrates that the ΦX genotype is stable through many generations in culture. This co-transformation system should allow the introduction and stable integration of virtually any defined gene into cultured cells. Ligation to either viral vectors or selectable biochemical markers is not required.

Introduction

Specific genes can be stably introduced into cultured cells by DNA-mediated gene transfer. The rare transformant is usually detected by biochemical selection. In this manner, we have isolated cells transformed with a variety of cellular and viral genes coding for selectable biochemical markers (Wigler et al., 1977, 1978, 1979). The isolation of cells transformed with

* Present address: Cold Spring Harbor Laboratory, Cold Spring Harbor, New York 11724.
† Division of Biology, California Institute of Technology, Pasadena, California 91125.

genes which do not code for selectable markers, however, is problematic since current transformation procedures are highly inefficient. This paper demonstrates the feasibility of co-transforming cells with two physically unlinked genes. Co-transformed cells can be identified and isolated when one of these genes codes for a selectable marker. We have used a viral thymidine kinase gene as a selectable marker to isolate mouse cell lines which contain the tk gene along with either bacteriophage ΦX174, plasmid pBR322 or the cloned rabbit β–globin gene sequences stably integrated into cellular DNA. The introduction of cloned eucaryotic genes into animal cells may provide a means for studying the functional consequences of DNA sequence organization.

Results

Experimental Design

The addition of the purified thymidine kinase (tk) gene from herpes simplex virus to mutant mouse cells lacking tk results in the appearance of stable transformants expressing the viral gene which can be selected by their ability to row in HAT (Maitland and McDougall, 1977; Wigler et al., 1977). To obtain co-transformants, cultures are exposed to the tk gene in the presence of a vast excess of a well defined DNA sequence for which hybridization probes are available. Tk+ transformants are isolated and scored for the co-transfer of additional DNA sequences by molecular hybridization.

Co-transformation of Mouse Cells with ΦX174 DNA

We initially used ΦX DNA in co-transformation experiments with the tk gene as the selectable marker. ΦX replicative form DNA was cleaved with Pst I, which recognizes a single site in the circular genome (Figure 1) (Sanger et al., 1977). 500 pg of the purified tk gene were mixed with 1–10 μg of Pst-cleaved ΦX replicative form DNA. This DNA was then added to mouse Ltk− cells using the transformation conditions previously described (Wigler et al., 1979). After 2 weeks in selective medium (HAT), tk+ transformants were observed at a frequency of one colony per 10^6 cells per 20 pg of purified gene. Clones were picked and grown into mass culture.

We then asked whether tk+ transformants also contained ΦX DNA sequences. High molecular weight DNA from the transformants was cleaved with the restriction endonuclease Eco RI, which recognizes no sites in the ΦX genome. The DNA was fractionated by agarose gel electrophoresis and transferred to nitrocellulose filters, and these filters were then annealed with nick-translated ^{32}P–ΦX DNA (blot hybridization) (Southern, 1975; Botchan, Topp and Sambrook, 1976; Pellicer et al., 1978). These annealing experi-

ϕX 174

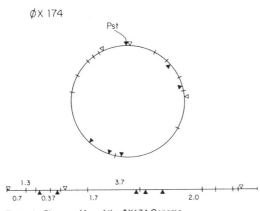

Figure 1. Cleavage Map of the ΦX174 Genome

Cleavage sites for the restriction endonucleases Pst I, Hpa I ($\triangledown$), Hpa II ($\blacktriangledown$) and Hae III (I) are shown for circular RFI and Pst I-linearized ΦX174 DNA (Sanger et al., 1977). The numbers above the line refer to the sizes of the internal Hpa I fragments in kbp, while those below the line refer to the sizes of the Hpa II fragments.

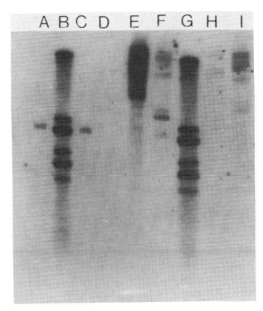

Figure 2. Identification of ΦX Sequences in Cells Transformed with ΦX174 DNA and the HSV tk Gene

Ltk⁻ aprt⁻ cells were transformed with ΦX174 DNA and the HSV tk gene using salmon sperm DNA as carrier. Tk⁺ transformants were selected by growth in HAT, cloned and grown into mass culture in HAT. High molecular weight DNA was extracted from seven independently isolated clones; 15 μg of DNA from each were digested with Eco RI, electrophoresed through 1% agarose gels, denatured in situ and transferred to nitrocellulose filters which were then annealed with ³²P-ΦX174 DNA (5 × 10⁶ cpm/μg) to identify co-transformants. Lanes B and G are Eco RI digests of ³²P-adenovirus 2 DNA; the six bands are 20.3, 4.2, 3.6, 2.6, 2.2 and 1.8 kbp. Lanes A, C, D, E, F, H and I are the seven independently isolated clones ΦX1–7, respectively. Only clone ΦX3 (lane D) lacks detectable ΦX sequences.

ments (Figure 2) demonstrate that six of the seven transformants had acquired bacteriophage sequences. Since the ΦX genome is not cut by the enzyme Eco RI, the number of bands observed reflects the minimum number of eucaryotic DNA fragments containing information homologous to ΦX. The clones contain variable amounts of ΦX sequences. Clones ΦX1 and ΦX2 (Figure 2, lanes A and C) reveal a single annealing fragment which is smaller than the ΦX genome. In these clones, therefore, only a portion of the transforming sequences persists. In lane D, we observe a tk⁺ transformant (clone ΦX3) with no detectable ΦX sequences. Clones ΦX4, 5, 6 and 7 (lanes E, F, H and I) reveal numerous high molecular weight bands which are too closely spaced to count, indicating that these clones contain multiple ΦX-specific fragments. These experiments demonstrate co-transformation of cultured mammalian cells with the viral tk gene and ΦX DNA.

Selection Is Necessary to Identify ΦX Transformants

We next asked whether transformation with ΦX DNA was restricted to the population of tk⁺ cells or whether a significant proportion of the original culture now contained ΦX sequences. Cultures were exposed to a mixture of the tk gene and ΦX DNA in a molar ratio of 1:2000 or 1:20,000. Half of the cultures were plated under selective conditions, while the other half were plated in neutral media at low density to facilitate cloning. Both selected (tk⁺) and unselected (tk⁻) colonies were picked, grown into mass culture and scored for the presence of ΦX sequences. In this series of experiments, eight of the nine tk⁺ selected colonies contained phage information (Figure 3). As

in the previous experiments, the clones contained varying amounts of ΦX DNA. In contrast, none of fifteen clones picked at random from neutral medium contained any ΦX information (data not shown). Thus the addition of a selectable marker facilitates the identification of those cells which contain ΦX DNA.

ΦX Sequences Are Integrated into Cellular DNA

Cleavage of DNA from ΦX transformants with Eco RI (Figure 2) generates a series of fragments which contain ΦX DNA sequences. These fragments may reflect multiple integration events. Alternatively, these fragments could result from tandem arrays of complete or partial ΦX sequences which are not integrated into cellular DNA. To distinguish between these possibilities, transformed cell DNA was cut with Bam HI or Eco RI, neither of which cleaves the ΦX genome. If the ΦX DNA sequences were not integrated, neither of these enzymes would cleave the ΦX fragments. Identical patterns would be generated from undigested DNA

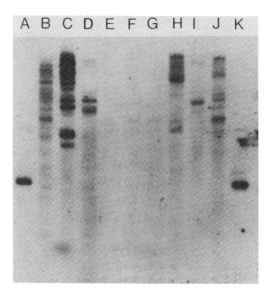

Figure 3. ΦX Sequences in tk⁺ Transformants

Cells were co-transformed as described in the legend to Figure 2, and half of the cultures were fed HAT, while the other half were replated under cloning conditions in DME. Nine colonies were selected in HAT and assayed for ΦX sequences as described (see Figure 2). Lanes A and K each contain 30 pg (2 gene equivalents) of Pst I-linearized ΦX174 DNA. Lanes B–J contain Eco RI-digested DNA from nine independently isolated tk⁺ transformants. Only one clone (lane E) does not contain ΦX sequences. None of fifteen clones isolated without selection contained ΦX sequences (blot not shown).

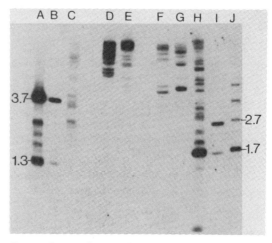

Figure 4. Extent of Sequence Representation in ΦX Co-transformants

High molecular weight DNA from co-transformant clones ΦX4 and ΦX5 was digested with either Eco RI, Bam HI, Hpa I or Hpa II and analyzed for the presence of ΦX sequences as described in the legend to Figure 2. (Lanes B and I) 50 pg (4 gene equivalents) of ΦX RFI DNA digested with Hpa I and Hpa II, respectively. (Lanes A, D, E and H) 15 μg of clone ΦX4 DNA digested with Hpa I, Eco RI, Bam HI and Hpa II, respectively, and analyzed for ΦX sequences by blot hybridization. (Lanes C, F, G and J) 15 μg of clone ΦX5 DNA digested with Hpa I, Eco RI, Bam HI or Hpa II, respectively.

and from DNA cleaved with either of these enzymes. If the sequences are integrated, then Bam HI and Eco RI should recognize different sites in the flanking cellular DNA and generate unique restriction patterns. DNA from clones ΦX4 and ΦX5 was cleaved with Bam HI or Eco RI and analyzed by Southern hybridization (Figure 4: clone 4, lanes D and E; clone 5, lanes F and G). In each instance, the annealing pattern with Eco RI fragments differed from that observed with the Bam HI fragments. Furthermore, the profile obtained with undigested DNA reveals annealing only in very high molecular weight regions with no discrete fragments observed (data not shown). Similar observations were made on clone ΦX1 (data not shown). Thus most of the ΦX sequences in these three clones are integrated into cellular DNA.

Intracellular Localization of the ΦX Sequences

The location of ΦX sequences in transformed cells was determined by subcellular fractionation. Nuclear and cytoplasmic fractions were prepared, and the ΦX DNA sequence content of each was assayed by blot hybridization. The data (not shown) indicate that 95% of the ΦX sequences are located in the nucleus. High and low molecular weight nuclear DNA was prepared by Hirt fractionation (Hirt, 1967). Hybridization with

DNA from these two fractions indicates that more than 95% of the ΦX information co-purifies with the high molecular weight DNA fraction. The small amount of hybridization observed in the supernatant fraction reveals a profile identical to that of the high molecular weight DNA, suggesting contamination of this fraction with high molecular weight DNA.

Extent of Sequence Representation of the ΦX Genome

The annealing profiles of DNA from transformed clones digested with enzymes that do not cleave the ΦX genome provide evidence that integration of ΦX sequences has occurred and allow us to estimate the number of ΦX sequences integrated. Annealing profiles of DNA from transformed clones digested with enzymes which cleave within the ΦX genome allow us to determine what proportion of the genome is present and how these sequences are arranged following integration. Cleavage of ΦX with the enzyme Hpa I generates three fragments for each integration event (see Figure 1): two "internal" fragments of 3.7 and 1.3 kb which together comprise 90% of the ΦX genome, and one "bridge" fragment of 0.5 kb which spans the Pst I cleavage site. The annealing profile observed when clone ΦX4 is digested with Hpa I is shown in Figure 4, lane A. Two intense bands are oberved at 3.7 and 1.3 kb. A less intense series of bands of higher molecular weight is also observed, some of which probably represent ΦX sequences adjacent to cellular DNA. These results indicate that at least 90% of the ΦX genome is present in these cells. It is worth noting that the internal 1.3 kb Hpa I fragment is bounded by an Hpa I site only 30 bp from the Pst I cleavage site. Comparison of the intensities of the internal bands with known quantities of Hpa I-cleaved ΦX DNA suggests that this clone contains approximately 100 copies of the ΦX genome (Figure 4, lanes A and B). The annealing patten of clone 5 DNA cleaved with Hpa I is more complex (Figure 4, lane C). If internal fragments are present, they are markedly reduced in intensity; instead, multiple bands of varying molecular weight are observed. The 0.5 kb Hpa I fragment which bridges the Pst I cleavage site is not observed for either clone ΦX 4 or clone ΦX5 (data not shown).

A similar analysis of clone ΦX4 and ΦX5 DNA was performed with the enzyme Hpa II. This enzyme cleaves the ΦX genome five times, thus generating four "internal" fragments of 1.7, 0.5, 0.5 and 0.2 kb, and a 2.6 kb "bridge" fragment which spans the Pst I cleavage site (Figure 1). The annealing patterns for Hpa II-cleaved DNA from ΦX clones 4 and 5 are shown in Figure 4 (clone ΦX4, lane H; clone ΦX5, lane J). In each clone an intense 1.7 kb band is observed, consistent with the retention of at least two internal Hpa II sites. The 0.5 kb internal fragments can also be observed, but they are not shown on this gel. Many additional fragments, mostly of higher molecular weight, are also present in each clone. These presumably reflect the multiple integration sites of ΦX DNA in the cellular genome. The 2.6 kb fragment bridging the Pst I cleavage site, however, is absent from clone ΦX4 (Figure 4, lane H). Reduced amounts of annealing fragments which co-migrate with the 2.6 kb Hpa II bridge fragment are observed in clone ΦX 5 (Figure 4, lane J). Similar observations were made in experiments with the enzyme Hae III. The annealing pattern of Hae III-digested DNA from these clones is shown in Figure 5 (clone ΦX4, lane B; clone ΦX5, lane C). In accord with our previous data, the 0.87 kb Hae III bridge fragment spanning the Pst site is absent or present in reduced amount in transformed cell DNA. Thus in general "internal" fragments of ΦX are found in these transformants, while "bridge" fragments which span the Pst I cleavage site are reduced or absent (see Discussion).

Stability of the Transformed Genotype

Our previous observations on the transfer of selectable biochemical markers indicate that the transformed phenotype remains stable for hundreds of generations if cells are maintained under selective pressure. If maintained in neutral medium, the transformed phenotype is lost at frequencies which range from <0.1 to as high as 30% per generation (Wigler et al., 1977,

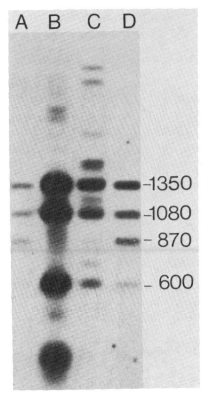

Figure 5. Annealing Pattern of DNA from Hae III-Digested ΦX Co-transformants

High molecular weight DNA from clones ΦX4 and ΦX5 was digested with Hae III, and the annealing profile was compared with that of Hae III-digested ΦXRFI DNA. Lanes A and D contain 30 and 50 pg (2 and 4 gene equivalents, respectively) of Hae III-digested ΦXRFI DNA. Lanes B and C contain 15 μg of Hae III-digested DNA from clones ΦX4 and ΦX5, respectively. The sizes of the prominent ΦX Hae III fragments in lanes A and D are 1350, 1080, 870 and 600 base pairs.

1979). The use of transformation to study the expression of foreign genes depends upon the stability of the transformed genotype. This is an important consideration with genes for which no selctive criteria are available. We assume that the presence of ΦX DNA in our transformants confers no selective advantage on the recipient cell. We therefore examined the stability of the ΦX genotype in the descendants of two clones after numerous generations in culture. Clones ΦX4 and ΦX5, both containing multiple copies of ΦX DNA, were subcloned and six independent subclones from each original clone were picked and grown into mass culture. DNA from each of these subclones was then digested with either Eco RI or Hpa I, and the annealing profiles of ΦX-containing fragments were compared with those of the original parental clone. The annealing pattern observed for four of the six ΦX4 subclones is virtually identical to that of the parent (Figure 6A). In

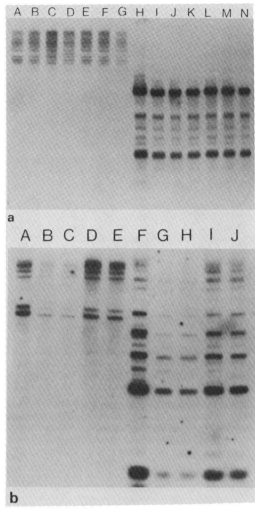

Figure 6. ΦX Sequences in Subclones of Co-transformants

(a) Annealing profiles of DNA from parental clone ΦX4 digested with Eco RI (lane A) and Hpa I (lane H) are compared with DNA from six independent subclones digested with either Eco RI (lanes B–G) or Hpa I (lanes I–N).

(b) High molecular weight DNA from four subclones of clone ΦX5 was isolated, cleaved with either Eco RI or Hpa II and compared with parental clone ΦX5 DNA. Lanes A and F contain clone ΦX5 DNA digested with Eco RI and Hpa II, respectively. DNA from four independently isolated subclones digested with either Eco RI (lanes B–E) or Hpa II (lanes G–J) was analyzed by blot hybridization.

two subclones, an additional Eco RI fragment appeared which is of identical molecular weight in both. This may have resulted from genotypic heterogeneity in the parental clone prior to subcloning. The patterns obtained for the subclones of ΦX5 are again virtually identical to the parental annealing profile (Figure 6B). These data indicate that ΦX DNA is maintained within the ten subclones examined for numerous generations

without significant loss or translocation of information.

Integration of pBR322 DNA into Mouse Cells

We have extended our observations on co-transformation to the EK2-approved bacterial vector, plasmid pBR322. pBR322 linearized with Bam HI was mixed with the purified viral tk gene in a molar ratio of 1000: 1. Tk⁺ transformants were selected and scored for the presence of pBR322 sequences. The Bgl I restriction map of Bam HI linearized pBR322 DNA is shown in Figure 7. Cleavage of this DNA with Bgl I generates two internal fragments of 2.4 and 0.3 kb. The sequence content of the pBR322 transformants was determined by digestion of transformed cell DNA with Bgl I followed by annealing with ^{32}P-labeled plasmid DNA. Four of five clones screened contained pBR sequences. Two of these clones contained the 2.4 kb internal fragment (Figure 8). The 0.3 kb fragment would not be detected on these gels. From the intensity of the 2.4 kb band in comparison with controls, we conclude that multiple copies of this fragment are present in these transformants. Other bands are observed which presumably represent the segments of pBR322 attached to cellular DNA.

Transformation of Mouse Cells with the Rabbit β–Globin Gene

Transformation with purified eucaryotic genes may provide a means for studying the expression of cloned genes in a heterologous host. We have therefore performed co-transformation experiments with the rabbit β major globin gene which was isolated from a cloned library of rabbit chromosomal DNA (Maniatis et al., 1978). One β–globin clone designated RβG-1 (Lacy et al., 1978) consists of a 15 kb rabbit DNA fragment carried on the bacteriophage λ cloning vector Charon 4a. Intact DNA from this clone (RβG-1) was mixed with the viral tk DNA at a molar ratio of 100:1, and tk⁺ transformants were isolated and examined for the presence of rabbit globin sequences. A restriction map of RβG-1 is shown in Figure 9. Cleavage of RβG-1 with the enzyme Kpn I generates a 4.7 kb fragment which contains the entire rabbit β–globin gene. This fragment was purified by gel electrophoresis and nick-translated to generate a probe for subsequent annealing experiments. The β–globin genes of mouse and rabbit are partially homologous, although we do not observe annealing of the rabbit β–globin probe with Kpn-cleaved mouse DNA under our experimental conditions (Figure 10, lanes C, D and G). In contrast, cleavage of rabbit liver DNA with Kpn I generates the expected 4.7 kb globin band (Figure 10, lane B). Cleavage of transformed cell DNA with the enzyme Kpn I generates a 4.7 kb fragment containing globin-specific information in six of the eight tk⁺ transformants examined (Figure 10). In two of the clones (Figure 10, lanes E and H), additional rabbit globin bands are observed which probably re-

PBR 322

Figure 7. Cleavage Map of pBR322

The Bgl I restriction endonuclease map for Bam HI-linearized pBR322 DNA is shown. The fragment sizes are in kbp, as determined by G. Sutcliffe (personal communication).

sult from the loss of at least one of the Kpn sites during transformation. The number of rabbit globin genes integrated in these transformants is variable. In comparison with control lanes (Figure 10, lanes A and L), some clones contain a single copy of the gene (lanes I, J and K), while others contain multiple copies of this heterologous gene (lanes E, F and H). These results demonstrate that cloned eucaryotic genes can be introduced into cultured mammalian cells by co-transformation.

Transformation Competence Is Not Stably Inherited

Our data suggest the existence of a subpopulation of transformation-competent cells within the total cell population. If competence is a stably inherited trait, then cells selected for transformation should be better recipients in subsequent gene transfer experiments than their parental cells. Two results indicate that as in procaryotes, competence is not stably heritable. In the first series of experiments, a double mutant, Ltk⁻ aprt⁻ (deficient in both tk and aprt), was transformed to either the tk⁺ aprt⁻ or the tk⁻ aprt⁺ phenotype using cellular DNA as donor (Wigler et al., 1978, 1979). These clones were then transformed to the tk⁺ aprt⁺ phenotype. The frequency of the second transformation was not significantly higher than the first. In another series of experiments, clones ΦX4 and ΦX5 were used as recipients for the transfer of a mutant folate reductase gene which renders recipient cells resistant to methotrexate (mtx). The cell line A29 Mtx^RIII contains a mutation in the structural gene for dihydrofolate reductase, reducing the affinity of this enzyme for methotrexate (Flintoff, Davidson and Siminovitch, 1976). Genomic DNA from this line was used to transform clones ΦX4 and ΦX5 and Ltk⁻ cells. The frequency of transformation to mtx resistance for the ΦX clones was identical to that observed with the parental Ltk⁻ cells. We conclude that competence is not a stably heritable trait and may therefore be a transient property of cells.

Discussion

In these studies, we have stably transformed mammalian cells with precisely defined procaryotic and eucaryotic genes for which no selective criteria exist. Our chosen experimental design derives from studies of transformation in bacteria which indicate that a

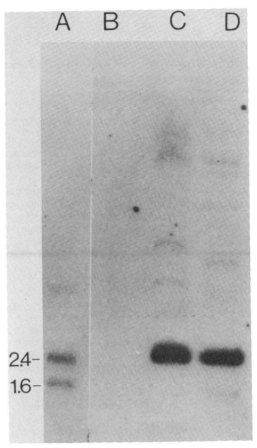

Figure 8. Physical Map of pBR322 Sequences in Co-transformants

Cells were exposed to pBR322 DNA and the viral tk gene and selected in HAT. High molecular weight DNA from three independent clones was digested with Bgl I and electrophoresed on a 1% agarose gel. The DNA was denatured in situ and transferred to nitrocellulose filters which were annealed with ³²P-pBR322 DNA. (Lane A) 5 pg of pBR322 DNA digested with Bgl I; (lanes B-D) 15 μg of DNA from three independent tk⁺ transformants.

small but selectable subpopulation of cells is competent in transformation (Thomas, 1955; Hotchkiss, 1959; Tomasz and Hotchkiss, 1964; Spizizen, Reilly and Evans, 1966). If this is also true for animal cells, then biochemical transformants will represent a subpopulation of competent cells which are likely to integrate other unlinked genes at frequencies higher than the general population. Thus, to identify transformants containing genes which provide no selectable trait, cultures were co-transformed with a physically unlinked gene which provided a selectable marker. This co-transformation system should allow the introduction and stable integration of virtually any defined gene into cultured cells. Ligation to either viral vectors or selectable biochemical markers is not required.

RβG-1

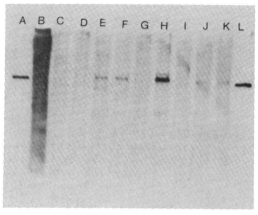

Figure 9. Physical Map of Rabbit β–Globin Phage Clone RβG-1

Cleavage sites for the restriction endonuclease Kpn I within RβG-1 are shown (Lacy et al., 1979). The numbers refer to the sizes of the fragments in kbp.

Co-transformation experiments were performed using the HSV tk gene as the selectable biochemical marker. The addition of this purified tk gene to mouse cells lacking thymidine kinase results in the appearance of stable transformants which can be selected by their ability to grow in HAT. Tk⁺ transformants were cloned and analyzed by blot hybridization for co-transfer of additional DNA sequences. In this manner, we have constructed mouse cell lines which contain multiple copies of ΦX, pBR322 and rabbit β–globin gene sequences.

The suggestion that these observations could result from contaminating procaryotic cells in our cultures is highly improbable. At least one of the rabbit β–globin mouse transformants expresses polyadenylated rabbit β–globin RNA sequences as a discrete 9S cytoplasmic species (B. Wold et al., manuscript in preparation). The elaborate processing events required to generate 9S globin RNA correctly are unlikely to occur in procaryotes.

The ΦX co-transformants were studied in greatest detail. The frequency of co-transformation is high: 14 of 16 tk⁺ transformants contain ΦX sequences. The ΦX sequences are integrated into high molecular weight nuclear DNA. The number of integration events varies from one to more than fifty in independent clones. The extent of the bacteriophage genome present within a given transformant is also variable; while some clones have lost up to half the genome, other clones contain over 90% of the ΦX sequences. Analysis of subclones demonstrates that the ΦX genotype is stable through many generations in culture. Similar conclusions are emerging from the characterization of the pBR322 and globin gene co-transformants.

Hybridization analysis of restriction endonuclease-cleaved transformed cell DNA allows us to make some preliminary statements on the nature of the integration intermediate. Only two ΦX clones have been examined in detail. In both clones, the donor DNA was Pst I-linearized ΦX DNA. We have attempted to distinguish between the integration of a linear or circular intermediate. If either precise circularization or the formation of linear concatamers had occurred at the Pst I cleavage site, and if integration occurred at random points along this DNA, we would expect cleavage maps of transformed cell DNA to mirror the circular ΦX map. The bridge fragment, however, is not observed or is present in reduced amounts in digests of

Figure 10. The Rabbit β–Globin Gene Is Present in Mouse DNA

Cells were exposed to RβG-1 DNA and the viral tk gene and selected in HAT. High molecular weight DNA from eight independent clones was digested with Kpn I and electrophoresed on a 1% agarose gel. The DNA was denatured in situ and transferred to nitrocellulose filters, which were then annealed with a ³²P-labeled 4.7 kbp fragment containing the rabbit β–globin gene. (Lanes A and L) 50 pg of the 4.7 kbp Kpn fragment of RβG-1; (lane B) 15 μg of rabbit liver DNA digested with Kpn; (lane C) 15 μg of Ltk⁻ aprt⁻ DNA; (lanes D–K) 15 μg of DNA from each of eight independently isolated tk⁺ transformants.

transformed cell DNA with three different restriction endonucleases. The fragments observed are in accord with a model in which ΦX DNA integrates as a linear molecule. Alternatively, it is possible that intramolecular recombination of ΦX DNA occurs, resulting in circularization with deletions at the Pst termini (Lai and Nathans, 1974). Random integration of this circular molecule would generate a restriction map similar to that observed for clones ΦX4 and ΦX5. Other more complex models of events occurring before, during or after integration can also be considered. Although variable amounts of DNA may be deleted from termini during transformation, most copies of integrated ΦX sequences in clone ΦX4 retain the Hpa I site, which is only 30 bp from the Pst I cleavage site. Whatever the mode of integration, it appears that cells can be stably transformed with long stretches of donor DNA. We have observed transformants containing contiguous stretches of donor DNA 50 kb long (B. Wold et al., unpublished studies).

We have attempted to identify cells transformed with ΦX sequences in the absence of selective pressure. Cultures were exposed to ΦX and tk DNA and cells were cloned under nonselective conditions. ΦX sequences were absent from all fifteen clones picked. In contrast, 14 of 16 clones selected for the tk⁺ phenotype contained ΦX DNA. The simplest interpretation is that a subpopulation of cells within the culture is competent in the uptake and integration of DNA. In this subpopulation of cells, two physically unlinked genes can be introduced into the same cell with high frequency. At present we can only speculate on the biological basis of competence. Competent cells may be genetic variants within the culture; however, our studies indicate that the competent phenotype is not stably inherited. If we extrapolate from studies in procaryotes, the phenomenon of competence is likely to be a complex and transient property reflecting the metabolic state of the cell.

Co-transformants contain at least one copy of the tk gene and variable amounts of ΦX DNA. Although transformation was performed with ΦX and tk sequences at a molar ratio of 1000:1, the sequence ratio observed in transformants never exceeded 100:1. There may be an upper limit to the number of integration events that a cell can tolerate, beyond which lethal mutations occur. Alternatively, it is possible that the efficiency of transformation may depend upon the nature of the transforming fragment. The tk gene may therefore represent a more efficient transforming agent than phage DNA.

The usefulness of the co-transformation method will depend to a large extent on its generality. To date, we have limited experience with other cell lines. The use of tk as a selectable marker restricts host cells to tk⁻ mutants. In unpublished studies, we have demonstrated the co-transfer of plasmid pBR 322DNA into Ltk⁻ aprt⁻ cells using aprt⁺ cellular DNA as donor and aprt as selectable marker. Furthermore, the use of dominant acting mutant genes which can confer drug resistance may extend the host range for co-transformation to virtually any cultured cell.

The stable transfer of ΦX DNA sequences to mammalian cells serves as a model system for the introduction of defined genes for which no selective criteria exist. We have used the tk co-transformation system to transform cells with the bacterial plasmid pBR322 and the cloned rabbit β–globin gene. Experiments which indicate that several of the pBR transformants contain an uninterrupted sequence which includes the replicative origin and the gene coding for ampicillin resistance (β–lactamase), suggest that DNA from pBR transformants may transfer ampicillin resistance to E. coli. Work in progress suggests that the rabbit β–globin gene is transcribed in at least one of the mouse transformants we have examined. Although preliminary, these studies indicate the potential value of co-transformation systems in the analysis of eucaryotic gene expression.

Experimental Procedures

Cell Culture

Ltk⁻ aprt⁻, a derivative of Ltk⁻ clone D (Kit et al., 1963), was obtained from R. Hughes and maintained in Dulbecco's modified Eagle's medium (DME) containing 10% calf serum and 50 µg/ml of diaminopurine (DAP). Prior to transformation, cells were washed and grown for three generations in the absence of DAP. A Chinese hamster cell line containing an altered dihydrofolate reductase (rendering it re-

sistant to methotrexate) A29 Mtx^RIII (Flintoff et al., 1976) was obtained from L. Siminovitch. These cells were propagated in DME supplemented with 3X nonessential amino acids, 10% calf serum and 1 µg/ml amethopterin.

Isolation of the HSV tk Gene

Intact herpes simplex virus (HSV) DNA was isolated from CV-1-infected cells as previously described (Pellicer et al., 1978). DNA was digested to completion with Kpn I (New England Biolabs) in a buffer containing 6 mM Tris (pH 7.9), 6 mM MgCl₂, 6 mM 2–mercaptoethanol, 6 mM NaCl and 200 µg/ml bovine serum albumin. The restricted DNA was fractionated by electrophoresis through 0.5% agarose gels (17 × 20 × 0.5 cm) for 24 hr at 70 V, and the 5.1 kb tk-containing fragment was extracted from the gel as described by Maxam and Gilbert (1977).

Source of DNAs

ΦX174 am3 RFI DNA was purchased from Bethesda Research Laboratories. Plasmid pBR322 DNA was grown in E. coli HB 101 and purified according to the method of Clewell (1972). The cloned rabbit β major globin gene in the λ Charon 4A derivative (RβG-1) was identified and isolated as previously described (Maniatis et al., 1978).

Co-transformation of Defined DNA Sequences and the HSV tk Gene

Ltk⁻ aprt⁻ mouse cells were transformed with either 1–10 µg of ΦX174, 1 µg of pBR322 or 1 µg of RβG-1 DNA in the presence of 1 ng of HSV-1 tk gene and 10–20 µg of salmon sperm carrier DNA, as previously described (Wigler et al., 1979). Tk⁺ transformants were selected in DME containing hypoxanthine, aminopterin and thymidine (HAT) and 10% calf serum. Isolated colonies were picked using cloning cylinders and grown into mass cultures.

Transformation

Methotrexate-resistant transformants of Ltk⁻ aprt⁻ cells were obtained following transformation with 20 µg of high molecular weight DNA from A29 Mtx^RIII cells and selection in DME containing 10% calf serum and 0.2 µg/ml amethopterin.

Transformation and selection for aprt⁺ transformants were as described by Wigler et al. (1979).

Isolation of Transformed Cell DNA

Cells were harvested by scraping into PBS and centrifuging at 1000 × g for 10 min. The pellet was resuspended in 40 vol of TNE [10 mM Tris–HCl (pH 8.0), 150 mM NaCl, 10 mM EDTA], and SDS and proteinase K were added to 0.2% and 100 µg/ml, respectively. The lysate was incubated at 37°C for 5–10 hr and then extracted sequentially with buffer-saturated phenol and CHCl₃. High molecular weight DNA was isolated by mixing the aqueous phase with 2 vol of cold ethanol and immediately removing the precipitate that formed. The DNA was washed with 70% ethanol and dissolved in 1 mM Tris, 0.1 mM EDTA.

Nuclei and cytoplasm from clones ΦX4 and ΦX5 were prepared as described by Ringold et al. (1977). The nuclear fraction was further fractionated into high and low molecular weight DNA as described by Hirt (1967).

Filter Hybridization

DNA from transformed cells was digested with various restriction endonucleases using the conditions specified by the supplier (New England Biolabs or Bethesda Research Laboratories). Digestions were performed at an enzyme to DNA ratio of 1.5 U/µg for 2 hr at 37°C. Reactions were terminated by the addition of EDTA, and the product was electrophoresed on horizontal agarose slab gels in 36 mM Tris, 30 mM NaH₂PO₄, 1 mM EDTA (pH 7.7). DNA fragments were transferred to nitrocellulose sheets, hybridized and washed as previously described (Weinstock et al., 1978) with two modifications. Two nitrocellulose filters were used during transfer (Jeffreys and Flavell, 1977b). The lower filter was discarded, and following hybridization the filter was washed 4 times for 20 min in 2 × SSC, 25 mM sodium phosphate, 1.5 mM $Na_4P_2O_7$, 0.05% SDS at 65°C and then successively in 1:1 and 1:5 dilutions of this buffer (Jeffreys and Flavell, 1977a).

Acknowledgments

We wish to thank Sharon Dana and Mary Chen for excellent technical assistance. This work was supported by grants from the NIH and the NSF.

The costs of publication of this article were defrayed in part by the payment of page charges. This article must therefore be hereby marked "*advertisement*" in accordance with 18 U.S.C. Section 1734 solely to indicate this fact.

Received January 8, 1979; revised January 29, 1979

References

Botchan, M., Topp, W. and Sambrook, J. (1976). Cell 9, 269–287.

Clewell, D. B. (1972). J. Bacteriol. 110, 667–676.

Flintoff, W. F., Davidson, S. V. and Siminovitch, L. (1976). Somatic Cell Genet. 2, 245–261.

Hirt, B. (1967). J. Mol. Biol. 26, 365–369.

Hotchkiss, R. (1959). Proc. Natl. Acad. Sci. USA 40, 49–55.

Jeffreys, A. J. and Flavell, R. A. (1977a). Cell 12, 429–439.

Jeffreys, A. J. and Flavell, R. A. (1977b). Cell 12, 1097–1108.

Kit, S., Dubbs, D., Piekarski, L. and Hsu, T. (1963). Exp. Cell Res. 31, 297–312.

Lacy, E., Lawn, R. M., Fritsch, D., Hardison, R. C., Parker, R. C. and Maniatis, T. (1979). In Cellular and Molecular Regulation of Hemoglobin Switching (New York: Grune & Stratton), in press.

Lai, C. J. and Nathans, D. (1974). Cold Spring Harbor Symp. Quant. Biol. 39, 53–60.

Maitland, N. J. and McDougall, J. K. (1977). Cell 11, 233–241.

Maniatis, T., Hardison, R. C., Lacy, E., Lauer, J., O'Connell, C., Quon, D., Sim, G. K. and Efstradiatis, A. (1978). Cell 15, 687–701.

Maxam, A. M. and Gilbert, W. (1977). Proc. Natl. Acad. Sci. USA 74, 560–564.

Pellicer, A., Wigler, M., Axel, R and Silverstein S. (1978). Cell 14, 133–141.

Ringold, G. M., Yamamoto, K. R., Shank, P. R. and Varmus, H. E. (1977). Cell 10, 19–26.

Sanger, F., Air, M., Barrell, B. G., Brown, N. L., Coulson, A. R., Fiddes, J. C., Hutchinson, C. A., Slocombe, P. M. and Smith, M. (1977). Nature 265, 687–695.

Southern, E. M. (1975). J. Mol. Biol. 98, 503–517.

Spizizen, J., Reilly, B. E. and Evans, A. H. (1966). Ann. Rev. Microbiol. 20, 371–400.

Thomas, R. (1955). Biochim. Biophys. Acta 18, 467–481.

Thomasz, A. and Hotchkiss, R. (1964). Proc. Nat. Acad. Sci. USA 51, 480–487.

Weinstock, R., Sweet, R., Weiss, M., Cedar, H. and Axel, R. (1978). Proc. Natl. Acad. Sci. USA 75, 1299–1303.

Wigler, M., Silverstein, S., Lee, L.-S., Pellicer, A., Cheng, Y.-C. and Axel, R. (1977). Cell 11, 223–232.

Wigler, M., Pellicer, A., Silverstein, S. and Axel, R. (1978). Cell 14, 725–731.

Wigler, M., Pellicer, A., Silverstein, S., Axel, R., Urlaub, G. and Chasin, L. (1979). Proc. Nat. Acad. Sci. USA, in press.

Supplementary Readings

Capecchi, M.R., High efficiency transformation by direct microinjection of DNA into cultured mammalian cells. *Cell* 22:479–488 (1980)

Hammer, R.E., Pursel, V.G., Rexroad, C.E., Jr., Wall, R.J., Bolt, D.J., Ebert, K.M., Palmiter, R.D. & Brinster, R.L., Production of transgenic rabbits, sheep and pigs by microinjection. *Nature* 315:680–683 (1985)

Maeda, S., Expression of foreign genes in insects using baculovirus vectors. *Annu. Rev. Entomol.* 34:351–372 (1989)

Maeda, S., Kawai, T., Obinata, M., Fujiwara, H., Horiuchi, T., Saeki, Y., Sato, Y. & Furusawa, M., Production of human α-interferon in silkworm using a baculovirus vector. *Nature* 315:592–594 (1985)

Mulligan, R.C. & Berg, P., Expression of a bacterial gene in mammalian cells. *Science* 209:1422–1427 (1980)

Sarver, N., Gruss, P., Law, M.-F., Khoury, G. & Howley, P.M., Bovine papilloma virus deoxyribonucleic acid: a novel eukaryotic cloning vector. *Mol. Cell. Biol.* 1:486–496 (1981)

Schaffner, W., Direct transfer of cloned genes from bacteria to mammalian cells. *Proc. Natl. Acad. Sci. U.S.A.* 77:2163–2167 (1980)

Simons, J.P., Wilmut, I., Clark, A.J., Archibald, A.L., Bishop, J.O. & Lathe, R., Gene transfer into sheep. *Bio/Technology* 6:179–183 (1988)

Wigler, M., Perucho, M., Kurtz, D., Dana, S., Pellicer, A., Axel, R. & Silverstein, S., Transformation of mammalian cells with an amplifiable dominant-acting gene. *Proc. Natl. Acad. Sci. U.S.A.* 77:3567–3570 (1980)

Diagnosis

One of the most visible products of recombinant DNA has been its application to diagnostic methodology. It is now possible, by rapid and highly sensitive procedures, to test for the presence of pathogens, identify genetic defects, characterize old and new species or genera. Classically, diagnostic bacteriology relied on metabolic reactions or the use of immunological tests employing polyclonal antibodies; the latter were also used in diagnostic virology. The advent of gene cloning provided new approaches to diagnosis.

If we consider the following scheme:

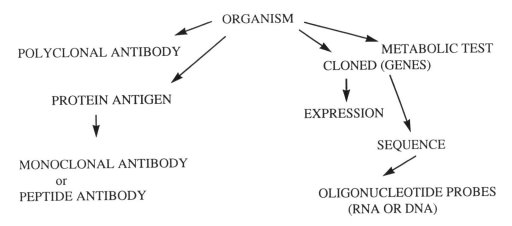

it is easy to see how the cloning and expression of genes for characteristic antigens could lead to the preparation of new and specific detection markers for any given organism. The process went hand in hand with extensive studies of the cloning of virus components and protein determinants of virulence in bacterial pathogens. Once a structural component of the pathogen had been cloned, one had a diagnostic test. This, combined with the power to produce monoclonal antibodies and the ability to sequence oligopeptides and nucleotides at will, has led to the development of epitope screening and nucleic acid probe diagnostics.

It must not be forgotten, however, that the discovery of restriction enzymes alone changed the approach to identifying carriers of hereditary diseases. Once the appropriate gene was cloned and identified on the chromosome, the use of endonuclease digestion to obtain restriction fragment length polymorphism patterns (RFLPs) permitted the detection of specific alterations in genes and new methods of prenatal diagnosis of hereditary diseases.

The use of the procedures described above has changed, and will continue to change, diagnostic and detection methodology; however the remarkable discovery of a simple approach to the amplification of the target sequence by the polymerase chain reaction

(PCR) now presents additional possibilities with respect to all aspects of DNA technology (cloning, site-directed mutagenesis, etc.) and all aspects of diagnostic studies. In bacterial diagnostics, the classical bacteriological test for the presence of tuberculosis due to *Mycobacterium tuberculosis* can take weeks or months. The use of PCR combined with specific oligonucleotide probes permits detection of the pathogen in sputum samples in a couple of days! Likewise, in food testing, it is possible to detect one bacterium of the food pathogen *Listeria monocytogenes* in 25 grams of a ripe camembert cheese—not an easy task without PCR! PCR amplification, which was developed by scientists in the biotechnology company Cetus, permits the enzymic synthesis *in vitro* of millions of copies of any specific DNA fragment. The fragment is recognized by flanking oligonucleotides which serve as primers to copy the DNA sequence chosen in a cyclic process. RNA (e.g. mRNA) can also be used as target molecule for amplification, as employed in generating antibody libraries. (See the Immunology section.) The PCR procedure has revolutionized molecular biology and is so commonly used that it has its own journal! In addition to the improvement of basic nucleic acid techniques, PCR has made available new approaches to molecular evolution and footprinting, and will be of value in the Human Genome Project.

In the field of diagnostic methodology, PCR will undoubtedly have an enormous impact. Identification of determinants for hereditary diseases will now be a question of screening for amplification of a specific sequence. Testing for genetic susceptibility to certain diseases and tissue typing are likely to become relatively simple and reliable procedures. Forensic use of PCR has already had an impact in extending the procedures pioneered by Jeffreys. PCR together with specific hybridization probes will be the most potent detective combination since Holmes and Watson, and molecular "whodunitry" is likely to replace conventional detection methods. How long will it take criminals to learn how to coat their persons with DNase?

An important aspect of diagnostic and detection methodology is how one measures the signal; what kinds of "reporter" systems can be used? Considerable progress has been made in the design of reporter enzymes that can amplify the assay sensitivity by several logs. For example fluorescent or light-producing enzyme reactions have been refined and are replacing radioactivity for a number of applications. For reasons such as this, of all applications of recombinant DNA technology, rapid and accurate automated diagnostic tests will probably have the most significant impact on medical practice.

Dystrophin: The Protein Product of the Duchene Muscular Dystrophy Locus

E.P. Hoffman, R.H. Brown and L.M. Kunkel

Summary

The protein product of the human Duchenne muscular dystrophy locus (DMD) and its mouse homolog (mDMD) have been identified by using polyclonal antibodies directed against fusion proteins containing two distinct regions of the mDMD cDNA. The DMD protein is shown to be approximately 400 kd and to represent approximately 0.002% of total striated muscle protein. This protein is also detected in smooth muscle (stomach). Muscle tissue isolated from both DMD-affected boys and mdx mice contained no detectable DMD protein, suggesting that these genetic disorders are homologous. Since mdx present no obvious clinical abnormalities, the identification of the mdx mouse as an animal model for DMD has important implications with regard to the etiology of the lethal DMD phenotype. We have named the protein dystrophin because of its identification via the isolation of the Duchenne muscular dystrophy locus.

Introduction

The muscular dystrophies are a heterogeneous group of both human and animal hereditary diseases whose primary manifestation is progressive muscle weakness due to intrinsic biochemical defects of muscle tissue (Mastaglia and Walton, 1982). The most common and devastating of the human muscular dystrophies is the X-linked recessive Duchenne muscular dystrophy (DMD), first described in the mid-1800s (Meryon, 1852; Duchenne, 1868). Affecting approximately 1 in 3,500 boys, this genetic disorder exhibits no obvious clinical manifestation until the age of 3 to 5 years, when proximal muscle weakness is first observed. The ensuing progressive loss of muscle strength usually leaves affected individuals wheelchair-bound by the age of 11, and results in early death due to respiratory failure. Both the typical histological pattern of widespread degeneration and regeneration of individual muscle fibers in most skeletal muscle groups (Dubowitz, 1985) and high concentrations of soluble muscle-specific enzymes in serum are present in affected individuals long before the clinical onset of the disease, and can often be found in female carriers (Emery and Holloway, 1977). Despite many years of intensive research, the primary biochemical defect responsible for this disorder has remained elusive, as have any rational therapies to slow the progression of the disease.

With no effective treatments available, an animal model for this disease has long been sought to test possible therapies. Despite the availability of numerous muscular dystrophies in many different species (Harris and Slater, 1980), the lack of any information concerning the biochemical defect involved in both DMD and the putative animal models has made it difficult to equate any specific animal muscular dystrophy with DMD. An X-linked murine muscular dystrophy, mdx, was fortuitously discovered during screening of normal mouse serum enzymes in preparation for a mutagenesis screen (Bulfield et al., 1984). Given the general conservation of the X chromosome within mammalian species (Ohno, Becak, and Becak, 1964), the chromosomal location of mdx in mouse suggested that the mdx mutation might indeed represent the same biochemical defect as that manifested in X-linked human DMD. However, homozygous mdx mice exhibit little, if any, clinically detectable phenotype. Homozygous strains are easily maintained, showing only slightly reduced fecundity (Torres and Duchen, 1987), and develop no obvious muscle weakness (Tanabe, Esaki, and Nomura, 1986). Histologically, the persistent degeneration/regeneration characteristic of human DMD muscle fibers appears very similar in the mdx mouse, though the extensive connective tissue proliferation (fibrosis) evident in human DMD muscle groups appears to be largely absent in mdx muscle (Bridges, 1986). Though it could be argued that the mdx mutation represents a mild allele of the mouse DMD homolog (such as the rarer Becker allele of Duchenne), two ethylnitrosourea-induced (ENU) alleles of the mdx locus exhibit similar phenotypes, making it unlikely that all three represent Becker-like mutations (Chapman, personal communication).

The differences between the mouse mdx and human DMD phenotypes have been used as evidence against the significance of the common X-linked nature of the mutations. In apparent agreement with the phenotypic differences, the mdx mutation has been recently shown to be more closely linked genetically to markers neighboring the less severe and less common human X-linked Emery–Dreifuss muscular dystrophy (Avner et al., 1987). These two human X-linked dystrophies are on opposite ends of the human X chromosome, yet recent studies using portions of the cloned DMD locus as genetic markers in the mouse have placed the mDMD locus in the same chromosomal region as the mdx mutation (Chapman et al., 1985; Heilig et al., 1987; Brockdorff et al., 1987; Chamberlain et al., 1987).

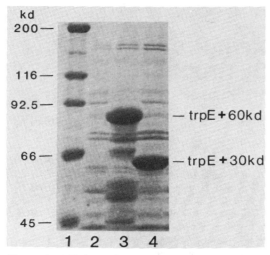

Figure 1. Insoluble Protein Fractions of Recombinant Bacterial Lysates

E. coli strain RR1 harboring either the parent plasmid vector (lane 2)(pATH2; Dieckmann and Tzagoloff, 1985) or one of its recombinants containing two different portions of the mDMD coding sequence (lane 3, trpE+60kd; lane 4, trpE+30kd) was induced with 3-B-indolacrylic acid and lysed, and insoluble proteins were isolated by centrifugation. Shown is 1% (~200 µg) of the insoluble protein obtained from a 100 ml culture of bacteria for each preparation. The trpE–mDMD fusion proteins are evident as the major protein species (~85% of total), at the expected molecular weights (trpE = ~33 kd). Protein size markers are indicated (lane 1).

A human cDNA clone representing a portion of the DMD transcript has been isolated on the basis of its conservation between mouse and man (Monaco et al., 1986). This partial cDNA was then used to isolate homologous cDNA sequences from mouse (Hoffman et al., 1987), and the entire 14 kb human coding sequence (Koenig et al., 1987). The mRNA product of this locus was detected only in muscle-containing tissues in both humans and mice (Monaco et al., 1986; Hoffman et al., 1987), and has been demonstrated to be expressed specifically in terminally differentiated myotubes in culture (Lev et al., 1987). In this paper we report the production of antibodies to the mDMD and DMD gene protein products. We describe the use of these antibodies to study the protein product of the Duchenne muscular dystrophy locus, called dystrophin, in tissues isolated from both normal and dystrophic mice and humans.

Results

Production of mDMD Fusion Proteins

The DNA and predicted amino acid sequences for 4.3 kb (30%) of the Duchenne muscular dystrophy (DMD) gene has been previously presented for cDNA clones isolated from human fetal skeletal muscle and for the mouse cardiac muscle homolog of the human DMD locus (mDMD; Hoffman et al., 1987; Koenig et al., 1987). These cDNAs were found to be highly conserved between mice and humans, exhibiting over 90% similarity at both the DNA

and amino acid levels. To study the mouse and human protein product of the DMD locus, polyclonal antibodies were produced against large regions of the mouse mDMD polypeptide.

Two different regions of the mouse heart DMD cDNA were fused to the 3' terminus of the E. coli trpE gene by using the expression vector pATH2 (Dieckmann and Tzagoloff, 1985). The two regions represent the majority of the mouse heart cDNA sequence previously described (Hoffman et al., 1987), but do not overlap. One construction resulted in the fusion of approximately 30 kd of mDMD protein to the 33 kd trpE protein, while the second employed roughly 60 kd of the mDMD protein. Since the trpE protein is insoluble, quantitative yields of induced fusion proteins were obtained simply by lysing of the cells and precipitation of insoluble proteins. As shown in Figure 1, novel insoluble fusion proteins of the expected size were produced; they were not present in lysates from bacteria containing the pATH2 vector alone.

Production of Antisera

Both fusion proteins were purified by preparative SDS–polyacrylamide gel electrophoresis (Laemmli, 1970) and used to immunize rabbits and sheep (see Experimental Procedures). Rabbits were immunized with electroeluted, "native" (free from SDS) insoluble antigen, while sheep were immunized with SDS–polyacrylamide gel slices containing denatured antigen. In order to monitor the titer and specificity of immune sera, dot blots of each antigen were made on nitrocellulose by using the insoluble fractions of bacterial lysates harboring either the parent plasmid vector (pATH2) or a recombinant fusion-protein plasmid. Each antigen solution was loaded such that the amount indicated refers to the amount of trpE protein in each dot (Figure 2).

The DMD portion of the smaller fusion protein (trpE+30kd) proved to be highly antigenic. Greater than 95% of rabbit antibodies (rabbit serum; Figure 2A) were found to be directed specifically against the DMD portion of the polypeptide, though this portion represented only half of the fusion protein. Similarly, approximately 90% of the sheep antibodies were directed specifically against the DMD portion of this fusion protein (data not shown). The larger trpE+60kd fusion protein also proved to be quite antigenic, with a 1:1000 dilution of primary sera having antibody titers nearly equal to the limits of the detection system (~1 ng antigen), though a much larger proportion of antibodies were directed against the trpE protein (sheep serum shown in Figure 2B; rabbit serum results similar but not shown).

To ensure that any protein species identified by the antisera was due to recognition by antibodies specific for he DMD portion of the fusion proteins, rabbit and sheep antibodies directed against the 30 kd antigen and sheep antibodies directed against the 60 kd antigen were affinity-purified (Figure 2). Affinity purification of the rabbit antibodies directed against the mDMD portion of the fusion protein was facilitated by the insolubility of the partially purified fusion protein. By simple resuspension of crude insoluble protein fractions (shown in Figure 1) in immune se-

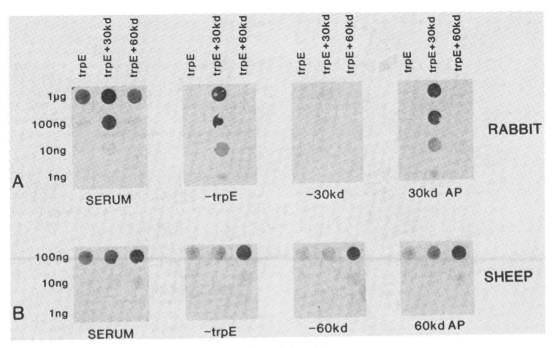

Figure 2. Specificity and Titer of Antisera Raised against the mDMD Fusion Proteins

Antigen dot blots were prepared on nitrocellulose by using the insoluble protein fractions shown in Figure 1. The amounts indicated refer to the relative amount of trpE protein in each dot.

(A) Immune serum from a rabbit immunized with the trpE+30kd antigen in insoluble form (see Experimental Procedures). The first filter shows the alkaline phosphatase staining after incubation with unprocessed immune serum diluted 1:1000, followed by incubation with alkaline phosphatase conjugated goat anti-rabbit second antibody (rabbit serum). Greater than 95% of the antibodies in the unprocessed serum are seen to be directed specifically against the DMD portion of the fusion protein. Antibodies specific for the trpE portion of the fusion peptide were then removed, with the resulting serum showing no apparent remaining reactivity for the trpE protein (−trpE). The immune serum with the antibodies against the 30 kd protein removed is shown to contain little remaining reactivity with the fusion protein (−30 kd). 30kd antigen–antibody complexes were disassociated, with the resulting supernatant showing a high titer of antibodies directed specifically against the DMD portion of the fusion peptide (30kd AP [affinity-purified]).

(B) The same affinity purification protocol as used in (A) was applied to immune serum from a sheep immunized with denatured trpE+60kd antigen.

rum, antibodies against the trpE protein were eliminated (−trpE; Figure 2A) and antibodies specific for the mDMD protein isolated (30 kd AP). The immune serum that had been absorbed with both the trpE and trpE+30kd antigens showed very little remaining reactivity with either of these antigens (−30kd; Figure 2A). The resulting affinity-purified antibody (30kd AP) had a titer above the limits of the detection system (1 ng) when a 1:1000 dilution was used (Figure 2A). The sheep antisera against this same fusion protein were affinity-purified in the same manner, with greater than 95% of the resulting affinity-purified immunoglobulins being directed specifically against the 30 kd DMD antigen (not shown). In the case of the sheep antiserum directed against the 60 kd antigen, the same affinity purification protocol was used but appeared to be much less efficient (Figure 2B; −trpE, −60kd, 60kd AP).

Identification of the Protein Product of the Duchenne Muscular Dystrophy Locus

Total protein samples were isolated from mouse (fresh) and human (frozen) tissues by direct solubilization of tissues in 10 volumes of gel loading buffer (100 mM Tris, pH

8.0; 10% SDS; 10 mM EDTA; 50 mM DTT). Alternatively, Triton X-100 insoluble fractions were isolated from human and mouse tissues by homogenization in 0.25% Triton X-100 using a Waring blender at full speed, and by pelleting of insoluble proteins. The protein concentrations in the Triton-insoluble fractions were quantitated by using the Bio-Rad protein assay, while the protein concentration of the directly SDS-solubilized tissues was estimated based on the starting mass of the tissue used. All protein samples (50 μg) were separated by electrophoresis on 3.5% to 12.5% gradient SDS–polyacrylamide gels (Laemmli, 1970) using a 3.0% stacking gel, and transferred to nitrocellulose (Towbin, Staehelin, and Gordon, 1979). Identical nitrocellulose blots of the separated proteins were incubated with affinity-purified rabbit antibodies directed against the 30 kd antigen (Figure 3A), affinity-purified sheep antibodies directed against the 60 kd antigen (Figure 3B), and affinity-purified sheep antibodies directed against the 30 kd antigen (Figure 3C), each at a 1:1000 dilution. Immune complexes were detected by using either ^{125}I-protein A (Figure 3A) or alkaline phosphatase conjugated donkey anti-sheep IgG second antibody (Figures

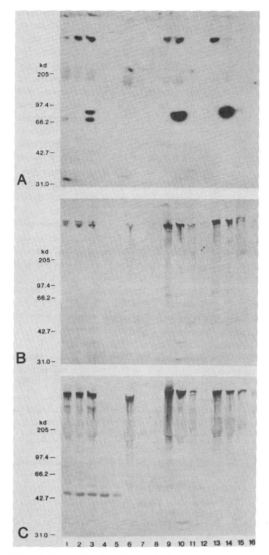

Figure 3. Identification of the Protein Product of the DMD Locus

Mouse (fresh) or human (frozen) tissues were either directly solubilized and denatured in gel loading buffer or were first homogenized with 0.25% Triton X-100, and Triton-insoluble proteins were isolated. Protein samples (50 μg) were fractionated on 3.5%–12.5% gradient SDS–poly-acrylamide gels (Laemmli, 1970), transferred to nitrocellulose (Towbin, Staehelin, and Gordon, 1979), and incubated with affinity-purified rabbit antibodies directed against the 30 kd cardiac mDMD (A), sheep antibodies directed against the 60 kd antigen (B), or sheep antibodies directed against the 30 kd antigen (C). Immune complexes were visualized by using either ^{125}I-protein A (A), or alkaline phosphatase conjugated donkey α-sheep IgG (B and C). Lanes are as follows: 1, human adult skeletal muscle; 2, human newborn cardiac muscle; 3, human newborn skeletal muscle (psoas); 4, human DMD-affected skeletal muscle (patient 1); 5, human DMD-affected skeletal muscle (patient 2); 6, Triton-insoluble extract of adult human skeletal muscle; 7, Triton-insoluble extract of DMD-affected skeletal muscle (patient 1); 8, Triton-insoluble extract of DMD-affected skeletal muscle (patient 2); 9, normal mouse heart; 10, normal mouse skeletal muscle; 11, normal mouse stomach; 12, normal mouse brain; 13, Triton-insoluble extract of normal mouse heart; 14, Triton-insoluble extract of normal mouse skeletal

3B and 3C) (sheep IgG binds very poorly to protein A). All antibodies detected a large molecular weight, apparently low abundance protein species calculated to be approximately 400 kd in total solubilized human and mouse skeletal and cardiac muscle (lanes 1–3, 9–10). The higher resolution of the alkaline phosphatase staining (Figures 3B and 3C) resolved this protein into doublets or triplets, though the slightly smaller bands most likely represent degradation products since there has been no evidence to date for alternatively spliced isoforms of the DMD mRNA (Koenig et al., 1987; Hoffman et al., 1987; Burghes et al., 1987). The 400 kd species was also clearly evident in mouse smooth muscle (stomach) (lane 11), though at a level substantially lower than that found in cardiac and skeletal muscle (lanes 9–10). The same apparent protein species was detectable in mouse brain at an extremely low level (Figures 3B and 3C, lane 12). Though transcriptional studies of the DMD gene in mice and humans (Monaco et al., 1986; Hoffman et al., 1987) were unable to identify DMD gene transcription in brain, the results presented here are completely compatible given the much greater sensitivity of the Western analysis used for protein detection relative to the Northern analysis used for mRNA detection of the large transcript. Further studies are required to determine whether the apparent low level of the DMD protein in brain is due to expression in smooth muscle or in other cell types.

The 400 kd protein species recognized by all antibodies was generally Triton-insoluble, though it appeared to be associated more strongly with the myofibrillar matrix fraction in cardiac muscle than it did in either skeletal muscle or smooth muscle (Figure 3, lanes 9–11, 13–15). This protein was not detectable in the SDS-solubilized (lanes 4–5) or Triton-insoluble (lanes 7–8) fractions of skeletal muscle samples from two boys affected with DMD. The deficiency of this protein in the DMD-affected boys is particularly evident in Figure 3C, where the presence of a cross-reactive 50 kd Triton-soluble protein serves to verify the equal protein content of the normal vs. DMD lanes (lanes 1–5).

Though sheep antibodies directed against the larger 60 kd DMD antigen (Figure 3B) recognized solely the 400 kd protein in both mouse and human tissues, additional protein species were clearly detected by rabbit and sheep antibodies raised against the 30 kd antigen (Figures 3A and 3C). However, none of the smaller proteins recognized by the sheep α-30kd antibodies were similarly recognized by the rabbit α-30kd antibodies, indicating that these smaller proteins represent cross-reactive protein species and are not themselves products of the DMD locus in either mice or humans. Though it appears that the amount of the DMD protein varies in different tissues depending on the anti-

muscle; 15, Triton-insoluble extract of normal mouse stomach; 16, Triton-insoluble extract of normal mouse brain.

Shown is the 400 kd, low abundance protein species recognized by all three antibodies in all normal muscle-containing tissues. Smaller, cross-reactive protein species are detected by antibodies raised against the 30 kd DMD antigen in either rabbit (A) or sheep (C). The size and location of myosin and biotinylated molecular weight markers are indicated.

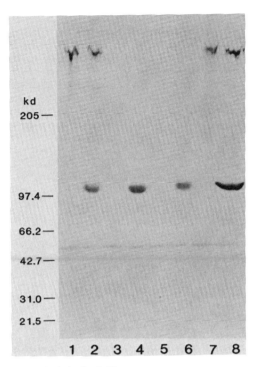

kd
205 —

97.4 —

66.2 —

42.7 —

31.0 —

21.5 —

1 2 3 4 5 6 7 8

Figure 4. Analysis of *mdx* Mice

Freshly dissected heart and skeletal muscle tissues from normal, *mdx*, *mdx*[467], and homozygous *Tr* mice were solubilized and denatured in gel loading buffer, and aliquots (50 μg) were fractionated on 3.5%–12.5% gradient SDS–polyacrylamide gels. Proteins were transferred to nitrocellulose and incubated with a mixture of rabbit (anti-30kd antigen) and sheep (anti-60kd antigen) anti-DMD antibodies, and the immune complexes were detected by using alkaline phosphatase conjugated second antibodies. Lanes are as follows: 1, homozygous *Tr* heart; 2, *Tr* skeletal muscle; 3, *mdx*[467] heart; 4, *mdx*[467] skeletal muscle; 5, *mdx* heart; 6, *mdx* skeletal muscle; 7, normal mouse heart; 8, normal mouse skeletal muscle.

The 400 kd protein species recognized by both sets of antibodies is evident in both normal and *Tr* mice, but is absent from both alleles of *mdx* mice. The more abundant 90 kd cross-reactive protein species recognized by rabbit antibodies directed against the 30 kd antigen (as shown in Figure 3A) is seen at equal levels in the skeletal muscle of all mice. This cross-reactive protein has an apparent molecular weight of 100 kd, which is probably a better representation of the size than the 90 kd weight calculated in Figure 3A. The size and location of myosin and biotinylated molecular weight markers are indicated.

body used (see Figure 3A vs. Figure 3B), this is assumed to be an artifact due to the increased contrast of the autoradiographic exposure of Figure 3A compared to the alkaline phosphatase staining of Figures 3B and 3C. Indeed, a comparison of Figures 3B and 3C, both of which employ immunochemical staining, indicates that antibodies raised against the two different antigens recognize the same 400 kd protein in equal relative abundances.

Analysis of *mdx* Mice

Skeletal and cardiac muscle was dissected from normal, *mdx* (Bulfield et al., 1984), *mdx*[467] (V. Chapman, personal communication), and homozygous *Tr* (a severe neuro-

pathological disorder; Falconer, 1951; Henry, Cowen, and Sidman, 1983) mice and solubilized directly in gel loading buffer as described above. Protein samples were separated by electrophoresis on 3.5% to 12.5% gradient polyacrylamide gels and analyzed as above, by using a cocktail of sheep α-60kd and rabbit α-30kd antibodies. As shown in Figure 4, the 400 kd protein species was present in the skeletal and cardiac muscle of both normal (lanes 7–8) and *Tr* (lanes 1–2) mice. The detected protein appeared the same with this cocktail of antisera as it did with each antiserum separately (shown in Figure 3), indicating that the two antibodies raised against different antigens recognized the same protein. Both antibodies failed to detect the 400 kd protein in muscle tissues isolated from mice harboring either allele of the *mdx* mutation (lanes 3–6).

A much smaller skeletal muscle–specific cross-reactive polypeptide recognized by the rabbit antibodies raised against the 30 kd antigen (Figure 3A) appeared at equal levels in all mice, and serves to verify the quantity and quality of protein loaded in the skeletal muscle lanes. On the gradient gel in this experiment this smaller protein species has an apparent molecular weight of 100 kd, which probably represents a more accurate determination than that shown in Figure 3A.

Relationship between the DMD Protein and Nebulin

Nebulin, a large molecular weight, high abundance myofibrillar protein (Wang, 1985), has recently been implicated as being a candidate for the primary product of the DMD gene (Wood et al., 1987). To compare nebulin levels to those of the DMD protein, tissue samples from normal and DMD-affected human individuals and from normal, *mdx*, and *Tr* mice were directly solubilized in gel loading buffer, the proteins fractionated on 3.5% SDS–polyacrylamide gels, and the gels processed as above. Identical nitrocellulose blots were incubated with affinity-purified rabbit antibody directed against the 30 kd antigen (Figure 5A), or with guinea pig anti-rabbit nebulin (Figure 5B), followed by incubation with ^{125}I-protein A. As expected from the previous experiments, the anti-DMD antibodies recognized a 400 kd protein species in normal human skeletal muscle (Figure 5A, lanes 6 and 9) and in normal and *Tr* mouse skeletal and cardiac muscle (lanes 1–2, 10–11). This protein species was not detectable in human DMD muscle biopsies (Figure 5A, lanes 7–8), in either allele of *mdx* mouse (lanes 4–5, 12–13), or (in this experiment) in normal mouse brain (lane 3). The cross-reactive 100 kd skeletal muscle–specific protein species normally detected by the rabbit α-30kd antibodies used (see Figures 3A and 4) was run off the 3.5% gels, and is therefore not seen.

The anti-nebulin antibodies detected the expected abundant, skeletal muscle–specific protein species of approximately 500 kd (Wang, 1985; Hu, Kimura, and Maruyama, 1986; Locker and Wild, 1986), though this represents the first reported immunological evidence for the apparent absence of nebulin in cardiac muscle. Comparison of the autoradiographic exposure times and the signal intensities of the 400 kd DMD protein to those of

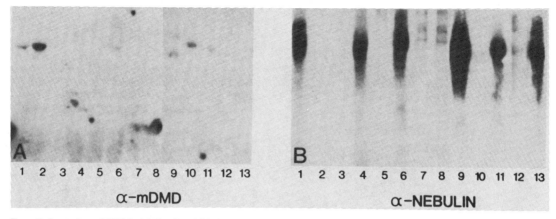

Figure 5. Comparison of DMD Protein Levels and Distribution with Those of Nebulin

Mouse (fresh) and human (frozen) tissues were solubilized in sample loading buffer, with aliquots (50 μg) fractionated on 3.5% SDS–polyacrylamide gels. Proteins were transferred to nitrocellulose and incubated with antibodies directed against the 30 kd DMD antigen (A) or anti-nebulin antibodies (B), followed by detection of bound IgG with ^{125}I-protein A. Lanes are as follows: 1, normal mouse skeletal muscle; 2, normal mouse heart; 3, normal mouse brain; 4, mdx skeletal muscle; 5, mdx heart; 6, normal human skeletal muscle; 7, DMD skeletal muscle (patient 1); 8, DMD skeletal muscle (patient 2); 9, normal human skeletal muscle; 10, homozygous Tr heart; 11, Tr skeletal muscle; 12, mdx[467] heart; 13, mdx[467] skeletal muscle. The primary anti-DMD protein antibody and the detection system used for (A) is the same as for Figure 3A, and the levels and size of the DMD protein appear similar. The apparent size difference of dystrophin between lanes 9 and 10 in (A) is due to the larger amount of protein loaded in lane 9, such that the migration of the DMD protein is distorted by nebulin (B). In contrast to the DMD protein, nebulin appears as a larger (500 kd; Wang, 1985), more abundant, skeletal muscle–specific protein species in all mouse and human skeletal muscle samples tested. Autoradiography in (A) was for 3 days; (B), for 2 hr. Comparison of (A) and (B) shows that the DMD protein is approximately one-thousandth the level of nebulin.

nebulin indicated that the DMD protein was approximately one-thousandth the level of nebulin (Figure 5). Since nebulin has been calculated to represent about 3% of myofibrillar protein (and thus 1% of total muscle protein) (Wang, 1985), the DMD protein can be estimated to represent approximately 0.001% of total muscle protein.

Nebulin is evident, though greatly reduced, in the human DMD muscle samples tested (Figure 5B, lanes 7–8). On the other hand, nebulin appeared at normal levels in both alleles of mdx mouse (lanes 4–5, 12–13). This immunological data provides conclusive evidence that nebulin and the DMD protein are indeed distinct proteins, and therefore indicates that nebulin cannot be the protein product of the DMD locus.

Relative Cellular Abundance of the DMD Protein

The protein product of the DMD locus has been previously calculated to be in very low abundance on the basis of mRNA levels in muscle tissue (Hoffman et al., 1987). In order to measure more directly the cellular abundance of the DMD protein, the amount of this protein in heart was quantitated. Known quantities of DMD fusion proteins were denatured and fractionated on 3.5%–12.5% gradient SDS–polyacrylamide alongside 100 μg of solubilized mouse heart. Proteins were transferred to nitrocellulose, incubated with sheep antibodies directed against the 60 kd DMD antigen, and then detected by using alkaline phosphatase conjugated second antibodies. As shown in Figure 6, the antiserum reacts only with the antigen to which it was raised and not with the 30 kd fusion protein, indicating that all immunostaining is due to antibodies specific for the DMD portion of the trpE+60kd fusion pro-

tein. The signal exhibited for the 400 kd DMD protein in 100 μg of total cardiac protein corresponds to approximately 2 ng of the partially purified antigen to which the antibody was raised (Figure 6, lanes 3–4). Thus, by this measurement, the DMD protein comprises approximately 0.002% of total muscle protein.

Discussion

Recent reports have substantiated the correlation of recently described human cDNA sequences to the Duchenne muscular dystrophy locus in humans (Monaco et al., 1986; Monaco and Kunkel, 1987; Burghes et al., 1987; Koenig et al., 1987). Indeed, the fact that a large proportion of affected individuals exhibit small deletions within the genomic locus covered by cloned cDNAs indicates that these sequences represent the human DMD gene. The mouse homolog of the human DMD locus has been shown to reside on the mouse X chromosome by both genomic DNA analysis (Monaco et al., 1986) and cDNA analysis (Hoffman et al., 1987; Chamberlain et al., 1987; Heilig et al., 1987; Brockdorff et al., 1987). The human and mouse DMD cDNAs have been shown to be greater than 90% homologous over the entire amino-terminal one-third of the protein (~130 kd), diverging only upstream of their common translation initiation codon (Hoffman et al., 1987; Koenig et al., 1987). The predicted amino acid sequences of this portion of the human and mouse DMD proteins indicated that the protein might serve a highly conserved structural role in the myofiber (Hoffman et al., 1987). By the raising of antibodies to in vitro engineered fusion proteins containing portions of the mDMD protein (30 kd and

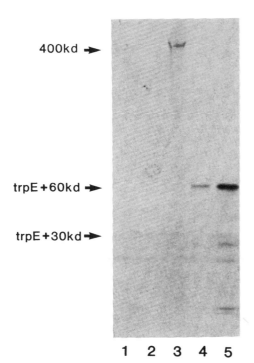

400kd →

trpE+60kd →

trpE+30kd →

1   2   3   4   5

Figure 6. Quantitation of DMD Protein Levels in Normal Heart

Aliquots of bacterial lysates containing DMD–trpE fusion proteins (see Figure 1) were quantitated, solubilized in gel loading buffer, and fractionated on 3.5%–12.5% SDS–polyacrylamide gels alongside 100 μg of solubilized mouse heart. Fractionated proteins were transferred to nitrocellulose and incubated with antibodies (sheep) directed against the 60 kd DMD antigen, followed by immune complex detection with alkaline phosphatase conjugated donkey anti-sheep second antibody. Lanes are as follows: 1, trpE+30kd (2 ng); 2, trpE+30kd (10 ng); 3, normal mouse heart (100 μg); 4, trpE+60kd (2 ng); 5, trpE+60kd (10 ng). The affinity-purified antibody directed against the trpE+60kd antigen is seen to react specifically with this antigen and with the 400 kd DMD protein product. The signal intensity of the 400 kd DMD protein is seen to correspond to approximately 2 ng of partially purified antigen, indicating that the DMD protein represents approximately 0.002% of total heart protein. This calculation has an inherent uncertainty due to the variable transfer efficiencies of proteins of different sizes, though the antigenicity of these proteins should be identical because of their identical amino acid sequence.

60 kd), the protein product of the DMD locus was identified in both mouse and human muscle. The protein identified by these antibodies fulfills the following requirements expected of the primary gene product disrupted in DMD. First, the mRNA product of the mouse and human DMD loci has been estimated to be 16 kb in length (Monaco et al., 1986; Hoffman et al., 1987). This estimated size was revised to 14 kb with the complete cloning of the human cDNA (Koenig et al., 1987). The protein species described in this paper has been estimated to be approximately 400 kd, a size that is in general agreement with the translation of a mRNA of 14 kb. Second, the mRNAs corresponding to both the human and mouse DMD loci have been found to represent 0.01–0.001% of total muscle mRNA, as evidenced by both clone frequency in cDNA libraries and

abundance relative to α-tubulin mRNA (Hoffman et al., 1987). The abundance of the identified protein agrees with the mRNA abundance, as evidenced by abundance relative to nebulin (Figure 5) and direct quantitation in mouse heart (Figure 6). Third, the expression pattern of the mDMD gene has been studied at the mRNA level, with the large DMD transcript detectable only in striated and possibly smooth muscle (Monaco et al., 1986; Hoffman et al., 1987; Burghes et al., 1987). The 400 kd protein was clearly detected in skeletal and cardiac muscle, with smaller amounts in stomach (smooth muscle) (Figure 3). Fourth, the primary amino acid sequence of the amino-terminal 30% of the DMD protein has been shown to exhibit features common to many structural proteins, being highly conserved and rich in α-helix (Hoffman et al., 1987; Koenig et al., 1987). In agreement with this hypothesis, the DMD protein was found to be largely Triton-insoluble, suggesting an association with the myofibrillar matrix (Figure 3, lanes 13–15). Fifth, muscle biopsies from two DMD-affected individuals contained no detectable 400 kd protein (Figures 3 and 5), consistent with the molecular analysis of the DMD gene which has shown that most DMD-affected individuals possess null mutations of the DMD locus (Koenig et al., 1987). Sixth, antibodies raised in both rabbit and sheep against fusion peptides encoded by two separate, distinct regions of the DMD cDNA recognize the same protein species, as evidenced by the identical size, abundance, and tissue distribution of the detected protein. Taken together, this evidence validates the specificity of the described antibodies for the DMD gene product, and thereby substantiates the identification of this protein as the primary biochemical defect in Duchenne muscular dystrophy. Since we know of no previously reported protein that shares the abundance, sequence, or size characteristics of the DMD protein, and since this protein was identified by molecular genetic studies of patients affected with Duchenne muscular dystrophy, we have named this protein dystrophin.

It is interesting to note the very small amount of dystrophin present in total mouse brain tissue. A 30% incidence of mental retardation has been observed in boys afflicted with DMD (Zellweger and Hanson, 1967). It is tempting to speculate that the observed mental retardation could be a direct consequence of dystrophin absence, though the variable penetrance of this phenotype would argue against this. It is also interesting that, to date, both the mRNA and the protein have been found only in terminally differentiated cells (Hoffman et al., 1987; Lev et al., 1987; this paper). Given the greater than 2 million base pair size of the genomic DMD locus in humans (Monaco and Kunkel, 1987; Koenig et al., 1987), it would take more than 24 hr for RNA polymerase II to transcribe a single mRNA molecule from the DMD gene (Ucker and Yamamoto, 1984). If it is assumed that DNA replication and mRNA transcription cannot take place simultaneously, then only cells that are mitotically inactive for longer than 24 hr would be capable of transcribing the DMD gene, thus limiting the production of the DMD protein to predominantly mitotically inactive cells.

Our results concerning the absence of dystrophin in tissues isolated from two alleles of *mdx* mice are particularly provocative. The lack of any detectable muscle weakness in *mdx* mice has led to the past hypotheses that the original *mdx* mutation represents either a mild allele of the mouse DMD homolog (Bulfield et al., 1984) or the homolog of the less severe human Emery–Dreyfuss dystrophy (Avner et al., 1987). We have found that two different mutant alleles of the *mdx* gene appear to lack the mDMD gene product, indicating that *mdx* and DMD most likely represent the same genetic disorder. Although the molecular defect in *mdx* has not been detected by DNA analysis (Chamberlain et al., 1987; L. M. Kunkel, unpublished data), the deficiency of dystrophin in *mdx* mice is presumed not to be a secondary consequence of a nonhomologous genetic disorder for the following reasons. First, the homology of the *mdx* and DMD loci is consistent with the linkage data (Bulfield et al., 1984; Chapman et al., 1985; Heilig et al., 1987; Brockdorff et al., 1987; Chamberlain et al., 1987). Second, the deficiency of dystrophin in *mdx* mice appears to be disease-specific, as muscle samples from homozygous Trembler mice, afflicted with a severe neuropathological disorder (Henry, Cowen, and Sidman, 1983), exhibit wild-type levels of this protein. Third, though it could be argued that the absence of the mDMD product represents a generalized degradation of muscle proteins in *mdx* muscle, nebulin, which is regarded as one of the more labile muscle proteins (Wang, 1985; Sugita et al., 1987), is detectable at wild-type levels in *mdx* skeletal muscle (Figure 5B). Indeed, the normal levels of nebulin protein observed in mouse *mdx* contrasts to the severely reduced levels observed in human DMD patient muscle (Wood et al., 1987; Sugita et al., 1987). Such differences might be indicative of a more active role of endogenous proteases in human DMD muscle fibers, and could possibly explain some of the differences in the clinical phenotype.

By equating the mouse *mdx* and human DMD loci, an animal model is now available for DMD. The differences in the clinical manifestation of the same primary biochemical defect in mice and in humans might be explained by differences in secondary biochemical effects or histological changes. Histologically, both DMD and *mdx* muscle exhibit nearly identical patterns of myofiber degeneration and regeneration (Dubowitz, 1985; Bridges, 1986), a process that is probably a direct consequence of dystrophin deficiency in myofibers. The *mdx* muscle, however, never develops the extensive connective tissue proliferation (endomysial fibrosis) that is characteristic of human DMD muscle tissue, especially in the later stages of the disease (Dubowitz, 1985; Tanabe, Esaki, and Nomura, 1986; Torres and Duchen, 1987). This indicates that the prominent fibrosis in human DMD muscle is probably an indirect or secondary consequence of dystrophin deficiency. Perhaps the extensive endomysial fibrosis in human DMD muscles results in the impairment of the ability of individual muscle fibers to regenerate. This would mean that muscle fiber number would progressively decrease as the connective tissue content of each muscle group increases— a process that is, in fact, observed (Cullen and Fulthorpe, 1975; Watkins and Cullen, 1985). Such a process could ul-

timately result in insufficient muscle fiber numbers for mobility and respiration. The muscle fibers of *mdx* mice, on the other hand, exhibit no such fibrosis and retain the ability to regenerate throughout the life of the mouse, posing no threat to either mobility or normal life span. Possible rational therapies for boys afflicted with DMD might therefore result from the ability to control the connective tissue proliferation within the muscle tissue. Alternatively, future medical research could address the primary biochemical defect responsible for the DMD and *mdx* phenotypes, namely the deficiency of the dystrophin protein leading to fiber degeneration. Possible chemical agents that might result in a slowing of fiber degeneration could then be tested on *mdx* mice.

Conclusion

Molecular biological techniques have led to the identification of the primary biochemical defect in an important hereditary human disease, Duchenne muscular dystrophy. The identification of this defect was based solely on the chromosomal location of the DMD locus. The antibodies produced against the DMD protein product, dystrophin, should prove useful in the diagnosis and characterization of this disorder. As more is understood about the role of dystrophin in normal muscle function, rational therapies for the many boys affected with this fatal disease will, we hope, emerge. Many of these therapies could be tested on the *mdx* mouse model for this disease.

Experimental Procedures

Plasmid Constructions

The predicted amino acid sequence has been determined from the cDNA sequence for the amino-terminal one-third of the Duchenne muscular dystrophy gene product in both mice and humans (Hoffman et al., 1987; Koenig et al., 1987). Two different regions of the mouse sequence were fused to the E. coli *trpE* gene as follows, with the predicted number of amino acids being deduced from the DNA sequence.

trpE+60kd

The mouse DMD cDNA (Hoffman et al., 1987) was restricted at the unique SpeI site, blunt-ended with Klenow, and then digested with HindIII in the 3' polylinker. The excised cDNA fragment of 1.4 kb was gel-purified and ligated to pATH2 (Dieckmann and Tzagoloff, 1985), which had been digested with SmaI and HindIII. Recombinants were identified by colony hybridization to random primer extended (^{32}P) insert (Feinberg and Vogelstein, 1983), and verified by subsequent plasmid DNA restriction analysis. The resulting plasmid construction fused the trpE protein (33 kd) to 410 amino acids (~60 kd) of the mDMD protein, and corresponds to position 1.3 kb to 2.7 kb on the equivalent human cDNA map (Koenig et al., 1987).

trpE+30kd

The most 3' end of the mouse cDNA currently available (Hoffman et al., 1987) was restricted at its unique nonmethylated XbaI site and at the BamHI site in the 3' polylinker. The excised 700 bp fragment was ligated to pATH2 digested with XbaI and BamHI as described above. This plasmid construction fused the trpE protein to 208 amino acids (~30 kd) of the mDMD protein, and corresponds to position 3.7 kb to 4.4 kb on the equivalent human cDNA map (Koenig et al., 1987).

Induction and Purification of Fusion Proteins

Plasmid constructions were maintained in E. coli RR1, which was grown as suggested by A. Tzagoloff (unpublished data) except that 200 μg/ml of tryptophan was used as a supplement to all media. Induction with 3-B-indolacrylic acid (IAA), harvesting, and initial purification of trpE fusion proteins was as described by Dieckmann and Tzagoloff

(1985). Between 15 and 25 mg of insoluble protein was obtained from 100 ml of induced bacterial culture, of which approximately 85% was estimated to be the desired fusion protein (Figure 1). Between 2 and 5 mg of insoluble protein was solubilized by boiling in SDS, and then size-fractionated on preparative SDS–polyacrylamide gels (Laemmli, 1970). The fusion proteins were visualized by rinsing of the gels in distilled water for 5 min followed by immersion in cold 0.25 M KCl, with the appropriate protein band then being excised.

For rabbit immunizations, fusion proteins were then purified by electroelution into dialysis sacs, followed by precipitation with 5 volumes of acetone to remove SDS. Protein pellets, which also contained coprecipitated glycine, were resuspended in sterile 10 mM Tris (pH 8.0), and the protein concentration was determined (Bio-Rad protein assay on extensively sonicated aliquots).

For sheep immunizations, gel slices containing SDS-denatured antigen were sent to Polyclonal Seralabs (Cambridge, MA).

Antibody Production
New Zealand white female rabbits were immunized according to the following schedules: 1, intravenous injection (10 μg) with weekly boosts; 2, intradermal using 10 μg of fusion protein emulsified with Freund's complete adjuvant, with 10 μg boosts every 3 weeks using Freund's incomplete adjuvant; 3, intradermal as above (2) using 50 μg of fusion protein, with 100 μg boosts.

One rabbit was immunized with each fusion peptide according to each of the schedules. The titers and specificity of the antibodies produced in each rabbit were constantly monitored by enzyme-linked immunoassays performed on nitrocellulose dot blots of insoluble protein fractions such as those shown in Figure 2. The best immune responses were obtained by using the trpE+30kd polypeptide with immunization protocols 1 and 3 above, with >95% of the antibodies produced being specifically against the mDMD portion of the fusion peptide, and with titers greater then the sensitivity of the ELISA assay system when using a 1:1000 dilution of crude serum 4 weeks after immunization. The trpE+60kd antigen took much longer (12 weeks) to evoke an immune response in rabbits, with the resulting sera showing a low specificity for the DMD portion of the fusion protein (not shown).

A single sheep was immunized with each antigen in the form of SDS-denatured protein in polyacrylamide gel slices. Approximately 1 mg of fusion protein was used per immunization. The initial immunization was with Freund's complete adjuvant, with boosts using incomplete adjuvant at days 14 and 28. Injections were at multiple sites both intramuscularly and subcutaneously in lymph node areas. Serum was collected at day 50.

Antibody Purification
Approximately 3 mg of partially purified trpE protein (insoluble fraction; see above) was reprecipitated, resuspended in 10 mM Tris (pH 8.0), and then precipitated again. The pellet was resuspended in 1.5 ml of immune serum, incubated on ice for 1 hr, centrifuged to pellet the trpE–antibody immune complexes, and discarded. The supernatant was then mixed with approximately 3 mg of partially purified fusion protein (insoluble fraction) that had been washed as above. After incubation on ice, the mDMD–antibody immune complexes were precipitated by centrifugation. The pellet was then resuspended in 500 μl of 0.2 M glycine (pH 2.3), incubated on ice for 5 min to disassociate the immune complexes, and centrifuged at 4°C to precipitate the insoluble antigen. The supernatant containing the purified immunoglobulins was neutralized with 50 μl Tris (pH 9.5), and either stabilized with BSA (fraction V)(5 mg/ml) or dialyzed extensively against phosphate-buffered saline (PBS).

Western Blotting
Mouse (fresh) or human (frozen) tissues were homogenized in 10 volumes of gel loading buffer (Sugita et al., 1987) by using a motorized Teflon tissue homogenizer. The protein concentration of the solubilized tissues was approximated based on the weight of the tissues used. Mouse skeletal muscle samples used were total hind limb muscle.

Triton X-100 insoluble proteins were prepared by homogenization of fresh or frozen tissues in a Waring blender at high speed for 30 sec in buffer consisting of 10 mM HEPES (pH 7.2), 5 mM EGTA, 1 mg/ml PMSF, 1 mM iodoacetamide, 1 mM benzamidine, 0.5 mg/ml aprotinin, 0.5 mg/ml leupeptin, 0.25 mg/ml pepstatin A, and 0.25% Triton X-100.

Triton-insoluble proteins were precipitated and then resuspended in buffer without Triton X-100, and the protein concentration was determined (Bio-Rad protein assay). Aliquots were diluted with gel loading buffer, and 50 μg was used per lane. Protein samples were heated to 95°C for 2 min, centrifuged, and electrophoretically fractionated on 0.75 mm SDS–polyacrylamide gels (Laemmli, 1970), by using a 3% stacking gel and either a 3.5% or 3.5%–12.5% gradient resolving gel.

Fractionated proteins were transferred to nitrocellulose (Towbin, Staehelin, and Gordon, 1979), and the filters were dried. Dried filters were blocked in 5% nonfat dry milk in TBST (10 mM Tris, pH 8.0; 500 mM NaCl; 0.05% Tween-20). All immunological reagent dilutions and filter washes were done in TBST. The affinity-purified anti-mDMD antibodies, affinity-purified second antibodies (Sigma), and guinea pig anti-rabbit nebulin antisera were diluted 1:1000 in TBST prior to use. Affinity-purified ^{125}I-protein A was from Amersham, and was used at 5 uCi per 20 ml of TBST. Biotinylated molecular weight markers were purchased from Bio-Rad, and were visualized by using protocols and reagents supplied by the manufacturer. Myosin was easily visualized on all filters by immunostaining ghosts due to the abundance of this protein.

Acknowledgments

We are grateful to Dr. A. Tzagoloff for the gift of the pATH vectors and protocols, Drs. Giovanni Salviati and Armand Miranda for the anti-nebulin antibodies, Dr. Vernon Chapman for supplying the mdx and mdx^{4cv} mice, Drs. C. Thomas Caskey and Jeffrey Chamberlain for the DMD and normal adult human muscle samples, and Dr. K. Abraham Chacko for human newborn heart and skeletal muscle. This work would not have been possible without the expert advice of Drs. Jerold Schwaber, Paul Rosenberg, Armand Miranda, Frost White, Rachael Neve, Paul Neumann, Anthony Monaco, Samuel Latt, and Nigel Flemming. We thank the Children's Hospital mouse facility supported by National Institutes of Health (NIH) grant NS20820 to Dr. Richard Sidman. E. P. Hoffman is the Harry Zimmerman post-doctoral fellow of the Muscular Dystrophy Association. This work was supported by NIH grants RO1 NS23740 (L. M. Kunkel) and NS00787-04 (R. H. Brown), the Muscular Dystrophy Association (L. M. Kunkel), and the Cecil B. Day Investment Company (R. H. Brown). L. M. Kunkel is an associate investigator of the Howard Hughes Medical Institute.

The costs of publication of this article were defrayed in part by the payment of page charges. This article must therefore be hereby marked "advertisment" in accordance with U.S.C. 18 Section 1734 solely to indicate this fact.

Received August 14, 1987; revised October 19, 1987.

References

Avner, P., Amar, L., Arnaud, D., Hanauer, A., and Cambrou, J. (1987). Detailed ordering of markers localizing to the Xq26-Xqter region of the human X chromosome by the use of an interspecific Mus spretus mouse cross. Proc. Natl. Acad. Sci. USA 84, 1629–1633.

Bridges, L. R. (1986). The association of cardiac muscle necrosis and inflammation with the degenerative and persistent myopathy of mdx mice. J. Neurol. Sci. 72, 147–157.

Brockdorff, N., Cross, G. S., Cananna, J. S., Fisher, E. M., Lyon, M. F., Davies, K. E., and Brown, S. D. M. (1987). The mapping of a cDNA from the human X-linked Duchenne muscular dystrophy gene to the mouse X chromosome. Nature 328, 166–168.

Bulfield, G., Siller, W. G., Wight, P. A., and Moore, K. J. (1984). X chromosome-linked muscular dystrophy (mdx) in the mouse. Proc. Natl. Acad. Sci. USA 81, 1189–1192.

Burghes, A. H. M., Logan, C., Hu, X., Belfall, B., Worton, R., and Ray, P. N. (1987). Isolation of a cDNA clone from the region of an X:21 translocation that breaks within the Duchenne/Becker muscular dystrophy gene. Nature 328, 434–437.

Chamberlain, J. S., Reeves, A. A., Caskey, C. T., Hoffman, E. P., Monaco, A. P., Kunkel, L. M., Grant, S. G., Mullins, L. J., Stephenson, D. A., and Chapman, V. M. (1987). Regional localization of the murine Duchenne muscular dystrophy gene on the mouse X chromosome. Somatic Cell Genet., in press.

Chapman, V. M., Murawski, M., Miller, D., and Swiatek, D. (1985). Mouse News Letter 72, 120.

Cullen, M. J., and Fulthorpe, J. J. (1975). Stages in fibre breakdown in Duchenne muscular dystrophy: an electron-microscopic study. J. Neurol. Sci. 24, 179–200.

Dieckmann, C. L., and Tzagoloff, A. (1985). Assembly of the mitochondrial membrane system. J. Biol. Chem. 260, 1513–1520.

Dubowitz, V. (1985). Muscle Biopsy: A Practical Approach (East Sussex, England: Balliere Tindall).

Duchenne, G. B. (1868). Recherches sur la paralysie musculaire pseudo-hypertophique ou paralysie myosclerosique. Arch. Gen. Med. 11, 5, 178, 305, 421, 552.

Emery, A. E. H., and Holloway, S. (1977). Use of normal daughters' and sisters' creatine kinase levels in estimating heterozygosity in Duchenne muscular dystrophy. Hum. Hered. 27, 118–126.

Falconer, D. S. (1951). Two new mutants, "Trembler" and "Reeler", with neurological actions in the house mouse (Mus musculus L.). J. Genet. 50, 192–201.

Feinberg, A. P., and Vogelstein, B. (1983). A technique for radiolabeling DNA restriction endonuclease fragments to high specific activity. Anal. Biochem. 132, 6–13.

Harris, J. B., and Slater, C. R. (1980). Animal models: what is their relevance to the pathogenesis of human muscular dystrophy? Br. Med. Bull. 36, 193–197.

Heilig, R., Lemaire, C., Mandel, J-L., Dandolo, L., Amar, L., and Avner, P. (1987). Localization of the region homologous to the Duchenne muscular dystrophy locus on the mouse X chromosome. Nature 328, 168–170.

Henry, E. W., Cowen, J. S., and Sidman, R. L. (1983). Comparison of Trembler and Trembler-J mouse phenotypes: varying severity of peripheral hypomyelination. J. Neuropath. Exp. Neurol. 42, 688–706.

Hoffman, E. H., Monaco, A. P., Feener, C. A., and Kunkel, L. M. (1987). Conservation of the Duchenne muscular dystrophy gene in mice and humans. Science 238, 347–350.

Hu, D. H., Kimura, S., and Maruyama, K. (1986). Sodium dodecyl sulfate gel electrophoresis studies of connectin-like high molecular weight proteins of various types of vertebrate and invertebrate muscles. J. Biochem. 99, 1485–1492.

Koenig, M., Hoffman, E. P., Bertelson, C. J., Monaco, A. P., Feener, C., and Kunkel, L. M. (1987). Complete cloning of the Duchenne muscular dystrophy (DMD) cDNA and preliminary genomic organization of the DMD gene in normal and affected individuals. Cell 50, 509–517.

Laemmli, U. K. (1970). Cleavage of structural proteins during the assembly of the head of bacteriophage T4. Nature 227, 680–685.

Lev, A., Feener, C., Kunkel, L. M., and Brown, R. H. (1987). Expression of the Duchenne's muscular dystrophy gene in cultured muscle cells. J. Biol. Chem., in press.

Locker, R. H., and Wild, D. J. C. (1986). A comparative study of high molecular weight proteins in various types of muscle across the animal kingdom. J. Biochem. 99, 1473–1484.

Mastaglia, F. L., and Walton, S. J. (1982) Skeletal Muscle Pathology (New York: Churchill Livingstone).

Meryon, E. (1852). On granular and fatty degeneration of the voluntary muscles. Medico-Chirurgical Trans. (London) 35, 73.

Monaco, A. P., and Kunkel, L. M. (1987). A giant locus for the Duchenne and Becker muscular dystrophy gene. Trends Genet. 3, 33–37.

Monaco, A. P., Neve, R. L., Colletti-Feener, C. A., Bertelson, C. J., Kurnit, D. M., and Kunkel, L. M. (1986). Isolation of candidate cDNAs for portions of the Duchenne muscular dystrophy gene. Nature 323, 646–650.

Ohno, S., Becak, W., and Becak, M. L. (1964). X-autosome ratio and the behavior pattern of individual X-chromosomes in placental mammals. Chromosoma 15, 14–30.

Sugita, H., Nonaka, I., Itoh, Y., Asakura, A., Hu, D. H., Kimura, S., and Maruyama, K. (1987). Is nebulin the product of Duchenne muscular dystrophy gene? Proc. Japan Acad. 63, 107–110.

Tanabe, Y., Esaki, K., and Nomura, T. (1986). Skeletal muscle pathology in X chromosome-linked muscular dystrophy (mdx) mouse. Acta Neuropath. 69, 91–95.

Torres, L. F., and Duchen, L. W. (1987). The mutant mdx: inherited myopathy in the mouse. Brain 110, 269–299.

Towbin, H., Staehelin, T., and Gordon, J. (1979). Electrophoretic transfer of proteins from polyacrylamide gels to nitrocellulose sheets: procedure and some applications. Proc. Natl. Acad. Sci. USA 76, 4350–4354.

Ucker, D. S., and Yamamoto, K. R. (1984). Early events in the stimulation of mammary tumor virus RNA synthesis by glucocorticoids. J. Biol. Chem. 259, 7416–7420.

Wang, K. (1985). Sarcomere-associated cytoskeletal lattices in striated muscle. Cell Muscle Mot. 6, 315–369.

Watkins, S. C., and Cullen, M. J. (1985). Histochemical fibre typing and ultrastructure of the small fibres in Duchenne muscular dystrophy. Neuropath. Appl. Neurobiol. 11, 447–460.

Wood, D. S., Zeviani, M., Prelle, A., Bonilla, E., Saviati, G., Miranda, A. F., diMauro, S., and Rowland, L. P. (1987). Is nebulin the defective gene product in Duchenne muscular dystrophy? N. Eng. J. Med. 316, 107–108.

Zellweger, H., and Hanson, J. W. (1967). Psychometric studies in muscular dystrophy type IIIa (Duchenne). Dev. Med. Child Neurol. 9, 576–581.

Hypervariable 'Minisatellite' Regions in Human DNA

A.J. Jeffreys, V. Wilson and S.L. Thein

The human genome contains many dispersed tandem-repetitive 'minisatellite' regions detected via a shared 10–15-base pair 'core' sequence similar to the generalized recombination signal (χ) of Escherichia coli. *Many minisatellites are highly polymorphic due to allelic variation in repeat copy number in the minisatellite. A probe based on a tandem-repeat of the core sequence can detect many highly variable loci simultaneously and can provide an individual-specific DNA 'fingerprint' of general use in human genetic analysis.*

DNA POLYMORPHISMS have revolutionized human genetic analysis and have found general use in antenatal diagnosis[1], mapping of human linkage groups[2,3], indirect localization of genetic disease loci by linkage[2,4,5] and analysis of the role of mitotic nondisjunction and recombination in inherited cancer[6-9]. Single-copy human DNA probes are used normally to detect restriction fragment length polymorphisms (RFLPs), most of which result from small-scale changes in DNA, usually base substitutions, which create or destroy specific restriction endonuclease cleavage sites[10,11]. As the mean heterozygosity of human DNA is low (~0.001 per base pair)[10-12], few if any restriction endonucleases will detect a RFLP at a given locus, although the probability of detection is improved for enzymes such as *Msp*I and *Taq*YI which contain the mutable CpG doublet in their recognition sequence[13]. Even when detected, most RFLPs are only dimorphic (presence or absence of a restriction endonuclease cleavage site) with a heterozygosity, determined by allele frequencies, which can never exceed 50% and which is usually much less. As a result, all such RFLPs will be uninformative in pedigree analysis whenever critical individuals are homozygous.

Genetic analysis in man could be simplified considerably by the availability of probes for hypervariable regions of human DNA showing multiallelic variation and correspondingly high heterozygosities. The first such region was isolated by chance by Wyman and White[14] from a library of random segments of human DNA. The structural basis for multiallelic variation at this locus is not yet known. Subsequently, and again by chance, several other highly variable regions have been discovered near the human insulin gene[15], α-related globin genes[16-18] and the c-Ha-*ras*-1 oncogene[19]. In each case, the variable region consists of tandem repeats of a short sequence (or 'minisatellite') and polymorphism results from allelic differences in the number of repeats, arising presumably by mitotic or meiotic unequal exchanges or by DNA slippage during replication. The resulting minisatellite length variation can be detected using any restriction endonuclease which does not cleave the repeat unit and provides for such loci a set of stably inherited genetic markers.

We have described previously a short minisatellite comprised of four tandem repeats of a 33-base pair (bp) sequence in an intron of the human myoglobin gene (ref. 20; Fig. 1). The 33-bp repeat showed some similarity in sequence to three other hypervariable human minisatellites characterized previously and on the basis that the myoglobin minisatellite was flanked by a 9-bp direct repeat characteristic of the target site duplications generated by transposable elements, we suggested that this minisatellite and some other hypervariable regions were related via transposition. We show here that the myoglobin 33-bp repeat is indeed capable of detecting other human minisatellites, some of which are highly polymorphic. These regions, however, are not related by transposition, but instead share a common short 'core' sequence in each repeat unit, which in turn provides a powerful probe for hypervariable regions.

Probe for variable human DNA

A pure repeat probe was prepared from the human myoglobin minisatellite by purification of a single 33-bp repeat element followed by head-to-tail ligation and cloning of the resulting polymer into pUC 13 (ref. 21; Fig. 1). Cleavage of one of the resulting recombinants, pAV33.7, with *Bam*HI plus *Eco*RI released a 767-bp DNA insert comprised almost entirely of 23 repeats of the 33-bp sequence.

Low stringency hybridization of this repetitive insert to human DNA digested with restriction endonuclease *Eco*RI detected numerous cross-hybridizing DNA fragments, some of which showed signs of polymorphic variation (data not shown). To improve the detection of polymorphisms, the hybridization was repeated with human DNA digested with *Hin*fI or *Hae*III, both of which cleave at a 4-bp sequence not present in the 33-bp repeat sequence and which should release minisatellites in relatively small DNA fragments whose size will reflect more closely the number of repeats per minisatellite. As shown in Fig. 2, the repetitive probe detected multiple DNA fragments in human DNA as well as the parent DNA fragment from the human myoglobin gene. The larger DNA fragments (in the range 2–6 kilobases (kb) and substantially larger than the mean DNA fragment size of ~0.3 kb in human DNA digested with *Hin*fI or *Hae*III) in particular showed variation between the three individuals examined; these variants were transmitted apparently in a Mendelian fashion, in that each polymorphic

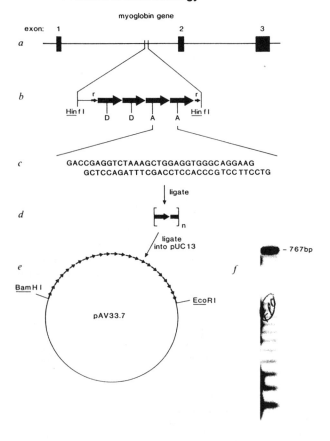

Fig. 1 Construction of a tandem-repetitive hybridization probe for 33-related DNA sequences. *a, b*, This probe was derived from a tandem repetitive segment of the human myoglobin gene[20]. This region, located in the first intron and comprising four repeats of a 33-bp sequence flanked by a 9-bp direct repeat (r), was isolated in a 169-bp *Hin*fI fragment, end-repaired and amplified by cloning into the *Sma*I site of pUC13 (ref. 21). *c*, A 33-bp repeat monomer was isolated by cleaving the third and fourth repeat with *Ava*II (A); a single base substitution in repeats 1 and 2 eliminates this site and creates instead a *Dde*I (D) cleavage site[20]. *d*, Ligation of the 33-bp monomer via the non-identical *Ava*II sticky ends produced a head-to-tail polymer. Polymers containing ⩾10 repeats were isolated by preparative agarose gel electrophoresis[33], end-repaired, ligated into the *Sma*I site of pUC13 and cloned in *E. coli* JM83 (ref. 21). *e*, The structure of clone pAV33.7 was confirmed by excision of the insert at the polylinker with *Bam*HI plus *Ec₂*RI, fill-in labelling with α-^{32}P-dCTP at the *Bam*HI site, and partial digestion with *Ava*II. *f*, Labelled partial digest products were resolved by electrophoresis on a 2% agarose gel. pAV33.7 contains 23 repeats of the 33-bp monomer contained in a 767-bp *Bam*HI/*Eco*RI fragment as shown (*e*).

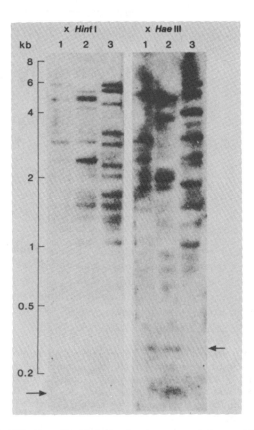

Fig. 2 Detection of multiple 33-related sequences in human DNA. 10 μg samples of DNA from individual 1 (daughter), 2 (mother) and 3 (father) were digested with *Hin*fI or *Hae*III, electrophoresed through a 1% agarose gel and transferred by blotting[35] to a Pall Biodyne membrane to prevent the loss of small DNA fragments. The membrane was hybridized in 1×SSC at 60° with dextran sulphate[36] to the insert from pAV33.7 (Fig. 1) labelled *in vitro* with [32]P (ref. 20). The arrowed fragment in each digest is derived from the minisatellite located in the myoglobin gene (Fig. 1); a survey of DNA from 12 individuals digested with *Hae*III showed that this myoglobin minisatellite is monomorphic (data not shown).

contains a minisatellite of 3–29 tandem copies of a repeat sequence whose length ranged from 16 bp in λ33.15 to 64 bp in λ33.4. Most minisatellites contained an integral number of repeats. In λ33.6, the 37-bp repeat consisted in turn of a diverged trimer of a basic 12-bp unit. Each λ33 recombinant represented a different region of the human genome, judged by the clone-specific DNA sequences flanking each minisatellite.

Highly polymorphic minisatellites

The eight cloned minisatellite regions were located in 0.5–2.2-kb *Hin*fI DNA fragments, smaller than the clearly polymorphic 2–6-kb DNA fragments detected by pAV33.7 in *Hin*fI digests of human DNA (Fig. 2). To determine whether any of the cloned regions were also polymorphic, ^{32}P-labelled single-stranded DNA probes were prepared from suitable M13 subclones of each minisatellite and hybridized at high stringency to a panel of 14 unrelated British caucasian DNAs digested with *Hin*fI. Typical hybridization patterns are shown in Fig. 5, showing that under these hybridization conditions each probe detects a unique region of the human genome. Alleles detected by each probe are summarized in Table 1 where in each case, the most common *Hin*fI allele corresponded in size to the *Hin*fI minisatellite fragment in clones λ33.1–15, suggesting that these regions have been isolated without major rearrangement.

In the limited population sample studied, four of the eight minisatellites showed polymorphic variation. Three of the regions were highly polymorphic with between five and eight resolvable *Hin*fI fragment-length alleles detected per locus. This variation almost certainly results from variation in the repeat

band in the daughter could be identified within one or other (but not both) parents. Variation was detectable in both *Hin*fI and *Hae*III digests of human DNA, consistent with polymorphism resulting from length variation of minisatellite regions.

Isolation of minisatellites

A human genomic library[20] of 10–20 kb *Sau*3A partials of human DNA cloned in phage λL47.1 (ref. 22) was screened by hybridization with the 33-bp repeat probe from pAV33.7. At least 40 strongly-to-weakly hybridizing plaques were identified in a library of 3×10⁵ recombinants, consistent with the complexity of the Southern blot hybridization (Fig. 2). A random selection of eight of these positive plaques was purified (λ33.1–15) and Southern blot analysis of phage DNA showed that in each recombinant the hybridizing DNA was localized in a unique short (0.2–2 kb) region of the recombinant. Sequence analysis (Fig. 3) showed that this region in each of the eight recombinants

Table 1 Allelic variation at individual cloned minisatellites

Clone λ33	Repeat length (bp)	Divergence %	Alleles Length (bp)	Alleles No. repeats	Alleles Frequency	$4N_e u$ A	$4N_e u$ B	$4N_e u$ C
1	62	0.2 ± 0.2	3,150	40	0.04	2	3	9
			2,600	31	0.11			
			2,350	27	0.04			
			*2,300	26	0.71			
			2,190	24	0.07			
			1,950	20	0.04			
3	32	14.1 ± 2.5	*450	6	1.00	0	0	0
4	64	6.9 ± 1.5	2,280	18	0.07	2	2	5
			2,140	16	0.11			
			*2,015	14	0.43			
			1,950	13	0.36			
			1,780	10	0.04			
5	17	9.2 ± 1.9	*1,660	14	1.00	0	0	0
6	37	0.7 ± 0.4	1,800	25†	0.04	4	6	20
			1,650	21†	0.07			
			1,570	19†	0.04			
			*1,535	18†	0.43			
			1,450	16†	0.04			
			1,400	15†	0.25			
			1,350	14†	0.04			
			1,280	12†	0.11			
10	41	5.9 ± 0.6	*1,460	5	1.00	0	0	0
11	33	0.0 ± 0.0	*990	3	1.00	0	0	0
15	16	1.1 ± 0.5	1,410	41	0.50	0.3	0.3	0.3
			*1,220	29	0.50			

DNA was prepared from white blood cells[10] from 14 unrelated British caucasians (seven male, seven female), digested with HinfI and Southern blot hybridized at very high stringency with probes from subcloned minisatellites as described in Fig. 5a, b. Cloned alleles whose sequences are shown in Fig. 3 are indicated (*). The divergence of each sequenced minisatellite from a hypothetical (consensus)$_n$ sequence is given as the mean percentage unique substitution divergence (±s.e.m.) to correct for variants which have diffused over more than one repeat. For example, λ33.15 shows 13 variants (10 substitutions and three deletions) but only five distinct variants over 29 repeats of a 16-bp sequence, giving a divergence from a (consensus)$_{29}$ sequence of $5/(29 \times 16) = 1.1\%$. The number of repeats in each allele was estimated from DNA fragment size; this estimate for alleles of λ33.6 is approximate (†) because of the trimeric nature of the λ33.6 repeat unit (Fig. 3). Approximate values of $\theta = 4N_e u$, where N_e is the effective population size and u is the rate of production of new length variants per locus per gamete, were estimated for each minisatellite from the number of different alleles (n_a) in the sample of 14 individuals by extensive population computer simulations designed to estimate the number of different, selectively equivalent alleles that could be maintained at steady-state in a population according to three recombinational models: A, random unequal crossing over between two alleles at meiosis, giving a new allele comprised of a random-length 5' segment of one allele fused to random length 3' segment of the second allele; B, constrained unequal exchange such that an allele mutates at random to gain or lose between one and three repeats; C, constrained slippage causing the gain or loss of only a single repeat. In each model, it is assumed as a first approximation that u is independent of the number of repeats in allele, as for the loci presented there is at most only a two-fold variation in allele length. Simulations were performed for values of θ from 0.1 to 100, using population sizes N_e from 50 to 500, and were continued for 10 N_e generations, steady state being achieved within ~N_e generations. Results from model A closely approximated the infinite allele model at $\theta < 2$, when the expected number of alleles n_a in a sample of i individuals is given by $n_a = \sum_1^{2i} (\theta/\theta + i - 1)$(ref. 31). Model C is the charge state model for which n_a has yet to be solved as a function of θ (ref. 32).

copy number in minisatellites, as alleles generally differed in length by an integral number of repeat units. In addition, longer alleles tended to hybridize with minisatellite probes more strongly than shorter alleles (see individual 1 in Fig. 5a), again suggesting that longer alleles contain more repeat sequence.

There is considerable variation in the level of repeat sequence homogeneity in each sequenced minisatellite region (Fig. 3). Some minisatellites (for example, λ33.3 and 5) show substantial repeat divergence, suggesting that these regions are not actively undergoing sequence homogenization by unequal exchange[23]; as expected, these regions show no polymorphic variation in repeat number (Table 1). Instead, the highly polymorphic minisatellites (λ33.1, 4 and 6) all show high repeat copy number together with substantial sequence homogeneity of repeats. In addition, base substitutions in the repeat units of hypervariable minisatellites tend to be present in more than one repeat (see, for example, λ33.15 in Fig. 3) which indicates that these minisatellites are actively and repeatedly engaging in unequal exchange, resulting in the diffusion of novel base substitutions across more than one repeat unit[23].

We used computer simulations to estimate the rate of unequal exchange needed to maintain the number of different (neutral) alleles (n_a) seen in our population sample (Table 1). Although space does not allow a detailed description of the population

simulations, we find that for a population of N_e diploid individuals starting with a monomorphic minisatellite containing say 30 repeats, n_a reaches a steady state in ~N_e generations, the mean value of n_a being determined both by the parameter $\theta = 4N_e u$, where u is the rate of production of new length alleles per locus per gamete, and by the model of unequal exchange used in the population simulation. Three models have been investigated: (1) random meiotic unequal exchange between minisatellite alleles; (2) constrained sister chromatid exchange; (3) DNA slippage causing the gain or loss of a single repeat (Table 1, models A, B and C respectively). We favour the constrained exchange model because of the tendency for minisatellite base substitution variants to diffuse to non-adjacent repeats (Fig. 3), together with the tendency of different length alleles of minisatellites to differ from each other by several rather than either one or many repeat units (Table 1). For the highly polymorphic minisatellites, we estimate θ to be 2–6. Given that the effective population size N_e for human populations has been estimated at ~10^4 (ref. 24), this gives values of u, the mutation rate to a new length allele, of ~$0.5–1.5 \times 10^{-4}$ per gamete for a minisatellite ~1 kb long.

This value is higher than the base substitution neutral mutation rate in man, which from studies of non-coding DNA sequence divergence in man and higher primates has been

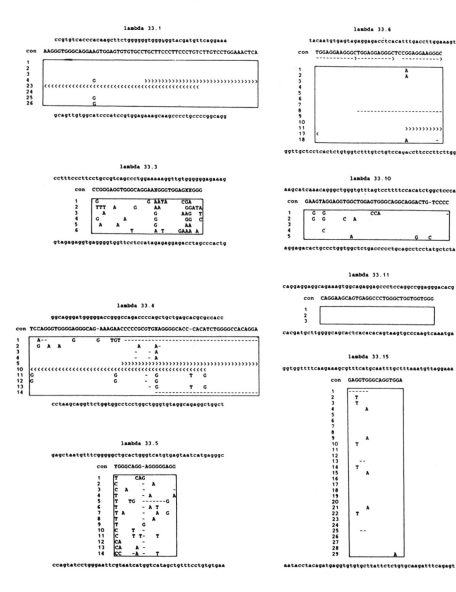

Fig. 3 Sequences of a selection of minisatellite regions detected by the myoglobin 33-repeat probe. The consensus sequence (con) of the tandem repetitive region in each of the genomic clones λ33.1–15 is shown, together with 50 bp of 5' and 3' flanking DNA (lower case). Differences from the consensus sequences are also shown for the individual numbered repeats (X, A or G; Y, C or T; −, missing nucleotide; ≫≪, region not sequenced although clearly a tandem repeat of the consensus sequence, or of a close derivative of the consensus, by inspection of sequencing autoradiographs).

Methods. A library of 10–20-kb human DNA fragments cloned in bacteriophage λL47.1 (refs 20, 22) was screened by hybridization with[32]P-labelled pAV33.7 insert as described in Fig. 2. A random selection of eight positive plaques was purified to give recombinants λ33.1–15. Each phage DNA was digested with *Hinf*I or *Hae*III, electrophoresed through a 1.5% agarose gel, and 33-repeat related sequences localized by Southern blot hybridization with pAV33.7 DNA (Fig. 2). Each recombinant gave a single positive *Hinf*I and *Hae*III fragment, except for λ33.4 and 11 which gave no detectable positive *Hae*III fragments (because of the presence of a *Hae*III cleavage site in the repeat regions in these clones; data not shown). Suitable positive *Hinf*I and *Hae*III fragments were isolated by preparative gel electrophoresis[33], end-repaired if necessary and blunt-end ligated into the *Sma*I site of M13mp8 (ref. 37). M13 recombinants were isolated after transformation into *E. coli* JM101 and sequenced by the dideoxynucleotide chain-termination method[38,39]. Each subcloned λ33 fragment contained a tandem repetitive region which in some cases could be sequenced directly. In other cases where the repeat region was too far from the sequencing primer site, the M13 inserts were shortened by cleavage with suitable restriction endonucleases and resequenced.

estimated at 1.0×10^{-9} substitution per nucleotide site per year[25–27]. Assuming that the generation time in man is 20 yr, this predicts a base substitution mutation rate of 2×10^{-5} per 1-kb minisatellite per gamete, lower than the estimated unequal exchange rate of 10^{-4} per gamete. This disparity in rates[23] is probably sufficient to maintain the amount of repeat sequence homogeneity seen in the hypervariable minisatellites λ33.1, 4 and 6.

The rate of unequal exchange can therefore be as high as 10^{-4} per kb minisatellite sequence and presumably is proportional to minisatellite length. In contrast, the rate of homologous recombination at meiosis in human DNA is ~1 centimorgan per 10^6 bp (ref. 2) or 10^{-5} per kb. The apparently very high rate of unequal exchange in minisatellites suggests either that they are hotspots for meiotic recombination, or that most exchanges are between sister chromatids at mitosis in the germline.

A χ sequence in minisatellites?

The length and sequence of the consensus minisatellite repeat sequences vary considerably; none of them are flanked by direct repeats (Fig. 3), in contrast to the repeat region in the myoglobin gene (Fig. 1). Thus, it is unlikely that these minisatellites are related by transposition of a common ancestral sequence. We therefore used dot-matrix comparisons[28] of each minisatellite repeat consensus with the myoglobin 33-bp repeat sequence to determine which region(s) of the 33-bp repeat probe were detecting each minisatellite. Remarkably, the consensus sequence of

each minisatellite repeat aligns with the myoglobin repeat specifically over a unique 10–15-bp core region of the 33-bp probe sequence (Fig. 4). This shared core region consists of an almost invariant sequence GGGCAGGAXG preceded by a 5-bp sequence common to most, but not all, repeats.

This core region in each cloned minisatellite suggests strongly that this sequence might help to generate minisatellites by promoting the initial tandem duplication of unique sequence DNA and/or by stimulating the subsequent unequal exchanges required to amplify the duplication into a minisatellite. As polymorphic minisatellites may also be recombination hotspots (see above), it might be significant that the core sequence is similar in length and in G content to the χ sequence, a signal for generalized recombination in *E. coli*[29] (Fig. 4a). Although the precise function of χ is unknown, current recombination models[30] suggest that this sequence binds the *rec*BC gene product, endonuclease V, which unwinds locally and nicks DNA to produce a single-stranded DNA projection required for the generation of Holliday junctions. In principle, DNA repair synthesis from the nicking site, followed by ligation to the single-stranded DNA projection, could generate a short tandem duplication with each duplicate containing a χ sequence capable of promoting unequal exchange and amplification of the duplicated region to produce a minisatellite (Fig. 4b). Although this model is highly speculative, it predicts that isolated core (or core-like) sequences may also be hotspots for initiating human chromosome recombination.

Probe for hypervariable regions

The repeat length of each minisatellite region is usually half (λ33.5 and 15), the same (λ33.3, 6, 10 and 11) or double the length (λ33.1 and 4) of the 33-bp probe from the human myoglobin gene (Fig. 3). This suggests that detection of minisatellites by pAV33.7 depends not only on the presence of a core sequence in each repeat but also on an in-phase alignment of cross-hybridizable core sequences in a heteroduplex between a minisatellite and the 33-bp repeat probe. If this is correct, then a probe consisting only of a tandem-repeated core sequence should be able to detect not only the human DNA fragments detectable by pAV33.7, but also additional minisatellites incapable of forming stable heteroduplexes with the 33-bp repeat probe. We therefore used the minisatellite from λ33.15, comprising 29 almost identical repeats of an almost perfect 16-bp core sequence, as a hybridization probe for additional minisatellites.

As shown in Fig. 5, this repeated core probe detects a complex profile of hybridizing fragments in human DNA digested with *Hin*fI, including most, but not all, of the bands detected previously with the 33-bp repeat probe from pAV33.7 (comparative data not shown). Only the largest (4–20 kb) *Hin*fI fragments can be resolved fully; these show extreme polymorphism to the extent that the hybridization profile provides an individual-specific DNA 'fingerprint'. Large fragment hyperpolymorphism is to be expected as, if the rate of unequal exchange is proportional to minisatellite length, then long minisatellites (~10 kb long) will have a greater unequal exchange rate ($u \sim 0.001$), raising both the heterozygosity and the number of alleles in a population.

Pedigree analysis

To establish that these large highly polymorphic fragments are stably inherited and segregate in a mendelian fashion, we analysed *Hin*fI digests of DNAs taken from an extensive Asian-Indian pedigree of Gujerati origin, including 54 individuals spanning four generations. The Southern blot/gel electrophoresis time was increased to improve the resolution of these large fragments. Typical examples of part of this pedigree are shown in Fig. 5d, e, confirming that the polymorphic variation is so great that all individuals, even in a single sibship of a first-cousin marriage (Fig. 5e), can be distinguished.

The families in Fig. 5d, e show that most of the large *Hin*fI fragments are transmitted from each parent to only some of the offspring, establishing that most of these fragments are present in the heterozygous state and that the heterozygosity for these large hypervariable fragments must be approaching 100%. Furthermore, inheritance is Mendelian in that these heterozygous bands are transmitted on average to 50% of the offspring; 48 clearly-resolved different heterozygous parental bands were scored in four sibships of 4–6 individuals and gave a total of 116 cases in which an offspring had inherited a given parental band, compared with 124 cases where the band had not been transmitted (data not shown). Conversely, all fragments in offspring can be traced back to one or other parent, then in turn to their parents (with one exception, see below) and therefore provide a set of stably inherited genetic markers. No band is specifically transmitted from father to son or father to daughter (Fig. 5d), eliminating Y and X linkage respectively and implying that these minisatellite fragments are mainly autosomal in origin. Although it is not yet known from where in the set of autosomes these DNA fragments originate, they are not derived from a single localized region of one autosome. Instead, pairs of parental fragments can be identified which segregate independently in the offspring (Fig. 5d). Precisely, a pair of bands AB in one parent (and absent from the other) cannot be allelic, nor linked closely in repulsion, if there is at least one AB or − − offspring; the presence of A− or −B recombinant progeny further establishes lack of tight linkage in coupling between A and B. Careful examination of the original autoradiograph of the family shown in Fig. 5d reveals, by these criteria, at least 10 resolvable bands in the mother, eight of which are mutually non-allelic and not

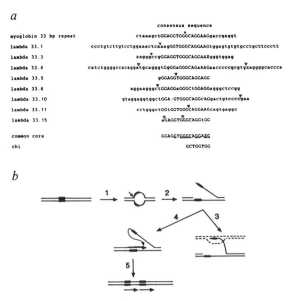

Fig. 4 A common χ-like core sequence shared by the repeat sequence of each minisatellite region. The sequence of each region was compared with the 33-bp tandem repetitive sequence in pAV33.7 and with its reverse complement, using dot-matrix analysis with variable windows and matching criteria[28]. In each case, only a single unambiguous region of sequence similarity was found between the myoglobin 33-bp repeat sequence and the λ33-repeat sequence. The same region was shared by the repeats of all eight λ33 clones. *a*, The aligned sequence of each repeat consensus, each of which is given as an arbitrary cyclic permutation of the consensus shown in Fig. 3. The common core sequence shared by all repeats is also shown and positions in each consensus which conform to the core sequence are identified by upper case letters. Invariant nucleotides in the canonical core sequence are underlined; the generalized recombination signal (χ) of E. coli is given also[29]. The beginning/end point of each repeat consensus (Fig. 3) is identified by ▼; in the case of λ33.4 and λ33.15, there is a non-integral number of repeats (Fig. 3) and the separate repeat beginning and end points are shown by ▽. *b*, Model for minisatellite generation promoted by χ-like sequences. (1) A χ-like region denoted by a box. RecBC enzyme binds to χ and unwinds DNA. (2) Nicking produces a single-strand projection (3), which can be assimilated into a homologous duplex to form the precursor of a Holliday junction[30]. (4) Alternatively, DNA repair synthesis followed by (5) ligation and segregation produces a tandem duplication of a χ-containing sequence which can amplify further by unequal exchange. The length of the tandem repeat is determined by the extent of repair synthesis.

linked closely. Two other bands may each be an allele of one of the eight unlinked fragments, in that only A− and −B progeny are observed in the limited number of offspring analysed, although such a small sample is insufficient to prove that such pairs of fragments are alleles of a single locus. We conclude that the core probe can give useful information simultaneously on at least several distinct unlinked hypervariable loci.

A new mutant allele

The extreme variability of the large *Hin*fI DNA fragments detected by the repeat core sequence suggests that the rate of generation of new length alleles must be very high and possibly amenable to direct measurement. There is a clear instance of a new mutation in the pedigree shown in Fig. 5e; individual 17 has a new fragment (arrowed), not present in either parent, which might have been derived by unequal exchange and slight expansion of a smaller maternal and paternal fragment present in the other offspring. In a survey of 27 individuals and their parents (data to be presented elsewhere), 240 clearly resolved offspring bands could be traced to one or other parent, the only exception being the band in individual 17. This gives a mutation rate u to a new allele for these hypervariable fragments of

Fig. 5 Polymorphic human DNA fragments detected by hybridization with individual λ33 probes. Southern blots of *Hin*fI digests of DNA from a random sample of British caucasians (1–6) and from selected members of a large British Asian-Indian pedigree (7–18) were hybridized with single-stranded ^{32}P-labelled hybridization probes prepared from suitable M13 recombinants containing minisatellite regions. The pair of bands (○) in individual 8 is an example of non-allelic fragments which are not tightly linked; the pair marked (●) illustrates possible allelism in that each of the five offspring inherits only one of the two fragments. The arrowed fragment in individual 17 is present in neither parent and is a new mutant. The correct paternity of individual 17 has been verified as described below.

Methods. 10 μg samples of DNA prepared from white blood cells[10] were digested with *Hin*fI, electrophoresed through a 20-cm long 1% agarose gel and transferred by blotting to a Sartorius nitrocellulose filter. Single-stranded ^{32}P-labelled hybridization probes were prepared from M13 recombinants as follows. Approximately 0.4 μg M13

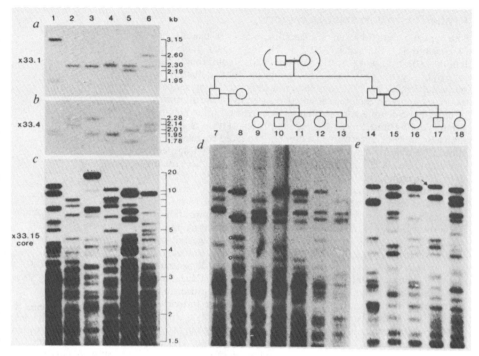

single-stranded DNA was annealed with 4 ng 17-mer sequencing primer[40] in 10 μl 10 mM MgCl$_2$, 10 mM Tris-HCl (pH 8.0) at 60 °C for 30 min. Primer extension was performed by adding 16 μl 80 μM dATP, 80 μM dGTP 80 μM TTP, 10 mM Tris-HCl (pH 8.0), 0.1 mM EDTA plus 3 μl (30 μCi)[α-^{32}P]dCTP (3,000 Ci mmol^{-1}) and 1 μl 5 units μl^{-1} Klenow fragment (Boehringer) and incubating at 37 °C for 15 min. Extension was completed by adding 2.5 μl 0.5 mM dCTP and chasing at 37 °C for a further 15 min. The DNA was cleaved at a suitable restriction endonuclease site either in the insert or in the M13 polylinker distal to the insert, denatured by adding 1/10 vol. 1.5 M NaOH, 0.1 M EDTA, and the ^{32}P-labelled single-stranded DNA fragment extending from the primer was recovered by electrophoresis through a 1.5% low melting point agarose gel (Sea Plaque). The excised band (specific activity > 10^9 c.p.m. μg^{-1} DNA) was melted at 100° C in the presence of 1 mg alkali-sheared carrier human placental DNA (sheared in 0.3 M NaOH, 20 mM EDTA at 100° C for 5 min) and added directly to a pre-warmed hybridization chamber; the carrier DNA also suppressed any subsequent hybridization to repetitive DNA sequences. The precise probes used were: 33.1, a 2,000-nucleotide subcloned *Hae*III fragment containing the minisatellite plus 350-nucleotide flanking human DNA; 33.4, a 695-nucleotide non-minisatellite *Eco*RI fragment on the primer-proximal side of the minisatellite contained in a 2,015-nucleotide *Hin*fI fragment; 33.15 core, a 592-nucleotide subcloned fragment containing the minisatellite sequence plus 128-nucleotide flanking human DNA. Hybridizations were performed as described elsewhere[36], except that dextran sulphate was replaced by 6% (w/v) polyethylene glycol 6,000 (Fisons) to reduce background labelling[41]. Filters A and B were hybridized overnight in 0.5 ×SSC at 65 °C and washed in 0.2 ×SSC at 65 °C. Filters C–E were hybridized and washed in 1 ×SSC at 65 °C. Filters were autoradiographed for 1–3 days at −80 °C using a fast tungstate intensifying screen. The correct paternity of individual 17 was established using a range of biochemical and blood group markers (haptoglobin, transferrin, red cell acid phosphatase, phosphoglucomutase I, adenylate kinase, adenosine deaminase, α$_1$-antitrypsin, G$_c$, Gm, esterase D, glyoxylase, phosphoglycollate phosphatase, C3, peptidase D, ABO, Rh and HLA; S.L.T., unpublished data), and confirmed further by rehybridizing this blot with the core minisatellite in λ33.6 to generate a second DNA 'fingerprint' in which all polymorphic bands in individual 17 could be traced back to one or other parent (data not shown).

~$\frac{1}{240}$ ≈ 0.004. This estimate is in reasonable agreement with the population genetic estimate of $u \sim 0.001$ for these very large hypervariable fragments (see above).

Conclusions

Here, we show not only that it is possible to design probes for the cloning of individual polymorphic minisatellite regions from human DNA, but also that the shared core sequence, which possibly serves as a recombination signal and promotes the formation of minisatellites, can be used for the simultaneous analysis of multiple hypervariable regions. We anticipate that these DNA 'fingerprints' will be of general use in human segregation analysis, in particular for detecting specific bands in close linkage with disease loci in large pedigrees and for studying marker loss in tumours. In addition, they provide a powerful method for paternity and maternity testing, can be used in forensic applications and might also be useful in detecting inbreeding between couples who have had an affected offspring possibly caused by an autosomal recessive gene carried by both parents.

The precise sequence of the core consensus shared by the repeat elements of the cloned minisatellites will be biased by the particular version of the core present in the myoglobin gene minisatellite used as the initial hybridization probe. Therefore, other variant (core)$_n$ probes might detect additional polymorphic loci not found by the λ33.15 repeated core sequence. Preliminary experiments have indeed shown that the core minisatellites in λ33.5 and λ33.6 also hybridize to multiple hypervariable loci, many of which are novel. We are attempting currently to clone large hypervariable regions to provide locus-specific probes for individual minisatellites.

The detection of new mutant-length alleles of minisatellites in human pedigrees will help the analysis of rates and mechanisms of unequal exchange and gene homogenization and can in principle be used to determine whether such exchanges occur by sister chromatid exchange or by recombination between homologous chromosomes at meiosis. In addition, if the core sequence is indeed a recombination signal, then its accurate definition could provide useful substrates for studying mechanisms of human recombination.

This work is the subject of a UK Patent Application. Enquiries should be addressed to National Research Development Corporation. We thank Alain Blanchetot for a human library, John F. Y. Brookfield and Robert Semeonoff for helpful discussions and Raymond Dalgleish for advice on hybridizations. We also thank Drs D. A. Hopkinson, P. Tippett and A. Ting for analysis of various biochemical, blood group and HLA markers. A.J.J. is a Lister Institute Research Fellow; this work was supported by a grant to A.J.J. and a Training Fellowship to S.L.T. from the MRC.

Prenatal Diagnosis of Alpha-thalassemia Clinical Application of Molecular Hybridization

W.Y. Kan, M.S. Golbus and A.M. Dozy

Abstract The technic of DNA-DNA hybridization was used for prenatal diagnosis of a pregnancy at risk for homozygous α-thalassemia. Fibroblasts were cultured from amniotic fluid, and the number of α-globin genes in the DNA was quantified by hybridization with radioactive DNA complementary to α-globin mRNA sequences. As compared to control studies of DNA from patients with α-thalassemia syndromes and from unaffected subjects, the results indicated that the fetus had α-thalassemia-1. The diagnosis was confirmed by umbilical-cord blood studies. (N Engl J Med 295:1165-1167, 1976)

I N the prenatal diagnosis of hereditary hemoglobin disorders[1,2] the method currently employed depends on the acquisition of fetal blood and the study of globin-chain synthesis. Homozygous β-thalassemia and sickle-cell anemia have been diagnosed in utero with this technic.[3-5] However, whereas the biochemical analyses, when carefully performed, have been accurate in predicting the homozygous state, the technics used in the acquisition of fetal blood need to be improved and made safer.

A different approach that does not require fetal blood can be used for the prenatal diagnosis of the α-thalassemia syndromes. This approach is based on the finding that the molecular defect in these syndromes is due to deletion of the gene that determines the structure of the α globin (α-globin structural gene).[6-9] The α-thalassemia syndromes have been best defined in Asian populations, in whom four clinical states of increasing severity are recognized (Table 1). A non-thalassemic person has four α-globin structural genes per diploid cell; in α-thalassemia-1 and hemoglobin-H disease, two and one α-globin genes are left intact respectively, and in homozygous α-thalassemia associated with hydrops fetalis, no α-globin genes remain intact.[10,11]* To measure the number of intact globin structural genes, the technic of molecular hybridization is now available.[6-11] Since nucleated cells, including cultured fibroblasts, contain the full complement of globin genes, cultured amniotic-fluid fibroblasts can be tested by the molecular hybridization method for the purpose of prenatal diagnosis of the α-thalassemia syndromes. This report describes the successful application of this approach to a pregnancy at risk for α-thalassemia.

From the Medical Service, San Francisco General Hospital, and the departments of Medicine, Obstetrics and Gynecology, Clinical Pathology and Laboratory Medicine, University of California, San Francisco (address reprint requests to Dr. Kan at San Francisco General Hospital, San Francisco, CA 94110).'

Supported by a grant (AM-16666) from the U.S. Public Health Service and by grants from the National Foundation–March of Dimes, and the John A. Hartford Foundation, Inc. (Dr. Kan is the recipient of a research career-development award [1-KO4-AM-70779] from the U.S. Public Health Service).

*In this paper, heterozygous α-thalassemia is referred to as α-thalassemia-1, and homozygous α-thalassemia as hydrops fetalis.

MATERIALS AND METHODS

A Chinese couple with α-thalassemia-1 had two previous pregnancies resulting in perinatal death due to hydrops fetalis. The parents have been found to have only two α-globin structural genes per diploid cell in their white blood cells. Their two living children are also affected by α-thalassemia-1. The couple requested prenatal diagnosis because the previous pregnancies resulting in hydrops were associated with maternal toxemia and because the mother did not wish to carry another unsuccessful pregnancy for 40 weeks. At the 15th week after the last menstrual period, amniocentesis was performed, and fibroblasts were cultured from the amniotic fluid.[12] After five weeks, 3×10^8 fibroblasts were obtained. DNA was extracted from the fibroblasts by previously described methods.[7] As a control, amniotic-fluid fibroblast DNA was similarly obtained from a pregnancy not at risk for thalassemia.

Table 1. The α-Thalassemia (α-Thal) Syndromes (Asian Populations).

SYNDROME	GENOTYPE	CLINICAL MANIFESTATION	NO. OF α-GLOBIN GENES*
Silent carrier state'	α-Thal-2	None	3
Heterozygous α-thalassemia	α-Thal-1	Microcytosis	2
Hemoglobin-H disease	α-Thal-1 $\times$ α-thal-2	Microcytosis, hemolysis	1
Homozygous α-thalassemia	α-Thal-1 $\times$ α-thal-1	Hydrops fetalis (fatal)	0

*No. of globin genes in α-thalassemia-2 is inferred. All others have been shown by hybridization.

†A variant is the carrier of hemoglobin Constant Spring, which is due to an elongated α chain produced in small quantities & which phenotypically resembles α-thalassemia-2.

As controls for normal and α-thalassemia, DNA was also extracted from either the placenta, bone marrow, spleen or liver of seven patients with hydrops fetalis, four patients with hemoglobin-H disease, six patients with α-thalassemia-1 and 12 non-thalassemic persons. The α-globin structural genes were quantified by the technic of molecular hybridization by methods previously described.[7,8] The cDNA probe used to assess percentage hybridization and thereby the number of α-globin genes in the fibroblast DNA was prepared with reverse transcriptase from ^{32}P-deoxycytidine triphosphate (New England Nuclear Corporation, 125 Ci per millimole).' The cDNA contained 75 to 80 per cent α sequences. The rest were mainly β sequences.

The α-cDNA was then annealed to cellular DNA's from these patients in duplicate 20-μl reaction mixtures, all of which contained 5 pg of cDNA, 100 μg of cellular DNA, 500 CPM of ^{3}H-deoxycytidine triphosphate-labeled unique-sequence HeLa DNA, 0.5 M sodium chloride, 0.002 M EDTA and 0.04 M Tris-hydrochloric acid (pH 7.4). The samples were incubated at 78°C for 78 hours. The unique-sequence DNA served as internal control for the extent of hybridization of the cellular DNA. The percentage of the cDNA annealed was assayed with batchwise elution with hydroxyapatite.[13]

†See accompanying editorial for further explanation.

RESULTS

The percentages of the α-cDNA annealed to the DNA of patients with various α-thalassemia syndromes and normal controls are shown in Figure 1. Twenty-five per cent of α-cDNA was annealed to hydrops DNA, probably owing to the presence of the β-DNA sequences in the α-cDNA preparation. Forty-five per cent, 58 per cent and 65 per cent of the α-cDNA was annealed to the DNA's from hemoglobin-H disease, α-thalassemia-1 and normal controls respectively. The differences between these groups are statistically significant (P<0.001 by Student's t-test). However, the differences between some individual cases of α-thalassemia-1 and normal are quite small. The DNA from the fibroblasts of the pregnancy studied annealed 56 per cent of the α-cDNA and was thus within the range for α-thalassemia-1. Fibroblasts grown from a pregnancy not at risk for α-thalassemia annealed 64 per cent of this α-cDNA. The ³H-labeled unique-sequence DNA used as internal controls was annealed to the same extent (80 to 83 per cent) with these DNA's. We therefore predicted that the fetus tested was probably affected by α-thalassemia-1. Although these results did not rule out the possibility that the fetus could be normal, it was clearly not af-

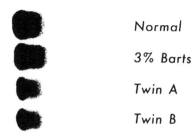

Figure 2. Cellulose Acetate Electrophoresis (pH 8.6) of the Hemoglobin from the Cord Blood of a Non-thalassemic Infant, One with 3 per Cent Hemoglobin Barts and the Twins Delivered in This Pregnancy.

Cathode is on the left. The bands from left to right are hemoglobin F, A and Barts.

fected by homozygous α-thalassemia. Chromosome analysis showed that the fetus tested was female.

About two months after this diagnosis was made, a twin gestation was discovered. Since at least one fetus was known to be unaffected, the parents had no hesitation about continuing the pregnancy. One male and one female infant were born, and neither was affected by hydrops fetalis. Analysis of umbilical-cord blood showed that the mean corpuscular volumes of the red blood cells of both fetuses were 91 fl and 87 fl, values that are microcytic for that gestational age (mean corpuscular volume of normal cord blood = 119±9.4 fl [mean±1 S.D.]). Hemoglobin electrophoresis demonstrated that both infants had about 8 per cent hemoglobin Barts (Fig. 2). These findings are diagnostic of α-thalassemia-1.

DISCUSSION

Our study shows that when the defect of a genetic disease is the result of a structural gene deletion, study of the amniotic-fluid fibroblasts by DNA-DNA hybridization is feasible. Furthermore, since amniocentesis for fibroblast culture is a well established technic, prenatal diagnosis of the α-thalassemia syndromes by this method is quite safe. Although α-thalassemia can also be detected by study of fetal blood,[14] fetal-blood sampling for prenatal diagnosis of this condition should not be used, since it is at present associated with a greater risk to the fetus.[15,16]

Variations in the purity and lengths of the α-cDNA's may affect the level of hybridization from one preparation to another. Therefore, in quantifying α-globin structural genes, both the DNA to be tested and the control DNA's from patients with well defined clinical syndromes should be tested with the same preparation of α-cDNA's.

In the pregnancy tested, the twin gestation was not discovered at the time of amniocentesis since ultrasound is not used routinely in this prenatal diagnosis center. Fortunately, the discovery of one unaffected fetus enabled the parents to continue the pregnancy without hesitation.

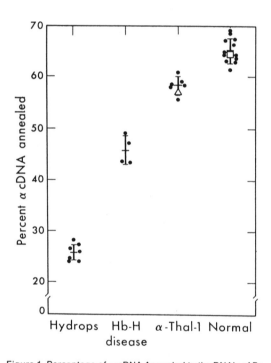

Figure 1. Percentage of α-cDNA Annealed to the DNA's of Patients with α-Thalassemia Syndromes and Non-thalassemic Controls (Solid Dots).

The means and 1 S.D. are shown. The △ denotes amniotic-fluid fibroblast DNA from the pregnancy tested, and the □ that from a pregnancy not at risk for α-thalassemia.

In prenatal diagnosis the finding of a female fetus with the same genotype as that of the mother always raises the possibility that the fibroblast tested could have been derived from the mother. However, the likelihood of this occurrence is extremely low. Our center has performed over 1200 prenatal diagnoses by amniocentesis; none of the fetuses predicted by chromosome analysis to be female have turned out to be male.

α-Thalassemia trait is a relatively common anomaly, especially in certain Asian countries. Because the homozygous state associated with hydrops fetalis is invariably fatal, the disease does not lead to the major medical and social problems caused by homozygous β-thalassemia. However, a safe and reliable method for prenatal diagnosis would be accepted by many parents, for pregnancies with hydropic fetuses are often associated with toxemia.[17] Also, many parents would choose not to continue a pregnancy doomed to an unsuccessful outcome. The methods described above can clearly distinguish the homozygous from the heterozygous state (α-thalassemia-1) and, in our hands, can also differentiate the homozygous state from hemoglobin-H disease, a relatively benign disorder in which three out of the four α genes are deleted. Difficulty may be encountered in distinguishing some cases of α-thalassemia-1 from the normal state. This separation is not clinically important since patients with α-thalassemia-1 are phenotypically normal.

At present, in addition to α-thalassemia, hereditary persistence of fetal hemoglobin and $\delta^0\beta^0$-thalassemia are the only other two diseases in which deletion of a globin structural gene has been demonstrated.[18,20] The homozygous state of both these disorders is rare, and the clinical manifestations of the former are mild.[21,22] Thus, prenatal diagnosis is not a practical consideration for these disorders. In the more common type of β-thalassemia associated with elevated hemoglobin A_2, the β-globin structural gene is intact,[8,23,24] and the prenatal diagnosis depends on fetal-blood studies.[1-3,15,16]

When cDNA's for other structural genes become available, molecular hybridization could conceivably be applied to prenatal diagnosis of those disorders where gene deletions are found.

We are indebted to Professor David Todd, of the University of Hong Kong, for supplying much of the material for the control group, to the Office of Program Resources and Logistics, Viral Cancer Program, Viral Oncology, Division of Cancer Cause and Prevention, National Cancer Institute, for the AMV DNA polymerase, to Ms. J.P. Holland for the mRNA preparation, to Ms. G. Abbo-Halbasch for the fibroblast culture and to Ms. K. Klaman for the cellulose acetate electrophoresis.

REFERENCES

1. Kan YW, Golbus MS, Klein P, et al: Successful application of prenatal diagnosis in a pregnancy at risk for homozygous β-thalassemia. N Engl J Med 292:1096-1099, 1975
2. Nathan DG, Alter BP, Frigoletto FD: Antenatal diagnosis of hemoglobinopathies: social and technical considerations. Semin Hematol 12:305-321, 1975
3. Kan YW, Golbus MS, Trecartin R, et al: Prenatal diagnosis of homozygous β-thalassemia. Lancet 2:790-792, 1975
4. Kan YW, Golbus MS, Trecartin R: Prenatal diagnosis of sickle-cell anemia. N Engl J Med 294:1039-1040, 1976
5. Alter BP, Friedman S, Hobbins JC, et al: Prenatal diagnosis of sickle-cell anemia and alpha G-Philadelphia. N Engl J Med 294:1040-1042, 1976
6. Ottolenghi S, Lanyon WG, Paul J, et al: Gene deletion as the cause of α-thalassaemia. Nature 251:389-392, 1974
7. Taylor JM, Dozy A, Kan YW, et al: Genetic lesion in homozygous alpha thalassaemia (hydrops fetalis). Nature 251:392-393, 1974
8. Ramirez F, Natta C, O'Donnell JV, et al: Relative numbers of human globin genes assayed with purified α and β complementary human DNA. Proc Natl Acad Sci USA 72:1550-1554, 1975
9. Gambino R, Kacian D, O'Donnell J, et al: A limited number of globin genes in human DNA. Proc Natl Acad Sci USA 71:3966-3970, 1974
10. Kan YW, Dozy AM, Varmus HE, et al: Deletion of α-globin genes in haemoglobin-H disease demonstrates multiple α-globin structural loci. Nature 255:255-256, 1975
11. Kan YW, Dozy AM, Varmus HE, et al: The molecular basis of the α-thalassaemia syndromes. Clin Res 23:398A, 1975
12. Golbus MS, Conte FA, Schneider EI, et al: Intrauterine diagnosis of genetic defects: results, problems, and follow-up of one hundred cases in a prenatal genetic detection center. Am J Obstet Gynecol 118:897-905, 1974
13. Garapin AC, Leong J, Fanshier L, et al: Identification of virus-specific RNA in cells infected with rous sarcoma virus. Biochem Biophys Res Commun 42:919-925, 1971
14. Kan YW, Bellevue R, Rieder RF, et al: Prenatal diagnosis of α thalassemia. Clin Res 22:374A, 1974
15. Kan YW, Golbus MS: Prenatal diagnosis of hereditary hemoglobin disorders: current developments. Pediatr Res 20:367, 1976
16. Alter BP, Sherman AS, Hobbins JC, et al: Prenatal diagnosis of hemoglobinopathies: experience in 17 cases. Clin Res 24:293A, 1976
17. Wasi P, Na-Nakorn S, Pootrakul S, et al: Alpha- and beta-thalassemia in Thailand. Ann NY Acad Sci 165:60-82, 1969
18. Kan YW, Holland JP, Dozy AM, et al: Deletion of the β-globin structure gene in hereditary persistence of foetal haemoglobin. Nature 258:162-163, 1975
19. Forget BG, Hillman DG, Lazarus H, et al: Absence of messenger RNA and gene DNA for β-globin chains in hereditary persistence of fetal hemoglobin. Cell 7:323-329, 1976
20. Ottolenghi S, Lanyon WG, Williamson R, et al: Human globin gene analysis for a patient with $\beta^0/\delta\beta^0$-thalassemia. Proc Natl Acad Sci USA 72:2294-2299, 1975
21. Weatherall DJ, Clegg JB: The Thalassaemia Syndromes. Oxford, Blackwell Scientific Publications, 1972
22. Stamatoyannopoulos G, Fessas Ph, Papayannopoulou Th: F-thalassemia: a study of thirty-one families with simple heterozygotes and combinations of F-thalassemia with A_2-thalassemia. Am J Med 47:194-208, 1969
23. Kan YW, Holland JP, Dozy AM, et al: Demonstration of non-functional β-globin mRNA in homozygous β^0-thalassemia. Proc Natl Acad Sci USA 72:5140-5144, 1975
24. Tolstoshev P, Mitchell J, Lanyon G, et al: Presence of gene for β-globin in homozygous β^0-thalassaemia. Nature 259:95-98, 1976

Enzymatic Amplification of β-globin Genomic Sequences and Restriction Site Analysis for Diagnosis of Sickle Cell Anemia

R.K. Saiki, S. Scharf, F. Faloona, K.B. Mullis, G.T. Horn, H.A. Erlich and N. Arnheim

Abstract. *Two new methods were used to establish a rapid and highly sensitive prenatal diagnostic test for sickle cell anemia. The first involves the primer-mediated enzymatic amplification of specific β-globin target sequences in genomic DNA, resulting in the exponential increase (220,000 times) of target DNA copies. In the second technique, the presence of the β^A and β^S alleles is determined by restriction endonuclease digestion of an end-labeled oligonucleotide probe hybridized in solution to the amplified β-globin sequences. The β-globin genotype can be determined in less than 1 day on samples containing significantly less than 1 microgram of genomic DNA.*

Recent advances in recombinant DNA technology have made possible the molecular analysis and prenatal diagnosis of several human genetic diseases. Fetal DNA obtained by aminocentesis or chorionic villus sampling can be analyzed by restriction enzyme digestion, with subsequent electrophoresis, Southern transfer, and specific hybridization to cloned gene or oligonucleotide probes. With polymorphic DNA markers linked genetically to a specific disease locus, segregation analysis must be carried out with restriction fragment length polymorphisms (RFLP's) found to be informative by examining DNA from family members (*1, 2*).

Many of the hemoglobinopathies, however, can be detected by more direct methods in which analysis of the fetus alone is sufficient for diagnosis. For example, the diagnosis of hydrops fetalis (homozygous α-thalassemia) can be made by documenting the absence of any α-globin genes by hybridization with an α-globin probe (*3–5*). Homozygosity for certain β-thalassemia alleles can be determined in Southern transfer experiments by using oligonucleotide probes that form stable duplexes with the normal β-globin gene sequence but form unstable hybrids with specific mutants (*6, 7*).

Sickle cell anemia can also be diagnosed by direct analysis of fetal DNA.

This disease results from homozygosity of the sickle-cell allele (β^S) at the β-globin gene locus. The S allele differs from the wild-type allele (β^A) by substitution of an A in the wild-type to a T at the second position of the sixth codon of the β chain gene, resulting in the replacement of a glutamic acid by a valine in the expressed protein. For the prenatal diagnosis of sickle cell anemia, DNA obtained by amniocentesis or chorionic villus sampling can be treated with a restriction endonuclease (for example, Dde I and Mst II) that recognizes a sequence altered by the β^S mutation (*8–11*). This generates β^A- and β^S-specific restriction fragments that can be resolved by Southern transfer and hybridization with a β-globin probe.

We have developed a procedure for the detection of the sickle cell mutation that is very rapid and is at least two orders of magnitude more sensitive than standard Southern blotting. There are two special features to this protocol. The first is a method for amplifying specific β-globin DNA sequences with the use of oligonucleotide primers and DNA polymerase (*12*). The second is the analysis of the β-globin genotype by solution hybridization of the amplified DNA with a specific oligonucleotide probe and subsequent digestion with a restriction endonuclease (*13*). These two techniques increase the speed and sensitivity, and

lessen the complexity of prenatal diagnosis for sickle cell anemia; they may also be generally applicable to the diagnosis of other genetic diseases and in the use of DNA probes for infectious disease diagnosis.

Sequence amplification by polymerase chain reaction. We use a two-step procedure for determining the β-globin genotype of human genomic DNA samples. First, a small portion of the β-globin gene sequence spanning the polymorphic Dde I restriction site diagnostic of the β^A allele is amplified. Next, the presence or absence of the Dde I restriction site in the amplified DNA sample is determined by solution hybridization with an end-labeled complementary oligomer followed by restriction endonuclease digestion, electrophoresis, and autoradiography.

The β-globin gene segment was amplified by the polymerase chain reaction (PCR) procedure of Mullis and Faloona (*12*) in which we used two 20-base oligonucleotide primers that flank the region to be amplified. One primer, PC04, is complementary to the (+)-strand and the other, PC03, is complementary to the (−)-strand (Fig. 1). The annealing of PC04 to the (+)-strand of denatured genomic DNA followed by extension with the Klenow fragment of *Escherichia coli* DNA polymerase I and deoxynucleotide triphosphates results in the synthesis of a (−)-strand fragment containing the target sequence. At the same time, a similar reaction occurs with PC03, creating a new (+)-strand. Since these newly synthesized DNA strands are themselves template for the PCR primers, repeated cycles of denaturation, primer annealing, and extension result in the exponential accumulation of the 110–base pair region defined by the primers.

An example of the degree of specific gene amplification achieved by the PCR method is shown in Fig. 2A. Samples of DNA (1 μg) were amplified for 20 cycles and a fraction of each sample, equivalent to 36 ng of the original DNA, was subjected to alkaline gel electrophoresis and transferred to a nylon filter. The filter was then hybridized with a ^{32}P-labeled 40-base oligonucleotide probe, RS06, which is complementary to the target sequence (Fig. 1A) but not to the PCR primers. The results, after a 2-hour autoradiographic exposure, show that a fragment hybridizing with the RS06 probe

The authors are in the Department of Human Genetics, Cetus Corporation, 1400 Fifty-Third Street, Emeryville, California 94608. The present address for N.A. is Department of Biological Sciences, University of Southern California, Los Angeles 90089-0371.

migrates at the position expected of the amplified target DNA segment (110 bases) (lanes 1 and 2). No hybridization with the RS06 probe could be detected in unamplified DNA (lane 4). When PCR amplification was performed on a DNA sample derived from an individual with hereditary persistence of fetal hemoglobin in which both the δ- and β-globin genes are deleted (14), again no 110-base fragment was detected (lane 3). To estimate the yield and efficiency of 20 cycles of PCR amplification, we prepared a Southern blot that contained 36 ng of an amplified genomic DNA sample and a dilution series consisting of various amounts of cloned β-globin sequence. The efficiency was calculated according to the formula: $(1 + X)^n = Y$, where X is the mean efficiency per cycle, n is the number of PCR cycles, and Y is the extent of amplification (yield) after n cycles (for example, a 200,000-fold increase after 20 cycles). The amounts of cloned β-globin sequences used in this experiment were calculated to represent efficiencies of 70 to 100 percent.

The reconstructions were prepared by digesting the β-globin plasmid, pBR328::β^A, with the restriction enzymes Hae III and Mae I. Both of these enzymes cleave the β-globin gene within or very near to the 20 base regions that hybridize to the PCR primers, generating a 103–base pair (bp) fragment that is almost identical in size and composition to the 110-bp segment created by PCR amplification. After hybridization with the RS06 probe and autoradiography, the amplified genomic sample was compared with the known standards, and the result indicated an overall efficiency of approximately 85 percent (Fig. 2B), representing an amplification of about 220,000 times (1.85^{20}).

Distinguishing the β^A and β^S alleles by the oligomer restriction method. We have previously described a rapid solution hybridization method that can indicate whether a genomic DNA sample contains a specific restriction enzyme site at, in principle, any chromosomal location (13). This method, called oligomer restriction (OR), involves the stringent hybridization of a ^32P end-labeled oligonucleotide probe to the specific segment of the denatured genomic DNA which spans the target restriction site. The ability of a mismatch within the restriction site to prevent cleavage of the duplex formed between the probe and the target genomic sequence is the basis for detecting allelic variants. The presence of the restriction site in the target DNA is revealed by the appearance of a specific

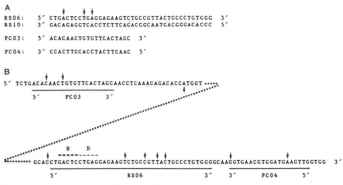

Fig. 1. Sequence of synthetic oligonucleotide primers and probe and their relation to the target β-globin region. (A) The primer PC03 is complementary to the (−)-strand and the primer PC04 is complementary to the (+)-strand of the β-globin gene. The probe RS06 is complementary to the (−)-strand of the wild-type (β^A) sequence of β-globin. RS10 is the "blocking oligomer", an oligomer complementary to the RS06 probe except for three nucleotides, indicated by the downward arrows. It is added before enzyme digestion to the OR reaction to anneal to the excess RS06 oligomer and prevent nonspecific cleavage products due to hybridization of RS06 to nontarget DNA (13). Because of the mismatches within the Dde I and Hinf I restriction sites, the RS06/RS10 duplex is not cleaved by Dde I and Hinf I digestion. (B) The relation between the primers, the probe, and the target β-globin sequence. The upward arrow indicates the β-globin initiation codon. The downward arrows indicate nucleotide differences between β- and δ-globin. The polymorphic Dde I site (CTCAG) is represented by a single horizontal dashed line (D), and the invariant Hinf I (GACTC) site is represented by double horizontal dashed lines (H).

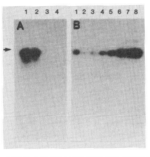

Fig. 2. Southern analysis of PCR amplified genomic DNA with the RS06 probe. (A) Samples (1 μg) of genomic DNA were dispensed in microcentrifuge tubes and adjusted to 100 μl in a buffer containing 10 mM tris, pH 7.5, 50 mM NaCl, 10 mM MgCl₂, 1.5 mM deoxynucleotide trisphosphate [(dNTP) each of all four was used], 1 μM PC03, and 1 μM PC04. After heating for 5 minutes at 95°C (to denature the genomic DNA), the tubes were centrifuged for 10 seconds in a microcentrifuge to remove the condensation. The samples were immediately transferred to a 30°C heat block for 2 minutes to allow the PC03 and PC04 primers to anneal to their target sequences. At the end of this period, 2 μl of the Klenow fragment of E. coli DNA polymerase I (Biolabs, 0.5 unit/μl in 10 mM tris, pH 7.5, 50 mM NaCl, 10 mM MgCl₂) was added, and the incubation was allowed to proceed for an additional 2 minutes at 30°C. This cycle—denaturation, centrifugation, hybridization, and extension—was repeated 19 more times, except that subsequent denaturations were done for 2 instead of 5 minutes. (The final volume after 20 cycles was 140 μl.) Thirty-six nanograms of the amplified genomic DNA (5 μl) were applied to a 4 percent Nusieve (FMC) alkaline agarose minigel and subjected to electrophoresis (50 V), for 2 hours until the bromcresol green dye front reached 4 cm. After neutralization and transfer to Genetrans nylon membrane (Plasco), the filter was "prehybridized" in 10 ml 3× SSPE (1× SSPE is 0.18M NaCl, 10 mM NaH₂PO₄, 1 mM EDTA, pH 7.4), 5× DET (1× DET is 0.02 percent each polyvinylpyrrolidone, Ficoll, and bovine serum albumin; 0.2 mM tris, 0.2 mM EDTA, pH 8.0), 0.5 percent SDS, and 30 percent formamide for 4 hours at 42°C. Hybridization with 1.0 pmol of phosphorylated (with [γ-³²P]ATP) RS06 (~5 μCi/pmol) in 10 ml of the same buffer was carried out for 18 hours at 42°C. The filter was washed twice in 2× SSPE, 0.1 percent sodium dodecyl sulfate (SDS) at room temperature for 30 minutes, and autoradiographed at −70°C for 2 hours with a single intensification screen. (Lanes 1 to 3) DNA's isolated from the cell lines Molt4, SC01, and GM2064, respectively. Molt4 is homozygous for the normal, wild-type allele of β-globin (β^Aβ^A), SC-1 is homozygous for the sickle cell allele (β^Sβ^S), and GM2064 is a cell line in which the β- and δ-globin genes have been deleted (ΔΔ) (13). (Lane 4) Contains 36 ng of Molt4 DNA that was not PCR amplified. The horizontal arrow indicates the position of a 114-base marker fragment obtained by digestion of pBR328 with Nar I. (B) Thirty-six nanograms of 20-cycle amplified Molt4 DNA (see above) was loaded onto a Nusieve gel along with measured amounts of Hae III–Mae I digested pBR328::β^A (13) calculated to represent the molar increase in β-globin target sequences at PCR efficiencies of 70, 75, 80, 85, 90, 95, and 100 percent (lanes 2 to 8, respectively). DNA was transferred to Genetrans and hybridized with the labeled RS06 probe as described above. (Lane 1) Molt4 DNA (36 ng); (lanes 2 to 8) 7.3×10^{-4} pmol, 1.3×10^{-3} pmol, 2.3×10^{-3} pmol, 4.0×10^{-3} pmol, 6.8×10^{-3} pmol, 1.1×10^{-2} pmol, and 1.9×10^{-2} pmol of pBR328::β^A, respectively (20).

labeled fragment generated by cleavage of the probe.

For the diagnosis of sickle cell anemia, the probe was designed to be complementary to a region of the β-globin gene locus surrounding the sixth codon. In the β^A allele, the nucleotide (nt) sequence at this position contains a Dde I restriction site, but due to the single base mutation, this site is absent in the β^S allele. Our strategy for generating specific probe cleavage products for each allele is shown in Fig. 3. It is based on the presence of an invariant Hinf I restriction enzyme site immediately adjacent to the polymorphic Dde I restriction site. Resolution of the labeled oligomer cleavage products produced by sequential digestion with Dde I and Hinf I allows us to distinguish between the two alleles. In an individual homozygous for the wild-type β-globin allele AA, Dde I digestion will produce a labeled octamer (8 nt) from the probe. Because of its short length, the 8-nt cleavage product will dissociate from the genomic target DNA and the subsequent digestion with Hinf I has no effect. In the case of SS homozygotes, however, Dde I digestion does not cleave the probe since a base pair mismatch exists in the recognition sequence formed between the probe and target DNA. The invariant Hinf I site will then be cleaved during Hinf I digestion, releasing a labeled trimer (3 nt). In an AS heterozygote, both a trimer and an oc-

tamer would be detected. The resolution of the intact 40-base probe, the 8-nt and the 3-nt cleavage products is achieved by polyacrylamide gel electrophoresis. Experiments testing the sequential digestion strategy with plasmids carrying the β^A and β^S alleles show that, in each case, the expected probe cleavage products were produced (Fig. 4).

Analysis of genomic DNA samples by PCR and OR. Eleven DNA samples derived from lymphoblastoid cell lines or white blood cells were analyzed for their β-globin genotype by standard Southern blotting and hybridization of the Mst II RFLP (10), identifying the genotypes of the samples as either AA, AS, or SS. Six of these samples (and one additional one) were then amplified by PCR for 20 cycles starting with 1 μg of DNA each. An aliquot of the amplified DNA sample (one-fourteenth of the original 1-μg sample) was hybridized to the RS06 probe and digested with Dde I and then Hinf I. A portion (one-tenth) of this oligomer restriction reaction was analyzed on a polyacrylamide gel to resolve the cleavage products, and the results obtained after 6 hours of autoradiography are shown Fig. 5. The high sensitivity achieved with the PCR and OR method is demonstrated by the strength of the autoradiographic signal derived from only 1/140 of the original 1-μg sample (7 ng). Each sample determined to be AA by RFLP analysis showed a strong 8-nt

fragment while those typed as SS showed a strong 3-nt fragment. Analysis of the known AS samples revealed both cleavage products.

In the analysis of the AA samples, a faint 3-nt could be detected in addition to the primary 8-nt signal. The reasons for this band remain unclear, although incomplete Dde I cleavage or the occasional failure of the 8-nt fragment to disassociate from the target DNA may contribute to the nonspecific 3-nt product generated by Hinf I digestion. In the analysis of the SS samples, a very faint 8-nt band was also observed in addition to the expected 3-nt signal. We have determined that the background 8-nt product detected in SS samples can be attributed to the δ-globin gene, which is highly homologous to β-globin. The nucleotide sequence of the two β-globin primers used for amplification is shown in Fig. 1. The downward pointing arrows indicate the differences between the β- and δ-globin genes. We hypothesized that the faint 8-nt signal observed in the SS samples was due to some amplification of the δ-globin gene by these primers and the subsequent cross-hybridization of the amplified δ sequences with the RS06 probe used in the OR procedure. δ-Globin has the same Dde I site as normal β-globin, and the duplex formed between an amplified δ gene segment and the RS06 probe would be expected to yield an 8-nt fragment on Dde I digestion even

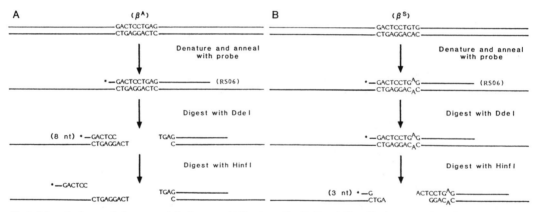

Fig. 3. Schematic diagram of oligomer restriction by sequential digestion to identify β^A- and β^S-specific cleavage products. The DNA sequences shown are the regions of the β-globin genomic DNA and the RS06 hybridization probe containing the invariant Hinf I site (GANTC, where N represents any nucleotide) and the polymorphic Dde I site (CTNAG). The remaining DNA sequences are represented as solid horizontal lines. The asterisk indicates the position of the radioactive ^{32}P label attached to the 5'-end of the RS06 probe with polynucleotide kinase. (A) Outline of the procedure and expected results when RS06 anneals to the normal β-globin gene (β^A). After denaturation of the genomic DNA and hybridization of the labeled RS06 probe to the complementary target sequence in the β^A gene, digestion of the probe-target hybrid with Dde I causes the release of a labeled (8-nt) cleavage product. Because of the relatively stringent conditions during Dde I digestion, the 8-nt cleavage product dissociates from the genomic DNA and the subsequent digestion with Hinf I has no effect. (B) Outline of Dde I and Hinf I digestion after hybridization of the RS06 probe to the sickle cell allele (β^S). As a consequence of the β^S mutation, the probe-target hybrid contains an A-A mismatch within the Dde I site and is not cleaved by the Dde I endonuclease. The Hinf I site, however, remains intact and digestion with that enzyme generates a labeled 3-nt product. Thus, the presence of the β^A allele is revealed by the release of a labeled 8-nt fragment, while the presence of β^S is indicated by a labeled 3-nt fragment.

though there are sequence differences (four mismatch out of 40 bases) between RS06 and δ-globin. It is likely that δ-globin sequences may be amplified to some extent and detected weakly with the RS06 probe in all DNA samples, but that its contribution to the total signal is very small and detectable only when the sample is SS and no 8-nt fragment from the β-globin gene is expected. We tested this hypothesis by treating an SS DNA sample before amplification with the enzyme Mbo I. Since there is a recognition site for this enzyme in the target DNA of the δ- but not the β-globin gene, cleavage of the δ gene between the regions that hybridize to the PCR primers should prevent its subsequent amplification (but not of β-globin). Our results showed that an SS DNA sample, first digested with Mbo I, gave only the 3-nt product but not the 8-nt product, this is consistent with the hypothesis of δ-globin amplification.

Effect of PCR cycle number on detection threshold. The strength of the autoradiograph signal detected by OR as a function of PCR cycle number and autoradiographic exposure was examined. The signal intensity after 20 cycles is at least 20 times as strong as that for 15 cycles and the determination of the β-globin genotype can be made with an autoradiographic exposure for only 2 hours (Fig. 6). The observed increase of ≥20-fold is consistent with our estimates of 85 percent efficiency per cycle, calculated from the data in Fig. 2B ($1.85^5 = 21.7$). Coupled with the time that it takes to actually carry out the PCR and OR procedures, a 20-cycle PCR allows a diagnosis to be made in less than 10 hours with a DNA sample of 1 μg.

Since all of the previous PCR experiments were done with 1 μg of genomic DNA, we explored the effect of using significantly smaller amounts of DNA as template for PCR amplification. The results obtained with 20 cycles of PCR amplification on 500, 100, 20, and 4 ng of DNA from an AS individual are shown in Fig. 7. After analysis of 1/40 of each sample by the OR procedure and a 20-hour autoradiographic exposure, the β-globin genotype could be easily determined on DNA samples of 20 ng or about 100 times less than is needed for a typical Southern transfer and hybridization experiment. In this experiment, only a small fraction (1/40) of the starting material was placed on the gel; therefore it should be possible to analyze samples of less than 20 ng of genomic DNA (20 ng is equivalent to approximately 6000 haploid genomes) if a larger proportion of the material was utilized in the OR and gel electrophoresis steps.

Diagnostic applications of the PCR-OR system. When currently available methods are used, the completion of a prenatal diagnosis for sickle cell anemia takes a period of several days after the DNA is isolated. With 1 μg of genomic DNA, the β-globin genotype can be determined by the PCR-OR method in less than 10 hours; 20 cycles of amplification requires about 2 hours (each full cycle takes 6 to 7 minutes in our protocol), the oligomer restriction procedure involving liquid hybridization and enzyme digestions require an additional 2 hours, and the electrophoresis takes about an hour. Autoradiographic exposure for 4 hours is sufficient to generate a strong signal.

Because this method includes a liquid hybridization protocol and involves the serial addition of reagents to a single tube, it is simpler to perform than the standard Southern transfer and hybridization procedure. Prior to electrophoresis, all of the reactions can be done in two small microcentrifuge tubes and could readily be automated.

The sensitivity, as well as the speed and simplicity, of this procedure is also important for clinical applications. Twenty nanograms of starting material can provide an easily detectable result in an overnight autoradiographic exposure. This sensitivity makes the PCR-OR method particularly valuable in cases

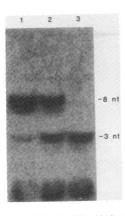

Fig. 4. Demonstration of OR sequential digestion with cloned β-globin genes. The sequential digestion strategy was demonstrated by annealing the RS06 probe to the β-globin plasmids pBR328::β^A and pBR328::β^S (13). The methods were similar to those described (13). Cloned β-globin DNA (45 ng; 0.01 pmol) was placed in a microcentrifuge tube, adjusted to 30 μl with TE buffer (10 mM tris, 0.1 mM EDTA, pH 8.0), overlaid with 0.1 ml of mineral oil. The DNA was denatured by heating for 5 to 10 minutes at 95°C. Ten microliters of 0.6M NaCl containing 0.02 pmol of phosphorylated (with [γ-^{32}P]ATP) RS06 probe oligomer (~5 μCi/pmol) was added and annealed for 60 minutes at 56°C. Unlabeled RS10 blocking oligomer (4 μl; 200 pmol/ml) (Fig. 1) (13) was then added, and the hybridization was continued for 5 to 10 minutes. Next, 5 μl of 100 mM MgCl₂ and 1 μl of Dde I (Biolabs, 10 units) was added and incubated for 20 minutes at 56°C; 1 μl of Hinf I (Biolabs, 10 units) was added and digestion was continued for 20 minutes at the same temperature. The reaction was terminated by the addition of 4 μl of 100 mM EDTA and 6 μl of tracking dye to a final volume of 61 μl; a portion (8 μl) (6 ng, 0.0013 pmol) was applied to a 0.75-mm thick, 30 percent polyacrylamide minigel (19 acrylamide:1 bis) in a Hoefer SE200 apparatus and subjected to electrophoresis (300 V) for 1 hour until the bromphenol blue dye front reached 3 cm. The top 1.5 cm of the gel, containing intact RS06, was cut off and discarded. The remaining gel was autoradiographed with a single intensification screen for 18 hours at −70°C. (Lane 1) six nanograms of pBR328::β^A; (lane 2) 3 ng of pBR28::β^A plus 3 ng of pBR328::β^S; and (lane 3) 6 ng of pBR328::β^S.

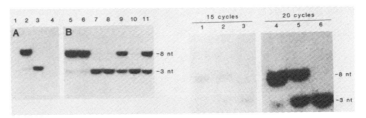

Fig. 5 (left). Determination of the β-globin genotype in human genomic DNA with PCR-OR. Samples (1 μg) of human genomic DNA were amplified for 20 cycles (as described in Fig. 2A). The amplified DNA's (71 ng) were hybridized to the RS06 probe and serially digested with Dde I and Hinf I (as described in Fig. 4). Each sample (6 μl) was analyzed by 30 percent polyacrylamide gel electrophoresis and autoradiographed for 6 hours at −70°C with one intensification screen. Each lane contains 7 ng of genomic DNA. (Lane 1) Unamplified Molt4 DNA (negative control); (lane 2) amplified Molt4 (β^Aβ^A); (lane 3) SC-1 (β^Sβ^S); (lane 4) GM2064 (ΔΔ); (lanes 5 to 11) clinical samples CH1 (β^Aβ^A), CH2 (β^Aβ^A), CH3 (β^Sβ^S), CH4 (β^Sβ^S), CH7 (β^Aβ^S), CH8 (β^Sβ^S), and CH12 (β^Aβ^S), respectively.   Fig. 6 (right). Effect of cycle number on signal strength. Genomic DNA (1 μg) from the clinical samples CH2 (β^Aβ^A), CH12 (β^Aβ^S), and CH5 (β^Sβ^S) were amplfied for 15 and 20 cycles and equivalent amounts of genomic DNA (80 ng) were analyzed by oligomer restriction. (Lanes 1 to 3) DNA (20 ng) from CH2, CH12, and CH5, respectively, amplified for 15 cycles; (lanes 4 to 6) DNA (20 ng) from CH2, CH12, and CH5, respectively, amplified for 20 cycles. Autoradiographic exposure was for 2.5 hours at −70°C with one intensification screen.

where poor DNA yields are obtained from prenatal samples. In addition, DNA samples of poor quality (very low average molecular weight) can give excellent results in the PCR-OR protocol.

The PCR method is likely to be generally applicable for specific gene amplification since a fragment encoding a portion of the HLA-DQα locus has recently been amplified with this procedure (15). We have carried out PCR amplification on a 110-bp β-globin sequence with an overall efficiency per cycle of about 85 percent. We have also amplified longer β-globin fragments (up to 267 bp), but the yield was lower under our standard conditions. Efficient amplification of a 267-bp fragment required some variation in the PCR procedure. In principle, increasing the number of PCR cycles should yield even greater amplification than that reported here (~220,000-fold after 20 cycles).

Our method for the diagnosis of sickle cell anemia involves the coupling of the PCR procedure with that of oligomer restriction. It was designed to distinguish between two alleles that differ by a polymorphic restriction site. The PCR-OR method is applicable as well to the diagnosis of other diseases where the lesion directly affects a restriction enzyme site or where the polymorphic site is in strong linkage disequilibrium with the disease causing locus. If the polymorphism is in linkage equilibrium with the disease, PCR-OR requires family studies to follow the inheritance of the disease locus.

In the case of the β-globin locus, the presence of the invariant Hinf I restriction site adjacent to the polymorphic Dde I site allows a sequential digestion procedure to identify both the β^A and β^S

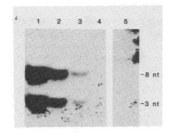

Fig. 7. Detection threshold for PCR-OR. Fivefold serial dilutions of genomic DNA (500, 100, 20, and 4 ng) from the clinical sample CH12 ($\beta^A\beta^S$) were amplifed by 20 cycles of PCR and one-tenth each reaction (50, 10, 2, and 0.4 ng) was analyzed by OR. The gel continued (lane 1) genomic DNA (12.5 ng); (lane 2) 2.5 ng; (lane 3) 0.5 ng; (lane 4) 0.1 ng; (lane 5) 12.5 ng genomic DNA from the globin deletion cell line GM2064. Autoradiographic exposure was for 20 hours at −70°C with an intensification screen.

alleles. In principle, this approach does not require that the two sites be immediately adjacent but only that the cleavage product generated by digestion at the polymorphic site dissociate from the target to prevent cutting at the invariant site. Since the restriction enzyme digestion conditions used here are fairly stringent for hybridization, we estimate that the polymorphic and invariant sites could perhaps be separated by as much as 20 bp.

The application of the PCR method to prenatal diagnosis does not necessarily depend on a polymorphic restriction site or on the use of radioactive probes. In fact, the significant amplification of target sequences achieved by the PCR method allows the use of nonisotopically labeled probes (16). Amplified target sequences could also be analyzed by a

number of other procedures including those involving the hybridization of small labeled oligomers which will form stable duplexes only if perfectly matched (6, 7, 17, 18) and the recently reported method based on the electrophoretic shifts of duplexes with base pair mismatches (19). The ability of the PCR procedure to amplify a target DNA segment in genomic DNA raises the possibility that its use may extend beyond that of prenatal diagnosis to other areas of molecular biology.

References and Notes

1. Y. W. Kan and A. M. Dozy, *Proc. Natl. Acad. U.S.A.* 75, 5631 (1978).
2. K. E. Davies and D. P. Ellis, *Biochem. J.* 226, 1 (1985).
3. Y. W. Kan, M. S. Golbus, A. M. Dozy, *N. Eng. J. Med.* 295, 1165 (1976).
4. A. M. Dozy *et al.*, *J. Am. Med. Assoc.* 241, 1610 (1979).
5. E. M. Rubin and Y. W. Kan, *Lancet* 1985-I, 75 (1985).
6. M. Pirastu *et al.*, *N. Engl. J. Med.* 309, 284 (1983).
7. S. H. Orkin, A. F. Markham, H. H. Kazazian, Jr., *J. Clin. Invest.* 71, 775 (1983).
8. R. F. Geever *et al.*, *Proc. Natl. Acad. Sci. U.S.A.* 78, 5081 (1981).
9. J. C. Chang and Y. W. Kan, *N. Eng. J. Med.* 307, 30 (1982).
10. S. H. Orkin *et al.*, *ibid.*, p. 32.
11. J. T. Wilson *et al.*, *Proc. Natl. Acad. Sci. U.S.A.* 79, 3628 (1982).
12. F. Faloona and K. Mullis, in preparation.
13. R. Saiki, N. Arnheim, H. Erlich, *Biotechnology* 3, 1008 (1985).
14. D. Tuan *et al.*, *Proc. Natl. Acad. Sci. U.S.A.* 80, 6937 (1983).
15. S. Scharf, unpublished data.
16. R. Saiki *et al.*, unpublished data.
17. B. J. Conner *et al.*, *Proc. Natl. Acad. Sci. U.S.A.* 80, 278 (1983).
18. V. J. Kidd, R. B. Wallace, K. Itakura, S. L. C. Woo, *Nature (London)* 304, 230 (1983).
19. R. M. Myers, N. Lumelsky, L. S. Lerman, T. Maniatis, *ibid.* 313, 495 (1985).
20. Calculation of the amounts of pBR328:β^A needed as standards to estimate PCR efficiency was done in the following way. If we assume that a human haploid genome size is 3×10^9 bp, 36 ng of DNA is equivalent to 1.8×10^{-8} pmol. The extent of amplification after 20 cycles at, for example, 85 percent efficiency is obtained with $(1.8 \times 10^{-8}) (1.85^{20})$ or 4.0×10^{-3} pmol of plasmid DNA (Fig. 2B, lane 5).

20 September 1985; accepted 15 November 1985

Evolution of Sickle Variant Gene

E. Solomon and W.F. Bodmer

SIR,—The 13 kilobase (kb) Hpa 1 restriction enzyme site polymorphism, close to the sickle variant of β-globin,[2,3] raises interesting evolutionary questions, as discussed by Dr Kurnit (Jan. 13, p. 104). He calculates that about 2785 generations would be needed to generate the 13% of sickle genes not associated with the 13 kb Hpa 1 polymorphic variant if the sickle mutation arose in a chromosome carrying this variant. This is inconsistent with the assumption that the sickle mutation occurred at the time agriculture and malaria were introduced into Africa, only 60–80 generations ago.[4] Kurnit therefore argues for a multicentric origin of the sickle gene. We should like to suggest an alternative interpretation of the data, assuming a much older origin of the sickle gene.

Though the increase in frequency of the sickle-cell gene may be traced to the introduction of malaria to Africa only 60–80 generations ago, this by no means implies that the mutation arose then. Indeed, it is more likely that the sickle mutation arose much earlier, in a founder with the 13 kb site, and that it achieved and was maintained at a low frequency due to random drift, until it was favoured by selection at the time of the introduction of malaria. At this time it would have already existed in coupling with both the 13 kb and the 7·6 kb sites, and both of these combinations would then have expanded due to selective pressure and so relatively quickly given rise to the situation today.

Using the standard population genetic random mating model for two linked loci and assuming, for simplicity, no change in the frequencies of either the sickle gene or the Hpa 1 site before the onset of malaria, we can calculate the number of generations (n) taken to generate the 13% recombinant chromosomes from the formula $(1-r)^n=(0.87-P)/(1-P)$, where P is the gene frequency of the Hpa 1 site, and r the recombination fraction between the sickle site and the Hpa 1 site. When P=0, implying that the Hpa 1 site is very rare and when r=1/20 000, n=2785 generations, as calculated by Kurnit. When P=0·03 (the present observed frequency of the 13 kb site in normal individuals, as given by Kan and Dozy[2]) n=2878. This number increases as the frequency of the Hpa 1 site increases. Rather than using the data as evidence of a multicentric origin we can use them to estimate the age for the sickle gene. This ranges from 2785 generations (or about 69 625 years) if the Hpa 1 site was very rare, to values as high as 6022 generations (150 550 years) if the 13 kb site was at one time more frequent than it is now. This analysis suggests that the origin of the sickle mutation might well predate the origin of the major human racial groups, although its striking increase in frequency was much more recent.

As mentioned by Kurnit, polymorphisms for restriction enzyme sites are likely to provide powerful new tools for evolutionary and genetic studies. These polymorphisms, which identify variation at the level of the gene itself, may provide a quantum jump in the range of available genetic markers for study. Kan and Dozy[2] have emphasised their application to the general problem of prenatal diagnosis of hereditary diseases, even where the specific biochemical defect is not known. If the level of polymorphism for these sites is as high as the initial studies suggest, and given the range of available restriction enzymes, one can envisage finding enough markers to cover systematically the whole human genome. Thus, only 200–300 suitably selected probes might be needed to provide a genetic marker for, say, every 10% recombination. Such a set of genetic markers could revolutionise our ability to study the genetic determination of complex attributes and to follow the inheritance of traits that are so far difficult or impossible to study at the cellular level. Association within families between a defined genetic marker and a trait whose inheritance is not clearly defined, provides the best evidence, through genetic linkage, for genetic determination. This powerful approach has so far been limited by the range of genetic markers available, but restriction-enzyme polymorphisms may soon solve this problem.

1. Schneider, R. G., Schmidt, R. M. in Abnormal Hemoglobins and Thalassemia edited by R. M. Schmidt; p. 33. New York, 1975.
2. Kan, Y. W., Dozy, A. M. Lancet, 1978, ii, 910.
3. Kan, Y. W., Dozy, A. M. Proc. natn. Acad. Sci. U.S.A., 1978, 75, 5631.
4. Weisenfeld, S. L. Science, 1967, 157, 1134.

Genetics Laboratory,
Department of Biochemistry,
University of Oxford, Oxford OX1 3QU

E. SOLOMON
W. F. BODMER

Supplementary Readings

Botstein, D., White, R.L., Skolnick, M. & Davis, R.W., Construction of genetic linkage map in man using restriction fragment length polymorphisms. *Am. J. Hum.Genet.* 32:314–331 (1980)

Brzustowicz, L.M., Lehner, T., Castilla, L.H., Penchaszadeh, G.K., Wilhelmsen, K.C., Daniels, R., Davies, K.E., Leppert, M., Ziter, F., Wood, D., Dubowitz, V., Zerres, K., Hausmanowa-Petrusewicz, I., Ott, J., Munsat, T.L. & Gilliam, T.C., Genetic mapping of chronic childhood-onset spinal muscular atrophy to chromosome 5q11.2-13.3. *Nature* 344:540–541 (1990)

Doolittle, R.F., Hunkapiller, M.W., Hood, L.E., Devare, S.G., Robbins, K.C., Aaronson, S.A. & Antoniades, H.N., Simian sarcoma *onc* gene, v-*sis*, is derived from the gene (or genes) encoding a platelet-derived growth factor. *Science* 221:275–277 (1983)

Friend, S.H., Bernards, R., Rogelj, S., Weinberg, R.A., Rapaport, J.M., Albert, D.M. & Dryja, T.P., A human DNA segment with properties of the gene that predisposes to retinoblastoma and osteosarcoma. *Nature* 323:643–646 (1986)

Gessler, M., Poustka, A., Cavenee, W., Neve, R.L., Orkin, S.H. & Bruns, G.A.P., Homozygous deletion in Wilms tumours of a zinc-finger gene identified by chromosome jumping. *Nature* 343:774–778 (1990)

Goate, A., Chartier-Harlin, M.-C., Mullan, M., Brown, J., Crawford, F., Fidani, L., Giuffra, L., Haynes, A, Irving, N., James, L., Mant, R., Newton, P., Rooke, K., Roques, P., Talbot, C., Pericak-Vance, M., Roses, A., Williamson, R., Rossor, M., Owen, M., & Hardy, J., Segregation of a missense mutation in the amyloid precursor protein gene with familial Alzheimer's disease. *Nature* 349:704–706 (1991)

Kerem, B., Rommens, J.M., Buchanan, J.A., Markeiwicz, D., Cox, T.K., Chakravarti, A., Buchwald, M. & Tsui, L.-C., Identification of the cystic fibrosis gene: Genetic analysis. *Science* 245:1073–1080 (1989)

Nigro, J.M., Baker, S.J., Preisinger, A.C., Jessup, J.M., Hostetter, R., Cleary, K., Bigner, S.H., Davidson, N., Baylin, S., Devilee, P., Glover, T., Collins, F.S., Weston, A., Modali, R., Harris, C.C. & Vogelstein, B., Mutations in the *p53* gene occur in diverse tumour types. *Nature* 342:705–708 (1989)

Riordan, J.R., Rommens, J.M., Kerem, B., Alon, N., Rozmahel, R., Grzelczak, Z., Zielenski, J., Lok, S., Plavsic, N., Chou, J.-L., Drumm, M.L., Iannuzzi, M.C., Collins, F.S. & Tsui, L.-C., Identification of the cystic fibrosis gene: Cloning and characterization of complementary DNA. *Science* 245:1066–1073 (1989)

Rommens, J.M., Iannuzzi, M.C., Kerem, B., Drumm, M.L., Melmer, G., Dean, M., Rozmahel, R., Cole, J.L., Kennedy, D., Hidaka, N., Zsiga, M., Buchwald, M., Riordan, J.R., Tsue, L.-C. & Collins, F.S., Identification of the cystic fibrosis gene: Chromosome walking and jumping. *Science* 245:1059–1065 (1989)

Waterfield, M.D., Scrace, G.T., Whittle, N., Stroobant, P., Johnsson, A., Wasteson, Å., Wastermark, B., Heldin, C.-H., Huang, J.S. and Deuel, T.F., Platelet-derived growth factor is structurally related to the putative transforming protein p28^{sis} of simian sarcoma virus. *Nature* 304:35–39 (1983)

Williams, J.G.K., Kubelik, A.R., Livak, K.J., Rafalski, J.A. & Tingey, S.V., DNA polymorphisms amplified by arbitrary primers are useful as genetic markers. *Nucleic Acids Res.* 18:6531–6535 (1990)

Vaccines

When gene cloning became "commercial" in the mid-1970s, it was obvious to everyone involved that the ability to produce quantities of purified antigens of viral, bacterial, or parasite origin by cloning the appropriate genes in hosts such as *E. coli* or *S. cerevisiae* could lead to an entirely new generation of vaccines (and diagnostic products). Why was this so? Conventional vaccines effective against infectious diseases were prepared by attenuation (usually by serial passage) to provide still-living avirulent forms of virus or bacterium, or inactivation by chemical and physical means leading to dead pathogens which were still capable of inducing a protective immune response. Such approaches have, of course, been highly successful and billions of people have been protected against polio, smallpox, tetanus, etc. Similar types of vaccines in the animal world have led to enormous economic gains in combatting diseases such as foot-and-mouth disease. The need for new approaches to vaccines and the potential for applications of recombinant DNA technology to provide them, came from the following arguments:

1) The production of vaccines by classical methods requires collection and work-up of large amounts of (often) highly infectious biological material. Hepatitis B vaccine production from blood is a typical example. Workers risk exposure to pathogenic materials during preparation.

2) The vaccine inactivation or attenuation process may not be 100% effective, or may even be reversed on subsequent use of the vaccine. Examples of this, especially in the case of polio vaccine, are well known.

3) The use of the vaccine, even though inactivated or attenuated, may cause serious side effects in a small percentage of recipients. This is known in the case of flu, pertussis, and other diseases; the side effects caused by vaccination against influenza attained political significance in a recent U.S. presidential election.

4) Vaccine preparation is often very expensive due to precautions required in protecting workers, and because vaccine use may require substantial liability insurance coverage to the producer as protection against litigation in the event of serious side-effects.

5) Conventional vaccine preparations may be unstable and often require storage at low temperature, which can limit their use in certain countries.

6) Effective vaccines are simply not available for many diseases, for example, malaria, other parasitic diseases, and certain virus infections.

7) Classical approaches to vaccine production do not provide any fundamental information on the pathogen and the infectious process.

Can the application of biotechnology solve all of the problems? In the first wave of enthusiasm during the development of small biotechnology companies, it was thought so! In 1991 we can look back on substantial successes, such as the case of the mammalian cell and yeast-produced hepatitis B vaccine, which may allow the World Health Organisation to mount a campaign for eradication of hepatitis similar to what was achieved with smallpox.

The development and introduction of an orally-administered rabies vaccine for animals in the wild is the first use of the vaccinia (smallpox) virus as a live vaccine carrier. The rabies virus protein was cloned into a vaccinia vector and is being used with the goal of eradicating the virus from its indigenous natural carrier, the common fox. If this reservoir can be depleted or eliminated, the chances are very good of removing the threat of rabies epidemics among the animal population. In addition, the use of the vaccinia-derived rabies vaccine in animals will be a good testing ground for a recombinant oral rabies vaccine for use in humans, and a model for other applications of such carriers which will permit more detailed analysis of humoral and cell-mediated immunity and the nature of the protective immune response induced by vaccines.

Nevertheless there have been failures, in the sense that despite much optimism and significant financial and research investment, we still do not have vaccines for malaria or human immunodeficiency syndrome (AIDS). In both of these cases, the lack of success is due to the complex life cycles of the pathogens rather than lack of effort by the scientific community. With foot and mouth disease (FMDV), on the order of 20 million dollars over a period of 7-10 years was invested by companies such as Biogen, Genentech, Molecular Genetics and Wellcome in efforts to develop a vaccine by cloning the viral surface antigen into a bacterial expression system to produce large quantities of pure coat protein for use as subunit vaccine. Although this was a good investment in fundamental research on the virus, it did not lead to an effective vaccine for foot-and-mouth disease. We know a great deal about the biology of the virus and how it infects cells, which will surely help in future developments, but the original goal has not been realised. None-theless, there are good reasons for optimism; improved and safer vaccines against *Bordetella* infections (pertussis) and other diseases should be available in the near future. Pertussis has been a particularly difficult problem since the determinants of virulence are complex, but several candidate genetically-engineered vaccines are currently under test in Sweden and Italy. We may cavil at the lack of immediate practical applications of all this effort, but in terms of an understanding of mechanisms of virulence of a number of pathogens, the advances have been spectacular. We have increased our knowledge of the pathology of infectious diseases enormously in the past few years, so that there is cause for continued optimism. Success in the future will come from using this accrued information to recognise the key immunogenic sequences and to obtain them in their native conformation. Other approaches to virus therapy are under intensive study: for example, the use of decoy receptor proteins to prevent virus attachment to target cells.

Although vaccine development has not been as successful as anticipated, it should be noted that the cloning and identification of a bacterial, viral, or parasite structural or virulence determinant provides the possibility of a sensitive detection method, by the use of either specific anti-pathogen antibodies or DNA probes. Developments in diagnostic techniques have been one of the most successful and quickly communicated applications of biotechnology. (See the Diagnostics section.) Hundreds of rapid, sensitive and reliable diagnostic methods for a variety of diseases are available and on the market.

Synthesis of Hepatitis B Surface and Core Antigens in *E. Coli*

J.C. Edman, R.A. Hallewell, P. Valenzuela, H.M. Goodman and W.J. Rutter

Hepatitis B virus (HBV) is the cause of a debilitating and potentially fatal disease, and possibly also of primary hepatocellular carcinoma[1]. A vaccine against hepatitis B would therefore be of considerable biomedical significance. One such vaccine, produced from the 22-nm particle form of hepatitis B surface antigen (HBsAg), has recently been found effective in clinical tests[2], but being derived from the sera of chronic carriers, it is expensive and its supply limited. The application of recombinant DNA methods can, in principle, provide a limitless source of vaccine. HBV DNA has been cloned in bacteria and the genes coding for HBsAg and hepatitis B core antigen (HBcAg) identified[3–5]. Small amounts of both antigens have been synthesized in *Escherichia coli*[3,6,7] and to enhance this synthesis we have constructed plasmids capable of expressing the genes for the antigens under the control of the efficient tryptophan (*trp*) operon regulatory region[8]. We describe here the construction of such recombinant plasmids which direct the synthesis of high levels of HBcAg and a β-lactamase: HBsAg fusion polypeptide.

The *trp* operon consists of a regulatory region and five coordinately expressed structural genes, *trp*E to *trp*A, respectively (see Fig. 1*a* and ref. 9). The leader peptide of the *trp* operon (*trp*L)[10] has a *Taq*I site between the Shine–Dalgarno sequence (SD)[11] and the initiator AUG codon (Fig. 1*b*). The location of this site facilitates the construction of a vector suitable for the creation of hybrid ribosome-binding sites, like those described previously for vectors based on the *lac* operon[12,13], containing the *trp*L SD sequence and the initiation codon of the foreign gene. Although the synthesis of the 14-amino acid *trp* leader peptide has not been demonstrated, an *in vivo* recombinant between the *trp*L SD sequence and the initiation codon of *trp*C has been shown to direct efficient *trp*C synthesis[14]. In addition, some *trp* operon deletion mutants between *trp*L and *trp*E have been shown to produce *trp*L:*trp*E fusion polypeptides, indicating that *trp*L does contain a functional ribosome-binding site[15].

Plasmid ptrpL1 was constructed from plasmid ptrpE2-1 (see ref. 16 and Fig. 1*c*) by first isolating the 32-base pair *Hpa*I–*Taq*I fragment of ptrpE2-1 by polyacrylamide gel electrophoresis (PAGE). This fragment contains part of the *trp* promoter-operator and the SD sequence of *trp*L[10]. The *Hpa*I–*Taq*I fragment was recloned into *Hpa*I–*Cla*I-digested ptrpE2-1. Because the *Taq*I sequence (T/CGA) of the *Hpa*I–*Taq*I fragment is preceded by an A residue on its 5' side, this construction resulted in the formation of a new *Cla*I site (AT/CGAT). The sequence of this junction is shown in Fig. 1*d*. The resulting plasmid, ptrpL1, has a unique *Cla*I site, three base pairs 3' to the SD sequence of *trp*L, which is useful because several different restriction fragments with 5' CG single-stranded termini can be cloned into it (for example, *Hpa*II, *Taq*I, *Acy*I).

Figure 2 outlines the procedure used to construct a series of plasmids that direct the synthesis of HBcAg sequences. A 1,005-base pair *Hha*I fragment of HBV DNA[5] containing all the HBcAg gene was purified by PAGE and treated with *Hpa*II methylase to prevent cleavage of the fragment during linker digestion with *Hpa*II. Our HBV DNA clone contains two potential initiation codons for HBcAg[5]. The start site for HBcAg is not known, but most of the evidence is consistent with translation initiation at the second codon[3]. This work utilizes the

first initiation codon which should result in the synthesis of a polypeptide containing the entire amino acid sequence for HBcAg and probably an additional 29 amino acids at the N terminus. The initiation codon for HBcAg lies 15 base pairs from one end of this fragment. This fragment was treated with T4 DNA polymerase in the absence of deoxynucleoside triphosphates (to allow the 3' exonuclease to function), in conditions where 0–10 nucleotides are removed from each 3' end, then with S₁ nuclease. *Bam*HI linkers (CCGGATCCGG, which also contain the *Hpa*II site, CCGG) were added and the fragments digested with *Hpa*II to generate the 5'CG cohesive end. After purification, these fragments were ligated with *Cla*I-digested and phosphatase-treated ptrpL1. The ligation mixture was used to transform HB101 and recombinants were screened for HBcAg by radioimmune assay (RIA). Seventeen out of 40 recombinants tested were HBcAg positive. Restriction enzyme analysis of the plasmids from HBcAg-positive clones showed that all contained HBcAg sequences and were in the proper orientation for *trp*-dependent expression of HBcAg. Plasmids with the HBcAg gene in the opposite orientation were HBcAg negative.

DNA sequence analysis of four positive clones confirmed that the initiation codon of HBcAg was in close proximity (13–16 base pairs) to the SD sequence of the leader peptide. An HBcAg standard was not available to measure the absolute levels of HBcAg produced in *E. coli*, but the relative levels varied fourfold. The plasmid responsible for the highest level of expression, pCA246, whose sequence is presented at the bottom of Fig. 2, was chosen for further characterization. The amount of HBcAg produced in a 20-min labelling period is 3.2% of the newly synthesized protein uncorrected for cysteine content (~150,000 molecules per cell) as estimated by densitometric scanning of the autoradiogram, see Fig. 3*A*, lane *d*. The level of HBcAg can be increased to 10% of newly synthesized protein by optimizing the concentration of the inducer, 3-β indolyl-acrylic acid (IA, unpublished results).

The efficacy of the *trp* promoter in enhancing expression of a linked coding sequence has been demonstrated by the significant production of a fused protein involving the *trp*D gene and human growth hormone coding sequences[17], and by the overproduction of a contiguous β-lactamase gene (Fig. 3*A*, lane *b*), which constitutes 12% of labelled protein in cells containing ptrpL1, as a result of partial derepression of *trp* promoters in cells grown in media lacking tryptophan. On addition of IA all *trp* promoters become fully derepressed[8,9] and the level of β-lactamase increases to 50% of labelled protein (Fig. 3*A*, lane *e*).

When cells containing pCA246 were treated with IA[9], labelled with ³⁵S-cysteine and the products run on SDS gels, an increased synthesis of a 22,000-molecular weight polypeptide not present in the control strain was observed (Fig. 3*A*, lanes *a*, *d*). The DNA sequence of the HBcAg gene predicts a protein of this molecular weight[5]. In such cells there is also a strong polar effect on β-lactamase production; the decreased level of β-lactamase may be due to a transcription termination site in the HBV DNA or inefficient transcription/translation of HBcAg sequences.

Anti-HBcAg serum precipitates (Fig. 3*B*, lanes *a*, *b*) a 22,000-MW polypeptide, corresponding to the HBcAg protein predicted by the constructed coding sequence, and a polypeptide of MW ~19,000 that is not evident in the labelled proteins resolved on SDS gels (Fig. 3*A*, lanes *a*, *d*). This latter polypeptide probably results from initiation of translation at the second methionine of the 22,000-MW polypeptide or degradation. The pattern of background bands precipitated with anti-HBcAg is different from that in the normal serum control. Sedimentation studies of HBcAg produced in *E. coli* demonstrate a sedimentation coefficient >100S, suggesting that HBcAg is present as some tightly associated complex. As HBcAg functions as a DNA-binding protein[3], it may be immunoprecipitated as a complex with DNA or RNA. This

* To whom enquiries concerning the *trp* vectors should be addressed.

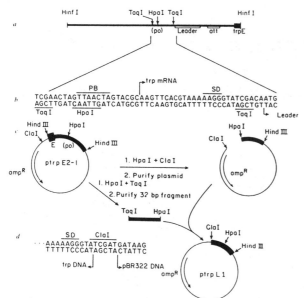

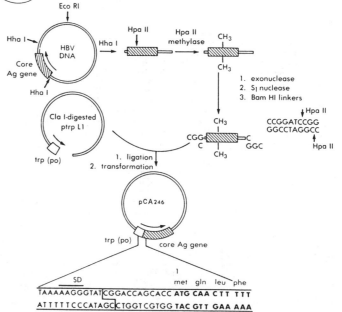

Fig. 1 Construction of expression vector ptrpL1. *a*, Partial restriction map of the 492-base pair *Hinf*I fragment of the *trp* operon of *E. coli*. The promoter–operator (po), leader region, attenuator (att) and *trp*E structural gene are shown in their approximate locations[10]. *b*, DNA sequence[24] of the promoter–operator and proximal portion of the leader region. Note that a *Taq*I site lies between the Shine–Dalgarno sequence (SD) and the initiation codon (ATG) of the leader region. The proposed Pribnow box (PB) and transcription initiation site are also shown. *c*, Construction of ptrpL1. Plasmid vector ptrpE2-1 (construction to be published elsewhere, see text) was digested with *Hpa*I and *Taq*I and the 32-base pair fragment of the *trp* operon purified by PAGE[25,26]. Separately, ptrpE2-1 was digested with *Cla*I and *Hpa*I and the plasmid purfied by agarose gel electrophoresis[26]. The plasmid vector (200 ng) was ligated to the 32-base pair fragment (5 ng) in 20 μl with T4 DNA ligase and the ligation mix used to transform HB101 (ref. 27). Ampicillin-resistant transformants were analysed for plasmids containing single *Cla*I and *Hind*III sites[28]. Several clones were analysed by restriction mapping and one by DNA sequence[29]. This plasmid had the expected sequence around the *Cla*I site and was designated ptrpL1. *d*, DNA sequence around the *Cla*I site of ptrpL1 showing the location of the *trp*L SD sequence. Enzymes were purchased from either New England Biolabs or Boehringer-Mannheim and reactions carried out according to manufacturer's instructions.

Fig. 2 Construction of plasmids expressing HBcAg. Cloned HBV DNA[5] was digested with *Hha*I and a 1,005-base pair fragment containing the HBcAg gene was isolated by PAGE[25,26]; 20 μg of this fragment were then treated with *Hpa*II methylase[30] (a gift from K. Agarwal). After phenol–chloroform extraction and ethanol precipitation the fragment was treated with 28 U of T4 DNA polymerase (prepared by J. Barry) in the absence of dNTPs (to allow the 3′ exonuclease activity to function) in 200 μl of 30mM Tris-acetate, pH 7.8, 67 mM KOAc, 10 mM MgOAc, 0.5 mM dithiothreitol and 100 μg ml⁻¹ bovine serum albumin for 30 s at 37 °C. The reaction was stopped with phenol, extracted with chloroform and precipitated with ethanol. The fragments were then treated with S₁ nuclease and *Bam*HI linkers (Collaborative Research) added using T4 DNA ligase[31]. The fragments were digested with *Hpa*II and purified from the digested linkers by PAGE and 150 ng ligated with 200 μg *Cla*I-digested and calf intestine alkaline phosphatase-treated ptrpL1 (10 U sigma type VII phosphatase incubated with 20 μg DNA at 68 °C for 30 min in 400 μl 10 mM Tris, pH 7.5, 0.5 mM EDTA) in 20 μl and the ligation mixture used to transform HB101 (ref. 32). Transformants were selected on L plates[33] containing 20 μg ml⁻¹ ampicillin and screened for recombinant plasmid DNA[34]. Forty such recombinants were then tested for HBcAg using a double antibody RIA with human anti-HBcAg IgG[35]; 17 were HBcAg positive. Analysis of DNA after digestion with restriction enzymes showed that all 17 contained HBcAg DNA sequences in a *trp*-dependent expression orientation. Four were chosen for DNA sequence analysis. The sequence of the plasmid producing the highest levels of HBcAg, pCA246, is shown. The sequence of the others was slightly longer having one or three additional nucleotides of HBV DNA.

would result in the precipitation of other proteins bound to nucleic acids and explain the character of the immunoprecipitates.

It has been suggested that the differential utilization of codons for particular amino acids between *E. coli* and certain eukaryotes[18] may account for the observation that some eukaryotic cDNAs are expressed at lower levels than expected in *E. coli*[17,19]. One mechanism proposed for this effect is that cognate tRNA molecules are in limiting concentrations for rarely used codons, thus preventing high-level expression of a gene containing several such codons. The HBcAg gene contains 25 arginine codons, 17 of which are rarely utilized in *E. coli*[5,18]. We have shown that HBcAg is produced in substantial quantities when placed under the auspices of ptrpL1. Thus at least in this instance, inclusion of many rarely utilized codons does not prevent high-level expression of a gene sequence in *E. coli*.

Plasmids similar to the HBcAg-expressing ones were constructed using the HBsAg gene. Cells transformed with these

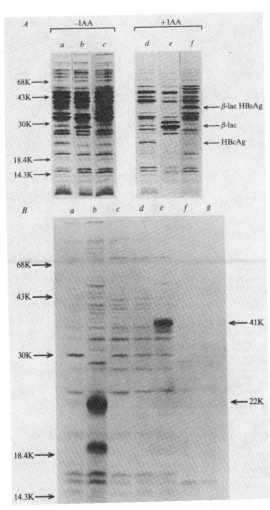

Fig. 3 A, autoradiography of SDS-polyacrylamide gel[36] of proteins produced in cells containing pCA246 and pSA4A. Overnight cultures were grown in M9 medium[33] containing 0.25% casamino acids, 0.5% glucose and 0.01% thiamine. The cultures were diluted 1:10 with fresh media, grown for 1 h at 30 °C, 15 μg ml⁻¹ of IA were added when induction of the *trp* operon was required, and cultures grown for another 2 h at 30 °C (ref. 8). The cultures (1 ml) were then labelled for 20 min at 30 °C with 10 μCi ml⁻¹ of ³⁵S-cysteine (NEN), labelled cells collected in 1.5-ml Eppendorf tubes and resuspended by boiling for 2 min in Laemmli sample buffer[36]. Samples were run on 12.5% Laemmli gels[36], the gels were dried and subjected to autoradiography. Lanes *a–c* are from uninduced cells containing pCA246 (HBcAg), ptrpL1 and pSA4A (β-lactamase: HBsAg fusion), respectively. Lanes *d–f* are samples induced with IA of cells containing pCA246, ptrpL1 and pSA4A, respectively. Molecular weight markers on the left: bovine serum albumin (68,000), ovalbumin (43,000), carbonic anhydrase (30,000), β-lactoglobulin (18,400) and lysozyme (14,300). Arrows on right of the bands denote the migration of the β-lactamase : HBsAg fusion (MW 41,000), β-lactamase (29,000) and HBcAg from pCA246 (22,000). B, autoradiogram of SDS-polyacrylamide gel of immunoprecipitates of HBcAg and β-lactamase : HBsAg fusion produced in *E. coli*. ³⁵S-cysteine-labelled samples were prepared as described above, except labelled cells were resuspended in phosphate-buffered saline containing 1 mM phenyl-methylsulphonylfluoride. Cells were sonicated and proteins immunoprecipitated using the SAC technique[17]. *a*, pCA246 + normal human IgG; *b*, pCA246 + human anti-HBcAg; *c*, pSA4A + normal guinea pig IgG; *d*, pSA4A + guinea pig anti-HBsAg and 2 μg unlabelled HBsAg; *e*, pSA4A + guinea pig anti-HBsAg; *f*, ptrpL1 + anti-HBcAg; and *g*, ptrpL1 + anti-HBsAg. One μg of each antibody (Merck) was added per tube. Markers on left as above. Numbers on right refer to the β-lactamase : HBsAg fusion (MW 41,000) and HBcAg (22,000) respectively.

plasmids were uniformly negative for HBsAg by RIA. Several clones were analysed by restriction mapping and two by DNA sequence. The distance separating the SD sequence and the initiation codon for HBsAg was 11 and 12 nucleotides. Neither of the clones accumulated polypeptides of the proper size (MW 25,398) on addition of IA, but the polar effect on β-lactamase was still present (data not shown). This result could have been due to rapid degradation or inefficient transcription/translation.

HBsAg contains a prominent hydrophobic region as well as an N-terminal region resembling a modified signal peptide. There are several studies of powerful peptide degradation systems in bacteria including an ATP-dependent system[20] probably operative at the cell membrane. We reasoned that elimination of part of the sequence coding for the N-terminal region and subsequent replacement with a nucleotide sequence coding for a stable protein fragment might enhance the stability of this molecule. Because β-lactamase is overproduced in cells containing ptrpL1, we elected to try to produce a β-lactamase: HBsAg fusion polypeptide. Figure 4 summarizes the construction of such plasmids. *Pst*I-digested ptrpL1 was 'tailed' with deoxyguanosine using terminal transferase in conditions where ~15 dG residues were added to each 3' end. A 744-base pair *Hinc*II fragment of HBV DNA containing the coding sequence for 204 amino acids (the amino terminal 22 are missing) of HBsAg was purified by PAGE, tailed with dC and hybridized with the tailed ptrpL1. The mixture was used to transform *E. coli* HB101.

The DNA sequence of this HBV DNA fragment predicts that when the HBsAg sequences are in the proper orientation and same phase as β-lactamase, a 43,000-MW polypeptide should be produced. This includes 183 amino acids of pre-β-lactamase (MW 20,100), 5–10 glycine residues from the G tail (MW 375–750), and 204 amino acids of HBsAg (22,600). If the pre-sequence is cleaved, the fusion protein would have a predicted MW of 41,000. Thirty-eight ampicillin-sensitive colonies were grown in the presence of IA, labelled, and the products run on SDS gels. Two clones producing a 41,000-MW polypeptide as a major product (Fig. 3A, lanes *c, f*) were chosen for DNA sequence. As expected, the HBsAg sequences were in the same phase as the amino-terminal β-lactamase sequences (Fig. 4). This fusion protein represents 8.5% of protein (uncorrected for cysteine content) synthesized in a 20-min period (~170,000 molecules per cell) as estimated by densitometric scanning of the autoradiogram.

The presumptive 41,000-MW fusion polypeptide is immuno-precipitated with anti-HBsAg IgG (Fig. 3B, lane *e*). Addition of cold HBsAg inhibits the precipitation completely (Fig. 3B, lane *d*). Thus, HBsAg determinants are present in the fusion polypeptide. The level of the 41,000-MW polypeptide varied widely in different experiments and in some cases was not detected by PAGE. However, in all cases a radioactive polypeptide of MW 41,000 was immunoprecipitated by anti-HBsAg IgG. The reason for this variability is not known, but may be due to either subtle variations in cellular growth conditions and/or shifting product stability.

We have presented here evidence that substantial quantities of hepatitis B surface and core antigen sequences can be synthesized in *E. coli*. These proteins may be useful in the development of an HBV vaccine. HBsAg is synthesized as a fusion polypeptide that contains an antigenic determinant found in native HBsAg; its effectiveness as an immunogen in generating protective antibodies is under investigation. The product will probably be less effective than the HBsAg particle itself because it has been demonstrated that dissociated HBsAg is much less immunogenic than the 22-nm HBsAg particle[21]. However, this disadvantage may not be decisive as HBsAg can be produced in bacteria in any desired quantity.

The availability of HBcAg synthesized in bacteria may answer some questions about the biology of HBV. For example, it has been suggested[22,23] that the cryptic 'e' antigen of HBV is a dissociated form of HBcAg protein, but the preparations used may have been contaminated by other antigens. As the HBcAg

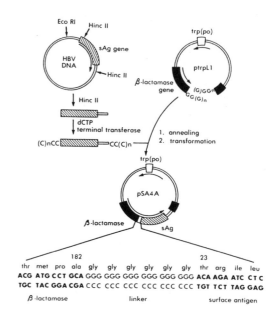

Fig. 4 Construction of plasmids expressing a β-lactamase:HBsAg fusion. Cloned HBV DNA was digested with *Hinc*II and a 744-base pair fragment containing part of the HBsAg gene was isolated by PAGE. This fragment was tailed with approximately 15 dC residues using terminal transferase[37]. Plasmid ptrpL1 was digested with *Pst*I and tailed with ~15 dG residues. The fragment (35 ng) was then hybridized with the tailed plasmid (200 ng)[38] and the mixture used to transform HB101. Transformants were selected on L plates containing 2.5 μg ml⁻¹ tetracycline[31] and tested for ampicillin resistance on plates containing 20 μg ml⁻¹ ampicillin. Samples were prepared from 38 ampicillin-sensitive transformants, induced with IA as described in Fig. 3 legend. Samples were run on 12.5% Laemmli gels[36], the gels were dried and subjected to autoradiography. Two clones, pSA4A and pSA7A, producing large amounts of a 41,000-MW polypeptide were found and these were chosen for further analysis. The DNA sequence of the fusion junction of pSA4A is shown at the bottom of the figure.

produced in bacteria has no other HBV gene products, this possibility can now be tested directly.

It is also possible through similar recombinant DNA methods to identify and express the remaining genes of HBV (putative e antigen, DNA polymerase) and evaluate their role in the cytopathology of HBV. These studies might elucidate the aetiology of the disease and suggest new strategies for its control.

Received 8 December 1980; accepted 20 March 1981.

1. Zuckerman, A. J. *J. Tox. envir. Hlth* **5**, 275–280 (1979).
2. Szmuness, W. *et al. New Engl. J. Med.* **303**, 833–841 (1980).
3. Pasek, M. *et al. Nature* **282**, 575–579 (1979).
4. Galibert, F. *et al. Nature* **281**, 646–650 (1979).
5. Valenzuela, P. *et al.* in *Animal Virus Genetics* (ed. Fields, B. N., Jaenisch, R. & Fox, C. F.) 57–71 (Academic, New York, 1980).
6. Burrel, C. J. *et al. Nature* **279**, 43–47 (1979).
7. Charnay, P. *et al. Nature* **286**, 893–895 (1980).
8. Hallewell, R. A. & Emtage, S. *Gene* **9**, 27–47 (1980).
9. Morse, D. E., Mosteller, R. D. & Yanofsky, C. *Cold Spring Harb. Symp. quant. Biol.* **34**, 725–740 (1969).
10. Lee, F., Bertrand, K., Bennett, G. & Yanofsky, C. *J. molec. Biol.* **121**, 193–227 (1978).
11. Shine, J. & Dalgarno, L. *Nature* **254**, 34–38 (1975).
12. Goeddel, D. *et al. Nature* **281**, 544–548 (1979).
13. Roberts, T. M. *et al. Proc. natn. Acad. Sci. U.S.A.* **76**, 5596–5600 (1979).
14. Christie, G. & Platt, T. *J. molec. Biol.* **143**, 335–341 (1980).
15. Miozzari, G. F. & Yanofsky, C. *J. Bact.* **133**, 1457–1466 (1978).
16. Tacon, W. *et al. Molec. gen. Genet.* **177**, 427–438 (1980).
17. Martial, J. A., Hallewell, R. A., Baxter, J. D. & Goodman, H. M. *Science* **205**, 602–607 (1979).
18. Grantham, R., Gautier, C., Gouy, M., Mercier, R. & Pavez, A. *Nucleic Acids Res.* **8**, 49–61 (1980).
19. Fraser, T. H. & Bruce, B. J. *Proc. natn. Acad. Sci. U.S.A.* **75**, 5936–5940 (1978).
20. Murakami, K., Voellmy, K. & Goldberg, A. L. *J. biol. Chem.* **254**, 8194–8200 (1979).
21. Cabral, G. A. *et al. J. gen. Virol.* **38**, 339–350 (1978).
22. Takahashi, K. *et al. J. Immun.* **122**, 275–279 (1979).
23. Ohori, H. *et al. Intervirology* **13**, 74–82 (1980).
24. Bennett, G. N., Schweingruber, M. E., Brown, K. D., Squires, C. & Yanofsky, C. *Proc. natn. Acad. Sci. U.S.A.* **73**, 2351–2355 (1976).
25. Maniatis, T., Jeffrey, A. & van de Sande, H. *Biochemistry* **14**, 3787–3794 (1975).
26. Smith, H. O. *Meth. Enzym.* **65**, 371–380 (1980).
27. Boyer, H. W. & Roulland-Dussoix, D. *J. molec. Biol.* **41**, 459–472 (1969).
28. Birnboim, H. C. & Doly, J. *Nucleic Acids Res.* **7**, 1513–1523 (1979).
29. Maxam, A. & Gilbert, W. *Proc. natn. Acad. Sci. U.S.A.* **74**, 560–564 (1977).
30. Yoo, J. O. & Agarwal, K. L. *J. biol. Chem.* **255**, 6445–6449 (1980).
31. Ullrich, A. *et al. Science* **196**, 1313–1319 (1977).
32. Cohen, S. N., Chang, A. C. Y. & Hsu, L. *Proc. natn. Acad. Sci. U.S.A.* **69**, 2110–2114 (1972).
33. Miller, J. H. in *Experiments in Molecular Genetics*, Appendix I (Cold Spring Harbor Laboratory, New York, 1972).
34. Barnes, W. M. *Science* **195**, 393–395 (1977).
35. Ling, C. M. & Overby, L. R. *J. Immun.* **109**, 834–841 (1972).
36. Laemmli, U.K. *Nature* **227**, 680–685 (1970).
37. Roychoudhury, R., Jay, E. & Wu, R. *Nucleic Acids Res.* **3**, 863–877 (1976).
38. Chang, A. C. Y. *et al. Nature* **275**, 617–624 (1978).

Influenza Antigenic Determinants are Expressed from Haemagglutinin Genes Cloned in *Escherichia Coli*

J.S. Emtage, W.C.A. Tacon, G.H. Catlin, B. Jenkins, A.G. Porter and N.H. Carey

A gene sequence for the fowl plague virus (FPV) haemagglutinin molecule has been inserted into a bacterial plasmid such that its transcription is under the control of a promoter derived from the tryptophan operon. Such plasmids direct the synthesis of a protein that reacts specifically with antisera to FPV haemagglutinin. Evidence is also presented that in some cases DNA inserted at the HindIII site of pBR322 is expressed.

INFLUENZA is still a major disease of man. It causes death primarily in the elderly, many of whom have chronic heart or vascular disease and other disorders. Severe infections of healthy individuals also occurs but in these cases morbidity, leading to incapacity to work, rather than mortality usually results. The morbidity figures however are of great economic importance. For example, the Hong Kong outbreak of 1968–70 resulted in 25 million lost working days in England[1], while in the US the effect of the Hong Kong variant of 1968–69 was costed at 3.8 billion dollars[2]. Thus, prevention of influenza epidemics and pandemics would be of value both economically and socially.

The main structure involved in immunity against influenza is the haemagglutinin (HA) surface glycoprotein[3,4]. The functional HA subunit, one of the spikes on the virus surface, is a triangular rod-shaped glycoprotein with a molecular weight of ~250,000 and is comprised of three HA monomers[5]. The HA monomers are synthesised as single polypeptide chains containing, in the case of FPV, an 18-amino acid precursor peptide at the N-terminus[6]. During maturation and virus assembly the pre-HA is further processed to HA1 and HA2 which remain linked by disulphide bridges[7,8].

Recent advances in genetic engineering and our knowledge of the structure and sequence of bacterial operons and control elements now allow the construction of new bacterial strains with the potential of synthesising large quantities of viral proteins (antigens) which may ultimately be useful as vaccines. In this article we describe the controlled production of haemagglutinin antigen from an HA gene cloned in *Escherichia coli*; this is the first step towards testing the feasibility of large scale antigen production by these means.

Experimental design

We have previously cloned at the *Hind*III site of pBR322 a DNA fragment containing the control region of the *E. coli* tryptophan operon, and coding for the ribosome binding site and first seven amino acids of *trpE*. From this, we have produced a series of vectors, pWT111, pWT121 and pWT131, capable of ensuring that any inserted DNA is read in the correct phase[9].

We recently cloned and sequenced the gene for the FPV haemagglutinin protein[6]. The gene was cloned using *Hind*III linkers and can therefore be transferred between vectors. From the nucleotide sequences of the FPV HA gene and the *Hind*III sites of the pWT series, it was clear that, if inserted at the *Hind*III site of pWT121 in the correct orientation, the HA gene would be translated by readthrough from the *trpE* fragment (Fig. 1A).

The protein resulting from initiation at the *trpE* AUG would be a hybrid consisting of the following fragments in order: (1) an N-terminus of 7 amino acids from anthranilate synthetase; (2) 6 amino acids specified by linker DNA; (3) 6 phenylalanine residues from the (T)$_{19}$ region of the FPV cloned DNA; (4) 7 amino acids from the normally non-translated 5'-portion of the HA gene; (5) 558 amino acids comprising the haemagglutinin protein and its prepeptide[6] and, finally; (6) 5 amino acids specified by *Hind*III linker at the C-terminus[9]. This is a total of 589 amino acids with a total MW of 69,000.

It was also clear that the HA gene should not be expressed from the initiator AUG of the *trpE* fragment in pWT111 or pWT131. Recognition of the ribosome binding site and initiator AUG of the HA was not anticipated. Similarly, we did not expect expression of the HA gene inserted at the *Hind*III site of pBR322 in either orientation.

Construction of expression plasmids

pWT121 was restricted with *Hind*III, treated with alkaline phosphatase, ligated to the purified HA gene and the mixture used to transform *E. coli* K12 HB101. A total of 47 ampicillin resistant transformants was obtained (pWT121 in the absence of the HA gene produced 13 colonies). Of these 47 colonies, 19 were tetracycline sensitive (Tc^S) and the remainder tetracycline resistant (Tc^R). We selected 12 colonies (3Tc^S and 9 Tc^R) for further characterisation. Analysis of plasmid DNA by gel electrophoresis after *Hind*III restriction showed that 5 of the 9 Tc^R colonies contained plasmids with a DNA insert. The other four contained the parent plasmid.

The orientation of the inserted DNA was determined by restriction enzyme analysis of the plasmids. We denote the two orientations as either R or L and define R and L orientation in terms of the direction of transcription required for HA

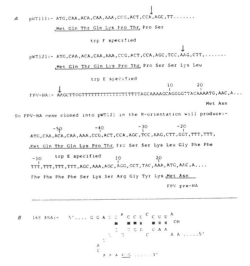

Fig. 1 *A*, Predicted nucleotide sequences around the *Hind*III sites of pWT111, pWT121 and the FPV·HA gene. Arrows indicate the position of the *Hind*III sites and, for clarity, only the sequence of the coding strand is shown. The protein sequence of the N-terminus of anthranilate synthetase has previously been reported by Lee *et al*[10]. The numbering system is bidirectional beginning with the 5'-nucleotide of the gene 4 complementary RNA. *B*, Potential base pairing between the 3'oligonucleotide of *E. coli* 16S RNA[11] and the pre-AUG region of the haemagglutinin mRNA.

EMTAGE, J.S., TACON, W.C.A., CATLIN, G.H., JENKINS, B., PORTER, A.G. and CAREY, N.H.
Influenza antigenic determinants are expressed from haemagglutinin genes cloned in *Escherichia coli*.
Reprinted by permission from *Nature* v. 283: pp. 171-174. Copyright 1980 Macmillan Magazines Ltd.

Fig. 2   A, Orientation of the FPV gene in pWT111, pWT121 and pBR322. pBR322, pWT111 or pWT121 (10 μg) was limit digested with HindIII. The resulting 5'-terminal phosphates of the vectors were removed by treatment with 20 μg bacterial alkaline phosphatase for 30 min at 37 °C in a 25 μl incubation containing 20 mM Tris-HCl pH 7.5 and 0.1% SDS. After phenol extraction and ethanol precipitation 0.2 μg of each DNA was ligated to ~40 ng FPV-HA gene in a 20 μl incubation containing 50 mM Tris-HCl pH 7.5, 10 mM MgCl$_2$, 20 mM dithiothreitol, 1 mM ATP and 0.02 units T4 DNA ligase (New England Biolabs). After overnight incubation at 15 °C the mixtures were diluted to 100 μl with TCM (10 mM Tris-HCl pH 7.5, 10 mM CaCl$_2$, 10 mM MgCl$_2$) and used to transfect 200 μl CaCl$_2$-treated E. coli K12 HB101 by a previously described procedure[12]. Transformants present after growth for 16 h at 37 °C on L-agar[13], plus 100 μg ml^{-1} carbenicillin (Pyopen) were tested for tetracycline resistance by picking onto agar plates containing M9 salts, glucose and casamino acids[13] plus carbenicillin and tetracycline (10 μg ml^{-1}). For plasmid isolation colonies were picked and 100-ml cultures grown in M9 salts, glucose, casamino acids medium supplemented with carbenicillin. At an A_{600} of ~0.6, choramphenicol was added to a final concentration of 150 μg ml^{-1} and incubation continued for 16 h at 37 °C. Cells were killed by the

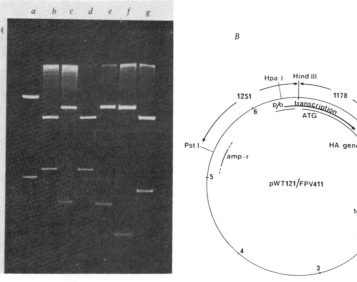

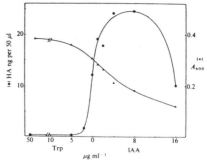

addition of diethylpyrocarbonate[14] to 0.4%, collected by centrifugation, washed with TE buffer (10 mM Tris-HCl pH 7.5, 1 mM EDTA) and suspended in 1.5 ml of iced 25% sucrose in 50 mM Tris-HCl pH 8. Cells were lysed by the addition of 0.5 ml lysozyme (10 mg ml^{-1}, 0.5 ml EDTA (0.5 M, pH 8) and 2.5 ml iced Triton solution (0.1% Triton X-100, 50 mM Tris-HCl pH 8, 50 mM EDTA) at 5-min intervals. The resulting lysates were cleared by spinning at 15,000 g for 20 min, extracted with phenol and then chloroform and finally precipitated by the addition of 0.1 vol 3M NaAc and 0.54 vol isopropanol. After 30 min at −20 °C the DNA was pelleted, washed with 70% ethanol, dried and dissolved in TE. PstI digests were carried out on 0.1 μg DNA in conditions recommended by the vendor (New England Biolabs) with the modification that RNase (50 μg ml^{-1}) was included in the incubation. Digests were fractionated on a 1.4% agarose slab gel[15] and bands visualised under UV light after staining with ethidium bromide. Lane a contains PM-2 DNA digested with HindIII. Bands seen are 5,400, 2,350 and 1,050 base pairs. The other lanes contain PstI digests of the following plasmids: b, pWT121/FPV 411(R); c, pWT121/FPV 412(L); d, pWT111/FPV 502(R); e, pWT111/FPV 503(L); f, pBR322/FPV 604(L); g, pBR322/FPV 605(R). B, Structure of pWT121/FPV 411(R). pWT121 contains a single HindIII site downstream from the trp promoter[9]. The plasmid shown contains the FPV-HA gene cloned at the HindIII site in the R orientation. The distances between the PstI and HindIII sites are shown as well as the direction of transcription from the trp promoter and the position of the initiator ATG. A full description of the construction of the pWT plasmid series has been published elsewhere[9]. Briefly, the 497-base pair HinfI fragment containing the trp promoter/operator region as well as the nucleotides specifying the leader sequence and first seven amino acids to trpE (ref. 10) was isolated and cloned, using HindIII linkers, into the HindIII site of pBR322. From this plasmid, pWT101, was isolated containing the trp promoter in the R orientation. In pWT101 tetracycline resistance is controlled from the trp promoter. Elimination of the HindIII site upstream of the trp promoter produced the vector pWT111 and phase changing at the HindIII site of pWT111 using DNA polymerase I and HindIII linkers produced pWT121 and pWT131.

production. That is, a gene in the R orientation required rightward (clockwise) transcription for expression; in the pWT plasmid series this is the orientation for expression from the trp promoter. For clarity we have included the letter (R) or (L) after the plasmid number to indicate its orientation. pWT121 contains a single PstI site, in the ampicillin gene[16], 1,251 base pairs from the HindIII site and the FPV HA gene contains a PstI site 1,178 base pairs from the 5' end of the coding strand[6]. Thus PstI digests of the recombinant plasmids should indicate which are correctly orientated with respect to the trp promoter. Representatives of the two FPV-gene containing groups are shown in Fig. 2A (b and c). The Tc^S plasmids gave rise to a band of 1,800 base pairs on PstI digestion while the Tc^R plasmids produced one of 2,450 base pairs. These are consistent with the fragments expected from the plasmid shown in Fig. 2B and indicate that all the Tc^S plasmids contained inserts in the L orientation while the Tc^R plasmids contained inserts in the R orientation.

In a similar way we re-cloned the HA gene into the HindIII site of pWT111 and pBR322. Again the pWT111/FPV plasmids could be divided into Tc^R and Tc^S groups and PstI restrictions (Fig. 2A, d and e) of representatives of these groups indicated that the HA gene was in the R orientation in the Tc^R group and in the L orientation in the Tc^S group. The pBR322/FPV plasmids however were all Tc^S but still contained L and R orientated genes as shown in Fig. 2A (f and g).

The correct phasing of the HA gene in pWT121/FPV 411(R) with the trpE AUG was confirmed by nucleotide sequencing. The HA gene contains an EcoRII restriction site at positions 36–40 (ref. 6). This site, however, is methylated and although not cleaved by EcoRII, it is cleaved by the EcoRII isoschizomer BstN1. Therefore we digested pWT121/FPV 411(R) with BstN1, labelled the 5' ends with ^{32}P, recut the DNA with HpaI and isolated the fragment containing the initiation region. The nucleotide sequence of this fragment was complementary to the sequence predicted in Fig. 1A, except for nucleotide −36 which

was an A instead of a G. This was also the case for the corresponding position in pWT111. In both plasmids this position corresponds to the 3'-terminal nucleotide of the HindIII linker. The reason for the change is not clear; it might be due to random mutation or, more likely, to micro-heterogeneity or incomplete deblocking after chemical synthesis at the 3' ends of the HindIII linker. In either event the effect in the mRNA strand is to have a serine codon (UCG) instead of a proline codon (CCG) but does not change the phase with respect to the trpE AUG.

Gene expression

E. coli colonies containing representatives of the plasmids described above were screened for FPV-HA antigen using a solid-phase immunological method[17]. Briefly, small cultures of individual colonies were grown, collected and lysed using lysozyme and Triton X-100. Any HA sequences present were

Fig. 3   Control of HA production. Overnight cultures of E. coli K12 HB 101 containing pWT121/FPV 411(R) were grown in M9 salts, glucose, casamino acids medium supplemented with carbenicillin (100 μg ml^{-1}). Aliquots of 100 μl were inoculated into 5 ml of the above medium containing either tryptophan or IAA as indicated. Cultures were grown for 2 h at 37 °C, the A_{600} determined and cells collected, lysed and assayed for HA content as described in Table 1.

Table 1 Radioimmunoassay of FPV-HA in lysates of bacteria containing influenza genes

Plasmid	Orientation of gene	Phenotype	HA content (ng per 50 μl)
pWT121/FPV 411	R	Tc^R	20.3
pWT121/FPV 412	L	Tc^S	5.0
pWT111/FPV 502	R	Tc^S	4.1
pWT111/FPV 503	L	Tc^S	2.1
pBR322/FPV 604	L	Tc^S	3.6
pBR322/FPV 605	R	Tc^S	0
pWT121	—	Tc^R	0
pBR322	—	Tc^R	0

Liquid cultures (5 ml) of individual colonies were grown in M9 medium. Cells were collected by centrifugation, and lysed, in a final volume of 1 ml, as described in Fig. 2A. This lysate was used for antigen assay without further treatment. HA-antigen assays were performed in polystyrene culture tubes (Nunc no. 1410) using an antibody sandwich technique[17]. Rabbit anti-FPV HA was purified by ammonium sulphate precipitation and DEAE-cellulose chromatography[18] before use and labelled with ^{125}I as described elsewhere[19]. Tubes were coated with 50 μl of 0.2M NaHCO$_3$ pH 9 containing 60 μg ml^{-1} purified, unlabelled IgG for 10 min at room temperature, washed with 3 × 1-ml aliquots of wash buffer (phosphate-buffered saline containing 0.1% bovine serum albumin, 0.1% NP-40 and 0.5% normal goat serum) and then incubated at room temperature either with known amounts (1–20 ng) of Sarkosyl-disrupted FPV or with the bacterial lysates. After 2 h, the tubes were washed with 4 × 1-ml aliquots of wash buffer and finally incubated at 4°C for 16 h with 50 μl wash buffer containing 200,000 c.p.m. ^{125}I-anti-HA. Tubes were again washed and counted in a Packard γ-counter. Antigen present was quantitated from standard curves relating the radioactivity bound to the amount of FPV present.

bound to a polystyrene tube coated with FPV-HA specific IgG. Bound antigen was then detected by incubating with high specific activity ^{125}I-anti FPV-HA. As indicated in Table 1, immune reactivity was detected from all colonies containing an FPV-HA gene inserted into the pWT plasmids and from pBR322 containing the HA gene in the L orientation. Neither of the parent plasmids nor pBR322 containing the HA gene in the R orientation produced any immune reacting material. Further, the positive reactions were abolished if the tubes were coated either with normal rabbit IgG or with anti A/Victoria—HA IgG instead of specific anti-FPV-HA IgG.

It was interesting to find expression of HA antigen from one orientation of pBR322/FPV. The HindIII site of pBR322 lies in the promoter region of the tetracycline gene and it is known that cloning at the HindIII site destroys this rightward transcribing promoter[20] unless the cloned DNA has its own promoter transcribing into the tetracycline gene(s)[21]. We therefore propose that a previously unknown promoter which transcribes in the leftward direction exists on the tetracycline gene side of the HindIII site of pBR322. It is necessary to postulate the existence of this promoter to explain the expression of the HA gene in pBR322/FPV 604(L), pWT121/FPV 412(L) and pWT111/FPV 503(L). The possibility that there is a pseudo-promoter sequence in the HA gene itself, for example the (T$_{19}$):(A$_{19}$) sequence, is ruled out as the plasmid pBR322/FPV 605(R) does not express its HA gene (Table 1).

This postulated promoter would explain the transcription of the HA gene in the L-orientated plasmids. How is translation explained? Inspection of the nucleotide sequence around the tetracycline–HA gene junction of pBR322/FPV 604(L) reveals nonsense triplets in all three translation phases[22]. Thus the bacterial translational system is probably recognising a nucleotide sequence on the HA mRNA itself and initiating protein synthesis at the AUG of the pre-peptide. Comparison of the sequence of the untranslated region of the HA gene with that of the 3' end of prokaryotic 16S ribosomal RNA[11] reveals surprising complementarity (Fig. 1B). Consistent with this interpretation is the expression of HA antigen in pWT111/FPV 502(R). In this case the HA gene is in the wrong phase to be translated from the trpE AUG; protein synthesis initiating at the trpE AUG is terminated by the now in phase UGA triplet at position 23–25 in the HA gene. Thus we conclude that initiation is from the natural HA AUG at position 22–24 (Fig. 1A). Assuming equal transcription from the trp promoter in pWT121/FPV 411(R) and pWT111/FPV 502(R) and that the products have equal antigenic properties, then from the data of Table 1 we conclude that the eukaryotic ribosome binding site is

recognised with an efficiency of 20% compared to the prokaryotic site.

The fivefold difference in HA expression between pWT121/FPV 411(R) and pWT111/FPV 502(R) could be due to differences in the plasmid complement of the respective cells. To exclude this possibility the relative copy numbers of the above plasmids were determined from agarose gels of lysed whole cells[9] and found to be the same (data not shown).

Hitherto the HindIII and EcoRI sites of pBR322 have been assumed to be non-expression sites because of their respective situations in and upstream of the tetracycline promoter[20,21]. This has been thought to be of value in assessing the containment categorisation when considering the safety of experiments in genetic engineering[23]. In view of the above findings it is clear that the validity of these assumptions should be reassessed. It seems likely that a potential ribosome binding site on the inserted gene will be of some importance.

Table 2 Control of HA expresion from the tryptophan promoter/operator region

Plasmid	Addition	HA* (ng per 50 μl)
pWT121/FPV 411(R)	Tryptophan	0.9
	IAA	59.1
pWT121/FPV 412(L)	Tryptophan	9.6
	IAA	4.9

Cultures (5 ml) of pWT121/FPV 411(R) and pWT121/FPV 412(L) were grown for 3 h at 37°C in M9 salts, glucose, casamino acids medium supplemented with carbenicillin (100 μg ml^{-1}) and either IAA (20 μg ml^{-1}) or tryptophan (100 μg ml^{-1}). Cells were collected, lysed and assayed for HA content as described in Table 1.
* Results normalised to same A_{600}.

Control of gene expression

In vivo the tryptophan operon is controlled by the level of tryptophan[24] and control is best obtained in culture by using tryptophan as repressor or β-indole acrylic acid (IAA) as inducer. The latter is a competitive inhibitor that prevents tryptophan binding and thereby effectively inactivates the repressor[25].

We selected a representative colony from each of the two expressing groups (in pWT121) to examine HA synthesis in the above conditions. Our initial observation was that pWT121/FPV 411(R) grew very slowly in the presence of 20 μg ml^{-1} IAA; there was a threefold difference compared to growth in the presence of tryptophan. pWT121/FPV 412(L) grew equally well in both sets of conditions. The results of radioimmunoassay of these cultures are shown in Table 2. Because of the differences in growth rate, the values are all normalised to the A_{600} of the culture containing pWT121/FPV 411(R) plus tryptophan. In pWT121/FPV 411(R), HA production is stimulated 65-fold when the trp operon is derepressed. In contrast, IAA decreased HA synthesis in pWT121/FPV 412(L). Thus increased transcription from the trp promoter is competing with transcription from the presumed promoter near or around the tetracycline gene and reducing expression in the leftward direction.

The effects of repression and induction on HA synthesis are also seen in Fig. 3. As little as 5 μg ml^{-1} tryptophan was sufficient for maximal repression. On the other hand, because of its effect on growth, the induction characteristics of IAA were more complex. Maximal induction occurred at IAA concentrations of 4–8 μg ml^{-1} and HA synthesis was stimulated 120-fold over the repressed level.

Finally, we examined the immunoprecipitable products in cells containing various plasmids after induction or repression of trp transcription. At 3 h after addition of either IAA or tryptophan to cultures of such cells, aliquots were pulse-labelled with ^{35}S-methionine and the immunoprecipitable proteins separated by SDS-polyacrylamide gel electrophoresis. No immunoprecipitable proteins were present in induced cells containing pWT121 or in repressed cells containing pWT121/FPV 411(R) (Fig. 4a and b, respectively). However, induced cells containing pWT121/FPV 411(R) or pWT111/FPV 502(R) and repressed cells containing

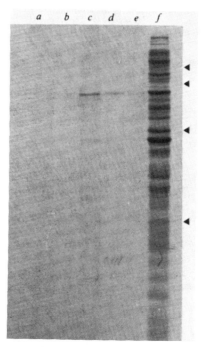

Fig. 4 Immunoprecipitation of the haemagglutinin-like protein. Cultures of *E. coli* K12 HB101 containing the plasmids pWT121, pWT121/FPV 411(R), pWT121/FPV 412(L) and pWT111/FPV 502(R) were grown in M9 salts, glucose, casamino acids medium containing carbenicillin (100 μg ml⁻¹) and either IAA (10 μg ml⁻¹) or tryptophan (40 μg ml⁻¹). After a 3-h period, 5-ml samples were removed and pulse-labelled for 10 min at 37 °C with 50 μCi of ³⁵S-methionine. The labelling was terminated by the addition of 20 ml cold M9 medium, the cells pelleted by centrifugation at 12,000g for 10 min at 4 °C and lysed, in a final volume of 1 ml, as described in Fig. 2A. Immunoprecipitations were performed as follows: to 400 μl *E. coli*. extract was added 2 μl 10% Nonidet P-40 (NP–40), 4 μl (2μg) normal rabbit IgG and the mixture incubated for 60 min at 20 °C. An aliquot of 10 μl of a 10% suspension of killed *Staphylococcus aureus*[31] (Cowan 1 strain) was then added and incubation continued for a further 15 min. The resulting immunoprecipitate was pelleted and discarded. To the supernatant was added 2 μg rabbit anti FPV-HA and, after 60 min at 20 °C, 10 μl of *S. aureus*. This immunoprecipitate was pelleted 15 min later and washed twice with a solution containing 0.15 M NaCl, 5 mM EDTA, 50 mM Tris-HCl pH 7.4 and 0.05% NP-40. Radioactive proteins were eluted from the complex with 30 μl SDS buffer, heated 90 °C for 2 min and electrophoresed on a 12.5% polyacrylamide-SDS gel[32] which was then dried and autoradiographed. The figure shows the immunoprecipitable proteins from cells containing the following plasmids: *a*, pWT121 + IAA; *b*, pWT121/FPV 411(R) + *trp*; *c*, pWT121/FPV 411(R) + IAA; *d*, pWT111/FPV 502(R) + IAA; *e*, pWT121/FPV 412(L) + *trp*; *f* contains an extract from induced cells containing pWT121/FPV 411(R). The positions of conalbumin (MW 76,000), bovine serum albumin (69,000), ovalbumin (43,000) and carbonic anhydrase (30,000) are indicated by the arrows.

pWT121/FPV 412(L) all produced a band of MW 61,000 (Fig. 4c, d and e respectively). Our experimental design predicted that the HA-like protein synthesised from pWT121/FPV 411(R) would have a MW of 69,168 and that the products from pWT121/FPV 502(R) and pWT121/FPV 412(R) would be 26 amino acids smaller at the N-terminus. As the three products are in fact the same size, it seems probable that the primary gene product has been processed and a portion of the polypeptide removed. The same results were obtained when the products of these plasmids were examined in minicells (unpublished results). Clearly the mechanism of the processing and the possible involvement of the pre-HA sequence need further investigation.

Conclusions

We have demonstrated the feasibility of producing controlled amounts of influenza antigenic determinants by genetic engineering. Obviously further analysis is necessary to characterise the protein product and a number of questions remain unanswered. First, for example, initiation of protein synthesis from the prokaryotic-like ribosome binding site on the HA will result in a polypeptide having a eukaryotic leader sequence. This sequence is thought to function in the transmembrane movement of the HA *in vivo*[26]; we may question whether this sequence is recognised by the regulatory systems for export in bacteria. Second, what effect do the $(T_{19}):(A_{19})$ region, resulting from the oligo(dT) primer and the poly(A)tail of the vRNA and situated between the AUG of *trpE* and the HA sequence, or the $(Phe)_6$ oligopeptide resulting from its translation, have on transcription and translation respectively? Homopolymeric nucleotide regions such as this are one of the signals implicated in transcription termination[27]; thus the $(T_{19}):(A_{19})$ region may have a polar effect on HA production. Similarly, the $(Phe)_6$ region may result in depletion of charged $tRNA_{Phe}$ and slow ribosome movement along the mRNA. Clearly examination of HA production by plasmids where this region has been removed is essential.

The absolute quantity of HA-like protein produced may also be estimated. The estimates based on immunological detection cannot be precise and are probably minimum values since if there are differences from the natural antigen it seems likely that the bacterial product would have fewer antigenic determinants or lower affinity for the antiserum. Based on our maximum yield of 60 ng per 50 μl bacterial lysate (Table 2) and a protein concentration of 160 μg ml⁻¹ for the same lysate we arrive at a figure of 0.75% of total protein. By comparison, the quantities of induced proteins can be determined by densitometry from autoradiographs of gels containing proteins from induced and repressed cells. This shows that the 61,000 MW protein is 2–3% of total protein synthesis (unpublished results). Neither of these values was as high as expected. Hallewell and Emtage[28] have recently cloned an *E. coli* HindIII fragment containing the *trp* promoter and the complete *trpE* gene in pBR322; in induced conditions this plasmid was capable of specifying up to 30% of total protein as anthranilate synthetase[28]. Our values however are similar to those reported for the synthesis of ovalbumin-like proteins under the control of the *lac* promoter[29,30]; that is, the level of expression of the eukaryotic genes is only about 10% of the expected. Whether this is due to limiting amounts of some tRNA species as proposed by Fraser and Bruce[30] or to the type of transcription termination discussed above remains to be seen.

We thank Drs J. Oxford and M. Gethin for gifts of rabbit anti-FPV HA and *Staphylococcus aureus* respectively and Dr A. Hale for providing research facilities. Cloning and expression experiments were performed in a CIII laboratory as recommended by GMAG.

Received 1 October; accepted 22 November 1979.

1. Schild, G. C. in *Chemoprophylaxis and Virus Infections of the Respiratory Tract* Vol. 1 (ed. Oxford, J.S.) 63–101 (CRC Press, Cleveland, 1977).
2. Kavet, J. thesis, Harvard Univ. (1972).
3. Drzeniek, R., Seto, J. T. & Rott, R. *Biochim. biophys. Acta* **128**, 547–558 (1966).
4. Laver, W. G. & Kilbourne, E. D. *Virology* **30**, 493–501 (1966).
5. Wiley, D. C., Skehel, J. J. & Waterfield, M. D. *Virology* **79**, 446–448 (1977).
6. Porter, A. G. *Nature* **282**, 471–477 (1979).
7. Skehel, J. J. & Schild, G. C. *Virology* **44**, 396–408 (1971).
8. Laver, W. G. *Virology* **45**, 275–288 (1971).
9. Tacon, W. A., Carey, N. H. & Emtage, J. S. *Molec. gen. Genet.* (in the press).
10. Lee, F., Bertrand, K., Bennett, G. & Yanofsky, C. *J. molec. Biol.* **121**, 193–217 (1978).
11. Steitz, J. A. & Jakes, K. *Proc. natn. Acad. Sci. U.S.A.* **72**, 4734–4738 (1975).
12. Glover, D. in *New Techniques in Biophysics and Cell Biology* Vol. 8 (eds Pain, R. H. & Smith, B. J.) 125–145 (Wiley, New York, 1976).
13. Miller, J. H. *Experiments in Molecular Genetics* (Cold Spring Harbor Laboratory, New York, 1976).
14. Weissman, C. & Boll, W. *Nature* **261**, 428–429 (1976).
15. Sharp, P. A., Sugden, B. & Sambrook, J. *Biochemistry* **12**, 3055–3063 (1973).
16. Sutcliffe, J. G. *Proc. natn. Acad. Sci. U.S.A.* **75**, 3737–3741 (1978).
17. Broome, S. & Gilbert, W. *Proc. natn. Acad. Sci. U.S.A.* **75**, 2746–2749 (1978).
18. Livingston, D. M. *Meth. Enzym.* **34**, 723–731 (1974).
19. Hunter, W. M. & Greenwood, F. C. *Biochem. J.* **91**, 43–46 (1964).
20. Rodriguez, R. L., West, R. W., Heyneker, H. L., Bolivar, F. & Boyer, H. W. *Nucleic Acids Res.* **6**, 3267–3287 (1978).
21. Widera, G., Gautier, F., Lindenmaier, W. & Collins, J. *Molec. gen. Genet.* **163**, 301–305 (1978).
22. Boyer, H. W. *et al.* in *Recombinant Molecules: Impact on Science and Society* (eds Beers, R. F. & Bassett, E. G.) 9–20 (Raven, New York, 1977).
23. Genetic Engineering: New Guidelines for U.K. *Nature* **276**, 104–108 (1978).
24. Platt, T. in *The Operon* (eds Miller, J. H. & Reznikoff, W. S.) 263–302 (Cold Spring Harbor Laboratory, New York, 1978).
25. Squires, C. L., Lee, F. D. & Yanofsky, C. *J. molec. Biol.* **92**, 93–111 (1975).
26. Waterfield, M. D., Espelie, K., Elder, K. & Skehel, J. J. *Br. med. Bull.* **35**, 57–63 (1979).
27. Adhya, S. & Gottesman, M. *A. Rev. Biochem.* **47**, 967–996 (1978).
28. Hallewell, R. A. & Emtage, J. S. *Gene* (in the press).
29. Mercereau-Puijalon, O. *et al. Nature* **275**, 505–510 (1978).
30. Fraser, T. H. & Bruce, B. J. *Proc. natn. Acad. Sci. U.S.A.* **75**, 5936–5940 (1978).
31. Kessler, S. W. *J. Immun.* **115**, 1617–1624 (1975).
32. Laemmli, U. K. *Nature* **227**, 680–685 (1970).

Vaccinia Virus: A Selectable Eukaryotic Cloning and Expression Vector

M. Mackett, G.L. Smith and B. Moss

ABSTRACT   Foreign DNA was inserted into two nonessential regions of the vaccinia virus genome by homologous recombination in cells infected with virus and transfected with plasmids containing the foreign DNA elements flanked by vaccinia virus DNA. Thymidine kinase-negative (TK$^-$) recombinants were selected after inserting foreign DNA into the coding region of the *TK* gene of wild-type vaccinia virus; TK$^+$ recombinants were selected after inserting the herpesvirus *TK* gene into TK$^-$ mutants of vaccinia virus. For TK$^+$ expression, it was necessary to insert a 275-base-pair DNA fragment containing the initiation site and sequences upstream of an early vaccinia virus transcript next to the coding sequences of the herpesvirus gene. The unique ability of the herpesvirus TK to phosphorylate ^{125}I-labeled deoxycytidine provided independent confirmation of gene expression. These studies demonstrate the use of vaccinia virus as a selectable cloning and expression vector, confirm the map location of the vaccinia virus *TK* gene, and provide initial information regarding the location of vaccinia virus transcriptional regulatory sequences.

Several virus groups, including the papovaviruses (1–3), papillomaviruses (4), adenoviruses (5, 6), and retroviruses (7, 8), have been employed as eukaryotic cloning and expression vectors. The relatively small sizes of these virus genomes have facilitated the *in vitro* construction of recombinant DNA molecules. Although genetic engineering of larger viruses is more difficult, such vectors have the advantage of greater capacity and potential of retaining complete infectivity in a wide range of host cells. For vaccinia virus, there is an added incentive of creating recombinants that may have value as live virus vaccines.

In considering the development of vaccinia virus as an expression vector, the following biological characteristics of this unique agent must be taken into account (9, 10): a large [180-kilobase (kb)] genome, a lack of infectivity of isolated viral DNA, the packaging of viral enzymes necessary for transcription within the infectious particle, the probability that vaccinia virus has evolved its own transcriptional regulatory sequences, and the cytoplasmic site of virus transcription and replication. Initially, the major technical problems involved insertion of DNA into the large genome, efficient expression of heterologous genes, and selection of recombinant virus.

Insertion of DNA into the vaccinia virus genome can be accomplished by homologous recombination *in vivo*. Because vaccinia virus DNA by itself is not infectious, intact virus and calcium phosphate-precipitated viral DNA (11, 12) or plasmids containing viral sequences (13) are added in succession. By using plasmids, it is possible to perform the majority of manipulations *in vitro* except for the final step of transfection.

Presumably, efficient expression of foreign genes will depend on the use of vaccinia virus promoters. Although vaccinia virus transcriptional signals have not been defined, the region upstream of one early gene was found to be extremely rich in adenine and thymine residues and differed substantially from prokaryotic or eukaryotic consensus sequences (14). We have now tested the possibility that this DNA segment contains vaccinia virus-specific transcriptional signals by inserting it next to the coding sequences of a foreign gene.

For selection of recombinant virus, advantage was taken of the recent localization of the vaccinia virus thymidine kinase (TK; EC 2.7.1.21) gene (13). Our plan was to construct plasmids containing foreign DNA inserted within the vaccinia virus *TK* gene and then use 5-bromodeoxyuridine to select *in vivo* recombinants on the basis of the resulting TK$^-$ phenotype. The success of this approach would also confirm the map location of this gene. As an alternative method of selection, the herpesvirus *TK* gene fused to a putative vaccinia promoter segment was added to cells infected with a TK$^-$ vaccinia virus mutant. TK$^+$ recombinants were selected by using amethopterin to inhibit thymidylate synthesis (15).

The studies described in this communication demonstrate the use of vaccinia virus as a selectable eukaryotic cloning and expression vector and provide information regarding the location of vaccinia virus transcriptional signals.

MATERIALS AND METHODS

Preparation of DNA. Recombinant plasmids were prepared from pBR328 (16) or pUC7 (a gift of J. Viera and J. Messing) and purified as described by Birnboim and Doly (17). DNA fragments were isolated from agarose gels by electrophoresis onto DEAE-paper (18) or by binding to powdered glass (19).

Marker Rescue. Two hours after infection of TK$^-$ 143 cells (20) with TK$^+$ or TK$^-$ vaccinia virus WR (0.01–0.05 plaque-forming unit/cell), calcium phosphate-precipitated plasmid DNA was added (13). For isolation of TK$^+$ recombinants, amethopterin-containing selective medium was added at 6 hr and cells were harvested at 48 hr after infection (13). For isolation of TK$^-$ mutants, selection with bromodeoxyuridine at 25 μg/ml was used only during the subsequent plaque assay.

Blot Hybridizations. For dot-blot hybridizations, monolayers in 16-mm-diameter wells were harvested 48 hr after infection, lysed by three freeze–thaw cycles, treated with trypsin at 0.125 mg/ml for 30 min at 37°C, and collected on nitrocellulose sheets by filtration using a microsample manifold (Schleicher & Schuell). The filter was washed with 100 mM NaCl/50 mM Tris·HCl, pH 7.5; blotted three times on successive Whatman 3 MM papers saturated with (*i*) 0.5 M NaOH, (*ii*) 1 M Tris·HCl, pH 7.5, and (*iii*) 2× NaCl/Cit (NaCl/Cit is 0.15 M NaCl/0.015 M sodium citrate); baked at 80°C for 2 hr; and then in-

Abbreviations: kb, kilobase(s); bp, base pair(s); TK, thymidine kinase; NaCl/Cit, 0.15 M NaCl/0.015 M sodium citrate.

MACKETT, M., SMITH, G.L. and MOSS, B.
Vaccinia virus: a selectable eukaryotic cloning and expression vector.
Proc. Natl. Acad. Sci. U.S.A. 79:7415-7419 (1982). Reprinted with the authors' permission.

cubated with 5× Denhardt's solution (21) supplemented with 0.1 mg of denatured salmon sperm DNA per ml in 4× NaCl/ Cit at 65°C for 4 hr. DNA labeled with ^{32}P by nick-translation (22) and sodium dodecyl sulfate at a final concentration of 0.1% were added and hybridization was continued for 12 hr. The filter was washed twice for 15 min at 65°C with 2× NaCl/Cit containing 0.1% sodium dodecyl sulfate and then with 0.2× NaCl/ Cit containing 0.1% sodium dodecyl sulfate.

Transfer of DNA restriction fragments from agarose gels was performed by a modification (23) of the Southern (24) blotting technique. The filters were hybridized to ^{32}P-labeled DNA and washed as described above.

Herpesvirus Pyrimidine Kinase Assay. A plaque autoradiography procedure involving ^{125}I-labeled deoxycytidine (^{125}I-deoxycytidine) (New England Nuclear) as a specific substrate for the herpesvirus pyrimidine kinase (25, 26) was used. Monolayers of TK$^-$ 143 cells were incubated with ^{125}I-deoxycytidine at 1 μCi/ml (1 Ci = 3.7 × 10^{10} becquerels) and tetrahydrouridine (Sigma) at 20 μg/ml to inhibit cytidine deaminase (27) between 14 and 48 hr after infection.

RESULTS

Insertion of Foreign DNA into the Vaccinia Virus *TK* Gene and Selection of Recombinants. Initial experiments were designed to develop a general method of inserting foreign DNA into a nonessential region of the vaccinia virus genome and selecting recombinants. The vaccinia virus *TK* gene seemed an ideal target because its inactivation would allow selection for the TK$^-$ phenotype. The *TK* gene was recently mapped, by marker rescue and by cell-free translation of hybrid-selected mRNA, within the *Hind*III J fragment of vaccinia virus (13). Additional data suggested that a unique *Eco*RI site was located within the body of the *TK* gene (28). To facilitate genetic manipulations, the *Hind*III J fragment was transferred to a derivative of pBR328 with an *Eco*RI site that had been eliminated by nuclease S1 digestion. A convenient 2.4-kb *Eco*RI E fragment of adenovirus type 18 DNA was inserted into the unique *Eco*RI site within the *TK* gene and the new plasmid containing vaccinia virus and adenovirus sequences was called pVJAd.

The next step involved the use of homologous recombination to transfer the adenovirus DNA flanked by vaccinia sequences into the vaccinia virus genome. TK$^-$ 143 cells were infected with wild-type TK$^+$ vaccinia virus and then transfected with calcium phosphate-precipitated pVJAd. At this stage, no selection was used. The yield of TK$^-$ virus, determined by plaque assay in

the presence of bromodeoxyuridine, was approximately 3 × 10^5 TK$^-$ plaques per μg of pVJAd added. This value was 5–20 times higher than the yield of spontaneous TK$^-$ mutants but about ⅓ to ⅕ the number obtained upon parallel transfection with a plasmid containing DNA from a previously isolated TK$^-$ vaccinia virus nonsense mutant (28).

The increase in number of TK$^-$ plaques after transfection with pVJAd suggested that insertion of adenovirus DNA into the vaccinia *TK* gene had occurred. To confirm this, and distinguish recombinants from spontaneous TK$^-$ mutants, a rapid dot-blot hybridization procedure was used. Of 40 plaque isolates screened, 70% clearly hybridized to an adenovirus probe and 100% hybridized to a vaccinia virus probe.

The site of integration of foreign DNA into the vaccinia virus genome was determined by analysis of restriction endonuclease fragments. Cleavage of pVJAd with *Eco*RI produced two fragments of about 2.4 and 10 kb. The smaller piece contained the entire adenovirus DNA insert, whereas the larger one contained both vaccinia and plasmid sequences. When the blot was probed with ^{32}P-labeled adenovirus DNA, we detected the 2.4-kb fragment from both pVJAd and the TK$^-$ recombinant virus but not from the wild-type virus control (Fig. 1A). Cleavage of pVJAd with *Hind*III resulted in the formation of three fragments, two of which (5.7 and 1.8 kb) contained adenovirus sequences. Fragments of these sizes from pVJAd and the TK$^-$ recombinant were detected by autoradiography (Fig. 1A). When DNA from the recombinant virus was not cleaved, the adenovirus DNA sequences were associated only with the high molecular weight virion DNA (Fig. 1A).

The above experiments demonstrated that the entire adenovirus DNA segment was inserted into the vaccinia virus genome. To identify the site of integration, restriction fragment blots were probed with ^{32}P-labeled *Hind*III J fragment (Fig. 1B) and ^{32}P-labeled total vaccinia virus DNA (Fig. 1C). It is evident that the 5.1-kb *Hind*III fragment of wild-type virus is absent from the TK$^-$ recombinant (Fig. 1 B and C). This fragment is replaced by a larger *Hind*III fragment of 5.7 kb and a smaller one of 1.8 kb containing both vaccinia virus and adenovirus DNA. Because of the small amount of vaccinia DNA in the 1.8-kb *Hind*III fragment, it was clearly visible only upon longer exposure. The identical patterns obtained when *Eco*RI fragments of wild-type and recombinant virus are probed with vaccinia DNA (Fig. 1 B and C) indicate that no additional genomic changes have occurred.

Collectively, these data demonstrate the site-specific insertion of a 2.4-kb adenovirus DNA segment into the *Hind*III J

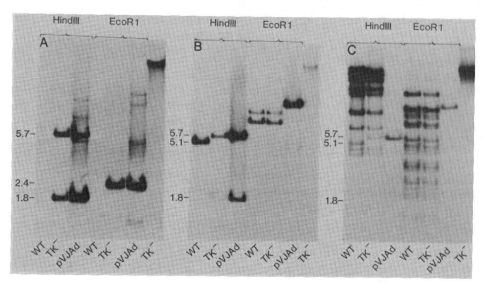

FIG. 1. Evidence for site-specific insertion of foreign DNA into the vaccinia virus genome. A TK$^-$ recombinant virus shown to contain adenovirus DNA by dot-blot hybridization was grown in HeLa cells and purified by sucrose gradient sedimentation. DNA from the recombinant (TK$^-$), purified wild-type virus (WT), and pVJAd was digested with *Hind*III or *Eco*RI, subjected to electrophoresis on a 1% agarose gel, blotted to nitrocellulose, and probed with a ^{32}P-labeled 2.4-kb *Eco*RI E fragment of adenovirus type 18 (A), a 5.1-kb *Hind*III J fragment of vaccinia virus (B), and total vaccinia virus DNA (C). The track at the extreme right of each panel contains DNA that was not digested with restriction endonuclease. An autoradiograph is shown. Fragment sizes are given in kb.

fragment of the vaccinia virus genome. Moreover, the TK⁻ phenotype of these recombinants indicates that the TK gene has been inactivated by insertion of foreign DNA.

Insertion and Expression of the Herpesvirus *TK* Gene Within the Vaccinia Virus Genome. As a second selection method, we inserted the herpesvirus *TK* gene into TK⁻ vaccinia virus mutants and isolated TK⁺ recombinants. Previous studies indicated that a 9-kb segment of the vaccinia virus genome, located proximal to the left inverted terminal repetition, was not essential for infectivity (29, 30). This seemed to be a suitable region for the insertion of foreign DNA. Accordingly, a 3-kb *Eco*RI/*Ava* I segment from the nonessential region was cloned

in pBR328 and the plasmid was called pMH5/1. A *Bam*HI fragment containing the entire herpesvirus *TK* gene (31) was then inserted between the *Bgl* II and *Bam*HI sites of pMH5/1 (Fig. 2). The resulting recombinant is called pVHTK1.

It was anticipated that vaccinia virus regulatory sequences would be needed to obtain efficient expression of heterologous DNA. Inspection of the nucleotide sequence of an early vaccinia virus gene encoding a 7.5-kilodalton polypeptide revealed a *Rsa* I site betwen the transcriptional and translational initiation sites (14). A *Hinc*II/*Rsa* I fragment of approximately 275 bp containing the transcriptional initiation site and upstream sequences was blunt-end ligated to *Hinc*II-cleaved pUC7 (Fig.

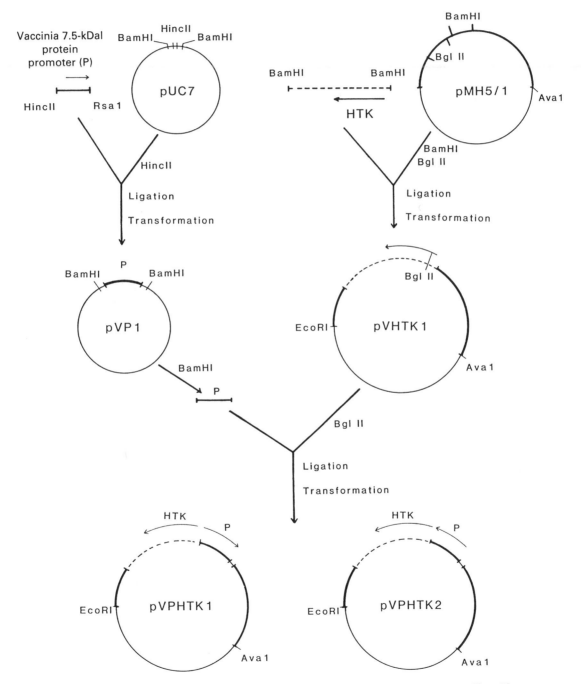

FIG. 2. Construction of plasmids containing the herpesvirus *TK* gene fused to a vaccinia virus promoter. Plasmid sequences are represented as a fine solid line, vaccinia sequences as a heavy solid line, and herpesvirus sequences as an interrupted line. Three plasmids, pVHTK1, pVPHTK1, and pVPHTK2, contain a herpesvirus DNA segment including the *TK* gene flanked by the same vaccinia virus DNA sequences. In pVPHTK1 and pVPHTK2, a 275-base-pair (bp) fragment containing a putative vaccinia virus promoter (P) for the gene encoding a 7.5-kilodalton polypeptide has been inserted in incorrect and correct orientations, respectively, adjacent to the herpesvirus TK (HTK) coding sequences.

2). By then excising the inserted vaccinia virus DNA with *Bam*HI, we effectively added restriction endonuclease linkers to the segment. The *Bam*HI fragment was then inserted at the unique *Bgl* II site of pVHTK1 (Fig. 2). Because the *Bgl* II site is located between the transcriptional and translational initiation sites of the herpesvirus *TK* gene (32, 33), this placed putative vaccinia regulatory sequences adjacent to herpesvirus TK coding sequences. In pVPHTK2, the regulatory and coding sequences are in proper orientation, whereas in pVPHTK1 they are opposite.

Transfection experiments were carried out by adding calcium phosphate-precipitated pVHTK1, pVPHTK1, or pVPHTK2 to TK⁻ cells infected with a TK⁻ mutant of vaccinia virus. The yield of TK⁺ virus was determined by plaque assay in selective medium (28). At the lowest dilution tested (1:100), no TK⁺ plaques were detected when the plasmid used for transfection contained the uninterrupted herpesvirus *TK* gene or the *TK* gene with vaccinia regulatory sequences in opposite orientation. However, when the two sequences were in correct orientation, 5,200 plaque-forming units per μg of plasmid was obtained. Of 23 TK⁺ plaque isolates tested, all hybridized to herpesvirus TK DNA (Fig. 3*A*). For subsequent experiments, virus was used that had been plaque purified twice in selective media and then amplified by successive passages in selective and nonselective media.

Site-specific integration of the herpesvirus *TK* gene into vaccinia virus DNA was demonstrated by blot hybridization of restriction fragments. Inspection of Fig. 2 reveals that homologous recombination of pVHTK2 with vaccinia virus DNA should lead to the deletion of vaccinia sequences between the *Bgl* II and *Bam*HI sites and insertion of herpesvirus DNA. Because the restriction endonuclease sites are not regenerated when *Bgl*

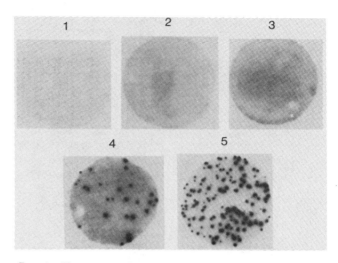

FIG. 4. Plaque autoradiograph demonstrating expression of herpesvirus TK. Monolayers of TK⁻ 143 cells mock-infected (*1*) or infected with TK⁻ vaccinia virus mutant (*2*), TK⁺ wild-type vaccinia virus (*3*), and vaccinia virus/herpesvirus TK recombinants at 10-fold different dilutions (*4*) and (*5*). Except for *2*, medium containing thymidine and amethopterin was added 2 hr after infection. At 14 hr after infection, cells were labeled for 20 hr with ¹²⁵I-deoxycytidine. After washing and fixation of the cell monolayers, autoradiographs were made.

II and *Bam*HI fragments are ligated, it was not possible to neatly excise the integrated herpesvirus DNA. The recombinant vaccinia DNA was therefore digested with *Hind*III or *Xho* I, both of which cut outside of the herpesvirus DNA segment. Hybridization of ³²P-labeled herpesvirus TK DNA to blots of electrophoretically separated restriction digests revealed bands of the predicted size (Fig. 3*B*). No hybridization of the probe to wild-type vaccinia virus DNA was detected.

A specific assay based on the ability of the herpesvirus TK enzyme to phosphorylate ¹²⁵I-deoxycytidine (25, 26) was used as a second measure of expression. As shown by autoradiography, recombinant vaccinia virus plaques incorporated the halogenated pyrimidine, whereas no incorporation was detected in visible plaques formed by wild-type TK⁺ vaccinia virus or by uninfected monolayers (Fig. 4). The number of recombinant plaques that formed in the presence of amethopterin correlated precisely with the number and position of the autoradiographic spots at two virus dilutions. Expression of the herpesvirus *TK* gene, as judged by growth in selective medium and by ¹²⁵I-deoxycytidine incorporation, was still obtained after six successive plaque purifications of the vaccinia virus recombinant. The absence of ¹²⁵I incorporation by wild-type vaccinia virus apparently reflects the stringent substrate specificity of the latter TK.

DISCUSSION

The directed insertion of foreign DNA into two nonessential regions of the vaccinia virus genome has been described. Selection was achieved either by interrupting the endogenous *TK* gene of wild-type vaccinia virus or by adding the herpesvirus *TK* gene to TK⁻ mutants. In the latter case, expression depended upon placement of a 275-bp fragment of known sequence (14) containing the transcriptional initiation site and upstream sequences of an early vaccinia virus gene next to herpesvirus TK coding sequences. The salutary effect of the correctly oriented vaccinia sequence implies that it contains transcriptional regulatory signals and was necessary for efficient early expression of the *TK* gene. Presumably, the same 275-bp vaccinia virus DNA fragment could be used to initiate transcription of prokaryotic and eukaryotic as well as other virus genes. Selection could be obtained by inserting those genes of

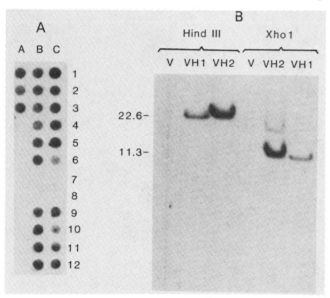

FIG. 3. Demonstration of herpesvirus DNA sequences in vaccinia virus recombinants. (*A*) Screening of plaque isolates by dot-blot hybridization. The filter was hybridized with ³²P-labeled herpesvirus *TK Bam*HI fragment. Samples A 1–3, B 1–6, B 9–12, C 1–6, and C 9–12 contain DNA from cells infected with independent TK⁺ plaque isolates. Samples B 7 and 8 contain DNA from cells infected with wild-type virus and samples C 7 and 8 are from mock-infected cells. An autoradiograph is shown. (*B*) Analysis of vaccinia virus recombinants by restriction endonuclease analysis. DNA was extracted from cells infected with wild-type vaccinia (V) and with independent TK⁺ recombinants (VH1 and VH2) and digested with *Hind*III or *Xho* I. After agarose gel electrophoresis, the DNA fragments were transferred to a nitrocellulose sheet and hybridized with ³²P-labeled herpesvirus TK DNA. An autoradiograph is shown.

interest in tandem with the herpesvirus *TK* gene or by insertion into and inactivation of the vaccinia virus *TK* gene.

Despite careful analysis of more than a dozen early vaccinia virus transcripts, no evidence of splicing has been obtained (14, 34–37). If, as seems likely, vaccinia virus lacks splicing enzymes, there may be a constraint on the use of the vector for expression of some eukaryotic genes. However, such difficulties may be avoided by the insertion of cDNA clones into the vaccinia virus genome.

Although the foreign DNA fragments inserted into the vaccinia virus genome were a few thousand nucleotides long, we suspect that the potential capacity is significantly greater. Vaccinia virus mutants containing many tandem repetitions of a 1,650-bp segment have been identified, indicating that even larger genomes can be packaged (38). Moreover, the capacity could be enhanced even further by using vaccinia virus (29, 30) or closely related rabbitpox virus (39) mutants with deletions of up to 30 kb as vectors. This should make it possible for a single recombinant virus to express many different genes.

In summary, we anticipate that the methods described here of inserting foreign genes into vaccinia virus and of obtaining expression and selection will be generally applicable. The successful use of vaccinia virus for immunization and eradication of smallpox raises the exciting possibility of employing vaccinia virus recombinants expressing antigens of pathogenic organisms to prevent and eliminate currently important infectious diseases of man and animals.

We thank M. Haffey, L. Enquist, and J. Janik for plasmids, N. Cooper for maintaining cell lines, and J. Carolan for typing the manuscript.

1. Hamer, D. H. & Leder, P. (1979) *Nature (London)* **281**, 35–40.
2. Gruss, P. & Khoury, G. (1981) *Proc. Natl. Acad. Sci. USA* **78**, 133–137.
3. Mulligan, R. C., Howard, B. H. & Berg, P. (1979) *Nature (London)* **277**, 108–114.
4. Sarver, N., Gruss, P., Law, M.-F., Khoury, G. & Howley, P. M. (1981) *Mol. Cell. Biol.* **1**, 486–496.
5. Thummel, C., Tjian, R. & Grodzicker, T. (1981) *Cell* **23**, 825–836.
6. Solnick, D. (1981) *Cell* **24**, 135–143.
7. Wei, C.-M., Gibson, M., Spear, P. G. & Scolnick, E. M. (1981) *J. Virol.* **39**, 935–944.
8. Shimotohno, K. & Temin, H. M. (1981) *Cell* **26**, 67–77.
9. Moss, B. (1974) in *Comprehensive Virology*, eds. Fraenkel-Conrat, H. & Wagner, R. R. (Plenum, New York), pp. 405–474.
10. Dales, S. & Pogo, B. G. T. (1981) *Virol. Monogr.* **18**, 1–109.
11. Sam, C. K. & Dumbell, K. R. (1981) *Ann. Virol. (Inst. Pasteur)* **132E**, 135–150.
12. Nakano, E., Panicalli, D. & Paoletti, E. (1982) *Proc. Natl. Acad. Sci. USA* **79**, 1593–1596.
13. Weir, J. P., Bajszar, G. & Moss, B. (1982) *Proc. Natl. Acad. Sci. USA* **79**, 1210–1214.
14. Venkatesan, S., Baroudy, B. M. & Moss, B. (1981) *Cell* **125**, 805–813.
15. Szybalska, E. H. & Szybalski, W. (1962) *Proc. Natl. Acad. Sci. USA* **48**, 2026–2034.
16. Bolivar, F., Rodriguez, R. L., Greene, P. J., Betlach, M. C., Heyneker, H. L. & Boyer, H. W. (1977) *Gene* **2**, 95–113.
17. Birnboim, H. C. & Doly, J. (1979) *Nucleic Acids Res.* **7**, 1513–1523.
18. Winberg, G. & Hammarskjold, M. L. (1980) *Nucleic Acids Res.* **8**, 253–264.
19. Vogelstein, B. & Gillespie, D. (1979) *Proc. Natl. Acad. Sci. USA* **76**, 615–619.
20. Rhim, J. S., Cho, H. Y. & Huebner, R. J. (1975) *Int. J. Cancer* **15**, 23–29.
21. Denhardt, D. T. (1966) *Biochem. Biophys. Res. Commun.* **23**, 641–646.
22. Rigby, P. W. J., Dieckmann, M., Rhodes, C. & Berg, P. (1977) *J. Mol. Biol.* **113**, 237–251.
23. Wahl, G. M., Stern, M. & Stark, G. R. (1979) *Proc. Natl. Acad. Sci. USA* **76**, 3683–3687.
24. Southern, E. M. (1975) *J. Mol. Biol.* **98**, 503–517.
25. Summers, W. C. & Summers, W. P. (1977) *J. Virol.* **24**, 314–318.
26. Smiley, J. R., Wagner, M. J., Summers, W. P. & Summers, W. C. (1980) *Virology* **102**, 83–93.
27. Camiener, G. W. (1968) *Biochem. Pharmacol.* **17**, 1981–1991.
28. Bajszar, G., Wittek, R., Weir, J. & Moss, B. (1983) *J. Virol.* **45**, in press.
29. Panicalli, D. L., Davis, S. W., Mercer, S. R. & Paoletti, E. (1981) *J. Virol.* **37**, 1000–1010.
30. Moss, B., Winters, E. & Cooper, J. (1981) *J. Virol.* **40**, 387–395.
31. Enquist, L. W., Vande Woude, G. F., Wagner, M., Smiley, J. R. & Summers, W. C. (1979) *Gene* **7**, 335–342.
32. McKnight, S. L. (1980) *Nucleic Acids Res.* **8**, 5949–5964.
33. Wagner, M. J., Sharp, J. A. & Summers, W. C. (1981) *Proc. Natl. Acad. Sci. USA* **78**, 1441–1445.
34. Wittek, R., Cooper, J. A., Barbosa, E. & Moss, B. (1980) *Cell* **21**, 487–493.
35. Cooper, J. A., Wittek, R. & Moss, B. (1981) *J. Virol.* **37**, 284–294.
36. Wittek, R., Cooper, J. A. & Moss, B. (1981) *J. Virol.* **39**, 722–737.
37. Wittek, R. & Moss, B. (1982) *J. Virol.* **42**, 447–455.
38. Moss, B., Winters, E. & Cooper, N. (1981) *Proc. Natl. Acad. Sci. USA* **78**, 1614–1618.
39. Moyer, R. W. & Rothe, C. T. (1980) *Virology* **102**, 119–132.

Human Hepatitis B Vaccine from Recombinant Yeast

W.J. McAleer, E.B. Buynak, R.Z. Maigetter, D.E. Wampler, W.J. Miller and M.R. Hilleman

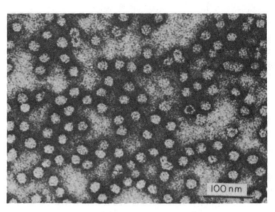

Fig. 1 Electron micrograph of HBsAg particles from recombinant yeast. Cells were grown in a 335-l fermentation vessel, collected by centrifugation, resuspended in an equal volume of 0.01 M sodium phosphate pH 7.5, containing 0.01% Triton X-100, and disrupted by rapid stirring with glass beads in a Dyno-Mill (Impandex; see ref. 23). The resulting extract was clarified by centrifugation for 90 min at 10,000g. The clarified yeast extract was applied to a column of Sepharose 4B to which had been attached goat antibody to human HBsAg. The column was developed at a flow rate of 2 column vol per h. Extraneous protein was washed away with 5 column vol of buffer A and the HBsAg was eluted with 3 M NH_4SCN. Fractions containing HBsAg were pooled and thiocyanate was removed by dialysis against 0.01 M sodium phosphate pH 6.8, containing 0.15 M NaCl. Dialysed antigen was diluted to 40 μg ml^{-1} and visualized by negative staining with 2% phosphotungstic acid.

The worldwide importance of human hepatitis B virus infection and the toll it takes in chronic liver disease, cirrhosis and hepatocarcinoma, make it imperative that a vaccine be developed for worldwide application[1]. Human hepatitis B vaccines[2-6] are presently prepared using hepatitis B surface antigen (HBsAg) that is purified from the plasma of human carriers of hepatitis B virus infection. The preparation of hepatitis B vaccine from a human source is restricted by the available supply of infected human plasma and by the need to apply stringent processes that purify the antigen and render it free of infectious hepatitis B virus and other possible living agents that might be present in the plasma. Joint efforts between our laboratories and those of Drs W. Rutter and B. Hall led to the preparation of vectors carrying the DNA sequence[7,8] for HBsAg and antigen expression in the yeast *Saccharomyces cerevisiae*[9]. Here we describe the development of hepatitis B vaccine of yeast cell origin. HBsAg of subtype adw was produced in recombinant yeast cell culture, and the purified antigen in alum formulation stimulated production of antibody in mice, grivet monkeys and chimpanzees. Vaccinated chimpanzees were totally protected when challenged intravenously with either homologous or heterologous subtype adr and ayw virus of human serum source. This is the first example of a vaccine produced from recombinant cells which is effective against a human viral infection.

Several alternative approaches to a hepatitis B vaccine are being developed. HBsAg has been expressed by several transformed mammalian cell lines, such as the human hepatoma line, PLC/PRF/5 (refs 10, 11), simian virus 40-infected monkey kidney cells[12] and mouse L cells[13]. These sources are of some concern, however, because the cell lines may be neoplastic. Although HBsAg has been cloned in bacteria[7,14], expression was very weak. Other laboratories[15-19] have described the synthesis of oligopeptides that carry antigenic determinants of HBsAg but their potency in animals is low and much work will need to be done to potentiate antigenicity. Smith and collaborators[20] have described the construction of a recombinant vaccinia virus which expresses HBsAg and have proposed its use as a live attenuated vaccine; its antigenic potency has been demonstrated but whether such a vaccine would be safe and effective in man is still unknown.

Valenzuela *et al.*[9] originally reported that yeast cells are able not only to express the HBsAg gene but also to assemble the polypeptides into particles that have much the same appearance as particles isolated from human plasma and which are immunogenic in mice. Since then, other laboratories[21,22] have shown that HBsAg produced in yeast is antigenic in rabbits and guinea pigs. With such progress, recombinant yeast has become an attractive alternative to human plasma as a source of antigen for hepatitis B vaccine.

For vaccine preparation, the HBsAg used was of subtype adw and was produced in fermentation cultures of *S. cerevisiae* carrying an expression vector using yeast alcohol dehydrogenase I as a promoter. The yeast strain used in these studies was obtained from G. Ammerer (University of Washington) and is similar to the strain described by Valenzuela *et al.*[9] in which the production of HBsAg in yeast was first reported.

Cells were collected by centrifugation and broken by homogenization with glass beads[23]. HBsAg particles were purified from the clarified extract by immune affinity chromatography using goat antibody to human HBsAg. Electron microscopy (Fig. 1) revealed a homogeneous array of particles free of extraneous morphological entities. The UV absorption pattern was the same as for the plasma antigen, with an $E^{1\%}$ of 45. SDS-polyacrylamide gel electrophoresis (Fig. 2) in reducing conditions revealed a major band at molecular

Table 1 Antigenic potency in mice of HBsAg purified from yeast and from human plasma

Vaccine source	Antigen dose per injection (μg protein)	Anti-HbsAg response after vaccination	
		No. positive/total	GMT
Human plasma (lot 799-2)	10	9/10	563
	2.5	10/10	2,235
	0.625	4/9	32
	0.156	0/10	4
ED$_{50}$	0.639		
Yeast (lot 81-4)	40	10/10	5,432
	10	10/10	3,400
	2.5	8/10	673
	0.625	8/10	967
ED$_{50}$	<0.625		

Groups of 10 5-week-old ICR/Ha mice propagated in our laboratories were given a single 1-ml injection intraperitoneally of serial fourfold dilutions of yeast or human plasma vaccine in alum diluent. The mice were bled individually and tested for serum antibody level 4 weeks later. Human plasma vaccine, lot 799-2, was prepared in these laboratories[2-4]. Yeast-derived vaccine, lot 81-4, was purified as described in Fig. 1 legend and adsorbed to alum. GMT, geometric mean titre, expressed in AUSAB units; ED$_{50}$, dose required to seroconvert 50% of the mice.

---

* To whom reprint requests should be addressed.

weight 23,000 (23K) corresponding to the non-glycosylated polypeptide which is the major polypeptide of the viral envelope. In this respect it differs from the plasma antigen which has, in addition to the 23K polypeptide, a glycosylated derivative which migrates at 27K. The yeast and plasma antigens differ also in their reactivity in the radioimmunoassay (RIA) (AUSRIA II, Abbott). RIA reactivity of purified yeast-derived HBsAg varied from preparation to preparation in the range 20–50% of the reference human antigen.

Because of this reduced radioimmune reactivity, and because the yeast antigen is not glycosylated, it was important to determine whether the antigen was immunogenic. To test both antigenicity and immunogenicity in animals, purified antigen was formulated into a vaccine by adsorbing on alum adjuvant to contain 40 μg HBsAg protein and 0.5 mg aluminium (hydroxide) per 1 ml dose.

Studies in mice (Table 1) showed the yeast-derived antigen to be at least as antigenic as the antigen purified from human plasma. Grivet monkeys also developed antibody following vaccination with the yeast-derived antigen (Table 2). A single injection of the vaccine at all dose levels resulted in seroconversion of all the animals in both vaccine groups. These results were important as they showed that high antibody titres were maintained for at least a year.

Protective efficacy was tested for by using susceptible chimpanzees. The four chimpanzees that received the recombinant vaccine developed antibody in substantial titre following vaccination (Table 3). Following challenge with infectious human plasma, all four vaccinated animals were protected. By contrast, all four unvaccinated animals developed hepatitis B virus infection with positive antigenaemia, antibody to hepatitis B core antigen (anti-HBcAg), elevation of serum glutamic oxalacetic transaminase (SGOT) and serum glutamic pyruvic transaminase (SGPT), and liver histopathology. It is important to note that the animals were protected against both subtype adr and ayw challenge even though the vaccine is of the adw subtype.

Yeast fermentation technology is well established and we have shown that HBsAg can be isolated from yeast extracts in a highly purified form by a single application of immune affinity chromatography. Vaccine made from this antigen is equally as potent as human plasma-derived vaccine in stimulating antibodies in mice, and is protective in challenge experiments in chimpanzees. Antibodies raised by yeast-derived vaccine persisted for at least a year in monkeys, showing no important deviation from that of the plasma vaccine.

Human HBsAg is composed of a sequence of 226 amino acids of which the a antigen determinant is dominant. Small differences in amino acid sequence may occur at several positions in the polypeptide chain and are responsible for the subtype specificities[24]. In previous studies, chimpanzees that were cross-challenged with heterologous subtypes of hepatitis B virus after recovery from infection or vaccination with human plasma-

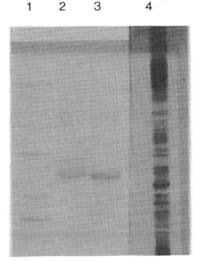

Fig. 2 SDS-polyacrylamide gel electrophoresis of cell culture and yeast-derived HBsAg. All samples were reduced, denatured and electrophoresed as described by Laemmli[30]. After electrophoresis, polypeptides were visualized with Coomassie brilliant blue (lanes 1–3) or with the silver stain procedure described by Morrissey[31] (lane 4). Lane 1, molecular weight standards (3 μg each): phosphorylase b (94K), bovine serum albumin (68K), ovalbumin (43K), carbonic anhydrase (30K), soybean trypsin inhibitor (21K) and lysozyme (14.3K). Lane 2, 30 μg of HBsAg from the human hepatoma cell line PLC/PRF/5 (ref. 10), also purified from yeast as described in Fig. 1 legend. Lane 3, 30 μg of HBsAg purified from yeast as described in Fig. 1. Lane 4, 10 μg of clarified yeast extract as described in Fig. 1 legend.

derived antigens, were solidly protected due to the common group specificity of the dominant a antigen that is present in all HBsAg subtypes[25]. A protective efficacy trial in man of subtype ad vaccine of human plasma origin has shown strong protection against the homologous subtype[26,27] and, most recently, against the heterologous subtype ay[28] in studies carried out on the staffs of renal dialysis centres where subtype ay hepatitis is most common. The positive cross-protection afforded against heterologous subtype ayw virus challenge in chimpanzee immunized with type adw vaccine of yeast origin, indicates that the a antigen remains dominant in the recombinant-produced antigen obtained from human plasma.

Table 2 Antigenic potency in grivet monkeys of HBsAg purified from yeast and from human plasma

Vaccine source	Antigen dose per injection (μg protein)	Week 4	Anti-HBsAg response after initial vaccine dose (geometric mean titre)		
			Week 8	Week 12	Week 52
Human plasma	10	36	213	170	127
(lot 86016)	2.5	343	6,227	17,348	9.924
	0.625	53	4,642	3,164	5,688
	0.156	15	128	83	358
Yeast	40	88	1,078	7,103	11,554
(lot 81-4)	10	184	877	8,489	4,984
	2.5	225	1,168	6,361	10,868
	0.625	109	925	518	313

A group of four initially seronegative grivet monkeys (*Cercopithecus aethiops*), weighing 3–5 kg, were each given two 1-ml intramuscular (i.m.) doses of yeast or human plasma vaccine 4 weeks apart. Dilutions of antigen were made in alum placebo of the same composition as the vaccine. Animals were bled at biweekly intervals for 1 yr and tested for antibody to HBsAg by using a commercial RIA kit (AUSAB, Abbott). Protein was measured by the method of Lowry[29]. Human plasma lot 86016 was prepared in these laboratories[2–4]

Table 3 Protective efficacy in chimpanzees of HBsAg purified from yeast and from human plasma

Injection	Chimp no.	Before challenge Anti-HBsAg titre (at 12 weeks)	Antigen subtype	After challenge (week of onset or weeks of duration) HBsAg Onset	Duration	Anti-HBcAg Onset	Duration	SGOT elevation Onset	Duration	SGPT elevation Onset	Duration	Liver pathology onset
Yeast vaccine	110	1,830	adr	–	–	–	–	–	–	–	–	–
(lot 81-4)	138	540	adr	–	–	–	–	–	–	–	–	–
	103	18,300	ayw	–	–	–	–	–	–	–	–	–
	120	7,200	ayw	–	–	–	–	–	–	–	–	–
Unvaccinated	111	<8	adr	10	10	15	9	17	3	17	6	20
controls	128	<8	adr	8	11	12	12	17	3	16	5	20
	127	<8	ayw	6	14	12	12	13	3	13	7	16
	130	<8	ayw	6	18	10	14	22	1	14	10	24

Eight chimpanzees, each weighing 40–60 kg, were selected for study based on negative findings in tests for HBsAg, anti-HBsAg, anti-HBcAg, elevation in transaminase, liver histopathology and tuberculin reaction. The animals were separated into two groups, four test animals and four controls. Each of the four test animals was given three 40-μg doses of yeast-derived HBsAg vaccine in 1 ml volume i.m. at 4-week intervals. All eight animals were then challenged by intravenous injection of 1,000 chimpanzee infectious doses of subtype adr or ayw virus in 1 ml of human hepatitis B plasma. Antigen and antibody titres were measured by commercial (Abbott) RIA kits (AUSRIA, AUSAB and CORAB for HBsAg, anti-HBsAg and anti-HBcAg, respectively). SGOT and SGPT assays were performed by the Sigma-Frankel (no. 505) and by the UV absorption (Boehringer-Mannheim) procedures, respectively. SGOT titres >40 and SGPT titres >30 were considered elevated. The subtype adr and ayw human plasmas used for challenge were obtained from Drs R. Gerety and E. Tabor of the Office of Biologics, US Food and Drug Administration; they were of measured viral infectiousness for chimpanzees and were subtyped serologically. The animals were bled at weekly intervals during the 36-week period of observation, covering 12 weeks before virus challenge and 24 weeks after. Liver biopsies were taken at 4-week intervals using a Menghini 16T needle. The tissues were fixed in 10% buffered formalin solution and the haematoxylin/eosin-stained sections were prepared by Dr A. Phelps of these laboratories under blind code number. The tests were carried out in animals that were held in isolation in the facilities of Dr William E. Greer at the Gulf South Research Institute, New Iberia, Louisiana. –, All remained negative.

We thank Dr C. E. Carty and F. X. Kovach for assistance in fermentation, B. J. Harder, N. Grason and J. Bailey for assistance in antigen purification, Dr B. Wolanski and R. Ziegler for electron microscopy, and H. E. Darmofal, J. T. Deviney, K. I. Guckert, R. R. Roehm and L. W. Stanton for assistance in the animal tests.

Received 11 July; accepted 16 November 1983.

1. Deinhardt, F. & Gust, I. D. *Bull. Wld Hlth Org.* **60**, 661–691 (1982).
2. Hilleman, M. R. *et al. Viral Hepatitis* (eds Vyas, G. N., Cohen, S. N. & Schmid, R.) 525–537 (The Franklin Institute Press, Philadelphia, 1978).
3. Hilleman, M. R. *et al.* in *Viral Hepatitis 1981 int. Symp.* (eds Szmuness, W., Alter, H. & Maynard, J.) 385–397 (The Franklin Institute Press, Philadelphia, 1982).
4. Buynak, E. B. *et al. J. Am. med. Ass.* **235**, 2832–2834 (1976).
5. Adamowicz, P. *et al. INSERM Symp.* No. 18 (eds Maupas, P. & Guesry, P.) 37–49 (Elsevier, Amsterdam, 1981).
6. Coutinho, R. A. *et al. Br. med. J.* **286**, 1305–1308 (1983).
7. Valenzuela, P. *et al. Nature* **280**, 815–819 (1979).
8. Edman, J. C., Hallewell, R. A., Valenzuela, P., Goodman, H. M. & Rutter, W. J. *Nature* **291**, 503–506 (1981).
9. Valenzuela, P., Medina, A., Rutter, W. J., Ammerer, G. & Hall, B. D. *Nature* **298**, 347–350 (1982).
10. Alexander, J. J., Bey, E. M., Geddes, E. W. & Lecatsas, G. *S. Afr. med. J.* **50**, 2124–2128 (1976).
11. Barin, F., Maupas, P., Coursaget, P., Goudeau, A. & Chiron, J. P. in *INSERM Symp.* No. 18 (eds Maupas, P. & Guesry, P.) 263–266 (Elsevier, Amsterdam, 1981).
12. Moriarty, A. M., Hoyer, B. H., Shih, J. W-K., Gerin, J. L. & Hamer, D. H. *Proc. natn. Acad. Sci. U.S.A.* **78**, 2606–2610 (1981).
13. Dubois, M. F., Pourcel, C., Rousset, S., Chany, C. & Tiollais, P. *Proc. natn. Acad. Sci. U.S.A.* **77**, 4549–4553 (1980).
14. Burrell, J. C., Mackay, P., Greenway, P. J., Hofschneider, P. H. & Murray, K. *Nature* **279**, 43–47 (1979).
15. Lerner, R. A. *et al. Proc. natn. Acad. Sci. U.S.A.* **78**, 3403–3407 (1981).
16. Dreesman, G. R. *et al. Nature* **295**, 158–160 (1982).
17. Bhatnagar, P. K. *et al. Proc. natn. Acad. Sci. U.S.A.* **79**, 4400–4404 (1982).
18. Prince, A. M., Ikram, H. & Hopp, T. P. *Proc. natn. Acad. Sci. U.S.A.* **79**, 579–582 (1982).
19. Gerin, J. L. *et al. Proc. natn. Acad. Sci. U.S.A.* **80**, 2365–2369 (1983).
20. Smith, G. L., Mackett, M. & Moss, B. *Nature* **302**, 490–495 (1983).
21. Miyanohara, A. *et al. Proc. natn. Acad. Sci. U.S.A.* **80**, 1–5 (1983).
22. Hitzeman, R. A. *et al. Nucleic Acids Res.* **11**, 2745–2763 (1983).
23. Deters, D., Muller, U. & Homberger, H. *Analyt. Biochem.* **70**, 263–267 (1976).
24. Ono, Y. *et al. Nucleic Acids Res.* **11**, 1747–1757 (1983).
25. Gerety, R. J., Tabor, E., Purcell, R. H. & Tyeryar, F. J. *J. infect. Dis.* **140**, 642–648 (1979).
26. Szmuness, W., Stevens, C. E., Zang, E. A., Harley, E. J. & Kellner, A. *Hepatology* **1**, 377–385 (1981).
27. Francis, D. P. *et al. Ann. intern. Med.* **97**, 362–366 (1982).
28. Szmuness, W. *et al. New Engl. J. Med.* **307**, 1481–1486 (1982).
29. Lowry, O. H., Rosebrough, N. J., Farr, A. L. & Randall, R. J. *biol. Chem.* **193**, 265–275 (1951).
30. Laemmli, U. K. *Nature* **227**, 680–685 (1970).
31. Morrissey, J. H. *Analyt. Biochem.* **117**, 307–310 (1981).

Construction of Poxviruses as Cloning Vectors: Insertion of the Thymidine Kinase Gene from Herpes Simplex Virus into the DNA of Infectious Vaccinia Virus

D. Panicali and E. Paoletti

ABSTRACT   We have constructed recombinant vaccinia viruses containing the thymidine kinase gene from herpes simplex virus. The gene was inserted into the genome of a variant of vaccinia virus that had undergone spontaneous deletion as well as into the 120-megadalton genome of the large prototypic vaccinia variant. This was accomplished via in vivo recombination by cotransfection of eukaryotic tissue culture cells with cloned BamHI-digested thymidine kinase gene from herpes simplex virus containing flanking vaccinia virus DNA sequences and infectious rescuing vaccinia virus. Pure populations of the recombinant viruses were obtained by replica filter techniques or by growth of the recombinant virus in biochemically selective medium. The herpes simplex virus thymidine kinase gene, as an insert in vaccinia virus, is transcribed in vivo and in vitro, and the fidelity of in vivo transcription into a functional gene product was detected by the phosphorylation of 5-[^{125}I]iodo-2'-deoxycytidine.

Various methods for the introduction of defined foreign DNA sequences into eukaryotic cells are currently under investigation. These include microinjection of DNA (1), fusion of liposomes containing DNA (2, 3), erythrocyte-mediated gene transfer (4), direct introduction into cells of calcium orthophosphate-precipitated DNA (5–8), and eukaryotic viral vectors (9–18).

Recently, we demonstrated that endogenous subgenomic elements can be inserted into infectious progeny vaccinia virus via recombination in vivo (19). The ability to integrate vaccinia virus DNA sequences into infectious vaccinia virus progeny suggested the possibility for insertion of foreign genetic elements into infectious vaccinia virus via similar protocols. To test this, we chose the thymidine kinase (TK) gene of herpes simplex virus (HSV). The HSV TK gene is available as a clone, and its sequence has been determined and is known to contain the gene's own regulatory signals (20, 21). Vaccinia recombinant viruses containing the HSV TK gene can be detected by nucleic acid hybridization, by selection in appropriate methotrexate-containing medium, or by a specific enzymatic assay.

We were able to insert the HSV TK gene into a number of vaccinia virus preparations and to obtain pure cultures of recombinant vaccinia virus expressing the herpes virus gene. These studies are reported in this communication.

MATERIALS AND METHODS

Cells and Viruses. TK⁻ human (line 143) cells derived by C. Croce and K. Huebner (Wistar Institute, Philadelphia, PA) were obtained from B. Moss (National Institutes of Health, Bethesda, MD). African green monkey kidney cells (CV-1) and baby hamster kidney cells (BHK-21) (C13) were maintained as monolayer cultures in Eagle's modified medium containing 10% fetal calf serum. The L and S variants of vaccinia virus have been described in detail (22). VTK⁻11, a TK⁻ mutant containing the L variant genome, was derived by R. Condit (State University of New York, Buffalo) and obtained through D. Hruby (University of Wisconsin, Madison). VTK⁻79, a TK⁻ mutant containing the S variant genome, was derived in our laboratory.

Construction of Plasmid Vectors. Plasmid pBR322 in which the BamHI Q fragment of HSV DNA (which includes the TK gene) had been inserted into the BamHI site (23) was provided by D. Hruby. The HindIII F fragment of vaccinia virus DNA was isolated from preparative agarose gels by binding to glass powder (24) as described (22). Approximately 200 ng of HindIII-cleaved pBR322 DNA was combined with 500 ng of isolated F fragment and ligated with phage T4 DNA ligase (New England BioLabs) in 50 mM Tris·HCl, pH 7.6/10 mM MgCl₂/10 mM dithiothreitol/1 mM ATP overnight at 10°C. Escherichia coli HB101 (25) cells were transformed (26) with the ligated DNA. Ampicillin-resistant tetracycline-sensitive recombinants were analyzed by restriction analysis by the method of Holmes and Quigley (27). A recombinant plasmid containing the HindIII F fragment was isolated and designated pDP3. Preparative amounts of pDP3 DNA were prepared by the cleared lysate procedure of Clewell and Helinski (28) after chloramphenicol amplification (29). Linearized pDP3 DNA from a partial BamHI digest was ligated with purified BamHI TK fragment and used to transform E. coli HB101. Transformants were screened for BamHI-produced TK inserts by colony hybridization (30).

Construction, Screening, and Recovery of Recombinant Vaccinia Viruses. Insertion of the HSV TK gene into infectious vaccinia virus followed the general protocol for the rescue of endogenous genetic markers described by Nakano et al. (19). Five to 10 μg of HindIII-cleaved pDP132 or pDP137 DNA was coprecipitated with 2–10 μg of vaccinia virus carrier DNA. The calcium orthophosphate-precipitated DNA (0.5 ml) was flooded onto either BHK 21 or CV-1 monolayers previously infected (1 hr) with infectious vaccinia virus. The cells were harvested after 24 hr, lysed by three cycles of freezing and thawing, and screened for recombinant vaccinia viruses. Three approaches were utilized for screening and isolating recombinant vaccinia viruses containing the HSV TK gene: (i) Infected cell monolayers on which the plaques were visualized by neutral red staining were imprinted onto nitrocellulose filters according to Villarreal and Berg (31) as described (19). A mirror image replica was imprinted onto a second nitrocellulose filter by contact with the primary filter. The primary nitrocellulose filter was used for in situ hybridization (31), using ³²P-labeled nick-translated (32) BamHI HSV TK DNA fragment as probe to detect viral plaques containing the HSV TK insert. Areas of the secondary nitrocellulose replica filter corresponding to those plaques showing

Abbreviations: TK, thymidine kinase; HSV, herpes simplex virus; MDal, megadalton(s).

PANICALI, D. and PAOLETTI, E.
Construction of poxviruses as cloning vectors: insertion of the thymidine kinase gene from herpes simplex virus into the DNA of infectious vaccinia virus. Proc. Natl. Acad. Sci. U.S.A. 79:4927-4931 (1982). Reprinted with the authors' permission.

positive hybridization were punched out and infectious recombinant viruses were recovered by inoculating monolayer cultures. (*ii*) Recombinant vaccinia virus expressing HSV TK activity on monolayers of TK⁻ human (line 143) cells were selected in the presence of methotrexate-containing medium (33) by using a modification of the procedure of Campione-Piccardo *et al.* (34). This cell line is viable in the presence of methotrexate-containing medium for up to 48 hr, long enough to allow detection of vaccinia virus plaques. (*iii*) Vaccinia virus recombinants containing functional HSV TK were detected by the phosphorylation of [^{125}I]iododeoxycytidine. Vaccinia virus was plated on monolayers of CV-1, and after 48 hr, 5 μCi (1 Ci = 3.7×10^{10} becquerels) of 5-[^{125}I]iodo-2′-deoxycytidine (New England Nuclear) was added in 1.5 ml of medium for overnight labeling. After staining with neutral red to visualize the plaques, the monolayers were washed three times with phosphate-buffered saline and imprinted onto nitrocellulose filters, and plaques expressing HSV TK were localized by radioautography. Infectious recombinant virus was recovered from punched-out areas of the nitrocellulose filter. All recombinant viruses were further purified by two consecutive cycles of plaque isolation.

DNA Restriction Analysis of Recombinant Vaccinia Virus. DNA from recombinant vaccinia virus-infected CV-1 monolayers was extracted by the procedure of Esposito *et al.* (35) or from virus purified from infected HeLa cells according to Joklik (36) as described (19, 22). *Hind*III-, *Bam*HI-, or *Sst* I-digested DNA was analyzed on agarose gels and transferred to nitrocellulose (Schleicher and Schuell, BA85) (37) for hybridization with ^{32}P-labeled nick-translated HSV *TK* probe as described (22, 38).

Analysis of RNA Transcripts. ^{32}P-Labeled recombinant vaccinia virus RNA was synthesized *in vitro* by using [α-^{32}P]GTP as described (22). The RNA was selected by hybridization and eluted from cloned HSV *TK* DNA bound to diazobenzyloxymethyl-paper as described (39). The selected RNA was hybridized to Southern blots of pDP137 digested with *Sst* I or with *Hind*III/*Bam*HI as described (22).

RESULTS

Strategy for the Construction of Recombinant Vaccinia Virus. The noninfectious nature of purified vaccinia virus DNA, its cytoplasmic site of replication, and its large genome present unique problems to approaches of genetic manipulation of the virus. A significant breakthrough demonstrating the feasibility of marker rescue experiments using defined DNA fragments has recently been reported (19, 40). The insertion of foreign DNA into infectious progeny vaccinia virus involves the formation of recombinant DNA containing the foreign gene flanked by contiguous vaccinia virus genomic sequences. This chimeric DNA is introduced into cells as a calcium orthophosphate precipitate. The cells are additionally infected with vaccinia virus. *In vivo* recombination occurs between the flanking vaccinia virus DNA sequences and homologous sequences present in the replicating vaccinia virus genome. The recombined DNA containing the foreign genetic insert replicates and is subsequently packaged into infectious virus.

Construction of Chimeric Plasmid Vectors. Insertion of foreign genes into vaccinia virus must occur at loci that do not disrupt essential gene functions. An obvious locus of nonessential DNA resides within the 5% of the genome deleted from the L variant vaccinia virus previously described (22). As part of another project to identify nonessential genetic loci, the single *Bam*HI site localized within the internal *Hind*III F fragment (22) was found not to disrupt essential genetic information and was utilized for the insertion of the HSV *TK* fragment. The construction of the recombinant DNA plasmids utilized in this study is outlined in Fig. 1. The 2.8-megadalton (MDal) pBR322

digested with *Hind*III was ligated with purified 8.6-MDal vaccinia virus *Hind*III F fragment. Full-length linearized plasmid from a partial *Bam*HI digest of the recombinant (11.3-MDal) plasmid pDP3 was isolated and ligated with the 2.3-MDal HSV *Bam*HI *TK* fragment. Plasmids containing the HSV *TK* insert were identified by colony hybridization using ^{32}P-labeled nick-translated HSV *TK* DNA probe. Two of these plasmids, pDP132 (13.6 MDal) and pDP137 (15.9 MDal), which contain the HSV *TK* gene inserted into the proper *Bam*HI site, were selected for further studies after analysis with *Hind*III and *Sst* I (*Sac* I). The presence of a single, asymmetric *Sst* I restriction site in both the *Hind*III F fragment and the *HSV TK* was used to determine the orientation and copy number of the HSV *TK* insert. *Sst* I digestion of pDP132 gave two fragments of 10.1 and 3.5 MDal, whereas *Sst* I digestion of pDP137 generated three fragments of 10.8, 2.8, and 2.3 MDal. These data demonstrated that the insertion of the HSV *TK* in pDP132 was in opposite orientation from the HSV *TK* insert in pDP137 and that two copies of the *TK* gene were present in pDP137 as a "head-to-tail" tandem. This was confirmed by additional digestion with *Bam*HI and densitometric analysis.

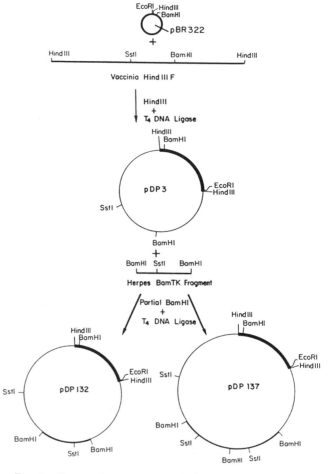

FIG. 1. Construction of chimeric vaccinia virus–herpesvirus plasmids. The purified 8.6-MDal vaccinia *Hind*III F fragment was combined with the *Hind*III-digested 2.8-MDal pBR322 plasmid and ligated with T4 DNA ligase. A partial *Bam*HI digest of the derivative pDP3 (11.3 MDal) plasmid was combined with the (2.3-MDal) HSV *Bam*HI *TK* fragment and ligated with T4 DNA ligase. After detection of colonies containing the HSV *TK* gene by colony hybridization and further additional restriction analysis, two plasmids, pDP132 (13.6 MDal) and pDP137 (15.9 MDal), were selected. pDP137 contains a head-to-tail tandem of the HSV *TK* in opposite orientation from the single-copy HSV *TK* gene in pDP132. *Bam*HI, *Sst* I, *Hind*III, and *Eco*RI sites are mapped as indicated and the plasmids are drawn essentially to scale.

Construction and Recovery of Recombinant Virus Containing the HSV _TK_ gene. The construction and analysis of six recombinant vaccinia virus populations, vP1–6, containing the HSV _TK_ gene in the single _Bam_HI site of the vaccinia virus _Hind_III F fragment will be described. The vaccinia virus recombinants vP1 and vP2 were constructed by employing the strategy of _in vivo_ recombination outlined above. Donor DNA consisted of a calcium orthophosphate precipitate of pDP132 for vP1 or pDP137 for vP2 and purified S variant vaccinia virus DNA as carrier. The recipient was infectious S variant vaccinia virus. _In vivo_ recombination and subsequent screening for vaccinia virus progeny containing the HSV _TK_ insert were performed on CV-1 cells with two approaches. The first consisted of screening replica nitrocellulose filters. Approximately 0.4% of the plaques generated from the progeny of _in vivo_ recombination contained the HSV _TK_ insert, as detected by _in situ_ hybridization. The second approach took advantage of the observation of Cooper (41) that the substrate specificity of the HSV TK allows phosphorylation of halogenated pyrimidine nucleosides, whereas cellular TK does not. This observation has been exploited by Summers and Summers (42), who used [125I]-iododeoxycytidine in combination with radioautography as a sensitive and specific assay for HSV TK. This assay can be used to detect vaccinia virus recombinants containing and expressing the HVS _TK_ even if vaccinia virus carries its own TK activity, because the vaccinia virus TK fails to phosphorylate iododeoxycytidine. With this assay, no plaques equivalent to vP1 generated with pDP132 donor DNA were detected, but a number of plaques equivalent to vP2 generated with pDP137 donor DNA were detected, suggesting that the HSV _TK_ was expressed in the vaccinia virus recombinant vP2 but not in recombinant vP1. The ability of recombinant vaccinia virus containing the HSV _TK_ to phosphorylate [125I]iododeoxycytidine is shown in Fig. 2. Only recombinant vaccinia virus vP2, containing the HSV _TK_ gene insert, utilized [125I]iododeoxycytidine. Uninfected, vaccinia virus-infected, or recombinant vaccinia virus vP1-infected CV-1 monolayers did not. The only difference between vP1 and vP2 is the orientation of HSV _TK_ in the donor DNA.

Additional vaccinia virus recombinants vP3 and vP4 were

generated by _in vivo_ recombination in BHK 21 cells by using pDP132 and pDP137 donor DNA, respectively. In this case, the carrier DNA used was purified from VTK⁻79, an S variant vaccinia deficient in TK activity. VTK⁻79 was also the rescuing virus. Because VTK⁻79 lacks endogenous TK activity, successful insertion and expression of the inserted HSV _TK_ gene was monitored in recombinant vaccinia virus progeny by growth on TK⁻ human (line 143) cells in the presence of methotrexate-containing medium, using modifications of the procedure described by Campione-Piccardo _et al._ (34). Plaques were detected in neutral red-stained monolayers 48 hr after infection when pDP137 was used as donor DNA to generate recombinant vaccinia virus vP4, suggesting successful insertion and expression of the HSV _TK_ gene into VTK⁻79 vaccinia virus. No viral plaques were detected when pDP132 was used as donor DNA. That this latter result was due to a failure to express the HSV _TK_ and not to a failure to insert the gene into the VTK⁻79 vaccinia virus, was subsequently demonstrated by finding vP3 recombinant viruses containing the HSV _TK_ insert by the _in situ_ hybridization and replica filter method described above.

Two additional virus recombinants containing the HSV _TK_ insert were derived by _in vivo_ recombination in BHK monolayers. Donor DNA was either pDP132 or pDP137. Carrier DNA was derived from VTK⁻11, an L variant of vaccinia virus deficient in TK activity. The rescuing infectious vaccinia virus was also VTK⁻11. Two recombinants of VTK⁻11, vP5 and vP6, were obtained by using pDP132 and pDP137, respectively. These two recombinants were identified by _in situ_ hybridization for the presence of the HSV _TK_ insert and infectious progeny derived from replica nitrocellulose filters. With the [125I]iododeoxycytidine assay vP6 was positive for HSV _TK_ expression but, as expected, vP5 was not.

Restriction Analysis of Recombinant Vaccinia Virions Modified by the Insertion of the HSV _TK_ Gene. DNA was extracted from S variant and from vP1 and vP2 recombinant vaccinia viruses, digested with _Hind_III, _Bam_HI, or _Sst_ I, and analyzed on agarose gels. As indicated in Fig. 3A, _Hind_III digestion of vP1 (lane 2) and vP2 (lane 3) generated a new 10.9-MDal fragment not found in the S variant (lane 1). The 10.9-MDal fragment is composed of the 8.6-MDal internal vaccinia virus _Hind_III F fragment and the 2.3-MDal HSV _TK_ insert. It should be noted that two moles of the _Hind_III F fragment of the S variant DNA are produced by digestion, one internal fragment and the left terminal fragment (22), but one mole of the F fragment is found in digests of vP1 and vP2, as expected from the insertion of the HSV _TK_ into the internal F fragment. Digestion of vP1 (lane 5) and vP2 (lane 6) with _Bam_HI gives an additional one mole of 2.3-MDal fragment not found in the _Bam_HI digestion of S variant DNA (lane 4), consistent with the size of the HSV _TK_ insert. Digestion of vP1 (lane 8) and vP2 (lane 9) with _Sst_ I gives new fragments not found in the S variant DNA (lane 7). _Sst_ I digestion of vP1 (lane 8) DNA gives fragments of 17.0 and 3.4 MDal while vP2 (lane 9) fragments of 17.7 and 2.7 MDal are detected. Because _Sst_ I cuts the HSV _TK_ gene once and asymmetrically, the digests of vP1 and vP2 confirm the insertion of the HSV _TK_ into the _Hind_III F fragment in opposite orientation. Additionally, the absence of a 2.3-MDal band from vP2 (lane 9) confirms the presence of a single HSV _TK_ insert in the recombinant virus. A Southern blot (37) of the digested DNA was prepared and ³²P-labeled nick-translated HSV _TK_ DNA was hybridized under conditions previously described (22). As expected, the labeled probe hybridizes to the novel DNA fragments in vP1 and vP2 containing the HSV _TK_ gene as described above (Fig. 3B, lanes 2, 3, 5, 6, 8, and 9). No hybridization of the probe to S variant digested DNA (lanes 1, 4, and 7) was detected. Reciprocal hybridizations using labeled vaccinia virus DNA as probe resulted in hybridization to all of the vaccinia

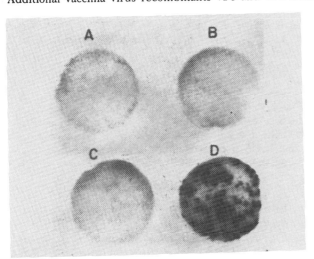

FIG. 2. [125I]Iododeoxycytidine assay for the expression of HSV _TK_ in recombinant vaccinia virus. Uninfected or 48-hr infected monolayers of CV-1 were incubated with 1.5 ml of liquid overlay medium containing 5 μCi of [125I]iododeoxycytidine for 16 hr and washed three times with phosphate-buffered saline, and the neutral red-stained monolayers were imprinted onto nitrocellulose filters. A radioautogram of uninfected (_A_), S variant vaccinia virus-infected (_B_), recombinant vaccinia virus vP1-infected (_C_), or recombinant vaccinia virus vP2-infected (_D_) CV-1 monolayers is shown.

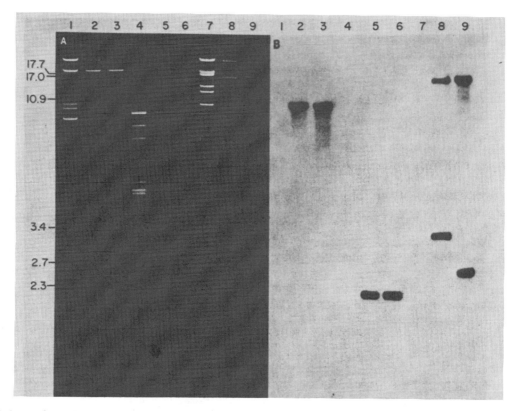

FIG. 3. Restriction analysis of recombinant vaccinia virus. DNAs of S variant and recombinant vP1 and vP2 vaccinia viruses were digested with *Hin*dIII, *Bam*HI, or *Sst* I and analyzed on an agarose gel. (*A*) *Hin*dIII digest of S variant (lane 1), vP1 (lane 2), and vP2 (lane 3); *Bam*HI digest of S variant (lane 4), vP1 (lane 5), and vP2 (lane 6); *Sst* I digest of S variant (lane 7), vP1 (lane 8), and vP2 (lane 9). (*B*) Radioautograph of the Southern blot to which ^{32}P-labeled nick-translated HSV *TK* DNA (1×10^5 cpm/ml) was hybridized. Hybridization to the 10.9-MDal *Hin*dIII fragment derived from vP1 (lane 2) and vP2 (lane 3), the 2.3-MDal *Bam*HI fragment derived from vP1 (lane 5) and vP2 (lane 6), and the 17.0- and 3.4-MDal fragments derived from vP1 (lane 8) and the 17.7- and 2.7-MDal fragments derived from vP2 (lane 9) by *Sst* I digestion are indicated. No hybridization to restricted S variant DNA (lanes 1, 4, and 7) was detected.

virus DNA-containing fragments but not to the *Bam*HI HSV *TK* fragment (data not shown).

Analysis of Transcription. Although regulatory signals are known to flank the body of the HSV *TK* gene (20, 21), it was not known whether the vaccinia virus transcriptional system would recognize these HSV *TK* regulatory signals. In order to

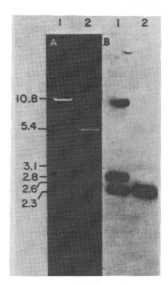

FIG. 4. Analysis of transcription of HSV *TK* in the vP2 vaccinia virus recombinant. (*A*) Agarose gel of *Sst* I- (lane 1) or *Hin*dIII/*Bam*HI- (lane 2) digested pDP137 DNA. (*B*) Radioautograph of hybridization of hybrid-selected ^{32}P-labeled RNA synthesized *in vitro* by using vP2 recombinant vaccinia virus.

gain information regarding the recognition of these signals by vaccinia virus, the recombinant vP2 was studied for transcription of the HSV *TK* gene, because this recombinant was shown to express HSV TK activity. ^{32}P-Labeled RNA was synthesized *in vitro* by using purified vP2 virus and was selected by hybridization to cloned HSV *TK*. The selected RNA was hybridized to a Southern blot of restriction endonuclease-digested pDP137 DNA. Fig. 4*A* shows the 10.8-, 2.8-, and 2.3-MDal fragments derived by *Sst* I digestion of pDP137 (lane 1) and the 5.4-, 3.1-, 2.6-, and 2.3-MDal fragments derived by *Hin*dIII/*Bam*HI digestion of pDP137 (lane 2). ^{32}P-Labeled selected RNA hybridized to all three *Sst* I fragments (Fig. 4*B*, lane 1) but only to the 2.3-MDal (HSV *TK*) *Hin*dIII/*Bam*HI fragment (lane 2). Identical results were observed with *in vivo* labeled RNA derived either at 2 hr after infection in the presence of cytosine arabinonucleoside, an inhibitor of DNA replication, or at 6 hr without metabolic inhibitors.

DISCUSSION

We have demonstrated the feasibility of using poxviruses as eukaryotic vectors by the insertion of the HSV *TK* into the genome of infectious vaccinia virus. Six unique recombinants have been constructed: vP1 and vP2 containing the HSV *TK* in wild-type S variant vaccinia virus; vP3 and vP4 containing the HSV *TK* in a TK$^-$ S variant vaccinia virus; and vP5 and vP6 containing the HSV *TK* in a TK$^-$ L variant genome. Expression of the HSV *TK* was observed in recombinants vP2, vP4, and vP6 but not in vP1, vP3, or vP5. Correct orientation appears to be essential for the expression of this gene at this site. Although the transcription analysis, indicating hybridization of *in vivo* and *in vitro* hybrid-selected RNA only to the HSV *TK* portion of the recombinant plasmid and not to the flanking vaccinia virus DNA sequences, might be interpreted as if the regulatory signals of the HSV *TK* gene are recognized, other (unpublished) data suggest that vaccinia signals may be operative for HSV *TK* expression.

The ability to insert foreign genetic material into the 120-MDal genome of the prototypic L variant vaccinia suggests that packaging of additional DNA is not prohibited, but the upper limit of packaged DNA remains to be determined. With the 6.3-MDal deletion of nonessential DNA found in the S variant (22) plus the additional 2.3-MDal HSV *TK* insert reported here in the L variant it appears that at least 8.6 MDal or approximately 2.5 simian virus 40 genome equivalents of foreign DNA can be readily handled by vaccinia virus. It is reasonable to assume that additional nonessential DNA can be deleted from the S variant, thus increasing the total amount of foreign DNA that can be inserted. This is one of the attractions of poxvirus vectors. In addition, unlike other established or potential eukaryotic viral vectors, such as simian virus 40, herpesvirus, adenovirus, retroviruses, or bovine papilloma virus, vaccinia virus is not considered to be oncogenic and therefore represents less of a risk at this level of consideration.

Although vaccinia virus recombinants vP2, vP4, and vP6 were constructed with the pDP137 recombinant plasmid containing a head-to-tail tandem of the HSV *TK* gene, no evidence for a tandem insertion was observed in these and two other additional vaccinia recombinants tested. The reason for this is not clear, but it may be loss of one copy by *in vivo* recombination.

We selected HSV *TK* for our first encounter with insertion of foreign DNA into vaccinia virus for a variety of reasons. The sequence of the gene is available and its regulatory signals are defined. The gene therefore offers a ready target for genetic manipulation. Vaccinia virus recombinants can be readily selected for TK expression and its unique substrate specificity, the ability to phosphorylate iododeoxycytidine, makes detection and assay simple. Splicing other foreign genes with non-biochemically selectable characteristics to the HSV *TK* followed by insertion into the vaccinia virus genome would facilitate detection and selection of additional recombinant viruses. Alternatively, splicing other foreign genes into the body of the HSV *TK* would render the recombinant virus TK⁻ and therefore selectable by its resistance to growth in the presence of bromodeoxyuridine. Some of these advantages enumerated for the HSV *TK* gene will also be available by analogous manipulations of the endogenous vaccinia TK gene.

The observations reported here open the door for exploring the genetic regulation of vaccinia virus by programmed manipulations of foreign genes and the possible production of biological reagents and gene replacement therapy via poxvirus vectors. Perhaps the most interesting aspect because of its more immediate potential for realization is the use of vaccinia virus vectors for the production of live vaccines. These could be produced by the insertion and expression of appropriate genes encoding specific antigens into vaccinia virus recombinants. Although vaccinia virus as an immunizing agent is not without its risks, vaccination has been used for several hundred years since its introduction by Jenner, and the medical community is well versed in its use. The success of using vaccinia virus to immunize against smallpox has been noted by the recent declaration of the World Health Organization that the world is free of smallpox. It is conceivable that other human or veterinary disease processes may be controlled or eliminated by utilizing live vaccines produced by recombinant vaccinia viruses. The construction of recombinant vaccinia viruses containing genes for the expression of pertinent antigens and the use of these recombinants as live vaccines will be presented in future communications.

We express our gratitude to S. Mercer, S. Davis, C. Whitkop, and B. Lipinskas, who participated in aspects of these studies, and to L. Bruno for preparing the manuscript. We thank B. Moss, D. Hruby, and R. Condit for supplying reagents. This work was supported in part by Grant GM23853 from the National Institutes of Health.

1. Mueller, C., Graessmann, A. & Graessmann, M. (1978) *Cell* **15**, 579–585.
2. Dimitriadis, G. J. (1978) *Nature (London)* **274**, 923–924.
3. Ostro, M. J., Giacomoni, D., Lavelle, D., Paxton, W. & Dray, S. (1978) *Nature (London)* **274**, 921–923.
4. Rechsteiner, M. (1978) *Natl. Cancer Inst. Monogr.* **48**, 57–64.
5. Graham, F. L. & van der Eb, A. J. (1973) *Virology* **52**, 456–467.
6. Mantei, N., Werner, B. & Weissmann, C. (1979) *Nature (London)* **281**, 40–46.
7. Wigler, M., Silverstein, S., Lee, L.-S., Pellicer, A., Cheng, Y.-C. & Axel, R. (1977) *Cell* **11**, 223–232.
8. Wold, B., Wigler, M., Lacy, E., Maniatis, T., Silverstein, S. & Axel, R. (1979) *Proc. Natl. Acad. Sci. USA* **76**, 5684–5688.
9. Gruss, P. & Khoury, G. (1981) *Proc. Natl. Acad. Sci. USA* **78**, 133–137.
10. Hamer, D. H. & Leder, P. (1979) *Nature (London)* **281**, 35–40.
11. Mulligan, R. C., Howard, B. H. & Berg, P. (1979) *Nature (London)* **277**, 108–114.
12. Sarver, N., Gruss, P., Law, M.-F., Khoury, G. & Howley, P. M. (1981) *Mol. Cell. Biol.* **1**, 486–496.
13. Sveda, M. M. & Lai, C. J. (1981) *Proc. Natl. Acad. Sci. USA* **78**, 5488–5492.
14. Shimotohno, K. & Temin, H. M. (1981) *Cell* **26**, 67–77.
15. Gething, M. J. & Sambrook, J. (1981) *Nature (London)* **293**, 620–625.
16. Gruss, P., Ellis, R. W., Shih, T. Y., Konig, M., Scolnick, E. M. & Khoury, G. (1981) *Nature (London)* **293**, 486–488.
17. Subramani, S., Mulligan, R. & Berg, P. (1981) *Mol. Cell. Biol.* **1**, 854–864.
18. Moriarty, A. M., Hoyer, B. H., Shih, J. W. K., Gerin, J. L. & Hamer, D. H. (1981) *Proc. Natl. Acad. Sci. USA* **78**, 2606–2610.
19. Nakano, E., Panicali, D. & Paoletti, E. (1982) *Proc. Natl. Acad. Sci. USA* **79**, 1593–1596.
20. McKnight, S. L. (1980) *Nucleic Acids Res.* **8**, 5949–5964.
21. Wagner, M. J., Sharp, J. A. & Summers, W. C. (1981) *Proc. Natl. Acad. Sci. USA* **78**, 1441–1445.
22. Panicali, D., Davis, S. W., Mercer, S. R. & Paoletti, E. (1981) *J. Virol.* **37**, 1000–1010.
23. Bolivar, F., Rodriguez, R. L., Greene, P. J., Betlach, M. C., Heyneker, H. L., Boyer, H. W., Crosa, J. H. & Falkow, S. (1977) *Gene* **2**, 95–113.
24. Vogelstein, B. & Gillespie, D. (1979) *Proc. Natl. Acad. Sci. USA* **76**, 615–619.
25. Boyer, H. W. & Roulland-Dussoix, D. (1969) *J. Mol. Biol.* **41**, 459–472.
26. Dagert, M. & Ehrlich, S. D. (1979) *Gene* **6**, 23–28.
27. Holmes, D. S. & Quigley, M. (1981) *Anal. Biochem.* **114**, 193–197.
28. Clewell, D. B. & Helinski, D. R. (1969) *Proc. Natl. Acad. Sci. USA* **62**, 1159–1166.
29. Clewell, D. B. (1972) *J. Bacteriol.* **110**, 667–676.
30. Hanahan, D. & Meselson, M. (1980) *Gene* **10**, 63–67.
31. Villarreal, L. P. & Berg, P. (1977) *Science* **196**, 183–185.
32. Rigby, P. W. J., Dieckmann, M., Rhodes, C. & Berg, P. (1977) *J. Mol. Biol.* **113**, 237–251.
33. Davis, D. B., Munyon, W., Buchsbaum, R. & Chawda, R. (1974) *J. Virol.* **13**, 140–145.
34. Campione-Piccardo, J., Rawls, W. E. & Bacchetti, S. (1979) *J. Virol.* **31**, 281–287.
35. Esposito, J., Condit, R. & Obijeski, J. (1981) *J. Virol. Methods* **2**, 175–179.
36. Joklik, W. K. (1962) *Virology* **18**, 9–18.
37. Southern, E. M. (1975) *J. Mol. Biol.* **98**, 503–517.
38. Wahl, G. M., Stern, M. & Stark, G. R. (1979) *Proc. Natl. Acad. Sci. USA* **76**, 3683–3687.
39. Whitkop, C., Lipinskas, B. R., Mercer, S., Panicali, D. & Paoletti, E. (1982) *J. Virol.* **42**, 734–741.
40. Sam, C. K. & Dumbell, K. R. (1981) *Ann. Virol. (Ann. Inst. Pasteur Paris)* **132E**, 135–150.
41. Cooper, G. M. (1973) *Proc. Natl. Acad. Sci. USA* **70**, 3788–3792.
42. Summers, W. C. & Summers, W. P. (1977) *J. Virol.* **24**, 314–318.

Protection from Rabies by a Vaccinia Virus Recombinant Containing the Rabies Virus Glycoprotein Gene

T.J. Wiktor, R.I. MacFarlan, K.J. Reagan, B. Dietzschold,
P.J. Curtis, W.H. Wunner, M.-P. Kieny, R. Lathe, J.-P. Lecocq,
M. Mackett, B. Moss and H. Koprowski

ABSTRACT   Inoculation of rabbits and mice with a vaccinia-rabies glycoprotein recombinant (V-RG) virus resulted in rapid induction of high concentrations of rabies virus-neutralizing antibodies and protection from severe intracerebral challenge with several strains of rabies virus. Protection from virus challenge also was achieved against the rabies-related Duvenhage virus but not against the Mokola virus. Effective immunization by V-RG depended on the expression of a rabies glycoprotein that registered proline rather than leucine as the eighth amino acid from its NH$_2$ terminus (V-RGpro8). A minimum dose required for effective immunization of mice was 10^4 plaque-forming units of V-RGpro8 virus. β-propiolactone-inactivated preparations of V-RGpro8 virus also induced high levels of rabies virus-neutralizing antibody and protected mice against intracerebral challenge with street rabies virus. V-RGpro8 virus was highly effective in priming mice to generate a secondary rabies virus-specific cytotoxic T-lymphocyte response following culture of lymphocytes with either ERA or PM strains of rabies virus.

Rabies is a disease of the central nervous system of major importance to human and veterinary medicine. The etiologic agent, rabies virus, is composed of five structural proteins and a linear, single-stranded RNA genome of negative sense. The rabies virus glycoprotein (G) forms surface projections through the viral lipid envelope and is the only protein capable of inducing and reacting with virus-neutralizing antibody (VNA) (1, 2). Several studies have established that the isolated G is capable of protecting animals against rabies (for review, see ref. 3).

Recently, cloned cDNA copies of the G mRNA from two rabies virus strains have been isolated and sequenced (4, 5). Expression of either G in bacterial hosts has so far failed to yield a product capable of immunizing animals against rabies (5–7). In order to provide post-translational modifications potentially necessary for production of authentic rabies virus G, a vector system allowing expression of rabies G in eukaryotic hosts was sought. To this end, successful expression, immunization, and protection has been reported with infectious vaccinia virus recombinants containing foreign genes such as hepatitis B surface antigen, influenza virus hemagglutinin, and herpes simplex virus glycoprotein D (8–12).

This study compares the biologic and protective properties of two vaccinia virus recombinants expressing ERA strain rabies G proteins differing at a single amino acid residue. We report that infection of cells with either vaccinia virus recombinant resulted in the expression of a novel rabies G; however, only one of these products induced virus-

neutralizing activity, cytotoxic T-cell memory, and protection against an intracerebral challenge with rabies virus.

MATERIALS AND METHODS

Cells and Viruses. Monolayer cultures of BHK-21 clone 13 cells (13) and NA neuroblastoma cells of A/J mouse origin (14) were grown at 37°C in Eagle's minimum essential medium supplemented with 10% fetal calf serum as described (15). The ERA strain of rabies virus (16) was propagated in BHK-21 cells. The PM strain of rabies virus, grown in Vero cells and inactivated by β-propiolactone (βPL), was a gift from the Institute Merieux (Bio Vero Lot S-1163). Challenge viruses included the MD5951 strain of street rabies virus (17), obtained from G. M. Baer of the Centers for Disease Control (Atlanta, GA); a human isolate (HI5) street rabies virus; and rabies-related Duvenhage virus (18) and Mokola strain IbAn 27377 (19) virus. A stock of each challenge virus was prepared from NA or BHK-21 cells and titrated by intracerebral inoculation into 5- to 6-wk-old ICR mice. Additional street rabies viruses used for testing the virus-neutralizing activity of antisera were isolated in 1983 from *Eptesicus fuscus* bat in Ontario, Canada, in 1974 from salivary glands of a red fox in France, and in 1956 from human brain in China (strain CTN-1) provided through the World Health Organization (Geneva); also used was rabies-related Lagos bat virus (19). Wild-type vaccinia viruses (strains WR and Copenhagen) were prepared in tissue culture and purified from cytoplasmic extracts by sucrose gradient centrifugation (20). Vaccinia recombinant viruses containing cloned rabies G cDNA were constructed by using methods previously described (21–23). The vaccinia–rabies glycoprotein recombinant (V-RG) virus containing the coding sequence for proline as the eighth amino acid of the rabies virion G is designated V-RGpro8 (VVTGgRAB in ref. 23), and that which codes for leucine as the eighth amino acid is designated V-RGleu8. Infectivity titers of wild-type vaccinia and V-RG recombinant viruses were determined by a plaque assay on CER cells as described for rabies virus (24).

Preparation of Inactivated V-RG Vaccines. BHK-21 cells at 80–90% confluence were infected with vaccinia (Copenhagen) or V-RGpro8 virus at an input multiplicity of 0.1 plaque-forming units (pfu) per cell. After a 1 hr adsorption at room

Abbreviations: V-RG, vaccinia-rabies glycoprotein recombinant; V-RGpro8, V-RG expressing proline at position 8; G, rabies virus glycoprotein; VNA, virus-neutralizing antibodies; βPL, β-propiolactone; V-RGleu8, V-RG expressing leucine at position 8; pfu, plaque-forming units; CTL, cytotoxic T lymphocyte(s); LU30, 30% lytic units.
§Present address: ARC-Animal Breeding Research Organization, King's Buildings, West Mains Road, Edinburgh, EH9 3JQ, U.K.
¶Present address: Paterson Laboratories, Christie Hospital and Holt Radium Institute, Manchester, M20 9BX, U.K.

WIKTOR, T.J., MACFARLAN, R.I., REAGAN, K.J., DIETZSCHOLD, B., CURTIS, P.J., WUNNER, W.H., KIENY, M.-P., LATHE, R., LECOCQ, J.-P., MACKETT, M., MOSS, B. and KOPROWSKI, H.
Protection from rabies by a vaccinia virus recombinant containing the rabies virus glycoprotein gene.
Proc. Natl. Acad. Sci. U.S.A. 81:7194-7198 (1984). Reprinted with the authors' permission.

temperature, cells were cultured at 37°C with Eagle's minimum essential medium supplemented with 0.2% bovine serum albumin until the cytopathic effect reached 50–75%. Cells were scraped from culture vessels, pelleted, washed once with phosphate-buffered saline (pH 7.4), and swelled for 15 min at 0°C in 10 mM Tris·HCl, pH 7.6/10 mM KCl/1.5 mM MgCl$_2$/2 mM phenylmethylsulfonyl fluoride. Cells were homogenized twice in a Dounce homogenizer, and the nuclei were pelleted. The supernatant represented a crude cell extract of vaccinia or V-RGpro8 viruses. A portion of this extract was inactivated by βPL (1:4000) as described elsewhere (25). The absence of live virus in these preparations was confirmed by infecting monolayers of BHK-21 cells and observing for virus-induced cytopathic effect. A blind passage of culture fluid from these cells onto fresh cultures of BHK-21 cells was performed 5 days after infection. No infectious virus could be detected. Part of the V-RGpro8 virus-infected cell extract was then used for isolation and purification of recombinant virus by sucrose gradient centrifugation and treated with βPL. The remaining extract was adjusted to 2% Triton X-100 and centrifuged for 1 hr at 23°C at 107,000 × g. The solubilized G was isolated from the supernatant by adsorption to an affinity column prepared with an anti-rabies virus-G monoclonal antibody (26). The eluted G was treated with βPL. Protein concentrations were determined with bovine serum albumin as the standard (27).

Animals. Female 5- to 6-wk-old ICR mice (Dominion Laboratories, Dublin, VA) and 5- to 8-wk-old A/J mice (The Jackson Laboratory) were used in these experiments. New Zealand White female rabbits were purchased from Hazleton Dutchland (Aberdeen, MD).

Immunization and Challenge Protocols. Rabbits were inoculated by intradermal injection of 2 × 10^8 pfu of V-RG virus distributed into three separate sites on the back. ICR mice were infected intradermally by scarification of tail skin or by injection into the footpad with either wild-type or recombinant vaccinia viruses (10^9 pfu/ml). When βPL-inactivated virus was used, mice were inoculated with two intraperitoneal injections (0.5 ml) 7 days apart. Immunized mice and rabbits were challenged with street rabies virus by intracerebral inoculation with 2400 and 24,000 mouse LD$_{50}$, respectively, and were observed for a minimum period of 3 mo.

Antibody Titrations. Rabies VNA titers were measured by a modified rapid fluorescent focus inhibition technique (28) against ERA strain rabies virus. Titers are expressed as the highest serum dilution that was capable of reducing the number of rabies virus-infected cells by 50%. A neutralization index was determined by comparing the number of infected cells in control cultures with the number of infected cells in cultures incubated in the presence of antibody-containing serum and expressed as the log$_{10}$ virus reduction per ml of undiluted serum (29). The virus neutralization titers for antibodies directed against vaccinia virus was determined by a plaque reduction assay with monolayers of CER cells (15).

Detection of Antigen by Immunofluorescence. Rabies G in V-RG virus-infected cells was visualized by indirect immunofluorescence in live or acetone-fixed cells using anti-G antiserum as described elsewhere (29).

Cytotoxic T-Lymphocyte (CTL) Responses. A/J mice were inoculated intraperitoneally with 10^7 pfu of ERA rabies virus or intravenously with 10^5 pfu of wild-type vaccinia or V-RGpro8 virus. Primary CTL responses were assayed at 6 days after infection (30), and secondary in vitro CTL responses at 4 wk after infection. To generate secondary CTL, spleen cells from primed mice were cultured at 2.5 × 10^6 cells per ml with dilutions of βPL-inactivated viruses in medium (31) containing 10% fetal calf serum. After incubation for 5 days at 37°C in 5% CO$_2$/95% air, the cells were washed in Eagle's minimum essential medium with 2% fetal calf se-

rum and titrated for cytotoxicity in a 6-hr ^{51}Cr-release assay as described (30). Infection of NA target cells with wild-type vaccinia or V-RGpro8 viruses was carried out as described for rabies virus (30) except that the infected cells were incubated in siliconized Petri dishes for 5–6 hr to allow expression of surface antigens. Cells were then labeled with ^{51}Cr and used as targets. Results are presented as 30% lytic units (LU30) and take into account the spontaneous release of ^{51}Cr into the medium (10–24%) and the maximum release in detergent. One LU30 is defined as the number of effector cells required to achieve 30% specific ^{51}Cr release. A large number of LU30 per culture indicates a potent CTL population.

RESULTS

Expression of Rabies G in Vaccinia Virus Vectors. The rabies-specific G cDNA isolated by Anilionis et al. (4) was inserted into the BamHI site of plasmid pGS20 so as to be controlled by an early vaccinia virus promoter translocated within the thymidine kinase gene (22). The chimeric gene formed in this manner contains the vaccinia RNA start site juxtaposed with the rabies translational initiation codon so as to avoid the production of a fusion protein. This plasmid construct was used to transfect vaccinia virus (strain WR)-infected cells to prepare a recombinant virus that contained the rabies G cDNA inserted into the thymidine kinase locus (V-RGleu8). Successful expression of a novel rabies G in V-RGleu8 virus-infected BHK-21 cells resulted in a protein that was metabolically labeled with [^{35}S]methionine and [^{3}H]glucosamine, was immunoprecipitable with polyclonal rabbit anti-G antibodies, but which migrated faster than rabies virion G in NaDodSO$_4$/polyacrylamide gel. In V-RGleu8 virus-infected BHK-21 cells, the pattern of immunofluorescence suggested that the protein expressed by V-RGleu8 virus was not in a native configuration (Fig. 1). First, the fluorescence that is characteristic of the rabies virus G on the surface of cells was weak in V-RGleu8 virus-infected cells, where the majority of antigen was detected within the cytoplasm (Fig. 1 A and B). Secondly, a panel of anti-G monoclonal antibodies that bind only to native rabies virus G failed to detect the V-RGleu8 virus-expressed antigen (not shown). Moreover, injection of V-RGleu8 virus into animals failed to induce rabies VNA (Table 1) and to protect against rabies.

Amino acid analysis of the NH$_2$ terminus of the rabies virus G (32) revealed a discrepancy at amino acid position 8 (proline) with the predicted sequence (leucine) of the original cDNA clone (4). By sequencing this entire viral G gene, this amino acid change and one other at position 399 (leucine to valine) were identified (ref. 33; data not shown). Assuming that the change near the NH$_2$ terminus might have a greater impact on the structure formation of nascent G, we modified the cDNA clone by site-directed mutagenesis to rectify the amino acid at position 8 (23). In addition, the guanosine tail originally introduced for cloning the cDNA was removed since it may impede expression (23). This modified DNA was inserted into plasmid pTG186 (23) and subsequently transferred into vaccinia virus to generate the recombinant designated V-RGpro8. The Copenhagen strain of vaccinia virus used for human vaccination was used as the vector. Infection of BHK-21 cells by V-RGpro8 virus resulted in expression of a rabies G on the cell surface and in cytoplasm detected by immunofluorescence (Fig. 1 C and D). The protein expressed by this recombinant virus reacted with a panel of anti-G monoclonal antibodies in a pattern identical with native rabies virus G (23).

Induction of VNA and Protection Against Rabies. Inoculation of rabbits and mice with V-RGpro8 virus resulted in a rapid induction of rabies VNA (Table 1). In rabbits, rabies

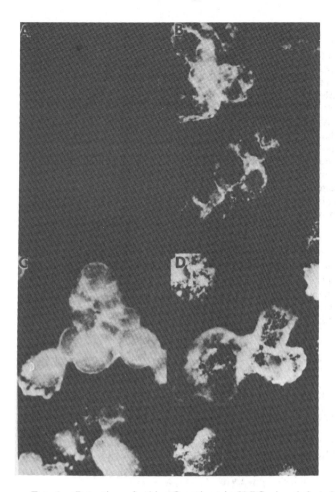

FIG. 1. Detection of rabies G antigen in V-RG virus-infected cells. Monolayers of BHK-21 cells were infected with 0.1 pfu per cell of virus and cultured for 16 hr. Antigen was visualized by indirect immunofluorescence using rabbit antirabies virus G antiserum unfixed (A and C) or acetone-fixed (B and D) cells infected with V-RGleu8 (A and B) or V-RGpro8 (C and D) viruses.

VNA titers at 5, 11, and 14 days after inoculation were 800, 10,000, and >30,000, respectively. Vaccinia VNA titers after 14 days were substantially lower. Rabbit serum (day 14) neutralized between $10^{5.3}$ and $10^{6.6}$ tissue culture ID_{50} of ERA rabies virus, and three street rabies virus isolates previously shown to differ from the ERA strain in their reactivity with a panel of anti-rabies virus G monoclonal antibodies. Neutralization indices against rabies-related Duvenhage, Lagos bat,

and Mokola viruses were $10^{6.2}$, $10^{3.1}$, and $10^{3.4}$, respectively. These results, which are comparable to those obtained with anti-ERA rabies virus antiserum, demonstrate that Duvenhage virus is more closely related to rabies than are rabies-related Lagos bat and Mokola viruses.

Three of four rabbits vaccinated with V-RGpro8 virus resisted a severe intracerebral challenge with 24,000 mouse LD_{50} of MD5951 rabies virus, whereas all five unvaccinated control rabbits died from rabies after 12–15 days (Table 1). The one vaccinated rabbit that died from rabies survived until 21 days after challenge.

Inoculation of mice with V-RGpro8 virus, by either scarification or injection into the footpad, resulted in rabies VNA titers of 30,000 or higher after 14 days. All mice were protected against challenge with either HI5 or MD5951 rabies viruses or with the rabies-related Duvenhage virus. No protection was seen after challenge with Mokola virus. Mice inoculated with wild-type vaccinia or V-RGleu8 virus did not develop rabies VNA and were not protected against rabies.

A minimum dose of V-RGpro8 virus capable of protecting 50% of recipient mice was 10^4 pfu (Fig. 2). In this experiment, mice were inoculated in the footpad and challenged intracerebrally with 2400 mouse LD_{50} of MD5951 rabies virus after 15 days. Levels of rabies VNA were determined at 7 and 14 days (Fig. 2).

Cellular Immune Response Induced by V-RGpro8 Virus. Rabies viruses induce a strong rabies-specific primary CTL response in A/J mice (31); in contrast, CTL induced by V-RGpro8 virus were predominantly directed against vaccinia virus-infected target cells (not shown). However, inoculation with V-RGpro8 virus effectively primed mice for a secondary CTL response after culture of lymphocytes with βPL-inactivated ERA or PM rabies viruses (Fig. 3 A and B). These CTL were specific only for target cells expressing rabies G (i.e., infected with ERA rabies virus or with V-RGpro8 virus). Target cells infected with V-RGpro8 virus were comparatively resistant to lysis, perhaps reflecting differences in density or display of rabies G. In contrast, lymphocytes from V-RGpro8-primed mice lysed only vaccinia or V-RGpro8 virus-infected cells after stimulation with inactivated, purified V-RGpro8 virus (Fig. 3C). Spleen cells from mice primed with vaccinia virus generated no CTL activity after stimulation with rabies viruses. In another experiment, lymphocytes from mice primed with ERA rabies virus generated a strong secondary CTL response after culture with either inactivated PM or ERA rabies virus (Fig. 3 D and E), whereas inactivated V-RGpro8 virus was ineffective at the dilutions tested (Fig. 3F) despite evidence that this preparation contained rabies G (see below).

Immunogenicity of Inactivated V-RGpro8 Virus. The abili-

Table 1. Induction of VNA and protection from rabies by vaccinia recombinant viruses

Animals/ inoculation route	Vaccine*	VNA titers					
		Rabies				Vaccinia	
		Day 0	Day 5	Day 11	Day 14	Day 14	Protection[†]
Rabbits/ intradermal	V-RGpro8	<10	800	10,000	>30,000	250	3/4
	V-RGleu8	<10	—	—	<10	—	—
	None	<10	—	—	<10	—	0/5
Mice/ intradermal	V-RGpro8	<10	—	—	>30,000	250	12/12
	V-RGleu8	<10	—	—	<10	—	0/12
	Vaccinia	<10	—	—	<10	250	0/12
Mice/ footpad	V-RGpro8	<10	—	—	>30,000	1250	12/12
	V-RGleu8	<10	—	—	<10	—	0/12
	Vaccinia	<10	—	—	<10	1250	0/12

*Vaccine was inoculated on day 0 using 2×10^8 pfu (intradermal) or 5×10^7 pfu (footpad) of virus.
[†]A challenge dose of 2400 or 24,000 mouse LD_{50} of MD5951 rabies virus was given on day 14 to mice and rabbits, respectively, by intracerebral inoculation.

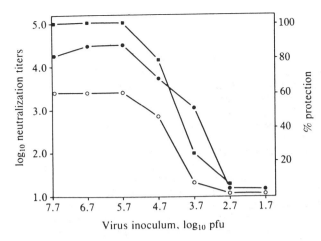

FIG. 2. Minimum protective dose of V-RGpro8 virus. Groups of 10 mice were inoculated in the footpad with serial 10-fold dilutions of V-RGpro8 virus. Levels of rabies VNA were determined 7 (○) and 14 (●) days after infection. On day 15, mice were challenged intracerebrally with 2400 mouse LD$_{50}$ of MD5951 rabies virus (■).

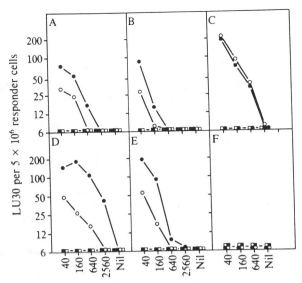

FIG. 3. Secondary CTL response stimulated by V-RGpro8 and rabies viruses *in vitro*. A/J mice were inoculated intravenously with 10^5 pfu of V-RGpro8 virus (*A*, *B*, and *C*) or intraperitoneally with 10^7 pfu of ERA rabies virus (*D*, *E*, and *F*). Four weeks later, spleen cells were cultured with dilutions of βPL-inactivated ERA (*B* and *E*) or PM (*A* and *D*) rabies virus or with βPL-inactivated V-RGpro8 virus (*C* and *F*). After 5 days, each culture was titrated for CTL activity, which is expressed as LU30 generated per 5 × 10^6 responder spleen cells. Target cells: uninfected (□), ERA rabies virus-infected (●), vaccinia virus-infected (■), or V-RGpro8 virus-infected (○) ^{51}Cr-labeled NA cells. Specific lysis due to anti-rabies antibody plus complement was 0%, 95.1%, 0%, and 93.2%, respectively.

ty of βPL-inactivated preparations of V-RGpro8 virus to induce an anti-rabies immune response was tested. Extracts of V-RGpro8 virus-infected cells, purified V-RGpro8 virus, and G isolated from V-RGpro8 virus-infected cell extracts by using an affinity column prepared with anti-rabies virus G antibody, were inactivated and inoculated intraperitoneally into mice. The mice were inoculated again after 7 days and challenged intracerebrally with 240 LD$_{50}$ of MD5951 rabies virus after a further 7 days. All three preparations induced high levels of rabies VNA and protected against rabies (Table 2).

DISCUSSION

The construction of vaccinia virus recombinants expressing genes derived from pathogenic agents has great potential for the production of vaccines. In this report we demonstrate the effectiveness of live and inactivated experimental rabies vaccines prepared by this technology. Initially, we constructed a WR strain vaccinia recombinant virus incorporating the ERA rabies virus G cDNA sequence described by Anilionis *et al.* (4), which codes for leucine at position 8 of the rabies virion G. However, direct amino acid sequencing of rabies virus G established that the eighth residue was proline. The difference in nucleotide sequence between the original cDNA clone and viron RNA, resulting in this amino acid substitution, could have arisen during the cloning procedure or in the transcription of virion RNA to mRNA. In any case, since we could not be sure whether the original cDNA coded

for a functional gene product, the cloned sequence was changed by site-directed mutagenesis to code for proline at position 8 and inserted into a second vaccinia vector derived from the Copenhagen vaccinia strain (23). Both recombinant viruses (V-RGleu8 and V-RGpro8) produced a protein of similar size that was detected in fixed preparations of infected cells by immunofluorescence using monospecific antiserum raised against rabies virus G. However, the V-RGpro8 virus-expressed antigen, but not the V-RGleu8 antigen, was detectable at high density on the surface of infected cells and reacted with monoclonal antibodies recognizing native rabies virus G. This information indicates that the V-RGleu8 virus rabies G has an altered antigenic structure. Remarkably, a single amino acid substitution near the NH$_2$ terminus evidently results in a generally altered conformation of G, which may affect either post-translational modification or

Table 2. Induction of VNA and protection from rabies by inactivated preparations from V-RGpro8 recombinant virus

Vaccine*	Titer before inactivation, log$_{10}$ pfu/ml	Protein concentration, μg per mouse†	Rabies VNA titers		
			Day 7	Day 14	Protection‡
V-RGpro8 virus-infected cell extract	7.5	140	80	8000	12/12
V-RGpro8 purified virus	8.6	9	270	4000	12/12
V-RGpro8 purified rabies G	<1.0§	50	120	15000	12/12
Vaccinia virus-infected cell extract	8.6	900	<10	<10	0/12
Unvaccinated controls	—	—	—	<10	0/12

*Vaccines were prepared from infected BHK-21 cells as described in *Materials and Methods* and inactivated with βPL.
†Total protein in two intraperitoneal inoculations given on days 0 and 7.
‡A challenge dose of 240 mouse LD$_{50}$ of MD5951 rabies virus was given on day 14 to mice by intracerebral inoculation.
§No infectivity detected in undiluted sample.

transport. The rabies G (Leu-8) was also defective when expressed in the Copenhagen vaccinia virus vector (not shown).

Inoculation of mice and rabbits with V-RGpro8 virus induced high levels of rabies VNA. The titers obtained with a single inoculation of this recombinant vaccinia virus were consistently higher than those seen after repeated immunization with inactivated rabies viral vaccines of the type currently used for vaccination (1, 33). V-RGpro8 virus effectively protected animals from rabies. Mice and rabbits survived intracerebral challenge with 2400 and 24,000 mouse LD_{50} of street rabies virus, respectively. This can be regarded as a severe test of immunity. These results of pre-exposure immunization experiments indicate that V-RGpro8 virus has potential as a vaccine for human and/or veterinary use.

In humans, rabies vaccination is used primarily for treatment after exposure to the virus. It has been postulated that not only VNA but also CTL responses are important in post-exposure protection (34, 35). Mice immunized with V-RGpro8 virus generated a substantial secondary cytotoxic response *in vitro* after re-exposure of lymphocytes to PM or ERA rabies viruses (Fig. 3) or *in vivo* after inoculation of V-RGpro8 virus-immunized mice with ERA rabies virus (unpublished data). The CTL generated were specific for rabies G and lysed target cells infected with V-RGpro8 or ERA rabies viruses. A similar priming effect also has been demonstrated after immunization with a vaccinia recombinant virus expressing the influenza virus hemagglutinin (36).

Despite the ability of V-RGpro8 virus to induce CTL memory specific for rabies G, primary rabies-specific CTL responses were weak. Since V-RGpro8 did induce a primary vaccinia-specific CTL response, this finding may reflect some form of immunodominance; however, the mechanisms involved are unclear (37).

Live vaccinia virus has a long history of safe use as a vaccine for humans, despite a low incidence of serious complications (38). Reintroduction of vaccinia virus-based vaccines may be controversial; therefore, we have evaluated the immunogenicity of purified inactivated V-RGpro8 virus and the rabies G isolated from V-RGpro8 virus-infected cells. Both preparations induced rabies VNA and protected mice against rabies. Induction of VNA by inactivated V-RGpro8 virus implies that the rabies G is closely associated with the V-RGpro8 virion. Immunoelectron microscopy should clarify whether the rabies G is a component of the viral membrane. However, these initial results suggest the possibility that inactivated V-RGpro8 virus could also be used as a vaccine against rabies.

This work was supported by Research Grants AI-09706 and AI-18883 from the National Institute of Allergy and Infectious Diseases.

1. Wiktor, T. J., Gyorgy, E., Schlumberger, H. D., Sokol, F. & Koprowski, H. (1973) *J. Immunol.* **110**, 269–276.
2. Cox, J. H., Dietzschold, B. & Schneider, L. G. (1977) *Infect. Immun.* **16**, 754–759.
3. Wunner, W. H., Dietzschold, B., Curtis, P. J. & Wiktor, T. J. (1983) *J. Gen. Virol.* **64**, 1649–1656.
4. Anilionis, A., Wunner, W. H. & Curtis, P. J. (1981) *Nature (London)* **294**, 275–278.
5. Yelverton, E., Norton, S., Obijeski, J. F. & Goeddel, D. V. (1983) *Science* **219**, 614–620.
6. Lathe, R. F., Kieny, M. P., Schmitt, D., Curtis, P. & Lecocq, J. P. (1984) *J. Mol. Appl. Genet.* **2**, 331–342.
7. Malek, L. T., Soostmeyer, G., Garvin, R. T. & James, E. (1984) in *Modern Approaches to Vaccines: Molecular and Chemical Basis of Virus Virulence and Immunogenicity*, eds. Channock, R. M. & Lerner, R. A. (Cold Spring Harbor Laboratory, Cold Spring Harbor, NY), Vol. 1, pp. 203–208.
8. Smith, G. L., Mackett, M. & Moss, B. (1983) *Nature (London)* **302**, 490–495.
9. Smith, G. L., Murphy, B. R. & Moss, B. (1983) *Proc. Natl. Acad. Sci. USA* **80**, 7155–7159.
10. Moss, B., Smith, G. L., Gerin, J. L. & Purcell, R. H. (1984) *Nature (London)* **311**, 67–69.
11. Panicali, D., Davis, S. W., Weinberg, R. L. & Paoletti, E. (1983) *Proc. Natl. Acad. Sci. USA* **80**, 5364–5368.
12. Paoletti, E., Lipinskas, B. R., Samsonoff, C., Mercer, S. & Panicali, D. (1984) *Proc. Natl. Acad. Sci. USA* **81**, 193–197.
13. Stoker, M. & MacPherson, I. (1964) *Nature (London)* **203**, 1355–1357.
14. Clark, H. F. (1980) *Infect. Immun.* **27**, 1012–1022.
15. Wiktor, T. J. (1973) in *Laboratory Techniques in Rabies*, World Health Organization Monograph No. 23, eds. Kaplan, M. & Koprowski, H. (World Health Organization, Geneva), pp. 101–123.
16. Clark, H. F. & Wiktor, T. J. (1972) in *Strains of Human Viruses*, ed. Plotkin, S. A. (Karger, Basel, Switzerland), pp. 177–182.
17. Smith, J. S., McClelland, C. L., Reid, F. L. & Baer, G. M. (1982) *Infect. Immun.* **35**, 213–221.
18. Meredith, C. D., Rossouw, A. P. & Van Praag Koch, H. (1971) *S. Afr. Med. J.* **45**, 767–769.
19. Shope, R. E., Murphy, F. A., Harrison, A. K., Causey, O. R., Kemp, G. E., Simpson, D. I. H. & Moore, D. L. (1970) *J. Virol.* **6**, 690–692.
20. Joklik, W. K. (1962) *Virology* **18**, 9–18.
21. Moss, B., Smith, G. L. & Mackett, M. (1983) in *Gene Amplification and Analysis*, eds. Papas, T. S., Rosenberg, M. & Chirikjian, J. K. (Elsevier/North-Holland, New York), Vol. 3, pp. 201–213.
22. Mackett, M., Smith, G. L. & Moss, B. (1984) *J. Virol.* **49**, 857–864.
23. Kieny, M. P., Lathe, R., Drillien, R., Spehner, D., Skory, S., Schmitt, D., Wiktor, T., Koprowski, H. & Lecocq, J.-P., *Nature (London)*, in press.
24. Lafon, M., Wiktor, T. J. & Macfarlan, R. I. (1983) *J. Gen. Virol.* **64**, 843–851.
25. Wiktor, T. J., Aaslestad, H. G. & Kaplan, M. M. (1972) *Appl. Microbiol.* **23**, 914–918.
26. Dietzschold, B., Wiktor, T. J., Wunner, W. H. & Varrichio, A. (1983) *Virology* **124**, 330–337.
27. Bramhall, S., Noack, N., Wu, M. & Loewenberg, J. R. (1969) *Anal. Biochem.* **31**, 146–148.
28. Reagan, K. J., Wunner, W. H., Wiktor, T. J. & Koprowski, H. (1983) *J. Virol.* **48**, 660–666.
29. Wiktor, T. J. & Koprowski, H. (1978) *Proc. Natl. Acad. Sci. USA* **75**, 3938–3942.
30. Macfarlan, R. I., Dietzschold, B., Wiktor, T. J., Kiel, M., Houghten, R., Lerner, R. A., Sutcliffe, J. G. & Koprowski, H. (1984) *J. Immunol.*, in press.
31. Wiktor, T. J., Doherty, P. C. & Koprowski, H. (1977) *Proc. Natl. Acad. Sci. USA* **74**, 334–338.
32. Dietzschold, B., Wiktor, T. J., Macfarlan, R. I. & Varrichio, A. (1982) *J. Virol.* **44**, 595–602.
33. Wunner, W. H., Smith, C. L., Lafon, M., Ideler, J. & Wiktor, T. J. (1983) in *Nonsegmented Negative Strand Viruses*, eds. Bishop, D. H. L. & Compans, R. W. (Academic, San Diego, CA), pp. 279–284.
34. Plotkin, S. A., Wiktor, T. J., Koprowski, H., Rosenoff, E. I. & Tint, H. (1976) *Am. J. Epidemiol.* **103**, 75–80.
35. Wiktor, T. J. (1978) *Dev. Biol. Stand.* **40**, 255–264.
36. Bennink, J. R., Yewdell, J. W., Smith, G. L., Moller, C. & Moss, B. (1984) *Nature (London)*, in press.
37. Wybier-Franqui, J., Gomard, E. & Levy, J. P. (1982) *Cell. Immunol.* **68**, 287–301.
38. Lane, J. M., Ruben, F. L., Neff, J. M. & Millar, J. D. (1970) *J. Infect. Dis.* **122**, 303–309.

Supplementary Readings

Berman, P.W., Gregory, T.J., Riddle, L., Nakamura, G.R., Champe, M.A., Porter, J.P., Wurm, F.M., Herschberg, R.D., Cobb, E.K. & Eichberg, J.W., Protection of chimpanzees from infection by HIV 1 after vaccination with recombinant glycoprotein gp120 but not gp160. *Nature* 345:622–624 (1990)

Brochier, B., Thomas, I., Bauduin, B., Leveau, T., Pastorett, P.-P., Languet, B., Chappuis, G., Desmettre, P., Blancou, J. & Artois, M., Use of a vaccinia-rabies recombinant virus for the oral vaccination of foxes against rabies. *Vaccine* 8:101–104 (1990)

Charfas, J., Malaria vaccines; the failed promise. *Science* 247:402–403 (1990)

Desrosiers, R.C., Wyard, M.S., Kodama, T., Ringler, D.J., Arthur, L.O., Sehgal, P.K., Letvin, N.L., King, N.W. & Daniel, M.D., Vaccine protection against simian immunodeficiency virus infection. *Proc. Natl. Acad. Sci. U.S.A.* 86:6353–6357 (1989)

Gething, M.J. & Sambrook, J., Cell-surface expression of influenza haemagglutinin from a cloned DNA copy of the RNA gene. *Nature* 293:620–625 (1981)

Hardy, K.H., Stahl, S. & Küpper, H., Production in *B. subtilis* of hepatitis B core antigen and of major antigen of foot and mouth disease virus. Nature 293:481–483 (1981)

Kieny, M.P., Lathe, R., Drillien, R., Spehner, D., Skory, S., Schmitt, D., Wiktor, T., Koprowski, H. & Lecocq, J.P., Expression of rabies virus glycoprotein from a recombinant vaccinia virus. *Nature* 312:163–166 (1984)

Marlin, S.D., Staunton, D.E., Sprintge, T.A., Stratowa, C., Sommergruber, W. & Merluzzi, V.J., A soluble form of intercellular adhesion molecule-1 inhibits rhinovirus infection. *Nature* 344:70–71 (1990)

Marshall, E., Malaria research - what next? *Science* 247:399–402 (1990)

Valenzuela, P., Medina, A., Rutter, W.J., Ammerer, G. & Hall, B.D., Synthesis and assembly of hepatitis B virus surface antigen particles in yeast. *Nature* 298:347–350 (1982)

Immunology

It is possible that no other area of biology has profited from the implementation of recombinant DNA methods to quite the same extent as immunology. In vaccine development the practical applications of antigens and antibodies for therapy, prophylaxis, and in diagnosis have been well appreciated for some time. However, in spite of great advances in the understanding of immunology before 1975, it was not always an "exact" experimental science in the sense that minute quantities of impure biologically active chemicals were of necessity assayed in crude mixtures containing many other biologically active molecules. Immunology suffered (as did other forms of biology) from the lack of pure materials that would enable the accurate identification of the ligands and receptors involved in the complex networks of immunological interactions. Even with respect to cell cultures, the lack of pure growth factors often complicated the interpretation of observations: which was the active molecule? This has now changed dramatically. The mechanism by which vast repertoires of antibodies could be generated in response to antigenic stimulation has been a major scientific problem and until recently was explained (largely) by hypotheses that had little chance of experimental verification.

The basic tools used in the study of immunological systems have changed as a result of advances in peptide synthesis technology, the ability to grow a variety of cell types in culture, the development of sensitive methods for labelling and detecting antibody/antigen interactions, and sensitive tissue distribution studies through the application of new types of imaging technology. Now, superimposed on these technical advances comes recombinant DNA technology.

The demonstration by Kohler and Milstein of the way in which monoclonal antibodies could be obtained, providing materials of high specificity, purity and in sufficient quantity, is not essentially recombinant DNA technology. But this work, which would have to be considered a milestone in the great scientific discoveries of all time, provided the impetus and technological basis for what was to follow in the application of recombinant DNA technology for the study and rapid development of knowledge of the immune system. Monoclonal antibodies are used in various types of immunoassays, immunocytopathology, and in flow cytometry. *In vivo* applications include diagnosis and therapy. In the latter case, the ability to target toxins to specific cells is an area of great promise.

It is not easy to select, from all of the beautiful work that has been done in this area, a few key papers that represent milestones in immunology, especially when we consider that cytokines and cell growth factors that were not known or suspected fifteen years ago are now available > 99% pure in handfuls! And on top of that their receptors have been cloned and expressed in convenient experimental systems! In addition, cloning and sequencing the immunoglobulin genes has provided critical knowledge of their structure and function which has been extended by protein X-ray crystallography.

The supplementary section contains important papers describing antigen receptor genes, the identification and molecular characterization of numerous cytokines, the deciphering of macro-molecular structures such as the major histocompatibility complex, and the studies which led to unravelling the molecular basis of antibody diversity. Our understanding of the networks of cell interactions involved in the immune response is increasing at an exponential rate, as is the potential for solving the mysteries of auto-immune diseases.

Two other papers (Ward *et al.* and Huse *et al.*) present genetic engineering of the production of antibodies in microbes. The prospects raised by these studies are extensive, provided that certain technical difficulties, such as yield, can be resolved. Since the discovery of monoclonal antibodies there have been many attempts to make human rather than mouse monoclonals, but this is still a long way from practicality. The expression of active antibody fragments in microbes is a procedure which lends itself directly to the production of "human" materials. Once again, *Escherichia coli* is the host of choice for antibody gene expression, permitting immunoglobulin fragments to be secreted into the bacterial periplasmic space in correctly folded, active form. (It is interesting to note that immunoglobulin genes were expressed in *E. coli* in the early days of recombinant DNA, but the proteins were not sufficiently active for practical use.)

The Ward and Huse papers use slightly different strategies but the end result is the same; they produce "libraries" of active antibody fragments from mRNA obtained from an immunized animal. In the Ward work the heavy chain variable genes (VH) were used, while the Huse group employed both light (VL) and heavy genes (VH). The net result is *E. coli* expression libraries carrying a wide repertoire of binding activities. In both cases, PCR amplification was the method for generating the antibody genes—a good example of the power of this technique. This is still early days in terms of antibody libraries, but their use in medicine and industry (development of abzymes) is, to say the least, a promising prospect. One difficulty raised by the generation of such huge libraries of varied binding capacities is that of screening; how does one pick out good candidate clones? Fortunately, highly sensitive and specific antibody screening methods have been developed recently, using *E. coli* bacteriophage M13 with a hybrid coat protein as detector. (See Scott *et al.* and Cwirla *et al.* in the supplementary references.)

Finally, although the preparation of antibody libraries on such a scale, mimicking the repertoire of the immune system, is a considerable advance, the use of *E. coli* constructions to obtain defined single antibodies is likely to be of great practical application. Perhaps they will replace traditional monoclonal antibodies. (See Plückthun in the supplementary readings.)

Generation of a Large Combinatorial Library of the Immunoglobulin Repertoire in Phage Lambda

W.D. Huse, L. Sastry, S.A. Iverson, A.S. Kang, M. Alting-Mees, D.R. Burton, S.J. Benkovic and R.A. Lerner

A novel bacteriophage lambda vector system was used to express in *Escherichia coli* a combinatorial library of Fab fragments of the mouse antibody repertoire. The system allows rapid and easy identification of monoclonal Fab fragments in a form suitable for genetic manipulation. It was possible to generate, in 2 weeks, large numbers of monoclonal Fab fragments against a transition state analog hapten. The methods described may supersede present-day hybridoma technology and facilitate the production of catalytic and other antibodies.

MONOCLONAL ANTIBODIES HAVE BEEN GENERATED THAT catalyze chemical transformations ranging from simple acyl transfer reactions to the energetically demanding hydrolysis of the peptide bond in the presence of metal cofactors (*1, 2–11*). Initially, it was widely held that antibodies would be most useful for catalysis where their predominant role was to overcome entropic barriers that occur along the reaction pathway. The basis of this hypothesis was that the chance occurrence of amino acid side chains capable of acid base catalysis in proximity to the reaction center was unlikely. However, for some reactions, study of the *p*H rate profile has revealed the participation of monobasic residues. Other studies have focused on placing appropriate charges on the antigen to induce specific binding interactions by complementary charged amino acid side chains on the antibody (*9, 12, 13*). Such functionalities might participate as a general acid, base, or nucleophile in the reaction under study.

Apart from the validity of the design of the mechanism based antigen, the probability of finding antibodies where particular amino acid side chains participate in catalysis also depends on the number of different antibodies assayed. Because current methods of generating monoclonal antibodies do not provide for an adequate survey of the available repertoire, we have been devising methods to clone the antibody repertoire in *Escherichia coli* and have described the preparation of a highly diverse immunoglobulin gene library (*14*). Given the difficulty of expressing both heavy and light chains together, we initially considered the construction and expression of libraries restricted to fragments of the variable region of the immunoglobulin (Ig) heavy chain V_H (*14*). In fact, a recent report describes the construction of a plasmid expression library in *E. coli* in which V_H fragments with affinity for keyhole limpet hemocyanin (KLH) and lysozyme have been isolated (*15*). However, the use of isolated V_H fragments as antibody mimics may be limited because (i) the available crystal structures of antibody-antigen complexes show considerable contact between antigen and V_L (light chain

W. D. Huse, L. Sastry, S. A. Iverson, and R. A. Lerner are with the Departments of Molecular Biology and Chemistry, Research Institute of Scripps Clinic, La Jolla, CA 92037. D. R. Burton and A. S. Kang are at the Departments of Molecular Biology and Chemistry, Research Institute of Scripps Clinic, La Jolla, CA 92037 and at the Krebs Institute, Department of Molecular Biology and Biotechnology, The University, Sheffield, United Kingdom. M. Alting-Mees is at Stratagene Inc., La Jolla, CA 92037. S. J. Benkovic is at the Department of Chemistry, Pennsylvania State University, University Park, PA 16802 and at the Departments of Molecular Biology and Chemistry, Research Institute of Scripps Clinic, La Jolla, CA 92037

HUSE, W.D., SASTRY, L., IVERSON, S.A., KANG, A.S., ALTING-MEES, M., BURTON, D.R., BENKOVIC, S.J. and LERNER, R.A.
Generation of a large combinatorial library of the immunoglobulin repertoire in phage lambda.
Science 246:1275-1281. Copyright 1989 by the AAAS.

variable) domain as well as V_H (*16*). More explicitly, in the case of a series of antibodies to dextran, the V_L domain provides contacts critical to antigen binding (*17*). Thus, it is unlikely that the affinity of isolated V_H fragments will generally match that of intact antibodies. (ii) The absence of the V_L domain leaves a large hydrophobic patch on one face of the V_H fragment, which will almost certainly lead to increased nonspecificity relative to whole antibodies (*15*). In contrast, Fab fragments (antigen binding) have been studied for more than 30 years. They behave as whole antibodies in terms of antigen recognition, and their affinity and specificity are well defined. Furthermore, for Fab, the combinatorial properties of heavy and light chains serve as an important source of diversity.

In that individual Fab molecules can be expressed and assembled in *E. coli* (*18*), the route to mimicking the diversity of the antibody system in vitro should lie in solving the problem of expressing the repertoires of heavy and light chains in combination. Accordingly, we used a novel system to enable the construction of bacteriophage lambda (λ) libraries expressing a population of functional antibody fragments (Fab's) with a potential diversity equal to or exceeding that of the parent animal.

Criteria for vector construction. To obtain a vector system for generating the largest number of Fab fragments that could be screened directly, we constructed the expression libraries in bacteriophage λ for the following reasons. First, in vitro packaging of phage DNA is the most efficient method of reintroducing DNA into host cells. Second, it is possible to detect protein expression at the level of single-phage plaques. Finally, in our experience, screens of phage libraries diminish the usual difficulties with nonspecific binding. The alternative, plasmid cloning vectors are only advantageous in the analysis of clones after they have been identified. This advantage is not lost in our system because we use λzap II and are able to excise a plasmid (*19*) containing the heavy chain, light chain, or Fab expressing inserts.

The vectors for expression of V_H, V_L, Fv (fragment of the variable region), and Fab sequences are diagrammed in Figs. 1 and 2. They were constructed by a modification of λzap II (*19*) in which we inserted synthetic oligonucleotides into the multiple cloning site. The vectors were designed to be antisymmetric with respect to the Not I and Eco RI restriction sites that flank the cloning and expression sequences. This antisymmetry in the placement of restriction sites in a linear vector such as bacteriophage allows a library expressing light chains to be combined with one expressing heavy chains in order to construct combinatorial Fab expression libraries. The vector λLc1 is designed to serve as a cloning vector for light

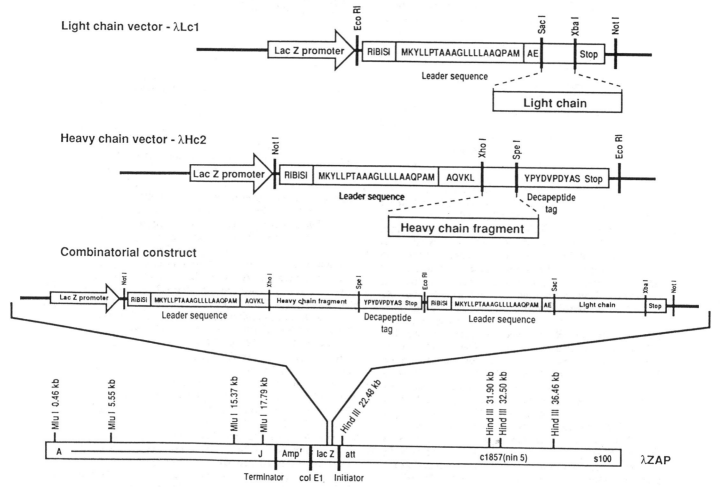

Fig. 1. Combinatorial bacteriophage λ vector system for expression of Fab antibody fragments. The λLc1 vector was constructed for the cloning of PCR amplified products of mRNA that code for light chain protein by inserting the nucleotide sequence shown depicted in Fig. 2A into the Sac I and Xho I sites of λzap II. The sequence was constructed from overlapping synthetic oligonucleotides varying in length from 25 to 50 nucleotides. The λHc2 vector was constructed for cloning PCR amplified products coding for heavy chain Fd sequences by inserting the nucleotide sequence (Fig. 2B) into the Not I and Xho I sites of λzap II. As with the light chain vector, the inserted sequence was constructed from overlapping synthetic oligonucleotides. The combinatorial constructs that can express Fab fragments are generated by cutting DNA isolated from light and heavy chain libraries at the antisymmetric Eco RI site of each vector, followed by re-ligation of the resulting arms. This generates constructs having random combination of light and heavy chains which can be expressed, upon induction with IPTG, from a dicistronic mRNA from the lac Z promoter.

chain fragments, and λHc2 is designed to serve as a cloning vector for heavy chain sequences in the initial step of library construction. These vectors are engineered to efficiently clone the products of PCR amplification with specific restriction sites incorporated at each end (*14, 15*). The sequence of the oligonucleotides used to construct these vectors include elements for construction, expression, and secretion of Fab fragments. These oligonucleotides introduce the antisymmetric Not I and Eco RI restriction sites; a leader peptide for the bacterial *pel* B gene, which has previously been successfully used in *E. coli* to secrete Fab fragments (*18*); a ribosome binding site at the optimal distance for expression of the cloned sequence; cloning sites for either the light or heavy chain PCR product; and, in λHc2, a decapeptide tag at the carboxyl terminus of the expressed heavy chain protein fragment. The sequence of the decapeptide tag was useful because of the availability of monoclonal antibodies to this peptide that were used for immunoaffinity purification of fusion proteins (*20*). The restriction endonuclease recognition sites included in the vectors were Sac I and Xba I in λLc1, and Xho I and Spe I in λHc2. The vectors were characterized by restriction digest analysis and DNA sequencing.

Choice of antigen and amplification of antibody fragments. We constructed the initial Fab expression library from mRNA isolated from a mouse that had been immunized with the KLH-coupled *p*-nitrophenyl phosphonamidate antigen **1** (NPN) (Fig. 3). This antigen was shown by Janda and co-workers (*7*) to be an effective one for the generation of catalytic antibodies. Also, the antibodies for the NPN reaction have been identified and therefore facilitate the implementation of assay systems. Finally, successful generation of catalytic antibodies generally requires binding to relatively small organic haptens, and it was necessary to test the suitability of our system for such molecules.

The PCR amplification of messenger RNA (mRNA) isolated from spleen cells or hybridomas with oligonucleotides that incorporate restriction sites into the ends of the amplified product can be used to clone and express heavy chain sequences (*14, 15*). This work is now extended to include the amplification of the Fd ($V_H - C_H1$) and κ chain sequences (Fig. 4) from mouse spleen cells. The oligonucleotide primers used for these amplifications (Tables 1 and 2) are analogous to those that have been successfully used for amplification of V_H sequences (*14*). The set of 5′ primers for heavy chain amplification was identical to those used to amplify V_H, and those for light chain amplification were chosen similarly (*14, 21*). The 3′ primers of heavy (IgG1) and light (κ) chain sequences included the cysteines involved in disulfide bond formation between heavy and light chains. At this stage no primer was constructed to amplify light (λ) chains since they constitute only a small fraction of murine antibodies (*22*). Restriction endonuclease recognition sequences were incorporated into the primers to allow for the cloning of the amplified fragment into a λ phage vector in a predetermined reading frame for expression.

A λLc1

```
         EcoRI                     Ribosome Binding Site
                                                        Met  Lys  Tyr
5'      TGAATTCTAAACTAGTCGCCAAGGAGACAGTCATAATGAAATA
3'  TCGAACTTAAGATTTGATCAGCGGTTCCTCTGTCAGTATTACTTTAT

   Pel B Leader
   Leu Leu Pro Thr Ala Ala Ala Gly Leu Leu Leu Leu Ala Ala Gln Pro
  CCTATTGCCTACGGCAGCCGCTGGATTGTTATTACTCGCTGCCCAAC
  GGATAACGGATGCCGTCGGCGACCTAACAATAATGAGCGACGGGTTG

   NcoI      SacI               XbaI          NotI
   Ala Met Ala Glu Leu               Stop     Stop
  CAGCCATGGCCGAGCTCGTCAGTTCTAGAGTTAAGCGGCCG        3'
  GTCGGTACCGGCTCGAGCAGTCAAGATCTCAATTCGCCGGCAGCT    5'
```

B λHc2

```
   NotI             Ribosome Binding Site
                                        Met Lys Tyr Leu Leu
  GGCCGCAAATTCTATTTCAAGGAGACAGTCATAATGAAATACCTATT
      CGTTTAAGATAAAGTTCCTCTGTCAGTATTACTTTATGGATAA

                                                      NcoI
   Pro Thr Ala Ala Ala Gly Leu Leu Leu Leu Ala Ala Gln Pro Ala Met
  GCCTACGGCAGCCGCTGGATTGTTATTACTCGCTGCCCAACCAGCCA
  CGGATGCCGTCGGCGACCTAACAATAATGAGCGACGGGTTCCTCGGT

                    XhoI          XbaI   SpeI
   Ala Gln Val Lys Leu Leu Glu                   Tyr Pro Tyr Asp
  TGGCCCAGGTGAAACTGCTCGAGATTTCTAGACTAGTTACCCGTACG
  ACCGGGTCCACTTTGACGAGCTCTAAAGATCTGATCAATGGGCATGC

                              EcoRI
   Val Pro Asp Tyr Gly Ser Stop
  ACGTTCCGGACTACGGTTCTTAATAGAATTCG
  TGCAAGGCCTGATGCCAAGAATTATCTTAAGCAGCT
```

Fig. 2. (**A**) The nucleotide sequence inserted into λzap II to construct λLc1. (**B**) The nucleotide sequence inserted into λzap II to construct λHc2.

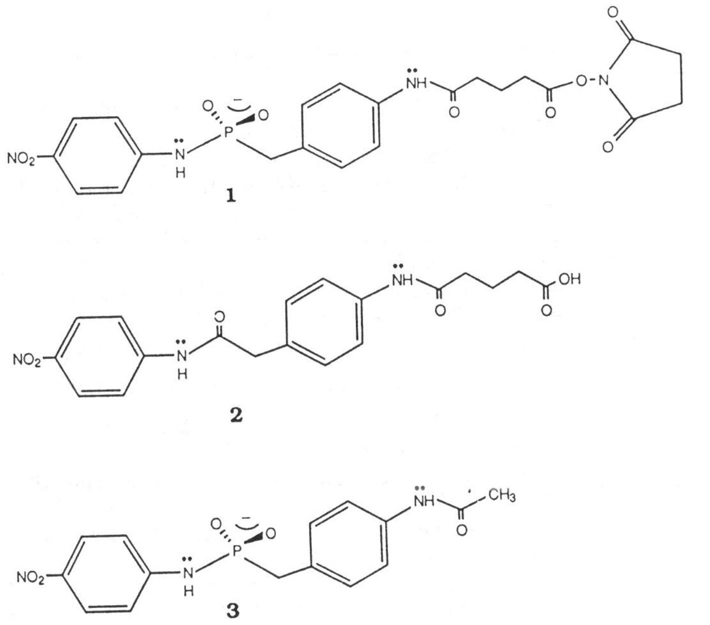

Fig. 3. The transition state analog **1**, which induces antibodies for hydrolyzing carboxamide substrate **2**. Compound **1** containing a glutaryl spacer and an *N*-hydroxysuccinimide–linker appendage is the form used to couple the hapten **1** to protein carriers KLH and BSA, while **3** is the inhibitor. The phosphonamidate functionality is a mimic of the stereoelectronic features of the transition state for hydrolysis of the amide bond.

Library construction. We constructed a combinatorial library in two steps. In the first step, separate heavy and light chain libraries were constructed in λHc2 and λLc1, respectively (Fig. 1). In the second step, these two libraries were combined at the antisymmetric Eco RI sites present in each vector. This resulted in a library of clones each of which potentially coexpresses a heavy and a light chain. The actual combinations are random and do not necessarily reflect the combinations present in the B cell population in the parent animal. The λHc2 expression vector has been used to create a library of heavy chain sequences from DNA obtained by PCR amplification of mRNA isolated from the spleen of a 129 G_{IX}^+ mouse previously immunized with NPN conjugated to KLH. This primary library contains 1.3×10^6 plaque-forming units (pfu) and has been screened for the expression of the decapeptide tag to determine the percentage of clones expressing Fd sequences. The sequence for this peptide is only in frame for expression after the genes for an Fd (or V_H) fragment have been cloned into the vector. At least 80 percent of the clones in the library express Fd fragments when assayed by immunodetection of the decapeptide tag.

The light chain library was constructed in the same way as the heavy chain and shown to contain 2.5×10^6 members. Plaque screening, with an antibody to κ chain, indicated that 60 percent of

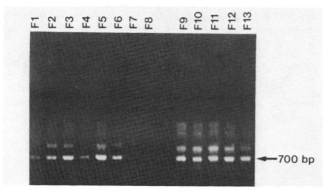

Fig. 4. PCR amplification of Fd and κ regions from the spleen mRNA of a mouse immunized with NPN. Amplification was performed as described (14) with RNA-cDNA hybrids obtained by the reverse transcription of the mRNA with primer specific for amplification of heavy chain sequences (12, Table 1) or light chain sequences (9, Table 2). Lanes F1 to F8 represent the product of heavy chain amplification reactions with one of each of the eight 5′ primers (primers 2 to 9, Table 1) and the 3′ primer (primer 12, Table 1). Light chain (κ) amplifications with the 5′ primers (primers 3 to 7, Table 2) and the appropriate 3′ primer (9, Table 2) are shown in lanes F9 through F13. A band of 700 base pairs is seen in all lanes indicating the successful amplification of Fd and κ regions.

Table 2. Primers used for amplification of κ light chain sequences for construction of Fab's. Amplification was performed in five separate reactions, each containing one of the 5′ primers (primers 3 to 7) and one of the 3′ primers (primer 9). The remaining 3′ primer (primer 8) has been used to construct Fv fragments. The underlined portion of the 5′ primers incorporate a Sac I restriction site and that of the 3′ primers an Xba I restriction site.

LIGHT CHAIN PRIMERS
1) 5′- CCA G TTCC <u>GAGCTC</u> G TTGTGACTCAGGAATCT -3′
2) 5′- CCA G TTCC <u>GAGCTC</u> G TGTTGACGCAGCCGCCC -3′
3) 5′- CCA G TTCC <u>GAGCTC</u> G TGCTC ACCCAGTCTCCA -3′
4) 5′- CCA G TTCC <u>GAGCTC</u> C CAG ATGACCCAGTCTCCA -3′
5) 5′- CCA GAT G T<u>GAGCTC</u> G TGATGACCCAGACTCCA -3′
6) 5′- CCA GAT G T<u>GAGCTC</u> G TCATGACCCAGTCTCCA -3′
7) 5′- CCA G TTCC <u>GAGCTC</u> G TGATGACACAGTCTCCA -3′
8) 5′- G CAGCATT<u>CTAGA</u>GTTTCAGCTCCAGCTTGCC -3′
9) 5′- GCGCCG T<u>CTAGA</u> ATTAAC ACTCATTCCTGTTGAA -3′

Table 1. Primers used for amplification of heavy chain Fd fragments for construction of Fab's. Amplification was performed in eight separate reactions, each containing one of the 5′ primers (primers 2 to 9) and one of the 3′ primers (primer 12). The remaining 5′ primers that were used for amplification in a single reaction are either a degenerate primer (primer 1) or a primer that incorporates inosine at four degenerate positions (primer 10). The remaining 3′ primer (primer 11) has been used to construct Fv fragments. The underlined portion of the 5′ primers incorporates an Xho I site and that of the 3′ primer on Spe I restriction site.

HEAVY CHAIN PRIMERS
1) 5′-AGG T CCAGCTG CTCGAGTC TGG-3′ (with GAA / CT and T/A variants)
2) 5′-AGGTCCAGCTGC<u>TCGAG</u>TCTGG-3′
3) 5′-AGGTCCAGCTGC<u>TCGAG</u>TCAGG-3′
4) 5′-AGGTCCAGCTT<u>CTCGAG</u>TCTGG-3′
5) 5′-AGGTCCAGCTT<u>CTCGAG</u>TCAGG-3′
6) 5′-AGGTCCAACTGC<u>TCGAG</u>TCTGG-3′
7) 5′-AGGTCCAACTGC<u>TCGAG</u>TCAGG-3′
8) 5′-AGGTCCAACTT<u>CTCGAG</u>TCTGG-3′
9) 5′-AGGTCCAACTT<u>CTCGAG</u>TCAGG-3′
10) 5′-AGGTIIAICTI<u>CTCGAG</u>TC TGG-3′ (with T/A variant)
11) 5′-CTATTA<u>ACTAGT</u>AACGGTAACAGT-GGTGCCTTGCCCCA-3′
12) 5′- AGGCTT<u>ACTAGT</u>ACA ATCCCTGG-GCACAAT-3′

the library contained expressed light chain inserts. This relatively small percentage of inserts probably resulted from incomplete dephosphorylation of the vector after cleavage with Sac I and Xba I.

Once obtained, the two libraries were used to construct a combinatorial library by crossing them at the Eco RI site as follows. DNA was first purified from each library. The light chain library was cleaved with Mlu I restriction endonuclease, the resulting 5′ ends were dephosphorylated, and the product was digested with Eco RI. This process cleaved the left arm of the vector into several pieces, but the right arm containing the light chain sequences remained intact. The DNA of heavy chain library was cleaved with Hind III, dephosphorylated, and then cleaved with Eco RI; this process destroyed the right arm, but the left arm containing the heavy chain sequences remained intact. The DNA's so prepared were then mixed and ligated. After ligation, only clones that resulted from combination of a right arm of light chain–containing clones and a left arm of heavy chain–containing clones reconstituted a viable phage. After ligation and packaging, 2.5×10^7 clones were obtained. This is the combinatorial Fab expression library that was screened to identify clones having affinity for NPN. For determining the frequency of the phage clones that coexpress the light and heavy chain fragments, we screened duplicate lifts of the combinatorial library for light and heavy chain expression. In our examination of approximately 500 recombinant phage, approximately 60 percent coexpressed light and heavy chain proteins.

Antigen binding. All three libraries, the light chain, the heavy chain, and Fab were screened to determine whether they contained recombinant phage that expressed antibody fragments binding NPN. In a typical procedure, 30,000 phage were plated and duplicate lifts with nitrocellulose screened for binding to NPN coupled to ^{125}I-labeled bovine serum albumin (BSA) (Fig. 5). Duplicate screens of 90,000 recombinant phage from the light chain library and a similar number from the heavy chain library did not identify any clones that bound the antigen. In contrast, the screen of a similar number of clones from the Fab expression library identified many phage plaques that bound NPN (Fig. 5). This observation indicates that, under conditions where many heavy chains in combination with light chains bind to antigen, heavy or light chains alone do not. Therefore, in the case of NPN, we expect that there are many heavy and light chains that only bind antigen when they are combined with specific light and heavy chains, respectively. This result supports our decision to screen large combinatorial Fab expression libraries. To assess our ability to screen large numbers of clones and obtain a more quantitative estimate of the frequency of antigen binding clones in the combinatorial library, we screened one million phage plaques and identified approximately 100 clones that bound to antigen. For six clones, a region of the plate containing the positive phage plaques and approximately 20 surrounding them was "cored," replated, and screened with duplicate lifts (Fig. 5). As expected, the expression products of approximately 1 in 20 of the phage specifically bind to antigen. Phage which were believed to be negative on the initial screen did not give positives on replating.

To determine the specificity of the antigen-antibody interaction, antigen-binding was subjected to competition with free unlabeled antigen (Fig. 6). These studies showed that individual clones could be distinguished on the basis of antigen affinity. The concentration of free haptens required for complete inhibition of binding varied between 10 to $100 \times 10^{-9} M$, suggesting that the expressed Fab fragments had binding constants in the nanomolar range.

In preparation for characterization of the protein products, a plasmid containing the heavy and light chain genes was excised with helper phage (Fig. 7). Mapping of the excised plasmid demonstrated a restriction pattern consistent with incorporation of heavy and light chain sequences. The protein products of one of the clones was analyzed by enzyme-linked immunosorbent assay (ELISA) and immunoblotting to establish the composition of the NPN binding protein. A bacterial supernatant after IPTG (isopropyl thiogalactoside) induction was concentrated and subjected to gel filtration. Fractions in the molecular size range 40 to 60 kD were pooled, concentrated, and subjected to a further gel filtration separation. ELISA analysis of the eluted fractions (Fig. 8) indicated that NPN binding was associated with a protein of a molecular size of about 50 kD, which contained both heavy and light chains. An immunoblot of a concentrated bacterial supernatant preparation under nonreducing conditions was developed with antibody to decapeptide. This revealed a 50-kD protein band. We have found that the antigen-binding protein can be purified to homogeneity from bacterial supernate in two steps involving affinity chromatography on protein G followed by gel filtration. SDS-PAGE analysis of the protein revealed a single band at ~50 kD under nonreducing conditions and a doublet at ~25 kD under reducing conditions. Taken together, these results are consistent with NPN-binding being a function of Fab fragments in which heavy and light chains are covalently linked by a disulfide bond.

Properties of the in vivo repertoire compared to the phage combinatorial library. Previously we constructed a highly diverse V_H library in *E. coli*. We have now combined heavy and light chain libraries to clone and express assembled and functional Fab fragments of immunoglobulin. A moderately restricted library was prepared because only a limited number of primers was used for polymerase chain reaction (PCR) amplification of Fd sequences. The library is expected to contain only clones expressing κ-γ1 sequences. However, this is not an inherent limitation of the method since the addition of more primers can amplify any antibody class or subclass. Despite this restriction we were able to isolate a large number of clones producing antigen binding proteins.

A central issue is how our phage library compares with the in vivo antibody repertoire in terms of size, characteristics of diversity, and ease of access.

The size of the mammalian antibody repertoire is difficult to judge, but a figure of the order of 10^6 to 10^8 different antigen specificities is often quoted. With some of the reservations discussed below, a phage library of this size or larger can readily be constructed by a modification of the method described. Once an initial combinatorial library has been constructed, heavy and light chains can be shuffled to obtain libraries of exceptionally large numbers.

In principle, the diversity characteristics of the naive (unimmunized) in vivo repertoire and corresponding phage library are expected to be similar in that both involve a random combination of heavy and light chains. However, different factors act to restrict the diversity expressed by an in vivo repertoire and phage library. For example, a physiological modification such as tolerance will restrict the expression of certain antigenic specificities from the in vivo repertoire, but these specificities may still appear in the phage library. However, bias in the cloning process may introduce restrictions into the diversity of the phage library. For example, the representation of mRNA for sequences expressed by stimulated B cells can be expected to predominate over those of unstimulated cells because of higher levels of expression. In addition, the resting repertoire might overrepresent spontaneously activated B cells whose immunoglobulins have been suggested to be less specific. In any event, methods exist to selectively exclude such populations of cells. Also, the fortuitous presence of restriction sites in the variable gene similar to those used for cloning and combination will cause them to be eliminated. We can circumvent some of these difficulties by making minor changes, such as introducing amber mutations in the vector system. Different source tissues (for example, peripheral blood, bone marrow, or regional lymph nodes) and different PCR primers (for example, those to amplify different antibody classes), may result in libraries with different diversity characteristics.

Another difference between in vivo repertoire and phage library is

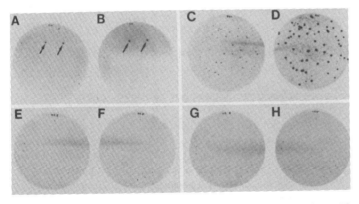

Fig. 5. Screening phage libraries for antigen-binding. Duplicate plaque lifts of Fab (filters **A** and **B**), heavy chain (filters **E** and **F**), and light chain (filters **G** and **H**) expression libraries were screened against ^{125}I-labeled BSA conjugated with NPN at a density of approximately 30,000 plaques per plate. Filters **C** and **D** illustrate the duplicate secondary screening of a cored positive from a primary filter A (arrows) as discussed in the text. Standard plaque lift methods were used in screening. Cells (XL1 blue) infected with phage were incubated on 150-mm plates for 4 hours at 37°C, protein expression was induced by overlay with nitrocellulose filters soaked in 1 m*M* IPTG, and the plates were incubated at 25°C for 8 hours. Duplicate filters were obtained during a second incubation under the same conditions. Filters were then blocked in a solution of 1 percent BSA in phosphate-buffered saline (PBS) for 1 hour before incubation (with rocking) at 25°C for 1 hour with a solution of ^{125}I-labeled BSA (at 0.1 μ*M*) conjugated to NPN (2×10^6 cpm/ml; approximately 15 NPN per BSA molecule), in 1 percent BSA in PBS. Background was reduced by preliminary centrifugation of stock ^{125}I-labeled BSA solution at 100,000*g* for 15 minutes and preliminary incubation of solutions with plaque lifts from plates containing bacteria infected with a phage having no insert. After labeling, filters were washed repeatedly with PBS containing 0.05 percent Tween 20 before the overnight development of autoradiographs.

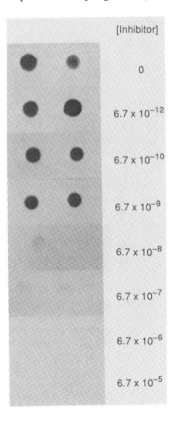

[Inhibitor]

0

6.7×10^{-12}

6.7×10^{-10}

6.7×10^{-9}

6.7×10^{-8}

6.7×10^{-7}

6.7×10^{-6}

6.7×10^{-5}

Fig. 6. Specificity of antigen binding shown by competitive inhibition. Filter lifts from positive plaques were exposed to ^{125}I-labeled BSA-NPN in the presence of increasing concentrations of the inhibitor NPN. A number of phages correlated with NPN-binding as in Fig. 5 were spotted in duplicate (about 100 particles per spot) directly onto a bacterial lawn. The plate was then overlaid with an IPTG-soaked filter and incubated for 19 hours at 25°C. The filters were then blocked in 1 percent BSA in PBS before incubation in ^{125}I-BSA–NPN as done previously with the inclusion of varying amounts of NPN in the labeling solution. Other conditions and procedures were as in Fig. 5. The results for a phage of moderate affinity are shown in duplicate in the figure. Similar results were obtained for four other phages with some differences in the effective inhibitor concentration ranges.

Fig. 7. A plasmid can be excised from λLc1, λHc2, and their combination because they are a modification of λzap II. M13mp8 was used as helper phage and the excised plasmid was infected into a F$^+$ derivative of MC1061. The excised plasmid contains the same constructs for antibody fragment expression as do the parent vectors (Fig. 1). These plasmid constructs are more conveniently analyzed for restriction pattern and protein expression of the λ phage clones identified and isolated on the basis of antigen binding. The plasmid also contains an f1 origin of replication which facilitates the preparation of single-stranded DNA for sequence analysis and in vitro mutagenesis.

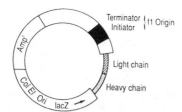

that antibodies isolated from the repertoire may have benefited from affinity maturation as a result of somatic mutations after combination of heavy and light chains whereas the phage library randomly combines the matured heavy and light chains. Given a large enough phage library derived from a particular in vivo repertoire, the original matured heavy and light chains will be recombined. However, since one of the potential benefits of this technology is to obviate the need for immunization by the generation of a single highly diverse "generic" phage library, it would be useful to have

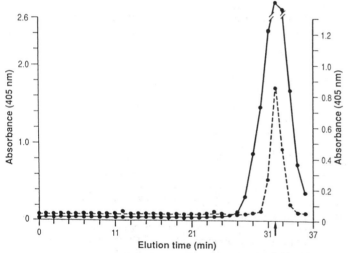

Fig. 8. Characterization of an antigen binding protein. The concentrated partially purified bacterial supernatant of an NPN binding clone was separated by gel filtration and samples from each fraction were applied to microtiter plates coated with BSA-NPN. Addition of either antibody to decapeptide (---) or antibody to κ chain (—, left-hand scale) conjugated with alkaline phosphatase was followed by color development. The arrow indicates the position of elution of a known Fab fragment. The results show that antigen binding is a property of a 50-kD protein containing both heavy and light chains. To permit protein characterization, a single plaque of a NPN-positive clone (Fig. 5) was picked, and the plasmid containing the heavy and light chain inserts (Fig. 7) was excised (19). Cultures (500 ml) in L broth were inoculated with 3 ml of a saturated culture of the clone and incubated for 4 hours at 37°C. Protein synthesis was induced by the addition of IPTG to a final concentration of 1 mM, and the cultures were incubated for 10 hours at 25°C. The supernatant from 200 ml of cells was concentrated to 2 ml and applied to a TSK-G4000 column. Samples (50 μl) from the eluted fractions were assayed by ELISA. Microtiter plates were coated with BSA-NPN at 1 μg/ml, 50-μl samples were mixed with 50 μl of PBS-Tween 20 (0.05 percent) BSA (0.1 percent) added, and the plates were incubated for 2 hours at 25°C. The plated material was then washed with PBS-Tween 20-BSA and 50 μl of appropriate concentrations of a rabbit antibody to decapeptide (20) or a goat antibody to mouse κ light chain (Southern Biotech) conjugated with alkaline phosphatase were added and incubated for 2 hours at 25°C. The plates were again washed, 50 μl of p-nitrophenyl phosphate (1 mg/ml in 0.1M tris, pH 9.5, containing 50 mM MgCl$_2$) was added, and the plates were incubated for 15 to 30 minutes and the absorbance was read at 405 nm.

methods to optimize sequences to compensate for the absence of somatic mutation and clonal selection. Three procedures are made readily available through the vector system presented. First, saturation mutagenesis may be performed on the complementarity-determining regions (CDR's) (23) and the resulting Fab's can be assayed for increased function. Second, a heavy or a light chain of a clone that binds antigen can be recombined with the entire light or heavy chain libraries, respectively, in a procedure identical to that used to construct the combinatorial library. Third, iterative cycles of the two above procedures can be performed to further optimize the affinity or catalytic properties of the immunoglobulin. The last two procedures are not permitted in B cell clonal selection, which suggests that the methods described here may actually increase our ability to identify optimal sequences.

Access is the third area where it is of interest to compare the in vivo antibody repertoire and phage library. In practical terms the phage library is much easier to access. The screening methods used have allowed one to survey the gene products of at least 50,000 clones per plate so that 10^6 to 10^7 antibodies can be readily examined in a day but the most powerful screening methods depend on selection. In the catalytic antibody system, this may be accomplished by incorporating into the antigen leaving groups necessary for replication of auxotrophic bacterial strains or toxic substituents susceptible to catalytic inactivation. Further advantages are related to the fact that the in vivo antibody repertoire can only be accessed via immunization, which is a selection on the basis of binding affinity. The phage library is not similarly restricted. For example, the only general method to identify antibodies with catalytic properties has been by preselection on the basis of affinity of the antibody to a transition state analog. Such restrictions do not apply to the in vitro library where catalysis can, in principle, be assayed directly. The ability to assay directly large numbers of antibodies for function may allow selection for catalysts in reactions where a mechanism is not well defined or synthesis of the transition state analog is difficult. Assaying for catalysis directly eliminates the bias of the screening procedure for reaction mechanisms limited to a particular synthetic analog; therefore, simultaneous exploration of multiple reaction pathways for a given chemical transformation are possible.

We have described procedures for the generation of Fab fragments that are clearly different in a number of important respects from antibodies. There is undoubtedly a loss of affinity in having monovalent Fab antigen binders, but it is possible to compensate for this by selection of suitably tight binders. For a number of applications such as diagnostics and biosensors, monovalent Fab fragments may be preferable. For applications requiring Fc effector functions, the technology already exists for extending the heavy chain gene and expressing the glycosylated whole antibody in mammalian cells.

Our data show that it is now possible to construct and screen at least three orders of magnitude more clones with monospecificity than previously possible. The data also invite speculation concerning the production of antibodies without the use of live animals.

REFERENCES AND NOTES

1. R. A. Lerner, in *Proceedings of the XVIIIth Solvay Conference* on Chemistry, G. Van Binst, Ed. (Springer-Verlag, Berlin, 1983), pp. 43–49.
2. A. Tramontano, K. D. Janda, R. A. Lerner, *Proc. Natl. Acad. Sci. U.S.A.* **83**, 6736, 1986; *Science* **234**, 1566 (1986); A. Tramontano, A. A. Ammann, R. A. Lerner, *J. Am. Chem. Soc.* **110**, 2282 (1988); S. J. Pollack and P. G. Schultz, *ibid.* **111**, 1929 (1989); S. J. Pollack, P. Hsiun, P. G. Schultz, *ibid.*, p. 5961.
3. S. J. Pollack, J. W. Jacobs, P. G. Schultz, *Science* **234**, 1570 (1986); J. Jacobs, P. G. Schultz, R. Sugasawara, M. J. Powell, *J. Am. Chem. Soc.* **109**, 2174 (1987).
4. S. J. Benkovic, A. D. Napper, R. A. Lerner, *Proc. Natl. Acad. Sci. U.S.A.* **85**, 5355 (1988); K. D. Janda, R. A. Lerner, A. Tramontano, *J. Am. Chem. Soc.* **110**, 4835 (1988); K. D. Janda, S. J. Benkovic, R. A. Lerner, *Science* **244**, 437 (1989).
5. A. D. Napper, S. J. Benkovic, A. Tramontano, R. A. Lerner, *Science* **237**, 1041 (1987).

6. D. Hilvert, S. H. Carpenter, K. D. Nared, M.-T. M. Auditor, *Proc. Natl. Acad. Sci. U.S.A.* **85**, 4953 (1988); D. Y. Jackson *et al.*, *J. Am. Chem. Soc.* **110**, 4841 (1988).
7. K. D. Janda, D. Schloeder, S. J. Benkovic, R. A. Lerner, *Science* **241**, 1188 (1988). We thank K. D. Janda for providing hapten 1 and inhibitor 3.
8. A. Cochran, R. Sugasawara, P. G. Schultz, *J. Am. Chem. Soc.* **110**, 7888 (1988).
9. K. M. Shokat *et al.*, *Nature* **338**, 269 (1989).
10. B. L. Iverson and R. A. Lerner, *Science* **243**, 1184 (1989).
11. D. Hilvert *et al.*, *J. Am. Chem. Soc.*, in press.
12. R. A. Lerner and S. J. Benkovic, *Bioassays* **9**, 107 (1988); K. D. Janda, M. I. Weinhouse, D. M. Schloeder, R. A. Lerner, S. J. Benkovic, unpublished results.
13. B. L. Iverson, K. Cameron, K. Jahangiri, D. Pasternak, unpublished results.
14. L. Sastry *et al.*, *Proc. Natl. Acad. Sci. U.S.A.* **86**, 5728 (1989).
15. E. S. Ward, D. Gussow, A. D. Griffiths, P. T. Jones, G. Winter, *Nature* **341**, 544 (1989); R. Orlandi, D. H. Gussow, P. T. Jones, G. Winter, *Proc. Natl. Acad. Sci. U.S.A.* **86**, 3833 (1989).
16. D. R. Davies, S. Sheriff, E. A. Padlan, *J. Biol. Chem.* **263**, 10541 (1988); J. N. Herron, X. He, M. L. Mason, E. W. Voss, Jr., A. B. Edmundson, *Proteins Struct. Funct. Genet.* **5**, 271 (1989); M. Whitlow, A. J. Howard, J. F. Wood, E. W. Voss, Jr., K. D. Hardman, unpublished results. In the structure of Whitlow *et al.*, the light chain in the fluorescein isothiocyanate conjugated Fab 4-4-20 crystal structure has major contributions to the binding of the hapten fluorescein. All of the hydrogen bonds to fluorescein are from the light chain. Three come from the light chain CDR 1 (residues His[27D], Tyr[32], and Arg[34], in Kabat's numbering system) and one from the light chain CDR 3 (residue Ser[91]). Fluorescein is bound in a deep pocket in the Fab 4-4-40 structure, so that only 10 percent (54 Å^2 of 521 Å^2) of its solvent accessible surface area is exposed to solvent; 38 percent of this pocket is lined with light chain residues, based on calculations of solvent-accessible surface area. Aromatic residues His[27D], Tyr[32], and Trp[96] from the light chain and Trp[33], Tyr[53], Tyr[100D], and Tyr[100E] from the heavy chain form this deep pocket for the fluorescein.
17. S. C. Wallick, E. A. Kabat, S. L. Morrison, *Immunology* **142**, 1235 (1989).
18. M. Better, C. P. Chang, R. R. Robinson, A. H. Horwitz, *Science* **240**, 1041 (1988); A. Skerra and A. Plückthun, *ibid.*, p. 1038.
19. J. M. Short, J. M. Fernandez, J. A. Sorge, W. D. Huse, *Nucleic Acids Res.* **16**, 7583 (1988).
20. J. Field *et al.*, *Mol. Cell. Biol.* **8**, 2159 (1988).
21. R. Orlandi, D. H. Gussow, P. T. Jones, G. Winter, *Proc. Natl. Acad. Sci. U.S.A.* **86**, 3833 (1989).
22. The Fv (variable region) fragments may be constructed with a 3' primer that is complementary to the mRNA in the J (joining) region (amino acid 128) and a set of 5' primers that are complementary to the first strand cDNA in the conserved amino-terminal region of the processed protein.
23. Amino acid and nucleotide sequences of immunoglobulins used in this paper were taken from E. A. Kabat, T. T. Wu, M. Reid-Miller, H. M. Perry, K. S. Gottesman, *Sequences of Proteins of Immunological Interest* (US Public Health Service, National Institutes of Health, Bethesda, MD, 1987).

24. We thank D. Schloeder, D. A. McLeod, R. Samodal, B. Hay for technical assistance; E. A. Kabat, N. Klinman, K. D. Janda, and B. L. Iverson for advice and comments. Supported by an NIH postdoctoral fellowship (S.A.I), the SERC Protein Engineering Club (A.K.), and a Jenner Fellowship of the Lister Institute of Preventive Medicine (D.R.B.).

26 October 1989; accepted 14 November 1989

Continuous Cultures of Fused Cells Secreting Antibody of Predefined Specificity

G. Köhler and C. Milstein

THE manufacture of predefined specific antibodies by means of permanent tissue culture cell lines is of general interest. There are at present a considerable number of permanent cultures of myeloma cells[1,2] and screening procedures have been used to reveal antibody activity in some of them. This, however, is not a satisfactory source of monoclonal antibodies of predefined specificity. We describe here the derivation of a number of tissue culture cell lines which secrete anti-sheep red blood cell (SRBC) antibodies. The cell lines are made by fusion of a mouse myeloma and mouse spleen cells from an immunised donor. To understand the expression and interactions of the Ig chains from the parental lines, fusion experiments between two known mouse myeloma lines were carried out.

Each immunoglobulin chain results from the integrated expression of one of several V and C genes coding respectively for its variable and constant sections. Each cell expresses only one of the two possible alleles (allelic exclusion; reviewed in ref. 3). When two antibody-producing cells are fused, the products of both parental lines are expressed[4,5], and although the light and heavy chains of both parental lines are randomly joined, no evidence of scrambling of V and C sections is observed[4]. These results, obtained in an heterologous system involving cells of rat and mouse origin, have now been confirmed by fusing two myeloma cells of the same mouse strain,

The protein secreted (MOPC 21) is an IgG1 (κ) which has been fully sequenced[7,8]. Equal numbers of cells from each parental line were fused using inactivated Sendai virus[9] and samples containing 2×10^5 cells were grown in selective medium in separate dishes. Four out of ten dishes showed growth in selective medium and these were taken as independent hybrid lines, probably derived from single fusion events. The karyotype of the hybrid cells after 5 months in culture was just under the sum of the two parental lines (Table 1). Figure 1 shows the isoelectric focusing[10] (IEF) pattern of the secreted products of different lines. The hybrid cells (samples c–h in Fig. 1) give a much more complex pattern than either parent (a and b) or a mixture of the parental lines (m). The important feature of the new pattern is the presence of extra bands (Fig. 1, arrows). These new bands, however, do not seem to be the result of differences in primary structure; this is indicated by the IEF pattern of the products after reduction to separate the heavy and light chains (Fig. 1B). The IEF pattern of chains of the hybrid clones (Fig. 1B, g) is equivalent to the sum of the IEF pattern (a and b) of chains of the parental clones with no evidence of extra products. We conclude that, as previously shown with interspecies hybrids[4,5], new Ig molecules are produced as a result of mixed association between heavy and light chains from the two parents. This process is intracellular as a mixed cell population does not give rise to such hybrid molecules (compare m and g, Fig. 1A). The individual cells must therefore be able to express both isotypes. This result shows that in hybrid cells the expression of one isotype and idiotype does not exclude the expression of another: both heavy chain

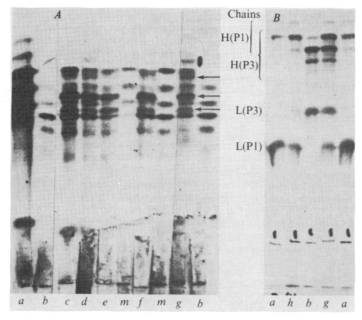

Fig. 1 Autoradiograph of labelled components secreted by the parental and hybrid cell lines analysed by IEF before (A) and after reduction (B). Cells were incubated in the presence of ^{14}C-lysine[14] and the supernatant applied on polyacrylamide slabs. A, pH range 6.0 (bottom) to 8.0 (top) in 4 M urea. B, pH range 5.0 (bottom) to 9.0 (top) in 6 M urea; the supernatant was incubated for 20 min at 37 °C in the presence of 8 M urea, 1.5 M mercaptoethanol and 0.1 M potassium phosphate pH 8.0 before being applied to the right slab. Supernatants from parental cell lines in: a, P1Bu1; b, P3-X67Ag8; and m, mixture of equal number of P1Bul and P3-X67Ag8 cells. Supernatants from two independently derived hybrid lines are shown: c–f, four subclones from Hy-3; g and h, two subclones from Hy-B. Fusion was carried out[4,9] using 10^6 cells of each parental line and 4,000 haemagglutination units inactivated Sendai virus (Searle). Cells were divided into ten equal samples and grown separately in selective medium (HAT medium, ref. 6). Medium was changed every 3 d. Successful hybrid lines were obtained in four of the cultures, and all gave similar IEF patterns. Hy-B and Hy-3 were further cloned in soft agar[14]. L, Light; H, heavy.

and provide the background for the derivation and understanding of antibody-secreting hybrid lines in which one of the parental cells is an antibody-producing spleen cell.

Two myeloma cell lines of BALB/c origin were used. P1Bul is resistant to 5-bromo-2'-deoxyuridine[4], does not grow in selective medium (HAT, ref. 6) and secretes a myeloma protein, Adj PC5, which is an IgG2A (κ), (ref. 1). Synthesis is not balanced and free light chains are also secreted. The second cell line, P3-X63Ag8, prepared from P3 cells[2], is resistant to 20 μg ml^{-1} 8-azaguanine and does not grow in HAT medium.

isotypes ($\gamma1$ and $\gamma2a$) and both V_H and both V_L regions (idiotypes) are expressed. There are no allotypic markers for the C_κ region to provide direct proof for the expression of both parental C_κ regions. But this is indicated by the phenotypic link between the V and C regions.

Figure 1A shows that clones derived from different hybridisation experiments and from subclones of one line are indistinguishable. This has also been observed in other experiments (data not shown). Variants were, however, found in a survey of 100 subclones. The difference is often associated with changes

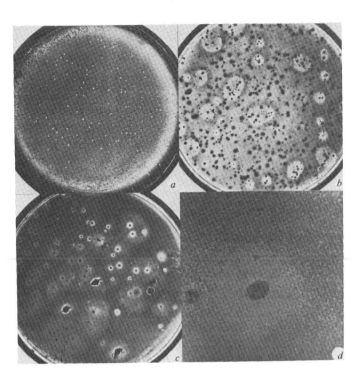

Fig. 2 Isolation of an anti-SRBC antibody-secreting cell clone. Activity was revealed by a halo of haemolysed SRBC. Direct plaques given by: *a*, 6,000 hybrid cells Sp-1; *b*, clones grown in soft agar from an inoculum of 2,000 Sp-1 cells; *c*, recloning of one of the positive clones Sp-1/7; *d*, higher magnification of a positive clone. Myeloma cells (10^7 P3-X67A g8) were fused to 10^8 spleen cells from an immunised BALB/c mouse. Mice were immunised by intraperitoneal injection of 0.2 ml packed SRBC diluted 1:10, boosted after 1 month and the spleens collected 4 d later. After fusion, cells (Sp-1) were grown for 8 d in HAT medium, changed at 1–3 d intervals. Cells were then grown in Dulbecco modified Eagle's medium, supplemented for 2 weeks with hypoxanthine and thymidine. Forty days after fusion the presence of anti-SRBC activity was revealed as shown in *a*. The ratio of plaque forming cells/total number of hybrid cells was 1/30. This hybrid cell population was cloned in soft agar (50% cloning efficiency). A modified plaque assay was used to reveal positive clones shown in *b*–*d* as follows. When cell clones had reached a suitable size, they were overlaid in sterile conditions with 2 ml 0.6% agarose in phosphate-buffered saline containing 25 µl packed SRBC and 0.2 ml fresh guinea pig serum (absorbed with SRBC) as source of complement. *b*, Taken after overnight incubation at 37 °C. The ratio of positive/total number of clones was 1/33. A suitable positive clone was picked out and grown in suspension. This clone was called Sp-1/7, and was recloned as shown in *c*; over 90% of the clones gave positive lysis. A second experiment in which 10^6 P3-X67Ag8 cells were fused with 10^8 spleen cells was the source of a clone giving rise to indirect plaques (clone Sp-2/3-3). Indirect plaques were produced by the addition of 1:20 sheep anti-MOPC 21 antibody to the agarose overlay.

in the ratios of the different chains and occasionally with the total disappearance of one or other of the chains. Such events are best visualised on IEF analysis of the separated chains (for example, Fig. 1*h*, in which the heavy chain of P3 is no longer observed). The important point that no new chains are detected by IEF complements a previous study[4] of a rat–mouse hybrid line in which scrambling of *V* and *C* regions from the light chains of rat and mouse was not observed. In this study, both light chains have identical C_κ regions and therefore scrambled V_L–C_L molecules would be undetected. On the other hand, the heavy chains are of different subclasses and we expect scrambled V_H–C_H to be detectable by IEF. They were not observed in the clones studied and if they occur must do so at a lower frequency. We conclude that in syngeneic cell hybrids (as well as in interspecies cell hybrids) *V*–*C* integration is not the result of cytoplasmic events. Integration as a result of DNA translocation or rearrangement during transcription is also suggested by the presence of integrated mRNA molecules[11] and by the existence of defective heavy chains in which a deletion of *V* and *C* sections seems to take place in already committed cells[12].

The cell line P3-X63Ag8 described above dies when exposed to HAT medium. Spleen cells from an immunised mouse also die in growth medium. When both cells are fused by Sendai virus and the resulting mixture is grown in HAT medium, surviving clones can be observed to grow and become established after a few weeks. We have used SRBC as immunogen, which enabled us, after culturing the fused lines, to determine the presence of specific antibody-producing cells by a plaque assay technique[13] (Fig. 2*a*). The hybrid cells were cloned in soft agar[14] and clones producing antibody were easily detected by an overlay of SRBC and complement (Fig. 2*b*). Individual clones were isolated and shown to retain their phenotype as almost all the clones of the derived purified line are capable of lysing SRBC (Fig. 2*c*). The clones were visible to the naked eye (for example, Fig. 2*d*). Both direct and indirect plaque

assays[13] have been used to detect specific clones and representative clones of both types have been characterised and studied.

The derived lines (Sp hybrids) are hybrid cell lines for the following reasons. They grow in selective medium. Their karyotype after 4 months in culture (Table 1) is a little smaller than the sum of the two parental lines but more than twice the chromosome number of normal BALB/c cells, indicating that the lines are not the result of fusion between spleen cells. In addition the lines contain a metacentric chromosome also present in the parental P3-X67Ag8. Finally, the secreted immunoglobulins contain MOPC 21 protein in addition to new, unknown components. The latter presumably represent the chains derived from the specific anti-SRBC antibody. Figure 3*A* shows the IEF pattern of the material secreted by two such Sp hybrid clones. The IEF bands derived from the parental P3 line are visible in the pattern of the hybrid cells, although obscured by the presence of a number of new bands. The pattern is very complex, but the complexity of hybrids of this type is likely to result from the random recombination of chains (see above, Fig. 1). Indeed, IEF patterns of the reduced material secreted by the spleen–P3 hybrid clones gave a simpler pattern of Ig chains. The heavy and light chains of the P3 parental line became prominent, and new bands were apparent.

The hybrid Sp-1 gave direct plaques and this suggested that it produces an IgM antibody. This is confirmed in Fig. 4 which shows the inhibition of SRBC lysis by a specific anti-IgM

Table 1 Number of chromosomes in parental and hybrid cell lines

Cell line	Number of chromosomes per cell	Mean
P3-X67Ag8	66,65,65,65,65	65
P1Bul	Ref. 4	55
Mouse spleen cells	—	40
Hy-B (P1–P3)	112,110,104,104,102	106
Sp-1/7-2	93,90,89,89,87	90
Sp-2/3-3	97,98,96,96,94,88	95

antibody. IEF techniques usually do not reveal 19S IgM molecules. IgM is therefore unlikely to be present in the unreduced sample *a* (Fig. 3*B*) but μ chains should contribute to the pattern obtained after reduction (sample *a*, Fig. 3*A*).

The above results show that cell fusion techniques are a powerful tool to produce specific antibody directed against a predetermined antigen. It further shows that it is possible to isolate hybrid lines producing different antibodies directed against the same antigen and carrying different effector functions (direct and indirect plaque).

The uncloned population of P3–spleen hybrid cells seems quite heterogeneous. Using suitable detection procedures it should be possible to isolate tissue culture cell lines making different classes of antibody. To facilitate our studies we have used a myeloma parental line which itself produced an Ig. Variants in which one of the parental chains is no longer expressed seem fairly common in the case of P1–P3 hybrids (Fig. 1*h*). Therefore selection of lines in which only the specific antibody chains are expressed seems reasonably simple. Alternatively, non-producing variants ￢f myeloma lines could be used for fusion.

We used SRBC as antigen. Three different fusion experiments were successful in producing a large number of antibody-producing cells. Three weeks after the initial fusion, 33/1,086

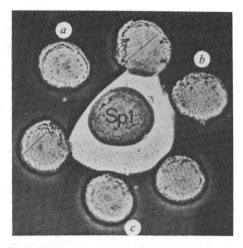

Fig. 4 Inhibition of haemolysis by antibody secreted by hybrid clone Sp-1/7-2. The reaction was in a 9-cm Petri dish with a layer of 5 ml 0.6% agarose in phosphate-buffered saline containing 1/80 (v/v) SRBC. Centre well contains 2.5 μl 20 times concentrated culture medium of clone Sp-1/7-2 and 2.5 μl mouse serum. *a*, Sheep specific anti-mouse macroglobulin (MOPC 104E, Dr Feinstein); *b*, sheep anti-MOPC 21 (P3) IgG1 absorbed with Adj PC-5; *c*, sheep anti-Adj PC-5 (IgG2a) absorbed with MOPC 21. After overnight incubation at room temperature the plate was developed with guinea pig serum diluted 1:10 in Dulbecco's medium without serum.

clones (3%) were positive by the direct plaque assay. The cloning efficiency in the experiment was 50%. In another experiment, however, the proportion of positive clones was considerably lower (about 0.2%). In a third experiment the hybrid population was studied by limiting dilution analysis. From 157 independent hybrids, as many as 15 had anti-SRBC activity. The proportion of positive over negative clones is remarkably high. It is possible that spleen cells which have been triggered during immunisation are particularly successful in giving rise to viable hybrids. It remains to be seen whether similar results can be obtained using other antigenes.

The cells used in this study are all of BALB/c origin and the hybrid clones can be injected into BALB/c mice to produce solid tumours and serum having anti-SRBC activity. It is possible to hybridise antibody-producing cells from different origins[4,5]. Such cells can be grown *in vitro* in massive cultures to provide specific antibody. Such cultures could be valuable for medical and industrial use.

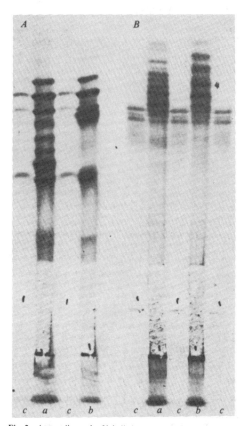

Fig. 3 Autoradiograph of labelled components secreted by anti-SRBC specific hybrid lines. Fractionation before (*B*) and after (*A*) reduction was by IEF. *p*H gradient was 5.0 (bottom) to 9.0 (top) in the presence of 6 M urea. Other conditions as in Fig. 1. Supernatants from: *a*, hybrid clone Sp-1/7-2; *b*, hybrid clone Sp-2/3-3; *c*, myeloma line P3-X67Ag8.

Cloning Immunoglobulin Variable Domains for Expression by the Polymerase Chain Reaction

R. Orlandi, D.H. Güssow, P.T. Jones and G. Winter

ABSTRACT   We have designed a set of oligonucleotide primers to amplify the cDNA of mouse immunoglobulin heavy and light chain variable domains by the polymerase chain reaction. The primers incorporate restriction sites that allow the cDNA of the variable domains to be force-cloned for sequencing and expression. Here we have applied the technique to clone and sequence the variable domains of five hybridoma antibodies and to express a mouse–human chimeric antibody that binds to the human mammary carcinoma line MCF-7. The technique should also lead to the cloning of antigen-binding specificities directly from immunoglobulin genes.

For serotherapy, human monoclonal antibodies (mAbs) would be preferred to mouse mAbs because the foreign immunoglobulin can elicit an anti-globulin response that may interfere with the therapy (1) or cause allergic or immune complex hypersensitivity (2). However, there are considerable difficulties in making human mAbs of the required specificity by hybridoma technology (3). Recently, protein engineering has been used to convert mouse mAbs into "human" mAbs by joining the entire immunoglobulin variable (V) domains from mouse mAbs to human constant (C) domains (4–6) or by transplanting the complementarity-determining regions (CDRs) of the mouse mAbs into human myeloma proteins (7–9). Thus, the rearranged immunoglobulin V genes provide the raw material for engineering and have been derived by cloning from genomic DNA (10, 11) or cDNA (12).

The polymerase chain reaction (PCR) (13) has been used recently for genomic (14) and cDNA cloning (15). It involves repeated rounds of extension from two primers specific for regions at each end of the gene. The primers need not match the gene sequence exactly (16), and restriction sites can be incorporated within the primers to allow the forced cloning of the amplified DNA (14). In principle, the restriction sites within the PCR primers could be devised to clone a gene directly for expression, although this has not been described. Therefore, we sought to apply the PCR to the cloning and expression of immunoglobulin V genes.

First, we identified conserved regions at each end of the nucleotide sequences encoding V domains of mouse immunoglobulin heavy chain (V_H) and κ light chain (V_κ). Second, we designed primers for the amplification, which incorporated restriction sites for forced cloning. Third, we constructed vectors that allow the amplified cDNA to be expressed, while retaining the amino acid sequence typical of V domains. We then applied the technique to clone and sequence cDNA encoding five mouse mAbs of therapeutic potential: MBr1 (17), BW 431/26 (18), BW 494/32 (19), BW 250/183 (18, 20), and BW 704/152. MBr1 has been raised against a human mammary carcinoma line MCF-7 and rec-

ognizes a saccharide epitope (21). The cDNA of the MBr1 V_H and V_κ domains (mouse μ and κ chains) was expressed as a simple chimeric antibody (human $\gamma1$ and κ chains). The human $\gamma1$ isotype was chosen, as it should mediate cell killing by both complement lysis and cell-mediated routes (9, 22).

METHODS

Comparison of Nucleotide Sequences. The aligned entries of the nucleotide sequences of the V genes were extracted from the Kabat data base (23): the beginning of the nucleotide sequences correspond to the mature N terminus of the protein, and nucleotides encoding the protein signal sequences are not included. To allow manipulation of the data with the computer program DBUTIL (24) and analysis with the FAMNS and PLOTD programs (R. Staden, personal communication), the entries were entered into a shotgun sequencing data base (24). By using DBUTIL, each V_H or V_κ block was trimmed to a "core alignment" by removing positions encoding extra amino acid residues (A to K in ref. 23) except for residues 82A, 82B, and 82C of the V_H domains. By using FAMNS, the frequency of the most common nucleotide was scored for each site, and the information was used to design the amplification primers (see *Results*). By using PLOTD, the 14 nucleotides at the 3′ end of the forward and back primers (see below) were matched with each entry in the data base. (Entries with nucleotides missing in this section of sequence were excluded from the comparison.) Likewise, PLOTD was used to match up the nucleotide sequences encoding CDR1 and CDR2 of MBr1 V_H and V_κ domains with each of the entries in the data base.

cDNA Synthesis and Amplification. RNA was prepared from about 5×10^8 hybridoma cells grown in roller bottles, and mRNA was selected on oligo(dT)-cellulose (25). First-strand cDNA synthesis was based on ref. 26. A 50-μl reaction mixture containing 10 μg of mRNA, 20 pmol of VH1FOR primer [5′-d(TGAGGAGACGGTGACCGTGGTCCCTTGG-CCCCAG)] or VK1FOR primer [5′-d(GTTAGATCTCCAG-CTTGGTCCC)], 250 μM of each dNTP, 10 mM dithiothreitol, 100 mM Tris·HCl (pH 8.3), 10 mM MgCl$_2$, and 140 mM KCl was heated at 70°C for 10 min and cooled. Reverse transcriptase (Anglian Biotec, Colchester, U.K.) was added (46 units) and incubated at 42°C for 1 hr. For amplification with a thermostable DNA polymerase (15), a 50-μl reaction mixture containing 5 μl of the cDNA·RNA hybrid, 25 pmol of primers VH1FOR or VK1FOR and VH1BACK [5′-d(AGGTSMARC-TGCAGSAGTCWGG) in which S = C or G, M = A or C, R = A or G, and W = A or T] or VK1BACK [5′-d(GA-CATTCAGCTGACCCAGTCTCCA)] as appropriate, 250 μM

Abbreviations: mAb, monoclonal antibody; V_H, heavy chain variable region; V_κ, κ light chain variable region; PCR, polymerase chain reaction; CDR, complementarity-determining region; C, constant; J, joining.
‡To whom reprint requests should be addressed.

ORLANDI, R., GÜSSOW, D.H., JONES, P.T. and WINTER, G.
Cloning immunoglobulin variable domains for expression by the polymerase chain reaction.
Proc. Natl. Acad. Sci. U.S.A. 86:3833-3837 (1989). Reprinted with the authors' permission.

of each dNTP, 67 mM Tris chloride (pH 8.8), 17 mM $(NH_4)_2SO_4$, 10 mM $MgCl_2$, 200 μg of gelatine per ml, and 2 units of thermus aquatics (Taq) polymerase (Cetus) was overlaid with paraffin oil and subjected to 25 rounds of temperature cycling with a Techne PHC-1 programable heating block. A typical cycle was 1 min at 95°C (denature), 1 min at 30°C (anneal), and 2 min at 72°C (elongate). The sample (and oil) was extracted twice with ether, once with phenol, and then with phenol/$CHCl_3$, followed by ethanol precipitation. The sample was taken up in 50 μl of water and frozen.

Vector Construction. To make the phage M13-VHPCR1 vector, a *Bst*EII site was introduced into the M13-HuVHNP vector (7) by site-directed mutagenesis (27, 28). To make the M13-VKPCR1 vector, *Pvu* II and *Bcl* I sites were likewise introduced into a version of M13-HuVKLYS (J. Foote and G.W., unpublished data) in which the *Pvu* II sites had been removed from the M13 backbone as follows: M13mp18 (29) was cut with *Pvu* II, and the vector backbone was blunt-ligated to a synthetic *Hin*dIII–*Bam*HI polylinker. The HuVKLYS gene was then introduced as a *Hin*dIII–*Bam*HI fragment.

Cloning of Amplified cDNA. The amplified cDNA was digested with the restriction enzymes *Pst* I and *Bst*EII for the V_H gene or *Pvu* II and *Bgl* II for the V_κ gene. The fragments were phenol-extracted, purified on 2% low-melting-point agarose gels, and force-cloned into M13-VHPCR1 (digested with *Pst* I and *Bst*EII) or into M13-VKPCR1 (digested with *Pvu* II and *Bcl* I). Note that there is a short [4 base pair (bp)] region of complementarity between the 3' ends of the VH1FOR and VH1BACK primers, and the two primers can prime on each other to give a short duplex, which is readily cloned (J. Gamble and G.W., unpublished data). Therefore, it is important to gel-purify the amplified cDNA. The vector backbone was purified on 0.8% agarose gels and, in the case of the M13-VKPCR1 vector, was also treated with calf intestinal phosphatase. The M13-VKPCR1 vector was pre-pared in *Escherichia coli* JM110 (29) to avoid Dam methylation at the *Bcl* I site. Clones containing V gene inserts were identified directly by sequencing (30) with primers based in the 3' noncoding region of the V gene in the M13-VHPCR1 and M13-VKPCR1 vectors.

Antibody Expression. The *Hin*dIII–*Bam*HI fragment carrying the MBr1 V_H gene in M13-VHPCR1 was recloned into a pSV-gpt vector with human immunoglobulin heavy chain $\gamma1$ C regions (9). Likewise, the MBr1 V_κ gene in M13-VKPCR1 was recloned as a *Hin*dIII–*Bam*HI fragment into pSV vector with a hygromycin-resistance marker and human κ chain C domains [pSV-hyg-HuCK] (J. Foote and G.W., unpublished data). Vectors were linearized with *Pvu* I and cotransfected into the nonsecreting mouse myeloma line NS0 (31) by electroporation (32). Cells were selected in the presence of mycophenolic acid at 0.3 μg/ml after 1 day and at 1 μg/ml after 7 days. After 14 days, four wells, each containing one or two major colonies, were screened by incorporation of [^{14}C]lysine (33), and the secreted antibody was detected after precipitation with protein A-Sepharose (Pharmacia) on Na-DodSO$_4$/PAGE (34). The gels were stained, fixed, soaked in fluorographic reagent Amplify (Amersham), dried, and autoradiographed on preflashed film at −70°C for 2 days. Supernatant was also tested for binding to the mammary carcinoma line MCF-7 and the colon carcinoma line HT-29 essentially as described (17), either by indirect immunofluorescence assay on cell suspensions (using fluorescein-labeled goat antihuman IgG; Amersham) or solid-phase RIA on monolayers of fixed cells (using ^{125}I-labeled protein A; Amersham). The MCF-7 and HT-29 cell lines were provided by J. Fogh.

RESULTS

Design of Amplification Primers and Vectors. The nucleotide sequences of V genes in ref. 23 were aligned as described

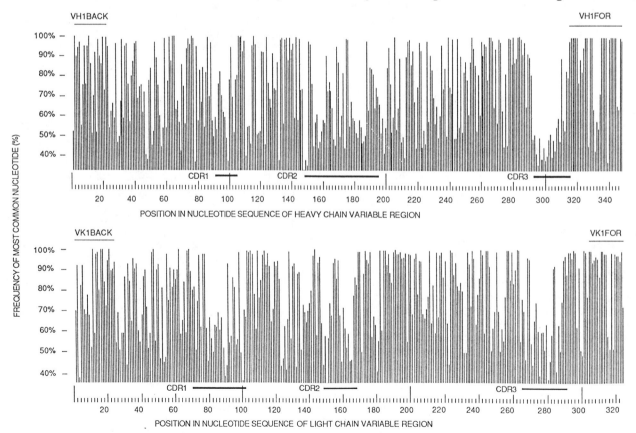

FIG. 1. Frequency of the most common nucleotides in V_H and V_κ gene sequences in ref. 23. CDR1, CDR2, and CDR3 are located, respectively, at nucleotide positions 91–105, 148–195, and 292–315 of V_H genes and positions 70–102, 148–168, and 265–291 of V_κ genes.

Table 1. Checking primers for mismatches with the data base entries

Primers	Entries with none, one, or two mismatches		
	0	1	2
VH1FOR	50/131	71/131	
VH1BACK	22/141	56/141	43/141
VK1FOR	38/61	20/61	
VK1BACK	19/115	54/115	26/115

For each primer, the number of entries with zero, one, or two mismatches with respect to the 14 nucleotides at the 3' end of each of the primers are given, as well as the total number of eligible entries. For the mixed VH1BACK primer, each of the possible variants was scored and the results summed.

in *Methods*. The frequency of the most common nucleotide was plotted for each position in the aligned V_H and V_κ gene domains (Fig. 1). As expected, the nucleotide sequences encoding the protein CDRs are variable, and those corresponding to the joining (J)-region segments are conserved. However, there are several other conserved regions, in particular those encoding the mature N terminus of both V_H and V_κ domains.

Amplification primers VH1FOR and VK1FOR were designed to be complementary to the mRNA in the J regions; and primers VH1BACK and VK1BACK, to be complementary to the first-strand cDNA encoding the conserved N-terminal region. VH1BACK primer has a mixed sequence. The restriction sites for forced cloning were incorporated toward the 5' end (the least conserved part) of the primer, and mismatches were minimized near the 3' end.

The degree of complementarity of the 3' end of each primer to each of the entries in the Kabat data base was checked with PLOTD. For most entries, this region of the primer proved to be perfectly matched with only one or two mismatches (Table 1). The 3' ends of the forward primers were also checked against each of the mouse heavy and light chain J regions, and the 3' ends of the back primers with consensus sequences derived from each of the mouse V_H and V_κ families (as listed in ref. 23). The primers match well to each J region or mouse V gene family, apart from the VK1BACK primer, which has several mismatches with the mouse V_κ II family.

The scheme for cloning of V_H and V_κ sequences from mouse Ig mRNA is shown in Fig. 2. The restriction sites at each end of the mature V domains allow the existing domain to be excised and the amplified cDNA to be inserted. At the

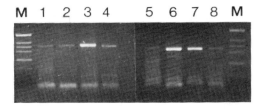

FIG. 3. Amplified cDNA from heavy and light chain V regions of four hybridomas. Each sample (5 μl) was checked on 2% agarose gels and stained with ethidium bromide. Lanes: M, 1 μg of pBR322 *Hin*fI fragments (517 and 506) bp, 396 bp, 344 bp, 298 bp, (221 and 220) bp, 154 bp, and 75 bp as markers; 1–4, V_H cDNA of BW 704/152, BW 250/183, BW 494/32, and BW 431/26, respectively; 5–8, separate gel for the corresponding V_κ cDNA.

cloning junction and at both ends of the V genes, the cloned cDNAs encode amino acid sequences that are typical of V_H or V_κ domains.

Amplification and Cloning. cDNA was prepared from each of five hybridoma lines: MBr1, BW 704/152, BW 250/183, BW 494/32, and BW 431/26 (17–20) by using the amplification primers. A typical result is shown in Fig. 3, in which the amplified cDNA gives a major band of the expected size (about 320–350 bp), although its intensity varies and there are also smaller bands. For the V_κ cDNA of MBr1, the band was very weak and was excised from the gel and reamplified in a second round (not shown). After digestion with restriction enzymes, the cDNA was gel-purified and cloned into the M13-VHPCR1 and M13-VKPCR1 vectors. The clones containing V_H or V_κ gene inserts were identified directly by sequencing, as the majority carried correct inserts. (There is an internal *Pst* I site in the V_H gene of BW 431/26, and the gene was therefore assembled in two steps: the 3' *Pst* I–*Bst*EII fragment was first cloned into M13-VHPCR1, followed in a second step by the 5' *Pst* I fragment.)

A comparison of the sequences of V_H and V_κ domains from the five antibodies with the Kabat protein data base (23) revealed that there were five different mouse V_H families (IA, IB, IIA, IIB and IIIA) and two different mouse V_K families (I and VI) (Table 2). To ensure accuracy, the sequences of the V genes for each antibody were determined from two independent amplifications from the cDNA·mRNA hybrid or, in the case of MBr1, from two independent batches of mRNA. With the exception of two single nucleotide discrepancies and a deletion of three nucleotides beyond the 3' end of the VK1BACK primer (presumably due to slippage of the

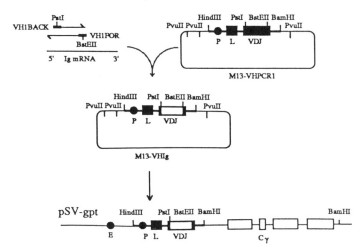

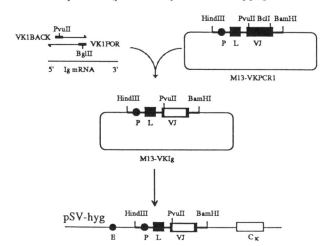

FIG. 2. Scheme for amplification of cDNA and cloning into phage M13 vectors to hook up V region genes for expression. The vectors M13-VHPCR1 and M13-VKPCR1, for cloning the amplified cDNA, contain introns: transcription is driven from the immunoglobulin heavy chain promoter (P), and the signal sequence (L) and leader intron are taken from the mouse V47 unrearranged V_H gene (12). The noncoding sequence to the 3' end of the V_H gene is described in ref. 12 and of the V_κ gene in ref. 9.

Table 2. Identification of mouse heavy or light chain V gene families cloned from hybridoma mRNA

Hybridoma	V_H family	V_κ family
MBr1	IIB	I
BW 431/26	IA	VI
BW 494/32	IIIA	VI
BW 250/183	IIA	VI
BW 704/152	IB	I

primer), the sequences of the independent clones were identical.

Sequencing and Expression of MBr1. The nucleotide sequence of the MBr1 heavy and light chain V genes is shown in Fig. 4 with part of the flanking regions of the M13-VHPCR1 and M13-VKPCR1 vectors. There are some unusual features of the MBr1 sequence—for example, CDR3 of the V_H domain is very short (35), with a diversity (D) segment encoding apparently only three amino acids. However, these features proved identical in two independent clones. The nucleotide sequences of CDR1 and CDR2 of MBr1 V genes were screened against each of the entries in the Kabat data base using PLOTD. This showed that the sequence of the MBr1 V_H gene was most closely related to that of 119.1, a member of the 4-hydroxy-3-nitrophenylacetyl (NP) major idiotypic family (36).

The V_H and V_κ gene regions of MBr1 were assembled together with human genes $C_{\gamma 1}$ and C_κ for expression of simple mouse–human chimeric antibodies in myeloma cells NS0 (31). The transfected cells were screened for secretion of antibody by precipitation of ^{14}C-labeled heavy and light chains with protein A-Sepharose: one of the four wells

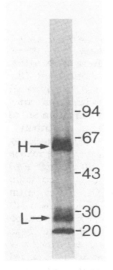

FIG. 5. Secretion of MBr1 chimeric antibody (human $C_{\gamma 1}$ and C_κ) from the myeloma NS0. Cells were labeled with [^{14}C]lysine, and the secreted antibody was subjected to NaDodSO$_4$/PAGE (10% acrylamide). The heavy (H) and light (L) chain bands are marked, and the molecular weights of the protein markers phosphorylase b, albumin, ovalbumin, carbonic anhydrase, and trypsin inhibitor are shown $\times 10^{-3}$.

secreted antibody (Fig. 5). The chimeric antibody in the supernatant, like the parent mouse MBr1 antibody, was found to bind to the mammary carcinoma MCF-7 but not to the colon carcinoma HT-29.

DISCUSSION

The sequencing and cloning of immunoglobulin V genes is the first and often rate-limiting step in making simple chimeric or reshaped human antibodies. We have devised a simple and rapid way of cloning these genes via the PCR. By making a systematic comparison of the aligned sequences of V_H and V_κ genes, we have identified nucleotide sequences at the 5' ends of both V_H and V_κ genes that are relatively conserved. Since the J_H and J_κ regions at the 3' ends of the genes are also conserved, we were able to design primers for PCR amplification based on these sequences and to include restriction sites for forced cloning. In general, the clones could be screened directly by sequencing, with almost all of the recombinants carrying the correct inserts.

From the five mouse hybridomas, we succeeded in preparing mRNA and amplified cDNA and in cloning heavy chain genes from five V_H gene families and light chain genes from two V_κ gene families. This suggests that our primers might amplify most immunoglobulin mRNA of the mouse repertoire. We have also used the same primers for amplification and cloning of immunoglobulin V_H and V_κ genes from mouse hybridoma mRNA and spleen genomic DNA (D.H.G. and G.W., unpublished data). Nevertheless, we might anticipate some problems—for example, the poor match of the mouse V_κ II gene family with VK1BACK primer. As an alternative to the "universal" amplification primers, we have also used mixtures of back primers based on the concensus sequences of each V_H and V_κ gene family and mixtures of forward primers based on individual J region sequences or mixtures of forward primers based in the CH$_1$ or C_κ gene regions (D.H.G. and G.W., unpublished data). However, the siting of both primers within the rearranged V genes is probably necessary for their cloning from genomic DNA.

Errors in the amplified V_H and V_κ genes were readily identified by sequencing clones from two independent amplifications. We observed only two single nucleotide discrepancies on comparing 10 pairs of genes of about 300 nucleotides. Thus, the overall error frequency (1/1500) for 25 cycles of PCR is in the same range as the previous estimates of 1/400 (15) and 1/4000 (37) for 30 cycles and is comparable to that expected for a single round of copying of cDNA from RNA by reverse transcriptase (38). As alternative to sequencing independent clones, the amplified cDNA could be sequenced directly (37).

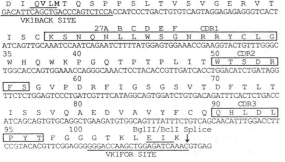

```
                Sequence of MBr1 VH

                                    Splice      -1
                                     ↓G  V  H  S
                                     AGGTGTCCACTCC
1          PstI              10                     20
Q  V  Q  L  Q  EQSPG  T  E  L  A  S  P  G  A  S  V  T  L
CAGGTCCAACTGCAGGAGTCAGGAACTGAGCTGGCGAGTCCTGGGGCATCAGTGACACTG
   VH1BACK SITE
                              30       CDR1
S  C  K  A  S  G  Y  T  F  T  D  H  I  I  N  W  V  K  K  R
TCCTGCAAGGCTTCTGGCTACACATTTACTGACCATATTATAAATTGGGTAAAAAAGAGG
                                   52a 53     CDR2
P  G  Q  G  L  E  W  I  G  R  I  Y  P  V  S  G  V  T  N  Y
CCTGGACAGGGCCTTGAGTGGATTGGAAGGATTTATCCAGTAAGTGGTGTAACTAACTAC
60     CDR2        65                70
N  Q  K  F  M  G  K  A  T  F  S  V  D  R  S  S  N  T  V  Y
AATCAAAAATTCATGGGCAAGGCCACATTCTCTGTAGACCGGTCCTCCAACACAGTGTAC
80     82A  B  C  83                90
M  V  L  N  S  L  T  S  E  D  P  A  V  Y  Y  C  G  R  G  F
ATGGTGTTGAACAGTCTGACATCTGAGGACCCTGCTGTCTATTACTGTGGAAGGGGCTTT
   CDR3    103             BstEII       Splice
D  F  D  Y  W  G  Q  G  T  T  T  V  L  T  V  S  S  ↓
GATTTTGACTACTGGGGCCAAGGGACCACGGTCACCGTCTCCTCAGGT......
                   VH1FOR SITE


                Sequence of MBr1 VK

                                    Splice      -1
                                     ↓G  V  H  S
                                     AGGTGTCCACTCC
1          PvuII              10                     20
D  I  Q  V  L  M  T  Q  S  P  P  S  L  T  V  S  V  G  E  R  V  T
GACATTCAGCTGACCCAGTCTCCACCATCCCTGACTGTGTCAGTAGGAGAGAGGGTCACT
   VK1BACK SITE
              27A  B  C  D  E  F        CDR1
I  S  C  K  S  N  Q  N  L  L  W  S  G  N  R  R  Y  C  L  G
ATCAGTTGCAAATCCAATCAGAATCTTTTATGGAGTGGGAAACCGAAGGTACTGTTTGGGC
35        40                  50        CDR2
W  H  Q  W  K  P  G  Q  T  P  T  P  L  I  T  W  T  S  D  R
TGGCACCAGTGGAAACCAGGGCAAACTCCTACACCGTTGATCACCTGGACATCTGATAGG
             60                      70
F  S  G  V  P  D  R  F  I  G  S  G  S  V  T  D  F  T  L  T
TTCTCTGGAGTCCCTGATCGTTCATAGGCAGTGGATCTGTGACAGATTTCACTCTGACC
              80                  90         CDR3
I  S  S  V  Q  A  E  D  V  A  V  Y  F  C  Q  Q  H  L  D  L
ATCAGCAGTGTGCAGGCTGAAGATGTGGCAGTTTATTTCTGTCAGCAACATTTGGACCTT
95        100         BglII/BclI   Splice
P  Y  T  F  G  G  G  T  K  L  E  I  K  ↓
CCGTACACGTTCGGAGGGGGGACCAAGCTGGAGATCAAACGTGAG
                   VK1FOR SITE
```

FIG. 4. Nucleotide sequences of MBr1 heavy and light chain V genes and adjacent flanking regions of the M13-VHPCR1 and M13-VKPCR1 vectors. Amino acids encoded within the amplification primers, which are likely to differ in MBr1, are marked in bold type.

Given our choice of priming sites, it is not possible to determine the exact sequence at both ends of the V genes, as it is dictated by the primer. However, to some extent, we can reconstruct the sequence in these regions (Fig. 4). For example, MBr1 uses heavy chain J_{H2} and light chain $J_{\kappa 2}$: assuming no somatic mutation, we would expect that residue 109 of the V_H MBr1 product is probably leucine and not valine. The V genes of MBr1 belong to the mouse V_H IIB and V_κ I families. For the MBr1 V_H gene, we might expect its product to have Gln-6 and Pro-7, rather than Glu-6 and Ser-7; for the MBr1 V_κ gene, we similarly might expect Val-3 and Met-4, rather than Gln-3 and Leu-4. Nevertheless, the chimeric MBr1 antibody binds to the antigen, and in any case, these uncertainties in the framework regions are unlikely to affect the antigenic specificity, since the specificity can be transplanted from one antibody to another by transfer of the CDRs only (7–9). For the construction of therapeutic antibodies, the sequence of the original mouse hybridoma antibody scarcely matters if the antigenic specificity is retained.

We envisage many other applications of this technique, ranging from the analysis of mouse immunoglobulin gene repertoire to the "rescue" of mAbs from unstable human hybridomas (3). (In view of the over-representation of the mouse immunoglobulin V genes in the Kabat nucleotide data base, these primers may not be optimal for the PCR amplification of human immunoglobulin genes.) However, the most exciting applications may come from the construction of antibody expression libraries. Our design of PCR primers allows the V genes to be joined in frame to a specialized vector, which directs the expression of antibody. Thus, from the mRNA of the mouse hybridoma MBr1, we readily derived a mouse–human chimeric antibody of the same specificity, which was expressed and secreted from the mouse myeloma NS0. In principle, the amplified cDNA could also be expressed and secreted as Fv or Fab fragments from *E. coli* (39, 40). Furthermore, we anticipate that antigenic specificities could be cloned directly from genomic DNA or the mRNA from stimulated spleen or peripheral blood lymphocytes, particularly if the sequence of either heavy or light chain partner is already known. For example, the products of the germ line V_H genes for the major mouse NP^b idiotypic family are associated with λ light chains (41). Alternatively, antigenic specificities might be derived directly from single cells, if both V_H and V_κ genes could be amplified (42). Thus, ultimately this technique could provide an alternative to hybridoma technology (43) for the cloning of antigenic specificities.

We thank R. Staden for providing the programmes PLOTD and FAMNS and advice in data analysis; J. Foote for the M13-HuVKLYS and pSV-hyg-HuCK vectors; M. Colnaghi, S. Canevari, and C. Milstein for their encouragement; and C. Milstein for invaluable comments on the manuscript. R.O. was supported by travel grants from the European Molecular Biology Organization and the Associazione Italiana Ricerca Cancro, and D.H.G., by Behringwerke, Marburg, F.R.G.

1. Miller, R. A., Oseroff, A. R., Stratte, P. T. & Levy, R. (1983) *Blood* **62**, 988–995.
2. Ratner, B. (1943) *Allergy Anaphylaxis and Immunotherapy* (Williams & Wilkins, Baltimore).
3. Carson, D. A. & Freimark, B. D. (1986) *Adv. Immun.* **38**, 275–311.
4. Morrison, S. L., Johnson, M. J., Herzenberg, L. A. & Oi, V. T. (1984) *Proc. Natl. Acad. Sci. USA* **81**, 6851–6855.
5. Boulianne, G. L., Hozumi, N. & Schulman, M. J. (1984) *Nature (London)* **312**, 643–646.
6. Neuberger, M., Williams, G. T., Mitchell, E. B., Jouhal, S. S., Flanagan, J. G. & Rabbitts, T. H. (1985) *Nature (London)* **314**, 268–270.
7. Jones, P. T., Dear, ·P. H., Foote, J., Neuberger, M. S. & Winter, G. (1986) *Nature (London)* **321**, 522–525.
8. Verhoeyen, M., Milstein, C. & Winter, G. (1988) *Science* **239**, 1534–1536.
9. Riechmann, L., Clark, M., Waldmann, H. & Winter, G. (1988) *Nature (London)* **332**, 323–327.
10. Oi, V. T., Morrison, S. L., Herzenberg, L. A. & Berg, P. (1983) *Proc. Natl. Acad. Sci. USA* **80**, 825–829.
11. Ochi, A., Hawley, R. G., Shulman, M. J. & Hozumi, N. (1983) *Nature (London)* **302**, 340–342.
12. Neuberger, M. S. (1983) *EMBO J.* **2**, 1373–1378.
13. Saiki, R. K., Scharf, S., Faloona, F., Mullis, K. B., Horn, G. T., Ehrlich, H. A. & Arnheim, N. (1985) *Science* **230**, 1350–1354.
14. Scharf, S. J., Horn, G. T. & Ehrlich, H. A. (1986) *Science* **233**, 1076–1078.
15. Saiki, R. K., Gelfand, D. H., Stoffel, S., Scharf, S. J., Higuchi, R., Horn, G. T., Mullis, K. B. & Ehrlich, H. A. (1988) *Science* **239**, 487–491.
16. Lee, C. C., Wu, X., Gibbs, R. A., Cook, R. G., Muzny, D. M. & Caskey, C. T. (1988) *Science* **239**, 1288–1291.
17. Menard, S., Tagliabue, E., Canevari, S., Fossati, G. & Colnaghi, M. I. (1983) *Cancer Res.* **43**, 1295–1300.
18. Bosslet, K., Steinstrasser, A., Schwarz, A., Harthus, H. P., Luben, G., Kuhlmann, L. & Sedlacek, H. H. (1988) *Eur. J. Nuclear Med.* **14**, 523–528.
19. Bosslet, K., Kern, H. F., Kanzy, E. J., Steinstraesser, A., Schwarz, A., Luben, G., Schorlemmer, H. U. & Sedlacek, H. H. (1986) *Cancer Immunol. Immunother.* **23**, 185–191.
20. Bosslet, K., Luben, G., Schwarz, A., Hundt, E., Harthus, E. P., Seiler, F. R., Murher, C., Kloppel, G., Kayser, K. & Sedlacek, H. H. (1985) *Int. J. Cancer* **36**, 75–84.
21. Bremer, E. G., Levery, S. B., Sonnino, S., Ghidoni, R., Canevari, S., Kannagi, R. & Hakamori, S. (1984) *J. Biol. Chem.* **259**, 14773–14777.
22. Brüggemann, M., Williams, G. T., Bindon, C. I., Clark, M. R., Walker, M. R., Jefferis, R., Waldmann, H. & Neuberger, M. S. (1987) *J. Exp. Med.* **166**, 1351–1361.
23. Kabat, E. A., Wu, T. T., Reid-Miller, M., Perry, H. M. & Gottesmann, K. S. (1987) *Sequences of Proteins of Immunological Interest* (U.S. Dept. of Health and Human Services, U.S. Government Printing Office).
24. Staden, R. (1986) *Nucleic Acids Res.* **14**, 217–231.
25. Griffiths, G. & Milstein, C. (1985) *Hybridoma Technology in the Biosciences and Medicine* (Plenum, New York), pp. 103–115.
26. Maniatis, T., Fritsch, E. F. & Sambrook, J. (1982) *Molecular Cloning: A Laboratory Manual* (Cold Spring Harbor Lab., Cold Spring Harbor, NY).
27. Zoller, M. & Smith, M. (1982) *Nucleic Acids Res.* **10**, 6457–6500.
28. Carter, P., Bedouelle, H. & Winter, G. (1985) *Nucleic Acids Res.* **13**, 4431–4443.
29. Yanisch-Perron, C., Vieira, J. & Messing, J. (1985) *Gene* **33**, 103–119.
30. Sanger, F., Nicklen, S. & Coulson, A. R. (1977) *Proc. Natl. Acad. Sci. USA* **74**, 5463–5467.
31. Kearney, J. F., Radbruch, A., Liesegang, B. Rajewski, K. (1979) *J. Immunol.* **123**, 1548–1550.
32. Potter, H., Weir, L. & Leder, P. (1984) *Proc. Natl. Acad. Sci. USA* **81**, 7161–7163.
33. Galfre, G. & Milstein, C. (1981) *Methods Enzymol.* **73**, 1–46.
34. Laemmli, U. K. (1970) *Nature (London)* **227**, 680–685.
35. Berek, C. & Milstein, C. (1988) *Immunol. Rev.* **105**, 5–26.
36. Maizels, N. & Bothwell, A. (1985) *Cell* **43**, 715–720.
37. Innis, M. A., Myambo, K. B., Gelfand, D. H. & Brow, M. A. D. (1988) *Proc. Natl. Acad. Sci. USA* **85**, 9436–9440.
38. Gopinathan, K. P., Weymouth, L. A., Kunkel, T. A. & Loeb, L. A. (1979) *Nature (London)* **278**, 857–859.
39. Skerra, A. & Pluckthun, A. (1988) *Science* **240**, 1038–1041.
40. Better, M., Chang, C. P., Robinson, R. & Horwitz, A. H. (1988) *Science* **240**, 1041–1043.
41. Bothwell, A. L. M., Paskind, M., Reth, M., Imanishi-Kari, T., Rajewsky, K. & Baltimore, D. (1981) *Cell* **24**, 625–637.
42. Li, H., Gyllensten, U. B., Ciu, X., Saiki, R. K., Ehrlich, H. A. & Arnheim, N. (1988) *Nature (London)* **335**, 414–417.
43. Kohler, G. & Milstein, C. (1975) *Nature (London)* **256**, 495–497.

Supplementary Readings

Bentley, G.A., Boulot, G., Riottot, M.M. & Poljak, R.J., Three-dimensional structure of an idiotope-anti-idiotope complex. *Nature* 2348:254–257 (1990)

Claesson, L., Larhammar, D., Rask, L. & Peterson, P.A., cDNA clone for the human invariant γchain of class II histocompatibility antigens and its implications for the protein structure. *Proc. Natl. Acad. Sci. U.S.A.* 80:7395–7399 (1983)

Collins, M.K.L., Goodfellow, P.N., Spurr, N.K., Solomon, E., Tanigawa, G., Tonegawa, S. & Owen, M.J., The human T-cell receptor α-chain gene maps to chromosome 14. *Nature* 314:273–274 (1985)

Cwirla, S.E., Peters, E.A., Barrett, R.W. & Dower, W.J., Peptides on phage: a vast library of peptides for identifying ligands. *Proc. Natl. Acad. Sci. U.S.A.* 87:6378-6382 (1990)

Davis, M.M. & Bjorkman, P.J., T-cell antigen receptor genes and T-cell recognition. *Nature* 334:395–402 (1988)

Davis, M.M., Chien, Y.H., Gascoigne, N.R. & Hedrick, S.M., A murine T-cell receptor gene complex: Isolation, structure and rearrangement. *Immunol. Rev.* 81:235–258 (1984)

Hatakeyama, M., Tsudo, M., Minamoto, S., Kono, T., Doi, T., Miyata, T., Miyasaka, M. & Taniguchi, T., Interleukin-2 receptor β chain gene: Generation of three receptor forms by cloned human α and β chain cDNA's. *Science* 244:551–556 (1989)

McCafferty, J., Griffiths, A.D., Winter, G. & Chiswell, D.J., Phage antibodies: filamentous phage displaying antibody variable domains. *Nature* 348:552–554 (1990)

Pennica, D., Hayflick, J.S., Bringman, T.S., Palladino, M.A., Jr. & Goeddal, D.V., Cloning and expression in *E. coli* of the cDNA for murine tumor necrosis factor. *Proc. Natl. Acad. Sci. U.S.A.* 82:6060–6064 (1985)

Plückthuhn, A., Antibody engineering: advances from the use of *Escherichia coli* expression systems. *Bio/Technology* 9:545–551 (1991)

Scott, J.K. & Smith, G.P., Searching for peptide ligands with an epitope library. *Science* 249:386–390 (1990)

Taniguchi, T., Matsui, H., Fujita, T., Takaoka, C., Kashima, N., Yoshimoto, R. & Hamuro, J., Structure and expression of a cloned cDNA for human interleukin-2. *Nature* 302:305–310 (1983)

Tonegawa, S., Somatic generation of antibody diversity. *Nature* 302:573 (1983)

Ward, E.S., Güssow, D., Griffiths, A.D., Jones, P.T. & Winter, G., Binding activities of a repertoire of single immunoglobulin variable domains secreted from *Escherichia coli*. *Nature* 341:544–546 (1989)

Patents and "Guidelines"

In the early 1970s it was unlikely that anyone could have predicted the extent of the legal and regulatory problems raised by the application of genetic engineering to biotechnology. The concerns voiced at the Asilomar conference and the subsequent establishment of the Recombinant Advisory Committee in the United States (and equivalent oversight committees in many other countries) begat rational and prudent guidelines for the manipulation and use of organisms by recombinant DNA methods— and a new series of acronyms (RAC, GMAG, etc.) However, with the prospect of commercial development, there followed the patenting of a recombinant organism and, with the formation of the fledgling modern biotechnology industry, myriads of processes and products: plasmids, promoters, hosts, secretion systems, etc., became the object of patent filings in the United States and other countries, with a corresponding explosion in Patent Office activity. The precursor of all of this was the patent covering the Cohen/ Boyer discovery of plasmid transformation of *E. coli* which provides the basis for the entire genetic engineering industry.

Only three patents are presented; they are considered to be milestones for several reasons. The "Chakrabarty" patent (4,259,444) represented the first application filed for a living organism and it created a great deal of controversy. The application was disputed by certain sectors of industry and was resolved finally by a U.S. Supreme Court ruling which upheld the patent on the basis that a genetically manipulated microbe is not naturally occurring. This was a landmark decision since it permitted the filing of patents for "anything under the sun that is made by man." The establishment of this concept of patentable subject matter was an important guideline for the U.S. Patent Office and has been followed (in most cases) by patent-granting authorities in other countries.

The Cohen and Boyer ("Stanford") patent (4, 237,224) is noteworthy for different reasons. The results of the research were publicly disclosed before the filing of the application, and the U.S. Patent Office deadline (one year of grace after disclosure) was met with little time to spare. As a result, this important invention is not protected outside of the United States; in Europe and Japan patents must be filed before any public disclosure, a restrictive measure for many since publication of research is oftentimes retarded. The "year of grace" is a favorable U.S. practice, but costly. Because this patent is not in force outside the United States, royalties accruing from its application are paid on U.S. sales only. The Cohen/Boyer patent is also of interest because of the fact that Stanford University granted licenses for the sum of $10,000 and rapidly accumulated a substantial war-chest to be used in the event of any litigation. Although there have been questions about its validity, the patent has not in fact been seriously challenged. It is enlightening to read the final claims section, the key part of any patent application, which provides a good example of how a very simple laboratory procedure can be protected. Can all the claims be justified based on the state-of-the-art? These are questions that must be considered by all patent examiners.

Finally, we include the Harvard "oncomouse" patent (4,736,866), which was the first filing for a man-manipulated animal. Following the Chakrabarty ruling, the filing of patents on animals was a natural progression. In this case the issue raised more emotional and ethical questions, but in spite of much public and legal commotion, the U.S. patent was awarded. In Europe the situation is unresolved, the European Community having failed to come to grips with the problems of animal patents. In this case also, you are urged to read the claims section, to see what property is given patent protection.

There are, of course, many other patents in biotechnology and several disputes over conflicting claims and priority have surfaced, leading to lengthy, complex (and expensive) litigations. Examples are the Genentech/Wellcome case on the rights to cloned tissue plasminogen activator and similar disputes over erythropoeitin (Amgen/Genetics Institute), Factor VIII (Scripps/Genentech), sandwich assays (Hybritech/Monoclonal Antibodies). Very high stakes are involved in questions of obviousness and infringement with respect to the protection and production of biotechnology products. (Who first cloned?, expressed?, had the idea!?) The acceptable scope and validity of biotechnology patents depends very much on the diligence of a skilled patent lawyer. This is not to say that all potential disputes have ended in legal battles; there have been numerous instances of judicious cross-licensing agreements that have benefitted both the parties involved and the consumer. A good example is that concerning recombinant alpha-interferon, where Schering-Plough and Hoffmann-Laroche reached an agreement in order to avoid a costly legal confrontation. Some cases have been resolved, but others, like the rights to commercialization of erythropoietin, have been appealed to the U.S. Supreme Court. The arguments presented in these cases provide fascinating reading.

It is not possible to complete this section without reference to legislation with respect to the use and release of genetically-engineered organisms; risk assessment has become a very hot topic and reams of governmental papers have been produced on the subject. Bear in mind that the "risks" are largely imaginary and stem from the public perception that taking genes from one living creature and putting them into another can lead to highly dangerous situations. As a result of a great deal of adverse, ill-informed and highly speculative publicity, any valuable applications of biotechnology have been held up or even stopped. In certain countries, activist groups have gained a very strong influence which, allied with the luddism of others, has created problems for academia, industry and government. Violent acts of destruction have occurred, with loss of life in some cases.

Increasing research effort is being put into studies of the properties of recombinant organisms in the environment and it is to be hoped that this will allay fear and permit the full benefits of modern biotechnological innovation to be developed. There is an interesting conceptual conflict for many scientists in this work. Most research is based on experiments which are designed to ask questions in which a *positive* response is obtained. In contrast many environmental release studies are designed to show that something does *not* happen, e.g. the organism does not survive, transfer of recombinant DNA is of no consequence, etc. How can one arrive at decisions based on considerations of risk and benefit when the risk cannot be stated in realistic terms? There is every reason to believe that biotechnology will contribute in a major way to finding solutions to both short term and long term problems in health, food, populations and the environment. To realise these potentials, ethical objections must be addressed intelligently based on existing knowledge and guidelines, set apart from the hue and cry of special interest groups. There are no milestone papers in this field, but it is of interest to read summaries of the Asilomar Conference in 1974 and see how current research guidelines have evolved from this meeting. By all means, avoid the popular press!

Microorganisms Having Multiple Compatible Degradative Energy-generating Plasmids and Preparation Thereof

A.M. Chakrabarty

[54] **MICROORGANISMS HAVING MULTIPLE COMPATIBLE DEGRADATIVE ENERGY-GENERATING PLASMIDS AND PREPARATION THEREOF**

[75] Inventor: **Ananda M. Chakrabarty,** Latham, N.Y.

[73] Assignee: **General Electric Company,** Schenectady, N.Y.

[21] Appl. No.: **260,563**

[22] Filed: **Jun. 7, 1972**

[51] Int. Cl.³ **C12N 15/00**
[52] U.S. Cl. **435/172; 435/253; 435/264; 435/281; 435/820; 435/875; 435/877**
[58] Field of Search .... 195/28 R, 1, 3 H, 3 R, 195/96, 78, 79, 112; 435/172, 253, 264, 820, 281, 875, 877

[56] **References Cited**

PUBLICATIONS

Annual Review of Microbiology vol. 26 Annual Review Inc. 1972 pp. 362–368.
Journal of Bacteriology vol. 106 pp. 468–478 (1971).
Bacteriological Reviews vol. 33 pp. 210–263 (1969).

Primary Examiner—R. B. Penland

Attorney, Agent, or Firm—Leo I. MaLossi; James C. Davis, Jr.

[57] **ABSTRACT**

Unique microorganisms have been developed by the application of genetic engineering techniques. These microorganisms contain at least two stable (compatible) energy-generating plasmids, these plasmids specifying separate degradative pathways. The techniques for preparing such multi-plasmid strains from bacteria of the genus Pseudomonas are described. Living cultures of two strains of Pseudomonas (*P. aeruginosa* [NRRL B-5472] and *P. putida* [NRRL B-5473]) have been deposited with the United States Department of Agriculture, Agricultural Research Service, Northern Marketing and Nutrient Research Division, Peoria, Ill. The *P. aeruginosa* NRRL B-5472 was derived from *Pseudomonas aeruginosa* strain 1c by the genetic transfer thereto, and containment therein, of camphor, octane, salicylate and naphthalene degradative pathways in the form of plasmids. The *P. putida* NRRL B-5473 was derived from *Pseudomonas putida* strain PpG1 by genetic transfer thereto, and containment therein, of camphor, salicylate and naphthalene degradative pathways and drug resistance factor RP-1, all in the form of plasmids.

18 Claims, 2 Drawing Figures

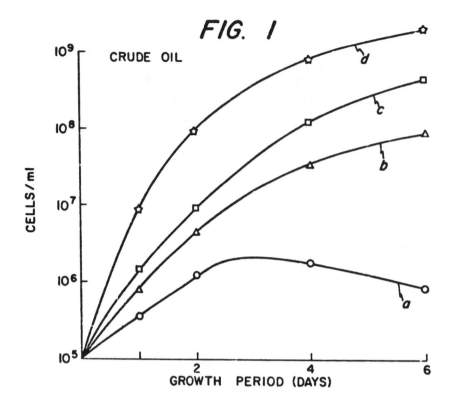

FIG. 1

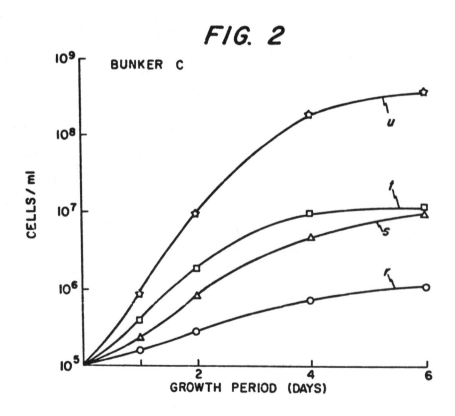

FIG. 2

1

MICROORGANISMS HAVING MULTIPLE COMPATIBLE DEGRADATIVE ENERGY-GENERATING PLASMIDS AND PREPARATION THEREOF

BACKGROUND OF THE INVENTION

The terminology of microbial genetics is sufficiently complicated that certain definitions will be particularly useful in the understanding of this invention: 10

Extrachromosomal element. . . a hereditary unit that is physically separate from the chromosome of the cell; the terms "extrachromosomal element" and "plasmid" are synonymous; when physically separated from the chromosome, some plasmids can be transmitted at high 15 frequency to other cells, the transfer being without associated chromosomal transfer;

Episome . . . a class of plasmids that can exist in a state of integration into the chromosome of their host cell or as an autonomous, independently replicating, cytoplas- 20 mic inclusion;

Transmissible plasmid . . . a plasmid that carries genetic determinants for its own intercell transfer via conjugation;

DNA . . . deoxytribonucleic acid; 25

Bacteriophage . . . a particle composed of a piece of DNA encoded and contained within a protein head portion and having a tail and tail fibers composed of protein;

Transducing phage . . . a bacteriophage that carries 30 fragments of bacterial chromosomal DNA and transfers this DNA on subsequent infection of another bacterium;

Conjugation . . . the process by which a bacterium establishes cellular contact with another bacterium and the transfer of genetic material occurs; 35

Curing . . . the process by which selective plasmids can be eliminated from the microorganism;

Curing agent . . . a chemical material or a physical treatment that enhances curing;

Genome . . . a combination of genes in some given 40 sequence;

Degradative pathway . . . a sequence of enzymatic reactions (e.g. 5 to 10 enzymes are produced by the microbe) converting the primary substrate to some simple common metabolite, a normal food substance for 45 microorganisms;

(Sole carbon source)⁻ . . . indicative of a mutant incapable of growing on the given sole carbon source;

(Plasmid)^{del} . . . indicative of cells from which the given plasmid has been completely driven out by curing 50 or in which no portion of the plasmid ever existed;

(Plasmid)⁻ . . . indicative of cells lacking in the given plasmid; or cells harboring a non-functional derivative of the given plasmid;

(Amino-acid)⁻ . . . indicative of a strain that cannot 55 manufacture the given amino acid;

(Vitamin)⁻ . . . indicative of a strain that cannot manufacture the given vitamin and

(Plasmid)⁺ . . . indicates that the cells contain the given plasmid. 60

Plasmids are believed to consist of double-stranded DNA molecules. The genetic organization of a plasmid is believed to include at least one replication site and a maintenance site for attachment thereof to a structural component of the host cell. Generally, plasmids are not 65 essential for cell viability.

Much work has been done supporting the existence, functions and genetic organization of plasmids. As is

2

reported in the review by Richard P. Novick "Extrachromosomal Inheritance in Bacteria" (Bacteriological Reviews, June 1969, pp. 210–263, [1969]) on page 229, "DNA corresponding to a number of different plasmids has been isolated by various methods from plasmid-positive cells, characterized physiochemically and in some cases examined in the electron microscope".

There is no recognition in the Novick review of the existence of energy-generating plasmids specifying degradative pathways. As reported on page 237 of the Novick review, of the known (non energy-generating) plasmids "Combinations of four or five different plasmids in a cell seem to be stable."

Plasmids may be compatible (i.e. they can reside stably in the same host cell) or incompatible (i.e. they are unable to reside stably in a single cell). Among the known plasmids, for example, are sex factor plasmids and drug-resistance plasmids.

Also, as stated on page 240 of the Novick review, "Cells provide specific maintenance systems or sites for plasmids. It is though that attachment of such sites is required for replication and for segregation of replicas. Each plasmid is matched to a particular maintenance site . . . ". Once a plasmid enters a given cell, if there is no maintenance site available, because of prior occupancy by another plasmid, these plasmids will be incompatible.

The biodegradation of aromatic hydrocarbons such as phenol, cresols and salicylate has been studied rather extensively with emphasis on the biochemistry of these processes, notably enzyme characterization, nature of intermediates involved and the regulatory aspects of the enzymic actions. The genetic basis of such biodegradation, on the other hand, has not been as thoroughly studied because of the lack of suitable transducing phages and other genetic tools.

The work of Chakrabarty and Gunsalus (Genetics, 68, No. 1, page S10, [1971]) has showed that the genes governing the synthesis of the enzymes responsible for the degradation of camphor constitute a plasmid. Similarly, this work has shown the plasmid nature of the octane-degradative pathway. However, attempts by the authors to provide a microorganism with both CAM and OCT plasmids were unsuccessful, these plasmids being incompatible.

In Escherichia coli artificial, transmissible plasmids (one per cell) have been made, each containing a degradative pathway. These plasmids, not naturally occurring, are F'lac and F'gal, wherein the lactose-and galactose-degrading genes were derived from the chromosome of the organism. Such plasmids are described in "F-prime Factor Formation in E. Coli K12" by J. Scaife (Genet. Res. Cambr. [1966], 8, pp. 189–196).

If the development of microorganisms containing multiple containing energy-generating plasmids specifying preselected degradative pathways could be made possible, the economic and environmental impact of such an invention would be vast. For example, there would be immediate application for such versatile microbes in the production of proteins from hydrocarbons ("Proteins from Petroleum"—Wang, Chemical Engineering, August 26, 1968, page 99); in cleaning up oil spills ("Oil Spills: An Environmental Threat"—Environmental Sciene and Technology, Volume 4, February 1970, page 97); and in the disposal of used automotive lubricating oils ("Waste Lube Oils Pose Disposal Di-

3

lemma", Environmental Science and Technology, Volume 6, page 25, January 1972).

SUMMARY OF THE INVENTION

A transmissible plasmid has been found that specifies 5 a degradative pathway for salicylate [SAL], an aromatic hydrocarbon. In addition, a plasmid has been identified that specifies a degradative pathway for naphthalene [NPL], a polynuclear aromatic hydrocarbon. The NPL plasmid is also transmissible. 10

Having established the existence of (and transmissibility of) plasmid-borne capabilities for specifying separate degradative pathways for salicylate and naphthalene, unique single-cell microbes have been developed containing various stable combinations of the [CAM], 15 [OCT], [SAL], and [NPL] plasmids. In addition, stable combinations in a single cell of the aforementioned plasmids together with a non energy-generating plasmid [drug resistance factor RP-1] have been achieved. The versatility of these novel microorganisms has been 20 demonstrated by the substantial extent to which degradation of such complex hydrocarbons as crude oil and Bunker C oil has been achieved thereby.

BRIEF DESCRIPTION OF THE DRAWING
25
The exact nature of the invention as well as objects and advantages thereof will be readily apparent from consideration of the following specification relating to the annexed drawing in which:

FIG. 1 shows the increase in growth rate in crude oil 30 of Pseudomonas strain bacteria provided with increasing numbers of energy-generating degradative plasmids by the practice of this invention and

FIG. 2 shows the increase in growth rate in Bunker C oil of Pseudomonas strain bacteria provided with in- 35 creasing numbers of energy-generating degradative plasmids by the practice of this invention.

DESCRIPTION OF THE PREFERRED EMBODIMENT
40
Microorganisms prepared by the genetic engineering processes described herein are exemplified by cultures now on deposit with the U.S. Department of Agriculture. These cultures are identified as follows:

Pseudomonas aeruginosa (NRRL B-5472) . . . derived 45 from *Pseudomonas aeruginosa* strain 1c (ATCC No. 15692) by genetic transfer thereto, and containment therein, of camphor, octane, salicylate and naphthalene degradative pathways in the form of plasmids.

Pseudomonas putida (NRRL B-5473) . . . derived from 50 *Pseudomonas putida* strain PpG1 (ATCC No. 17453) by genetic transfer thereto, and containment therein, of camphor, salicylate and naphthalene degradative pathways and a drug resistance factor RP-1, all in the form of plasmids. The drug resistance factor is responsible for 55 resistance to neomycin/kanamycin, carbenicillin and tetracycline.

A sub-culture of each of these strains can be obtained from the permanent collection of the Northern Marketing and Nutrient Research Division, Agricultural Ser- 60 vice, U.S. Department of Agriculture, Peoria, IL, U.S.A.

Morphological observations in various media, growth in various media, general group characterization tests, utilization of sugars and optimum growth 65 conditions for the strains from which the above-identified organisms were derived are set forth in "The Aerobic Pseudomonads: A Taxonomic Study" by Stanier, R.

4

Y. et al [Journal of General Microbiology 43, pp. 159–271 (1966)]. The taxonomic properties of the above-identified organisms remain the same as those of the parent strains.

P. aeruginosa strain 1c (ATCC No. 15692) is the same as strain 131 (ATCC No. 17503) in the Stanier et al study. Later the designation for this strain was changed to *P. aeruginosa* PAO [Holloway, B. W. "Genetics of Pseudomonas", Bacteriological Reviews, 33, 419–443 (1969)]. *P. putida* strain PpG1 (ATCC No. 17453) is the same as strain 77 (ATCC No. 17453) in the Stanier et al study

As will be described in more detail hereinbelow, these organisms thrive on a very wide range of hydrocarbons including crude oil and Bunker C oil. These organisms are non-pathogenic as is the general case with laboratory strains of Pseudomonas.

In brief, the process for preparing microbes containing multiple compatible energy-generating plasmids specifying separate degradative pathways is as follows:

(1) selecting the complex or mixture to be degraded;

(2) identifying the plurality of degradative pathways required in a single cell to degrade the several components of the complex or mixture therewith;

(3) isolating a strain of some given microorganism on one particular selective substrate identical or similar to one of the several components (the selection of the microorganism is generally on the basis of a demonstrated superior growth capability);

(4) determining whether the capability of the given strain to degrade the selective substrate is plasmid-borne;

(5) attempting to transfer this first degradative pathway by conjugation to other strains of the same organism (or to the same strain which has been cured of the pathway) and then verifying the transmissible nature of the plasmid;

(6) purifying the conjugatants (recipients of the plasmids by conjugation) and checking for distinctive characteristics of the recipient to insure that the recipient did, in fact, receive the degradative pathway;

(7) repeating the process so as to introduce a second plasmid to the conjugatants;

(8) rendering the first and second plasmids compatible, if necessary, by fusion of the plasmids and

(9) repeating the process as outlined above until the full complement of degradative pathways desired in a single cell has been accomplished by plasmid transfer (and fusion, when required).

In the first reported instance (Chakrabarty et al article mentioned hereinabove) in which the attempt was made to locate more than one energy-generating degradative pathway in the same cell, it was found that CAM and OCT plasmids cannot exist stably under these conditions. In spite of the implication from these results that multiple energy-generating plasmid content in a single cell could be achieved but not maintained, it was decided to attempt to discover some way in which to overcome this problem of plasmid incompatibility As noted hereinabove and described more fully hereinbelow with specific reference to energy-generating plasmid transfer in the genus Pseudomonas, the problem of plasmid instability has now been solved by bringing about fusion of the plasmids in the recipient cell.

The development of single cell capability for the degradation and conversion of complex hydrocarbons was selected as the immediate beneficial application with particular emphasis on the genetic control of oil

5

spills by the way of a single strain of Pseudomonas. In order to be able to cope with crude oil and Bunker C oil spills it was decided that the single cells of Pseudomonas derivate produced by this invention should possess degradative pathways for linear aliphatic, cyclic ali- 5
phatic, aromatic and polynuclear aromatic hydrocarbons. *Pseudomonas aeruginosa* (NRRL B-5472) strain, which displays these degradative capabilities was thereupon eventually developed.

Massive oil spills that are not promptly contained and 10
cleaned up have a catastrophic effect on aquatic lives. Microbial strains are known that can decompose individual components of crude oil (thus, various yeasts can degrade aliphatic straight-chain hydrocarbons, but not most of the aromatic and polynuclear hydrocarbons). 15
Pseudomonas and other bacteria species are known to degrade the aliphatic, aromatic and polynuclear aromatic hydrocarbon compounds, but unfortunately any given strain can degrade only a particular component. For this reason, prior to the instant invention, biological 20
control of oil spills had involved the use of a mixture of bacterial strains, each capable of degrading a single component of the oil complex on the theory that the cumulative degradative actions would consume the oil and convert it to cell mass. This cell mass in turn serves 25
as food for aquatic life. However, since bacterial strains differ from one another in (a) their rates of growth on the various hydrocarbon components, (b) nutritional requirements, production of antibiotics or other toxic material, and (c) requisite pH, temperature and mineral 30
salts, the use of a mixed culture leads to the ultimate survival of but a portion of the initial collection of bacterial strains. As a result, when a mixed culture of hydrocarbon-degrading bacteria are deposited on an oil spill the bulk of the oil often remains unattacked for a 35
long period of time (weeks) and is free to spread or sink.

By establishing that SAL and NPL degradative pathways are specified by genes borne by transmissible plasmids in Pseudomonas and by the discovery that plasmids can be rendered stable (e.g. CAM and OCT) by 40
fusion of the plasmids it has been made possible, for the first time, to genetically engineer a strain of Pseudomonas having the single cell capability for multiple separate degradative pathways. Such a strain of microbes equipped to simultaneously degrade several compo- 45
nents of crude oil can degrade an oil spill much more quickly (days) than a mixed culture meanwhile bringing about coalescence of the remaining portions into large drops. This action quickly removes the opportunity for spreading of the oil thereby enhancing recovery of the 50
coalesced residue.

Preparation of *P. aeruginosa* [NRRL B-5472]

The compositions of the synthetic mineral media for growth of the cultures were the same for all the Pseudo- 55
monas species employed. The mineral medium was prepared from:

6

PA Concentrate . . .
  100 ml of 1 Molar K_2HPO_4
  50 ml of 1 Molar KH_2PO_4
  160 ml of 1 Molar NH_4Cl
$100 \times$ Salts . . .
  19.5 gm $MgSO_4$
  5.0 gm $MnSO_4.H_2O$
  5.0 gm $FeSO_4.7H_2O$
  0.3 gm $CaCl_2.2H_2O$
  1.0 gm Ascorbic acid
  1 liter H_2O

Each of the above (PA Concentrate and $100 \times$ Salts) was sterilized by autoclaving. Thereafter, one liter of the mineral medium was prepared as follows:

PA Concentrate	77.5 ml
100 X Salts	10.0 ml
Agar	15.0 gm
H_2O	to one liter (The pH is adjusted to 6.8–7.0).

All experiments were carried out at 32° C. unless otherwise stated.

It was decided that a very useful hydrocarbon degradation capability would be attained in a single *Pseudomonas aeruginosa* cell, if the degradative pathways for linear aliphatic, cyclic aliphatic, aromatic and polynuclear aromatic hydrocarbons could be transferred thereto. *Pseudomonas aeruginosa* PAO was selected because of its high growth rate even at temperatures as high as 45° C. Four strains of Pseudomonas were selected having the individual capabilities for degrading n-octane (a linear aliphatic hydrocarbon), camphor (a cyclic aliphatic hydrocarbon), salicylate (an aromatic hydrocarbon) and naphthalene (a polynuclear aromatic hydrocarbon).

The specific strains of Pseudomonas able to degrade these hydrocarbons were then treated with curing agent to verify the plasmid-nature of each of these degradative pathways. Of the known curing agents (e.g. sodium dodecyl sulfate, urea, acriflavin, rifampicin, ethidium bromide, high temperature, mitomycin C, acridine orange etc.) most were unable to cure any of the degradative pathways. However, it was found (Table I) that the degradative pathways of the several species could be cured with mitomycin C. Each of the Pseudomonas strains bearing the specified degradative pathways are known in the art:

(a)	CAM$^+$ *P. putida* PpG1	Proc. Nat. Acad. Sci. (U.S.A.), 60, 168 (1968)
(b)	OCT$^+$ *P. oleovorans*	J. Biol. Chem. 242, 4334 (1967)
(c)	SAL$^+$ *P. putida* R-1	Bacteriological Proceedings 1972 p. 60
(d)	NPL$^+$ *P. aeruginosa*	Biochem. J. 91, 251 (1964)

TABLE I

Strain	Degradative Pathway	Mitomycin C Concentration ($\mu g/ml$)	Frequency of Curing (Percent)
CAM · *P. putida* PpG1	cyclic aliphatic hydrocarbon (camphor)	0 10 20	<0.01 5 95
OCT · *P. oleovorans*	aliphatic hydrocarbon (n-octane)	0 10 20	<0.1 1.0 3.0
SAL · *P. putida* R-1	aromatic hydrocarbon	0	<0.1

7

8

TABLE I-continued

Strain	Degradative Pathway	Mitomycin C Concentration (μg/ml)	Frequency of Curing (Percent)
	(salicylate)	5	0.7
		10	3.0
		15	4.0
NPL+ *P. aeruginosa*	polynuclear aromatic hydrocarbon (naphthalene)	0	<0.1
		5	0.5
		10	1.8

Curing degradative pathways from each strain with mitomycin C was accomplished by preparing several test tubes of L broth [Lennox E.S. (1955), *Virology*, 1, 190] containing varying concentrations of mitomycin C and inoculating these tubes with suitable dilutions of early stationary phase cells of the given strain to give concentrations 10^4 to 10^5 cells/ml. These tubes were incubated on a shaker at 32° C. for 2–3 days. Aliquots from tubes that showed some growth were then diluted and plated on glucose minimal plates. After growth at 32° C. for 24 hours, individual colonies were split and respotted on glucose-minimal and degradative pathway-minimal plates to give the proportion of CAM−, OCT−, SAL− and NPL− in order to determine the frequency of curing. It was, therefore, shown that in each instance the degradative pathway genes are plasmid-borne.

Transductional studies with a number of point mutants in the camphor and salicylate pathways has suggested that the cured segments lost either the entire or the major portion the plasmid genes. The plasmid nature of the degradative pathways was also confirmed from evidence of their transmissibility by conjugation from one strain to another (Table II). Although the frequency of plasmid transfer varies widely with individual plasmids and although OCT plasmid cannot be transferred from *P. oleovorans* to *P. aeruginosa* PAO at any detectable frequency, most of the plasmids can nevertheless be transferred from one strain to another by conjugation.

The plasmid transfers, instead of being made to other strains could have been made to organisms of the same strain, that had been cured of the given pathway with mitomycin C, acridine orange or other curing agent.

Pseudomonas putida U has been described in the article by Feist et al [J. Bacteriology 100, p. 869–877 (1969)].

The auxotropic mutants (mutants that require a food source containing a particular amino acid or vitamin for growth) shown in Table II as donors were each grown in a complex nutrient medium (e.g. L broth) to a population density of at least about 10^8 cells/ml without shaking in a period of from 6 to 24 hours. The prototropic (cells capable of growing on some given minimal source of carbon) recipients to which degradative pathway transfer was desired were grown separately in the same complex nutrient medium to a population density of at least about 10^8 cells/ml with shaking in a period of from 4 to 26 hours. For each degradative pathway transfer these cultures were mixed in equal volumes, kept for 15 minutes to 2 hours at 32° C. without shaking (to permit conjugation to occur) and then plated on minimal plates containing the particular substrate as the sole source of carbon. This procedure for cell growth of donor and recipient and the mixing thereof is typical of the manner in which conjugation and plasmid transfer is encouraged in the laboratory, this procedure being designed to provide a very efficient transfer system. Temperature is not critical, but the preferred temperature range is 30°–37° C. Reduction in the population density of either donor or recipient below about 1,000,000 cells/ml or any change in the optimal growth conditions (stationary growth of donor, agitated growth of recipient, growth in high nutrient content medium, harvest of recipient cells at log phase) will drastically reduce the frequency of plasmid transfer.

The details for preparing and isolating auxotropic mutants is described in the textbook, "The Genetics of Bacteria and Their Viruses" by William Hays [John Wiley & Sons, Inc. (1965)].

TABLE II

Donor	Recipient	Degradative Pathway	Frequency of Transfer
Trp−CAM+	*P. aeruginosa* PAO	CAM	10^{-3}
P. putida PpG1	CAM^{del} *P. putida*	CAM	10^{-2}
Met−OCT+	*P. aeruginosa* PAO	OCT	$<10^{-9}$
P. oleovorans	*P. putida* PpG1	OCT	10^{-9}
	P. putida U	OCT	10^{-7}
His−SAL+	*P. aeruginosa* PAO	SAL	10^{-7}
P. putida R-1	*P. putida* PpG1	SAL	10^{-6}
Trp−NPL+	*P. putida* PpG1	NPL	10^{-7}
P. aeruginosa	NPL^{del} *P. aeruginosa* PAO	NPL	10^{-5}

Abbreviations:
Trp - tryptophane
Met - methionine
His - histidine

Control cultures of donors and recipients were also plated individually on minimal plates containing the requisite substrate in each instance as the sole source of carbon, to determine the reversion frequency of donor and recipient cells.

All plates (including controls) were incubated at 30°–37° C. for several days. In each instance in which colonies appeared in numbers exceeding the colony growth on the reversion plates, it was established that degradative pathway transfer had occurred between the donors and recipients. Such conjugatants were than purified by a series of single colony isolation cultures and checked for growth rates or other distinctive characteristics of the recipient to insure that the recipient actually received the given degradative pathway.

Having determined that the degradative pathways were plasmid-borne and transmissible, the task of transferring the multiplicity of plasmids to a single cell *P. aeruginosa* PAO was undertaken. Prior work (referred to hereinabove) had established that OCT placmids could not be transferred from *P. oleovorans* to *P. aeruginosa* PAO. Therefore, the first task was to discover how (if at all) the OCT and CAM plasmids could be rendered compatible.

9

The CAM plasmid was transferred to a Met⁻ mutant of OCT⁺ *P. oleovorans* strain from a CAM⁺ *P. putida* strain. The conjugatant is, of course, unstable and will segregate either CAM or OCT at an appreciable rate. Therefore, the conjugatant was alternately grown in camphor and then octane as sole sources of carbon to isolate those cells in which both of these degradative pathways were present, even though unstable. The surviving cells were centrifuged, suspended in 0.9% saline solution and irradiated with UV rays (3 General Electric FS-5 lamps providing a total of about 24 watts). Aliquots were drawn from the suspension as follows: one aliquot was removed before UV treatment, one aliquot after UV exposure for 30 seconds and one aliquot after UV exposure for 60 seconds. These aliquots of irradiated cells were grown in the absence of light for 3 hours in L broth and were then used as donors for the transfer of plasmids to the *P. aeruginosa* PAO strain as recipient, selection being made for the OCT plasmid on an octane minimal plate.

As is shown in Table III aliquots of similarly irradiated suspensions for Met⁻OCT⁺CAM*del* *P. oleovorans* and Met⁻CAM⁺OCT*del* *P. oleovorans were prepared and used as plasmid donors to*

P. aeruginosa PAO, selection being made for the plasmids shown. The Met⁻CAM⁺OCT*del* strain was prepared by introducing CAM plasmids into Met⁻OCT⁺ mutant of *P. oleovorans* and selecting for CAM⁺ conjugatants, which have lost the OCT plasmid. The Met⁻OCT⁺CAM*del* *P. oleovorans* is the Met⁻ mutant of wild type *P. oleovorans*.

The failure to secure determinable transfer of OCT plasmids from Met⁻OCT⁺ *P. oleovorans* to the recipient and the success in securing transfer of CAM plasmids from Met⁻CAM⁻OCT*del* *P. oleovorans* to the recipient are shown. These results support the theory that the successful transfer of OCT plasmids from the

10

mutant of CAM⁺OCT⁺ *P. aeruginosa* PAO that had been provided with its multiple plasmids by the methods described herein for plasmid transfer and plasmid fusion was used as the donor. After conjugation between the donor and OCT*del* CAM*del* *P. putida* PpGl, the resulting culture was plated on minimal plates containing camphor and also on minimal plates containing n-octane. Part of each of 132 colonies growing on the CAM minimal plates were transferred to OCT minimal plates and part of each of 219 colonies growing on the OCT minimal plates were transferred to CAM minimal plates. Each of these transferred portions grew, which tends to establish that (a) both CAM and OCT plasmids had been transferred to the conjugatant, (b) the transfer had been on a one-for-one basis and, therefore, (c) the CAM and OCT plasmids were fused together.

Similar plasmid transfer was carried out between the Trp⁻CAM⁺OCT⁺*P. aeruginosa* PAO donor and OCT*del* CAM*del* *P. aeruginosa* PAO and similar selection procedures were employed. The results further reinforced the above position as to the fused nature of the transferred CAM and OCT plasmids. When the CAM and OCT plasmids have been subjected to UV radiation as disclosed, if either CAM or OCT plasmid is transferred, the other plasmid will always be associated with it regardless of which plasmid is selected first. If either plasmid of the fused pair is cured from the cell, both plasmids are lost simultaneously. Thus, the conjugatants were treated with mitomycin C and the resultant CAM*del* segregants were examined. Invariably all CAM*del* segregants were found to have lost the OCT plasmid as well. Thus, the facts of simultaneous curing of the two plasmids and the co-transfer thereof strongly suggest that incompatible plasmids treated with means for cleaving the DNA of the plasmids results in fusion of the DNA segments to become part of the same replicon.

TABLE IV

Donor	Recipient	Selected Plasmid	Non-selected Plasmid	Total OCT⁺/CAM⁺
Trp⁻CAM⁺OCT⁺ *P. aeruginosa* PAO	OCT*del*CAM*del* *P. putida* PpG1	CAM OCT	OCT CAM	132/132 219/219
	OCT*del*CAM*del* *P. aeruginosa* PAO	CAM OCT	OCT CAM	107/107 96/96

Met⁻CAM⁻OCT⁺ *P. oleovorans* (that had been irradiated for 30 seconds with UV rays) to *P. aeruginosa* PAO had been made possible by the fusion of the CAM and OCT plasmids in the *P. oleovorans* by the UV exposure and the subsequent transfer of CAM/OCT plasmids in combination (with separate degradative pathways), to the recipient.

Having successfully overcome all obstacles to the formation of a stable CAM⁺OCT⁺SAL⁺NPL⁺ Pseudomonas the several energy-generating degradative plasmids were transferred to a single cell as is shown in Table V by the conjugation techniques described hereinabove. The initial *P. aeruginosa* strain used is referred to herein as *P. aeruginosa* PAO, formerly known as *P.*

TABLE III

Donor	Recipient	Selected Plasmid	Period of UV-Irradiation (Sec)	Transfer of Frequency
Met⁻OCT⁺ *P. oleovorans*	*P. aeruginosa* PAO	OCT	0 30 60	<10⁻⁹ <10⁻⁹ <10⁻⁹
Met⁻CAM⁻OCT*del* *P. oleovorans*	*P. aeruginosa* PAO	CAM	0 30 60	10⁻⁴ 10⁻⁵ 10⁻⁷
Met⁻CAM⁻OCT⁺ *P. oleovorans*	*P. aeruginosa* PAO	OCT	0 30 60	<10⁻⁹ 10⁻⁸ <10⁻⁹

Table IV presents verification of this theory of co-transfer of CAM and OCT fused plasmids. A Trp⁻

aeruginosa strain 1c available as ATCC No. 15692 and-

11

/ro ATCC No. 17503. This strain of *P. aeruginosa* does not contain any known energy-generating plasmid. The CAM and OCT plasmids exist in the fused state, are individually and simultaneously functional and appear perfectly compatible with the individual compatible SAL and NPL plasmids. Tests for compatibility of obth CAM+OCT+SAL+ *P. aeruginosa* PAO and CAM-+OCT+SAL+NPL+ *P. aeruginosa* PAO revealed that there is no segregation of the plasmids in excess of that found in the donor. Plasmids will be accepted and maintained by *P. acidovorans*, *P. alcaligenes* and *P. fluorescens*. All of these plasmids should be transferable to and maintainable in these and many other species of Pseudomonas, such as *P. putida*, *P. oleovorans*, *P. multivorans*, etc.

Superstrains such as the CAM+OCT+SAL+NPL+ strain of *P. aeruginosa* PAO can grow on a minimal plate of any of camphor, n-octane, salicylate, naphthalene and, because of the phenomenon of relaxed specificity, on compounds similar thereto. Thus, the effectiveness of a given degradative plasmid does not appear to be diminished in its ability to function singly by the presence of other degradative plasmids in the same cell.

12

strains of *P. aeruginosa* PAO. Curve a shows the cell growth as a function of time of

P. aeruginosa without any plasmid-borne energy-generating degradative pathways. Curve b shows greater cell growth as a function of time for SAL+ *P. aeruginosa*. Curve c shows still greater cell growth as a function of time for SAL+NPL+ *P. aeruginosa*. Curve d shows cell growth that is significantly greater still as a function of time for the CAM+OCT+SAL+NPL+ superstrain of *P. aeruginosa*. These results clearly establish that cells artifically provided by the practice of this invention with the genetic capability for degrading different hydrocarbons can grow at a faster rate and better on crude oil as the plasmid-borne degradative pathways are increased in number and variety, because of the facility of these degradative pathways to simultaneously function at full capacity.

Similar results are shown in FIG. 2 displaying the growth capabilities of this same series of organisms utilizing Bunker C oil as the sole source of carbon. Bunker C is (or is prepared from) the residuum remaining after the more commercially useful components have been removed from crude oil. This residuum is

TABLE V

Donor	Recipient	Selected Plasmid	Phenotype of the Conjugatant
Trp⁻CAM+OCT+ *P. aeruginosa* PAO	*P. aeruginosa* PAO	CAM	CAM+OCT+ *P. aeruginosa* PAO
His⁻SAL+ *P. putida* R-1	CAM+OCT+ *P. aeruginosa* PAO	SAL	CAM+OCT+SAL+ *P. aeruginosa* PAO
Trp⁻NPL+ *P. aeruginosa*	CAM+OCT+SAL+ *P. aeruginosa* PAO	NPL	CAM+OCT+SAL+NPL+ *P. aeruginosa* PAO

Indication of the capability of all degradative plasmids to function simultaneously in energy generation is provided by tests in which CAM+OCT+SAL+NPL+ *P. aeruginosa* PAO superstrain was added to separate broth samples each of which contained 1 millimolar (mM) of nutrient (a suboptimal concentration), one set of samples containing camphor, a second set of samples containing n-octane, a third set of samples containing salicylate and a fourth set of samples containing naphthalene, these being the sole sources of carbon in each instance. The superstrain grew very slowly in the separate sole carbon source samples. However, when the superstrain was added to samples containing all four sources of carbon present together in the same (1 mM)

very thick and sticky and without significant use, per se. A small amount of volatile hydrocarbons is often added thereto to lower the viscosity. Curve r reflects the cell growth as a function of time of *P. aeruginosa* cells not having any plasmid-borne energy-degradative pathways. Curve s shows increased cell growth as a function of time for SAL+ *P. aeruginosa*. Curve t shows further increase in cell growth as a function of time for SAL+NPL+ *P. aeruginosa*. Curve u shows still more significant cell growth as a function of time for CAM-+OCT+SAL+NPL+ *P. aeruginosa*.

The SAL+ *P. aeruginosa* and SAL+NPL+ *P. aeruginosa* cultures were prepared as shown in Table VI below:

TABLE VI

Donor	Recipient	Selected Plasmid	Conjugatant
His⁻ SAL+ *P. putida* R-1	*P. aeruginosa* PAO	SAL	SAL+ *P. aeruginosa* PAO
Trp⁻NPL+ *P. aeruginosa*	SAL+ *P. aeruginosa* PAO	NPL	SAL+NPL+ *P. aeruginosa* PAO

concentration of 4 mM, the rate of growth increased considerably establishing that simultaneous utilization of all four sources of carbon had occurred.

Next, the ability of such superstrains to degrade crude oil was demonstrated. Crude oils, of course, vary greatly (depending upon source, period of activity of the well, etc.) in the relative amounts of linear aliphatic, cyclic aliphatic, aromatic and polynuclear hydrocarbons present, although some of each of these classes of hydrocarbons is typically present in some amount in the chemical make up of all crude oils from producing wells.

FIG. 1 shows the difference in growth capabilities in crude oil as the sole source of carbon of four single cell

The experiments providing the data for FIGS. 1 and 2 were conducted in 250 ml Erlenmeyer flasks. To each flask was added 50 ml of mineral medium (described hereinabove) with pH adjusted to 6.8–7.0; 2.5 ml of the sole carbon source (crude oil or Bunker C) and $5 \times 10^6 - 1 \times 10^7$ cells. Growth was conducted at 32° C. with shaking. At daily intervals 5 ml aliquots were taken. The optical densities of these aliquots were determined at 660 nm in a Bausch & Lomb, Inc. colorimeter to determine organism density. Also, viable cell counts were determined by diluting portions of the aliquots and plating on L-agar (L-broth containing agar) plates. The colonies were counted after 24 hours of incubation at 32° C. and these counts were used to construct FIGS. 1

13

and 2. Also, the cells were submitted to protein analysis, to be discussed hereinbelow.

The 2.5 ml of crude oil or Bunker C appears to have initially offered an essentially unlimited food supply, but the results shown may well represent less than the full capability of the superstrain, because the relative amounts of the various hydrocarbons (degradable by the CAM+, OCT+, SAL+ and NPL+ plasmids) present in the carbon sources had not been ascertained and after a couple of days the food supply for one or more plasmids may have been limited.

A very significant aspect of the growth of the superstrain in crude and Bunker C oils is the fact that the components, which would spread the quickest on the water's surface from spills of these oils, disappear within 2-3 days and the remaining components of the oil co-

14

tion of stable plasmids to which the newly introduced plasmid can be fused.

Preparation of *P. putida* [NRRL B-5473]

The mineral medium and the technique for fostering conjugation was the same as described above. A culture of antibiotic-sensitive *P. putida* PpGl was cured of its CAM plasmids with mitomycin C and was used as the initial recipient. This strain of *P. putida* is sensitive to small (e.g. 25 micrograms/ml) concentrations of neomycin/kanamycin, carbenicillin and tetracycline. As is shown in Table VII below, all the donor strains are auxotropic mutants, because the use of auxotropic mutant donors facilitates counterselection of the conjugatants due to the ease of selecting against such donors.

TABLE VII

Donor	Recipient	Selected Plasmid	Phenotype of the Conjugatant
Trp CAM⁺ *P. putida* PpGl	CAMᵈᵉˡ *P. putida* PpGl	CAM	CAM⁺ *P. putida* PpGl
His SAL⁺ *P. putida* R-1	CAM⁺ *P. putida* PpGl	SAL	CAM⁺SAL⁺ *P. putida* PpGl
Trp NPL⁺ *P. aeruginosa*	CAM⁺SAL⁺ *P. putida* PpGl	NPL	CAM⁺SAL⁺NPL⁺ *P. putida* PpGl
Met *P. aeruginosa* Strain 1822 (RP-1)	CAM⁺SAL⁺NPL⁺ *P. putida* PpGl	RP-1	CAM⁺SAL⁺NPL⁺RP-1⁺ *P. putida* PpGl

alesce to form large droplets that cannot spread out. These droplets can be removed more easily by mechanical recovery techniques as the microbes continue to consume these remaining components.

In practice an inoculum of dry (or lyophilized) powders of these genetically engineered microbes will be dispersed over (e.g. from overhead) an oil spill as soon as possible to control spreading of the oil, which is so destructive of marine flora and fauna and the microbes will degrade as much of the oil as possible to reduce the amount that need be recovered mechanically, when equipment has reached the scene and has been rendered operative. A particularly beneficial manner of depositing the inoculum on the oil spill is to impregnate straw with the inoculum and drop the inoculated straw on the oil spill where both components will be put to use—the inoculum (mass of microbes) to degrade the oil and the straw to act as a carrier for the microbes and also to function as an oil absorbent. Other absorbent materials may be used, if desired, but straw is the most practical. No special care need be taken in the preparation and storage of the dried inoculum or straw (or other absorbent material) coated with inoculum. No additional nutrient or mineral content need be supplied. Also, although culture from the logarithmic growth phase is preferred, culture from either the early stationary or logarithmic growth phases can be used.

It is reasonable to expect that a vast number of plasmid-borne hydrocarbon degradative pathways remain undiscovered. Hopefully, now that a method for controlled genetic additions to the natural degradative capabilities of microbes has been demonstrated by this invention, still more new and useful single cell organisms can be prepared able to degrade even more of the large number of hydrocarbons in crude oil, whether or not the plasmids yet to be found are compatible with each other or with those plasmids present in superstrains NRRL B-5472 and NRRL B-5473.

Both of these superstrains can be used as recipients for more plasmids. The capability for utilizing fusion (by UV irradiation or X-ray exposure) to render additional plasmids compatible is actually increased in a multiplasmid conjugatant, because of the larger selec-

The *P. aeruginosa* RP-1 strain is disclosed in the Sykes et al article [Nature 226, 952 (1970)]. Selection for the RP-1 plasmid was accomplished on a neomycin/kanamycin plate. Further, CAM+SAL+NPL+RP-1+ *P. putida* PpGl has been determined to be resistant to carbenicillin and tetracycline establishing that the RP-1 plasmid is actually present and that the organisms that survived the selection process were not merely the results of mutant development. Also, the plasmids of this superstrain can be transferred and can be cured. The rate of segregation (spontaneous loss) of plasmids from the superstrain has been found to be the same as in the donors.

Both superstrains can, of course, be used as a source of plasmids in addition to those sources disclosed herein. For example, to transfer CAM, SAL or NPL plasmids from CAM+SAL+NPL+RP-1+ *P. putida* PpGl to a given Pseudomonas recipient, the procedures for cell growth of donor and recipient and the mixing thereof for optimized conjugation is the same as described hereinabove. These plasmids will have different frequencies of transfer at different times. The order of diminishing frequency of transfer is CAM, NPL, SAL. For the transfer of CAM plasmid, after conjugation, selection is made for CAM. Surviving colonies are subdivided and selection is made for SAL, NPL and CAM plasmids from each colony. Those portions surviving only on camphor as the sole source of carbon will have received the CAM plasmid free of the SAL or NPL plasmids. The same procedure can be followed for the individual transfer of SAL or NPL plasmids.

In addition to the previously discussed capability for improved treatment of oil spills, considerable improvement is now possible in the microbial single-cell synthesis of proteins from carbon-containing substrates. The restriction of having to employ substantially single-component substrates, e.g. alkanes, paraffins, carbohydrates, etc. has now been removed, simultaneously providing the opportunity for increases of 50–100 fold in the amount of cell mass that may be produced by a single cell in a given time period, when the given single cell has been provided with multiple energy-generating

15

plasmids. Also, being able to optimize the protein production of bacteria is of particular interest since bacterial cell mass has a much greater protein content and most bacteria have greater tolerance for heat than yeasts. This latter aspect is of importance since less refrigeration is necessary to remove the heat generated by the oxidative degradation of the substrate.

The general process and apparatus for single cell production of protein is set forth in the Wang article (incorporated by reference) referred to hereinabove. One particular advantage of the multi-plasmid single cell organism of this invention is that after the cell mass has been harvested it can be subjected to a subsequent incubation period in a mineral medium free of any carbon source for a sufficient period of time to insure the metabolism of residual intra-cellular hydrocarbons, e.g. polynuclear aromatics, which are frequently carcinogenic. Presently, treatment of cell mass to remove unattacked hydrocarbons often leads to reduction in the quality of the protein product.

The economics of protein production by single-cell organisms will be further improved by the practice of this invention, because of the reduced cost of substrate (e.g. oil refinery residue, waste lubricating oil, crude oil) utilizable by organisms provided with preselected plasmid content.

Cell mass growth in crude oil using NRRL B-5472 was harvested by centrifugation, washed two times in water and dried by blowing air (55° C.) over the mass overnight. The dried mass was hydrolyzed and analyzed for amino acid content by the technique described "High Recovery of Tryptophane from Acid Hydrolysis of Proteins"-Matsubara et al [Biochem. and Biophys. Res. Comm. 35 No. 2, 175–181 (1969)]. The amino acid analysis showed that the amino acid distribution of superstrain cell mass grown in crude oil is comparable to beef in threonine, valine, cystine, methionine, isoleucine, leucine, phenylalanine and tryptophane content and significantly superior to yeast in methionine content.

Continued capacity for increasing the degrading capability of the superstrains now on deposit has been made possible by the practice of this invention as more plasmid-borne degradative pathways are discovered. To date P. aeruginosa strain 1822 has been provided with all four known hydrocarbon degradative pathways (OCT, CAM, SAL, NPL) plus the drug-resistance factor RP-1 found therein. If there is an upper limit to the number of energy-generating plasmids that will be received and maintained in a single cell, this limit is yet to be reached. Attempts to integrate plasmids (CAM, OCT, SAL) with the cell chromosome have been unsuccessful as indicated by failure to mobilize the chromosome. Such results have so far verified the extra-chromosomal nature of the energy-generating and drug-resistance plasmids. There is, of course, no reason to expect that the only plasmids are those that specify degradative pathways for hydrocarbons. Conceivably plasmids may be discovered that will provide requisite enzyme series for the degradation of environmental pollutants such as insecticides, pesticides, plastics and other inert compounds.

Energy-generating plasmids in general are known to have broad inducer and substrate specificity [i.e. enzymes will be formed and will act on a variety of structurally similar compounds]. For example, the CAM plasmid is known to have a very relaxed inducer and substrate specificity [Gunsalus et al-Israel J. Med. Sci.,

16

1, 1099–1119 (1965) and Hartline et al-Journal of Bacteriology, 106, 468–478 (1971)]. Similarly, the OCT plasmid has broad inducer and substrate specificity [Peterson et al-J. Biol. Chem. 242, 4334 (1967)]. In the practice of the instant invention it has been demonstrated that plasmids display the same degree of relaxed specificity in the conjugatant as in the donor.

Thus, by the practice of this invention new facility and capability for growth has been embodied in useful single-cell organisms by the manipulation of phenomena that had been previously undiscovered (i.e. the plasmid-borne nature of the degradative pathways for salicylate and naphthalene) and/or had been previously unsuccessfully applied (i.e. rendering stable a plurality of previously incompatible plasmids in the same single cell).

Filed concurrently herewith is U.S. Application Ser. No. 260,488-Chakrabarty, filed June 7, 1972 now U.S. Pat. No. 3,814,474 and assigned to the assignee of the instant invention.

What I claim as new and desire to secure by Letters Patent of the United States is:

1. A bacterium from the genus Pseudomonas containing therein at least two stable energy-generating plasmids, each of said plasmids providing a separate hydrocarbon degradative pathway.

2. The Pseudomonas bacterium of claim 1 wherein the hydrocarbon degradative pathways are selected from the group consisting of linear aliphatic, cyclic aliphatic, aromatic and polynuclear aromatic.

3. The Pseudomonas bacterium of claim 1, said bacterium being of the specie P. aeruginosa.

4. The P. aeruginosa bacterium of claim 3 wherein the bacterium contains CAM, OCT, SAL and NPL plasmids.

5. The Pseudomonas bacterium of claim 1, said bacterium being of the specie P. putida.

6. The P. putida bacterium of claim 5 wherein the bacterium contains CAM, SAL, NPL and RP-1 plasmids.

7. An inoculum for the degradation of a preselected substrate comprising a complex or mixture of hydrocarbons, said inoculum consisting essentially of bacteria of the genus Pseudomonas at least some of which contain at least two stable energy-generating plasmids, each of said plasmids providing a separate hydrocarbon degradative pathway.

8. The inoculum of claim 7 wherein the hydrocarbon degradative pathways are selected from the group consisting of linear aliphatic, cyclic aliphatic, aromatic and polynuclear aromatic.

9. The inoculum of claim 8 wherein the bacteria having multiple energy-generating plasmids are of the specie P. aeruginosa.

10. The inoculum of claim 8 wherein the bacteria having multiple energy-generating plasmids are of the specie P. putida.

11. In the process in which a first energy-generating plasmid specifying a degradative pathway is transferred by conjugation from a donor Pseudomonas bacterium to a recipient Pseudomonas bacterium containing at least one energy-generating plasmid that is incompatible with said first plasmid, said transfer occurring in the quiescent state after the mixing of substantially equal volumes of cultures of said donor and said recipient, each culture presenting the respective organisms in a complex nutrient liquid medium at a population density of at least about 1,000,000 cells/ml, the improvement

17

wherein after conjugation has occurred, the multi-plasmid conjugatant bacteria are subjected to DNA-cleaving radiation in a dosage sufficient to fuse the first plasmid and the plasmid incompatible therewith located in the same cell.

12. The improvement of claim 11 wherein the DNA-cleaving radiation is UV radiation.

13. The improvement of claim 12 wherein the first plasmid provides the degradative pathway for camphor and the recipient Pseudomonas contains the degradative pathway for n-octane.

14. An inoculated medium for the degradation of liquid hydrocarbon substrate material floating on water, said inoculated medium comprising a carrier material able to float on water and bacteria *from the genus Pseudomonas* carried thereby, at least some of said bacteria

18

each containing at least two stable energy-generating plasmids, each of said plasmids providing a separate hydrocarbon degradative pathway and said carrier material being able to absorb said hydrocarbon material.

15. The inoculated medium of claim 14 wherein the carrier material is straw.

16. The inoculated medium of claim 14 wherein the hydrocarbon degradative pathways are selected from the group consisting of linear aliphatic, cyclic aliphatic, aromatic and polynuclear aromatic.

17. The inoculated medium of claim 14 wherein the bacteria are of the specie *P. aeruginosa.*

18. The inoculated medium of claim 14 wherein the bacteria are of the specie *P. putida.*

* * * * *

Process for Producing Biologically Functional Molecular Chimeras

S.N. Cohen and H.W. Boyer

[54] **PROCESS FOR PRODUCING BIOLOGICALLY FUNCTIONAL MOLECULAR CHIMERAS**

[75] Inventors: **Stanley N. Cohen**, Portola Valley; **Herbert W. Boyer**, Mill Valley, both of Calif.

[73] Assignee: **Board of Trustees of the Leland Stanford Jr. University**, Stanford, Calif.

[21] Appl. No.: **1,021**

[22] Filed: **Jan. 4, 1979**

Related U.S. Application Data

[63] Continuation-in-part of Ser. No. 959,288, Nov. 9, 1978, which is a continuation-in-part of Ser. No. 687,430, May 17, 1976, abandoned, which is a continuation-in-part of Ser. No. 520,691, Nov. 4, 1974.

[51] Int. Cl.^{3} C12P 21/00
[52] U.S. Cl. .... 435/68; 435/172; 435/231; 435/183; 435/317; 435/849; 435/820; 435/91; 435/207; 260/112.5 S; 260/27R; 435/212
[58] Field of Search .... 195/1, 28 N, 28 R, 112, 195/78, 79; 435/68, 172, 231, 183

[56] **References Cited**

U.S. PATENT DOCUMENTS

3,813,316 5/1974 Chakrabarty .... 195/28 R

OTHER PUBLICATIONS

Morrow et al., Proc. Nat. Acad. Sci. USA, vol. 69, pp. 3365–3369, Nov. 1972.
Morrow et al., Proc. Nat. Acad. Sci. USA, vol. 71, pp. 1743–1747, May 1974.
Hershfield et al., Proc. Nat. Acad. Sci. USA, vol. 71, pp. 3455 et seq. (1974).
Jackson et al., Proc. Nat. Acad. Sci. USA, vol. 69, pp. 2904–2909, Oct. 1972.
Mertz et al., Proc. Nat. Acad. Sci. USA, vol. 69, pp. 3370–3374, Nov. 1972.
Cohen, et al., Proc. Nat. Acad. Sci. USA, vol. 70, pp. 1293–1297, May 1973.
Cohen et al., Proc. Nat. Acad. Sci. USA, vol. 70, pp. 3240–3244, Nov. 1973.
Chang et al., Proc. Nat. Acad. Sci, USA, vol. 71, pp. 1030–1034, Apr. 1974.
Ullrich et al., Science vol. 196, pp. 1313–1319, Jun. 1977.
Singer et al., Science vol. 181, p. 1114 (1973).
Itakura et al., Science vol. 198, pp. 1056–1063 Dec. 1977.
Komaroff et al., Proc. Nat. Acad. Sci. USA, vol. 75, pp. 3727–3731, Aug. 1978.
Chemical and Engineering News, p. 4, May 30, 1977.
Chemical and Engineering News, p. 6, Sep. 11, 1978.

Primary Examiner—Alvin E. Tanenholtz
Attorney, Agent, or Firm—Bertram I. Rowland

[57]   **ABSTRACT**

Method and compositions are provided for replication and expression of exogenous genes in microorganisms. Plasmids or virus DNA are cleaved to provide linear DNA having ligatable termini to which is inserted a gene having complementary termini, to provide a biologically functional replicon with a desired phenotypical property. The replicon is inserted into a microorganism cell by transformation. Isolation of the transformants provides cells for replication and expression of the DNA molecules present in the modified plasmid. The method provides a convenient and efficient way to introduce genetic capability into microorganisms for the production of nucleic acids and proteins, such as medically or commercially useful enzymes, which may have direct usefulness, or may find expression in the production of drugs, such as hormones, antibiotics, or the like, fixation of nitrogen, fermentation, utilization of specific feedstocks, or the like.

14 Claims, No Drawings

COHEN, S.N. and BOYER, H.W.
Process for producing biologically functional molecular chimeras.
U.S. Patent 4,237,224. Filed Jan. 4, 1979, awarded Dec. 2, 1980.

1

PROCESS FOR PRODUCING BIOLOGICALLY FUNCTIONAL MOLECULAR CHIMERAS

The invention was supported by generous grants of NIH, NSF and the American Cancer Society.

CROSS-REFERENCE TO RELATED APPLICATIONS

This application is a continuatin-in-part of applicatin Ser. No. 959,288, filed Nov. 9, 1978, which is a continuation of application Ser. No. 687,430 filed May 17, 1976, now abandoned, which was a continuation-in-part of application Ser. No. 520,691, filed Nov. 4, 1974, now abandoned.

BACKGROUND OF THE INVENTION

1. Field of the Invention

Although transfer of plasmids among strains of *E. coli* and other Enterobacteriaceae has long been accomplished by conjugation and/or transduction, it has not been previously possible to selectively introduce particular species of plasmid DNA into these bacterial hosts or other microorganisms. Since microorganisms that have been transformed with plasmid DNA contain autonomously replicating extrachromosomal DNA species having the genetic and molecular characteristics of the parent plasmid, transformation has enabled the selective cloning and amplification of particular plasmid genes.

The ability of genes derived from totally different biological classes to replicate and be expressed in a particular microorganism permits the attainment of interspecies genetic recombination. Thus, it becomes practical to introduce into a particular microorganism, genes specifying such metabolic or synthetic functions as nitrogen fixation, photosynthesis, antibiotic production, hormone synthesis, protein synthesis, e.g. enzymes or antibodies, or the like—functions which are indigenous to other classes of organisms—by linking the foreign genes to a particular plasmid or viral replicon.

BRIEF DESCRIPTION OF THE PRIOR ART

References which relate to the subject invention are Cohen, et al., Proc. Nat. Acad, Sci., USA, 69, 2110 (1972); ibid, 70, 1293 (1973); ibid, 70, 3240 (1973); ibid, 71, 1030 (1974); Morrow, et al., Proc. Nat. Acad. Sci., 71, 1743 (1974); Novick, Bacteriological Rev., 33, 210 (1969); and Hershfeld, et al., Proc. Nat. Acad. Sci., in press; Jackson, et al., ibid, 69, 2904 (1972);

SUMMARY OF THE INVENTION

Methods and compositions are provided for genetically transforming microorganisms, particularly bacteria, to provide diverse genotypical capability and producing recombinant plasmids. A plasmid or viral DNA is modified to form a linear segment having ligatable termini which is joined to DNA having at least one intact gene and complementary ligatable termini. The termini are then bound together to form a "hybrid" plasmid molecule which is used to transform susceptible and compatible microorganisms. After transformation, the cells are grown and the transformants harvested. The newly functionalized microorganisms may then be employed to carry out their new function; for example, by producing proteins which are the desired end product, or metabolities of enzymic conversion, or be lysed and the desired nucleic acids or proteins recovered.

2

DESCRIPTION OF THE SPECIFIC EMBODIMENTS

The process of this invention employs novel plasmids, which are formed by inserting DNA having one or more intact genes into a plasmid in such a location as to permit retention of an intact replicator locus and system (replicon) to provide a recombinant plasmid molecule. The recombinant plasmid molecule will be referred to as a "hybrid" plasmid or plasmid "chimera." The plasmid chimera contains genes that are capable of expressing at least one phenotypical property. The plasmid chimera is used to transform a susceptible and competent microorganism under conditions where transformation occurs. The microorganism is then grown under conditions which allow for separation and harvesting of transformants that contain the plasmid chimera.

The process of this invention will be divided into the following stages:

I. preparation of the recombinant plasmid or plasmid chimera;

II. transformation or preparation of transformants; and

III. replication and transcription of the recombinant plasmid in transformed bacteria.

Preparation of Plasmid Chimera

In order to prepare the plasmid chimera, it is necessary to have a DNA vector, such as a plasmid or phage, which can be cleaved to provide an intact replicator locus and system (replicon), where the linear segment has ligatable termini or is capable of being modified to introduce ligatable termini. Of particular interest are those plasmids which have a phenotypical property, which allow for ready separation of transformants from the parent microorganism. The plasmid will be capable of replicating in a microorganism, particularly a bacterium which is susceptible to transformation. Various unicellular microorganisms can be transformed, such as bacteria, fungii and algae. That is, those unicellular organisms which are capable of being grown in cultures of fermentation. Since bacteria are for the most part the most convenient organisms to work with, bacteria will be hereinafter referred to as exemplary of the other unicellular organisms. Bacteria, which are susceptible to transformation, include members of the Enterobacteriaceae, such as strains of *Escherichia coli;* Salmonella; Bacillaceae, such as *Bacillus subtilis;* Pneumococcus; Streptococcus, and *Haemophilus influenzae.*

A wide variety of plasmids may be employed of greatly varying molecular weight. Normally, the plasmids employed will have molecular weights in the range of about 1×10^6 to 50×10^6d, more usually from about 1 to 20×10^6d, and preferably, from about 1 to 10×10^6d. The desirable plasmid size is determined by a number of factors. First, the plasmid must be able to accommodate a replicator locus and one or more genes that are capable of allowing replication of the plasmid. Secondly, the plasmid should be of a size which provides for a reasonable probability of recircularization with the foreign gene(s) to form the recombinant plasmid chimera. Desirably, a restriction enzyme should be available, which will cleave the plasmid without inactivating the replicator locus and system associated with the replicator locus. Also, means must be provided for providing ligatable termini for the plasmid, which are

3

complementary to the termini of the foreign gene(s) to allow fusion of the two DNA segments.

Another consideration for the recombinant plasmid is that it be compatible with the bacterium to be transformed. Therefore, the original plasmid will usually be derived from a member of the family to which the bacterium belongs.

The original plasmid should desirably have a phenotypical property which allows for the separation of transformant bacteria from parent bacteria. Particularly useful is a gene, which provides for survival selection. Survival selection can be achieved by providing resistance to a growth inhibiting substance or providing a growth factor capability to a bacterium deficient in such capability.

Conveniently, genes are available, which provide for antibiotic or heavy metal resistance or polypeptide resistance, e.g. colicin. Therefore, by growing the bacteria on a medium containing a bacteriostatic or bacteriocidal substance, such as an antibiotic, only the transformants having the antibiotic resistance will survive. Illustrative antibiotics include tetracycline, streptomycin, sulfa drugs, such as sulfonamide, kanamycin, neomycin, penicillin, chloramphenicol, or the like.

Growth factors include the synthesis of amino acids, the isomerization of substrates to forms which can be metabolized or the like. By growing the bacteria on a medium which lacks the appropriate growth factor, only the bacteria which have been transformed and have the growth factor capability will clone.

One plasmid of interest derived from *E. coli* is referred to as pSC101 and is described in Cohen, et al., Proc. Nat. Acad. Sci., USA, 70, 1293 (1972), (referred to in that article as Tc6-5). Further description of this particular plasmid and its use is found in the other articles previously referred to.

The plasmid pSC101 has a molecular weight of about 5.8×10^6d and provides tetracycline resistance.

Another plasmid of interest is colicinogenic factor EI (ColE1), which has a molecular weight of 4.2×10^6d, and is also derived from *E. coli.* The plasmid has a single EcoRI substrate site and provides immunity to colicin E1.

In preparing the plasmid for joining with the exogenous gene, a wide variety of techniques can be provided, including the formation of or introduction of cohesive termini. Flush ends can be joined. Alternatively, the plasmid and gene may be cleaved in such a manner that the two chains are cleaved at different sites to leave extensions at each end which serve as cohesive termini. Cohesive termini may also be introduced by removing nucleic acids from the opposite ends of the two chains or alternatively, introducing nucleic acids at opposite ends of the two chains.

To illustrate, a plasmid can be cleaved with a restriction endonuclease or other DNA cleaving enzyme. The restriction enzyme can provide square ends, which are then modified to provide cohesive termini or can cleave in a staggered manner at different, but adjacent, sites on the two strands, so as to provide cohesive termini directly.

Where square ends are formed such as, for example, by HIN (Haemophilus influenzae RII) or pancreatic DNAse, one can ligate the square ends or alternatively one can modify the square ends by chewing back, adding particular nucleic acids, or a combination of the two. For example, one can employ appropriate transferases to add a nucleic acid to the 5' and 3' ends of the

4

DNA. Alternatively, one can chew back with an enzyme, such as a λ-exonuclease, and it is found that there is a high probability that cohesive termini will be achieved in this manner.

An alternative way to achieve a linear segment of the plasmid with cohesive termini is to employ an endonuclease such as EcoRI. The endonuclease cleaves the two strands at different adjacent sites providing cohesive termini directly.

With flush ended molecules, a T4 ligase may be employed for linking the termini. See, for example, Scaramella and Khorana, J. Mol. Biol. 72: 427–444 (1972) and Scaramella, DNAS 69: 3389 (1972), whose disclosure is incorporated herein by reference.

Another way to provide ligatable termini is to leave employing DNAse and Mn^{++} as reported by Lai and Nathans, J. Mol. Biol, 89: 179 (1975).

The plasmid, which has the replicator locus, and serves as the vehicle for introduction of a foreign gene into the bacterial cell, will hereafter be referred to as "the plasmid vehicle."

It is not necessary to use plasmid, but any molecule capable of replication in bacteria can be employed. Therefore, instead of plasmid, viruses may be employed, which will be treated in substantially the same manner as the plasmid, to provide the ligatable termini for joining to the foreign gene.

If production of cohesive termini is by restriction endonuclease cleavage, the DNA containing the foreign gene(s) to be bound to the plasmid vehicle will be cleaved in the same manner as the plasmid vehicle. If the cohesive termini are produced by a different technique, an analogous technique will normally be employed with the foreign gene. (By foreign gene is intended a gene derived from a source other than the transformant strain.) In this way, the foreign gene(s) will have ligatable termini, so as to be able to covalently bonded to the termini of the plasmid vehicle. One can carry out the cleavage or digest of the plasmids together in the same medium or separately, combine the plasmids and recircularize the plasmids to form the plasmid chimera in the absence of active restriction enzyme capable of cleaving the plasmids.

Descriptions of methods of cleavage with restriction enzymes may be found in the following articles: Greene, et al., *Methods in Molecular Biology,* Vol. 9, ed. Wickner, R. B., (Marcel Dekker, Inc., New York), "DNA Replication and Biosynthesis"; Mertz and Davis, 69, Proc. Nat. Acad. Sci., USA, 69, 3370 (1972);

The cleavage and non-covalent joining of the plasmid vehicle and the foreign DNA can be readily carried out with a restriction endonuclease, with the plasmid vehicle and foreign DNA in the same or different vessels. Depending on the number of fragments, which are obtained from the DNA endonuclease digestion, as well as the genetic properties of the various fragments, digestion of the foreign DNA may be carried out separately and the fragments separated by centrifugation in an appropriate gradient. Where the desired DNA fragment has a phenotypical property, which allows for the ready isolation of its transformant, a separation step can usually be avoided.

Endonuclease digestion will normally be carried out at moderate temperatures, normally in the range of 10° to 40° C. in an appropriately buffered aqueous medium, usually at a pH of about 6.5 to 8.5. Weight percent of total DNA in the reaction mixture will generally be about 1 to 20 weight percent. Time for the reaction will

5

vary, generally being from 0.1 to 2 hours. The amount of endonuclease employed is normally in excess of that required, normally being from about 1 to 5 units per 10 μg of DNA.

Where cleavage into a plurality of DNA fragments results, the course of the reaction can be readily followed by electrophoresis. Once the digestion has gone to the desired degree, the endonuclease is inactivated by heating above about 60° C. for five minutes. The digestion mixture may be worked up by dialysis, gradient separation, or the like, or used directly.

After preparation of the two double stranded DNA sequences, the foreign gene and vector are combined for annealing and/or ligation to provide for a functional recombinant DNA structure. With plasmids, the annealing involves the hydrogen bonding together of the cohesive ends of the vector and the foreign gene to form a circular plasmid which has cleavage sites. The cleavage sites are then normally ligated to form the complete closed and circularized plasmid.

The annealing, and as appropriate, recircularization can be performed in whole or in part in vitro or in vivo. Preferably, the annealing is performed in vitro. The annealing requires an appropriate buffered medium containing the DNA fragments. The temperature employed initially for annealing will be about 40° to 70° C., followed by a period at lower temperature, generaly from about 10° to 30° C. The molar ratio of the two segments will generally be in the range of about 1–5:-5–1. The particular temperature for annealing will depend upon the binding strength of the cohesive termi. While 0.5 hr to 2 or more days may be employed for annealing, it is believed that a period of 0.5 to 6 hrs may be sufficient. The time employed for the annealing will vary with the temperature employed, the nature of the salt solution, as well as the nature of the cohesive termini.

The ligation, when in vitro, can be achieved in conventional ways employing DNA ligase. Ligation is conveniently carried out in an aqueous solution (pH 6–8) at temperatures in the range of about 5° to 40° C. The concentration of the DNA will generally be from about 10 to 100 g/ml. A sufficient amount of the DNA ligase or other ligating agent e.g. T4 ligase, is employed to provide a convenient rate of reaction, generally ranging from about 5 to 50 U/ml. A small amount of a protein e.g. albumin, may be added at concentrations of about 10 to 200 g/ml. The ligation with DNA ligase is carried out in the presence of magnesium at about 1–10 mM.

At the completion of the annealing or ligation, the solution may be chilled and is ready for use in transformation.

It is not necessary to ligate the recircularized plasmid prior to transformation, since it is found that this function can be performed by the bacterial host. However, in some situations ligation prior to transformation may be desirable.

The foreign DNA can be derived from a wide variety of sources. The DNA may be derived from eukaryotic or prokaryotic cells, viruses, and bacteriophage. The fragments employed will generally have molecular weights in the range of about 0.5 to 20×10⁶d, usually in the range of 1 to 10×10⁶d. The DNA fragment may include one or more genes or one or more operons.

Desirably, if the plasmid vehicle does not have a phenotypical property which allows for isolation of the transformants, the foreign DNA fragment should have

6

such property. Also, an intact promoter and base sequences coding for initiation and termination sites should be present for gene expression.

In accordance with the subject invention, plasmids may be prepared which have replicons and genes which could be present in bacteria as a result of normal mating of bacteria. However, the subject invention provides a technique, whereby a replicon and gene can coexist in a plasmid, which is capable of being introduced into a unicellular organism, which could not exist in nature. The first type of plasmid which cannot exist in nature is a plasmid which derives its replicon from one organism and the exogenous gene from another organism, where the two organisms do not exchange genetic information. In this situation, the two organisms will either be eukaryotic or prokaryotic. Those organisms which are able to exchange genetic information by mating are well known. Thus, prior to this invention, plasmids having a replicon and one or more genes from two sources which do not exchange genetic information would not have existed in nature. This is true, even in the event of mutations, and induced combinations of genes from different strains of the same species. For the natural formation of plasmids formed from a replicon and genes from different microorganisms it is necessary that the microorganisms be capable of mating and exchanging genetic information.

In the situation, where the replicon comes from a eukaryotic or prokaryotic cell, and at least one gene comes from the other type of cell, this plasmid heretofore could not have existed in nature. Thus, the subject invention provides new plasmids which cannot naturally occur and can be used for transformation of unicellular organisms to introduce genes from other unicellular organisms, where the replicon and gene could not previously naturally coexist in a plasmid.

Besides naturally occurring genes, it is feasible to provide synthetic genes, where fragments of DNA may be joined by various techniques known in the art. Thus, the exogenous gene may be obtained from natural sources or from synthetic sources.

The plasmid chimera contains a replicon which is compatible with a bacterium susceptible of transformation and at least one foreign gene which is directly or indirectly bonded through deoxynucleotides to the replicon to form the circularized plasmid structure. As indicated previously, the foreign gene normally provides a phenotypical property, which is absent in the parent bacterium. The foreign gene may come from another bacterial strain, species or family, or from a plant or animal cell. The original plasmid chimera will have been formed by in vitro covalent bonding between the replicon and foreign gene. Once the originally formed plasmid chimera has been used to prepare transformants, the plasmid chimera will be replicated by the bacterial cell and cloned in vivo by growing the bacteria in an appropriate growth medium. The bacterial cells may be lysed and the DNA isolated by conventional means or the bacteria continually reproduced and allowed to express the genotypical property of the foreign DNA.

Once a bacterium has been transformed, it is no longer necessary to repeat the in vitro preparation of the plasmid chimera or isolate the plasmid chimera from the transformant progeny. Bacterial cells can be repeatedly multiplied which will express the genotypical property of the foreign gene.

7

One method of distinguishing between a plasmid which originates in vivo from a plasmid chimera which originates in vitro is the formation of homoduplexes between an in vitro prepared plasmid chimera and the plasmid formed in vivo. It will be an extremely rare event where a plasmid which originates in vivo will be the same as a plasmid chimera and will form homoduplexes with plasmid chimeras. For a discussion of homoduplexes, see Sharp, Cohen and Davidson, J. Mol. Biol., 75, 235 (1973), and Sharp, et al, ibid, 71, 471 (1972).

The plasmid derived from molecular cloning need not homoduplex with the in vitro plasmid originally employed for transformation of the bacterium. The bacterium may carry out modification processes, which will not affect the portion of the replicon introduced which is necessary for replication nor the portion of the exogenous DNA which contains the gene providing the genotypical trait. Thus, nucleotides may be introduced or excised and, in accordance with naturally occurring mating and transduction, additional genes may be introduced. In addition, for one or more reasons, the plasmids may be modified in vitro by techniques which are known in the art. However, the plasmids obtained by molecular cloning will homoduplex as to those parts which relate to the original replicon and the exogenous gene.

II. Transformation

After the recombinant plasmid or plasmid chimera has been prepared, it may then be used for the transformation of bacteria. It should be noted that the annealing and ligation process not only results in the formation of the recombinant plasmid, but also in the recircularization of the plasmid vehicle. Therefore, a mixture is obtained of the original plasmid, the recombinant plasmid, and the foreign DNA. Only the original plasmid and the DNA chimera consisting of the plasmid vehicle and linked foreign DNA will normally be capable of replication. When the mixture is employed for transformation of the bacteria, replication of both the plasmid vehicle genotype and the foreign genotype will occur with both genotypes being replicated in those cells having the recombinant plasmid.

Various techniques exist for transformation of a bacterial cell with plasmid DNA. A technique, which is particularly useful with Escherichia coli, is described in Cohen, et al., ibid, 69, 2110 (1972). The bacterial cells are grown in an appropriate medium to a predetermined optical density. For example, with E. coli strain C600, the optical density was 0.85 at 590 nm. The cells are concentrated by chilling, sedimentation and washing with a dilute salt solution. After centrifugation, the cells are resuspended in a calcium chloride solution at reduced temperatures (approx. 5°–15° C.), sedimented, resuspended in a smaller volume of a calcium chloride solution and the cells combined with the DNA in an appropriately buffered calcium chloride solution and incubated at reduced temperatures. The concentration of Ca^{++} will generally be about 0.01 to 0.1 M. After a sufficient incubation period, generally from about 0.5–3.0 hours, the bacteria are subjected to a heat pulse generally in the range of 35° to 45° C. for a short period of time; namely from about 0.5 to 5 minutes. The transformed cells are then chilled and may be transferred to a growth medium, whereby the transformed cells having the foreign genotype may be isolated.

8

An alternative transformation technique may be found in Lederberg and Cohen, J. Bacteriol., 119, 1072 (1974), whose disclosure is incorporated herein by reference.

III. Replication and Transcription of the Plasmid

The bacterial cells, which are employed, will be of such species as to allow replication of the plasmid vehicle. A number of different bacteria which can be employed, have been indicated previously. Strains which lack indigenous modification and restriction enzymes are particularly desirable for the cloning of DNA derived from foreign sources.

The transformation of the bacterial cells will result in a mixture of bacterial cells, the dominant proportion of which will not be transformed. Of the fraction of cells which are transformed, some significant proportion, but normally a minor proportion, will have been transformed by recombinant plasmid. Therefore, only a very small fraction of the total number of cells which are present will have the desired phenotypical characteristics.

In order to enhance the ability to separate the desired bacterial clones, the bacterial cells, which have been subjected to transformation, will first be grown in a solution medium, so as to amplify the absolute number of the desired cells. The bacterial cells may then be harvested and streaked on an appropriate agar medium. Where the recombinant plasmid has a phenotype, which allows for ready separation of the transformed cells from the parent cells, this will aid in the ready separation of the two types of cells. As previously indicated, where the genotype provides resistance to a growth inhibiting material, such as an antibiotic or heavy metal, the cells can be grown on an agar medium containing the growth inhibiting substance. Only available cells having the resistant genotype will survive. If the foreign gene does not provide a phenotypical property, which allows for distinction between the cells transformed by the plasmid vehicle and the cells transformed by the plasmid chimera, a further step is necessary to isolate the replicated plasmid chimera from the replicated plasmid vehicle. The steps include lysing of the cells and isolation and separation of the DNA by conventional means or random selection of transformed bacteria and characterization of DNA from such transformants to determine which cells contain molecular chimeras. This is accomplished by physically characterizing the DNA by electrophoresis, gradient centrifugation or electron microscopy.

Cells from various clones may be harvested and the plasmid DNA isolated from these transformants. The plasmid DNA may then be analyzed in a variety of ways. One way is to treat the plasmid with an appropriate restriction enzyme and analyze the resulting fragments for the presence of the foreign gene. Other techniques have been indicated above.

Once the recombinant plasmid has been replicated in a cell and isolated, the cells may be grown and multiplied and the recombinant plasmid employed for transformation of the same or different bacterial strain.

The subject process provides a technique for introducing into a bacterial strain a foreign capability which is genetically mediated. A wide variety of genes may be employed as the foreign genes from a wide variety of sources. Any intact gene may be employed which can be bonded to the plasmid vehicle. The source of the gene can be other bacterial cells, mammalian cells, plant

9

cells, etc. The process is generally applicable to bacterial cells capable of transformation. A plasmid must be available, which can be cleaved to provide a linear segment having ligatable termini, and an interact replicator locus and system, preferably a system including a gene which provides a phenotypical property which allows for easy separation of the transformants. The linear segment may then be annealed with a linear segment of DNA having one or more genes and the resulting recombinant plasmid employed for transformation of the bacteria.

By introducing one or more exogenous genes into a unicellular organism, the organism will be able to produce polypeptides and proteins ("poly(amino acids)") which the organism could not previously produce. In some instances the poly(amino acids) will have utility in themselves, while in other situations, particularly with enzymes, the enzymatic product(s) will either be useful in itself or useful to produce a desirable product.

One group of poly(amino acids) which are directly useful are hormones. Illustrative hormones include parathyroid hormone, growth hormone, gonadotropins (FSH, luteinizing hormone, chorionogonadatropin, and glycoproteins), insulin, ACTH, somatostatin, prolactin, placental lactogen, melanocyte stimulating hormone, thyrotropin, parathyroid hormone, calcitonin, enkephalin, and angiotensin.

Other poly(amino acids) of interest include serum proteins, fibrinogin, prothrombin, thromboplastin, globulin e.g. gamma-globulins or antibodies, heparin, antihemophilia protein, oxytocin, albumins, actin, myosin, hemoglobin, ferritin, cytochrome, myoglobin, lactoglobulin, histones, avidin, thyroglobulin, interferin, kinins and transcortin.

Where the genes or genes produce one or more enzymes, the enzymes may be used for fulfilling a wide variety of functions. Included in these functions are nitrogen fixation, production of amino acids, e.g. polyiodothyronine, particularly thyroxine, vitamins, both water and fat soluble vitamins, antimicrobial drugs, chemotheropeutic agents e.g. antitumor drugs, polypeptides and proteins e.g. enzymes from apoenzymes and hormones from prohormones, diagnostic reagents, energy producing combinations e.g. photosynthesis and hydrogen production, prostaglandins, steroids, cardiac glycosides, coenzymes, and the like.

The enzymes may be individually useful as agents separate from the cell for commercial applications, e.g. in detergents, synthetic transformations, diagnostic agents and the like. Enzymes are classified by the I.U.B. under the classifications as I. Oxidoreductases; II. Transferases; III. Hydrolases; IV. Lyases; V. Isomerases; and VI. Ligases.

EXPERIMENTAL

In order to demonstrate the subject invention, the following experiments were carried out with a variety of foreign genes.

(All temperatures not otherwise indicated are Centigrade. All percents not otherwise indicated are percents by weight.)

EXAMPLE A

A. Preparation of pSC101 Plasmid

Covalently closed R6-5 DNA was sheared with a Virtis stainless steel microshaft in a one milliliter cup. The R6-5 DNA was sheared at 2,000 r.p.m. for 30 minutes in TEN buffer solution (0.02 M Tris-HCl (pH 8.0)-1

10

mM EDTA (pH 8.0)-0.02 M NaCl), while chilled at 0°-4°.

The sheared DNA sample was subjected to sucrose gradient sedimentation at 39,500 r.p.m. in a Spinco SW 50.1 rotor at 20°. A 0.12 mil fraction was collected on a 2.3 cm diameter circle of Whatman No. 3 filter paper, dried for 20 minutes and precipitated by immersion of the disc in cold 5% trichloroacetic acid, containing 100 μg/ml thymidine. The precipitate was filtered and then washed once with 5% trichloroacetic acid, twice with 99% ethanol and dried. pSC101 was the 27S species having a calculated molecular weight of 5.8×10^6 d.

B. Generalized Transformation Procedure

E. coli strain C600 was grown at 37° in H1 medium to an optical density of 0.85 at 590 nm. At this point the cells were chilled quickly, sedimented and washed once in 0.5 volume 10 nM NaCl. After centrifugation, the bacteria was resuspended in half the original volume of chilled 0.03 M calcium chloride, kept at 0° for 20 minutes, sedimented, and then resuspended in 0.1 of the original volume of 0.03 M of calcium chloride solution. Chilled DNA samples in TEN buffer were supplemented with 0.1 M calcium chloride to a final concentration of 0.03 M.

0.2 ml of competent cells treated with calcium chloride was added to 0.1 ml of DNA solution with chilled pipets and an additional incubation was done for 60 minutes at 0°. The bacteria were then subjected to a heat pulse at 42° for two minutes, chilled, and then either placed directly onto nutrient agar containing appropriate antibiotics or, where indicated, diluted 10 times in L-broth and incubated at 37° before plating. The cell survival is greater than 50% after calcium chloride treatment and heat pulse. Drug resistance was assayed on nutrient agar plates with the antibiotics indicated in specific experiments.

EXAMPLE I: Construction of Biologically Functional Bacterial Plasmids in vitro

A. Covalently closed R6-5 plasmid DNA was cleaved by incubation at 37° for 15 minutes in a 0.2 ml reaction mixture containing DNA (40 μg/ml, 100 mM Tris.HCl (pH 7.4)), 5 mM MgCl$_2$, 50 mM NaCl, and excess (2 U) EcoRI endonuclease in 1 μl volume. An additional incubation at 60° for 5 minutes was employed to inactivate the endonuclease.

The resulting mixture of plasmid fragments was employed for transformation of E. coli strain C600 in accordance with the procedure previously described. A single clone was examined further which was selected for resistance to kanamycin and was also found to carry resistance to neomycin and sulfonamide, but not to tetracycline, chloramphenicol, or streptomycin after transformation of E. coli by EcoRI generated DNA fragments of R6-5. Closed circular DNA obtained from this isolate (plasmid designation pSC102) by CsCl-ethidium bromide gradient centrifugation had an S value of 39.5 in neutral surcrose gradients.

Treatment of pSC102 plasmid DNA with EcoRI restriction endonuclease in accordance with the above-described procedure resulted in the formation of 3 fragments that were separable by electrophoresis in agarose gels. Intact pSC102 plasmid DNA and pSC101 plasmid DNA, which had been separately purified by dye-buoyant density centrifugation, were treated with EcoRI endonuclease followed by annealing at 0°-2° for

11

about six hours. The mixture was then subjected to ligation with pSC101 and pSC102 in a ratio of 1:1 respectively, by ligating for 6 hours at 14° in 0.2 ml reaction mixtures containing 5 mM MgCl₂, 0.1 mM NAD, 100 µg/ml of bovine-serum albumin (BSA), 10 mM ammonium sulphate (pH 7.0), and 18 U/ml of DNA ligase. (J. Mertz and Davis, Proc. Nat. Acad. Sci., USA, 69, 3370 (1972); and Modrich, et al., J. Biol. Chem., 248, 7495 (1973). Ligated mixtures were incubated at 37° for 5 minutes and then chilled in ice water. Aliquots containing 3.3–6.5 µg/ml of total DNA were used directly for transformation.

Transformation of E. coli strain C600 was carried out as previously described. For comparison purposes, transformation was also carried out with a mixture of pSC101 and pSC102 plasmid DNA, which had been subjected to EcoRI endonuclease, but not DNA ligase. The antibiotics used for selection were tetracycline (10 µg/ml) and kanamycin (25 µg/ml). The results are reported as transformants per microgram of DNA. The following table indicates the results.

TABLE I

Transformation of E. coli C600 by a mixture of pSC101 and pSC102 DNA

Treatment of DNA	Transformation frequency for antibiotic resistence markers		
	Tetracycline	Kanamycin	Tetracycline + kanamycin
None	2 × 10⁵	1 × 10⁵	2 × 10²
EcoRI	1 × 10⁴	1.1 × 10³	7 × 10¹
EcoRI + DNA ligase	1.2 × 10⁴	1.3 × 10³	5.7 × 10²

Kanamycin resistance in the R65 plasmid is a result of the presence of the enzyme kanamycin monophosphotransferase. The enzyme can be isolated from the bacteria by known procedures and employed in an assay for kanamycin in accordance with the procedure described in Smith, et al., New England J. Medicine, 286, 583 (1972).

In the preparation for the enzyme extracts, the E. coli are grown in ML-broth and harvested in a late logarithm phase of growth. The cells are osmotically shocked (see Nossal, et al., J. Biol. Chem. 241, 3055 (1966), washed twice at room temperature with 10 ml 0.01 M Tris and 0.03 M NaCl, pH 7.3, and the pellet suspended in 10 ml 20% sucrose, 3 × 10³ M EDTA and 0.033 M Tris (pH 7.5), stirred for 10 minues at room temperature and centrifuged at 16,000 g for 5 minutes. The pellet is then suspended in 2 ml of cold 5 × 10⁻⁴ M MCl₂, stirred for 10 minutes at 2° and centrifuged at 26,000 g for 10 minutes to yield a supernatant fluid referred to as the osmotic shockate. The solution should be stored at −20° or lower. (See Benveneste, et al., FEBS Letters, 14 293 (1971).

The osmotic shockate may then be used in accordance with the procedure of Smith, et al., supra.

EXAMPLE II: Genome Construction between Bacterial Species in vitro: Replication and Expression of Staphylococcus Plasmid Genes in E. coli

S. aureus strain 8325 contains the plasmid pI258, which expresses resistance to penicillin, erythromycin, cadmium and mercury. (Lindberg, et al., J. Bacteriol., 115, 139 (1973)). Covalently closed circular pSC101 and pI258 plasmid DNA were separately cleaved by incubation at 37° for 15 minutes in 0.2 ml reaction mixtures by EcoRI endonuclease in accordance with the procedure

12

described previously. Aliquots of the two cleaved species were mixed in a ratio of 3 µg of pI258:1 µg of pSC101 and annealed at 2°–4° for 48 hours. Subsequent ligation was carried out for six hours at 14° as described previously and aliquots containing 3.3–6.5 µg/ml of total DNA were used directly in the transformation as described previously.

Other transformations were carried out employing the two plasmids independently and a mixture of the two plasmids. Selection of transformants was carried out at antibiotic concentrations for tetracycline (Tc, 25 µg/ml) or pencillin (Pc, 25 OU/ml). The transformation was carried out with E. coli strain C600 r$_K$⁻ m$_K$⁻. The following table indicates the results.

TABLE III

Transformation of C600 r$_K$⁻ m$_K$⁻ by pSC101 and pI258 Plasmid DNA

DNA	Transformants/µg DNA	
	Tc	Pc
PSC101 closed circular	1 × 10⁶	<3
pI258 closed circular	<3.6	<3.6
pSC101 + pI258 untreated	9.1 × 10⁵	<5
pSC101 + pI258 EcoRI-treated	4.7 × 10³	10

The above table demonstrates that bacteria can be formed which have both tetracycline resistance and penicillin resistance. Thus, one can provide the phenotypical property penicillin resistance in bacteria from DNA, which is indigenous to another biological organism. One can thus use E. coli for the production of the enzyme, which imparts penicillin resistance to bacteria, and assay for penicillin in a manner similar to that employed for kanamycin. Penicillinase is used for destroying penicillin in blood serum of patients treated with penicillin in order to determine whether pathogenic organisms whose growth is inhibited by penicillin may be present.

EXAMPLE III: Replication and Transcription of Eukaryotic DNA in E. coli

The amplified ribosomal DNA (rDNA) codeing for 18S and 28S ribosomal RNA of the South African toad, Xenopus laevis was used as a source of eukaryotic DNA for these experiments. Dawid, et al., J. Mol. Biol., 51, 341 (1970). E. coli-X. laevis recombinant plasmids were constructed in vitro as follows:

The reaction mixture (60 µl) contained 100 mM Tris.HCl (pH 7.5) 50 mM NaCl, 5 mM MgCl₂, 1.0 µg of pSC101 plasmid DNA and 2.5 µg of X. laevis rDNA, and excess EcoRI restriction endonuclease (1 µl, 2 U). After a 15 minute incubation at 37°, the reaction mixture was placed at 63° for 5 minutes to inactivate EcoRI endonuclease. The product was then refrigerated at 0.5° for 24 hours, to allow association of the short cohesive termini.

The reaction mixture for ligation of phosphodiester bonds was adjusted to a total volume of 100 µl and contained in addition to the components of the endonuclease reaction, 30 mM Tris.HCl (pH 8.1), 1 mM sodium EDTA, 5 mM MgCl₂, 3.2 nM NAD, 10 mM ammonium sulphate, 5 µg BSA, and 9 U of E. coli DNA ligase. All components were chilled to 5° before their addition to the reaction mixture. The ligase reaction mixture was incubated at 14° for 45 minutes, and then at 0.5° for 48 hours. Additional NAD and ligase were added and the mixture incubated at 15° for 30 minutes and then for 15 minutes at 37°. The ligated DNA was used directly in

13

the plasmid transformation procedure previously described. The DNA was used to transform *E. coli* strain C600 $r_K^- m_K^-$ and tetracycline resistant transformants (3.3×10^3/μg of pSC101 DNA) were selected and numbered consecutively CD1, CD2, etc. Plasmid DNA was isolated from a number of the transformants.

^{32}P-labeled 18 S and 28 S *X. laevis* rRNA were hybridized with DNA obtained from the plasmids CD4, CD18, CD30, and CD42. CD4 DNA annealed almost equally with both the 18 S and 28 S rRNA species. CD18 plasmid DNA hybridized principally with 28 S *X. laevis* rRNA, while the DNA of plasmids CD30 and CD42 annealed primarily with 18 S rRNA. These data indicate that portions of the *X. laevis* rDNA were, in fact, incorporated into a plasmid recombinant with pSC101, which was capable of transforming *E. coli*, so as to be capable of replicating X. laevis rDNA.

Transcription of *X. laevis* DNA was also carried out in *E. coli* minicells. The minicell producing *E. coli* strain P678-54 was transformed with plasmid DNA isolated from *E. coli* strain C600 $r_K^- m_K^-$ containing CD4, CD18, or CD42. Many cells containing the plasmids were isolated and incubated with [^{3}H] uridine; RNA purified from such minicells was hybridized with *X. laevis* rDNA immobilized on nitrocellulose membranes in order to determine whether the *X. laevis* rDNA linked to the pSC101 replicon is transcribed in *E. coli*. The results in the following table show that RNA species capable of annealing with purified *X. laevis* rDNA are synthesized in *E. coli* minicells carrying the recombinant plasmids, CD4, CD18, and CD42, but not by minicells carrying the pSC101 plasmid alone.

Minicells containing plasmids were isolated as described by Cohen, et al., Nature New Biol., 231, 249 (1971). They were incubated with [^{3}H] uridine (50 μCi/ml, 30 Ci/mol) as described by Roozen, et al., J. Bacteriol., 107, 21 (1971) for 10 minutes at 37°. Minicells collected by centrifugation were resuspended in Tris.HCl (20 mM, pH 7.5)-5 mM MgCl$_2$-1 mM EDTA pH 8.0 and rapidly frozen and thawed 3 times. RNA was extracted as described in Cohen, et al., J. Mol. Biol., 37, 387 (1968). Hybridization assays were carried out in nitrocellulose membranes as described in Cohen, et al., ibid, at saturating levels of pSC101 DNA. Hybridizations involving *X. laevis* DNA were not performed at DNA excess. Counts bound to blank filters (5–10 c.p.m.) were substracted from experimentally determined values. ^{3}H count eluted from filters containing *X. laevis* DNA were rendered acid soluble by ribonuclease A 20 μg/ml, 0.30 M NaCl-0.030 M sodium citrate, 1 hour, 37°. The following table indicates the results.

TABLE III

Plasmid carried by minicells	Input cpm	[^{3}H] RNA synthesized by *E. coli* minicells		
		[^{3}H] RNA counts hybridized to		
		X. laevis rDNA		pSC101 DNA
		0.2μg	0.4μg	18μg
CD42	4810	905 (19%)	1436 (30%)	961 (20%)
CD18	3780	389 (10%)	—	1277 (34%)
CD4	5220	789 (15%)	—	1015 (19%)
pSC101	4170	0 (0%)	—	1500 (36%)

EXAMPLE IV: Plasmid ColEl as a Molecular Vehicle for Cloning and Amplification of Trp Operon

In a volume of 200 μl (100 mM Tris.HCl (pH 7.5)-5 mM MgCl$_2$-50 mM NaCl), 5.7 μg of ColEl (*E. coli* JC411Thy$^-$/ColEl) (Clewell, et al., Proc. Nat. Acad.

14

Sci., USA, 62, 1159 (1969) and 6.0 μg DNA from bacteriophage φ80pt190 (Deeb, et al., Virology, 31, 289 (1967) were digested to completion with homogeneously purified EcoRI endonuclease, monitoring the digestion by electrophoresis of the fragments in an agarose gel. The endonuclease was inactivated by heating at 65° for 5 minutes, the digest dialyzed overnight against 5 mM Tris.HCl, pH 7.5, and the sample concentrated to 50 μl. The fragments were ligated as described in Dugaiczyk, et al., Biochemistry, 13, 503 (1974) at a concentration of 75 pmoles/ml of fragments.

Transformation was carried out as previously described except that the cells were grown to $A_{590} = 0.600$ and following exposure to DNA were incubated in L-broth for 90 minutes. The cells were collected and resuspended in 10 mM NaCl before plating. Cells employed as recipients for the transformations were *E. coli* strains C600 trpR$^-$, ΔtrpE5(MV1), C600 trpR$^-$ trpE 10220 recA(MV2), C600 ΔtrpE5(MV10) and C600 ΔtrpE5 recA(MV12). (trpR$^-$ is the structural gene for the trp repressor and ΔtrpE5 is a trp operon deletion entirely within trpE and removing most of the gene.) Approximately 2 μg of the DNA was used to transform the cells.

Cultures were plated on Vogel-Bonner agar supplemented with 50 μg/ml of the non-selective amino acids, 0.2% glucose and 5 μg/ml of required vitamins. Transformants to colicin immunity were initially selected on a lawn of a culture of a mutant strain carrying ColE1. Clones were then selected for their ability to grow in the absence of tryptophan. Cells capable of producing tryptophan were isolated, which could be used for the production of exogenous tryptophan. The subject example demonstrates the introduction of a complete operon from foreign DNA to provide a transformant capable of replicating the operon and transcribing and translating to produce enzymes capable of producing an aromatic amino acid.

EX. V: Cloning of Synthetic Somatostatin Gene

The deoxyribonucleotide sequence for the somatostatin gene was prepared in accordance with conventional procedures. (Itakura et al, Science, 198 1056 (1977)). To prepare the recombinant plasmid, plasmid pBR 322 was digested with Eco RI. The reaction was terminated by extraction with a mixture of phenol and chloroform, the DNA precipitated with ethanol and resuspended in 50 μl of T$_4$ DNA polymerase buffer. The reaction was started by the addition of 2 units of T$_4$ DNA polymerase. After 30 min at 37°, the mixture was extracted with phenol and chloroform and the DNA precipitated with ethanol. The λplac5 DNA (3 μg) was digested with the endonuclease Hae III and the digested pBR 322 DNA blunt end ligated with the Hae III-digested λplac5 DNA in a final volume of 30 μl with T$_4$ DNA ligase (hydroxylopatite fraction) in 20 mM tris-HCl pH 7.6), 10 mM MgCl$_2$, 10 mM dithiothreitol and 0.5 mM ATP for 12 hrs at 12°. The ligated DNA mixture was dialyzed against 10 mM tris-HCl (pH 7.6) and used to transform *E. coli* strain RR1. Transformants were selected for tetracycline resistance and ampicillin resistance on antibiotic (20 μg/ml) X-gal (40 μg/ml) medium. Colonies constitutive for the synthesis of β-galactosiodase were identified by their blue color and of 45 colonies so identified, 3 of them were found to contain plasmids with 2 Eco RI sites separated by ~200 base pairs.

15

The plasmid so obtained pBH10 was modified to eliminate the Eco R1 site distal to the iac operator and plasmid pBH20 was obtained.

Plasmid pBH20 (10 μg) was digested with endonucleases Eco R1 and Bam HI and treated with bacterial alkaline phosphatase (0.1 unit of BAPF, Worthington) and incubation was continued for 10 min at 65°. After extract with a phenol-chloroform mixture, the DNA was precipiated with ethanol. Somatostatin DNA (50 μl containing 4 μg/ml) was ligated with the Bam HI-Eco R1 alkaline phosphatase=treated pBH20 DNA in a total volume of 50 μl with 4 units of T4 DNA ligase for 2 hrs at 22° and the recombinant plasmid used to transform E. coli RR1. Of the Tc^r transformants isolated (10), four plasmids has Eco R1 and Bam HI sites. Base sequence analysis indicated that the plasmid pSOM1 had the desired somatostatin DNA fragment inserted. Because of the failure to detect somatostatin activity from cultures carrying plasmid pSOM1, a plasmid was constructed in which the somatostatin gene could be located at the COOH-terminus of the β-galactosidase gene, keeping the translation in phase. For the construction of such a plasmid, pSOM1 (50 μg) was digested with restriction enzymes Eco R1 and Pst I. A preparative 5 percent polyacrylamide gel was used to separate the large Pst I-Eco R1 fragment that carries the somatostatin gene from the small fragment carrying the lac control elements (12). In a similar way plasmid pBR322 DNA (50 μg) was digested with Pst I and Eco R1 restriction endonucleases, and the two resulting DNA fragments were purified by preparative electrophoresis on a 5 percent polyacrylamide gel. The small Pst I-Eco R1 fragment from pBR322 (1 μg) was ligated with the large PstI-Eco R1 DNA fragment (5 μg) from pSOM1. The ligated mixture was used to transform E. coli RR1, and transformants were selected for Ap^r on X-gal medium. Almost all the Ap^r transformants (95 percent) gave white colonies (no lac operator) on X-gal indicator plates. The resulting plasmid, pSOM11, was used in the construction of plasmid pSOM11-3. A mixture of 5 μg of pSOM11 DNA and 5 μg of λplac5 DNA was digested with Eco R1. The DNA was extracted with a mixture of phenol and chloroform; the extract was precipitated by ethanol, and the precipitate was resuspended in T4 DNA ligase buffer (50 μl) in the presence of T4 DNA ligase (1 unit). The ligated mixture was used to transform E. coli strain RR1. Transformants were selected for Ap^r on X-gal plates containing ampicillin aidn screened for constitutive β-galactosidase production. Approximately 2 percent of the colonies were blue (such as pSOM11-1 and 11-2). Restriction enzyme analysis of plasmid DNA obtained from these clones revealed that all the plasmids carried a new Eco R1 fragment of approximately 4.4 megadaltons, which carries the lac operon control sites and most of the β-galactosidase gene (13, 14). Two orientations of the Eco R1 fragment are possible, and the asymmetric location of a Hind III restriction in this fragment can indicate which plasmids had transcription proceeding into the somatostatin gene. The clones carrying plasmids SOM11-3, pSOM11-5, pSOM11-6, and pSOM11-7 contained the Eco R1 fragment in this orientation.

It is evident from the above results, that both DNA from a eukaryotic source and RNA transcribed from the eukaryotic DNA can be formed in a bacterial cell and isolated. Thus, the subject process provides a simple technique for producing large amounts of eukaryotic DNA and/or RNA without requiring the repro-

16

duction and maintenance of the eukaryotic organism or cells. The employment of DNA for production of ribosomal RNA is merely illustrative of using a genome from a eukaryotic cell for formation of a recombinant plasmid for replication in a bacteria. Genomes from a eukaryotic cell for formation of genotypical properties, such as the production of enzymes, could have equivalently been used. As evidenced by the transformation with DNA from a bacteriophage, and entire operon can be introduced into a bacterial cell and the cell becomes capable of its transcription, translation, and production of a functional gene product. Thus, a wide variety of auxotrophic properties can be introduced into a bacterial cell.

In accordance with the subject invention, DNA vehicles are provided, which are covalently closed circular extrachromosomal replicons or genetic elements, including plasmids and viral DNA. The vehicles generally will have molecular weights in the range of about 1 to 20×10^6 and are characterized by having an intact replicon, which includes a replicator locus and gene. The vehicle is capable of cleavage by a restriction enzyme to provide a linear segment having an intact replicon and cohesive termini, which may be directly obtained by the cleavage or by subsequent modification of the termini of the linear segment. The vehicle will be capable of transforming a bacterial cell and to that extent is compatible with the cell which will provide replication and translation. Preferably, the vehicle will have a phenotypical property which will allow for segregation of the transformant cells. Phenotypical properties include resistance to growth inhibiting materials, such as antibiotics, peptides and heavy metals, morphological properties, color, or the like, and production of growth factors, e.g. amino acids.

The vehicle is combined with DNA indigenous to a biological organism other than the cell which provides replication and provides a genotypical or phenotypical property which is alien to the cell. The source of the DNA can be prokaryotic or eukaryotic, thus including bacteria, fungi, vertebrates, e.g. mammals, and the like.

The plasmid vehicle and the alien DNA having complementary cohesive termini can be annealed together and covalently linked to provide a recombinant plasmid, which is capable of transforming a bacterial cell, so as to be capable of replication, transcription, and translation. As a result, a wide variety of unique capabilities can be readily introduced into bacteria, so as to provide convenient ways to obtain nucleic acids and to study nucleic acids from a foreign host. Thus, the method provides the ability to obtain large amounts of a foreign nucleic acid from bacteria in order to be able to study the function and nature of the nucleic acid. In addition, the subject method provides means for preparing enzymes and enzymic products from bacteria where the natural host is not as convenient or efficient a source of such product. Particularly, bacteria may allow for more ready isolation of particular enzymes, uncontaminated by undersirable contaminants, which are present in the original host. In addition, the products of the enzymic reactions may be more readily isolated and more efficiently produced by a transformant than by the original host. Besides enzymes, other proteins can be produced such as antibodies, antigens, albumins, globulins, glycoproteins, and the like.

Although the foregoing invention has been described in some detail by way of illustration and example for purposes of clarity of understanding, it will be obvious

17

that certain changes and modifications may be practiced within the scope of the appended claims.

We claim:

1. A method for replicating a biologically functional DNA, which comprises:

transforming under transforming conditions compatible unicellular organisms with biologically functional DNA to form transformants; said biologically functional DNA prepared in vitro by the method of:

(a) cleaving a viral or circular plasmid DNA compatible with said unicellular organism to provide a first linear segment having an intact replicon and termini of a predetermined character;

(b) combining said first linear segment with a second linear DNA segment, having at least one intact gene and foreign to said unicellular organism and having termini ligatable to said termini of said first linear segment, wherein at least one of said first and second linear DNA segments has a gene for a phenotypical trait, under joining conditions where the termini of said first and second segments join to provide a functional DNA capable of replication and transcription in said unicellular organism;

growing said unicellular organisms under appropriate nutrient conditions; and

isolating said transformants from parent unicellular organisms by means of said phenotypical trait imparted by said biologically functional DNA.

2. A method according to claim 1, wherein said unicellular organisms are bacteria.

3. A method according to claim 2, wherein said transformation is carried out in the presence of calcium chloride.

4. A method according to claim 3, wherein said phenotypical trait is resistance to growth inhibiting substance, and said growth is carried out in the presence of a sufficient amount of said growth inhibiting substance to inhibit the growth of parent unicellular organisms, but insufficient to inhibit the growth of transformants.

5. A method according to claim 1, wherein said unicellular organism is *E. coli.*

18

6. A method according to claim 1, wherein said predetermined termini are staggered and cohesive.

7. A method according to claim 6, wherein said joining conditions includes enzymatic ligation.

8. A method according to claim 6, wherein said cohesive ends are formed by staggered cleavage of said viral or circular plasmid DNA and a source of said second segment with a restriction enzyme.

9. A method according to claim 6 wherein said cohesive termini are formed by addition of nucleotides.

10. A method according to claim 1, wherein said predetermined termini are blunt end and said joining conditions include enzymatic ligation.

11. A method for replicating a biologically functional DNA comprising, a replicon compatible with a host unicellular organism joined to a gene derived from a source which does not exchange genetic information with said host organism, said method comprising:

isolating said biologically functional DNA from transformants prepared in accordance with claim 1;

transforming unicellular microorganisms with which said replicon is compatible with said isolated DNA to provide second transformants; and

growing said second transformants under appropriate nutrient conditions to replicate said biologically functional DNA.

12. A method for producing a protein foreign to a unicellular organism by means of expression of a gene by said unicellular organism, wherein said gene is derived from a source which does not exchange genetic information with said organism, said method comprising:

growing transformants prepared in accordance with any of claims 1 and 11 under appropriate nutrient conditions, whereby said organism expresses said foreign gene and produces said protein.

13. A method according to claim 12, wherein said protein is an enzyme.

14. A method according to claim 11, wherein said method is repeated substituting said biologically functional DNA from transformants prepared in accordance with claim 1 with second or subsequent transformants to produce additional transformants.

* * * * *

Transgenic Non-human Mammals

H. Leder and T.A. Stewart

[54] TRANSGENIC NON-HUMAN MAMMALS

[75] Inventors: **Philip Leder**, Chestnut Hill, Mass.; **Timothy A. Stewart**, San Francisco, Calif.

[73] Assignee: **President and Fellows of Harvard College**, Cambridge, Mass.

[21] Appl. No.: **623,774**

[22] Filed: **Jun. 22, 1984**

[51] Int. Cl.⁴ C12N 1/00; C12Q 1/68; C12N 15/00; C12N 5/00

[52] U.S. Cl. 800/1; 435/6; 435/172.3; 435/240.1; 435/240.2; 435/320; 435/317.1; 935/32; 935/59; 935/70; 935/76; 935/111

[58] Field of Search 435/6, 172.3, 240, 317, 435/320, 240.1, 240.2; 935/70, 76, 59, 111, 32; 800/1

[56] **References Cited**

U.S. PATENT DOCUMENTS

4,535,058 8/1985 Weinberg et al. 435/91
4,579,821 4/1986 Palmiter et al. 435/240

OTHER PUBLICATIONS

Ucker et al, Cell 27:257–266, Dec. 1981.
Ellis et al, Nature 292:506–511, Aug. 1981.
Goldfarb et al, Nature 296:404–409, Apr. 1981.
Huang et al, Cell 27:245–255, Dec. 1981.
Blair et al, Science 212:941–943, 1981.
Der et al, Proc. Natl. Acad. Sci. USA 79:3637–3640, Jun. 1982.
Shih et al, Cell 29:161–169, 1982.
Gorman et al, Proc. Natl. Acad. Sci. USA 79:6777–6781, Nov. 1982.
Schwab et al, EPA–600/9–82–013, Sym: Carcinogen, Polynucl. Aromat. Hydrocarbons Mar. Environ., 212–32 (1982).
Wagner et al. (1981) Proc. Natl. Acad. Sci USA 78, 5016–5020.
Stewart et al. (1982) Science 217, 1046–8.
Costantini et al. (1981) Nature 294, 92–94.
Lacy et al. (1983) Cell 34, 343–358.
McKnight et al. (1983) Cell 34, 335.
Binster et al. (1983) Nature 306, 332–336.
Palmiter et al. (1982) Nature 300, 611–615.
Palmiter et al. (1983) Science 222, 814.
Palmiter et al. (1982) Cell 29, 701–710.

Primary Examiner—Alvin E. Tanenholtz
Attorney, Agent, or Firm—Paul T. Clark

[57]   **ABSTRACT**

A transgenic non-human eukaryotic animal whose germ cells and somatic cells contain an activated oncogene sequence introduced into the animal, or an ancestor of the animal, at an embryonic stage.

12 Claims, 2 Drawing Sheets

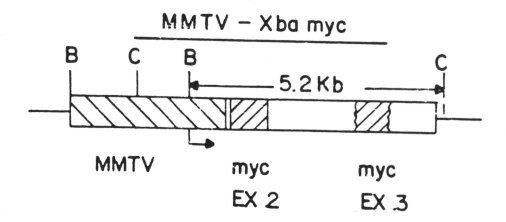

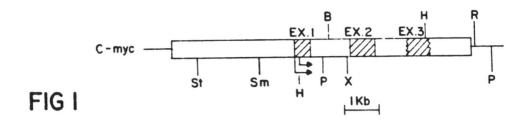

FIG 1

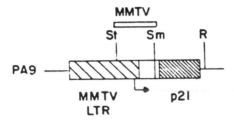

FIG 2

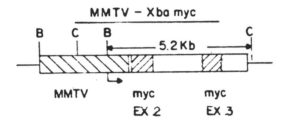

FIG 3

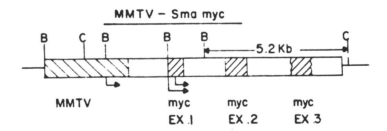

FIG 4

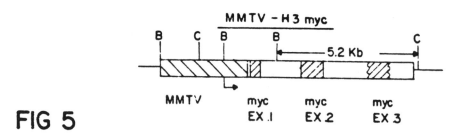

FIG 5

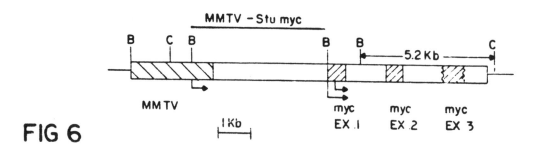

FIG 6

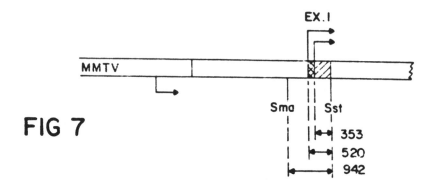

FIG 7

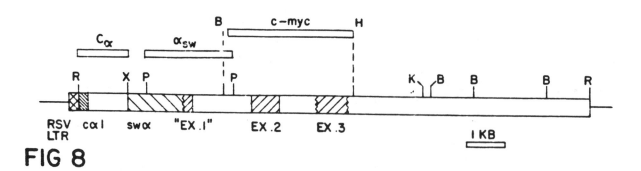

FIG 8

1

TRANSGENIC NON-HUMAN MAMMALS

BACKGROUND OF THE INVENTION

This invention relates to transgenic animals.

Transgenic animals carry a gene which has been introduced into the germline of the animal, or an ancestor of the animal, at an early (usually one-cell) developmental stage. Wagner et al. (1981) *P.N.A.S. U.S.A.* 78, 5016; and Stewart et al. (1982) *Science* 217, 1046 describe transgenic mice containing human globin genes. Constantini et al. (1981) *Nature* 294, 92; and Lacy et al. (1983) *Cell* 34, 343 describe transgenic mice containing rabbit globin genes. McKnight et al. (1983) *Cell* 34, 335 describes transgenic mice containing the chicken transferrin gene. Brinster et al. (1983) *Nature* 306, 332 describes transgenic mice containing a functionally rearranged immunoglobulin gene. Palmiter et al. (1982) *Nature* 300, 611 describes transgenic mice containing the rat growth hormone gene fused to a heavy metal-inducible metalothionein promoter sequence. Palmiter et al. (1982) *Cell* 29, 701 describes transgenic mice containing a thymidine kinase gene fused to a metalothionein promoter sequence. Palmiter et al. (1983) *Science* 222, 809 describes transgenic mice containing the human growth hormone gene fused to a metalothionein promoter sequence.

SUMMARY OF THE INVENTION

In general, the invention features a transgenic non-human eukaryotic animal (preferably a rodent such as a mouse) whose germ cells and somatic cells contain an activated oncogene sequence introduced into the animal, or an ancestor of the animal, at an embryonic stage (preferably the one-cell, or fertilized oocyte, stage, and generally not later than about the 8-cell stage). An activated oncogene sequence, as the term is used herein, means an oncogene which, when incorporated into the genome of the animal, increases the probability of the development of neoplasms (particularly malignant tumors) in the animal. There are several means by which an oncogene can be introduced into an animal embryo so as to be chromosomally incorporated in an activated state. One method is to transfect the embryo with the gene as it occurs naturally, and select transgenic animals in which the gene has integrated into the chromosome at a locus which results in activation. Other activation methods involve modifying the oncogene or its control sequences prior to introduction into the embryo. One such method is to transfect the embryo using a vector containing an already translocated oncogene. Other methods are to use an oncogene whose transcription is under the control of a synthetic or viral activating promoter, or to use an oncogene activated by one or more base pair substitutions, deletions, or additions.

In a preferred embodiment, the chromosome of the transgenic animal includes an endogenous coding sequence (most preferably the c-myc gene, hereinafter the myc gene), which is substantially the same as the oncogene sequence, and transcription of the oncogene sequence is under the control of a promoter sequence different from the promoter sequence controlling transcription of the endogenous coding sequence. The oncogene sequence can also be under the control of a synthetic promoter sequence. Preferably, the promoter sequence controlling transcription of the oncogene sequence is inducible.

2

Introduction of the oncogene sequence at the fertilized oocyte stage ensures that the oncogene sequence will be present in all of the germ cells and somatic cells of the transgenic animal. The presence of the oncogene sequence in the germ cells of the transgenic "founder" animal in turn means that all of the founder animal's descendants will carry the activated oncogene sequence in all of their germ cells and somatic cells. Introduction of the oncogene sequence at a later embryonic stage might result in the oncogene's absence from some somatic cells of the founder animal, but the descendants of such an animal that inherit the gene will carry the activated oncogene in all of their germ cells and somatic cells.

Any oncogene or effective sequence thereof can be used to produce the transgenic mice of the invention. Table 1, below, lists some known viral and cellular oncogenes, many of which are homologous to DNA sequences endogenous to mice and/or humans, as indicated. The term "oncogene" encompasses both the viral sequences and the homologous endogenous sequences.

TABLE 1

Abbreviation	Virus
src	Rous Sarcoma Virus (Chicken)
yes	Y73 Sarcoma Virus (Chicken)
fps	Fujinami (St Feline) Sarcoma Virus (Chicken, Cat)
abl	Abelson Murine Leukemia Virus (Mouse)
ros	Rochester-2 Sarcoma Virus (Chicken)
fgr	Gardner-Rasheed Feline Sarcoma Virus (Cat)
erbB	Avian Erythroblastosis Virus (Chicken)
fms	McDonough Feline Sarcoma Virus (Cat)
mos	Moloney Murine Sarcoma Virus (Mouse)
raf	3611 Murine Sarcoma+ Virus (Mouse)
Ha-ras-1	Harvey Murine Sarcoma Virus (Rat) (Balb/c mouse; 2 loci)
Ki-ras 2	Kirsten Murine Sarcoma Virus (Rat)
Ki-ras 1	Kirsten Murine Sarcoma Virus (Rat)
myc	Avian MC29 Myelocytomatosis Virus (Chicken)
myt	Avian Myelo Blastomas (Chicken)
fos	FBJ Osteosarcoma Virus (Mouse)
ski	Avian SKV T10 Virus (Chicken)
rel	Reticuloendotheliosis Virus (Turkey)
sis	Simian Sarcoma Virus (Woolly Monkey)
N-myc	Neuroblastomas (Human)
N-ras	Neuro astoma, Leukemia Sarcoma Virus (Human)
Blym	Bursal Lymphomas (Chicken)
mam	Mammary Carcinoma (Human)
neu	Neuro, Glioblastoma (Rat)
ertAI	Chicken AEV (Chicken)
ra-ras	Rasheed Sarcoma Virus

3

TABLE 1-continued

Abbreviation	Virus
	(Rat)
mmt-myc	Carcinoma Virus MH2 (Chicken)
myc	Myelocytomatosis OK10 (Chicken)
myb-ets	Avian myeloblastosis/ erythroblastosis Virus E26 (Chicken)
raf-2	3611-MSV (Mouse)
raf-1	3611-MSV (Mouse)
Ha-ras-2	Ki-MSV (Rat)
erbB	Erythroblastosis virus (Chicken)

The animals of the invention can be used to test a material suspected of being a carcinogen, by exposing the animal to the material and determining neoplastic growth as an indicator of carcinogenicity. This test can be extremely sensitive because of the propensity of the transgenic animals to develop tumors. This sensitivity will permit suspect materials to be tested in much smaller amounts than the amounts used in current animal carcinogenicity studies, and thus will minimize one source of criticism of current methods, that their validity is questionable because the amounts of the tested material used is greatly in excess of amounts to which humans are likley to be exposed. Furthermore, the animals will be expected to develop tumors much sooner because they already contain an activated oncogene. The animals are also preferable, as a test system, to bacteria (used, e.g., in the Ames test) because they, like humans, are vertebrates, and because carcinogenicity, rather than mutogenicity, is measured.

The animals of the invention can also be used as tester animals for materials, e.g. antioxidants such as beta-carotine or Vitamin E, thought to confer protection against the development of neoplasms. An animal is treated with the material, and a reduced incidence of neoplasm development, compared to untreated animals, is detected as an indication of protection. The method can further include exposing treated and untreated animals to a carcinogen prior to, after, or simultaneously with treatment with the protective material.

The animals of the invention can also be used as a source of cells for cell culture. Cells from the animals may advantageously exhibit desirable properties of both normal and transformed cultured cells; i.e., they will be normal or nearly normal morphologically and physiologically, but can, like cells such as NIH 3T3 cells, be cultured for long, and perhaps indefinite, periods of time. Further, where the promoter sequence controlling transcription of the oncogene sequence is inducible, cell growth rate and other culture characteristics can be controlled by adding or eliminating the inducing factor.

Other features and advantages of the invention will be apparent from the description of the preferred embodiments, and from the claims.

DESCRIPTION OF THE PREFERRED EMBODIMENTS

The drawings will first briefly be described.

DRAWINGS

FIG. 1 is a diagrammatic representation of a region of a plasmid bearing the mouse myc gene and flanking regions.

4

FIG. 2 is a diagrammatic represenation of a region of a plasmid, pA9, bearing the mouse mammary tumor virus long terminal repeat (MMTV LTR) sequences.

FIGS. 3–6 and 8 are diagrammatic representations of activated oncogene fusions.

FIG. 7 is a diagrammatic representation of a probe useful for detecting activated myc fusions.

MMTV-MYC FUSED GENES

Gene fusions were made using the mouse myc gene and the MMTV LTR. The myc gene is known to be an activatable oncogene. (For example, Leder et al. (1983) *Science* 222, 765 explains how chromosomal translocations that characterize Burkitt's Lymphoma and mouse plasmacytomas result in a juxtaposition of the myc gene and one of the immunoglobulin constant regions; amplification of the myc gene has also been observed in transformed cell lines.) FIG. 1 illustrates the subclone of the mouse myc gene which provided the myc regions.

The required MMTV functions were provided by the pA9 plasmid (FIG. 2) that demonstrated hormone inducibility of the p21 protein; this plasmid is described in Huang et al. (1981) *Cell* 27, 245. The MMTV functions on pA9 include the region required for glucocorticoid control, the MMTV promoter, and the cap site.

The above plasmids were used to construct the four fusion gene contructions illustrated in FIGS. 3–6. The constructions were made by deleting from pA9 the Sma-EcoRI region that included the p21 protein coding sequences, and replacing it with the four myc regions shown in the Figures. Procedures were the conventional techniques described in Maniatis et al. (1982) *Molecular Cloning: A Laboratory Manual* (Cold Spring Harbor Laboratory). The restriction sites shown in FIG. 1 are StuI (St), SmaI (Sm), EcoRI (R), HindIII (H), PvuI (P), BamHI (B), XbaI (X), and ClaI (C). The solid arrows below the constructions represent the promoter in the MMTV LTR and in the myc gene. The size (in Kb) of the major fragment, produced by digestion with BamHI and ClaI, that will hybridize to the myc probe, is shown for each construction.

MMTV-H3 myc (FIG. 5) was constructed in two steps: Firstly, the 4.7 Kb Hind III myc fragment which contains most of the myc sequences was made blunt with Klenow polymerase and ligated to the pA9 Smal-EcoRI vector that had been similarly treated. This construction is missing the normal 3' end of the myc gene. In order to introduce the 3' end of the myc gene, the PvuI-PvuI fragment extending from the middle of the first myc intron to the pBR322 PvuI site in the truncated MMTV-H3 myc was replaced by the related PvuI-PvuI fragment from the mouse myc subclone.

The MMTV-Xba myc construction (FIG. 3) was produced by first digesting the MMTV-Sma myc plasmid with SmaI and XbaI. The XbaI end was then made blunt with Klenow polymerase and the linear molecule recircularized with T4 DNA ligase. The MMTV-Stu myc (FIG. 6) and the MMTV-Sma myc (FIG. 4) constructions were formed by replacing the p21 protein coding sequences with, respectively, the StuI-EcoRI or Sma-EcoRI myc fragments (the EcoRI site is within the pBR322 sequences of the myc subclone). As shown in FIG. 1, there is only one StuI site within the myc gene. As there is more than one SmaI site within the myc gene (FIG. 4), a partial SmaI digestion was carried out to generate a number of MMTV-Sma myc plasmids; the plasmid illustrated in FIG. 4 was selected as not showing rearrangements and also including a suffi-

5

ciently long region 5' of the myc promoter (approximately 1 Kb) to include myc proximal controlling regions.

The constructions of FIGS. 4 and 6 contain the two promoters naturally preceding the unactivated myc gene. The contruction of FIG. 5 has lost both myc promoters but retains the cap site of the shorter tranript. The construction of FIG. 3 does not include the first myc exon but does include the entire protein coding sequence. The 3' end of the myc sequence in all of the illustrated constructions is located at the HindIII site approximately 1 Kb 3' to the myc polyA addition site.

These constructions were all checked by multiple restriction enzyme digestions and were free of detectable rearrangements.

PRODUCTION OF TRANSGENIC MICE CONTAINING MMTV-MYC FUSIONS

The above MMTV-myc plasmids were digested with SalI and EcoRI (each of which cleaves once within the pBR322 sequence) and separately injected into the male pronuclei of fertilized one-cell mouse eggs; this resulted in about 500 copies of linearized plasmid per pronucleus. The injected eggs were then transferred to pseudo-pregnant foster females as described in Wagner et al. (1981) P.N.A.S. U.S.A. 78, 5016. The eggs were derived from a CD-1 X C57Bl/6J mating. Mice were obtained from the Charles River Laboratories (CD^R-1-Ha/Icr (CD-1), an albino outbred mouse) and Jackson Laboratories (C57Bl/6J), and were housed in an environmentally controlled facility maintained on a 10 hour dark: 14 hour light cycle. The eggs in the foster females were allowed to develop to term.

ANALYSIS OF TRANSGENIC MICE

At four weeks of age, each pup born was analyzed using DNA taken from the tail in a Southern hybridization, using a ^{32}P DNA probe (labeled by nick-translation). In each case, DNA from the tail was digested with BamHI and ClaI and probed with the ^{32}P-labeled BamHI/HindIII probe from the normal myc gene (FIG. 1).

The DNA for analysis was extracted from 0.1–1.5 cm sections of tail, by the method described in Davis et al. (1980) in Methods in Enzymology, Grossman et al., eds., 65, 404, except that one chloroform extraction was performed prior to ethanol precipitation. The resulting nucleic acid pellet was washed once in 20% ethanol, dried, and resuspended in 300 μl of 1.0 mM Tris, pH 7.4, 0.1 mM EDTA.

Ten μl of the tail DNA preparation (approximately 10 μg DNA) were digested to completion, electrophoresed through 0.8% agarose gels, and transferred to nitrocellulose, as described in Southern (1975) J. Mol. Biol. 98, 503. Filters were hybridized overnight to probes in the presence of 10% dextran sulfate and washed twice in 2 X SSC, 0.1% SDS at room temperature and four times in 0.1 X SSC, 0.1% SDS at 64° C.

The Southern hybridizations indicated that ten founder mice had retained an injected MMTV-myc fusion. Two founder animals had integrated the myc gene at two different loci, yielding two genetically distinct lines of transgenic mice. Another mouse yielded two polymorphic forms of the integrated myc gene and thus yielded two genetically distinct offspring, each of which carried a different polymorphic form of the gene.

6

Thus, the 10 founder animals yielded 13 lines of transgenic offspring.

The founder animals were mated to uninjected animals and DNA of the resulting thirteen lines of transgenic offspring analyzed; this analysis indicated that in every case the injected genes were transmitted through the germline. Eleven of the thirteen lines also expressed the newly acquired MMTV-myc genes in at least one somatic tissue; the tissue in which expression was most prevalent was salivary gland.

Transcription of the newly acquired genes in tissues was determined by extracting RNA from the tissues and assaying the RNA in an S1 nuclease protection procedure, as follows. The excised tissue was rinsed in 5.0 ml cold Hank's buffered saline and total RNA was isolated by the method of Chrigwin et al. (1979) Biochemistry 18, 5294, using the CsCl gradient modification. RNA pellets were washed twice by reprecipitation in ethanol and quantitated by absorbance at 260 nm. An appropriate single stranded, uniformly labeled DNA probe was prepared as described by Ley et al. (1982) PNAS USA 79, 4775. To test for transcription of the MMTV-Stu myc fusion of FIG. 6, for example, the probe illustrated in FIG. 7 was used. This probe extends from a SmaI site 5' to the first myc exon to an SstI site at the 3' end of the first myc exon. Transcription from the endogenous myc promoters will produce RNA that will protect fragments of the probe 353 and 520 base pairs long; transcription from the MMTV promoter will completely protect the probe and be revealed as a band 942 base pairs long, in the following hybridization procedure.

Labelled, single-stranded probe fragments were isolated on 8M urea 5% acrylamide gels, electroeluted, and hybridized to total RNA in a modification of the procedure of Berk et al. (1977) Cell 12, 721. The hybridization mixture contained 50,000 cpm to 100,000 cpm of probe (SA = 10^8 cpm/μg), 10 μg total cellular RNA, 75% formamide, 500 mM NaCl, 20 mM Tris pH 7.5, 1 mM EDTA, as described in Battey et al. (1983) Cell 34, 779. Hybridization temperatures were varied according to the GC content in the region of the probe expected to hybridize to mRNA. The hybridizations were terminated by the addition of 1500 units of S1 nuclease (Boehringer Mannheim). S1 nuclease digestions were carried out at 37° C. for 1 hour. The samples were then ethanol-precipitated and electrophoresed on thin 8M urea 5% acrylamide gels.

Northern hybridization analysis was also carried out, as follows. Total RNA was electrophoresed through 1% formaldehyde 0.8% agarose gels, blotted to nitrocellulose filters (Lehrach et al. (1979) Biochemistry 16, 4743), and hybridized to nick-translated probes as described in Taub et al. (1982) PNAS USA 79, 7837. The tissues analyzed were thymus, pancreas, spleen, kidney, testes, liver, heart, lung, skeletal muscle, brain, salivary gland, and preputial gland.

Both lines of mice which had integrated and were transmitting to the next generation the MMTV-Stu myc fusion (FIG. 6) exhibited transcription of the fusion in salivary gland, but in no other tissue.

One of two lines of mice found to carry the MMTV-Sma myc fusion (FIG. 4) expressed the gene fusion in all tissues examined, with the level of expression being particularly high in salivary gland. The other line expressed the gene fusion only in salivary gland, spleen, testes, lung, brain, and preputial gland.

Four lines of mice carried the MMTV-H3 myc fusion (FIG. 5). In one, the fusion was transcribed in testes,

7

lung, salivary gland, and brain; in a second, the fusion was transcribed only in salivary gland; in a third, the fusion was transcribed in none of the somatic tissues tested; and in a fourth, the fusion was transcribed in salivary gland and intestinal tissue.

In two mouses lines found to carry the MMTV-Xba myc fusion, the fusion was transcribed in testes and salivary gland.

RSV-MYC FUSED GENES

Referring to FIG. 8, the plasmid designated RSV-S107 was generated by inserting the EcoRI fragment of the S107 plasmacytoma myc gene, (Kirsch et al. (1981) *Nature* 293, 585) into a derivative of the Rous Sarcoma Virus (RSV) enhancer-containing plasmid (pRSVcat) described in Gorman et al. (1982) *PNAS USA* 79, 6777, at the EcoRI site 3' to the RSV enhancer sequence, using standard recombinant DNA techniques. All chloramphenicol acetyl transferase and SV40 sequences are replaced in this vector by the myc gene; the RSV promoter sequence is deleted when the EcoRI fragments are replaced, leaving the RSV enhancer otherwise intact. The original translocation of the myc gene in the S107 plasmacytoma deleted the two normal myc promoters as well as a major portion of the untranslated first myc exon, and juxtaposed, 5' to 5', the truncated myc gene next to the α immunoglobulin heavy chain switch sequence.

The illustrated (FIG. 8) regions of plasmid RSV-S107 are: crosshatched, RSV sequences; fine-hatched, alpha 1 coding sequences; left-hatched, immunoglobulin alpha switch sequences; right-hatched, myc exons. The thin lines flanking the RSV-S107 myc exon represent pBR322 sequences. The marked restriction enzyme sites are: R, EcoRI; X, Xbal; P, Pst 1; K, Kpn 1; H, HindIII; B, BamHI. The sequences used for three probes used in assays described herein (C-α, α-sw and c-myc) are marked.

PRODUCTION OF TRANSGENIC MICE

Approximately 500 copies of the RSV-S107 myc plasmid (linearized at the unique Kpn-1 site 3' to the myc gene) were injected into the male pronucleus of eggs derived from a C57BL/6J x CD-1 mating. Mice were obtained from Charles River Laboratories (CD-1, an albino outbred mouse) and from Jackson Laboratories (C57BL/6J). These injected eggs were transferred into pseudopregnant foster females, allowed to develop to term, and at four weeks of age the animals born were tested for retention of the injected sequences by Southern blot analysis of DNA extracted from the tail, as described above. Of 28 mice analyzed, two males were found to have retained the new genes and both subsequently transmitted these sequences through the germline in a ratio consistent with Mendelian inheritance of single locus.

First generation transgenic offspring of each of these founder males were analyzed for expression of the rearranged myc genes by assaying RNA extracted from the major internal tissues and organs in an S1 nuclease protection assay, as described above. The hearts of the offspring of one line showed aberrant myc expression; the other 13 tissues did not.

Backcrossing (to C57Bl/6J) and in-breeding matings produced some transgenic mice which did not demonstrate the same restriction site patterns on Southern blot analysis as either their transgenic siblings or their parents. In the first generation progeny derived from a

8

mating between the founder male and C57BL/6J females, 34 F1 animals were analyzed and of these, 19 inherited the newly introduced gene, a result consistent with the founder being a heterozygote at one locus. However, of the 19 transgenic mice analyzed, there were three qualitatively different patterns with respect to the more minor myc hybridizing fragments.

In order to test the possibility that these heterogenous genotypes arose as a consequence of multiple insertions and/or germline mosaicism in the founder, two F1 mice (one carrying the 7.8 and 12 Kb BamHI bands, and the other carrying only the 7.8 Kb BamHI band) were mated and the F2 animals analyzed. One male born to the mating of these two appeared to have sufficient copies of the RSV-S107 myc gene to be considered as a candidate for having inherited the two alleles; this male was backcrossed with a wild-type female. All 23 of 23 backcross offspring analyzed inherited the RSV-S107 myc genes, strongly suggesting that the F2 male mouse had inherited two alleles at one locus. Further, as expected, the high molecular weight fragment (12 Kb) segregated as a single allele.

To determine whether, in addition to the polymorphisms arising at the DNA level, the level of aberrant myc expression was also altered, heart mRNA was analyzed in eight animals derived from the mating of the above double heterozygote to a wild-type female. All eight exhibited elevated myc mRNA, with the amount appearing to vary between animals; the lower levels of expression segregated with the presence of the 12 Kb myc hybridizing band. The level of myc mRNA in the hearts of transgenic mice in a second backcross generation also varied. An F1 female was backcrossed to a C57Bl/6J male to produce a litter of seven pups, six of which inherited the RSV-S107 myc genes. All seven of these mice were analyzed for expression. Three of the six transgenic mice had elevated levels of myc mRNA in the hearts whereas in the other three the level of myc mRNA in the hearts was indistinguishable from the one mouse that did not carry the RSV-S107 myc gene. This result suggests that in addition to the one polymorphic RSV-S107 myc locus from which high levels of heart-restricted myc mRNA were transcribed, there may have been another segregating RSV-S107 myc locus that was transcriptionally silent.

CARCINOGENICITY TESTING

The animals of the invention can be used to test a material suspected of being a carcinogen, as follows. If the animals are to be used to test materials thought to be only weakly carcinogenic, the transgenic mice most susceptible of developing tumors are selected, by exposing the mice to a low dosage of a known carcinogen and selecting those which first develop tumors. The selected animals and their descendants are used as test animals by exposing them to the material suspected of being a carcinogen and determining neoplastic growth as an indicator of carcinogenicity. Less sensitive animals are used to test more strongly carcinogenic materials. Animals of the desired sensitivity can be selected by varying the type and concentration of known carcinogen used in the selection process. When extreme sensitivity is desired, the selected test mice can consist of those which spontaneously develop tumors.

TESTING FOR CANCER PROTECTION

The animals of the invention can be used to test materials for the ability to confer protection against the

9

development of neoplasms. An animal is treated with the material, in parallel with an untreated control transgenic animal. A comparatively lower incidence of neoplasm development in the treated animal is detected as an indication of protection.

TISSUE CULTURE

The transgenic animals of the invention can be used as a source of cells for cell culture. Tissues of transgenic mice are analyzed for the presence of the activated oncogene, either by directly analyzing DNA or RNA, or by assaying the tissue for the protein expressed by the gene. Cells of tissues carrying the gene can be cultured, using standard tissue culture techniques, and used, e.g., to study the functioning of cells from normally difficult to culture tissues such as heart tissue.

DEPOSITS

Plasmids bearing the fusion genes shown in FIGS. 3, 4, 5, 6. and 8 have been deposited in the American Type Culture Collection, Rockville, Md., and given, respectively, ATCC Accession Nos. 39745, 39746, 39747, 39748, and 39749.

OTHER EMBODIMENTS

Other embodiments are within the following claims. For example, any species of transgenic animal can be employed. In some circumstances, for instance, it may be desirable to use a species, e.g., a primate such as the rhesus monkey, which is evolutionarily closer to humans than mice.

We claim:

1. A transgenic non-human mammal all of whose germ cells and somatic cells contain a recombinant activated oncogene sequence introduced into said mam-

10

mal, or an ancestor of said mammal, at an embryonic stage.

2. The mammal of claim 1, a chromosome of said mammal including an endogenous coding sequence substantially the same as a coding sequence of said oncogene sequence.

3. The mammal of claim 2, said oncogene sequence being integrated into a chromosome of said mammal at a site different from the location of said endogenous coding sequence.

4. The mammal of claim 2 wherein transcription of said oncogene sequence is under the control of a promoter sequence different from the promoter sequence controlling the transcription of said endogenous coding sequence.

5. The mammal of claim 4 wherein said promoter sequence controlling transcription of said oncogene sequence is inducible.

6. The mammal of claim 1 wherein said oncogene sequence comprises a coding sequence of a c-myc gene.

7. The mammal of claim 1 wherein transcription of said oncogene sequence is under the control of a viral promoter sequence.

8. The mammal of claim 7 wherein said viral promoter sequence comprises a sequence of an MMTV promoter.

9. The mammal of claim 7 wherein said viral promoter sequence comprises a sequence of an RSV promoter.

10. The mammal of claim 1 wherein transcription of said oncogene sequence is under the control of a synthetic promoter sequence.

11. The mammal of claim 1, said mammal being a rodent.

12. The mammal of claim 11, said rodent being a mouse.

* * * * *